CHAPTER 12 Reflex and Voluntary Control of Posture & Movement 253

vestibulospinal tracts) regulate proximal muscles and posture. The lateral corticospinal and rubrospinal tracts control distal limb muscles for fine motor control and skilled voluntary movements.

- Decerebrate rigidity leads to hyperactivity in extensor muscles in all four extremities; it is actually spasticity due to facilitation of the stretch reflex. It resembles what is seen with uncal herniation due to a supratentorial lesion. Decorticate rigidity is flexion of the upper extremities at the elbow and extensor hyperactivity in the lower extremities. It occurs on the hemiplegic side after hemorrhage or thrombosis in the internal capsule.
- The basal ganglia include the caudate nucleus, putamen, globus pallidus, subthalamic nucleus, and substantia nigra. The connections between the parts of the basal ganglia include a dopaminergic nigrostriatal projection from the substantia nigra to the striatum and a GABAergic projection from the striatum to substantia nigra.
- Parkinson disease is due to degeneration of the nigrostriatal dopaminergic neurons and is characterized by akinesia, bradykinesia, cogwheel rigidity, and tremor at rest. Huntington disease is characterized by choreiform movements due to the loss of the GABAergic inhibitory pathway to the globus pallidus.
- The cerebellar cortex contains five types of neurons: Purkinje, granule, basket, stellate, and Golgi cells. The two main inputs to the cerebellar cortex are climbing fibers and mossy fibers. Purkinje cells are the only output from the cerebellar cortex, and they generally project to the deep nuclei. Damage to the cerebellum leads to several characteristic abnormalities, including hypotonia, ataxia, and intention tremor.

MULTIPLE-CHOICE QUESTIONS

For all questions, select the single best answer unless otherwise directed.

1. When dynamic γ-motor neurons are activated at the same time as α-motor neurons to muscle,
   A. prompt inhibition of discharge in spindle Ia afferents takes place.
   B. clonus is likely to occur.
   C. the muscle will not contract.
   D. the number of impulses in spindle Ia afferents is smaller than when α discharge alone is increased.
   E. the number of impulses in spindle Ia afferents is greater than when α discharge alone is increased.
2. The inverse stretch reflex
   A. occurs when Ia spindle afferents are inhibited.
   B. is a monosynaptic reflex initiated by activation of the Golgi tendon organ.
   C. is a disynaptic reflex with a single interneuron inserted between the afferent and efferent limbs.
   D. is a polysynaptic reflex with many interneurons inserted between the afferent and efferent limbs.
   E. uses type II afferent fibers from the Golgi tendon organ.
3. Withdrawal reflexes are *not*
   A. initiated by nociceptive stimuli.
   B. prepotent.
   C. prolonged if the stimulus is strong.
   D. an example of a flexor reflex.
   E. accompanied by the same response on both sides of the body.
4. While exercising, a 42-year-old female developed sudden onset of tingling in her right leg and an inability to control movement in that limb. A neurological exam showed a hyperactive knee jerk reflex and a positive Babinski sign. Which of the following is *not* characteristic of a reflex?
   A. Reflexes can be modified by impulses from various parts of the CNS.
   B. Reflexes may involve simultaneous contraction of some muscles and relaxation of others.
   C. Reflexes are chronically suppressed after spinal cord transection.
   D. Reflexes involve transmission across at least one synapse.
   E. Reflexes often occur without conscious perception.
5. Increased neural activity before a skilled voluntary movement is *first* seen in the
   A. spinal motor neurons.
   B. precentral motor cortex.
   C. midbrain.
   D. cerebellum.
   E. cortical association areas.
6. A 58-year-old woman was brought to the emergency room of her local hospital because of a sudden change of consciousness. All four limbs were extended, suggestive of decerebrate rigidity. A brain CT showed a rostral pontine hemorrhage. Which of the following describes components of the central pathway responsible for control of posture?
   A. The tectospinal pathway terminates on neurons in the dorsolateral area of the spinal ventral horn that innervate limb muscles.
   B. The medullary reticulospinal pathway terminates on neurons in the ventromedial area of the spinal ventral horn that innervate axial and proximal muscles.
   C. The pontine reticulospinal pathway terminates on neurons in the dorsomedial area of the spinal ventral horn that innervate limb muscles.
   D. The medial vestibular pathway terminates on neurons in the dorsomedial area of the spinal ventral horn that innervate axial and proximal muscles.
   E. The lateral vestibular pathway terminates on neurons in the dorsolateral area of the spinal ventral horn that innervate axial and proximal muscles.
7. A 38-year-old female had been diagnosed with a metastatic brain tumor. She was brought to the emergency room of her local hospital because of irregular breathing and progressive loss of consciousness. She also showed signs of decerebrate posturing. Which of the following is *not* true about decerebrate rigidity?
   A. It involves hyperactivity in extensor muscles of all four limbs.
   B. The excitatory input from the reticulospinal pathway activates γ-motor neurons which indirectly activate α-motor neurons.
   C. It is actually a type of spasticity due to inhibition of the stretch reflex.
   D. It resembles what ensues after uncal herniation.
   E. Lower extremities are extended with toes pointed inward.

End-of-chapter review questions help you assess your comprehension

246 SECTION II Central and Peripheral Neurophysiology

CLINICAL BOX 12–7

Basal Ganglia Diseases

The initial detectable damage in **Huntington disease** is to medium spiny neurons in the striatum. This loss of this GABAergic pathway to the globus pallidus external segment releases inhibition, permitting the hyperkinetic features of the disease to develop. An early sign is a jerky trajectory of the hand when reaching to touch a spot, especially toward the end of the reach. Later, hyperkinetic **choreiform movements** appear and gradually increase until they incapacitate the patient. Speech becomes slurred and then incomprehensible, and a progressive dementia is followed by death, usually within 10–15 years after the onset of symptoms. Huntington disease affects 5 out of 100,000 people worldwide. It is inherited as an autosomal dominant disorder, and its onset is usually between the ages of 30 and 50. The abnormal gene responsible for the disease is located near the end of the short arm of chromosome 4. It normally contains 11–34 cytosine-adenine-guanine (CAG) repeats, each coding for glutamine. In patients with Huntington disease, this number is increased to 42–86 or more copies, and the greater the number of repeats, the earlier the age of onset and the more rapid the progression of the disease. The gene codes for **huntingtin,** a protein of unknown function. Poorly soluble protein aggregates, which are toxic, form in cell nuclei and elsewhere. However, the correlation between aggregates and symptoms is less than perfect. It appears that a loss of the function of huntingtin occurs that is proportional to the size of the CAG insert. In animal models of the disease, intrastriatal grafting of fetal striatal tissue improves cognitive performance. In addition, tissue caspase-1 activity is increased in the brains of humans and animals with the disease, and in mice in which the gene for this apoptosis-regulating enzyme has been knocked out, progression of the disease is slowed.

Another basal ganglia disorder is **Wilson disease** (or **hepatolenticular degeneration**), which is a rare disorder of copper metabolism which has an onset between 6 and 25 years of age, affecting about four times as many females as males. Wilson disease affects about 30,000 people worldwide. It is a genetic autosomal recessive disorder due to a mutation on the long arm of chromosome 13q. It affects the copper-transporting ATPase gene (*ATP7B*) in the liver, leading to an accumulation of copper in the liver and resultant progressive liver damage. About 1% of the population carries a single abnormal copy of this gene but does not develop any symptoms. A child who inherits the gene from both parents may develop the disease. In affected individuals, copper accumulates in the periphery of the cornea in the eye accounting for the characteristic yellow **Kayser–Fleischer rings.** The dominant neuronal pathology is degeneration of the putamen, a part of the **lenticular nucleus.** Motor disturbances include "wing-beating" tremor or **asterixis, dysarthria,** unsteady gait, and rigidity.

Another disease commonly referred to as a disease of the basal ganglia is **tardive dyskinesia.** This disease indeed involves the basal ganglia, but it is caused by medical treatment of another disorder with **neuroleptic drugs** such as phenothiazides or haloperidol. Therefore, tardive dyskinesia is iatrogenic in origin. Long-term use of these drugs may produce biochemical abnormalities in the striatum. The motor disturbances include either temporary or permanent uncontrolled involuntary movements of the face and tongue and cogwheel rigidity. The neuroleptic drugs act via blockade of dopaminergic transmission. Prolonged drug use leads to hypersensitivity of D_3 dopaminergic receptors and an imbalance in nigrostriatal influences on motor control.

THERAPEUTIC HIGHLIGHTS

Treatment for Huntington disease is directed at treating the symptoms and maintaining quality of life as there is no cure. In general, drugs used to treat the symptoms of this disease have side effects such as fatigue, nausea, and restlessness. In August 2008, the U.S. Food and Drug Administration approved the use of **tetrabenazine** to reduce choreiform movements that characterize the disease. This drug binds reversibly to vesicular monoamine transporters (VMAT) and thus inhibits the uptake of monoamines into synaptic vesicles. It also acts as a dopamine receptor antagonist. Tetrabenazine is the first drug to receive approval for individuals with Huntington disease. It is also used to treat other hyperkinetic movement disorders such as tardive dyskinesia. **Chelating agents** (eg, **penicillamine, trienthine**) are used to reduce the copper in the body in individuals with Wilson disease. Tardive dyskinesia has proven to be difficult to treat. Treatment in patients with psychiatric disorders is often directed at prescribing a neuroleptic with less likelihood of causing the disorder. **Clozapine** is an example of an atypical neuroleptic drug that has been an effective substitute for traditional neuroleptic drugs but with less risk for development of tardive dyskinesia.

Clinical cases add real-world relevance to the text

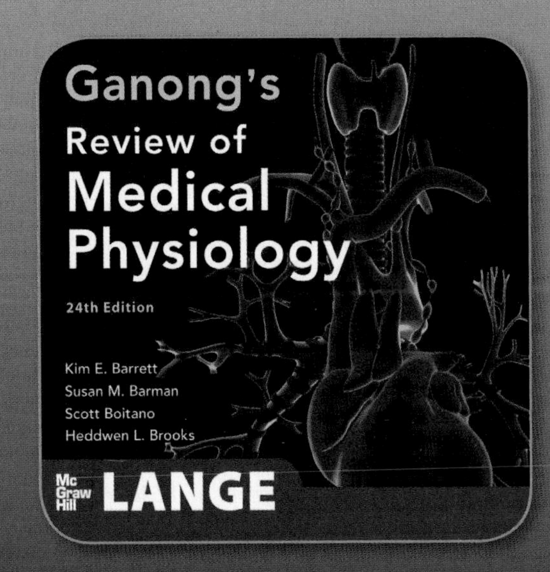

Also available on the iPad through Inkling

- Vivid full-color and annotated diagrams
- Interactive quizzes—with constructive feedback
- Videos and animations demonstrating the function of human systems

About the Authors

KIM E. BARRETT

Kim Barrett received her PhD in biological chemistry from University College London in 1982. Following postdoctoral training at the National Institutes of Health, she joined the faculty at the University of California, San Diego, School of Medicine in 1985, rising to her current rank of Professor of Medicine in 1996. Since 2006, she has also served the University as Dean of Graduate Studies. Her research interests focus on the physiology and pathophysiology of the intestinal epithelium, and how its function is altered by commensal, probiotics, and pathogenic bacteria as well as in specific disease states, such as inflammatory bowel diseases. She has published more than 200 articles, chapters, and reviews, and has received several honors for her research accomplishments including the Bowditch and Davenport Lectureships from the American Physiological Society and the degree of Doctor of Medical Sciences, honoris causa, from Queens University, Belfast. She has also been very active in scholarly editing, serving currently as the Deputy Editor-in-Chief of the *Journal of Physiology*. She is also a dedicated and award-winning instructor of medical, pharmacy, and graduate students, and has taught various topics in medical and systems physiology to these groups for more than 20 years. Her efforts as a teacher and mentor will be recognized with the Bodil M. Schmidt-Nielson Distinguished Mentor and Scientist Award from the American Physiological Society in 2012. Her teaching experiences led her to author a prior volume (Gastrointestinal Physiology, McGraw-Hill, 2005) and she was honored to have been invited to take over the helm of Ganong in 2007 for the 23rd edition, and to have guided this new edition.

SUSAN M. BARMAN

Susan Barman received her PhD in physiology from Loyola University School of Medicine in Maywood, Illinois. Afterward she went to Michigan State University (MSU) where she is currently a Professor in the Department of Pharmacology/Toxicology and the Neuroscience Program. Dr Barman has had a career-long interest in neural control of cardiorespiratory function with an emphasis on the characterization and origin of the naturally occurring discharges of sympathetic and phrenic nerves. She was a recipient of a prestigious National Institutes of Health MERIT (Method to Extend Research in Time) Award. She is also a recipient of an Outstanding University Woman Faculty Award from the MSU Faculty Professional Women's Association and an MSU College of Human Medicine Distinguished Faculty Award. She has been very active in the American Physiological Society (APS) and was recently elected to serve as its 85th President. She has also served as a Councillor as well as Chair of the Central Nervous System Section of APS, Women in Physiology Committee and Section Advisory Committee of APS. In her spare time, she enjoys daily walks, aerobic exercising, and mind-challenging activities like puzzles of various sorts.

SCOTT BOITANO

Scott Boitano received his PhD in genetics and cell biology from Washington State University in Pullman, Washington, where he acquired an interest in cellular signaling. He fostered this interest at University of California, Los Angeles, where he focused his research on second messengers and cellular physiology of the lung epithelium. He continued to foster these research interests at the University of Wyoming and at his current positions with the Department of Physiology and the Arizona Respiratory Center, both at the University of Arizona.

HEDDWEN L. BROOKS

Heddwen Brooks received her PhD from Imperial College, University of London and is an Associate Professor in the Department of Physiology at the University of Arizona (UA). Dr Brooks is a renal physiologist and is best known for her development of microarray technology to address in vivo signaling pathways involved in the hormonal regulation of renal function. Dr Brooks' many awards include the American Physiological Society (APS) Lazaro J. Mandel Young Investigator Award, which is for an individual demonstrating outstanding promise in epithelial or renal physiology. In 2009, she received the APS Renal Young Investigator Award at the annual meeting of the Federation of American Societies for Experimental Biology. Dr Brooks is currently Chair of the APS Renal Section Steering Committee. She serves on the Editorial Board of the *American Journal of Physiology-Renal Physiology* (since 2001), and has served on study sections of the National Institutes of Health and the American Heart Association. She is a current member of the Merit Review Board for the Department of Veterans' Affairs.

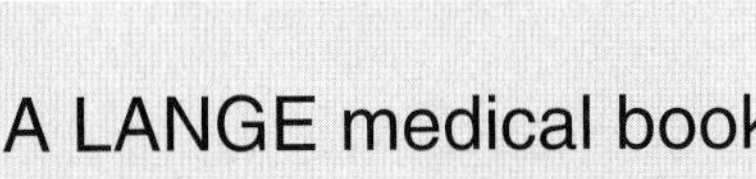

Ganong's Review of Medical Physiology

TWENTY-FOURTH EDITION

Kim E. Barrett, PhD
Professor, Department of Medicine
Dean of Graduate Studies
University of California, San Diego
La Jolla, California

Susan M. Barman, PhD
Professor, Department of Pharmacology/Toxicology
Michigan State University
East Lansing, Michigan

Scott Boitano, PhD
Associate Professor, Physiology
Arizona Respiratory Center
Bio5 Collaborative Research Institute
University of Arizona
Tucson, Arizona

Heddwen L. Brooks, PhD
Associate Professor, Physiology
College of Medicine
Bio5 Collaborative Research Institute
University of Arizona
Tucson, Arizona

New York Chicago San Francisco Lisbon London Madrid Mexico City
Milan New Delhi San Juan Seoul Singapore Sydney Toronto

The McGraw-Hill Companies

Ganong's Review of Medical Physiology, Twenty-Fourth Edition

1 2 3 4 5 6 7 8 9 0 CTP/CTP 17 16 15 14 13 12

ISBN 978-0-07-178003-2
MHID 0-07-178003-3
ISSN 0892-1253

Notice

Medicine is an ever-changing science. As new research and clinical experience broaden our knowledge, changes in treatment and drug therapy are required. The authors and the publisher of this work have checked with sources believed to be reliable in their efforts to provide information that is complete and generally in accord with the standards accepted at the time of publication. However, in view of the possibility of human error or changes in medical sciences, neither the authors nor the publisher nor any other party who has been involved in the preparation or publication of this work warrants that the information contained herein is in every respect accurate or complete, and they disclaim all responsibility for any errors or omissions or for the results obtained from use of the information contained in this work. Readers are encouraged to confirm the information contained herein with other sources. For example and in particular, readers are advised to check the product information sheet included in the package of each drug they plan to administer to be certain that the information contained in this work is accurate and that changes have not been made in the recommended dose or in the contraindications for administration. This recommendation is of particular importance in connection with new or infrequently used drugs.

This book was set in Minion Pro by Thomson Digital.
The editors were Michael Weitz and Brian Kearns.
The production supervisor was Catherine H. Saggese.
Project managment was provided by Aakriti Kathuria, Thomson Digital.
The designer was Elise Lansdon.
China Translation & Printing, Ltd. was printer and binder.

This book is printed on acid-free paper.

Dedication to

William Francis Ganong

William Francis ("Fran") Ganong was an outstanding scientist, educator, and writer. He was completely dedicated to the field of physiology and medical education in general. Chairman of the Department of Physiology at the University of California, San Francisco, for many years, he received numerous teaching awards and loved working with medical students.

Over the course of 40 years and some 22 editions, he was the sole author of the best selling *Review of Medical Physiology*, and a co-author of 5 editions of *Pathophysiology of Disease: An Introduction to Clinical Medicine.* He was one of the "deans" of the Lange group of authors who produced concise medical text and review books that to this day remain extraordinarily popular in print and now in digital formats. Dr. Ganong made a gigantic impact on the education of countless medical students and clinicians.

A general physiologist par excellence and a neuroendocrine physiologist by subspecialty, Fran developed and maintained a rare understanding of the entire field of physiology. This allowed him to write each new edition (every 2 years!) of the *Review of Medical Physiology* as a sole author, a feat remarked on and admired whenever the book came up for discussion among physiologists. He was an excellent writer and far ahead of his time with his objective of distilling a complex subject into a concise presentation. Like his good friend, Dr. Jack Lange, founder of the Lange series of books, Fran took great pride in the many different translations of the *Review of Medical Physiology* and was always delighted to receive a copy of the new edition in any language.

He was a model author, organized, dedicated, and enthusiastic. His book was his pride and joy and like other best-selling authors, he would work on the next edition seemingly every day, updating references, rewriting as needed, and always ready and on time when the next edition was due to the publisher. He did the same with his other book, *Pathophysiology of Disease: An Introduction to Clinical Medicine*, a book that he worked on meticulously in the years following his formal retirement and appointment as an emeritus professor at UCSF.

Fran Ganong will always have a seat at the head table of the greats of the art of medical science education and communication. He died on December 23, 2007. All of us who knew him and worked with him miss him greatly.

Key Features of the 24th Edition of *Ganong's Review of Medical Physiology*

- **Provides concise coverage of every important topic without sacrificing comprehensiveness or readability**
- **Reflects the latest research and developments in the areas of chronic pain, reproductive physiology, and acid-base homeostasis**
- **Incorporates examples from clinical medicine to illustrate important physiologic concepts**
- **NEW: Section introductions help you build a solid foundation on the given topic**
- **NEW: Introductory materials that cover overarching principles of endocrine regulation in physiology**
- **NEW: Detailed explanations of incorrect answer choices**
- **NEW: More clinical cases and flow charts than ever**
- **NEW: Expanded legends for each illustration—so you don't have to refer back to the text**

More than 600 full-color illustrations

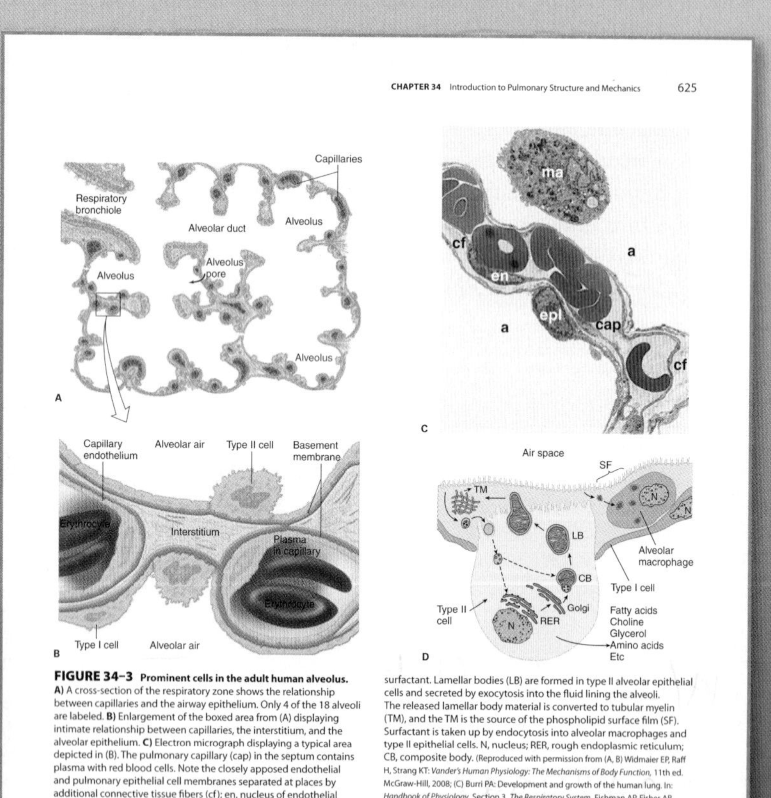

FIGURE 34–3 Prominent cells in the adult human alveolus. **A)** A cross-section of the respiratory zone shows the relationship between capillaries and the airway epithelium. Only 4 of the 18 alveoli are labeled. **B)** Enlargement of the boxed area from (A) displaying intimate relationship between capillaries, the interstitium, and the alveolar epithelium. **C)** Electron micrograph displaying a typical area depicted in (B). The pulmonary capillary (cap) in the septum contains plasma with red blood cells. Note the closely apposed endothelial and pulmonary epithelial cell membranes separated at places by additional connective tissue fibers (cf); en, nucleus of endothelial cell; epl, nucleus of type I alveolar epithelial cell; a, alveolar space; ma, alveolar macrophage. **D)** Type II cell formation and metabolism of surfactant. Lamellar bodies (LB) are formed in type II alveolar epithelial cells and secreted by exocytosis into the fluid lining the alveoli. The released lamellar body material is converted to tubular myelin (TM), and the TM is the source of the phospholipid surface film (SF). Surfactant is taken up by endocytosis into alveolar macrophages and type II epithelial cells. N, nucleus; RER, rough endoplasmic reticulum; CB, composite body. (Reproduced with permission from (A, B) Widmaier EP, Raff H, Strang KT: *Vander's Human Physiology: The Mechanisms of Body Function*, 11th ed. McGraw-Hill, 2008; (C) Burri PA: Development and growth of the human lung. In: *Handbook of Physiology*, Section 3, *The Respiratory System*. Fishman AP, Fisher AB [editors]. American Physiological Society, 1985; and (D) Wright JR: Metabolism and turnover of lung surfactant. Am Rev Respir Dis 136:426, 1987.)

0.5 (μm apart (Figure 34–3). The alveoli also contain other specialized cells, including pulmonary alveolar macrophages (PAMs, or AMs), lymphocytes, plasma cells, neuroendocrine cells, and mast cells. PAMs are an important component of the pulmonary defense system. Like other macrophages, these cells come originally from the bone marrow. PAMs are actively phagocytic and ingest small particles that evade the mucociliary escalator and reach the alveoli. They also help process inhaled antigens for immunologic attack, and they secrete substances that attract granulocytes to the lungs as well as substances that stimulate granulocyte and monocyte formation in the bone marrow. PAM function can also be detrimental—when they ingest large amounts of the substances in cigarette smoke or other irritants, they may release lysosomal products into the extracellular space to cause inflammation.

Contents

Preface ix

SECTION I Cellular and Molecular Basis for Medical Physiology 1

1 General Principles & Energy Production in Medical Physiology 3

2 Overview of Cellular Physiology in Medical Physiology 35

3 Immunity, Infection, & Inflammation 67

4 Excitable Tissue: Nerve 83

5 Excitable Tissue: Muscle 97

6 Synaptic & Junctional Transmission 119

7 Neurotransmitters & Neuromodulators 135

SECTION II Central and Peripheral Neurophysiology 155

8 Somatosensory Neurotransmission: Touch, Pain, and Temperature 157

9 Vision 177

10 Hearing & Equilibrium 199

11 Smell & Taste 217

12 Reflex and Voluntary Control of Posture & Movement 227

13 Autonomic Nervous System 255

14 Electrical Activity of the Brain, Sleep–Wake States, & Circadian Rhythms 269

15 Learning, Memory, Language, & Speech 283

SECTION III Endocrine and Reproductive Physiology 297

16 Basic Concepts of Endocrine Regulation 299

17 Hypothalamic Regulation of Hormonal Functions 307

18 The Pituitary Gland 323

19 The Thyroid Gland 339

20 The Adrenal Medulla & Adrenal Cortex 353

21 Hormonal Control of Calcium & Phosphate Metabolism & the Physiology of Bone 377

22 Reproductive Development & Function of the Female Reproductive System 391

23 Function of the Male Reproductive System 419

24 Endocrine Functions of the Pancreas & Regulation of Carbohydrate Metabolism 431

Gastrointestinal Physiology 453

25 Overview of Gastrointestinal Function & Regulation 455

26 Digestion, Absorption, & Nutritional Principles 477

27 Gastrointestinal Motility 497

28 Transport & Metabolic Functions of the Liver 509

Cardiovascular Physiology 519

29 Origin of the Heartbeat & the Electrical Activity of the Heart 521

30 The Heart as a Pump 539

31 Blood as a Circulatory Fluid & the Dynamics of Blood & Lymph Flow 555

32 Cardiovascular Regulatory Mechanisms 587

33 Circulation Through Special Regions 601

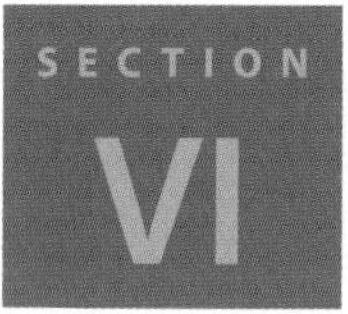

Respiratory Physiology 619

34 Introduction to Pulmonary Structure and Mechanics 621

35 Gas Transport & pH 641

36 Regulation of Respiration 657

Renal Physiology 671

37 Renal Function & Micturition 673

38 Regulation of Extracellular Fluid Composition & Volume 697

39 Acidification of the Urine & Bicarbonate Excretion 711

Answers to Multiple Choice Questions 721

Index 723

Preface

FROM THE AUTHORS

We are very pleased to launch the 24th edition of *Ganong's Review of Medical Physiology*. The current authors have attempted to maintain the highest standards of excellence, accuracy, and pedagogy developed by Fran Ganong over the 46 years in which he educated countless students worldwide with this textbook.

We were pleased with the reaction to the 23rd edition, our first at the helm. However, recognizing that improvement is always possible, and that medical knowledge is constantly evolving, we convened panels both of expert colleagues and of students to give us feedback on style, content, level and organizational issues. Based on this input, we have thoroughly reorganized the text and redoubled our efforts to ensure that the book presents state of the art knowledge. We have also increased clinical content, particularly related to the burden of disease states that arise from abnormal physiology of the systems we discuss.

We remain grateful to many colleagues and students who contact us with suggestions for clarifications and new material. This input helps us to ensure that the text is as useful as possible. We hope that you enjoy the fruits of our labors, and the new material in the 24th Edition.

This edition is a revision of the original works of Dr. Francis Ganong.

NEW THERAPEUTIC HIGHLIGHTS

- Recognizing the critical links between physiology and therapeutics, the boxed clinical cases, also now include succinct summaries of modern pharmacological approaches to the treatment or management of the condition discussed.

NEW—GANONG'S REVIEW OF MEDICAL PHYSIOLOGY 24/E COMES TO LIFE WITH THE GANONG IPAD DIGITAL VERSION!

- Integrated assessment
- Engaging interactivity and gorgeous, high resolution illustrations.
- Concepts are brought to life with movies, and integrates them right into the book
- Opportunity to buy individual chapters

NEW! GANONG ONLINE LEARNING CENTER

WWW.LANGETEXTBOOKS.COM/BARRETT

This dedicated Ganong website will include the following:

- Movies and Animations for both students and professors to access. See concepts come to life!
- PowerPoint of all images and tables for the instructor
- Review Questions for students to test themselves

NEW TO THIS EDITION!

Each section will now have an introduction

- Information on the burden of disease associated with each organ system
- New introductory materials covering overarching principles of endocrine regulation in physiology
- Answers to the review questions in the book, with additional explanations to incorrect questions will now be included
- Additional Flow Charts—Students expressed how helpful flow charts are in tying concepts together and seeing the big picture!
- Chapter summaries tied to chapter objectives
- Expanded legends—This will help students understand figures, without necessarily referring back to the text
- Increased number of clinical cases

SECTION I

Cellular and Molecular Basis for Medical Physiology

The detailed study of physiological system structure and function has its foundations in physical and chemical laws and the molecular and cellular makeup of each tissue and organ system. This first section provides an overview to the basic building blocks that provide the important framework for human physiology. It is important to note here that these initial sections are not meant to provide an exhaustive understanding of biophysics, biochemistry, or cellular and molecular physiology, rather they are to serve as a reminder of how the basic principals from these disciplines contribute to medical physiology discussed in later sections.

In the first part of this section, the basic building blocks: electrolytes; carbohydrates, lipids, and fatty acids; amino acids and proteins; and nucleic acids; are introduced and discussed. The students are reminded of some of the basic principles and building blocks of biophysics and biochemistry and how they fit into the physiological environment. Examples of direct clinical applications are provided in the Clinical Boxes to help bridge the gap between building blocks, basic principles and human physiology. These basic principles are followed up with a discussion of the generic cell and its components. It is important to realize the cell is the basic unit within the body, and it is the collection and fine-tuned interactions among and between these fundamental units that allow for proper tissue, organ, and organism function.

In the second part of this introductory section, we take a cellular approach to lay a groundwork of understanding groups of cells that interact with many of the systems discussed in future chapters. The first group of cells presented are those that contribute to inflammatory reactions in the body. These individual players, their coordinated behavior and the net effects of the "open system" of inflammation in the body are discussed in detail. The second group of cells discussed are responsible for the excitatory responses in human physiology and include both neuronal and muscle cells. A fundamental understanding of the inner workings of these cells, and how they are controlled by their neighboring cells helps the student to understand their eventual integration into individual systems discussed in later sections.

In the end, this first section serves as an introduction, refresher, and quick source of material to best understand systems physiology presented in the later sections. For detailed understanding of any of the chapters within this section, several excellent and current text books that provide a more in depth review of principles of biochemistry, biophysics, cell physiology, muscle and neuronal physiology are provided as resources at the end of each individual chapter. Students who are intrigued by the overview provided in this first section are encouraged to visit these texts for a more thorough understanding of these basic principles.

General Principles & Energy Production in Medical Physiology

CHAPTER 1

OBJECTIVES

After studying this chapter, you should be able to:

- Define units used in measuring physiological properties.
- Define pH and buffering.
- Understand electrolytes and define diffusion, osmosis, and tonicity.
- Define and explain the significance of resting membrane potential.
- Understand in general terms the basic building blocks of the cell: nucleotides, amino acids, carbohydrates, and fatty acids.
- Understand higher-order structures of the basic building blocks: DNA, RNA, proteins, and lipids.
- Understand the basic contributions of the basic building blocks to cell structure, function, and energy balance.

INTRODUCTION

In unicellular organisms, all vital processes occur in a single cell. As the evolution of multicellular organisms has progressed, various cell groups organized into tissues and organs have taken over particular functions. In humans and other vertebrate animals, the specialized cell groups include a gastrointestinal system to digest and absorb food; a respiratory system to take up O_2 and eliminate CO_2; a urinary system to remove wastes; a cardiovascular system to distribute nutrients, O_2, and the products of metabolism; a reproductive system to perpetuate the species; and nervous and endocrine systems to coordinate and integrate the functions of the other systems. This book is concerned with the way these systems function and the way each contributes to the functions of the body as a whole. This first chapter focuses on a review of basic biophysical and biochemical principles and the introduction of the molecular building blocks that contribute to cellular physiology.

GENERAL PRINCIPLES

THE BODY AS ORGANIZED "SOLUTIONS"

The cells that make up the bodies of all but the simplest multicellular animals, both aquatic and terrestrial, exist in an "internal sea" of **extracellular fluid (ECF)** enclosed within the integument of the animal. From this fluid, the cells take up O_2 and nutrients; into it, they discharge metabolic waste products. The ECF is more dilute than present-day seawater, but its composition closely resembles that of the primordial oceans in which, presumably, all life originated.

In animals with a closed vascular system, the ECF is divided into the **interstitial fluid** the circulating **blood plasma** and **the lymph fluid that bridges these two domains.** The plasma and the cellular elements of the blood, principally red blood cells, fill the vascular system, and together they constitute the **total blood volume.** The interstitial fluid is that part of the ECF that is outside the vascular and lymph systems, bathing the cells. About a third of the **total body water** is extracellular; the remaining two thirds is intracellular **(intracellular fluid).** Inappropriate compartmentalization of the body fluids can result in edema (Clinical Box 1–1). In the average young adult male, 18% of the body weight is protein and related substances, 7% is mineral, and 15% is fat.

CLINICAL BOX 1–1

Edema

Edema is the build up of body fluids within tissues. The increased fluid is related to an increased leak from the blood and/or reduced removal by the lymph system. Edema is often observed in the feet, ankles, and legs, but can happen in many areas of the body in response to disease, including those of the heart, lung, liver, kidney, or thyroid.

THERAPEUTIC HIGHLIGHTS

The best treatment for edema includes reversing the underlying disorder. Thus, proper diagnosis of the cause of edema is the primary first step in therapy. More general treatments include restricting dietary sodium to minimize fluid retention, and employing appropriate diuretic therapy.

The remaining 60% is water. The distribution of this water is shown in Figure 1–1A.

The intracellular component of the body water accounts for about 40% of body weight and the extracellular component for about 20%. Approximately 25% of the extracellular component is in the vascular system (plasma = 5% of body weight) and 75% outside the blood vessels (interstitial fluid = 15% of body weight). The total blood volume is about 8% of body weight. Flow between these compartments is tightly regulated.

UNITS FOR MEASURING CONCENTRATION OF SOLUTES

In considering the effects of various physiologically important substances and the interactions between them, the number of molecules, electric charges, or particles of a substance per unit volume of a particular body fluid are often more meaningful than simply the weight of the substance per unit volume. For this reason, physiological concentrations are frequently expressed in moles, equivalents, or osmoles.

Moles

A mole is the gram-molecular weight of a substance, that is, the molecular weight of the substance in grams. Each mole (mol) consists of 6×10^{23} molecules. The millimole (mmol) is 1/1000 of a mole, and the micromole (μmol) is 1/1,000,000 of a mole. Thus, 1 mol of NaCl = 23 g + 35.5 g = 58.5 g and 1 mmol = 58.5 mg. The mole is the standard unit for expressing the amount of substances in the SI unit system.

The molecular weight of a substance is the ratio of the mass of one molecule of the substance to the mass of one twelfth the mass of an atom of carbon-12. Because molecular weight is a ratio, it is dimensionless. The dalton (Da) is a unit of mass equal to one twelfth the mass of an atom of carbon-12. The kilodalton (kDa = 1000 Da) is a useful unit for expressing the molecular mass of proteins. Thus, for example, one can speak of a 64-kDa protein or state that the molecular mass of the protein is 64,000 Da. However, because molecular weight is a dimensionless ratio, it is incorrect to say that the molecular weight of the protein is 64 kDa.

Equivalents

The concept of electrical equivalence is important in physiology because many of the solutes in the body are in the form of charged particles. One equivalent (eq) is 1 mol of an ionized substance divided by its valence. One mole of NaCl dissociates into 1 eq of Na^+ and 1 eq of Cl^-. One equivalent of Na^+ = 23 g, but 1 eq of Ca^{2+} = 40 g/2 = 20 g. The milliequivalent (meq) is 1/1000 of 1 eq.

Electrical equivalence is not necessarily the same as chemical equivalence. A gram equivalent is the weight of a substance that is chemically equivalent to 8.000 g of oxygen. The normality (N) of a solution is the number of gram equivalents in 1 L. A 1 N solution of hydrochloric acid contains both H^+ (1 g) and Cl^- (35.5 g) equivalents, = (1 g + 35.5 g)/L = 36.5 g/L.

WATER, ELECTROLYTES, & ACID/BASE

The water molecule (H_2O) is an ideal solvent for physiological reactions. H_2O has a **dipole moment** where oxygen slightly pulls away electrons from the hydrogen atoms and creates a charge separation that makes the molecule **polar.** This allows water to dissolve a variety of charged atoms and molecules. It also allows the H_2O molecule to interact with other H_2O molecules via hydrogen bonding. The resulting hydrogen bond network in water allows for several key properties relevant to physiology: (1) water has a high surface tension, (2) water has a high heat of vaporization and heat capacity, and (3) water has a high dielectric constant. In layman's terms, H_2O is an excellent biological fluid that serves as a solute; it provides optimal heat transfer and conduction of current.

Electrolytes (eg, NaCl) are molecules that dissociate in water to their cation (Na^+) and anion (Cl^-) equivalents. Because of the net charge on water molecules, these electrolytes tend not to reassociate in water. There are many important electrolytes in physiology, notably Na^+, K^+, Ca^{2+}, Mg^{2+}, Cl^-, and HCO_3^-. It is important to note that electrolytes and other charged compounds (eg, proteins) are unevenly distributed in the body fluids (Figure 1–1B). These separations play an important role in physiology.

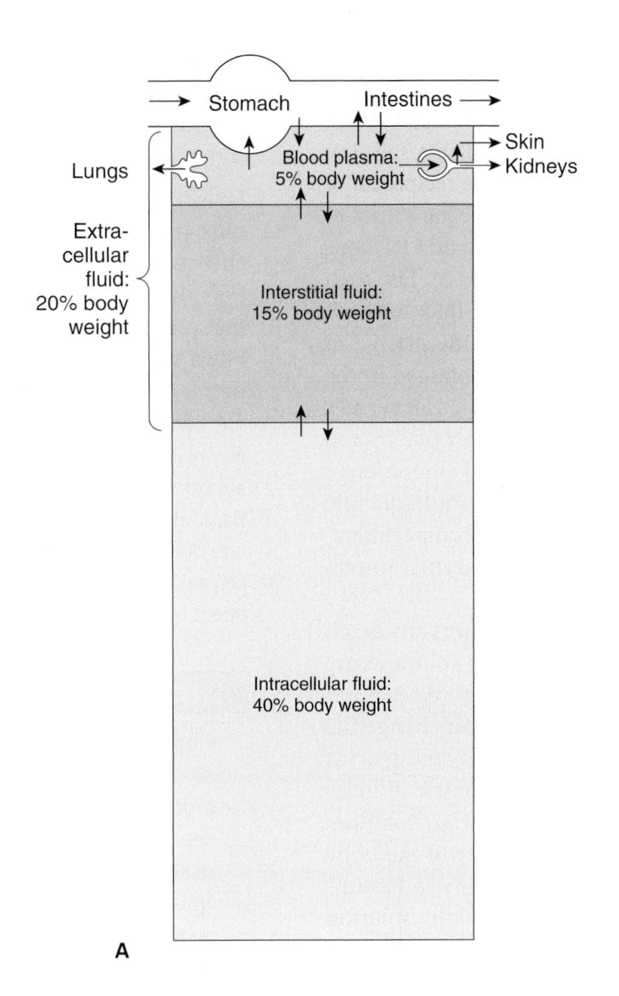

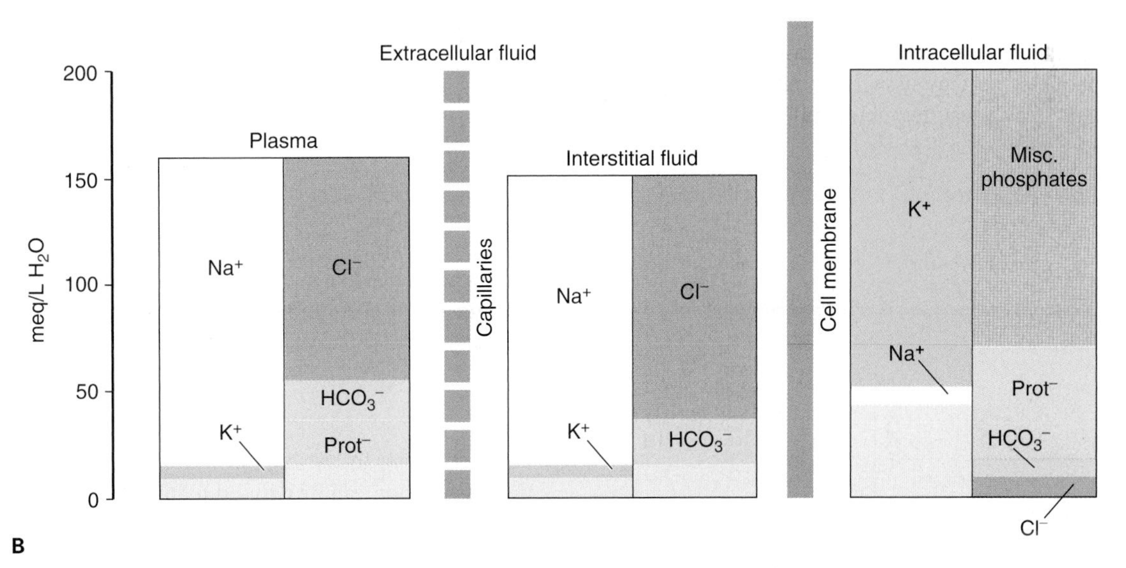

FIGURE 1–1 Organization of body fluids and electrolytes into compartments. A) Body fluids can be divided into Intracellular and Extracellular fluid compartments (ICF and ECF, respectively). Their contribution to percentage body weight (based on a healthy young adult male; slight variations exist with age and gender) emphasizes the dominance of fluid makeup of the body. Transcellular fluids, which constitute a very small percentage of total body fluids, are not shown. Arrows represent fluid movement between compartments. **B)** Electrolytes and proteins are unequally distributed among the body fluids. This uneven distribution is crucial to physiology. $Prot^-$, protein, which tends to have a negative charge at physiologic pH.

pH AND BUFFERING

The maintenance of a stable hydrogen ion concentration ($[H^+]$) in body fluids is essential to life. The **pH** of a solution is defined as the logarithm to the base 10 of the reciprocal of the H^+ concentration ($[H^+]$), that is, the negative logarithm of the $[H^+]$. The pH of water at 25°C, in which H^+ and OH^- ions are present in equal numbers, is 7.0 (Figure 1–2). For each pH unit less than 7.0, the $[H^+]$ is increased 10-fold; for each pH unit above 7.0, it is decreased 10-fold. In the plasma of healthy individuals, pH is slightly alkaline, maintained in the narrow range of 7.35–7.45 (Clinical Box 1–2). Conversely, gastric fluid pH can be quite acidic (on the order of 3.0) and pancreatic secretions can be quite alkaline (on the order of 8.0). Enzymatic activity and protein structure are frequently sensitive to pH; in any given body or cellular compartment, pH is maintained to allow for maximal enzyme/protein efficiency.

Molecules that act as H^+ donors in solution are considered acids, while those that tend to remove H^+ from solutions are considered bases. Strong acids (eg, HCl) or bases (eg, NaOH) dissociate completely in water and thus can most change the $[H^+]$ in solution. In physiological compounds, most acids or bases are considered "weak," that is, they contribute relatively few H^+ or take away relatively few H^+ from solution. Body pH is stabilized by the **buffering capacity** of the body fluids. A **buffer** is a substance that has the ability to bind or release H^+ in solution, thus keeping the pH of the solution relatively constant despite the addition of considerable quantities of acid or base. Of course there are a number of buffers at work in biological fluids at any given time. All buffer pairs in a homogenous solution are in equilibrium with the same $[H^+]$; this is known as the **isohydric principle.** One outcome of this principle is that by assaying a single buffer system, we can understand a great deal about all of the biological buffers in that system.

CLINICAL BOX 1–2

Acid–Base Disorders

Excesses of acid (acidosis) or base (alkalosis) exist when the blood is outside the normal pH range (7.35–7.45). Such changes impare the delivery of O_2 to and removal of CO_2 from tissues. There are a variety of conditions and diseases that can interfere with pH control in the body and cause blood pH to fall outside of healthy limits. Acid–base disorders that result from respiration to alter CO_2 concentration are called respiratory acidosis and respiratory alkalosis. Nonrespiratory disorders that affect HCO_3^- concentration are refereed to as metabolic acidosis and metabolic alkalosis. Metabolic acidosis/alkalosis can be caused by electrolyte disturbances, severe vomiting or diarrhea, ingestion of certain drugs and toxins, kidney disease, and diseases that affect normal metabolism (eg, diabetes).

THERAPEUTIC HIGHLIGHTS

Proper treatments for acid–base disorders are dependent on correctly identifying the underlying causal process(es). This is especially true when mixed disorders are encountered. Treatment of respiratory acidosis should be initially targeted at restoring ventilation, whereas treatment for respiratory alkalosis is focused on the reversal of the root cause. Bicarbonate is typically used as a treatment for acute metabolic acidosis. An adequate amount of a chloride salt can restore acid–base balance to normal over a matter of days for patients with a chloride-responsive metabolic alkalosis whereas chloride-resistant metabolic alkalosis requires treatment of the underlying disease.

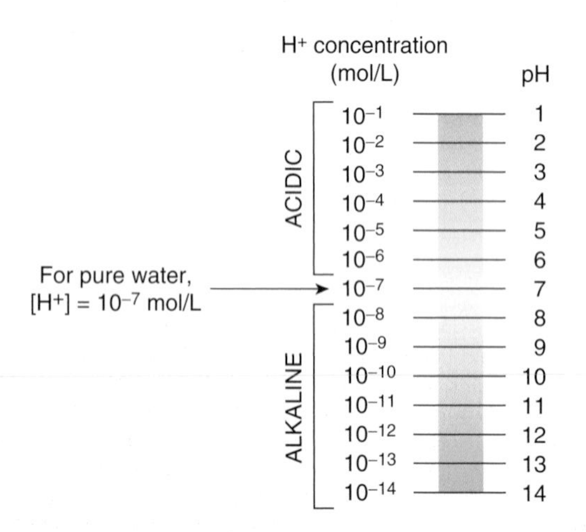

FIGURE 1–2 Proton concentration and pH. Relative proton (H^+) concentrations for solutions on a pH scale are shown.

When acids are placed into solution, there is dissociation of some of the component acid (HA) into its proton (H^+) and free acid (A^-). This is frequently written as an equation:

$$HA \rightleftarrows H^+ + A^-.$$

According to the laws of mass action, a relationship for the dissociation can be defined mathematically as:

$$K_a = [H^+][A^-]/[HA]$$

where K_a is a constant, and the brackets represent concentrations of the individual species. In layman's terms, the product of the proton concentration ($[H^+]$) times the free acid concentration ($[A^-]$) divided by the bound acid concentration ($[HA]$) is a defined constant (K). This can be rearranged to read:

$$[H^+] = K_a [HA]/[A^-]$$

If the logarithm of each side is taken:

$$\log[H^+] = \log K_a + \log[HA]/[A^-]$$

Both sides can be multiplied by −1 to yield:

$$-\log[H^+] = -\log K_a + \log[A^-]/[HA]$$

This can be written in a more conventional form known as the **Henderson Hasselbalch equation:**

$$pH = pK_a + \log[A^-]/[HA]$$

This relatively simple equation is quite powerful. One thing that we can discern right away is that the buffering capacity of a particular weak acid is best when the pK_a of that acid is equal to the pH of the solution, or when:

$$[A^-] = [HA], pH = pK_a$$

Similar equations can be set up for weak bases. An important buffer in the body is carbonic acid. Carbonic acid is a weak acid, and thus is only partly dissociated into H^+ and bicarbonate:

$$H_2CO_3 \leftrightarrow H^+ + HCO_3^-$$

If H^+ is added to a solution of carbonic acid, the equilibrium shifts to the left and most of the added H^+ is removed from solution. If OH^- is added, H^+ and OH^- combine, taking H^+ out of solution. However, the decrease is countered by more dissociation of H_2CO_3, and the decline in H^+ concentration is minimized. A unique feature of bicarbonate is the linkage between its buffering ability and the ability for the lungs to remove carbon dioxide from the body. Other important biological buffers include phosphates and proteins.

DIFFUSION

Diffusion is the process by which a gas or a substance in a solution expands, because of the motion of its particles, to fill all the available volume. The particles (molecules or atoms) of a substance dissolved in a solvent are in continuous random movement. A given particle is equally likely to move into or out of an area in which it is present in high concentration. However, because there are more particles in the area of high concentration, the total number of particles moving to areas of lower concentration is greater; that is, there is a **net flux** of solute particles from areas of high to areas of low concentration. The time required for equilibrium by diffusion is proportional to the square of the diffusion distance. The magnitude of the diffusing tendency from one region to another is directly proportional to the cross-sectional area across which diffusion is taking place and the **concentration gradient,** or **chemical gradient,** which is the difference in concentration of the diffusing substance divided by the thickness of the boundary **(Fick's law of diffusion).** Thus,

$$J = -DA\frac{\Delta c}{\Delta x}$$

where J is the net rate of diffusion, D is the diffusion coefficient, A is the area, and $\Delta c/\Delta x$ is the concentration gradient. The minus sign indicates the direction of diffusion. When considering movement of molecules from a higher to a lower concentration, $\Delta c/\Delta x$ is negative, so multiplying by −DA gives a positive value. The permeabilities of the boundaries across which diffusion occurs in the body vary, but diffusion is still a major force affecting the distribution of water and solutes.

OSMOSIS

When a substance is dissolved in water, the concentration of water molecules in the solution is less than that in pure water, because the addition of solute to water results in a solution that occupies a greater volume than does the water alone. If the solution is placed on one side of a membrane that is permeable to water but not to the solute, and an equal volume of water is placed on the other, water molecules diffuse down their concentration (chemical) gradient into the solution (Figure 1–3). This process—the diffusion of **solvent** molecules into a region in which there is a higher concentration of a **solute** to which the membrane is impermeable—is called **osmosis.** It is an important factor in physiologic processes. The tendency for movement of solvent molecules to a region of greater solute concentration can be prevented by applying pressure to the more concentrated solution. The pressure necessary to prevent solvent migration is the **osmotic pressure** of the solution.

Osmotic pressure—like vapor pressure lowering, freezing-point depression, and boiling-point elevation—depends on the number rather than the type of particles in a solution; that is, it is a fundamental colligative property of solutions. In an **ideal solution,** osmotic pressure (P) is related to temperature and volume in the same way as the pressure of a gas:

$$P = \frac{nRT}{V}$$

where n is the number of particles, R is the gas constant, T is the absolute temperature, and V is the volume. If T is held

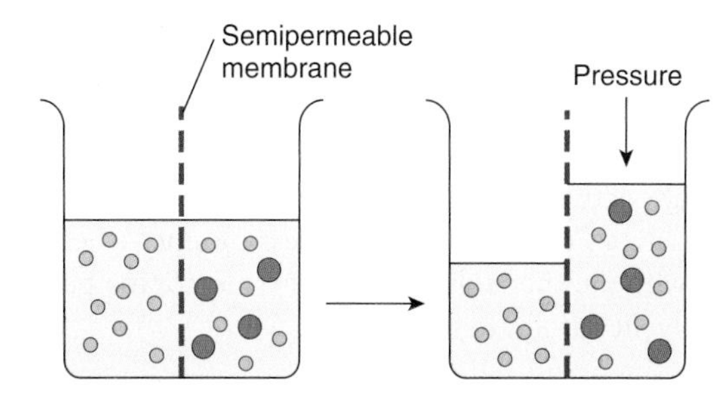

FIGURE 1–3 Diagrammatic representation of osmosis. Water molecules are represented by small open circles, and solute molecules by large solid circles. In the diagram on the left, water is placed on one side of a membrane permeable to water but not to solute, and an equal volume of a solution of the solute is placed on the other. Water molecules move down their concentration (chemical) gradient into the solution, and, as shown in the diagram on the right, the volume of the solution increases. As indicated by the arrow on the right, the osmotic pressure is the pressure that would have to be applied to prevent the movement of the water molecules.

constant, it is clear that the osmotic pressure is proportional to the number of particles in solution per unit volume of solution. For this reason, the concentration of osmotically active particles is usually expressed in **osmoles.** One osmole (Osm) equals the gram-molecular weight of a substance divided by the number of freely moving particles that each molecule liberates in solution. For biological solutions, the milliosmole (mOsm; 1/1000 of 1 Osm) is more commonly used.

If a solute is a nonionizing compound such as glucose, the osmotic pressure is a function of the number of glucose molecules present. If the solute ionizes and forms an ideal solution, each ion is an osmotically active particle. For example, NaCl would dissociate into Na^+ and Cl^- ions, so that each mole in solution would supply 2 Osm. One mole of Na_2SO_4 would dissociate into Na^+, Na^+, and SO_4^{2-} supplying 3 Osm. However, the body fluids are not ideal solutions, and although the dissociation of strong electrolytes is complete, the number of particles free to exert an osmotic effect is reduced owing to interactions between the ions. Thus, it is actually the effective concentration **(activity)** in the body fluids rather than the number of equivalents of an electrolyte in solution that determines its osmotic capacity. This is why, for example, 1 mmol of NaCl per liter in the body fluids contributes somewhat less than 2 mOsm of osmotically active particles per liter. The more concentrated the solution, the greater the deviation from an ideal solution.

The osmolal concentration of a substance in a fluid is measured by the degree to which it depresses the freezing point, with 1 mol of an ideal solution depressing the freezing point by 1.86°C. The number of milliosmoles per liter in a solution equals the freezing point depression divided by 0.00186. The **osmolarity** is the number of osmoles per liter of solution (eg, plasma), whereas the **osmolality** is the number of osmoles per kilogram of solvent. Therefore, osmolarity is affected by the volume of the various solutes in the solution and the temperature, while the osmolality is not. Osmotically active substances in the body are dissolved in water, and the density of water is 1, so osmolal concentrations can be expressed as osmoles per liter (Osm/L) of water. In this book, osmolal (rather than osmolar) concentrations are considered, and osmolality is expressed in milliosmoles per liter (of water).

Note that although a homogeneous solution contains osmotically active particles and can be said to have an osmotic pressure, it can exert an osmotic pressure only when it is in contact with another solution across a membrane permeable to the solvent but not to the solute.

OSMOLAL CONCENTRATION OF PLASMA: TONICITY

The freezing point of normal human plasma averages −0.54°C, which corresponds to an osmolal concentration in plasma of 290 mOsm/L. This is equivalent to an osmotic pressure against pure water of 7.3 atm. The osmolality might be expected to be higher than this, because the sum of all the cation and anion equivalents in plasma is over 300. It is not this high because plasma is not an ideal solution and ionic interactions reduce the number of particles free to exert an osmotic effect. Except when there has been insufficient time after a sudden change in composition for equilibrium to occur, all fluid compartments of the body are in (or nearly in) osmotic equilibrium. The term **tonicity** is used to describe the osmolality of a solution relative to plasma. Solutions that have the same osmolality as plasma are said to be **isotonic;** those with greater osmolality are **hypertonic;** and those with lesser osmolality are **hypotonic.** All solutions that are initially isosmotic with plasma (ie, that have the same actual osmotic pressure or freezing-point depression as plasma) would remain isotonic if it were not for the fact that some solutes diffuse into cells and others are metabolized. Thus, a 0.9% saline solution remains isotonic because there is no net movement of the osmotically active particles in the solution into cells and the particles are not metabolized. On the other hand, a 5% glucose solution is isotonic when initially infused intravenously, but glucose is metabolized, so the net effect is that of infusing a hypotonic solution.

It is important to note the relative contributions of the various plasma components to the total osmolal concentration of plasma. All but about 20 of the 290 mOsm in each liter of normal plasma are contributed by Na^+ and its accompanying anions, principally Cl^- and HCO_3^-. Other cations and anions make a relatively small contribution. Although the concentration of the plasma proteins is large when expressed in grams per liter, they normally contribute less than 2 mOsm/L because of their very high molecular weights. The major nonelectrolytes of plasma are glucose and urea, which in the steady state are in equilibrium with cells. Their contributions to osmolality are normally about 5 mOsm/L each but can become quite large in hyperglycemia or uremia. The total plasma osmolality is important in assessing dehydration, overhydration, and other fluid and electrolyte abnormalities (Clinical Box 1–3).

NONIONIC DIFFUSION

Some weak acids and bases are quite soluble in cell membranes in the undissociated form, whereas they cannot cross membranes in the charged (ie, dissociated) form. Consequently, if molecules of the undissociated substance diffuse from one side of the membrane to the other and then dissociate, there is appreciable net movement of the undissociated substance from one side of the membrane to the other. This phenomenon is called **nonionic diffusion.**

DONNAN EFFECT

When an ion on one side of a membrane cannot diffuse through the membrane, the distribution of other ions to which the membrane is permeable is affected in a predictable way. For example, the negative charge of a nondiffusible anion hinders

CLINICAL BOX 1–3

Plasma Osmolality & Disease

Unlike plant cells, which have rigid walls, animal cell membranes are flexible. Therefore, animal cells swell when exposed to extracellular hypotonicity and shrink when exposed to extracellular hypertonicity. Cells contain ion channels and pumps that can be activated to offset moderate changes in osmolality; however, these can be overwhelmed under certain pathologies. Hyperosmolality can cause coma (hyperosmolar coma). Because of the predominant role of the major solutes and the deviation of plasma from an ideal solution, one can ordinarily approximate the plasma osmolality within a few mosm/liter by using the following formula, in which the constants convert the clinical units to millimoles of solute per liter:

$$\text{Osmolality (mOsm/L)} = 2[\text{Na+}]\ (\text{mEq/L}) + 0.055[\text{Glucose}]\ (\text{mg/dL}) + 0.36[\text{BUN}]\ (\text{mg/dL})$$

BUN is the blood urea nitrogen. The formula is also useful in calling attention to abnormally high concentrations of other solutes. An observed plasma osmolality (measured by freezing-point depression) that greatly exceeds the value predicted by this formula probably indicates the presence of a foreign substance such as ethanol, mannitol (sometimes injected to shrink swollen cells osmotically), or poisons such as ethylene glycol (component of antifreeze) or methanol (alternative automotive fuel).

diffusion of the diffusible cations and favors diffusion of the diffusible anions. Consider the following situation,

X	m	Y
K^+	|	K^+
Cl^-	|	Cl^+
$Prot^-$	|	

in which the membrane (m) between compartments X and Y is impermeable to charged proteins ($Prot^-$) but freely permeable to K^+ and Cl^-. Assume that the concentrations of the anions and of the cations on the two sides are initially equal. Cl^- diffuses down its concentration gradient from Y to X, and some K^+ moves with the negatively charged Cl^- because of its opposite charge. Therefore

$$[K^+_X] > [K^+_Y]$$

Furthermore,

$$[K^+_X] + [Cl^-_X] + [Prot^-_X] > [K^+_Y] + [Cl^-_Y]$$

that is, more osmotically active particles are on side X than on side Y.

Donnan and Gibbs showed that in the presence of a nondiffusible ion, the diffusible ions distribute themselves so that at equilibrium their concentration ratios are equal:

$$\frac{[K^+_X]}{[K^+_Y]} = \frac{[Cl^-_Y]}{[Cl^-_X]}$$

Cross-multiplying,

$$[K^+_X] + [Cl^-_X] = [K^+_Y] + [Cl^-_Y]$$

This is the **Gibbs–Donnan equation.** It holds for any pair of cations and anions of the same valence.

The Donnan effect on the distribution of ions has three effects in the body introduced here and discussed below. First, because of charged proteins ($Prot^-$) in cells, there are more osmotically active particles in cells than in interstitial fluid, and because animal cells have flexible walls, osmosis would make them swell and eventually rupture if it were not for **Na, K ATPase** pumping ions back out of cells. Thus, normal cell volume and pressure depend on Na, K ATPase. Second, because at equilibrium the distribution of permeant ions across the membrane (m in the example used here) is asymmetric, an electrical difference exists across the membrane whose magnitude can be determined by the **Nernst equation.** In the example used here, side X will be negative relative to side Y. The charges line up along the membrane, with the concentration gradient for Cl^- exactly balanced by the oppositely directed electrical gradient, and the same holds true for K^+. Third, because there are more proteins in plasma than in interstitial fluid, there is a Donnan effect on ion movement across the capillary wall.

FORCES ACTING ON IONS

The forces acting across the cell membrane on each ion can be analyzed mathematically. Chloride ions (Cl^-) are present in higher concentration in the ECF than in the cell interior, and they tend to diffuse along this **concentration gradient** into the cell. The interior of the cell is negative relative to the exterior, and chloride ions are pushed out of the cell along this **electrical gradient.** An equilibrium is reached between Cl^- influx and Cl^- efflux. The membrane potential at which this equilibrium exists is the **equilibrium potential.** Its magnitude can be calculated from the Nernst equation, as follows:

$$E_{Cl} = \frac{RT}{FZ_{Cl}} \ln \frac{[Cl_o^-]}{[Cl_i^-]}$$

where

E_{Cl} = equilibrium potential for Cl^-
R = gas constant
T = absolute temperature
F = the Faraday number (number of coulombs per mole of charge)
Z_{Cl} = valence of Cl^- (−1)
$[Cl_o^-]$ = Cl^- concentration outside the cell
$[Cl_i^-]$ = Cl^- concentration inside the cell

Converting from the natural log to the base 10 log and replacing some of the constants with numerical values holding temperature at 37°C, the equation becomes:

$$E_{Cl} = 61.5 \log \frac{[Cl_i^-]}{[Cl_o^-]} \quad \text{(at 37°C)}.$$

Note that in converting to the simplified expression the concentration ratio is reversed because the −1 valence of Cl^- has been removed from the expression.

The equilibrium potential for Cl^- (E_{Cl}) in the mammalian spinal neuron, calculated from the standard values listed in Table 1–1, is −70 mV, a value identical to the typical measured resting membrane potential of −70 mV. Therefore, no forces other than those represented by the chemical and electrical gradients need be invoked to explain the distribution of Cl^- across the membrane.

A similar equilibrium potential can be calculated for K^+ (E_K; again, at 37°C):

$$E_K = \frac{RT}{FZ_K} \ln \frac{[K_o^+]}{[K_i^+]} = 61.5 \log \frac{[K_o^+]}{[K_i^+]} \quad \text{(at 37°C)}$$

where

E_K = equilibrium potential for K^+

Z_K = valence of K^+ (+1)

$[K_o^+]$ = K^+ concentration outside the cell

$[K_i^+]$ = K^+ concentration inside the cell

R, T, and F as above

In this case, the concentration gradient is outward and the electrical gradient inward. In mammalian spinal motor neurons E_K is −90 mV (Table 1–1). Because the resting membrane potential is −70 mV, there is somewhat more K^+ in the neurons that can be accounted for by the electrical and chemical gradients.

The situation for Na^+ in the mammalian spinal motor neuron is quite different from that for K^+ or Cl^-. The direction of the chemical gradient for Na^+ is inward, to the area where it is in lesser concentration, and the electrical gradient is in the same direction. E_{Na} is +60 mV (Table 1–1). Because neither E_K nor E_{Na} is equal to the membrane potential, one would expect the cell to gradually gain Na^+ and lose K^+ if only passive electrical and chemical forces were acting across the membrane. However, the intracellular concentration of Na^+ and K^+ remain constant because selective permeability and because of the action of the Na, K ATPase that actively transports Na^+ out of the cell and K^+ into the cell (against their respective electrochemical gradients).

TABLE 1–1 Concentration of some ions inside and outside mammalian spinal motor neurons.

	Concentration (mmol/L of H_2O)		
Ion	Inside Cell	Outside Cell	Equilibrium Potential (mV)
Na+	15.0	150.0	+60
K+	150.0	5.5	−90
Cl−	9.0	125.0	−70

Resting membrane potential = −70 mV

GENESIS OF THE MEMBRANE POTENTIAL

The distribution of ions across the cell membrane and the nature of this membrane provide the explanation for the membrane potential. The concentration gradient for K^+ facilitates its movement out of the cell via K^+ channels, but its electrical gradient is in the opposite (inward) direction. Consequently, an equilibrium is reached in which the tendency of K^+ to move out of the cell is balanced by its tendency to move into the cell, and at that equilibrium there is a slight excess of cations on the outside and anions on the inside. This condition is maintained by Na, K ATPase, which uses the energy of ATP to pump K^+ back into the cell and keeps the intracellular concentration of Na^+ low. Because the Na, K ATPase moves three Na^+ out of the cell for every two K^+ moved in, it also contributes to the membrane potential, and thus is termed an **electrogenic** pump. It should be emphasized that the number of ions responsible for the membrane potential is a minute fraction of the total number present and that the total concentrations of positive and negative ions are equal everywhere except along the membrane.

ENERGY PRODUCTION

ENERGY TRANSFER

Energy used in cellular processes is primarily stored in bonds between phosphoric acid residues and certain organic compounds. Because the energy of bond formation in some of these phosphates is particularly high, relatively large amounts of energy (10–12 kcal/mol) are released when the bond is hydrolyzed. Compounds containing such bonds are called **high-energy phosphate compounds.** Not all organic phosphates are of the high-energy type. Many, like glucose 6-phosphate, are low-energy phosphates that on hydrolysis liberate 2–3 kcal/mol. Some of the intermediates formed in carbohydrate metabolism are high-energy phosphates, but the most important high-energy phosphate compound is **adenosine triphosphate (ATP).** This ubiquitous molecule (Figure 1–4) is the energy storehouse of the body. On hydrolysis to adenosine diphosphate (ADP), it liberates energy directly to such processes as muscle contraction, active transport, and the synthesis of many chemical compounds. Loss of another phosphate to form adenosine monophosphate (AMP) releases more energy.

Another group of high-energy compounds are the thioesters, the acyl derivatives of mercaptans. **Coenzyme A (CoA)** is a widely distributed mercaptan-containing adenine, ribose, pantothenic acid, and thioethanolamine (Figure 1–5).

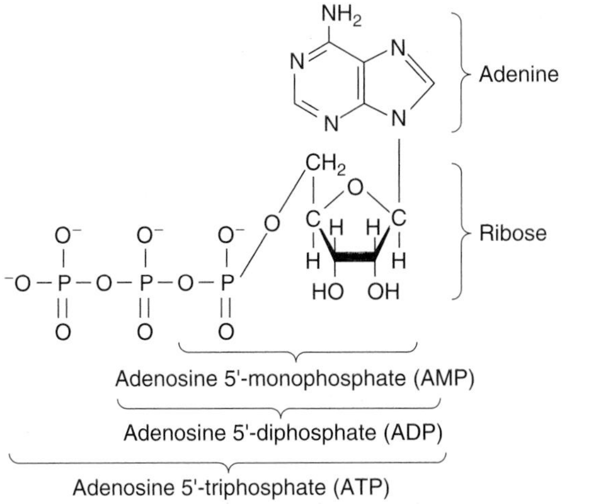

FIGURE 1–4 Energy-rich adenosine derivatives. Adenosine triphosphate is broken down into its backbone purine base and sugar (at right) as well as its high energy phosphate derivatives (across bottom). (Reproduced, with permission, from Murray RK et al: *Harper's Biochemistry*, 26th ed. McGraw-Hill, 2003.)

Reduced CoA (usually abbreviated HS–CoA) reacts with acyl groups (R–CO–) to form R–CO–S–CoA derivatives. A prime example is the reaction of HS-CoA with acetic acid to form acetylcoenzyme A (acetyl-CoA), a compound of pivotal importance in intermediary metabolism. Because acetyl-CoA has a much higher energy content than acetic acid, it combines readily with substances in reactions that would otherwise require outside energy. Acetyl-CoA is therefore often called "active acetate." From the point of view of energetics, formation of 1 mol of any acyl-CoA compound is equivalent to the formation of 1 mol of ATP.

BIOLOGIC OXIDATIONS

Oxidation is the combination of a substance with O_2, or loss of hydrogen, or loss of electrons. The corresponding reverse processes are called **reduction.** Biologic oxidations are catalyzed by specific enzymes. Cofactors (simple ions) or coenzymes (organic, nonprotein substances) are accessory substances that usually act as carriers for products of the reaction. Unlike the enzymes, the coenzymes may catalyze a variety of reactions.

A number of coenzymes serve as hydrogen acceptors. One common form of biologic oxidation is removal of hydrogen from an R–OH group, forming R=O. In such dehydrogenation reactions, nicotinamide adenine dinucleotide (NAD^+) and dihydronicotinamide adenine dinucleotide phosphate ($NADP^+$) pick up hydrogen, forming dihydronicotinamide adenine dinucleotide (NADH) and dihydronicotinamide adenine dinucleotide phosphate (NADPH) (Figure 1–6). The hydrogen is then transferred to the flavoprotein–cytochrome system, reoxidizing the NAD^+ and $NADP^+$. Flavin adenine dinucleotide (FAD) is formed when riboflavin is phosphorylated, forming flavin mononucleotide (FMN). FMN then combines with AMP, forming the dinucleotide. FAD can accept hydrogens in a similar fashion, forming its hydro (FADH) and dihydro ($FADH_2$) derivatives.

The flavoprotein–cytochrome system is a chain of enzymes that transfers hydrogen to oxygen, forming water. This process occurs in the mitochondria. Each enzyme in the chain is reduced and then reoxidized as the hydrogen is passed down the line. Each of the enzymes is a protein with an attached nonprotein prosthetic group. The final enzyme in the chain is cytochrome c oxidase, which transfers hydrogens to O_2, forming H_2O. It contains two atoms of Fe and three of Cu and has 13 subunits.

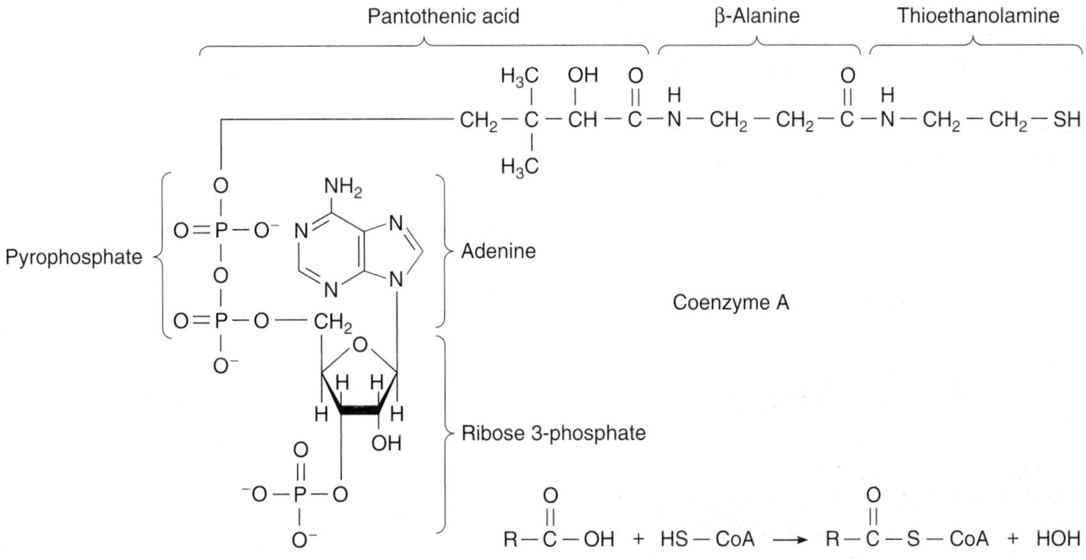

FIGURE 1–5 Coenzyme A (CoA) and its derivatives. Left: Formula of reduced coenzyme A (HS-CoA) with its components highlighted. **Right:** Formula for reaction of CoA with biologically important compounds to form thioesters. R, remainder of molecule.

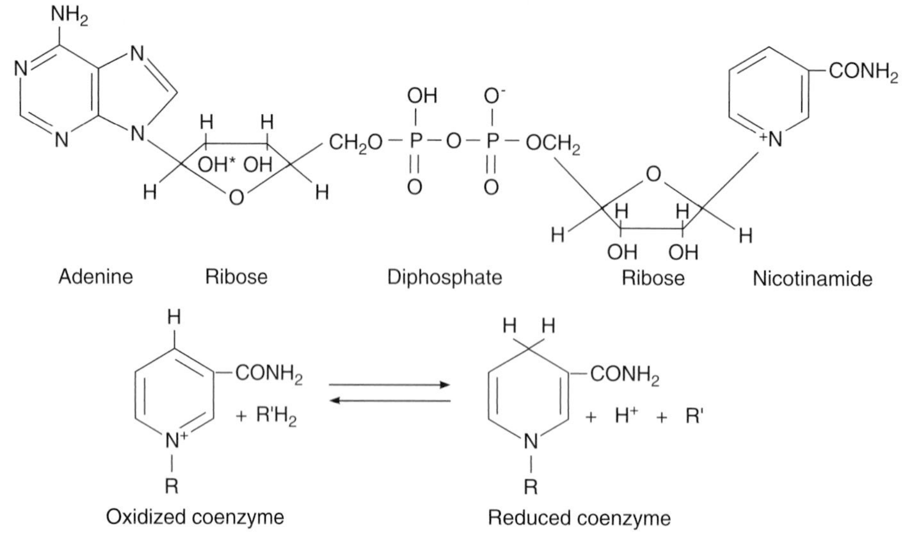

FIGURE 1–6 Structures of molecules important in oxidation reduction reactions to produce energy. Top: Formula of the oxidized form of nicotinamide adenine dinucleotide (NAD^+). Nicotinamide adenine dinucleotide phosphate ($NADP^+$) has an additional phosphate group at the location marked by the asterisk. **Bottom:** Reaction by which NAD^+ and $NADP^+$ become reduced to form NADH and NADPH. R, remainder of molecule; R′, hydrogen donor.

The principal process by which ATP is formed in the body is **oxidative phosphorylation.** This process harnesses the energy from a proton gradient across the mitochondrial membrane to produce the high-energy bond of ATP and is briefly outlined in Figure 1–7. Ninety per cent of the O_2 consumption in the basal state is mitochondrial, and 80% of this is coupled to ATP synthesis. ATP is utilized throughout the cell, with the bulk used in a handful of processes: approximately 27% is used for protein synthesis, 24% by Na, K ATPase to help set membrane potential, 9% by gluconeogenesis, 6% by Ca^{2+} ATPase, 5% by myosin ATPase, and 3% by ureagenesis.

MOLECULAR BUILDING BLOCKS

NUCLEOSIDES, NUCLEOTIDES, & NUCLEIC ACIDS

Nucleosides contain a sugar linked to a nitrogen-containing base. The physiologically important bases, **purines** and **pyrimidines,** have ring structures (Figure 1–8). These structures are bound to ribose or 2-deoxyribose to complete the nucleoside.

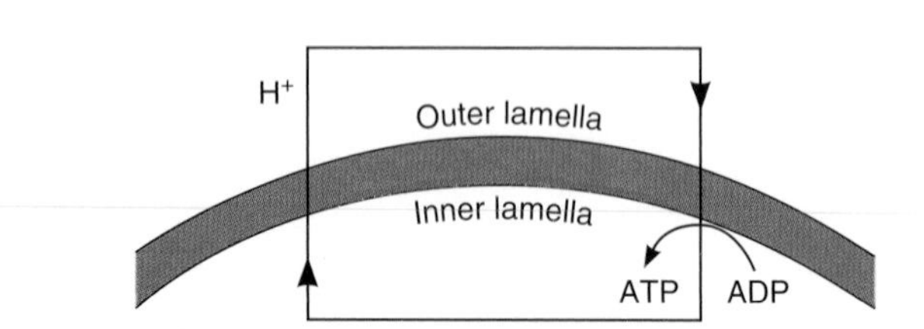

FIGURE 1–7 Simplified diagram of transport of protons across the inner and outer lamellas of the inner mitochondrial membrane. The electron transport system (flavoprotein-cytochrome system) helps create H^+ movement from the inner to the outer lamella. Return movement of protons down the proton gradient generates ATP.

When inorganic phosphate is added to the nucleoside, a **nucleotide** is formed. Nucleosides and nucleotides form the backbone for RNA and DNA, as well as a variety of coenzymes and regulatory molecules of physiological importance (eg, NAD^+, $NADP^+$, and ATP; Table 1–2). Nucleic acids in the diet are digested and their constituent purines and pyrimidines absorbed, but most of the purines and pyrimidines are synthesized from amino acids, principally in the liver. The nucleotides and RNA and DNA are then synthesized. RNA is in dynamic equilibrium with the amino acid pool, but DNA, once formed, is metabolically stable throughout life. The purines and pyrimidines released by the breakdown of nucleotides may be reused or catabolized. Minor amounts are excreted unchanged in the urine.

Purine nucleus

Adenine:	6-Aminopurine
Guanine:	1-Amino-6-oxypurine
Hypoxanthine:	6-Oxypurine
Xanthine:	2,6-Dioxypurine

Pyrimidine nucleus

Cytosine:	4-Amino-2-oxypyrimidine
Uracil:	2,4-Dioxypyrimidine
Thymine:	5-Methyl-2,4-dioxypyrimidine

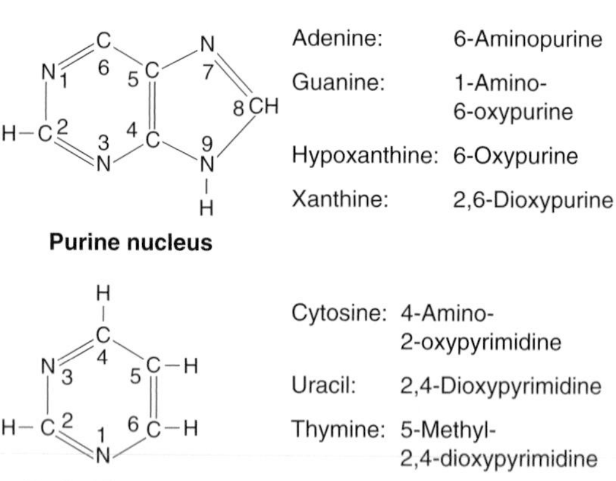

FIGURE 1–8 Principal physiologically important purines and pyrimidines. Purine and pyrimidine structures are shown next to representative molecules from each group. Oxypurines and oxypyrimidines may form enol derivatives (hydroxypurines and hydroxypyrimidines) by migration of hydrogen to the oxygen substituents.

TABLE 1–2 Purine- and pyrimidine-containing compounds.

Type of Compound	Components
Nucleoside	Purine or pyrimidine plus ribose or 2-deoxyribose
Nucleotide (mononucleotide)	Nucleoside plus phosphoric acid residue
Nucleic acid	Many nucleotides forming double-helical structures of two polynucleotide chains
Nucleoprotein	Nucleic acid plus one or more simple basic proteins
Contain ribose	Ribonucleic acids (RNA)
Contain 2-deoxyribose	Deoxyribonucleic acids (DNA)

The pyrimidines are catabolized to the **β-amino acids,** β-alanine and β-aminoisobutyrate. These amino acids have their amino group on β-carbon, rather than the α-carbon typical to physiologically active amino acids. Because β-aminoisobutyrate is a product of thymine degradation, it can serve as a measure of DNA turnover. The β-amino acids are further degraded to CO_2 and NH_3.

Uric acid is formed by the breakdown of purines and by direct synthesis from 5-phosphoribosyl pyrophosphate (5-PRPP) and glutamine (Figure 1–9). In humans, uric acid is excreted in the urine, but in other mammals, uric acid is further oxidized to allantoin before excretion. The normal blood uric acid level in humans is approximately 4 mg/dL (0.24 mmol/L). In the kidney, uric acid is filtered, reabsorbed, and secreted. Normally, 98% of the filtered uric acid is reabsorbed and the remaining 2% makes up approximately 20% of the amount excreted. The remaining 80% comes from the tubular secretion. The uric acid excretion on a purine-free diet is about 0.5 g/24 h and on a regular diet about 1 g/24 h. Excess uric acid in the blood or urine is a characteristic of gout (Clinical Box 1–4).

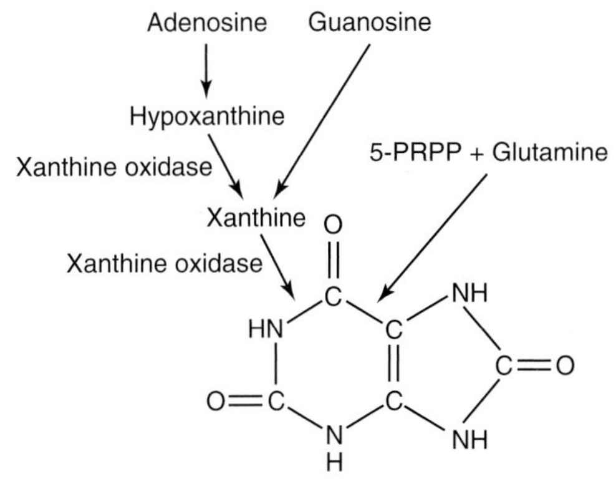

FIGURE 1–9 Synthesis and breakdown of uric acid. Adenosine is converted to hypoxanthine, which is then converted to xanthine, and xanthine is converted to uric acid. The latter two reactions are both catalyzed by xanthine oxidase. Guanosine is converted directly to xanthine, while 5-PRPP and glutamine can be converted to uric acid. An additional oxidation of uric acid to allantoin occurs in some mammals.

CLINICAL BOX 1–4

Gout

Gout is a disease characterized by recurrent attacks of arthritis; urate deposits in the joints, kidneys, and other tissues; and elevated blood and urine uric acid levels. The joint most commonly affected initially is the metatarsophalangeal joint of the great toe. There are two forms of "primary" gout. In one, uric acid production is increased because of various enzyme abnormalities. In the other, there is a selective deficit in renal tubular transport of uric acid. In "secondary" gout, the uric acid levels in the body fluids are elevated as a result of decreased excretion or increased production secondary to some other disease process. For example, excretion is decreased in patients treated with thiazide diuretics and those with renal disease. Production is increased in leukemia and pneumonia because of increased breakdown of uric acid-rich white blood cells.

THERAPEUTIC HIGHLIGHTS

The treatment of gout is aimed at relieving the acute arthritis with drugs such as colchicine or nonsteroidal anti-inflammatory agents and decreasing the uric acid level in the blood. Colchicine does not affect uric acid metabolism, and it apparently relieves gouty attacks by inhibiting the phagocytosis of uric acid crystals by leukocytes, a process that in some way produces the joint symptoms. Phenylbutazone and probenecid inhibit uric acid reabsorption in the renal tubules. Allopurinol, which directly inhibits xanthine oxidase in the purine degradation pathway, is one of the drugs used to decrease uric acid production.

DNA

Deoxyribonucleic acid (DNA) is found in bacteria, in the nuclei of eukaryotic cells, and in mitochondria. It is made up of two extremely long nucleotide chains containing the bases adenine (A), guanine (G), thymine (T), and cytosine (C) (Figure 1–10). The chains are bound together by hydrogen bonding between the bases, with adenine bonding to thymine and guanine to cytosine. This stable association forms a double-helical structure (Figure 1–11). The double helical structure of DNA is compacted in the cell by association with **histones,** and further compacted into **chromosomes.** A diploid human cell contains 46 chromosomes.

A fundamental unit of DNA, or a **gene,** can be defined as the sequence of DNA nucleotides that contain the information for the production of an ordered amino acid sequence for a single polypeptide chain. Interestingly, the protein encoded by a single gene may be subsequently divided into several different physiologically active proteins. Information is accumulating at an accelerating rate about the structure of genes and their regulation. The basic structure of a typical eukaryotic gene is shown in diagrammatic form in Figure 1–12. It is made up of a strand of DNA that includes coding and noncoding regions. In eukaryotes, unlike prokaryotes, the portions of the genes that dictate the formation of proteins are usually broken into several segments **(exons)** separated by segments that are not translated **(introns).** Near the transcription start site of the gene is a **promoter,** which is the site at which RNA polymerase and its cofactors bind. It often includes a thymidine–adenine–thymidine–adenine (TATA) sequence **(TATA box),** which ensures that transcription starts at the proper point. Farther out in the 5' region are **regulatory elements,** which include enhancer and silencer sequences. It has been estimated that each gene has an average of five regulatory sites. Regulatory sequences are sometimes found in the 3'-flanking region as well. In a diploid cell each gene will have two **alleles,** or versions of that gene. Each allele occupies the same position on the homologous chromosome. Individual alleles can confer slightly different properties of the gene when fully transcribed. It is interesting to note that changes in single nucleotides within or outside coding regions of a gene **(single nucleotide polymorphisms; SNPs)** can have great consequences for gene function. The study of SNPs in human disease is a growing and exciting area of genetic research.

Gene mutations occur when the base sequence in the DNA is altered from its original sequence. Alterations can be through insertions, deletions, or duplications. Such alterations can affect protein structure and be passed on to daughter cells after cell division. **Point mutations** are single base substitutions. A variety of chemical modifications (eg, alkylating or intercalating agents, or ionizing radiation) can lead to changes in DNA sequences and mutations. The collection of genes within the full expression of DNA from an organism is termed its **genome.** An indication of the complexity of DNA in the human haploid genome (the total genetic message) is its size; it is made up of 3×10^9 base pairs that can code for approximately 30,000 genes. This genetic message is the blueprint for the heritable characteristics of the cell and its descendants. The proteins formed from the DNA blueprint include all the enzymes, and these in turn control the metabolism of the cell.

Each nucleated somatic cell in the body contains the full genetic message, yet there is great differentiation and specialization in the functions of the various types of adult cells. Only small parts of the message are normally transcribed. Thus, the genetic message is normally maintained in a repressed state. However, genes are controlled both spatially and temporally. The double helix requires highly regulated interaction by proteins to unravel for **replication, transcription,** or both.

REPLICATION: MITOSIS & MEIOSIS

At the time of each somatic cell division **(mitosis),** the two DNA chains separate, each serving as a template for the synthesis of a new complementary chain. DNA polymerase catalyzes this reaction. One of the double helices thus formed goes to one daughter cell and one goes to the other, so the amount of DNA in each daughter cell is the same as that in the parent cell. The life cycle of the cell that begins after mitosis is highly regulated and is termed the **cell cycle** (Figure 1–13). The G_1 (or Gap 1) phase represents a period of cell growth and divides the end of mitosis from the DNA synthesis (or S) phase. Following DNA synthesis, the cell enters another period of cell growth, the G_2 (Gap 2) phase. The ending of this stage is marked by chromosome condensation and the beginning of mitosis (M stage).

In germ cells, reductive division **(meiosis)** takes place during maturation. The net result is that one of each pair of chromosomes ends up in each mature germ cell; consequently, each mature germ cell contains half the amount of chromosomal material found in somatic cells. Therefore, when a sperm unites with an ovum, the resulting zygote has the full complement of DNA, half of which came from the father and half from the mother. The term "ploidy" is sometimes used to refer to the number of chromosomes in cells. Normal resting diploid cells are **euploid** and become **tetraploid** just before division. **Aneuploidy** is the condition in which a cell contains other than the haploid number of chromosomes or an exact multiple of it, and this condition is common in cancerous cells.

RNA

The strands of the DNA double helix not only replicate themselves, but also serve as templates by lining up complementary bases for the formation in the nucleus of **ribonucleic acids (RNA).** RNA differs from DNA in that it is single-stranded, has **uracil** in place of thymine, and its sugar moiety is ribose rather than 2'-deoxyribose (Figure 1–10). The production of

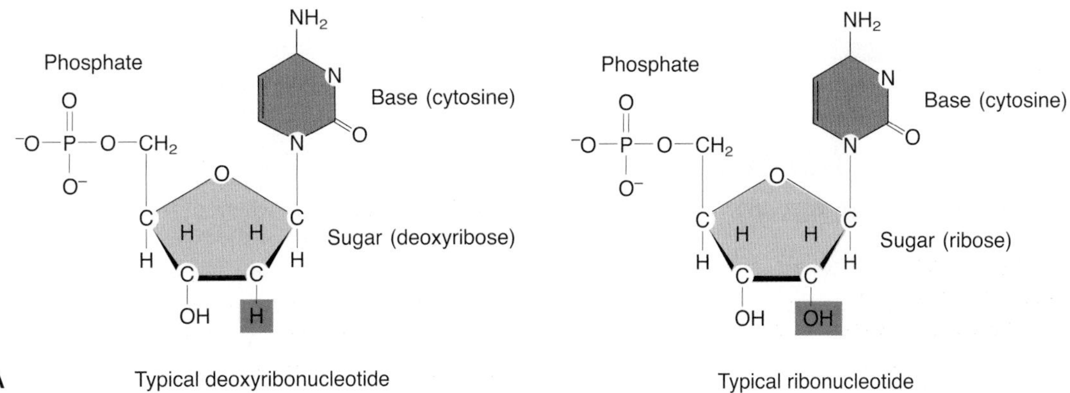

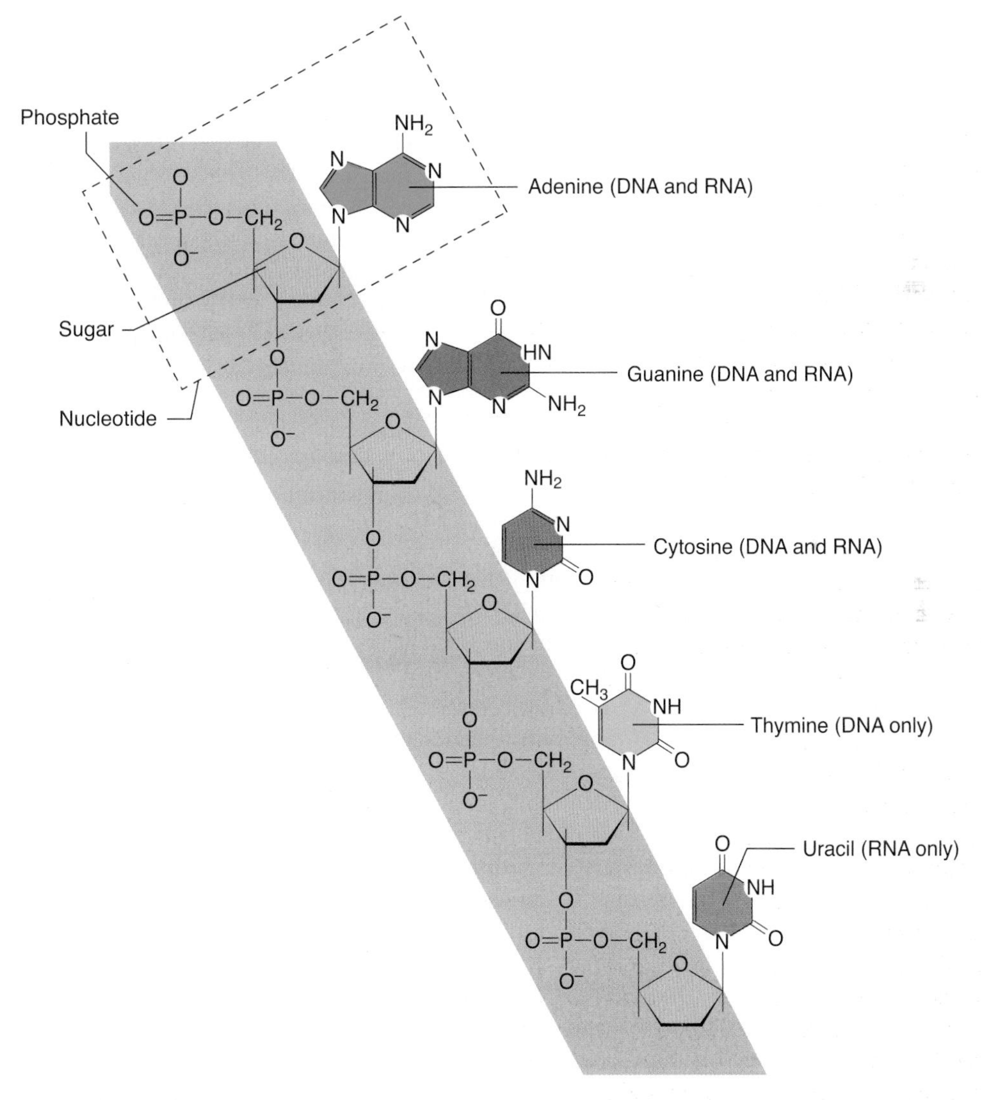

FIGURE 1–10 Basic structure of nucleotides and nucleic acids. A) At left, the nucleotide cytosine is shown with deoxyribose and at right with ribose as the principal sugar. **B)** Purine bases adenine and guanine are bound to each other or to pyrimidine bases, cytosine, thymine, or uracil via a phosphodiester backbone between 2′-deoxyribosyl moieties attached to the nucleobases by an N-glycosidic bond. Note that the backbone has a polarity (ie, a 5′ and a 3′ direction). Thymine is only found in DNA, while the uracil is only found in RNA.

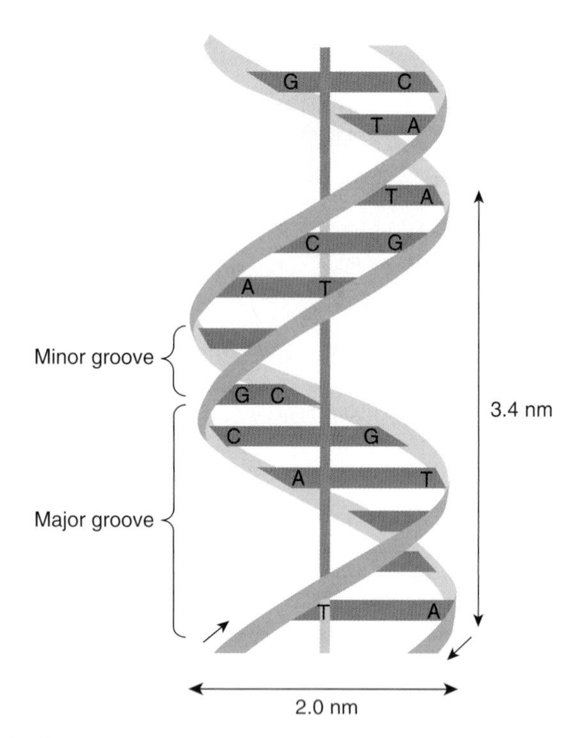

FIGURE 1–11 Double-helical structure of DNA. The compact structure has an approximately 2.0 nm thickness and 3.4 nm between full turns of the helix that contains both major and minor grooves. The structure is maintained in the double helix by hydrogen bonding between purines and pyrimidines across individual strands of DNA. Adenine (A) is bound to thymine (T) and cytosine (C) to guanine (G). (Reproduced with permission from Murray RK et al: *Harper's Biochemistry,* 28th ed. McGraw-Hill, 2009.)

RNA from DNA is called **transcription.** Transcription can lead to several types of RNA including: **messenger RNA (mRNA), transfer RNA (tRNA), ribosomal RNA (rRNA),** and other RNAs. Transcription is catalyzed by various forms of **RNA polymerase.**

Typical transcription of an mRNA is shown in Figure 1–14. When suitably activated, transcription of the gene into a pre-mRNA starts at the **cap site** and ends about 20 bases beyond the AATAAA sequence. The RNA transcript is capped in the nucleus by addition of 7-methylguanosine triphosphate to the 5' end; this cap is necessary for proper binding to the ribosome. A **poly(A) tail** of about 100 bases is added to the untranslated segment at the 3' end to help maintain the stability of the mRNA. The pre-mRNA formed by capping and addition of the poly(A) tail is then processed by elimination of the introns, and once this posttranscriptional modification is complete, the mature mRNA moves to the cytoplasm. Posttranscriptional modification of the pre-mRNA is a regulated process where differential splicing can occur to form more than one mRNA from a single pre-mRNA. The introns of some genes are eliminated by **spliceosomes,** complex units that are made up of small RNAs and proteins. Other introns are eliminated by **self-splicing** by the RNA they contain. Because of introns and splicing, more than one mRNA can be formed from the same gene.

Most forms of RNA in the cell are involved in **translation,** or protein synthesis. A brief outline of the transition from transcription to translation is shown in Figure 1–15. In the cytoplasm, ribosomes provide a template for tRNA to deliver specific amino acids to a growing polypeptide chain based on specific sequences in mRNA. The mRNA molecules are smaller than the DNA molecules, and each represents a transcript of a small segment of the DNA chain. For comparison, the molecules of tRNA contain only 70–80 nitrogenous bases, compared with hundreds in mRNA and 3 billion in DNA. A newer class of RNA, **microRNAs,** have recently been reported. MicroRNAs are small RNAs, approximately 21–25-nucleotides in length, that have been shown to negatively regulate gene expression at the posttranscriptional level. It is expected that roles for these small RNAs will continue to expand as research into their function continues.

AMINO ACIDS & PROTEINS

AMINO ACIDS

Amino acids that form the basic building blocks for proteins are identified in Table 1–3. These amino acids are often referred to by their corresponding three-letter, or single-letter abbreviations. Various other important amino acids such as ornithine, 5-hydroxytryptophan, L-dopa, taurine, and thyroxine (T_4) occur in the body but are not found in proteins. In higher animals, the L isomers of the amino

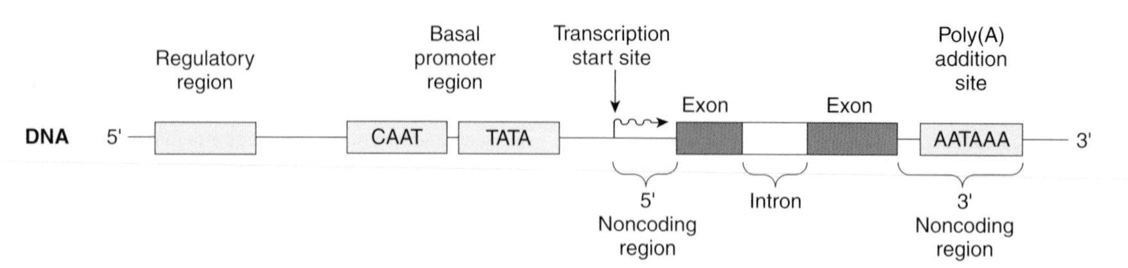

FIGURE 1–12 Diagram of the components of a typical eukaryotic gene. The region that produces introns and exons is flanked by noncoding regions. The 5′-flanking region contains stretches of DNA that interact with proteins to facilitate or inhibit transcription. The 3′-flanking region contains the poly(A) addition site. (Modified from Murray RK et al: *Harper's Biochemistry,* 26th ed. McGraw-Hill, 2003.)

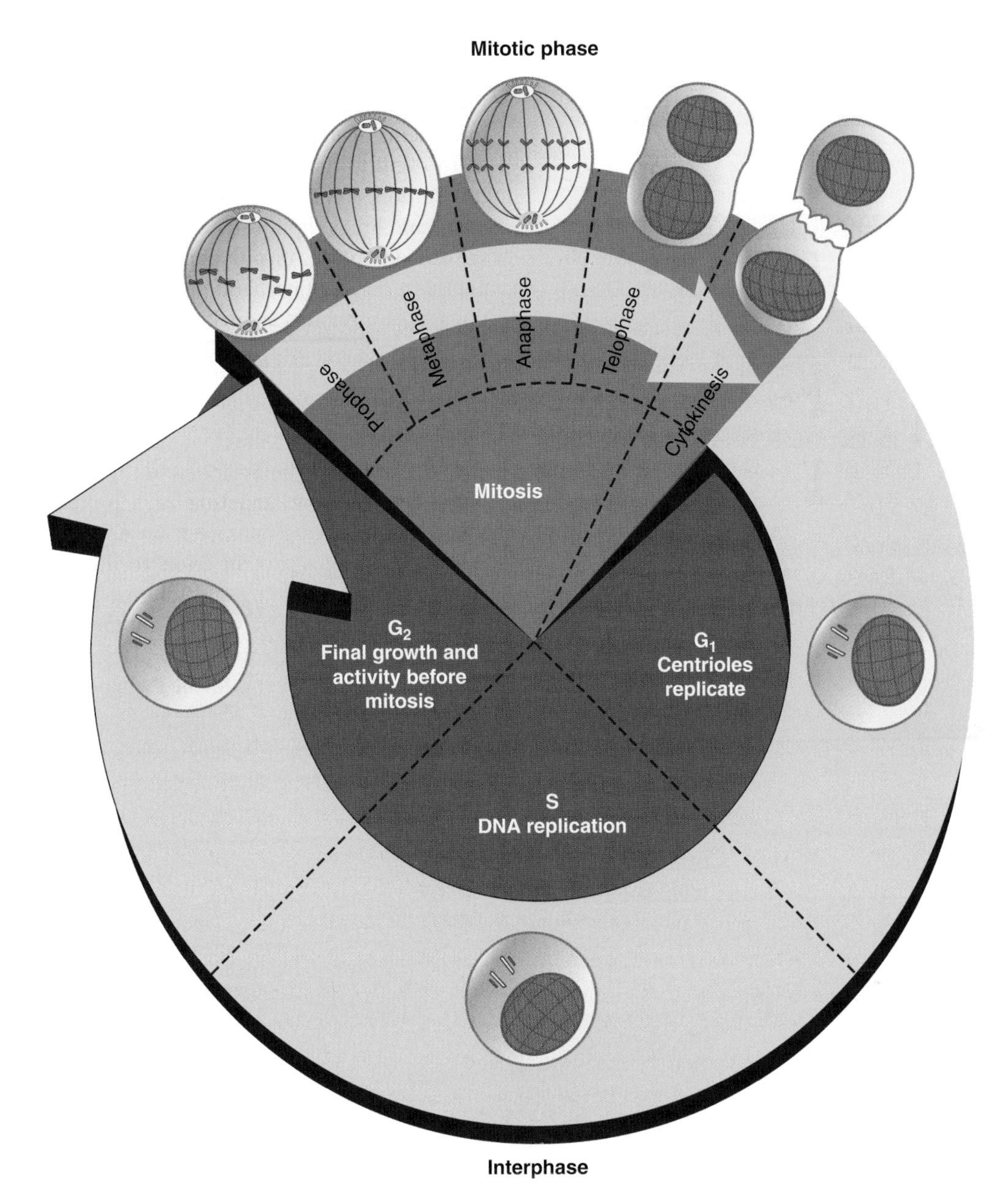

FIGURE 1–13 Sequence of events during the cell cycle. Immediately following mitosis (M) the cell enters a gap phase (G1) before a DNA synthesis phase (S) a second gap phase (G2) and back to mitosis. Collectively G1, S, and G2 phases are referred to as interphase (I).

acids are the only naturally occurring forms in proteins. The L isomers of hormones such as thyroxine are much more active than the D isomers. The amino acids are acidic, neutral, or basic, depending on the relative proportions of free acidic (–COOH) or basic ($–NH_2$) groups in the molecule. Some of the amino acids are **nutritionally essential amino acids,** that is, they must be obtained in the diet, because they cannot be made in the body. Arginine and histidine must be provided through diet during times of rapid growth or recovery from illness and are termed **conditionally essential.** All others are **nonessential amino acids** in the sense that they can be synthesized in vivo in amounts sufficient to meet metabolic needs.

THE AMINO ACID POOL

Although small amounts of proteins are absorbed from the gastrointestinal tract and some peptides are also absorbed, most ingested proteins are digested into their constituent amino acids before absorption. The body's proteins are being continuously hydrolyzed to amino acids and resynthesized. The turnover rate of endogenous proteins averages 80–100 g/d, being highest in the intestinal mucosa and practically nil in the extracellular structural protein, collagen. The amino acids formed by endogenous protein breakdown are identical to those derived from ingested protein. Together they form a common **amino acid pool** that supplies the needs of the body (Figure 1–16).

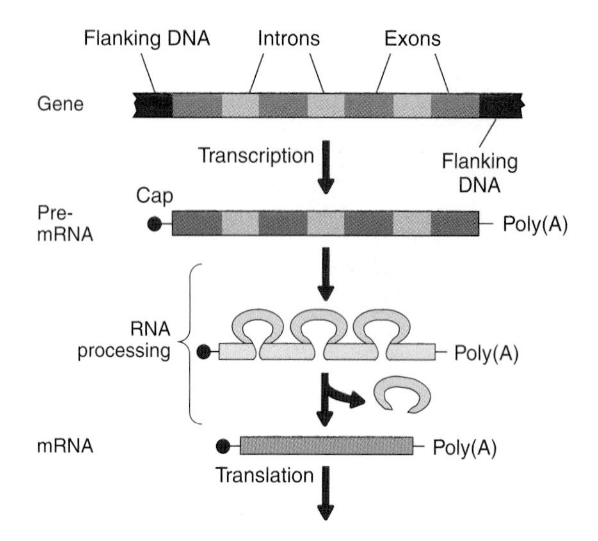

FIGURE 1–14 Transcription of a typical mRNA. Steps in transcription from a typical gene to a processed mRNA are shown. Cap, cap site. (Adapted from Nienhuis AW, et al: Thalassemia major: molecular and clinical aspects. NIH Conference Ann Intern Med 1979 Dec;91(6):883–897.)

PROTEINS

Proteins are made up of large numbers of amino acids linked into chains by **peptide bonds** joining the amino group of one amino acid to the carboxyl group of the next (Figure 1–17). In addition, some proteins contain carbohydrates (glycoproteins) and lipids (lipoproteins). Smaller chains of amino acids are called **peptides** or **polypeptides.** The boundaries between peptides, polypeptides, and proteins are not well defined. For this text, amino acid chains containing 2–10 amino acid residues are called peptides, chains containing more than 10 but fewer than 100 amino acid residues are called polypeptides, and chains containing 100 or more amino acid residues are called proteins.

The order of the amino acids in the peptide chains is called the **primary structure** of a protein. The chains are twisted and folded in complex ways, and the term **secondary structure** of a protein refers to the spatial arrangement produced by the twisting and folding. A common secondary structure is a regular coil with 3.7 amino acid residues per turn (α-helix). Another common secondary structure is a β-sheet. An anti-parallel β-sheet is formed when extended polypeptide

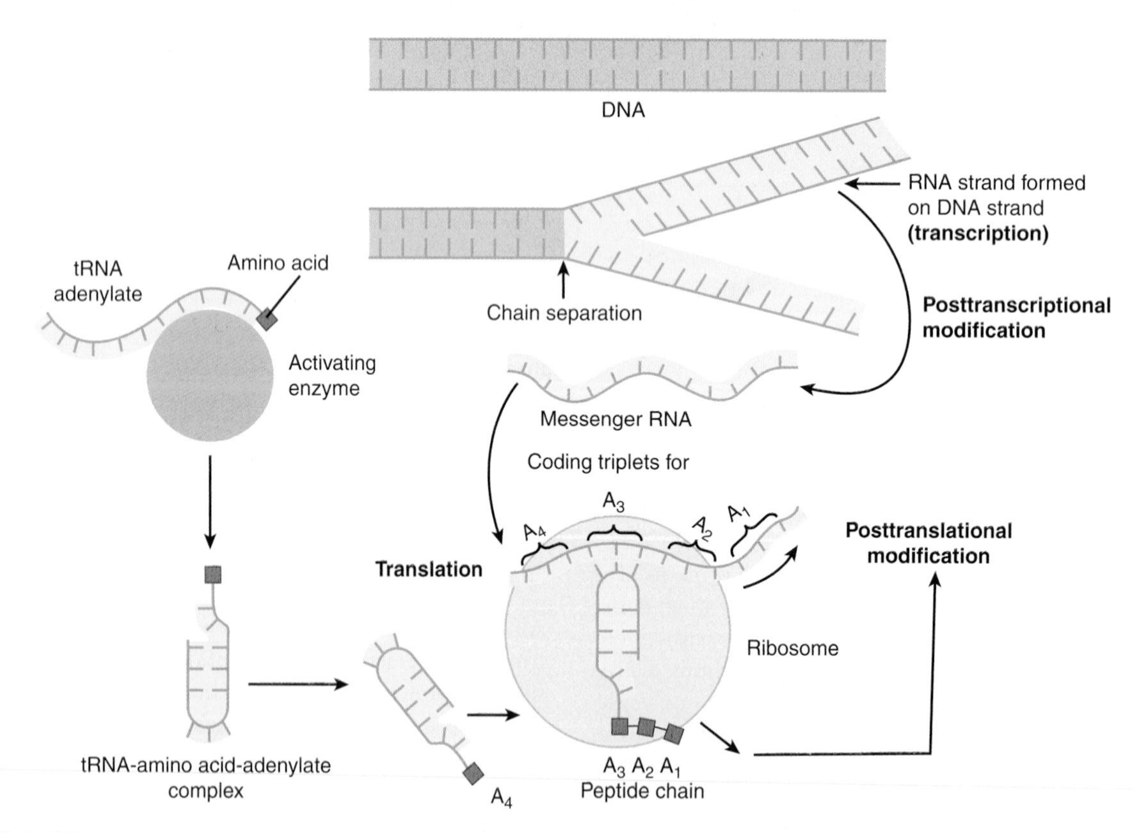

FIGURE 1–15 Diagrammatic outline of transcription to translation. From the DNA molecule, a messenger RNA is produced and presented to the ribosome. It is at the ribosome where charged tRNA match up with their complementary codons of mRNA to position the amino acid for growth of the polypeptide chain. DNA and RNA are represented as lines with multiple short projections representing the individual bases. Small boxes labeled A represent individual amino acids.

TABLE 1–3 Amino acids found in proteins.

Amino acids with aliphatic side chains	Amino acids with acidic side chains, or their amides
Alanine (Ala, A)	Aspartic acid (Asp, D)
Valine (Val, V)	Asparagine (Asn, N)
Leucine (Leu, L)	Glutamine (Gln, Q)
Isoleucine (Ile, I)	Glutamic acid (Glu, E)
Hydroxyl-substituted amino acids	γ-Carboxyglutamic acid[b] (Gla)
Serine (Ser, S)	Amino acids with side chains containing basic groups
Threonine (Thr, T)	**Arginine**[c] (Arg, R)
Sulfur-containing amino acids	**Lysine** (Lys, K)
Cysteine (Cys, C)	Hydroxylysine[b] (Hyl)
Methionine (Met, M)	**Histidine**[c] (His, H)
Selenocysteine[a]	Imino acids (contain imino group but no amino group)
Amino acids with aromatic ring side chains	Proline (Pro, P)
Phenylalanine (Phe, F)	4-Hydroxyproline[b] (Hyp)
Tyrosine (Tyr, Y)	3-Hydroxyproline[b]
Tryptophan (Trp, W)	

Those in bold type are the nutritionally essential amino acids. The generally accepted three-letter and one-letter abbreviations for the amino acids are shown in parentheses.

[a]Selenocysteine is a rare amino acid in which the sulfur of cysteine is replaced by selenium. The codon UGA is usually a stop codon, but in certain situations it codes for selenocysteine.

[b]There are no tRNAs for these four amino acids; they are formed by posttranslational modification of the corresponding unmodified amino acid in peptide linkage. There are tRNAs for selenocysteine and the remaining 20 amino acids, and they are incorporated into peptides and proteins under direct genetic control.

[c]Arginine and histidine are sometimes called "conditionally essential"—they are not necessary for maintenance of nitrogen balance, but are needed for normal growth.

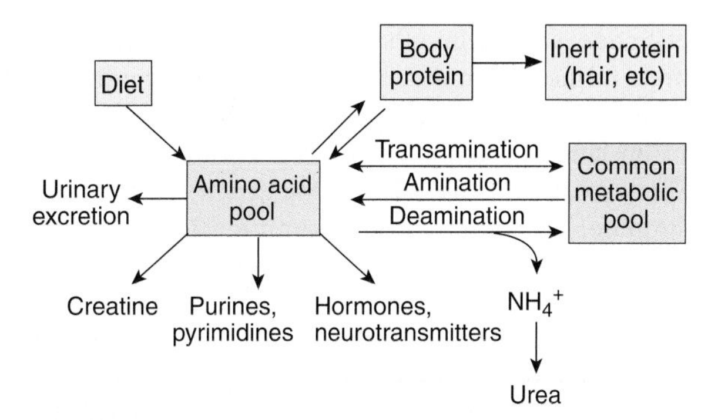

FIGURE 1–16 Amino acids in the body. There is an extensive network of amino acid turnover in the body. Boxes represent large pools of amino acids and some of the common interchanges are represented by arrows. Note that most amino acids come from the diet and end up in protein; however, a large portion of amino acids are interconverted and can feed into and out of a common metabolic pool through amination reactions.

chains fold back and forth on one another and hydrogen bonding occurs between the peptide bonds on neighboring chains. Parallel β-sheets between polypeptide chains also occur. The **tertiary structure** of a protein is the arrangement of the twisted chains into layers, crystals, or fibers. Many protein molecules are made of several proteins, or subunits (eg, hemoglobin), and the term **quaternary structure** is used to refer to the arrangement of the subunits into a functional structure.

PROTEIN SYNTHESIS

The process of protein synthesis, **translation,** is the conversion of information encoded in mRNA to a protein (Figure 1–15). As described previously, when a definitive mRNA reaches a ribosome in the cytoplasm, it dictates the formation of a polypeptide chain. Amino acids in the cytoplasm are activated by combination with an enzyme and adenosine monophosphate (adenylate), and each **activated amino acid** then combines with a specific molecule of tRNA. There is at least one tRNA for each of the 20 unmodified amino acids found in large quantities in the body proteins of animals, but some amino acids have more than one tRNA. The tRNA–amino acid–adenylate complex is next attached to the mRNA template, a process that occurs in the ribosomes. The tRNA "recognizes" the proper spot to attach on the mRNA template because it has on its active end a set of three bases that are complementary to a set of three bases in a particular spot on the mRNA chain. The

Amino acid   Polypeptide chain

FIGURE 1–17 Amino acid structure and formation of peptide bonds. The dashed line shows where peptide bonds are formed between two amino acids. The highlighted area is released as H_2O. R, remainder of the amino acid. For example, in glycine, R = H; in glutamate, R = —$(CH_2)_2$—COO^-.

genetic code is made up of such triplets **(codons),** sequences of three purine, pyrimidine, or purine and pyrimidine bases; each codon stands for a particular amino acid.

Translation typically starts in the ribosomes with an AUG (transcribed from ATG in the gene), which codes for methionine. The amino terminal amino acid is then added, and the chain is lengthened one amino acid at a time. The mRNA attaches to the 40S subunit of the ribosome during protein synthesis, the polypeptide chain being formed attaches to the 60S subunit, and the tRNA attaches to both. As the amino acids are added in the order dictated by the codon, the ribosome moves along the mRNA molecule like a bead on a string. Translation stops at one of three stop, or nonsense, codons (UGA, UAA, or UAG), and the polypeptide chain is released. The tRNA molecules are used again. The mRNA molecules are typically reused approximately 10 times before being replaced. It is common to have more than one ribosome on a given mRNA chain at a time. The mRNA chain plus its collection of ribosomes is visible under the electron microscope as an aggregation of ribosomes called a **polyribosome.**

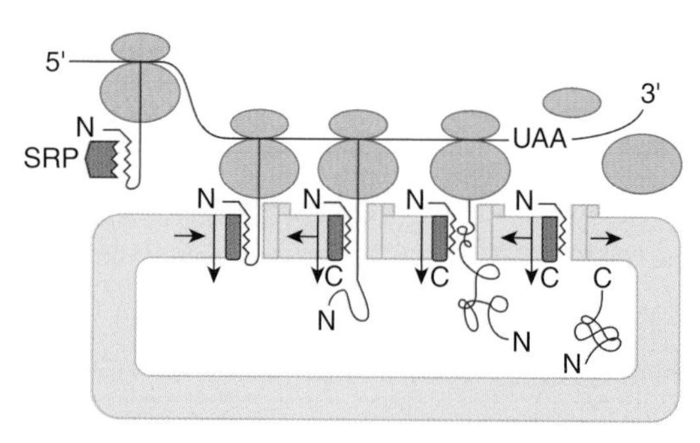

FIGURE 1–18 Translation of protein into the endoplasmic reticulum according to the signal hypothesis. The ribosomes synthesizing a protein move along the mRNA from the 5′ to the 3′ end. When the signal peptide of a protein destined for secretion, the cell membrane, or lysosomes emerges from the large unit of the ribosome, it binds to a signal recognition particle (SRP), and this arrests further translation until it binds to the translocon on the endoplasmic reticulum. N, amino end of protein; C, carboxyl end of protein. (Reproduced, with permission, from Perara E, Lingappa VR: Transport of proteins into and across the endoplasmic reticulum membrane. In: *Protein Transfer and Organelle Biogenesis.* Das RC, Robbins PW (editors). Academic Press, 1988.)

POSTTRANSLATIONAL MODIFICATION

After the polypeptide chain is formed, it "folds" into its biological form and can be further modified to the final protein by one or more of a combination of reactions that include hydroxylation, carboxylation, glycosylation, or phosphorylation of amino acid residues; cleavage of peptide bonds that converts a larger polypeptide to a smaller form; and the further folding, packaging, or folding and packaging of the protein into its ultimate, often complex configuration. Protein folding is a complex process that is dictated primarily by the sequence of the amino acids in the polypeptide chain. In some instances, however, nascent proteins associate with other proteins called **chaperones,** which prevent inappropriate contacts with other proteins and ensure that the final "proper" conformation of the nascent protein is reached.

Proteins also contain information that helps to direct them to individual cell compartments. Many proteins that are destined to be secreted or stored in organelles and most transmembrane proteins have at their amino terminal a **signal peptide (leader sequence)** that guides them into the endoplasmic reticulum. The sequence is made up of 15–30 predominantly hydrophobic amino acid residues. The signal peptide, once synthesized, binds to a **signal recognition particle (SRP),** a complex molecule made up of six polypeptides and 7S RNA, one of the small RNAs. The SRP stops translation until it binds to a **translocon,** a pore in the endoplasmic reticulum that is a heterotrimeric structure made up of Sec 61 proteins. The ribosome also binds, and the signal peptide leads the growing peptide chain into the cavity of the endoplasmic reticulum (Figure 1–18). The signal peptide is next cleaved from the rest of the peptide by a signal peptidase while the rest of the peptide chain is still being synthesized. SRPs are not the only signals that help to direct proteins to their proper place in or out of the cell; other signal sequences, posttranslational modifications, or both (eg, glycosylation) can serve this function.

UBIQITINATION AND PROTEIN DEGRADATION

Like protein synthesis, protein degradation is a carefully regulated, complex process. It has been estimated that overall, up to 30% of newly produced proteins are abnormal, such as can occur during improper folding. Aged normal proteins also need to be removed as they are replaced. Conjugation of proteins to the 74-amino-acid polypeptide **ubiquitin** marks them for degradation. This polypeptide is highly conserved and is present in species ranging from bacteria to humans. The process of binding ubiquitin is called **ubiquitination,** and in some instances, multiple ubiquitin molecules bind **(polyubiquitination).** Ubiquitination of cytoplasmic proteins, including integral proteins of the endoplasmic reticulum, can mark the proteins for degradation in multisubunit proteolytic particles, or **proteasomes.** Ubiquitination of membrane proteins, such as the growth hormone receptors, also marks them for degradation; however these can be degraded in lysosomes as well as via the proteasomes. Alteration of proteins by ubiquitin or the small ubiquitin-related modifier **(SUMO),** however, does not necessarily lead to degradation. More recently it has been shown that these posttranslational modifications can play important roles in protein–protein interactions and various cellular signaling pathways.

There is an obvious balance between the rate of production of a protein and its destruction, so ubiquitin conjugation is of major importance in cellular physiology. The rates at which individual proteins are metabolized vary, and the body has mechanisms by which abnormal proteins are recognized and degraded more rapidly than normal body constituents. For example, abnormal hemoglobins are metabolized rapidly in individuals with congenital hemoglobinopathies (see Chapter 31).

CATABOLISM OF AMINO ACIDS

The short-chain fragments produced by amino acid, carbohydrate, and fat catabolism are very similar (see below). From this **common metabolic pool** of intermediates, carbohydrates, proteins, and fats can be synthesized. These fragments can enter the citric acid cycle, a final common pathway of catabolism, in which they are broken down to hydrogen atoms and CO_2. Interconversion of amino acids involves transfer, removal, or formation of amino groups. **Transamination** reactions, conversion of one amino acid to the corresponding keto acid with simultaneous conversion of another keto acid to an amino acid, occur in many tissues:

$$\text{Alanine} + \alpha - \text{Ketoglutarate} \rightleftarrows \text{Pyruvate} + \text{Glutamate}$$

Oxidative deamination of amino acids occurs in the liver. An imino acid is formed by dehydrogenation, and this compound is hydrolyzed to the corresponding keto acid, with production of NH_4^+:

$$\text{Amino acid} + NAD^+ \rightarrow \text{Imino acid} + NADH + H^+$$

$$\text{Imino acid} + H_2O \rightarrow \text{Keto acid} + NH_4^+$$

Interconversions between the amino acid pool and the common metabolic pool are summarized in Figure 1–19. Leucine, isoleucine, phenylalanine, and tyrosine are said to be **ketogenic** because they are converted to the ketone body acetoacetate (see below). Alanine and many other amino acids are **glucogenic** or **gluconeogenic;** that is, they give rise to compounds that can readily be converted to glucose.

UREA FORMATION

Most of the NH_4^+ formed by deamination of amino acids in the liver is converted to urea, and the urea is excreted in the urine. The NH_4^+ forms carbamoyl phosphate, and in the mitochondria it is transferred to ornithine, forming citrulline. The enzyme involved is ornithine carbamoyltransferase. Citrulline is converted to arginine, after which urea is split off and ornithine is regenerated (urea cycle; Figure 1–20). The overall reaction in the urea cycle consumes 3 ATP (not shown) and thus requires significant energy. Most of the urea is formed in the liver, and in severe liver disease the blood urea nitrogen

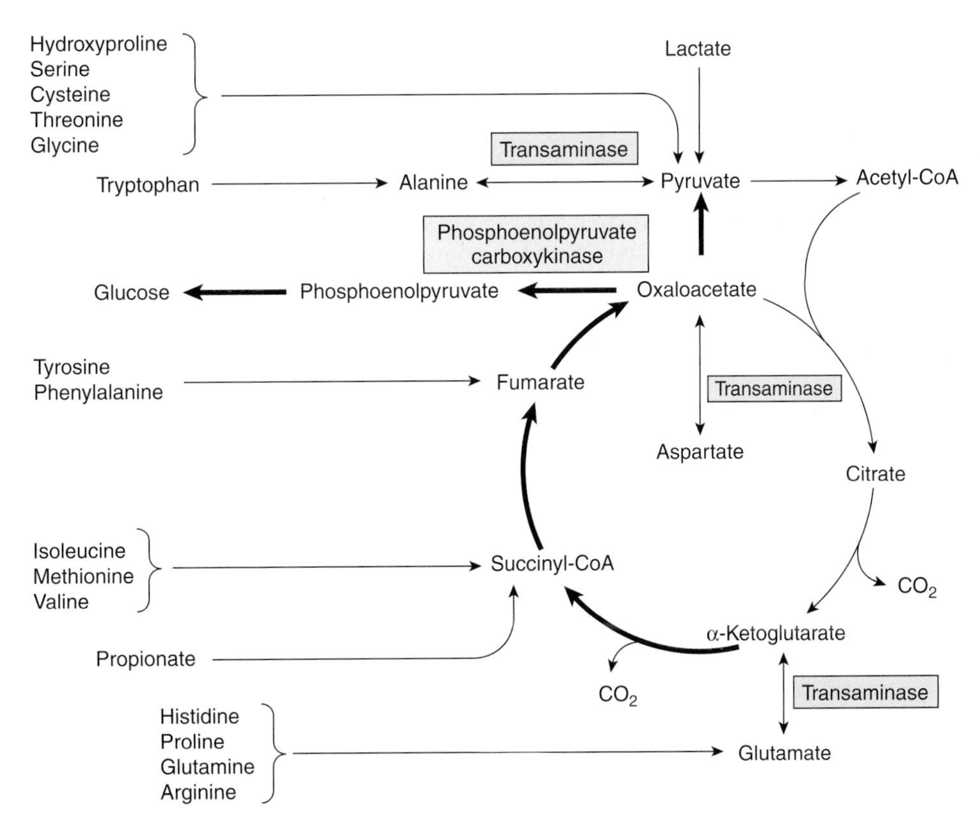

FIGURE 1–19 Involvement of the citric acid cycle in transamination and gluconeogenesis. The bold arrows indicate the main pathway of gluconeogenesis. Note the many entry positions for groups of amino acids into the citric acid cycle. (Reproduced with permission from Murray RK et al: *Harper's Biochemistry,* 26th ed. McGraw-Hill, 2003.)

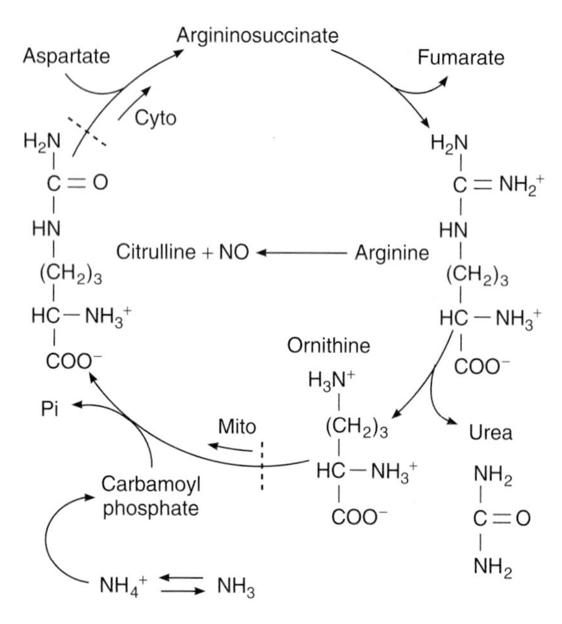

FIGURE 1–20 Urea cycle. The processing of NH_3 to urea for excretion contains several coordinative steps in both the cytoplasm (Cyto) and the mitochondria (Mito). The production of carbamoyl phosphate and its conversion to citrulline occurs in the mitochondria, whereas other processes are in the cytoplasm.

(BUN) falls and blood NH_3 rises (see Chapter 28). Congenital deficiency of ornithine carbamoyltransferase can also lead to NH_3 intoxication.

METABOLIC FUNCTIONS OF AMINO ACIDS

In addition to providing the basic building blocks for proteins, amino acids also have metabolic functions. Thyroid hormones, catecholamines, histamine, serotonin, melatonin, and intermediates in the urea cycle are formed from specific amino acids. Methionine and cysteine provide the sulfur contained in proteins, CoA, taurine, and other biologically important compounds. Methionine is converted into S-adenosylmethionine, which is the active methylating agent in the synthesis of compounds such as epinephrine.

CARBOHYDRATES

Carbohydrates are organic molecules made of equal amounts of carbon and H_2O. The simple sugars, or **monosaccharides,** including **pentoses** (five carbons; eg, ribose) and **hexoses** (six carbons; eg, glucose) perform both structural (eg, as part of nucleotides discussed previously) and functional roles (eg, inositol 1,4,5 trisphosphate acts as a cellular signaling molecules) in the body. Monosaccharides can be linked together to form disaccharides (eg, sucrose), or polysaccharides (eg, glycogen). The placement of sugar moieties onto proteins (glycoproteins) aids in cellular targeting, and in the case of

D-Glucose	D-Galactose	D-Fructose
H–C=O	H–C=O	CH_2OH
H–C–OH	H–C–OH	C=O
HO–C–H	HO–C–H	HO–C–H
H–C–OH	HO–C–H	H–C–OH
H–C–OH	H–C–OH	H–C–OH
CH_2OH	CH_2OH	CH_2OH

FIGURE 1–21 Structures of principal dietary hexoses. Glucose, galactose, and fructose are shown in their naturally occurring D isomers.

some receptors, recognition of signaling molecules. In this section, we will discuss a major role for carbohydrates in physiology, the production and storage of energy.

Dietary carbohydrates are for the most part polymers of hexoses, of which the most important are glucose, galactose, and fructose (Figure 1–21). Most of the monosaccharides occurring in the body are the D isomers. The principal product of carbohydrate digestion and the principal circulating sugar is glucose. The normal fasting level of plasma glucose in peripheral venous blood is 70–110 mg/dL (3.9–6.1 mmol/L). In arterial blood, the plasma glucose level is 15–30 mg/ dL higher than in venous blood.

Once it enters cells, glucose is normally phosphorylated to form glucose 6-phosphate. The enzyme that catalyzes this reaction is **hexokinase.** In the liver, there is an additional enzyme, **glucokinase,** which has greater specificity for glucose and which, unlike hexokinase, is increased by insulin and decreased in starvation and diabetes. The glucose 6-phosphate is either polymerized into glycogen or catabolized. The process of glycogen formation is called **glycogenesis,** and glycogen breakdown is called **glycogenolysis.** Glycogen, the storage form of glucose, is present in most body tissues, but the major supplies are in the liver and skeletal muscle. The breakdown of glucose to pyruvate or lactate (or both) is called **glycolysis.** Glucose catabolism proceeds via cleavage through fructose to trioses or via oxidation and decarboxylation to pentoses. The pathway to pyruvate through the trioses is the **Embden–Meyerhof pathway,** and that through 6-phosphogluconate and the pentoses is the **direct oxidative pathway (hexose monophosphate shunt).** Pyruvate is converted to acetyl-CoA. Interconversions between carbohydrate, fat, and protein include conversion of the glycerol from fats to dihydroxyacetone phosphate and conversion of a number of amino acids with carbon skeletons resembling intermediates in the Embden–Meyerhof pathway and citric acid cycle to these intermediates by deamination. In this way, and by conversion of lactate to glucose, nonglucose molecules can be converted to glucose **(gluconeogenesis).** Glucose can be converted to fats through acetyl-CoA, but because the conversion of pyruvate to acetyl-CoA, unlike most reactions in glycolysis, is irreversible, fats are not converted to glucose via this pathway. There is therefore very little net conversion of fats to carbohydrates in the body because, except for the quantitatively unimportant production from glycerol, there is no pathway for conversion.

CITRIC ACID CYCLE

The **citric acid cycle** (Krebs cycle, tricarboxylic acid cycle) is a sequence of reactions in which acetyl-CoA is metabolized to CO_2 and H atoms. Acetyl-CoA is first condensed with the anion of a four-carbon acid, oxaloacetate, to form citrate and HS-CoA. In a series of seven subsequent reactions, $2CO_2$ molecules are split off, regenerating oxaloacetate (Figure 1–22). Four pairs of H atoms are transferred to the flavoprotein–cytochrome chain, producing 12ATP and $4H_2O$, of which $2H_2O$ is used in the cycle. The citric acid cycle is the common pathway for oxidation to CO_2 and H_2O of carbohydrate, fat, and some amino acids. The major entry into it is through acetyl CoA, but a number of amino acids can be converted to citric acid cycle intermediates by deamination. The citric acid cycle requires O_2 and does not function under anaerobic conditions.

ENERGY PRODUCTION

The net production of energy-rich phosphate compounds during the metabolism of glucose and glycogen to pyruvate depends on whether metabolism occurs via the Embden–Meyerhof pathway or the hexose monophosphate shunt. By oxidation at the substrate level, the conversion of 1 mol of phosphoglyceraldehyde to phosphoglycerate generates 1 mol of ATP, and the conversion of 1 mol of phosphoenolpyruvate to pyruvate generates another. Because 1 mol of glucose 6-phosphate produces, via the Embden–Meyerhof pathway, 2 mol of phosphoglyceraldehyde, 4 mol of ATP is generated per mole of glucose metabolized to pyruvate. All these reactions occur in the absence of O_2 and consequently represent anaerobic production of energy. However, 1 mol of ATP is used in forming fructose 1,6-diphosphate from fructose 6-phosphate and 1 mol in phosphorylating glucose when it enters the cell. Consequently, when pyruvate is formed anaerobically from glycogen, there is a *net* production of 3 mol of ATP per mole of glucose 6-phosphate; however, when pyruvate is formed from 1 mol of blood glucose, the net gain is only 2 mol of ATP.

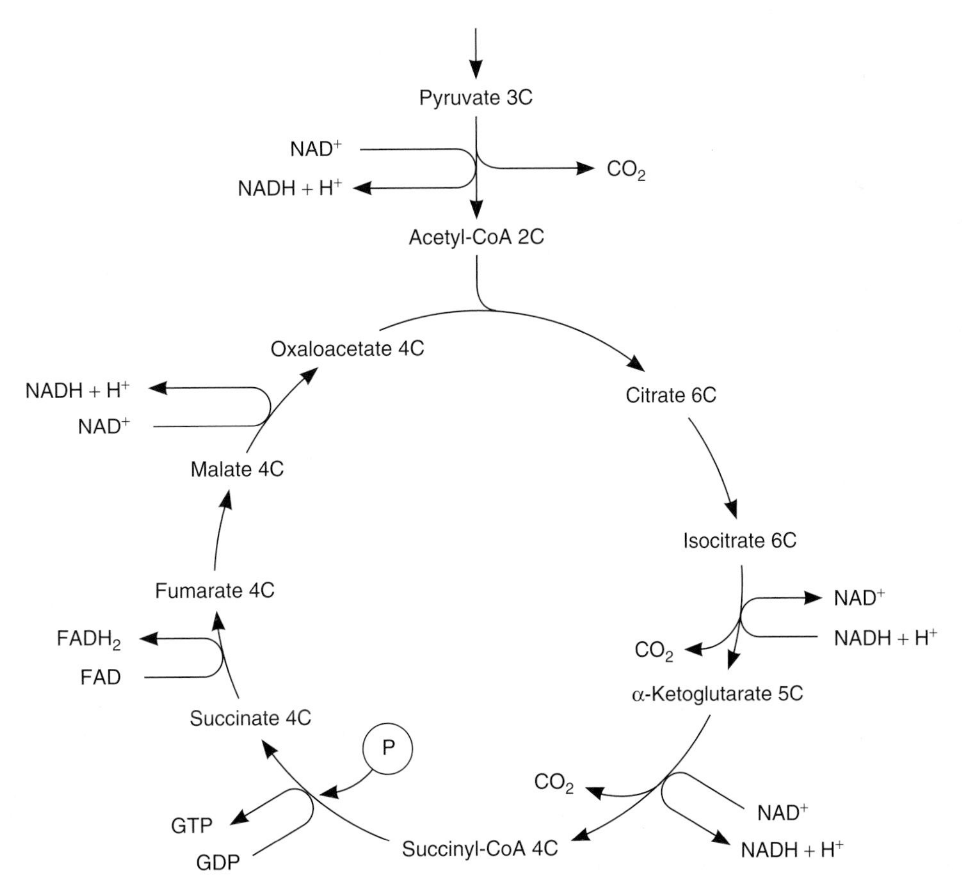

FIGURE 1–22 Citric acid cycle. The numbers (6C, 5C, etc) indicate the number of carbon atoms in each of the intermediates. The conversion of pyruvate to acetyl-CoA and each turn of the cycle provide four NADH and one $FADH_2$ for oxidation via the flavoprotein-cytochrome chain plus formation of one GTP that is readily converted to ATP.

A supply of NAD^+ is necessary for the conversion of phosphoglyceraldehyde to phosphoglycerate. Under anaerobic conditions (anaerobic glycolysis), a block of glycolysis at the phosphoglyceraldehyde conversion step might be expected to develop as soon as the available NAD^+ is converted to NADH. However, pyruvate can accept hydrogen from NADH, forming NAD^+ and lactate:

$$\text{Pyruvate} + \text{NADH} \leftrightarrow \text{Lactate} + \text{NAD}^+$$

In this way, glucose metabolism and energy production can continue for a while without O_2. The lactate that accumulates is converted back to pyruvate when the O_2 supply is restored, with NADH transferring its hydrogen to the flavoprotein–cytochrome chain.

During aerobic glycolysis, the net production of ATP is 19 times greater than the two ATPs formed under anaerobic conditions. Six ATPs are formed by oxidation, via the flavoprotein–cytochrome chain, of the two NADHs produced when 2 mol of phosphoglyceraldehyde is converted to phosphoglycerate (Figure 1–22), six ATPs are formed from the two NADHs produced when 2 mol of pyruvate is converted to acetyl-CoA, and 24 ATPs are formed during the subsequent two turns of the citric acid cycle. Of these, 18 are formed by oxidation of six NADHs, 4 by oxidation of two $FADH_2$s, and two by oxidation at the substrate level, when succinyl-CoA is converted to succinate (this reaction actually produces GTP, but the GTP is converted to ATP). Thus, the net production of ATP per mol of blood glucose metabolized aerobically via the Embden–Meyerhof pathway and citric acid cycle is $2 + [2 \times 3] + [2 \times 3] + [2 \times 12] = 38$.

Glucose oxidation via the hexose monophosphate shunt generates large amounts of NADPH. A supply of this reduced coenzyme is essential for many metabolic processes. The pentoses formed in the process are building blocks for nucleotides (see below). The amount of ATP generated depends on the amount of NADPH converted to NADH and then oxidized.

"DIRECTIONAL-FLOW VALVES" IN METABOLISM

Metabolism is regulated by a variety of hormones and other factors. To bring about any net change in a particular metabolic process, regulatory factors obviously must drive a chemical reaction in one direction. Most of the reactions in intermediary metabolism are freely reversible, but there are a number of "directional-flow valves," that is, reactions that proceed in one direction under the influence of one enzyme or transport mechanism and in the opposite direction under the influence of another. Five examples in the intermediary metabolism of carbohydrate are shown in Figure 1–23. The different pathways for fatty acid synthesis and catabolism (see below) are another example. Regulatory factors exert their influence on metabolism by acting directly or indirectly at these directional-flow valves.

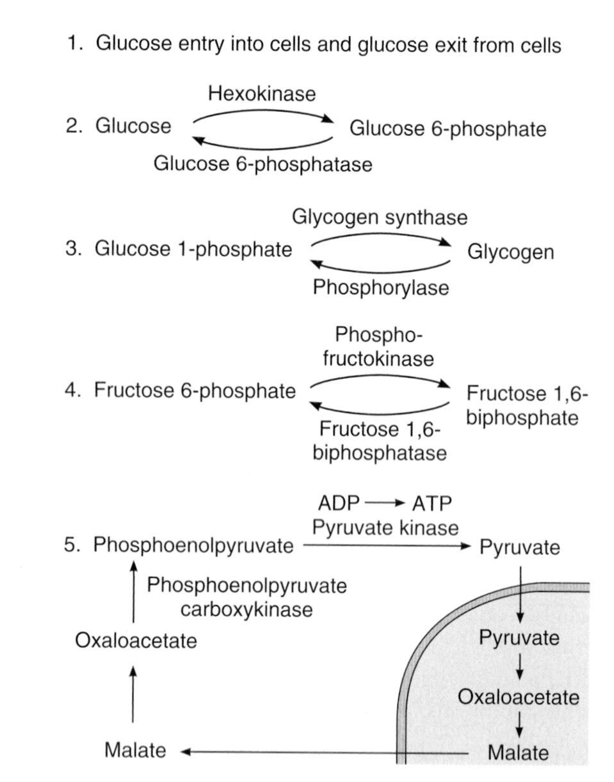

FIGURE 1–23 Directional flow valves in energy production reactions. In carbohydrate metabolism there are several reactions that proceed in one direction by one mechanism and in the other direction by a different mechanism, termed "directional-flow valves." Five examples of these reactions are illustrated (numbered at left). The double line in example 5 represents the mitochondrial membrane. Pyruvate is converted to malate in mitochondria, and the malate diffuses out of the mitochondria to the cytosol, where it is converted to phosphoenolpyruvate.

GLYCOGEN SYNTHESIS & BREAKDOWN

Glycogen is a branched glucose polymer with two types of glycoside linkages: 1:4α and 1:6α (Figure 1–24). It is synthesized on **glycogenin,** a protein primer, from glucose 1-phosphate via uridine diphosphoglucose (UDPG). The enzyme **glycogen synthase** catalyses the final synthetic step. The availability of glycogenin is one of the factors determining the amount of glycogen synthesized. The breakdown of glycogen in 1:4α linkage is catalyzed by phosphorylase, whereas another enzyme catalyzes the breakdown of glycogen in 1:6α linkage.

FACTORS DETERMINING THE PLASMA GLUCOSE LEVEL

The plasma glucose level at any given time is determined by the balance between the amount of glucose entering the bloodstream and the amount of glucose leaving the bloodstream. The principal determinants are therefore the dietary intake; the rate of entry into the cells of muscle, adipose tissue,

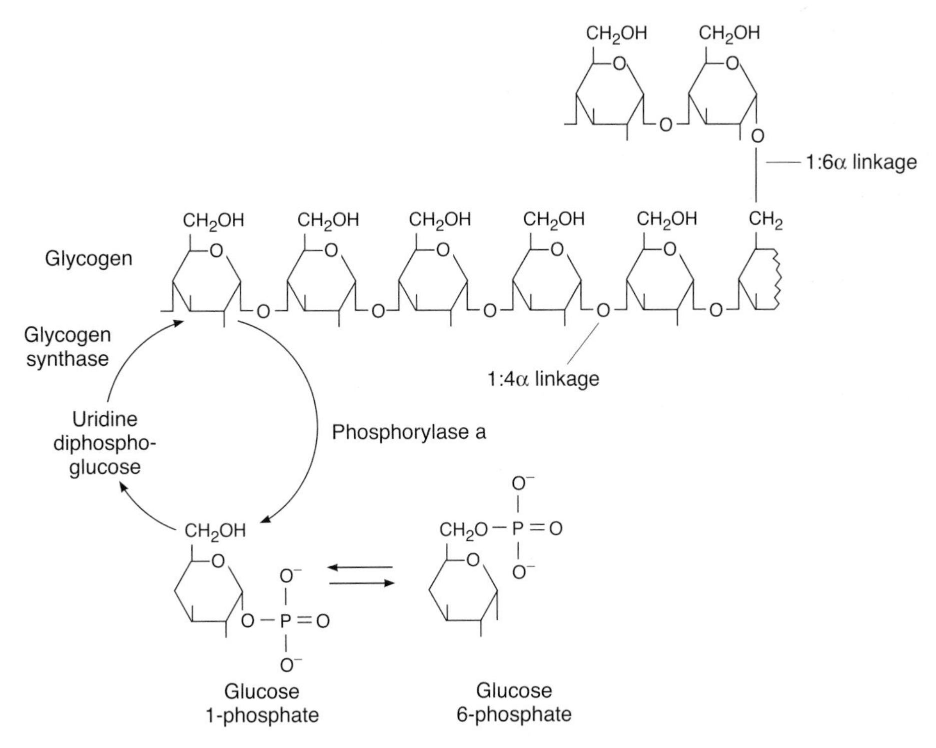

FIGURE 1–24 Glycogen formation and breakdown. Glycogen is the main storage for glucose in the cell. It is cycled: built up from glucose 6-phosphate when energy is stored and broken down to glucose 6-phosphate when energy is required. Note the intermediate glucose 1-phosphate and enzymatic control by phosphorylase a and glycogen kinase.

and other organs; and the glucostatic activity of the liver (Figure 1–25). Five per cent of ingested glucose is promptly converted into glycogen in the liver, and 30–40% is converted into fat. The remainder is metabolized in muscle and other tissues. During fasting, liver glycogen is broken down and the liver adds glucose to the bloodstream. With more prolonged fasting, glycogen is depleted and there is increased gluconeogenesis from amino acids and glycerol in the liver. Plasma glucose declines modestly to about 60 mg/dL during prolonged starvation in normal individuals, but symptoms of hypoglycemia do not occur because gluconeogenesis prevents any further fall.

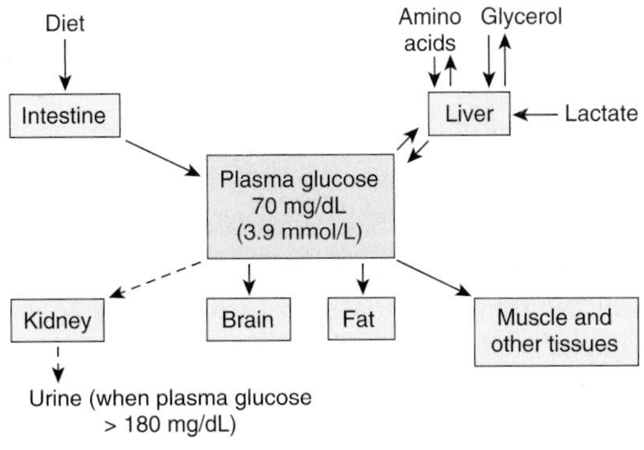

Diet
Intestine
Amino acids
Glycerol
Liver
Lactate
Plasma glucose 70 mg/dL (3.9 mmol/L)
Kidney
Brain
Fat
Muscle and other tissues
Urine (when plasma glucose > 180 mg/dL)

FIGURE 1–25 Plasma glucose homeostasis. Note the glucostatic function of the liver, as well as the loss of glucose in the urine when the renal threshold is exceeded (dashed arrows).

METABOLISM OF HEXOSES OTHER THAN GLUCOSE

Other hexoses that are absorbed from the intestine include galactose, which is liberated by the digestion of lactose and converted to glucose in the body; and fructose, part of which is ingested and part produced by hydrolysis of sucrose. After phosphorylation, galactose reacts with UDPG to form uridine diphosphogalactose. The uridine diphosphogalactose is converted back to UDPG, and the UDPG functions in glycogen synthesis. This reaction is reversible, and conversion of UDPG to uridine diphosphogalactose provides the galactose necessary for formation of glycolipids and mucoproteins when dietary galactose intake is inadequate. The utilization of galactose, like that of glucose, depends on insulin. The inability to make UDPG can have serious health consequences (Clinical Box 1–5).

Fructose is converted in part to fructose 6-phosphate and then metabolized via fructose 1,6-diphosphate. The enzyme catalyzing the formation of fructose 6-phosphate is hexokinase, the same enzyme that catalyzes the conversion of glucose to glucose 6-phosphate. However, much more fructose is converted to fructose 1-phosphate in a reaction catalyzed

CLINICAL BOX 1–5

Galactosemia

In the inborn error of metabolism known as **galactosemia,** there is a congenital deficiency of galactose 1-phosphate uridyl transferase, the enzyme responsible for the reaction between galactose 1-phosphate and UDPG, so that ingested galactose accumulates in the circulation; serious disturbances of growth and development result.

THERAPEUTIC HIGHLIGHTS

Treatment with galactose-free diets improves galactosemia without leading to galactose deficiency. This occurs because the enzyme necessary for the formation of uridine diphosphogalactose from UDPG is present.

by fructokinase. Most of the fructose 1-phosphate is then split into dihydroxyacetone phosphate and glyceraldehyde. The glyceraldehyde is phosphorylated, and it and the dihydroxyacetone phosphate enter the pathways for glucose metabolism. Because the reactions proceeding through phosphorylation of fructose in the 1 position can occur at a normal rate in the absence of insulin, it has been recommended that fructose be given to diabetics to replenish their carbohydrate stores. However, most of the fructose is metabolized in the intestines and liver, so its value in replenishing carbohydrate elsewhere in the body is limited.

Fructose 6-phosphate can also be phosphorylated in the 2 position, forming fructose 2,6-diphosphate. This compound is an important regulator of hepatic gluconeogenesis. When the fructose 2,6-diphosphate level is high, conversion of fructose 6-phosphate to fructose 1,6-diphosphate is facilitated, and thus breakdown of glucose to pyruvate is increased. A decreased level of fructose 2,6-diphosphate facilitates the reverse reaction and consequently aids gluconeogenesis.

TABLE 1–4 Lipids.

Typical fatty acids:

Palmitic acid:	$CH_3(CH_2)_{14}—C(=O)—OH$
Stearic acid:	$CH_3(CH_2)_{16}—C(=O)—OH$
Oleic acid:	$CH_3(CH_2)_7CH{=}CH(CH_2)_7—C(=O)—OH$ (Unsaturated)

Triglycerides (triacylglycerols): Esters of glycerol and three fatty acids.

$$\begin{matrix} CH_2—O—C(=O)—R \\ | \\ CH_2—O—C(=O)—R \\ | \\ CH_2—O—C(=O)—R \\ \text{Triglyceride} \end{matrix} + 3H_2O \rightleftarrows \begin{matrix} CH_2OH \\ | \\ CHOH \\ | \\ CH_2OH \\ \text{Glycerol} \end{matrix} + 3HO—C(=O)—R$$

R = Aliphatic chain of various lengths and degrees of saturation.

Phospholipids:

A. Esters of glycerol, two fatty acids, and

1. Phosphate = phosphatidic acid
2. Phosphate plus inositol = phosphatidylinositol
3. Phosphate plus choline = phosphatidylcholine (lecithin)
4. Phosphate plus ethanolamine = phosphatidyl-ethanolamine (cephalin)
5. Phosphate plus serine = phosphatidylserine

B. Other phosphate-containing derivatives of glycerol

C. Sphingomyelins: Esters of fatty acid, phosphate, choline, and the amino alcohol sphingosine.

Cerebrosides: Compounds containing galactose, fatty acid, and sphingosine.

Sterols: Cholesterol and its derivatives, including steroid hormones, bile acids, and various vitamins.

FATTY ACIDS & LIPIDS

The biologically important lipids are the fatty acids and their derivatives, the neutral fats (triglycerides), the phospholipids and related compounds, and the sterols. The triglycerides are made up of three fatty acids bound to glycerol (Table 1–4). Naturally occurring fatty acids contain an even number of carbon atoms. They may be saturated (no double bonds) or unsaturated (dehydrogenated, with various numbers of double bonds). The phospholipids are constituents of cell membranes and provide structural components of the cell membrane, as well as an important source of intra- and intercellular signaling molecules. Fatty acids also are an important source of energy in the body.

FATTY ACID OXIDATION & SYNTHESIS

In the body, fatty acids are broken down to acetyl-CoA, which enters the citric acid cycle. The main breakdown occurs in the mitochondria by β-oxidation. Fatty acid oxidation begins with activation (formation of the CoA derivative) of the fatty acid, a reaction that occurs both inside and outside the mitochondria. Medium- and short-chain fatty acids can enter the mitochondria without difficulty, but long-chain fatty acids must be bound to **carnitine** in ester linkage before they can cross the inner mitochondrial membrane.

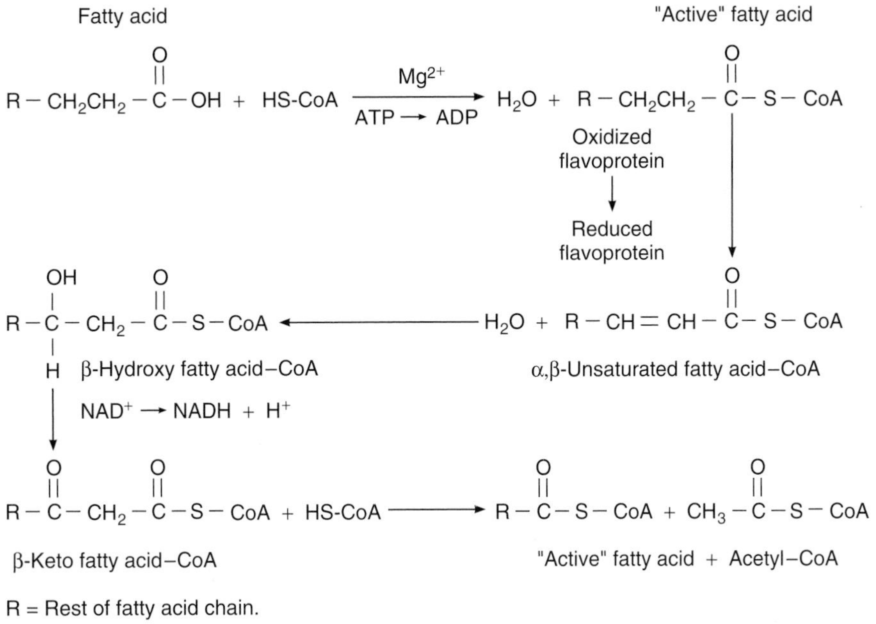

Fatty acid → "Active" fatty acid

$$R-CH_2CH_2-\overset{O}{\overset{\|}{C}}-OH + HS\text{-}CoA \xrightarrow[ATP \rightarrow ADP]{Mg^{2+}} H_2O + R-CH_2CH_2-\overset{O}{\overset{\|}{C}}-S-CoA$$

Oxidized flavoprotein → Reduced flavoprotein

$$H_2O + R-CH=CH-\overset{O}{\overset{\|}{C}}-S-CoA$$

α,β-Unsaturated fatty acid–CoA

$$\longrightarrow R-\underset{H}{\overset{OH}{C}}-CH_2-\overset{O}{\overset{\|}{C}}-S-CoA$$

β-Hydroxy fatty acid–CoA

$NAD^+ \rightarrow NADH + H^+$

$$R-\overset{O}{\overset{\|}{C}}-CH_2-\overset{O}{\overset{\|}{C}}-S-CoA + HS\text{-}CoA \longrightarrow R-\overset{O}{\overset{\|}{C}}-S-CoA + CH_3-\overset{O}{\overset{\|}{C}}-S-CoA$$

β-Keto fatty acid–CoA → "Active" fatty acid + Acetyl–CoA

R = Rest of fatty acid chain.

FIGURE 1–26 Fatty acid oxidation. This process, splitting off two carbon fragments at a time, is repeated to the end of the chain.

Carnitine is β-hydroxy-γ-trimethylammonium butyrate, and it is synthesized in the body from lysine and methionine. A translocase moves the fatty acid–carnitine ester into the matrix space. The ester is hydrolyzed, and the carnitine recycles. β-oxidation proceeds by serial removal of two carbon fragments from the fatty acid (Figure 1–26). The energy yield of this process is large. For example, catabolism of 1 mol of a six-carbon fatty acid through the citric acid cycle to CO_2 and H_2O generates 44 mol of ATP, compared with the 38 mol generated by catabolism of 1 mol of the six-carbon carbohydrate glucose.

KETONE BODIES

In many tissues, acetyl-CoA units condense to form acetoacetyl-CoA (Figure 1–27). In the liver, which (unlike other tissues) contains a deacylase, free acetoacetate is formed. This β-keto acid is converted to β-hydroxybutyrate and acetone, and because these compounds are metabolized with difficulty in the liver, they diffuse into the circulation. Acetoacetate is also formed in the liver via the formation of 3-hydroxy-3-methylglutaryl-CoA, and this pathway is quantitatively more important than deacylation. Acetoacetate, β-hydroxybutyrate, and acetone are called **ketone bodies.** Tissues other than liver transfer CoA from succinyl-CoA to acetoacetate and metabolize the "active" acetoacetate to CO_2 and H_2O via the citric acid cycle. Ketone bodies are also metabolized via other pathways. Acetone is discharged in the urine and expired air. An imbalance of ketone bodies can lead to serious health problems (Clinical Box 1–6).

CELLULAR LIPIDS

The lipids in cells are of two main types: **structural lipids,** which are an inherent part of the membranes and can serve as progenitors for cellular signaling molecules; and **neutral fat,** stored in the adipose cells of the fat depots. Neutral fat is mobilized during starvation, but structural lipid is preserved. The fat depots obviously vary in size, but in nonobese individuals they make up about 15% of body weight in men and 21% in women. They are not the inert structures they were once thought to be but, rather, active dynamic tissues undergoing continuous breakdown and resynthesis. In the depots, glucose is metabolized to fatty acids, and neutral fats are synthesized. Neutral fat is also broken down, and free fatty acids (FFAs) are released into the circulation.

A third, special type of lipid is **brown fat,** which makes up a small percentage of total body fat. Brown fat, which is somewhat more abundant in infants but is present in adults as well, is located between the scapulas, at the nape of the neck, along the great vessels in the thorax and abdomen, and in other scattered locations in the body. In brown fat depots, the fat cells as well as the blood vessels have an extensive sympathetic innervation. This is in contrast to white fat depots, in which some fat cells may be innervated but the principal sympathetic innervation is solely on blood vessels. In addition, ordinary lipocytes have only a single large droplet of white fat, whereas brown fat cells contain several small droplets of fat. Brown fat cells also contain many mitochondria. In these mitochondria, an inward proton conductance that generates ATP takes places as usual, but in addition there is a second proton conductance that does not generate ATP. This "short-circuit" conductance

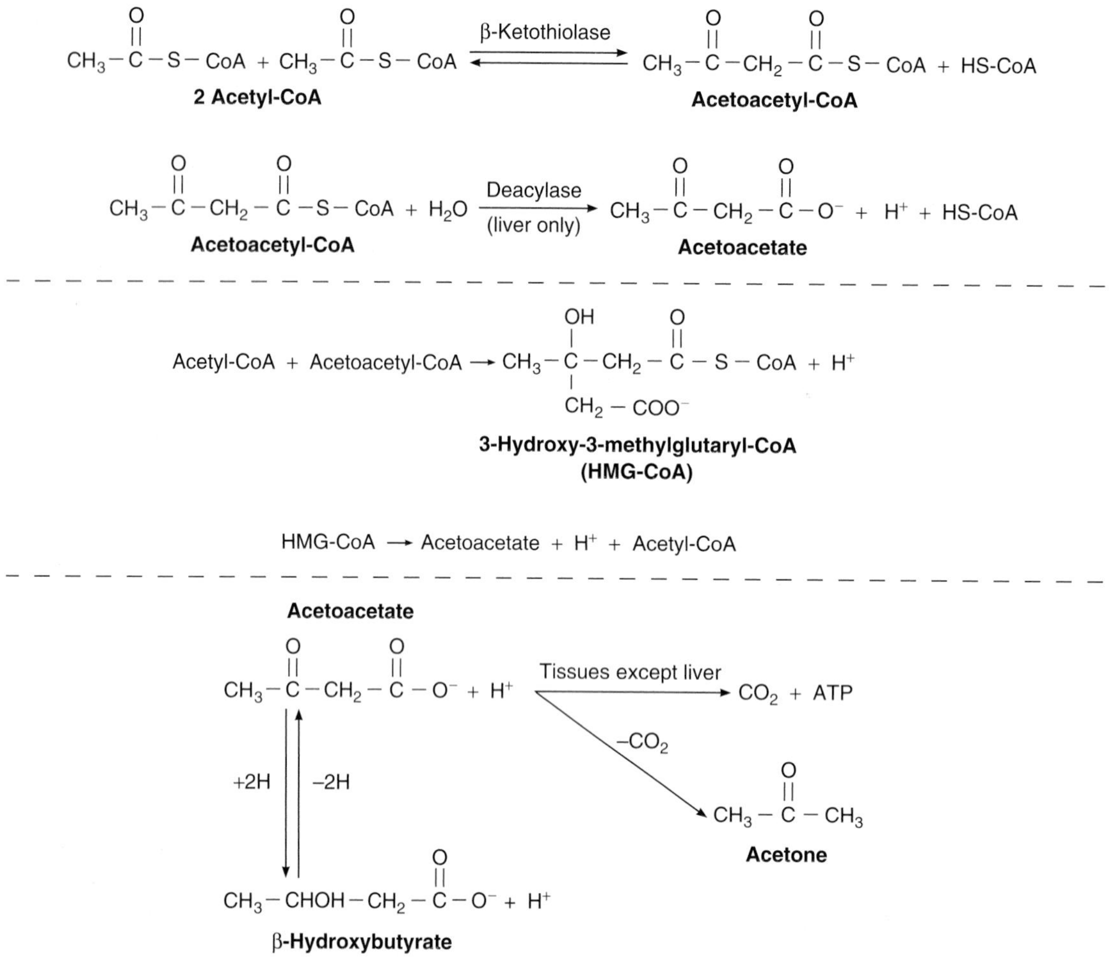

$$CH_3-\overset{O}{\overset{\|}{C}}-S-CoA + CH_3-\overset{O}{\overset{\|}{C}}-S-CoA \underset{}{\overset{\beta\text{-Ketothiolase}}{\rightleftharpoons}} CH_3-\overset{O}{\overset{\|}{C}}-CH_2-\overset{O}{\overset{\|}{C}}-S-CoA + HS\text{-}CoA$$

2 Acetyl-CoA → **Acetoacetyl-CoA**

$$CH_3-\overset{O}{\overset{\|}{C}}-CH_2-\overset{O}{\overset{\|}{C}}-S-CoA + H_2O \xrightarrow[\text{(liver only)}]{\text{Deacylase}} CH_3-\overset{O}{\overset{\|}{C}}-CH_2-\overset{O}{\overset{\|}{C}}-O^- + H^+ + HS\text{-}CoA$$

Acetoacetyl-CoA → **Acetoacetate**

$$\text{Acetyl-CoA} + \text{Acetoacetyl-CoA} \rightarrow CH_3-\underset{CH_2-COO^-}{\overset{OH}{C}}-CH_2-\overset{O}{\overset{\|}{C}}-S-CoA + H^+$$

3-Hydroxy-3-methylglutaryl-CoA (HMG-CoA)

$$\text{HMG-CoA} \rightarrow \text{Acetoacetate} + H^+ + \text{Acetyl-CoA}$$

Acetoacetate

$$CH_3-\overset{O}{\overset{\|}{C}}-CH_2-\overset{O}{\overset{\|}{C}}-O^- + H^+ \xrightarrow{\text{Tissues except liver}} CO_2 + ATP$$

$$\text{Acetoacetate} \xrightarrow{-CO_2} CH_3-\overset{O}{\overset{\|}{C}}-CH_3 \quad \textbf{Acetone}$$

$$\text{Acetoacetate} \underset{-2H}{\overset{+2H}{\rightleftharpoons}} CH_3-CHOH-CH_2-\overset{O}{\overset{\|}{C}}-O^- + H^+$$

β-Hydroxybutyrate

FIGURE 1–27 Formation and metabolism of ketone bodies. Note the two pathways for the formation of acetoacetate.

CLINICAL BOX 1–6

Diseases Associated with Imbalance of β-oxidation of Fatty Acids

Ketoacidosis

The normal blood ketone level in humans is low (about 1 mg/dL) and less than 1 mg is excreted per 24 h, because the ketones are normally metabolized as rapidly as they are formed. However, if the entry of acetyl-CoA into the citric acid cycle is depressed because of a decreased supply of the products of glucose metabolism, or if the entry does not increase when the supply of acetyl-CoA increases, acetyl-CoA accumulates, the rate of condensation to acetoacetyl-CoA increases, and more acetoacetate is formed in the liver. The ability of the tissues to oxidize the ketones is soon exceeded, and they accumulate in the bloodstream (ketosis). Two of the three ketone bodies, acetoacetate and β-hydroxybutyrate, are anions of the moderately strong acids acetoacetic acid and β-hydroxybutyric acid. Many of their protons are buffered, reducing the decline in pH that would otherwise occur. However, the buffering capacity can be exceeded, and the metabolic acidosis that develops in conditions such as diabetic ketosis can be severe and even fatal. Three conditions lead to deficient intracellular glucose supplies, and hence to ketoacidosis: starvation; diabetes mellitus; and a high-fat, low-carbohydrate diet. The acetone odor on the breath of children who have been vomiting is due to the ketosis of starvation. Parenteral administration of relatively small amounts of glucose abolishes the ketosis, and it is for this reason that carbohydrate is said to be antiketogenic.

Carnitine Deficiency

Deficient β-oxidation of fatty acids can be produced by carnitine deficiency or genetic defects in the translocase or other enzymes involved in the transfer of long-chain fatty acids into the mitochondria. This causes cardiomyopathy. In addition, it causes **hypoketonemic hypoglycemia** with coma, a serious and often fatal condition triggered by fasting, in which glucose stores are used up because of the lack of fatty acid oxidation to provide energy. Ketone bodies are not formed in normal amounts because of the lack of adequate CoA in the liver.

depends on a 32-kDa uncoupling protein (UCP1). It causes uncoupling of metabolism and generation of ATP, so that more heat is produced.

PLASMA LIPIDS & LIPID TRANSPORT

The major lipids are relatively insoluble in aqueous solutions and do not circulate in the free form. **FFAs** are bound to albumin, whereas cholesterol, triglycerides, and phospholipids are transported in the form of **lipoprotein** complexes. The complexes greatly increase the solubility of the lipids. The six families of lipoproteins (Table 1–5) are graded in size and lipid content. The density of these lipoproteins is inversely proportionate to their lipid content. In general, the lipoproteins consist of a hydrophobic core of triglycerides and cholesteryl esters surrounded by phospholipids and protein. These lipoproteins can be transported from the intestine to the liver via an **exogenous pathway,** and between other tissues via an **endogenous pathway.**

Dietary lipids are processed by several pancreatic lipases in the intestine to form mixed micelles of predominantly FFA, **2-monoacylglycerols,** and cholesterol derivatives (see Chapter 26). These micelles additionally can contain important water-insoluble molecules such as **vitamins A, D, E, and K.** These mixed micelles are taken up into cells of the intestinal mucosa where large lipoprotein complexes, **chylomicrons,** are formed. The chylomicrons and their remnants constitute a transport system for ingested exogenous lipids (exogenous pathway). Chylomicrons can enter the circulation via the lymphatic ducts. The chylomicrons are cleared from the circulation by the action of **lipoprotein lipase,** which is located on the surface of the endothelium of the capillaries. The enzyme catalyzes the breakdown of the triglyceride in the chylomicrons to FFA and glycerol, which then enter adipose cells and are reesterified. Alternatively, the FFA can remain in the circulation bound to albumin. Lipoprotein lipase, which requires heparin as a cofactor, also removes triglycerides from circulating **very low density lipoproteins (VLDL).** Chylomicrons depleted of their triglyceride remain in the circulation as cholesterol-rich lipoproteins called **chylomicron remnants,** which are 30–80 nm in diameter. The remnants are carried to the liver, where they are internalized and degraded.

The endogenous system, made up of VLDL, **intermediate-density lipoproteins (IDL), low-density lipoproteins (LDL),** and **high-density lipoproteins (HDL),** also transports triglycerides and cholesterol throughout the body. VLDL are formed in the liver and transport triglycerides formed from fatty acids and carbohydrates in the liver to extrahepatic tissues. After their triglyceride is largely removed by the action of lipoprotein lipase, they become IDL. The IDL give up phospholipids and, through the action of the plasma enzyme **lecithin-cholesterol acyltransferase (LCAT),** pick up cholesteryl esters formed from cholesterol in the HDL. Some IDL are taken up by the liver. The remaining IDL then lose more triglyceride and protein, probably in the sinusoids of the liver, and become LDL. LDL provide cholesterol to the tissues. The cholesterol is an essential constituent in cell membranes and is used by gland cells to make steroid hormones.

FREE FATTY ACID METABOLISM

In addition to the exogenous and endogenous pathways described above, FFA are also synthesized in the fat depots in which they are stored. They can circulate as lipoproteins bound to albumin and are a major source of energy for many organs. They are used extensively in the heart, but probably all tissues can oxidize FFA to CO_2 and H_2O.

The supply of FFA to the tissues is regulated by two lipases. As noted above, lipoprotein lipase on the surface of the endothelium of the capillaries hydrolyzes the triglycerides in chylomicrons and VLDL, providing FFA and glycerol, which are reassembled into new triglycerides in the fat cells. The intracellular **hormone-sensitive lipase** of adipose tissue catalyzes the breakdown of stored triglycerides into glycerol and fatty acids, with the latter entering the circulation. Hormone-sensitive

TABLE 1–5 The principal lipoproteins.[a]

		Composition (%)					
Lipoprotein	**Size (nm)**	**Protein**	**Free Cholesteryl**	**Cholesterol Esters**	**Triglyceride**	**Phospholipid**	**Origin**
Chylomicrons	75–1000	2	2	3	90	3	Intestine
Chylomicron remnants	30–80	...	...	...	...	...	Capillaries
Very low density lipoproteins (VLDL)	30–80	8	4	16	55	17	Liver and intestine
Intermediate-density lipo-proteins (IDL)	25–40	10	5	25	40	20	VLDL
Low-density lipoproteins (LDL)	20	20	7	46	6	21	IDL
High-density lipoproteins (HDL)	7.5–10	50	4	16	5	25	Liver and intestine

[a]The plasma lipids include these components plus free fatty acids from adipose tissue, which circulate bound to albumin.

lipase is increased by fasting and stress and decreased by feeding and insulin. Conversely, feeding increases and fasting and stress decrease the activity of lipoprotein lipase.

CHOLESTEROL METABOLISM

Cholesterol is the precursor of the steroid hormones and bile acids and is an essential constituent of cell membranes. It is found only in animals. Related sterols occur in plants, but plant sterols are poorly absorbed from the gastrointestinal tract. Most of the dietary cholesterol is contained in egg yolks and animal fat.

Cholesterol is absorbed from the intestine and incorporated into the chylomicrons formed in the intestinal mucosa. After the chylomicrons discharge their triglyceride in adipose tissue, the chylomicron remnants bring cholesterol to the liver. The liver and other tissues also synthesize cholesterol. Some of the cholesterol in the liver is excreted in the bile, both in the free form and as bile acids. Some of the biliary cholesterol is reabsorbed from the intestine. Most of the cholesterol in the liver is incorporated into VLDL and circulates in lipoprotein complexes.

The biosynthesis of cholesterol from acetate is summarized in Figure 1–28. Cholesterol feeds back to inhibit its own synthesis by inhibiting **HMG-CoA reductase,** the enzyme that converts 3-hydroxy-3-methylglutaryl-coenzyme A (HMG-CoA) to mevalonic acid. Thus, when dietary cholesterol intake is high, hepatic cholesterol synthesis is decreased, and vice versa. However, the feedback compensation is incomplete, because a diet that is low in cholesterol and saturated fat leads to only a modest decline in circulating plasma cholesterol. The most effective and most commonly used cholesterol-lowering drugs are lovastatin and other **statins,** which reduce cholesterol synthesis by inhibiting HMG-CoA. The relationship between cholesterol and vascular disease is discussed in Clinical Box 1–7.

ESSENTIAL FATTY ACIDS

Animals fed a fat-free diet fail to grow, develop skin and kidney lesions, and become infertile. Adding linolenic, linoleic, and arachidonic acids to the diet cures all the deficiency symptoms. These three acids are polyunsaturated fatty acids and because of their action are called **essential fatty acids.** Similar deficiency symptoms have not been unequivocally demonstrated in humans, but there is reason to believe that some unsaturated fats are essential dietary constituents, especially for children. Dehydrogenation of fats is known to occur in the body, but there does not appear to be any synthesis of carbon chains with the arrangement of double bonds found in the essential fatty acids.

EICOSANOIDS

One of the reasons that essential fatty acids are necessary for health is that they are the precursors of prostaglandins, prostacyclin, thromboxanes, lipoxins, leukotrienes, and related compounds. These substances are called **eicosanoids,** reflecting their origin from the 20-carbon (eicosa-) polyunsaturated fatty acid **arachidonic acid (arachidonate)** and the 20-carbon derivatives of linoleic and linolenic acids.

The **prostaglandins** are a series of 20-carbon unsaturated fatty acids containing a cyclopentane ring. They were first isolated from semen but are synthesized in most and possibly in all organs in the body. Prostaglandin H_2 (PGH_2) is the precursor for various other prostaglandins, thromboxanes, and prostacyclin. Arachidonic acid is formed from tissue phospholipids by **phospholipase A_2.** It is converted to prostaglandin H_2 (PGH_2) by **prostaglandin G/H synthases** 1 and 2. These are bifunctional enzymes that have both cyclooxygenase and peroxidase activity, but they are more commonly known by the names cyclooxygenase 1 **(COX1)** and cyclooxygenase 2 **(COX2).** Their structures are very similar, but COX1 is constitutive whereas COX2 is induced by growth factors, cytokines, and

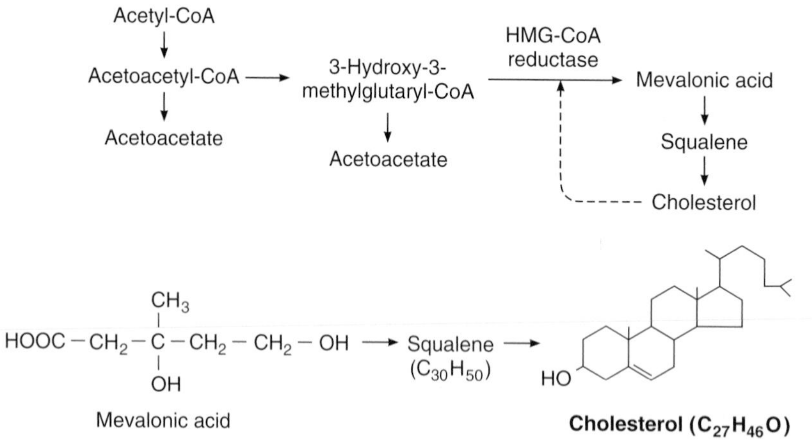

FIGURE 1–28 Biosynthesis of cholesterol. Six mevalonic acid molecules condense to form squalene, which is then hydroxylated to cholesterol. The dashed arrow indicates feedback inhibition by cholesterol of HMG-CoA reductase, the enzyme that catalyzes mevalonic acid formation.

CLINICAL BOX 1–7

Cholesterol & Atherosclerosis

The interest in cholesterol-lowering drugs stems from the role of cholesterol in the etiology and course of **atherosclerosis.** This extremely widespread disease predisposes to myocardial infarction, cerebral thrombosis, ischemic gangrene of the extremities, and other serious illnesses. It is characterized by infiltration of cholesterol and oxidized cholesterol into macrophages, converting them into foam cells in lesions of the arterial walls. This is followed by a complex sequence of changes involving platelets, macrophages, smooth muscle cells, growth factors, and inflammatory mediators that produces proliferative lesions which eventually ulcerate and may calcify. The lesions distort the vessels and make them rigid. In individuals with elevated plasma cholesterol levels, the incidence of atherosclerosis and its complications is increased. The normal range for plasma cholesterol is said to be 120–200 mg/dL, but in men, there is a clear, tight, positive correlation between the death rate from ischemic heart disease and plasma cholesterol levels above 180 mg/dL. Furthermore, it is now clear that lowering plasma cholesterol by diet and drugs slows and may even reverse the progression of atherosclerotic lesions and the complications they cause.

In evaluating plasma cholesterol levels in relation to atherosclerosis, it is important to analyze the LDL and HDL levels as well. LDL delivers cholesterol to peripheral tissues, including atheromatous lesions, and the LDL plasma concentration correlates positively with myocardial infarctions and ischemic strokes. On the other hand, HDL picks up cholesterol from peripheral tissues and transports it to the liver, thus lowering plasma cholesterol. It is interesting that women, who have a lower incidence of myocardial infarction than men, have higher HDL levels. In addition, HDL levels are increased in individuals who exercise and those who drink one or two alcoholic drinks per day, whereas they are decreased in individuals who smoke, are obese, or live sedentary lives. Moderate drinking decreases the incidence of myocardial infarction, and obesity and smoking are risk factors that increase it. Plasma cholesterol and the incidence of cardiovascular diseases are increased in **familial hypercholesterolemia,** due to various loss-of-function mutations in the genes for LDL receptors.

THERAPEUTIC HIGHLIGHTS

Although atherosclerosis is a progressive disease, it is also preventable in many cases by limiting risk factors, including lowering "bad" cholesterol through a healthy diet and exercise. Drug treatments for high cholesterol, including the statins among others, provide additional relief that can complement a healthy diet and exercise. If atherosclerosis is advanced, invasive techniques, such as angioplasty and stenting, can be used to unblock arteries.

tumor promoters. PGH_2 is converted to prostacyclin, thromboxanes, and prostaglandins by various tissue isomerases. The effects of prostaglandins are multitudinous and varied. They are particularly important in the female reproductive cycle, in parturition, in the cardiovascular system, in inflammatory responses, and in the causation of pain. Drugs that target production of prostaglandins are among the most common over the counter drugs available (Clinical Box 1–8).

Arachidonic acid also serves as a substrate for the production of several physiologically important **leukotrienes** and **lipoxins.** The leukotrienes, thromboxanes, lipoxins, and prostaglandins have been called local hormones. They

CLINICAL BOX 1–8

Pharmacology of Prostaglandins

Because prostaglandins play a prominent role in the genesis of pain, inflammation, and fever, pharmacologists have long sought drugs to inhibit their synthesis. Glucocorticoids inhibit phospholipase A_2 and thus inhibit the formation of all eicosanoids. A variety of nonsteroidal anti-inflammatory drugs (NSAIDs) inhibit both cyclooxygenases, inhibiting the production of PGH_2 and its derivatives. Aspirin is the best-known of these, but ibuprofen, indomethacin, and others are also used. However, there is evidence that prostaglandins synthesized by COX2 are more involved in the production of pain and inflammation, and prostaglandins synthesized by COX1 are more involved in protecting the gastrointestinal mucosa from ulceration. Drugs such as celecoxib and rofecoxib that selectively inhibit COX2 have been developed, and in clinical use they relieve pain and inflammation, possibly with a significantly lower incidence of gastrointestinal ulceration and its complications than is seen with nonspecific NSAIDs. However, rofecoxib has been withdrawn from the market in the United States because of a reported increase of strokes and heart attacks in individuals using it. More research is underway to better understand all the effects of the COX enzymes, their products, and their inhibitors.

have short half-lives and are inactivated in many different tissues. They undoubtedly act mainly in the tissues at sites in which they are produced. The leukotrienes are mediators of allergic responses and inflammation. Their release is provoked when specific allergens combine with IgE antibodies on the surfaces of mast cells (see Chapter 3). They produce bronchoconstriction, constrict arterioles, increase vascular permeability, and attract neutrophils and eosinophils to inflammatory sites. Diseases in which they may be involved include asthma, psoriasis, adult respiratory distress syndrome, allergic rhinitis, rheumatoid arthritis, Crohn's disease, and ulcerative colitis.

CHAPTER SUMMARY

- Cells contain approximately two thirds of the body fluids, while the remaining extracellular fluid is found between cells (interstitial fluid) or in the circulating lymph and blood plasma.
- The number of molecules, electrical charges, and particles of substances in solution are important in physiology.
- Biological buffers including bicarbonate, proteins, and phosphates can bind or release protons in solution to help maintain pH. Biological buffering capacity of a weak acid or base is greatest when pK_a = pH.
- Although the osmolality of solutions can be similar across a plasma membrane, the distribution of individual molecules and distribution of charge across the plasma membrane can be quite different. The separation of concentrations of charged species sets up an electrical gradient at the plasma membrane (inside negative). The electro-chemical gradient is in large part maintained by the Na, K ATPase. These are affected by the Gibbs–Donnan equilibrium and can be calculated using the Nernst potential equation.
- Cellular energy can be stored in high-energy phosphate compounds, including adenosine triphosphate (ATP). Coordinated oxidation–reduction reactions allow for the production of a proton gradient at the inner mitochondrial membrane that ultimately yields to the production of ATP in the cell.
- Nucleotides made from purine or pyrimidine bases linked to ribose or 2-deoxyribose sugars with inorganic phosphates are the basic building blocks for nucleic acids, DNA, and RNA. The fundamental unit of DNA is the gene, which encodes information to make proteins in the cell. Genes are transcribed into messenger RNA, and with the help of ribosomal RNA and transfer RNAs, translated into proteins.
- Amino acids are the basic building blocks for proteins in the cell and can also serve as sources for several biologically active molecules. Translation is the process of protein synthesis. After synthesis, proteins can undergo a variety of posttranslational modifications prior to obtaining their fully functional cell state.
- Carbohydrates are organic molecules that contain equal amounts of C and H_2O. Carbohydrates can be attached to proteins (glycoproteins) or fatty acids (glycolipids) and are critically important for the production and storage of cellular and body energy. The breakdown of glucose to generate energy, or glycolysis, can occur in the presence or absence of O_2 (aerobic or anaerobically). The net production of ATP during aerobic glycolysis is 19 times higher than anaerobic glycolysis.
- Fatty acids are carboxylic acids with extended hydrocarbon chains. They are an important energy source for cells and fatty acid derivatives—including triglycerides, phospholipids and sterols—have additional important cellular applications.

MULTIPLE-CHOICE QUESTIONS

For all questions, select the single best answer unless otherwise directed.

1. The membrane potential of a particular cell is at the K^+ equilibrium. The intracellular concentration for K^+ is at 150 mmol/L and the extracellular concentration for K^+ is at 5.5 mmol/L. What is the resting potential?
   A. −70 mv
   B. −90 mv
   C. +70 mv
   D. +90 mv
2. The difference in concentration of H^+ in a solution of pH 2.0 compared with one of pH 7.0 is
   A. 5-fold
   B. 1/5 as much
   C. 10^5 fold
   D. 10^{-5} as much
3. Transcription refers to
   A. the process where an mRNA is used as a template for protein production.
   B. the process where a DNA sequence is copied into RNA for the purpose of gene expression.
   C. the process where DNA wraps around histones to form a nucleosome.
   D. the process of replication of DNA prior to cell division.
4. The primary structure of a protein refers to
   A. the twist, folds, or twist and folds of the amino acid sequence into stabilized structures within the protein (ie, α-helices and β-sheets).
   B. the arrangement of subunits to form a functional structure.
   C. the amino acid sequence of the protein.
   D. the arrangement of twisted chains and folds within a protein into a stable structure.
5. Fill in the blanks: Glycogen is a storage form of glucose. ________ refers to the process of making glycogen and ________ refers to the process of breakdown of glycogen.
   A. Glycogenolysis, glycogenesis
   B. Glycolysis, glycogenolysis
   C. Glycogenesis, glycogenolysis
   D. Glycogenolysis, glycolysis
6. The major lipoprotein source of the cholesterol used in cells is
   A. chylomicrons
   B. intermediate-density lipoproteins (IDLs)
   C. albumin-bound free fatty acids
   D. LDL
   E. HDL

7. Which of the following produces the most high-energy phosphate compounds?
   A. aerobic metabolism of 1 mol of glucose
   B. anaerobic metabolism of 1 mol of glucose
   C. metabolism of 1 mol of galactose
   D. metabolism of 1 mol of amino acid
   E. metabolism of 1 mol of long-chain fatty acid
8. When LDL enters cells by receptor-mediated endocytosis, which of the following does not occur?
   A. Decrease in the formation of cholesterol from mevalonic acid.
   B. Increase in the intracellular concentration of cholesteryl esters.
   C. Increase in the transfer of cholesterol from the cell to HDL.
   D. Decrease in the rate of synthesis of LDL receptors.
   E. Decrease in the cholesterol in endosomes.

CHAPTER RESOURCES

Alberts B, Johnson A, Lewis J, et al: *Molecular Biology of the Cell*, 5th ed. Garland Science, 2008.

Hille B: *Ionic Channels of Excitable Membranes*, 3rd ed. Sinauer Associates, 2001.

Kandel ER, Schwartz JH, Jessell TM: *Principles of Neural Science*, 4th ed. McGraw-Hill, 2000.

Macdonald RG, Chaney WG: *USMLE Road Map, Biochemistry.* McGraw-Hill, 2007.

Murray RK, Bender DA, Botham KM, et al: *Harper's Biochemistry*, 28th ed. McGraw-Hill, 2009.

Pollard TD, Earnshaw WC: *Cell Biology*, 2nd ed. Saunders, Elsevier, 2008.

Sack GH, Jr: *USMLE Road Map, Genetics.* McGraw Hill, 2008.

Scriver CR, Beaudet AL, Sly WS, et al (editors): *The Metabolic and Molecular Bases of Inherited Disease*, 8th ed. McGraw-Hill, 2001.

Overview of Cellular Physiology in Medical Physiology

CHAPTER 2

OBJECTIVES

After studying this chapter, you should be able to:

- Name the prominent cellular organelles and state their functions in cells.
- Name the building blocks of the cellular cytoskeleton and state their contributions to cell structure and function.
- Name the intercellular and cellular to extracellular connections.
- Define the processes of exocytosis and endocytosis, and describe the contribution of each to normal cell function.
- Define proteins that contribute to membrane permeability and transport.
- Recognize various forms of intercellular communication and describe ways in which chemical messengers (including second messengers) affect cellular physiology.

INTRODUCTION

The cell is the fundamental working unit of all organisms. In humans, cells can be highly specialized in both structure and function; alternatively, cells from different organs can share features and function. In the previous chapter, we examined some basic principles of biophysics and the catabolism and metabolism of building blocks found in the cell. In some of those discussions, we examined how the building blocks could contribute to basic cellular physiology (eg, DNA replication, transcription, and translation). In this chapter, we will briefly review more of the fundamental aspects of cellular and molecular physiology. Additional aspects that concern specialization of cellular and molecular physiology are considered in the next chapters concerning immune function and excitable cells, and, within the sections that highlight each physiological system.

FUNCTIONAL MORPHOLOGY OF THE CELL

A basic knowledge of cell biology is essential to an understanding of the organ systems and the way they function in the body. A key tool for examining cellular constituents is the microscope. A light microscope can resolve structures as close as 0.2 μm, while an electron microscope can resolve structures as close as 0.002 μm. Although cell dimensions are quite variable, this resolution can give us a good look at the inner workings of the cell. The advent of common access to phase contrast, fluorescent, confocal, and many other microscopy techniques along with specialized probes for both static and dynamic cellular structures further expanded the examination of cell structure and function. Equally revolutionary advances in modern biophysical, biochemical, and molecular biological techniques have also greatly contributed to our knowledge of the cell.

The specialization of the cells in the various organs is considerable, and no cell can be called "typical" of all cells in the body. However, a number of structures **(organelles)** are common to most cells. These structures are shown in Figure 2–1. Many of them can be isolated by ultracentrifugation combined with other techniques. When cells are homogenized and the resulting suspension is centrifuged, the nuclei sediment first, followed by the mitochondria. High-speed centrifugation that

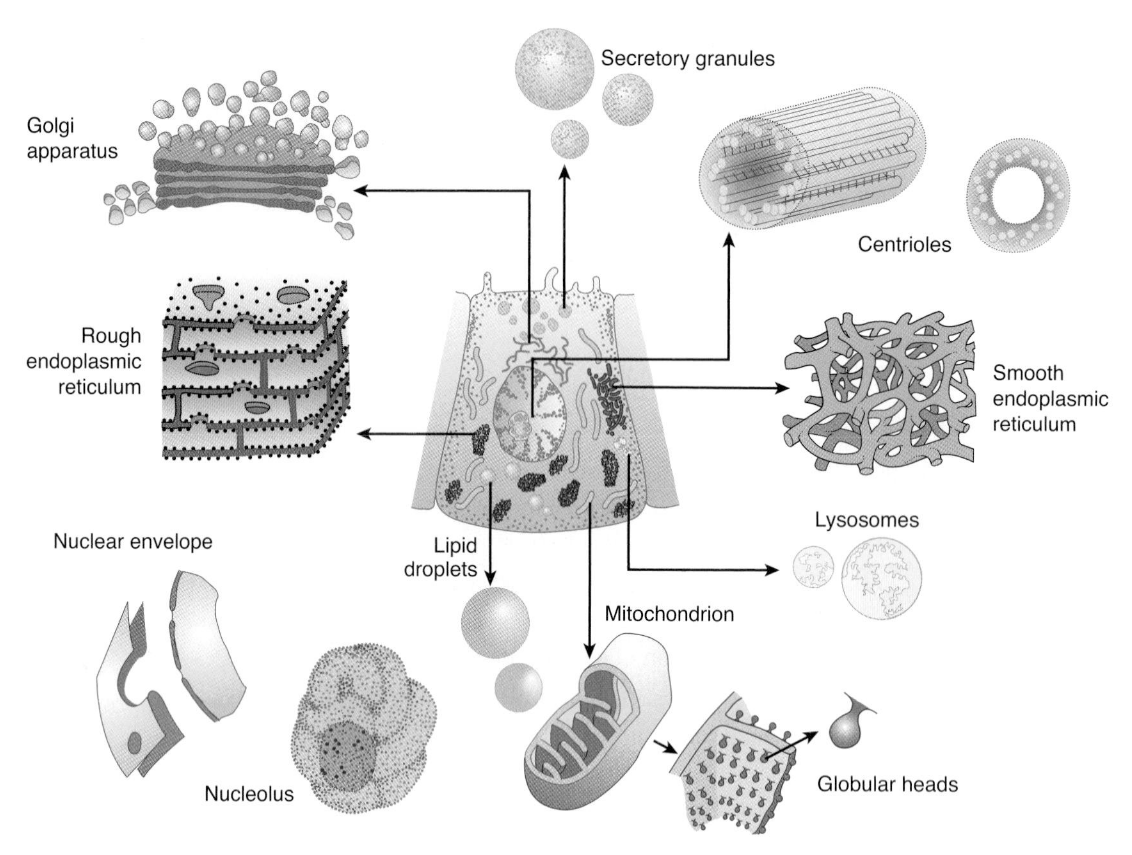

FIGURE 2–1 Diagram showing a hypothetical cell in the center as seen with the light microscope. Individual organelles are expanded for closer examination. (Adapted from Bloom and Fawcett. Reproduced with permission from Junqueira LC, Carneiro J, Kelley RO: *Basic Histology*, 9th ed. McGraw-Hill, 1998.)

generates forces of 100,000 times gravity or more causes a fraction made up of granules called the **microsomes** to sediment. This fraction includes organelles such as the **ribosomes** and **peroxisomes.**

CELL MEMBRANES

The membrane that surrounds the cell is a remarkable structure. It is made up of lipids and proteins and is semipermeable, allowing some substances to pass through it and excluding others. However, its permeability can also be varied because it contains numerous regulated ion channels and other transport proteins that can change the amounts of substances moving across it. It is generally referred to as the **plasma membrane.** The nucleus and other organelles in the cell are bound by similar membranous structures.

Although the chemical structures of membranes and their properties vary considerably from one location to another, they have certain common features. They are generally about 7.5 nm (75 Å) thick. The major lipids are phospholipids such as phosphatidylcholine, phosphotidylserine, and phosphatidylethanolamine. The shape of the phospholipid molecule reflects its solubility properties: the "head" end of the molecule contains the phosphate portion and is relatively soluble in water (polar, **hydrophilic**) and the "tail" ends are relatively insoluble (nonpolar, **hydrophobic**). The possession of both hydrophilic and hydrophobic properties makes the lipid an **amphipathic** molecule. In the membrane, the hydrophilic ends of the molecules are exposed to the aqueous environment that bathes the exterior of the cells and the aqueous cytoplasm; the hydrophobic ends meet in the water-poor interior of the membrane (Figure 2–2). In **prokaryotes** (ie, bacteria in which there is no nucleus), the membranes are relatively simple, but in **eukaryotes** (cells containing nuclei), cell membranes contain various glycosphingolipids, sphingomyelin, and cholesterol in addition to phospholipids and phosphatidylcholine.

Many different proteins are embedded in the membrane. They exist as separate globular units and many pass through or are embedded in one leaflet of the membrane **(integral proteins),** whereas others **(peripheral proteins)** are associated with the inside or outside of the membrane (Figure 2–2). The amount of protein varies significantly with the function of the membrane but makes up on average 50% of the mass of the membrane; that is, there is about one protein molecule per 50 of the much smaller phospholipid molecules. The proteins in the membrane carry out many functions. Some are **cell adhesion molecules (CAMs)** that anchor cells to their neighbors or to basal laminas. Some proteins function

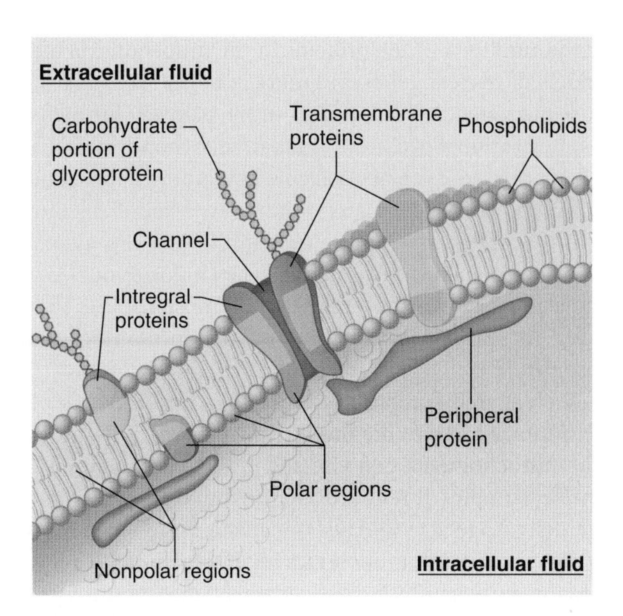

FIGURE 2–2 Organization of the phospholipid bilayer and associated proteins in a biological membrane. The phospholipid molecules each have two fatty acid chains (wavy lines) attached to a phosphate head (open circle). Proteins are shown as irregular colored globules. Many are integral proteins, which extend into the membrane, but peripheral proteins are attached to the inside or outside (not shown) of the membrane. Specific protein attachments and cholesterol commonly found in the bilayer are omitted for clarity. (Reproduced with permission from Widmaier EP, Raff H, Strang K: *Vander's Human Physiology: The Mechanisms of Body Function,* 11th ed. McGraw-Hill, 2008.)

as **pumps,** actively transporting ions across the membrane. Other proteins function as **carriers,** transporting substances down electrochemical gradients by facilitated diffusion. Still others are **ion channels,** which, when activated, permit the passage of ions into or out of the cell. The role of the pumps, carriers, and ion channels in transport across the cell membrane is discussed below. Proteins in another group function as **receptors** that bind **ligands** or messenger molecules, initiating physiologic changes inside the cell. Proteins also function as **enzymes,** catalyzing reactions at the surfaces of the membrane. Examples from each of these groups are discussed later in this chapter.

The uncharged, hydrophobic portions of the proteins are usually located in the interior of the membrane, whereas the charged, hydrophilic portions are located on the surfaces. Peripheral proteins are attached to the surfaces of the membrane in various ways. One common way is attachment to glycosylated forms of phosphatidylinositol. Proteins held by these **glycosylphosphatidylinositol (GPI) anchors** (Figure 2–3) include enzymes such as alkaline phosphatase, various antigens, a number of CAMs, and three proteins that combat cell lysis by complement. Over 45 GPI-linked cell surface proteins have now been described in humans. Other proteins are **lipidated,** that is, they have specific lipids attached to them (Figure 2–3). Proteins may be **myristoylated, palmitoylated,** or **prenylated** (ie, attached to geranylgeranyl or farnesyl groups).

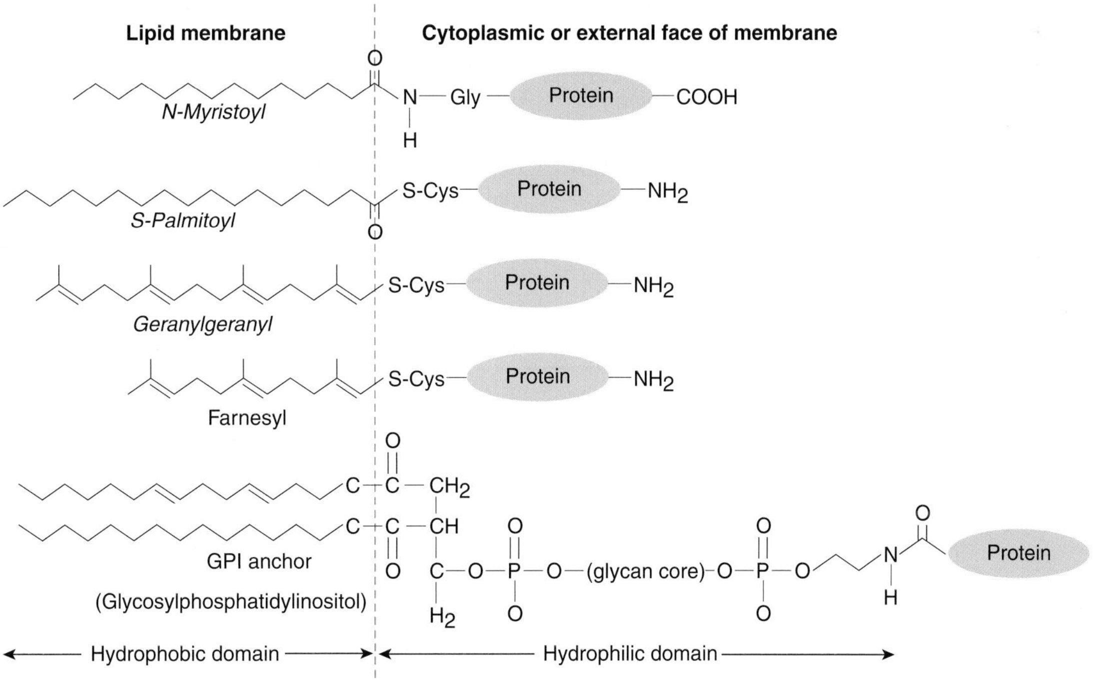

Lipid membrane
Cytoplasmic or external face of membrane
N-Myristoyl — N(H)—Gly—Protein—COOH
S-Palmitoyl — S-Cys—Protein—NH_2
Geranylgeranyl — S-Cys—Protein—NH_2
Farnesyl — S-Cys—Protein—NH_2
GPI anchor (Glycosylphosphatidylinositol) — O—P—O—(glycan core)—O—P—O—N(H)—Protein
Hydrophobic domain
Hydrophilic domain

FIGURE 2–3 Protein linkages to membrane lipids. Some are linked by their amino terminals, others by their carboxyl terminals. Many are attached via glycosylated forms of phosphatidylinositol (GPI anchors). (Adapted with permission from Fuller GM, Shields D: *Molecular Basis of Medical Cell Biology.* McGraw-Hill, 1998.)

The protein structure—and particularly the enzyme content—of biologic membranes varies not only from cell to cell, but also within the same cell. For example, some of the enzymes embedded in cell membranes are different from those in mitochondrial membranes. In epithelial cells, the enzymes in the cell membrane on the mucosal surface differ from those in the cell membrane on the basal and lateral margins of the cells; that is, the cells are **polarized.** Such polarization makes directional transport across epithelia possible. The membranes are dynamic structures, and their constituents are being constantly renewed at different rates. Some proteins are anchored to the cytoskeleton, but others move laterally in the membrane.

Underlying most cells is a thin, "fuzzy" layer plus some fibrils that collectively make up the **basement membrane** or, more properly, the **basal lamina.** The basal lamina and, more generally, the extracellular matrix are made up of many proteins that hold cells together, regulate their development, and determine their growth. These include collagens, laminins, fibronectin, tenascin, and various proteoglycans.

MITOCHONDRIA

Over a billion years ago, aerobic bacteria were engulfed by eukaryotic cells and evolved into **mitochondria,** providing the eukaryotic cells with the ability to form the energy-rich compound ATP by **oxidative phosphorylation.** Mitochondria perform other functions, including a role in the regulation of **apoptosis** (programmed cell death), but oxidative phosphorylation is the most crucial. Each eukaryotic cell can have hundreds to thousands of mitochondria. In mammals, they are generally depicted as sausage-shaped organelles (Figure 2–1), but their shape can be quite dynamic. Each has an outer membrane, an intermembrane space, an inner membrane, which is folded to form shelves **(cristae),** and a central matrix space. The enzyme complexes responsible for oxidative phosphorylation are lined up on the cristae (Figure 2–4).

Consistent with their origin from aerobic bacteria, the mitochondria have their own genome. There is much less DNA in the mitochondrial genome than in the nuclear genome, and 99% of the proteins in the mitochondria are the products of nuclear genes, but mitochondrial DNA is responsible for certain key components of the pathway for oxidative phosphorylation. Specifically, human mitochondrial DNA is a double-stranded circular molecule containing approximately 16,500 base pairs (compared with over a billion in nuclear DNA). It codes for 13 protein subunits that are associated with proteins encoded by nuclear genes to form four enzyme complexes plus two ribosomal and 22 transfer RNAs that are needed for protein production by the intramitochondrial ribosomes.

The enzyme complexes responsible for oxidative phosphorylation illustrate the interactions between the products of the mitochondrial genome and the nuclear genome. For example, complex I, reduced nicotinamide adenine dinucleotide dehydrogenase (NADH), is made up of seven protein subunits coded by mitochondrial DNA and 39 subunits coded by nuclear DNA. The origin of the subunits in the other complexes is shown in Figure 2–4. Complex II, succinate dehydrogenase-ubiquinone oxidoreductase; complex III, ubiquinonecytochrome c oxidoreductase; and complex IV, cytochrome c oxidase, act with complex I, coenzyme Q, and cytochrome c to convert metabolites to CO_2 and water. Complexes I, III, and IV pump protons (H^+) into the intermembrane space during this electron transfer. The protons then flow down their electrochemical gradient through complex V, ATP synthase, which harnesses this energy to generate ATP.

As zygote mitochondria are derived from the ovum, their inheritance is maternal. This maternal inheritance has been used as a tool to track evolutionary descent. Mitochondria have an ineffective DNA repair system, and the mutation rate for mitochondrial DNA is over 10 times the rate for nuclear DNA. A large number of relatively rare diseases have now been traced to mutations in mitochondrial DNA. These include disorders of tissues with high metabolic rates in which energy production is defective as a result of abnormalities in the production of ATP, as well as other disorders (Clinical Box 2–1).

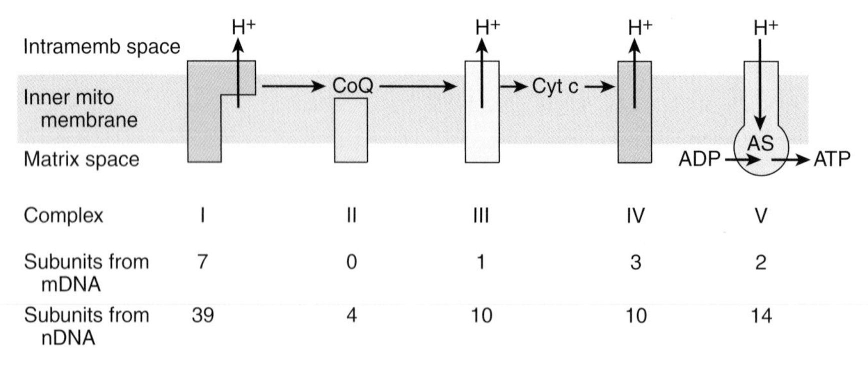

FIGURE 2–4 Components involved in oxidative phosphorylation in mitochondria and their origins. As enzyme complexes I through IV convert 2-carbon metabolic fragments to CO_2 and H_2O, protons (H^+) are pumped into the intermembrane space. The proteins diffuse back to the matrix space via complex V, ATP synthase (AS), in which ADP is converted to ATP. The enzyme complexes are made up of subunits coded by mitochondrial DNA (mDNA) and nuclear DNA (nDNA), and the figures document the contribution of each DNA to the complexes.

CLINICAL BOX 2–1

Mitochondrial Diseases

Mitochondrial diseases encompass at least 40 diverse disorders that are grouped because of their links to mitochondrial failure. These diseases can occur following inheritance or spontaneous mutations in mitochondrial or nuclear DNA that lead to altered functions of the mitochondrial proteins (or RNA). Depending on the target cell and/or tissues affected, symptoms resulting from mitochondrial diseases may include altered motor control, altered muscle output, gastrointestinal dysfunction, altered growth, diabetes, seizures, visual/hearing problems, lactic acidosis, developmental delays, susceptibility to infection or, cardiac, liver and respiratory disease. Although there is evidence for tissue-specific isoforms of mitochondrial proteins, mutations in these proteins do not fully explain the highly variable patterns or targeted organ systems observed with mitochondrial diseases.

THERAPEUTIC HIGHLIGHTS

With the diversity of disease types and the overall importance of mitochondria in energy production, it is not surprising that there is no single cure for mitochondrial diseases and focus remains on treating the symptoms when possible. For example, in some mitochondrial myopathies (ie, mitochondrial diseases associated with neuromuscular function), physical therapy may help to extend the range of movement of muscles and improve dexterity.

LYSOSOMES

In the cytoplasm of the cell there are large, somewhat irregular structures surrounded by membranes. The interior of these structures, which are called **lysosomes,** is more acidic than the rest of the cytoplasm, and external material such as endocytosed bacteria, as well as worn-out cell components, are digested in them. The interior is kept acidic by the action of a **proton pump,** or **H^+ ATPase.** This integral membrane protein uses the energy of ATP to move protons from the cytosol up their electrochemical gradient and keep the lysosome relatively acidic, near pH 5.0. Lysosomes can contain over 40 types of hydrolytic enzymes, some of which are listed in Table 2–1. Not surprisingly, these enzymes are all acid hydrolases, in that they function best at the acidic pH of the lysosomal compartment. This can be a safety feature for the cell; if the lysosomes were to break open and release their contents, the enzymes would not be efficient at the near neutral cytosolic pH (7.2), and thus would be unable to digest cytosolic enzymes they may encounter. Diseases associated with lysosomal dysfunction are discussed in Clinical Box 2–2.

TABLE 2–1 Some of the enzymes found in lysosomes and the cell components that are their substrates.

Enzyme	Substrate
Ribonuclease	RNA
Deoxyribonuclease	DNA
Phosphatase	Phosphate esters
Glycosidases	Complex carbohydrates; glycosides and polysaccharides
Arylsulfatases	Sulfate esters
Collagenase	Collagens
Cathepsins	Proteins

CLINICAL BOX 2–2

Lysosomal Diseases

When a lysosomal enzyme is congenitally absent, the lysosomes become engorged with the material the enzyme normally degrades. This eventually leads to one of the **lysosomal diseases** (also called lysosomal storage diseases). There are over 50 such diseases currently recognized. For example, Fabry disease is caused by a deficiency in α-galactosidase; Gaucher disease is caused by a deficiency in β-galactocerebrosidase and Tay–Sachs disease, which causes mental retardation and blindness, is caused by the loss of hexosaminidase A, a lysosomal enzyme that catalyzes the biodegradation of gangliosides (fatty acid derivatives). Such individual lysosomal diseases are rare, but they are serious and can be fatal.

THERAPEUTIC HIGHLIGHTS

Since there are many different lysosomal disorders, treatments vary considerably and "cures" remain elusive for most of these diseases. Much of the care is focused on managing symptoms of each specific disorder. Enzyme replacement therapy has shown to be effective for certain lysosomal diseases, including Gaucher's disease and Fabry's disease. However, the long-term effectiveness and the tissue specific effects of many of the enzyme replacement treatments have not yet been established. Alternative approaches include bone marrow or stem cell transplantation. Again, these are limited in use and medical advances are necessary to fully combat this group of diseases.

PEROXISOMES

Peroxisomes are 0.5 μm in diameter, are surrounded by a membrane, and contain enzymes that can either produce H_2O_2 **(oxidases)** or break it down **(catalases).** Proteins are directed to the peroxisome by a unique signal sequence with the help of protein chaperones, **peroxins.** The peroxisome membrane contains a number of peroxisome-specific proteins that are concerned with transport of substances into and out of the matrix of the peroxisome. The matrix contains more than 40 enzymes, which operate in concert with enzymes outside the peroxisome to catalyze a variety of anabolic and catabolic reactions (eg, breakdown of lipids). Peroxisomes can form by budding of the endoplasmic reticulum, or by division. A number of synthetic compounds were found to cause proliferation of peroxisomes by acting on receptors in the nuclei of cells. These **peroxisome proliferator-activated receptors (PPARs)** are members of the nuclear receptor superfamily. When activated, they bind to DNA, producing changes in the production of mRNAs. The known effects for PPARs are extensive and can affect most tissues and organs.

CYTOSKELETON

All cells have a **cytoskeleton,** a system of fibers that not only maintains the structure of the cell but also permits it to change shape and move. The cytoskeleton is made up primarily of **microtubules, intermediate filaments,** and **microfilaments** (Figure 2–5), along with proteins that anchor them and tie them together. In addition, proteins and organelles move along microtubules and microfilaments from one part of the cell to another, propelled by molecular motors.

Cytoskeletal filaments	Diameter (nm)	Protein subunit
Microfilament	7	Actin
Intermediate filament	10	Several proteins
Microtubule	25	Tubulin

FIGURE 2–5 Cytoskeletal elements of the cell. Artistic impressions that depict the major cytoskeletal elements are shown on the left, with basic properties of these elements on the right. (Reproduced with permission from Widmaier EP, Raff H, Strang KT: *Vander's Human Physiology: The Mechanisms of Body Function,* 11th ed. McGraw-Hill, 2008.)

Microtubules (Figure 2–5 and Figure 2–6) are long, hollow structures with 5-nm walls surrounding a cavity 15 nm in diameter. They are made up of two globular protein subunits: α- and β-tubulin. A third subunit, γ-tubulin, is associated with the production of microtubules by the centrosomes. The α and β subunits form heterodimers, which aggregate to form long tubes made up of stacked rings, with each ring usually containing 13 subunits. The tubules interact with GTP to facilitate their formation. Although microtubule subunits can be added to either end, microtubules are polar with assembly predominating at the "+" end and disassembly predominating at the "–" end. Both processes occur simultaneously in vitro. The growth of microtubules is temperature sensitive (disassembly is favored under cold conditions) as well as under the control of a variety of cellular factors that can directly interact with microtubules in the cell.

Because of their constant assembly and disassembly, microtubules are a dynamic portion of the cytoskeleton. They provide the tracks along which several different molecular motors move transport vesicles, organelles such as secretory granules, and mitochondria from one part of the cell to another. They also form the spindle, which moves the chromosomes in mitosis. Cargo can be transported in either direction on microtubules.

There are several drugs available that disrupt cellular function through interaction with microtubules. Microtubule assembly is prevented by colchicine and vinblastine. The

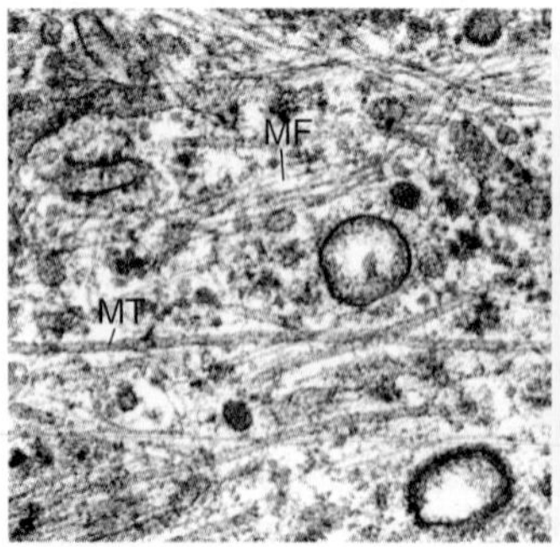

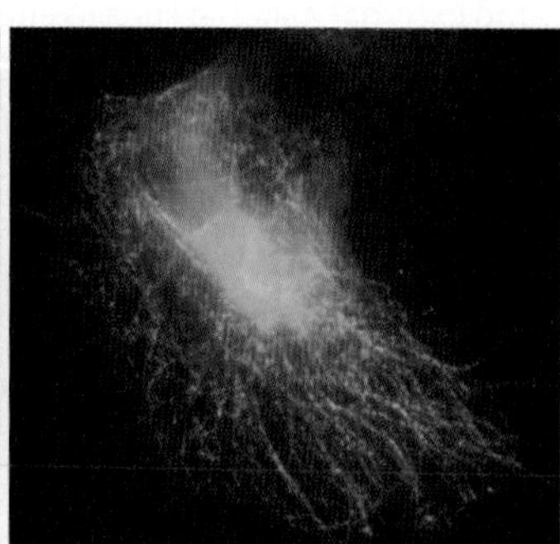

FIGURE 2–6 Microfilaments and microtubules. Electron-micrograph **(Left)** of the cytoplasm of a fibroblast, displaying actin microfilaments (MF) and microtubules (MT). Fluorescent micrographs of airway epithelial cells displaying actin microfilaments stained with phalloidin **(Middle)** and microtubules visualized with an antibody to β-tubulin **(Right).** Both fluorescent micrographs are counterstained with Hoechst dye (blue) to visualize nuclei. Note the distinct differences in cytoskeletal structure. (For left; Courtesy of E Katchburian.)

anticancer drug **paclitaxel (Taxol)** binds to microtubules and makes them so stable that organelles cannot move. Mitotic spindles cannot form, and the cells die.

Intermediate filaments (Figure 2–5) are 8–14 nm in diameter and are made up of various subunits. Some of these filaments connect the nuclear membrane to the cell membrane. They form a flexible scaffolding for the cell and help it resist external pressure. In their absence, cells rupture more easily, and when they are abnormal in humans, blistering of the skin is common. The proteins that make up intermediate filaments are cell-type specific, and are thus frequently used as cellular markers. For example, vimentin is a major intermediate filament in fibroblasts, whereas cytokeratin is expressed in epithelial cells.

Microfilaments (Figures 2–5 and 2–6) are long solid fibers with a 4–6 nm diameter that are made up of **actin.** Although actin is most often associated with muscle contraction, it is present in all types of cells. It is the most abundant protein in mammalian cells, sometimes accounting for as much as 15% of the total protein in the cell. Its structure is highly conserved; for example, 88% of the amino acid sequences in yeast and rabbit actin are identical. Actin filaments polymerize and depolymerize in vivo, and it is not uncommon to find polymerization occurring at one end of the filament while depolymerization is occurring at the other end. **Filamentous (F) actin** refers to intact microfilaments and **globular (G) actin** refers to the unpolymerized protein actin subunits. F-actin fibers attach to various parts of the cytoskeleton and can interact directly or indirectly with membrane-bound proteins. They reach to the tips of the microvilli on the epithelial cells of the intestinal mucosa. They are also abundant in the lamellipodia that cells put out when they crawl along surfaces. The actin filaments interact with integrin receptors and form **focal adhesion complexes,** which serve as points of traction with the surface over which the cell pulls itself. In addition, some molecular motors use microfilaments as tracks.

MOLECULAR MOTORS

The molecular motors that move proteins, organelles, and other cell parts (collectively referred to as "cargo") to all parts of the cell are 100–500 kDa ATPases. They attach to their cargo at one end of the molecule and to microtubules or actin polymers with the other end, sometimes referred to as the "head." They convert the energy of ATP into movement along the cytoskeleton, taking their cargo with them. There are three super families of molecular motors: **kinesin, dynein,** and **myosin.** Examples of individual proteins from each superfamily are shown in Figure 2–7. It is important to note that there is extensive variation among superfamily members, allowing for the specialization of function (eg, choice of cargo, cytoskeletal filament type, and/or direction of movement).

The conventional form of **kinesin** is a doubleheaded molecule that tends to move its cargo toward the "+" ends of microtubules. One head binds to the microtubule and then bends its neck while the other head swings forward and binds, producing almost continuous movement. Some kinesins are associated with mitosis and meiosis. Other kinesins perform different functions, including, in some instances, moving cargo to the "–" end of microtubules. **Dyneins** have two heads, with their neck pieces embedded in a complex of proteins. **Cytoplasmic dyneins** have a function like that of conventional kinesin, except they tend to move particles and membranes to the "–" end of the microtubules. The multiple forms of **myosin** in the body are divided into 18 classes. The heads of myosin molecules bind to actin and produce motion by bending their neck

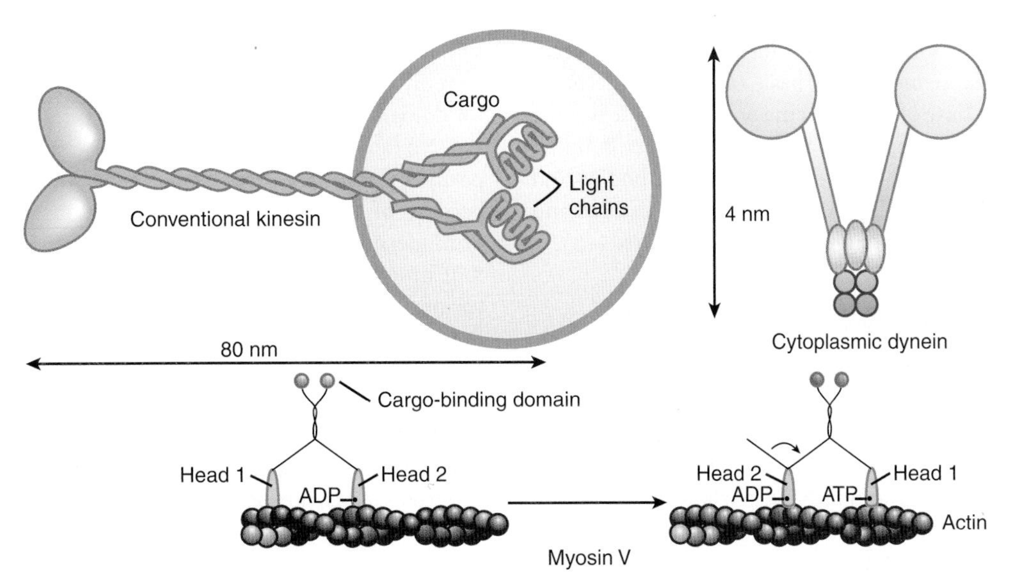

FIGURE 2–7 Three examples of molecular motors. Conventional kinesin is shown attached to cargo, in this case a membrane-bound organelle. The way that myosin V "walks" along a microtubule is also shown. Note that the heads of the motors hydrolyze ATP and use the energy to produce motion.

regions (myosin II) or walking along microfilaments, one head after the other (myosin V). In these ways, they perform functions as diverse as contraction of muscle and cell migration.

CENTROSOMES

Near the nucleus in the cytoplasm of eukaryotic animal cells is a **centrosome.** The centrosome is made up of two **centrioles** and surrounding amorphous **pericentriolar material.** The centrioles are short cylinders arranged so that they are at right angles to each other. Microtubules in groups of three run longitudinally in the walls of each centriole (Figure 2–1). Nine of these triplets are spaced at regular intervals around the circumference.

The centrosomes are **microtubule-organizing centers (MTOCs)** that contain γ-tubulin. The microtubules grow out of this γ-tubulin in the pericentriolar material. When a cell divides, the centrosomes duplicate themselves, and the pairs move apart to the poles of the mitotic spindle, where they monitor the steps in cell division. In multinucleate cells, a centrosome is near each nucleus.

CILIA

Cilia are specialized cellular projections that are used by unicellular organisms to propel themselves through liquid and by multicellular organisms to propel mucus and other substances over the surface of various epithelia. Additionally, virtually all cells in the human body contain a primary cilium that emanates from the surface. The primary cilium serves as a sensory organelle that receives both mechanical and chemical signals from other cells and the environment. Cilia are functionally indistinct from the eukaryotic flagella of sperm cells. Within the cilium there is an **axoneme** that comprises a unique arrangement of nine outer microtubule doublets and two inner microtubules ("9+2" arrangement). Along this cytoskeleton is **axonemal dynein.** Coordinated dynein–microtubule interactions within the axoneme are the basis of ciliary and sperm movement. At the base of the axoneme and just inside lies the **basal body.** It has nine circumferential triplet microtubules, like a centriole, and there is evidence that basal bodies and centrioles are interconvertible. A wide variety of diseases and disorders arise from dysfunctional cilia (Clinical Box 2–3).

CLINICAL BOX 2–3

Ciliary Diseases

Primary ciliary dyskinesia refers to a set of inherited disorders that limit ciliary structure and/or function. Disorders associated with ciliary dysfunction have long been recognized in the conducting airway. Altered ciliary function in the conducting airway can slow the mucociliary escalator and result in airway obstruction and increased infection. Dysregulation of ciliary function in sperm cells has also been well characterized to result in loss of motility and infertility. Ciliary defects in the function or structure of primary cilia have been shown to have effects on a variety of tissues/organs. As would be expected, such diseases are quite varied in their presentation, largely due to the affected tissue and include mental retardation, retinal blindness, obesity, polycystic kidney disease, liver fibrosis, ataxia, and some forms of cancer.

THERAPEUTIC HIGHLIGHTS

The severity in ciliary disorders can vary widely, and treatments targeted to individual organs also vary. Treatment of ciliary dyskinesia in the conducting airway is focused on keeping the airways clear and free of infection. Strategies include routine washing and suctioning of the sinus cavities and ear canals and liberal use of antibiotics. Other treatments that keep the airway from being obstructed (eg, bronchodilators, mucolytics, and steroids) are also commonly used.

CELL ADHESION MOLECULES

Cells are attached to the basal lamina and to each other by **CAMs** that are prominent parts of the intercellular connections described below. These adhesion proteins have attracted great attention in recent years because of their unique structural and signaling functions found to be important in embryonic development and formation of the nervous system and other tissues, in holding tissues together in adults, in inflammation and wound healing, and in the metastasis of tumors. Many CAMs pass through the cell membrane and are anchored to the cytoskeleton inside the cell. Some bind to like molecules on other cells (homophilic binding), whereas others bind to nonself molecules (heterophilic binding). Many bind to **laminins,** a family of large cross-shaped molecules with multiple receptor domains in the extracellular matrix.

Nomenclature in the CAM field is somewhat chaotic, partly because the field is growing so rapidly and partly because of the extensive use of acronyms, as in other areas of modern biology. However, the CAMs can be divided into four broad families: (1) **integrins,** heterodimers that bind to various receptors; (2) adhesion molecules of the **IgG superfamily** of immunoglobulins; (3) **cadherins,** Ca^{2+}-dependent molecules that mediate cell-to-cell adhesion by homophilic reactions; and (4) **selectins,** which have lectin-like domains that bind carbohydrates. Specific functions of some of these molecules are addressed in later chapters.

The CAMs not only fasten cells to their neighbors, but they also transmit signals into and out of the cell. For example, cells that lose their contact with the extracellular matrix via integrins have a higher rate of apoptosis than anchored cells, and interactions between integrins and the cytoskeleton are involved in cell movement.

INTERCELLULAR CONNECTIONS

Intercellular junctions that form between the cells in tissues can be broadly split into two groups: junctions that fasten the cells to one another and to surrounding tissues, and junctions that permit transfer of ions and other molecules from one cell to another. The types of junctions that tie cells together and endow tissues with strength and stability include **tight junctions,** which are also known as the **zonula occludens** (Figure 2–8). The **desmosome** and **zonula adherens** also help to hold cells together, and the **hemidesmosome** and **focal adhesions** attach cells to their basal laminas. The **gap junction** forms a cytoplasmic "tunnel" for diffusion of small molecules (< 1000 Da) between two neighboring cells.

Tight junctions characteristically surround the apical margins of the cells in epithelia such as the intestinal mucosa, the walls of the renal tubules, and the choroid plexus. They are also important to endothelial barrier function. They are made up of ridges—half from one cell and half from the other—which adhere so strongly at cell junctions that they almost obliterate the space between the cells. There are three main families of transmembrane proteins that contribute to tight junctions: **occludin, junctional adhesion molecules (JAMs),** and **claudins;** and several more proteins that interact from the cytosolic side. Tight junctions permit the passage of some ions and solute in between adjacent cells (**paracellular pathway**) and the degree of this "leakiness" varies, depending in part on the protein makeup of the tight junction. Extracellular fluxes of ions and solute across epithelia at these junctions are a significant part of overall ion and solute flux. In addition, tight junctions prevent the movement of proteins in the plane of the membrane, helping to maintain the different distribution of transporters and channels in the apical and basolateral cell membranes that make transport across epithelia possible.

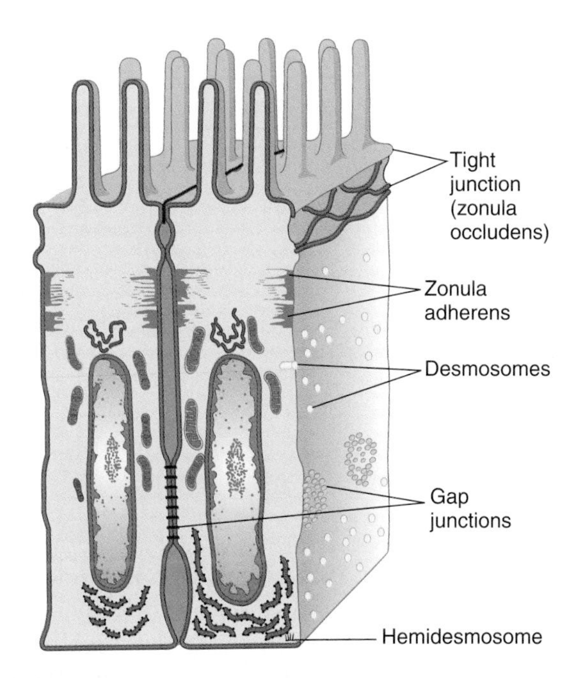

FIGURE 2–8 Intercellular junctions in the mucosa of the small intestine. Tight junctions (zonula occludens), adherens junctions (zonula adherens), desmosomes, gap junctions, and hemidesmosomes are all shown in relative positions in a polarized epithelial cell.

In epithelial cells, each zonula adherens is usually a continuous structure on the basal side of the zonula occludens, and it is a major site of attachment for intracellular microfilaments. It contains cadherins.

Desmosomes are patches characterized by apposed thickenings of the membranes of two adjacent cells. Attached to the thickened area in each cell are intermediate filaments, some running parallel to the membrane and others radiating away from it. Between the two membrane thickenings the intercellular space contains filamentous material that includes cadherins and the extracellular portions of several other transmembrane proteins.

Hemidesmosomes look like half-desmosomes that attach cells to the underlying basal lamina and are connected intracellularly to intermediate filaments. However, they contain integrins rather than cadherins. Focal adhesions also attach cells to their basal laminas. As noted previously, they are labile structures associated with actin filaments inside the cell, and they play an important role in cell movement.

GAP JUNCTIONS

At gap junctions, the intercellular space narrows from 25 to 3 nm, and units called **connexons** in the membrane of each cell are lined up with one another to form the dodecameric gap junction (Figure 2–9). Each connexon is made up of six protein subunits called **connexins.** They surround a channel that, when lined up with the channel in the corresponding connexon in the adjacent cell, permits substances to pass between the cells without entering the ECF. The diameter of the channel is normally about 2 nm, which permits the passage of ions, sugars, amino acids, and other solutes with molecular weights up to about 1000. Gap junctions thus permit the rapid propagation of electrical activity from cell to cell, as well as the exchange of various chemical messengers. However, the gap junction channels are not simply passive, nonspecific conduits. At least 20 different genes code for connexins in humans, and mutations in these genes can lead to diseases that are highly selective in terms of the tissues involved and the type of communication between cells produced (Clinical Box 2–4). Experiments in which particular connexins are deleted by gene manipulation or replaced with different connexins confirm that the particular connexin subunits that make up connexons determine their permeability and selectivity. It should be noted that connexons can also provide a conduit for regulated passage of small molecules between the cytoplasm and the ECF. Such movement can allow additional signaling pathways between and among cells in a tissue.

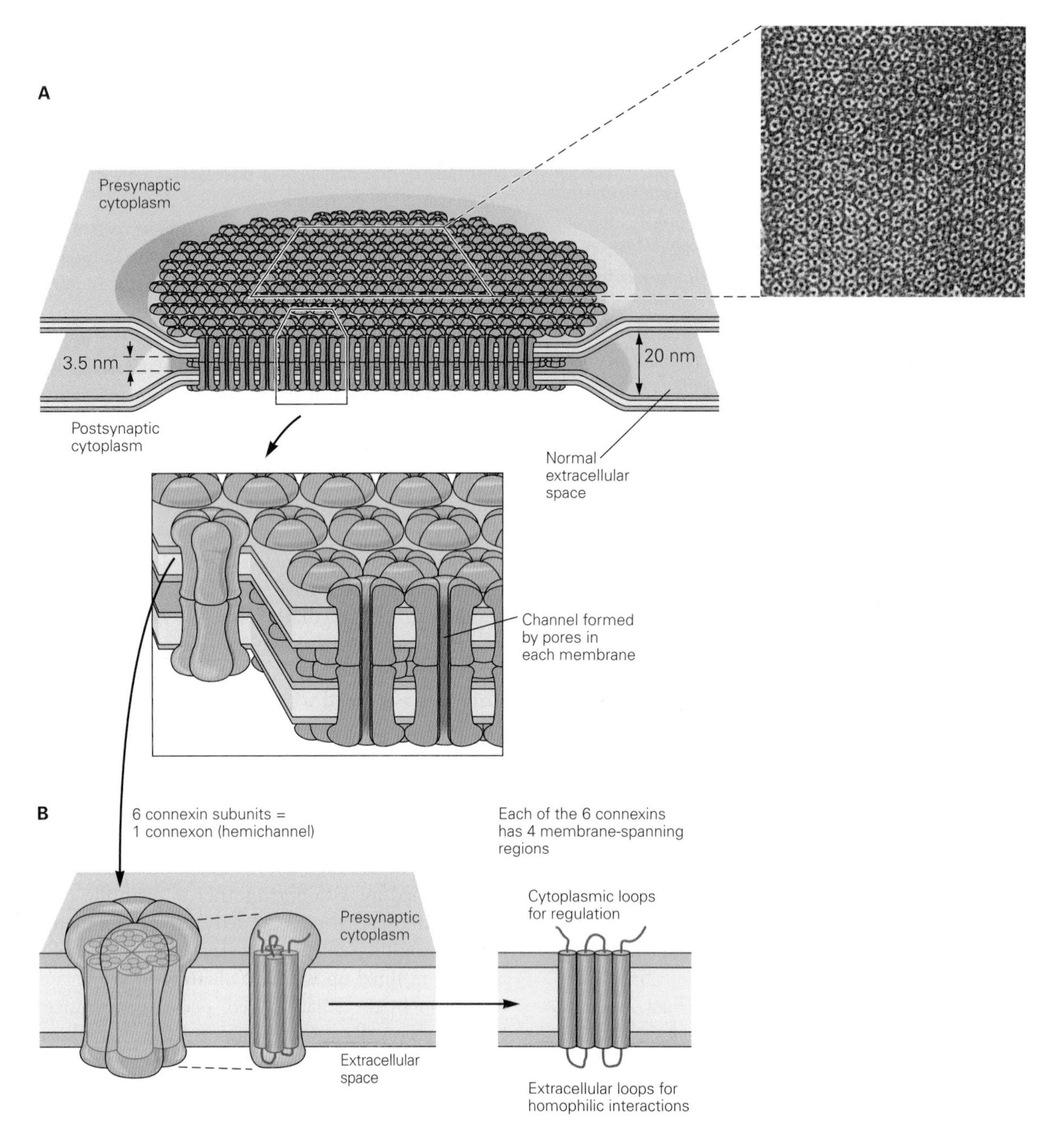

FIGURE 2–9 Gap junction connecting the cytoplasm of two cells. A) A gap junction plaque, or collection of individual gap junctions, is shown to form multiple pores between cells that allow for the transfer of small molecules. Inset is electronmicrograph from rat liver (N. Gilula). **B)** Topographical depiction of individual connexon and corresponding six connexin proteins that traverse the membrane. Note that each connexin traverses the membrane four times. (Reproduced with permission from Kandel ER, Schwartz JH, Jessell TM (editors): *Principles of Neural Science*, 4th ed. McGraw-Hill, 2000.)

NUCLEUS & RELATED STRUCTURES

A nucleus is present in all eukaryotic cells that divide. The nucleus is made up in large part of the **chromosomes,** the structures in the nucleus that carry a complete blueprint for all the heritable species and individual characteristics of the animal. Except in germ cells, the chromosomes occur in pairs, one originally from each parent. Each chromosome is made up of a giant molecule of **DNA.** The DNA strand is about 2 m long, but it can fit in the nucleus because at intervals it is wrapped around a core of histone proteins to form a **nucleosome.** There are about 25 million nucleosomes in each nucleus. Thus, the structure of the chromosomes has been likened to a string of beads. The beads are the nucleosomes, and the linker DNA between them is the string. The whole complex of DNA and proteins is called **chromatin.** During cell division, the coiling around histones is loosened, probably by acetylation of

CLINICAL BOX 2–4

Connexins in Disease

In recent years, there has been an explosion of information related to the in vivo functions of connexins, growing out of work on connexin knock-outs in mice and the analysis of mutations in human connexins. The mouse knock-outs demonstrated that connexin deletions lead to electrophysiological defects in the heart and predisposition to sudden cardiac death, female sterility, abnormal bone development, abnormal growth in the liver, cataracts, hearing loss, and a host of other abnormalities. Information from these and other studies has allowed for the identification of several connexin mutations now known to be responsible for almost 20 different human diseases. These diseases include several skin disorders such as Clouston Syndrome (a connexin 30 (Cx30) defect) and erythrokeratoderma variabilis (Cx30.3 and Cx31); inherited deafness (Cx26, Cx30, and Cx31); predisposition to myoclonic epilepsy (Cx36), predisposition to arteriosclerosis (Cx37); cataract (Cx46 and Cx50); idiopathic atrial fibrillation (Cx40); and X-linked Charcot-Marie-Tooth disease (Cx32). It is interesting to note that each of these target tissues for disease contain other connexins that do not fully compensate for loss of the crucial connexins in disease development. Understanding how loss of individual connexins alters cell physiology to contribute to these and other human diseases is an area of intense research.

the histones, and pairs of chromosomes become visible, but between cell divisions only clumps of chromatin can be discerned in the nucleus. The ultimate units of heredity are the **genes** on the chromosomes). As discussed in Chapter 1, each gene is a portion of the DNA molecule.

The nucleus of most cells contains a **nucleolus** (Figure 2–1), a patchwork of granules rich in **RNA.** In some cells, the nucleus contains several of these structures. Nucleoli are most prominent and numerous in growing cells. They are the site of synthesis of ribosomes, the structures in the cytoplasm in which proteins are synthesized.

The interior of the nucleus has a skeleton of fine filaments that are attached to the **nuclear membrane,** or **envelope** (Figure 2–1), which surrounds the nucleus. This membrane is a double membrane, and spaces between the two folds are called **perinuclear cisterns.** The membrane is permeable only to small molecules. However, it contains **nuclear pore complexes.** Each complex has eightfold symmetry and is made up of about 100 proteins organized to form a tunnel through which transport of proteins and mRNA occurs. There are many transport pathways; many proteins that participate in these pathways, including **importins** and **exportins** have been isolated and characterized. Much current research is focused on transport into and out of the nucleus, and a more detailed understanding of these processes should emerge in the near future.

ENDOPLASMIC RETICULUM

The **endoplasmic reticulum** is a complex series of tubules in the cytoplasm of the cell (Figure 2–1; Figure 2–10; and Figure 2–11). The inner limb of its membrane is continuous with a segment of the nuclear membrane, so in effect this part of the nuclear membrane is a cistern of the endoplasmic reticulum. The tubule walls are made up of membrane. In **rough,** or **granular, endoplasmic reticulum,** ribosomes are attached to the cytoplasmic side of the membrane, whereas in **smooth,** or **agranular, endoplasmic reticulum,** ribosomes are absent. Free ribosomes are also found in the cytoplasm. The granular endoplasmic reticulum is concerned with protein synthesis

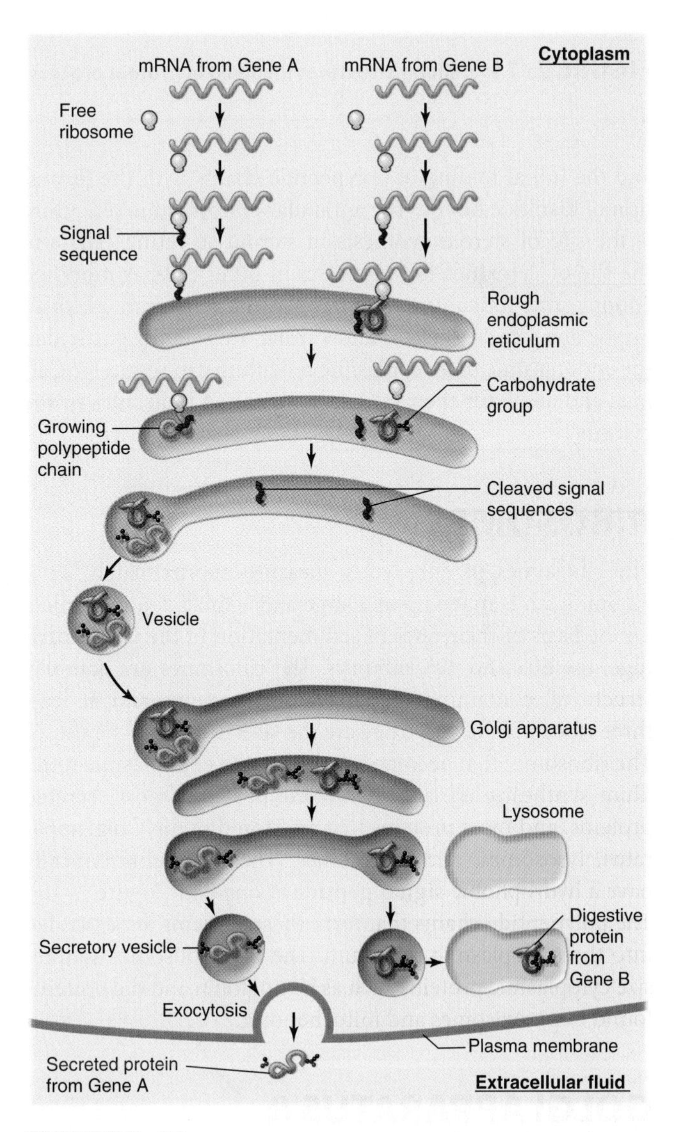

FIGURE 2–10 Rough endoplasmic reticulum and protein translation. Messenger RNA and ribosomes meet up in the cytosol for translation. Proteins that have appropriate signal peptides begin translation, and then associate with the endoplasmic reticulum (ER) to complete translation. The association of ribosomes is what gives the ER its "rough" appearance. (Reproduced with permission from Widmaier EP, Raff H, Strang KT: *Vander's Human Physiology: The Mechanisms of Body Function*, 11th ed. McGraw-Hill, 2008.)

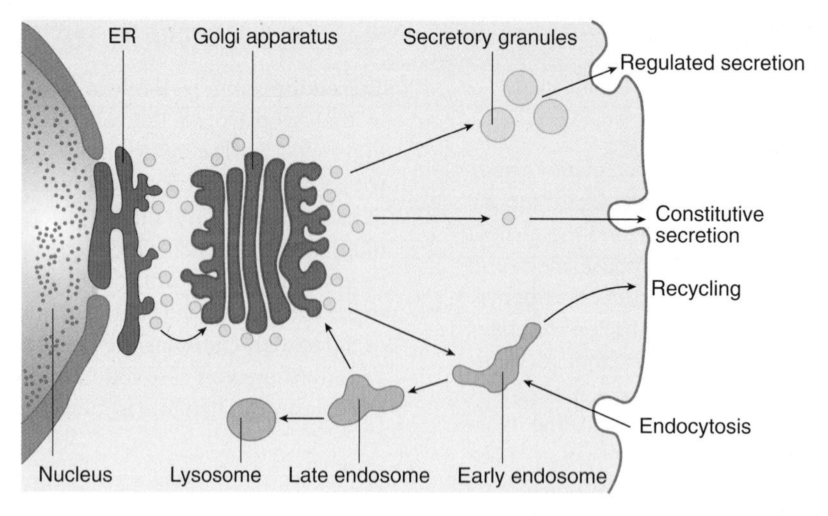

FIGURE 2–11 Cellular structures involved in protein processing. See text for details.

and the initial folding of polypeptide chains with the formation of disulfide bonds. The agranular endoplasmic reticulum is the site of steroid synthesis in steroid-secreting cells and the site of detoxification processes in other cells. A modified endoplasmic reticulum, the sarcoplasmic reticulum, plays an important role in skeletal and cardiac muscle. In particular, the endoplasmic or sarcoplasmic reticulum can sequester Ca^{2+} ions and allow for their release as signaling molecules in the cytosol.

RIBOSOMES

The ribosomes in eukaryotes measure approximately 22 × 32 nm. Each is made up of a large and a small subunit called, on the basis of their rates of sedimentation in the ultracentrifuge, the 60S and 40S subunits. The ribosomes are complex structures, containing many different proteins and at least three ribosomal RNAs. They are the sites of protein synthesis. The ribosomes that become attached to the endoplasmic reticulum synthesize all transmembrane proteins, most secreted proteins, and most proteins that are stored in the Golgi apparatus, lysosomes, and endosomes. These proteins typically have a hydrophobic **signal peptide** at one end (Figure 2–10). The polypeptide chains that form these proteins are extruded into the endoplasmic reticulum. The free ribosomes synthesize cytoplasmic proteins such as hemoglobin and the proteins found in peroxisomes and mitochondria.

GOLGI APPARATUS & VESICULAR TRAFFIC

The Golgi apparatus is a collection of membrane-enclosed sacs (cisternae) that are stacked like dinner plates (Figure 2–1). One or more Golgi apparati are present in all eukaryotic cells, usually near the nucleus. Much of the organization of the Golgi is directed at proper glycosylation of proteins and lipids. There are more than 200 enzymes that function to add, remove, or modify sugars from proteins and lipids in the Golgi apparatus.

The Golgi apparatus is a polarized structure, with cis and trans sides (Figures 2–1; 2–10; 2–11). Membranous vesicles containing newly synthesized proteins bud off from the granular endoplasmic reticulum and fuse with the cistern on the cis side of the apparatus. The proteins are then passed via other vesicles to the middle cisterns and finally to the cistern on the trans side, from which vesicles branch off into the cytoplasm. From the trans Golgi, vesicles shuttle to the lysosomes and to the cell exterior via constitutive and nonconstitutive pathways, both involving **exocytosis.** Conversely, vesicles are pinched off from the cell membrane by **endocytosis** and pass to endosomes. From there, they are recycled.

Vesicular traffic in the Golgi, and between other membranous compartments in the cell, is regulated by a combination of common mechanisms along with special mechanisms that determine where inside the cell they will go. One prominent feature is the involvement of a series of regulatory proteins controlled by GTP or GDP binding **(small G proteins)** associated with vesicle assembly and delivery. A second prominent feature is the presence of proteins called SNAREs (for soluble N-ethylmaleimide-sensitive factor attachment receptor). The v- (for vesicle) SNAREs on vesicle membranes interact in a lock-and-key fashion with t- (for target) SNAREs. Individual vesicles also contain structural protein or lipids in their membrane that help to target them for specific membrane compartments (eg, Golgi sacs, cell membranes).

QUALITY CONTROL

The processes involved in protein synthesis, folding, and migration to the various parts of the cell are so complex that it is remarkable that more errors and abnormalities do not occur. The fact that these processes work as well as they do is because of mechanisms at each level that are responsible for "quality control." Damaged DNA is detected and repaired or bypassed. The various RNAs are also checked during the translation process. Finally, when the protein chains are in the endoplasmic reticulum and Golgi apparatus, defective structures are detected and the abnormal proteins are degraded in lysosomes and proteasomes. The net result is a remarkable accuracy in the production of the proteins needed for normal body function.

APOPTOSIS

In addition to dividing and growing under genetic control, cells can die and be absorbed under genetic control. This process is called **programmed cell death,** or **apoptosis** (Gr. apo "away" + ptosis "fall"). It can be called "cell suicide" in the sense that the cell's own genes play an active role in its demise. It should be distinguished from necrosis ("cell murder"), in which healthy cells are destroyed by external processes such as inflammation.

Apoptosis is a very common process during development and in adulthood. In the central nervous system (CNS), large numbers of neurons are produced and then die during the remodeling that occurs during development and synapse formation. In the immune system, apoptosis gets rid of inappropriate clones of immunocytes and is responsible for the lytic effects of glucocorticoids on lymphocytes. Apoptosis is also an important factor in processes such as removal of the webs between the fingers in fetal life and regression of duct systems in the course of sexual development in the fetus. In adults, it participates in the cyclic breakdown of the endometrium that leads to menstruation. In epithelia, cells that lose their connections to the basal lamina and neighboring cells undergo apoptosis. This is responsible for the death of the enterocytes sloughed off the tips of intestinal villi. Abnormal apoptosis probably occurs in autoimmune diseases, neurodegenerative diseases, and cancer. It is interesting that apoptosis occurs in invertebrates, including nematodes and insects. However, its molecular mechanism is much more complex than that in vertebrates.

One final common pathway bringing about apoptosis is activation of **caspases,** a group of cysteine proteases. Many of these have been characterized to date in mammals; 11 have been found in humans. They exist in cells as inactive proenzymes until activated by the cellular machinery. The net result is DNA fragmentation, cytoplasmic and chromatin condensation, and eventually membrane bleb formation, with cell breakup and removal of the debris by phagocytes (see Clinical Box 2–5).

TRANSPORT ACROSS CELL MEMBRANES

There are several mechanisms of transport across cellular membranes. Primary pathways include exocytosis, endocytosis, movement through ion channels, and primary and secondary active transport. Each of these are discussed below.

CLINICAL BOX 2–5

Molecular Medicine

Fundamental research on molecular aspects of genetics, regulation of gene expression, and protein synthesis has been paying off in clinical medicine at a rapidly accelerating rate.

One early dividend was an understanding of the mechanisms by which antibiotics exert their effects. Almost all act by inhibiting protein synthesis at one or another of the steps described previously. Antiviral drugs act in a similar way; for example, acyclovir and ganciclovir act by inhibiting DNA polymerase. Some of these drugs have this effect primarily in bacteria, but others inhibit protein synthesis in the cells of other animals, including mammals. This fact makes antibiotics of great value for research as well as for treatment of infections.

Single genetic abnormalities that cause over 600 human diseases have now been identified. Many of the diseases are rare, but others are more common and some cause conditions that are severe and eventually fatal. Examples include the defectively regulated Cl^- channel in cystic fibrosis and the unstable **trinucleotide repeats** in various parts of the genome that cause Huntington's disease, the fragile X syndrome, and several other neurologic diseases. Abnormalities in mitochondrial DNA can also cause human diseases such as Leber's hereditary optic neuropathy and some forms of cardiomyopathy. Not surprisingly, genetic aspects of cancer are probably receiving the greatest current attention. Some cancers are caused by **oncogenes,** genes that are carried in the genomes of cancer cells and are responsible for producing their malignant properties. These genes are derived by somatic mutation from closely related **proto-oncogenes,** which are normal genes that control growth. Over 100 oncogenes have been described. Another group of genes produce proteins that suppress tumors, and more than 10 of these **tumor suppressor genes** have been described. The most studied of these is the p53 gene on human chromosome 17. The p53 protein produced by this gene triggers apoptosis. It is also a nuclear transcription factor that appears to increase production of a 21-kDa protein that blocks two cell cycle enzymes, slowing the cycle and permitting repair of mutations and other defects in DNA. The p53 gene is mutated in up to 50% of human cancers, with the production of p53 proteins that fail to slow the cell cycle and permit other mutations in DNA to persist. The accumulated mutations eventually cause cancer.

EXOCYTOSIS

Vesicles containing material for export are targeted to the cell membrane (Figure 2–11), where they bond in a similar manner to that discussed in vesicular traffic between Golgi stacks, via the v-SNARE/t-SNARE arrangement. The area of fusion then breaks down, leaving the contents of the vesicle outside and the cell membrane intact. This is the Ca^{2+}-dependent process of **exocytosis** (Figure 2–12).

Note that secretion from the cell occurs via two pathways (Figure 2–11). In the **nonconstitutive pathway,** proteins from the Golgi apparatus initially enter secretory granules, where processing of prohormones to the mature hormones occurs before exocytosis. The other pathway, the **constitutive pathway,** involves the prompt transport of proteins to the cell membrane in vesicles, with little or no processing or storage. The nonconstitutive pathway is sometimes called the **regulated pathway,** but this term is misleading because the output of proteins by the constitutive pathway is also regulated.

ENDOCYTOSIS

Endocytosis is the reverse of exocytosis. There are various types of endocytosis named for the size of particles being ingested as well as the regulatory requirements for the particular process. These include **phagocytosis, pinocytosis, clathrin-mediated endocytosis, caveolae-dependent uptake,** and **nonclathrin/noncaveolae endocytosis.**

Phagocytosis ("cell eating") is the process by which bacteria, dead tissue, or other bits of microscopic material are engulfed by cells such as the polymorphonuclear leukocytes of the blood. The material makes contact with the cell membrane, which then invaginates. The invagination is pinched off, leaving the engulfed material in the membrane-enclosed vacuole and the cell membrane intact. **Pinocytosis** ("cell drinking") is a similar process with the vesicles much smaller in size and the substances ingested are in solution. The small size membrane that is ingested with each event should not be misconstrued; cells undergoing active pinocytosis (eg, macrophages) can ingest the equivalent of their entire cell membrane in just 1 h.

Clathrin-mediated endocytosis occurs at membrane indentations where the protein **clathrin** accumulates. Clathrin molecules have the shape of triskelions, with three "legs" radiating from a central hub (Figure 2–13). As endocytosis progresses, the clathrin molecules form a geometric array that surrounds the endocytotic vesicle. At the neck of the vesicle, the GTP binding protein **dynamin** is involved, either directly or indirectly, in pinching off the vesicle. Once the complete vesicle is formed, the clathrin falls off and the three-legged proteins recycle to form another vesicle. The vesicle fuses with and dumps its contents into an **early endosome** (Figure 2–11). From the early

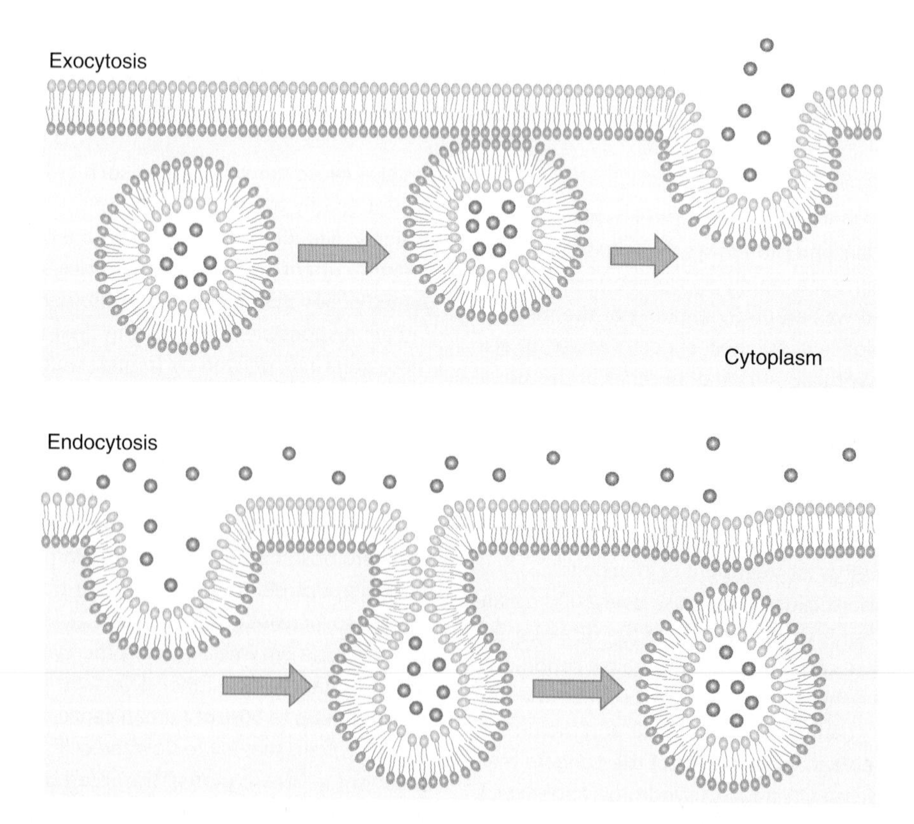

FIGURE 2–12 Exocytosis and endocytosis. Note that in exocytosis the cytoplasmic sides of two membranes fuse, whereas in endocytosis two noncytoplasmic sides fuse.

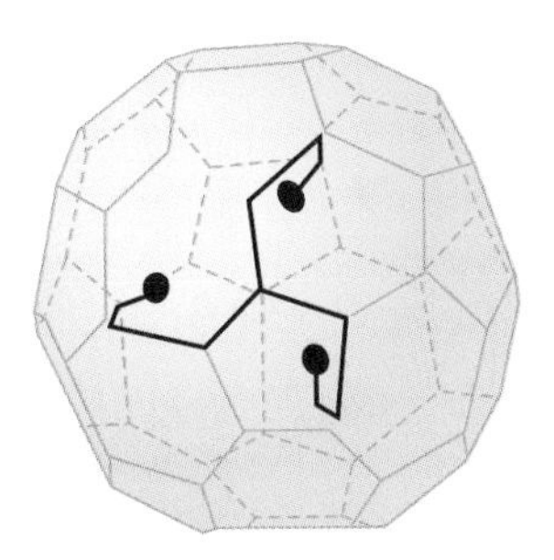

FIGURE 2–13 Clathrin molecule on the surface of an endocytotic vesicle. Note the characteristic triskelion shape and the fact that with other clathrin molecules it forms a net supporting the vesicle.

endosome, a new vesicle can bud off and return to the cell membrane. Alternatively, the early endosome can become a **late endosome** and fuse with a lysosome (Figure 2–11) in which the contents are digested by the lysosomal proteases. Clathrin-mediated endocytosis is responsible for the internalization of many receptors and the ligands bound to them—including, for example, nerve growth factor (NGF) and low-density lipoproteins. It also plays a major role in synaptic function.

It is apparent that exocytosis adds to the total amount of membrane surrounding the cell, and if membrane were not removed elsewhere at an equivalent rate, the cell would enlarge. However, removal of cell membrane occurs by endocytosis, and such exocytosis–endocytosis coupling maintains the surface area of the cell at its normal size.

RAFTS & CAVEOLAE

Some areas of the cell membrane are especially rich in cholesterol and sphingolipids and have been called **rafts.** These rafts are probably the precursors of flask-shaped membrane depressions called **caveolae** (little caves) when their walls become infiltrated with a protein called **caveolin** that resembles clathrin. There is considerable debate about the functions of rafts and caveolae, with evidence that they are involved in cholesterol regulation and transcytosis. It is clear, however, that cholesterol can interact directly with caveolin, effectively limiting the protein's ability to move around in the membrane. Internalization via caveolae involves binding of cargo to caveolin and regulation by dynamin. Caveolae are prominent in endothelial cells, where they help in the uptake of nutrients from the blood.

COATS & VESICLE TRANSPORT

It now appears that all vesicles involved in transport have protein coats. In humans, over 50 coat complex subunits have been identified. Vesicles that transport proteins from the trans Golgi to lysosomes have **assembly protein 1 (AP-1)** clathrin coats, and endocytotic vesicles that transport to endosomes have AP-2 clathrin coats. Vesicles that transport between the endoplasmic reticulum and the Golgi have coat proteins I and II (COPI and COPII). Certain amino acid sequences or attached groups on the transported proteins target the proteins for particular locations. For example, the amino acid sequence Asn–Pro–any amino acid–Tyr targets transport from the cell surface to the endosomes, and mannose-6-phosphate groups target transfer from the Golgi to mannose-6-phosphate receptors (MPR) on the lysosomes.

Various small G proteins of the Rab family are especially important in vesicular traffic. They appear to guide and facilitate orderly attachments of these vesicles. To illustrate the complexity of directing vesicular traffic, humans have 60 Rab proteins and 35 SNARE proteins.

MEMBRANE PERMEABILITY & MEMBRANE TRANSPORT PROTEINS

An important technique that has permitted major advances in our knowledge about transport proteins is **patch clamping.** A micropipette is placed on the membrane of a cell and forms a tight seal to the membrane. The patch of membrane under the pipette tip usually contains only a few transport proteins, allowing for their detailed biophysical study (Figure 2–14). The cell can be left intact **(cell-attached patch clamp).** Alternatively, the patch can be pulled loose from the cell, forming an **inside-out patch.** A third alternative is to suck out the patch with the micropipette still attached to the rest of the cell membrane, providing direct access to the interior of the cell **(whole cell recording).**

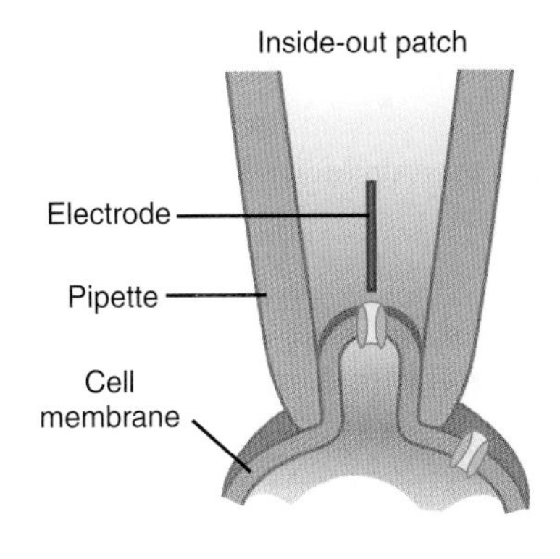

FIGURE 2–14 Patch clamp to investigate transport. In a patch clamp experiment, a small pipette is carefully maneuvered to seal off a portion of a cell membrane. The pipette has an electrode bathed in an appropriate solution that allows for recording of electrical changes through any pore in the membrane (shown below). The illustrated setup is termed an "inside-out patch" because of the orientation of the membrane with reference to the electrode. Other configurations include cell attached, whole cell, and outside-out patches. (Modified from Ackerman MJ, Clapham DE: Ion channels: Basic science and clinical disease. *N Engl J Med* 1997;336:1575.)

Small, nonpolar molecules (including O_2 and N_2) and small uncharged polar molecules such as CO_2 diffuse across the lipid membranes of cells. However, the membranes have very limited permeability to other substances. Instead, they cross the membranes by endocytosis and exocytosis and by passage through highly specific **transport proteins,** transmembrane proteins that form channels for ions or transport substances such as glucose, urea, and amino acids. The limited permeability applies even to water, with simple diffusion being supplemented throughout the body with various water channels **(aquaporins).** For reference, the sizes of ions and other biologically important substances are summarized in Table 2–2.

Some transport proteins are simple aqueous **ion channels,** though many of these have special features that make them selective for a given substance such as Ca^{2+} or, in the case of aquaporins, for water. These membrane-spanning proteins (or collections of proteins) have tightly regulated pores that can be **gated** opened or closed in response to local changes (Figure 2–15). Some are gated by alterations in membrane potential **(voltage-gated),** whereas others are opened or closed in response to a ligand **(ligand-gated).** The ligand is often external (eg, a neurotransmitter or a hormone). However, it can also be internal; intracellular Ca^{2+}, cAMP, lipids, or one of the G proteins produced in cells can bind directly to channels and activate them. Some channels are also opened by mechanical stretch, and these mechanosensitive channels play an important role in cell movement.

Other transport proteins are **carriers** that bind ions and other molecules and then change their configuration, moving the bound molecule from one side of the cell membrane to the other. Molecules move from areas of high concentration to areas of low concentration (down their **chemical gradient**),

TABLE 2–2 Size of hydrated ions and other substances of biologic interest.

Substance	Atomic or Molecular Weight	Radius (nm)
Cl^-	35	0.12
K^+	39	0.12
H_2O	18	0.12
Ca^{2+}	40	0.15
Na^+	23	0.18
Urea	60	0.23
Li^+	7	0.24
Glucose	180	0.38
Sucrose	342	0.48
Inulin	5000	0.75
Albumin	69,000	7.50

Data from Moore EW: *Physiology of Intestinal Water and Electrolyte Absorption.* American Gastroenterological Association, 1976.

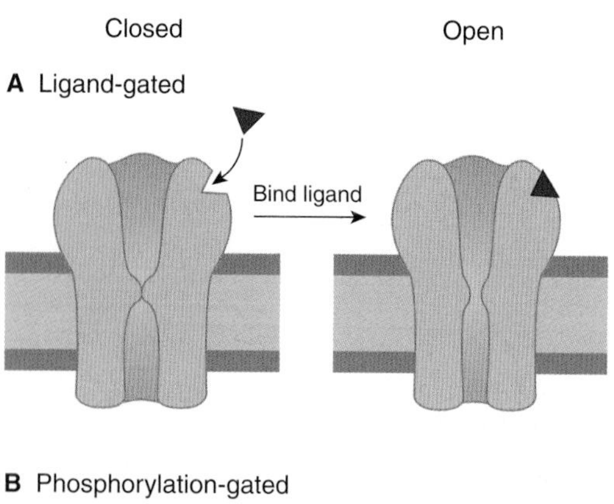

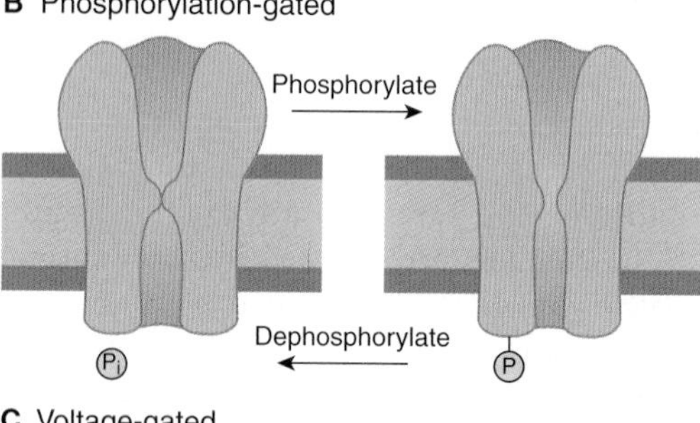

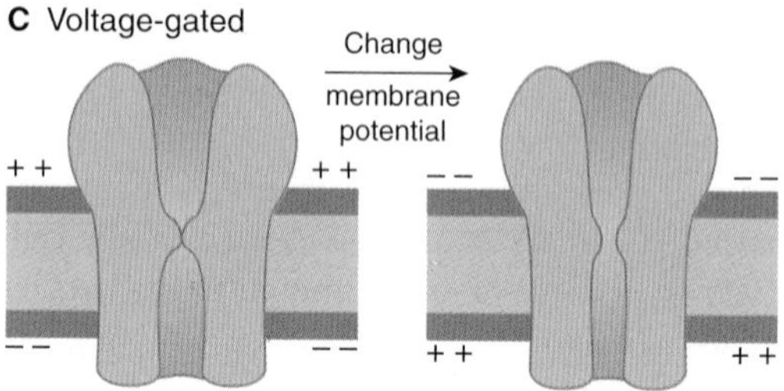

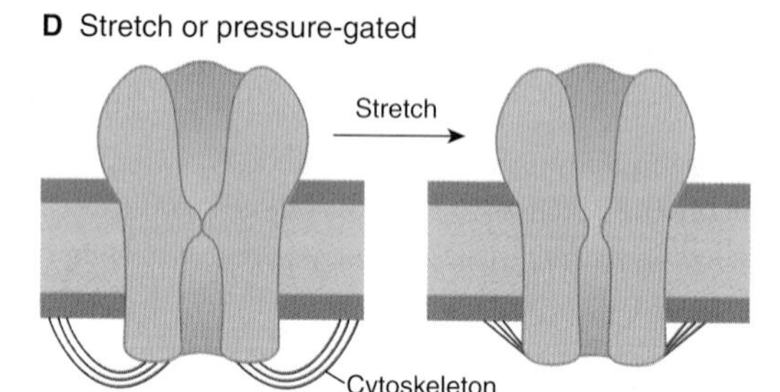

FIGURE 2–15 Regulation of gating in ion channels. Several types of gating are shown for ion channels. **A)** Ligand-gated channels open in response to ligand binding. **B)** Protein phosphorylation or dephosphorylation regulate opening and closing of some ion channels. **C)** Changes in membrane potential alter channel openings. **D)** Mechanical stretch of the membrane results in channel opening. (Reproduced with permission from Kandel ER, Schwartz JH, Jessell TM (editors): *Principles of Neural Science,* 4th ed. McGraw-Hill, 2000.)

and cations move to negatively charged areas whereas anions move to positively charged areas (down their **electrical gradient**). When carrier proteins move substances in the direction of their chemical or electrical gradients, no energy input is required and the process is called **facilitated diffusion.** A typical example is glucose transport by the glucose transporter, which moves glucose down its concentration gradient from the ECF to the cytoplasm of the cell. Other carriers transport substances against their electrical and chemical gradients. This form of transport requires energy and is called **active transport.** In animal cells, the energy is provided almost exclusively by hydrolysis of ATP. Not surprisingly, therefore, many carrier molecules are ATPases, enzymes that catalyze the hydrolysis of ATP. One of these ATPases is **sodium–potassium adenosine triphosphatase (Na, K ATPase),** which is also known as the **Na, K pump.** There are also H, K ATPases in the gastric mucosa and the renal tubules. Ca^{2+} ATPase pumps Ca^{2+} out of cells. Proton ATPases acidify many intracellular organelles, including parts of the Golgi complex and lysosomes.

Some of the transport proteins are called **uniports** because they transport only one substance. Others are called **symports** because transport requires the binding of more than one substance to the transport protein and the substances are transported across the membrane together. An example is the symport in the intestinal mucosa that is responsible for the cotransport of Na^+ and glucose from the intestinal lumen into mucosal cells. Other transporters are called **antiports** because they exchange one substance for another.

ION CHANNELS

There are ion channels specific for K^+, Na^+, Ca^{2+}, and Cl^-, as well as channels that are nonselective for cations or anions. Each type of channel exists in multiple forms with diverse properties. Most are made up of identical or very similar subunits. Figure 2–16 shows the multiunit structure of various channels in diagrammatic cross-section.

Most K^+ channels are tetramers, with each of the four subunits forming part of the pore through which K^+ ions pass. Structural analysis of a bacterial voltage-gated K^+ channel indicates that each of the four subunits have a paddle-like extension containing four charges. When the channel is closed, these extensions are near the negatively charged interior of the cell. When the membrane potential is reduced, the paddles containing the charges bend through the membrane to its exterior surface, causing the channel to open. The bacterial K^+ channel is very similar to the voltage-gated K^+ channels in a wide variety of species, including mammals. In the acetylcholine ion channel and other ligand-gated cation or anion channels, five subunits make up the pore. Members of the ClC family of Cl^- channels are dimers, but they have two pores, one in each subunit. Finally, aquaporins are tetramers with a water pore in each of the subunits. Recently, a number of ion channels with intrinsic enzyme activity have been cloned. More than 30 different voltage-gated or cyclic nucleotide-gated Na^+ and Ca^{2+} channels of this type have been described. Representative Na^+, Ca^{2+}, and K^+ channels are shown in extended diagrammatic form in Figure 2–17.

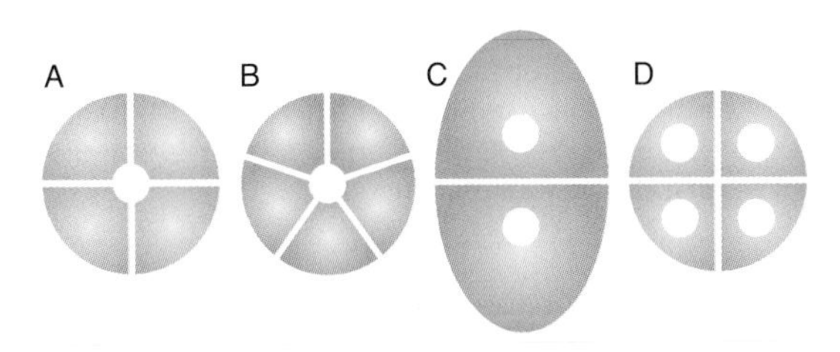

FIGURE 2–16 Different ways in which ion channels form pores. Many K^+ channels are tetramers **A),** with each protein subunit forming part of the channel. In ligand-gated cation and anion channels **B)** such as the acetylcholine receptor, five identical or very similar subunits form the channel. Cl^- channels from the ClC family are dimers **C),** with an intracellular pore in each subunit. Aquaporin water channels **(D)** are tetramers with an intracellular channel in each subunit. (Reproduced with permission from Jentsch TJ: Chloride channels are different. Nature 2002;415:276.)

Another family of Na^+ channels with a different structure has been found in the apical membranes of epithelial cells in the kidneys, colon, lungs, and brain. The **epithelial sodium channels (ENaCs)** are made up of three subunits encoded by three different genes. Each of the subunits probably spans the membrane twice, and the amino terminal and carboxyl terminal are located inside the cell. The α subunit transports Na^+, whereas the β and γ subunits do not. However, the addition of the β and γ subunits increases Na^+ transport through the α subunit. ENaCs are inhibited by the diuretic amiloride, which binds to the α subunit, and they used to be called **amiloride inhibitable Na^+ channels.** The ENaCs in the kidney play an important role in the regulation of ECF volume by aldosterone. ENaC knockout mice are born alive but promptly die because they cannot move Na^+, and hence water, out of their lungs.

Humans have several types of Cl^- channels. The ClC dimeric channels are found in plants, bacteria, and animals, and there are nine different ClC genes in humans. Other Cl^- channels have the same pentameric form as the acetylcholine receptor; examples include the γ-aminobutyric acid A ($GABA_A$) and glycine receptors in the CNS. The cystic fibrosis transmembrane conductance regulator (CFTR) that is mutated in cystic fibrosis is also a Cl^- channel. Ion channel mutations cause a variety of **channelopathies**—diseases that mostly affect muscle and brain tissue and produce episodic paralyses or convulsions, but are also observed in nonexcitable tissues (Clinical Box 2–6).

Na, K ATPase

As noted previously, Na, K ATPase catalyzes the hydrolysis of ATP to adenosine diphosphate (ADP) and uses the energy to extrude three Na^+ from the cell and take two K^+ into the cell for each molecule of ATP hydrolyzed. It is an **electrogenic pump** in that it moves three positive charges out of the cell for each two that it moves in, and it is therefore said to have a **coupling ratio** of 3:2. It is found in all parts of the body. Its activity is inhibited by ouabain and related digitalis glycosides used in the treatment of heart failure. It is a heterodimer made up of an α subunit with a molecular weight of approximately 100,000 and a β subunit with a molecular weight of approximately 55,000. Both extend through the cell membrane

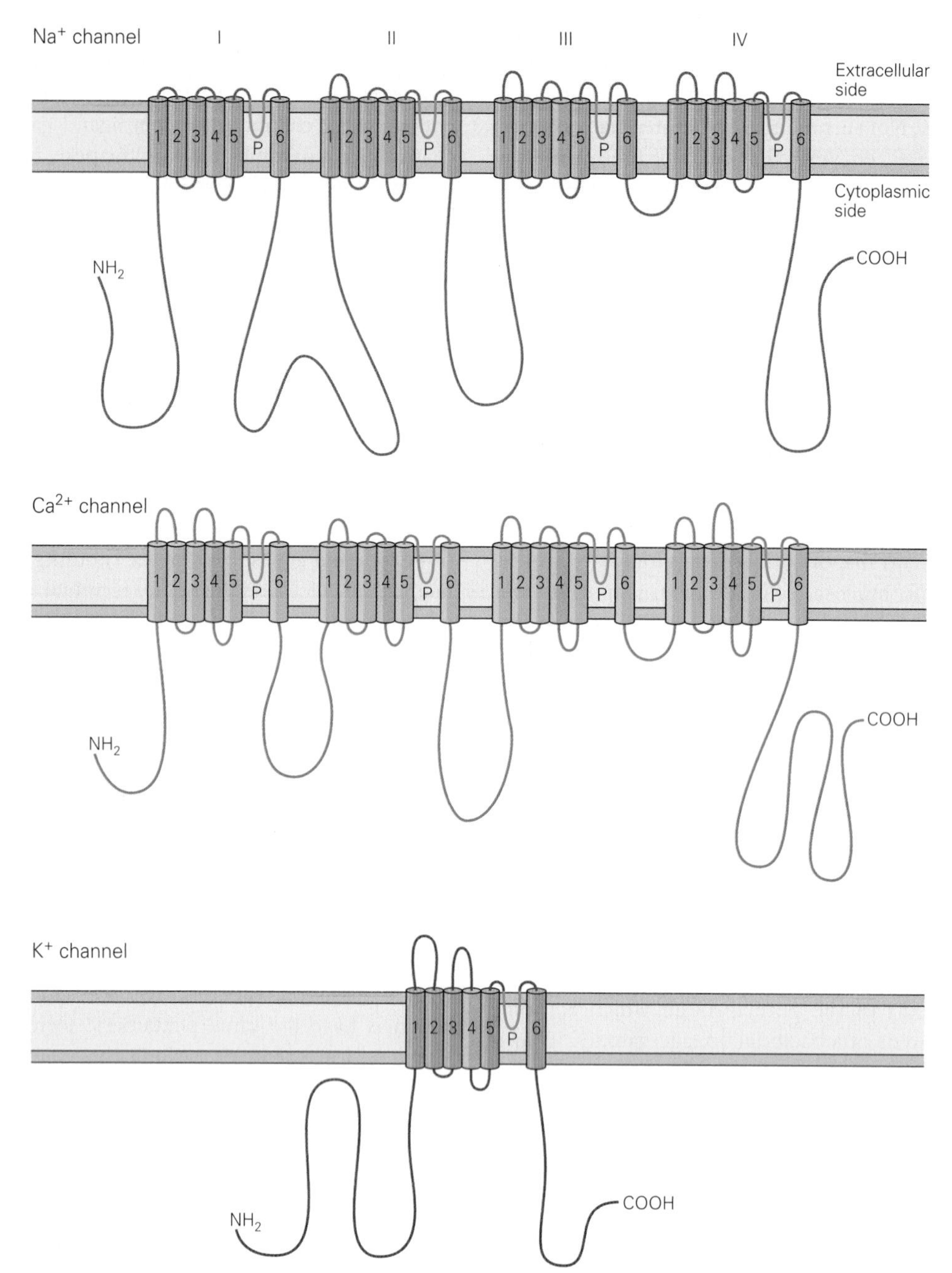

FIGURE 2–17 Diagrammatic representation of the pore-forming subunits of three ion channels. The α subunit of the Na^+ and Ca^{2+} channels traverse the membrane 24 times in four repeats of six membrane-spanning units. Each repeat has a "P" loop between membrane spans 5 and 6 that does not traverse the membrane. These P loops are thought to form the pore. Note that span 4 of each repeat is colored in red, representing its net "+" charge. The K^+ channel has only a single repeat of the six spanning regions and P loop. Four K^+ subunits are assembled for a functional K^+ channel. (Reproduced with permission from Kandel ER, Schwartz JH, Jessell TM (editors): *Principles of Neural Science,* 4th ed. McGraw-Hill, 2000.)

(Figure 2–18). Separation of the subunits eliminates activity. The β subunit is a glycoprotein, whereas Na^+ and K^+ transport occur through the α subunit. The β subunit has a single membrane-spanning domain and three extracellular glycosylation sites, all of which appear to have attached carbohydrate residues. These residues account for one third of its molecular weight. The α subunit probably spans the cell membrane 10 times, with the amino and carboxyl terminals both located intracellularly. This subunit has intracellular Na^+- and ATP-binding sites and a phosphorylation site; it also has extracellular binding sites for K^+ and ouabain. The endogenous ligand of the ouabain-binding site is unsettled. When Na^+ binds to the α subunit, ATP also binds and is converted to ADP, with a phosphate being transferred to Asp 376, the phosphorylation site. This causes a change in the configuration of the protein, extruding Na^+ into the ECF. K^+ then binds extracellularly,

CLINICAL BOX 2–6

Channelopathies

Channelopathies include a wide range of diseases that can affect both excitable (eg, neurons and muscle) and non-excitable cells. Using molecular genetic tools, many of the pathological defects in channelopathies have been traced to mutations in single ion channels. Examples of channelopathies in excitable cells include periodic paralysis (eg, Kir2.1, a K^+ channel subunit, or $Na_v2.1$, a Na^+ channel subunit), myasthenia (eg, nicotinic Acetyl Choline Receptor, a ligand gated nonspecific cation channel), myotonia (eg, Kir1.1, a K^+ channel subunit), malignant hypothermia (Ryanodine Receptor, a Ca^{2+} channel), long QT syndrome (both Na^+ and K^+ channel subunit examples) and several other disorders. Examples of channelopathies in nonexcitable cells include the underlying cause for Cystic fibrosis (CFTR, a Cl^- channel) and a form of Bartter's syndrome (Kir1.1, a K^+ channel subunit). Importantly, advances in treatment of these disorders can come from the understanding of the basic defect and tailoring drugs that act to alter the mutated properties of the affected channel.

dephosphorylating the α subunit, which returns to its previous conformation, releasing K^+ into the cytoplasm.

The α and β subunits are heterogeneous, with α_1, α_2, and α_3 subunits and β_1, β_2, and β_3 subunits described so far. The α_1 isoform is found in the membranes of most cells, whereas α_2 is present in muscle, heart, adipose tissue, and brain, and α_3 is present in heart and brain. The β_1 subunit is widely distributed but is absent in certain astrocytes, vestibular cells of the inner ear, and glycolytic fast-twitch muscles. The fast-twitch muscles contain only β_2 subunits. The different α and β subunit structures of Na, K ATPase in various tissues probably represent specialization for specific tissue functions.

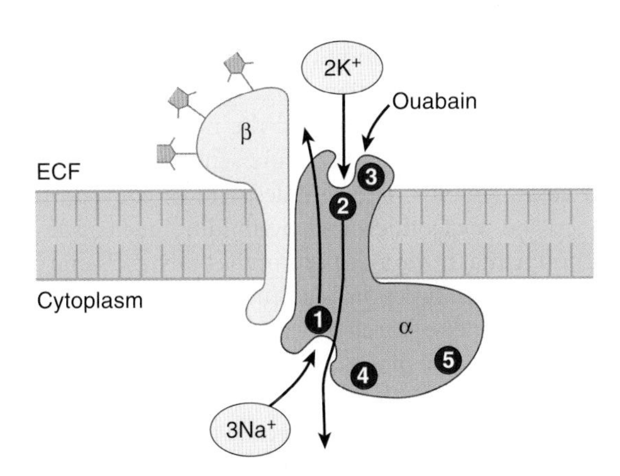

FIGURE 2–18 Na, K ATPase. The intracellular portion of the α subunit has a Na^+-binding site **(1)**, a phosphorylation site **(4)**, and an ATP-binding site **(5)**. The extracellular portion has a K^+-binding site **(2)** and an ouabain-binding site **(3)**. (From Horisberger J-D et al: Structure–function relationship of Na, K ATPase. Annu Rev Physiol 1991;53:565. Reproduced with permission from the *Annual Review of Physiology*, vol. 53. Copyright 1991 by Annual Reviews.)

REGULATION OF Na, K ATPase

The amount of Na^+ normally found in cells is not enough to saturate the pump, so if the Na^+ increases, more is pumped out. Pump activity is affected by second messenger molecules (eg, cAMP and diacylglycerol [DAG]). The magnitude and direction of the altered pump effects vary with the experimental conditions. Thyroid hormones increase pump activity by a genomic action to increase the formation of Na, K ATPase molecules. Aldosterone also increases the number of pumps, although this effect is probably secondary. Dopamine in the kidney inhibits the pump by phosphorylating it, causing a natriuresis. Insulin increases pump activity, probably by a variety of different mechanisms.

SECONDARY ACTIVE TRANSPORT

In many situations, the active transport of Na^+ is coupled to the transport of other substances **(secondary active transport).** For example, the luminal membranes of mucosal cells in the small intestine contain a symport that transports glucose into the cell only if Na^+ binds to the protein and is transported into the cell at the same time. From the cells, the glucose enters the blood. The electrochemical gradient for Na^+ is maintained by the active transport of Na^+ out of the mucosal cell into ECF. Other examples are shown in Figure 2–19. In the heart, Na, K ATPase indirectly affects Ca^{2+} transport. An antiport in the membranes of cardiac muscle cells normally exchanges intracellular Ca^{2+} for extracellular Na^+.

Active transport of Na^+ and K^+ is one of the major energy-using processes in the body. On the average, it accounts for about 24% of the energy utilized by cells, and in neurons it accounts for 70%. Thus, it accounts for a large part of the basal metabolism. A major payoff for this energy use is the establishment of the electrochemical gradient in cells.

TRANSPORT ACROSS EPITHELIA

In the gastrointestinal tract, the pulmonary airways, the renal tubules, and other structures lined with polarized epithelial cells, substances enter one side of a cell and exit another, producing movement of the substance from one side of the epithelium to the other. For transepithelial transport to occur, the cells need to be bound by tight junctions and, obviously, have different ion channels and transport proteins in different parts of their membranes. Most of the instances of secondary active transport cited in the preceding paragraph involve transepithelial movement of ions and other molecules.

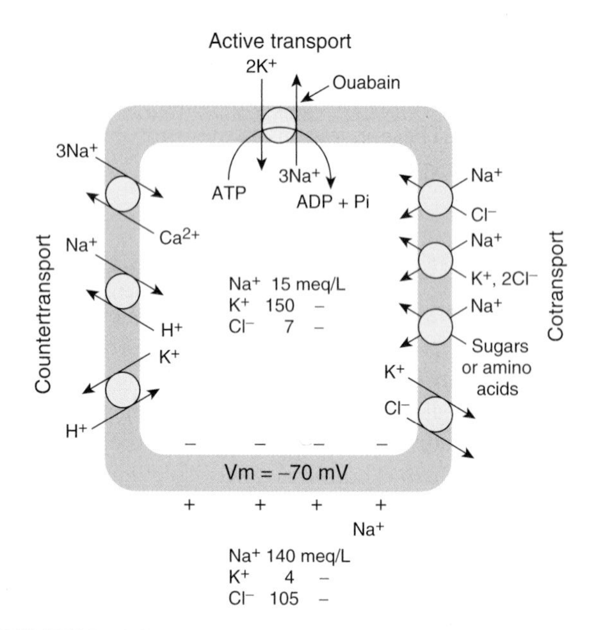

FIGURE 2–19 Composite diagram of main secondary effects of active transport of Na^+ and K^+. Na,K ATPase converts the chemical energy of ATP hydrolysis into maintenance of an inward gradient for Na^+ and an outward gradient for K^+. The energy of the gradients is used for countertransport, cotransport, and maintenance of the membrane potential. Some examples of cotransport and countertransport that use these gradients are shown. (Reproduced with permission from Skou JC: The Na–K pump. News Physiol Sci 1992;7:95.)

SPECIALIZED TRANSPORT ACROSS THE CAPILLARY WALL

The capillary wall separating plasma from interstitial fluid is different from the cell membranes separating interstitial fluid from intracellular fluid because the pressure difference across it makes **filtration** a significant factor in producing movement of water and solute. By definition, filtration is the process by which fluid is forced through a membrane or other barrier because of a difference in pressure on the two sides.

The structure of the capillary wall varies from one vascular bed to another. However, near skeletal muscle and many other organs, water and relatively small solutes are the only substances that cross the wall with ease. The apertures in the junctions between the endothelial cells are too small to permit plasma proteins and other colloids to pass through in significant quantities. The colloids have a high molecular weight but are present in large amounts. Small amounts cross the capillary wall by vesicular transport, but their effect is slight. Therefore, the capillary wall behaves like a membrane impermeable to colloids, and these exert an osmotic pressure of about 25 mm Hg. The colloid osmotic pressure due to the plasma colloids is called the **oncotic pressure.** Filtration across the capillary membrane as a result of the hydrostatic pressure head in the vascular system is opposed by the oncotic pressure. The way the balance between the hydrostatic and oncotic pressures controls exchanges across the capillary wall is considered in detail in Chapter 31.

TRANSCYTOSIS

Vesicles are present in the cytoplasm of endothelial cells, and tagged protein molecules injected into the bloodstream have been found in the vesicles and in the interstitium. This indicates that small amounts of protein are transported out of capillaries across endothelial cells by endocytosis on the capillary side followed by exocytosis on the interstitial side of the cells. The transport mechanism makes use of coated vesicles that appear to be coated with caveolin and is called **transcytosis, vesicular transport,** or **cytopempsis.**

INTERCELLULAR COMMUNICATION

Cells communicate with one another via chemical messengers. Within a given tissue, some messengers move from cell to cell via gap junctions without entering the ECF. In addition, cells are affected by chemical messengers secreted into the ECF, or by direct cell–cell contacts. Chemical messengers typically bind to protein receptors on the surface of the cell or, in some instances, in the cytoplasm or the nucleus, triggering sequences of intracellular changes that produce their physiologic effects. Three general types of intercellular communication are mediated by messengers in the ECF: (1) **neural communication,** in which neurotransmitters are released at synaptic junctions from nerve cells and act across a narrow synaptic cleft on a postsynaptic cell; (2) **endocrine communication,** in which hormones and growth factors reach cells via the circulating blood or the lymph; and (3) **paracrine communication,** in which the products of cells diffuse in the ECF to affect neighboring cells that may be some distance away (Figure 2–20). In addition, cells secrete chemical messengers that in some situations bind to receptors on the same cell, that is, the cell that secreted the messenger **(autocrine communication).** The chemical messengers include amines, amino acids, steroids, polypeptides, and in some instances, lipids, purine nucleotides, and pyrimidine nucleotides. It is worth noting that in various parts of the body, the same chemical messenger can function as a neurotransmitter, a paracrine mediator, a hormone secreted by neurons into the blood (neural hormone), and a hormone secreted by gland cells into the blood.

An additional form of intercellular communication is called **juxtacrine communication.** Some cells express multiple repeats of growth factors such as **transforming growth factor alpha (TGFα)** extracellularly on transmembrane proteins that provide an anchor to the cell. Other cells have TGFα receptors. Consequently, TGFα anchored to a cell can bind to a TGFα receptor on another cell, linking the two. This could be important in producing local foci of growth in tissues.

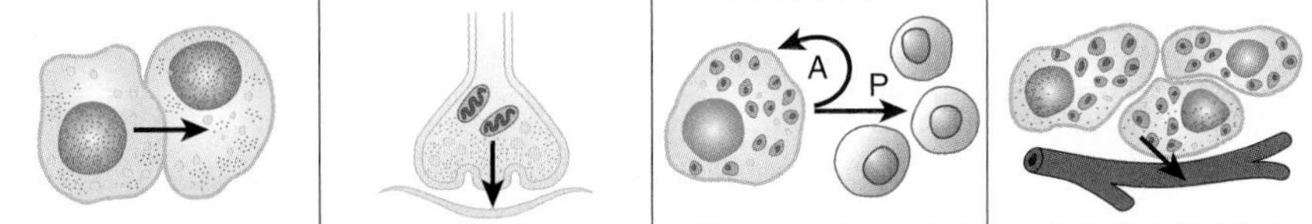

	GAP JUNCTIONS	SYNAPTIC	PARACRINE AND AUTOCRINE	ENDOCRINE
Message transmission	Directly from cell to cell	Across synaptic cleft	By diffusion in interstitial fluid	By circulating body fluids
Local or general	Local	Local	Locally diffuse	General
Specificity depends on	Anatomic location	Anatomic location and receptors	Receptors	Receptors

FIGURE 2–20 Intercellular communication by chemical mediators. A, autocrine; P, paracrine.

RECEPTORS FOR CHEMICAL MESSENGERS

The recognition of chemical messengers by cells typically begins by interaction with a receptor at that cell. There have been over 20 families of receptors for chemical messengers characterized. These proteins are not static components of the cell, but their numbers increase and decrease in response to various stimuli, and their properties change with changes in physiological conditions. When a hormone or neurotransmitter is present in excess, the number of active receptors generally decreases **(down-regulation),** whereas in the presence of a deficiency of the chemical messenger, there is an increase in the number of active receptors **(up-regulation).** In its actions on the adrenal cortex, angiotensin II is an exception; it increases rather than decreases the number of its receptors in the adrenal. In the case of receptors in the membrane, receptor-mediated endocytosis is responsible for down-regulation in some instances; ligands bind to their receptors, and the ligand–receptor complexes move laterally in the membrane to coated pits, where they are taken into the cell by endocytosis **(internalization).** This decreases the number of receptors in the membrane. Some receptors are recycled after internalization, whereas others are replaced by de novo synthesis in the cell. Another type of down-regulation is **desensitization,** in which receptors are chemically modified in ways that make them less responsive.

MECHANISMS BY WHICH CHEMICAL MESSENGERS ACT

Receptor–ligand interaction is usually just the beginning of the cell response. This event is transduced into secondary responses within the cell that can be divided into four broad categories: (1) ion channel activation, (2) **G-protein** activation, (3) activation of enzyme activity within the cell, or (4) direct activation of transcription. Within each of these groups, responses can be quite varied. Some of the common mechanisms by which chemical messengers exert their intracellular effects are summarized in Table 2–3. Ligands such as acetylcholine bind directly to ion channels in the cell membrane, changing their conductance. Thyroid and steroid hormones, 1,25-dihydroxycholecalciferol, and retinoids enter cells and act on one or another member of a family of structurally related cytoplasmic or nuclear receptors. The activated receptor binds to DNA and increases transcription of selected mRNAs. Many other ligands in the ECF bind to receptors on the surface of cells and trigger the release of intracellular mediators such as cAMP, IP_3, and DAG that initiate changes in cell function. Consequently, the extracellular ligands are called **"first messengers"** and the intracellular mediators are called **"second messengers."** Second messengers bring about many short-term changes in cell function by altering enzyme function, triggering exocytosis, and so on, but they also can lead to the alteration of transcription of various genes. A variety of enzymatic changes, protein–protein interactions, or second messenger changes can be activated within a cell in an orderly fashion following receptor recognition of the primary messenger. The resulting **cell signaling pathway** provides amplification of the primary signal and distribution of the signal to appropriate targets within the cell. Extensive cell signaling pathways also provide opportunities for feedback and regulation that can fine-tune the signal for the correct physiological response by the cell.

The most predominant posttranslation modification of proteins, phosphorylation, is a common theme in cell signaling pathways. Cellular phosphorylation is under the control of two groups of proteins: **kinases,** enzymes that catalyze the phosphorylation of tyrosine or serine and threonine residues in proteins (or in some cases, in lipids); and **phosphatases,** proteins that remove phosphates from proteins (or lipids). Some of the larger receptor families are themselves kinases. Tyrosine kinase receptors initiate phosphorylation on tyrosine residues on complementary receptors following ligand binding. Serine/threonine kinase receptors initiate phosphorylation on serines or threonines in complementary receptors following ligand binding. Cytokine receptors are directly

TABLE 2–3 Common mechanisms by which chemical messengers in the ECF bring about changes in cell function.

Mechanism	Examples
Open or close ion channels in cell membrane	Acetylcholine on nicotinic cholinergic receptor; norepinephrine on K^+ channel in the heart
Act via cytoplasmic or nuclear receptors to increase transcription of selected mRNAs	Thyroid hormones, retinoic acid, steroid hormones
Activate phospholipase C with intracellular production of DAG, IP_3, and other inositol phosphates	Angiotensin II, norepinephrine via α_1-adrenergic receptor, vasopressin via V_1 receptor
Activate or inhibit adenylyl cyclase, causing increased or decreased intracellular production of cAMP	Norepinephrine via β_1-adrenergic receptor (increased cAMP); norepinephrine via α_2-adrenergic receptor (decreased cAMP)
Increase cGMP in cell	Atrial natriuretic peptide; nitric oxide
Increase tyrosine kinase activity of cytoplasmic portions of transmembrane receptors	Insulin, epidermal growth factor (EGF), platelet-derived growth factor (PDGF), monocyte colony-stimulating factor (M-CSF)
Increase serine or threonine kinase activity	TGFβ, activin, inhibin

TABLE 2–4 Sample protein kinases.

Phosphorylate serine or threonine residues, or both
Calmodulin-dependent
Myosin light-chain kinase
Phosphorylase kinase
Ca^{2+}/calmodulin kinase I
Ca^{2+}/calmodulin kinase II
Ca^{2+}/calmodulin kinase III
Calcium-phospholipid-dependent
Protein kinase C (seven subspecies)
Cyclic nucleotide-dependent
cAMP-dependent kinase (protein kinase A; two subspecies)
cGMP-dependent kinase
Phosphorylate tyrosine residues
Insulin receptor, EGF receptor, PDGF receptor, and M-CSF receptor

associated with a group of protein kinases that are activated following cytokine binding. Alternatively, second messenger changes can lead to phosphorylation further downstream in the signaling pathway. More than 500 protein kinases have been described. Some of the principal ones that are important in mammalian cell signaling are summarized in Table 2–4. In general, addition of phosphate groups changes the conformation of the proteins, altering their functions and consequently the functions of the cell. The close relationship between phosphorylation and dephosphorylation of cellular proteins allows for a temporal control of activation of cell signaling pathways. This is sometimes referred to as a **"phosphate timer."** The dysregulation of the phosphate timer and subsequent cellular signaling in a cell can lead to disease (Clinical Box 2–7).

STIMULATION OF TRANSCRIPTION

The activation of transcription, and subsequent translation, is a common outcome of cellular signaling. There are three distinct pathways for primary messengers to alter transcription of cells. First, as is the case with steroid or thyroid hormones, the primary messenger is able to cross the cell membrane and bind to a nuclear receptor, which then can directly interact with DNA to alter gene expression. A second pathway to gene transcription is the activation of cytoplasmic protein kinases that can move to the nucleus to phosphorylate a latent transcription factor for activation. This pathway is a common endpoint of signals that go through the **mitogen activated protein (MAP) kinase** cascade. MAP kinases can be activated following a variety of receptor–ligand interactions through second messenger signaling. They comprise a series of three kinases that coordinate a stepwise phosphorylation to activate each protein in series in the cytosol. Phosphorylation of the last MAP kinase in series allows it to migrate to the nucleus where it phosphorylates a latent transcription factor. A third common pathway is the activation of a latent transcription factor in the cytosol, which then migrates to the nucleus and alters transcription. This pathway is shared by a diverse set of transcription factors that include **nuclear factor kappa B (NFκB;** activated following tumor necrosis family receptor binding and others), and **signal transducers of activated transcription (STATs;** activated following cytokine receptor binding). In all cases, the binding of the activated transcription factor to DNA increases (or in some cases, decreases) the transcription of mRNAs encoded by the gene to which it binds. The mRNAs are translated in the ribosomes, with the production of increased quantities of proteins that alter cell function.

INTRACELLULAR Ca^{2+} AS A SECOND MESSENGER

Ca^{2+} regulates a very large number of physiological processes that are as diverse as proliferation, neural signaling, learning, contraction, secretion, and fertilization, so regulation of intracellular Ca^{2+} is of great importance. The free Ca^{2+} concentration in the cytoplasm at rest is maintained at about 100 nmol/L. The Ca^{2+} concentration in the interstitial fluid is about 12,000 times the cytoplasmic concentration (ie, 1,200,000 nmol/L), so there is a marked inwardly directed concentration

CLINICAL BOX 2–7

Kinases in Cancer: Chronic Myeloid Leukemia

Kinases frequently play important roles in regulating cellular physiology outcomes, including cell growth and cell death. Dysregulation of cell proliferation or cell death is a hallmark of cancer. Although cancer can have many causes, a role for kinase dysregulation is exemplified in Chronic myeloid leukemia (CML). CML is a pluripotent hematopoietic stem cell disorder characterized by the Philadelphia (Ph) chromosome translocation. The Ph chromosome is formed following a translocation of chromosomes 9 and 22. The resultant shortened chromosome 22 (Ph chromosome). At the point of fusion a novel gene (BCR-ABL) encoding the active tyrosine kinase domain from a gene on chromosome 9 (Abelson tyrosine kinase; c-Abl) is fused to novel regulatory region of a separate gene on chromosome 22 (breakpoint cluster region; bcr). The BCR-ABL fusion gene encodes a cytoplasmic protein with constitutively active tyrosine kinase. The dysregulated kinase activity in BCR-ABL protein effectively limits white blood cell death signaling pathways while promoting cell proliferation and genetic instability. Experimental models have shown that translocation to produce the fusion BCR-ABL protein is sufficient to produce CML in animal models.

THERAPEUTIC HIGHLIGHTS

The identification of BCR-ABL as the initial transforming event in CML provided an ideal target for drug discovery. The drug imatinib (Gleevac) was developed to specifically block the tyrosine kinase activity of the BCR-ABL protein. Imatinib has proven to be an effective agent for treating chronic phase CML.

gradient as well as an inwardly directed electrical gradient. Much of the intracellular Ca^{2+} is stored at relatively high concentrations in the endoplasmic reticulum and other organelles (Figure 2–21), and these organelles provide a store from which Ca^{2+} can be mobilized via ligand-gated channels to increase the concentration of free Ca^{2+} in the cytoplasm. Increased cytoplasmic Ca^{2+} binds to and activates calcium-binding proteins. These proteins can have direct effects in cellular physiology, or can activate other proteins, commonly protein kinases, to further cell signaling pathways.

Ca^{2+} can enter the cell from the extracellular fluid, down its electrochemical gradient, through many different Ca^{2+} channels. Some of these are ligand-gated and others are voltage-gated. Stretch-activated channels exist in some cells as well.

Many second messengers act by increasing the cytoplasmic Ca^{2+} concentration. The increase is produced by releasing Ca^{2+} from intracellular stores—primarily the endoplasmic reticulum—or by increasing the entry of Ca^{2+} into cells, or by

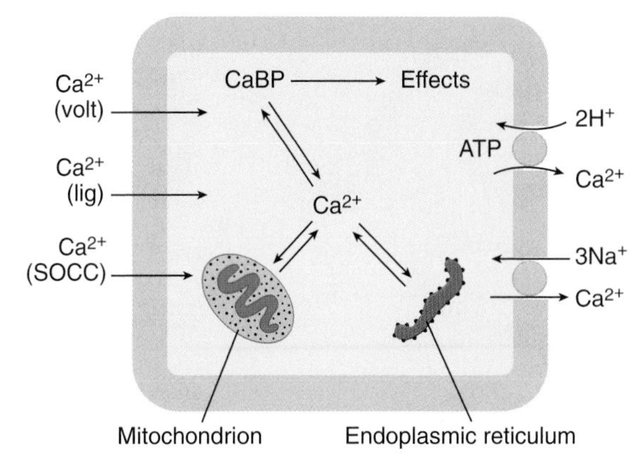

FIGURE 2–21 Ca^{2+} handling in mammalian cells. Ca^{2+} is stored in the endoplasmic reticulum and, to a lesser extent, mitochondria and can be released from them to replenish cytoplasmic Ca^{2+}. Calcium-binding proteins (CaBP) bind cytoplasmic Ca^{2+} and, when activated in this fashion, bring about a variety of physiologic effects. Ca^{2+} enters the cells via voltage-gated (volt) and ligand-gated (lig) Ca^{2+} channels and store-operated calcium channels (SOCCs). It is transported out of the cell by Ca, Mg ATPases (not shown), Ca, H ATPase and a Na, Ca antiport. It is also transported into the ER by Ca ATPases.

both mechanisms. IP_3 is the major second messenger that causes Ca^{2+} release from the endoplasmic reticulum through the direct activation of a ligand-gated channel, the IP_3 receptor. In effect, the generation of one second messenger (IP_3) can lead to the release of another second messenger (Ca^{2+}). In many tissues, transient release of Ca^{2+} from internal stores into the cytoplasm triggers opening of a population of Ca^{2+} channels in the cell membrane **(store-operated Ca^{2+} channels; SOCCs).** The resulting Ca^{2+} influx replenishes the total intracellular Ca^{2+} supply and refills the endoplasmic reticulum. Recent research has identified the physical relationships between SOCCs and regulatory interactions of proteins from the endoplasmic reticulum that gate these channels.

As with other second messenger molecules, the increase in Ca^{2+} within the cytosol is rapid, and is followed by a rapid decrease. Because the movement of Ca^{2+} outside of the cytosol (ie, across the plasma membrane or the membrane of the internal store) requires that it move up its electrochemical gradient, it requires energy. Ca^{2+} movement out of the cell is facilitated by the plasma membrane Ca^{2+} ATPase. Alternatively, it can be transported by an antiport that exchanges three Na^+ for each Ca^{2+} driven by the energy stored in the Na^+ electrochemical gradient. Ca^{2+} movement into the internal stores is through the action of the **sarcoplasmic or endoplasmic reticulum Ca^{2+} ATPase,** also known as the **SERCA pump.**

CALCIUM-BINDING PROTEINS

Many different Ca^{2+}-binding proteins have been described, including **troponin, calmodulin,** and **calbindin.** Troponin is the Ca^{2+}-binding protein involved in contraction of skeletal muscle (Chapter 5). Calmodulin contains 148 amino acid

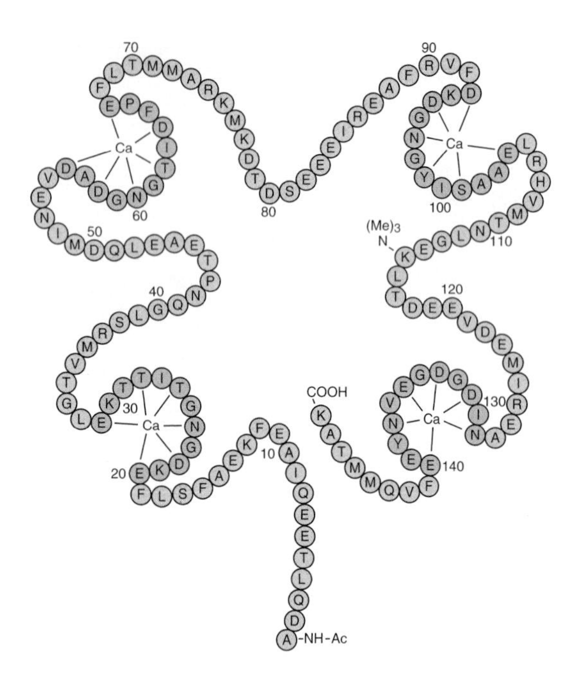

FIGURE 2–22 Secondary structure of calmodulin from bovine brain. Single-letter abbreviations are used for the amino acid residues. Note the four calcium domains (purple residues) flanked on either side by stretches of amino acids that form α-helices in tertiary structure. (Reproduced with permission from Cheung WY: Calmodulin: An overview. Fed Proc 1982;41:2253.)

residues (Figure 2–22) and has four Ca^{2+}-binding domains. It is unique in that amino acid residue 115 is trimethylated, and it is extensively conserved, being found in plants as well as animals. When calmodulin binds Ca^{2+}, it is capable of activating five different calmodulin-dependent kinases (CaMKs; Table 2–4), among other proteins. One of the kinases is **myosin light-chain kinase,** which phosphorylates myosin. This brings about contraction in smooth muscle. CaMKI and CaMKII are concerned with synaptic function, and CaMKIII is concerned with protein synthesis. Another calmodulin-activated protein is **calcineurin,** a phosphatase that inactivates Ca^{2+} channels by dephosphorylating them. It also plays a prominent role in activating T cells and is inhibited by some immunosuppressants.

MECHANISMS OF DIVERSITY OF Ca^{2+} ACTIONS

It may seem difficult to understand how intracellular Ca^{2+} can have so many varied effects as a second messenger. Part of the explanation is that Ca^{2+} may have different effects at low and at high concentrations. The ion may be at high concentration at the site of its release from an organelle or a channel (**Ca^{2+} sparks**) and at a subsequent lower concentration after it diffuses throughout the cell. Some of the changes it produces can outlast the rise in intracellular Ca^{2+} concentration because of the way it binds to some of the Ca^{2+}-binding proteins. In addition, once released, intracellular Ca^{2+} concentrations frequently oscillate at regular intervals, and there is evidence that the frequency and, to a lesser extent, the amplitude of those oscillations codes information for effector mechanisms. Finally, increases in intracellular Ca^{2+} concentration can spread from cell to cell in waves, producing coordinated events such as the rhythmic beating of cilia in airway epithelial cells.

G PROTEINS

A common way to translate a signal to a biologic effect inside cells is by way of nucleotide regulatory proteins that are activated after binding GTP (**G proteins).** When an activating signal reaches a G protein, the protein exchanges GDP for GTP. The GTP–protein complex brings about the activating effect of the G protein. The inherent GTPase activity of the protein then converts GTP to GDP, restoring the G protein to an inactive resting state. G proteins can be divided into two principal groups involved in cell signaling: **small G proteins** and **heterotrimeric G proteins.** Other groups that have similar regulation and are also important to cell physiology include elongation factors, dynamin, and translocation GTPases.

There are six different families of small G proteins (or **small GTPases**) that are all highly regulated. **GTPase activating proteins (GAPs)** tend to inactivate small G proteins by encouraging hydrolysis of GTP to GDP in the central binding site. **Guanine exchange factors (GEFs)** tend to activate small G proteins by encouraging exchange of GDP for GTP in the active site. Some of the small G proteins contain lipid modifications that help to anchor them to membranes, while others are free to diffuse throughout the cytosol. Small G proteins are involved in many cellular functions. Members of the Rab family regulate the rate of vesicle traffic between the endoplasmic reticulum, the Golgi apparatus, lysosomes, endosomes, and the cell membrane. Another family of small GTP-binding proteins, the Rho/Rac family, mediates interactions between the cytoskeleton and cell membrane; and a third family, the Ras family, regulates growth by transmitting signals from the cell membrane to the nucleus.

Another family of G proteins, the larger **heterotrimeric G proteins,** couple cell surface receptors to catalytic units that catalyze the intracellular formation of second messengers or couple the receptors directly to ion channels. Despite the knowledge of the small G proteins described above, the heteromeric G proteins are frequently referred to in the shortened "G protein" form because they were the first to be identified. Heterotrimeric G proteins are made up of three subunits designated α, β, and γ (Figure 2–23). Both the α and the γ subunits have lipid modifications that anchor these proteins to the plasma membrane. The α subunit is bound to GDP. When a ligand binds to a G protein-coupled receptor (GPCR, discussed below), this GDP is exchanged for GTP and the α subunit separates from the combined β and γ subunits. The separated α subunit brings about many biologic effects. The β and γ subunits are tightly bound in the cell and together

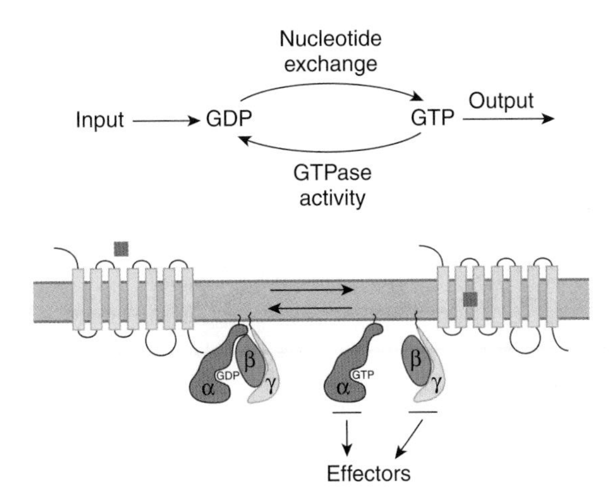

FIGURE 2–23 Heterotrimeric G proteins. Top: Summary of overall reaction that occurs in the Gα subunit. **Bottom:** When the ligand (square) binds to the G protein-coupled receptor in the cell membrane, GTP replaces GDP on the α subunit. GTP-α separates from the βγ subunit and GTP-α and βγ both activate various effectors, producing physiologic effects. The intrinsic GTPase activity of GTP-α then converts GTP to GDP, and the α, β, and γ subunits reassociate.

form a signaling molecule that can also activate a variety of effectors. The intrinsic GTPase activity of the α subunit then converts GTP to GDP, and this leads to re-association of the α with the βγ subunit and termination of effector activation. The GTPase activity of the α subunit can be accelerated by a family of **regulators of G protein signaling (RGS).**

Heterotrimeric G proteins relay signals from over 1000 GPCRs, and their effectors in the cells include ion channels and enzymes. There are 20 α, 6 β, and 12 γ genes, which allow for over 1400 α, β, and γ combinations. Not all combinations occur in the cell, but over 20 different heterotrimeric G proteins have been well documented in cell signaling. They can be divided into five families, each with a relatively characteristic set of effectors.

G PROTEIN-COUPLED RECEPTORS

All the **GPCRs** that have been characterized to date are proteins that span the cell membrane seven times. Because of this structure they are alternatively referred to as **seven-helix receptors** or **serpentine receptors.** A very large number have been cloned, and their functions are multiple and diverse. This is emphasized by the extensive variety of ligands that target GPCRs (Table 2–5). The structures of four GPCRs are shown in Figure 2–24. These receptors assemble into a barrel-like structure. Upon ligand binding, a conformational change activates a resting heterotrimeric G protein associated with the cytoplasmic leaf of the plasma membrane. Activation of a single receptor can result in 1, 10, or more active heterotrimeric G proteins, providing amplification as well as transduction of the first messenger. Bound receptors can be inactivated to limit the amount of cellular signaling. This frequently occurs through phosphorylation of the cytoplasmic side of the receptor. Because of their diversity and importance in cellular signaling pathways, GPCRs are prime targets for drug discovery (Clinical Box 2–8).

TABLE 2–5 Examples of ligands for G-protein coupled receptors.

Class	Ligand
Neurotransmitters	Epinephrine Norepinephrine Dopamine 5-Hydroxytryptamine Histamine Acetylcholine Adenosine Opioids
Tachykinins	Substance P Neurokinin A Neuropeptide K
Other peptides	Angiotensin II Arginine vasopressin Oxytocin VIP, GRP, TRH, PTH
Glycoprotein hormones	TSH, FSH, LH, hCG
Arachidonic acid derivatives	Thromboxane A_2
Other	Odorants Tastants Endothelins Platelet-activating factor Cannabinoids Light

INOSITOL TRISPHOSPHATE & DIACYLGLYCEROL AS SECOND MESSENGERS

The link between membrane binding of a ligand that acts via Ca^{2+} and the prompt increase in the cytoplasmic Ca^{2+} concentration is often **inositol trisphosphate (inositol 1,4,5-trisphosphate; IP_3).** When one of these ligands binds to its receptor, activation of the receptor produces activation of phospholipase C (PLC) on the inner surface of the membrane. Ligands bound to GPCR can do this through the G_q heterotrimeric G proteins, while ligands bound to tyrosine kinase receptors can do this through other cell signaling pathways. PLC has at least eight isoforms; PLC_β is activated by heterotrimeric G proteins, while PLCγ forms are activated through tyrosine kinase receptors. PLC isoforms can catalyze the hydrolysis

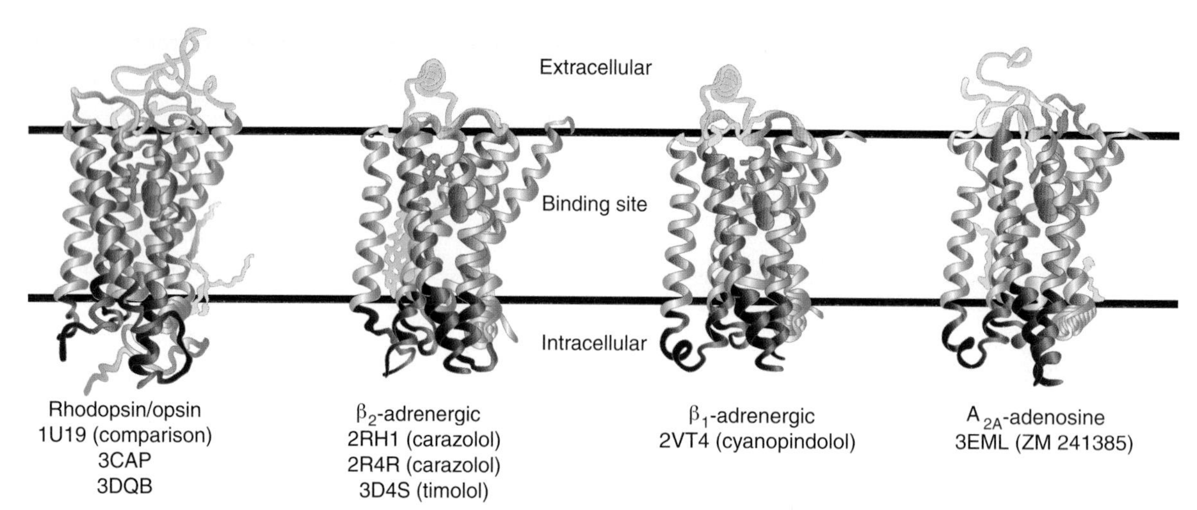

FIGURE 2–24 Representative structures of four G protein-coupled receptors from solved crystal structures. Each group of receptors is represented by one structure, all rendered with the same orientation and color scheme: transmembrane helices are colored light blue, intracellular regions are colored darker blue, and extracellular regions are brown. Each ligand is colored orange and rendered as sticks, bound lipids are colored yellow, and the conserved tryptophan residue is rendered as spheres and colored green. This figure highlights the observed differences seen in the extracellular and intracellular domains as well as the small differences seen in the ligand binding orientations among the four GPCRs various ligands. (Reproduced with permission from Hanson MA, Stevens RC: Discovery of new GPCR biology: one receptor structure at a time. Structure 1988 Jan 14;17(1):8–14.)

of the membrane lipid phosphatidylinositol 4,5-diphosphate (PIP_2) to form IP_3 and **DAG** (Figure 2–25). The IP_3 diffuses to the endoplasmic reticulum, where it triggers the release of Ca^{2+} into the cytoplasm by binding the IP_3 receptor, a ligand-gated Ca^{2+} channel (Figure 2–26). DAG is also a second messenger; it stays in the cell membrane, where it activates one of several isoforms of **protein kinase C.**

CYCLIC AMP

Another important second messenger is cyclic adenosine 3',5'-monophosphate **(cyclic AMP** or **cAMP;** Figure 2–27). Cyclic AMP is formed from ATP by the action of the enzyme **adenylyl cyclase** and converted to physiologically inactive 5'AMP by the action of the enzyme **phosphodiesterase.** Some of the

CLINICAL BOX 2–8

Drug Development: Targeting the G-Protein Coupled Receptors (GPCRs)

GPCRs are among the most heavily investigated drug targets in the pharmaceutical industry, representing approximately 40% of all the drugs in the marketplace today. These proteins are active in just about every organ system and present a wide range of opportunities as therapeutic targets in areas including cancer, cardiac dysfunction, diabetes, central nervous system disorders, obesity, inflammation, and pain. Features of GPCRs that allow them to be drug targets are their specificity in recognizing extracellular ligands to initiate cellular response, the cell surface location of GPCRs that make them accessible to novel ligands or drugs, and their prevalence in leading to human pathology and disease.

Specific examples of successful GPCR drug targets are noted with two types of **Histamine Receptors.**

Histamine-1 Receptor (H1-Receptor) antagonists: allergy therapy. Allergens can trigger local mast cells or basophils to release histamine in the airway. A primary target for histamine is the H1-Receptor in several airway cell types and this can lead to transient itching, sneezing, rhinorrhea, and nasal congestion. There are a variety of drugs with improved peripheral H1 receptor selectivity that are currently used to block histamine activation of the H1-Receptor and thus limit allergen effects in the upper airway. Current H1-Receptor antagonists on the market today include loratadine, fexofenadine, cetirizine, and desloratadine. These "second" an "third" generation anti H1-Receptor drugs have improved specificity and reduced adverse side effects (eg, drowsiness and central nervous system dysfunction) associated with some of the "first" generation drugs first introduced in the late 1930's and widely developed over the next 40 years.

Histamine-2 Receptor (H2-Receptor) antagonists: treating excess stomach acid. Excess stomach acid can result in gastroesophageal reflux disease or even peptic ulcer symptoms. The parietal cell in the stomach can be stimulated to produce acid via histamine action at the H2-Receptor. Excess stomach acid results in heartburn. Antagonists or H2-Receptor blockers, reduce acid production by preventing H2-Receptor signaling that leads to production of stomach acid. There are several drugs (eg, ranitidine, famotidine, cimetidine, and nizatidine) that specifically block the H2-receptor and thus reduce excess acid production.

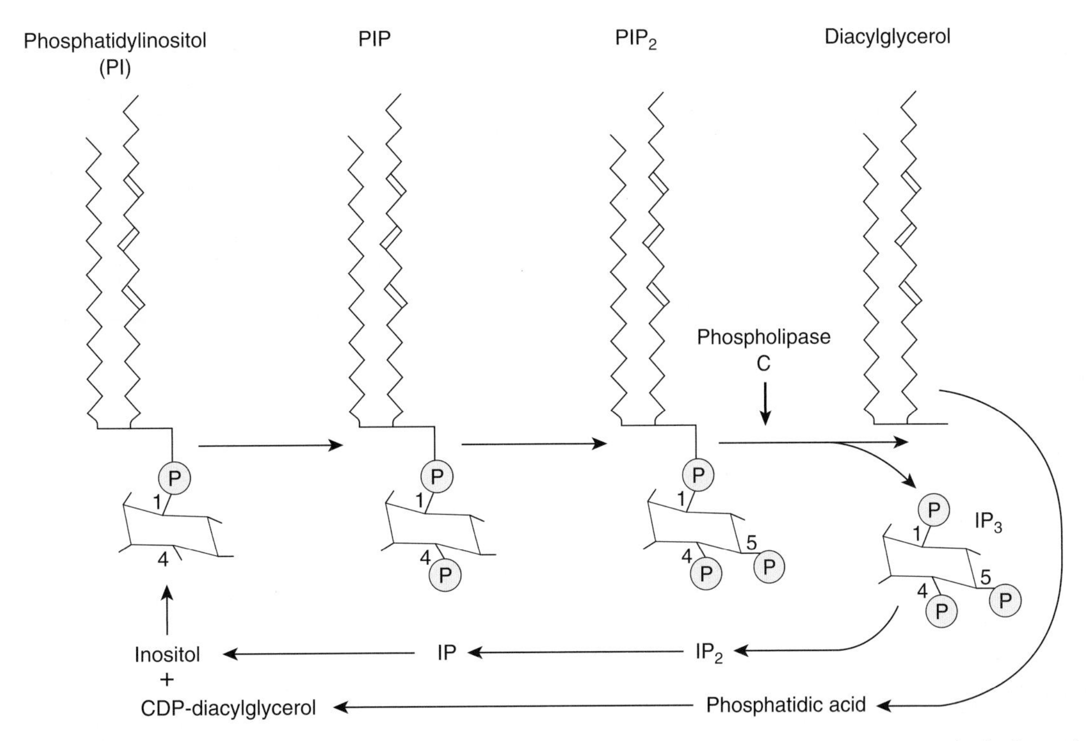

FIGURE 2–25 Metabolism of phosphatidylinositol in cell membranes. Phosphatidylinositol is successively phosphorylated to form phosphatidylinositol 4-phosphate (PIP), then phosphatidylinositol 4,5-bisphosphate (PIP_2). Phospholipase C_β and phospholipase Cγ catalyze the breakdown of PIP_2 to inositol 1,4,5-trisphosphate (IP_3) and diacylglycerol. Other inositol phosphates and phosphatidylinositol derivatives can also be formed. IP_3 is dephosphorylated to inositol, and diacylglycerol is metabolized to cytosine diphosphate (CDP)-diacylglycerol. CDP-diacylglycerol and inositol then combine to form phosphatidylinositol, completing the cycle. (Modified from Berridge MJ: Inositol triphosphate and diacylglycerol as second messengers. Biochem J 1984;220:345.)

phosphodiesterase isoforms that break down cAMP are inhibited by methylxanthines such as caffeine and theophylline. Consequently, these compounds can augment hormonal and transmitter effects mediated via cAMP. Cyclic AMP activates one of the cyclic nucleotide-dependent protein kinases **(protein kinase A, PKA)** that, like protein kinase C, catalyzes the phosphorylation of proteins, changing their conformation and altering their activity. In addition, the active catalytic subunit of PKA moves to the nucleus and phosphorylates the **cAMP-responsive element-binding protein (CREB).** This transcription factor then binds to DNA and alters transcription of a number of genes.

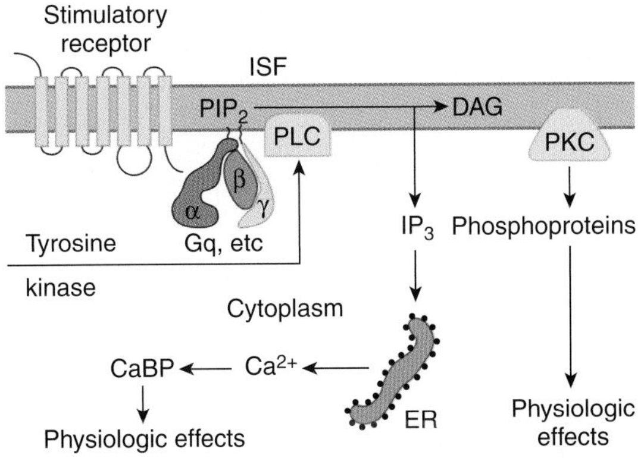

FIGURE 2–26 Diagrammatic representation of release of inositol triphosphate (IP_3) and diacylglycerol (DAG) as second messengers. Binding of ligand to G protein-coupled receptor activates phospholipase C $(PLC)_\beta$. Alternatively, activation of receptors with intracellular tyrosine kinase domains can activate PLCγ. The resulting hydrolysis of phosphatidylinositol 4,5-diphosphate (PIP_2) produces IP_3, which releases Ca^{2+} from the endoplasmic reticulum (ER), and DAG, which activates protein kinase C (PKC). CaBP, Ca^{2+}-binding proteins; ISF, interstitial fluid.

PRODUCTION OF cAMP BY ADENLYL CYCLASE

Adenylyl cyclase is a membrane bound protein with 12 transmembrane regions. Ten isoforms of this enzyme have been described and each can have distinct regulatory properties, permitting the cAMP pathway to be customized to specific tissue needs. Notably, stimulatory heterotrimeric G proteins (G_s) activate, while inhibitory heterotrimeric G proteins (G_i) inactivate adenylyl cyclase (Figure 2–28). When the appropriate ligand binds to a stimulatory receptor, a G_s α subunit activates one of the adenylyl cyclases. Conversely, when the appropriate ligand binds to an inhibitory receptor, a G_i α sub-unit inhibits adenylyl cyclase. The receptors are specific, responding at low threshold to only one or a select group of related ligands. However, heterotrimeric G proteins mediate the stimulatory and inhibitory effects produced by many different ligands. In addition, cross-talk occurs between the

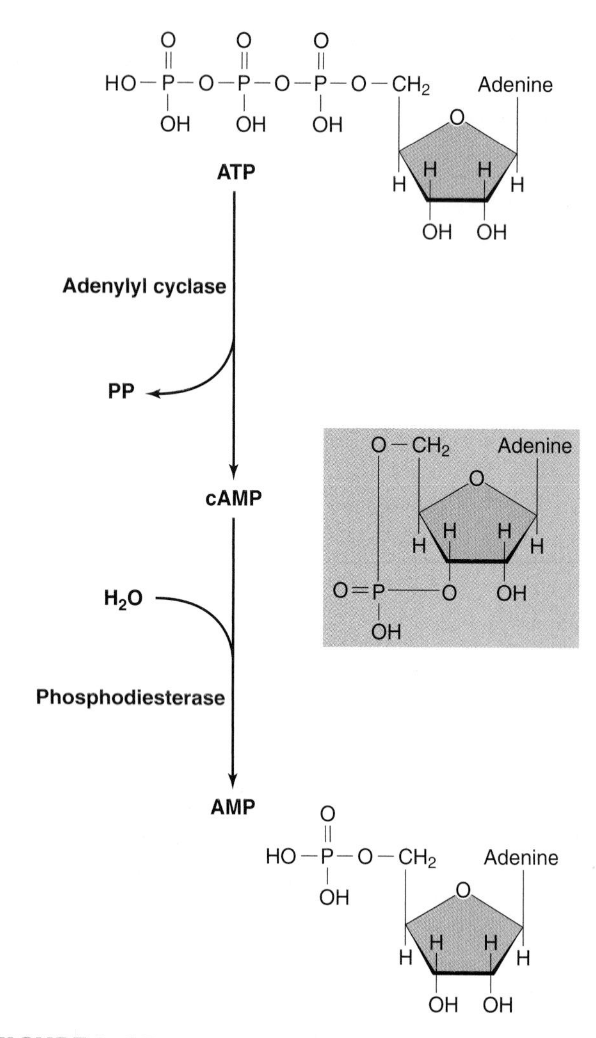

FIGURE 2–27 Formation and metabolism of cAMP. The second messenger cAMP is made from ATP by adenylyl cyclase and broken down into AMP by phosphodiesterase.

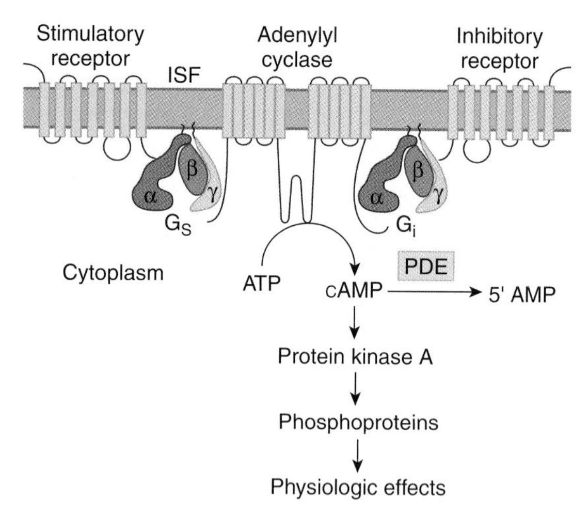

FIGURE 2–28 The cAMP system. Activation of adenylyl cyclase catalyzes the conversion of ATP to cAMP. Cyclic AMP activates protein kinase A, which phosphorylates proteins, producing physiologic effects. Stimulatory ligands bind to stimulatory receptors and activate adenylyl cyclase via G_s. Inhibitory ligands inhibit adenylyl cyclase via inhibitory receptors and G_i. ISF, interstitial fluid.

phospholipase C system and the adenylyl cyclase system, as several of the isoforms of adenylyl cyclase are stimulated by calmodulin. Finally, the effects of protein kinase A and protein kinase C are very widespread and can also affect directly, or indirectly, the activity at adenylyl cyclase. The close relationship between activation of G proteins and adenylyl cyclases also allows for spatial regulation of cAMP production. All of these events, and others, allow for fine-tuning the cAMP response for a particular physiological outcome in the cell.

Two bacterial toxins have important effects on adenylyl cyclase that are mediated by G proteins. The A subunit of **cholera toxin** catalyzes the transfer of ADP ribose to an arginine residue in the middle of the α subunit of G_s. This inhibits its GTPase activity, producing prolonged stimulation of adenylyl cyclase. **Pertussis toxin** catalyzes ADP-ribosylation of a cysteine residue near the carboxyl terminal of the α subunit of G_i. This inhibits the function of G_i. In addition to the implications of these alterations in disease, both toxins are used for fundamental research on G protein function. The drug forskolin also stimulates adenylyl cyclase activity by a direct action on the enzyme.

GUANYLYL CYCLASE

Another cyclic nucleotide of physiologic importance is **cyclic guanosine monophosphate (cyclic GMP** or **cGMP).** Cyclic GMP is important in vision in both rod and cone cells. In addition, there are cGMP-regulated ion channels, and cGMP activates cGMP-dependent kinase, producing a number of physiologic effects.

Guanylyl cyclases are a family of enzymes that catalyze the formation of cGMP. They exist in two forms (Figure 2–29). One form has an extracellular amino terminal domain that is a receptor, a single transmembrane domain, and a cytoplasmic portion with guanylyl cyclase catalytic activity. Several such guanylyl cyclases have been characterized. Two are receptors for atrial natriuretic peptide (ANP; also known as atrial natriuretic factor), and a third binds an *Escherichia coli* enterotoxin and the gastrointestinal polypeptide guanylin. The other form of guanylyl cyclase is soluble, contains heme, and is not bound to the membrane. There appear to be several isoforms of the intracellular enzyme. They are activated by nitric oxide (NO) and NO-containing compounds.

GROWTH FACTORS

Growth factors have become increasingly important in many different aspects of physiology. They are polypeptides and proteins that are conveniently divided into three groups. One group is made up of agents that foster the multiplication or development of various types of cells; NGF, insulin-like growth factor I (IGF-I), activins and inhibins, and epidermal growth

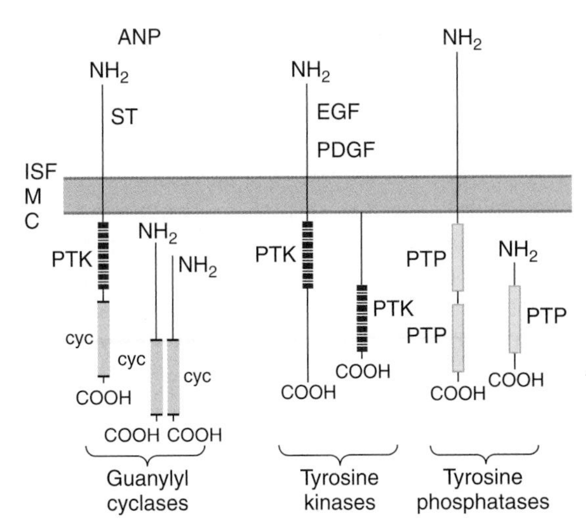

FIGURE 2–29 Diagrammatic representation of guanylyl cyclases, tyrosine kinases, and tyrosine phosphatases. ANP, atrial natriuretic peptide; C, cytoplasm; cyc, guanylyl cyclase domain; EGF, epidermal growth factor; ISF, interstitial fluid; M, cell membrane; PDGF, platelet-derived growth factor; PTK, tyrosine kinase domain; PTP, tyrosine phosphatase domain; ST, *E. coli* enterotoxin. (Modified from Koesling D, Böhme E, Schultz G: Guanylyl cyclases, a growing family of signal transducing enzymes. FASEB J 1991;5:2785.)

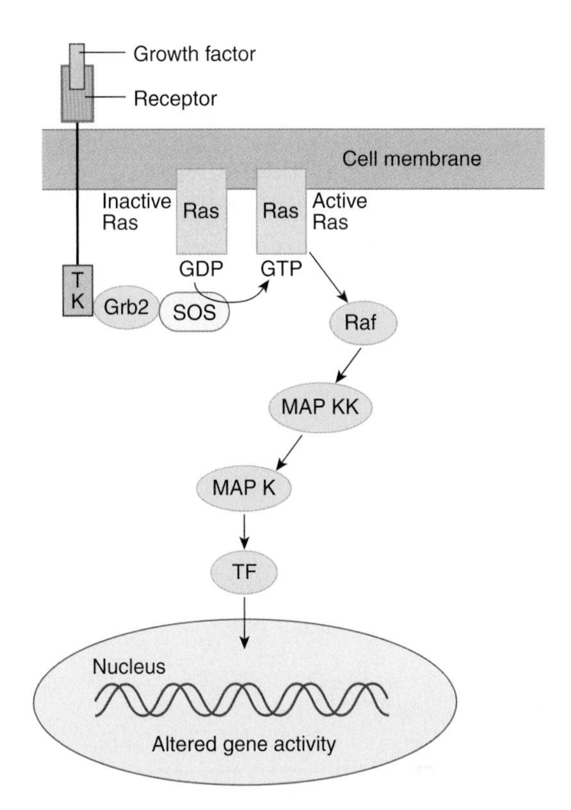

FIGURE 2–30 One of the direct pathways by which growth factors alter gene activity. TK, tyrosine kinase domain; Grb2, Ras activator controller; Sos, Ras activator; Ras, product of the ras gene; MAP K, mitogen-activated protein kinase; MAP KK, MAP kinase kinase; TF, transcription factors. There is a cross-talk between this pathway and the cAMP pathway, as well as a cross-talk with the IP_3–DAG pathway.

factor (EGF) are examples. More than 20 have been described. The cytokines are a second group. These factors are produced by macrophages and lymphocytes, as well as other cells, and are important in regulation of the immune system (see Chapter 3). Again, more than 20 have been described. The third group is made up of the colony-stimulating factors that regulate proliferation and maturation of red and white blood cells.

Receptors for EGF, platelet-derived growth factor (PDGF), and many of the other factors that foster cell multiplication and growth have a single membrane-spanning domain with an intracellular tyrosine kinase domain (Figure 2–29). When ligand binds to a tyrosine kinase receptor, it first causes a dimerization of two similar receptors. The dimerization results in partial activation of the intracellular tyrosine kinase domains and a cross-phosphorylation to fully activate each other. One of the pathways activated by phosphorylation leads, through the small G protein Ras, to MAP kinases, and eventually to the production of transcription factors in the nucleus that alter gene expression (Figure 2–30).

Receptors for cytokines and colony-stimulating factors differ from the other growth factors in that most of them do not have tyrosine kinase domains in their cytoplasmic portions and some have little or no cytoplasmic tail. However, they initiate tyrosine kinase activity in the cytoplasm. In particular, they activate the so-called Janus tyrosine kinases **(JAKs)** in the cytoplasm (Figure 2–31). These in turn phosphorylate **STAT** proteins. The phosphorylated STATs form homo- and heterodimers and move to the nucleus, where they act as transcription factors. There are four known mammalian JAKs and seven known STATs. Interestingly, the JAK–STAT pathway can also be activated by growth hormone and is another important direct path from the cell surface to the nucleus. However, it should be emphasized that both the Ras and the JAK–STAT pathways are complex and there is cross-talk between them and other signaling pathways discussed previously.

Finally, note that the whole subject of second messengers and intracellular signaling has become immensely complex, with multiple pathways and interactions. It is only possible in a book such as this to list highlights and present general themes that will aid the reader in understanding the rest of physiology (see Clinical Box 2–9).

HOMEOSTASIS

The actual environment of the cells of the body is the interstitial component of the ECF. Because normal cell function depends on the constancy of this fluid, it is not surprising that in multicellular animals, an immense number of regulatory mechanisms have evolved to maintain it. To describe "the various physiologic arrangements which serve to restore the normal state, once it has been disturbed," W.B. Cannon coined the term **homeostasis.** The buffering properties of the body fluids and the renal and respiratory adjustments to the presence of excess acid or alkali are examples of homeostatic mechanisms. There are countless other examples, and a large

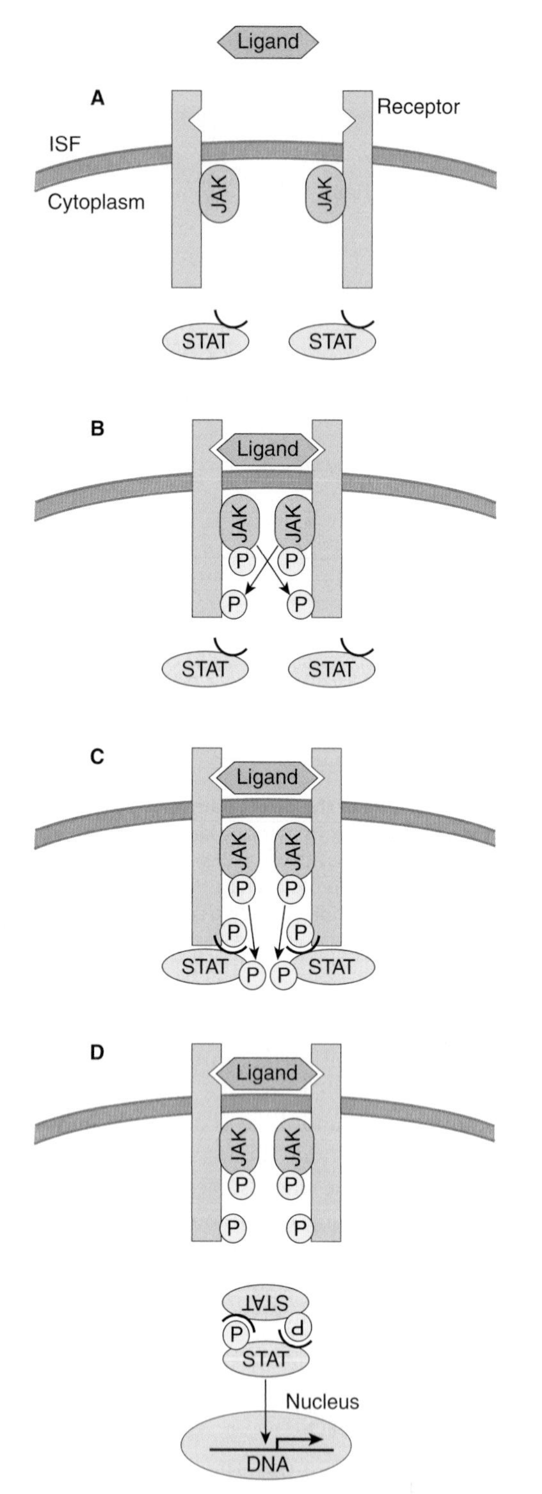

FIGURE 2–31 Signal transduction via the JAK–STAT pathway. A) Ligand binding leads to dimerization of receptor. **B)** Activation and tyrosine phosphorylation of JAKs. **C)** JAKs phosphorylate STATs. **D)** STATs dimerize and move to nucleus, where they bind to response elements on DNA. (Modified from Takeda K, Kishimoto T, Akira S: STAT6: Its role in interleukin 4-mediated biological functions. J Mol Med 1997;75:317.)

CLINICAL BOX 2–9

Receptor & G Protein Diseases

Many diseases are being traced to mutations in the genes for receptors. For example, loss-of-function receptor mutations that cause disease have been reported for the 1,25-dihydroxycholecalciferol receptor and the insulin receptor. Certain other diseases are caused by production of antibodies against receptors. Thus, antibodies against thyroid-stimulating hormone (TSH) receptors cause Graves' disease, and antibodies against nicotinic acetylcholine receptors cause myasthenia gravis.

An example of loss of function of a receptor is the type of nephrogenic diabetes insipidus that is due to loss of the ability of mutated V2 vasopressin receptors to mediate concentration of the urine. Mutant receptors can gain as well as lose function. A gain-of-function mutation of the Ca^{2+} receptor causes excess inhibition of parathyroid hormone secretion and familial hypercalciuric hypocalcemia. G proteins can also undergo loss-of-function or gain-of-function mutations that cause disease (Table 2–6). In one form of pseudohypoparathyroidism, a mutated $G_s\alpha$ fails to respond to parathyroid hormone, producing the symptoms of hypoparathyroidism without any decline in circulating parathyroid hormone. Testotoxicosis is an interesting disease that combines gain and loss of function. In this condition, an activating mutation of $G_s\alpha$ causes excess testosterone secretion and prepubertal sexual maturation. However, this mutation is temperature-sensitive and is active only at the relatively low temperature of the testes (33°C). At 37°C, the normal temperature of the rest of the body, it is replaced by loss of function, with the production of hypoparathyroidism and decreased responsiveness to TSH. A different activating mutation in $G_s\alpha$ is associated with the rough-bordered areas of skin pigmentation and hypercortisolism of the McCune–Albright syndrome. This mutation occurs during fetal development, creating a mosaic of normal and abnormal cells. A third mutation in $G_s\alpha$ reduces its intrinsic GTPase activity. As a result, it is much more active than normal, and excess cAMP is produced. This causes hyperplasia and eventually neoplasia in somatotrope cells of the anterior pituitary. Forty per cent of somatotrope tumors causing acromegaly have cells containing a somatic mutation of this type.

part of physiology is concerned with regulatory mechanisms that act to maintain the constancy of the internal environment. Many of these regulatory mechanisms operate on the principle of negative feedback; deviations from a given normal set point are detected by a sensor, and signals from the sensor trigger compensatory changes that continue until the set point is again reached.

TABLE 2–6 Examples of abnormalities caused by loss- or gain-of-function mutations of heterotrimeric G protein-coupled receptors and G proteins.

Site	Type of Mutation	Disease
Receptor		
Cone opsins	Loss	Color blindness
Rhodopsin	Loss	Congenital night blindness; two forms of retinitis pigmentosa
V_2 vasopressin	Loss	X-linked nephrogenic diabetes insipidus
ACTH	Loss	Familial glucocorticoid deficiency
LH	Gain	Familial male precocious puberty
TSH	Gain	Familial nonautoimmune hyperthyroidism
TSH	Loss	Familial hypothyroidism
Ca^{2+}	Gain	Familial hypercalciuric hypocalcemia
Thromboxane A_2	Loss	Congenital bleeding
Endothelin B	Loss	Hirschsprung disease
G protein		
$G_s\alpha$	Loss	Pseudohypothyroidism type 1a
$G_s\alpha$	Gain/loss	Testotoxicosis
$G_s\alpha$	Gain (mosaic)	McCune–Albright syndrome
$G_s\alpha$	Gain	Somatotroph adenomas with acromegaly
$G_i\alpha$	Gain	Ovarian and adrenocortical tumors

CHAPTER SUMMARY

- The cell and the intracellular organelles are surrounded by semipermeable membranes. Biological membranes have a lipid bilayer core that is populated by structural and functional proteins. These proteins contribute greatly to the semipermeable properties of biological membrane.
- Cells contain a variety of organelles that perform specialized cell functions. The nucleus is an organelle that contains the cellular DNA and is the site of transcription. The endoplasmic reticulum and the Golgi apparatus are important in protein processing and the targeting of proteins to correct compartments within the cell. Lysosomes and peroxisomes are membrane-bound organelles that contribute to protein and lipid processing. Mitochondria are organelles that allow for oxidative phosphorylation in eukaryotic cells and also are important in specialized cellular signaling.
- The cytoskeleton is a network of three types of filaments that provide structural integrity to the cell as well as a means for trafficking of organelles and other structures around the cell. Actin filaments are important in cellular contraction, migration, and signaling. Actin filaments also provide the backbone for muscle contraction. Intermediate filaments are primarily structural. Microtubules provide a dynamic structure in cells that allows for the movement of cellular components around the cell.
- There are three superfamilies of molecular motor proteins in the cell that use the energy of ATP to generate force, movement, or both. Myosin is the force generator for muscle cell contraction. Cellular myosins can also interact with the cytoskeleton (primarily thin filaments) to participate in contraction as well as movement of cell contents. Kinesins and cellular dyneins are motor proteins that primarily interact with microtubules to move cargo around the cells.
- Cellular adhesion molecules aid in tethering cells to each other or to the extracellular matrix as well as providing for initiation of cellular signaling. There are four main families of these proteins: integrins, immunoglobulins, cadherins, and selectins.
- Cells contain distinct protein complexes that serve as cellular connections to other cells or the extracellular matrix. Tight junctions provide intercellular connections that link cells into a regulated tissue barrier and also provide a barrier to movement of proteins in the cell membrane. Gap junctions provide contacts between cells that allow for direct passage of small molecules between two cells. Desmosomes and adherens junctions are specialized structures that hold cells together. Hemidesmosomes and focal adhesions attach cells to their basal lamina.
- Exocytosis and endocytosis are vesicular fusion events that allow for movement of proteins and lipids between the cell interior, the plasma membrane, and the cell exterior. Exocytosis can be constitutive or nonconstitutive; both are regulated processes that require specialized proteins for vesicular fusion. Endocytosis is the formation of vesicles at the plasma membrane to take material from the extracellular space into the cell interior.
- Cells can communicate with one another via chemical messengers. Individual messengers (or ligands) typically bind to a plasma membrane receptor to initiate intracellular changes that lead to physiologic changes. Plasma membrane receptor families include ion channels, G protein-coupled receptors, or a variety of enzyme-linked receptors (eg, tyrosine kinase receptors). There are additional cytosolic receptors (eg, steroid receptors) that can bind membrane-permeant compounds. Activation of receptors leads to cellular changes that include changes in membrane potential, activation of heterotrimeric G proteins, increase in second messenger molecules, or initiation of transcription.
- Second messengers are molecules that undergo a rapid concentration changes in the cell following primary messenger recognition. Common second messenger molecules include Ca^{2+}, cyclic adenosine monophosphate (cAMP), cyclic guanine monophosphate (cGMP), inositol trisphosphate (IP_3), and nitric oxide (NO).

MULTIPLE-CHOICE QUESTIONS

For all questions, select the single best answer unless otherwise directed.

1. The electrogenic Na, K ATPase plays a critical role in cellular physiology by
   A. using the energy in ATP to extrude 3 Na^+ out of the cell in exchange for taking two K^+ into the cell.
   B. using the energy in ATP to extrude 3 K^+ out of the cell in exchange for taking two Na^+ into the cell.
   C. using the energy in moving Na^+ into the cell or K^+ outside the cell to make ATP.
   D. using the energy in moving Na^+ outside of the cell or K^+ inside the cell to make ATP.
2. Cell membranes
   A. contain relatively few protein molecules.
   B. contain many carbohydrate molecules.
   C. are freely permeable to electrolytes but not to proteins.
   D. have variable protein and lipid contents depending on their location in the cell.
   E. have a stable composition throughout the life of the cell.
3. Second messengers
   A. are substances that interact with first messengers outside cells.
   B. are substances that bind to first messengers in the cell membrane.
   C. are hormones secreted by cells in response to stimulation by another hormone.
   D. mediate the intracellular responses to many different hormones and neurotransmitters.
   E. are not formed in the brain.
4. The Golgi complex
   A. is an organelle that participates in the breakdown of proteins and lipids.
   B. is an organelle that participates in posttranslational processing of proteins.
   C. is an organelle that participates in energy production.
   D. is an organelle that participates in transcription and translation.
   E. is a subcellular compartment that stores proteins for trafficking to the nucleus.
5. Endocytosis
   A. includes phagocytosis and pinocytosis, but not clathrin-mediated or caveolae-dependent uptake of extracellular contents.
   B. refers to the merging of an intracellular vesicle with the plasma membrane to deliver intracellular contents to the extracellular milieu.
   C. refers to the invagination of the plasma membrane to uptake extracellular contents into the cell.
   D. refers to vesicular trafficking between Golgi stacks.
6. G protein-coupled receptors
   A. are intracellular membrane proteins that help to regulate movement within the cell.
   B. are plasma membrane proteins that couple the extracellular binding of primary signaling molecules to exocytosis.
   C. are plasma membrane proteins that couple the extracellular binding of primary signaling molecules to the activation of heterotrimeric G proteins.
   D. are intracellular proteins that couple the binding of primary messenger molecules with transcription.
7. Gap junctions are intercellular connections that
   A. primarily serve to keep cells separated and allow for transport across a tissue barrier.
   B. serve as a regulated cytoplasmic bridge for sharing of small molecules between cells.
   C. serve as a barrier to prevent protein movement within the cellular membrane.
   D. are cellular components for constitutive exocytosis that occurs between adjacent cells.
8. F-actin is a component of the cellular cytoskeleton that
   A. provides a structural component for cell movement.
   B. is defined as the "functional" form of actin in the cell.
   C. refers to the actin subunits that provide the molecular building blocks of the extended actin molecules found in the cell.
   D. provide the molecular architecture for cell to cell communication.

CHAPTER RESOURCES

Alberts B, Johnson A, Lewis J, et al: *Molecular Biology of the Cell*, 5th ed. Garland Science, 2008.
Cannon WB: *The Wisdom of the Body.* Norton, 1932.
Junqueira LC, Carneiro J, Kelley RO: *Basic Histology*, 9th ed. McGraw-Hill, 1998.
Kandel ER, Schwartz JH, Jessell TM (editors): *Principles of Neural Science*, 4th ed. McGraw-Hill, 2000.
Pollard TD, Earnshaw WC: *Cell Biology*, 2nd ed. Saunders, Elsevier, 2008.
Sperelakis N (editor): *Cell Physiology Sourcebook*, 3rd ed. Academic Press, 2001.

CHAPTER

3

Immunity, Infection, & Inflammation

OBJECTIVES

After studying this chapter, you should be able to:

- Understand the significance of immunity, particularly with respect to defending the body against microbial invaders.
- Define the circulating and tissue cell types that contribute to immune and inflammatory responses.
- Describe how phagocytes are able to kill internalized bacteria.
- Identify the functions of hematopoietic growth factors, cytokines, and chemokines.
- Delineate the roles and mechanisms of innate, acquired, humoral, and cellular immunity.
- Understand the basis of inflammatory responses and wound healing.

INTRODUCTION

As an open system, the body is continuously called upon to defend itself from potentially harmful invaders such as bacteria, viruses, and other microbes. This is accomplished by the immune system, which is subdivided into innate and adaptive (or acquired) branches. The immune system is composed of specialized effector cells that sense and respond to foreign antigens and other molecular patterns not found in human tissues. Likewise, the immune system clears the body's own cells that have become senescent or abnormal, such as cancer cells. Finally, normal host tissues occasionally become the subject of inappropriate immune attack, such as in autoimmune diseases or in settings where normal cells are harmed as innocent bystanders when the immune system mounts an inflammatory response to an invader. It is beyond the scope of this volume to provide a full treatment of all aspects of modern immunology. Nevertheless, the student of physiology should have a working knowledge of immune functions and their regulation, due to a growing appreciation for the ways in which the immune system can contribute to normal physiological regulation in a variety of tissues, as well as contributions of immune effectors to pathophysiology.

IMMUNE EFFECTOR CELLS

Many immune effector cells circulate in the blood as the white blood cells. In addition, the blood is the conduit for the precursor cells that eventually develop into the immune cells of the tissues. The circulating immunologic cells include **granulocytes (polymorphonuclear leukocytes, PMNs)**, comprising **neutrophils, eosinophils,** and **basophils; lymphocytes;** and **monocytes.** Immune responses in the tissues are further amplified by these cells following their extravascular migration, as well as tissue **macrophages** (derived from monocytes) and **mast cells** (related to basophils). Acting together, these cells provide the body with powerful defenses against tumors and viral, bacterial, and parasitic infections.

GRANULOCYTES

All granulocytes have cytoplasmic granules that contain biologically active substances involved in inflammatory and allergic reactions.

The average half-life of a neutrophil in the circulation is 6 h. To maintain the normal circulating blood level, it is therefore necessary to produce over 100 billion neutrophils per day. Many neutrophils enter the tissues, particularly if triggered to do so by an infection or by inflammatory cytokines. They are attracted to the endothelial surface by cell adhesion molecules known as selectins, and they roll along it. They then bind firmly to neutrophil adhesion molecules of the integrin family. They next insinuate themselves through the walls of the capillaries between endothelial cells by a process called **diapedesis.** Many of those that leave the circulation enter the gastrointestinal tract and are eventually lost from the body.

Invasion of the body by bacteria triggers the **inflammatory response.** The bone marrow is stimulated to produce and release large numbers of neutrophils. Bacterial products interact with plasma factors and cells to produce agents that attract neutrophils to the infected area **(chemotaxis).** The chemotactic agents, which are part of a large and expanding family of **chemokines** (see following text), include a component of the complement system (C5a); leukotrienes; and polypeptides from lymphocytes, mast cells, and basophils. Other plasma factors act on the bacteria to make them "tasty" to the phagocytes **(opsonization).** The principal opsonins that coat the bacteria are immunoglobulins of a particular class (IgG) and complement proteins (see following text). The coated bacteria then bind to G protein-coupled receptors on the neutrophil cell membrane. This triggers increased motor activity of the cell, exocytosis, and the so-called respiratory burst. The increased motor activity leads to prompt ingestion of the bacteria by endocytosis **(phagocytosis).** By **exocytosis,** neutrophil granules discharge their contents into the phagocytic vacuoles containing the bacteria and also into the interstitial space **(degranulation).** The granules contain various proteases plus antimicrobial proteins called **defensins.** In addition, the cell membrane-bound enzyme **NADPH oxidase** is activated, with the production of toxic oxygen metabolites. The combination of the toxic oxygen metabolites and the proteolytic enzymes from the granules makes the neutrophil a very effective killing machine.

Activation of NADPH oxidase is associated with a sharp increase in O_2 uptake and metabolism in the neutrophil (the **respiratory burst**) and generation of O_2^- by the following reaction:

$$NADPH + H^+ + 2O_2 + \rightarrow NADP^+ + 2H^+ + 2O_2^-$$

O_2^- is a **free radical** formed by the addition of one electron to O_2. Two O_2^- react with two H^+ to form H_2O_2 in a reaction catalyzed by the cytoplasmic form of superoxide dismutase (SOD-1):

$$O_2^- + O_2^- + H^+ \xrightarrow{SOD\text{-}1} \rightarrow H_2O_2 + O_2$$

O_2^- and H_2O_2 are both oxidants that are effective bactericidal agents, but H_2O_2 is converted to H_2O and O_2 by the enzyme **catalase.** The cytoplasmic form of SOD contains both Zn and Cu. It is found in many parts of the body. It is defective as a result of genetic mutation in a familial form of **amyotrophic lateral sclerosis** (ALS; see Chapter 15). Therefore, it may be that O_2^- accumulates in motor neurons and kills them in at least one form of this progressive, fatal disease. Two other forms of SOD encoded by at least one different gene are also found in humans.

Neutrophils also discharge the enzyme **myeloperoxidase,** which catalyzes the conversion of Cl^-, Br^-, I^-, and SCN^- to the corresponding acids (HOCl, HOBr, etc). These acids are also potent oxidants. Because Cl^- is present in greatest abundance in body fluids, the principal product is HOCl.

In addition to myeloperoxidase and defensins, neutrophil granules contain elastase, metalloproteinases that attack collagen, and a variety of other proteases that help destroy invading organisms. These enzymes act in a cooperative fashion with O_2^-, H_2O_2, and HOCl to produce a killing zone around the activated neutrophil. This zone is effective in killing invading organisms, but in certain diseases (eg, rheumatoid arthritis) the neutrophils may also cause local destruction of host tissue.

Like neutrophils, **eosinophils** have a short half-life in the circulation, are attracted to the surface of endothelial cells by selectins, bind to integrins that attach them to the vessel wall, and enter the tissues by diapedesis. Like neutrophils, they release proteins, cytokines, and chemokines that produce inflammation but are capable of killing invading organisms. However, eosinophils have some selectivity in the way in which they respond and in the killing molecules they secrete. Their maturation and activation in tissues is particularly stimulated by IL-3, IL-5, and GM-CSF (see below). They are especially abundant in the mucosa of the gastrointestinal tract, where they defend against parasites, and in the mucosa of the respiratory and urinary tracts. Circulating eosinophils are increased in allergic diseases such as asthma and in various other respiratory and gastrointestinal diseases.

Basophils also enter tissues and release proteins and cytokines. They resemble but are not identical to mast cells, and like mast cells they contain histamine (see below). They release histamine and other inflammatory mediators when activated by binding of specific antigens to cell-fixed IgE molecules, and participate in immediate-type hypersensitivity (allergic) reactions. These range from mild urticaria and rhinitis to severe anaphylactic shock. The antigens that trigger IgE formation and basophil (and mast cell) activation are innocuous to most individuals, and are referred to as allergens.

MAST CELLS

Mast cells are heavily granulated cells of the connective tissue that are abundant in tissues that come into contact with the external environment, such as beneath epithelial surfaces. Their granules contain proteoglycans, histamine, and many proteases. Like basophils, they degranulate when allergens bind to cell-bound IgE molecules directed against them. They are involved in inflammatory responses initiated by immunoglobulins IgE and IgG (see below). The inflammation combats invading parasites. In addition to this involvement in acquired immunity, they release TNF-α in response to bacterial products by an antibody-independent mechanism, thus participating in the nonspecific **innate immunity** that combats infections prior to the development of an adaptive immune response (see following text). Marked mast cell degranulation produces clinical manifestations of allergy up to and including anaphylaxis.

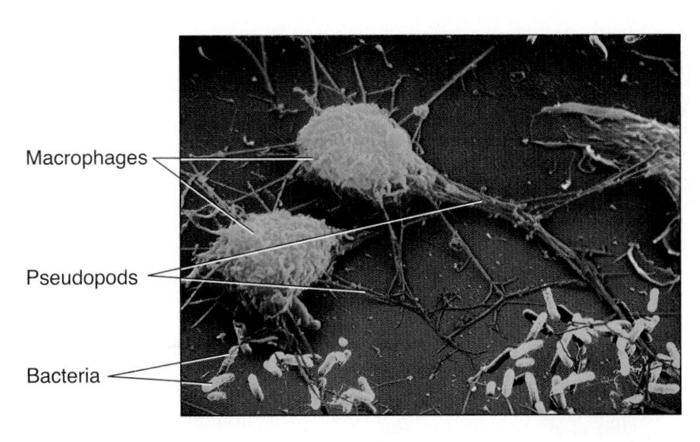

FIGURE 3–1 Macrophages contacting bacteria and preparing to engulf them. Figure is a colorized version of a scanning electronmicrograph.

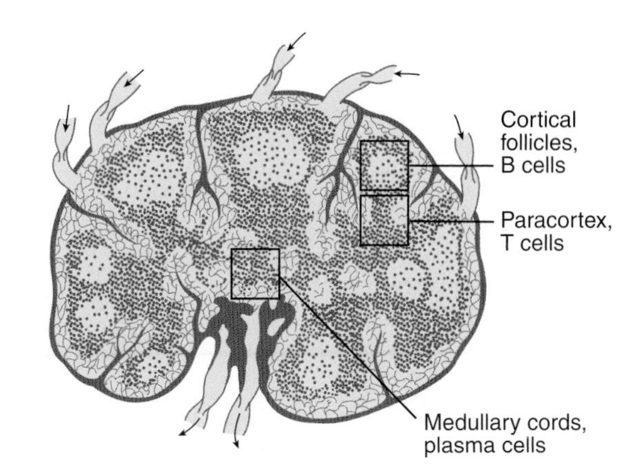

FIGURE 3–2 Anatomy of a normal lymph node. (After Chandrasoma. Reproduced with permission from McPhee SJ, Lingappa VR, Ganong WF [editors]: *Pathophysiology of Disease*, 4th ed. McGraw-Hill, 2003.)

MONOCYTES

Monocytes enter the blood from the bone marrow and circulate for about 72 h. They then enter the tissues and become **tissue macrophages** (Figure 3–1). Their life span in the tissues is unknown, but bone marrow transplantation data in humans suggest that they persist for about 3 months. It appears that they do not reenter the circulation. Some may end up as the multinucleated giant cells seen in chronic inflammatory diseases such as tuberculosis. The tissue macrophages include the Kupffer cells of the liver, pulmonary alveolar macrophages (see Chapter 34), and microglia in the brain, all of which come originally from the circulation. In the past, they have been called the **reticuloendothelial system,** but the general term **tissue macrophage system** seems more appropriate.

Macrophages are activated by cytokines released from T lymphocytes, among others. Activated macrophages migrate in response to chemotactic stimuli and engulf and kill bacteria by processes generally similar to those occurring in neutrophils. They play a key role in innate immunity (see below). They also secrete up to 100 different substances, including factors that affect lymphocytes and other cells, prostaglandins of the E series, and clot-promoting factors.

LYMPHOCYTES

Lymphocytes are key elements in the production of acquired immunity (see below). After birth, some lymphocytes are formed in the bone marrow. However, most are formed in the lymph nodes (Figure 3–2), thymus, and spleen from precursor cells that originally came from the bone marrow and were processed in the thymus (T cells) or bursal equivalent (B cells, see below). Lymphocytes enter the bloodstream for the most part via the lymphatics. At any given time, only about 2% of the body lymphocytes are in the peripheral blood. Most of the rest are in the lymphoid organs. It has been calculated that in humans, 3.5×10^{10} lymphocytes per day enter the circulation via the thoracic duct alone; however, this count includes cells that reenter the lymphatics and thus traverse the thoracic duct more than once. The effects of adrenocortical hormones on the lymphoid organs, the circulating lymphocytes, and the granulocytes are discussed in Chapter 20.

During fetal development, and to a much lesser extent during adult life, lymphocyte precursors come from the bone marrow. Those that populate the thymus (Figure 3–3) become transformed by the environment in this organ into T lymphocytes. In birds, the precursors that populate the bursa of Fabricius, a lymphoid structure near the cloaca, become transformed into B lymphocytes. There is no bursa in mammals, and the transformation to B lymphocytes occurs in **bursal equivalents,** that is, the fetal liver and, after birth, the bone marrow. After residence in the thymus or liver, many of the T and B lymphocytes migrate to the lymph nodes.

T and B lymphocytes are morphologically indistinguishable but can be identified by markers on their cell membranes. B cells differentiate into **plasma cells** and **memory B cells.** There are three major types of T cells: **cytotoxic T cells, helper T cells,** and **memory T cells.** There are two subtypes of helper T cells: T helper 1 (TH1) cells secrete IL-2 and γ-interferon and are concerned primarily with cellular immunity; T helper 2 (TH2) cells secrete IL-4 and IL-5 and interact primarily with B cells in relation to humoral immunity. Cytotoxic T cells destroy transplanted and other foreign cells, with their development aided and directed by helper T cells. Markers on the surface of lymphocytes are assigned CD (clusters of differentiation) numbers on the basis of their reactions to a panel of monoclonal antibodies. Most cytotoxic T cells display the glycoprotein CD8, and helper T cells display the glycoprotein CD4. These proteins are closely associated with the T cell receptors and may function as coreceptors. On the basis of differences in their receptors and functions, cytotoxic T cells are divided into αβ and γδ types (see below). Natural killer (NK) cells (see above) are also cytotoxic lymphocytes, though they are not T cells. Thus, there are three main types of cytotoxic lymphocytes in the body: αβ T cells, γδ T cells, and NK cells.

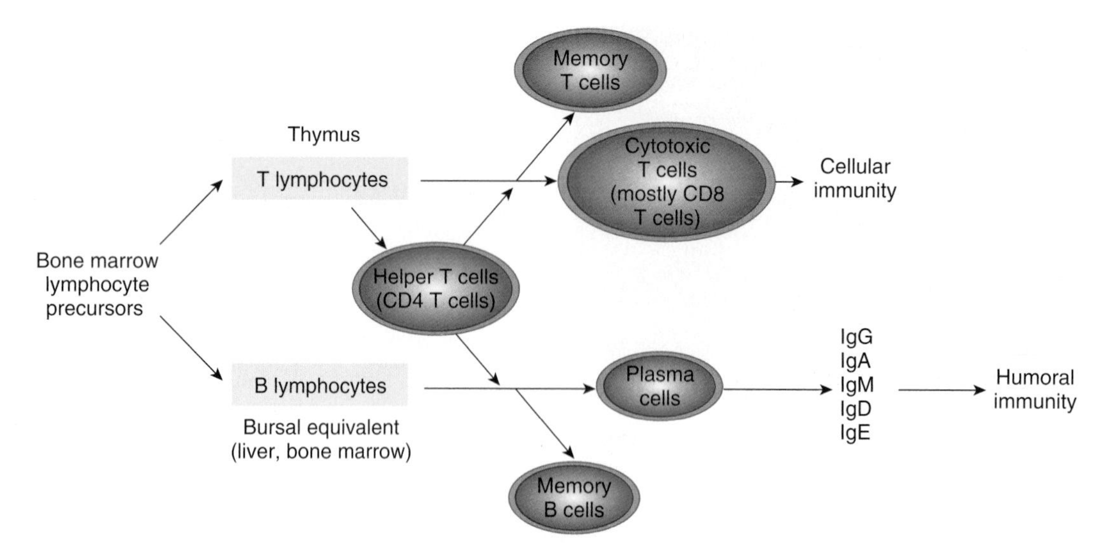

FIGURE 3–3 Development of the system mediating acquired immunity.

MEMORY B CELLS & T CELLS

After exposure to a given antigen, a small number of activated B and T cells persist as memory B and T cells. These cells are readily converted to effector cells by a later encounter with the same antigen. This ability to produce an accelerated response to a second exposure to an antigen is a key characteristic of acquired immunity. The ability persists for long periods of time, and in some instances (eg, immunity to measles) it can be life-long.

After activation in lymph nodes, lymphocytes disperse widely throughout the body and are especially plentiful in areas where invading organisms enter the body, for example, the mucosa of the respiratory and gastrointestinal tracts. This puts memory cells close to sites of reinfection and may account in part for the rapidity and strength of their response. Chemokines are involved in guiding activated lymphocytes to these locations.

GRANULOCYTE & MACROPHAGE COLONY-STIMULATING FACTORS

The production of white blood cells is regulated with great precision in healthy individuals, and the production of granulocytes is rapidly and dramatically increased in infections. The proliferation and self-renewal of hematopoietic stem cells (HSCs) depends on **stem cell factor (SCF).** Other factors specify particular lineages. The proliferation and maturation of the cells that enter the blood from the marrow are regulated by growth factors that cause cells in one or more of the committed cell lines to proliferate and mature (Table 3–1). The regulation of erythrocyte production by **erythropoietin** is discussed in Chapter 38. Three additional factors are called **colony-stimulating factors (CSFs),** because they cause appropriate single stem cells to proliferate in soft agar, forming colonies. The factors stimulating the production of committed stem cells include **granulocyte–macrophage CSF (GM-CSF), granulocyte CSF (G-CSF),** and **macrophage CSF (M-CSF).** Interleukins **IL-1** and **IL-6** followed by **IL-3** (Table 3–1) act in sequence to convert pluripotential uncommitted stem cells to committed progenitor cells. IL-3 is also known as **multi-CSF.** Each of the CSFs has a predominant action, but all the CSFs and interleukins also have other overlapping actions. In addition, they activate and sustain mature blood cells. It is interesting in this regard that the genes for many of these factors are located together on the long arm of chromosome 5 and may have originated by duplication of an ancestral gene. It is also interesting that basal hematopoiesis is normal in mice in which the GM-CSF gene is knocked out indicating that loss of one factor can be compensated for by others. On the other hand, the absence of GM-CSF causes accumulation of surfactant in the lungs (see Chapter 34).

As noted in Chapter 38, erythropoietin is produced in part by kidney cells and is a circulating hormone. The other factors are produced by macrophages, activated T cells, fibroblasts, and endothelial cells. For the most part, the factors act locally in the bone marrow (Clinical Box 3–1).

IMMUNITY

OVERVIEW

Insects and other invertebrates have only **innate immunity.** This system is triggered by receptors that bind sequences of sugars, lipids, amino acids, or nucleic acids that are common on bacteria and other microorganisms, but are not found in eukaryotic cells. These receptors, in turn, activate various defense mechanisms. The receptors are coded in the germ line, and their fundamental structure is not modified by exposure to antigen. The activated defenses include, in various species, release of interferons, phagocytosis, production of antibacterial peptides, activation of the complement system, and several proteolytic cascades. Even plants release antibacterial peptides

TABLE 3–1 Hematopoietic growth factors.

Cytokine	Cell Lines Stimulated	Cytokine Source
IL-1	Erythrocyte Granulocyte Megakaryocyte Monocyte	Multiple cell types
IL-3	Erythrocyte Granulocyte Megakaryocyte Monocyte	T lymphocytes
IL-4	Basophil	T lymphocytes
IL-5	Eosinophil	T lymphocytes
IL-6	Erythrocyte Granulocyte Megakaryocyte Monocyte	Endothelial cells Fibroblasts Macrophages
IL-11	Erythrocyte Granulocyte Megakaryocyte	Fibroblasts Osteoblasts
Erythropoietin	Erythrocyte	Kidney Kupffer cells of liver
SCF	Erythrocyte Granulocyte Megakaryocyte Monocyte	Multiple cell types
G-CSF	Granulocyte	Endothelial cells Fibroblasts Monocytes
GM-CSF	Erythrocyte Granulocyte Megakaryocyte	Endothelial cells Fibroblasts Monocytes T lymphocytes
M-CSF	Monocyte	Endothelial cells Fibroblasts Monocytes
Thrombopoietin	Megakaryocyte	Liver, kidney

Key: CSF, colony stimulating factor; G, granulocyte; IL, interleukin; M, macrophage; SCF, stem cell factor.

Reproduced with permission from McPhee SJ, Lingappa VR, Ganong WF (editors): *Pathophysiology of Disease*, 6th ed. McGraw-Hill, 2010.

CLINICAL BOX 3–1

Disorders of Phagocytic Function

More than 15 primary defects in neutrophil function have been described, along with at least 30 other conditions in which there is a secondary depression of the function of neutrophils. Patients with these diseases are prone to infections that are relatively mild when only the neutrophil system is involved, but which can be severe when the monocyte-tissue macrophage system is also involved. In one syndrome (neutrophil hypomotility), actin in the neutrophils does not polymerize normally, and the neutrophils move slowly. In another, there is a congenital deficiency of leukocyte integrins. In a more serious disease (chronic granulomatous disease), there is a failure to generate O_2^- in both neutrophils and monocytes and consequent inability to kill many phagocytosed bacteria. In severe congenital glucose 6-phosphate dehydrogenase deficiency, there are multiple infections because of failure to generate the NADPH necessary for O_2^- production. In congenital myeloperoxidase deficiency, microbial killing power is reduced because hypochlorous acid is not formed.

THERAPEUTIC HIGHLIGHTS

The cornerstones of treatment in disorders of phagocytic function include scrupulous efforts to avoid exposure to infectious agents, and antibiotic and antifungal prophylaxis. Antimicrobial therapies must also be implemented aggressively if infections occur. Sometimes, surgery is needed to excise and/or drain abscesses and relieve obstructions. Bone marrow transplantation may offer the hope of a definitive cure for severe conditions, such as chronic granulomatous disease. Sufferers of this condition have a significantly reduced life expectancy due to recurrent infections and their complications, and so the risks of bone marrow transplantation may be deemed acceptable. Gene therapy, on the other hand, remains a distant goal.

in response to infection. This primitive immune system is also important in vertebrates, particularly in the early response to infection. However, in vertebrates, innate immunity is also complemented by **adaptive** or **acquired immunity,** a system in which T and B lymphocytes are activated by specific antigens. T cells bear receptors related to antibody molecules, but which remain cell-bound. When these receptors encounter their cognate antigen, the T cell is stimulated to proliferate and produce cytokines that orchestrate the immune response, including that of B cells. Activated B lymphocytes form clones that produce secreted antibodies, which attack foreign proteins. After the invasion is repelled, small numbers of lymphocytes persist as memory cells so that a second exposure to the same antigen provokes a prompt and magnified immune attack. The genetic event that led to acquired immunity occurred 450 million years ago in the ancestors of jawed vertebrates and was probably insertion of a transposon into the genome in a way that made possible the generation of the immense repertoire of T cell receptors and antibodies that can be produced by the body.

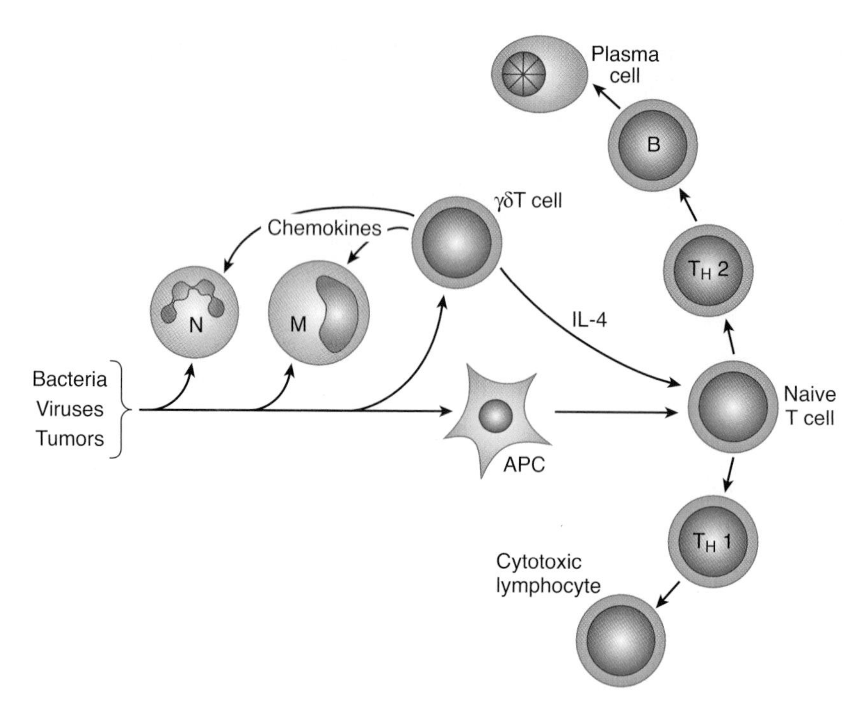

FIGURE 3–4 How bacteria, viruses, and tumors trigger innate immunity and initiate the acquired immune response. Arrows indicate mediators/cytokines that act on the target cell shown and/or pathways of differentiation. APC, antigen-presenting cell; M, monocyte; N, neutrophil; TH1 and TH2, helper T cells type 1 and type 2, respectively.

In vertebrates, including humans, innate immunity provides the first line of defense against infections, but it also triggers the slower but more specific acquired immune response (Figure 3–4). In vertebrates, natural and acquired immune mechanisms also attack tumors and tissue transplanted from other animals.

Once activated, immune cells communicate by means of cytokines and chemokines. They kill viruses, bacteria, and other foreign cells by secreting other cytokines and activating the complement system.

CYTOKINES

Cytokines are hormone-like molecules that act—generally in a paracrine fashion—to regulate immune responses. They are secreted not only by lymphocytes and macrophages but by endothelial cells, neurons, glial cells, and other types of cells. Most of the cytokines were initially named for their actions, for example, B cell-differentiating factor, or B cell-stimulating factor 2. However, the nomenclature has since been rationalized by international agreement to that of the **interleukins.** For example, the name of B cell-differentiating factor was changed to interleukin-4. A number of cytokines selected for their biological and clinical relevance are listed in Table 3–2, but it would be beyond the scope of this text to list all cytokines, which now number more than 100.

Many of the receptors for cytokines and hematopoietic growth factors (see above), as well as the receptors for prolactin (see Chapter 22), and growth hormone (see Chapter 18) are members of a cytokine-receptor superfamily that has three subfamilies (Figure 3–5). The members of subfamily 1, which includes the receptors for IL-4 and IL-7, are homodimers. The members of subfamily 2, which includes the receptors for IL-3, IL-5, and IL-6, are heterodimers. The receptor for IL-2 (and several other cytokines) consists of a heterodimer plus an unrelated protein, the so-called Tac antigen. The other members of subfamily 3 have the same γ chain as IL-2R. The extracellular domain of the homodimer and heterodimer subunits all contain four conserved cysteine residues plus a conserved Trp-Ser-X-Trp-Ser domain, and although the intracellular portions do not contain tyrosine kinase catalytic domains, they activate cytoplasmic tyrosine kinases when ligand binds to the receptors.

The effects of the principal cytokines are listed in Table 3–2. Some of them have systemic as well as local paracrine effects. For example, IL-1, IL-6, and tumor necrosis factor α cause fever, and IL-1 increases slow-wave sleep and reduces appetite.

Another superfamily of cytokines is the **chemokine** family. Chemokines are substances that attract neutrophils (see previous text) and other white blood cells to areas of inflammation or immune response. Over 40 have now been identified, and it is clear that they also play a role in the regulation of cell growth and angiogenesis. The chemokine receptors are G protein-coupled receptors that cause, among other things, extension of pseudopodia with migration of the cell toward the source of the chemokine.

THE COMPLEMENT SYSTEM

The cell-killing effects of innate and acquired immunity are mediated in part by a system of more than 30 plasma proteins originally named the **complement system** because they "complemented" the effects of antibodies. Three different pathways or enzyme

TABLE 3–2 Examples of cytokines and their clinical relevance.

Cytokine	Cellular Sources	Major Activities	Clinical Relevance
Interleukin-1	Macrophages	Activation of T cells and macrophages; promotion of inflammation	Implicated in the pathogenesis of septic shock, rheumatoid arthritis, and atherosclerosis
Interleukin-2	Type 1 (TH1) helper T cells	Activation of lymphocytes, natural killer cells, and macrophages	Used to induce lymphokine-activated killer cells; used in the treatment of metastatic renal-cell carcinoma, melanoma, and various other tumors
Interleukin-4	Type 2 (TH2) helper T cells, mast cells, basophils, and eosinophils	Activation of lymphocytes, monocytes, and IgE class switching	As a result of its ability to stimulate IgE production, plays a part in mast-cell sensitization and thus in allergy and in defense against nematode infections
Interleukin-5	Type 2 (TH2) helper T cells, mast cells, and eosinophils	Differentiation of eosinophils	Monoclonal antibody against interleukin-5 used to inhibit the antigen-induced late-phase eosinophilia in animal models of allergy
Interleukin-6	Type 2 (TH2) helper T cells and macrophages	Activation of lymphocytes; differentiation of B cells; stimulation of the production of acute-phase proteins	Overproduced in Castleman's disease; acts as an autocrine growth factor in myeloma and in mesangial proliferative glomerulonephritis
Interleukin-8	T cells and macrophages	Chemotaxis of neutrophils, basophils, and T cells	Levels are increased in diseases accompanied by neutrophilia, making it a potentially useful marker of disease activity
Interleukin-11	Bone marrow stromal cells	Stimulation of the production of acute-phase proteins	Used to reduce chemotherapy-induced thrombocytopenia in patients with cancer
Interleukin-12	Macrophages and B cells	Stimulation of the production of interferon γ by type 1 (TH1) helper T cells and by natural killer cells; induction of type 1 (TH1) helper T cells	May be useful as an adjuvant for vaccines
Tumor necrosis factor α	Macrophages, natural killer cells, T cells, B cells, and mast cells	Promotion of inflammation	Treatment with antibodies against tumor necrosis factor α beneficial in rheumatoid arthritis and Crohn's disease
Lymphotoxin (tumor necrosis factor β)	Type 1 (TH1) helper T cells and B cells	Promotion of inflammation	Implicated in the pathogenesis of multiple sclerosis and insulin-dependent diabetes mellitus
Transforming growth factor β	T cells, macrophages, B cells, and mast cells	Immunosuppression	May be useful therapeutic agent in multiple sclerosis and myasthenia gravis
Granulocyte-macrophage colony-stimulating factor	T cells, macrophages, natural killer cells, and B cells	Promotion of the growth of granulocytes and monocytes	Used to reduce neutropenia after chemotherapy for tumors and in ganciclovir-treated patients with AIDS; used to stimulate cell production after bone marrow transplantation
Interferon-α	Virally infected cells	Induction of resistance of cells to viral infection	Used to treat AIDS-related Kaposi sarcoma, melanoma, chronic hepatitis B infection, and chronic hepatitis C infection
Interferon-β	Virally infected cells	Induction of resistance of cells to viral infection	Used to reduce the frequency and severity of relapses in multiple sclerosis
Interferon-γ	Type 1 (TH1) helper T cells and natural killer cells	Activation of macrophages; inhibition of type 2 (TH2) helper T cells	Used to enhance the killing of phagocytosed bacteria in chronic granulomatous disease

Reproduced with permission from Delves PJ, Roitt IM: The immune system. First of two parts. *N Engl J Med* 2000;343:37.

cascades activate the system: the **classic pathway,** triggered by immune complexes; the **mannose-binding lectin pathway,** triggered when this lectin binds mannose groups in bacteria; and the **alternative** or **properdin pathway,** triggered by contact with various viruses, bacteria, fungi, and tumor cells. The proteins that are produced have three functions: they help kill invading organisms by opsonization, chemotaxis, and eventual lysis of the cells; they serve in part as a bridge from innate to acquired immunity by activating B cells and aiding immune memory; and they help dispose of waste products after apoptosis. Cell lysis, one of the principal ways the complement system kills cells, is brought about by inserting proteins called **perforins** into their cell membranes. These create holes, which permit free flow of ions and thus disruption of membrane polarity.

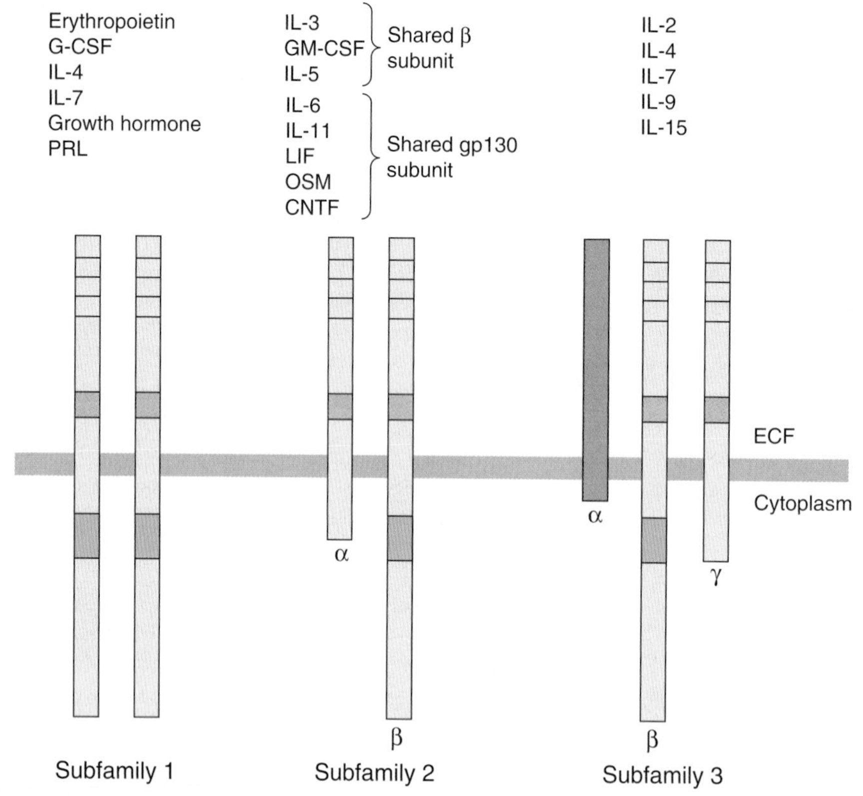

FIGURE 3–5 Members of one of the cytokine receptor superfamilies, showing shared structural elements. Note that all the subunits except the α subunit in subfamily 3 have four conserved cysteine residues (open boxes at top) and a Trp-Ser-X-Trp-Ser motif (pink). Many subunits also contain a critical regulatory domain in their cytoplasmic portions (green). CNTF, ciliary neurotrophic factor; LIF, leukemia inhibitory factor; OSM, oncostatin M; PRL, prolactin. (Modified from D'Andrea AD: Cytokine receptors in congenital hematopoietic disease. *N Engl J Med* 1994;330:839.)

INNATE IMMUNITY

The cells that mediate innate immunity include neutrophils, macrophages, and **natural killer cells,** large cytotoxic lymphocytes distinct from both T and B cells . All these cells respond to molecular patterns produced by bacteria and to other substances characteristic of viruses, tumor, and transplant cells. Many cells that are not professional immunocytes may nevertheless also contribute to innate immune responses, such as endothelial and epithelial cells. The activated cells produce their effects via the release of cytokines, as well as, in some cases, complement and other systems.

Innate immunity in *Drosophila* centers around a receptor protein named **toll,** which binds fungal antigens and triggers activation of genes coding for antifungal proteins. An expanding list of toll-like receptors (TLRs) have now been identified in humans and other vertebrates. One of these, TLR4, binds bacterial lipopolysaccharide and a protein called CD14, and this initiates intracellular events that activate transcription of genes for a variety of proteins involved in innate immune responses. This is important because bacterial lipopolysaccharide produced by gram-negative organisms is the cause of septic shock. TLR2 mediates the response to microbial lipoproteins, TLR6 cooperates with TLR2 in recognizing certain peptidoglycans, TLR5 recognizes a molecule known as flagellin in bacterial flagellae, and TLR9 recognizes bacterial DNA. TLRs are referred to as **pattern recognition receptors (PRRs),** because they recognize and respond to the molecular patterns expressed by pathogens. Other PRRs may be intracellular, such as the so-called NOD proteins. One NOD protein, NOD2, has received attention as a candidate gene leading to the intestinal inflammatory condition, Crohn's disease (Clinical Box 3–2).

ACQUIRED IMMUNITY

As noted previously, the key to acquired immunity is the ability of lymphocytes to produce antibodies (in the case of B cells) or cell-surface receptors (in the case of T cells) that are specific for one of the many millions of foreign agents that may invade the body. The antigens stimulating production of T cell receptors or antibodies are usually proteins and polypeptides, but antibodies can also be formed against nucleic acids and lipids if these are presented as nucleoproteins and lipoproteins. Antibodies to small molecules can also be produced experimentally if the molecules are bound to protein. Acquired immunity has two components: humoral immunity and cellular immunity. **Humoral immunity** is mediated by circulating immunoglobulin antibodies in the γ-globulin fraction of the plasma proteins. Immunoglobulins are produced by differentiated forms of B lymphocytes known as **plasma cells,** and they activate the complement system and attack and neutralize antigens. Humoral immunity is a major defense against bacterial infections. **Cellular immunity** is mediated by T lymphocytes. It is responsible for delayed allergic reactions and rejection of transplants of foreign tissue. Cytotoxic T cells attack and destroy cells bearing the antigen that activated them. They kill by inserting perforins (see above) and by initiating apoptosis. Cellular immunity constitutes a major defense against infections due to viruses, fungi, and a few bacteria such as the tubercle bacillus. It also helps defend against tumors.

CLINICAL BOX 3–2

Crohn's Disease

Crohn's disease is a chronic, relapsing, and remitting disease that involves transmural inflammation of the intestine that can occur at any point along the length of the gastrointestinal tract, but most commonly is confined to the distal small intestine and colon. Patients with this condition suffer from changes in bowel habits, bloody diarrhea, severe abdominal pain, weight loss, and malnutrition. Evidence is accumulating that the disease reflects a failure to down-regulate inflammatory responses to the normal gut commensal microbiota. In genetically-susceptible individuals, mutations in genes controlling innate immune responses (eg, NOD2) or regulators of acquired immunity appear to predispose to disease when individuals are exposed to appropriate environmental factors, which can include a change in the microbiota or stress

THERAPEUTIC HIGHLIGHTS

During flares of Crohn's disease, the mainstay of treatment remains high-dose corticosteroids to suppress inflammation nonspecifically. Surgery is often required to treat complications such as strictures, fistulas, and abscesses. Some patients with severe disease also benefit from ongoing treatment with immunosuppressive drugs, or from treatment with antibodies targeted against tumor necrosis factor-α. Probiotics, therapeutic microorganisms designed to restore a "healthy" microbiota, may have some role in prophylaxis. The pathogenesis of Crohn's disease, as well as the related inflammatory bowel disease, ulcerative colitis, remains the subject of intense investigation, and therapies that target specific facets of the inflammatory cascade that may be selectively implicated in individual patients with differing genetic backgrounds are under development.

ANTIGEN RECOGNITION

The number of different antigens recognized by lymphocytes in the body is extremely large. The repertoire develops initially without exposure to the antigen. Stem cells differentiate into many million different T and B lymphocytes, each with the ability to respond to a particular antigen. When the antigen first enters the body, it can bind directly to the appropriate receptors on B cells. However, a full antibody response requires that the B cells contact helper T cells. In the case of T cells, the antigen is taken up by an antigen-presenting cell (APC) and partially digested. A peptide fragment of it is presented to the appropriate receptors on T cells. In either case, the cells are stimulated to divide, forming **clones** of cells that respond to this antigen **(clonal selection).** Effector cells are also subject to **negative selection,** during which lymphocyte precursors that are reactive with self-antigens are normally deleted. This results in immune **tolerance.** It is this latter process that presumably goes awry in autoimmune diseases, where the body reacts to and destroys cells expressing normal proteins, with accompanying inflammation that may lead to tissue destruction.

ANTIGEN PRESENTATION

APCs include specialized cells called **dendritic cells** in the lymph nodes and spleen and the Langerhans dendritic cells in the skin. Macrophages and B cells themselves, and likely many other cell types, can also function as APCs. For example, in the intestine, the epithelial cells that line the tract are likely important in presenting antigens derived from commensal bacteria. In APCs, polypeptide products of antigen digestion are coupled to protein products of the **major histocompatibility complex (MHC)** genes and presented on the surface of the cell. The products of the MHC genes are called human leukocyte antigens (HLA).

The genes of the MHC, which are located on the short arm of human chromosome 6, encode glycoproteins and are divided into two classes on the basis of structure and function. Class I antigens are composed of a 45-kDa heavy chain associated noncovalently with β_2-microglobulin encoded by a gene outside the MHC (Figure 3–6). They are found on all nucleated cells. Class II antigens are heterodimers made up of a 29–34-kDa α chain associated noncovalently with a 25–28-kDa β chain. They are present in "professional" APCs, including B cells, and in activated T cells.

The class I MHC proteins (MHC-I proteins) are coupled primarily to peptide fragments generated from proteins synthesized within cells. Peptides to which the host is not tolerant (eg, those from mutant or viral proteins) are recognized by T cells. The digestion of these proteins occurs in complexes of proteolytic enzymes known as **proteasomes,** and the peptide fragments bind to MHC proteins in the endoplasmic reticulum. The class II MHC proteins (MHC-II proteins) are concerned primarily with peptide products of extracellular antigens, such as bacteria, that enter the cell by endocytosis and are digested in the late endosomes.

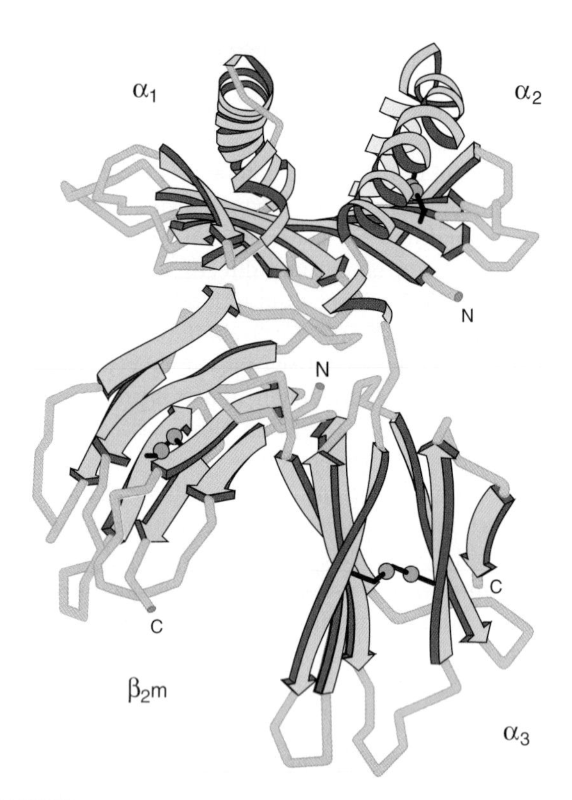

FIGURE 3–6 Structure of human histocompatibility antigen HLA-A2. The antigen-binding pocket is at the top and is formed by the α_1 and α_2 parts of the molecule. The α_3 portion and the associated β_2-microglobulin (β_2m) are close to the membrane. The extension of the C terminal from α 3 that provides the transmembrane domain and the small cytoplasmic portion of the molecule have been omitted. (Reproduced with permission from Bjorkman PJ, et al: Structure of the human histocompatibility antigen HLA-A2. Nature 1987;329:506.)

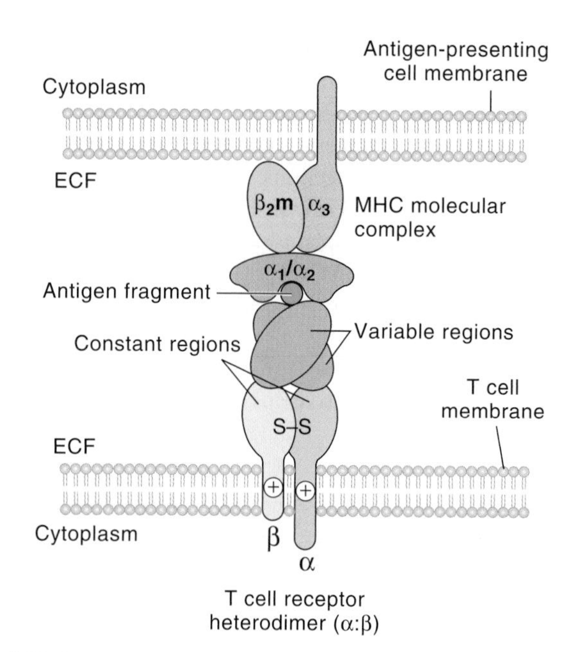

FIGURE 3–7 Interaction between antigen-presenting cell (top) and αβ T lymphocyte (bottom). The MHC proteins (in this case, MHC-I) and their peptide antigen fragment bind to the α and β units that combine to form the T cell receptor.

T CELL RECEPTORS

The MHC protein–peptide complexes on the surface of the APCs bind to appropriate T cells. Therefore, receptors on the T cells must recognize a very wide variety of complexes. Most of the receptors on circulating T cells are made up of two polypeptide units designated α and β. They form heterodimers that recognize the MHC proteins and the antigen fragments with which they are combined (Figure 3–7). These cells are called αβ T cells. On the other hand, about 10% of circulating T cells have two different polypeptides designated γ and δ in their receptors, and they are called γδ T cells. These T cells are prominent in the mucosa of the gastrointestinal tract, and there is evidence that they form a link between the innate and acquired immune systems by way of the cytokines they secrete (Figure 3–3).

CD8 occurs on the surface of cytotoxic T cells that bind MHC-I proteins, and CD4 occurs on the surface of helper T cells that bind MHC-II proteins (Figure 3–8). The CD8 and CD4 proteins facilitate the binding of the MHC proteins to the T cell receptors, and they also foster lymphocyte development. The activated CD8 cytotoxic T cells kill their targets directly, whereas the activated CD4 helper T cells secrete cytokines that activate other lymphocytes.

The T cell receptors are surrounded by adhesion molecules and proteins that bind to complementary proteins in the APC when the two cells transiently join to form the "immunologic synapse" that permits T cell activation to occur (Figure 3–7). It is now generally accepted that two signals are necessary to produce activation. One is produced by the binding of the digested antigen to the T cell receptor. The other is produced by the joining of the surrounding proteins in the "synapse."

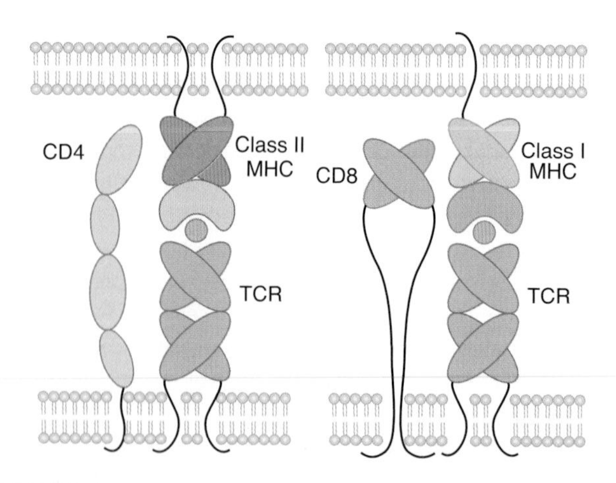

FIGURE 3–8 Diagrammatic summary of the structure of CD4 and CD8, and their relation to MHC-I and MHC-II proteins. Note that CD4 is a single protein, whereas CD8 is a heterodimer.

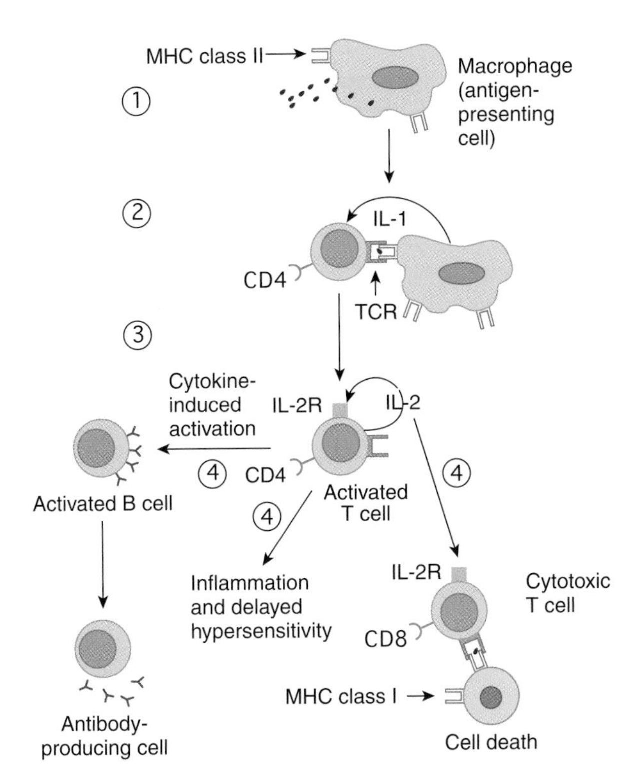

FIGURE 3–9 Summary of acquired immunity. (1) An antigen-presenting cell ingests and partially digests an antigen, then presents part of the antigen along with MHC peptides (in this case, MHC II peptides on the cell surface). **(2)** An "immune synapse" forms with a naive CD4 T cell, which is activated to produce IL-2. **(3)** IL-2 acts in an auto-crine fashion to cause the cell to multiply, forming a clone. **(4)** The activated CD4 cell may promote B cell activation and production of plasma cells or it may activate a cytotoxic CD8 cell. The CD8 cell can also be activated by forming a synapse with an MCH I antigen-presenting cell. (Reproduced with permission from McPhee SJ, Lingappa VR, Ganong WF [editors]: *Pathophysiology of Disease,* 6th ed. McGraw-Hill, 2010.)

If the first signal occurs but the second does not, the T cell is inactivated and becomes unresponsive.

B CELLS

As noted above, B cells can bind antigens directly, but they must contact helper T cells to produce full activation and antibody formation. It is the TH2 subtype that is mainly involved. Helper T cells develop along the TH2 lineage in response to IL-4 (see below). On the other hand, IL-12 promotes the TH1 phenotype. IL-2 acts in an autocrine fashion to cause activated T cells to proliferate. The role of various cytokines in B cell and T cell activation is summarized in Figure 3–9.

The activated B cells proliferate and transform into **memory B cells** (see above) and **plasma cells.** The plasma cells secrete large quantities of antibodies into the general circulation. The antibodies circulate in the globulin fraction of the plasma and, like antibodies elsewhere, are called **immunoglobulins.** The immunoglobulins are actually the secreted form of antigen-binding receptors on the B cell membrane.

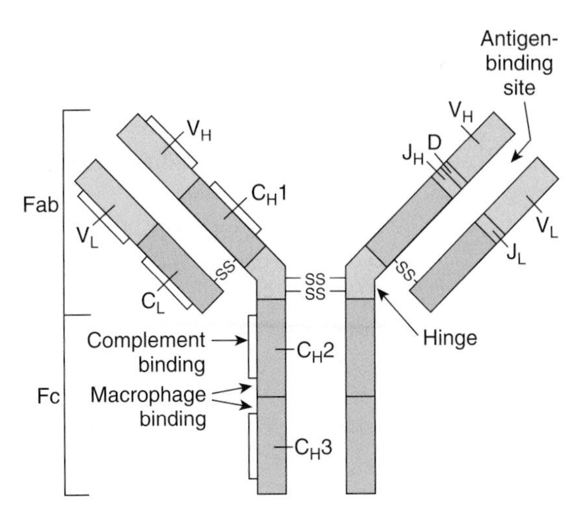

FIGURE 3–10 Typical immunoglobulin G molecule. Fab, portion of the molecule that is concerned with antigen binding; Fc, effector portion of the molecule. The constant regions are pink and purple, and the variable regions are orange. The constant segment of the heavy chain is subdivided into CH1, CH2, and CH3. SS lines indicate intersegmental disulfide bonds. On the right side, the C labels are omitted to show regions J_H, D, and J_L.

IMMUNOGLOBULINS

Circulating antibodies protect their host by binding to and neutralizing some protein toxins, by blocking the attachment of some viruses and bacteria to cells, by opsonizing bacteria (see above), and by activating complement. Five general types of immunoglobulin antibodies are produced by plasma cells. The basic component of each is a symmetric unit containing four polypeptide chains (Figure 3–10). The two long chains are called **heavy chains,** whereas the two short chains are called **light chains.** There are two types of light chains, k and λ, and eight types of heavy chains. The chains are joined by disulfide bridges that permit mobility, and there are intrachain disulfide bridges as well. In addition, the heavy chains are flexible in a region called the hinge. Each heavy chain has a variable (V) segment in which the amino acid sequence is highly variable, a diversity (D) segment in which the amino acid segment is also highly variable, a joining (J) segment in which the sequence is moderately variable, and a constant (C) segment in which the sequence is constant. Each light chain has a V, J, and C segment. The V segments form part of the antigen-binding sites (Fab portion of the molecule [Figure 3–10]). The Fc portion of the molecule is the effector portion, which mediates the reactions initiated by antibodies.

Two of the classes of immunoglobulins contain additional polypeptide components (Table 3–3). In IgM, five of the basic immunoglobulin units join around a polypeptide called the J chain to form a pentamer. In IgA, the **secretory immunoglobulin,** the immunoglobulin units form dimers and trimers around a J chain and a polypeptide that comes from epithelial cells, the secretory component (SC).

TABLE 3–3 Human immunoglobulins.[a]

Immunoglobulin	Function	Heavy Chain	Additional Chain	Structure	Plasma Concentration (mg/dL)
IgG	Complement activation	$\gamma_1, \gamma_2, \gamma_3, \gamma_4$		Monomer	1000
IgA	Localized protection in external secretions (tears, intestinal secretions, etc)	α_1, α_2	J, SC	Monomer; dimer with J or SC chain; trimer with J chain	200
IgM	Complement activation	μ	J	Pentamer with J chain	120
IgD	Antigen recognition by B cells	Δ	Monomer	3	
IgE	Reagin activity; releases histamine from basophils and mast cells	ε	Monomer	0.05	

[a]In all instances, the light chains are k or γ.

In the intestine, bacterial and viral antigens are taken up by M cells (see Chapter 26) and passed on to underlying aggregates of lymphoid tissue **(Peyer's patches),** where they activate naive T cells. These lymphocytes then form B cells that infiltrate mucosa of the gastrointestinal, respiratory, genitourinary, and female reproductive tracts and the breast. There they secrete large amounts of IgA when exposed again to the antigen that initially stimulated them. The epithelial cells produce the SC, which acts as a receptor for, and binds to, IgA. The resulting secretory immunoglobulin passes through the epithelial cell and is secreted by exocytosis. This system of **secretory immunity** is an important and effective defense mechanism at all mucosal surfaces. It also accounts for the immune protection that is conferred by breast feeding of infants whose immune systems are otherwise immature, because IgA is secreted into the breast milk.

GENETIC BASIS OF DIVERSITY IN THE IMMUNE SYSTEM

The genetic mechanism for the production of the immensely large number of different configurations of immunoglobulins produced by human B cells, as well as T cell receptors, is a fascinating biologic problem. Diversity is brought about in part by the fact that in immunoglobulin molecules there are two kinds of light chains and eight kinds of heavy chains. As noted previously, there are areas of great variability **(hypervariable regions)** in each chain. The variable portion of the heavy chains consists of the V, D, and J segments. In the gene family responsible for this region, there are several hundred different coding regions for the V segment, about 20 for the D segment, and four for the J segment. During B cell development, one V coding region, one D coding region, and one J coding region are selected at random and recombined to form the gene that produces that particular variable portion. A similar variable recombination takes place in the coding regions responsible for the two variable segments (V and J) in the light chain. In addition, the J segments are variable because the gene segments join in an imprecise and variable fashion (junctional site diversity) and nucleotides are sometimes added (junctional insertion diversity). It has been calculated that these mechanisms permit the production of about 10^{15} different immunoglobulin molecules. Additional variability is added by somatic mutation.

Similar gene rearrangement and joining mechanisms operate to produce the diversity in T cell receptors. In humans, the α subunit has a V region encoded by 1 of about 50 different genes and a J region encoded by 1 of another 50 different genes. The β subunits have a V region encoded by 1 of about 50 genes, a D region encoded by 1 of 2 genes, and a J region encoded by 1 of 13 genes. These variable regions permit the generation of up to an estimated 10^{15} different T cell receptors (Clinical Box 3–3 and Clinical Box 3–4).

A variety of immunodeficiency states can arise from defects in these various stages of B and T lymphocyte maturation. These are summarized in Figure 3–11.

PLATELETS

Platelets are circulating cells that are important mediators of hemostasis. While not immune cells, per se, they often participate in the response to tissue injury in cooperation with inflammatory cell types (see below). They have a ring of microtubules around their periphery and an extensively invaginated membrane with an intricate canalicular system in contact with the ECF. Their membranes contain receptors for collagen, ADP, vessel wall von Willebrand factor (see below), and fibrinogen. Their cytoplasm contains actin, myosin, glycogen, lysosomes, and two types of granules: (1) dense granules, which contain the nonprotein substances that are secreted in response to platelet activation, including serotonin, ADP, and other adenine nucleotides; and (2) α-granules, which contain secreted proteins. These proteins include clotting factors and **platelet-derived growth factor (PDGF).** PDGF is also produced by macrophages and endothelial cells. It is a dimer made up of A and B subunit polypeptides. Homodimers (AA and BB), as well as the heterodimer (AB), are produced.

CLINICAL BOX 3–3

Autoimmunity

Sometimes the processes that eliminate antibodies against self-antigens fail and a variety of different **autoimmune diseases** are produced. These can be B cell- or T cell-mediated and can be organ-specific or systemic. They include type 1 diabetes mellitus (antibodies against pancreatic islet B cells), myasthenia gravis (antibodies against nicotinic cholinergic receptors), and multiple sclerosis (antibodies against myelin basic protein and several other components of myelin). In some instances, the antibodies are against receptors and are capable of activating those receptors; for example, antibodies against TSH receptors increase thyroid activity and cause Graves' disease (see Chapter 19). Other conditions are due to the production of antibodies against invading organisms that cross-react with normal body constituents **(molecular mimicry).** An example is rheumatic fever following a streptococcal infection; a portion of cardiac myosin resembles a portion of the streptococcal M protein, and antibodies induced by the latter attack the former and damage the heart. Some conditions may be due to **bystander effects,** in which inflammation sensitizes T cells in the neighborhood, causing them to become activated when otherwise they would not respond.

THERAPEUTIC HIGHLIGHTS

The therapy of autoimmune disorders rests on efforts to replace or restore the damaged function (eg, provision of exogenous insulin in type 1 diabetes) as well as nonspecific efforts to reduce inflammation (using corticosteroids) or to suppress immunity. Recently, agents that deplete or dampen the function of B cells have been shown to have some efficacy in a range of autoimmune disorders, including rheumatoid arthritis, most likely by interrupting the production of autoantibodies that contribute to disease pathogenesis.

CLINICAL BOX 3–4

Tissue Transplantation

The T lymphocyte system is responsible for the rejection of transplanted tissue. When tissues such as skin and kidneys are transplanted from a donor to a recipient of the same species, the transplants "take" and function for a while but then become necrotic and are "rejected" because the recipient develops an immune response to the transplanted tissue. This is generally true even if the donor and recipient are close relatives, and the only transplants that are never rejected are those from an identical twin. Nevertheless, organ transplantation remains the only viable option in a number of end stage diseases.

THERAPEUTIC HIGHLIGHTS

A number of treatments have been developed to overcome the rejection of transplanted organs in humans. The goal of treatment is to stop rejection without leaving the patient vulnerable to massive infections. One approach is to kill T lymphocytes by killing all rapidly dividing cells with drugs such as azathioprine, a purine antimetabolite, but this makes patients susceptible to infections and cancer. Another is to administer glucocorticoids, which inhibit cytotoxic T cell proliferation by inhibiting production of IL-2, but these cause osteoporosis, mental changes, and the other facets of Cushing syndrome (see Chapter 20). More recently, immunosuppressive drugs such as **cyclosporine** or **tacrolimus (FK-506)** have found favor. Activation of the T cell receptor normally increases intracellular Ca^{2+}, which acts via calmodulin to activate calcineurin. Calcineurin dephosphorylates the transcription factor NF-AT, which moves to the nucleus and increases the activity of genes coding for IL-2 and related stimulatory cytokines. Cyclosporine and tacrolimus prevent the dephosphorylation of NF-AT. However, these drugs inhibit all T cell-mediated immune responses, and cyclosporine causes kidney damage and cancer. A new and promising approach to transplant rejection is the production of T cell unresponsiveness by using drugs that block the costimulation that is required for normal activation (see text). Clinically effective drugs that act in this fashion could be of great value to transplant surgeons.

PDGF stimulates wound healing and is a potent mitogen for vascular smooth muscle. Blood vessel walls as well as platelets contain von Willebrand factor, which, in addition to its role in adhesion, regulates circulating levels of factor VIII (see below).

When a blood vessel wall is injured, platelets adhere to the exposed collagen and **von Willebrand factor** in the wall via receptors on the platelet membrane. Von Willebrand factor is a very large circulating molecule that is produced by endothelial cells. Binding produces platelet activations, which release the contents of their granules. The released ADP acts on the ADP receptors in the platelet membranes to produce further accumulation of more platelets **(platelet aggregation).** Humans have at least three different types of platelet ADP receptors: $P2Y_1$, $P2Y_2$, and $P2X_1$. These are obviously attractive targets for drug development, and several new inhibitors have shown

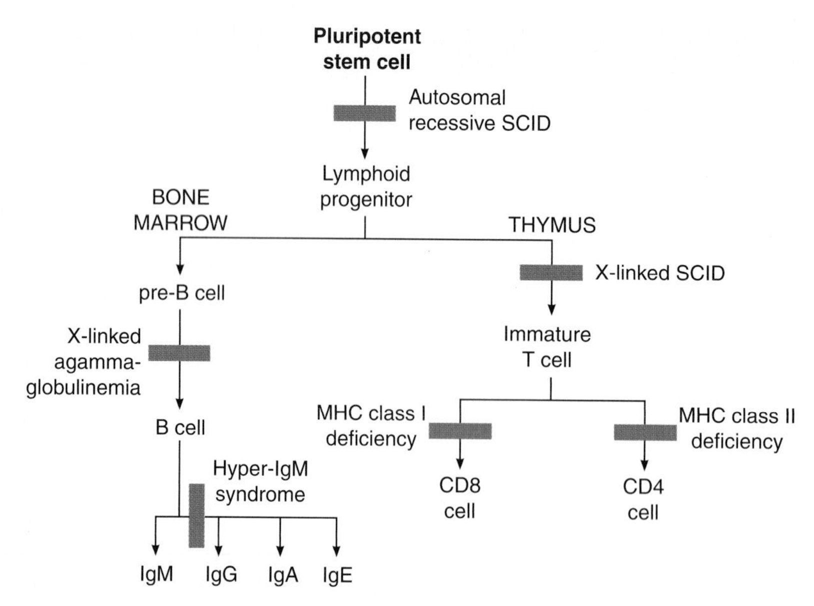

FIGURE 3–11 Sites of congenital blockade of B and T lymphocyte maturation in various immunodeficiency states. SCID, severe combined immune deficiency. (Modified from Rosen FS, Cooper MD, Wedgwood RJP: The primary immunodeficiencies. *N Engl J Med* 1995;333:431.)

promise in the prevention of heart attacks and strokes. Aggregation is also fostered by **platelet-activating factor (PAF),** a cytokine secreted by neutrophils and monocytes as well as platelets. This compound also has inflammatory activity. It is an ether phospholipid, 1-alkyl-2-acetylglyceryl-3-phosphorylcholine, which is produced from membrane lipids. It acts via a G protein-coupled receptor to increase the production of arachidonic acid derivatives, including thromboxane A_2. The role of this compound in the balance between clotting and anticlotting activity at the site of vascular injury is discussed in Chapter 31.

Platelet production is regulated by the CSFs that control the production of the platelet precursors in the bone marrow, known as megakaryocytes, plus **thrombopoietin,** a circulating protein factor. This factor, which facilitates megakaryocyte maturation, is produced constitutively by the liver and kidneys, and there are thrombopoietin receptors on platelets. Consequently, when the number of platelets is low, less is bound and more is available to stimulate production of platelets. Conversely, when the number of platelets is high, more is bound and less is available, producing a form of feedback control of platelet production. The amino terminal portion of the thrombopoietin molecule has the platelet-stimulating activity, whereas the carboxyl terminal portion contains many carbohydrate residues and is concerned with the bioavailability of the molecule.

When the platelet count is low, clot retraction is deficient and there is poor constriction of ruptured vessels. The resulting clinical syndrome **(thrombocytopenic purpura)** is characterized by easy bruisability and multiple subcutaneous hemorrhages. Purpura may also occur when the platelet count is normal, and in some of these cases, the circulating platelets are abnormal **(thrombasthenic purpura).** Individuals with thrombocytosis are predisposed to thrombotic events.

INFLAMMATION & WOUND HEALING

LOCAL INJURY

Inflammation is a complex localized response to foreign substances such as bacteria or in some instances to internally produced substances. It includes a sequence of reactions initially involving cytokines, neutrophils, adhesion molecules, complement, and IgG. PAF, an agent with potent inflammatory effects, also plays a role. Later, monocytes and lymphocytes are involved. Arterioles in the inflamed area dilate, and capillary permeability is increased (see Chapters 32 and 33). When the inflammation occurs in or just under the skin (Figure 3–12), it is characterized by redness, swelling, tenderness, and pain. Elsewhere, it is a key component of asthma, ulcerative colitis, Crohn's disease, rheumatoid arthritis, and many other diseases (Clinical Box 3–2).

Evidence is accumulating that a transcription factor, **nuclear factor-κB,** plays a key role in the inflammatory response. NF-κB is a heterodimer that normally exists in the cytoplasm of cells bound to IκBα, which renders it inactive. Stimuli such as cytokines, viruses, and oxidants induce signals that allow NF-κB to dissociate from IκBα, which is then degraded. NF-κB moves to the nucleus, where it binds to the DNA of the genes for numerous inflammatory mediators, resulting in their increased production and secretion. Glucocorticoids inhibit the activation of NF-κB by increasing the production of IκBα, and this is probably the main basis of their anti-inflammatory action (see Chapter 20).

SYSTEMIC RESPONSE TO INJURY

Cytokines produced in response to inflammation and other injuries, as well as disseminated infection, also produce systemic responses. These include alterations in plasma **acute**

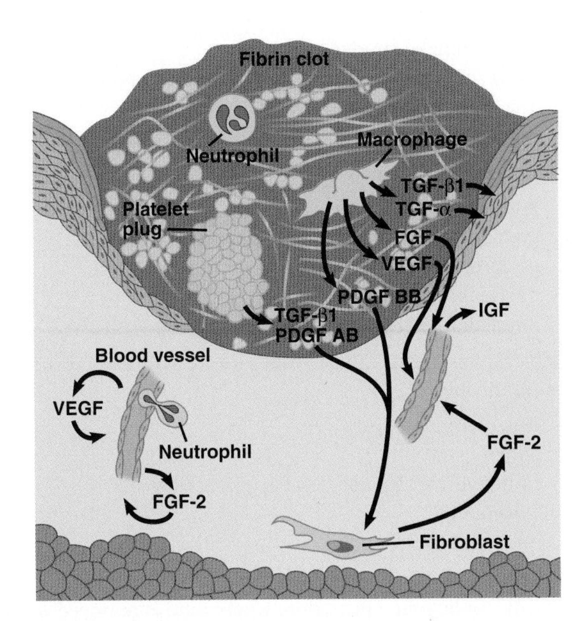

FIGURE 3–12 Cutaneous wound 3 days after injury, showing the multiple cytokines and growth factors affecting the repair process. VEGF, vascular endothelial growth factor. For other abbreviations, see Appendix. Note the epidermis growing down under the fibrin clot, restoring skin continuity. (Modified from Singer AJ, Clark RAF: Cutaneous wound healing. *N Engl J Med* 1999;341:738.)

phase proteins, defined as proteins whose concentration is increased or decreased by at least 25% following injury. Many of the proteins are of hepatic origin. A number of them are shown in Figure 3–13. The causes of the changes in concentration are incompletely understood, but it can be said that many of the changes make homeostatic sense. Thus, for example, an increase in C-reactive protein activates monocytes and causes further production of cytokines. Other changes that occur in response to injury include somnolence, negative nitrogen balance, and fever.

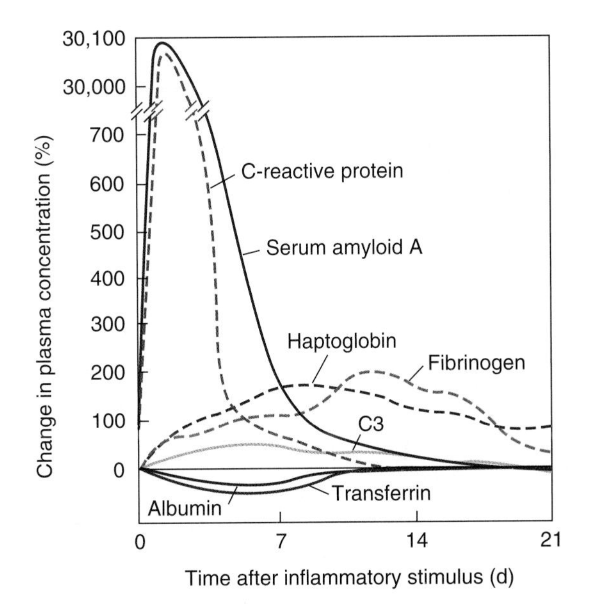

FIGURE 3–13 Time course of changes in some major acute phase proteins. C3, C3 component of complement. (Modified and reproduced with permission from McAdam KP, Elin RJ, Sipe JD, Wolff SM: Changes in human serum amyloid A and C-reactive protein after etiocholanolone-induced inflammation. J Clin Invest, 1978 Feb;61(2):390–394.)

WOUND HEALING

When tissue is damaged, platelets adhere to exposed matrix via integrins that bind to collagen and laminin (Figure 3–12). Blood coagulation produces thrombin, which promotes platelet aggregation and granule release. The platelet granules generate an inflammatory response. White blood cells are attracted by selectins and bind to integrins on endothelial cells, leading to their extravasation through the blood vessel walls. Cytokines released by the white blood cells and platelets up-regulate integrins on macrophages, which migrate to the area of injury, and on fibroblasts and epithelial cells, which mediate wound healing and scar formation. Plasmin aids healing by removing excess fibrin. This aids the migration of keratinocytes into the wound to restore the epithelium under the scab. Collagensynthesis is upregulated, producing the scar. Wounds gain 20% of their ultimate strength in 3 weeks and later gain more strength, but they never reach more than about 70% of the strength of normal skin.

CHAPTER SUMMARY

- Immune and inflammatory responses are mediated by several different cell types—granulocytes, lymphocytes, monocytes, mast cells, tissue macrophages, and antigen-presenting cells—that arise predominantly from the bone marrow and may circulate or reside in connective tissues.
- Granulocytes mount phagocytic responses that engulf and destroy bacteria. These are accompanied by the release of reactive oxygen species and other mediators into adjacent tissues that may cause tissue injury.
- Mast cells and basophils underpin allergic reactions to substances that would be treated as innocuous by nonallergic individuals.
- A variety of soluble mediators orchestrate the development of immunologic effector cells and their subsequent immune and inflammatory reactions.
- Innate immunity represents an evolutionarily conserved, primitive response to stereotypical microbial components.
- Acquired immunity is slower to develop than innate immunity, but long-lasting and more effective.
- Genetic rearrangements endow B and T lymphocytes with a vast array of receptors capable of recognizing billions of foreign antigens.
- Self-reactive lymphocytes are normally deleted; a failure of this process leads to autoimmune disease. Disease can also result from abnormal function or development of granulocytes and lymphocytes. In these latter cases, deficient immune responses to microbial threats usually result.

- Inflammatory responses occur in response to infection or injury, and serve to resolve the threat, although they may cause damage to otherwise healthy tissue. A number of chronic diseases reflect excessive inflammatory responses that persist even once the threat is controlled, or are triggered by stimuli that healthy individuals would not respond to.

MULTIPLE-CHOICE QUESTIONS

For all questions, select the single best answer unless otherwise directed.

1. In an experiment, a scientist treats a group of mice with an antiserum that substantially depletes the number of circulating neutrophils. Compared with untreated control animals, the mice with reduced numbers of neutrophils were found to be significantly more susceptible to death induced by bacterial inoculation. The increased mortality can be ascribed to a relative deficit in which of the following?
   A. Acquired immunity
   B. Oxidants
   C. Platelets
   D. Granulocyte/macrophage colony stimulating factor (GM-CSF)
   E. Integrins
2. A 20-year-old college student comes to the student health center in April complaining of runny nose and congestion, itchy eyes, and wheezing. She reports that similar symptoms have occurred at the same time each year, and that she obtains some relief from over-the-counter antihistamine drugs, although they make her too drowsy to study. Her symptoms can most likely be attributed to inappropriate synthesis of which of the following antibodies specific for tree pollen?
   A. IgA
   B. IgD
   C. IgE
   D. IgG
   E. IgM
3. If a nasal biopsy were performed on the patient described in Question 2 while symptomatic, histologic examination of the tissue would most likely reveal degranulation of which of the following cell types?
   A. Dendritic cells
   B. Lymphocytes
   C. Neutrophils
   D. Monocytes
   E. Mast cells
4. A biotechnology company is working to design a new therapeutic strategy for cancer that involves triggering an enhanced immune response to cellular proteins that are mutated in the disease. Which of the following immune cells or processes will most likely **not** be required for a successful therapy?
   A. Cytotoxic T cells
   B. Antigen presentation in the context of MHC-II
   C. Proteosomal degradation
   D. Gene rearrangements producing T cell receptors
   E. The immune synapse
5. The ability of the blood to phagocytose pathogens and mount a respiratory burst is increased by
   A. interleukin-2 (IL-2)
   B. granulocyte colony-stimulating factor (G-CSF)
   C. erythropoietin
   D. interleukin-4 (IL-4)
   E. interleukin-5 (IL-5)
6. Cells responsible for innate immunity are activated most commonly by
   A. glucocorticoids
   B. pollen
   C. carbohydrate sequences in bacterial cell walls
   D. eosinophils
   E. thrombopoietin
7. A patient suffering from an acute flare in his rheumatoid arthritis undergoes a procedure where fluid is removed from his swollen and inflamed knee joint. Biochemical analysis of the inflammatory cells recovered from the removed fluid would most likely reveal a decrease in which of the following proteins?
   A. Interleukin 1
   B. Tumor necrosis factor-α
   C. Nuclear factor-κB
   D. IκBα
   E. von Willbrand factor

CHAPTER RESOURCES

Delibro G: The Robin Hood of antigen presentation. Science 2004;302:485.

Delves PJ, Roitt IM: The immune system. (Two parts.) N Engl J Med 2000;343:37,108.

Dhainaut J-K, Thijs LG, Park G (editors): *Septic Shock.* WB Saunders, 2000.

Ganz T: Defensins and host defense. Science 1999;286:420.

Karin M, Ben-Neriah Y: Phosphorylation meets ubiquitination: the control of NF-κB activity. Annu Rev Immunol 2000; 18:621.

Samstein B, Emond JC: Liver transplant from living related donors. Annu Rev Med 2001;52:147.

Singer AJ, Clark RAF: Cutaneous wound healing. N Engl J Med 1999;341:738

Tedder TF, Steeber DA, Chen A, et al: The selectins: Vascular adhesion molecules. FASEB J 1995;9:866.

Tilney NL: *Transplant: From Myth to Reality.* Yale University Press, 2003.

Walport MJ: Complement. (Two parts.) N Engl J Med 2001;344:1058, 1140.

CHAPTER

4

Excitable Tissue: Nerve

OBJECTIVES

After studying this chapter, you should be able to:

- Name the various types of glia and their functions.
- Name the parts of a neuron and their functions.
- Describe the chemical nature of myelin, and summarize the differences in the ways in which unmyelinated and myelinated neurons conduct impulses.
- Describe orthograde and retrograde axonal transport.
- Describe the changes in ionic channels that underlie the action potential.
- List the various nerve fiber types found in the mammalian nervous system.
- Describe the function of neurotrophins.

INTRODUCTION

The human central nervous system (CNS) contains about 10^{11} (100 billion) **neurons.** It also contains 10–50 times this number of **glial cells.** The CNS is a complex organ; it has been calculated that 40% of the human genes participate, at least to a degree, in its formation. The neurons, the basic building blocks of the nervous system, have evolved from primitive neuroeffector cells that respond to various stimuli by contracting. In more complex animals, contraction has become the specialized function of muscle cells, whereas integration and transmission of nerve impulses have become the specialized functions of neurons. Neurons and glial cells along with brain capillaries form a functional unit that is required for normal brain function, including synaptic activity, extracellular fluid homeostasis, energy metabolism, and neural protection. Disturbances in the interaction of these elements are the pathophysiological basis for many neurological disorders (eg, cerebral ischemia, seizures, neurodegenerative diseases, and cerebral edema). This chapter describes the cellular components of the CNS and the excitability of neurons, which involves the genesis of electrical signals that enable neurons to integrate and transmit impulses (eg, action potentials, receptor potentials, and synaptic potentials).

CELLULAR ELEMENTS IN THE CNS

GLIAL CELLS

For many years following their discovery, glial cells (or glia) were viewed as CNS connective tissue. In fact, the word *glia* is Greek for *glue.* However, today theses cells are recognized for their role in communication within the CNS in partnership with neurons. Unlike neurons, glial cells continue to undergo cell division in adulthood and their ability to proliferate is particularly noticeable after brain injury (eg, stroke).

There are two major types of glial cells in the vertebrate nervous system: **microglia** and **macroglia.** Microglia are scavenger cells that resemble tissue macrophages and remove debris resulting from injury, infection, and disease (eg, multiple sclerosis, AIDS-related dementia, Parkinson disease, and Alzheimer disease). Microglia arise from macrophages outside of the nervous system and are physiologically and embryologically unrelated to other neural cell types.

There are three types of macroglia: oligodendrocytes, Schwann cells, and astrocytes (Figure 4–1). **Oligodendrocytes** and **Schwann cells** are involved in myelin formation around axons in the CNS and peripheral nervous system, respectively. **Astrocytes,** which are found throughout the brain, are of two subtypes. **Fibrous astrocytes,** which contain many intermediate filaments, are found primarily in white matter.

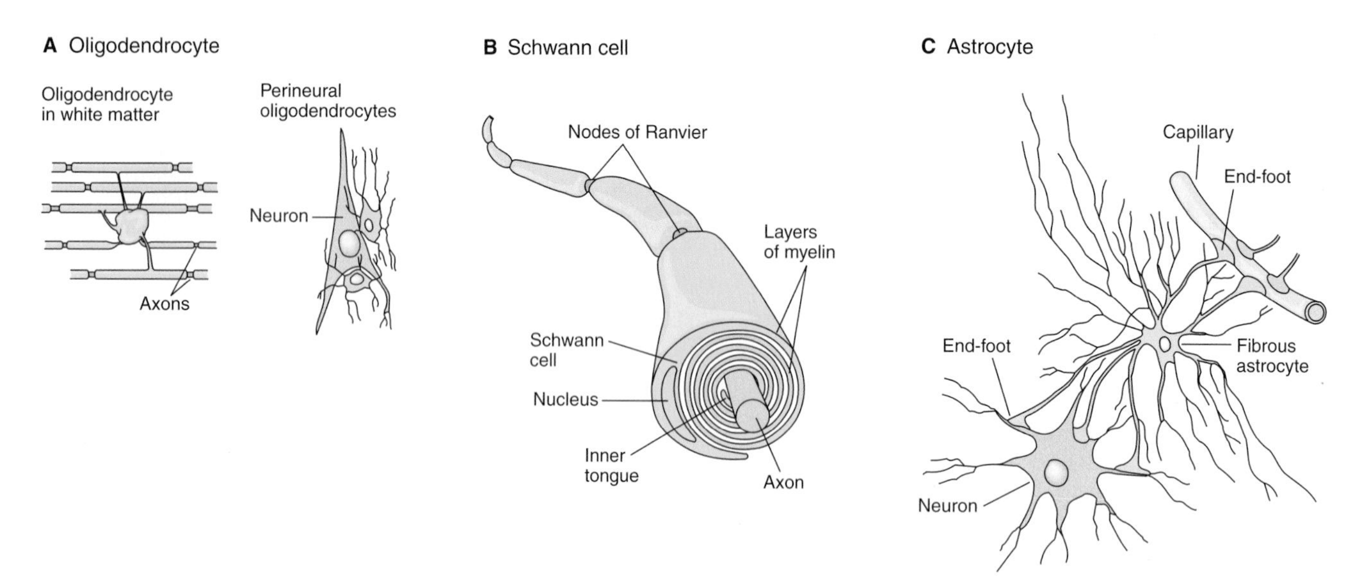

FIGURE 4–1 The principal types of macroglia in the nervous system. A) Oligodendrocytes are small with relatively few processes. Those in the white matter provide myelin, and those in the gray matter support neurons. **B)** Schwann cells provide myelin to the peripheral nervous system. Each cell forms a segment of myelin sheath about 1 mm long; the sheath assumes its form as the inner tongue of the Schwann cell turns around the axon several times, wrapping in concentric layers. Intervals between segments of myelin are the nodes of Ranvier. **C)** Astrocytes are the most common glia in the CNS and are characterized by their starlike shape. They contact both capillaries and neurons and are thought to have a nutritive function. They are also involved in forming the blood–brain barrier. (From Kandel ER, Schwartz JH, Jessell TM (editors): *Principles of Neural Science,* 4th ed. McGraw-Hill, 2000.)

Protoplasmic astrocytes are found in gray matter and have a granular cytoplasm. Both types send processes to blood vessels, where they induce capillaries to form the tight junctions making up the **blood–brain barrier.** They also send processes that envelop synapses and the surface of nerve cells. Protoplasmic astrocytes have a membrane potential that varies with the external K^+ concentration but do not generate propagated potentials. They produce substances that are tropic to neurons, and they help maintain the appropriate concentration of ions and neurotransmitters by taking up K^+ and the neurotransmitters glutamate and γ-aminobutyrate (GABA).

NEURONS

Neurons in the mammalian CNS come in many different shapes and sizes. Most have the same parts as the typical spinal motor neuron illustrated in Figure 4–2. The cell body **(soma)** contains the nucleus and is the metabolic center of the neuron. Neurons have several processes called **dendrites** that extend outward from the cell body and arborize extensively. Particularly in the cerebral and cerebellar cortex, the dendrites have small knobby projections called **dendritic spines.** A typical neuron also has a long fibrous **axon** that originates from a somewhat thickened area of the cell body, the **axon hillock.** The first portion

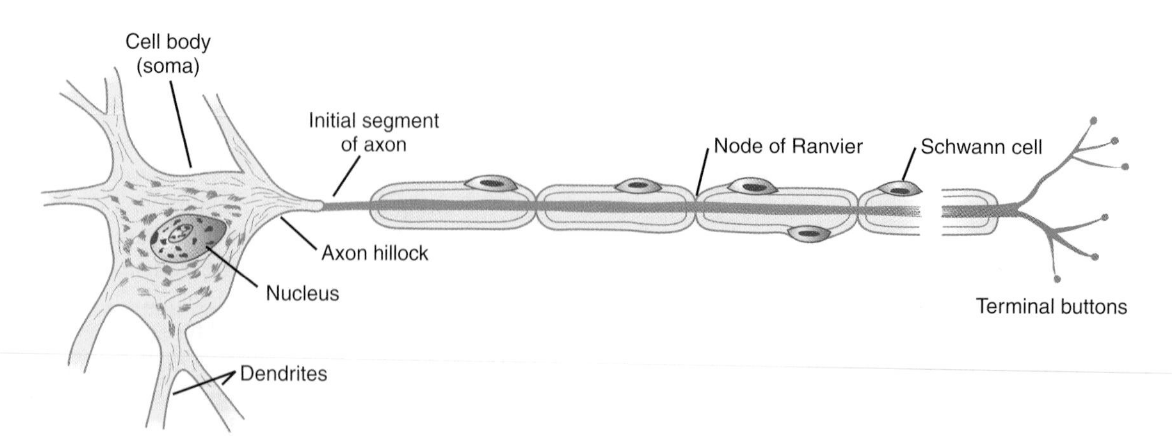

FIGURE 4–2 Motor neuron with a myelinated axon. A motor neuron is comprised of a cell body (soma) with a nucleus, several processes called dendrites, and a long fibrous axon that originates from the axon hillock. The first portion of the axon is called the initial segment. A myelin sheath forms from Schwann cells and surrounds the axon except at its ending and at the nodes of Ranvier. Terminal buttons (boutons) are located at the terminal endings.

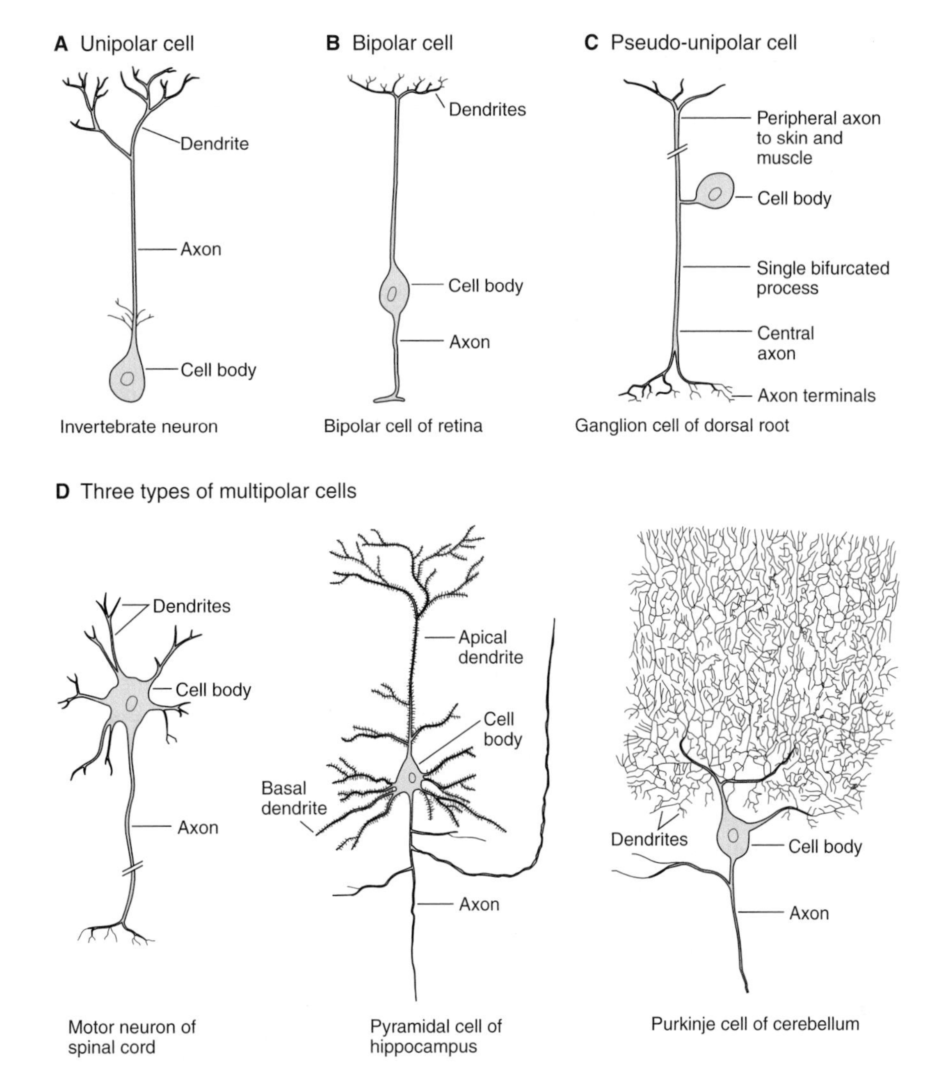

FIGURE 4–3 Some of the types of neurons in the mammalian nervous system. A) Unipolar neurons have one process, with different segments serving as receptive surfaces and releasing terminals. **B)** Bipolar neurons have two specialized processes: a dendrite that carries information to the cell and an axon that transmits information from the cell. **C)** Some sensory neurons are in a subclass of bipolar cells called pseudo-unipolar cells. As the cell develops, a single process splits into two, both of which function as axons—one going to skin or muscle and another to the spinal cord. **D)** Multipolar cells have one axon and many dendrites. Examples include motor neurons, hippocampal pyramidal cells with dendrites in the apex and base, and cerebellar Purkinje cells with an extensive dendritic tree in a single plane. (From Kandel ER, Schwartz JH, Jessell TM (editors): *Principles of Neural Science,* 4th ed. McGraw-Hill, 2000.)

of the axon is called the **initial segment.** The axon divides into **presynaptic terminals,** each ending in a number of **synaptic knobs** which are also called **terminal buttons** or **boutons.** They contain granules or vesicles in which the synaptic transmitters secreted by the nerves are stored. Based on the number of processes that emanate from their cell body, neurons can be classified as **unipolar, bipolar,** and **multipolar** (Figure 4–3).

The conventional terminology used for the parts of a neuron works well enough for spinal motor neurons and interneurons, but there are problems in terms of "dendrites" and "axons" when it is applied to other types of neurons found in the nervous system. From a functional point of view, neurons generally have four important zones: (1) a receptor, or dendritic zone, where multiple local potential changes generated by synaptic connections are integrated; (2) a site where propagated action potentials are generated (the initial segment in spinal motor neurons, the initial node of Ranvier in cutaneous sensory neurons); (3) an axonal process that transmits propagated impulses to the nerve endings; and (4) the nerve endings, where action potentials cause the release of synaptic transmitters. The cell body is often located at the dendritic zone end of the axon, but it can be within the axon (eg, auditory neurons) or attached to the side of the axon (eg, cutaneous neurons). Its location makes no difference as far as the receptor function of the dendritic zone and the transmission function of the axon are concerned.

The axons of many neurons are myelinated, that is, they acquire a sheath of **myelin,** a protein–lipid complex that is wrapped around the axon (Figure 4–1B). In the peripheral nervous system, myelin forms when a Schwann cell wraps its membrane around an axon up to 100 times. The myelin is then

CLINICAL BOX 4–1

Demyelinating Diseases

Normal conduction of action potentials relies on the insulating properties of **myelin.** Thus, defects in myelin can have major adverse neurological consequences. One example is **multiple sclerosis (MS),** an autoimmune disease that affects over 3 million people worldwide, usually striking between the ages of 20 and 50 and affecting women about twice as often as men. The cause of MS appears to include both genetic and environmental factors. It is most common among Caucasians living in countries with temperate climates including Europe, southern Canada, northern United States, and southeastern Australia. Environmental triggers include early exposure to viruses such as Epstein-Barr virus and those that cause measles, herpes, chicken pox, or influenza. In MS, antibodies and white blood cells in the immune system attack myelin, causing inflammation and injury to the sheath and eventually the nerves that it surrounds. Loss of myelin leads to leakage of K^+ through voltage-gated channels, hyperpolarization, and failure to conduct action potentials. Initial presentation commonly includes reports of **paraparesis** (weakness in lower extremities) that may be accompanied by mild spasticity and hyperreflexia; **paresthesia;** numbness; urinary incontinence; and heat intolerance. Clinical assessment often reports **optic neuritis,** characterized by blurred vision, a change in color perception, visual field defect (**central scotoma**), and pain with eye movements; **dysarthria;** and **dysphagia.** Symptoms are often exacerbated by increased body temperature or ambient temperature. Progression of the disease is quite variable. In the most common form called **relapsing-remitting MS,** transient episodes appear suddenly, last a few weeks or months, and then gradually disappear. Subsequent episodes can appear years later, and eventually full recovery does not occur. Many of these individuals later develop a steadily worsening course with only minor periods of remission (**secondary-progressive MS**). Others have a progressive form of the disease in which there are no periods of remission (**primary-progressive MS**). Diagnosing MS is very difficult and generally is delayed until multiple episodes occur with deficits separated in time and space. **Nerve conduction tests** can detect slowed conduction in motor and sensory pathways. Cerebral spinal fluid analysis can detect the presence of **oligoclonal** bands indicative of an abnormal immune reaction against myelin. The most definitive assessment is **magnetic resonance imaging (MRI)** to visualize multiple scarred (sclerotic) areas or plaques in the brain. These plaques often appear in the periventricular regions of the cerebral hemispheres.

THERAPEUTIC HIGHLIGHTS

Although there is no cure for MS, **corticosteroids** (eg, **prednisone**) are the most common treatment used to reduce the inflammation that is accentuated during a relapse. Some drug treatments are designed to modify the course of the disease. For example, daily injections of **β-interferons** suppress the immune response to reduce the severity and slow the progression of the disease. **Glatiramer acetate** may block the immune system's attack on the myelin. **Natalizumab** interferes with the ability of potentially damaging immune cells to move from the bloodstream to the CNS. A recent clinical trial using B cell–depleting therapy with **rituximab,** an anti-CD20 monoclonal antibody, showed that the progression of the disease was slowed in patients under the age of 51 who were diagnosed with the primary-progressive form of MS. Another recent clinical trial has shown that oral administration of **fingolimod** slowed the progression of the relapsing-remitting form of MS. This immunosuppressive drug acts by sequestering lymphocytes in the lymph nodes, thereby limiting their access to the CNS.

compacted when the extracellular portions of a membrane protein called protein zero (P_0) lock to the extracellular portions of P_0 in the apposing membrane. Various mutations in the gene for P_0 cause peripheral neuropathies; 29 different mutations have been described that cause symptoms ranging from mild to severe. The myelin sheath envelops the axon except at its ending and at the **nodes of Ranvier,** periodic 1-μm constrictions that are about 1 mm apart (Figure 4–2). The insulating function of myelin is discussed later in this chapter. Not all neurons are myelinated; some are **unmyelinated,** that is, simply surrounded by Schwann cells without the wrapping of the Schwann cell membrane that produces myelin around the axon.

In the CNS of mammals, most neurons are myelinated, but the cells that form the myelin are oligodendrocytes rather than Schwann cells (Figure 4–1). Unlike the Schwann cell, which forms the myelin between two nodes of Ranvier on a single neuron, oligodendrocytes emit multiple processes that form myelin on many neighboring axons. In multiple sclerosis, a crippling autoimmune disease, patchy destruction of myelin occurs in the CNS (see Clinical Box 4–1). The loss of myelin is associated with delayed or blocked conduction in the demyelinated axons.

AXONAL TRANSPORT

Neurons are secretory cells, but they differ from other secretory cells in that the secretory zone is generally at the end of the axon, far removed from the cell body. The apparatus for protein synthesis is located for the most part in the cell body, with transport of proteins and polypeptides to the axonal ending by

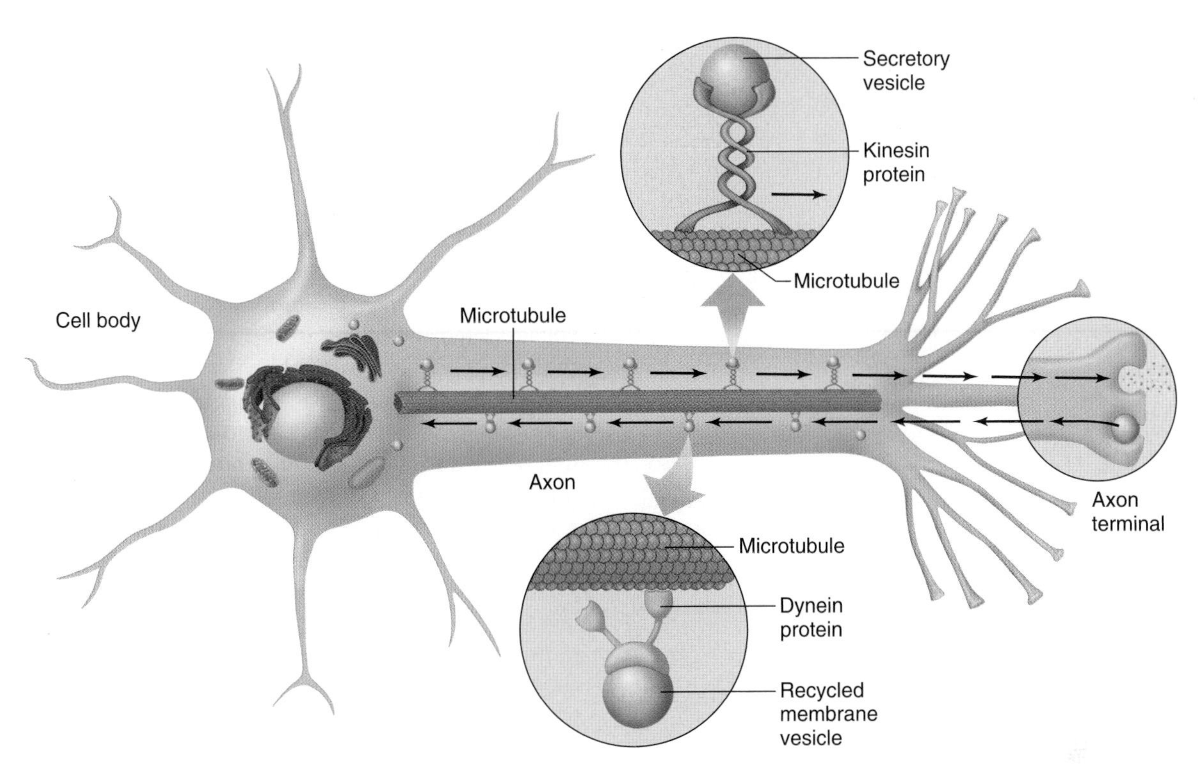

FIGURE 4–4 Axonal transport along microtubules by dynein and kinesin. Fast (400 mm/day) and slow (0.5–10 mm/day) axonal orthograde transport occurs along microtubules that run along the length of the axon from the cell body to the terminal. Retrograde transport (200 mm/day) occurs from the terminal to the cell body. (From Widmaier EP, Raff H, Strang KT: *Vander's Human Physiology.* McGraw-Hill, 2008.)

axoplasmic flow. Thus, the cell body maintains the functional and anatomic integrity of the axon; if the axon is cut, the part distal to the cut degenerates **(wallerian degeneration).**

Orthograde transport occurs along microtubules that run along the length of the axon and requires two molecular motors, dynein and kinesin (Figure 4–4). Orthograde transport moves from the cell body toward the axon terminals. It has both fast and slow components; **fast axonal transport** occurs at about 400 mm/day, and **slow axonal transport** occurs at 0.5 to 10 mm/day. **Retrograde transport,** which is in the opposite direction (from the nerve ending to the cell body), occurs along microtubules at about 200 mm/day. Synaptic vesicles recycle in the membrane, but some used vesicles are carried back to the cell body and deposited in lysosomes. Some materials taken up at the ending by endocytosis, including **nerve growth factor (NGF)** and some viruses, are also transported back to the cell body. A potentially important exception to these principles seems to occur in some dendrites. In them, single strands of mRNA transported from the cell body make contact with appropriate ribosomes, and protein synthesis appears to create local protein domains.

EXCITATION & CONDUCTION

A hallmark of nerve cells is their excitable membrane. Nerve cells respond to electrical, chemical, or mechanical stimuli. Two types of physicochemical disturbances are produced: local, nonpropagated potentials called, depending on their location, **synaptic, generator,** or **electrotonic potentials;** and propagated potentials, the **action potentials** (or **nerve impulses**). Action potentials are the primary electrical responses of neurons and other excitable tissues, and they are the main form of communication within the nervous system. They are due to changes in the conduction of ions across the cell membrane. The electrical events in neurons are rapid, being measured in **milliseconds (ms);** and the potential changes are small, being measured in **millivolts (mV).**

The impulse is normally transmitted **(conducted)** along the axon to its termination. Nerves are not "telephone wires" that transmit impulses passively; conduction of nerve impulses, although rapid, is much slower than that of electricity. Nerve tissue is in fact a relatively poor passive conductor, and it would take a potential of many volts to produce a signal of a fraction of a volt at the other end of a meter-long axon in the absence of active processes in the nerve. Instead, conduction is an active, self-propagating process, and the impulse moves along the nerve at a constant amplitude and velocity. The process is often compared to what happens when a match is applied to one end of a trail of gunpowder; by igniting the powder particles immediately in front of it, the flame moves steadily down the trail to its end as it is extinguished in its wake.

RESTING MEMBRANE POTENTIAL

When two electrodes are connected through a suitable amplifier and placed on the surface of a single axon, no potential difference is observed. However, if one electrode is inserted

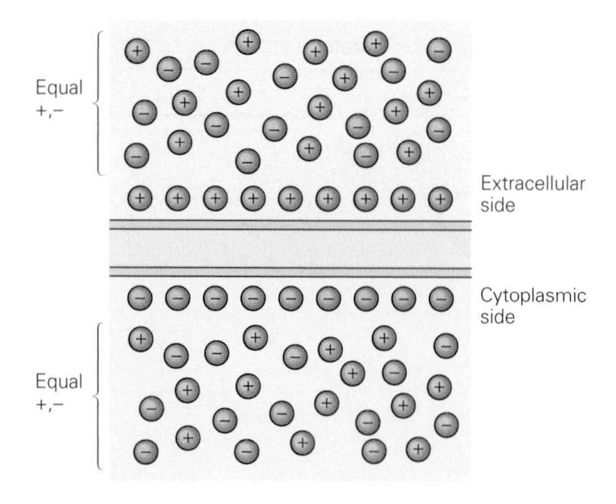

FIGURE 4–5 A membrane potential results from separation of positive and negative charges across the cell membrane. The excess of positive charges (red circles) outside the cell and negative charges (blue circles) inside the cell at rest represents a small fraction of the total number of ions present. (From Kandel ER, Schwartz JH, Jessell TM (editors): *Principles of Neural Science*, 4th ed. McGraw-Hill, 2000.).

into the interior of the cell, a constant **potential difference** is observed, with the inside negative relative to the outside of the cell at rest. A membrane potential results from separation of positive and negative charges across the cell membrane (Figure 4–5).

In order for a potential difference to be present across a membrane lipid bilayer, two conditions must be met. First, there must be an unequal distribution of ions of one or more species across the membrane (ie, a concentration gradient). Second, the membrane must be permeable to one or more of these ion species. The permeability is provided by the existence of channels or pores in the bilayer; these channels are usually permeable to a single species of ions. The resting membrane potential represents an equilibrium situation at which the driving force for the membrane-permeant ions down their concentration gradients across the membrane is equal and opposite to the driving force for these ions down their electrical gradients.

In neurons, the concentration of K^+ is much higher inside than outside the cell, while the reverse is the case for Na^+. This concentration difference is established by Na, K ATPase. The outward K^+ concentration gradient results in passive movement of K^+ out of the cell when K^+-selective channels are open. Similarly, the inward Na^+ concentration gradient results in passive movement of Na^+ into the cell when Na^+-selective channels are open.

In neurons, the **resting membrane potential** is usually about –70 mV, which is close to the equilibrium potential for K^+ (step 1 in Figure 4–6). Because there are more open K^+ channels than Na^+ channels at rest, the membrane permeability to K^+ is greater. Consequently, the intracellular and extracellular K^+ concentrations are the prime determinants of the resting membrane potential, which is therefore close to the equilibrium potential for K^+. Steady ion leaks cannot continue forever without eventually dissipating the ion gradients. This is prevented by the Na, K ATPase, which actively moves Na^+ and K^+ against their electrochemical gradients.

IONIC FLUXES DURING THE ACTION POTENTIAL

The cell membranes of nerves, like those of other cells, contain many different types of ion channels. Some of these are voltage-gated and others are ligand-gated. It is the behavior of these channels, and particularly Na^+ and K^+ channels, which explains the electrical events in neurons.

The changes in membrane conductance of Na^+ and K^+ that occur during the action potentials are shown by steps 1 through 7 in Figure 4–6. The conductance of an ion is the reciprocal of its electrical resistance in the membrane and is a measure of the membrane permeability to that ion. In response to a depolarizing stimulus, some of the voltage-gated Na^+ channels open and Na^+ enters the cell and the membrane is brought to its **threshold potential** (step 2) and the voltage-gated Na^+ channels overwhelm the K^+ and other channels. The entry of Na^+ causes the opening of more voltage-gated Na^+ channels and further depolarization, setting up a **positive feedback loop.** The rapid upstroke in the membrane potential ensues (step 3). The membrane potential moves toward the equilibrium potential for Na^+ (+60 mV) but does not reach it during the action potential (step 4), primarily because the increase in Na^+ conductance is short-lived. The Na^+ channels rapidly enter a closed state called the **inactivated state** and remain in this state for a few milliseconds before returning to the resting state, when they again can be activated. In addition, the direction of the electrical gradient for Na^+ is reversed during the **overshoot** because the membrane potential is reversed, and this limits Na^+ influx; also the voltage-gated K^+ channels open. These factors contribute to **repolarization.** The opening of voltage-gated K^+ channels is slower and more prolonged than the opening of the Na^+ channels, and consequently, much of the increase in K^+ conductance comes after the increase in Na^+ conductance (step 5). The net movement of positive charge out of the cell due to K^+ efflux at this time helps complete the process of repolarization. The slow return of the K^+ channels to the closed state also explains the **after-hyperpolarization** (step 6), followed by a return to the resting membrane potential (step 7). Thus, voltage-gated K^+ channels bring the action potential to an end and cause closure of their gates through a **negative feedback process.** Figure 4–7 shows the sequential feedback control in voltage-gated K^+ and Na^+ channels during the action potential.

Decreasing the external Na^+ concentration reduces the size of the action potential but has little effect on the resting membrane potential. The lack of much effect on the resting membrane potential would be predicted, since the permeability of the membrane to Na^+ at rest is relatively low. In contrast, since the resting membrane potential is close to the equilibrium potential for K^+, changes in the changes in the external concentration of this ion can have major effects on the resting membrane potential. If the extracellular level of K^+ is

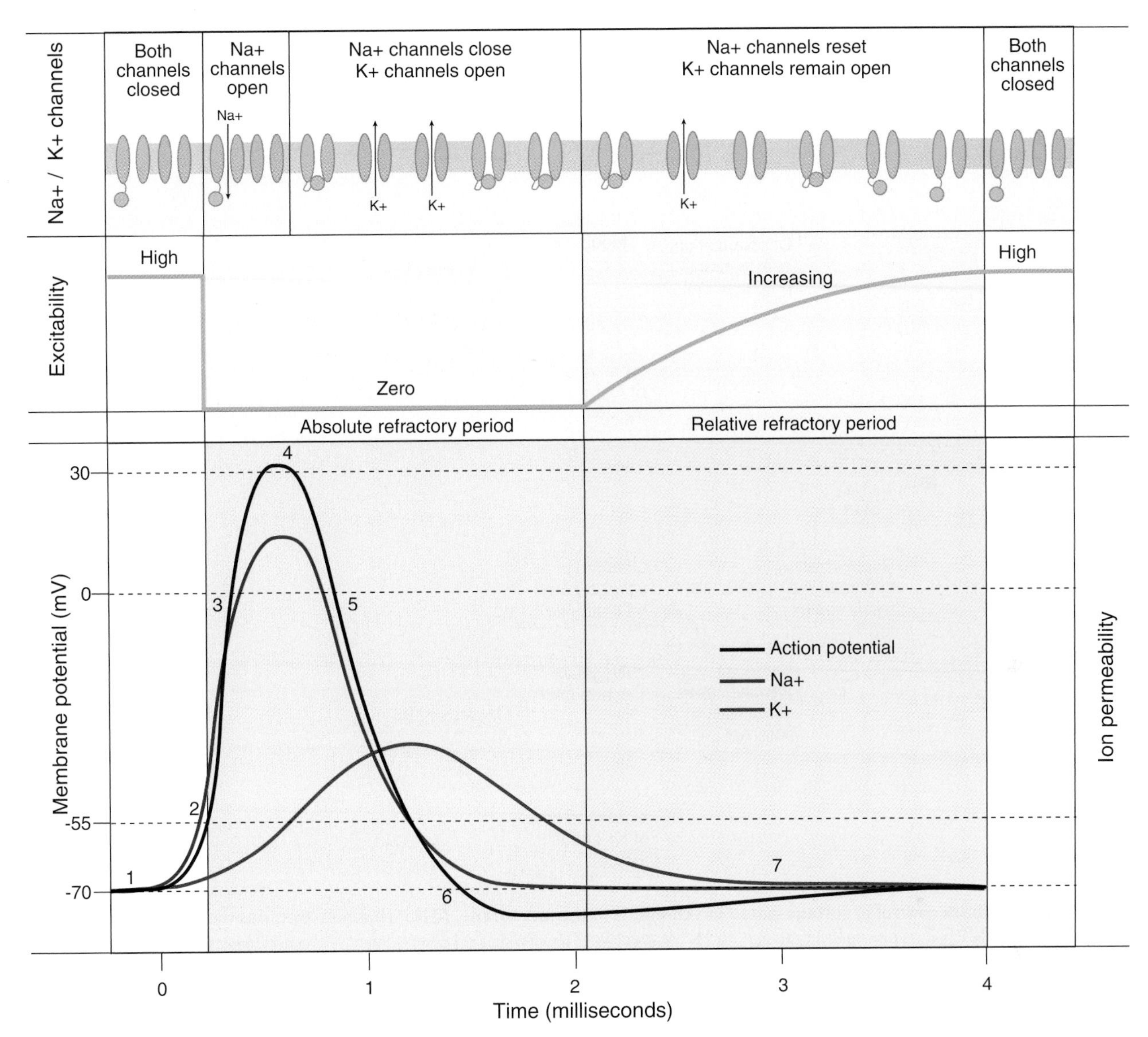

FIGURE 4–6 Changes in membrane potential and relative membrane permeability to Na⁺ and K⁺ during an action potential. Steps 1 through 7 are detailed in the text. These changes in threshold for activation (excitability) are correlated with the phases of the action potential. (Modified from Silverthorn DU: *Human Physiology: An Integrated Approach,* 5th ed. Pearson, 2010.)

increased **(hyperkalemia),** the resting potential moves closer to the threshold for eliciting an action potential, thus the neuron becomes more excitable. If the extracellular level of K^+ is decreased **(hypokalemia),** the membrane potential is reduced and the neuron is hyperpolarized.

Although Na^+ enters the nerve cell and K^+ leaves it during the action potential, very few ions actually move across the membrane. It has been estimated that only 1 in 100,000 K^+ ions cross the membrane to change the membrane potential from +30 mV (peak of the action potential) to –70 mV (resting potential). Significant differences in ion concentrations can be measured only after prolonged, repeated stimulation.

Other ions, notably Ca^{2+}, can affect the membrane potential through both channel movement and membrane interactions. A decrease in extracellular Ca^{2+} concentration increases the excitability of nerve and muscle cells by decreasing the amount of depolarization necessary to initiate the changes in the Na^+ and K^+ conductance that produce the action potential. Conversely, an increase in extracellular Ca^{2+} concentration can stabilize the membrane by decreasing excitability.

ALL-OR-NONE ACTION POTENTIALS

It is possible to determine the minimal intensity of stimulating current **(threshold intensity)** that, acting for a given duration, will just produce an action potential. The threshold intensity varies with the duration; with weak stimuli it is long, and with strong stimuli it is short. The relation between the strength and the duration of a threshold stimulus is called the **strength–duration curve.** Slowly rising currents fail to fire the nerve because the nerve adapts to the applied stimulus, a process called **adaptation.**

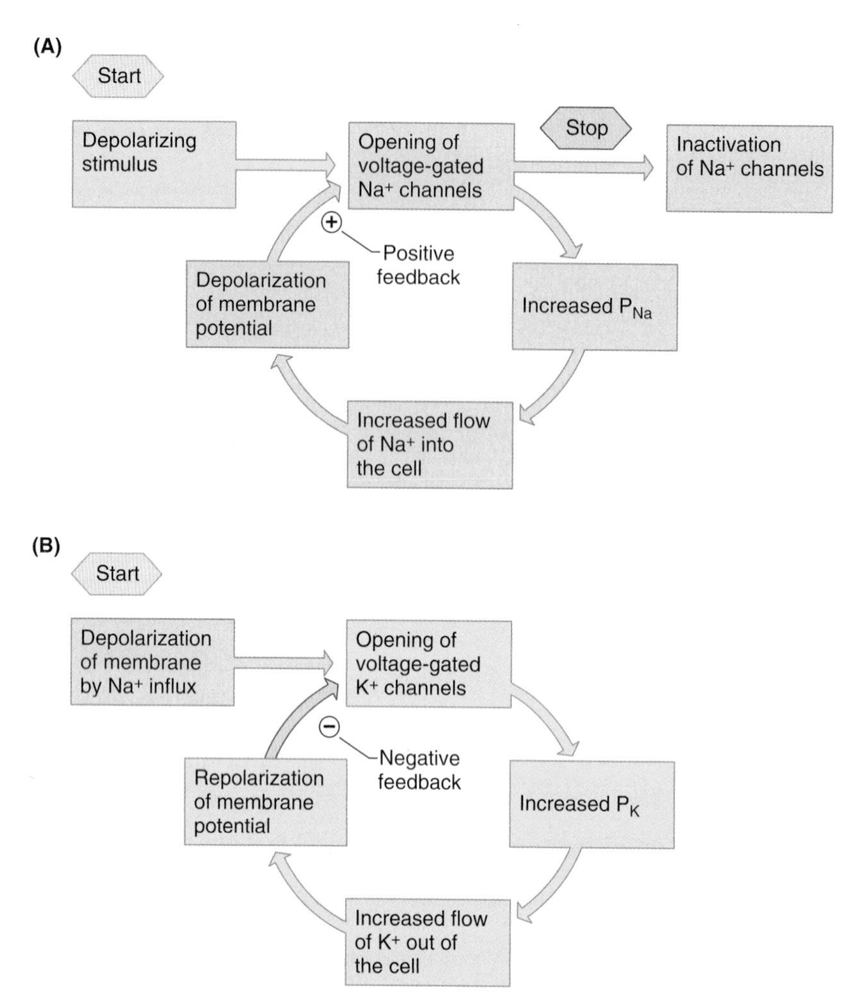

FIGURE 4–7 Feedback control in voltage-gated ion channels in the membrane. A) Na^+ channels exert positive feedback. **B)** K^+ channels exert negative feedback. P_{Na}, P_K is permeability to Na^+ and K^+, respectively. (From Widmaier EP, Raff H, Strang KT: *Vander's Human Physiology*. McGraw-Hill, 2008.)

Once threshold intensity is reached, a full-fledged action potential is produced. Further increases in the intensity of a stimulus produce no increment or other change in the action potential as long as the other experimental conditions remain constant. The action potential fails to occur if the stimulus is subthreshold in magnitude, and it occurs with constant amplitude and form regardless of the strength of the stimulus if the stimulus is at or above threshold intensity. The action potential is therefore **all-or-none** in character.

ELECTROTONIC POTENTIALS, LOCAL RESPONSE, & FIRING LEVEL

Although subthreshold stimuli do not produce an action potential, they do have an effect on the membrane potential. This can be demonstrated by placing recording electrodes within a few millimeters of a stimulating electrode and applying subthreshold stimuli of fixed duration. Application of such currents leads to a localized depolarizing potential change that rises sharply and decays exponentially with time. The magnitude of this response drops off rapidly as the distance between the stimulating and recording electrodes is increased. Conversely, an anodal current produces a hyperpolarizing potential change of similar duration. These potential changes are called **electrotonic potentials.** As the strength of the current is increased, the response is greater due to the increasing addition of a **local response** of the membrane (Figure 4–8). Finally, at 7–15 mV of depolarization (potential of –55 mV), the **firing level** (threshold potential) is reached and an action potential occurs.

CHANGES IN EXCITABILITY DURING ELECTROTONIC POTENTIALS & THE ACTION POTENTIAL

During the action potential, as well as during electrotonic potentials and the local response, the threshold of the neuron to stimulation changes (Figure 4–6). Hyperpolarizing responses elevate the threshold, and depolarizing potentials

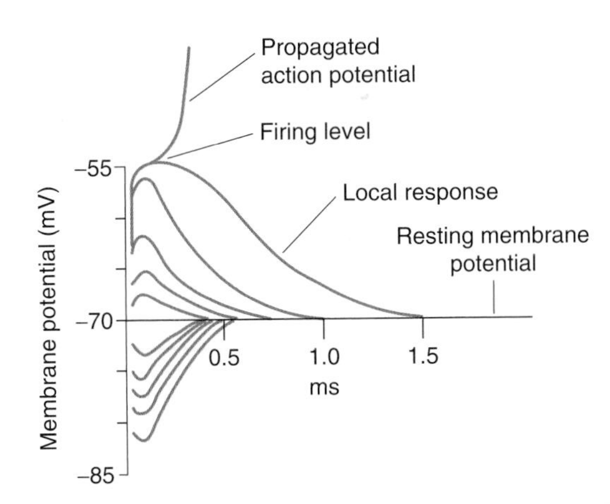

FIGURE 4–8 Electrotonic potentials and local response. The changes in the membrane potential of a neuron following application of stimuli of 0.2, 0.4, 0.6, 0.8, and 1.0 times threshold intensity are shown superimposed on the same time scale. The responses below the horizontal line are those recorded near the anode, and the responses above the line are those recorded near the cathode. The stimulus of threshold intensity was repeated twice. Once it caused a propagated action potential (top line), and once it did not.

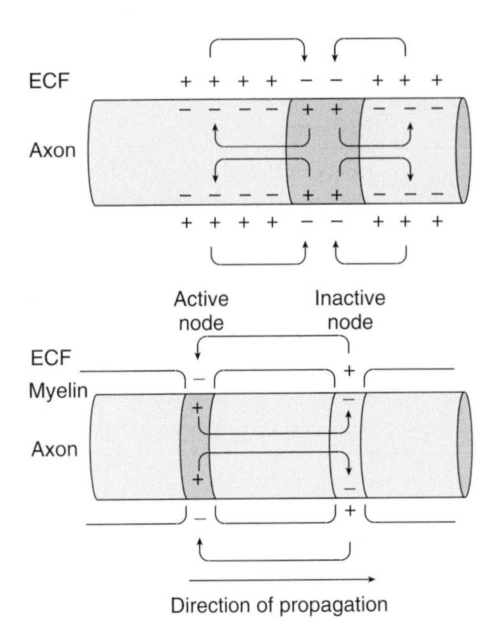

FIGURE 4–9 Local current flow (movement of positive charges) around an impulse in an axon. Top: Unmyelinated axon. **Bottom:** Myelinated axon. Positive charges from the membrane ahead of and behind the action potential flow into the area of negativity represented by the action potential ("current sink"). In myelinated axons, depolarization appears to "jump" from one node of Ranvier to the next (saltatory conduction).

lower it as they move the membrane potential closer to the firing level. During the local response, the threshold is lowered, but during the rising and much of the falling phases of the spike potential, the neuron is refractory to stimulation. This **refractory period** is divided into an **absolute refractory period,** corresponding to the period from the time the firing level is reached until repolarization is about one-third complete, and a **relative refractory period,** lasting from this point to the start of after-depolarization. During the absolute refractory period, no stimulus, no matter how strong, will excite the nerve, but during the relative refractory period, stronger than normal stimuli can cause excitation. These changes in threshold are correlated with the phases of the action potential in Figure 4–6.

CONDUCTION OF THE ACTION POTENTIAL

The nerve cell membrane is polarized at rest, with positive charges lined up along the outside of the membrane and negative charges along the inside. During the action potential, this polarity is abolished and for a brief period is actually reversed (Figure 4–9). Positive charges from the membrane ahead of and behind the action potential flow into the area of negativity represented by the action potential ("current sink"). By drawing off positive charges, this flow decreases the polarity of the membrane ahead of the action potential. Such electrotonic depolarization initiates a local response, and when the firing level is reached, a propagated response occurs that in turn electrotonically depolarizes the membrane in front of it.

The spatial distribution of ion channels along the axon plays a key role in the initiation and regulation of the action potential. Voltage-gated Na^+ channels are highly concentrated in the nodes of Ranvier and the initial segment in myelinated neurons. The number of Na^+ channels per square micrometer of membrane in myelinated mammalian neurons has been estimated to be 50–75 in the cell body, 350–500 in the initial segment, less than 25 on the surface of the myelin, 2000–12,000 at the nodes of Ranvier, and 20–75 at the axon terminals. Along the axons of unmyelinated neurons, the number is about 110. In many myelinated neurons, the Na^+ channels are flanked by K^+ channels that are involved in repolarization.

Conduction in myelinated axons depends on a similar pattern of circular current flow as described above. However, myelin is an effective insulator, and current flow through it is negligible. Instead, depolarization in myelinated axons travels from one node of Ranvier to the next, with the current sink at the active node serving to electrotonically depolarize the node ahead of the action potential to the firing level (Figure 4–9). This "jumping" of depolarization from node to node is called **saltatory conduction.** It is a rapid process that allows myelinated axons to conduct up to 50 times faster than the fastest unmyelinated fibers.

ORTHODROMIC & ANTIDROMIC CONDUCTION

An axon can conduct in either direction. When an action potential is initiated in the middle of the axon, two impulses traveling in opposite directions are set up by electrotonic

TABLE 4–1 Types of mammalian nerve fibers.

Fiber Type	Function	Fiber Diameter (μm)	Conduction Velocity (m/s)	Spike Duration (ms)	Absolute Refractory Period (ms)
Aα	Proprioception; somatic motor	12–20	70–120		
Aβ	Touch, pressure	5–12	30–70	0.4–0.5	0.4–1
Aγ	Motor to muscle spindles	3–6	15–30		
Aδ	Pain, temperature	2–5	12–30		
B	Preganglionic autonomic	<3	3–15	1.2	1.2
C, Dorsal root	Pain, temperature	0.4–1.2	0.5–2	2	2
C, Sympathetic	Postganglionic sympathetic	0.3–1.3	0.7–2.3	2	2

depolarization on either side of the initial current sink. In the natural situation, impulses pass in one direction only, ie, from synaptic junctions or receptors along axons to their termination. Such conduction is called **orthodromic.** Conduction in the opposite direction is called **antidromic.** Because synapses, unlike axons, permit conduction in one direction only, an antidromic impulse will fail to pass the first synapse they encounter and die out at that point.

PROPERTIES OF MIXED NERVES

Peripheral nerves in mammals are made up of many axons bound together in a fibrous envelope called the **epineurium.** Potential changes recorded extracellularly from such nerves therefore represent an algebraic summation of the all-or-none action potentials of many axons. The thresholds of the individual axons in the nerve and their distance from the stimulating electrodes vary. With subthreshold stimuli, none of the axons are stimulated and no response occurs. When the stimuli are of threshold intensity, axons with low thresholds fire and a small potential change is observed. As the intensity of the stimulating current is increased, the axons with higher thresholds are also discharged. The electrical response increases proportionately until the stimulus is strong enough to excite all of the axons in the nerve. The stimulus that produces excitation of all the axons is the **maximal stimulus,** and application of greater, supramaximal stimuli produces no further increase in the size of the observed potential.

After a stimulus is applied to a nerve, there is a **latent period** before the start of the action potential. This interval corresponds to the time it takes the impulse to travel along the axon from the site of stimulation to the recording electrodes. Its duration is proportionate to the distance between the stimulating and recording electrodes and inversely proportionate to the speed of conduction. If the duration of the latent period and the distance between the stimulating and recording electrodes are known, **axonal conduction velocity** can be calculated.

NERVE FIBER TYPES & FUNCTION

Erlanger and Gasser divided mammalian nerve fibers into A, B, and C groups, further subdividing the A group into α, β, γ, and δ fibers. In Table 4–1, the various fiber types are listed with their diameters, electrical characteristics, and functions. By comparing the neurologic deficits produced by careful dorsal root section and other nerve-cutting experiments with the histologic changes in the nerves, the functions and histologic characteristics of each of the families of axons responsible for the various peaks of the compound action potential have been established. In general, the greater the diameter of a given nerve fiber, the greater is its speed of conduction. The large axons are concerned primarily with proprioceptive sensation, somatic motor function, conscious touch, and pressure, while the smaller axons subserve pain and temperature sensations and autonomic function.

Further research has shown that not all the classically described lettered components are homogeneous, and a numerical system (Ia, Ib, II, III, and IV) has been used by some physiologists to classify sensory fibers. Unfortunately, this has led to confusion. A comparison of the number system and the letter system is shown in Table 4–2.

TABLE 4–2 Numerical classification of sensory nerve fibers.

Number	Origin	Fiber Type
Ia	Muscle spindle, annulo-spiral ending	Aα
Ib	Golgi tendon organ	Aα
II	Muscle spindle, flower-spray ending; touch, pressure	Aβ
III	Pain and cold receptors; some touch receptors	Aδ
IV	Pain, temperature, and other receptors	Dorsal root C

TABLE 4–3 Relative susceptibility of mammalian A, B, and C nerve fibers to conduction block produced by various agents.

Susceptibility To:	Most Susceptible	Intermediate	Least Susceptible
Hypoxia	B	A	C
Pressure	A	B	C
Local anesthetics	C	B	A

TABLE 4–4 Neurotrophins.

Neurotrophin	Receptor
Nerve growth factor (NGF)	Trk A
Brain-derived neurotrophic factor (BDNF)	Trk B
Neurotrophin 3 (NT-3)	Trk C, less on Trk A and Trk B
Neurotrophin 4/5 (NT-4/5)	Trk B

In addition to variations in speed of conduction and fiber diameter, the various classes of fibers in peripheral nerves differ in their sensitivity to hypoxia and anesthetics (Table 4–3). This fact has clinical as well as physiologic significance. Local anesthetics depress transmission in the group C fibers before they affect group A touch fibers (see Clinical Box 4–2). Conversely, pressure on a nerve can cause loss of conduction in large-diameter motor, touch, and pressure fibers while pain sensation remains relatively intact. Patterns of this type are sometimes seen in individuals who sleep with their arms under their heads for long periods, causing compression of the nerves in the arms. Because of the association of deep sleep with alcoholic intoxication, the syndrome is most common on weekends and has acquired the interesting name Saturday night or Sunday morning paralysis.

NEUROTROPHINS

A number of proteins necessary for survival and growth of neurons have been isolated and studied. Some of these **neurotrophins** are products of the muscles or other structures that the neurons innervate, but many in the CNS are produced by astrocytes. These proteins bind to receptors at the endings of a neuron. They are internalized and then transported by retrograde transport to the neuronal cell body, where they foster the production of proteins associated with neuronal development, growth, and survival. Other neurotrophins are produced in neurons and transported in an anterograde fashion to the nerve ending, where they maintain the integrity of the postsynaptic neuron.

RECEPTORS

Four established neurotrophins and their three high-affinity **tyrosine kinase associated (Trk) receptors** are listed in Table 4–4. Each of these Trk receptors dimerizes, and this initiates autophosphorylation in the cytoplasmic tyrosine kinase domains of the receptors. An additional low-affinity NGF receptor that is a 75-kDa protein is called **p75^{NTR}**. This receptor binds all four of the listed neurotrophins with equal affinity. There is some evidence that it can form a heterodimer with Trk A monomer and that the dimer has increased affinity and specificity

CLINICAL BOX 4–2

Local Anesthesia

Local or regional anesthesia is used to block the conduction of action potentials in sensory and motor nerve fibers. This usually occurs as a result of blockade of voltage-gated Na^+ channels on the nerve cell membrane. This causes a gradual increase in the threshold for electrical excitability of the nerve, a reduction in the rate of rise of the action potential, and a slowing of axonal conduction velocity. There are two major categories of local anesthetics: **ester-linked** (eg, **cocaine, procaine, tetracaine**) or **amide-linked** (eg, **lidocaine, bupivacaine**). In addition to either the ester or amide, all local anesthetics contain an aromatic and an amine group. The structure of the aromatic group determines the drug's hydrophobic characteristics, and the amine group determines its latency to onset of action and its potency. Application of these drugs into the vicinity of a central (eg, **epidural, spinal anesthesia**) or peripheral nerve can lead to rapid, temporary, and near complete interruption of neural traffic to allow a surgical or other potentially noxious procedure to be done without eliciting pain. Cocaine (from the coca shrub, *Erythroxylan coca*) was the first chemical to be identified as having local anesthetic properties and remains the only naturally occurring local anesthetic. In 1860, Albert Niemann isolated the chemical, tasted it, and reported a numbing effect on his tongue. The first clinical use of cocaine as a local anesthetic was in 1886 when Carl Koller used it as a topical ophthalmic anesthetic. Its addictive and toxic properties prompted the development of other local anesthetics. In 1905, procaine was synthesized as the first suitable substitute for cocaine. Nociceptive fibers (unmyelinated C fibers) are the most sensitive to the blocking effect of local anesthetics. This is followed by sequential loss of sensitivity to temperature, touch, and deep pressure. Motor nerve fibers are the most resistant to the actions of local anesthetics.

CLINICAL BOX 4–3

Axonal Regeneration

Peripheral nerve damage is often reversible. Although the axon will degenerate distal to the damage, connective elements of the so-called **distal stump** often survive. **Axonal sprouting** occurs from the proximal stump, growing toward the nerve ending. This results from **growth-promoting factors** secreted by **Schwann cells** that attract axons toward the distal stump. Adhesion molecules of the immunoglobulin superfamily (eg, NgCAM/L1) promote axon growth along cell membranes and extracellular matrices. Inhibitory molecules in the perineurium assure that the regenerating axons grow in a correct trajectory. Denervated distal stumps are able to upregulate production of **neurotrophins** that promote growth. Once the regenerated axon reaches its target, a new functional connection (eg, neuromuscular junction) is formed. Regeneration allows for considerable, although not full, recovery. For example, fine motor control may be permanently impaired because some motor neurons are guided to an inappropriate motor fiber. Nonetheless, recovery of peripheral nerves from damage far surpasses that of central nerve pathways. The proximal stump of a damaged axon in the CNS will form short sprouts, but distant stump recovery is rare, and the damaged axons are unlikely to form new synapses. This is in part because CNS neurons do not have the growth-promoting chemicals needed for regeneration. In fact, CNS myelin is a potent inhibitor of axonal growth. In addition, following CNS injury several events—**astrocytic proliferation, activation of microglia, scar formation, inflammation,** and **invasion of immune cells**—provide an inappropriate environment for regeneration. Thus, treatment of brain and spinal cord injuries frequently focuses on rehabilitation rather than reversing the nerve damage. New research is aiming to identify ways to initiate and maintain axonal growth, to direct regenerating axons to reconnect with their target neurons, and to reconstitute original neuronal circuitry.

THERAPEUTIC HIGHLIGHTS

There is evidence showing that the use of **nonsteroidal anti-inflammatory drugs** (NSAIDs) like ibuprofen can overcome the factors that inhibit axonal growth following injury. This effect is thought to be mediated by the ability of NSAIDs to inhibit RhoA, a small GTPase protein that normally prevents repair of neural pathways and axons. Growth cone collapse in response to myelin-associated inhibitors after nerve injury is prevented by drugs such as **pertussis toxin,** which interfere with signal transduction via trimeric G protein. Experimental drugs that inhibit the **phosphoinositide 3-kinase (PI3) pathway** or the **inositol triphosphate (IP_3) receptor** have also been shown to promote regeneration after nerve injury.

for NGF. However, it now appears that $p75^{NTR}$ receptors can form homodimers that in the absence of Trk receptors cause apoptosis, an effect opposite to the usual growth-promoting and nurturing effects of neurotrophins. Research is ongoing to characterize the distinct roles of $p75^{NTR}$ and Trk receptors and factors that influence their expression in neurons.

FUNCTION OF NEUROTROPHINS

The first neurotrophin to be characterized was NGF, a protein growth factor that is necessary for the growth and maintenance of sympathetic neurons and some sensory neurons. It is present in a broad spectrum of animal species, including humans, and is found in many different tissues. In male mice, there is a particularly high concentration in the submandibular salivary glands, and the level is reduced by castration to that seen in females. The factor is made up of two α, two β, and two γ subunits. The β subunits, each of which has a molecular mass of 13,200 Da, have all the nerve growth-promoting activity, the α subunits have trypsin-like activity, and the γ subunits are serine proteases. The function of the proteases is unknown. The structure of the β subunit of NGF resembles that of insulin.

NGF is picked up by neurons and is transported in retrograde fashion from the endings of the neurons to their cell bodies. It is also present in the brain and appears to be responsible for the growth and maintenance of cholinergic neurons in the basal forebrain and the striatum. Injection of antiserum against NGF in newborn animals leads to almost total destruction of the sympathetic ganglia; it thus produces an **immunosympathectomy.** There is evidence that the maintenance of neurons by NGF is due to a reduction in apoptosis.

Brain-derived neurotrophic factor (BDNF), neurotrophin 3 (NT-3), NT-4/5, and NGF each maintain a different pattern of neurons, although there is some overlap. Disruption of NT-3 by gene knockout causes a marked loss of cutaneous mechanoreceptors, even in heterozygotes. BDNF acts rapidly and can actually depolarize neurons. BDNF-deficient mice lose peripheral sensory neurons and have severe degenerative changes in their vestibular ganglia and blunted long-term potentiation.

OTHER FACTORS AFFECTING NEURONAL GROWTH

The regulation of neuronal growth is a complex process. Schwann cells and astrocytes produce **ciliary neurotrophic factor (CNTF).** This factor promotes the survival of damaged and embryonic spinal cord neurons and may prove to be of value in treating human diseases in which motor neurons degenerate. **Glial cell line-derived neurotrophic factor (GDNF)** maintains midbrain dopaminergic neurons in vitro. However, GDNF knockouts have dopaminergic neurons that appear normal, but they have no kidneys and fail to develop an enteric nervous system. Another factor that enhances the growth of neurons is **leukemia inhibitory factor (LIF).** In addition, neurons as well as other cells respond to **insulin-like growth factor I (IGF-I)** and the various forms of **transforming growth factor (TGF), fibroblast growth factor (FGF),** and **platelet-derived growth factor (PDGF).**

Clinical Box 4–3 compares the ability to regenerate neurons after central and peripheral nerve injury.

CHAPTER SUMMARY

- There are two main types of glia: microglia and macroglia. Microglia are scavenger cells. Macroglia include oligodendrocytes, Schwann cells, and astrocytes. The first two are involved in myelin formation; astrocytes produce substances that are tropic to neurons, and they help maintain the appropriate concentration of ions and neurotransmitters.
- Neurons are composed of a cell body (soma) that is the metabolic center of the neuron, dendrites that extend outward from the cell body and arborize extensively, and a long fibrous axon that originates from a somewhat thickened area of the cell body, the axon hillock.
- The axons of many neurons acquire a sheath of myelin, a protein–lipid complex that is wrapped around the axon. Myelin is an effective insulator, and depolarization in myelinated axons travels from one node of Ranvier to the next, with the current sink at the active node serving to electrotonically depolarize to the firing level the node ahead of the action potential.
- Orthograde transport occurs along microtubules that run the length of the axon and requires two molecular motors: dynein and kinesin. It moves from the cell body toward the axon terminals and has both fast (400 mm/day) and slow (0.5–10 mm/day) components. Retrograde transport, which is in the opposite direction (from the nerve ending to the cell body), occurs along microtubules at about 200 mm/day.
- In response to a depolarizing stimulus, voltage-gated Na^+ channels become active, and when the threshold potential is reached, an action potential results. The membrane potential moves toward the equilibrium potential for Na^+. The Na^+ channels rapidly enter a closed state (inactivated state) before returning to the resting state. The direction of the electrical gradient for Na^+ is reversed during the overshoot because the membrane potential is reversed, and this limits Na^+ influx. Voltage-gated K^+ channels open and the net movement of positive charge out of the cell helps complete the process of repolarization. The slow return of the K^+ channels to the closed state explains after-hyperpolarization, followed by a return to the resting membrane potential.
- Nerve fibers are divided into different categories (A, B, and C) based on axonal diameter, conduction velocity, and function. A numerical classification (Ia, Ib, II, III, and IV) is also used for sensory afferent fibers.
- Neurotrophins such as NGF are carried by retrograde transport to the neuronal cell body, where they foster the production of proteins associated with neuronal development, growth, and survival.

MULTIPLE-CHOICE QUESTIONS

For all questions, select the single best answer unless otherwise directed.

1. Which of the following statements about glia is true?
   A. Microglia arise from macrophages outside of the nervous system and are physiologically and embryologically similar to other neural cell types.
   B. Glia do not undergo proliferation.
   C. Protoplasmic astrocytes produce substances that are tropic to neurons to help maintain the appropriate concentration of ions and neurotransmitters by taking up K^+ and the neurotransmitters glutamate and GABA.
   D. Oligodendrocytes and Schwann cells are involved in myelin formation around axons in the peripheral and central nervous systems, respectively.
   E. Macroglia are scavenger cells that resemble tissue macrophages and remove debris resulting from injury, infection, and disease.
2. A 13-year-old girl was being seen by her physician because of experiencing frequent episodes of red, painful, warm extremities. She was diagnosed with primary erythromelalgia, which may be due to a peripheral nerve sodium channelopathy. Which part of a neuron has the highest concentration of Na^+ channels per square micrometer of cell membrane?
   A. dendrites
   B. cell body near dendrites
   C. initial segment
   D. axonal membrane under myelin
   E. none of Ranvier
3. A 45-year-old female office worker had been experiencing tingling in her index and middle fingers and thumb of her right hand. Recently, her wrist and hand had become weak. Her physician ordered a nerve conduction test to evaluate her for carpal tunnel syndrome. Which one of the following nerves has the slowest conduction velocity?
   A. Aα fibers
   B. Aβ fibers
   C. Aγ fibers
   D. B fibers
   E. C fibers
4. Which of the following is *not* correctly paired?
   A. Synaptic transmission: Antidromic conduction
   B. Molecular motors: Dynein and kinesin
   C. Fast axonal transport: ~400 mm/day
   D. Slow axonal transport: 0.5–10 mm/day
   E. Nerve growth factor: Retrograde transport
5. A 32-year-old female received an injection of a local anesthetic for a tooth extraction. Within 2 h, she noted palpitations, diaphoresis, and dizziness. Which of the following ionic changes is correctly matched with a component of the action potential?
   A. Opening of voltage-gated K^+ channels: After-hyperpolarization
   B. A decrease in extracellular Ca^{2+}: Repolarization
   C. Opening of voltage-gated Na^+ channels: Depolarization
   D. Rapid closure of voltage-gated Na^+ channels: Resting membrane potential
   E. Rapid closure of voltage-gated K^+ channels: Relative refractory period

6. A man falls into a deep sleep with one arm under his head. This arm is paralyzed when he awakens, but it tingles, and pain sensation in it is still intact. The reason for the loss of motor function without loss of pain sensation is that in the nerves to his arm,
   A. A fibers are more susceptible to hypoxia than B fibers.
   B. A fibers are more sensitive to pressure than C fibers.
   C. C fibers are more sensitive to pressure than A fibers.
   D. Motor nerves are more affected by sleep than sensory nerves.
   E. Sensory nerves are nearer the bone than motor nerves and hence are less affected by pressure.
7. Which of the following statements about nerve growth factor is *not* true?
   A. It is made up of three polypeptide subunits.
   B. It is responsible for the growth and maintenance of adrenergic neurons in the basal forebrain and the striatum.
   C. It is necessary for the growth and development of the sympathetic nervous system.
   D. It is picked up by nerves from the organs they innervate.
   E. It can express both $p75^{NTR}$ and Trk A receptors.
8. A 20-year old female student awakens one morning with severe pain and blurry vision in her left eye; the symptoms abate over several days. About 6 months later, on a morning after playing volleyball with friends, she notices weakness but not pain in her right leg; the symptoms intensify while taking a hot shower. Which of the following is most likely to be the case?
   A. The two episodes described are not likely to be related.
   B. She may have primary-progressive multiple sclerosis.
   C. She may have relapsing-remitting multiple sclerosis.
   D. She may have a lumbar disk rupture.
   E. She may have Guillain–Barre syndrome.

CHAPTER RESOURCES

Aidley DJ: *The Physiology of Excitable Cells*, 4th ed. Cambridge University Press, 1998.

Benarroch EE: Neuron-astrocyte interactions: Partnership for normal function and disease. Mayo Clin Proc 2005;80:1326.

Boron WF, Boulpaep EL: *Medical Physiology*, 2nd ed. Elsevier, 2009.

Bradbury EJ, McMahon SB: Spinal cord repair strategies: Why do they work? Nat Rev Neurosci 2006;7:644.

Brunton L, Chabner B, Knollman B (editors): *Goodman and Gilman's The Pharmacological Basis of Therapeutics*, 12th ed. McGraw-Hill, 2010.

Catterall WA: Structure and function of voltage-sensitive ion channels. Science 1988; 242:649.

Golan DE, Tashjian AH, Armstrong EJ, Armstrong AW (editors): *Principles of Pharmacology: The Pathophysiological Basis of Drug Therapy*, 2nd ed. Lippincott Williams & Wilkins, 2008.

Hille B: *Ionic Channels of Excitable Membranes*, 3rd ed. Sinauer Associates, 2001.

Kandel ER, Schwartz JH, Jessell TM (editors): *Principles of Neural Science*, 4th ed. McGraw-Hill, 2000.

Nicholls JG, Martin AR, Wallace BG: *From Neuron to Brain: A Cellular and Molecular Approach to the Function of the Nervous System*, 4th ed. Sinauer Associates, 2001.

Thuret S, Moon LDF, Gage FH: Therapeutic interventions after spinal cord injury. Nat Rev Neurosci 2006;7:628.

Volterra A, Meldolesi J: Astrocytes, from brain glue to communication elements: The revolution continues. Nat Rev Neurosci 2005;6:626.

Widmaier EP, Raff H, Strang KT: *Vander's Human Physiology.* McGraw-Hill, 2008.

CHAPTER 5

Excitable Tissue: Muscle

OBJECTIVES

After studying this chapter, you should be able to:

- Differentiate the major classes of muscle in the body.
- Describe the molecular and electrical makeup of muscle cell excitation–contraction coupling.
- Define elements of the sarcomere that underlie striated muscle contraction.
- Differentiate the role(s) for Ca^{2+} in skeletal, cardiac, and smooth muscle contraction.
- Appreciate muscle cell diversity and function.

INTRODUCTION

Muscle cells, like neurons, can be excited chemically, electrically, and mechanically to produce an action potential that is transmitted along their cell membranes. Unlike neurons, they respond to stimuli by activating a contractile mechanism. The contractile protein myosin and the cytoskeletal protein actin are abundant in muscle, where they are the primary structural components that bring about contraction.

Muscle is generally divided into three types: **skeletal, cardiac,** and **smooth,** although smooth muscle is not a homogeneous single category. Skeletal muscle makes up the great mass of the somatic musculature. It has well-developed cross-striations, does not normally contract in the absence of nervous stimulation, lacks anatomic and functional connections between individual muscle fibers, and is generally under voluntary control. Cardiac muscle also has cross-striations, but it is functionally syncytial and, although it can be modulated via the autonomic nervous system, it can contract rhythmically in the absence of external innervation owing to the presence in the myocardium of pacemaker cells that discharge spontaneously (see Chapter 29). Smooth muscle lacks cross-striations and can be further subdivided into two broad types: unitary (or visceral) smooth muscle and multiunit smooth muscle. The type found in most hollow viscera is functionally syncytial and contains pacemakers that discharge irregularly. The multiunit type found in the eye and in some other locations is not spontaneously active and resembles skeletal muscle in graded contractile ability.

SKELETAL MUSCLE MORPHOLOGY

ORGANIZATION

Skeletal muscle is made up of individual muscle fibers that are the "building blocks" of the muscular system in the same sense that the neurons are the building blocks of the nervous system. Most skeletal muscles begin and end in tendons, and the muscle fibers are arranged in parallel between the tendinous ends, so that the force of contraction of the units is additive. Each muscle fiber is a single cell that is multinucleated, long, cylindrical, and surrounded by a cell membrane, the **sarcolemma** (Figure 5–1). There are no syncytial bridges between cells. The muscle fibers are made up of myofibrils, which are divisible into individual filaments. These myofilaments contain several proteins that together make up the contractile machinery of the skeletal muscle.

The contractile mechanism in skeletal muscle largely depends on the proteins **myosin-II, actin, tropomyosin,** and **troponin.** Troponin is made up of three subunits: **troponin I, troponin T,** and **troponin C.** Other important proteins in muscle are involved in maintaining the proteins that participate in contraction in appropriate structural relation to one another and to the extracellular matrix.

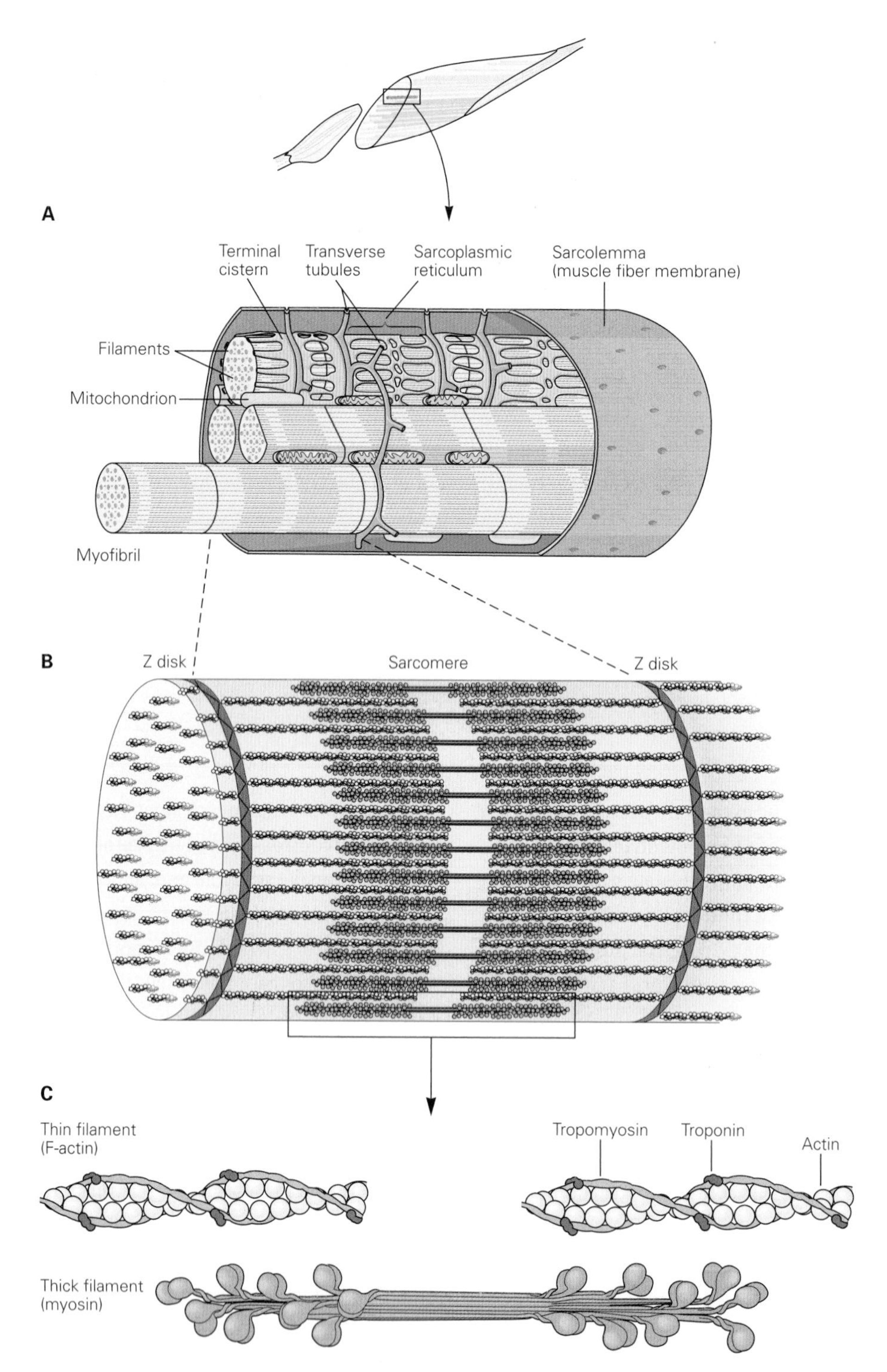

FIGURE 5–1 Mammalian skeletal muscle. A single muscle fiber surrounded by its sarcolemma has been cut away to show individual myofibrils. The cut surface of the myofibrils shows the arrays of thick and thin filaments. The sarcoplasmic reticulum with its transverse (T) tubules and terminal cisterns surrounds each myofibril. The T tubules invaginate from the sarcolemma and contact the myofibrils twice in every sarcomere. Mitochondria are found between the myofibrils and a basal lamina surrounds the sarcolemma.

STRIATIONS

Differences in the refractive indexes of the various parts of the muscle fiber are responsible for the characteristic cross-striations seen in skeletal muscle when viewed under the microscope. The parts of the cross-striations are frequently identified by letters (Figure 5–2). The light I band is divided by the dark Z line, and the dark A band has the lighter H band in its center. A transverse M line is seen in the middle of the H band, and this line plus the narrow light areas on either side of

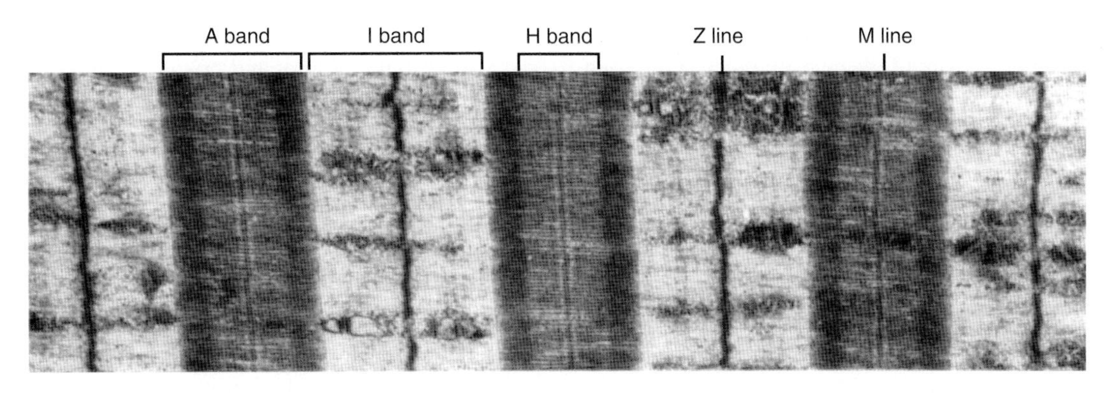

FIGURE 5–2 Electronmicrograph of human gastrocnemius muscle. The various bands and lines are identified at the top (× 13,500). (Courtesy of Walker SM, Schrodt GR.)

it are sometimes called the pseudo-H zone. The area between two adjacent Z lines is called a **sarcomere.** The orderly arrangement of actin, myosin, and related proteins that produces this pattern is shown in Figure 5–3. The thick filaments, which are about twice the diameter of the thin filaments, are made up of myosin; the thin filaments are made up of actin, tropomyosin, and troponin. The thick filaments are lined up to form the A bands, whereas the array of thin filaments extends out of the A band and into the less dense staining I bands. The lighter H bands in the center of the A bands are the regions where, when the muscle is relaxed, the thin filaments do not overlap the thick filaments. The Z lines allow for anchoring of the thin filaments. If a transverse section through the A band is examined under the electron microscope, each thick filament is seen to be surrounded by six thin filaments in a regular hexagonal pattern.

The form of myosin found in muscle is myosin-II, with two globular heads and a long tail. The heads of the myosin molecules form cross-bridges with actin. Myosin contains heavy chains and light chains, and its heads are made up of the light chains and the amino terminal portions of the heavy chains. These heads contain an actin-binding site and a catalytic site that hydrolyzes ATP. The myosin molecules are arranged symmetrically on either side of the center of the sarcomere, and it is this arrangement that

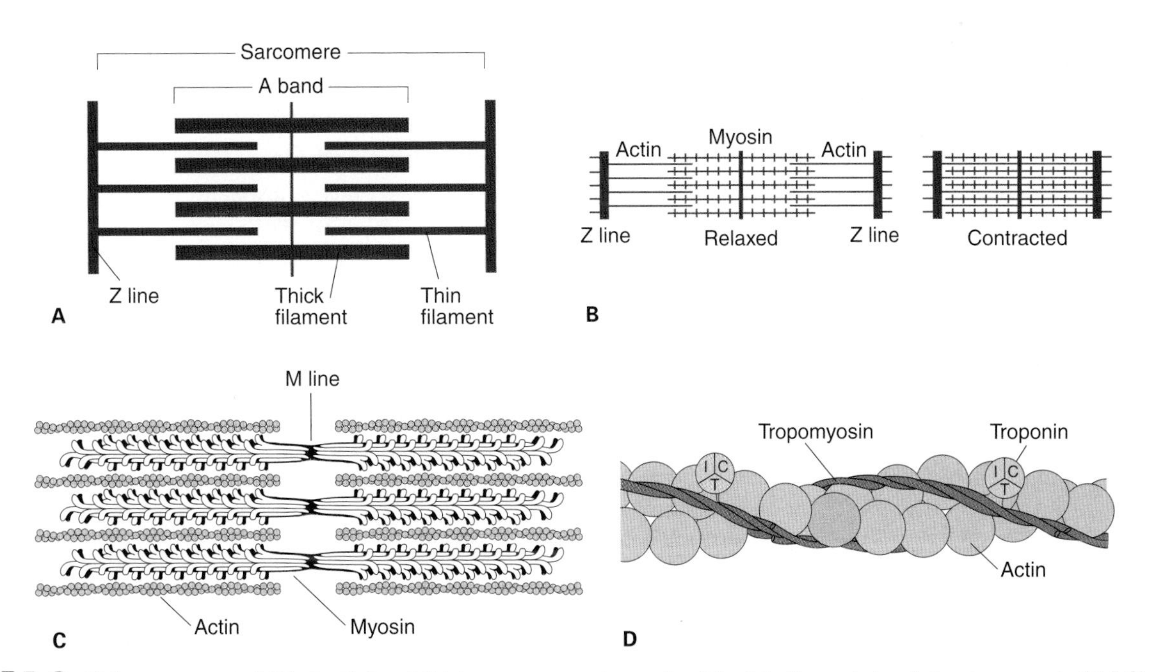

FIGURE 5–3 A) Arrangement of thin (actin) and thick (myosin) filaments in skeletal muscle (compare to Figure 5–2). B) Sliding of actin on myosin during contraction so that Z lines move closer together. **C)** Detail of relation of myosin to actin in an individual sarcomere, the functional unit of the muscle. **D)** Diagrammatic representation of the arrangement of actin, tropomyosin, and troponin of the thin filaments in relation to a myosin thick filament. The globular heads of myosin interact with the thin filaments to create the contraction. Note that myosin thick filaments reverse polarity at the M line in the middle of the sarcomere, allowing for contraction. (C and D are modified with permision from Kandel ER, Schwartz JH, Jessell TM [editors]: *Principles of Neural Science,* 4th ed. McGraw-Hill, 2000.)

creates the light areas in the pseudo-H zone. The M line is the site of the reversal of polarity of the myosin molecules in each of the thick filaments. At these points, there are slender cross-connections that hold the thick filaments in proper array. Each thick filament contains several hundred myosin molecules.

The thin filaments are polymers made up of two chains of actin that form a long double helix. Tropomyosin molecules are long filaments located in the groove between the two chains in the actin (Figure 5–3). Each thin filament contains 300–400 actin molecules and 40–60 tropomyosin molecules. Troponin molecules are small globular units located at intervals along the tropomyosin molecules. Each of the three troponin subunits has a unique function: Troponin T binds the troponin components to tropomyosin; troponin I inhibits the interaction of myosin with actin; and troponin C contains the binding sites for the Ca^{2+} that helps to initiate contraction.

Some additional structural proteins that are important in skeletal muscle function include **actinin, titin,** and **desmin.** Actinin binds actin to the Z lines. Titin, the largest known protein (with a molecular mass near 3,000,000 Da), connects the Z lines to the M lines and provides scaffolding for the sarcomere. It contains two kinds of folded domains that provide muscle with its elasticity. At first when the muscle is stretched there is relatively little resistance as the domains unfold, but with further stretch there is a rapid increase in resistance that protects the structure of the sarcomere. Desmin adds structure to the Z lines in part by binding the Z lines to the plasma membrane. Some muscle disorders associated these structural components are described in Clinical Box 5–1. It should be noted that although these proteins are important in muscle structure/function, by no means do they represent an exhaustive list.

SARCOTUBULAR SYSTEM

The muscle fibrils are surrounded by structures made up of membranes that appear in electronmicrographs as vesicles and tubules. These structures form the **sarcotubular system,** which is made up of a **T system** and a **sarcoplasmic reticulum.** The

CLINICAL BOX 5–1

Structural and Metabolic Disorders in Muscle Disease

The term **muscular dystrophy** is applied to diseases that cause progressive weakness of skeletal muscle. About 50 such diseases have been described, some of which include cardiac as well as skeletal muscle. They range from mild to severe and some are eventually fatal. They have multiple causes, but mutations in the genes for the various components of the dystrophin–glycoprotein complex are a prominent cause. The dystrophin gene is one of the largest in the body, and mutations can occur at many different sites in it. **Duchenne muscular dystrophy** is a serious form of dystrophy in which the dystrophin protein is absent from muscle. It is X-linked and usually fatal by the age of 30. In a milder form of the disease, **Becker muscular dystrophy,** dystrophin is present but altered or reduced in amount. Limb-girdle muscular dystrophies of various types are associated with mutations of the genes coding for the sarcoglycans or other components of the dystrophin–glycoprotein complex.

Due to its enormous size and structural role in the sarcomere, **titin** is a prominent target for mutations that give rise to muscle disease. Mutations that encode for shorter titin structure have been associated with dilated cardiomyopathy, while other mutations have been associated with hypertrophic cardiomyopathy. The skeletal muscle-associated tibialis muscular dystrophy is a genetic muscle disease of titin that is predicted to destabilize the folded state of the protein. Interestingly, many of the titin mutations identified thus far are in regions of titin that are expressed in all striated muscles, yet, not all muscles are affected in the same way. Such muscle type–specific phenotypes underscore the need to study titin's multiple functions in different muscles, under both normal and pathological conditions.

Desmin-related myopathies are a very rare heterogeneous group of muscle disorders that typically result in cellular aggregates of desmin. Common symptoms of these diseases are failing and wasting in the distal muscles of the lower limbs that can later be identified in other body areas. Studies in desmin knockout mice have revealed defects in skeletal, smooth, and cardiac muscle, notably in the diaphragm and heart.

Metabolic Myopathies

Mutations in genes that code for enzymes involved in the metabolism of carbohydrates, fats, and proteins to CO_2 and H_2O in muscle and the production of ATP can cause **metabolic myopathies** (eg, McArdle syndrome). Metabolic myopathies all have in common exercise intolerance and the possibility of muscle breakdown due to accumulation of toxic metabolites.

THERAPEUTIC HIGHLIGHTS

Although acute muscle pain and soreness can be treated with anti-inflammatory drugs and rest, the genetic dysfunctions described above are not as easily addressed. The overall goals are to slow muscle function/structure loss and, when possible relieve symptoms associated with the disease. Extensive monitoring, physical therapy and appropriate drugs including corticosteroids can aid to slow disease progression. Assistive devices and surgery are not uncommon as the diseases progress.

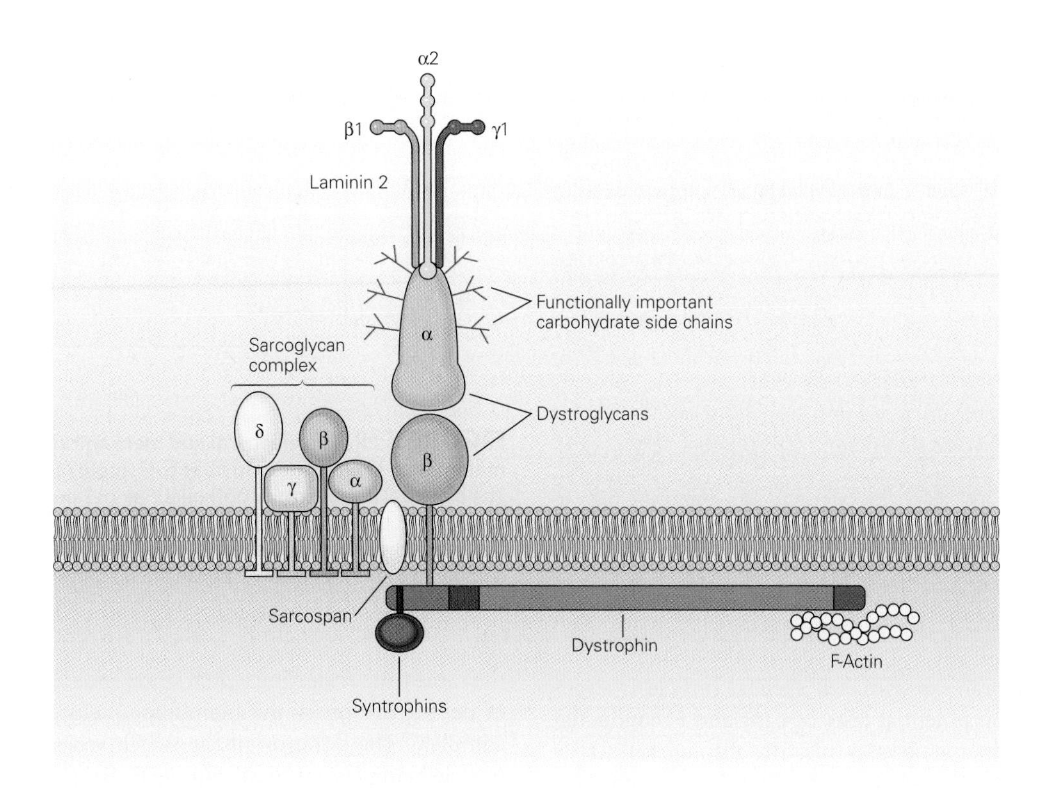

FIGURE 5–4 The dystrophin–glycoprotein complex. Dystrophin connects F-actin to the two members of the dystroglycan (DG) complex, α and β-dystroglycan, and these in turn connect to the merosin subunit of laminin 211 in the extracellular matrix. The sarcoglycan complex of four glycoproteins, α-, β-, γ-, and γ-sarcoglycan, sarcospan, and syntropins are all associated with the dystroglycan complex. There are muscle disorders associated with loss, abnormalities, or both of the sarcoglycans and merosin. (This diagram was adapted from diagrams by Justin Fallon and Kevin Campbell.)

T system of transverse tubules, which is continuous with the sarcolemma of the muscle fiber, forms a grid perforated by the individual muscle fibrils (Figure 5–1). The space between the two layers of the T system is an extension of the extracellular space. The sarcoplasmic reticulum, which forms an irregular curtain around each of the fibrils, has enlarged **terminal cisterns** in close contact with the T system at the junctions between the A and I bands. At these points of contact, the arrangement of the central T system with a cistern of the sarcoplasmic reticulum on either side has led to the use of the term **triads** to describe the system. The T system, which is continuous with the sarcolemma, provides a path for the rapid transmission of the action potential from the cell membrane to all the fibrils in the muscle. The sarcoplasmic reticulum is an important store of Ca^{2+} and also participates in muscle metabolism.

DYSTROPHIN–GLYCOPROTEIN COMPLEX

The large **dystrophin** protein (molecular mass 427,000 Da) forms a rod that connects the thin actin filaments to the transmembrane protein **β-dystroglycan** in the sarcolemma by smaller proteins in the cytoplasm, **syntrophins.** β-dystroglycan is connected to **merosin** (merosin refers to laminins that contain the α2 subunit in their trimeric makeup) in the extracellular matrix by **α-dystroglycan** (Figure 5–4). The dystroglycans are in turn associated with a complex of four transmembrane glycoproteins: α-, β-, γ-, and δ-**sarcoglycan.** This **dystrophin–glycoprotein complex** adds strength to the muscle by providing a scaffolding for the fibrils and connecting them to the extracellular environment. Disruption of these important structural features can result in several different muscular dystrophies (see Clinical Box 5–1).

ELECTRICAL PHENOMENA & IONIC FLUXES

ELECTRICAL CHARACTERISTICS OF SKELETAL MUSCLE

The electrical events in skeletal muscle and the ionic fluxes that underlie them share distinct similarities to those in nerve, with quantitative differences in timing and magnitude.

TABLE 5–1 Steady-state distribution of ions in the intracellular and extracellular compartments of mammalian skeletal muscle, and the equilibrium potentials for these ions.

	Concentration (mmol/L)		
Ion[a]	Intracellular Fluid	Extracellular Fluid	Equilibrium Potential (mV)
Na^+	12	145	+65
K^+	155	4	−95
H^+	13×10^{-5}	3.8×10^{-5}	−32
Cl^-	3.8	120	−90
HCO_3^-	8	27	−32
A^-	155	0	...
Membrane potential = −90 mV			

[a]A^- represents organic anions. The value for intracellular Cl^- is calculated from the membrane potential, using the Nernst equation.

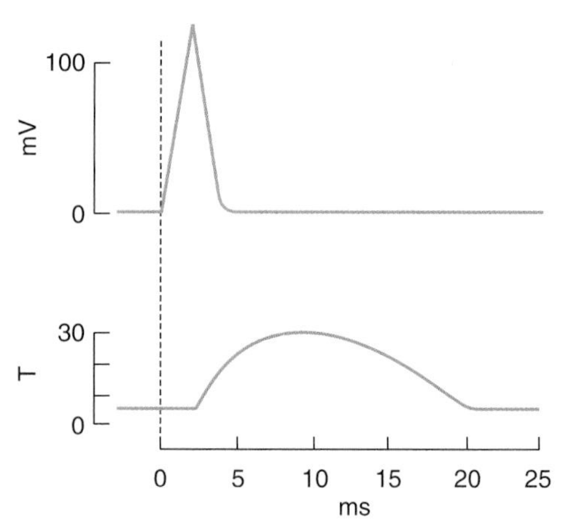

FIGURE 5–5 The electrical and mechanical responses of a mammalian skeletal muscle fiber to a single maximal stimulus. The electrical response (mV potential change) and the mechanical response (T, tension in arbitrary units) are plotted on the same abscissa (time). The mechanical response is relatively long-lived compared to the electrical response that initiates contraction.

The resting membrane potential of skeletal muscle is about −90 mV. The action potential lasts 2–4 ms and is conducted along the muscle fiber at about 5 m/s. The absolute refractory period is 1–3 ms long, and the after-polarizations, with their related changes in threshold to electrical stimulation, are relatively prolonged. The initiation of impulses at the myoneural junction is discussed in the next chapter.

ION DISTRIBUTION & FLUXES

The distribution of ions across the muscle fiber membrane is similar to that across the nerve cell membrane. Approximate values for the various ions and their equilibrium potentials are shown in Table 5–1. As in nerves, depolarization is largely a manifestation of Na^+ influx, and repolarization is largely a manifestation of K^+ efflux.

CONTRACTILE RESPONSES

It is important to distinguish between the electrical and mechanical events in skeletal muscle. Although one response does not normally occur without the other, their physiologic bases and characteristics are different. Muscle fiber membrane depolarization normally starts at the motor end plate, the specialized structure under the motor nerve ending. The action potential is transmitted along the muscle fiber and initiates the contractile response.

THE MUSCLE TWITCH

A single action potential causes a brief contraction followed by relaxation. This response is called a **muscle twitch.** In Figure 5–5, the action potential and the twitch are plotted on the same time scale. The twitch starts about 2 ms after the start of depolarization of the membrane, before repolarization is complete. The duration of the twitch varies with the type of muscle being tested. "Fast" muscle fibers, primarily those concerned with fine, rapid, precise movement, have twitch durations as short as 7.5 ms. "Slow" muscle fibers, principally those involved in strong, gross, sustained movements, have twitch durations up to 100 ms.

MOLECULAR BASIS OF CONTRACTION

The process by which the contraction of muscle is brought about is a sliding of the thin filaments over the thick filaments. Note that this shortening is not due to changes in the actual lengths of the thick and thin filaments, rather, by their increased overlap within the muscle cell. The width of the A bands is constant, whereas the Z lines move closer together when the muscle contracts and farther apart when it relaxes (Figure 5–3).

The sliding during muscle contraction occurs when the myosin heads bind firmly to actin, bend at the junction of the head with the neck, and then detach. This "power stroke" depends on the simultaneous hydrolysis of ATP. Myosin-II molecules are dimers that have two heads, but only one attaches to actin at any given time. The probable sequence of events of the power stroke is outlined in Figure 5–6. In resting muscle, troponin I is bound to actin and tropomyosin and covers the sites where myosin heads interact with actin. Also at rest, the myosin head contains tightly bound ADP. Following an action potential, cytosolic Ca^{2+} is increased and free Ca^{2+} binds to troponin C. This binding results in a weakening of the troponin I interaction with actin and exposes the actin binding site for myosin to allow for formation of myosin/actin cross-bridges.

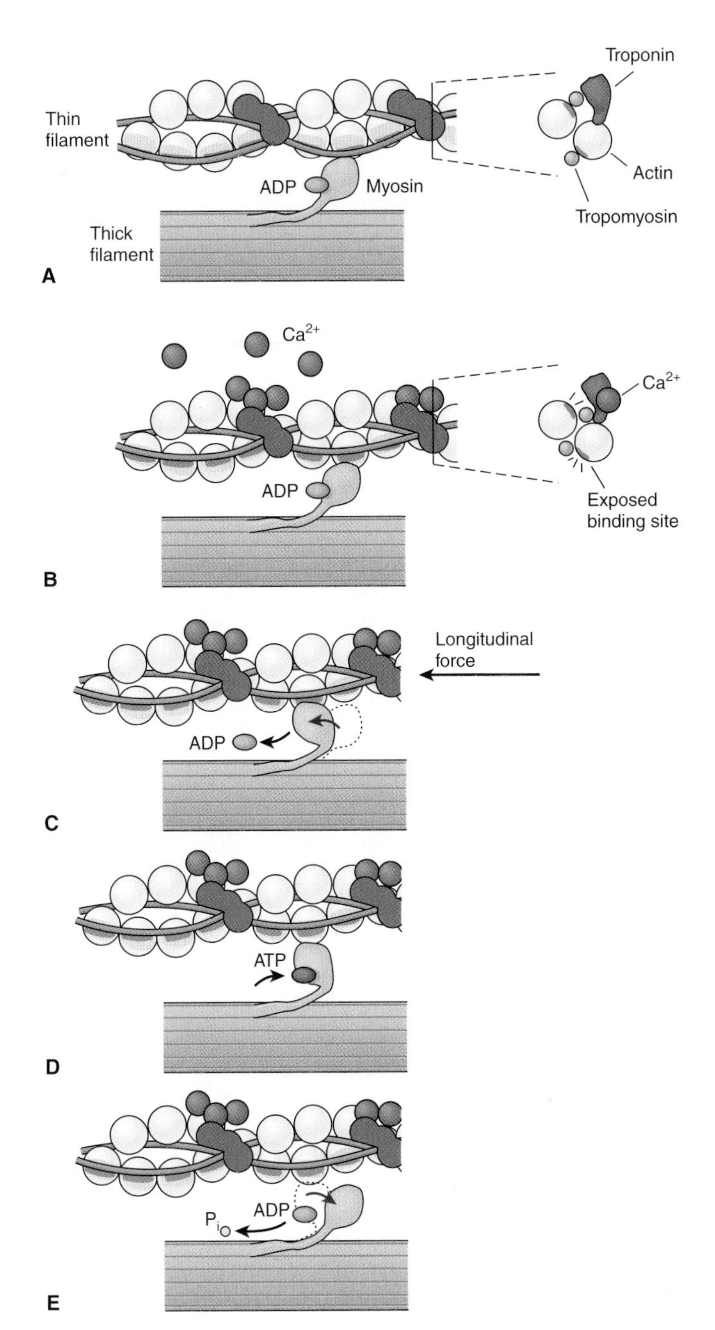

FIGURE 5–6 Power stroke of myosin in skeletal muscle. A) At rest, myosin heads are bound to adenosine diphosphate and are said to be in a "cocked" position in relation to the thin filament, which does not have Ca^{2+} bound to the troponin–tropomyosin complex. **B)** Ca^{2+} bound to the troponin–tropomyosin complex induces a conformational change in the thin filament that allows for myosin heads to cross-bridge with thin filament actin. **C)** Myosin heads rotate, move the attached actin and shorten the muscle fiber, forming the power stroke. **D)** At the end of the power stroke, ATP binds to a now exposed site, and causes a detachment from the actin filament. **E)** ATP is hydrolyzed into ADP and inorganic phosphate (P_i) and this chemical energy is used to "re-cock" the myosin head. (Based on Huxley AF, Simmons RM: Proposed mechanism of force generation in striated muscle. Nature Oct 22;233(5321):533–538, 1971 and Squire JM: Molecular mechanisms in muscular contraction. Trends Neurosci 6:409–413, 1093.)

Upon formation of the cross-bridge, ADP is released, causing a conformational change in the myosin head that moves the thin filament relative to the thick filament, comprising the cross-bridge "power stroke." ATP quickly binds to the free site on the myosin, which leads to a detachment of the myosin head from the thin filament. ATP is hydrolyzed and inorganic phosphate (P_i) released, causing a "re-cocking" of the myosin head and completing the cycle. As long as Ca^{2+} remains elevated and sufficient ATP is available, this cycle repeats. Many myosin heads cycle at or near the same time, and they cycle repeatedly, producing gross muscle contraction. Each power stroke shortens the sarcomere about 10 nm. Each thick filament has about 500 myosin heads, and each head cycles about five times per second during a rapid contraction.

The process by which depolarization of the muscle fiber initiates contraction is called **excitation–contraction coupling.** The action potential is transmitted to all the fibrils in the fiber via the T system (Figure 5–7). It triggers the release of Ca^{2+} from the terminal cisterns, the lateral sacs of the sarcoplasmic reticulum next to the T system. Depolarization of the T tubule membrane activates the sarcoplasmic reticulum via **dihydropyridine receptors (DHPR),** named for the drug dihydropyridine, which blocks them (Figure 5–8). DHPR are voltage-gated Ca^{2+} channels in the T tubule membrane. In cardiac muscle, influx of Ca^{2+} via these channels triggers the release of Ca^{2+} stored in the sarcoplasmic reticulum (calcium-induced calcium release) by activating the **ryanodine receptor (RyR).** The RyR is named after the plant alkaloid ryanodine that was used in its discovery. The RyR is a ligand-gated Ca^{2+} channel with Ca^{2+} as its natural ligand. In skeletal muscle, Ca^{2+} entry from the extracellular fluid (ECF) by this route is not required for Ca^{2+} release. Instead, the DHPR that serves as the voltage sensor unlocks release of Ca^{2+} from the nearby sarcoplasmic reticulum via physical interaction with the RyR. The released Ca^{2+} is quickly amplified through calcium-induced calcium release. Ca^{2+} is reduced in the muscle cell by the sarcoplasmic or endoplasmic reticulum Ca^{2+} ATPase (SERCA). The SERCA pump uses energy from ATP hydrolysis to remove Ca^{2+} from the cytosol back into the terminal cisterns, where it is stored until released by the next action potential. Once the Ca^{2+} concentration outside the reticulum has been lowered sufficiently, chemical interaction between myosin and actin ceases and the muscle relaxes. Note that ATP provides the energy for both contraction (at the myosin head) and relaxation (via SERCA). If transport of Ca^{2+} into the reticulum is inhibited, relaxation does not occur even though there are no more action potentials; the resulting sustained contraction is called a **contracture.** Alterations in the excitable response in muscle underscore many different pathologies (Clinical Box 5–2).

TYPES OF CONTRACTION

Muscular contraction involves shortening of the contractile elements, but because muscles have elastic and viscous elements in series with the contractile mechanism, it is possible

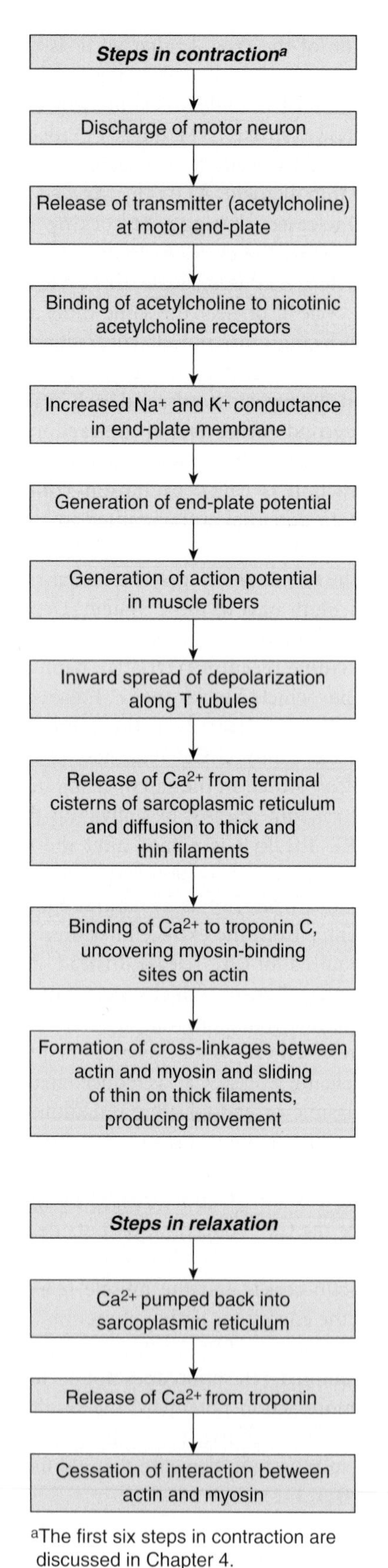

FIGURE 5–7 Flow of information that leads to muscle contraction.

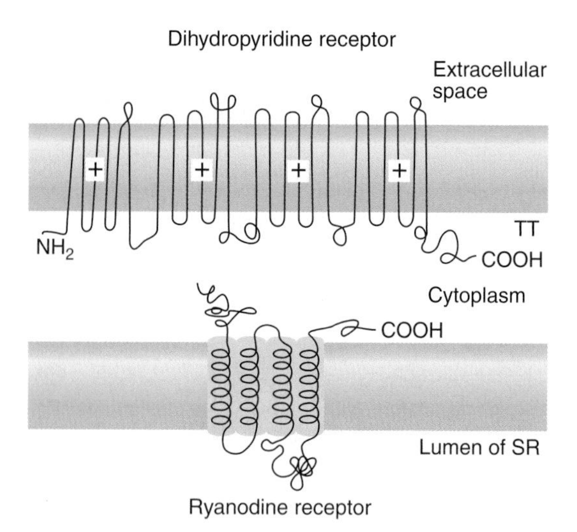

FIGURE 5–8 Relation of the T tubule (TT) to the sarcoplasmic reticulum in Ca^{2+} transport. In skeletal muscle, the voltage-gated dihydropyridine receptor in the T tubule triggers Ca^{2+} release from the sarcoplasmic reticulum (SR) via the ryanodine receptor (RyR). Upon sensing a voltage change, there is a physical interaction between the sarcolemmal-bound DHPR and the SR-bound RyR. This interaction gates the RyR and allows for Ca^{2+} release from the SR.

for contraction to occur without an appreciable decrease in the length of the whole muscle (Figure 5–9). Such a contraction is called **isometric** ("same measure" or length). Contraction against a constant load with a decrease in muscle length is **isotonic** ("same tension"). Note that because work is the product of force times distance, isotonic contractions do work, whereas isometric contractions do not. In other situations, muscle can do negative work while lengthening against a constant weight.

SUMMATION OF CONTRACTIONS

The electrical response of a muscle fiber to repeated stimulation is like that of nerve. The fiber is electrically refractory only during the rising phase and part of the falling phase of the spike potential. At this time, the contraction initiated by the first stimulus is just beginning. However, because the contractile mechanism does not have a refractory period, repeated stimulation before relaxation has occurred produces additional activation of the contractile elements and a response that is added to the contraction already present. This phenomenon is known as **summation of contractions.** The tension developed during summation is considerably greater than that during the single muscle twitch. With rapidly repeated stimulation, activation of the contractile mechanism occurs repeatedly before any relaxation has occurred, and the individual responses fuse into one continuous contraction. Such a response is called a **tetanus (tetanic contraction).** It is a **complete tetanus** when no relaxation occurs between stimuli and an **incomplete tetanus** when periods of incomplete relaxation take place between the summated stimuli. During a complete tetanus, the tension developed is about four times that developed by the individual

CLINICAL BOX 5–2

Muscle Channelopathies

Channelopathies are diseases that have as their underlying feature mutations or dysregulation of ion channels. Such diseases are frequently associated with excitable cells, including muscle. In the various forms of clinical **myotonia**, muscle relaxation is prolonged after voluntary contraction. The molecular bases of myotonias are due to dysfunction of channels that shape the action potential. Myotonia dystrophy is caused by an autosomal dominant mutation that leads to over-expression of a K^+ channel (although the mutation is *not* at the K^+ channel). A variety of myotonias are associated with mutations in Na^+ channels (eg, hyperkalemic periodic paralysis, paramyotonia congenita, or Na^+ channel congenita) or Cl^- channels (eg, dominant or recessive myotonia congenita). **Myasthenia,** defined as abnormal muscle weakness or disease, can also be related to loss of ion channel function in the muscle. In **congenital myasthenia,** the patient has an inheritable disorder of one of a group of ion channels necessary for the transmission of neuronal signaling to muscle response. Mutations in Ca^{2+} channels that allow for neuronal transmitter release or in the acetylcholine receptor nonspecific cation channels, important in recognition of neuronal transmitters, have both been shown to cause congenital myasthenia. Alterations of channel functions can also occur via autoimmune disease, such as that observed in **myasthenia gravis.** In this disease, antibodies to the nicotinic acetylcholine receptor can reduce its functional presence at the muscle membrane by up to 80%, and thus limit muscle response to neuronal transmitter release.

Channelopathies can also occur in the Ca^{2+} release channels in muscle (ryanodine receptors) that amplify the Ca^{2+} response within the cell. Such mutations can cause **malignant hyperthermia.** Patients with this conditions display normal muscle function under normal conditions. However, certain anesthetic agents, or in rare cases exposure to high environmental heat or strenuous exercise, can trigger abnormal release of Ca^{2+} from the sarcoplasmic reticulum in the muscle cell, resulting in sustained muscle contraction and heat production. In severe cases, fatality can occur.

THERAPEUTIC HIGHLIGHTS

Although the symptoms associated with each individual channelopathy may be similar, treatments for the individual diseases include a wide variety of drugs that are targeted to the individual ion channel (or proteins associated with ion channel) defect. Appropriate drug therapy helps to improve symptoms and maintain acceptable muscle function. Further interventions related to individual diseases are to avoid muscle movements that exacerbate the disease.

twitch contractions. The development of an incomplete and a complete tetanus in response to stimuli of increasing frequency is shown in Figure 5–10.

The stimulation frequency at which summation of contractions occurs is determined by the twitch duration of the particular muscle being studied. For example, if the twitch duration is 10 ms, frequencies less than 1/10 ms (100/s) cause discrete responses interrupted by complete relaxation, and frequencies greater than 100/s cause summation.

RELATION BETWEEN MUSCLE LENGTH, TENSION & VELOCITY OF CONTRACTION

Both the tension that a muscle develops when stimulated to contract isometrically (the **total tension**) and the **passive tension** exerted by the unstimulated muscle vary with the length of the muscle fiber. This relationship can be studied in a whole skeletal muscle preparation such as that shown in Figure 5–9. The length of the muscle can be varied by changing the distance between its two attachments. At each length, the passive tension is measured, the muscle is then stimulated electrically, and the total tension is measured. The difference between the two values at any length is the amount of tension actually generated by the contractile process, the **active tension.** The records obtained by plotting passive tension and total tension against muscle length are shown in Figure 5–11. Similar curves are obtained when single muscle fibers are studied. The length of the muscle at which the active tension is maximal is usually called its **resting length.** The term comes originally from experiments demonstrating that the length of many of the muscles in the body at rest is the length at which they develop maximal tension.

The observed length–tension relation in skeletal muscle can be explained by the sliding filament mechanism of muscle contraction. When the muscle fiber contracts isometrically, the tension developed is proportional to the number of cross-bridges between the actin and the myosin molecules. When muscle is stretched, the overlap between actin and myosin is reduced and the number of cross-linkages is therefore reduced. Conversely, when the muscle is appreciably shorter than resting length, the distance the thin filaments can move is reduced.

The velocity of muscle contraction varies inversely with the load on the muscle. At a given load, the velocity is maximal at the resting length and declines if the muscle is shorter or longer than this length.

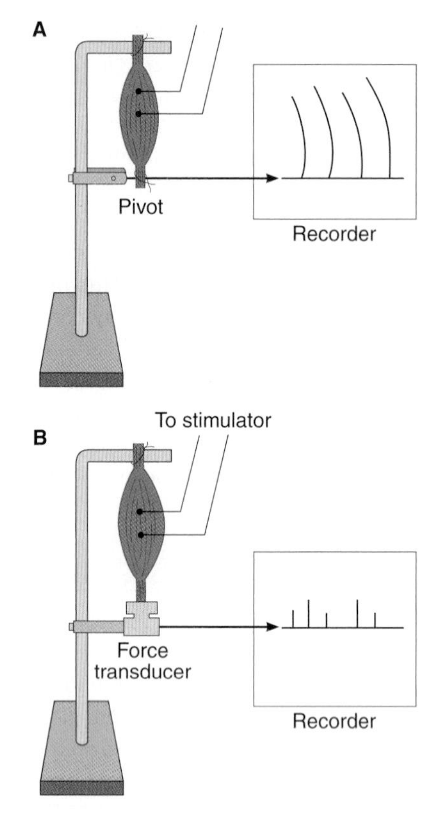

FIGURE 5–9 **A)** Muscle preparation arranged for recording isotonic contractions. **B)** Preparation arranged for recording isometric contractions. In **A,** the muscle is fastened to a writing lever that swings on a pivot. In **B,** it is attached to an electronic transducer that measures the force generated without permitting the muscle to shorten.

FIBER TYPES

Although skeletal muscle fibers resemble one another in a general way, skeletal muscle is a heterogeneous tissue made up of fibers that vary in myosin ATPase activity, contractile speed, and other properties. Muscles are frequently classified into two types, "slow" and "fast." These muscles can contain a mixture of three fiber types: type I (or SO for slow-oxidative); type IIA (FOG for fast-oxidative-glycolytic); or type IIB (FG for fast glycolytic). Some of the properties associated with type I, type IIA, and type IIB fibers are summarized in Table 5–2. Although these classification schemes are valid for muscles across many mammalian species, there are significant variations of fibers within and between muscles. For example, type I fibers in a given muscle can be larger than type IIA fibers from a different muscle in the same animal. Many of the differences in the fibers that make up muscles stem from differences in the proteins within them. Most of these are encoded by multigene families. Ten different **isoforms** of the myosin heavy chains (MHCs) have been characterized. Each of the two types of light chains also have isoforms. It appears that there is only one form of actin, but multiple isoforms of tropomyosin and all three components of troponin.

ENERGY SOURCES & METABOLISM

Muscle contraction requires energy, and muscle has been called "a machine for converting chemical energy into mechanical work." The immediate source of this energy is ATP, and this is formed by the metabolism of carbohydrates and lipids.

PHOSPHORYLCREATINE

ATP is resynthesized from ADP by the addition of a phosphate group. Some of the energy for this endothermic reaction is supplied by the breakdown of glucose to CO_2 and H_2O, but there also exists in muscle another energy-rich phosphate compound that can supply this energy for short periods. This compound is **phosphorylcreatine,** which is hydrolyzed to creatine and phosphate groups with the release of considerable energy (Figure 5–12). At rest, some ATP in the mitochondria transfers its phosphate to creatine, so that a phosphorylcreatine store is built up. During exercise, the phosphorylcreatine is hydrolyzed at the junction between the myosin heads and actin, forming ATP from ADP and thus permitting contraction to continue.

CARBOHYDRATE & LIPID BREAKDOWN

At rest and during light exercise, muscles utilize lipids in the form of free fatty acids as their energy source. As the intensity of exercise increases, lipids alone cannot supply energy fast enough and so use of carbohydrate becomes the predominant component in the muscle fuel mixture. Thus, during exercise,

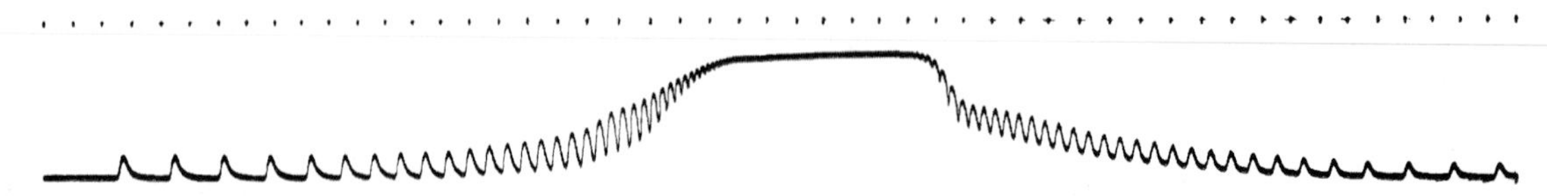

FIGURE 5–10 **Tetanus.** Isometric tension of a single muscle fiber during continuously increasing and decreasing stimulation frequency. Dots at the top are at intervals of 0.2 s. Note the development of incomplete and then complete tetanus as stimulation is increased, and the return of incomplete tetanus, then full response, as stimulation frequency is decreased.

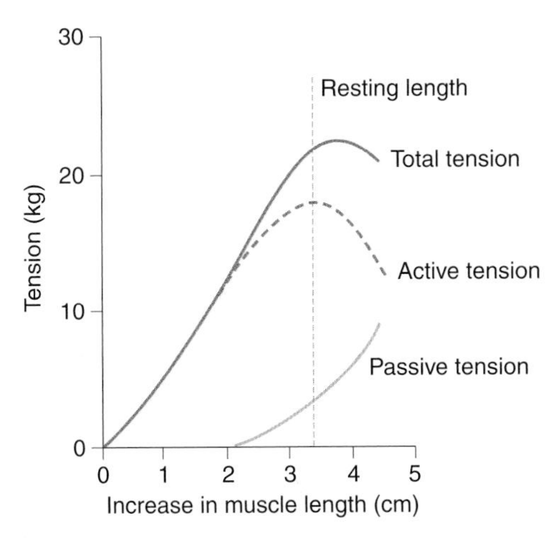

FIGURE 5–11 Length-tension relationship for the human triceps muscle. The passive tension curve measures the tension exerted by this skeletal muscle at each length when it is not stimulated. The total tension curve represents the tension developed when the muscle contracts isometrically in response to a maximal stimulus. The active tension is the difference between the two.

much of the energy for phosphorylcreatine and ATP resynthesis comes from the breakdown of glucose to CO_2 and H_2O. Glucose in the bloodstream enters cells, where it is degraded through a series of chemical reactions to pyruvate. Another source of intracellular glucose, and consequently of pyruvate, is glycogen, the carbohydrate polymer that is especially abundant in liver and skeletal muscle. When adequate O_2 is

TABLE 5–2 Classification of fiber types in skeletal muscles.

	Type 1	Type IIA	Type IIB
Other names	Slow, Oxidative (SO)	Fast, Oxidative, Glycolytic (FOG)	Fast, Glycolytic (FG)
Color	Red	Red	White
Myosin ATPase activity	Slow	Fast	Fast
Ca^{2+}-pumping capacity of sarcoplasmic reticulum	Moderate	High	High
Diameter	Small	Large	Large
Glycolytic capacity	Moderate	High	High
Oxidative capacity	High	Moderate	Low
Associated Motor Unit Type	Slow (S)	Fast Resistant to Fatigue (FR)	Fast Fatigable (FF)
Membrane potential = −90 mV			
Oxidative capacity	High	Moderate	Low

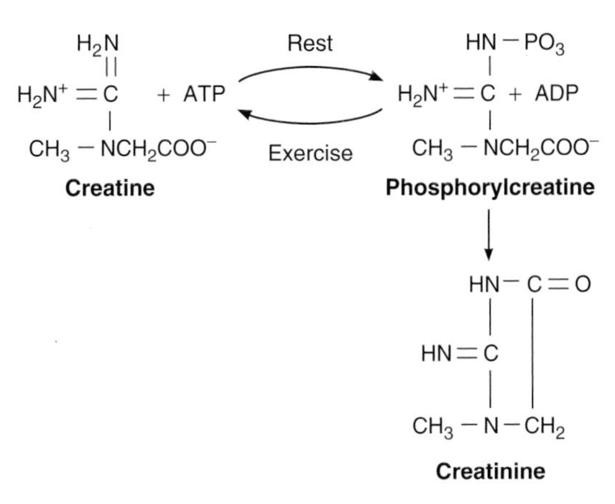

FIGURE 5–12 Creatine, phosphorylcreatine, and creatinine cycling in muscle. During periods of high activity, cycling of phosphorylcreatine allows for quick release of ATP to sustain muscle activity.

present, pyruvate enters the citric acid cycle and is metabolized—through this cycle and the so-called respiratory enzyme pathway—to CO_2 and H_2O. This process is called **aerobic glycolysis.** The metabolism of glucose or glycogen to CO_2 and H_2O forms large quantities of ATP from ADP. If O_2 supplies are insufficient, the pyruvate formed from glucose does not enter the tricarboxylic acid cycle but is reduced to lactate. This process of **anaerobic glycolysis** is associated with the net production of much smaller quantities of energy-rich phosphate bonds, but it does not require the presence of O_2. A brief overview of the various reactions involved in supplying energy to skeletal muscle is shown in Figure 5–13.

THE OXYGEN DEBT MECHANISM

During exercise, the muscle blood vessels dilate and blood flow is increased so that the available O_2 supply is increased. Up to a point, the increase in O_2 consumption is proportional to the

$$ATP + H_2O \rightarrow ADP + H_3PO_4 + 7.3\text{ kcal}$$

$$\text{Phosphorylcreatine} + ADP \rightleftharpoons \text{Creatine} + ATP$$

$$\text{Glucose} + 2\,ATP\ (\text{or glycogen} + 1\,ATP) \xrightarrow{\text{Anaerobic}} 2\ \text{Lactic acid} + 4\,ATP$$

$$\text{Glucose} + 2\,ATP\ (\text{or glycogen} + 1\,ATP) \xrightarrow{\text{Oxygen}} 6\,CO_2 + 6\,H_2O + 40\,ATP$$

$$FFA \xrightarrow{\text{Oxygen}} CO_2 + H_2O + ATP$$

FIGURE 5–13 ATP turnover in muscle cells. Energy released by hydrolysis of 1 mol of ATP and reactions responsible for resynthesis of ATP. The amount of ATP formed per mole of free fatty acid (FFA) oxidized is large but varies with the size of the FFA. For example, complete oxidation of 1 mol of palmitic acid generates 140 mol of ATP.

energy expended, and all the energy needs are met by aerobic processes. However, when muscular exertion is very great, aerobic resynthesis of energy stores cannot keep pace with their utilization. Under these conditions, phosphorylcreatine is still used to resynthesize ATP. In addition, some ATP synthesis is accomplished by using the energy released by the anaerobic breakdown of glucose to lactate. Use of the anaerobic pathway is self-limiting because in spite of rapid diffusion of lactate into the bloodstream, enough accumulates in the muscles to eventually exceed the capacity of the tissue buffers and produce an enzyme-inhibiting decline in pH. However, for short periods, the presence of an anaerobic pathway for glucose breakdown permits muscular exertion of a far greater magnitude than would be possible without it. For example, in a 100-m dash that takes 10 s, 85% of the energy consumed is derived anaerobically; in a 2-mile race that takes 10 min, 20% of the energy is derived anaerobically; and in a long-distance race that takes 60 min, only 5% of the energy comes from anaerobic metabolism.

After a period of exertion is over, extra O_2 is consumed to remove the excess lactate, replenish the ATP and phosphorylcreatine stores, and replace the small amounts of O_2 that were released by myoglobin. Without replenishment of ATP, muscles enter a state of rigor (Clinical Box 5–3). The amount of extra O_2 consumed is proportional to the extent to which the energy demands during exertion exceeded the capacity for the aerobic synthesis of energy stores, that is, the extent to which an **oxygen debt** was incurred. The O_2 debt is measured experimentally by determining O_2 consumption after exercise until a constant, basal consumption is reached and subtracting the basal consumption from the total. The amount of this debt may be six times the basal O_2 consumption, which indicates that the subject is capable of six times the exertion that would have been possible without it.

HEAT PRODUCTION IN MUSCLE

Thermodynamically, the energy supplied to a muscle must equal its energy output. The energy output appears in work done by the muscle, in energy-rich phosphate bonds formed for later use, and in heat. The overall mechanical efficiency of skeletal muscle (work done/total energy expenditure) ranges up to 50% while lifting a weight during isotonic contraction and is essentially 0% during isometric contraction. Energy storage in phosphate bonds is a small factor. Consequently, heat production is considerable. The heat produced in muscle can be measured accurately with suitable thermocouples.

Resting heat, the heat given off at rest, is the external manifestation of basal metabolic processes. The heat produced in excess of resting heat during contraction is called the **initial heat.** This is made up of **activation heat,** the heat that muscle produces whenever it is contracting, and **shortening heat,** which is proportional in amount to the distance the muscle shortens. Shortening heat is apparently due to some change in the structure of the muscle during shortening.

Following contraction, heat production in excess of resting heat continues for as long as 30 min. This **recovery heat** is the heat liberated by the metabolic processes that restore the muscle to its precontraction state. The recovery heat of muscle is approximately equal to the initial heat; that is, the heat produced during recovery is equal to the heat produced during contraction.

If a muscle that has contracted isotonically is restored to its previous length, extra heat in addition to recovery heat is produced **(relaxation heat).** External work must be done on the muscle to return it to its previous length, and relaxation heat is mainly a manifestation of this work.

CLINICAL BOX 5–3

Muscle Rigor

When muscle fibers are completely depleted of ATP and phosphorylcreatine, they develop a state of rigidity called rigor. When this occurs after death, the condition is called **rigor mortis.** In rigor, almost all of the myosin heads attach to actin but in an abnormal, fixed, and resistant way. The muscles effectively are locked into place and become quite stiff to the touch.

PROPERTIES OF SKELETAL MUSCLES IN THE INTACT ORGANISM

THE MOTOR UNIT

Innervation of muscle fibers is critical to muscle function (Clinical Box 5–4). Because the axons of the spinal motor neurons supplying skeletal muscle each branch to innervate

CLINICAL BOX 5–4

Denervation of Muscle

In the intact animal healthy skeletal muscle does not contract except in response to stimulation of its motor nerve supply. Destruction of this nerve supply causes muscle atrophy. It also leads to abnormal excitability of the muscle and increases its sensitivity to circulating acetylcholine (**denervation hypersensitivity;** see Chapter 6). Fine, irregular contraction of individual fibers (**fibrillations**) appears. This is the classic picture of a **lower motor neuron lesion.** If the motor nerve regenerates, the fibrillations disappear. Usually, the contractions are not visible grossly, and they should not be confused with **fasciculations,** which are jerky, visible contractions of groups of muscle fibers that occur as a result of pathologic discharge of spinal motor neurons.

several muscle fibers, the smallest possible amount of muscle that can contract in response to the excitation of a single motor neuron is not one muscle fiber but all the fibers supplied by the neuron. Each single motor neuron and the muscle fibers it innervates constitute a **motor unit.** The number of muscle fibers in a motor unit varies. In muscles such as those of the hand and those concerned with motion of the eye (ie, muscles concerned with fine, graded, precise movement), each motor unit innervates very few (on the order of three to six) muscle fibers. On the other hand, values of 600 muscle fibers per motor unit can occur in human leg muscles. The group of muscle fibers that contribute to a motor unit can be intermixed within a muscle. That is, although they contract as a unit, they are not necessarily "neighboring" fibers within the muscle.

Each spinal motor neuron innervates only one kind of muscle fiber, so that all the muscle fibers in a motor unit are of the same type. On the basis of the type of muscle fiber they innervate, and thus on the basis of the duration of their twitch contraction, motor units are divided into S (slow), FR (fast, resistant to fatigue), and FF (fast, fatigable) units. Interestingly, there is also a gradation of innervation of these fibers, with S fibers tending to have a low innervation ratio (ie, small units) and FF fibers tending to have a high innervation ratio (ie, large units). The recruitment of motor units during muscle contraction is not random; rather it follows a general scheme, the **size principle.** In general, a specific muscle action is developed first by the recruitment of S muscle units that contract relatively slowly to produce controlled contraction. Next, FR muscle units are recruited, resulting in more powerful response over a shorter period of time. Lastly, FF muscle units are recruited for the most demanding tasks. For example, in muscles of the leg, the small, slow units are first recruited for standing. As walking motion is initiated, their recruitment of FR units increases. As this motion turns to running or jumping, the FF units are recruited. Of course, there is overlap in recruitment, but, in general, this principle holds true.

The differences between types of muscle units are not inherent but are determined by, among other things, their activity. When the nerve to a slow muscle is cut and the nerve to a fast muscle is spliced to the cut end, the fast nerve grows and innervates the previously slow muscle. However, the muscle becomes fast and corresponding changes take place in its muscle protein isoforms and myosin ATPase activity. This change is due to changes in the pattern of activity of the muscle; in stimulation experiments, changes in the expression of MHC genes and consequently of MHC isoforms can be produced by changes in the pattern of electrical activity used to stimulate the muscle. More commonly, muscle fibers can be altered by a change in activity initiated through exercise (or lack thereof). Increased activity can lead to muscle cell hypertrophy, which allows for increase in contractile strength. Type IIA and IIB fibers are most susceptible to these changes. Alternatively, inactivity can lead to muscle cell atrophy and a loss of contractile strength. Type I fibers—that is, the ones used most often—are most susceptible to these changes.

ELECTROMYOGRAPHY

Activation of motor units can be studied by electromyography, the process of recording the electrical activity of muscle. This may be done in unanaesthetized humans by using small metal disks on the skin overlying the muscle as the pick-up electrodes or by using needle or fine wire electrodes inserted into the muscle. The record obtained with such electrodes is the **electromyogram (EMG).** With needle or fine wire electrodes, it is usually possible to pick up the activity of single muscle fibers. The measured EMG depicts the potential difference between the two electrodes, which is altered by the activation of muscles in between the electrodes. A typical EMG is shown in Figure 5–14.

It has been shown by electromyography that little if any spontaneous activity occurs in the skeletal muscles of normal individuals at rest. With minimal voluntary activity a few motor units discharge, and with increasing voluntary effort, more and more are brought into play to monitor the **recruitment of motor units.** Gradation of muscle response is therefore in part a function of the number of motor units activated. In addition, the frequency of discharge in the individual nerve fibers plays a role, the tension developed during a tetanic contraction being greater than that during individual twitches. The length of the muscle is also a factor. Finally, the motor units fire asynchronously, that is, out of phase with one another. This asynchronous firing causes the individual muscle fiber responses to merge into a smooth contraction of the whole muscle. In summary, EMGs can be used to quickly (and roughly) monitor abnormal electrical activity associated with muscle responses.

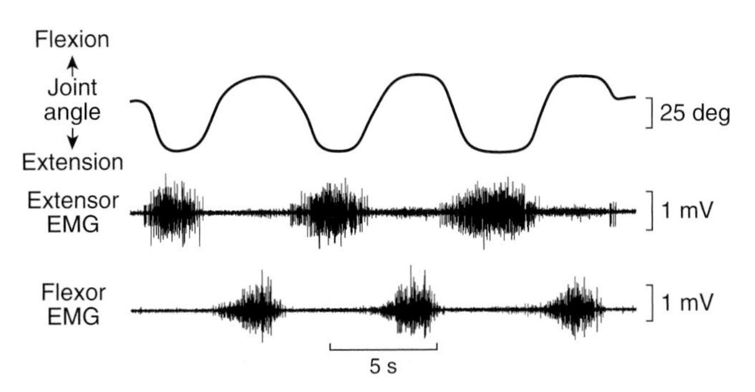

FIGURE 5–14 Relative joint angle and electromyographic tracings from human extensor pollicis longus and flexor pollicis longus during alternate flexion and extension of the distal joint of the thumb. The extensor pollicis longus and flexor pollicis longus extend and flex the distal joint of the thumb, respectively. The distal thumb joint angle (top) is superimposed over the extensor pollicis longus (middle) and flexor pollicus longus (bottom) EMGs. Note the alternate activation and rest patterns as one muscle is used for extension and the other for flexion. (Courtesy of Andrew J. Fuglevand.)

THE STRENGTH OF SKELETAL MUSCLES

Human skeletal muscle can exert 3–4 kg of tension per square centimeter of cross-sectional area. This figure is about the same as that obtained in a variety of experimental animals and seems to be constant for mammalian species. Because many of the muscles in humans have a relatively large cross-sectional area, the tension they can develop is quite large. The gastrocnemius, for example, not only supports the weight of the whole body during climbing but resists a force several times this great when the foot hits the ground during running or jumping. An even more striking example is the gluteus maximus, which can exert a tension of 1200 kg. The total tension that could be developed if all muscles in the body of an adult man pulled together is approximately 22,000 kg (nearly 25 tons).

BODY MECHANICS

Body movements are generally organized in such a way that they take maximal advantage of the physiologic principles outlined above. For example, the attachments of the muscles in the body are such that many of them are normally at or near their resting length when they start to contract. In muscles that extend over more than one joint, movement at one joint may compensate for movement at another in such a way that relatively little shortening of the muscle occurs during contraction. Nearly isometric contractions of this type permit development of maximal tension per contraction. The hamstring muscles extend from the pelvis over the hip joint and the knee joint to the tibia and fibula. Hamstring contraction produces flexion of the leg on the thigh. If the thigh is flexed on the pelvis at the same time, the lengthening of the hamstrings across the hip joint tends to compensate for the shortening across the knee joint. In the course of various activities, the body moves in a way that takes advantage of this. Such factors as momentum and balance are integrated into body movement in ways that make possible maximal motion with minimal muscular exertion. One net effect is that the stress put on tendons and bones rarely exceeds 50% of their failure strength, protecting them from damage.

In walking, each limb passes rhythmically through a support or stance phase when the foot is on the ground and a swing phase when the foot is off the ground. The support phases of the two legs overlap, so that two periods of double support occur during each cycle. There is a brief burst of activity in the leg flexors at the start of each step, and then the leg is swung forward with little more active muscular contraction. Therefore, the muscles are active for only a fraction of each step, and walking for long periods causes relatively little fatigue.

A young adult walking at a comfortable pace moves at a velocity of about 80 m/min and generates a power output of 150–175 W per step. A group of young adults asked to walk at their most comfortable rate selected a velocity close to 80 m/min, and it was found that they had selected the velocity at which their energy output was minimal. Walking more rapidly or more slowly took more energy.

CARDIAC MUSCLE MORPHOLOGY

The striations in cardiac muscle are similar to those in skeletal muscle, and Z lines are present. Large numbers of elongated mitochondria are in close contact with the muscle fibrils. The muscle fibers branch and interdigitate, but each is a complete unit surrounded by a cell membrane. Where the end of one muscle fiber abuts on another, the membranes of both fibers parallel each other through an extensive series of folds. These areas, which always occur at Z lines, are called **intercalated disks** (Figure 5–15). They provide a strong union between fibers, maintaining cell-to-cell cohesion, so that the pull of one contractile cell can be transmitted along its axis to the next. Along the sides of the muscle fibers next to the disks, the cell membranes of adjacent fibers fuse for considerable distances, forming gap junctions. These junctions provide low-resistance bridges for the spread of excitation from one fiber to another. They permit cardiac muscle to function as if it were a syncytium, even though no protoplasmic bridges are present between cells. The T system in cardiac muscle is located at the Z lines rather than at the A–I junction, where it is located in mammalian skeletal muscle.

ELECTRICAL PROPERTIES

RESTING MEMBRANE & ACTION POTENTIALS

The resting membrane potential of individual mammalian cardiac muscle cells is about –80 mV. Stimulation produces a propagated action potential that is responsible for initiating contraction. Although action potentials vary among the cardiomyocytes in different regions of the heart (discussed in Chapter 29), the action potential of a typical ventricular cardiomyocyte can be used as an example (Figure 5–16). Depolarization proceeds rapidly and an overshoot of the zero potential is present, as in skeletal muscle and nerve, but this is followed by a plateau before the membrane potential returns to the baseline. In mammalian hearts, depolarization lasts about 2 ms, but the plateau phase and repolarization last 200 ms or more. Repolarization is therefore not complete until the contraction is half over.

As in other excitable tissues, changes in the external K^+ concentration affect the resting membrane potential of cardiac muscle, whereas changes in the external Na^+ concentration affect the magnitude of the action potential. The initial rapid depolarization and the overshoot (phase 0) are due to opening of voltage-gated Na^+ channels similar to that occurring in nerve and skeletal muscle (Figure 5–17). The initial rapid repolarization (phase 1) is due to closure of Na^+ channels and opening of one type of K^+ channel. The subsequent prolonged plateau (phase 2) is due to a slower but prolonged opening of voltage-gated Ca^{2+} channels. Final repolarization (phase 3) to the resting membrane potential (phase 4) is due

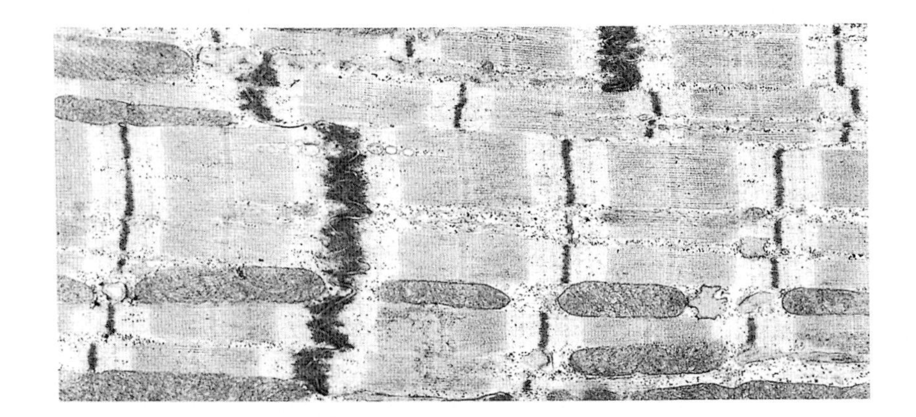

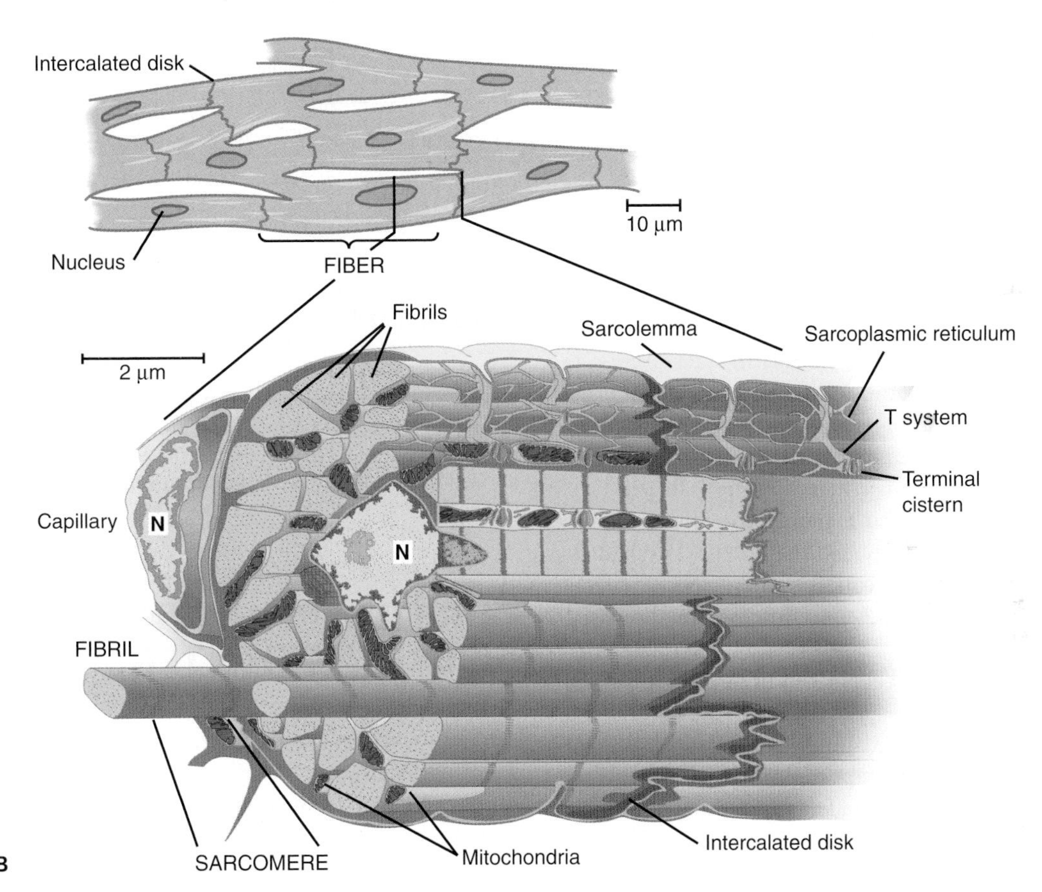

FIGURE 5–15 Cardiac muscle. A) Electronmicrograph of cardiac muscle. Note the similarity of the A-I regions seen in the skeletal muscle EM of Figure 3–2. The fuzzy thick lines are intercalated disks and function similarly to the Z-lines but occur at cell membranes (× 12,000). (Reproduced with permission from Bloom W, Fawcett DW: *A Textbook of Histology,* 10th ed. Saunders, 1975.) **B)** Artist interpretation of cardiac muscle as seen under the light microscope (top) and the electron microscope (bottom). Again, note the similarity to skeletal muscle structure. N, nucleus. (Reproduced with permission from Braunwald E, Ross J, Sonnenblick EH: Mechanisms of contraction of the normal and failing heart. *N Engl J Med* 1967;277:794.)

to closure of the Ca^{2+} channels and a slow, delayed increase of K^+ efflux through various types of K^+ channels. Cardiac myocytes contain at least two types of Ca^{2+} channels (T- and L-types), but the Ca^{2+} current is mostly due to opening of the slower L-type Ca^{2+} channels. Mutations or dysfunction in any of these channels lead to serious pathologies of the heart (eg, Clinical Box 5–5).

MECHANICAL PROPERTIES

CONTRACTILE RESPONSE

The contractile response of cardiac muscle begins just after the start of depolarization and lasts about 1.5 times as long as the action potential (Figure 5–16). The role of Ca^{2+} in excitation–

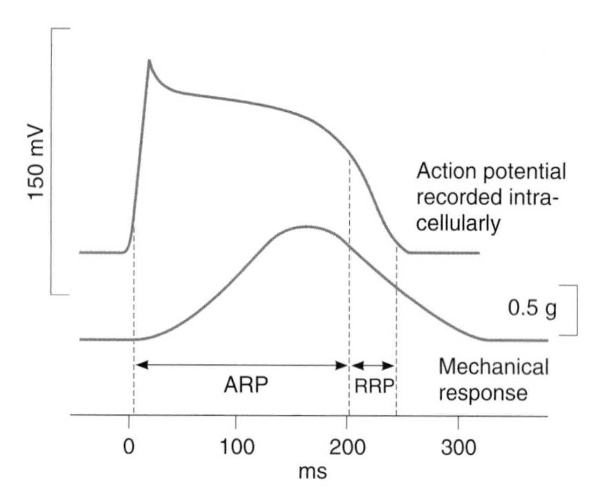

FIGURE 5–16 Comparison of action potentials and contractile response of a mammalian cardiac muscle fiber in a typical ventricular cell. In the top trace, the intracellular recording of the action potential shows the quick depolarization and extended recovery. In the bottom trace, the mechanical response is matched to the extracellular and intracellular electrical activities. Note that in the absolute refractory period (ARP), the cardiac myocyte cannot be excited, whereas in the relative refractory period (RRP) minimal excitation can occur.

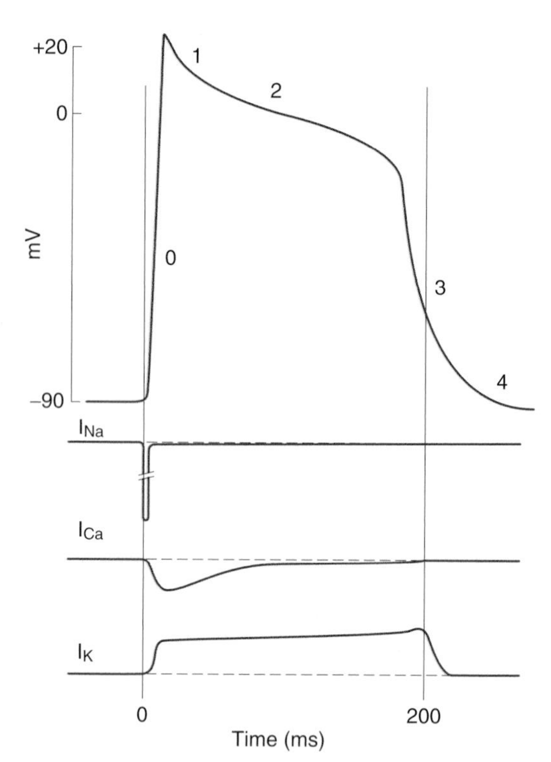

FIGURE 5–17 Dissection of the cardiac action potential. Top: The action potential of a cardiac muscle fiber can be broken down into several phases: 0, depolarization; 1, initial rapid repolarization; 2, plateau phase; 3, late rapid repolarization; 4, baseline. Bottom: Diagrammatic summary of Na^+, Ca^{2+}, and cumulative K+ currents during the action potential. As is convention, inward currents are downward, and outward currents are upward.

contraction coupling is similar to its role in skeletal muscle (see above). However, it is the influx of extracellular Ca^{2+} through the voltage-sensitive DHPR in the T system that triggers calcium-induced calcium release through the RyR at the sarcoplasmic reticulum. Because there is a net influx of Ca^{2+} during activation, there is also a more prominent role for plasma membrane Ca^{2+} ATPases and the Na^+/Ca^{2+} exchanger in recovery of intracellular Ca^{2+} concentrations. Specific effects of drugs that indirectly alter Ca^{2+} concentrations are discussed in Clinical Box 5–6.

During phases 0 to 2 and about half of phase 3 (until the membrane potential reaches approximately –50 mV during repolarization), cardiac muscle cannot be excited again; that is, it is in its **absolute refractory period.** It remains relatively refractory until phase 4. Therefore, tetanus of the type seen in skeletal muscle cannot occur. Of course, tetanization of cardiac muscle for any length of time would have lethal consequences, and in this sense, the fact that cardiac muscle cannot be tetanized is a safety feature.

ISOFORMS

Cardiac muscle is generally slow and has relatively low ATPase activity. Its fibers are dependent on oxidative metabolism and hence on a continuous supply of O_2. The human heart contains both the α and the β isoforms of the myosin heavy chain (α MHC and β MHC). β MHC has lower myosin ATPase activity than α MHC. Both are present in the atria, with the α isoform predominating, whereas the β isoform predominates in the ventricle. The spatial differences in expression contribute to the well-coordinated contraction of the heart.

CORRELATION BETWEEN MUSCLE FIBER LENGTH & TENSION

The relation between initial fiber length and total tension in cardiac muscle is similar to that in skeletal muscle; there is a resting length at which the tension developed on stimulation is maximal. In the body, the initial length of the fibers is determined by the degree of diastolic filling of the heart, and the pressure developed in the ventricle is proportional to the volume of the ventricle at the end of the filling phase **(Starling's law of the heart).** The developed tension (Figure 5–18) increases as the diastolic volume increases until it reaches a maximum, then tends to decrease. However, unlike skeletal muscle, the decrease in developed tension at high degrees of stretch is not due to a decrease in the number of cross-bridges between actin and myosin, because even severely dilated hearts are not stretched to this degree. The decrease is instead due to beginning disruption of the myocardial fibers.

The force of contraction of cardiac muscle can be also increased by catecholamines, and this increase occurs without a change in muscle length. This positive ionotropic effect of catecholamines is mediated via innervated β_1-adrenergic receptors, cyclic AMP, and their effects on Ca^{2+} homeostasis. The heart also contains noninnervated β_2-adrenergic receptors,

CLINICAL BOX 5–5

Long QT Syndrome

Long QT syndrome (LQTS) is defined as a prolongation of the QT interval observed on an electrocardiogram. LQTS can lead to irregular heartbeats and subsequent fainting, seizure, cardiac arrest, or even death. Although certain medications can lead to LQTS, it is more frequently associated with genetic mutations in a variety of cardiac-expressed ion channels. Mutations in cardiac-expressed voltage gated K^+ channel genes (KCNQ1 or KCNH2) account for most of the mutation-based cases of LQTS (~90%). Mutations in cardiac-expressed voltage-gated Na^+ channels (eg, SCN5A) or cardiac-expressed Ca^{2+} channels (eg, CACNA1C) have also been associated with the disease. The fact that mutations in diverse channels all can result in the prolongation of the QT interval and subsequent pathology underlies the intricate interplay of these channels in shaping the heart's electrical response.

THERAPEUTIC HIGHLIGHTS

Patients with long QT syndrome (LQTS) should avoid drugs that prolong the QT interval or reduce their serum K^+ or Mg^{2+} levels; any K^+ or Mg^{2+} deficiencies should be corrected. Drug interventions in asymptomatic patients remain somewhat controversial, although patients with congenital defects that lead to LQTS are considered candidates for intervention independent of symptoms. In general, β-blockers have been used for LQTS to reduce the risk of cardiac arrhythmias. More specific and effective treatments can be introduced once the underlying cause of LQTS is identified.

which also act via cyclic AMP, but their ionotropic effect is smaller and is maximal in the atria. Cyclic AMP activates protein kinase A, and this leads to phosphorylation of the voltage-dependent Ca^{2+} channels, causing them to spend more time in the open state. Cyclic AMP also increases the active transport of Ca^{2+} to the sarcoplasmic reticulum, thus accelerating relaxation and consequently shortening systole. This is important when the cardiac rate is increased because it permits adequate diastolic filling (see Chapter 30).

METABOLISM

Mammalian hearts have an abundant blood supply, numerous mitochondria, and a high content of myoglobin, a muscle pigment that can function as an O_2 storage mechanism. Normally, less than 1% of the total energy liberated is provided

CLINICAL BOX 5–6

Glycolysidic Drugs & Cardiac Contractions

Oubain and other digitalis glycosides are commonly used to treat failing hearts. These drugs have the effect of increasing the strength of cardiac contractions. Although there is discussion as to full mechanisms, a working hypothesis is based on the ability of these drugs to inhibit the Na, K ATPase in cell membranes of the cardiomyocytes. The block of the Na, K ATPase in cardiomyocytes would result in an increased intracellular Na^+ concentration. Such an increase would result in a decreased Na^+ influx and hence Ca^{2+} efflux via the Na^+-Ca^{2+} exchange antiport during the Ca^{2+} recovery period. The resulting increase in intracellular Ca^{2+} concentration in turn increases the strength of contraction of the cardiac muscle. With this mechanism in mind, these drugs can also be quite toxic. Overinhibition of the Na, K ATPase would result in a depolarized cell that could slow conduction, or even spontaneously activate. Alternatively, an overly increased Ca^{2+} concentration could also have ill effects on cardiomyocyte physiology.

by anaerobic metabolism. During hypoxia, this figure may increase to nearly 10%; but under totally anaerobic conditions, the energy liberated is inadequate to sustain ventricular contractions. Under basal conditions, 35% of the caloric needs of the human heart are provided by carbohydrate, 5% by ketones and amino acids, and 60% by fat. However, the proportions of substrates utilized vary greatly with the nutritional state. After ingestion of large amounts of glucose, more lactate

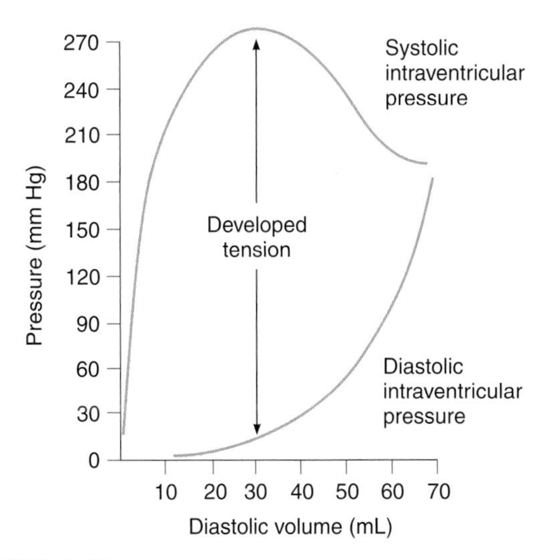

FIGURE 5–18 Length-tension relationship for cardiac muscle. Comparison of the systolic intraventricular pressure (top trace) and diastolic intraventricular pressure (bottom trace) display the developed tension in the cardiomyocyte. Values shown are for canine heart.

and pyruvate are used; during prolonged starvation, more fat is used. Circulating free fatty acids normally account for almost 50% of the lipid utilized. In untreated diabetics, the carbohydrate utilization of cardiac muscle is reduced and that of fat is increased.

SMOOTH MUSCLE MORPHOLOGY

Smooth muscle is distinguished anatomically from skeletal and cardiac muscle because it lacks visible cross-striations. Actin and myosin-II are present, and they slide on each other to produce contraction. However, they are not arranged in regular arrays, as in skeletal and cardiac muscle, and so the striations are absent. Instead of Z lines, there are **dense bodies** in the cytoplasm and attached to the cell membrane, and these are bound by α-actinin to actin filaments. Smooth muscle also contains tropomyosin, but troponin appears to be absent. The isoforms of actin and myosin differ from those in skeletal muscle. A sarcoplasmic reticulum is present, but it is less extensive than those observed in skeletal or cardiac muscle. In general, smooth muscles contain few mitochondria and depend, to a large extent, on glycolysis for their metabolic needs.

TYPES

There is considerable variation in the structure and function of smooth muscle in different parts of the body. In general, smooth muscle can be divided into **unitary** (or **visceral**) **smooth muscle** and **multiunit smooth muscle.** Unitary smooth muscle occurs in large sheets, has many low-resistance gap junctional connections between individual muscle cells, and functions in a syncytial fashion. Unitary smooth muscle is found primarily in the walls of hollow viscera. The musculature of the intestine, the uterus, and the ureters are examples. Multiunit smooth muscle is made up of individual units with few (or no) gap junctional bridges. It is found in structures such as the iris of the eye, in which fine, graded contractions occur. It is not under voluntary control, but it has many functional similarities to skeletal muscle. Each multiunit smooth muscle cell has en passant endings of nerve fibers, but in unitary smooth muscle there are en passant junctions on fewer cells, with excitation spreading to other cells by gap junctions. In addition, these cells respond to hormones and other circulating substances. Blood vessels have both unitary and multiunit smooth muscle in their walls.

ELECTRICAL & MECHANICAL ACTIVITY

Unitary smooth muscle is characterized by the instability of its membrane potential and by the fact that it shows continuous, irregular contractions that are independent of its nerve supply. This maintained state of partial contraction is called **tonus,** or **tone.** The membrane potential has no true "resting" value, being relatively low when the tissue is active and higher when it is inhibited, but in periods of relative quiescence values for resting potential are on the order of –20 to –65 mV. Smooth muscle cells can display divergent electrical activity (eg, Figure 5–19). There are slow sine wave-like fluctuations a few millivolts in magnitude and spikes that sometimes overshoot the zero potential line and sometimes do not. In many tissues, the spikes have a duration of about 50 ms, whereas in some tissues the action potentials have a prolonged plateau during repolarization, like the action potentials in cardiac muscle. As in the other muscle types, there are significant contributions of K^+, Na^+, and Ca^{2+} channels and Na, K ATPase to this electrical activity. However, discussion of contributions to individual smooth muscle types is beyond the scope of this text.

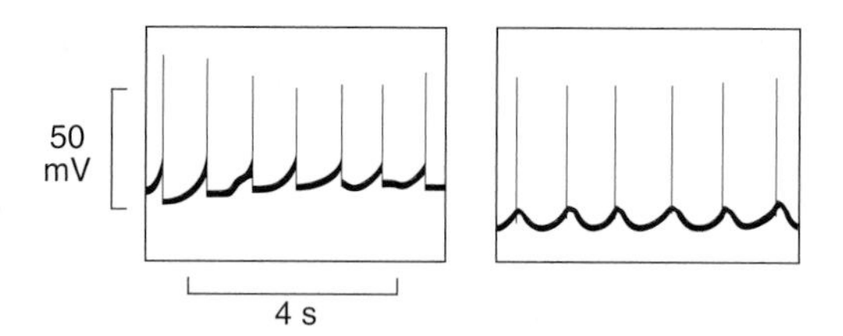

FIGURE 5–19 Electrical activity of individual smooth muscle cells in the guinea pig taenia coli. Left: Pacemaker-like activity with spikes firing at each peak. **Right:** Sinusoidal fluctuation of membrane potential with firing on the rising phase of each wave. In other fibers, spikes can occur on the falling phase of sinusoidal fluctuations and there can be mixtures of sinusoidal and pacemaker potentials in the same fiber.

Because of the continuous activity, it is difficult to study the relation between the electrical and mechanical events in unitary smooth muscle, but in some relatively inactive preparations, a single spike can be generated. In such preparations, the excitation–contraction coupling in unitary smooth muscle can occur with as much as a 500 ms delay. Thus, it is a very slow process compared with that in skeletal and cardiac muscle, in which the time from initial depolarization to initiation of contraction is less than 10 ms. Unlike unitary smooth muscle, multiunit smooth muscle is nonsyncytial and contractions do not spread widely through it. Because of this, the contractions of multiunit smooth muscle are more discrete, fine, and localized than those of unitary smooth muscle.

MOLECULAR BASIS OF CONTRACTION

As in skeletal and cardiac muscle, Ca^{2+} plays a prominent role in the initiation of contraction of smooth muscle. However, the source of Ca^{2+} increase can be quite different in unitary smooth muscle. Depending on the activating stimulus, Ca^{2+} increase can be due to influx through voltage- or ligand-gated plasma membrane channels, efflux from intracellular stores through the RyR, efflux from intracellular stores through the **inositol trisphosphate receptor (IP_3R)** Ca^{2+} channel, or via a combination of these channels. In addition, the lack of troponin in smooth muscle prevents Ca^{2+} activation via troponin binding. Rather, myosin in smooth muscle must be phosphorylated for activation of the myosin ATPase. Phosphorylation and dephosphorylation of myosin also occur in skeletal muscle, but phosphorylation is not necessary for activation of the ATPase. In smooth muscle, Ca^{2+} binds to calmodulin, and the resulting complex activates **calmodulin-dependent myosin light chain kinase.** This enzyme catalyzes the phosphorylation of the myosin light chain on serine at position 19, increasing its ATPase activity.

Myosin is dephosphorylated by **myosin light chain phosphatase** in the cell. However, dephosphorylation of myosin light chain kinase does not necessarily lead to relaxation of the smooth muscle. Various mechanisms are involved. One appears to be a latch bridge mechanism by which myosin cross-bridges remain attached to actin for some time after the cytoplasmic Ca^{2+} concentration falls. This produces sustained contraction with little expenditure of energy, which is especially important in vascular smooth muscle. Relaxation of the muscle presumably occurs when the Ca^{2+}-calmodulin complex finally dissociates or when some other mechanism comes into play. The events leading to contraction and relaxation of unitary smooth muscle are summarized in Figure 5–20. The events in multiunit smooth muscle are generally similar.

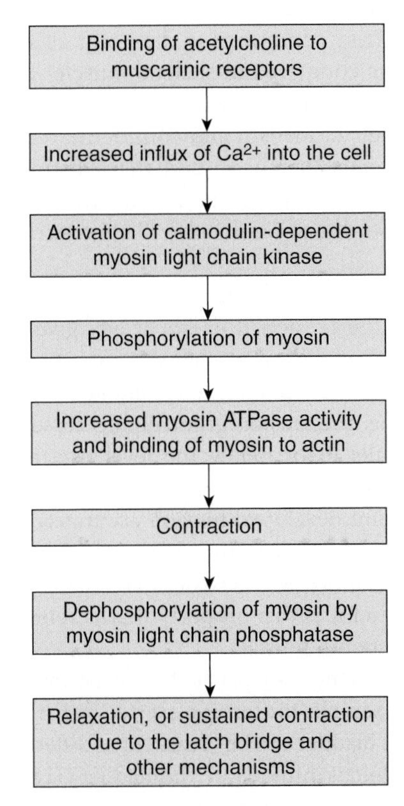

FIGURE 5–20 Sequence of events in contraction and relaxation of smooth muscle. Flow chart illustrates many of the molecular changes that occur from the initiation of contraction to its relaxation. Note the distinct differences from skeletal and cardiac muscle excitation.

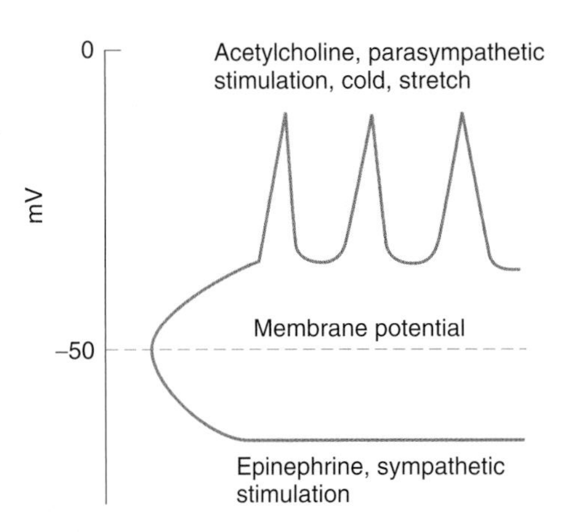

FIGURE 5–21 Effects of various agents on the membrane potential of intestinal smooth muscle. Drugs and hormones can alter firing of smooth muscle action potentials by raising (top trace) or lowering (bottom trace) resting membrane potential.

Unitary smooth muscle is unique in that, unlike other types of muscle, it contracts when stretched in the absence of any extrinsic innervation. Stretch is followed by a decline in membrane potential, an increase in the frequency of spikes, and a general increase in tone.

If epinephrine or norepinephrine is added to a preparation of intestinal smooth muscle arranged for recording of intracellular potentials in vitro, the membrane potential usually becomes larger, the spikes decrease in frequency, and the muscle relaxes (Figure 5–21). Norepinephrine is the chemical mediator released at noradrenergic nerve endings, and stimulation of the noradrenergic nerves to the preparation produces inhibitory potentials. Acetylcholine has an effect opposite to that of norepinephrine on the membrane potential and contractile activity of intestinal smooth muscle. If acetylcholine is added to the fluid bathing a smooth muscle preparation in vitro, the membrane potential decreases and the spikes become more frequent. The muscle becomes more active, with an increase in tonic tension and the number of rhythmic contractions. The effect is mediated by phospholipase C, which produces IP_3 and allows for Ca^{2+} release through IP_3 receptors. In the intact animal, stimulation of cholinergic nerves causes release of acetylcholine, excitatory potentials, and increased intestinal contractions.

Like unitary smooth muscle, multiunit smooth muscle is very sensitive to circulating chemical substances and is normally activated by chemical mediators (acetylcholine and norepinephrine) released at the endings of its motor nerves. Norepinephrine in particular tends to persist in the muscle and to cause repeated firing of the muscle after a single stimulus rather than a single action potential. Therefore, the contractile response produced is usually an irregular tetanus rather than a single twitch. When a single twitch response is obtained, it resembles the twitch contraction of skeletal muscle except that its duration is 10 times as long.

CLINICAL BOX 5–7

Common Drugs That Act on Smooth Muscle

Overexcitation of smooth muscle in the airways, such as that observed during an asthma attack, can lead to bronchoconstriction. Inhalers that deliver drugs to the conducting airway are commonly used to offset this smooth muscle bronchoconstriction, as well as other symptoms in the asthmatic airways. The rapid effects of drugs in inhalers are related to smooth muscle relaxation. Rapid response inhaler drugs (eg, ventolin, albuterol, sambuterol) frequently target β-adrenergic receptors in the airway smooth muscle to elicit a relaxation. Although these β-adrenergic receptor agonists targeting the smooth muscle do not treat all symptoms associated with asthma (eg, inflammation and increased mucus), they act rapidly and frequently allow for sufficient opening of the conducting airway to restore airflow, and thus allow for other treatments to reduce airway obstruction.

Smooth muscle is also a target for drugs developed to increase blood flow. As discussed in the text, NO is a natural signaling molecule that relaxes smooth muscle by raising cGMP. This signaling pathway is naturally down-regulated by the action of **phosphodiesterase (PDE)**, which transforms cGMP into a nonsignaling form, GMP. The drugs sildenafil, tadalafil, and vardenafil are all specific inhibitors of PDE V, an isoform found mainly in the smooth muscle in the corpus cavernosum of the penis (see Chapters 25 and 32). Thus, oral administration of these drugs can block the action of PDE V, increasing blood flow in a very limited region in the body and offsetting erectile dysfunction.

RELAXATION

In addition to cellular mechanisms that increase contraction of smooth muscle, there are cellular mechanisms that lead to its relaxation (Clinical Box 5–7). This is especially important in smooth muscle that surrounds the blood vessels to increase blood flow. It was long known that endothelial cells that line the inside of blood cells could release a substance that relaxed smooth muscle **(endothelial derived relaxing factor, EDRF).** EDRF was later identified as the gaseous second messenger molecule, **nitric oxide (NO).** NO produced in endothelial cells is free to diffuse into the smooth muscle for its effects. Once in muscle, NO directly activates a soluble guanylate cyclase to produce another second messenger molecule, **cyclic guanosine monophosphate (cGMP).** This molecule can activate cGMP-specific protein kinases that can affect ion channels, Ca^{2+} homeostasis, or phosphatases, or all of those mentioned, leading to smooth muscle relaxation (see Chapters 7 and 32).

FUNCTION OF THE NERVE SUPPLY TO SMOOTH MUSCLE

The effects of acetylcholine and norepinephrine on unitary smooth muscle serve to emphasize two of its important properties: (1) its spontaneous activity in the absence of nervous stimulation and (2) its sensitivity to chemical agents released from nerves locally or brought to it in the circulation. In mammals, unitary muscle usually has a dual nerve supply from the two divisions of the autonomic nervous system. The function of the nerve supply is not to initiate activity in the muscle but rather to modify it. Stimulation of one division of the autonomic nervous system usually increases smooth muscle activity, whereas stimulation of the other decreases it. In some organs, noradrenergic stimulation increases and cholinergic stimulation decreases smooth muscle activity; in others, the reverse is true.

FORCE GENERATION & PLASTICITY OF SMOOTH MUSCLE

Smooth muscle displays a unique economy when compared to skeletal muscle. Despite approximately 20% of the myosin content and a 100-fold difference in ATP use when compared with skeletal muscle, they can generate similar force per cross-sectional area. One of the tradeoffs of obtaining force under these conditions is the noticeably slower contractions when compared to skeletal muscle. There are several known reasons for these noticeable changes, including unique isoforms of myosin and contractile-related proteins expressed in smooth muscle and their distinct regulation (discussed above). The unique architecture of the smooth cell and its coordinated units also likely contribute to these changes.

Another special characteristic of smooth muscle is the variability of the tension it exerts at any given length. If a unitary smooth muscle is stretched, it first exerts increased tension. However, if the muscle is held at the greater length after stretching, the tension gradually decreases. Sometimes the tension falls to or below the level exerted before the muscle was stretched. It is consequently impossible to correlate length and developed tension accurately, and no resting length can be assigned. In some ways, therefore, smooth muscle behaves more like a viscous mass than a rigidly structured tissue, and it is this property that is referred to as the **plasticity** of smooth muscle.

The consequences of plasticity can be demonstrated in humans. For example, the tension exerted by the smooth muscle walls of the bladder can be measured at different degrees of distention as fluid is infused into the bladder via a catheter. Initially, tension increases relatively little as volume is increased because of the plasticity of the bladder wall. However, a point is eventually reached at which the bladder contracts forcefully (see Chapter 37).

CHAPTER SUMMARY

- There are three main types of muscle cells: skeletal, cardiac, and smooth.
- Skeletal muscle is a true syncytium under voluntary control. Skeletal muscles receive electrical stimuli from neurons to elicit contraction: "excitation–contraction coupling." Action potentials in muscle cells are developed largely through coordination of Na^+, K^+, and Ca^{2+} channels. Contraction in skeletal muscle cells is coordinated through Ca^{2+} regulation of the actomyosin system that gives the muscle its classic striated pattern under the microscope.
- There are several different types of skeletal muscle fibers (I, IIA, IIB) that have distinct properties in terms of protein makeup and force generation. Skeletal muscle fibers are arranged into motor units of like fibers within a muscle. Skeletal motor units are recruited in a specific pattern as the need for more force is increased.
- Cardiac muscle is a collection of individual cells (cardiomyocytes) that are linked as a syncytium by gap junctional communication. Cardiac muscle cells also undergo excitation–contraction coupling. Pacemaker cells in the heart can initiate propagated action potentials. Cardiac muscle cells also have a striated, actomyosin system that underlies contraction.
- Smooth muscle exists as individual cells and are frequently under control of the autonomic nervous system.
- There are two broad categories of smooth muscle cells: unitary and multiunit. Unitary smooth muscle contraction is synchronized by gap junctional communication to coordinate contraction among many cells. Multiunit smooth muscle contraction is coordinated by motor units, functionally similar to skeletal muscle.
- Smooth muscle cells contract through an actomyosin system, but do not have well-organized striations. Unlike skeletal and cardiac muscle, Ca^{2+} regulation of contraction is primarily through phosphorylation–dephosphorylation reactions.

MULTIPLE-CHOICE QUESTIONS

For all questions, select the single best answer unless otherwise directed.

1. The action potential of skeletal muscle
   A. has a prolonged plateau phase.
   B. spreads inward to all parts of the muscle via the T tubules.
   C. causes the immediate uptake of Ca^{2+} into the lateral sacs of the sarcoplasmic reticulum.
   D. is longer than the action potential of cardiac muscle.
   E. is not essential for contraction.
2. The functions of tropomyosin in skeletal muscle include
   A. sliding on actin to produce shortening.
   B. releasing Ca^{2+} after initiation of contraction.
   C. binding to myosin during contraction.
   D. acting as a "relaxing protein" at rest by covering up the sites where myosin binds to actin.
   E. generating ATP, which it passes to the contractile mechanism.
3. The cross-bridges of the sarcomere in skeletal muscle are made up of
   A. actin.
   B. myosin.
   C. troponin.
   D. tropomyosin.
   E. myelin.
4. The contractile response in skeletal muscle
   A. starts after the action potential is over.
   B. does not last as long as the action potential.
   C. produces more tension when the muscle contracts isometrically than when the muscle contracts isotonically.
   D. produces more work when the muscle contracts isometrically than when the muscle contracts isotonically.
   E. decreases in magnitude with repeated stimulation.
5. Gap junctions
   A. are absent in cardiac muscle.
   B. are present but of little functional importance in cardiac muscle.
   C. are present and provide the pathway for rapid spread of excitation from one cardiac muscle fiber to another.
   D. are absent in smooth muscle.
   E. connect the sarcotubular system to individual skeletal muscle cells.

CHAPTER RESOURCES

Alberts B, Johnson A, Lewis J, et al: *Molecular Biology of the Cell,* 5th ed. Garland Science, 2007.
Fung YC: *Biomechanics*, 2nd ed. Springer, 1993.
Hille B: *Ionic Channels of Excitable Membranes*, 3rd ed. Sinaver Associates, 2001.
Horowitz A: Mechanisms of smooth muscle contraction. Physiol Rev 1996;76:967.
Kandel ER, Schwartz JH, Jessell TM (editors): *Principles of Neural Science*, 4th ed. McGraw-Hill, 2000.
Katz AM: *Phyysiology of the Heart*, 4th ed. Raven Press, 2006.
Sperelakis N (editor): *Cell Physiology Sourcebook,* 3rd ed. Academic Press, 2001.

CHAPTER 6

Synaptic & Junctional Transmission

OBJECTIVES

After studying this chapter, you should be able to:

- Describe the main morphologic features of synapses.
- Distinguish between chemical and electrical transmission at synapses.
- Describe fast and slow excitatory and inhibitory postsynaptic potentials, outline the ionic fluxes that underlie them, and explain how the potentials interact to generate action potentials.
- Define and give examples of direct inhibition, indirect inhibition, presynaptic inhibition, and postsynaptic inhibition.
- Describe the neuromuscular junction, and explain how action potentials in the motor neuron at the junction lead to contraction of the skeletal muscle.
- Define denervation hypersensitivity.

INTRODUCTION

The "all-or-none" type of conduction seen in axons and skeletal muscle has been discussed in Chapters 4 and 5. Impulses are transmitted from one nerve cell to another cell at **synapses.** These are the junctions where the axon or some other portion of one cell (the **presynaptic cell**) terminates on the dendrites, soma, or axon of another neuron (Figure 6–1) or, in some cases, a muscle or gland cell (the **postsynaptic cell**). Cell-to-cell communication occurs across either a **chemical** or **electrical synapse.** At chemical synapses, a **synaptic cleft** separates the terminal of the presynaptic cell from the postsynaptic cell. An impulse in the presynaptic axon causes secretion of a chemical that diffuses across the synaptic cleft and binds to receptors on the surface of the postsynaptic cell. This triggers events that open or close channels in the membrane of the postsynaptic cell. In electrical synapses, the membranes of the presynaptic and postsynaptic neurons come close together, and gap junctions form between the cells (see Chapter 2). Like the intercellular junctions in other tissues, these junctions form low-resistance bridges through which ions can pass with relative ease. There are also a few conjoint synapses in which transmission is both electrical and chemical.

Regardless of the type of synapse, transmission is not a simple transmission of an action potential from the presynaptic to the postsynaptic cell. The effects of discharge at individual synaptic endings can be excitatory or inhibitory, and when the postsynaptic cell is a neuron, the summation of all the excitatory and inhibitory effects determines whether an action potential is generated. Thus, synaptic transmission is a complex process that permits the grading and adjustment of neural activity necessary for normal function. Because most synaptic transmission is chemical, consideration in this chapter is limited to chemical transmission unless otherwise specified.

Transmission from nerve to muscle resembles chemical synaptic transmission from one neuron to another. The **neuromuscular junction,** the specialized area where a motor nerve terminates on a skeletal muscle fiber, is the site of a stereotyped transmission process. The contacts between autonomic neurons and smooth and cardiac muscle are less specialized, and transmission in these locations is a more diffuse process. These forms of transmission are also considered in this chapter.

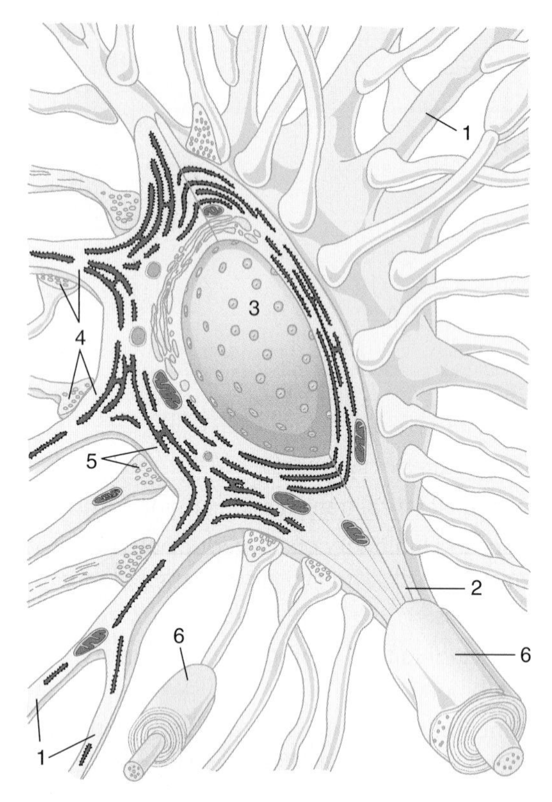

FIGURE 6–1 Synapses on a typical motor neuron. The neuron has dendrites **(1)**, an axon **(2)**, and a prominent nucleus **(3)**. Note that rough endoplasmic reticulum extends into the dendrites but not into the axon. Many different axons converge on the neuron, and their terminal boutons form axodendritic **(4)** and axosomatic **(5)** synapses. **(6)** Myelin sheath. (Reproduced with permission from Krstic RV: *Ultrastructure of the Mammalian Cell.* Springer, 1979.)

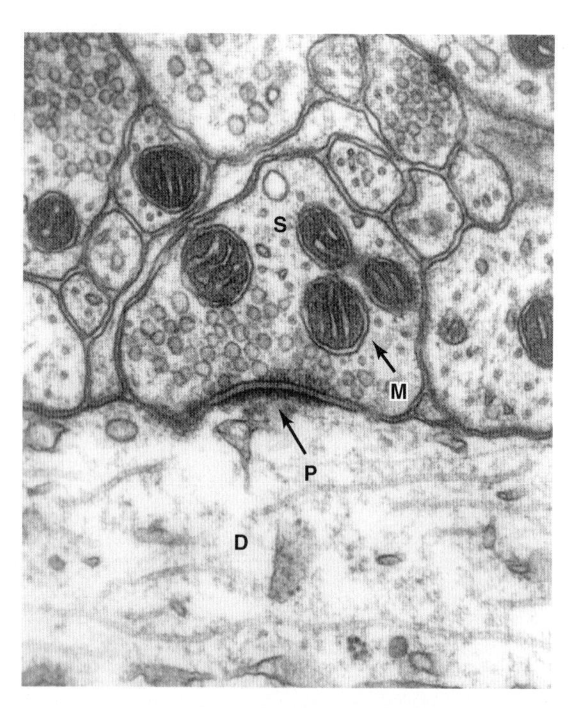

FIGURE 6–2 Electronmicrograph of synaptic knob (S) ending on the shaft of a dendrite (D) in the central nervous system. P, postsynaptic density; M, mitochondrion. (×56,000). (Courtesy of DM McDonald.)

SYNAPTIC TRANSMISSION: FUNCTIONAL ANATOMY

The anatomic structure of synapses varies considerably in the different parts of the mammalian nervous system. The ends of the presynaptic fibers are generally enlarged to form **terminal boutons or synaptic knobs** (Figure 6–2). In the cerebral and cerebellar cortex, endings are commonly located on dendrites and frequently on **dendritic spines,** which are small knobs projecting from dendrites (Figure 6–3). In some instances, the terminal branches of the axon of the presynaptic neuron form a basket or net around the soma of the postsynaptic cell (eg, basket cells of the cerebellum). In other locations, they intertwine with the dendrites of the postsynaptic cell (eg, climbing fibers of the cerebellum) or end on the dendrites directly (eg, apical dendrites of cortical pyramidal cells). Some end on axons of postsynaptic neurons (axoaxonal endings). On average, each neuron divides to form over 2000 synaptic endings, and because the human central nervous system (CNS) has 10^{11} neurons, it follows that there are about 2×10^{14} synapses. Obviously, therefore, communication between neurons is extremely complex. Synapses are dynamic structures, increasing and decreasing in complexity and number with use and experience.

It has been calculated that in the cerebral cortex, 98% of the synapses are on dendrites and only 2% are on cell bodies. In the spinal cord, the proportion of endings on dendrites is less; there are about 8000 endings on the dendrites of a typical spinal neuron and about 2000 on the cell body, making the soma appear encrusted with endings.

FUNCTIONS OF SYNAPTIC ELEMENTS

Each presynaptic terminal of a chemical synapse is separated from the postsynaptic structure by a synaptic cleft that is 20–40 nm wide. Across the synaptic cleft are many neurotransmitter receptors in the postsynaptic membrane, and usually a postsynaptic thickening called the **postsynaptic density** (Figures 6–2 and 6–3). The postsynaptic density is an ordered complex of specific receptors, binding proteins, and enzymes induced by postsynaptic effects.

Inside the presynaptic terminal are many mitochondria, as well as many membrane-enclosed vesicles, which contain neurotransmitters. There are three kinds of **synaptic vesicles:** small, clear synaptic vesicles that contain acetylcholine, glycine, GABA, or glutamate; small vesicles with a dense core that contain catecholamines; and large vesicles with a dense core that contain neuropeptides. The vesicles and the proteins contained in their walls are synthesized in the neuronal cell body and transported along the axon to the endings by fast axoplasmic transport. The neuropeptides in the large

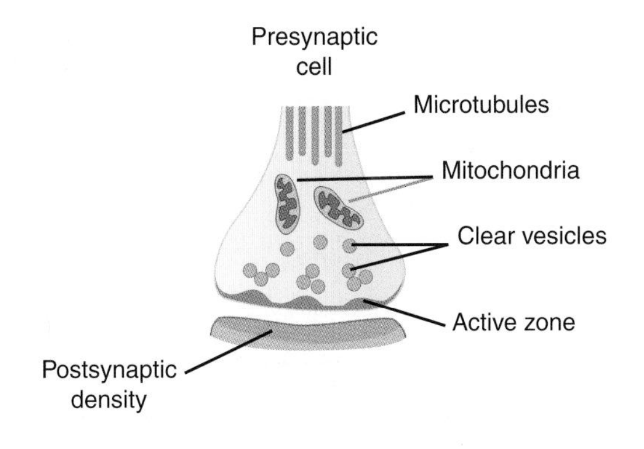

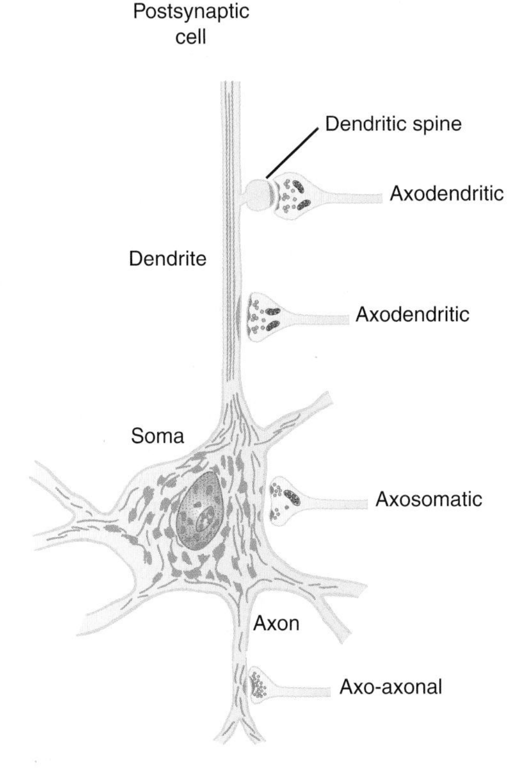

FIGURE 6–3 Axodendritic, axoaxonal, and axosomatic synapses. Many presynaptic neurons terminate on dendritic spines, as shown at the top, but some also end directly on the shafts of dendrites. Note the presence of clear and granulated synaptic vesicles in endings and clustering of clear vesicles at active zones.

dense-core vesicles must also be produced by the protein-synthesizing machinery in the cell body. However, the small clear vesicles and the small dense-core vesicles recycle in the nerve ending. These vesicles fuse with the cell membrane and release transmitters through exocytosis and are then recovered by endocytosis to be refilled locally. In some instances, they enter endosomes and are budded off the endosome and refilled, starting the cycle over again. The steps involved are shown in Figure 6–4. More commonly, however, the synaptic vesicle discharges its contents through a small hole in the cell membrane, then the opening reseals rapidly and the main vesicle stays inside the cell **(kiss-and-run discharge).** In this way, the full endocytotic process is short-circuited.

The large dense-core vesicles are located throughout the presynaptic terminals that contain them and release their neuropeptide contents by exocytosis from all parts of the terminal. On the other hand, the small vesicles are located near the synaptic cleft and fuse to the membrane, discharging their contents very rapidly into the cleft at areas of membrane thickening called **active zones** (Figure 6–3). The active zones contain many proteins and rows of Ca^{2+} channels.

The Ca^{2+} that triggers exocytosis of transmitters enters the presynaptic neurons, and transmitter release starts within 200 μs. Therefore, it is not surprising that the voltage-gated Ca^{2+} channels are very close to the release sites at the active zones. In addition, the transmitter must be released close to the postsynaptic receptors to be effective on the postsynaptic neuron. This orderly organization of the synapse depends in part on **neurexins,** proteins bound to the membrane of the presynaptic neuron that bind neurexin receptors in the membrane of the postsynaptic neuron. In many vertebrates, neurexins are produced by a single gene that codes for the α isoform. However, in mice and humans they are encoded by three genes, and both α and β isoforms are produced. Each of the genes has two regulatory regions and extensive alternative splicing of their mRNAs. In this way, over 1000 different neurexins are produced. This raises the possibility that the neurexins not only hold synapses together, but also provide a mechanism for the production of synaptic specificity.

As noted in Chapter 2, vesicle budding, fusion, and discharge of contents with subsequent retrieval of vesicle membrane are fundamental processes occurring in most, if not all, cells. Thus, neurotransmitter secretion at synapses and the accompanying membrane retrieval are specialized forms of the general processes of exocytosis and endocytosis. The details of the processes by which synaptic vesicles fuse with the cell membrane are still being worked out. They involve the **v-snare** protein **synaptobrevin** in the vesicle membrane locking with the **t-snare** protein **syntaxin** in the cell membrane; a multiprotein complex regulated by small GTPases such as Rab3 is also involved in the process (Figure 6–5). The one-way gate at the synapses is necessary for orderly neural function.

Several deadly toxins that block neurotransmitter release are zinc endopeptidases that cleave and hence inactivate proteins in the fusion-exocytosis complex. Clinical Box 6–1 describes how neurotoxins from bacteria called *Clostridium tetani* and *Clostridium botulinum* can disrupt neurotransmitter release in either the CNS or at the neuromuscular junction.

ELECTRICAL EVENTS IN POSTSYNAPTIC NEURONS

EXCITATORY & INHIBITORY POSTSYNAPTIC POTENTIALS

Penetration of an α-motor neuron is a good example of a technique used to study postsynaptic electrical activity. It is achieved by advancing a microelectrode through the ventral portion of the spinal cord. Puncture of a cell membrane is signaled by the

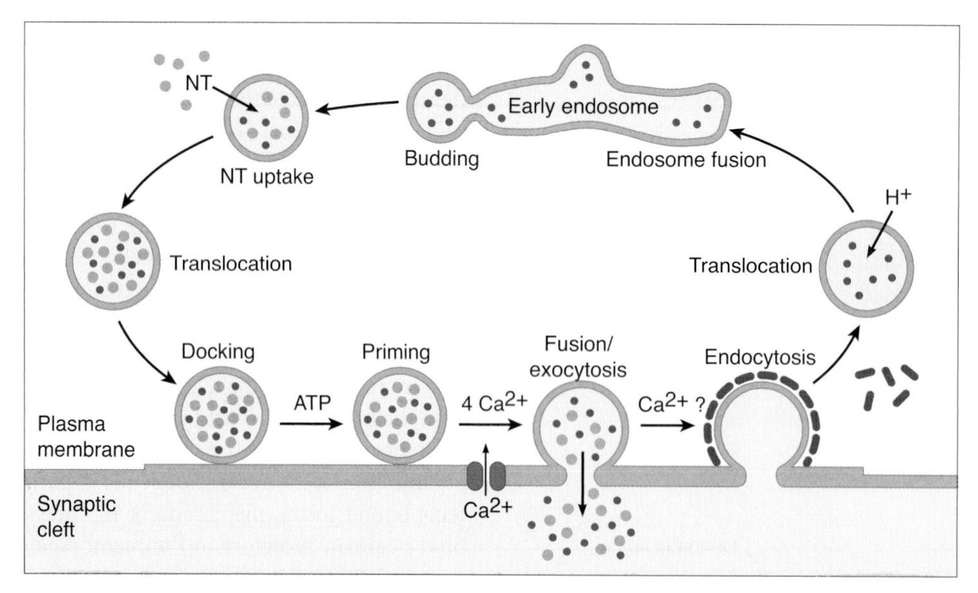

FIGURE 6–4 Small synaptic vesicle cycle in presynaptic nerve terminals. Vesicles bud off the early endosome and then fill with neurotransmitter (NT; top left). They then move to the plasma membrane, dock, and become primed. Upon arrival of an action potential at the ending, Ca^{2+} influx triggers fusion and exocytosis of the granule contents to the synaptic cleft. The vesicle wall is then coated with clathrin and taken up by endocytosis. In the cytoplasm, it fuses with the early endosome, and the cycle is ready to repeat. (Reproduced with permission from Südhof TC: The synaptic vesicle cycle: A cascade of protein-protein interactions. Nature 1995;375:645.)

appearance of a steady 70 mV potential difference between the microelectrode and an electrode outside the cell. The cell can be identified as a spinal motor neuron by stimulating the appropriate ventral root and observing the electrical activity of the cell. Such stimulation initiates an antidromic impulse (see Chapter 4) that is conducted to the soma and stops at that point. Therefore, the presence of an action potential in the cell after antidromic stimulation indicates that the cell that has been penetrated is an α-motor neuron. Stimulation of a dorsal root afferent (sensory neuron) can be used to study both excitatory and inhibitory events in α-motor neurons (Figure 6–6).

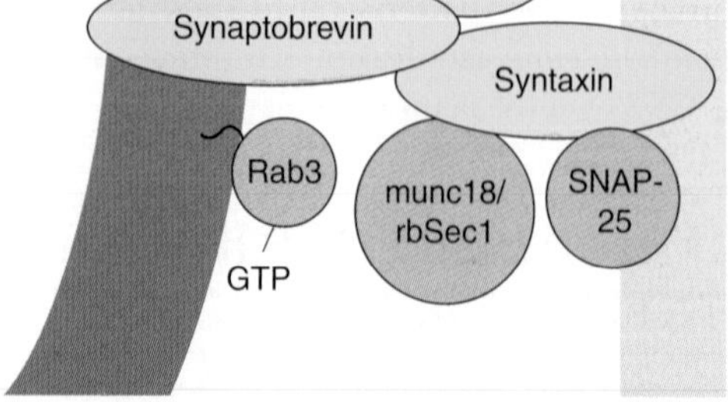

Once an impulse reaches the presynaptic terminals, a response can be obtained in the postsynaptic neuron after a **synaptic delay.** The delay is due to the time it takes for the synaptic mediator to be released and to act on the receptors on the membrane of the postsynaptic cell. Because of it, conduction along a chain of neurons is slower if there are many synapses compared to if there are only a few synapses. Because the minimum time for transmission across one synapse is 0.5 ms, it is also possible to determine whether a given reflex pathway is **monosynaptic** or **polysynaptic** (contains more than one synapse) by measuring the synaptic delay.

A single stimulus applied to the sensory nerves characteristically does not lead to the formation of a propagated action potential in the postsynaptic neuron. Instead, the stimulation produces either a transient partial depolarization or a transient hyperpolarization. The initial depolarizing response produced by a single stimulus to the proper input begins about 0.5 ms after the afferent impulse enters the spinal cord. It reaches its peak 11.5 ms later and then declines exponentially. During this potential, the excitability of the neuron to other stimuli is increased, and consequently the potential is called an **excitatory postsynaptic potential (EPSP)** (Figure 6–6).

The EPSP is produced by depolarization of the postsynaptic cell membrane immediately under the presynaptic ending. The excitatory transmitter opens Na^{+} or Ca^{2+} channels in the postsynaptic membrane, producing an inward current. The area of current flow thus created is so small that it does

FIGURE 6–5 Main proteins that interact to produce synaptic vesicle docking and fusion in nerve endings. The processes by which synaptic vesicles fuse with the cell involve the v-snare protein synaptobrevin in the vesicle membrane locking with the t-snare protein syntaxin in the cell membrane; a multiprotein complex regulated by small GTPases such as Rab3 is also involved in the process. (Reproduced with permission from Ferro-Novick S, John R: Vesicle fusion from yeast to man. Nature 1994;370:191.)

CLINICAL BOX 6–1

Botulinum and Tetanus Toxins

Clostridia are gram-positive bacteria. Two varieties, *Clostridium tetani* and *Clostridium botulinum,* produce some of the most potent biological toxins **(tetanus toxin** and **botulinum toxin)** known to affect humans. These neurotoxins act by preventing the release of neurotransmitters in the CNS and at the neuromuscular junction. Tetanus toxin binds irreversibly to the presynaptic membrane of the neuromuscular junction and uses retrograde axonal transport to travel to the cell body of the motor neuron in the spinal cord. From there it is picked up by the terminals of presynaptic inhibitory interneurons. The toxin attaches to **gangliosides** in these terminals and blocks the release of glycine and GABA. As a result, the activity of motor neurons is markedly increased. Clinically, tetanus toxin causes spastic paralysis; the characteristic symptom of "lockjaw" involves spasms of the masseter muscle. Botulism can result from ingestion of contaminated food, colonization of the gastrointestinal tract in an infant, or wound infection. Botulinum toxins are actually a family of seven neurotoxins, but it is mainly botulinum toxins A, B, and E that are toxic to humans. Botulinum toxins A and E cleave synaptosome-associated protein **(SNAP-25).** This is a presynaptic membrane protein needed for fusion of synaptic vesicles containing acetylcholine to the terminal membrane, an important step in transmitter release. Botulinum toxin B cleaves **synaptobrevin,** a vesicle-associated membrane protein **(VAMP).** By blocking acetylcholine release at the neuromuscular junction, these toxins cause flaccid paralysis. Symptoms can include ptosis, diplopia, dysarthria, dysphonia, and dysphagia.

THERAPEUTIC HIGHLIGHTS

Tetanus can be prevented by treatment with **tetanus toxoid vaccine.** The widespread use of this vaccine in the U.S. beginning in the mid 1940s has led to a marked decline in the incidence of tetanus toxicity. The incidence of botulinum toxicity is also low (about 100 cases per year in the U.S.), but in those individuals that are affected, the fatality rate is 5–10%. An antitoxin is available for treatment, and those who are at risk for respiratory failure are placed on a ventilator. On the positive side, local injection of small doses of botulinum toxin (**botox**) has proven to be effective in the treatment of a wide variety of conditions characterized by muscle hyperactivity. Examples include injection into the lower esophageal sphincter to relieve achalasia and injection into facial muscles to remove wrinkles.

not drain off enough positive charge to depolarize the whole membrane. Instead, an EPSP is inscribed. The EPSP due to activity in one synaptic knob is small, but the depolarizations produced by each of the active knobs summate.

EPSPs are produced by stimulation of some inputs, but stimulation of other inputs produces hyperpolarizing responses. Like the EPSPs, they peak 11.5 ms after the stimulus and decrease exponentially. During this potential, the excitability of the neuron to other stimuli is decreased; consequently, it is called an **inhibitory postsynaptic potential (IPSP)** (Figure 6–6).

An IPSP can be produced by a localized increase in Cl^- transport. When an inhibitory synaptic knob becomes active, the released transmitter triggers the opening of Cl^- channels in the area of the postsynaptic cell membrane under the knob. Cl^- moves down its concentration gradient. The net effect is the transfer of negative charge into the cell, so that the membrane potential increases.

The decreased excitability of the nerve cell during the IPSP is due to movement of the membrane potential away from the firing level. Consequently, more excitatory (depolarizing) activity is necessary to reach the firing level. The fact that an IPSP is mediated by Cl^- can be demonstrated by repeating the stimulus while varying the resting membrane potential of the postsynaptic cell. When the membrane potential is at the equilibrium potential for chloride (E_{Cl}), the postsynaptic potential disappears (Figure 6–7), and at more negative membrane potentials, it becomes positive **(reversal potential).**

Because IPSPs are net hyperpolarizations, they can be produced by alterations in other ion channels in the neuron. For example, they can be produced by opening of K^+ channels, with movement of K^+ out of the postsynaptic cell, or by closure of Na^+ or Ca^{2+} channels.

SLOW POSTSYNAPTIC POTENTIALS

In addition to the EPSPs and IPSPs described previously, slow EPSPs and IPSPs have been described in autonomic ganglia, cardiac and smooth muscle, and cortical neurons. These postsynaptic potentials have a latency of 100–500 ms and last several seconds. The slow EPSPs are generally due to decreases in K^+ conductance, and the slow IPSPs are due to increases in K^+ conductance.

ELECTRICAL TRANSMISSION

At synaptic junctions where transmission is electrical, the impulse reaching the presynaptic terminal generates an EPSP in the postsynaptic cell that, because of the low-resistance bridge between the two, has a much shorter latency than the EPSP at a synapse where transmission is chemical. In conjoint synapses, both a short-latency response and a longer-latency, chemically mediated postsynaptic response can occur.

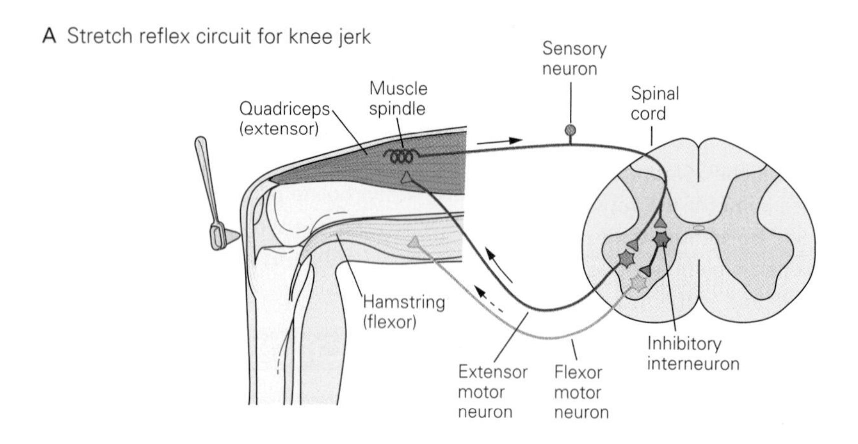

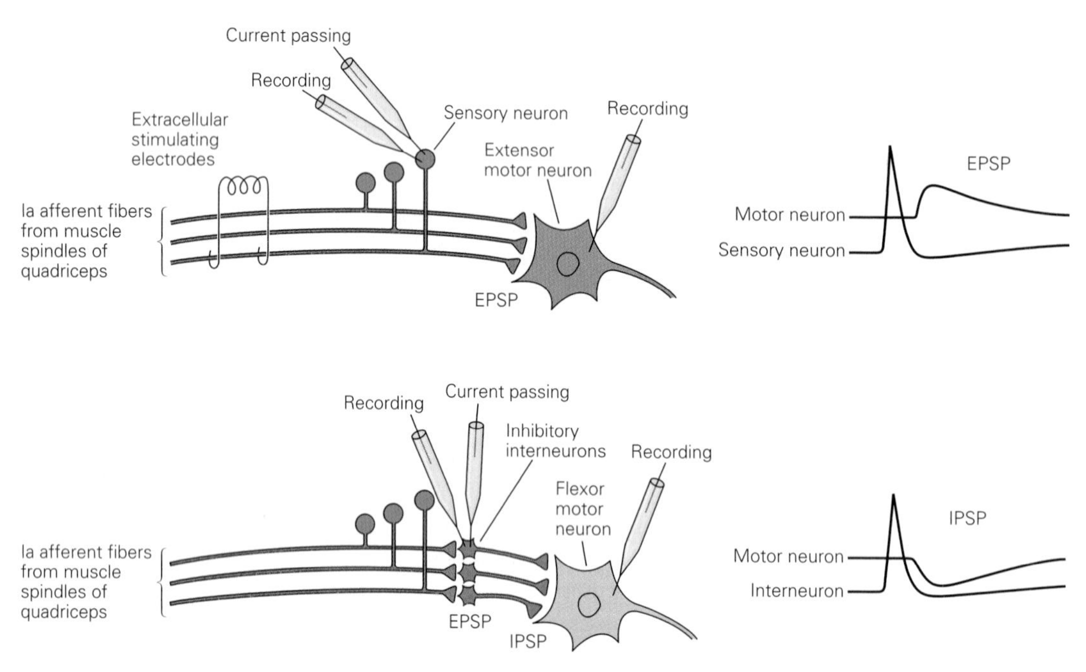

FIGURE 6–6 Excitatory and inhibitory synaptic connections mediating the stretch reflex provide an example of typical circuits within the CNS. A) The stretch receptor sensory neuron of the quadriceps muscle makes an excitatory connection with the extensor motor neuron of the same muscle and an inhibitory interneuron projecting to flexor motor neurons supplying the antagonistic hamstring muscle. **B)** Experimental setup to study excitation and inhibition of the extensor motor neuron. Top panel shows two approaches to elicit an excitatory (depolarizing) postsynaptic potential or EPSP in the extensor motor neuron–electrical stimulation of the whole Ia afferent nerve using extracellular electrodes and intracellular current passing through an electrode inserted into the cell body of a sensory neuron. Bottom panel shows that current passing through an inhibitory interneuron elicits an inhibitory (hyperpolarizing) postsynaptic potential or IPSP in the flexor motor neuron. (From Kandel ER, Schwartz JH, Jessell TM [editors]: *Principles of Neural Science,* 4th ed. McGraw-Hill, 2000.)

GENERATION OF AN ACTION POTENTIAL IN THE POSTSYNAPTIC NEURON

The constant interplay of excitatory and inhibitory activity on the postsynaptic neuron produces a fluctuating membrane potential that is the algebraic sum of the hyperpolarizing and depolarizing activities. The soma of the neuron thus acts as an integrator. When the level of depolarization reaches the threshold voltage, a propagated action potential will occur. However, the discharge of the neuron is slightly more complicated than this. In motor neurons, the portion of the cell with the lowest threshold for the production of an action potential is the **initial segment,** the portion of the axon at and just beyond the axon hillock. This unmyelinated segment is depolarized or hyperpolarized electrotonically by the current sinks and sources under the excitatory and inhibitory synaptic knobs. It is the first part of the neuron to fire, and its discharge

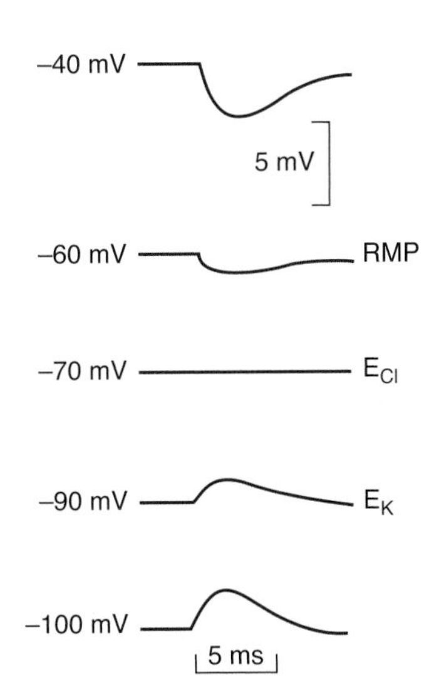

FIGURE 6–7 IPSP is due to increased Cl^- influx during stimulation. This can be demonstrated by repeating the stimulus while varying the resting membrane potential (RMP) of the postsynaptic cell. When the membrane potential is at E_{Cl}, the potential disappears, and at more negative membrane potentials (eg, E_K and below), it becomes positive (reversal potential).

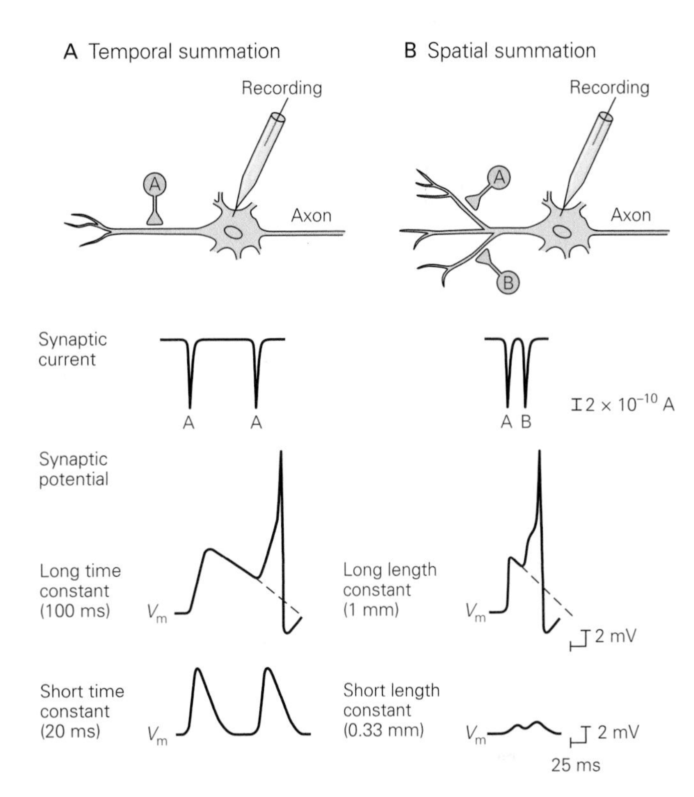

FIGURE 6–8 Central neurons integrate a variety of synaptic inputs through temporal and spatial summation. A) The time constant of the postsynaptic neuron affects the amplitude of the depolarization caused by consecutive EPSPs produced by a single presynaptic neuron. In cases of a long time constant, if a second EPSP is elicited before the first EPSP decays, the two potentials summate to induce an action potential. **B)** The length constant of a postsynaptic cell affects the amplitude of two EPSPs produced by two presynaptic neurons, A and B. If the length constant is long, the depolarization induced at two points on the neuron can spread to the trigger zone with minimal decrement so that the two potentials summate and an action potential is elicited. (From Kandel ER, Schwartz JH, Jessell TM [editors]: *Principles of Neural Science,* 4th ed. McGraw-Hill, 2000.)

is propagated in two directions: down the axon and back into the soma. Retrograde firing of the soma in this fashion probably has value in wiping the slate clean for subsequent renewal of the interplay of excitatory and inhibitory activity on the cell.

TEMPORAL & SPATIAL SUMMATION OF POSTSYNAPTIC POTENTIALS

Two passive membrane properties of a neuron affect the ability of postsynaptic potentials to summate to elicit an action potential (Figure 6–8). The **time constant** of a neuron determines the time course of the synaptic potential, and the **length constant** of a neuron determines the degree to which a depolarizing current is reduced as it spreads passively. Figure 6–8 also shows how the time constant of the postsynaptic neuron can affect the amplitude of the depolarization caused by consecutive EPSPs produced by a single presynaptic neuron. The longer the time constant, the greater is the chance for two potentials to summate to induce an action potential. If a second EPSP is elicited before the first EPSP decays, the two potentials summate and, as in this example, their additive effects are sufficient to induce an action potential in the postsynaptic neuron **(temporal summation).** Figure 6–8 also shows how the length constant of a postsynaptic neuron can affect the amplitude of two EPSPs produced by different presynaptic neurons in a process called **spatial summation.** If a neuron has a long length constant, the membrane depolarization induced by input arriving at two points on the neuron can spread to the trigger zone of the neuron with minimal decrement. The two potentials can summate and induce an action potential.

FUNCTION OF THE DENDRITES

For many years, the standard view was that dendrites were simply the sites of current sources or sinks that electrotonically change the membrane potential at the initial segment; that is, they were regarded merely as extensions of the soma that expand the area available for integration. When the dendritic tree of a neuron is extensive and has multiple presynaptic knobs ending on it, there is room for a great interplay of inhibitory and excitatory activity.

It is now well established that dendrites contribute to neural function in more complex ways. Action potentials can be recorded in dendrites. In many instances, these are initiated in the initial segment and conducted in a retrograde fashion, but propagated action potentials are initiated in some dendrites. Further research has demonstrated the malleability of dendritic spines. Dendritic spines appear, change, and even disappear over a time scale of minutes and hours, not days and months. Also, although protein synthesis occurs mainly in the soma with its nucleus, strands of mRNA migrate into the dendrites. There, each can become associated with a single ribosome in a dendritic spine and produce proteins, which alters the effects of input from individual synapses on the spine. These changes in dendritic spines have been implicated in motivation, learning, and long-term memory.

INHIBITION & FACILITATION AT SYNAPSES

Inhibition in the CNS can be postsynaptic or presynaptic. The neurons responsible for postsynaptic and presynaptic inhibition are compared in Figure 6–9. **Postsynaptic inhibition** during the course of an IPSP is called **direct inhibition** because it is not a consequence of previous discharges of the postsynaptic neuron. There are various forms of **indirect inhibition,** which is inhibition due to the effects of previous postsynaptic neuron discharge. For example, the postsynaptic cell can be refractory to excitation because it has just fired and is in its refractory period. During after-hyperpolarization it is also less excitable. In spinal neurons, especially after repeated firing, this after-hyperpolarization may be large and prolonged.

POSTSYNAPTIC INHIBITION

Postsynaptic inhibition occurs when an inhibitory transmitter (eg, glycine, GABA) is released from a presynaptic nerve terminal onto the postsynaptic neuron. Various pathways in the nervous system are known to mediate postsynaptic inhibition, and one illustrative example is presented here. Afferent fibers from the muscle spindles (stretch receptors) in skeletal muscle project directly to the spinal motor neurons of the motor units supplying the same muscle (Figure 6–6). Impulses in this afferent fiber cause EPSPs and, with summation, propagated responses in the postsynaptic motor neurons. At the same time, IPSPs are produced in motor neurons supplying the antagonistic muscles which have an inhibitory interneuron interposed between the afferent fiber and the motor neuron. Therefore, activity in the afferent fibers from the muscle spindles excites the motor neurons supplying the muscle from which the impulses come, and inhibits the motor neurons supplying its antagonists **(reciprocal innervation).** These reflexes are considered in more detail in Chapter 12.

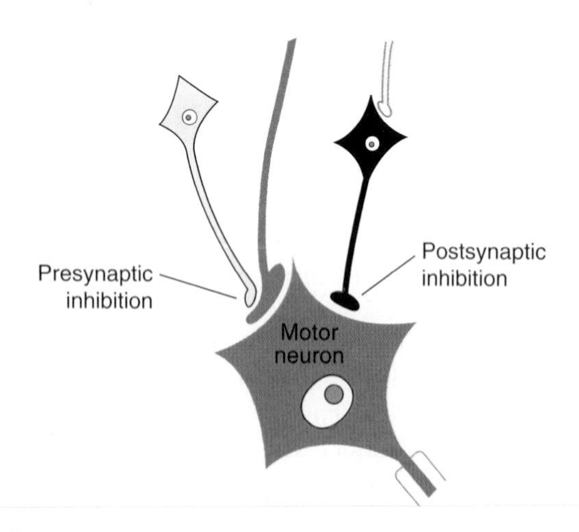

FIGURE 6–9 Comparison of neurons producing presynaptic and postsynaptic inhibition. Presynaptic inhibition is a process mediated by neurons whose terminals are on excitatory nerve endings, forming axoaxonal synapses, and reducing transmitter release form the excitatory neuron. Postsynaptic inhibition occurs when an inhibitory transmitter (eg, glycine, GABA) is released from a presynaptic nerve terminal onto the postsynaptic neuron.

PRESYNAPTIC INHIBITION & FACILITATION

Another type of inhibition occurring in the CNS is **presynaptic inhibition,** a process mediated by neurons whose terminals are on excitatory endings, forming **axoaxonal synapses** (Figure 6–3). Three mechanisms of presynaptic inhibition have been described. First, activation of the presynaptic receptors increases Cl^- conductance, and this has been shown to decrease the size of the action potentials reaching the excitatory ending (Figure 6–10). This in turn reduces Ca^{2+} entry and consequently the amount of excitatory transmitter released. Voltage-gated K^+ channels are also opened, and the resulting K^+ efflux also causes a decrease in Ca^{2+} influx. Finally, there is evidence for direct inhibition of transmitter release independent of Ca^{2+} influx into the excitatory ending.

The first transmitter shown to produce presynaptic inhibition was GABA. Acting via $GABA_A$ receptors, GABA increases Cl^- conductance. $GABA_B$ receptors are also present in the spinal cord and appear to mediate presynaptic inhibition via a G protein that produces an increase in K^+ conductance. Baclofen, a $GABA_B$ agonist, is effective in the treatment of the spasticity of spinal cord injury and multiple sclerosis, particularly when administered intrathecally via an implanted pump. Other transmitters also mediate presynaptic inhibition by G protein-mediated effects on Ca^{2+} channels and K^+ channels.

Conversely, **presynaptic facilitation** is produced when the action potential is prolonged (Figure 6–10) and the Ca^{2+} channels are open for a longer period. The molecular events responsible for the production of presynaptic facilitation mediated by serotonin in the sea snail *Aplysia* have been worked out in detail. Serotonin released at an axoaxonal ending increases intraneuronal cAMP levels, and the resulting phosphorylation of one group of K^+ channels closes the channels, slowing repolarization and prolonging the action potential.

ORGANIZATION OF INHIBITORY SYSTEMS

Presynaptic inhibition and postsynaptic inhibition are usually produced by stimulation of certain systems converging on a given postsynaptic neuron. Neurons may also inhibit themselves in a negative feedback fashion (negative feedback

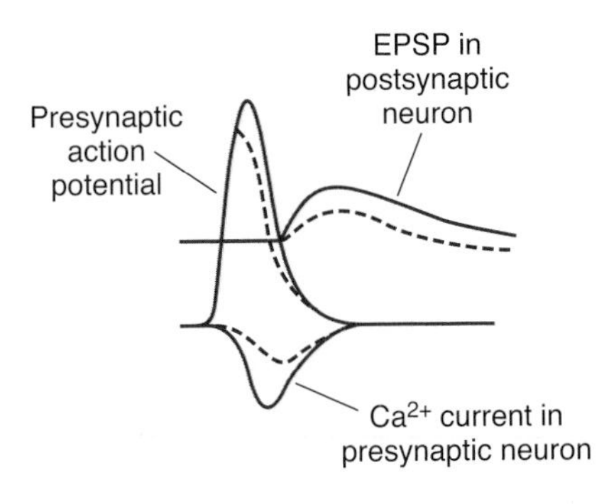

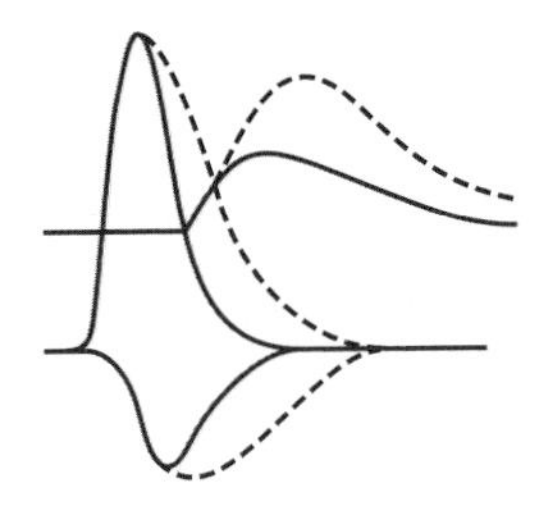

FIGURE 6–10 Effects of presynaptic inhibition and facilitation on the action potential and the Ca^{2+} current in the presynaptic neuron and the EPSP in the postsynaptic neuron. In each case, the solid lines are the controls and the dashed lines the records obtained during inhibition or facilitation. Presynaptic inhibition occurs when activation of presynaptic receptors increases Cl^- conductance which decreases the size of the action potential. This reduces Ca^{2+} entry and thus the amount of excitatory transmitter released. Presynaptic facilitation is produced when the action potential is prolonged and the Ca^{2+} channels are open for a longer duration. (Modified from Kandel ER, Schwartz JH, Jessell TM [editors]: *Principles of Neural Science*, 4th ed. McGraw-Hill, 2000.)

inhibition). For instance, a spinal motor neuron emits a recurrent collateral that synapses with an inhibitory interneuron, which then terminates on the cell body of the spinal neuron and other spinal motor neurons (Figure 6–11). This particular inhibitory neuron is sometimes called a **Renshaw cell** after its discoverer. Impulses generated in the motor neuron activate the inhibitory interneuron to secrete the inhibitory neurotransmitter **glycine,** and this reduces or stops the discharge of the motor neuron. Similar inhibition via recurrent collaterals is seen in the cerebral cortex and limbic system. Presynaptic inhibition due to descending pathways that terminate on afferent pathways in the dorsal horn may be involved in the gating of pain transmission.

Another type of inhibition is seen in the cerebellum. In this part of the brain, stimulation of basket cells produces IPSPs in the Purkinje cells. However, the basket cells and the Purkinje cells are excited by the same parallel-fiber excitatory input (see Chapter 12). This arrangement, which has been called feed-forward inhibition, presumably limits the duration of the excitation produced by any given afferent volley.

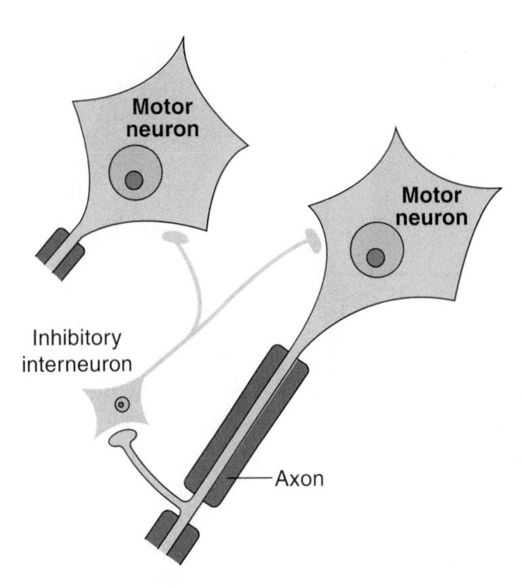

FIGURE 6–11 Negative feedback inhibition of a spinal motor neuron via an inhibitory interneuron. The axon of a spinal motor neuron has a recurrent collateral that synapses on an inhibitory interneuron that terminates on the cell body of the same and other motor neurons. The inhibitory interneuron is called a Renshaw cell and its neurotransmitter is glycine.

NEUROMUSCULAR TRANSMISSION

NEUROMUSCULAR JUNCTION

As the axon supplying a skeletal muscle fiber approaches its termination, it loses its myelin sheath and divides into a number of terminal boutons (Figure 6–12). The terminal contains many small, clear vesicles that contain acetylcholine, the transmitter at these junctions. The endings fit into **junctional folds,** which are depressions in the **motor end plate,** the thickened portion of the muscle membrane at the junction. The space between the nerve and the thickened muscle membrane is comparable to the synaptic cleft at neuron-to-neuron synapses. The whole structure is known as the **neuromuscular junction.** Only one nerve fiber ends on each end plate, with no convergence of multiple inputs.

SEQUENCE OF EVENTS DURING TRANSMISSION

The events occurring during transmission of impulses from the motor nerve to the muscle are somewhat similar to those occurring at neuron-to-neuron synapses (Figure 6–13). The impulse arriving in the end of the motor neuron increases the permeability of its endings to Ca^{2+}. Ca^{2+} enters the endings and triggers a marked increase in exocytosis of the acetylcholine-containing synaptic vesicles. The acetylcholine diffuses to nicotinic cholinergic (N_M) receptors that are concentrated at the tops of the junctional folds of the membrane of the motor end plate. Binding of acetylcholine to these receptors increases

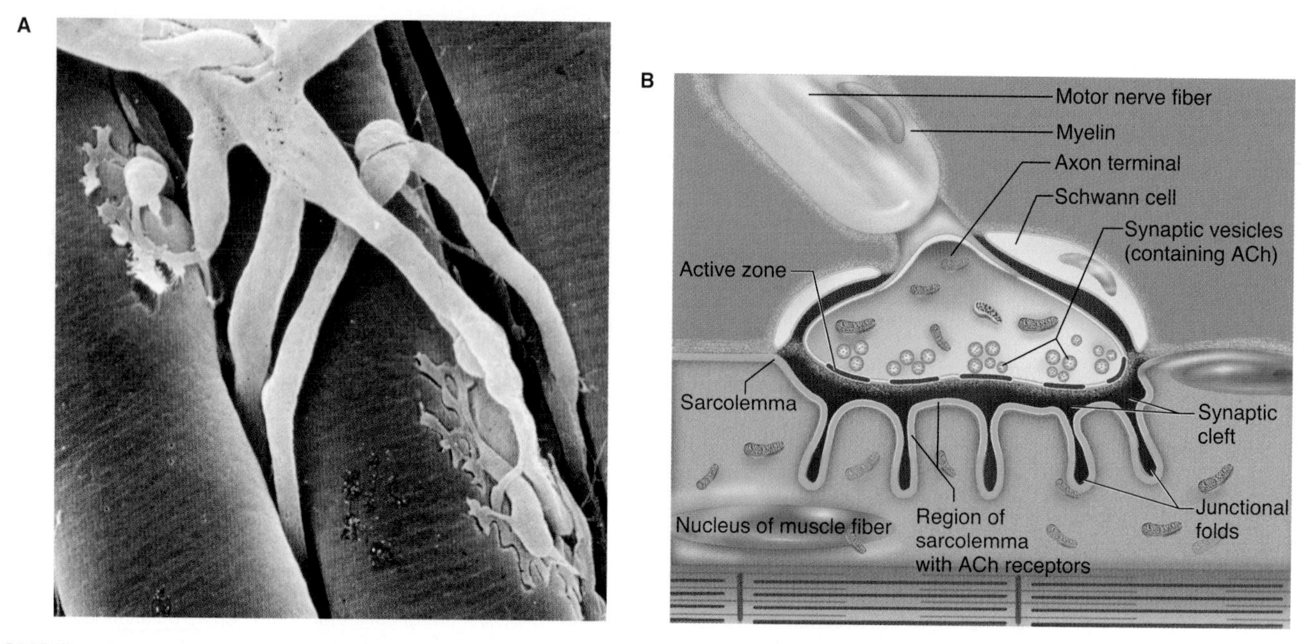

FIGURE 6–12 The neuromuscular junction. A) Scanning electronmicrograph showing branching of motor axons with terminals embedded in grooves in the muscle fiber's surface. **B)** Structure of a neuromuscular junction. (From Widmaier EP, Raff H, Strang KT: *Vanders Human Physiology*. McGraw-Hill, 2008.)

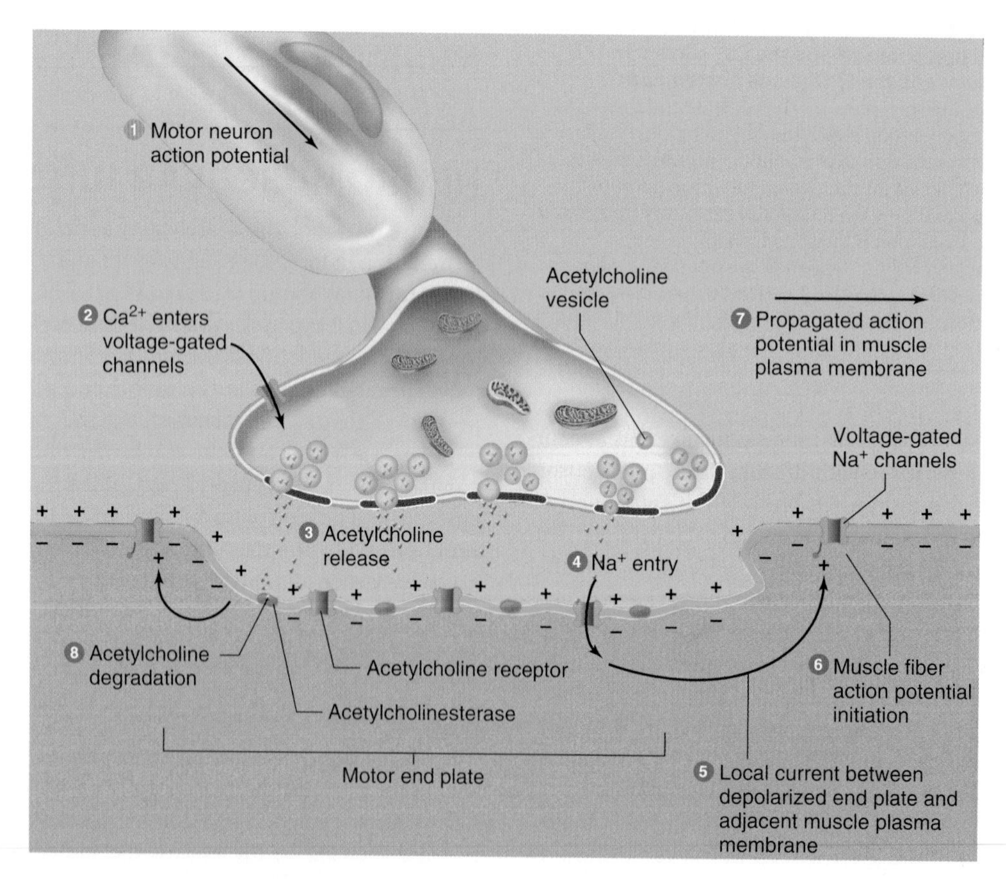

FIGURE 6–13 Events at the neuromuscular junction that lead to an action potential in the muscle fiber plasma membrane. The impulse arriving in the end of the motor neuron increases the permeability of its endings to Ca^{2+} which enters the endings and triggers exocytosis of the acetylcholine (ACh)-containing synaptic vesicles. ACh diffuses and binds to nicotinic cholinergic (N_M) receptors in the motor end plate which increases Na^+ and K^+ conductance. The resultant influx of Na^+ produces the end plate potential. The current sink created by this local potential depolarizes the adjacent muscle membrane to its firing level. Action potentials are generated on either side of the end plate and are conducted away from the end plate in both directions along the muscle fiber and the muscle contracts. ACh is then removed from the synaptic cleft by acetylcholinesterase. (From Widmaier EP, Raff H, Strang KT: *Vanders Human Physiology*. McGraw-Hill, 2008.)

the Na^+ and K^+ conductance, and the resultant influx of Na^+ produces a depolarizing potential, the **end plate potential.** The current sink created by this local potential depolarizes the adjacent muscle membrane to its firing level. Action potentials are generated on either side of the end plate and are conducted away from the end plate in both directions along the muscle fiber. The muscle action potential, in turn, initiates muscle contraction, as described in Chapter 5. Acetylcholine is then removed from the synaptic cleft by acetylcholinesterase, which is present in high concentration at the neuromuscular junction.

An average human end plate contains about 15–40 million acetylcholine receptors. Each nerve impulse releases acetylcholine from about 60 synaptic vesicles, and each vesicle contains about 10,000 molecules of the neurotransmitter. This amount is enough to activate about 10 times the number of N_M receptors needed to produce a full end plate potential. Therefore, a propagated action potential in the muscle is regularly produced, and this large response obscures the end plate potential. However, the end plate potential can be seen if the 10-fold safety factor is overcome and the potential is reduced to a size that is insufficient to activate the adjacent muscle membrane. This can be accomplished by administration of small doses of curare, a drug that competes with acetylcholine for binding to N_M receptors. The response is then recorded only at the end plate region and decreases exponentially away from it. Under these conditions, end plate potentials can be shown to undergo temporal summation.

QUANTAL RELEASE OF TRANSMITTER

Small quanta (packets) of acetylcholine are released randomly from the nerve cell membrane at rest. Each produces a minute depolarizing spike called a **miniature end plate potential,** which is about 0.5 mV in amplitude. The size of the quanta of acetylcholine released in this way varies directly with the Ca^{2+} concentration and inversely with the Mg^{2+} concentration at the end plate. When a nerve impulse reaches the ending, the number of quanta released increases by several orders of magnitude, and the result is the large end plate potential that exceeds the firing level of the muscle fiber. Quantal release of acetylcholine similar to that seen at the myoneural junction has been observed at other cholinergic synapses, and quantal release of other transmitters occurs at noradrenergic, glutamatergic, and other synaptic junctions. Two diseases of the neuromuscular junction, myasthenia gravis and Lambert-Eaton syndrome, are described in Clinical Box 6–2 and Clinical Box 6–3, respectively.

CLINICAL BOX 6–2

Myasthenia Gravis

Myasthenia gravis is a serious and sometimes fatal disease in which skeletal muscles are weak and tire easily. It occurs in 25 to 125 of every 1 million people worldwide and can occur at any age but seems to have a bimodal distribution, with peak occurrences in individuals in their 20s (mainly women) and 60s (mainly men). It is caused by the formation of circulating antibodies to the muscle type of **nicotinic cholinergic receptors.** These antibodies destroy some of the receptors and bind others to neighboring receptors, triggering their removal by endocytosis. Normally, the number of quanta released from the motor nerve terminal declines with successive repetitive stimuli. In myasthenia gravis, neuromuscular transmission fails at these low levels of quantal release. This leads to the major clinical feature of the disease, muscle fatigue with sustained or repeated activity. There are two major forms of the disease. In one form, the extraocular muscles are primarily affected. In the second form, there is a generalized skeletal muscle weakness. In severe cases, all muscles, including the diaphragm, can become weak and respiratory failure and death can ensue. The major structural abnormality in myasthenia gravis is the appearance of sparse, shallow, and abnormally wide or absent synaptic clefts in the motor end plate. Studies show that the postsynaptic membrane has a reduced response to acetylcholine and a 70–90% decrease in the number of receptors per end plate in affected muscles. Patients with mysathenia gravis have a greater than normal tendency to also have rheumatoid arthritis, systemic lupus erythematosus, and polymyositis. About 30% of mysathenia gravis patients have a maternal relative with an autoimmune disorder. These associations suggest that individuals with myasthenia gravis share a genetic predisposition to autoimmune disease. The thymus may play a role in the pathogenesis of the disease by supplying helper T cells sensitized against thymic proteins that cross-react with acetylcholine receptors. In most patients, the thymus is hyperplastic; and 10–15% have a thymoma.

THERAPEUTIC HIGHLIGHTS

Muscle weakness due to myasthenia gravis improves after a period of rest or after administration of an **acetylcholinesterase inhibitor** such as **neostigmine** or **pyridostigmine.** Cholinesterase inhibitors prevent metabolism of acetylcholine and can thus compensate for the normal decline in released neurotransmitters during repeated stimulation. **Immunosuppressive drugs** (eg, **prednisone, azathioprine,** or **cyclosporine**) can suppress antibody production and have been shown to improve muscle strength in some patients with myasthenia gravis. **Thymectomy** is indicated especially if a thymoma is suspected in the development of myasthenia gravis. Even in those without thymoma, thymectomy induces remission in 35% and improves symptoms in another 45% of patients.

CLINICAL BOX 6–3

Lambert–Eaton Syndrome

In a relatively rare condition called **Lambert–Eaton Syndrome (LEMS),** muscle weakness is caused by an autoimmune attack against one of the voltage-gated Ca^{2+} channels in the nerve endings at the neuromuscular junction. This decreases the normal Ca^{2+} influx that causes acetylcholine release. The incidence of LEMS in the U.S. is about 1 case per 100,000 people; it is usually an adult-onset disease that appears to have a similar occurrence in men and women. Proximal muscles of the lower extremities are primarily affected, producing a waddling gait and difficulty raising the arms. Repetitive stimulation of the motor nerve facilitates accumulation of Ca^{2+} in the nerve terminal and increases acetylcholine release, leading to an increase in muscle strength. This is in contrast to myasthenia gravis in which symptoms are exacerbated by repetitive stimulation. About 40% of patients with LEMS also have cancer, especially small cell cancer of the lung. One theory is that antibodies that have been produced to attack the cancer cells may also attack Ca^{2+} channels, leading to LEMS. LEMS has also been associated with lymphosarcoma, malignant thymoma, and cancer of the breast, stomach, colon, prostate, bladder, kidney, or gall bladder. Clinical signs usually precede the diagnosis of cancer. A syndrome similar to LEMS can occur after the use of **aminoglycoside antibiotics,** which also impair Ca^{2+} channel function.

THERAPEUTIC HIGHLIGHTS

Since there is a high comorbidity with small cell lung cancer, the first treatment strategy is to determine whether the individual also has cancer and, if so, to treat that appropriately. In patients without cancer, **immunotherapy** is initiated. **Prednisone** administration, **plasmapheresis,** and **intravenous immunoglobulin** are some examples of effective therapies for LEMS. Also, the use of **aminopyridines** facilitates the release of acetylcholine in the neuromuscular junction and can improve muscle strength in LEMS patients. This class of drugs causes blockade of presynaptic K^+ channels and promote activation of voltage-gated Ca^{2+} channels. Acetylcholinesterase inhibitors can be used but often do not ameliorate the symptoms of LEMS.

NERVE ENDINGS IN SMOOTH & CARDIAC MUSCLE

The postganglionic neurons in the various smooth muscles that have been studied in detail branch extensively and come in close contact with the muscle cells (Figure 6–14). Some of these nerve fibers contain clear vesicles and are cholinergic, whereas others contain the characteristic dense-core vesicles that contain norepinephrine. There are no recognizable end plates or other postsynaptic specializations. The nerve fibers run along the membranes of the muscle cells and sometimes groove their surfaces. The multiple branches of the noradrenergic and, presumably, the cholinergic neurons are beaded with enlargements **(varicosities)** and contain synaptic vesicles (Figure 6–14). In noradrenergic neurons, the varicosities are about 5 μm apart, with up to 20,000 varicosities per neuron. Transmitter is apparently liberated at each varicosity, that is, at many locations along each axon. This arrangement permits one neuron to innervate many effector cells. The type of contact in which a neuron forms a synapse on the surface of another neuron or a smooth muscle cell and then passes on to make similar contacts with other cells is called a **synapse en passant.**

In the heart, cholinergic and noradrenergic nerve fibers end on the sinoatrial node, the atrioventricular node, and the bundle of His (see Chapter 29). Noradrenergic fibers also innervate the ventricular muscle. The exact nature of the endings on nodal tissue is not known. In the ventricle, the contacts between the noradrenergic fibers and the cardiac muscle fibers resemble those found in smooth muscle.

JUNCTIONAL POTENTIALS

In smooth muscles in which noradrenergic discharge is excitatory, stimulation of the noradrenergic nerves produces discrete partial depolarizations that look like small end plate potentials and are called **excitatory junction potentials (EJPs).** These potentials summate with repeated stimuli. Similar EJPs are seen in tissues excited by cholinergic discharges. In tissues inhibited by noradrenergic stimuli, hyperpolarizing **inhibitory junction potentials (IJPs)** are produced by stimulation of the noradrenergic nerves. Junctional potentials spread electrotonically.

DENERVATION SUPERSENSITIVITY

When the motor nerve to skeletal muscle is cut and allowed to degenerate, the muscle gradually becomes extremely sensitive to acetylcholine. This is called **denervation hypersensitivity** or **supersensitivity.** Normally nicotinic receptors are located only in the vicinity of the motor end plate where the axon of the motor nerve terminates. When the motor nerve is severed,

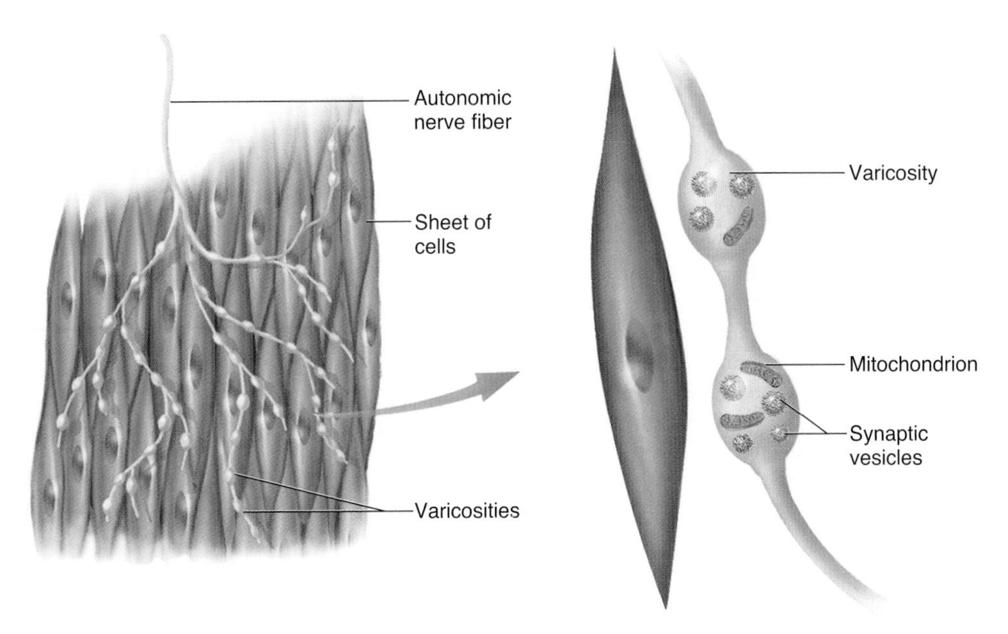

FIGURE 6–14 Endings of postganglionic autonomic neurons on smooth muscle. The nerve fibers run along the membranes of the smooth muscle cells and sometimes groove their surfaces. The multiple branches of postganglionic neurons are beaded with enlargements (varicosities) and contain synaptic vesicles. Neurotransmitter is released from the varicosities and diffuses to receptors on smooth muscle cell plasma membranes. (From Widmaier EP, Raff H, Strang KT: *Vanders Human Physiology.* McGraw-Hill, 2008.)

there is a marked proliferation of nicotinic receptors over a wide region of the neuromuscular junction. Denervation supersensitivity also occurs at autonomic junctions. Smooth muscle, unlike skeletal muscle, does not atrophy when denervated, but it becomes hyperresponsive to the chemical mediator that normally activates it. This hyperresponsiveness can be demonstrated by using pharmacological tools rather than actual nerve section. Prolonged use of a drug such as reserpine can be used to deplete transmitter stores and prevent the target organ from being exposed to norepinephrine for an extended period. Once the drug usage is stopped, smooth muscle and cardiac muscle will be supersensitive to subsequent release of the neurotransmitter.

The reactions triggered by section of an axon are summarized in Figure 6–15. Hypersensitivity of the postsynaptic structure to the transmitter previously secreted by the axon endings is a general phenomenon, largely due to the synthesis or activation of more receptors. Both orthograde degeneration **(wallerian degeneration)** and retrograde degeneration of the axon stump to the nearest collateral **(sustaining collateral)** will occur. There are a series of changes in the cell body that leads to a decrease in Nissl substance **(chromatolysis).** The nerve then starts to regrow, with multiple small branches projecting along the path the axon previously followed **(regenerative sprouting).** Axons sometimes grow back to their original targets, especially in locations like the neuromuscular junction. However, nerve regeneration is generally limited because axons often become entangled in the area of tissue damage at the site where they were disrupted. This difficulty has been reduced by administration of **neurotrophins** (see Chapter 4).

Denervation hypersensitivity has multiple causes. As noted in Chapter 2, a deficiency of a given chemical messenger generally produces an upregulation of its receptors. Another factor is a lack of reuptake of secreted neurotransmitters.

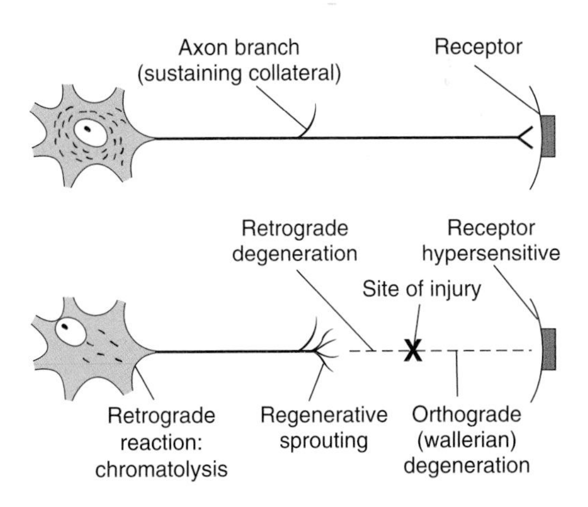

FIGURE 6–15 Summary of changes occurring in a neuron and the structure it innervates when its axon is crushed or cut at the point marked X. Hypersensitivity of the postsynaptic structure to the transmitter previously secreted by the axon occurs largely due to the synthesis or activation of more receptors. There is both orthograde (wallerian) degeneration from the point of damage to the terminal and retrograde degeneration of the axon stump to the nearest collateral (sustaining collateral). Changes also occur in the cell body, including chromatolysis. The nerve starts to regrow, with multiple small branches projecting along the path the axon previously followed (regenerative sprouting).

CHAPTER SUMMARY

- The terminals of the presynaptic fibers have enlargements called terminal boutons or synaptic knobs. The presynaptic terminal is separated from the postsynaptic structure by a synaptic cleft. The postsynaptic membrane contains neurotransmitter receptors and usually a postsynaptic thickening called the postsynaptic density.
- At chemical synapses, an impulse in the presynaptic axon causes secretion of a neurotransmitter that diffuses across the synaptic cleft and binds to postsynaptic receptors, triggering events that open or close channels in the membrane of the postsynaptic cell. At electrical synapses, the membranes of the presynaptic and postsynaptic neurons come close together, and gap junctions form low-resistance bridges through which ions pass with relative ease from one neuron to the next.
- An EPSP is produced by depolarization of the postsynaptic cell after a latency of 0.5 ms; the excitatory transmitter opens Na^+ or Ca^{2+} ion channels in the postsynaptic membrane, producing an inward current. An IPSP is produced by a hyperpolarization of the postsynaptic cell; it can be produced by a localized increase in Cl^- transport. Slow EPSPs and IPSPs occur after a latency of 100–500 ms in autonomic ganglia, cardiac, and smooth muscle, and cortical neurons. The slow EPSPs are due to decreases in K^+ conductance, and the slow IPSPs are due to increases in K^+ conductance.
- Postsynaptic inhibition during the course of an IPSP is called direct inhibition. Indirect inhibition is due to the effects of previous postsynaptic neuron discharge; for example, the postsynaptic cell cannot be activated during its refractory period. Presynaptic inhibition is a process mediated by neurons whose terminals are on excitatory endings, forming axoaxonal synapses; in response to activation of the presynaptic terminal. Activation of the presynaptic receptors can increase Cl^- conductance, decreasing the size of the action potentials reaching the excitatory ending, and reducing Ca^{2+} entry and the amount of excitatory transmitter released.
- The axon terminal of motor neurons synapses on the motor end plate on the skeletal muscle membrane to form the neuromuscular junction. The impulse arriving in the motor nerve terminal leads to the entry of Ca^{2+} which triggers the exocytosis of the acetylcholine-containing synaptic vesicles. The acetylcholine diffuses and binds to nicotinic cholinergic receptors on the motor end plate, causing an increase in Na^+ and K^+ conductance; the influx of Na^+ induces the end plate potential and subsequent depolarization of the adjacent muscle membrane. Action potentials are generated and conducted along the muscle fiber, leading in turn to muscle contraction.
- When a nerve is damaged and then degenerates, the postsynaptic structure gradually becomes extremely sensitive to the transmitter released by the nerve. This is called denervation hypersensitivity or supersensitivity.

MULTIPLE-CHOICE QUESTIONS

For all questions, select the single best answer unless otherwise directed.

1. Which of the following electrophysiological events is correctly paired with the change in ionic currents causing the event?
   A. Fast inhibitory postsynaptic potentials (IPSPs) and closing of Cl^- channels.
   B. Fast excitatory postsynaptic potentials (EPSPs) and an increase in Ca^{2+} conductance.
   C. End plate potential and an increase in Na^+ conductance.
   D. Presynaptic inhibition and closure of voltage-gated K^+ channels.
   E. Slow EPSPs and an increase in K^+ conductance.
2. Which of the following physiological processes is not correctly paired with a structure?
   A. Electrical transmission : gap junction
   B. Negative feedback inhibition : Renshaw cell
   C. Synaptic vesicle docking and fusion : presynaptic nerve terminal
   D. End plate potential : muscarinic cholinergic receptor
   E. Action potential generation : initial segment
3. Initiation of an action potential in skeletal muscle
   A. requires spatial facilitation.
   B. requires temporal facilitation.
   C. is inhibited by a high concentration of Ca^{2+} at the neuromuscular junction.
   D. requires the release of norepinephrine.
   E. requires the release of acetylcholine.
4. A 35-year-old woman sees her physician to report muscle weakness in the extraocular eye muscles and muscles of the extremities. She states that she feels fine when she gets up in the morning, but the weakness begins soon after she becomes active. The weakness is improved by rest. Sensation appears normal. The physician treats her with an anticholinesterase inhibitor, and she notes immediate return of muscle strength. Her physician diagnoses her with
   A. Lambert–Eaton syndrome.
   B. myasthenia gravis.
   C. multiple sclerosis.
   D. Parkinson disease.
   E. muscular dystrophy.
5. A 55-year-old female had an autonomic neuropathy which disrupted the sympathetic nerve supply to the pupillary dilator muscle of her right eye. While having her eyes examined, the ophthalmologist placed phenylephrine in her eyes. The right eye became much more dilated than the left eye. This suggests that
   A. the sympathetic nerve to the right eye had regenerated.
   B. the parasympathetic nerve supply to the right eye remained intact and compensated for the loss of the sympathetic nerve.
   C. phenylephrine blocked the pupillary constrictor muscle of the right eye.
   D. denervation supersensitivity had developed.
   E. the left eye also had nerve damage and so was not responding as expected.

6. A 47-year-old female was admitted to the hospital after reporting that she had been experiencing nausea and vomiting for about two days and then developed severe muscle weakness and neurological symptoms including ptosis and dysphagia. She indicated she had eaten at a restaurant the evening before the symptoms began. Lab tests were positive for *Clostridium botulinum.* Neurotoxins
   A. block the reuptake of neurotransmitters into presynaptic terminals.
   B. such as tetanus toxin bind reversibly to the presynaptic membrane at the neuromuscular junction.
   C. reach the cell body of the motor neuron by diffusion into the spinal cord.
   D. exert all of their adverse effects by acting centrally rather than peripherally.
   E. such as botulinum toxin prevent the release of acetylcholine from motor neurons due to cleavage of either synaptosome-associated proteins or vesicle-associated membrane proteins.

CHAPTER RESOURCES

Di Maoi V: Regulation of information passing by synaptic transmission: A short review. Brain Res 2008;1225:26.

Hille B: *Ionic Channels of Excitable Membranes,* 3rd ed. Sinauer Associates, 2001.

Magee JC: Dendritic integration of excitatory synaptic input. Nature Rev Neurosci 2000;1:181.

Sabatini B, Regehr WG: Timing of synaptic transmission. Annu Rev Physiol 1999;61:521.

Van der Kloot W, Molg J: Quantal acetylcholine release at the vertebrate neuromuscular junction. Physiol Rev 1994;74:899.

WuH, Xiong WC, Mei L: To build a synapse: signaling pathways in neuromuscular junction assembly. Development 2010;137:1017.

CHAPTER 7

Neurotransmitters & Neuromodulators

OBJECTIVES

After studying this chapter, you should be able to:

- List the major types of neurotransmitters.
- Summarize the steps involved in the biosynthesis, release, action, and removal from the synaptic cleft of the major neurotransmitters.
- Describe the various types of receptors for amino acids, acetylcholine, monoamines, ATP, opioids, nitric oxide, and cannabinoids.
- Identify the endogenous opioid peptides, their receptors, and their functions.

INTRODUCTION

Nerve endings have been called biological transducers that convert electrical energy into chemical energy. An observation made by Otto Loewi, a German pharmacologist, in 1920 serves as the foundation for the concept of chemical neurotransmission and his receipt of the Nobel Prize in Physiology and Medicine. He provided the first decisive evidence that a chemical messenger was released by the vagus nerve supplying the heart to reduce heart rate. The experimental design came to him in a dream on Easter Sunday of that year. He awoke from the dream, jotted down notes, but the next morning they were indecipherable. The next night, the dream recurred and he went to his laboratory at 3:00 AM to conduct a simple experiment on a frog heart. He isolated the hearts from two frogs, one with and one without its innervation. Both hearts were attached to cannulas filled with a saline solution. The vagus nerve of the first heart was stimulated, and then the saline solution from that heart was transferred to the noninnervated heart. The rate of its contractions slowed as if its vagus nerve had been stimulated. Loewi called the chemical released by the vagus nerve *vagusstoff*. Not long after, it was identified chemically to be acetylcholine. Loewi also showed that when the sympathetic nerve of the first heart was stimulated and its effluent was passed to the second heart, the rate of contraction of the "donor" heart increased as if its sympathetic fibers had been stimulated. These results proved that nerve terminals release chemicals that cause the modifications of cardiac function that occur in response to stimulation of its nerve supply.

CHEMICAL TRANSMISSION OF SYNAPTIC ACTIVITY

Regardless of the type of chemical mediator involved, several common steps comprise the process of transmission at a chemical synapse. The first steps are the synthesis of the **neurotransmitter** usually within the nerve terminal and its storage within **synaptic vesicles.** This is followed by release of the chemical into the **synaptic cleft** in response to nerve impulses. The secreted neurotransmitter can then act on **receptors** on the membrane of the postsynaptic neuron, effector organ (eg, muscle or gland), or even on the presynaptic nerve terminal. The final steps in the process lead to termination of the actions of the neurotransmitter and include diffusion away from the synaptic cleft, reuptake into the nerve terminal, and enzymatic degradation. All of these processes, plus the events in the postsynaptic neuron, are regulated by many physiologic factors and can be altered by drugs. Therefore, pharmacologists (in theory) should be able to develop drugs that regulate not only somatic and visceral motor activity but also emotions, behavior, and all the other complex functions of the brain. Some chemicals released by neurons have little or no direct effects on their own but can modify the effects of neurotransmitters. These chemicals are called **neuromodulators.**

CHEMISTRY OF TRANSMITTERS

Many neurotransmitters and the enzymes involved in their synthesis and catabolism have been localized in nerve endings by **immunohistochemistry,** a technique in which antibodies to a given substance are labeled and applied to brain and other tissues. The antibodies bind to the substance, and the location of the substance is then determined by locating the label with the light or electron microscope. **In situ hybridization histochemistry,** which permits localization of the mRNAs for particular synthesizing enzymes or receptors, has also been a valuable tool.

There are two main classes of chemical substances that serve as neurotransmitters and neuromodulators: small-molecule transmitters and large-molecule transmitters. Small-molecule transmitters include amino acids (eg, **glutamate, GABA,** and **glycine**), **acetylcholine,** monoamines (eg, **norepinephrine, epinephrine, dopamine,** and **serotonin**), and **adenosine triphosphate (ATP).** Large-molecule transmitters include neuropeptides such as **substance P, enkephalin, vasopressin,** and a host of others. In general, neuropeptides are colocalized with one of the small-molecule neurotransmitters (Table 7–1).

Figure 7–1 shows the biosynthesis of some common small-molecule transmitters released by neurons in the central nervous system (CNS) or peripheral nervous system. Figure 7–2 shows the location of major groups of neurons that contain norepinephrine, epinephrine, dopamine, and acetylcholine. These are some of the major central neuromodulatory systems.

TABLE 7–1 Examples of co-localization of small-molecule transmitters with neuropeptides.

Small Molecule Transmitter	Neuropeptide
Glutamate	Substance P
GABA	Cholecystokinin, enkephalin, somatostatin, substance P, thyrotropin-releasing hormone
Glycine	Neurotensin
Acetylcholine	Calcitonin gene-related protein, enkephalin, galanin, gonadotropin-releasing hormone, neurotensin, somatostatin, substance P, vasoactive intestinal polypeptide
Dopamine	Cholecystokinin, enkephalin, neurotensin
Norepinephrine	Enkephalin, neuropeptide Y, neurotensin, somatostatin, vasopressin
Epinephrine	Enkephalin, neuropeptide Y, neurotensin, substance P
Serotonin	Cholecystokinin, enkephalin, neuropeptide Y, substance P, vasoactive intestinal polypeptide

RECEPTORS

The action of a chemical mediator on its target structure is more dependent on the type of receptor on which it acts than on the properties of the mediator per se. Cloning and other molecular biology techniques have permitted spectacular advances in knowledge about the structure and function of receptors for neurotransmitters and other chemical messengers. The individual receptors, along with their ligands (the molecules that bind to them), are discussed in the following parts of this chapter. However, five themes have emerged that should be mentioned in this introductory discussion.

First, in every instance studied in detail to date, each chemical mediator has the potential to act on many subtypes of receptors. Thus, for example, norepinephrine acts on α_1, α_2, β_1, β_2, and β_3 adrenergic receptors. Obviously, this multiplies the possible effects of a given ligand and makes its effects in a given cell more selective.

Second, there are receptors on the presynaptic as well as the postsynaptic elements for many secreted transmitters. One type of presynaptic receptor called an **autoreceptor** often inhibits further secretion of the transmitter, providing feedback control. For example, norepinephrine acts on α_2 presynaptic receptors to inhibit additional norepinephrine secretion. A second type of presynaptic receptor is called a **heteroreceptor** whose ligand is a chemical other than the transmitter released by the nerve ending on which the receptor is located. For example, norepinephrine acts on a heteroreceptor on a cholinergic nerve terminal to inhibit the release of acetylcholine. In some cases, presynaptic receptors facilitate the release of neurotransmitters.

Third, although there are many neurotransmitters and many subtypes of receptors for each ligand, the receptors tend to group in two large families in terms of structure and function: **ligand-gated channels** (also known as **ionotropic receptors**) and **metabotropic receptors.** In the case of ionotropic receptors, a membrane channel is opened when a ligand binds to the receptor; and activation of the channel usually elicits a brief (few to tens of milliseconds) increase in ionic conductance. Thus, these receptors are important for fast synaptic transmission. Metabotropic receptors are 7-transmembrane **G protein-coupled receptors** (GPCR), and binding of a neurotransmitter to these receptors initiates the production of a second messenger that modulates the voltage-gated channels on neuronal membranes. The receptors for some neurotransmitters and neuromodulators are listed in Table 7–2, along with their principal second messengers and, where established, their net effect on ion channels. It should be noted that this table is an over-simplification. For example, activation of α_2-adrenergic receptors decreases intracellular cAMP concentrations, but there is evidence that the G protein activated by α_2-adrenergic presynaptic receptors also acts directly on Ca^{2+} channels to inhibit norepinephrine release.

Fourth, receptors are concentrated in clusters on the postsynaptic membrane close to the endings of neurons that secrete the neurotransmitters specific for them. This is generally due to the presence of specific binding proteins for them.

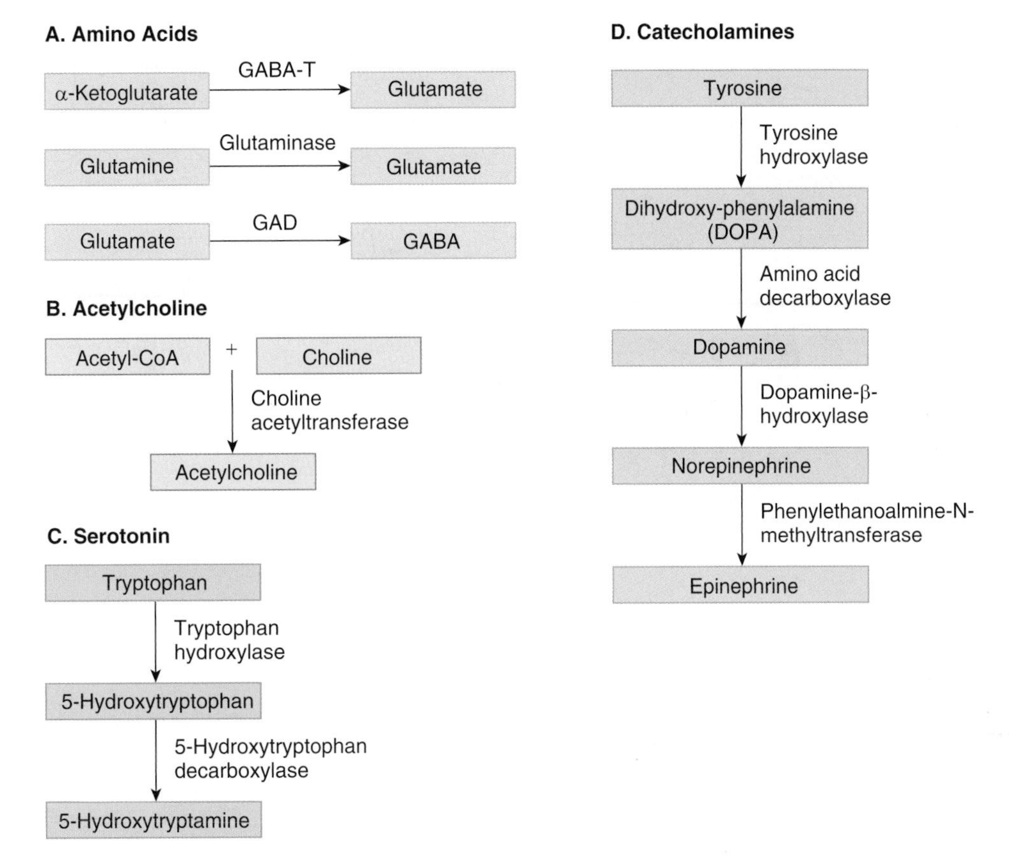

FIGURE 7–1 Biosynthesis of some common small molecule neurotransmitters. A) Glutamate is synthesized in the Krebs cycle by the conversion of α-ketoglutarate to the amino acid via the enzyme γ-aminobutyric acid transferase (GABA-T) or in nerve terminals by the hydrolysis of glutamine by the enzyme glutaminase. GABA is synthesized by the conversion of glutamate by the enzyme glutamic acid decarboxylase (GAD). **B)** Acetylcholine is synthesized in the cytoplasm of a nerve terminal from acetyl-Co-A and choline by the enzyme choline acetyltransferase. **C)** Serotonin is synthesized from the amino acid tryptophan in a two-step process: the enzymatic hydroxylation of tryptophan to 5-hydroxytryptophan and the enzymatic decarboxylation of this intermediate to form 5-hydroxytryptamine (also called serotonin). **D)** Catecholamines are synthesized from the amino acid tyrosine by a multi-step process. Tyrosine is oxidized to dihydroxy-phenylalanine (DOPA) by the enzyme tyrosine hydroxylase in the cytoplasm of the neuron; DOPA is then decarboxylated to dopamine. In dopaminergic neurons the process stops there. In noradrenergic neurons, the dopamine is transported into synaptic vesicles where it is converted to norepinephrine by dopamine-β-hydroxylase. In neurons that also contain the enzyme phenylethanolamine-N-methyltransferase, norepinephrine is converted to epinephrine.

Fifth, in response to prolonged exposure to their ligands, most receptors become unresponsive; that is, they undergo **desensitization.** This can be of two types: **homologous desensitization,** with loss of responsiveness only to the particular ligand and maintained responsiveness of the cell to other ligands; and **heterologous desensitization,** in which the cell becomes unresponsive to other ligands as well.

REUPTAKE

Neurotransmitters are rapidly transported from the synaptic cleft back into the cytoplasm of the neurons that secreted them via a process called **reuptake,** which involves a high-affinity, Na^+-dependent membrane transporter. Figure 7–3 illustrates the principle of reuptake of norepinephrine released from a sympathetic postganglionic nerve. After release of norepinephrine into the synaptic cleft, it is rapidly routed back into the sympathetic nerve terminal by a **norepinephrine transporter (NET).** A portion of the norepinephrine that re-enters the neuron is sequestered into the synaptic vesicles through the **vesicular monoamine transporter (VMAT).** There are analogous membrane and vesicular transporters for other small-molecule neurotransmitters released at other synapses in the CNS and peripheral nervous system.

Reuptake is a major factor in terminating the action of transmitters, and when it is inhibited, the effects of transmitter release are increased and prolonged. This has clinical consequences. For example, several effective antidepressant drugs are inhibitors of the reuptake of amine transmitters, and cocaine may inhibit dopamine reuptake. Glutamate uptake into neurons and glia is important because glutamate is an excitotoxin that can kill cells by overstimulating them (see Clinical Box 7–1). There is evidence that during ischemia

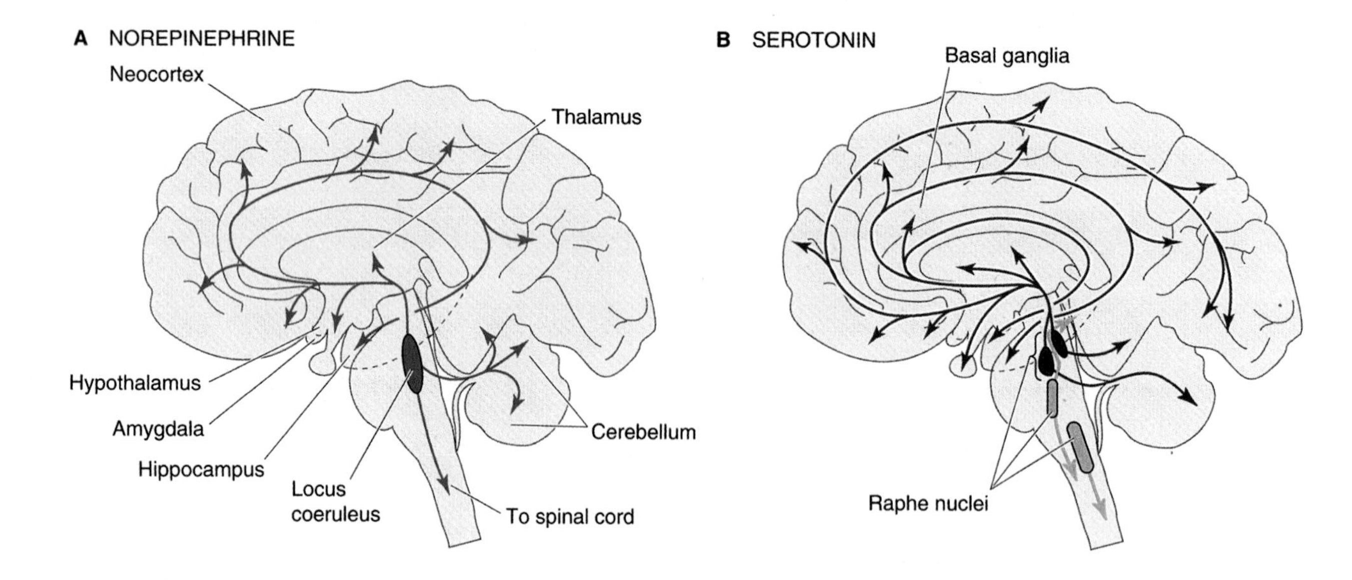

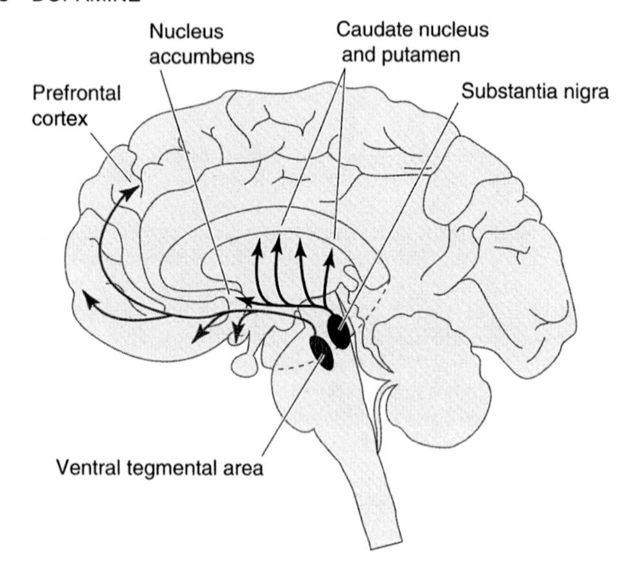

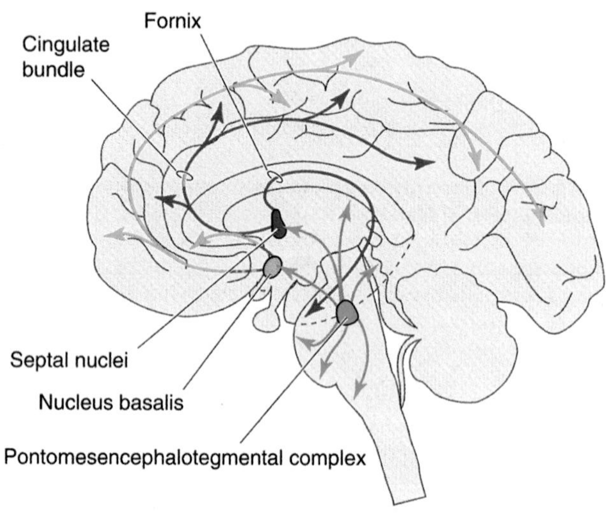

FIGURE 7–2 Four diffusely connected systems of central neuromodulators. A) Noradrenergic neurons in the locus coeruleus innervate the spinal cord, cerebellum, several nuclei of the hypothalamus, thalamus, basal telencephalon, and neocortex. **B)** Serotonergic neurons in the raphé nuclei project to the hypothalamus, limbic system, neocortex, cerebellum, and spinal cord. **C)** Dopaminergic neurons in the substantia nigra project to the striatum and those in the ventral tegmental area of the midbrain project to the prefrontal cortex the limbic system. **D)** Cholinergic neurons in the basal forebrain complex project to the hippocampus and the neocortex and those in the pontomesencephalotegmental cholinergic complex project to the dorsal thalamus and the forebrain. (Reproduced with permission from Boron WF, Boulpaep EL: *Medical Physiology.* Elsevier, 2005.)

and anoxia, loss of neurons is increased because glutamate reuptake is inhibited.

SMALL-MOLECULE TRANSMITTERS

Synaptic physiology is a rapidly expanding, complex field that cannot be covered in detail in this book. However, it is appropriate to summarize information about the principal neurotransmitters and their receptors.

EXCITATORY & INHIBITORY AMINO ACIDS

Glutamate

Glutamate is the main excitatory transmitter in the brain and spinal cord and has been calculated to be responsible for 75% of the excitatory transmission in the CNS. There are two distinct pathways involved in the synthesis of glutamate (Figure 7–1). In one pathway, **α-ketoglutarate** produced by the Krebs cycle is converted to glutamate by the enzyme **GABA transaminase**

TABLE 7–2 Pharmacology of a selection of receptors for some small molecule neurotransmitters.

Neurotransmitter	Receptor	Second Messenger	Net Channel Effects	Agonists	Antagonists
Glutamate	AMPA		↑Na^+, K^+	AMPA	CNQX, DNQX
	Kainate		↑Na^+, K^+	Kainate	CNQX, DNQX
	NMDA		↑Na^+, K^+,Ca^{2+}	NMDA	AP5, AP7
	$mGluR_1$	↑cAMP, ↑IP_3, DAG	↓K^+, ↑Ca^{2+}	DHPG	
	$mGluR_5$	↑IP_3, DAG	↓K^+, ↑Ca^{2+}	Quisqualate	
	$mGluR_2$, $mGluR_3$	↓cAMP	↑K^+, ↓Ca^{2+}	DCG-IV	
	$mGluR_4$, $mGluR_{6\text{-}7}$	↓cAMP	↓Ca^{2+}	L-AP4	
GABA	$GABA_A$		↑Cl^-	Muscimol	Bicuculline, Gabazine, Picrotoxin
	$GABA_B$	↑IP_3, DAG	↑K^+,↓Ca^{2+}	Baclofen	Saclofen
Glycine	Glycine		↑Cl^-	Taurine, β-alanine	Strychnine
Acetylcholine	N_M		↑Na^+, K^+	Nicotine	Tubocurarine, Gallamine triethiodide
	N_N		↑Na^+, K^+	Nicotine, lobeline	Trimethaphan
	M_1, M_3, M_5	↑IP_3, DAG	↑Ca^{2+}	Muscarine, Bethanechol, Oxotremorine (M_1)	Atropine, Pirenzipine (M_1)
	M_2, M_4	↓cAMP	↑K^+	Muscarine, Bethanechol (M_2)	Atropine, Tropicamide (M_4)
Norepinephrine	α_1	↑IP_3, DAG	↓K^+	Phenylephrine	Prazosin, Tamsulosin
	α_2	↓cAMP	↑K^+,↓Ca^{2+}	Clonidine	Yohimbine
	β_1	↑cAMP	↓K^+	Isoproterenol, Dobutamine	Atenolol, Esmolol
	β_2	↑cAMP		Albuterol	Butoxamine
Serotonin	$5HT_{1A}$	↓cAMP	↑K^+	8-OH-DPAT	Metergoline, Spiperone
	$5HT_{1B}$	↓cAMP		Sumatriptan	
	$5HT_{1D}$	↓cAMP	↓K^+	Sumatriptan	
	$5HT_{2A}$	↑IP_3, DAG	↓K^+	Dobutamine	Ketanserin
	$5HT_{2C}$	↑IP_3, DAG		α-Methly-5-HT	
	$5HT_3$		↑Na^+	α-Methly-5-HT	Ondansetron
	$5HT_4$	↑cAMP	↓K^+	5-Methoxytryptamine	

8-OH-DPAT, 8-hydroxy-N, N-dipropyl-2-aminotetralin; AMPA, α-amino-3-hydroxyl-5-methyl-4-isoxazole-propionate; DAG, diacylglycerol; DCG-IV, 2-(2, 3-dicarboxycyclopropyl) glycine; DHPG, 3,5-Dihydroxyphenylglycine ; IP_3, inositol triphosphate; L-AP4, 2-amino-4-phosphonobutyrate; NMDA, N-methyl D-aspartate.

(GABA-T). In the second pathway, glutamate is released from the nerve terminal into the synaptic cleft by Ca^{2+}-dependent exocytosis and transported via a **glutamate reuptake transporter** into glia, where it is converted to **glutamine** by the enzyme **glutamine synthetase** (Figure 7–4). Glutamine then diffuses back into the nerve terminal where it is hydrolyzed back to glutamate by the enzyme **glutaminase.** In addition to uptake of released glutamate into glia, the membrane transporters also return glutamate directly into the nerve terminal. Within glutamatergic neurons, glutamate is highly concentrated in synaptic vesicles by a **vesicular glutamate transporter.**

Glutamate Receptors

Glutamate acts on both ionotropic and metabotropic receptors in the CNS (Figure 7–4). There are three subtypes of ionotropic glutamate receptors, each named for its relatively specific agonist. These are the **AMPA** (α-amino-3-hydroxy-

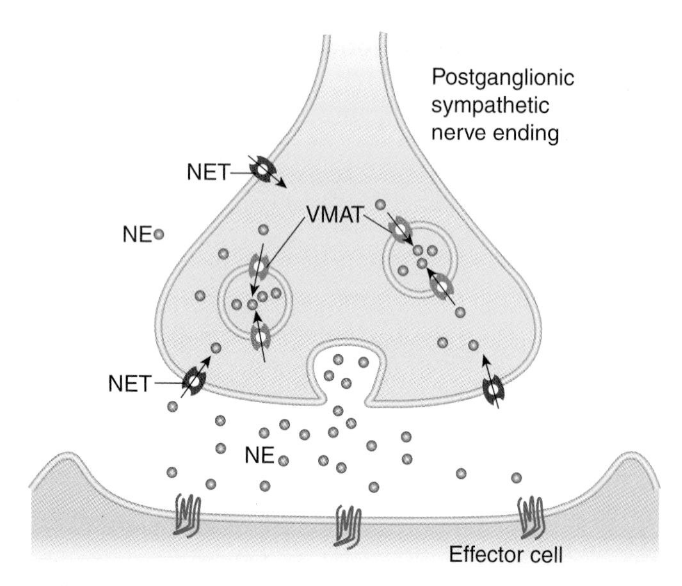

FIGURE 7–3 Fate of monoamines secreted at synaptic junctions. In each monoamine-secreting neuron, the monoamine is synthesized in the cytoplasm and the secretory granules and its concentration in secretory granules is maintained by the two vesicular monoamine transporters (VMAT). The monoamine is secreted by exocytosis of the granules, and it acts on G protein-coupled receptors. In this example, the monoamine is norepinephrine acting on adrenoceptors. Many of these receptors are postsynaptic, but some are presynaptic and some are located on glia. In addition, there is extensive reuptake of the monoamine into the cytoplasm of the presynaptic terminal via a monoamine transporter, in this case the norepinephrine transporter (NET). (Modified with permission from Katzung BG, Masters SB, Trevor AJ: *Basic and Clinical Pharmacology,* 11th ed. McGraw-Hill, 2009.)

5-methylisoxazole-4-propionate), **kainate** (kainate is an acid isolated from seaweed), and **NMDA** (N-methyl-D-aspartate) **receptors.** Table 7–2 summarizes some of the major properties of these receptors. Ionotropic glutamate receptors are tetramers composed of different subunits whose helical domains span the membranes three times and a short sequence that forms the channel pore. Four AMPA (GluR1 – GluR4), five kainite (GluR5 – GluR7, KA1, KA2), and six NMDA (NR1, NR2A – NR2D) subunits have been identified, and each is coded by a different gene.

The release of glutamate and its binding to AMPA or kainate receptors primarily permits the influx of Na^+ and the efflux of K^+, accounting for fast excitatory postsynaptic responses (EPSPs). Most AMPA receptors have low Ca^{2+} permeability, but the absence of certain subunits in the receptor complex at some sites allows for the influx of Ca^{2+}, which may contribute to the excitotoxic effect of glutamate (see Clinical Box 7–1).

Activation of the NMDA receptor permits the influx of relatively large amounts of Ca^{2+} along with Na^+. When glutamate is in excess in the synaptic cleft, the NMDA receptor-induced influx of Ca^{2+} into neurons is the major basis for the excitotoxic actions of glutamate. The NMDA receptor is unique in several ways (Figure 7–5). First, glycine facilitates its function by binding to the receptor. In fact, glycine binding is essential for the receptor to respond to glutamate. Second, when glutamate binds to the NMDA receptor, it opens, but at normal membrane potentials, the channel is blocked by extracellular Mg^{2+}. This block is removed only when the neuron containing the receptor is partially depolarized by the activation of

CLINICAL BOX 7–1

Excitotoxins

Glutamate is usually cleared from the brain's extracellular fluid by Na^+-dependent uptake systems in neurons and glia, keeping only micromolar levels of the chemical in the extracellular fluid despite millimolar levels inside neurons. However, excessive levels of glutamate occur in response to ischemia, anoxia, hypoglycemia, or trauma. Glutamate and some of its synthetic agonists are unique in that when they act on neuronal cell bodies, they can produce so much Ca^{2+} influx that the neurons die. This is the reason why microinjection of these **excitotoxins** is used in research to produce discrete lesions that destroy neuronal cell bodies without affecting neighboring axons. Evidence is accumulating that excitotoxins play a significant role in the damage done to the brain by a **stroke.** When a cerebral artery is occluded, the cells in the severely ischemic area die. Surrounding partially ischemic cells may survive but lose their ability to maintain the transmembrane Na^+ gradient. The elevated levels of intracellular Na^+ prevent the ability of **astrocytes** to remove glutamate from the brain's extracellular fluid. Therefore, glutamate accumulates to the point that excitotoxic damage and cell death occurs in the **penumbra,** the region around the completely infarcted area. In addition, excessive glutamate receptor activation may contribute to the pathophysiology of some neurodegenerative diseases such as **amyotrophic lateral sclerosis (ALS), Parkinson disease,** and **Alzheimer disease.**

THERAPEUTIC HIGHLIGHTS

Riluzole is a voltage-gated channel blocker that may antagonize NMDA receptors. It has been shown to slow the progression of impairment and modestly improve the life expectancy of patients with ALS. Another NMDA receptor antagonist **memantine** has been used to slow the progressive decline in patients with Alzheimer disease. A third NMDA receptor antagonist, **amantadine,** in conjunction with **levodopa,** has been shown to improve function in patients with Parkinson disease.

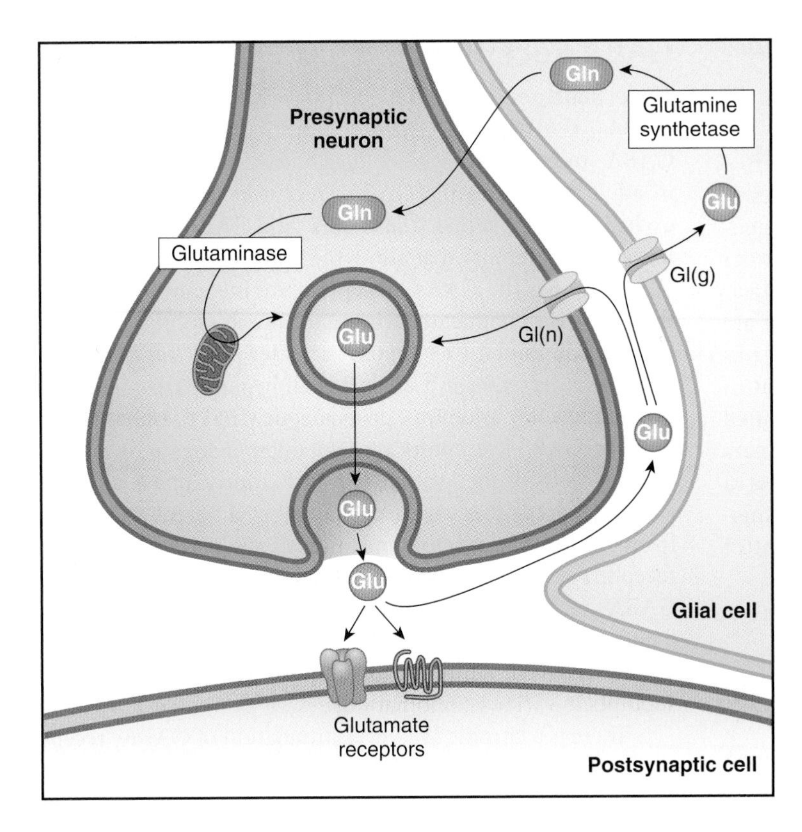

FIGURE 7–4 Biochemical events at a glutamatergic synapse. Glutamate (Glu) released into the synaptic cleft by Ca^{2+}-dependent exocytosis. Released Glu can act on ionotropic and G protein-coupled receptors on the postsynaptic neuron. Synaptic transmission is terminated by the active transport of Glu via by a Na^+-dependent glutamate transporters located on membranes of the presynaptic terminal [Gt(n)] and glia [Gt(g)]. In glia, Glu is converted to glutamine (Gln) by the enzyme glutamine synthetase; Gln then diffuses into the nerve terminal where it is hydrolyzed back to Glu by the enzyme glutaminase. In the nerve terminal, Glu is highly concentrated in synaptic vesicles by a vesicular glutamate transporter.

adjacent AMPA and kainite receptors. Third, the excitatory postsynaptic potential induced by activation of NMDA receptors is slower than that elicited by activation of the AMPA and kainite receptors.

Essentially all neurons in the CNS have both AMPA and NMDA receptors. Kainate receptors are located presynaptically on GABA-secreting nerve endings and postsynaptically at various sites, most notably in the hippocampus, cerebellum, and spinal cord. Kainate and AMPA receptors are found in glia as well as neurons, but NMDA receptors occur only in neurons. The concentration of NMDA receptors in the hippocampus is high, and blockade of these receptors prevents **long-term potentiation,** a long-lasting facilitation of transmission in neural pathways following a brief period of high-frequency stimulation. Thus, these receptors may well be involved in memory and learning (see Chapter 15).

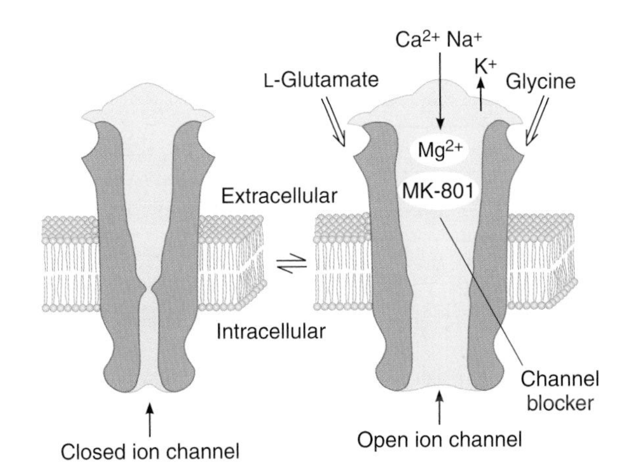

FIGURE 7–5 Diagrammatic representation of the NMDA receptor. When glycine and glutamate bind to the receptor, the closed ion channel (left) opens, but at the resting membrane potential, the channel is blocked by Mg^{2+} (right). This block is removed if partial depolarization is produced by other inputs to the neuron containing the receptor, and Ca^{2+} and Na^+ enter the neuron. Blockade can also be produced by the drug dizocilpine maleate (MK-801).

Activation of the metabotropic glutamate receptors (mGluR) leads to either an increase in intracellular inositol 1,4,5-triphosphate (IP_3) and diacylglycerol (DAG) levels or a decrease in intracellular cyclic adenosine monophosphate (cAMP) levels (Table 7–2). There are eight known subtypes of mGLuR. These receptors are located at both presynaptic ($mGluR_{2\text{-}4,\ 6\text{-}8}$) and postsynaptic sites ($mGluR_{1,\ 5}$), and they are widely distributed in the brain. They appear to be involved in the production of **synaptic plasticity,** particularly in the hippocampus and the cerebellum. Activation of presynaptic mGluR autoreceptors on neurons in the hippocampus limits the release of glutamate from these neurons. Knockout of the gene mGluR1 causes severe motor incoordination and deficits in spatial learning.

A characteristic of an excitatory synapse is the presence of a thickened area called a **postsynaptic density (PSD)** on the membrane of the postsynaptic neuron. This is a complex structure containing ionotropic glutamate receptors and signaling, scaffolding, and cytoskeletal proteins. The mGLuR are located adjacent to the PSD.

Pharmacology of Glutamate Synapses

Table 7–2 shows some of the pharmacological properties of various types of glutamate receptors, examples of the agonists that bind to these receptors, and some of the antagonists that prevent activation of the receptors. The clinical applications of drugs that modulate glutamatergic transmission are still in their infancy. This is because the role of glutamate as a neurotransmitter was discovered much later than that of most other small-molecule transmitters. It was not identified as a neurotransmitter until the 1970's, more than 50 years after the discovery of chemical neurotransmission. One area of drug development is intraspinal or extradural administration of NMDA receptor antagonists for the treatment of chronic pain.

GABA

GABA is the major inhibitory mediator in the brain and mediates both presynaptic and postsynaptic inhibition. GABA, which exists as β-aminobutyrate in the body fluids, is formed by decarboxylation of glutamate (Figure 7–1) by the enzyme **glutamate decarboxylase (GAD),** which is present in nerve endings in many parts of the brain. GABA is metabolized primarily by transamination to succinic semialdehyde and then to succinate in the citric acid cycle. **GABA-T** is the enzyme that catalyzes the transamination. In addition, there is an active reuptake of GABA via the GABA transporter. A **vesicular GABA transporter (VGAT)** transports GABA and glycine into secretory vesicles.

GABA Receptors

Three subtypes of GABA receptors have been identified: $GABA_A$, $GABA_B$, and $GABA_C$ (Table 7–2). The $GABA_A$ and $GABA_B$ receptors are widely distributed in the CNS, whereas in adult vertebrates the $GABA_C$ receptors are found almost exclusively in the retina. The $GABA_A$ and $GABA_C$ receptors are ionotropic receptors that allow the entry of Cl^- into neurons (Figure 7–6). The $GABA_B$ receptors are metabotropic GPCR that increase conductance in K^+ channels, inhibit adenylyl cyclase, and inhibit Ca^{2+} influx. Increases in Cl^- influx and K^+ efflux and decreases in Ca^{2+} influx all hyperpolarize neurons, producing a fast inhibitory postsynaptic (IPSP) response.

The $GABA_A$ receptors are pentamers made up of various combinations of six α subunits, four β, four γ, one δ, and one ε. This endows them with considerably different properties from one location to another. However, most synaptic $GABA_A$ receptors have two α, two β, and one γ subunit (Figure 7–6). $GABA_A$ receptors on dendrites, axons, or somas often contain δ and ε subunits in place of the γ subunit. The $GABA_C$ receptors are relatively simple in that they are pentamers of three ρ subunits in various combinations.

There is a chronic low-level stimulation of $GABA_A$ receptors in the CNS that is aided by GABA in the interstitial fluid. This background stimulation cuts down on the "noise" caused by incidental discharge of the billions of neural units and greatly improves the signal-to-noise ratio in the brain.

Pharmacology of GABA Synapses

Table 7–2 shows some of the pharmacological properties of GABA receptors, including examples of agonists that bind to receptors and some of the antagonists that prevent activation of

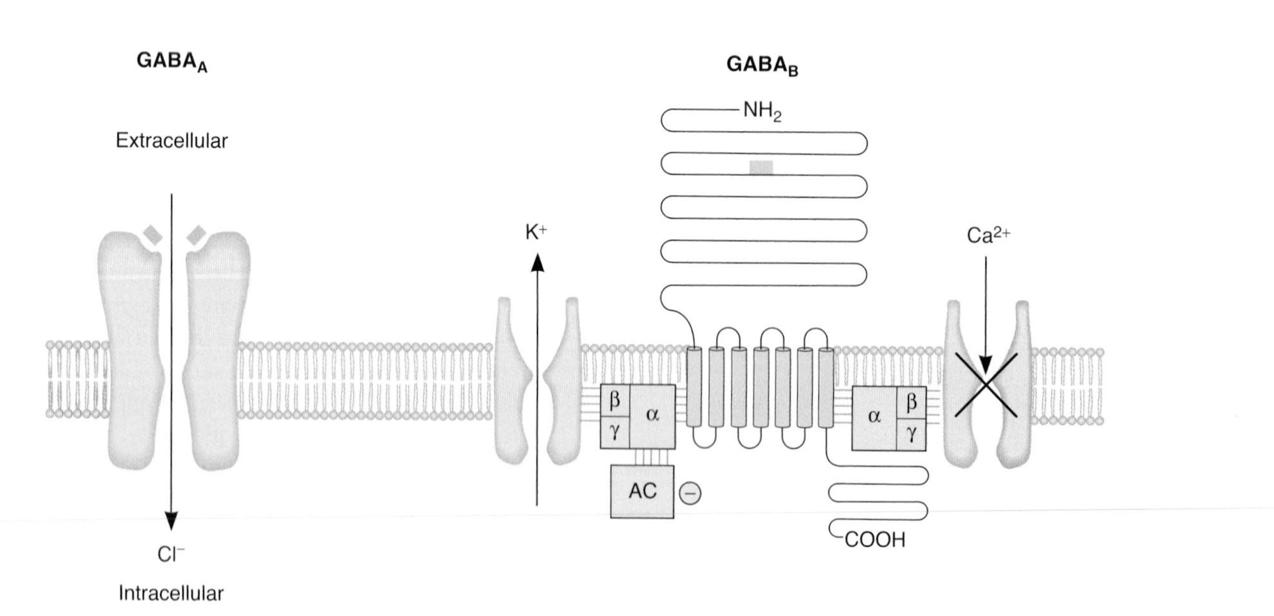

FIGURE 7–6 Diagram of $GABA_A$ and $GABA_B$ receptors, showing their principal actions. Two molecules of GABA (*squares*) bind to the $GABA_A$ receptor to allow an influx of Cl^-. One molecule of GABA binds to the $GABA_B$ receptor, which couples to the α-subunit of the G protein. G_i inhibits adenylyl cyclase (AC) to open a K^+ channel; G_o delays the opening of a Ca^{2+} channel. (Reproduced with permission from Bowery NG, Brown DA: The cloning of GABAB receptors. Nature 1997;386:223.)

these receptors. The increase in Cl^- conductance produced by $GABA_A$ receptors is potentiated by **benzodiazepines** (eg, **diazepam**). Thus, these are examples of neuromodulators. These drugs have marked anti-anxiety activity and are also effective muscle relaxants, anticonvulsants, and sedatives. Benzodiazepines bind to α subunits of $GABA_A$ receptors. **Barbiturates** such as **phenobarbital** are effective anticonvulsants because they enhance $GABA_A$ receptor-mediated inhibition as well as suppress AMPA receptor-mediated excitation. The anesthetic actions of barbiturates (**thiopental, pentobarbital,** and **methoxital**) result from their actions as agonists at $GABA_A$ receptors as well as by acting as neuromodulators of GABA transmission. Regional variation in anesthetic actions in the brain seems to parallel the variation in subtypes of $GABA_A$ receptors. Other inhaled anesthetics do not act by increasing GABA receptor activity; rather, they act by inhibiting NMDA and AMPA receptors.

Glycine

Glycine has both excitatory and inhibitory effects in the CNS. When it binds to NMDA receptors, it makes them more sensitive to the actions of glutamate. Glycine may spill over from synaptic junctions into the interstitial fluid, and in the spinal cord; for example, it may facilitate pain transmission by NMDA receptors in the dorsal horn. However, glycine is also responsible in part for direct inhibition, primarily in the brain stem and spinal cord. Like GABA, it acts by increasing Cl^- conductance. Its action is antagonized by strychnine. The clinical picture of convulsions and muscular hyperactivity produced by strychnine emphasizes the importance of postsynaptic inhibition in normal neural function. The glycine receptor responsible for inhibition is a Cl^- channel. It is a pentamer made up of two subunits: the ligand-binding α subunit and the structural β subunit. There are three kinds of neurons that are responsible for direct inhibition in the spinal cord: neurons that secrete glycine, neurons that secrete GABA, and neurons that secrete both. Neurons that secrete only glycine have the glycine transporter GLYT2, those that secrete only GABA have GAD, and those that secrete glycine and GABA have both. This third type of neuron is of special interest because the neurons seem to have glycine and GABA in the same vesicles.

Acetylcholine

Acetylcholine is the transmitter at the neuromuscular junction, in autonomic ganglia, and in postganglionic parasympathetic nerve-target organ junctions and some postganglionic sympathetic nerve-target junctions (see Chapter 13). In fact, acetylcholine is the transmitter released by all neurons that exit the CNS (cranial nerves, motor neurons, and preganglionic neurons). Acetylcholine is also found in the basal forebrain complex (septal nuclei and nucleus basalis), which projects to the hippocampus and neocortex, and the pontomesencephalic cholinergic complex, which projects to the dorsal thalamus and forebrain (Figure 7–2). These systems may be involved in regulation of sleep-wake states, learning, and memory (see Chapters 14 and 15).

Acetylcholine is largely enclosed in small, clear synaptic vesicles in high concentration in the terminals of **cholinergic neurons**. It is synthesized in the nerve terminal from choline and acetyl-CoA by the enzyme **choline acetyltransferase (ChAT)** (Figure 7–1 and Figure 7–7). Choline used in the synthesis of acetylcholine is transported from the extracellular space into the nerve terminal via a Na^+-dependent **choline transporter**

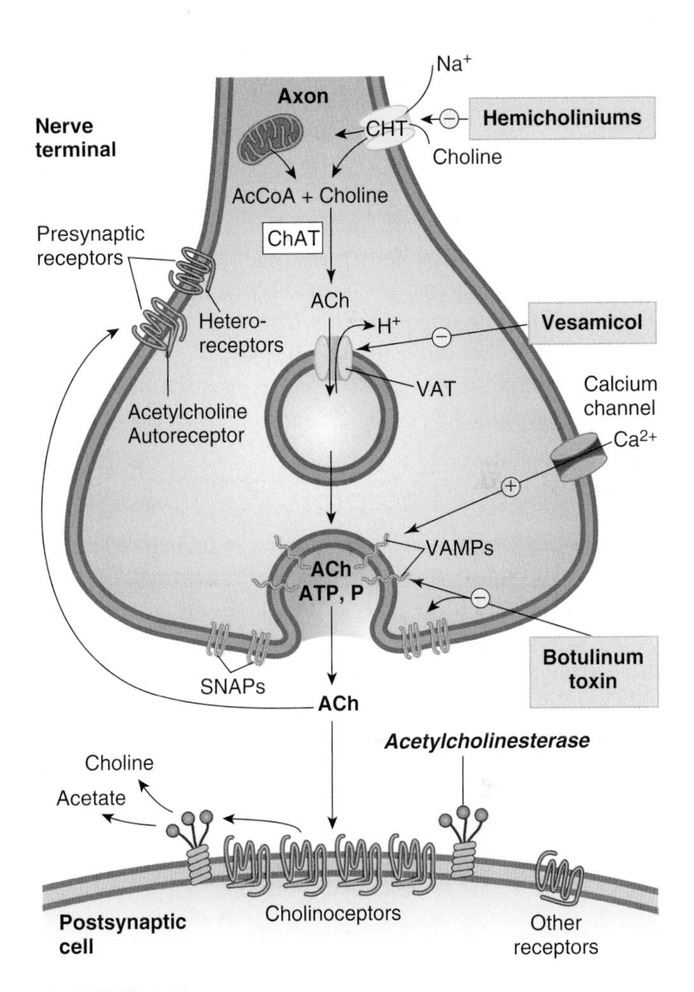

FIGURE 7–7 Biochemical events at a cholinergic synapse. Choline is transported into the presynaptic nerve terminal by a Na^+-dependent choline transporter (CHT), which can be blocked by the drug hemicholinium. Acetylcholine (ACh) is synthesized from choline and acetyl Co-A (AcCoA) by the enzyme choline acetyltransferase (ChAT) in the cytoplasm. ACh is then transported from the cytoplasm into vesicles by the vesicle-associated transporter (VAT) along with peptides (P) and adenosine triphosphate (ATP). This step can be blocked by the drug vesamicol. ACh is released from the nerve terminal when voltage-sensitive Ca^{2+} channels open, allowing an influx of Ca^{2+}, which leads to fusion of vesicles with the surface membrane and expulsion of ACh and co-transmitters into the synaptic cleft. This process involves synaptosome-associated proteins (SNAPs) and vesicle-associated membrane proteins (VAMPs) and can be prevented by the drug botulinum toxin. The released ACh can act on muscarinic G protein-coupled receptors on the postsynaptic target (eg, smooth muscle) or on nicotinic ionotropic receptors in autonomic ganglia or the endplate of skeletal muscle (not shown). In the synaptic junction, ACh is readily metabolized by the enzyme acetylcholinesterase. Autoreceptors and heteroreceptors on the presynaptic nerve ending modulate neurotransmitter release.

(CHT). Following its synthesis, acetylcholine is transported from the cytoplasm into vesicles by a **vesicle-associated transporter (VAT).** Acetylcholine is released when a nerve impulse triggers the influx of Ca^{2+} into the nerve terminal.

Acetylcholine must be rapidly removed from the synapse if repolarization is to occur. The removal occurs by way of hydrolysis of acetylcholine to choline and acetate, a reaction catalyzed by the enzyme **acetylcholinesterase** in the synaptic cleft. This enzyme is also called **true** or **specific cholinesterase.** Its greatest affinity is for acetylcholine, but it also hydrolyzes other choline esters. Acetylcholinesterase molecules are clustered in the postsynaptic membrane of cholinergic synapses. Hydrolysis of acetylcholine by this enzyme is rapid enough to explain the observed changes in Na^+ conductance and electrical activity during synaptic transmission. There are a variety of cholinesterases in the body that are not specific for acetylcholine. One found in plasma is capable of hydrolyzing acetylcholine but has different properties from acetylcholinesterase. It is called **pseudocholinesterase.** The plasma moiety is partly under endocrine control and is affected by variations in liver function.

Acetylcholine Receptors

Acetylcholine receptors are divided into two main types on the basis of their pharmacologic properties. Muscarine, the alkaloid responsible for the toxicity of toadstools, mimics the stimulatory action of acetylcholine on smooth muscle and glands. These actions of acetylcholine are called **muscarinic actions,** and the receptors involved are **muscarinic cholinergic receptors.** In sympathetic ganglia and skeletal muscle, nicotine mimics the stimulatory actions of acetylcholine. These actions of acetylcholine are called **nicotinic actions,** and the receptors involved are **nicotinic cholinergic receptors.** Nicotinic receptors are subdivided into those found in muscle at the neuromuscular junction (N_M) and those found in the CNS and autonomic ganglia (N_N). Both muscarinic and nicotinic acetylcholine receptors are also found within the brain.

The nicotinic acetylcholine receptors are members of a superfamily of ligand-gated ion channels (ionotropic receptors) that also includes the $GABA_A$ and glycine receptors and some of the glutamate receptors. Each nicotinic cholinergic receptor is made up of five subunits that form a central channel which, when the receptor is activated, permits the passage of Na^+ and other cations. The five subunits come from several types designated as α, β, γ, δ, and ε that are each coded by different genes. The N_M receptor is comprised of two α, one β, one δ, and either one γ or one ε subunit (Figure 7–8). The N_N receptors are comprised of only α and β subunits. Each α subunit has a binding site for acetylcholine, and binding of an acetylcholine molecule to each of them induces a conformational change in the protein so that the channel opens. This increases the conductance of Na^+, and the resulting influx of Na^+ produces a depolarizing potential. A prominent feature of neuronal nicotinic cholinergic receptors is their high permeability to Ca^{2+}. Many of the nicotinic cholinergic receptors in the brain are located presynaptically on glutamate-secreting axon terminals, and they facilitate the release of this transmitter.

There are five types of muscarinic cholinergic receptors (M_1–M_5), which are encoded by five separate genes. These are metabotropic receptors that are coupled via G proteins to adenylyl cyclase, K^+ channels, and/or phospholipase C

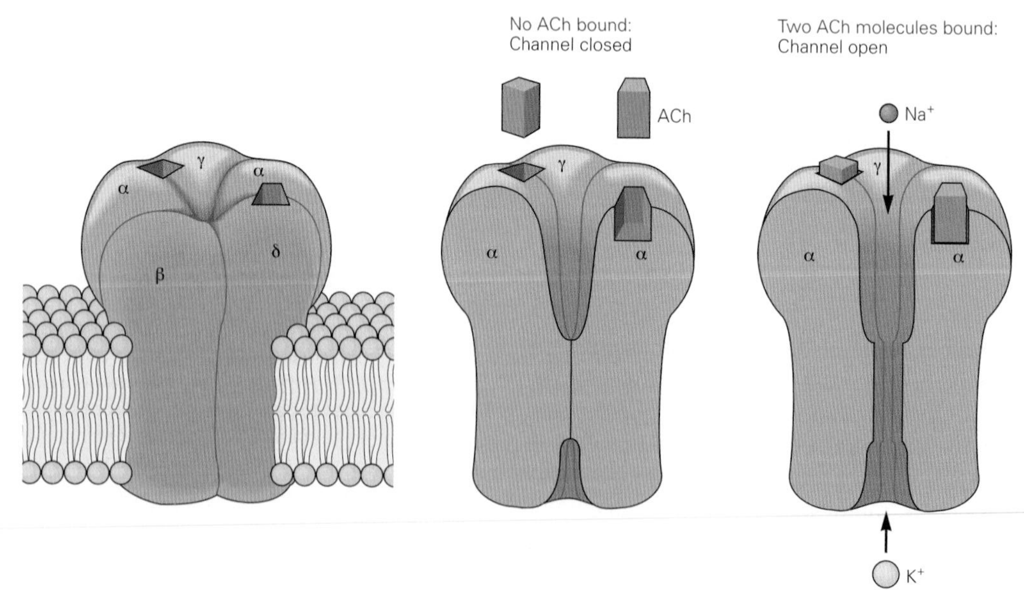

FIGURE 7–8 Three-dimensional model of the nicotinic acetylcholine-gated ion channel. The receptor–channel complex consists of five subunits, all of which contribute to forming the pore. When two molecules of acetylcholine bind to portions of the α-subunits exposed to the membrane surface, the receptor–channel changes conformation. This opens the pore in the portion of the channel embedded in the lipid bilayer, and both K^+ and Na^+ flow through the open channel down their electrochemical gradient. (From Kandel ER, Schwartz JH, Jessell TM [editors]: *Principles of Neural Science,* 4th ed. McGraw-Hill, 2000.)

(Table 7–2). M_1, M_4, and M_5 receptors are located in the CNS; M_2 receptors are in the heart, M_3 are on glands and smooth muscle. M_1 receptors are also located on autonomic ganglia where they can modulate neurotransmission.

Pharmacology of Cholinergic Synapses

Table 7–2 shows some of the major agonists that bind to cholinergic receptors as well as some of the cholinergic receptor antagonists. Figure 7–7 also shows the site of action of various drugs that alter cholinergic transmission. For example, **hemicholinium** blocks the choline transporter that moves choline into the nerve terminal, and **vesamicol** blocks the VAT that moves acetylcholine into the synaptic vesicle. Also, **botulinum toxin** prevents the release of acetylcholine from the nerve terminal.

MONOAMINES

Norepinephrine & Epinephrine

The chemical transmitter present at most sympathetic postganglionic endings is norepinephrine. It is stored in the synaptic knobs of the neurons that secrete it in characteristic small vesicles that have a dense core (granulated vesicles). Norepinephrine and its methyl derivative, epinephrine, are also secreted by the adrenal medulla (see Chapter 20), but epinephrine is not a mediator at postganglionic sympathetic endings. As discussed in Chapter 6, each sympathetic postganglionic neuron has multiple varicosities along its course, and each of these varicosities appears to be a site at which norepinephrine is secreted.

There are also norepinephrine-secreting and epinephrine-secreting neurons in the brain. Norepinephrine-secreting neurons are properly called **noradrenergic neurons,** although the term **adrenergic neurons** is also applied. However, the latter term should be reserved for epinephrine-secreting neurons. The cell bodies of the norepinephrine-containing neurons are located in the locus coeruleus and other medullary and pontine nuclei (Figure 7–2). From the locus coeruleus, the axons of the noradrenergic neurons descend into the spinal cord, enter the cerebellum, and ascend to innervate the paraventricular, supraoptic, and periventricular nuclei of the hypothalamus, the thalamus, the basal telencephalon, and the entire neocortex. The action of norepinephrine in these regions is primarily as a neuromodulator.

Biosynthesis & Release of Catecholamines

The principal **catecholamines** found in the body (norepinephrine, epinephrine, and dopamine) are formed by hydroxylation and decarboxylation of the amino acid tyrosine (Figure 7–1 and Figure 7–9). Some of the tyrosine is formed from phenylalanine, but most is of dietary origin. **Phenylalanine hydroxylase** is found primarily in the liver (see Clinical Box 7–2). Tyrosine is transported into catecholamine-secreting neurons via a Na^+-dependent carrier. It is converted to dihydroxy-phenylalanine (DOPA) and then to dopamine in the cytoplasm of the cells by **tyrosine hydroxylase** and **DOPA decarboxylase,** respectively. The decarboxylase is also called **amino acid decarboxylase.** The rate-limiting step in synthesis of catecholamines is the conversion of tyrosine to DOPA. Tyrosine hydroxylase is subject to feedback inhibition by dopamine and norepinephrine, thus providing internal control of the synthetic process.

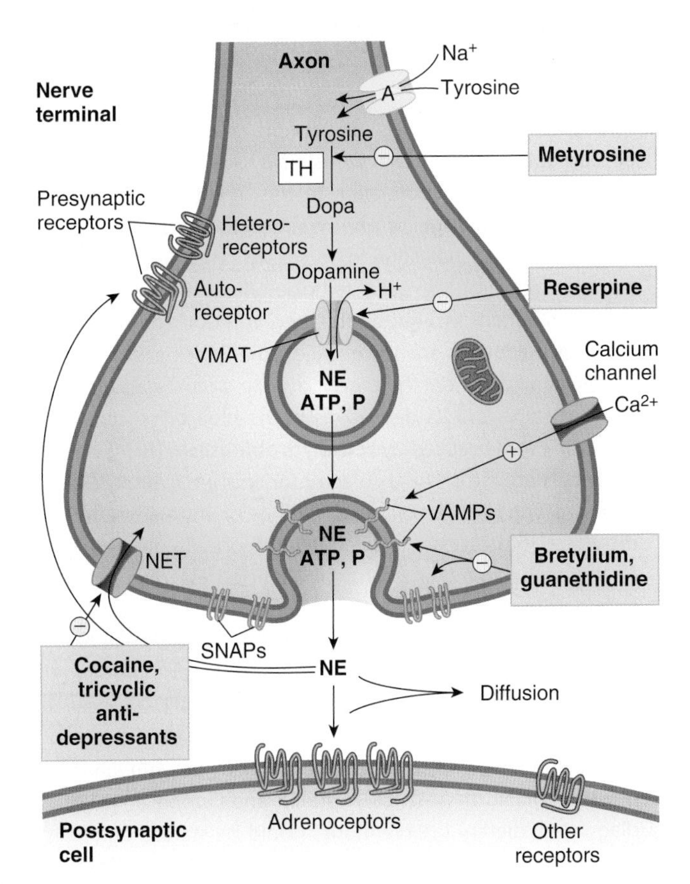

FIGURE 7–9 Biochemical events at a noradrenergic synapse. Tyrosine (Tyr) is transported into the noradrenergic nerve terminal by a Na^+-dependent carrier (A). The steps involved in the conversion of Tyr to dopamine and dopamine to norepinephrine (NE) are described in Figure 7–1. Dopamine is transported from the cytoplasm into the vesicle by the vesicular monoamine transporter (VMAT), which can be blocked by the drug reserpine. NE and other amines can also be carried by VMAT. Dopamine is converted to NE in the vesicle. An action potential opens voltage-sensitive Ca^{2+} channels to allow an influx of Ca^{2+}, and the vesicles then fuse with the surface membrane to trigger expulsion of NE along with peptides (P) and adenosine triphosphate (ATP). This process involves synaptosome-associated proteins (SNAPs) and vesicle-associated membrane proteins (VAMPs); it can be blocked by drugs such as guanethidine and bretylium. NE released into the nerve terminal can act on G protein-coupled receptors on the postsynaptic neuron or neuroeffector organ (eg, blood vessels). NE can also diffuse out of the cleft or be transported back into the nerve terminal by the norepinephrine transporter (NET). NET can be blocked by cocaine and tricyclic antidepressants. Autoreceptors and heteroreceptors on the presynaptic nerve ending modulate neurotransmitter release.

CLINICAL BOX 7–2

Phenylketonuria

Phenylketonuria (PKU) is an example of an inborn error of metabolism. PKU is characterized by severe mental deficiency and the accumulation in the blood, tissues, and urine of large amounts of **phenylalanine** and its keto acid derivatives. It is usually due to decreased function resulting from mutation of the gene for **phenylalanine hydroxylase.** This gene is located on the long arm of chromosome 12. Catecholamines are still formed from tyrosine, and the cognitive impairment is largely due to accumulation of phenylalanine and its derivatives in the blood. The condition can also be caused by **tetrahydrobiopterin (BH4) deficiency.** Because BH4 is a cofactor for tyrosine hydroxylase and tryptophan hydroxylase, as well as phenylalanine hydroxylase, PKU cases due to tetrahydrobiopterin deficiency have catecholamine and serotonin deficiencies in addition to hyperphenylalaninemia. These cause hypotonia, inactivity, and developmental problems. BH4 is also essential for the synthesis of nitric oxide (NO) by nitric oxide synthase. Severe BH4 deficiency can lead to impairment of NO formation, and the CNS may be subjected to increased oxidative stress. Blood phenylalanine levels are usually determined in newborns in North America, Australia, and Europe; if PKU is diagnosed, dietary interventions should be started before the age of 3 weeks to prevent the development of mental retardation.

THERAPEUTIC HIGHLIGHTS

PKU can usually be treated successfully by markedly reducing the amount of phenylalanine in the diet. This means restricting the intake of high-protein foods such as milk, eggs, cheese, meats, and nuts. In individuals with a BH4 deficiency, treatment can include tetrahydrobiopterin, **levodopa,** and **5-hydroxytryptophan** in addition to a low-phenylalanine diet. The US Food and Drug Administration approved the drug **sapropterin,** a synthetic BH4, for the treatment of some people with PKU.

Once dopamine is synthesized, it is transported into the vesicle by the **VMAT.** Here the dopamine is converted to norepinephrine by **dopamine β-hydroxylase (DBH).** Norepinephrine is the only small-molecule transmitter that is synthesized in synaptic vesicles instead of being transported into the vesicle after its synthesis.

Some neurons in the CNS and adrenal medullary cells also contain the cytoplasmic enzyme **phenylethanolamine-N-methyltransferase (PNMT),** which catalyzes the conversion of norepinephrine to epinephrine. In these cells, norepinephrine leaves the vesicles, is converted to epinephrine in the cytoplasm, and then enters other vesicles for storage until it is released by exocytosis.

Catabolism of Catecholamines

Norepinephrine, like other amine and amino acid transmitters, is removed from the synaptic cleft by binding to postsynaptic receptors, binding to presynaptic receptors, reuptake into the presynaptic neurons, or catabolism. Reuptake via a **NET** is a major mechanism to terminate the actions of norepinephrine (Figure 7–3), and the hypersensitivity of sympathetically denervated structures is explained in part on this basis. After the noradrenergic neurons are cut, their endings degenerate with loss of NET to remove norepinephrine from the synaptic cleft. Consequently, more norepinephrine from other sources is available to stimulate the receptors on the autonomic effectors.

Epinephrine and norepinephrine are metabolized to biologically inactive products by oxidation and methylation. The former reaction is catalyzed by **monoamine oxidase (MAO)** and the latter by **catechol-O-methyltransferase (COMT).** MAO is located on the outer surface of the mitochondria. MAO is widely distributed, being particularly plentiful in the nerve endings at which catecholamines are secreted. COMT is also widely distributed, particularly in the liver, kidneys, and smooth muscles. In the brain, it is present in glial cells, and small amounts are found in postsynaptic neurons, but none is found in presynaptic noradrenergic neurons. Consequently, catecholamine metabolism has two different patterns.

Extracellular epinephrine and norepinephrine are for the most part O-methylated, and measurement of the concentrations of the O-methylated derivatives normetanephrine and metanephrine in the urine is a good index of the rate of secretion of norepinephrine and epinephrine. The O-methylated derivatives that are not excreted are largely oxidized, and vanillylmandelic acid (VMA) is the most plentiful catecholamine metabolite in the urine.

In the noradrenergic nerve terminals, some of the norepinephrine is constantly being converted by intracellular MAO to the physiologically inactive deaminated derivatives, 3, 4-dihydroxymandelic acid (DOMA) and its corresponding glycol (DHPG). These are subsequently converted to their corresponding O-methyl derivatives, VMA and 3-methoxy-4-hydroxyphenylglycol (MHPG).

α- & β-Adrenoceptors

Epinephrine and norepinephrine both act on α-and β-adrenergic receptors **(adrenoceptors),** with norepinephrine having a greater affinity for α-adrenoceptors and epinephrine for β-adrenoceptors. These receptors are metabotropic GPCR, and each has multiple subtypes (α_{1A}, α_{1B}, α_{1D}, α_{2A}, α_{2B}, α_{2C}, and β_1 – β_3). Most α_1-adrenoceptors are coupled via G_q proteins to phospholipase C, leading to the formation of IP_3 and DAG, which mobilizes intracellular Ca^{2+} stores and activates protein kinase C, respectively. Thus, at many synapses, activation of α_1-adrenoceptors is excitatory to the postsynaptic target. In contrast, α_2-adrenoceptors activate G_i inhibitory proteins to inhibit adenylyl cyclase and decrease cAMP. Other actions of α_2-adrenoceptors are to activate G protein coupled inward rectifier K^+ channels to cause membrane hyperpolarization and to inhibit neuronal Ca^{2+} channels. Thus, at many synapses, activation of α_2-adrenoceptors inhibits the postsynaptic target. Presynaptic α_2-adrenoceptors are autoreceptors which, when activated, inhibit further release of norepinephrine from postganglionic sympathetic nerve terminals. β-adrenoceptors activate a stimulatory G_S protein to activate adenylyl cyclase to increase cAMP.

α_1-Adrenoceptors are located on smooth muscle and the heart, and α2-adrenoceptors are located in the CNS and on pancreatic islets cells and nerve terminals. β_1-adrenoceptors are located in the heart and renal juxtaglomerular cells. β_2-adrenoceptors are located in bronchial smooth muscle and skeletal muscle. β_3-adrenoceptors are located in adipose tissue.

Pharmacology of Noradrenergic Synapses

Table 7–2 shows some of the common agonists that bind to adrenoceptors as well as some of the common adrenoceptor antagonists. Figure 7–9 also shows the site of action of various drugs that alter noradrenergic transmission. For example, **metyrosine** blocks the action of tyrosine hydroxylase, the rate limiting step in the synthetic pathway for catecholamine production the nerve terminal. **Reserpine** blocks the VMAT that moves dopamine into the synaptic vesicle. Also, **bretylium** and **guanethidine** prevent the release of norepinephrine from the nerve terminal. **Cocaine** and **tricyclic antidepressants** block the NET. In addition to the agonists listed in Table 7–2, some drugs mimic the actions of norepinephrine by releasing stored transmitter from the noradrenergic endings. These are called **sympathomimetics** and include **amphetamines** and **ephedrine.**

Dopamine

In some parts of the brain, catecholamine synthesis stops at dopamine (Figure 7–1), which can then be secreted into the synaptic cleft. Active reuptake of dopamine occurs via a Na^+ and Cl^--dependent **dopamine transporter**. Dopamine is metabolized to inactive compounds by MAO and COMT in a manner analogous to the inactivation of norepinephrine. 3, 4-Dihydroxyphenylacetic acid (DOPAC) and homovanillic acid (HVA) are conjugated, primarily to sulfate.

Dopaminergic neurons are located in several brain regions (Figure 7–2). One region is the **nigrostriatal system,** which projects from the midbrain substantia nigra to the striatum in the basal ganglia and is involved in motor control. Another dopaminergic system is the **mesocortical system,** which arises primarily in the ventral tegmental area which projects to the nucleus accumbens and limbic subcortical areas; it is involved in reward behavior and addiction and in psychiatric disorders such as schizophrenia (Clinical Box 7–3). Studies

CLINICAL BOX 7–3

Schizophrenia

Schizophrenia is an illness involving deficits of multiple brain systems that alter an individual's inner thoughts as well as their interactions with others. Individuals with schizophrenia suffer from hallucinations, delusions, and racing thoughts (positive symptoms); and they experience apathy, difficulty dealing with novel situations, and little spontaneity or motivation (negative symptoms). Worldwide, about 1–2% of the population lives with schizophrenia. A combination of genetic, biological, cultural, and psychological factors contributes to the illness. A large amount of evidence indicates that a defect in the **mesocortical system** is responsible for the development of at least some of the symptoms of schizophrenia. Attention was initially focused on overstimulation of limbic **D_2 dopamine receptors. Amphetamine,** which causes release of dopamine as well as norepinephrine in the brain, causes a schizophrenia-like psychosis; brain levels of D_2 receptors are said to be elevated in schizophrenics; and there is a clear positive correlation between the anti-schizophrenic activity of many drugs and their ability to block D_2 receptors. However, several recently developed drugs are effective antipsychotic agents but bind to D_2 receptors to a limited degree. Instead, they bind to D_4 receptors, and there is active ongoing research into the possibility that these receptors are abnormal in individuals with schizophrenia.

THERAPEUTIC HIGHLIGHTS

Since the mid-1950s numerous antipsychotic drugs (eg, **chlorpromazine, haloperidol, perphenazine,** and **fluphenazine**) have been used to treat schizophrenia. In the 1990s, new "atypical" antipsychotics were developed. These include **clozapine,** which reduces psychotic symptoms, hallucinations, and breaks with reality. However a potential adverse side effect is agranulocytosis (a loss of the white blood cells), which impairs the ability to fight infections. Other atypical antipsychotics do not cause agranulocytosis, including **risperidone, olanzapine, quetiapine, ziprasidone, aripiprazole,** and **paliperidone.**

by **positive emission tomography (PET)** scanning in normal humans show that a steady loss of dopamine receptors occurs in the basal ganglia with age. The loss is greater in men than in women.

Dopamine Receptors

Five dopamine receptors have been cloned, but they fall into two major categories: D_1-like (D_1 and D_5) and D_2-like (D_2, D_3, and D_4). All dopamine receptors are metabotropic GPCR. Activation of D_1-type receptors leads to an increase in cAMP, whereas activation of D_2-like receptors reduces cAMP levels. Overstimulation of D_2 receptors may contribute to the pathophysiology of schizophrenia (Clinical Box 7–3). D_3 receptors are highly localized, especially to the nucleus accumbens (Figure 7–2). D_4 receptors have a greater affinity than the other dopamine receptors for the "atypical" antipsychotic drug **clozapine,** which is effective in schizophrenia but produces fewer extrapyramidal side effects than the other major tranquilizers do.

Serotonin

Serotonin (5-hydroxytryptamine; 5-HT) is present in highest concentration in blood platelets and in the gastrointestinal tract, where it is found in the enterochromaffin cells and the myenteric plexus. It is also found within the brain stem in the midline raphé nuclei, which project to a wide portion of the CNS including the hypothalamus, limbic system, neocortex, cerebellum, and spinal cord (Figure 7–2).

Serotonin is synthesized from the essential amino acid **tryptophan** (Figure 7–1 and Figure 7–10). The rate-limiting step is the conversion of the amino acid to **5-hydroxytrptophan** by **tryptophan hydroxylase.** This is then converted to serotonin by the **aromatic L-amino acid decarboxylase.** Serotonin is transported into the vesicles by the VMAT. After release from serotonergic neurons, much of the released serotonin is recaptured by the relatively selective **serotonin transporter (SERT).** Once serotonin is returned to the nerve terminal it is either taken back into the vesicles or is inactivated by MAO to form 5-hydroxyindoleacetic acid (5-HIAA). This substance is the principal urinary metabolite of serotonin, and urinary output of 5-HIAA is used as an index of the rate of serotonin metabolism in the body.

Tryptophan hydroxylase in the CNS is slightly different from the tryptophan hydroxylase in peripheral tissues, and is coded by a different gene. This is presumably why knockout of the *TPH1* gene, which codes for tryptophan hydroxylase in peripheral tissues, has much less effect on brain serotonin production than on peripheral serotonin production.

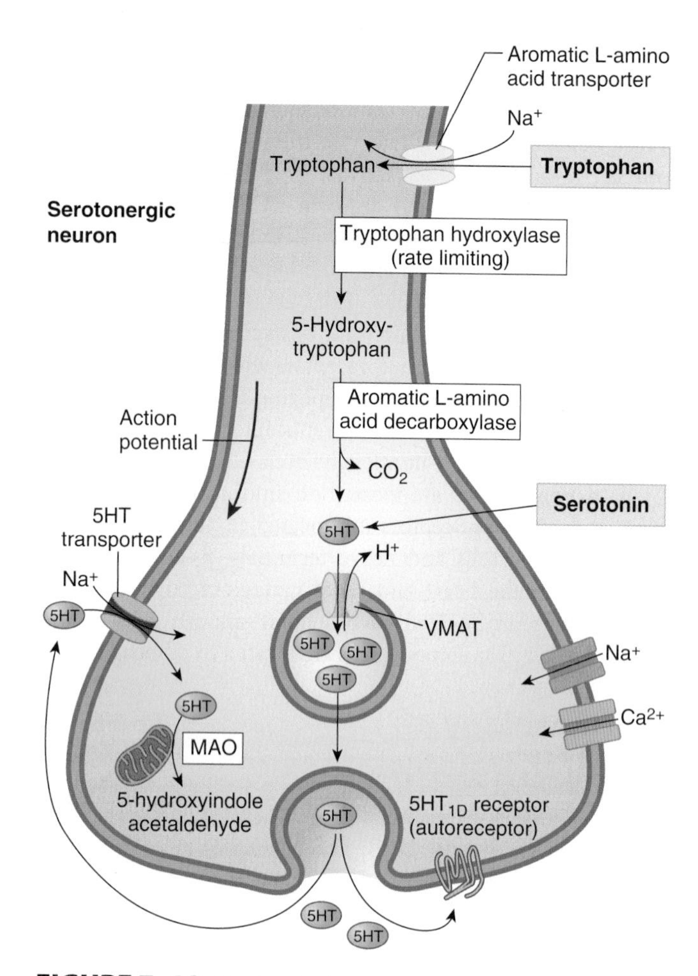

FIGURE 7–10 Biochemical events at a serotonergic synapse. Tryptophan is transported into the serotonergic nerve terminal by a Na^+-dependent aromatic L-amino acid transporter. The steps involved in the conversion of tryptophan to serotonin (5-hydroxytryptamine, 5-HT) are described in Figure 7–1. 5-HT is transported from the cytoplasm into vesicles by the vesicular monoamine transporter (VMAT). 5-HT release occurs when an action potential opens voltage-sensitive Ca^{2+} channels to allow an influx of Ca^{2+} and fusion of vesicles with the surface membrane. 5-HT released into the nerve terminal can act on G protein-coupled receptors on the postsynaptic neuron (not shown). 5-HT can also diffuse out of the cleft or be transported back into the nerve terminal by the 5-HT transporter. 5-HT can act on presynaptic autoreceptors to inhibit further neurotransmitter release. Cytoplasmic 5-HT is either sequestered in vesicles as described or metabolized to 5-hydroxyindole acetaldehyde by mitochondrial monoamine oxidase (MAO).

Serotonergic Receptors

There are seven classes of 5-HT receptors (from 5-HT_1 through 5-HT_7 receptors), and all except one (5-HT_3) are GPCR and affect adenylyl cyclase or phospholipase C (Table 7–2). Within the 5-HT_1 group are the 5-HT_{1A}, 5-HT_{1B}, 5-HT_{1D}, 5-HT_{1E}, and 5-HT_{1F} subtypes. Within the 5-HT_2 group there are 5-HT_{2A}, 5-HT_{2B}, and 5-HT_{2C} subtypes. There are two 5-HT_5 sub-types: 5-HT_{5A} and 5-HT_{5B}. Some of the serotonin receptors are presynaptic, and others are postsynaptic.

5-HT_{2A} receptors mediate platelet aggregation and smooth muscle contraction. Mice in which the gene for 5-HT_{2C} receptors has been knocked out are obese as a result of increased food intake despite normal responses to leptin (see Chapter 26), and they are prone to fatal seizures. 5-HT_3

CLINICAL BOX 7–4

Major Depression

According to the National Institutes of Mental Health, nearly 21 million Americans over the age of 18 have a mood disorder that includes **major depressive disorder**, **dysthymia**, and **bipolar disease.** The largest group is those diagnosed with major depression. Major depression has a median age of onset of 32 years and is more prevalent in women than men. Symptoms of major depression include depressed mood, anhedonia, loss of appetite, insomnia or hypersomnia, restlessness, fatigue, feelings of worthlessness, diminished ability to think or concentrate, and recurrent thoughts of suicide. **Typical depression** is characterized by feelings of sadness, early-morning awakenings, decreased appetite, restlessness, and anhedonia. Symptoms of **atypical depression** include pleasure-seeking behavior and hypersomnia.

The precise cause of depression is unknown, but genetic factors likely contribute. There is strong evidence for a role of central monoamines, including norepinephrine, serotonin, and dopamine. The hallucinogenic agent **lysergic acid diethylamide (LSD)** is a central 5-HT_2 receptor agonist. The transient hallucinations produced by this drug were discovered when the chemist who synthesized it inhaled some by accident. Its discovery drew attention to the correlation between behavior and variations in brain serotonin content. **Psilocin,** a substance found in certain mushrooms, and **N, N-dimethyltryptamine (DMT)** are also hallucinogenic and, like serotonin, are derivatives of tryptamine. **2, 5-Dimethoxy-4-methyl-amphetamine (DOM)** and **mescaline** and other true hallucinogens are phenylethylamines. However, each of these may exert their effects by binding to 5-HT_2 receptors. **3, 4-Methylenedioxymethamphetamine** (MDMA or **ecstasy**) is a popular drug of abuse that produces euphoria followed by difficulty in concentrating and depression. The drug causes release of serotonin followed by serotonin depletion; the euphoria may be due to the release and the later symptoms to the depletion.

THERAPEUTIC HIGHLIGHTS

In cases of typical depression, drugs such as **fluoxetine (Prozac),** which are selective **serotonin reuptake inhibitors (SSRIs),** are effective as antidepressants. SSRIs are also used to treat **anxiety disorders.** In atypical depression, SSRIs are often ineffective. Instead, **monoamine oxidase inhibitors (MAOIs)** such as **phenelzine** and **selegiline** have been shown to be effective as antidepressants. However, they have adverse consequences including hypertensive crisis if the patient ingests large quantities of products high in **tyramine,** which include aged cheese, processed meats, avocados, dried fruits, and red wines (especially Chianti). Based on evidence that atypical depression may result from a decrease in both serotonin and dopamine, drugs acting more generally on monoamines have been developed. These drugs, called **atypical antidepressants,** include **bupropion,** which resembles amphetamine and increases both serotonin and dopamine levels in the brain. Bupropion is also used as **smoking cessation therapy.**

receptors are present in the gastrointestinal tract and the area postrema and are related to vomiting. 5-HT_4 receptors are also present in the gastrointestinal tract, where they facilitate secretion and peristalsis, and in the brain. 5-HT_6 and 5-HT_7 receptors in the brain are distributed throughout the limbic system, and the 5-HT_6 receptors have a high affinity for antidepressant drugs.

Pharmacology of Serotonergic Synapses

Table 7–2 shows some of the common agonists that bind to 5-HT receptors as well as some of the common 5-HT receptor antagonists. In addition, tricyclic antidepressants inhibit the uptake of serotonin by block of SERT, similar to what was described for their actions at noradrenergic synapses. **Selective serotonin uptake inhibitors (SSRIs)** such as **fluoxetine** are widely used in the treatment of depression (see Clinical Box 7–4).

Histamine

Histaminergic neurons have their cell bodies in the tuberomammillary nucleus of the posterior hypothalamus, and their axons project to all parts of the brain, including the cerebral cortex and the spinal cord. Histamine is also found in cells in the gastric mucosa and in heparin-containing cells called **mast cells** that are plentiful in the anterior and posterior lobes of the pituitary gland as well as at body surfaces. Histamine is formed by decarboxylation of the amino acid histidine. The three well-characterized types of histamine receptors (H_1, H_2, and H_3) are all found in both peripheral tissues and the brain. Most, if not all, of the H_3 receptors are presynaptic, and they mediate inhibition of the release of histamine and other transmitters via a G protein. H_1 receptors activate phospholipase C, and H_2 receptors increase intracellular cAMP. The function of this diffuse histaminergic system is unknown, but evidence links brain histamine to arousal, sexual behavior, blood pressure, drinking, pain thresholds, and regulation of the secretion

release and bind to presynaptic CB_1 receptors to inhibit further transmitter release. A CB_2 receptor, which also couples to G proteins, has also been cloned; it is located primarily in the periphery. Agonists of this class of receptor do not induce the euphoric effects of activation of CB_1 receptors and they may have potential for use in the treatment of chronic pain.

CHAPTER SUMMARY

- Neurotransmitters and neuromodulators are divided into two major categories: small-molecule transmitters and large-molecule transmitters (neuropeptides). Usually neuropeptides are colocalized with one of the small-molecule neurotransmitters.
- Rapid removal of the chemical transmitter from the synaptic cleft occurs by diffusion, metabolism, and, in many instances, reuptake into the presynaptic neuron.
- Major neurotransmitters include glutamate, GABA, and glycine, acetylcholine, norepinephrine, serotonin, and opioids. ATP, NO, and cannabinoids can also act as neurotransmitters or neuromodulators.
- The amino acid glutamate is the main excitatory transmitter in the CNS. There are two major types of glutamate receptors: metabotropic (GPCR) and ionotropic (ligand-gated ion channels receptors, including kainite, AMPA, and NMDA).
- GABA is the major inhibitory mediator in the brain. Three subtypes of GABA receptors have been identified: GABAA and GABAC (ligand-gated ion channel) and GABAB (G protein-coupled). The GABAA and GABAB receptors are widely distributed in the CNS.
- Acetylcholine is found at the neuromuscular junction, in autonomic ganglia, and in postganglionic parasympathetic nerve-target organ junctions and a few postganglionic sympathetic nerve-target junctions. It is also found in the basal forebrain complex and pontomesencephalic cholinergic complex. There are two major types of cholinergic receptors: muscarinic (GPCR) and nicotinic (ligand-gated ion channel receptors).
- Norepinephrine-containing neurons are in the locus coeruleus and other medullary and pontine nuclei. Some neurons also contain PNMT, which catalyzes the conversion of norepinephrine to epinephrine. Epinephrine and norepinephrine act on α- and β-adrenoceptors, with norepinephrine having a greater affinity for α-adrenoceptors and epinephrine for β-adrenoceptors. They are GPCR, and each has multiple forms.
- Serotonin (5-HT) is found within the brain stem in the midline raphé nuclei which project to portions of the hypothalamus, the limbic system, the neocortex, the cerebellum, and the spinal cord. There are at least seven types of 5-HT receptors, and many of these contain subtypes. Most are GPCR.
- The three types of opioid receptors (μ, κ, and δ) are GPCR that differ in physiological effects, distribution in the brain and elsewhere, and affinity for various opioid peptides.

MULTIPLE-CHOICE QUESTIONS

For all questions, select the single best answer unless otherwise directed.

1. Which of the following statements about neurotransmitters is true?
   A. All neurotransmitters are derived from amino acid precursors.
   B. Small molecule neurotransmitters include dopamine, histamine, ATP, glycine, enkephalin, and norepinephrine.
   C. Large molecule transmitters include ATP, cannabinoids, substance P, and vasopressin.
   D. Norepinephrine can act as a neurotransmitter in the periphery and a neuromodulator in the CNS.
   E. Nitrous oxide is a neurotransmitter in the CNS.
2. Which of the following statements is *not* true?
   A. Neuronal glutamate is synthesized in glia by the enzymatic conversion from glutamine and then diffuses into the neuronal terminal where it is sequestered into vesicles until released by an influx of Ca^{2+} into the cytoplasm after an action potential reaches the nerve terminal.
   B. After release of serotonin into the synaptic cleft, its actions are terminated by reuptake into the presynaptic nerve terminal, an action that can be blocked by tricyclic antidepressants.
   C. Norepinephrine is the only small-molecule transmitter that is synthesized in synaptic vesicles instead of being transported into the vesicle after its synthesis.
   D. Each nicotinic cholinergic receptor is made up of five subunits that form a central channel which, when the receptor is activated, permits the passage of Na^+ and other cations.
   E. GABA transaminase converts glutamate to GABA; the vesicular GABA transporter transports both GABA and glycine into synaptic vesicles.
3. Which of the following receptors is correctly identified as an ionotropic or a G protein-coupled receptor (GPCR)?
   A. Neurokinin receptor: ionotropic
   B. Nicotinic receptor: GPCR
   C. $GABA_A$ receptor: ionotropic
   D. NMDA receptor: GPCR
   E. Glycine: GPCR
4. A 27-year-old male was brought to the emergency room and presented with symptoms of opioid intoxication. He was given an intravenous dose of naloxone. Endogenous opioids
   A. bind to both ionotropic receptors and GPCR.
   B. include morphine, endorphins, and dynorphins.
   C. show the following order of affinity for δ receptors: dynorphins > > endorphins.
   D. show the following order of affinity for μ receptors: dynorphins > endorphins.
   E. show the following order of affinity for κ receptors: endorphins > > enkephalins.

5. A 38-year-old woman was sent to a psychiatrist after she reported to her primary care physician that she had difficulty sleeping (awakening at 4 am frequently for the past few months) and a lack of appetite causing a weight loss of over 20 lbs. She also said she no longer enjoyed going out with her friends or doing volunteer service for underprivileged children. What type of drug is her doctor most likely to suggest as an initial step in her therapy?
   A. A serotonergic receptor antagonist.
   B. An inhibitor of neuronal uptake of serotonin.
   C. An inhibitor of monoamine oxidase.
   D. An amphetamine-like drug.
   E. A drug that causes an increase in both serotonin and dopamine.
6. A 55-year-old woman had been on long-term treatment with phenelzine for her depression. One night she was at a party where she consumed Chianti wine, aged cheddar cheese, processed meats, and dried fruits. She then developed a severe headache, chest pain, rapid heartbeat, enlarged pupils, increased sensitivity to light, and nausea. What is the most likely cause of these symptoms?
   A. The foods were contaminated with botulinum toxin.
   B. She had a myocardial infarction.
   C. She experienced a migraine headache.
   D. She had an adverse reaction to the mixture of alcohol with her antidepressant.
   E. She had a hypertensive crisis from eating foods high in tyramine while taking a monoamine oxidase inhibitor for her depression.

CHAPTER RESOURCES

Cooper JR, Bloom FE, Roth RH: *The Biochemical Basis of Neuropharmacology*, 8th ed. Oxford University Press, 2002.

Fink KB, Göthert M: 5-HT receptor regulation of neurotransmitter release. Pharmacol Rev 2007;59:360.

Jacob TJ, Moss SJ, Jurd R: $GABA_A$ receptor trafficking and its role in the dynamic modulation of neuronal inhibition. Nat Rev Neurosci 2008;9:331.

Katzung BG, Masters SB, Trevor AJ: *Basic and Clinical Pharmacology*, 11th ed. McGraw-Hill, 2009.

Madden DR: The structure and function of glutamate receptor ion channels. Nat Rev Neurosci 2002;3:91.

Monaghan DT, Bridges RJ, Cotman CW: The excitatory amino acid receptors: Their classes, pharmacology, and distinct properties in the function of the central nervous system. Annu Rev Pharmacol Toxicol 1989;29:365.

Olsen RW: The molecular mechanism of action of general anesthetics: Structural aspects of interactions with $GABA_A$ receptors. Toxicol Lett 1998;100:193.

Owens DF, Kriegstein AR: Is there more to GABA than synaptic inhibition? Nat Rev Neurosci 2002;3:715.

Roth BL: *The Serotonin Receptors: From Molecular Pharmacology to Human Therapeutics,* Humana Press, 2006.

Small KM, McGraw DW, Liggett SB: Pharmacology and physiology of human adrenergic receptor polymorphisms. Annu Rev Pharmacol Toxicol 2003;43:381.

Snyder SH, Pasternak GW: Historical review: Opioid receptors. Trends Pharamcol Sci 2003;24:198.

SECTION II Central and Peripheral Neurophysiology

INTRODUCTION TO NEUROPHYSIOLOGY

The central nervous system (CNS) can be likened to a computer processor that is the command center for most if not all of the functions of the body. The peripheral nervous system is like a set of cables that transfers critical data from the CNS to the body and then feeds back information from the body to the CNS. This "computer system" is very sophisticated and is designed to continually make appropriate adjustments to its inputs and outputs in order to allow one to react and adapt to changes in the external and internal environment (sensory systems), to maintain posture, permit locomotion, and use the fine motor control in our hands to create pieces of art (somatomotor system), to maintain homeostasis (autonomic nervous system), to regulate the transitions between sleep and wakefulness (consciousness), and to allow us to recall past events and to communicate with the outside world (higher cortical functions). This section on Neurophysiology will describe the fundamental properties and integrative capabilities of neural systems that allow for the exquisite control of this vast array of physiological functions. Medical fields such as neurology, neurosurgery, and clinical psychology build on the foundation of neurophysiology.

One of the most common reasons that an individual seeks the advice of a physician is because they are in pain. Severe chronic pain involves the rewiring of neural circuits that can result in an unpleasant sensation from even a simple touch to the skin. Chronic pain is a devastating health problem that is estimated to affect nearly one in 10 Americans (more than 25 million people). Within the past decade or so there have been considerable advancements made in understanding how activity is altered in these individuals and in identifying receptors types that are unique to nociceptive pathways. These findings have led to an expanding research effort to develop novel therapies that specifically target synaptic transmission in central nociceptive pathways and in peripheral sensory transduction. This is welcomed by the many individuals who do not get pain relief from nonsteroidal anti-inflammatory agents or even morphine. These kinds of research breakthroughs would not be possible without a thorough understanding of how the brain and body communicate with each other.

In addition to chronic pain, there are over 600 known neurological disorders. Nearly 50 million people in the United States alone and an estimated 1 billion people worldwide suffer from the effects of damage to the central or peripheral nervous system. Nearly 7 million people die annually as a result of a neurological disorder. Neurological disorders include genetic disorders (eg, Huntington disease), demyelinating diseases (eg, multiple sclerosis), developmental disorders (eg, cerebral palsy), degenerative diseases that target specific types of neurons (eg, Parkinson disease and Alzheimer disease), an imbalance of neurotransmitters (eg, depression, anxiety, and eating disorders), trauma (eg, spinal cord and head injury), and convulsive disorders (eg, epilepsy). In addition, there are neurological complications associated with cerebrovascular problems (eg, stroke) and exposure to neurotoxic chemicals (eg, nerve gases, mushroom poisoning, and pesticides).

Advances in stem cell biology and brain imaging techniques, a greater understanding of the basis for synaptic plasticity of the brain, a wealth of new knowledge about the regulation of receptors and the release of neurotransmitters, and the detection of genetic and molecular defects that lead to neurological problems have all contributed to advancements in identifying the pathophysiological basis for neurological disorders. They have also set the stage for identifying better therapies to prevent, reverse, or stabilize the physiological deficits that the more than 600 neurological disorders cause.

Somatosensory Neurotransmission: Touch, Pain, and Temperature

CHAPTER 8

OBJECTIVES

After studying this chapter, you should be able to:

- Name the types of touch and pressure receptors found in the skin.
- Describe the receptors that mediate the sensations of pain and temperature.
- Define generator potential.
- Explain the basic elements of sensory coding.
- Explain the differences between pain and nociception, first and second pain, acute and chronic pain, hyperalgesia and allodynia.
- Describe and explain visceral and referred pain.
- Compare the pathway that mediates sensory input from touch, proprioceptive, and vibratory senses to that mediating information from nociceptors and thermoreceptors.
- Describe processes involved in modulation of transmission in pain pathways.
- List some drugs that have been used for relief of pain and give the rationale for their use and their clinical effectiveness.

INTRODUCTION

We learn in elementary school that there are "five senses" (touch, sight, hearing, smell, and taste); but this dictum takes into account only those senses that reach our consciousness. There are many **sensory receptors** that relay information about the internal and external environment to the central nervous system (CNS) but do not reach consciousness. For example, the muscle spindles provide information about muscle length, and other receptors provide information about arterial blood pressure, the levels of oxygen and carbon dioxide in the blood, and the pH of the cerebrospinal fluid. The list of sensory modalities listed in Table 8–1 is overly simplified. The rods and cones, for example, respond maximally to light of different wavelengths, and three different types of cones are present, one for each of the three primary colors. There are five different modalities of taste: sweet, salt, sour, bitter, and umami. Sounds of different pitches are heard primarily because different groups of hair cells in the cochlea are activated maximally by sound waves of different frequencies.

Sensory receptors can be thought of as transducers that convert various forms of energy in the environment into action potentials in sensory neurons. The cutaneous receptors for touch and pressure are **mechanoreceptors. Proprioceptors** are located in muscles, tendons, and joints and relay information about muscle length and tension. **Thermoreceptors** detect the sensations of warmth and cold. Potentially harmful stimuli such as pain, extreme heat, and extreme cold are mediated by **nociceptors.** The term **chemoreceptor** refers to receptors stimulated by a change in the chemical composition of the environment in which they are located. These include receptors for taste and smell as well as visceral receptors such as those sensitive to changes in the plasma level of O_2, pH, and osmolality. **Photoreceptors** are those in the rods and cones in the retina that respond to light.

This chapter describes primarily the characteristics of cutaneous receptors that mediate the sensations of touch, pressure, pain, and temperature, the way they generate impulses in afferent neurons, and the central pathways that mediate or modulate information from these receptors. Since pain is one of the main reasons an individual seeks the advice of a physician, this topic gets considerable attention in this chapter. Receptors involved in the somatosensory modality of proprioception are described in Chapter 12 as they play key roles in the control of balance, posture, and limb movement.

SENSE RECEPTORS & SENSE ORGANS

CUTANEOUS MECHANORECEPTORS

Sensory receptors can be specialized dendritic endings of afferent nerve fibers, and they are often associated with non-neural cells that surround them forming a **sense organ.** Touch and pressure are sensed by four types of mechanoreceptors (Figure 8–1). **Meissner's corpuscles** are dendrites encapsulated in connective tissue and respond to changes in texture and slow vibrations. **Merkel cells** are expanded dendritic endings, and they respond to sustained pressure and touch. **Ruffini corpuscles** are enlarged dendritic endings with elongated capsules, and they respond to sustained pressure. **Pacinian corpuscles** consist of unmyelinated dendritic endings of a sensory nerve fiber, 2 μm in diameter, encapsulated by concentric lamellae of connective tissue that give the organ the appearance of a cocktail onion. Theses receptors respond to deep pressure and fast vibration. The sensory nerves from these mechanoreceptors are large myelinated Aα and Aβ fibers whose conduction velocities range from ~70–120 to ~40–75 m/s, respectively.

NOCICEPTORS

Some cutaneous sensory receptors are not specialized organs but rather they are free nerve endings. Pain and temperature sensations arise from unmyelinated dendrites of sensory neurons located throughout the glabrous and hairy skin as well as deep tissue. Nociceptors can be separated into several types. **Mechanical nociceptors** respond to strong pressure (eg, from a sharp object). **Thermal nociceptors** are activated by skin temperatures above 42°C or by severe cold. **Chemically sensitive nociceptors** respond to various chemicals like bradykinin, histamine, high acidity, and environmental irritants. **Polymodal nociceptors** respond to combinations of these stimuli.

TABLE 8–1 Principle sensory modalities.

Sensory System	Modality	Stimulus Energy	Receptor Class	Receptor Cell Types
Somatosensory	Touch	Tap, flutter 5–40 Hz	Cutaneous mechanoreceptor	Meissner corpuscles
Somatosensory	Touch	Motion	Cutaneous mechanoreceptor	Hair follicle receptors
Somatosensory	Touch	Deep pressure, vibration 60–300 Hz	Cutaneous mechanoreceptor	Pacinian corpuscles
Somatosensory	Touch	Touch, pressure	Cutaneous mechanoreceptor	Merkel cells
Somatosensory	Touch	Sustained pressure	Cutaneous mechanoreceptor	Ruffini corpuscles
Somatosensory	Proprioception	Stretch	Mechanoreceptor	Muscle spindles
Somatosensory	Proprioception	Tension	Mechanoreceptor	Golgi tendon organ
Somatosensory	Temperature	Thermal	Thermoreceptor	Cold and warm receptors
Somatosensory	Pain	Chemical, thermal, and mechanical	Chemoreceptor, thermoreceptor, and mechanoreceptor	Polymodal receptors or chemical, thermal, and mechanical nociceptors
Somatosensory	Itch	Chemical	Chemoreceptor	Chemical nociceptor
Visual	Vision	Light	Photoreceptor	Rods, cones
Auditory	Hearing	Sound	Mechanoreceptor	Hair cells (cochlea)
Vestibular	Balance	Angular acceleration	Mechanoreceptor	Hair cells (semicircular canals)
Vestibular	Balance	Linear acceleration, gravity	Mechanoreceptor	Hair cells (otolith organs)
Olfactory	Smell	Chemical	Chemoreceptor	Olfactory sensory neuron
Gustatory	Taste	Chemical	Chemoreceptor	Taste buds

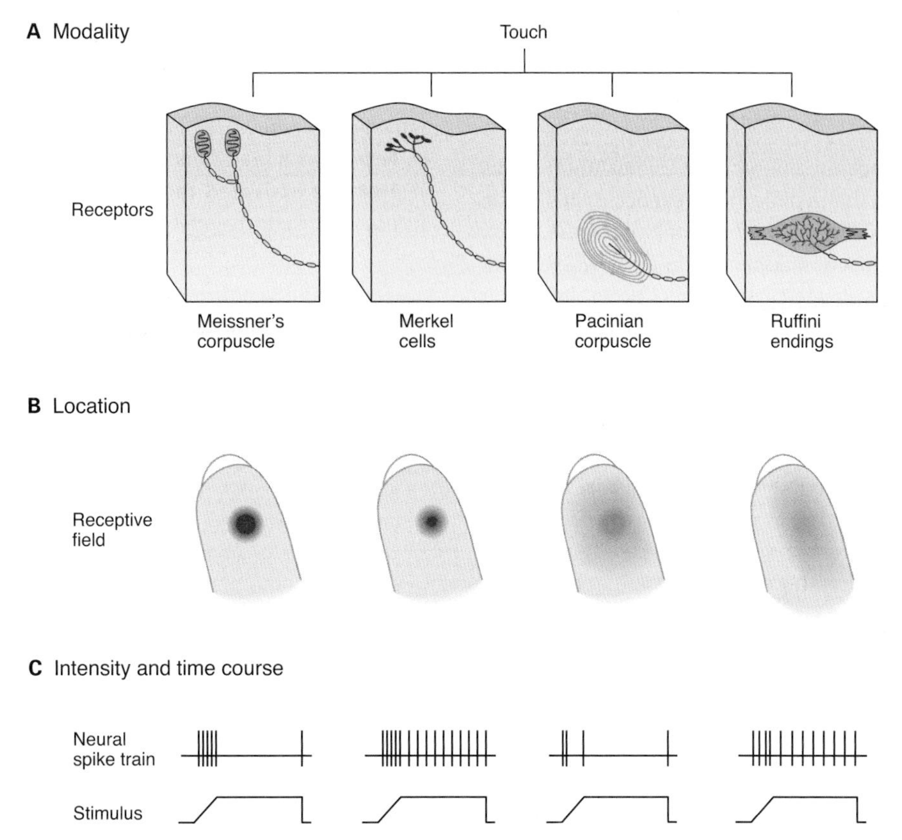

FIGURE 8–1 Sensory systems encode four elementary attributes of stimuli: modality, location (receptive field), intensity, and duration (timing). A) The human hand has four types of mechanoreceptors; their combined activation produces the sensation of contact with an object. Selective activation of Merkel cells and Ruffini endings causes sensation of steady pressure; selective activation of Meissner's and Pacinian corpuscles causes tingling and vibratory sensation. **B)** Location of a stimulus is encoded by spatial distribution of the population of receptors activated. A receptor fires only when the skin close to its sensory terminals is touched. These receptive fields of mechanoreceptors (shown as red areas on fingertips) differ in size and response to touch. Merkel cells and Meissner's corpuscles provide the most precise localization as they have the smallest receptive fields and are most sensitive to pressure applied by a small probe. **C)** Stimulus intensity is signaled by firing rates of individual receptors; duration of stimulus is signaled by time course of firing. The spike trains indicate action potentials elicited by pressure from a small probe at the center of each receptive field. Meissner's and Pacinian corpuscles adapt rapidly, the others adapt slowly. (From Kandel ER, Schwartz JH, Jessell TM [editors]: *Principles of Neural Science,* 4th ed. McGraw-Hill, 2000.)

Impulses from nociceptors are transmitted via two fiber types, thinly myelinated Aδ fibers (2–5 μm in diameter) that conduct at rates of ~12–35 m/s and unmyelinated C fibers (0.4–1.2 μm in diameter) that conduct at low rates of ~0.5–2 m/s. Activation of Aδ fibers, which release **glutamate,** is responsible for **first pain** (also called **fast pain** or **epicritic pain)** which is a rapid response and mediates the discriminative aspect of pain or the ability to localize the site and intensity of the noxious stimulus. Activation of C fibers, which release a combination of glutamate and **substance P,** is responsible for the delayed **second pain** (also called **slow pain** or **protopathic pain**) which is the dull, intense, diffuse, and unpleasant feeling associated with a noxious stimulus. **Itch** and **tickle** are also related to pain sensation (see Clinical Box 8–1).

There are a variety of receptors located on the endings of nociceptive sensory nerves that respond to noxious thermal, mechanical, or chemical stimuli (Figure 8–2). Many of these are part of a family of nonselective cation channels called **transient receptor potential** (TRP) channels. This includes **TRPV1** receptors (the V refers to a group of chemicals called **vanilloids**) that are activated by intense heat, acids, and chemicals such as **capsaicin** (the active principle of hot peppers and an example of a vanilloid). TRPV1 receptors can also be activated indirectly by initial activation of TRPV3 receptors in keratinocytes in the skin. Noxious mechanical, cold, and chemical stimuli may activate **TRPA1** receptors (A, for **ankyrin**) on sensory nerve terminals. Sensory nerve endings also have **acid sensing ion channel (ASIC) receptors** that are activated by pH changes within a physiological range and may be the dominant receptors mediating acid-induced pain. In addition to direct activation of receptors on nerve endings, some nociceptive stimuli release intermediate molecules that then activate receptors on the nerve ending. For example, nociceptive mechanical stimuli cause the release of ATP that acts on **purinergic receptors** (eg, P2X, an ionotropic receptor and P2Y, a G protein-coupled receptor). **Tyrosine receptor kinase A (TrkA)** is activated by **nerve growth factor (NGF)** that is released as a result of tissue damage.

CLINICAL BOX 8–1

Itch & Tickle

Itching (**pruritus**) is not a major problem for healthy individuals, but severe itching that is difficult to treat occurs in diseases such as chronic renal failure, some forms of liver disease, atopic dermatitis, and HIV infection. Especially in areas where many free endings of unmyelinated nerve fibers occur, itch spots can be identified on the skin by careful mapping. In addition, itch-specific fibers have been demonstrated in the ventrolateral spinothalamic tract. This and other evidence implicate the existence of an itch-specific path. Relatively mild stimulation, especially if produced by something that moves across the skin, produces itch and tickle. It is interesting that a tickling sensation is usually regarded as pleasurable, whereas itching is annoying and pain is unpleasant. Itching can be produced not only by repeated local mechanical stimulation of the skin but also by a variety of chemical agents including **histamine** and kinins such as **bradykinin** which are released in the skin in response to tissue damage. Kinins exert their effects by activation of two types of G protein-coupled receptors, B_1 and B_2. Activation of bradykinin B_2 receptors is a downstream event in **protease-activated receptor-2 (PAR-2)** activation, which induces both a nociceptive and a pruritogenic response.

THERAPEUTIC HIGHLIGHTS

Simple scratching relieves itching because it activates large, fast-conducting afferents that gate transmission in the dorsal horn in a manner analogous to the inhibition of pain by stimulation of similar afferents. **Antihistamines** are primarily effective in reducing pruritis associated with an allergic reaction. In a mouse model exhibiting scratching behavior in response to activation of PAR-2, treatment with a B_2 receptor antagonist reduced the scratching behavior. B_2 receptor antagonists may be a useful therapy for treating pruriginous conditions.

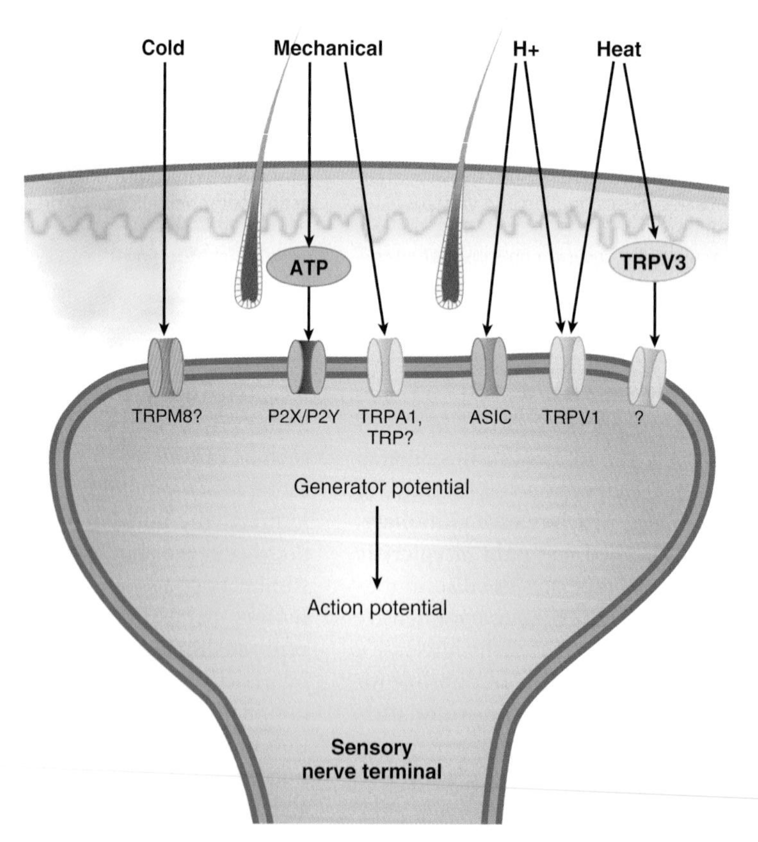

FIGURE 8–2 Receptors on nociceptive unmyelinated nerve terminals in the skin. Nociceptive stimuli (eg, heat) can activate some receptors directly due to transduction of the stimulus energy by receptors (eg, transient receptor potential (TRP) channel TRPV1) or indirectly by activation of TRP channels on keratinocytes (eg, TRPV3). Nociceptors (eg, mechanoreceptors) can also be activated by the release of intermediate molecules (eg, ATP). ASIC, acid-sensitive ion channel; P2X, ionotropic purinoceptor; P2Y, G protein-coupled purinergic receptor.

Nerve endings also have a variety of receptors that respond to immune mediators that are released in response to tissue injury. These include B_1 and B_2 receptors (bradykinin), prostanoid receptors (prostaglandins), and cytokine receptors (interleukins). These receptors mediate **inflammatory pain.**

THERMORECEPTORS

Innocuous **cold receptors** or **cool receptors** are on dendritic endings of Aδ fibers and C fibers, whereas innocuous **warmth receptors** are on C fibers. Mapping experiments show that the skin has discrete cold-sensitive and heat-sensitive spots. There are 4–10 times as many cold-sensitive as heat-sensitive spots.

The threshold for activation of warmth receptors is 30°C, and they increase their firing rate as the skin temperature increases to 46°C. Cold receptors are inactive at temperatures of 40°C, but then steadily increase their firing rate as skin temperature falls to about 24°C. As skin temperature further decreases, the firing rate of cold receptors decreases until the temperature reaches 10°C. Below that temperature, they are inactive and cold becomes an effective local anesthetic.

The receptor that is activated by moderate cold is **TRPM8.** The M refers to **menthol,** the ingredient in mint that gives it its "cool" taste. **TRPV4** receptors are activated by warm temperatures up to 34°C; **TRPV3** receptors respond to slightly higher temperatures of 35–39°C.

GENERATION OF IMPULSES IN CUTANEOUS RECEPTORS

The way that sensory receptors generate action potentials in the nerves that innervate them varies based on the complexity of the sense organ. In the skin, the Pacinian corpuscle has been studied in some detail. The myelin sheath of the sensory nerve begins inside the corpuscle (Figure 8–3). The first node of Ranvier is also located inside; the second is usually near the point at which the nerve fiber leaves the corpuscle.

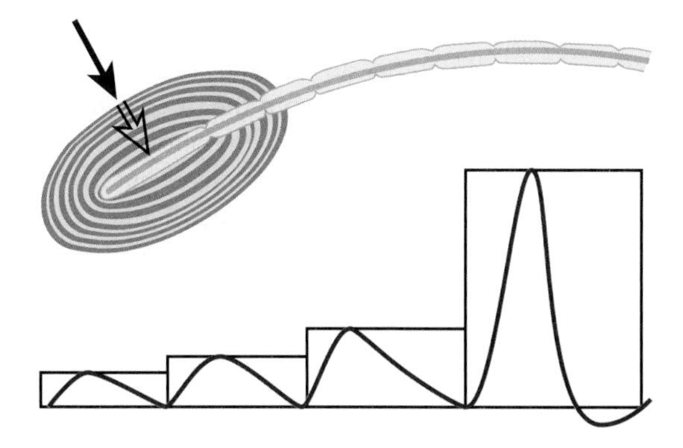

FIGURE 8–3 Demonstration that the generator potential in a Pacinian corpuscle originates in the unmyelinated nerve terminal. The electrical responses to pressures (black arrow) of 1, 2×, 3×, and 4× are shown. The strongest stimulus produced an action potential in the sensory nerve, originating in the center of the corpuscle. (From Waxman SG: *Clinical Neuroanatomy,* 26th ed. McGraw-Hill, 2010.)

GENERATOR POTENTIALS

When a small amount of pressure is applied to the Pacinian corpuscle, a nonpropagated depolarizing potential resembling an excitatory postsynaptic potential (EPSP) is recorded. This is called the **generator potential** or **receptor potential** (Figure 8–3). As the pressure is increased, the magnitude of the receptor potential is increased. The receptor therefore converts mechanical energy into an electrical response, the magnitude of which is proportional to the intensity of the stimulus. Thus, the responses are described as **graded potentials** rather than all-or-none as is the case for an **action potential.** When the magnitude of the generator potential reaches about 10 mV, an action potential is produced at the first node of Ranvier. The nerve then repolarizes. If the generator potential is great enough, the neuron fires again as soon as it repolarizes, and it continues to fire as long as the generator potential is large enough to bring the membrane potential of the node to the firing level. Thus, the node converts the graded response of the receptor into action potentials, the frequency of which is proportional to the magnitude of the applied stimulus.

SENSORY CODING

Converting a receptor stimulus to a recognizable sensation is termed **sensory coding.** All sensory systems code for four elementary attributes of a stimulus: modality, location, intensity, and duration. **Modality** is the type of energy transmitted by the stimulus. **Location** is the site on the body or space where the stimulus originated. **Intensity** is signaled by the response amplitude or frequency of action potential generation. **Duration** refers to the time from start to end of a response in the receptor. These attributes of sensory coding are shown for the modality of touch in Figure 8–1.

When the nerve from a particular sensory receptor is stimulated, the sensation evoked is that for which the receptor is specialized no matter how or where along the nerve the activity is initiated. This principle, first enunciated by Johannes Müller in 1835, has been called the **law of specific nerve energies.** For example, if the sensory nerve from a Pacinian corpuscle in the hand is stimulated by pressure at the elbow or by irritation from a tumor in the brachial plexus, the sensation evoked is touch. The general principle of specific nerve energies remains one of the cornerstones of sensory physiology.

MODALITY

Humans have four basic classes of receptors based on their sensitivity to one predominant form of energy: mechanical, thermal, electromagnetic, or chemical. The particular form of energy to which a receptor is most sensitive is called its **adequate stimulus.** The adequate stimulus for the rods and cones in the eye, for example, is light (an example of electromagnetic energy). Receptors do respond to forms of energy other than their adequate stimuli, but the threshold for these nonspecific responses is much higher. Pressure on the eyeball will stimulate the rods and cones, for example, but the threshold of these receptors to pressure is much higher than the threshold of the pressure receptors in the skin.

LOCATION

The term **sensory unit** refers to a single sensory axon and all of its peripheral branches. These branches vary in number but may be numerous, especially in the cutaneous senses. The **receptive field** of a sensory unit is the spatial distribution from which a stimulus produces a response in that unit (Figure 8–1). Representation of the senses in the skin is punctate. If the skin is carefully mapped, millimeter by millimeter, with a fine hair, a sensation of touch is evoked from spots overlying these touch receptors. None is evoked from the intervening areas. Similarly, temperature sensations and pain are produced by stimulation of the skin only over the spots where the receptors for these modalities are located. In the cornea and adjacent sclera of the eye, the surface area supplied by a single sensory unit is 50–200 mm^2. The area supplied by one sensory unit usually overlaps and interdigitates with the areas supplied by others.

One of the most important mechanisms that enable localization of a stimulus site is **lateral inhibition.** Information from sensory neurons whose receptors are at the peripheral edge of the stimulus is inhibited compared to information from the sensory neurons at the center of the stimulus. Thus, lateral inhibition enhances the contrast between the center and periphery of a stimulated area and increases the ability of the brain to localize a sensory input. Lateral inhibition underlies **two-point discrimination** (see Clinical Box 8–2).

INTENSITY

The intensity of sensation is determined by the amplitude of the stimulus applied to the receptor. This is illustrated in Figure 8–4. As a greater pressure is applied to the skin, the receptor potential in the mechanoreceptor increases (not shown), and the frequency of the action potentials in a single axon transmitting information to the CNS is also increased. In addition to increasing the firing rate in a single axon, the greater intensity of stimulation also will recruit more receptors into the receptive field.

CLINICAL BOX 8–2

Neurological Exam

The size of the receptive fields for light touch can be measured by the **two-point threshold test.** In this procedure, the two points on a pair of calipers are simultaneously positioned on the skin and one determines the minimum distance between the two caliper points that can be perceived as separate points of stimulation. This is called the **two-point discrimination threshold.** If the distance is very small, each caliper point is touching the receptive field of only one sensory neuron. If the distance between stimulation points is less than this threshold, only one point of stimulation can be felt. Thus, the two-point discrimination threshold is a measure of **tactile acuity.** The magnitude of two-point discrimination thresholds varies from place to place on the body and is smallest where touch receptors are most abundant. Stimulus points on the back, for instance, must be separated by at least 65 mm before they can be distinguished as separate, whereas on the fingertips two stimuli are recognized if they are separated by as little as 2 mm. Blind individuals benefit from the tactile acuity of fingertips to facilitate the ability to read Braille; the dots forming Braille symbols are separated by 2.5 mm. Two-point discrimination is used to test the integrity of the **dorsal column (medial lemniscus) system,** the central pathway for touch and proprioception.

Vibratory sensibility is tested by applying a vibrating (128-Hz) tuning fork to the skin on the fingertip, tip of the toe, or bony prominences of the toes. The normal response is a "buzzing" sensation. The sensation is most marked over bones. The term **pallesthesia** is also used to describe this ability to feel mechanical vibrations. The receptors involved are the receptors for touch, especially **Pacinian corpuscles,** but a time factor is also necessary. A pattern of rhythmic pressure stimuli is interpreted as vibration. The impulses responsible for the vibrating sensation are carried in the dorsal columns. Degeneration of this part of the spinal cord occurs in poorly controlled diabetes, pernicious anemia, vitamin B_{12} deficiencies, or early tabes dorsalis. Elevation of the threshold for vibratory stimuli is an early symptom of this degeneration. Vibratory sensation and proprioception are closely related; when one is diminished, so is the other.

Stereognosis is the perception of the form and nature of an object without looking at it. Normal persons can readily identify objects such as keys and coins of various denominations. This ability depends on relatively intact touch and pressure sensation and is compromised when the dorsal columns are damaged. The inability to identify an object by touch is called **tactile agnosia.** It also has a large cortical component; impaired stereognosis is an early sign of damage to the cerebral cortex and sometimes occurs in the absence of any detectable defect in touch and pressure sensation when there is a lesion in the primary sensory cortex. Stereoagnosia can also be expressed by the failure to identify an object by sight **(visual agnosia),** the inability to identify sounds or words **(auditory agnosia)** or color **(color agnosia),** or the inability to identify the location or position of an extremity **(position agnosia).**

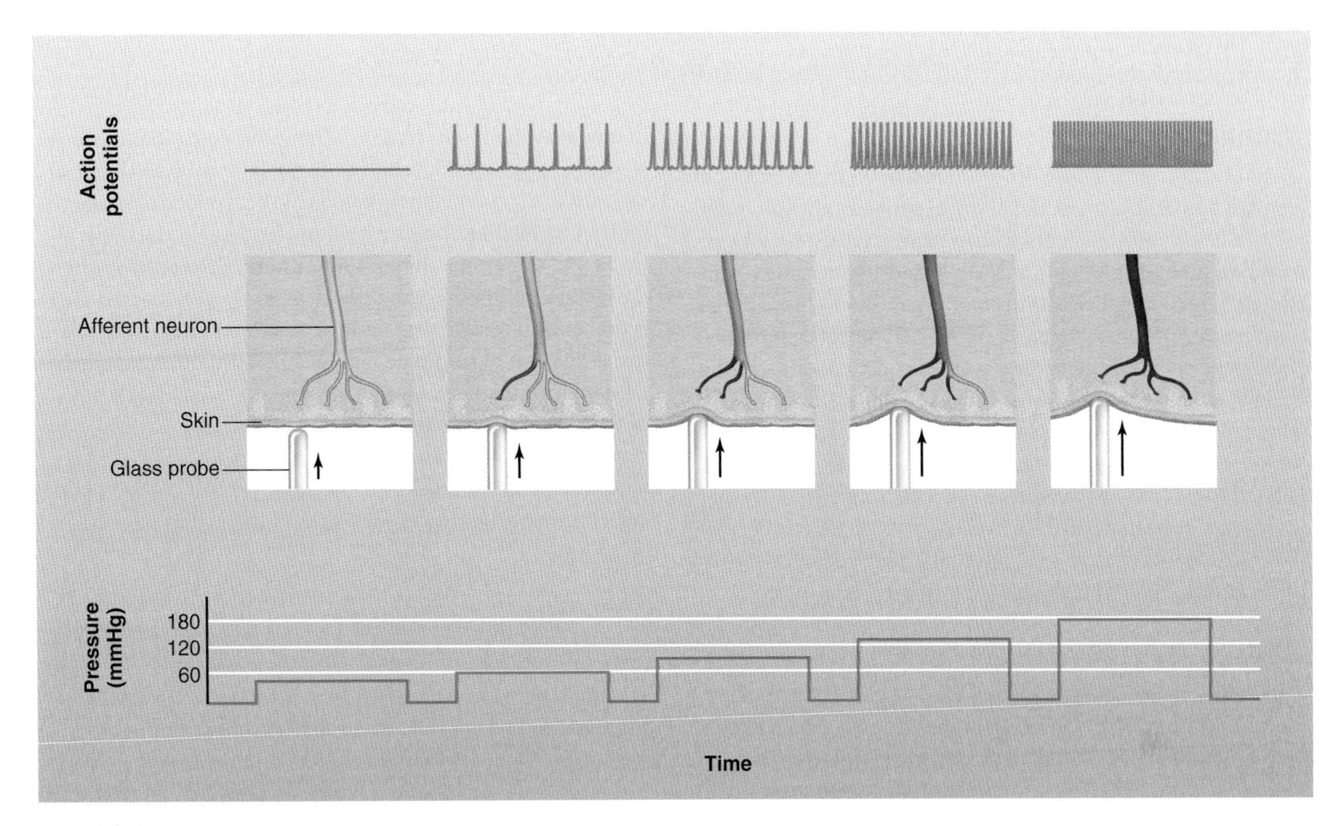

FIGURE 8–4 Relationship between stimulus and impulse frequency in an afferent fiber. Action potentials in an afferent fiber from a mechanoreceptor of a single sensory unit increase in frequency as branches of the afferent neuron are stimulated by pressure of increasing magnitude. (From Widmaier EP, Raff H, Strang KT: *Vander's Human Physiology.* McGraw-Hill, 2008.)

As the strength of a stimulus is increased, it tends to spread over a large area and generally not only activates the sense organs immediately in contact with it but also "recruits" those in the surrounding area. Furthermore, weak stimuli activate the receptors with the lowest thresholds, and stronger stimuli also activate those with higher thresholds. Some of the receptors activated are part of the same sensory unit, and impulse frequency in the unit therefore increases. Because of overlap and interdigitation of one unit with another, however, receptors of other units are also stimulated, and consequently more units fire. In this way, more afferent pathways are activated, which is interpreted in the brain as an increase in intensity of the sensation.

DURATION

If a stimulus of constant strength is maintained on a sensory receptor, the frequency of the action potentials in its sensory nerve declines over time. This phenomenon is known as **receptor adaptation** or **desensitization.** The degree to which adaptation occurs varies from one sense to another. Receptors can be classified into **rapidly adapting (phasic) receptors** and **slowly adapting (tonic) receptors.** This is illustrated for different types of touch receptors in Figure 8–1. Meissner and Pacinian corpuscles are examples of rapidly adapting receptors, and Merkel cells and Ruffini endings are examples of slowly adapting receptors. Other examples of slowly adapting receptors are muscle spindles and nociceptors. Different types of sensory adaptation likely have some value to the individual. Light touch would be distracting if it were persistent; and, conversely, slow adaptation of spindle input is needed to maintain posture. Similarly, input from nociceptors provides a warning that it would lose its value if it is adapted and disappeared.

NEUROLOGICAL EXAM

The sensory component of a neurological exam includes an assessment of various sensory modalities including touch, proprioception, vibratory sense, and pain. Cortical sensory function can be tested by placing familiar objects in a patient's hands and asking him or her to identify it with the eyes closed. Clinical Box 8–2 describes some of the common assessments made in a neurological exam.

PAIN

One of the most common reasons an individual seeks the advice of a physician is because he or she is in pain. Pain was called by Sherrington, "the physical adjunct of an imperative protective reflex." Painful stimuli generally initiate potent withdrawal and avoidance responses. Pain differs from other sensations in that it sounds a warning that something is wrong, preempts other signals, and is associated with an unpleasant affect. It is

CLINICAL BOX 8–3

Chronic Pain

A 2009 report in *Scientific American* indicated that 10–20% of the US and European populations experience **chronic pain;** 59% of these individuals are women. Based on a survey of primary care physicians, only 15% indicated that they felt comfortable treating patients with chronic pain; and 41% said they waited until patients specifically requested narcotic pain killers before prescribing them. Nearly 20% of adults with chronic pain indicated that they have visited an alternative medicine therapist. Risk factors for chronic neck and back pain include aging, being female, anxiety, repetitive work, obesity, depression, heavy lifting, and nicotine use. One example of chronic pain is **neuropathic pain** that may occur when nerve fibers are injured. Nerve damage can cause an inflammatory response due to activation of microglia in the spinal cord. Commonly, it is excruciating and a difficult condition to treat. For example, in **causalgia,** a spontaneous burning pain occurs long after seemingly trivial injuries. The pain is often accompanied by **hyperalgesia** and **allodynia. Reflex sympathetic dystrophy** is often present as well. In this condition, the skin in the affected area is thin and shiny, and there is increased hair growth. This may result because of sprouting and eventual overgrowth of noradrenergic sympathetic nerve fibers into the dorsal root ganglia of the sensory nerves from the injured area. Sympathetic discharge then brings on pain. Thus, it appears that the periphery has been short-circuited and that the relevant altered fibers are being stimulated by norepinephrine at the dorsal root ganglion level.

THERAPEUTIC HIGHLIGHTS

Chronic pain is often refractory to most conventional therapies such as **NSAIDs** and even **opioids.** In new efforts to treat chronic pain, some therapies focus on synaptic transmission in nociceptive pathways and peripheral sensory transduction. **TRPV1,** a capsaicin receptor, is activated by noxious stimuli such as heat, protons, and products of inflammation. **Capsaicin transdermal patches** or creams reduce pain by exhausting the supply of substance P in nerves. Nav1.8 (a tetrodotoxin-resistant voltage-gated sodium channel) is uniquely associated with nociceptive neurons in dorsal root ganglia. **Lidocaine** and **mexiletine** are useful in some cases of chronic pain and may act by blocking this channel. **Ralfinamide,** a Na^+ channel blocker, is under development for potential treatment of neuropathic pain. **Ziconotide,** a voltage-gated N-type Ca^{2+} channel blocker, has been approved for intrathecal analgesia in patients with refractory chronic pain. **Gabapentin** is an anticonvulsant drug that is an analog of GABA; it has been shown to be effective in treatment of neuropathic and **inflammatory pain** by acting on voltage-gated Ca^{2+} channels. **Topiramate,** a Na^+ channel blocker, is another example of an anticonvulsant drug that can be used to treat **migraine** headaches. **NMDA receptor antagonists** can be co-administered with an opioid to reduce tolerance to an opioid. Endogenous **cannabinoids** have analgesic actions in addition to their euphopric effects. Drugs that act on CB_2 receptors which are devoid of euphoric effects are under development for the treatment of neuropathic pain.

immensely complex because when tissue is damaged, central nociceptive pathways are sensitized and reorganized which leads to persistent or **chronic pain** (see Clinical Box 8–3).

CLASSIFICATION OF PAIN

For scientific and clinical purposes, **pain** is defined by the International Association for the Study of Pain (IASP) as, "an unpleasant sensory and emotional experience associated with actual or potential tissue damage, or described in terms of such damage." This is to be distinguished from the term **nociception** which the IASP defines as the unconscious activity induced by a harmful stimulus applied to sense receptors.

Pain is frequently classified as **physiologic** or **acute pain** and **pathologic** or **chronic pain,** which includes **inflammatory pain** and **neuropathic pain.** Acute pain typically has a sudden onset and recedes during the healing process; it can be regarded as "good pain" as it serves an important protective mechanism. The withdrawal reflex is an example of the expression of this protective role of pain.

Chronic pain can be considered "bad pain" because it persists long after recovery from an injury and is often refractory to common analgesic agents, including **nonsteroidal anti-inflammatory drugs (NSAIDs)** and **opioids.** Chronic pain can result from nerve injury (neuropathic pain) including diabetic neuropathy, toxin-induced nerve damage, and ischemia. **Causalgia** is a type of neuropathic pain (see Clinical Box 8–3).

HYPERALGESIA AND ALLODYNIA

Pain is often accompanied by **hyperalgesia** and **allodynia.** Hyperalgesia is an exaggerated response to a noxious stimulus, and allodynia is a sensation of pain in response to a normally innocuous stimulus. An example of the latter is the painful

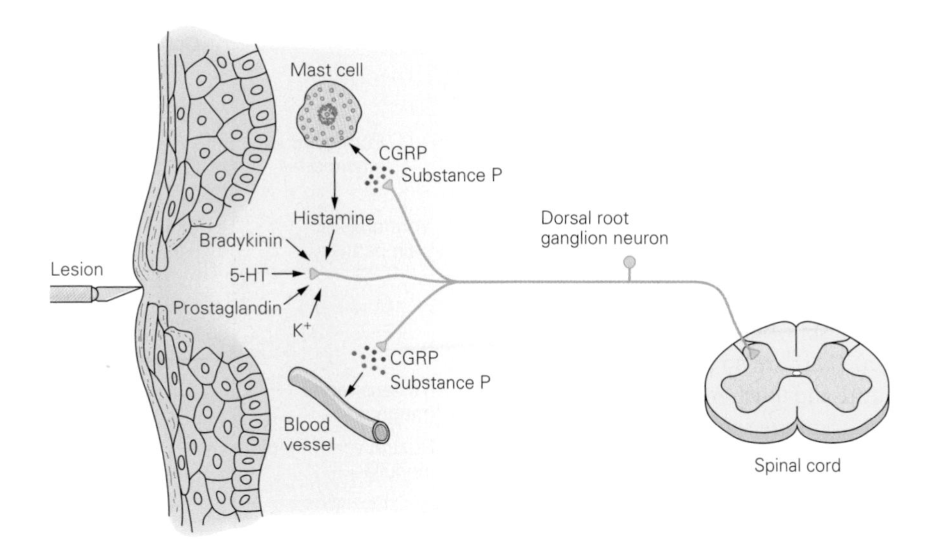

FIGURE 8–5 Chemical mediators are released in response to tissue damage and can sensitize or directly activate nociceptors. These factors contribute to hyperalgesia and allodynia. Tissue injury releases bradykinin and prostaglandins that sensitize or activate nociceptors, which in turn releases substance P and calcitonin gene-related peptide (CGRP). Substance P acts on mast cells to cause degranulation and release histamine, which activates nociceptors. Substance P causes plasma extravasation and CGRP dilates blood vessels; the resulting edema causes additional release of bradykinin. Serotonin (5-HT) is released from platelets and activates nociceptors. (From Lembeck F:CIBA Foundation Symposium, London: Pitman Medical; Summit, NJ, 1981.)

sensation from a warm shower when the skin is damaged by sunburn.

Hyperalgesia and allodynia signify increased sensitivity of nociceptive afferent fibers. Figure 8–5 shows how chemicals released at the site of injury can further directly activate receptors on sensory nerve endings leading to inflammatory pain. Injured cells also release chemicals such as K^+ that directly depolarize nerve terminals, making nociceptors more responsive **(sensitization).** Injured cells also release bradykinin and substance P, which can further sensitize nociceptive terminals. Histamine is released from mast cells, serotonin (5-HT) from platelets, and prostaglandins from cell membranes, all contributing to the inflammatory process and they activate or sensitize the nociceptors. Some released substances act by releasing another one (eg, bradykinin activates both Aδ and C nerve endings and increases synthesis and release of prostaglandins). Prostaglandin E_2 (a cyclooxygenase metabolite of arachidonic acid) is released from damaged cells and produces hyperalgesia. This is why aspirin and other NSAIDs (inhibitors of cyclooxygenase) alleviate pain.

In addition to sensitization of nerve endings by chemical mediators, several other changes occur within the periphery and CNS that can contribute to the chronic pain. The NGF released by tissue damage is picked up by nerve terminals and transported retrogradely to cell bodies in dorsal root ganglia where it can alter gene expression. Transport may be facilitated by the activation of TrkA receptors on the nerve endings. In the dorsal root ganglia, NGF increases production of substance P and converts nonnociceptive neurons to nociceptive neurons (a phenotypic change). NGF also influences expression of a tetrodotoxin-resistant sodium channel (**Nav1.8**) on dorsal root ganglia, further increasing activity.

Damaged nerve fibers undergo sprouting, so fibers from touch receptors synapse on spinal dorsal horn neurons that normally receive only nociceptive input (see below). This can explain why innocuous stimuli can induce pain after injury. The combined release of substance P and glutamate from nociceptive afferents in the spinal cord causes excessive activation of **NMDA (n-methyl-D-aspartate) receptors** on spinal neurons, a phenomenon called "wind-up" that leads to increased activity in pain transmitting pathways. Another change in the spinal cord is due to the activation of **microglia** near afferent nerve terminals in the spinal cord by the release of transmitters from sensory afferents. This, in turn, leads to the release of pro-inflammatory cytokines and chemokines that modulate pain processing by affecting presynaptic release of neurotransmitters and postsynaptic excitability. There are P2X receptors on microglia; antagonists of these receptors may be a useful therapy for treatment of chronic pain.

DEEP AND VISCERAL PAIN

The main difference between superficial and deep or visceral pain is the nature of the pain evoked by noxious stimuli. This is probably due to a relative deficiency of Aδ nerve fibers in deep structures, so there is little rapid, sharp pain. In addition, deep pain and visceral pain are poorly localized, nauseating, and frequently are accompanied by sweating and changes in blood pressure. Pain can be elicited experimentally from the periosteum and ligaments by injecting hypertonic saline into them. The pain produced in this fashion initiates reflex contraction of nearby skeletal muscles. This reflex contraction is similar to the muscle spasm associated with injuries to bones, tendons, and joints. The steadily contracting muscles become ischemic, and ischemia stimulates the pain receptors in the muscles. The pain in turn initiates more spasm, setting up a vicious cycle.

In addition to being poorly localized, unpleasant, and associated with nausea and autonomic symptoms, visceral pain often radiates or is referred to other areas. The autonomic nervous system, like the somatic, has afferent components, central integrating stations, and effector pathways. The receptors for pain and the other sensory modalities present in the viscera are similar to those in skin, but there are marked differences in their distribution. There are no proprioceptors in the viscera, and few temperature and touch receptors. Nociceptors are present, although they are more sparsely distributed than in somatic structures.

Afferent fibers from visceral structures reach the CNS via sympathetic and parasympathetic nerves. Their cell bodies are located in the dorsal root ganglia and the homologous cranial nerve ganglia. Specifically, there are visceral afferents in the facial, glossopharyngeal, and vagus nerves; in the thoracic and upper lumbar dorsal roots; and in the sacral dorsal roots.

As almost everyone knows from personal experience, visceral pain can be very severe. The receptors in the walls of the hollow viscera are especially sensitive to distention of these organs. Such distention can be produced experimentally in the gastrointestinal tract by inflation of a swallowed balloon attached to a tube. This produces pain that waxes and wanes (**intestinal colic**) as the intestine contracts and relaxes on the balloon. Similar colic is produced in intestinal obstruction by the contractions of the dilated intestine above the obstruction. When a visceral organ is inflamed or hyperemic, relatively minor stimuli cause severe pain, a form of hyperalgesia.

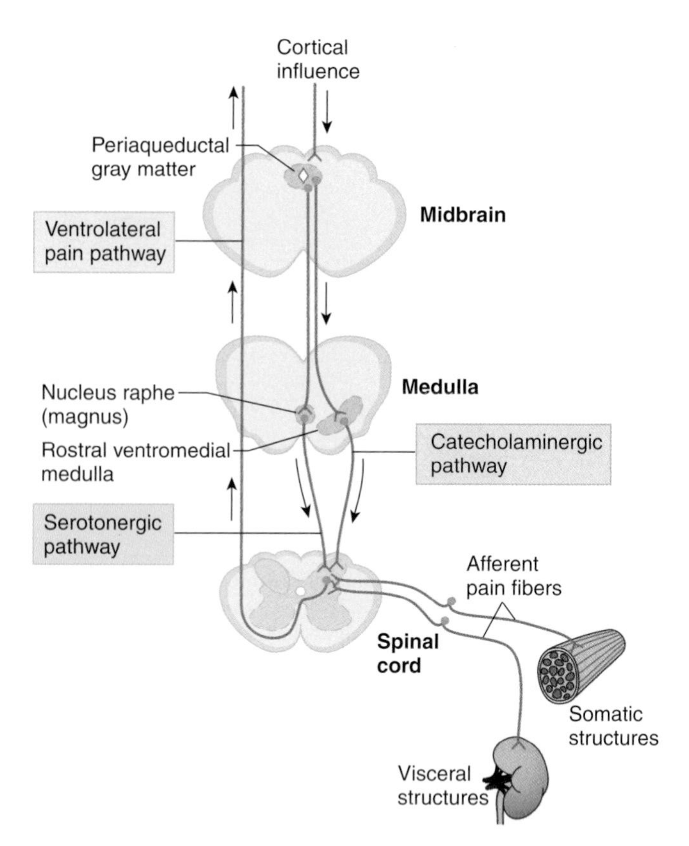

FIGURE 8–6 Schematic illustration of the convergence-projection theory for referred pain and descending pathways involved in pain control. The basis for referred pain may be convergence of somatic and visceral pain fibers on the same second-order neurons in the dorsal horn of the spinal cord that project higher brain regions. The periaqueductal gray (PAG) is a part of a descending pathway that includes serotonergic neurons in the nucleus raphé magnus and catecholaminergic neurons in the rostral ventromedial medulla to modulate pain transmission by inhibition of primary afferent transmission in the dorsal horn. (Courtesy of Al Basbaum.)

REFERRED PAIN

Irritation of a visceral organ frequently produces pain that is felt not at that site but in a somatic structure that may be some distance away. Such pain is said to be referred to the somatic structure **(referred pain).** Knowledge of the common sites of pain referral from each of the visceral organs is of importance to a physician. One of the best-known examples is referral of cardiac pain to the inner aspect of the left arm. Other examples include pain in the tip of the shoulder caused by irritation of the central portion of the diaphragm and pain in the testicle due to distention of the ureter. Additional instances abound in the practices of medicine, surgery, and dentistry. However, sites of reference are not stereotyped, and unusual reference sites occur with considerable frequency. Cardiac pain, for instance, may be referred to the right arm, the abdominal region, or even the back, neck, or jaw.

When pain is referred, it is usually to a structure that developed from the same embryonic segment or dermatome as the structure in which the pain originates. For example, the heart and the arm have the same segmental origin, and the testicle migrated with its nerve supply from the primitive urogenital ridge from which the kidney and ureter also developed.

The basis for referred pain may be convergence of somatic and visceral pain fibers on the same second-order neurons in the dorsal horn that project to the thalamus and then to the somatosensory cortex (Figure 8–6). This is called the **convergence–projection theory.** Somatic and visceral neurons converge in the ipsilateral dorsal horn. The somatic nociceptive fibers normally do not activate the second-order neurons, but when the visceral stimulus is prolonged, facilitation of the somatic fiber endings occurs. They now stimulate the second-order neurons, and of course the brain cannot determine whether the stimulus came from the viscera or from the area of referral.

SOMATOSENSORY PATHWAYS

The sensation evoked by impulses generated in a sensory receptor depends in part on the specific part of the brain they ultimately activate. The ascending pathways from sensory receptors

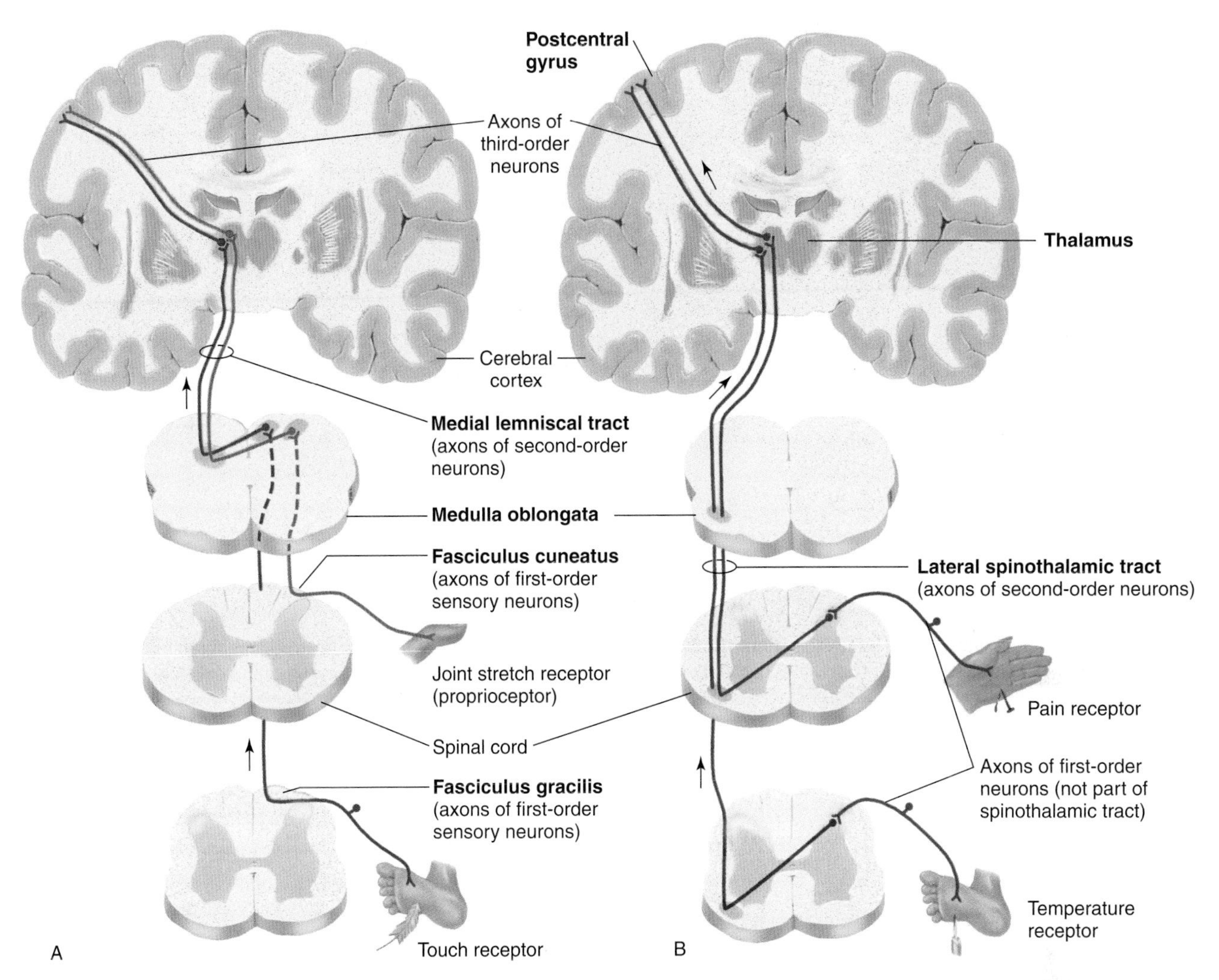

FIGURE 8–7 Ascending tracts carrying sensory information from peripheral receptors to the cerebral cortex. A) Dorsal column pathway mediates touch, vibratory sense, and proprioception. Sensory fibers ascend ipsilaterally via the spinal dorsal columns to medullary gracilus and cuneate nuclei; from there the fibers cross the midline and ascend in the medial lemniscus to the contralateral thalamic ventral posterior lateral (VPL) and then to the primary somatosensory cortex. **B)** Ventrolateral spinothalamic tract mediates pain and temperature. These sensory fibers terminate in the dorsal horn and projections from there cross the midline and ascend in the ventrolateral quadrant of the spinal cord to the VPL and then to the primary somatosensory cortex. (From Fox SI, *Human Physiology*. McGraw-Hill, 2008.)

to the cortex are different for the various sensations. Below is a comparison of the ascending sensory pathway that mediates touch, vibratory sense, and proprioception **(dorsal column-medial lemniscal pathway)** and that which mediates pain and temperature **(ventrolateral spinothalamic pathway).**

DORSAL COLUMN PATHWAY

The principal pathways to the cerebral cortex for touch, vibratory sense, and proprioception are shown in Figure 8–7. Fibers mediating these sensations ascend ipsilaterally in the dorsal columns of the spinal cord to the medulla, where they synapse in the **gracilus** and **cuneate nuclei.** The second-order neurons from these nuclei cross the midline and ascend in the **medial lemniscus** to end in the contralateral **ventral posterior lateral (VPL) nucleus** and related specific sensory relay nuclei of the thalamus. This ascending system is called the dorsal column or medial lemniscal system. The fibers within the dorsal column pathway are joined in the brain stem by fibers mediating sensation from the head. Touch and proprioception from the head are relayed mostly via the main sensory and mesencephalic nuclei of the trigeminal nerve.

Somatotopic Organization

Within the dorsal columns, fibers arising from different levels of the cord are somatotopically organized (Figure 8–7). Specifically, fibers from the sacral cord are positioned most medially and those from the cervical cord are positioned most laterally. This arrangement continues in the medulla with

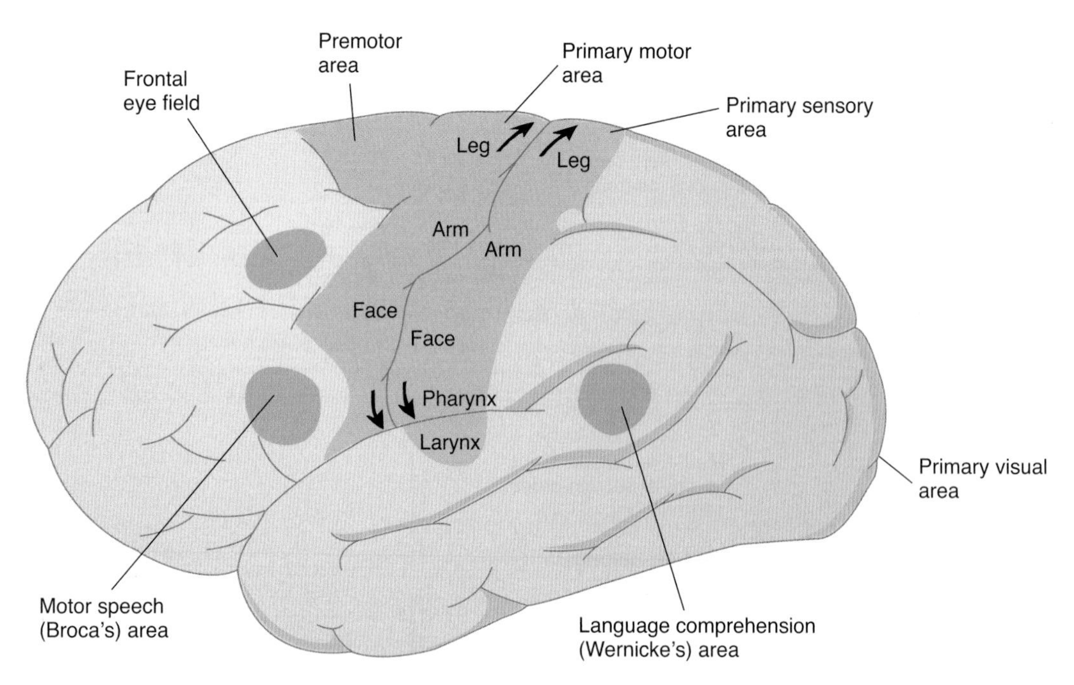

FIGURE 8–8 A lateral view of the left hemisphere showing some principal cortical areas and their functional correlates in the human brain. The primary somatosensory area is in the postcentral gyrus of the parietal lobe, and the primary motor cortex is in the precentral gyrus. (From Waxman SG: *Clinical Neuroanatomy,* 26th ed. McGraw-Hill, 2010.)

lower body (eg, foot) representation in the gracilus nucleus and upper body (eg, finger) representation in cuneate nucleus. The medial lemniscus is organized dorsal to ventral representing from neck to foot.

Somatotopic organization continues through the thalamus and cortex. VPL thalamic neurons carrying sensory information project in a highly specific way to the **primary somatosensory cortex** in the **postcentral gyrus** of the **parietal lobe** (Figure 8–8). The arrangement of projections to this region is such that the parts of the body are represented in order along the postcentral gyrus, with the legs on top and the head at the foot of the gyrus. Not only is there detailed localization of the fibers from the various parts of the body in the postcentral gyrus, but also the size of the cortical receiving area for impulses from a particular part of the body is proportional to the use of the part. The relative sizes of the cortical receiving areas are shown dramatically in Figure 8–9, in which the proportions of the **homunculus** have been distorted to correspond to the size of the cortical receiving areas for each. Note that the cortical areas for sensation from the trunk and back are small, whereas very large areas are concerned with impulses from the hand and the parts of the mouth concerned with speech.

Studies of the sensory receiving area emphasize the very discrete nature of the point-for-point localization of peripheral areas in the cortex and provide further evidence for the general validity of the law of specific nerve energies. Stimulation of the various parts of the postcentral gyrus gives rise to sensations projected to appropriate parts of the body. The sensations produced are usually numbness, tingling, or a sense

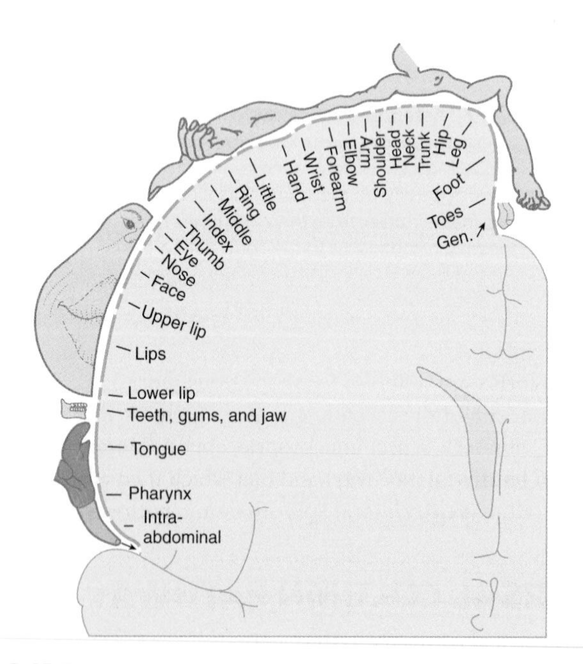

FIGURE 8–9 Sensory homunculus, drawn overlying a coronal section through the postcentral gyrus. The parts of the body are represented in order along the postcentral gyrus, with the legs on top and the head at the foot of the gyrus. The size of the cortical receiving area for impulses from a particular part of the body is proportionate to the use of the part. Gen., genitalia. (Reproduced with permission from Penfield W, Rasmussen G: *The Cerebral Cortex of Man.* Macmillan, 1950.)

of movement, but with fine enough electrodes it has been possible to produce relatively pure sensations of touch, warmth, and cold. The cells in the postcentral gyrus are organized in vertical columns. The cells in a given column are all activated by afferents from a given part of the body, and all respond to the same sensory modality.

In addition to the primary somatosensory cortex, there are two other cortical regions that contribute to the integration of sensory information. The **sensory association area** is located in the parietal cortex and the **secondary somatosensory cortex** is located in the wall of the **lateral fissure** (also called **sylvian fissure**) that separates the temporal from the frontal and parietal lobes. These regions receive input from the primary somatosensory cortex.

Conscious awareness of the positions of the various parts of the body in space depends in part on impulses from sensory receptors in and around the joints. Impulses from these receptors, from touch receptors in the skin and other tissues, and from muscle spindles are synthesized in the cortex into a conscious picture of the position of the body in space.

VENTROLATERAL SPINOTHALAMIC TRACT

Fibers from nociceptors and thermoreceptors synapse on neurons in the dorsal horn of the spinal cord. The axons from these dorsal horn neurons cross the midline and ascend in the ventrolateral quadrant of the spinal cord, where they form the ventrolateral spinothalamic pathway (Figure 8–7). Fibers within this tract synapse in the VPL. Some dorsal horn neurons that receive nociceptive input synapse in the reticular formation of the brain stem **(spinoreticular pathway)** and then project to the centrolateral nucleus of the thalamus.

Positron emission tomographic (PET) and functional magnetic resonance imaging (fMRI) studies in normal humans indicate that pain activates the primary and secondary somatosensory cortex and the cingulate gyrus on the side opposite the stimulus. In addition, the amygdala, frontal lobe, and the insular cortex are activated. These technologies were important in distinguishing two components of pain pathways. Researchers found that noxious stimuli that did not induce a change in affect caused an increased metabolism in the primary somatosensory cortex, whereas stimuli that elicited motivational-affective responses activated a larger portion of the cortex. This showed that the pathway to the primary somatosensory cortex is responsible for the discriminative aspect of pain. In contrast, the pathway that includes synapses in the brain stem reticular formation and centrolateral thalamic nucleus projects to the frontal lobe, limbic system, and insular cortex. This pathway mediates the motivational-affective component of pain.

Visceral sensation travels along the same central pathways as somatic sensation in the spinothalamic tracts and thalamic radiations, and the cortical receiving areas for visceral sensation are intermixed with the somatic receiving areas.

CORTICAL PLASTICITY

It is now clear that the extensive neuronal connections described above are not innate and immutable but can be changed relatively rapidly by experience to reflect the use of the represented area. Clinical Box 8–4 describes remarkable changes in cortical and thalamic organization that occur in response to limb amputation to lead to the phenomenon of **phantom limb pain.**

CLINICAL BOX 8–4

Phantom Limb Pain

In 1551, a military surgeon, Ambroise Pare, wrote "...the patients, long after the amputation is made, say they still feel pain in the amputated part. Of this they complain strongly, a thing worthy of wonder and almost incredible to people who have not experienced this." This is perhaps the earliest description of **phantom limb pain.** Between 50 and 80% of amputees experience phantom sensations, usually pain, in the region of their amputated limb. Phantom sensations may also occur after the removal of body parts other than the limbs, for example, after amputation of the breast, extraction of a tooth **(phantom tooth pain),** or removal of an eye **(phantom eye syndrome).** Numerous theories have been evoked to explain this phenomenon. The current theory is based on evidence that the brain can reorganize if sensory input is cut off. The **ventral posterior thalamic nucleus** is one example where this change can occur. In patients who have had their leg amputated, single neuron recordings show that the thalamic region that once received input from the leg and foot now respond to stimulation of the stump (thigh). Others have demonstrated remapping of the somatosensory cortex. For example, in some individuals who have had an arm amputated, stroking different parts of the face can lead to the feeling of being touched in the area of the missing limb.

THERAPEUTIC HIGHLIGHTS

There is some evidence that the use of **epidural anesthesia** during the amputation surgery can prevent the acute pain associated with the surgery, thereby reducing the need for opioid therapy in the immediate postoperative period. They also reported a reduced incidence of phantom pain following this anesthetic procedure. **Spinal cord stimulation** has been shown to be an effective therapy for phantom pain. Electric current is passed through an electrode that is placed next to the spinal cord to stimulate spinal pathways. This interferes with the impulses ascending to the brain and lessens the pain felt in the phantom limb. Instead, amputees feel a tingling sensation in the phantom limb.

Numerous animal studies point to dramatic reorganization of cortical structures. If a digit is amputated in a monkey, the cortical representation of the neighboring digits spreads into the cortical area that was formerly occupied by the representation of the amputated digit. Conversely, if the cortical area representing a digit is removed, the somatosensory map of the digit moves to the surrounding cortex. Extensive, long-term deafferentation of limbs leads to even more dramatic shifts in somatosensory representation in the cortex, with, for example, the hand cortical area responding to touching the face. The explanation of these shifts appears to be that cortical connections of sensory units to the cortex have extensive convergence and divergence, with connections that can become weak with disuse and strong with use.

Plasticity of this type occurs not only with input from cutaneous receptors but also with input in other sensory systems. For example, in cats with small lesions of the retina, the cortical area for the blind spot begins to respond to light striking other areas of the retina. Development of the adult pattern of retinal projections to the visual cortex is another example of this plasticity. At a more extreme level, experimentally routing visual input to the auditory cortex during development creates visual receptive fields in the auditory system.

PET scanning in humans also documents plastic changes, sometimes from one sensory modality to another. Thus, for example, tactile and auditory stimuli increase metabolic activity in the visual cortex in blind individuals. Conversely, deaf individuals respond faster and more accurately than normal individuals to moving stimuli in the visual periphery. Plasticity also occurs in the motor cortex. These findings illustrate the malleability of the brain and its ability to adapt.

EFFECTS OF CNS LESIONS

Clinical Box 8–2 describes some of the deficits noted after damage within the somatosensory pathways. Clinical Box 8–5 describes the characteristic changes in sensory and motor functions that occur in response to spinal hemisection.

Damage to the dorsal columns leads to ipsilateral loss of the ability to detect light touch, vibration, and proprioception from body structures represented caudal to the level of damage. Damage to the ventrolateral spinothalamic pathway leads to contralateral loss of pain and temperature sensation below the level of the lesion. Such spinal damage could occur with a penetrating wound or a tumor.

Lesions of the primary somatosensory cortex do not abolish somatic sensation. Irritation of this region causes **paresthesia** or an abnormal sensation of numbness and tingling on the contralateral side of the body. Destructive lesions impair the ability to localize noxious stimuli in time, space, and intensity. Damage to the cingulate cortex impairs the recognition of the aversive nature of a noxious stimulus.

An infarct in the thalamus can lead to a loss of sensation. **Thalamic pain syndrome** is sometimes seen during recovery from a thalamic infarct. The syndrome is characterized by chronic pain on the side of the body contralateral to the stroke.

CLINICAL BOX 8–5

Brown-Séquard Syndrome

A functional hemisection of the spinal cord causes a characteristic and easily recognized clinical picture that reflects damage to ascending sensory (dorsal-column pathway, ventrolateral spinothalamic tract) and descending motor (corticospinal tract) pathways, which is called the **Brown-Séquard syndrome.** The lesion to fasciculus gracilus or fasciculus cuneatus leads to ipsilateral loss of discriminative touch, vibration, and proprioception below the level of the lesion. The loss of the spinothalamic tract leads to contralateral loss of pain and temperature sensation beginning one or two segments below the lesion. Damage to the corticospinal tract produces weakness and spasticity in certain muscle groups on the same side of the body. Although a precise spinal hemisection is rare, the syndrome is fairly common because it can be caused by a spinal cord tumor, spinal cord trauma, degenerative disc disease, and ischemia.

THERAPEUTIC HIGHLIGHTS

Drug treatments for Brown-Séquard syndrome are based on the etiology and time since onset. High doses of **corticosteriods** have been shown to be of value particularly if administered soon after the onset of such a spinal cord injury. Steroids decrease the inflammation by suppressing polymorphonuclear leukocytes and reverse the increase in capillary permeability.

MODULATION OF PAIN TRANSMISSION

PROCESSING INFORMATION IN THE DORSAL HORN

Transmission in nociceptive pathways can be interrupted by actions within the dorsal horn of the spinal cord at the site of sensory afferent termination. Many people have learned from practical experience that rubbing or shaking an injured area decreases the pain due to the injury. The relief may be due to the simultaneous activation of innocuous cutaneous mechanoreceptors whose afferents emit collaterals that terminate in the dorsal horn. The activity of these cutaneous mechanosensitive afferents may reduce the responsiveness of dorsal horn neurons to their input from nociceptive afferent terminals. This is called the **gate-control mechanism** of pain modulation and it serves as the rationale behind the use of **transcutaneous electrical nerve stimulation (TENS)** for pain relief. This method uses electrodes to activate Aα and Aβ fibers in the vicinity of the injury.

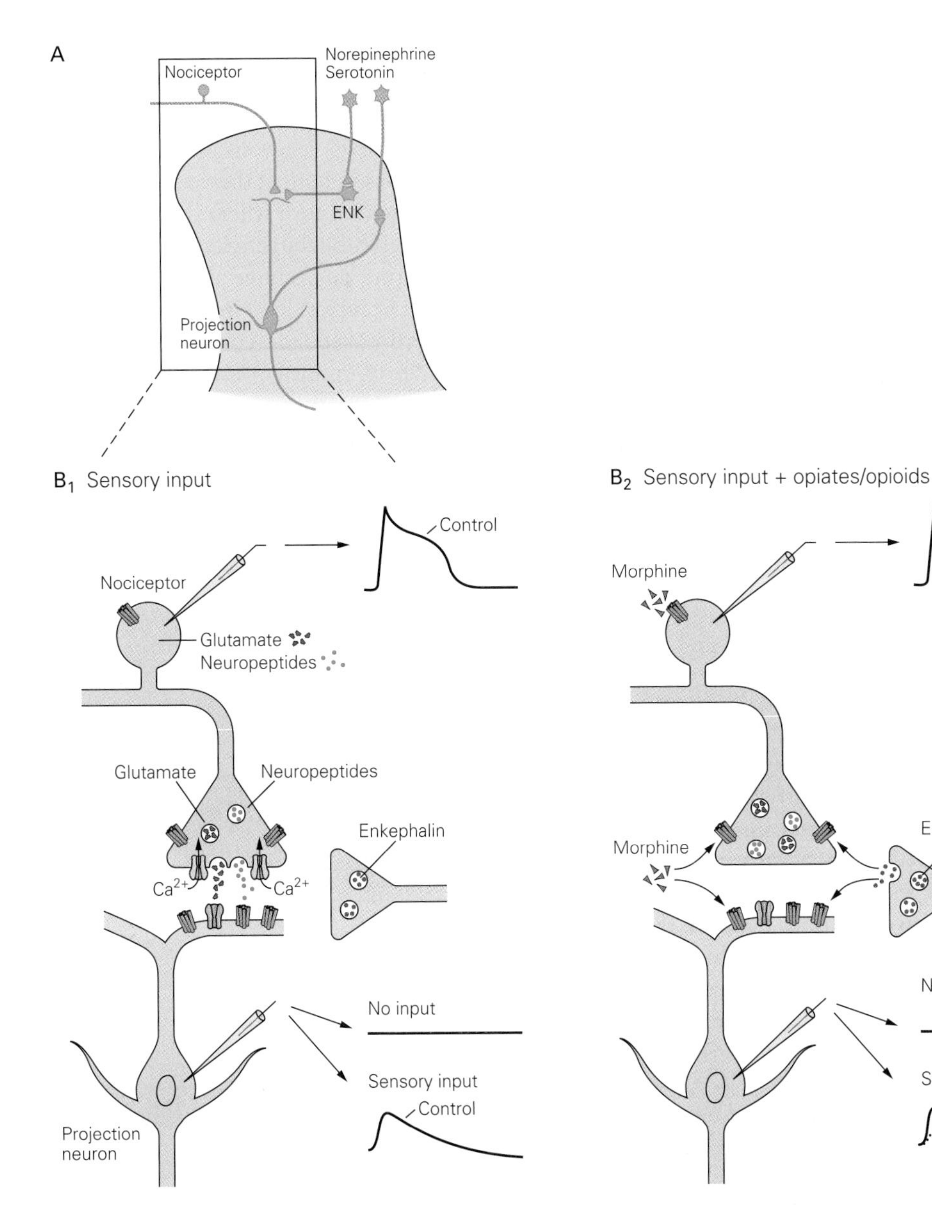

FIGURE 8–10 Local-circuit interneurons in the superficial dorsal horn of the spinal cord integrate descending and afferent pathways. A) Interactions of nociceptive afferent fibers, interneurons, and descending fibers in the dorsal horn. Nociceptive fibers terminate on spinothalamic projection neurons. Enkephalin (ENK)-containing interneurons exert both presynaptic and postsynaptic inhibitory actions. Serotonergic and noradrenergic neurons in the brainstem activate ENK interneurons and suppress the activity of spinothalamic projection neurons. **B_1)** Activation of nociceptors releases glutamate and neuropeptides from sensory terminals, depolarizing, and activating projection neurons. **B_2)** Opioids decrease Ca^{2+} influx leading to a decrease in the duration of nociceptor action potentials and a decreased release of transmitter. Also, opioids hyperpolarize the membrane of dorsal horn neurons by activating K^+ conductance and decrease the amplitude of the EPSP produced by stimulation of nociceptors. (From Kandel ER, Schwartz JH, Jessell TM [editors]: *Principles of Neural Science*, 4th ed. McGraw-Hill, 2000.)

Opioids are a commonly used analgesic that can exert their effects at various places in the CNS, including in the spinal cord and dorsal root ganglia. Figure 8–10 shows some of the various modes of action of opioids to decrease nociceptive transmission. There are interneurons in the superficial regions of the dorsal horn that contain endogenous opioid peptides (**enkephalin** and **dynorphin**). These interneurons terminate in the region of the dorsal horn where nociceptive afferents terminate. Opioid receptors (OR) are located on the terminals of nociceptive fibers and on dendrites of dorsal horn neurons, allowing for both presynaptic and postsynaptic sites of actions for opioids. Activation of the postsynaptic OR hyperpolarizes the dorsal horn interneuron by causing an increase in K^+ conductance. Activation of the presynaptic OR leads to a decrease in Ca^{2+} influx, resulting in a decrease in release of glutamate and substance P. Together these actions reduce the duration of the EPSP in the dorsal horn neuron. Activation of OR on dorsal root ganglia cell bodies also contributes to reduced transmission from nociceptive afferents.

Chronic use of morphine to relieve pain can cause patients to develop resistance to the drug, requiring progressively higher doses for pain relief. This **acquired tolerance** is different from **addiction,** which refers to a psychological craving. Psychological addiction rarely occurs when morphine is used to treat chronic pain, provided the patient does not have a history of drug abuse. Clinical Box 8–6 describes mechanisms involved in motivation and addiction.

ROLES OF PERIAQUEDUCTAL GRAY & BRAINSTEM

Another site of action for morphine and endogenous opioid peptides is the mesencephalic **periaqueductal gray (PAG).** An injection of opioids into the PAG induces analgesia. The PAG is a part of a descending pathway that modulates pain transmission by inhibition of primary afferent transmission in the dorsal horn (Figure 8–6). These PAG neurons project directly to and activate two groups of neurons in the brainstem: serotonergic neurons in the **nucleus raphé magnus** and catecholaminergic neurons in the **rostral ventromedial medulla.** Neurons in both of these regions project to the dorsal horn of the spinal cord where the released serotonin and norepinephrine inhibit the activity of dorsal horn neurons that receive input from nociceptive afferent fibers (Figure 8–10). This inhibition occurs, at least in part, due to the activation of the dorsal horn enkephalin-containing interneurons. There is also a group of brainstem catecholaminergic neurons in the **locus coeruleus** that are elements of this descending pain modulating pathway. These pontine neurons also exert their analgesic effect by the release of norepinephrine in the dorsal horn.

The analgesic effect of **electroacupuncture** may involve the release of endogenous opioids and activation of this descending pain modulatory pathway. Electroacupuncture

CLINICAL BOX 8–6

Motivation & Addiction

Neurons in the forebrain **ventral tegmental area** and **nucleus acumbens** are involved in motivated behaviors such as reward, laughter, pleasure, addiction, and fear. These areas have been referred to as the brain's **reward center or pleasure center**. The **mesocortical dopaminergic neurons** that project from the midbrain to the **nucleus accumbens** and the frontal cortex are also involved. **Addiction**, defined as the repeated compulsive use of a substance despite negative health consequences, can be produced by a variety of different drugs. According to the World Health Organization, over 76 million people worldwide suffer from alcohol abuse, and over 15 million suffer from drug abuse. Not surprisingly, alcohol and drug addiction are associated with the reward system. The best studied addictive drugs are opioids (eg, morphine and heroin); others include cocaine, amphetamine, alcohol, cannabinoids, and nicotine. These drugs affect the brain in different ways, but all have in common the fact that they increase the amount of dopamine available to act on **D_3 receptors** in the nucleus accumbens. Thus, acutely they stimulate the reward system of the brain. Long-term addiction involves the development of **tolerance,** which is the need for increasing amounts of a drug to produce a high. Also, **withdrawal** produces psychologic and physical symptoms. One of the characteristics of addiction is the tendency of addicts to relapse after treatment. For opioid addicts, the relapse rate in the first year is about 80%. Relapse often occurs on exposure to sights, sounds, and situations that were previously associated with drug use. Even a single dose of an addictive drug facilitates release of excitatory neurotransmitters in brain areas concerned with memory. The medial frontal cortex, hippocampus, and amygdala are concerned with memory, and they all project via excitatory glutamatergic pathways to the nucleus accumbens. Despite intensive study, relatively little is known about the brain mechanisms that cause tolerance and dependence. However, the two can be separated. Absence of **β-arrestin-2** blocks tolerance but has no effect on dependence. β-Arrestin-2 is a member of a family of proteins that inhibit heterotrimeric G proteins by phosphorylating them.

THERAPEUTIC HIGHLIGHTS

Withdrawal symptoms and cravings associated with addiction to opioids can be reversed by treatment with various drugs that act on the same CNS receptors as morphine and heroin. These include **methadone** and **buprenorphine.** The U.S. Federal Drug Administration has approved the use of three drugs for treatment of alcohol abuse: **naltrexone, acamprosate,** and **disulfiram.** Naltrexone is an opioid receptor antagonist that blocks the reward system and the craving for alcohol. Acamprosate may reduce the withdrawal effects associated with alcohol abuse. Disulfiram causes an accumulation of acetaldehyde by preventing the full degradation of alcohol. This leads to an unpleasant reaction to alcohol ingestion (eg, flushing, nausea, and palpitations). **Topiramate,** a Na^+ channel blocker, is showing promise in clinical trials of alcohol addiction. This is the same drug that has shown to be effective in treatment of migraine headaches.

activates ascending sensory pathways that emit collaterals in the PAG and in the brainstem serotonergic and catecholaminergic regions. The analgesic effect of electroacupuncture is prevented by administration of naloxone, an OR antagonist.

STRESS-INDUCED ANALGESIA

It is well known that soldiers wounded in the heat of battle often feel no pain until the battle is over. This is an example of **stress-induced analgesia** that can also be exemplified by reduced pain sensitivity when being attacked by a predator or other stressful events. Release of norepinephrine, perhaps from brainstem catecholaminergic neurons, in the amygdala may contribute to this phenomenon. As described above, the amygdala is a part of the limbic system that is involved in mediating the motivational-affective responses to pain.

The release of endogenous **cannabinoids** such as 2-arachidonoylglycerol (2AG) and anandamide may also contribute to stress-induced analgesia. These chemicals can act on at least two types of G protein-coupled receptors (CB_1 and CB_2). CB_1 receptors are located in many brain regions, and activation of these receptors accounts for the euphoric actions of cannabinoids. CB_2 receptors are expressed in activated microglia under various pathologies that are associated with chronic neuropathic pain (see Clinical Box 8–3). Binding of an agonist to CB_2 receptors on microglia reduces the inflammatory response and has an analgesic effect. Work is underway to develop selective CB_2 agonists for therapeutic treatment of neuropathic pain.

CHAPTER SUMMARY

- Touch and pressure are sensed by four types of mechanoreceptors that are innervated by rapidly conducting Aα and Aβ sensory afferents. They are rapidly adapting Meissner's corpuscles (respond to changes in texture and slow vibrations), slowly adapting Merkel's cells (respond to sustained pressure and touch), slowly adapting Ruffini corpuscles (respond to sustained pressure), and rapidly adapting Pacinian corpuscles (respond to deep pressure and fast vibrations).
- Nociceptors and thermoreceptors are free nerve endings on unmyelinated C fibers or lightly myelinated Aδ fibers in hairy and glaborous skin and deep tissues. These nerve endings have various types of receptors that are activated by noxious chemical (eg, TRPV1, ASIC), mechanical (eg, P2X, P2Y, TRPA1), and thermal (eg, TRPV1) stimuli. In addition, chemical mediators (eg, bradykinin, prostaglandin, serotonin, histamine) released in response to tissue injury directly activate or sensitize nociceptors.
- The generator or receptor potential is the nonpropagated depolarizing potential recorded in a sensory organ after an adequate stimulus is applied. As the stimulus is increased, the magnitude of the receptor potential is also increased. When it reaches a critical threshold, an action potential is generated in the sensory nerve.
- Converting a receptor stimulus to a recognizable sensation is termed sensory coding. All sensory systems code for four elementary attributes of a stimulus: modality, location, intensity, and duration.
- Pain is an unpleasant sensory and emotional experience associated with actual or potential tissue damage, or described in terms of such damage, whereas nociception is the unconscious activity induced by a harmful stimulus applied to sense receptors. First pain is mediated by Aδ fibers and causes a sharp, localized sensation. Second pain is mediated by C fibers and causes a dull, intense, diffuse, and unpleasant feeling. Acute pain has a sudden onset, recedes during the healing process, and serves as an important protective mechanism. Chronic pain is persistent and caused by nerve damage; it is often associated with hyperalgesia (an exaggerated response to a noxious stimulus) and allodynia (a sensation of pain in response to an innocuous stimulus). Chronic pain is often refractory to NSAIDs and opioids.
- Visceral pain is poorly localized, unpleasant, and associated with nausea and autonomic symptoms. It often radiates (or is referred) to other somatic structures perhaps due to convergence of somatic and visceral nociceptive afferent fibers on the same second-order neurons in the spinal dorsal horn that project to the thalamus and then to the primary somatosensory cortex.
- Discriminative touch, proprioception, and vibratory sensations are relayed via the dorsal column (medial lemniscus) pathway to the VPL in the thalamus and then to the primary somatosensory cortex. Pain and temperature sensations are mediated via the ventrolateral spinothalamic tract, which projects to the VPL and then to cortex. The discriminative aspect of pain results from activation of the primary somatosensory cortex; the motivational-affective component of pain is from activation of the frontal lobe, limbic system, and insular cortex.
- Transmission in pain pathways is modulated by endogenous opioids that can act in the PAG, brainstem, spinal cord, and dorsal root ganglia. Descending pain modulating pathways include neurons in the PAG, nucleus raphé magnus, rostral ventromedial medulla, and locus coeruleus.
- New pain therapies focus on synaptic transmission in nociception and peripheral sensory transduction. Capsaicin transdermal patches or creams reduce pain by exhausting the supply of substance P in nerves and by acting on TRPV1 receptors in the skin. Lidocaine and mexiletine are useful in some cases of chronic pain and act by blocking Nav1.8, which is uniquely associated with nociceptive neurons in dorsal root ganglia. Ziconotide, a voltage-gated N-type Ca^{2+} channel blocker, is used for intrathecal analgesia in patients with refractory chronic pain. Gabapentin, an anticonvulsant drug, is effective in treatment of neuropathic and inflammatory pain by acting on voltage-gated Ca^{2+} channels. Topiramate, a Na^+ channel blocker, is another anticonvulsant drug that can be used to treat migraines. NMDA receptor antagonists can be co-administered with an opioid to reduce tolerance to an opioid.

MULTIPLE-CHOICE QUESTIONS

For all questions, select the single best answer unless otherwise directed.

1. A 28-year-old male was seen by a neurologist because he had experienced prolonged episodes of tingling and numbness in his right arm. He underwent a neurological exam to evaluate his sensory nervous system. Which of the following receptors is correctly paired with the type of stimulus to which it is most apt to respond?
   A. Pacinian corpuscle and motion.
   B. Meissner's corpuscle and deep pressure.
   C. Merkel cells and warmth.
   D. Ruffini corpuscles and sustained pressure.
   E. Muscle spindle and tension.
2. Nociceptors
   A. are activated by strong pressure, severe cold, severe heat, and chemicals.
   B. are absent in visceral organs.
   C. are specialized structures located in the skin and joints.
   D. are innervated by group II afferents.
   E. are involved in acute but not chronic pain.
3. A generator potential
   A. always leads to an action potential.
   B. increases in amplitude as a more intense stimulus is applied.
   C. is an all-or-none phenomenon.
   D. is unchanged when a given stimulus is applied repeatedly over time.
   E. all of the above.
4. Sensory systems code for the following attributes of a stimulus:
   A. modality, location, intensity, and duration
   B. threshold, receptive field, adaptation, and discrimination
   C. touch, taste, hearing, and smell
   D. threshold, laterality, sensation, and duration
   E. sensitization, discrimination, energy, and projection
5. Which of the following are correctly paired?
   A. Neuropathic pain and withdrawal reflex
   B. First pain and dull, intense, diffuse, and unpleasant feeling
   C. Physiological pain and allodynia
   D. Second pain and C fibers
   E. Nociceptive pain and nerve damage
6. A 32-year-old female experienced the sudden onset of a severe cramping pain in the abdominal region. She also became nauseated. Visceral pain
   A. shows relatively rapid adaptation.
   B. is mediated by B fibers in the dorsal roots of the spinal nerves.
   C. is poorly localized.
   D. resembles "fast pain" produced by noxious stimulation of the skin.
   E. causes relaxation of nearby skeletal muscles.
7. A ventrolateral cordotomy is performed that produces relief of pain in the right leg. It is effective because it interrupts the
   A. left dorsal column.
   B. left ventrolateral spinothalamic tract.
   C. right ventrolateral spinothalamic tract.
   D. right medial lemniscal pathway.
   E. a direct projection to the primary somatosensory cortex.
8. Which of the following CNS regions is *not* correctly paired with a neurotransmitter or a chemical involved in pain modulation?
   A. Periaqueductal gray matter and morphine
   B. Nucleus raphé magnus and norepinephrine
   C. Spinal dorsal horn and enkephalin
   D. Dorsal root ganglion and opioids
   E. Spinal dorsal horn and serotonin
9. A 47-year-old female experienced migraine headaches that were not relived by her current pain medications. Her doctor prescribed one of the newer analgesic agents that exert their effects by targeting synaptic transmission in nociception and peripheral sensory transduction. Which of the following drugs is correctly paired with the type of receptor it acts on to exert its antinociceptive effects?
   A. Topiramate and Na^+ channel
   B. Ziconotide and NMDA receptors
   C. Naloxone and opioid receptors
   D. Lidocaine and TRPVI channels
   E. Gabapentin and Nav1.8
10. A 40-year-old man loses his right hand in a farm accident. Four years later, he has episodes of severe pain in the missing hand (phantom limb pain). A detailed PET scan study of his cerebral cortex might be expected to show
    A. expansion of the right hand area in his right primary somatosensory cortex.
    B. expansion of the right-hand area in his left primary somatosensory cortex.
    C. a metabolically inactive spot where his hand area in his left primary somatosensory cortex would normally be.
    D. projection of fibers from neighboring sensory areas into the right-hand area of his right primary somatosensory cortex.
    E. projection of fibers from neighboring sensory areas into the right-hand area of his left primary somatosensory cortex.
11. A 50-year-old woman undergoes a neurological exam that indicates loss of pain and temperature sensitivity, vibratory sense, and proprioception in the left leg. These symptoms could be explained by
    A. a tumor on the right medial lemniscal pathway in the sacral spinal cord.
    B. a peripheral neuropathy.
    C. a tumor on the left medial lemniscal pathway in the sacral spinal cord.
    D. a tumor affecting the right posterior paracentral gyrus.
    E. a large tumor in the right lumbar ventrolateral spinal cord.

CHAPTER RESOURCES

Baron R, Maier C: Phantom limb pain: Are cutaneous nociceptors and spinothalamic neurons involved in the signaling and maintenance of spontaneous and touch-evoked pain? A case report. Pain 1995;60:223.

Bell J, Bolanowski S, Holmes MH: The structure and function of Pacinian corpuscles: A review. Prog Neurobiol 1994;42:79.

Blumenfeld H: *Neuroanatomy Through Clinical Cases*. Sinauer Associates, 2002.

Brownjohn PW, Ashton JC. Novel targets in pain research: The case for CB_2 receptors as a biorational pain target. Current Anesth Critical Care 2009;20:198.

Craig AD: How do you feel? Interoception: The sense of the physiological condition of the body. Nat Rev Neurosci 2002;3:655.
Fields RD: New culprits in chronic pain. Scientific American 2009;301:50.
Garry EM, Jones E, Fleetwood-Walker SM: Nociception in vertebrates: Key receptors participating in spinal mechanisms of chronic pain in animals. Brain Res Rev 2004;46: 216.
Herman J: Phantom limb: From medical knowledge to folk wisdom and back. Ann Int Med 1998;128:76.
Hopkins K: Show me where it hurts: Tracing the pathways of pain. J NIH Res 1997;9:37.
Marchand F, Perretti M, McMahon SB: Role of the immune system in chronic pain. Nat Rev Neurosci 2005;6:521.
Mendell JR, Sahenk Z: Painful sensory neuropathy. N Engl J Med 2003;348:1243.
Mountcastle VB: *Perceptual Neuroscience.* Harvard University Press, 1999.
Willis WD: The somatosensory system, with emphasis on structures important for pain. Brain Res Rev 2007;55:297.
Wu MC, David SV, Gallant JL: Complete functional characterization of sensory neurons by system identification. Annu Rev Neurosci 2006;29:477.

CHAPTER

9

Vision

OBJECTIVES

After studying this chapter, you should be able to:

- Describe the various parts of the eye and list the functions of each.
- Describe the organization of the retina.
- Explain how light rays in the environment are brought to a focus on the retina and the role of accommodation in this process.
- Define hyperopia, myopia, astigmatism, presbyopia, and strabismus.
- Describe the electrical responses produced by rods and cones, and explain how these responses are produced.
- Describe the electrical responses and function of bipolar, horizontal, amacrine, and ganglion cells.
- Trace the neural pathways that transmit visual information from the rods and cones to the visual cortex.
- Describe the responses of cells in the visual cortex and the functional organization of the dorsal and ventral pathways to the parietal cortex.
- Define and explain dark adaptation and visual acuity.
- Describe the neural pathways involved in color vision.
- Identify the muscles involved in eye movements.
- Name the four types of eye movements and the function of each.

INTRODUCTION

The eyes are complex sense organs that have evolved from primitive light-sensitive spots on the surface of invertebrates. They gather information about the environment; and the brain interprets this information to form an image of what appears within the field of vision. The eye is often compared to a camera, with the cornea acting as the lens, the pupillary diameter functioning like the aperture of the camera, and the retina serving as the film. However the eye, especially the retina, is far more sophisticated than even the most expensive camera. Within its protective casing, each eye has a layer of photoreceptors that respond to light, a lens system that focuses the light on these receptors, and a system of nerves that conducts impulses from the receptors to the brain. A great deal of work has been done on the neurophysiology of vision; in fact it is said to be the most studied and perhaps the best understood sensory system. The way the components of the visual system operate to set up conscious visual images is the subject of this chapter.

ANATOMY OF THE EYE

The principal structures of the eye are shown in Figure 9–1. The outer protective layer of the eyeball is the **sclera** or the "white of the eye" through which no light can pass. It is modified anteriorly to form the transparent **cornea,** through which light rays enter the eye. The lateral margin of the cornea is contiguous with the **conjunctiva,** a clear mucous membrane that covers the sclera. Just inside the sclera is the **choroid,** which is a vascular layer that provides oxygen and nutrients

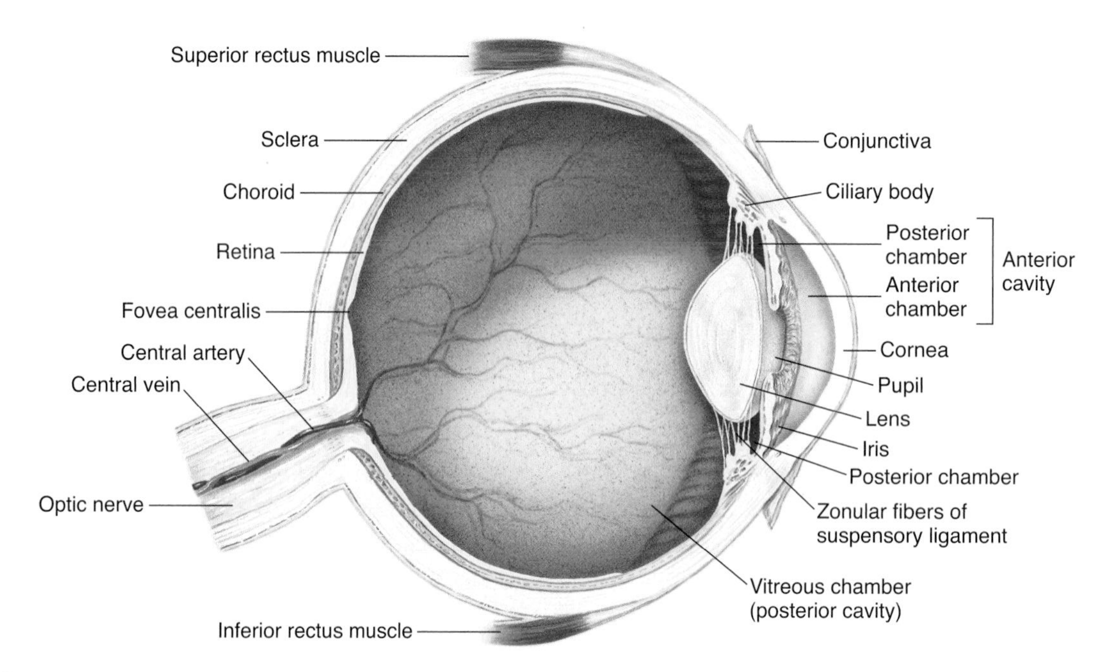

FIGURE 9–1 A schematic of the anatomy of the eye. (From Fox SI, *Human Physiology.* McGraw-Hill, 2008.)

to the structures in the eye. Lining the posterior two thirds of the choroid is the **retina,** the neural tissue containing the photoreceptors.

The **crystalline lens** is a transparent structure held in place by a circular **lens suspensary ligament (zonule).** The zonule is attached to the **ciliary body,** which contains circular muscle fibers and longitudinal muscle fibers that attach near the corneoscleral junction. In front of the lens is the pigmented and opaque **iris,** the colored portion of the eye. The iris, ciliary body, and choroid are collectively called the **uvea.** The iris contains circular muscle fibers that constrict and radial fibers that dilate the **pupil.** Variations in the diameter of the pupil can produce up to a fivefold change in the amount of light reaching the retina.

The **aqueous humor** is a clear protein-free liquid that nourishes the cornea and iris; it is produced in the ciliary body by diffusion and active transport from plasma. It flows through the pupil and fills the **anterior chamber** of the eye. It is normally reabsorbed through a network of trabeculae into the **canal of Schlemm,** which is a venous channel at the junction between the iris and the cornea (**anterior chamber angle**). Obstruction of this outlet leads to increased **intraocular pressure,** a critical risk factor for glaucoma (see Clinical Box 9–1).

The **posterior chamber** is a narrow aqueous-containing space between the iris, zonule, and the lens. The **vitreous chamber** is the space between the lens and the retina that is filled primarily with a clear gelatinous material called the **vitreous (vitreous humor).**

The eye is well protected from injury by the bony walls of the orbit. The cornea is moistened and kept clear by tears that course from the **lacrimal gland** in the upper portion of each orbit across the surface of the eye to empty via the **lacrimal duct** into the nose. Blinking helps keep the cornea moist.

RETINA

The retina extends anteriorly almost to the ciliary body. It is organized into layers containing different types of cells and neural processes (Figure 9–2). The outer nuclear layer contains the **photoreceptors,** the **rods,** and **cones.** The inner nuclear layer contains the cell bodies of various types of excitatory and inhibitory interneurons including **bipolar cells, horizontal cells,** and **amacrine cells.** The ganglion cell layer contains various types of **ganglion cells** that can be distinguished on the basis of morphology, projections, and functions. Ganglion cells are the only output neuron of the retina; their axons form the **optic nerve.** The outer plexiform layer is interposed between the outer and inner nuclear layers; the inner plexiform layer is interposed between the inner nuclear and ganglion cell layers. The neural elements of the retina are bound together by a type of glia called **Müller cells,** which form the **inner limiting membrane,** the boundary between the retina and the vitreous chamber. The elongated processes of these cells extend the entire thickness of the retina. The **outer limiting membrane** separates the inner segment portion of the rods and cones from their cell bodies.

The rods and cones, which are next to the choroid, synapse with bipolar cells, and the bipolar cells synapse with ganglion cells. There are various types of bipolar cells that differ in terms of morphology and function. Horizontal cells connect photoreceptor cells to the other photoreceptor cells in the outer plexiform layer. Amacrine cells connect ganglion cells to one another in the inner plexiform layer via processes of varying length and patterns. Amacrine cells also make connections on the terminals of bipolar cells. At least 29 types of amacrine cells have been described on the basis of their connections. Gap junctions also connect retinal neurons to one another.

CLINICAL BOX 9–1

Glaucoma

Increased intraocular pressure (IOP) is not the only cause of **glaucoma,** a degenerative disease in which there is loss of retinal ganglia cells; however, it is a critical risk factor. In a substantial minority of the patients with this disease, IOP is normal (10–20 mm Hg); however, increased IOP makes glaucoma worse, and treatment is aimed at lowering the pressure. Indeed, elevations in IOP due to injury or surgery can cause glaucoma. Glaucoma is caused by poor drainage of the aqueous humor through the filtration angle formed between the iris and the cornea. **Open-angle glaucoma,** a chronic disease, is caused by decreased permeability through the trabeculae into the canal of Schlemm, which leads to an increase in IOP. In some cases this type of glaucoma is due to a genetic defect. **Closed-angle glaucoma** results from a forward ballooning of the iris so that it reaches the back of the cornea and obliterates the filtration angle, thus reducing the outflow of aqueous humor. If left untreated, glaucoma can lead to blindness.

THERAPEUTIC HIGHLIGHTS

Glaucoma can be treated with agents that decrease the secretion or production of aqueous humor or with drugs that increase outflow of the aqueous humor. β-adrenergic blocking drugs such as **timolol** decrease the secretion of aqueous fluid. Carbonic anhydrase inhibitors (eg, **dorzolamide, acetozolamide**) also exert their beneficial effects by decreasing the secretion of aqueous humor. Glaucoma can also be treated with cholinergic agonists (eg, **pilocarpine, carbachol, physostigmine**) that increase aqueous outflow by causing ciliary muscle contraction. Aqueous outflow is also increased by **prostaglandins.** Prolonged use of **corticosteroids** can lead to glaucoma and increase the risk of occurrence of ocular infections due to fungi or viruses.

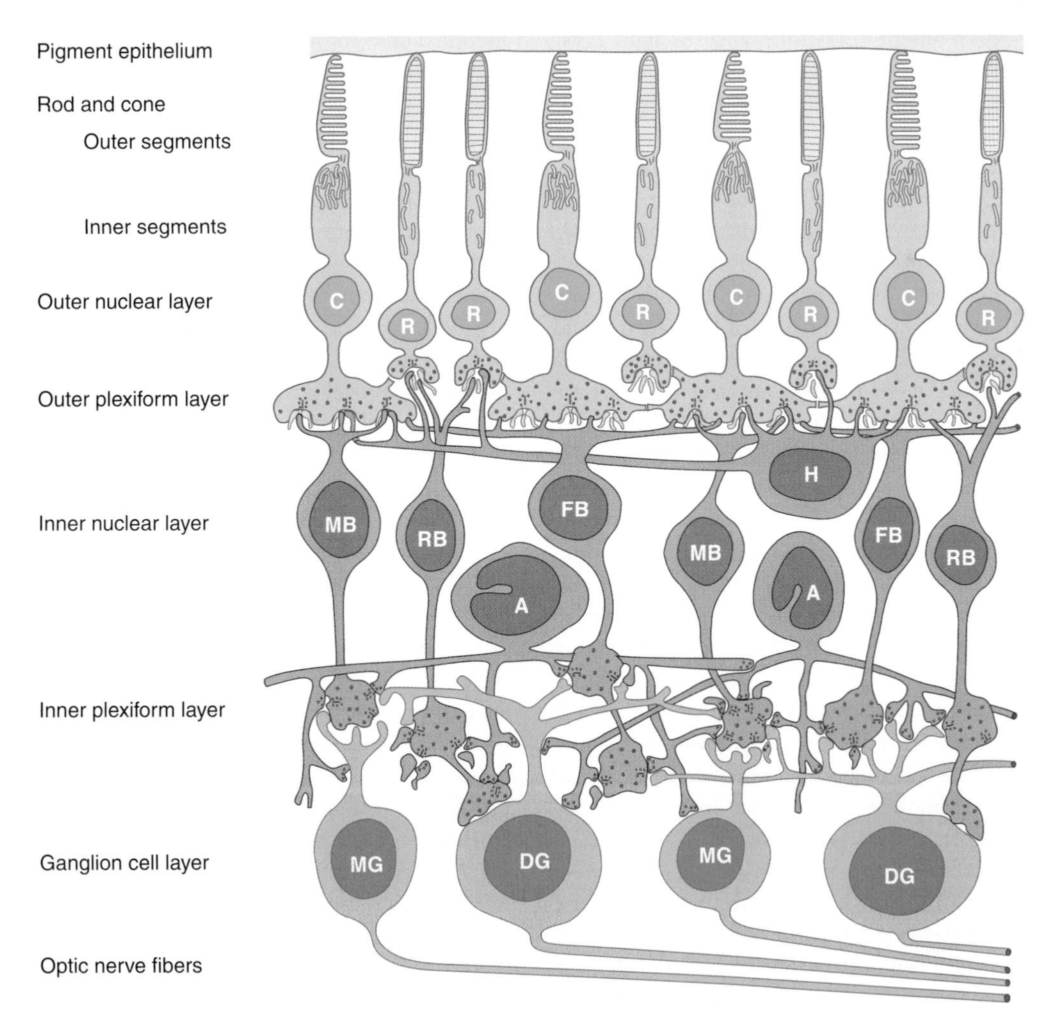

FIGURE 9–2 Neural components of the extrafoveal portion of the retina. C, cone; R, rod; MB, RB, and FB, midget, rod, and flat bipolar cells; DG and MG, diffuse and midget ganglion cells; H, horizontal cells; A, amacrine cells. (Modified from Dowling JE, Boycott BB: Organization of the primate retina: Electron microscopy. Proc R Soc Lond Ser B [Biol] 1966;166:80.)

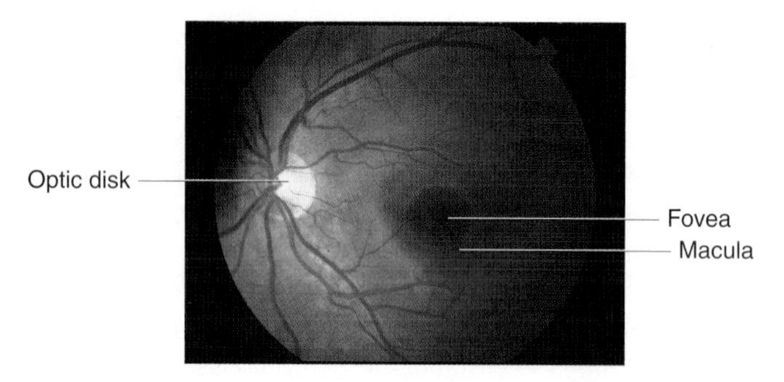

FIGURE 9–3 The fundus of the eye in a healthy human as seen through the ophthalmoscope. The fundus of the eye refers to the interior surface of the eye, opposite the lens, and includes the retina, optic disc, macula and fovea, and posterior pole. Optic nerve fibers leave the eyeball at the optic disc to form the optic nerve. The arteries, arterioles, and veins in the superficial layers of the retina near its vitreous surface can be seen through the ophthalmoscope. (Courtesy of Dr AJ Weber, Michigan State University.)

Because the receptor layer of the retina rests on the **pigment epithelium** next to the choroid, light rays must pass through the ganglion cell and bipolar cell layers to reach the rods and cones. The pigment epithelium absorbs light rays, preventing the reflection of rays back through the retina. Such reflection would otherwise produce blurring of the visual images.

The optic nerve leaves the eye at a point 3 mm medial to and slightly above the posterior pole of the globe. This region is visible through the ophthalmoscope as the **optic disk** (Figure 9–3). Since there are no visual receptors over the disk, this area of the retina does not respond to light and is known as the **blind spot.** Near the posterior pole of the eye, there is a yellowish pigmented spot called the **macula.** The **fovea** is in the center of the macula; it is a thinned-out, rod-free portion of the retina in humans and other primates. In it, the cones are densely packed, and each synapses on a single bipolar cell, which, in turn, synapses on a single ganglion cell, providing a direct pathway to the brain. There are very few overlying cells and no blood vessels. Consequently, the fovea is the point where **visual acuity** is greatest. When attention is attracted to or fixed on an object, the eyes are normally moved so that light rays coming from the object fall on the fovea. **Age-related macular degeneration** is a disease in which sharp, central vision is gradually destroyed (Clinical Box 9–2).

An ophthalmoscope is used to view the **fundus** of the eye, which is the interior surface of the eye, opposite to the lens; it includes the retina, optic disc, macula and fovea, and posterior pole (Figure 9–3). The arteries, arterioles, and veins in the superficial layers of the retina near its vitreous surface can be examined. Because this is the one place in the body where arterioles are readily visible, ophthalmoscopic examination is of great value in the diagnosis and evaluation of diabetes mellitus, hypertension, and other diseases that affect blood vessels. The retinal vessels supply the bipolar and ganglion cells, but the receptors are nourished, for the most part, by the capillary plexus in the choroid. This is why retinal detachment is so damaging to the receptor cells.

Glaucoma (Clinical Box 9–1) causes changes in the appearance of the fundus of the eye as seen through an ophthalmoscope (Figure 9–4). The photograph on the left is from a primate with a normal eye and shows an optic disc with a uniform "pinkish" color. The blood vessels appear relatively flat as they cross the margin of the disc. This is because there are a normal number of ganglion cell fibers, and the blood vessels have intact support tissue around them. The photograph on the right is from a primate with glaucoma that was experimentally induced by causing a chronic elevation in intraocular pressure. As is characteristic of glaucomatous optic neuropathy, the disc is pale, especially in the center. The retinal blood vessels are distorted, especially at the disc margin, due to a lack of support tissue; and there is increased "cupping" of the disc.

PHOTORECEPTORS

Each rod and cone photoreceptor is divided into an outer segment, an inner segment that includes a nuclear region, and a synaptic terminal zone (Figure 9–5). The outer segments are modified **cilia** comprised of regular stacks of flattened **saccules** or membranous **disks.** The inner segments are rich in mitochondria; this is the region that synthesizes the photosensitive compounds. The inner and outer segments are connected by a ciliary stalk through which the photosensitive compounds travel from the inner segment to the outer segment of the rods and cones.

The rods are named for the thin, rod-like appearance of their outer segments. Each rod contains a stack of disk membranes that are flattened membrane-bound intracellular organelles that have detached from the outer membrane, and are thus free floating. Cones generally have thick inner segments and conical outer segments, although their morphology varies from place to place in the retina. The saccules of the cones are formed by infolding of the membrane of the outer segment. The saccules and disks contain the photosensitive compounds that react to light, initiating action potentials in the visual pathways.

Rod outer segments are being constantly renewed by the formation of new disks at the inner edge of the segment and phagocytosis of old disks from the outer tip by cells of the pigment epithelium. Cone renewal is a more diffuse process and appears to occur at multiple sites in the outer segments.

In the extrafoveal portions of the retina, rods predominate (Figure 9–6), and there is a good deal of convergence. *Flat* bipolar cells (Figure 9–2) make synaptic contact with several cones, and *rod* bipolar cells make synaptic contact with several rods. Because there are approximately 6 million cones and 120 million rods in each human eye but only 1.2 million nerve fibers in each optic nerve, the overall convergence of receptors through bipolar cells on ganglion cells is about 105:1. However, there is divergence from this point on. For example, in the visual cortex the number of neurons concerned with vision is 1000 times the number of fibers in the optic nerves.

CLINICAL BOX 9–2

Visual Acuity and Age-Related Macular Degeneration

Visual acuity is the degree to which the details and contours of objects are perceived, and it is usually defined in terms of the shortest distance by which two lines can be separated and still be perceived as two lines. Clinically, visual acuity is often determined by the use of the familiar **Snellen letter charts** viewed at a distance of 20 ft (6 m). The individual being tested reads aloud the smallest line distinguishable. The results are expressed as a fraction. The numerator of the fraction is 20, the distance at which the subject reads the chart. The denominator is the greatest distance from the chart at which a normal individual can read the smallest line. Normal visual acuity is 20/20; a subject with 20/15 visual acuity has better than normal vision (not farsightedness); and one with 20/100 visual acuity has subnormal vision. Visual acuity is a complex phenomenon and is influenced by many factors, including optical factors (eg, the state of the image-forming mechanisms of the eye), retinal factors (eg, the state of the cones), and stimulus factors (eg, illumination, brightness of the stimulus, contrast between the stimulus and the background, length of time the subject is exposed to the stimulus). Many drugs can also have adverse side effects on visual acuity. Many patients treated with the anti-arrhythmic drug **amiodarone** report corneal changes (**kerotopathy**) including complaints of blurred vision, glare and halos around lights or light sensitivity. **Aspirin** and other anti-coagulants can cause conjunctival or retinal hemorrhaging which can impair vision. **Maculopathy** is a risk factor for those treated with **tamoxifen** for breast cancer. Anti-psychotic therapies such as **thioridazine** can cause pigmentary changes, which can affect visual acuity, color vision, and dark adaptation.

There are over 20 million individuals in the United States and Europe with **age-related macular degeneration (AMD),** which is a deterioration of central visual acuity. Nearly 30% of those aged 75 or older have this disorder, and it is the most common cause of visual loss in those aged 50 or older. Women are at greater risk than men for developing AMD; also Caucasians have a greater risk than blacks. There are two types: wet and dry. Wet AMD occurs when fragile blood vessels begin to form under the macula. Blood and fluid leak from these vessels and rapidly damage the macula. Vascular endothelial growth factors (VEGF) may contribute to the growth of these blood vessels. Dry AMD occurs when the cones in the macula slowly break down, causing a gradual loss of central vision.

THERAPEUTIC HIGHLIGHTS

The U.S. Food and Drug Administration has approved the use **ranibizumab** (Lucentis) to treat wet AMD. It acts by inhibiting VEGF. Another drug approved for the treatment of wet AMD is **pegaptanib sodium** (Macugen), which attacks VEGF. **Photodynamic therapy** uses a drug called **visudyne,** which is injected into the vein in an arm and is activated by a laser light, which produces a chemical reaction that destroys abnormal blood vessels. **Laser surgery** can be done to repair damaged blood vessels if they are at a distance from the fovea. However, new vessels may form after the surgery, and vision loss may progress.

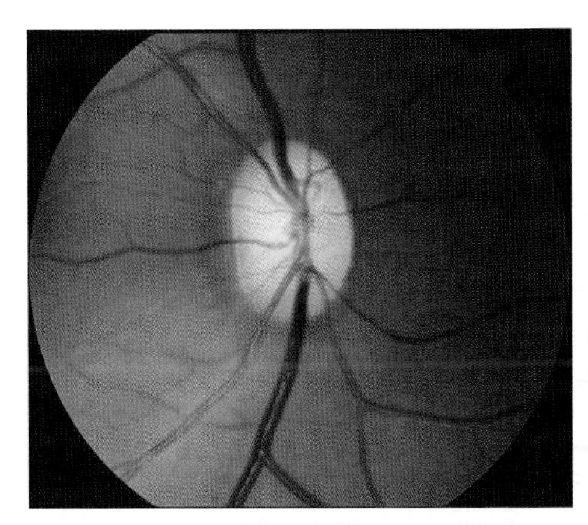

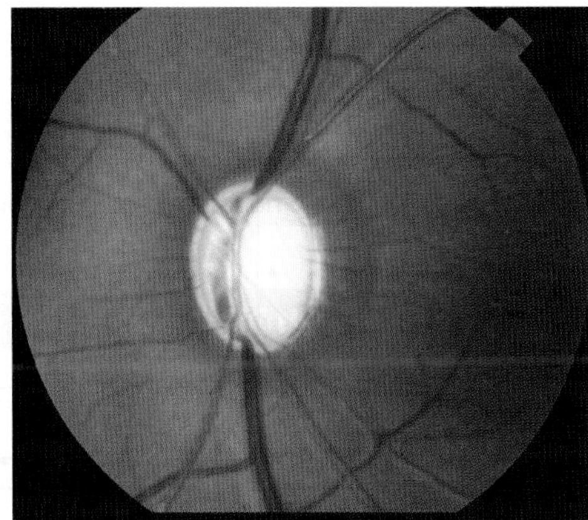

FIGURE 9–4 The fundus of the eye in a normal primate (left) and in a primate with experimentally induced glaucoma (right) as seen through an ophthalmoscope. Normal: uniform "pinkish" color, vessels appear relatively flat crossing the margin of disc due to a normal number of ganglion cell fibers and since they have intact support tissue around them. Glaucomatous: disc is pale, especially in center, vessels are distorted, especially at the disc margin due to lack of support tissue and increased "cupping" of the disc. (Courtesy of Dr AJ Weber, Michigan State University.)

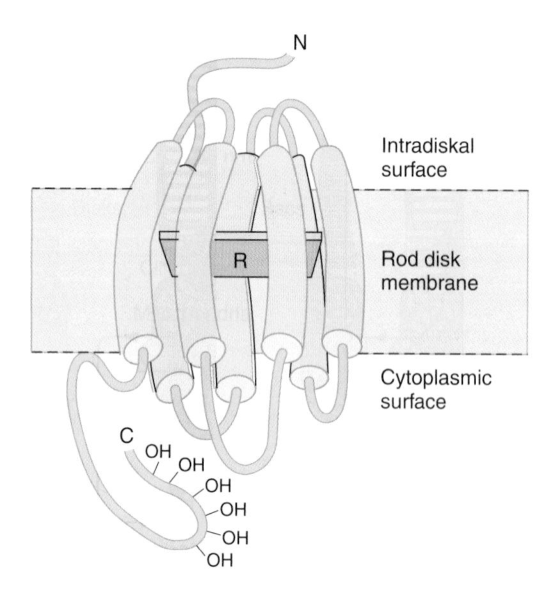

FIGURE 9–8 Diagrammatic representation of the structure of rhodopsin, showing the position of retinal in the rod disk membrane. Retinal (R) is located in parallel to the surface of the membrane and is attached to a lysine residue at position 296 in the 7th transmembrane domain.

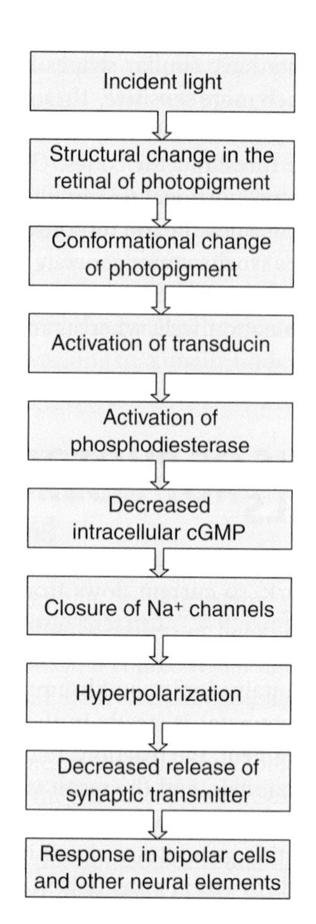

FIGURE 9–9 Sequence of events involved in phototransduction in rods and cones.

The sequence of events in photoreceptors by which incident light leads to production of a signal in the next succeeding neural unit in the retina is summarized in Figure 9–9. In the dark, the retinal in rhodopsin is in the 11-*cis* configuration. The only action of light is to change the shape of the retinal, converting it to the all-*trans* isomer. This, in turn, alters the configuration of the opsin, and the opsin change activates its associated heterotrimeric G protein, which in this case is called **transducin,** which has several subunits Tα, Gβ1, and Gγ1. After 11-*cis* retinal is converted to the all-*trans* configuration, it separates from the opsin in a process called bleaching. This changes the color from the rosy red of rhodopsin to the pale yellow of opsin.

Some of the all-*trans* retinal is converted back to the 11-*cis* retinal by retinal isomerase, and then again associates with scotopsin, replenishing the rhodopsin supply. Some 11-*cis* retinal is also synthesized from vitamin A. All of these reactions, except the formation of the all-*trans* isomer of retinal, are independent of the light intensity, proceeding equally well in light or darkness. The amount of rhodopsin in the receptors therefore varies inversely with the incident light level.

The G protein transducin exchanges GDP for GTP, and the α subunit separates. This subunit remains active until its intrinsic GTPase activity hydrolyzes the GTP. Termination of the activity of transducin is also accelerated by its binding of β-arrestin. The α subunit activates cGMP phosphodiesterase, which converts cGMP to 5′-GMP. cGMP normally acts directly on Na^+ channels to maintain them in the open position, so the decline in the cytoplasmic cGMP concentration causes some Na^+ channels to close. This produces the hyperpolarizing potential. This cascade of reactions occurs very rapidly and amplifies the light signal. The amplification helps explain the remarkable sensitivity of rod photoreceptors; these receptors are capable of producing a detectable response to as little as one photon of light.

Light reduces the concentration of Ca^{2+} as well as that of Na^+ in photoreceptors. The resulting decrease in Ca^{2+} concentration activates guanylyl cyclase, which generates more cGMP. It also inhibits the light-activated phosphodiesterase. Both actions speed recovery, restoring the Na^+ channels to their open position.

CONE PIGMENTS

Primates have three different kinds of cones. These receptors subserve color vision and respond maximally to light at wavelengths of 440, 535, and 565 nm. Each contains retinal and an opsin. The opsin resembles rhodopsin and spans the cone membrane seven times but has a characteristic structure in each type of cone. The details of the responses of cones to light are probably similar to those in rods. Light activates retinal, and this activates a cone transducin, a G protein that differs somewhat from rod transducin. Cone transducin in turn activates phosphodiesterase, catalyzing the conversion of cGMP to 5′-GMP. This causes closure of Na^+ channels between the extracellular fluid and the cone cytoplasm, a decrease in intracellular Na^+ concentration, and hyperpolarization of the cone synaptic terminals.

MELANOPSIN

A few retinal ganglion cells contain **melanopsin** rather than rhodopsin or cone pigments. The axons of these neurons project to the suprachiasmatic nuclei of the hypothalamus, where they form connections that synchronize a variety of endocrine and other circadian rhythms with the light–dark cycle (Chapter 14). When the gene for melanopsin is knocked out, circadian photo-entrainment is abolished. The pupillary light reflex (described below) is also reduced, and it is abolished when the rods and cones are also inactivated. Thus, a part of the pupillary responses and all the circadian entrainment responses to light–dark changes are controlled by a system distinct from the rods and cones.

PROCESSING OF VISUAL INFORMATION IN THE RETINA

In a sense, the processing of visual information in the retina involves the formation of three images. The first image, formed by the action of light on the photoreceptors, is changed to a second image in the bipolar cells, and this in turn is converted to a third image in the ganglion cells. In the formation of the second image, the signal is altered by the horizontal cells, and in the formation of the third, it is altered by the amacrine cells. There is little change in the impulse pattern in the lateral geniculate bodies, so the third image reaches the occipital cortex.

A characteristic of the bipolar and ganglion cells (as well as the lateral geniculate cells and the cells in layer 4 of the visual cortex) is that they respond best to a small, circular stimulus and that, within their receptive field, an annulus of light around the center (surround illumination) antagonizes the response to the central spot (Figure 9–10). The center can be excitatory with an inhibitory surround (an **"on-center" cell**) or inhibitory with an excitatory surround (an **"off-center" cell**). The inhibition of the center response by the surround is probably due to inhibitory feedback from one photoreceptor to another mediated via horizontal cells. Thus, activation of nearby photoreceptors by addition of the annulus triggers horizontal cell hyperpolarization, which in turn inhibits the response of the centrally activated photoreceptors. The inhibition of the response to central illumination by an increase in surrounding illumination is an example of **lateral inhibition**—that form of inhibition in which activation of a particular neural unit is associated with inhibition of the activity of nearby units. It is a general phenomenon in mammalian sensory systems and helps to sharpen the edges of a stimulus and improve discrimination.

A remarkable degree of processing of visual input occurs in the retina, largely via amacrine cells. For example, movement of an object within the visual field is separated from movement of the background caused by changes in posture and movement of the eyes. This was demonstrated by recording from optic neurons. When an object moved at a different speed or in a different direction than the background, an impulse was generated. However, when the object moved like the background, inhibition occurred and no optic nerve signal was generated.

THE IMAGE-FORMING MECHANISM

The eyes convert energy in the visible spectrum into action potentials in the optic nerve. The wavelength of visible light ranges from approximately 397 to 723 nm. The images of objects in the environment are focused on the retina. The light rays striking the retina generate potentials in the rods and cones. Impulses initiated in the retina are conducted to the cerebral cortex, where they produce the sensation of vision.

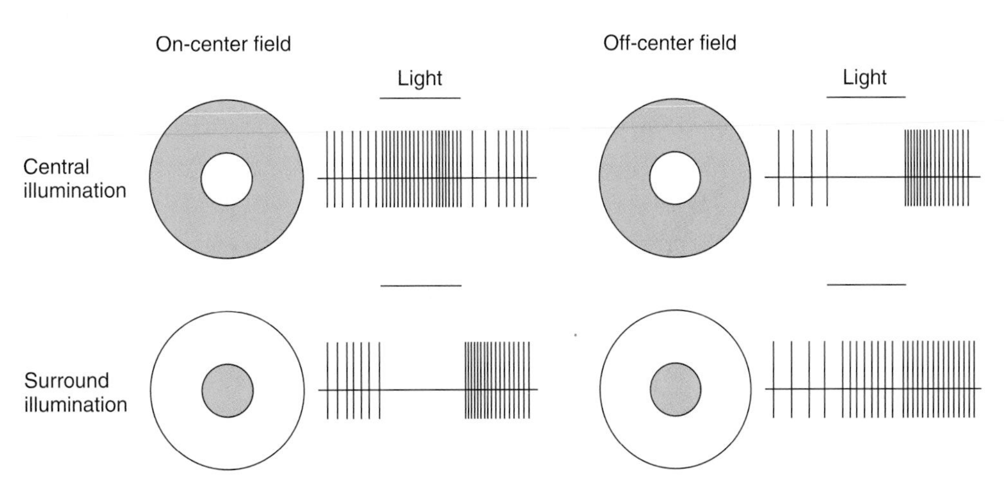

FIGURE 9–10 Responses of retinal ganglion cells to light focused on the portions of their receptive fields indicated in white. To the right of each receptive field diagram is a representation of the action potentials recorded from a ganglion cell in response to the light being turned on or off. Note that in three of the four situations, there is increased discharge when the light is turned off. (Adapted from Kuffler SW: Discharge patterns and functional organization of mammalian retina. J Neurophysiol 1953 Jan;16(1):37–68.)

PRINCIPLES OF OPTICS

Light rays are bent when they pass from a medium of one density into a medium of a different density, except when they strike perpendicular to the interface (Figure 9–11). The bending of light rays is called **refraction** and is the mechanism that allows one to focus an accurate image onto the retina. Parallel light rays striking a biconvex lens are refracted to a point **(principal focus)** behind the lens. The principal focus is on a line passing through the centers of curvature of the lens, the **principal axis.** The distance between the lens and the principal focus is the **principal focal distance.** For practical purposes, light rays from an object that strike a lens more than 6 m (20 ft) away are considered to be parallel. The rays from an object closer than 6 m are diverging and are therefore brought to a focus farther back on the principal axis than the principal focus. Biconcave lenses cause light rays to diverge.

Refractive power is greatest when the curvature of a lens is greatest. The refractive power of a lens is conveniently measured in **diopters,** the number of diopters being the reciprocal of the principal focal distance in meters. For example, a lens with a principal focal distance of 0.25 m has a refractive power of 1/0.25, or 4 diopters. The human eye has a refractive power of approximately 60 diopters at rest.

In the eye, light is actually refracted at the anterior surface of the cornea and at the anterior and posterior surfaces of the lens. The process of refraction can be represented diagrammatically, however, without introducing any appreciable error, by drawing the rays of light as if all refraction occurs at the anterior surface of the cornea (Figure 9–11). It should be noted that the retinal image is inverted. The connections of the retinal receptors are such that from birth any inverted image on the retina is viewed right side up and projected to the visual field on the side opposite to the retinal area stimulated. This perception is present in infants and is innate. If retinal images are turned right side up by means of special lenses, the objects viewed look as if they are upside down.

COMMON DEFECTS OF THE IMAGE-FORMING MECHANISM

In some individuals, the eyeball is shorter than normal and the parallel rays of light are brought to a focus behind the retina. This abnormality is called **hyperopia** or farsightedness

A)

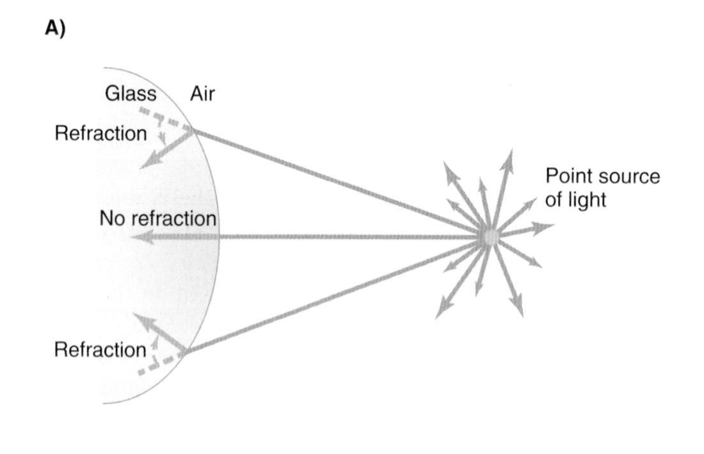

B)

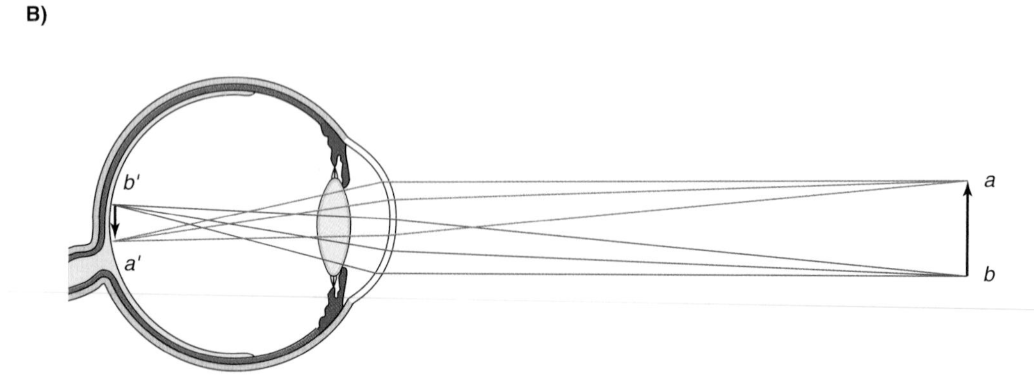

FIGURE 9–11 Focusing point sources of light. A) When diverging light rays enter a dense medium at an angle to its convex surface, refraction bends them inward. **B)** Refraction of light by the lens system. For simplicity, refraction is shown only at the corneal surface (site of greatest refraction) although it also occurs in the lens and elsewhere. Incoming light from *a* (above) and *b* (below) is bent in opposite directions, resulting in *b′* being above *a′* on the retina. (From Widmaier EP, Raff H, Strang KT: *Vander's Human Physiology,* 11th ed. McGraw-Hill, 2008.)

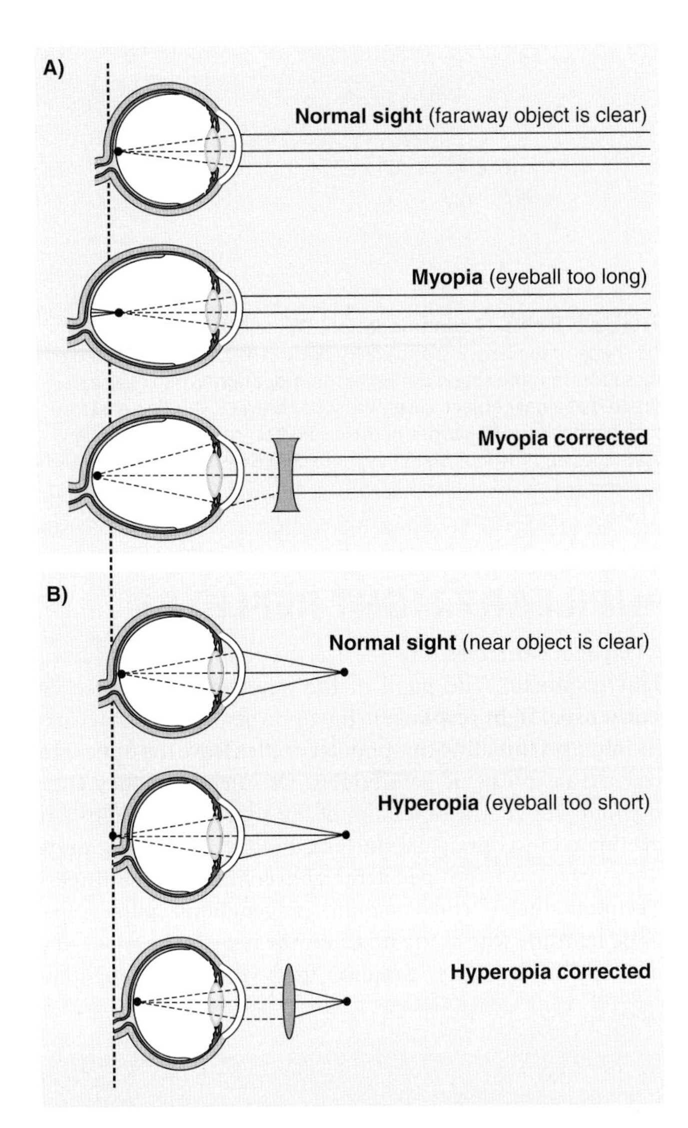

FIGURE 9–12 Common defects of the optical system of the eye. A) In myopia (nearsightedness), the eyeball is too long and light rays focus in front of the retina. Placing a biconcave lens in front of the eye causes the light rays to diverge slightly before striking the eye, so that they are brought to a focus on the retina. **B)** In hyperopia (farsightedness), the eyeball is too short and light rays come to a focus behind the retina. A biconvex lens corrects this by adding to the refractive power of the lens of the eye. (From Widmaier EP, Raff H, Strang KT: Vander's Human Physiology, 11th ed. McGraw-Hill, 2008.).

(Figure 9–12). Sustained accommodation, even when viewing distant objects, can partially compensate for the defect, but the prolonged muscular effort is tiring and may cause headaches and blurring of vision. The prolonged convergence of the visual axes associated with the accommodation may lead eventually to **strabismus** (Clinical Box 9–4). The defect can be corrected by using glasses with convex lenses, which aid the refractive power of the eye in shortening the focal distance.

In **myopia** (nearsightedness), the anteroposterior diameter of the eyeball is too long (Figure 9–12). Myopia is said to be genetic in origin. However, there is a positive correlation between sleeping in a lighted room before the age of 2 and the subsequent

CLINICAL BOX 9–4

Strabismus & Amblyopia

Strabismus is a misalignment of the eyes and one of the most common eye problems in children, affecting about 4% of children under 6 years of age. It is characterized by one or both eyes turning inward (**esotropia**), outward (**exotropia**), upward, or downward. In some cases, more than one of these conditions is present. Strabismus is also commonly called "wandering eye" or "crossed-eyes." It results in visual images that do not fall on corresponding retinal points. When visual images chronically fall on noncorresponding points in the two retinas in young children, one is eventually suppressed **(suppression scotoma).** This suppression is a cortical phenomenon, and it usually does not develop in adults. It is important to institute treatment before age 6 in affected children, because if the suppression persists, the loss of visual acuity in the eye generating the suppressed image is permanent. A similar suppression with subsequent permanent loss of visual acuity can occur in children in whom vision in one eye is blurred or distorted owing to a refractive error. The loss of vision in these cases is called **amblyopia ex anopsia,** a term that refers to uncorrectable loss of visual acuity that is not directly due to organic disease of the eye. Typically, an affected child has one weak eye with poor vision and one strong eye with normal vision. It affects about 3% of the general population. Amblyopia is also referred to as "lazy eye," and it often co-exists with strabismus.

THERAPEUTIC HIGHLIGHTS

Atropine (a cholinergic muscarinic receptor antagonist) and **miotics** such as **echothiophate iodide** can be administered in the eye to correct strabismus and ambylopia. Atropine will blur the vision in the good eye to force the individual to use the weaker eye. Eye muscle training through **optometric vision therapy** has also proven to be useful, even in patients older than 17 years of age. Some types of strabismus can be corrected by surgical shortening of some of the eye muscles, by eye muscle training exercises, and by the use of glasses with prisms that bend the light rays sufficiently to compensate for the abnormal position of the eyeball. However, subtle defects in **depth perception** persist. It has been suggested that congenital abnormalities of the visual tracking mechanisms may cause both strabismus and the defective depth perception. In infant monkeys, covering one eye with a patch for 3 months causes a loss of ocular dominance columns; input from the remaining eye spreads to take over all the cortical cells, and the patched eye becomes functionally blind. Comparable changes may occur in children with strabismus.

development of myopia. Thus, the shape of the eye appears to be determined in part by the refraction presented to it. In young adult humans the extensive close work involved in activities such as studying accelerates the development of myopia. This defect can be corrected by glasses with biconcave lenses, which make parallel light rays diverge slightly before they strike the eye.

Astigmatism is a common condition in which the curvature of the cornea is not uniform. When the curvature in one meridian is different from that in others, light rays in that meridian are refracted to a different focus, so that part of the retinal image is blurred. A similar defect may be produced if the lens is pushed out of alignment or the curvature of the lens is not uniform, but these conditions are rare. Astigmatism can usually be corrected with cylindric lenses placed in such a way that they equalize the refraction in all meridians.

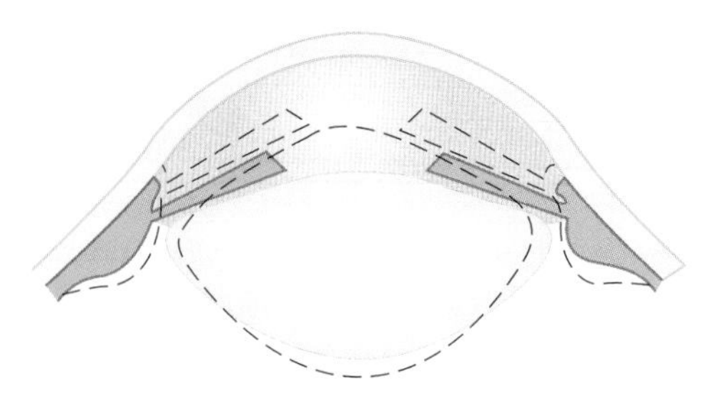

FIGURE 9–13 Accommodation. The solid lines represent the shape of the lens, iris, and ciliary body at rest, and the dashed lines represent the shape during accommodation. When gaze is directed at a near object, ciliary muscles contract. This decreases the distance between the edges of the ciliary body and relaxes the lens ligaments, and the lens becomes more convex. (From Waxman SG: *Clinical Neuroanatomy*, 26th ed. McGraw-Hill, 2010.)

ACCOMMODATION

When the ciliary muscle is relaxed, parallel light rays striking the optically normal **(emmetropic)** eye are brought to a focus on the retina. As long as this relaxation is maintained, rays from objects closer than 6 m from the observer are brought to a focus behind the retina, and consequently the objects appear blurred. The problem of bringing diverging rays from close objects to a focus on the retina can be solved by increasing the distance between the lens and the retina or by increasing the curvature or refractive power of the lens. In bony fish, the problem is solved by increasing the length of the eyeball, a solution analogous to the manner in which the images of objects closer than 6 m are focused on the film of a camera by moving the lens away from the film. In mammals, the problem is solved by increasing the curvature of the lens.

The process by which the curvature of the lens is increased is called **accommodation.** At rest, the lens is held under tension by the lens ligaments. Because the lens substance is malleable and the lens capsule has considerable elasticity, the lens is pulled into a flattened shape. If the gaze is directed at a near object, the ciliary muscle contracts. This decreases the distance between the edges of the ciliary body and relaxes the lens ligaments, so that the lens springs into a more convex shape (Figure 9–13). The change is greatest in the anterior surface of the lens. In young individuals, the change in shape may add as many as 12 diopters to the refractive power of the eye. The relaxation of the lens ligaments produced by contraction of the ciliary muscle is due partly to the sphincterlike action of the circular muscle fibers in the ciliary body and partly to the contraction of longitudinal muscle fibers that attach anteriorly, near the corneoscleral junction. As these fibers contract, they pull the whole ciliary body forward and inward. This motion brings the edges of the ciliary body closer together. Changes in accommodation with age are described in Clinical Box 9–5.

In addition to accommodation, the visual axes converge and the pupil constricts when an individual looks at a near object. This three-part response—accommodation, convergence of the visual axes, and pupillary constriction—is called the **near response.**

PUPILLARY LIGHT REFLEXES

When light is directed into one eye, the pupil constricts **(direct light response).** The pupil of the other eye also constricts **(consensual light response).** The optic nerve fibers that carry the impulses initiating this pupillary reflex leave the optic tract near the lateral geniculate bodies. On each side, they enter the midbrain via the brachium of the superior colliculus and terminate in the pretectal nucleus. From this nucleus, nerve fibers project to the ipsilateral and contralateral **Edinger–Westphal nuclei** which contain preganglionic parasympathetic neurons within the **oculomotor nerve.** These neurons terminate in the ciliary ganglion from which postganglionic

CLINICAL BOX 9–5

Accommodation & Aging

Accommodation is an active process, requiring muscular effort, and can therefore be tiring. Indeed, the ciliary muscle is one of the most used muscles in the body. The degree to which the lens curvature can be increased is limited, and light rays from an object very near the individual cannot be brought to a focus on the retina, even with the greatest of effort. The nearest point to the eye at which an object can be brought into clear focus by accommodation is called the **near point of vision.** The near point recedes throughout life, slowly at first and then rapidly with advancing age, from approximately 9 cm at age 10 to approximately 83 cm at age 60. This recession is due principally to increasing hardness of the lens, with a resulting loss of accommodation due to the steady decrease in the degree to which the curvature of the lens can be increased. By the time a normal individual reaches age 40–45, the loss of accommodation is usually sufficient to make reading and close work difficult. This condition, which is known as **presbyopia,** can be corrected by wearing glasses with convex lenses.

nerves project to the ciliary muscle. This pathway is dorsal to the pathway for the near response. Consequently, the light response is sometimes lost while the response to accommodation remains intact **(Argyll Robertson pupil).** One cause of this abnormality is CNS syphilis, but the Argyll Robertson pupil is also seen in other diseases producing selective lesions in the midbrain.

RESPONSES IN THE VISUAL PATHWAYS & CORTEX

NEURAL PATHWAYS

The axons of the ganglion cells pass caudally in the optic nerve and **optic tract** to end in the **lateral geniculate body** in the thalamus (Figure 9–14). The fibers from each nasal hemiretina decussate in the **optic chiasm.** In the geniculate body, the fibers from the nasal half of one retina and the temporal half of the other synapse on the cells whose axons form the **geniculocalcarine tract.** This tract passes to the occipital lobe of the cerebral cortex. The effects of lesions in these pathways on visual function are discussed below.

Some ganglion cell axons bypass the lateral geniculate nucleus (LGN) to project directly to the pretectal area; this pathway mediates the pupillary light reflex and eye movements. The frontal cortex is also concerned with eye movement, and especially its refinement. The bilateral **frontal eye fields** in this part of the cortex are concerned with control of saccades, and an area just anterior to these fields is concerned with vergence and the near response.

The brain areas activated by visual stimuli have been investigated in monkeys and humans by positron emission tomography (PET) and other imaging techniques. Activation occurs not only in the occipital lobe but also in parts of the inferior temporal cortex, the posteroinferior parietal cortex, portions of the frontal lobe, and the amygdala. The subcortical structures activated in addition to the lateral geniculate body

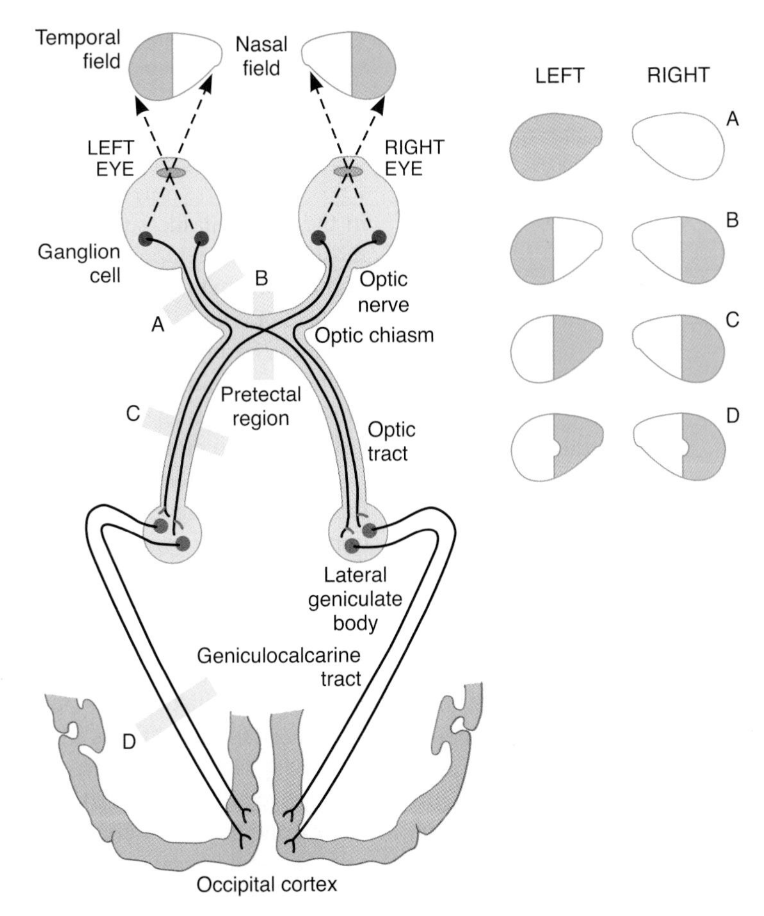

FIGURE 9–14 Visual pathways. Transection of the pathways at the locations indicated by the letters causes the visual field defects shown in the diagrams on the right. The fibers from the nasal half of each retina decussate in the optic chiasm, so that the fibers in the optic tracts are those from the temporal half of one retina and the nasal half of the other. A lesion that interrupts one optic nerve causes blindness in that eye (A). A lesion in one optic tract causes blindness in half of the visual field (C) and is called homonymous (same side of both visual fields) hemianopia (half-blindness). Lesions affecting the optic chiasm destroy fibers from both nasal hemiretinas and produce a heteronymous (opposite sides of the visual fields) hemianopia (B). Occipital lesions may spare the fibers from the macula (as in D) because of the separation in the brain of these fibers from the others subserving vision.

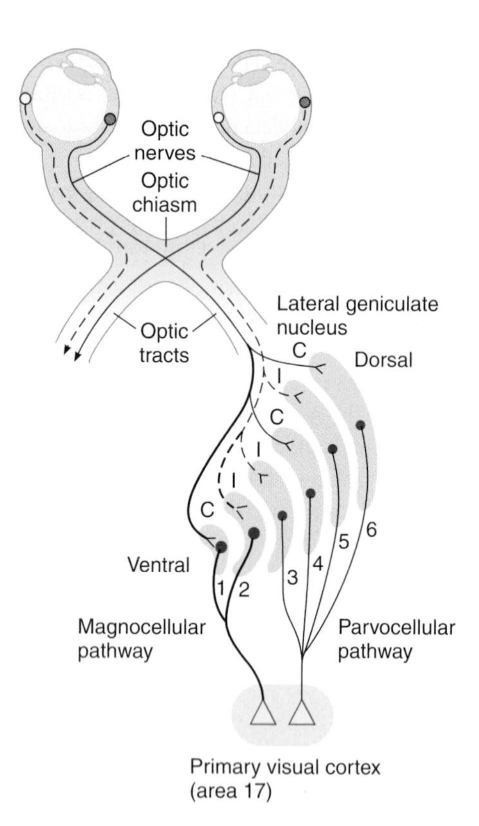

FIGURE 9–15 Ganglion cell projections from the right hemiretina of each eye to the right lateral geniculate body and from this nucleus to the right primary visual cortex. Note the six layers of the geniculate. P ganglion cells project to layers 3–6, and M ganglion cells project to layers 1 and 2. The ipsilateral (I) and contralateral (C) eyes project to alternate layers. Not shown are the interlaminar area cells, which project via a separate component of the P pathway to blobs in the visual cortex. (Modified from Kandel ER, Schwartz JH, Jessell TM [editors]: *Principles of Neural Science*, 4th ed. McGraw-Hill, 2000.)

include the superior colliculus, pulvinar, caudate nucleus, putamen, and claustrum.

The axons of retinal ganglion cells project a detailed spatial representation of the retina on the lateral geniculate body. Each geniculate body contains six well-defined layers (Figure 9–15). Layers 3–6 have small cells and are called parvocellular, whereas layers 1 and 2 have large cells and are called magnocellular. On each side, layers 1, 4, and 6 receive input from the contralateral eye, whereas layers 2, 3, and 5 receive input from the ipsilateral eye. In each layer, there is a precise point-for-point representation of the retina, and all six layers are in register so that along a line perpendicular to the layers, the receptive fields of the cells in each layer are almost identical. It is worth noting that only 10–20% of the input to the LGN comes from the retina. Major inputs also occur from the visual cortex and other brain regions. The feedback pathway from the visual cortex has been shown to be involved in visual processing related to the perception of orientation and motion.

There are several types of retinal ganglion cells. These include large ganglion cells (magno, or M cells), which add responses from different kinds of cones and are concerned with movement and stereopsis. Another type is the small ganglion cells (parvo, or P cells), which subtract input from one type of cone from input from another and are concerned with color, texture, and shape. The M ganglion cells project to the magnocellular portion of the lateral geniculate, whereas the P ganglion cells project to the parvocellular portion. From the LGN, a magnocellular pathway and a parvocellular pathway project to the visual cortex. The magnocellular pathway, from layers 1 and 2, carries signals for detection of movement, depth, and flicker. The parvocellular pathway, from layers 3–6, carries signals for color vision, texture, shape, and fine detail. The small-field bistratified ganglion cells may be involved in color vision and carry the short (blue) wavelength information to the intralaminar zones of the LGN.

Cells in the interlaminar region of the LGN also receive input from P ganglion cells, probably via dendrites of interlaminar cells that penetrate the parvocellular layers. They project via a separate component of the P pathway to the blobs in the visual cortex.

EFFECT OF LESIONS IN THE OPTIC PATHWAYS

Lesions along the visual pathways can be localized with a high degree of accuracy by the effects they produce in the visual fields. The fibers from the nasal half of each retina decussate in the optic chiasm, so that the fibers in the optic tracts are those from the temporal half of one retina and the nasal half of the other. In other words, each optic tract subserves half of the field of vision. Therefore, a lesion that interrupts one optic nerve causes blindness in that eye, but a lesion in one optic tract causes blindness in half of the visual field (Figure 9–14). This defect is classified as a **homonymous** (same side of both visual fields) **hemianopia** (half-blindness).

Lesions affecting the optic chiasm, such as pituitary tumors expanding out of the sella turcica, cause destruction of the fibers from both nasal hemiretinas and produce a **heteronymous** (opposite sides of the visual fields) **hemianopia.** Because the fibers from the macula are located posteriorly in the optic chiasm, hemianopic scotomas develop before vision in the two hemiretinas is completely lost. Selective visual field defects are further classified as bitemporal, binasal, and right or left.

The optic nerve fibers from the upper retinal quadrants subserving vision in the lower half of the visual field terminate in the medial half of the lateral geniculate body, whereas the fibers from the lower retinal quadrants terminate in the lateral half. The geniculocalcarine fibers from the medial half of the lateral geniculate terminate on the superior lip of the calcarine fissure, and those from the lateral half terminate on the inferior lip. Furthermore, the fibers from the lateral geniculate body that subserve macular vision separate from those that subserve peripheral vision and end more posteriorly on the lips of the calcarine fissure (Figure 9–16). Because of this anatomic arrangement, occipital lobe lesions may produce

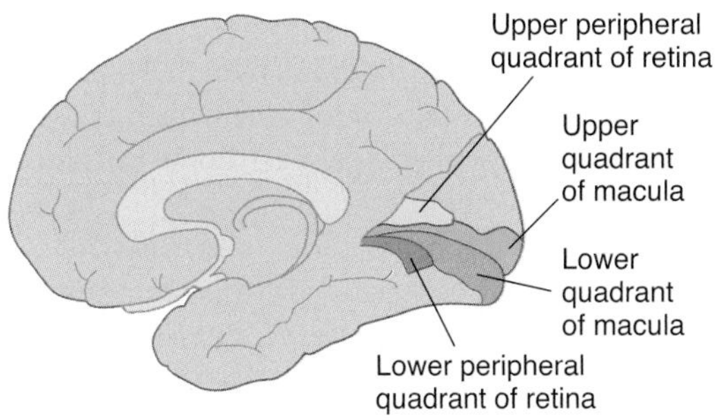

Upper peripheral quadrant of retina
Upper quadrant of macula
Lower quadrant of macula
Lower peripheral quadrant of retina

FIGURE 9–16 Medial view of the human right cerebral hemisphere showing projection of the retina on the primary visual cortex in the occipital cortex around the calcarine fissure. The geniculocalcarine fibers from the medial half of the lateral geniculate terminate on the superior lip of the calcarine fissure, and those from the lateral half terminate on the inferior lip. Also, the fibers from the lateral geniculate body that relay macular vision separate from those that relay peripheral vision and end more posteriorly on the lips of the calcarine fissure.

discrete quadrantic visual field defects (upper and lower quadrants of each half visual field).

Macular sparing, that is, loss of peripheral vision with intact macular vision, is also common with occipital lesions (Figure 9–14) because the macular representation is separate from that of the peripheral fields and very large relative to that of the peripheral fields. Therefore, occipital lesions must extend considerable distances to destroy macular as well as peripheral vision. Bilateral destruction of the occipital cortex in humans causes subjective blindness. However, there is appreciable **blind-sight,** that is, residual responses to visual stimuli even though they do not reach consciousness. For example, when these individuals are asked to guess where a stimulus is located during perimetry, they respond with much more accuracy than can be explained by chance. They are also capable of considerable discrimination of movement, flicker, orientation, and even color. Similar biasing of responses can be produced by stimuli in the blind areas in patients with hemianopia due to lesions in the visual cortex.

The fibers to the pretectal region that subserve the pupillary reflex produced by shining a light into the eye leave the optic tracts near the geniculate bodies. Therefore, blindness with preservation of the pupillary light reflex is usually due to bilateral lesions caudal to the optic tract.

PRIMARY VISUAL CORTEX

The primary visual receiving area (**primary visual cortex;** also known as V1) is located principally on the sides of the calcarine fissure (Figure 9–16). Just as the ganglion cell axons project a detailed spatial representation of the retina on the lateral geniculate body, the lateral geniculate body projects a similar point-for-point representation on the primary visual cortex. In the visual cortex, many nerve cells are associated with each incoming fiber. Like the rest of the neocortex, the visual cortex has six layers. The axons from the LGN that form the magnocellular pathway end in layer 4, specifically in its deepest part, layer 4C. Many of the axons that form the parvocellular pathway also end in layer 4C. However, the axons from the interlaminar region end in layers 2 and 3.

Layers 2 and 3 of the cortex contain clusters of cells about 0.2 mm in diameter that, unlike the neighboring cells, contain a high concentration of the mitochondrial enzyme cytochrome oxidase. The clusters have been named **blobs.** They are arranged in a mosaic in the visual cortex and are concerned with color vision. However, the parvocellular pathway also carries color opponent data to the deep part of layer 4.

Like the ganglion cells, the lateral geniculate neurons and the neurons in layer 4 of the visual cortex respond to stimuli in their receptive fields with on centers and inhibitory surrounds or off centers and excitatory surrounds. A bar of light covering the center is an effective stimulus for them because it stimulates the entire center and relatively little of the surround. However, the bar has no preferred orientation and, as a stimulus, is equally effective at any angle.

The responses of the neurons in other layers of the visual cortex are strikingly different. So-called **simple cells** respond to bars of light, lines, or edges, but only when they have a particular orientation. When, for example, a bar of light is rotated as little as 10° from the preferred orientation, the firing rate of the simple cell is usually decreased, and if the stimulus is rotated much more, the response disappears. There are also **complex cells,** which resemble simple cells in requiring a preferred orientation of a linear stimulus but are less dependent upon the location of a stimulus in the visual field than the simple cells and the cells in layer 4. They often respond maximally when a linear stimulus is moved laterally without a change in its orientation. They probably receive input from the simple cells.

The visual cortex, like the somatosensory cortex, is arranged in vertical columns that are concerned with orientation (**orientation columns**). Each is about 1 mm in diameter. However, the orientation preferences of neighboring columns differ in a systematic way; as one moves from column to column across the cortex, sequential changes occur in orientation preference of 5–10°. Thus, it seems likely that for each ganglion cell receptive field in the visual field, there is a collection of columns in a small area of visual cortex representing the possible preferred orientations at small intervals throughout the full 360°. The simple and complex cells have been called **feature detectors** because they respond to and analyze certain features of the stimulus. Feature detectors are also found in the cortical areas for other sensory modalities.

The orientation columns can be mapped with the aid of radioactive 2-deoxyglucose. The uptake of this glucose derivative is proportional to the activity of neurons. When this technique is employed in animals exposed to uniformly oriented visual stimuli such as vertical lines, the brain shows a remarkable array of intricately curved but evenly spaced orientation columns over a large area of the visual cortex.

Another feature of the visual cortex is the presence of **ocular dominance columns.** The geniculate cells and the cells in

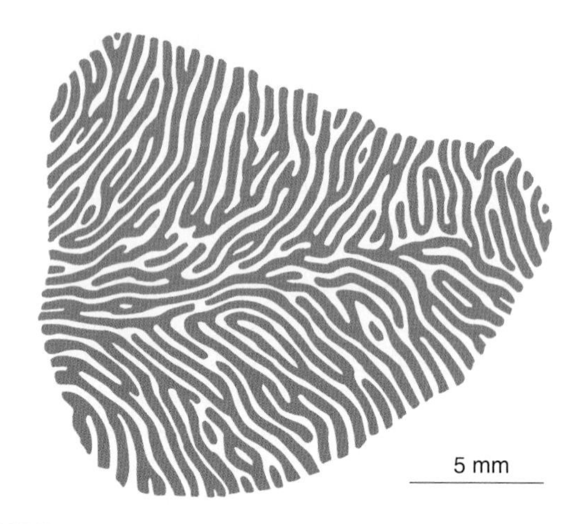

FIGURE 9–17 Reconstruction of ocular dominance columns in a subdivision of layer 4 of a portion of the right visual cortex of a rhesus monkey. Dark stripes represent one eye, light stripes the other. (Reproduced with permission from LeVay S, Hubel DH, Wiesel TN: The pattern of ocular dominance columns in macaque visual cortex revealed by a reduced silver stain. J Comp Neurol 1975;159:559.)

layer 4 receive input from only one eye, and the layer 4 cells alternate with cells receiving input from the other eye. If a large amount of a radioactive amino acid is injected into one eye, the amino acid is incorporated into protein and transported by axoplasmic flow to the ganglion cell terminals, across the geniculate synapses, and along the geniculocalcarine fibers to the visual cortex. In layer 4, labeled endings from the injected eye alternate with unlabeled endings from the uninjected eye. The result, when viewed from above, is a vivid pattern of stripes that covers much of the visual cortex (Figure 9–17) and is separate from and independent of the grid of orientation columns.

About half the simple and complex cells receive an input from both eyes. The inputs are identical or nearly so in terms of the portion of the visual field involved and the preferred orientation. However, they differ in strength, so that between the cells to which the input comes totally from the ipsilateral or the contralateral eye, there is a spectrum of cells influenced to different degrees by both eyes.

Thus, the primary visual cortex segregates information about color from that concerned with form and movement, combines the input from the two eyes, and converts the visual world into short line segments of various orientations.

OTHER CORTICAL AREAS CONCERNED WITH VISION

As mentioned above, the primary visual cortex (V1) projects to many other parts of the occipital lobes and other parts of the brain. These are often identified by number (V2, V3, etc) or by letters (LO, MT, etc). The distribution of some of these in the human brain is shown in Figure 9–18, and their putative functions are listed in Table 9–1. Studies of these areas have been carried out in monkeys trained to do various tasks and then fitted with implanted microelectrodes. In addition, the availability of PET and functional magnetic resonance imaging (fMRI) scanning has made it possible to conduct sophisticated experiments on visual cognition and other cortical visual functions in normal, conscious humans. The visual projections from V1 can be divided roughly into a **dorsal** or **parietal pathway,** concerned primarily with motion, and a **ventral** or **temporal pathway,** concerned with shape and recognition of forms and faces. In addition, connections to the sensory areas are important. For example, in the occipital cortex, visual responses to an object are better if the object is felt at the same time. There are many other relevant connections to other systems.

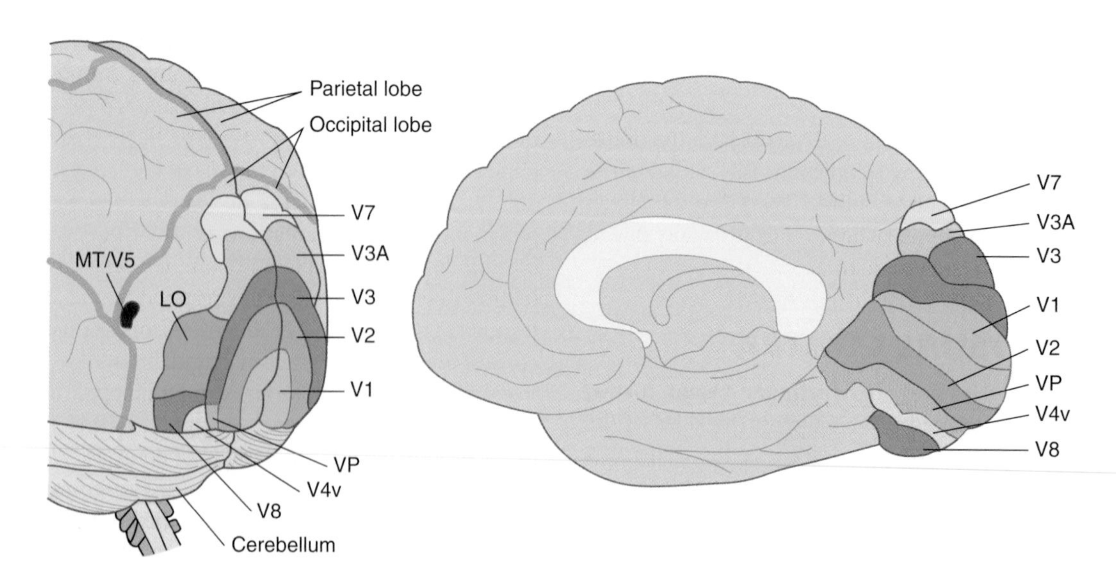

FIGURE 9–18 Some of the main areas to which the primary visual cortex (V1) projects in the human brain. Lateral and medial views. LO, lateral occipital; MT, medial temporal; VP, ventral parietal. See also Table 9–1. (Modified from Logothetis N: Vision: A window on consciousness. Sci Am [Nov] 1999;281:69–75.)

TABLE 9–1 Functions of visual projection areas in the human brain.

V1	Primary visual cortex; receives input from lateral geniculate nucleus, begins processing in terms of orientation, edges, etc
V2, V3, VP	Continued processing, larger visual fields
V3A	Motion
V4v	Unknown
MT/V5	Motion; control of movement
LO	Recognition of large objects
V7	Unknown
V8	Color vision

LO, lateral occipital; MT, medial temporal; VP, ventral parietal.

Modified from Logothetis N: Vision: a window on consciousness. Sci Am (Nov) 1999;281:69–75.

It is apparent from the preceding paragraphs that parallel processing of visual information occurs along multiple paths. In some as yet unknown way, all the information is eventually pulled together into what we experience as a conscious visual image.

COLOR VISION

Colors have three attributes: **hue, intensity,** and **saturation** (degree of freedom from dilution with white). For any color there is a **complementary color** that, when properly mixed with it, produces a sensation of white. Black is the sensation produced by the absence of light, but it is probably a positive sensation because the blind eye does not "see black;" rather, it "sees nothing."

Another observation of basic importance is the demonstration that the sensation of white, any spectral color, and even the extraspectral color, purple, can be produced by mixing various proportions of red light (wavelength 723–647 nm), green light (575–492 nm), and blue light (492–450 nm). Red, green, and blue are therefore called the **primary colors.** A third important point is that the color perceived depends in part on the color of other objects in the visual field. Thus, for example, a red object is seen as red if the field is illuminated with green or blue light, but as pale pink or white if the field is illuminated with red light. Clinical Box 9–6 describes color blindness.

CLINICAL BOX 9–6

Color Blindness

The most common test for **color blindness** uses the **Ishihara charts,** which are plates containing figures made up of colored spots on a background of similarly shaped colored spots. The figures are intentionally made up of colors that are liable to look the same as the background to an individual who is color blind. Some color-blind individuals are unable to distinguish certain colors, whereas others have only a color weakness. The prefixes "prot-," "deuter-," and "trit-" refer to defects of the red, green, and blue cone systems, respectively. Individuals with normal color vision are called **trichromats. Dichromats** are individuals with only two cone systems; they may have protanopia, deuteranopia, or tritanopia. **Monochromats** have only one cone system. Dichromats can match their color spectrum by mixing only two primary colors, and monochromats match theirs by varying the intensity of only one. Abnormal color vision is present as an inherited abnormality in Caucasian populations in about 8% of the males and 0.4% of the females. Tritanopia is rare and shows no sexual selectivity. However, about 2% of the colorblind males are dichromats who have protanopia or deuteranopia, and about 6% are anomalous trichromats in whom the red-sensitive or the green-sensitive pigment is shifted in its spectral sensitivity. These abnormalities are inherited as recessive and X-linked characteristics. Color blindness is present in males if the X chromosome has the abnormal gene. Females show a defect only when both X chromosomes contain the abnormal gene. However, female children of a man with X-linked color blindness are carriers of the color blindness and pass the defect on to half of their sons. Therefore, X-linked color blindness skips generations and appears in males of every second generation. Color blindness can also occur in individuals with lesions of area V8 of the visual cortex since this region appears to be uniquely concerned with color vision in humans. This deficit is called **achromatopsia.** Transient blue–green color weakness occurs as a side effect in individuals taking sildenafil (Viagra) for the treatment of erectile dysfunction because the drug inhibits the retinal as well as the penile form of phosphodiesterase.

RETINAL MECHANISMS

The **Young–Helmholtz theory** of color vision in humans postulates the existence of three kinds of cones, each containing a different photopigment and that are maximally sensitive to one of the three primary colors, with the sensation of any given color being determined by the relative frequency of the impulses from each of these cone systems. The correctness of this theory has been demonstrated by the identification and chemical characterization of each of the three pigments (Figure 9–19). One pigment (the blue-sensitive, or short-wave, pigment) absorbs light maximally in the blue-violet portion of the spectrum. Another (the green-sensitive, or middle-wave, pigment) absorbs maximally in the green portion. The third (the red-sensitive, or long-wave, pigment) absorbs maximally in the yellow portion. Blue, green, and red are the primary colors, but the cones with their maximal sensitivity in the yellow portion of the spectrum are sensitive enough in the red portion to respond to red light at a

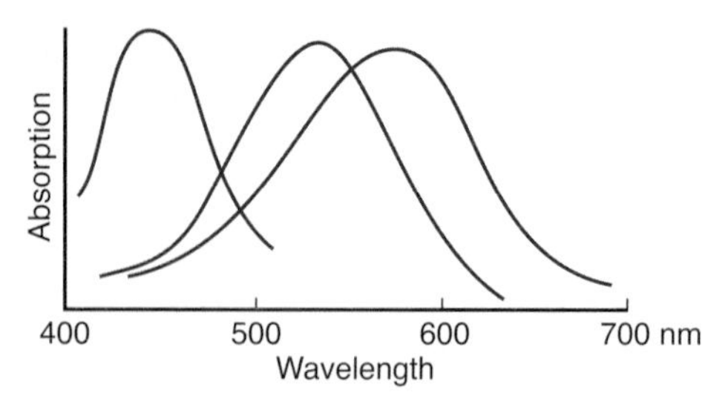

FIGURE 9–19 Absorption spectra of the three cone pigments in the human retina. The S pigment that peaks at 440 nm senses blue, and the M pigment that peaks at 535 nm senses green. The remaining L pigment peaks in the yellow portion of the spectrum, at 565 nm, but its spectrum extends far enough into the long wavelengths to sense red. (Reproduced with permission from Michael CR: Color vision. N Engl J Med 1973;288:724.)

lower threshold than green. This is all the Young–Helmholtz theory requires.

The gene for human rhodopsin is on chromosome 3, and the gene for the blue-sensitive S cone pigment is on chromosome 7. The other two cone pigments are encoded by genes arranged in tandem on the q arm of the X chromosome. The green-sensitive M and red-sensitive L pigments are very similar in structure; their opsins show 96% homology of amino acid sequences, whereas each of these pigments has only about 43% homology with the opsin of blue-sensitive pigment, and all three have about 41% homology with rhodopsin. Many mammals are **dichromats;** that is, they have only two cone pigments, a short-wave and a long-wave pigment. Old World monkeys, apes, and humans are **trichromats,** with separate middle- and long-wave pigments—in all probability because there was duplication of the ancestral long-wave gene followed by divergence.

There are variations in the red, long-wave pigment in humans. It has been known for some time that responses to the **Rayleigh match,** the amounts of red and green light that a subject mixes to match a monochromatic orange, are bimodal. This correlates with new evidence that 62% of otherwise color-normal individuals have serine at site 180 of their long-wave cone opsin, whereas 38% have alanine. The absorption curve of the subjects with serine at position 180 peaks at 556.7 nm, and they are more sensitive to red light, whereas the absorption curve of the subjects with alanine at position 180 peaks at 552.4 nm.

NEURAL MECHANISMS

Color is mediated by ganglion cells that subtract or add input from one type of cone to input from another type. Processing in the ganglion cells and the LGN produces impulses that pass along three types of neural pathways that project to V1: a red–green pathway that signals differences between Land M-cone responses, a blue–yellow pathway that signals differences between S-cone and the sum of L- and M-cone responses, and a luminance pathway that signals the sum of L- and M-cone responses. These pathways project to the blobs and the deep portion of layer 4C of V1. From the blobs and layer 4, color information is projected to V8. However, it is not known how V8 converts color input into the sensation of color.

OTHER ASPECTS OF VISUAL FUNCTION

DARK ADAPTATION

If a person spends a considerable length of time in brightly lighted surroundings and then moves to a dimly lighted environment, the retinas slowly become more sensitive to light as the individual becomes "accustomed to the dark." This decline in visual threshold is known as **dark adaptation.** It is nearly maximal in about 20 min, although some further decline occurs over longer periods. On the other hand, when one passes suddenly from a dim to a brightly lighted environment, the light seems intensely and even uncomfortably bright until the eyes adapt to the increased illumination and the visual threshold rises. This adaptation occurs over a period of about 5 min and is called **light adaptation,** although, strictly speaking, it is merely the disappearance of dark adaptation.

The dark adaptation response actually has two components. The first drop in visual threshold, rapid but small in magnitude, is known to be due to dark adaptation of the cones because when only the foveal, rod-free portion of the retina is tested, the decline proceeds no further. In the peripheral portions of the retina, a further drop occurs as a result of adaptation of the rods. The total change in threshold between the light-adapted and the fully dark-adapted eye is very great.

Radiologists, aircraft pilots, and others who need maximal visual sensitivity in dim light can avoid having to wait 20 min in the dark to become dark-adapted if they wear red goggles when in bright light. Light wavelengths in the red end of the spectrum stimulate the rods to only a slight degree while permitting the cones to function reasonably well. Therefore, a person wearing red glasses can see in bright light during the time it takes for the rods to become dark-adapted.

The time required for dark adaptation is determined in part by the time required to build up the rhodopsin stores. In bright light, much of the pigment is continuously being broken down, and some time is required in dim light for accumulation of the amounts necessary for optimal rod function. However, dark adaptation also occurs in the cones, and additional factors are undoubtedly involved.

CRITICAL FUSION FREQUENCY

The time-resolving ability of the eye is determined by measuring the **critical fusion frequency (CFF),** the rate at which stimuli can be presented and still be perceived as separate stimuli. Stimuli presented at a higher rate than the CFF are perceived as continuous stimuli. Motion pictures move

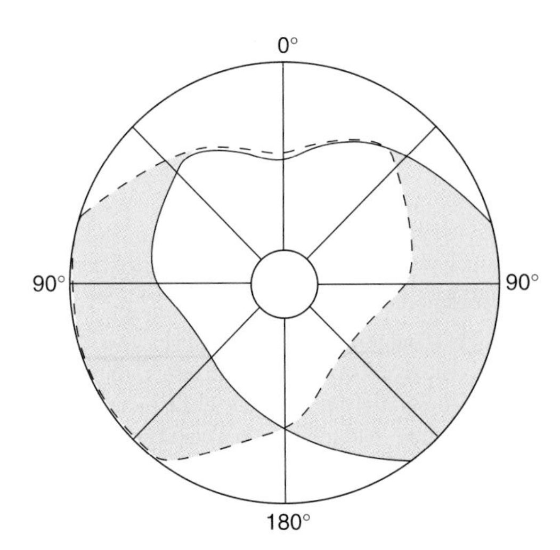

FIGURE 9–20 Monocular and binocular visual fields. The dashed line encloses the visual field of the left eye; the solid line, that of the right eye. The common area (heart-shaped clear zone in the center) is viewed with binocular vision. The colored areas are viewed with monocular vision.

because the frames are presented at a rate above the CFF, and movies begin to flicker when the projector slows down.

VISUAL FIELDS & BINOCULAR VISION

The visual field of each eye is the portion of the external world visible out of that eye. Theoretically, it should be circular, but actually it is cut off medially by the nose and superiorly by the roof of the orbit (Figure 9–20). Mapping the visual fields is important in neurologic diagnosis. The peripheral portions of the visual fields are mapped with an instrument called a **perimeter,** and the process is referred to as **perimetry.** One eye is covered while the other is fixed on a central point. A small target is moved toward this central point along selected meridians, and, along each, the location where the target first becomes visible is plotted in degrees of arc away from the central point (Figure 9–20). The central visual fields are mapped with a **tangent screen,** a black felt screen across which a white target is moved. By noting the locations where the target disappears and reappears, the blind spot and any **objective scotomas** (blind spots due to disease) can be outlined.

The central parts of the visual fields of the two eyes coincide; therefore, anything in this portion of the field is viewed with **binocular vision.** The impulses set up in the two retinas by light rays from an object are fused at the cortical level into a single image **(fusion).** The term **corresponding points** is used to describe the points on the retina on which the image of an object must fall if it is to be seen binocularly as a single object. If one eye is gently pushed out of alignment while staring fixedly at an object in the center of the visual field, double vision **(diplopia)** results; the image on the retina of the eye that is displaced no longer falls on the corresponding point. When visual images no longer fall on corresponding retinal points, strabismus occurs (see Clinical Box 9–4).

Binocular vision has an important role in the perception of depth. However, depth perception also has numerous monocular components, such as the relative sizes of objects, the degree one looks down at them, their shadows, and, for moving objects, their movement relative to one another (**movement parallax**).

EYE MOVEMENTS

The eye is moved within the orbit by six ocular muscles that are innervated by the oculomotor, trochlear, and abducens nerves. Figure 9–21 shows the movements produced by the

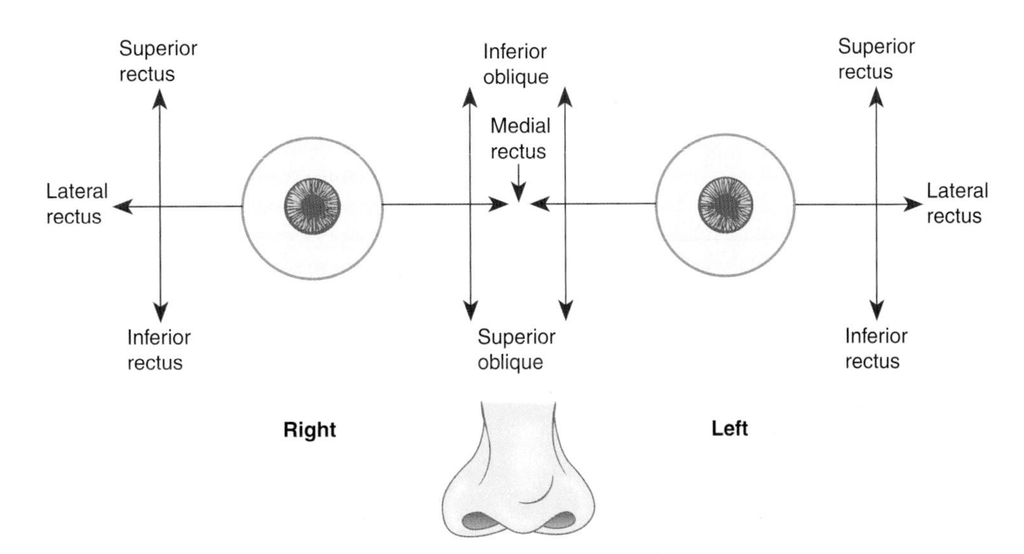

FIGURE 9–21 Diagram of eye muscle actions. The eye is adducted by the medial rectus and abducted by the lateral rectus. The adducted eye is elevated by the inferior oblique and depressed by the superior oblique; the abducted eye is elevated by the superior rectus and depressed by the inferior rectus. (From Waxman SG: *Clinical Neuroanatomy,* 26th ed. McGraw-Hill, 2010.)

six pairs of muscles. Because the oblique muscles pull medially, their actions vary with the position of the eye. When the eye is turned nasally, the inferior oblique elevates it and the superior oblique depresses it. When it is turned laterally, the superior rectus elevates it and the inferior rectus depresses it.

Because much of the visual field is binocular, it is clear that a very high order of coordination of the movements of the two eyes is necessary if visual images are to fall at all times on corresponding points in the two retinas and diplopia is to be avoided.

There are four types of eye movements, each controlled by a different neural system but sharing the same final common path, the motor neurons that supply the external ocular muscles. **Saccades,** sudden jerky movements, occur as the gaze shifts from one object to another. They bring new objects of interest onto the fovea and reduce adaptation in the visual pathway that would occur if gaze were fixed on a single object for long periods. **Smooth pursuit movements** are tracking movements of the eyes as they follow moving objects. **Vestibular movements,** adjustments that occur in response to stimuli initiated in the semicircular canals, maintain visual fixation as the head moves. **Convergence movements** bring the visual axes toward each other as attention is focused on objects near the observer. The similarity to a human-made tracking system on an unstable platform such as a ship is apparent: saccadic movements seek out visual targets, pursuit movements follow them as they move about, and vestibular movements stabilize the tracking device as the platform on which the device is mounted (ie, the head) moves about. In primates, these eye movements depend on an intact visual cortex. Saccades are programmed in the frontal cortex and the superior colliculi and pursuit movements in the cerebellum.

SUPERIOR COLLICULI

The superior colliculi, which regulate saccades, are innervated by M fibers from the retina. They also receive extensive innervation from the cerebral cortex. Each superior colliculus has a map of visual space plus a map of the body surface and a map for sound in space. A motor map projects to the regions of the brain stem that control eye movements. There are also projections via the tectopontine tract to the cerebellum and via the tectospinal tract to areas concerned with reflex movements of the head and neck. The superior colliculi are constantly active positioning the eyes, and they have one of the highest rates of blood flow and metabolism of any region in the brain.

CHAPTER SUMMARY

- The major parts of the eye are the sclera (protective covering), cornea (transfer light rays), choroid (nourishment), retina (receptor cells), lens, and iris.
- The retina is organized into several layers: the outer nuclear layer contains the photoreceptors (rods and cones), the inner nuclear layer contains bipolar cells, horizontal cells, and amacrine cells, and the ganglion cell layer contains the only output neuron of the retina.
- The bending of light rays (refraction) allows one to focus an accurate image onto the retina. Light is refracted at the anterior surface of the cornea and at the anterior and posterior surfaces of the lens. To bring diverging rays from close objects to a focus on the retina, the curvature of the lens is increased, a process called accommodation.
- In hyperopia (farsightedness), the eyeball is too short and light rays come to a focus behind the retina. In myopia (nearsightedness), the anteroposterior diameter of the eyeball is too long. Astigmatism is a common condition in which the curvature of the cornea is not uniform. Presbyopia is the loss of accommodation for near vision. Strabismus is a misalignment of the eyes; it is also known as "crossed eyes." Eyes can be deviated outward (exotropia) or inward (esotropia).
- Na^+ channels in the outer segments of the rods and cones are open in the dark, so current flows from the inner to the outer segment. When light strikes the outer segment, some of the Na^+ channels are closed and the cells are hyperpolarized.
- In response to light, horizontal cells are hyperpolarized, bipolar cells are either hyperpolarized or depolarized, and amacrine cells are depolarized and develop spikes that may act as generator potentials for the propagated spikes produced in the ganglion cells.
- The visual pathway is from the rods and cones to bipolar cells to ganglion cells then via the optic tract to the thalamic lateral geniculate body to the occipital lobe of the cerebral cortex. The fibers from each nasal hemiretina decussate in the optic chiasm; the fibers from the nasal half of one retina and the temporal half of the other synapse on the cells whose axons form the geniculocalcarine tract.
- Neurons in layer 4 of the visual cortex respond to stimuli in their receptive fields with on centers and inhibitory surrounds or off centers and excitatory surrounds. Neurons in other layers are called simple cells if they respond to bars of light, lines, or edges, but only when they have a particular orientation. Complex cells also require a preferred orientation of a linear stimulus but are less dependent on the location of a stimulus in the visual field. Projections from area V1 can be divided into a dorsal or parietal pathway (concerned primarily with motion) and a ventral or temporal pathway (concerned with shape and recognition of forms and faces).
- The decline in visual threshold after spending long periods of time in a dimly lit room is called dark adaptation. The fovea in the center of the retina is the point where visual acuity is greatest.
- The Young–Helmholtz theory of color vision postulates the existence of three kinds of cones, each containing a different photopigment and that are maximally sensitive to one of the three primary colors, with the sensation of any given color being determined by the relative frequency of the impulses from each of these cone systems.
- Eye movement is controlled by six ocular muscles innervated by the oculomotor, trochlear, and abducens nerves.

The inferior oblique muscle turns the eye upward and outward; the superior oblique turns it downward and outward. The superior rectus muscle turns the eye upward and inward; the inferior rectus turns it downward and inward. The medial rectus muscle turns the eye inward; the lateral rectus turns it outward.

- Saccades (sudden jerky movements) occur as the gaze shifts from one object to another, and they reduce adaptation in the visual pathway that would occur if gaze were fixed on a single object for long periods. Smooth pursuit movements are tracking movements of the eyes as they follow moving objects. Vestibular movements occur in response to stimuli in the semicircular canals to maintain visual fixation as the head moves. Convergence movements bring the visual axes toward each other as attention is focused on objects near the observer.

MULTIPLE-CHOICE QUESTIONS

For all questions, select the single best answer unless otherwise directed.

1. A visual exam in an 80-year-old man shows he has a reduced ability to see objects in the upper and lower quadrants of the left visual fields of both eyes but some vision remains in the central regions of the visual field. The diagnosis is
   A. central scotoma.
   B. heteronymous hemianopia with macular sparing.
   C. lesion of the optic chiasm.
   D. homonymous hemianopia with macular sparing.
   E. retinopathy.
2. A 45-year-old female who had never needed to wear glasses experienced difficulty reading a menu in a dimly-lit restaurant. She then recalled that as of late she needed to have the newspaper closer to her eyes in order to read it. A friend recommended she purchase reading glasses. Visual accommodation involves
   A. increased tension on the lens ligaments.
   B. a decrease in the curvature of the lens.
   C. relaxation of the sphincter muscle of the iris.
   D. contraction of the ciliary muscle.
   E. increased intraocular pressure.
3. A 28-year-old male with severe myopia made an appointment to see his ophthalmologist when he began to notice flashing lights and floaters in his visual field. He was diagnosed with a retinal detachment. The retina
   A. is epithelial tissue that contains photoreceptors.
   B. lines the anterior one-third of the choroid.
   C. has an inner nuclear layer that contains bipolar cells, horizontal cells, and amacrine cells.
   D. contains ganglion cells whose axons form the oculomotor nerve.
   E. contains an optic disk where visual acuity is greatest.
4. A 62-year-old Caucasian woman experienced a rapid onset of blurry vision along with loss of central vision. A comprehensive eye exam showed that she had wet age-related macular degeneration. The fovea of the eye
   A. has the lowest light threshold.
   B. is the region of highest visual acuity.
   C. contains only red and green cones.
   D. contains only rods.
   E. is situated over the head of the optic nerve.
5. Which of the following parts of the eye has the greatest concentration of rods?
   A. Ciliary body
   B. Iris
   C. Optic disk
   D. Fovea
   E. Parafoveal region
6. Which of the following is *not* correctly paired?
   A. Rhodopsin: retinal and opsin
   B. Obstruction of the canal of Schlemm: elevated intraocular pressure
   C. Myopia: convex lenses
   D. Astigmatism: nonuniform curvature of the cornea
   E. Inner segments of rods and cones: synthesis of the photosensitive compounds
7. The correct sequence of events involved in phototransduction in rods and cones in response to light is:
   A. activation of transducin, decreased release of glutamate, structural changes in rhodopsin, closure of Na^+ channels, and decrease in intracellular cGMP.
   B. decreased release of glutamate, activation of transducin, closure of Na^+ channels, decrease in intracellular cGMP, and structural changes in rhodopsin.
   C. structural changes in rhodopsin, decrease in intracellular cGMP, decreased release of glutamate, closure of Na^+ channels, and activation of transducin.
   D. structural changes in rhodopsin, activation of transducin, decrease in intracellular cGMP, closure of Na^+ channels, and decreased release of glutamate.
   E. activation of transducin, structural changes in rhodopsin, closure of Na^+ channels, decrease in intracellular cGMP, and decreased release of glutamate.
8. A 25-year-old medical student spent a summer volunteering in the sub-Saharan region of Africa. There he noted a high incidence of people reporting difficulty with night vision due to a lack of vitamin A in their diet. Vitamin A is a precursor for the synthesis of
   A. rods and cones.
   B. retinal.
   C. rod transducin.
   D. opsin.
   E. cone transducin.
9. An 11-year-old male was having difficulty reading the graphs that his teacher was showing at the front of classroom. His teacher recommended he be seen by an ophthalmologist. Not only was he asked to look at a Snellen letter chart for visual acuity but he was also asked to identify numbers in an Ishihara chart. He responded that he merely saw a bunch of dots. Abnormal color vision is 20 times more common in men than women because most cases are caused by an abnormal
   A. dominant gene on the Y chromosome.
   B. recessive gene on the Y chromosome.
   C. dominant gene on the X chromosome.
   D. recessive gene on the X chromosome.
   E. recessive gene on chromosome 22.

10. Which of the following is *not* involved in color vision?
    A. Activation of a pathway that signals differences between S cone responses and the sum of L and M cone responses
    B. Geniculate layers 3–6
    C. P pathway
    D. Area V3A of visual cortex
    E. Area V8 of visual cortex

11. A 56-year-old female was diagnosed with a tumor near the base of the skull, impinging on her optic tract. Which of the following statements about the central visual pathway is correct?
    A. The fibers from each temporal hemiretina decussate in the optic chiasm, so that the fibers in the optic tracts are those from the temporal half of one retina and the nasal half of the other.
    B. In the geniculate body, the fibers from the nasal half of one retina and the temporal half of the other synapse on the cells whose axons form the geniculocalcarine tract.
    C. Layers 2 and 3 of the visual cortex contain clusters of cells called globs that contain a high concentration of cytochrome oxidase.
    D. Complex cells have a preferred orientation of a linear stimulus and, compared to simple cells, are more dependent on the location of the stimulus within the visual field.
    E. The visual cortex is arranged in horizontal columns that are concerned with orientation.

CHAPTER RESOURCES

Baccus SA: Timing and computation in inner retinal circuitry. Annu Rev Physiol 2007; 69:271.

Chiu C, Weliky M: Synaptic modification by vision. Science 2003;300:1890.

Gegenfurtner KR, Kiper DC: Color vision. Annu Rev Neurosci 2003;26:181.

Logothetis N: Vision: A window on consciousness. Sci Am 1999;281:99.

Masland RH: The fundamental plan of the retina. Nat Neurosci 2001;4:877.

Oyster CW: *The Human Eye: Structure and Function*. Sinauer, 1999.

Pugh EN, Nikonov S, Lamb TD: Molecular mechanisms of vertebrate photoreceptor light adaptation. Curr Opin Neurobiol 1999;9:410.

Tobimatsu S, Celesia GG, Haug BA, Onofri M, Sartucci F, Porciatti V: Recent advances in clinical neurophysiology of vision. Suppl Clin Neurophysiol 2000;53:312.

Wässle H, Boycott BB: Functional architecture of the mammalian retina. Physiol Rev 1991;71:447.

Wu S: Synaptic Organization of the Vertebrate Retina: General Principles and Species-Specific Variations (The Friedenwald Lecture). Invest Ophthalmol Vis Sci 2010;51:1264.

CHAPTER 10

Hearing & Equilibrium

OBJECTIVES

After studying this chapter, you should be able to:

- Describe the components and functions of the external, middle, and inner ear.
- Describe the way that movements of molecules in the air are converted into impulses generated in hair cells in the cochlea.
- Explain the roles of the tympanic membrane, the auditory ossicles (malleus, incus, and stapes), and scala vestibule in sound transmission.
- Explain how auditory impulses travel from the cochlear hair cells to the auditory cortex.
- Explain how pitch, loudness, and timbre are coded in the auditory pathways.
- Describe the various forms of deafness and the tests used to distinguish between them.
- Explain how the receptors in the semicircular canals detect rotational acceleration and how the receptors in the saccule and utricle detect linear acceleration.
- List the major sensory inputs that provide the information that is synthesized in the brain into the sense of position in space.

INTRODUCTION

Our ears not only let us detect sounds, but they also help us maintain balance. Receptors for two sensory modalities (hearing and equilibrium) are housed in the ear. The external ear, the middle ear, and the cochlea of the inner ear are concerned with hearing. The semicircular canals, the utricle, and the saccule of the inner ear are concerned with equilibrium. Both hearing and equilibrium rely on a very specialized type of receptor called a hair cell. There are six groups of hair cells in each inner ear: one in each of the three semicircular canals, one in the utricle, one in the saccule, and one in the cochlea. Receptors in the semicircular canals detect rotational acceleration, receptors in the utricle detect linear acceleration in the horizontal direction, and receptors in the saccule detect linear acceleration in the vertical direction.

STRUCTURE AND FUNCTION OF THE EAR

EXTERNAL & MIDDLE EAR

The external ear funnels sound waves to the **external auditory meatus** (Figure 10–1). In some animals, the ears can be moved like radar antennas to seek out sound. From the external auditory meatus, sound waves pass inward to the **tympanic membrane** (eardrum).

The middle ear is an air-filled cavity in the temporal bone that opens via the **eustachian (auditory) tube** into the nasopharynx and through the nasopharynx to the exterior. The tube is usually closed, but during swallowing, chewing, and yawning it opens, keeping the air pressure on the two sides of the eardrum equalized. The three **auditory ossicles,** the **malleus, incus,** and **stapes,** are located in the middle ear (Figure 10–2). The **manubrium** (handle of the malleus) is attached to the back of the tympanic membrane. Its head is attached to the wall of the middle ear, and its short process

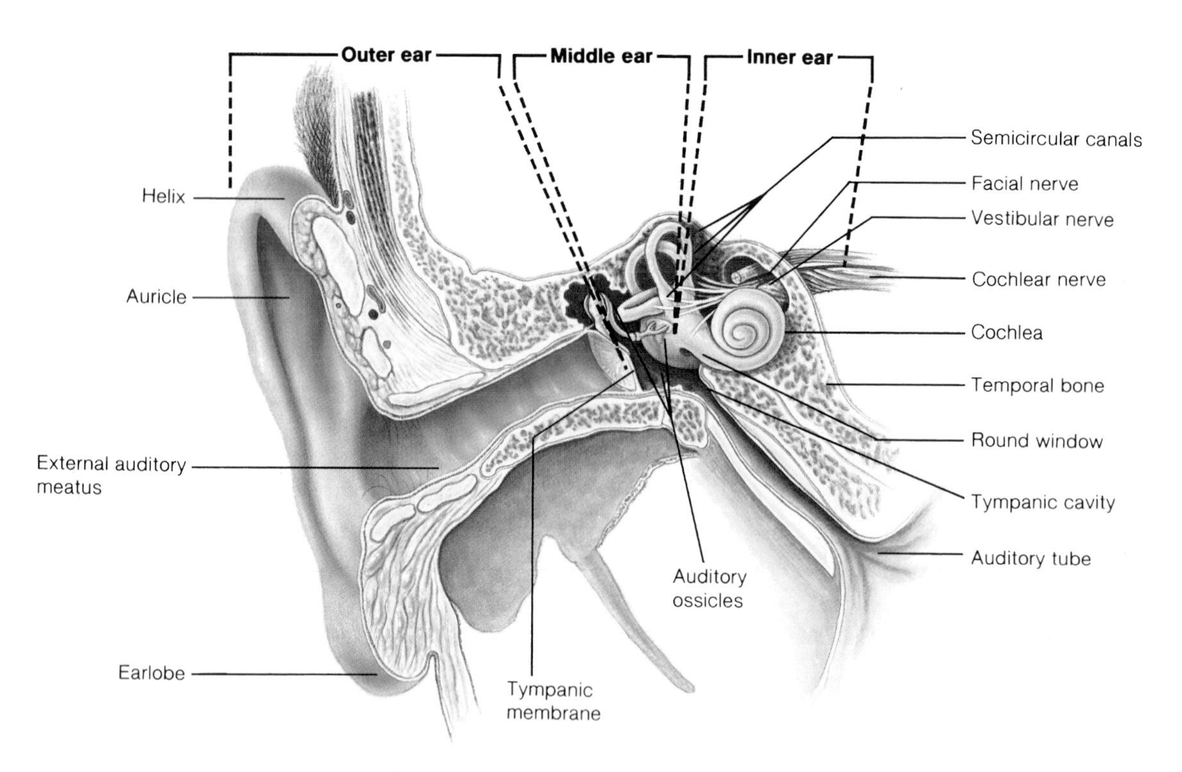

FIGURE 10–1 The structures of the external, middle, and inner portions of the human ear. Sound waves travel from the external ear to the tympanic membrane via the external auditory meatus. The middle ear is an air-filled cavity in the temporal bone; it contains the auditory ossicles. The inner ear is comprised of the bony and membranous labyrinths. To make the relationships clear, the cochlea has been turned slightly and the middle ear muscles have been omitted. (From Fox SI, *Human Physiology*. McGraw-Hill, 2008.)

is attached to the incus, which in turn articulates with the head of the stapes. The stapes is named for its resemblance to a stirrup. Its **foot plate** is attached by an annular ligament to the walls of the **oval window.** Two small skeletal muscles, the **tensor tympani** and the **stapedius,** are also located in the middle ear. Contraction of the former pulls the manubrium of the malleus medially and decreases the vibrations of the tympanic membrane; contraction of the latter pulls the foot plate of the stapes out of the oval window. The functions of the ossicles and the muscles are considered in more detail below.

INNER EAR

The inner ear **(labyrinth)** is made up of two parts, one within the other. The **bony labyrinth** is a series of channels in the petrous portion of the **temporal bone** and is filled with a fluid called **perilymph,** which has a relatively low concentration of K^+, similar to that of plasma or the cerebral spinal fluid. Inside these bony channels, surrounded by the perilymph, is the **membranous labyrinth.** The membranous labyrinth more or less duplicates the shape of the bony channels and is filled with a K^+-rich fluid called **endolymph.** The labyrinth has three components: the **cochlea** (containing receptors for hearing), **semicircular canals** (containing receptors that respond to head rotation), and the **otolith organs** (containing receptors that respond to gravity and head tilt).

The cochlea is a coiled tube that, in humans, is 35 mm long and makes two and three quarter turns (Figure 10–3). The basilar membrane and Reissner's membrane divide it into three chambers or **scalae** (Figure 10–4). The upper **scala vestibuli** and the lower **scala tympani** contain perilymph and communicate with each other at the apex of the cochlea through a small opening called the **helicotrema.** At the base of the cochlea, the scala vestibuli ends at the oval window, which is closed by the footplate of the stapes. The scala tympani end at the **round window,** a foramen on the medial wall of the middle ear that is closed by the flexible **secondary tympanic membrane.** The **scala media,** the middle cochlear chamber, is continuous with the membranous labyrinth and does not communicate with the other two scalae.

The **organ of Corti** on the basilar membrane extends from the apex to the base of the cochlea and thus has a spiral shape. This structure contains the highly specialized auditory receptors (hair cells) whose processes pierce the tough, membrane-like **reticular lamina** that is supported by the **pillar cells** or **rods of Corti** (Figure 10–4). The hair cells are arranged in four rows: three rows of **outer hair cells** lateral to the tunnel formed by the rods of Corti, and one row of **inner hair cells** medial to the tunnel. There are 20,000 outer hair cells and 3500 inner hair cells in each human cochlea. Covering the rows of hair cells is a thin, viscous, but elastic **tectorial membrane** in which the tips of the hairs of the outer but not the inner hair cells are embedded. The cell bodies of the sensory neurons that

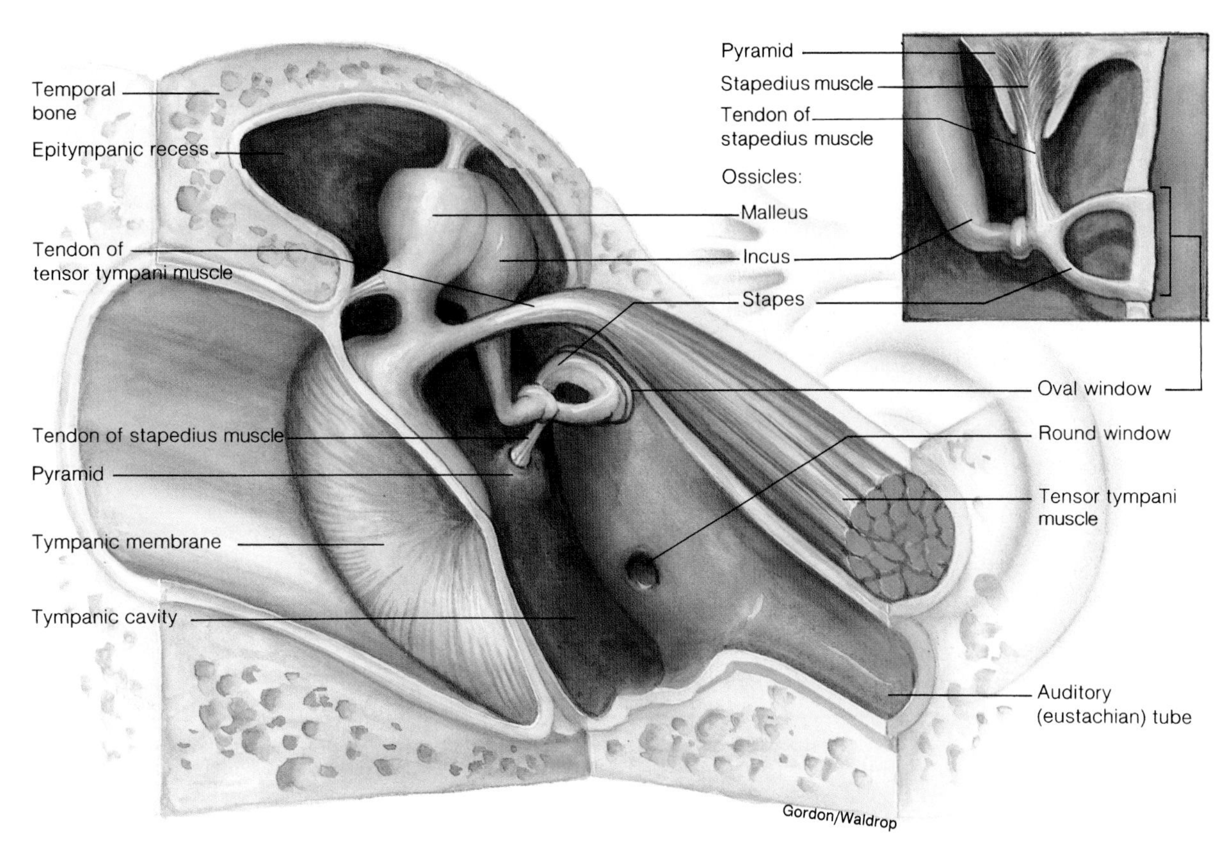

FIGURE 10–2 The medial view of the middle ear containing three auditory ossicles malleus, incus, and stapes) and two small skeletal muscles (tensor tympani muscle and stapedius). The manubrium (handle of the malleus) is attached to the back of the tympanic membrane. Its head is attached to the wall of the middle ear, and its short process is attached to the incus, which in turn articulates with the head of the stapes. The foot plate of the stapes is attached by an annular ligament to the walls of the oval window. Contraction of the tensor tympani muscle pulls the manubrium medially and decreases the vibrations of the tympanic membrane; contraction of the stapedius muscle pulls the foot plate of the stapes out of the oval window. (From Fox SI, *Human Physiology*. McGraw-Hill, 2008.)

arborize around the bases of the hair cells are located in the **spiral ganglion** within the **modiolus,** the bony core around which the cochlea is wound. Ninety to 95% of these sensory neurons innervate the inner hair cells; only 5–10% innervates the more numerous outer hair cells, and each sensory neuron innervates several outer hair cells. By contrast, most of the efferent fibers in the auditory nerve terminate on the outer rather than inner hair cells. The axons of the afferent neurons that innervate the hair cells form the **auditory (cochlear) division** of the eighth cranial nerve.

In the cochlea, tight junctions between the hair cells and the adjacent phalangeal cells prevent endolymph from reaching the bases of the cells. However, the basilar membrane is relatively permeable to perilymph in the scala tympani, and consequently, the tunnel of the organ of Corti and the bases of the hair cells are bathed in perilymph. Because of similar tight junctions, the arrangement is similar for the hair cells in other parts of the inner ear; that is, the processes of the hair cells are bathed in endolymph, whereas their bases are bathed in perilymph.

On each side of the head, the semicircular canals are perpendicular to each other, so that they are oriented in the three planes of space. A receptor structure, the **crista ampullaris,** is located in the expanded end **(ampulla)** of each of the membranous canals. Each crista consists of hair cells and supporting (sustentacular) cells surmounted by a gelatinous partition **(cupula)** that closes off the ampulla (Figure 10–3). The processes of the hair cells are embedded in the cupula, and the bases of the hair cells are in close contact with the afferent fibers of the **vestibular division** of the eighth cranial nerve.

A pair of otolith organs, the **saccule** and **utricle,** are located near the center of the membranous labyrinth. The sensory epithelium of these organs is called the **macula.** The maculae are vertically oriented in the saccule and horizontally located in the utricle when the head is upright. The maculae contain supporting cells and hair cells, surrounded by an otolithic membrane in which are embedded crystals of calcium carbonate, the **otoliths** (Figure 10–3). The otoliths, which are also called **otoconia** or **ear dust,** range from 3 to 19 μm in length in humans. The processes of the hair cells are embedded in the membrane. The nerve fibers from the hair cells join those from the cristae in the vestibular division of the eighth cranial nerve.

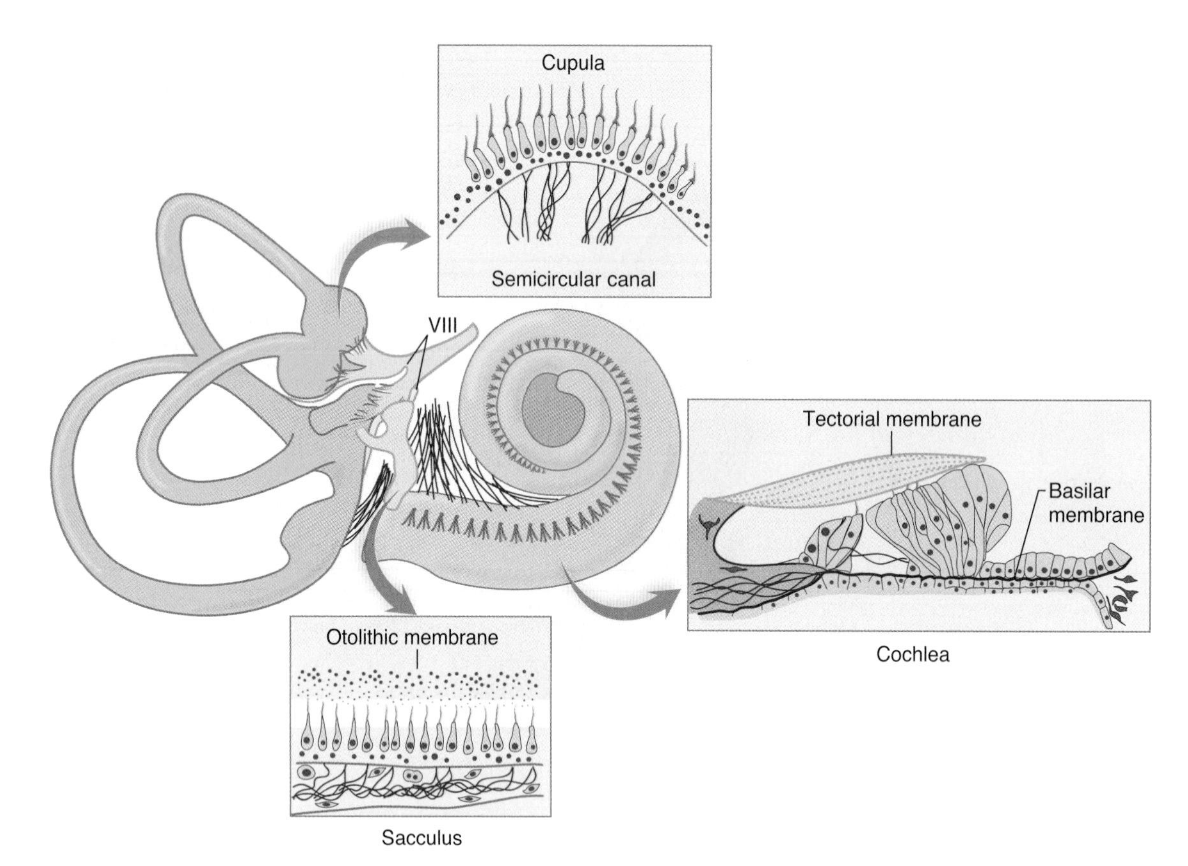

FIGURE 10–3 Schematic of the human inner ear showing the membranous labyrinth with enlargements of the structures in which hair cells are embedded. The membranous labyrinth is suspended in perilymph and filled with K^+-rich endolymph, which bathes the receptors. Hair cells (darkened for emphasis) occur in different arrays characteristic of the receptor organs. The three semicircular canals are sensitive to angular accelerations that deflect the gelatinous cupula and associated hair cells. In the cochlea, hair cells spiral along the basilar membrane in the organ of Corti. Airborne sounds set the eardrum in motion, which is conveyed to the cochlea by bones of the middle ear. This flexes the membrane up and down. Hair cells in the organ of Corti are stimulated by shearing motion. The otolithic organs (saccule and utricle) are sensitive to linear acceleration in vertical and horizontal planes. Hair cells are attached to the otolithic membrane. VIII, eighth cranial nerve, with auditory and vestibular divisions. (Adapted with permission from Hudspeth AJ: How the ear's works work. Nature 1989;341(6241):397–404.)

SENSORY RECEPTORS IN THE EAR: HAIR CELLS

The specialized sensory receptors in the ear consist of six patches of hair cells in the membranous labyrinth. These are examples of **mechanoreceptors.** The hair cells in the organ of Corti signal hearing; the hair cells in the utricle signal horizontal acceleration; the hair cells in the saccule signal vertical acceleration; and a patch in each of the three semicircular canals signal rotational acceleration. These hair cells have a common structure (Figure 10–5). Each is embedded in an epithelium made up of supporting cells, with the basal end in close contact with afferent neurons. Projecting from the apical end are 30–150 rod-shaped processes, or hairs. Except in the cochlea, one of these, the **kinocilium,** is a true but nonmotile cilium with nine pairs of microtubules around its circumference and a central pair of microtubules. It is one of the largest processes and has a clubbed end. The kinocilium is lost from the hair cells of the cochlea in adult mammals. However, the other processes, which are called **stereocilia,** are present in all hair cells. They have cores composed of parallel filaments of actin. The actin is coated with various isoforms of myosin. Within the clump of processes on each cell there is an orderly structure. Along an axis toward the kinocilium, the stereocilia increase progressively in height; along the perpendicular axis, all the stereocilia are the same height.

ELECTRICAL RESPONSES

Very fine processes called **tip links** (Figure 10–6) tie the tip of each stereocilium to the side of its higher neighbor, and at the junction are mechanically-sensitive cation channels in the taller process. When the shorter stereocilia are pushed toward the taller ones, the open time of these channels is increased. K^+—the most abundant cation in endolymph—and Ca^{2+} enter via the channel and produce depolarization. A myosin-based molecular motor in the taller neighbor then moves the channel toward the base, releasing tension in the tip link (Figure 10–6). This causes the channel to close and permits restoration of the resting state. Depolarization of hair cells causes them to release a neurotransmitter, probably glutamate, which initiates depolarization of neighboring afferent neurons.

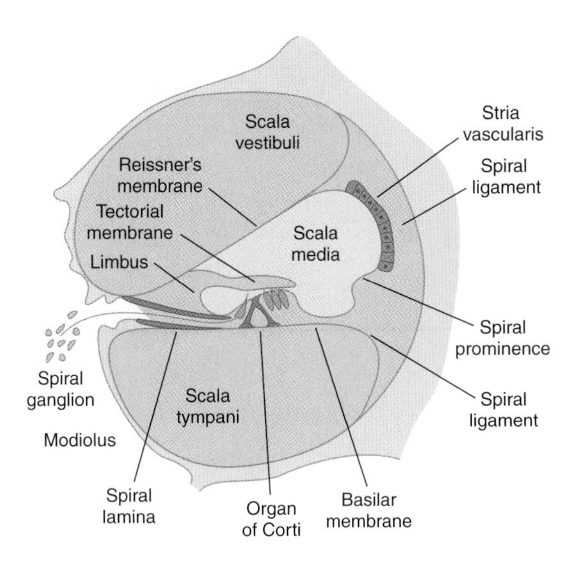

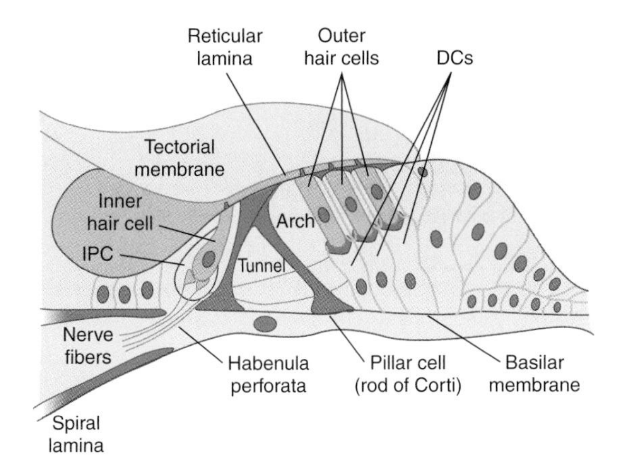

FIGURE 10–4 Schematic of the cochlea and organ of Corti in the membranous labyrinth of the inner ear. Top: The cross-section of the cochlea shows the organ of Corti and the three scalae of the cochlea. **Bottom:** This shows the structure of the organ of Corti as it appears in the basal turn of the cochlea. DC, outer phalangeal cells (Deiters' cells) supporting outer hair cells; IPC, inner phalangeal cell supporting inner hair cell. (Reproduced with permission from Pickels JO: *An Introduction to the Physiology of Hearing,* 2nd ed. Academic Press, 1988.)

The K^+ that enters hair cells via the mechanically-sensitive cation channels is recycled (Figure 10–7). It enters supporting cells and then passes on to other supporting cells by way of tight junctions. In the cochlea, it eventually reaches the stria vascularis and is secreted back into the endolymph, completing the cycle.

As described above, the processes of the hair cells project into the endolymph whereas the bases are bathed in perilymph. This arrangement is necessary for the normal production of receptor potentials. The perilymph is formed mainly from plasma. On the other hand, endolymph is formed in the scala media by the stria vascularis and has a high concentration of K^+ and a low concentration of Na^+ (Figure 10–7). Cells in the stria vascularis have a high concentration of Na, K ATPase. In addition, it appears that a unique electrogenic K^+ pump in the stria vascularis accounts for the fact that the scala media is electrically positive by 85 mV relative to the scala vestibuli and scala tympani.

The resting membrane potential of the hair cells is about –60 mV. When the stereocilia are pushed toward the kinocilium, the membrane potential is decreased to about –50 mV. When the bundle of processes is pushed in the opposite direction, the cell is hyperpolarized. Displacing the processes in a direction perpendicular to this axis provides no change in membrane potential, and displacing the processes in directions that are intermediate between these two directions produces depolarization or hyperpolarization that is proportional to the degree to which the direction is toward or away from the kinocilium. Thus, the hair processes provide a mechanism for generating changes in membrane potential proportional to the direction and distance the hair moves.

HEARING

SOUND WAVES

Sound is the sensation produced when longitudinal vibrations of the molecules in the external environment—that is, alternate phases of condensation and rarefaction of the molecules—strike the tympanic membrane. A plot of these movements as changes in pressure on the tympanic membrane per unit of time is a series of waves (Figure 10–8); such movements in the environment are generally called **sound waves.** The waves travel through air at a speed of approximately 344 m/s (770 mph) at 20°C at sea level. The speed of sound increases with temperature and with altitude. Other media can also conduct sound waves, but at a different speed. For example, the speed of sound is 1450 m/s at 20°C in fresh water and is even greater in salt water. It is said that the whistle of the blue whale is as loud as 188 dB and is audible for 500 miles.

In general, the **loudness** of a sound is directly correlated with the **amplitude** of a sound wave. The **pitch** of a sound is directly correlated with the **frequency** (number of waves per unit of time) of the sound wave. Sound waves that have repeating patterns, even though the individual waves are complex, are perceived as musical sounds; aperiodic nonrepeating vibrations cause a sensation of noise. Most musical sounds are made up of a wave with a primary frequency that determines the pitch of the sound plus a number of harmonic vibrations **(overtones)** that give the sound its characteristic **timbre** (quality). Variations in timbre permit us to identify the sounds of the various musical instruments even though they are playing notes of the same pitch.

Although the pitch of a sound depends primarily on the frequency of the sound wave, loudness also plays a part; low tones (below 500 Hz) seem lower and high tones (above 4000 Hz) seem higher as their loudness increases. Duration also affects pitch to a minor degree. The pitch of a tone cannot be perceived unless it lasts for more than 0.01 s, and with durations between 0.01 and 0.1 s, pitch rises as duration increases.

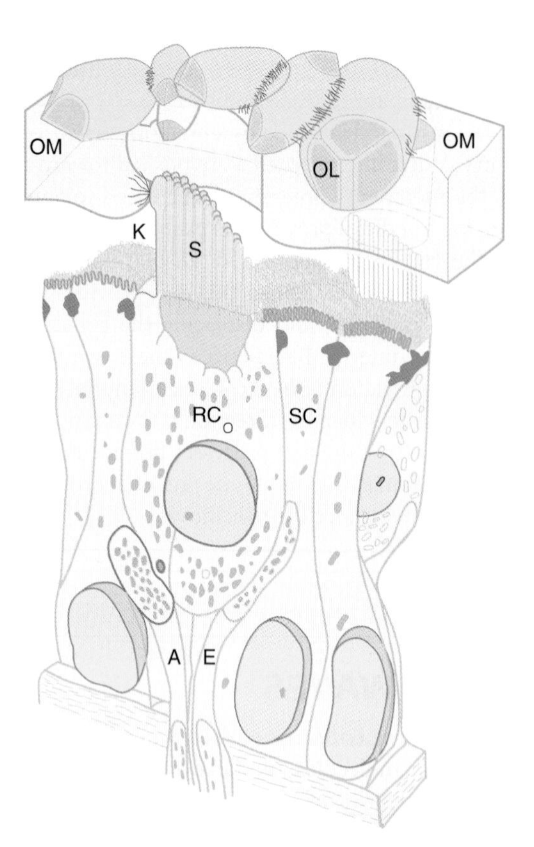

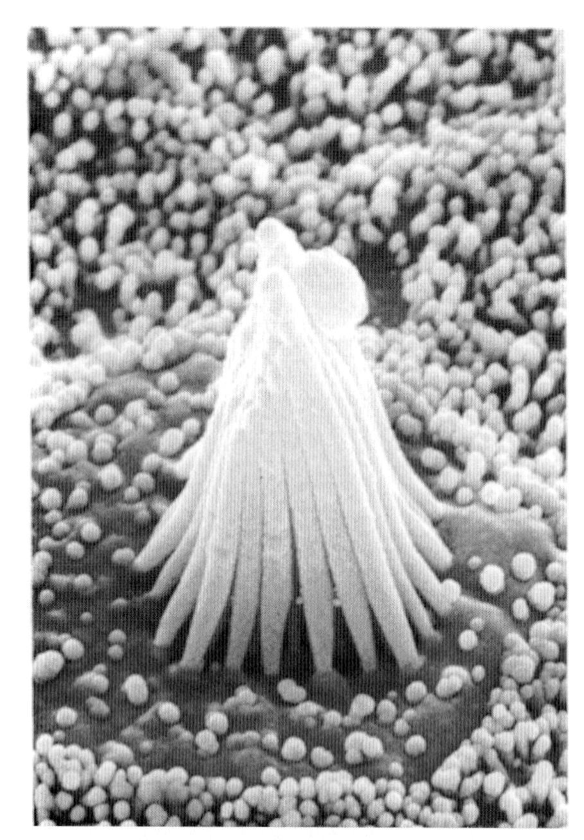

FIGURE 10–5 Structure of hair cell in the saccule. Left: Hair cells in the membranous labyrinth of the ear have a common structure, and each is within an epithelium of supporting cells (SC) surmounted by an otolithic membrane (OM) embedded with crystals of calcium carbonate, the otoliths (OL). Projecting from the apical end are rod-shaped processes, or hair cells (RC), in contact with afferent (A) and efferent (E) nerve fibers. Except in the cochlea, one of these, **kinocilium** (K), is a true but nonmotile cilium with nine pairs of microtubules around its circumference and a central pair of microtubules. The other processes, **stereocilia** (S), are found in all hair cells; they have cores of actin filaments coated with isoforms of myosin. Within the clump of processes on each cell there is an orderly structure. Along an axis toward the kinocilium, the stereocilia increase progressively in height; along the perpendicular axis, all the stereocilia are the same height. (Reproduced with permission from Hillman DE: Morphology of peripheral and central vestibular systems. In: Llinas R, Precht W [editors]: *Frog Neurobiology*. Springer, 1976.) **Right:** Scanning electronmicrograph of processes on a hair cell in the saccule. The otolithic membrane has been removed. The small projections around the hair cell are microvilli on supporting cells. (Courtesy of AJ Hudspeth.)

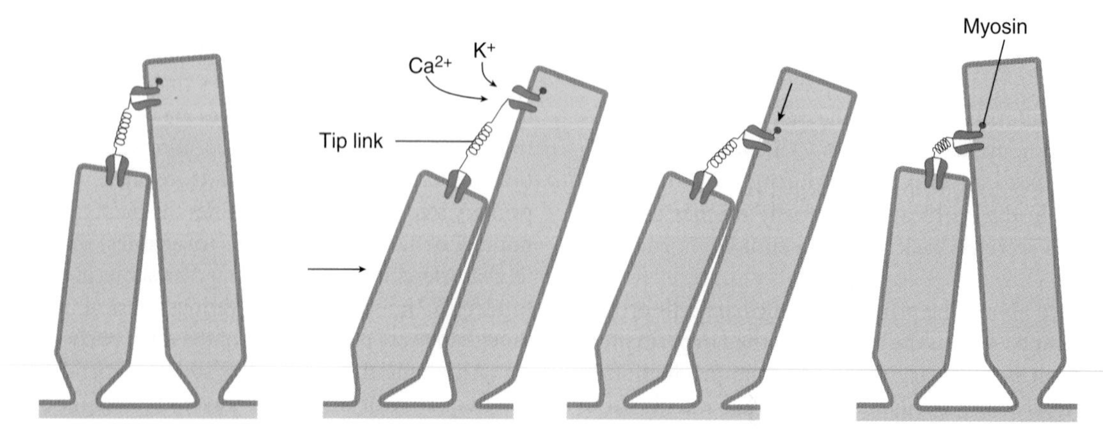

FIGURE 10–6 Schematic representation of the role of tip links in the responses of hair cells. When a stereocilium is pushed toward a taller stereocilium, the tip link is stretched and opens an ion channel in its taller neighbor. The channel next is presumably moved down the taller stereocilium by a molecular motor, so the tension on the tip link is released. When the hairs return to the resting position, the motor moves back up the stereocilium. (Reproduced with permission from Hudspeth AJ, Gillespie PG: Pulling springs to tune transduction: adaptation by hair cells. Neuron 1944 Jan;12(1):1–9.)

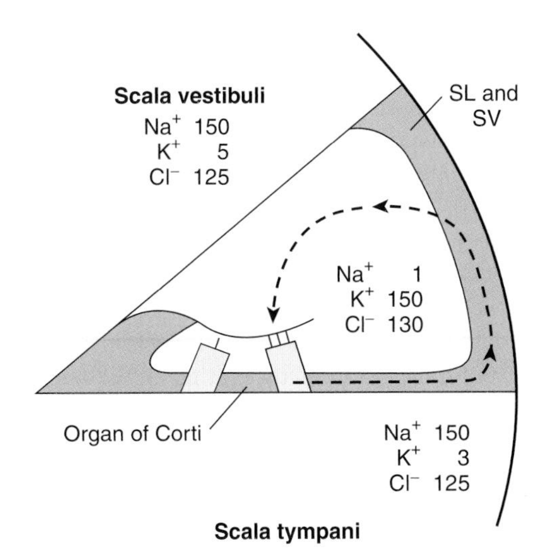

FIGURE 10–7 Ionic composition of perilymph in the scala vestibuli, endolymph in the scala media, and perilymph in the scala tympani. SL, spiral ligament. SV, stria vascularis. The dashed arrow indicates the path by which K^+ recycles from the hair cells to the supporting cells to the spiral ligament and is then secreted back into the endolymph by cells in the stria vascularis.

Finally, the pitch of complex sounds that include harmonics of a given frequency is still perceived even when the primary frequency (missing fundamental) is absent.

The amplitude of a sound wave can be expressed in terms of the maximum pressure change at the eardrum, but a relative scale is more convenient. The **decibel scale** is such a scale. The intensity of a sound in **bels** is the logarithm of the ratio of the intensity of that sound and a standard sound. A decibel (dB) is 0.1 bel. The standard sound reference level adopted by the Acoustical Society of America corresponds to 0 dB at a pressure level of 0.000204 × dyne/cm^2, a value that is just at the auditory threshold for the average human. A value of 0 dB does not mean the absence of sound but a sound level of an intensity equal to that of the standard. The 0–140 dB range from threshold pressure to a pressure that is potentially damaging to the organ of Corti actually represents a 10^7 (10 million)-fold variation in sound pressure. Put another way, atmospheric pressure at sea level is 15 lb/in^2 or 1 bar, and the range from the threshold of hearing to potential damage to the cochlea is 0.0002–2000 μbar.

A range of 120–160 dB (eg, firearms, jackhammer, jet plane on take off) is classified as painful; 90–110 dB (eg, subway, bass drum, chain saw, lawn mower) is classified as extremely high; 60–80 dB (eg, alarm clock, busy traffic, dishwasher, conversation) is classified as very loud; 40–50 dB (eg, moderate rainfall, normal room noise) is moderate; and 30 dB (eg, whisper, library) is faint. Prolonged or frequent exposure to sounds greater than 85 dB can cause hearing loss.

The sound frequencies audible to humans range from about 20 to a maximum of 20,000 cycles per second (cps, Hz). In bats and dogs, much higher frequencies are audible. The threshold of the human ear varies with the pitch of the sound (Figure 10–9), the greatest sensitivity being in the 1000- to 4000-Hz range. The pitch of the average male voice in conversation is about 120 Hz and that of the average female voice about 250 Hz. The number of pitches that can be distinguished by an average individual is about 2000, but trained musicians can improve on this figure considerably. Pitch discrimination is best in the 1000- to 3000-Hz range and is poor at high and low pitches.

The presence of one sound decreases an individual's ability to hear other sounds, a phenomenon known as **masking.** It is believed to be due to the relative or absolute refractoriness of previously stimulated auditory receptors and nerve fibers to other stimuli. The degree to which a given tone masks others is

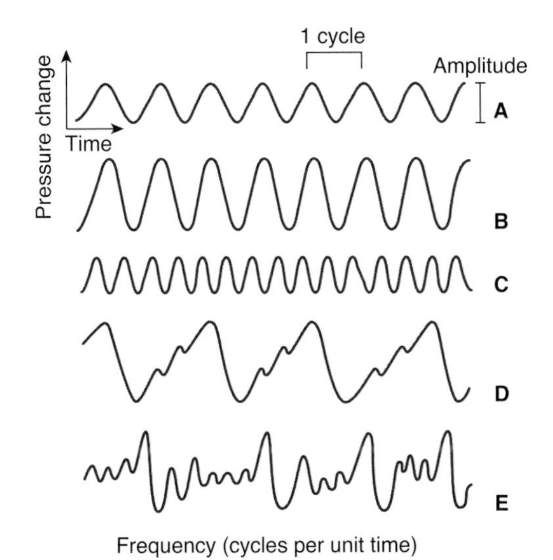

FIGURE 10–8 Characteristics of sound waves. A is the record of a pure tone. **B** has a greater amplitude and is louder than **A. C** has the same amplitude as **A** but a greater frequency, and its pitch is higher. **D** is a complex wave form that is regularly repeated. Such patterns are perceived as musical sounds, whereas waves like that shown in **E,** which have no regular pattern, are perceived as noise.

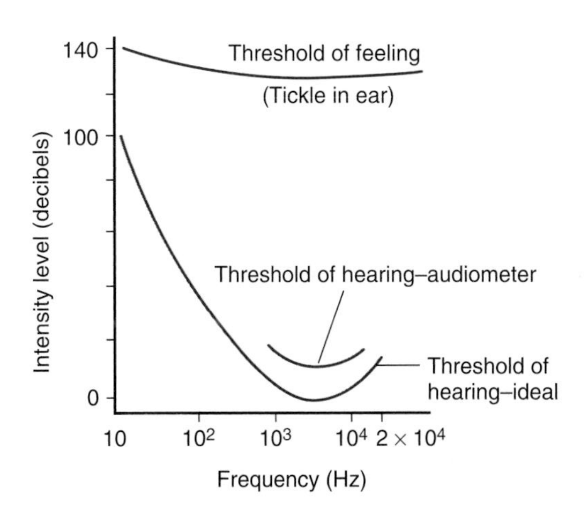

FIGURE 10–9 Human audibility curve. The middle curve is that obtained by audiometry under the usual conditions. The lower curve is that obtained under ideal conditions. At about 140 db (top curve), sounds are felt as well as heard.

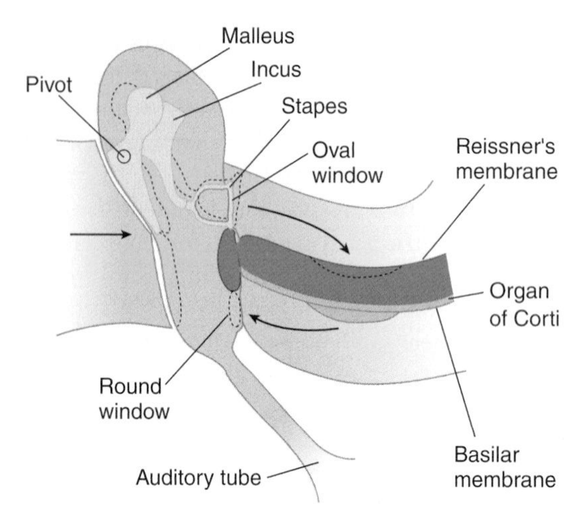

FIGURE 10–10 Schematic representation of the auditory ossicles and the way their movement translates movements of the tympanic membrane into a wave in the fluid of the inner ear. The wave is dissipated at the round window. The movements of the ossicles, the membranous labyrinth, and the round window are indicated by dashed lines. The waves are transformed by the eardrum and auditory ossicles into movements of the foot plate of the stapes. These movements set up waves in the fluid of the inner ear. In response to the pressure changes produced by sound waves on its external surface, the tympanic membrane moves in and out to function as a resonator that reproduces the vibrations of the sound source. The motions of the tympanic membrane are imparted to the manubrium of the malleus, which rocks on an axis through the junction of its long and short processes, so that the short process transmits the vibrations of the manubrium to the incus. The incus moves so that the vibrations are transmitted to the head of the stapes. Movements of the head of the stapes swing its foot plate.

related to its pitch. The masking effect of the background noise in all but the most carefully soundproofed environments raises the auditory threshold by a definite and measurable amount.

SOUND TRANSMISSION

The ear converts sound waves in the external environment into action potentials in the auditory nerves. The waves are transformed by the eardrum and auditory ossicles into movements of the foot plate of the stapes. These movements set up waves in the fluid of the inner ear (Figure 10–10). The action of the waves on the organ of Corti generates action potentials in the nerve fibers.

In response to the pressure changes produced by sound waves on its external surface, the tympanic membrane moves in and out. The membrane therefore functions as a **resonator** that reproduces the vibrations of the sound source. It stops vibrating almost immediately when the sound wave stops. The motions of the tympanic membrane are imparted to the manubrium of the malleus. The malleus rocks on an axis through the junction of its long and short processes, so that the short process transmits the vibrations of the manubrium to the incus. The incus moves in such a way that the vibrations are transmitted to the head of the stapes. Movements of the head of the stapes swing its foot plate to and fro like a door hinged at the posterior edge of the oval window. The auditory ossicles thus function as a lever system that converts the resonant vibrations of the tympanic membrane into movements of the stapes against the perilymph-filled scala vestibuli of the cochlea (Figure 10–10). This system increases the sound pressure that arrives at the oval window, because the lever action of the malleus and incus multiplies the force 1.3 times and the area of the tympanic membrane is much greater than the area of the foot plate of the stapes. Some sound energy is lost as a result of resistance, but it has been calculated that at frequencies below 3000 Hz, 60% of the sound energy incident on the tympanic membrane is transmitted to the fluid in the cochlea.

Contraction of the tensor tympani and stapedius muscles of the middle ear cause the manubrium of the malleus to be pulled inward and the footplate of the stapes to be pulled outward (Figure 10–2). This decreases sound transmission. Loud sounds initiate a reflex contraction of these muscles called the **tympanic reflex.** Its function is protective, preventing strong sound waves from causing excessive stimulation of the auditory receptors. However, the reaction time for the reflex is 40–160 ms, so it does not protect against brief intense stimulation such as that produced by gunshots.

BONE & AIR CONDUCTION

Conduction of sound waves to the fluid of the inner ear via the tympanic membrane and the auditory ossicles, the main pathway for normal hearing, is called **ossicular conduction.** Sound waves also initiate vibrations of the secondary tympanic membrane that closes the round window. This process, unimportant in normal hearing, is **air conduction.** A third type of conduction, **bone conduction,** is the transmission of vibrations of the bones of the skull to the fluid of the inner ear. Considerable bone conduction occurs when tuning forks or other vibrating bodies are applied directly to the skull. This route also plays a role in transmission of extremely loud sounds.

TRAVELING WAVES

The movements of the foot plate of the stapes set up a series of traveling waves in the perilymph of the scala vestibuli. A diagram of such a wave is shown in Figure 10–11. As the wave moves up the cochlea, its height increases to a maximum and then drops off rapidly. The distance from the stapes to this point of maximum height varies with the frequency of the vibrations initiating the wave. High-pitched sounds generate waves that reach maximum height near the base of the cochlea; low-pitched sounds generate waves that peak near the apex. The bony walls of the scala vestibuli are rigid, but Reissner's membrane is flexible. The basilar membrane is not under tension, and it is also readily depressed into the scala tympani by the peaks of waves in the scala vestibuli. Displacements of the fluid in the scala tympani are dissipated into air at the round window. Therefore, sound produces distortion of the basilar membrane, and the site at which this distortion is

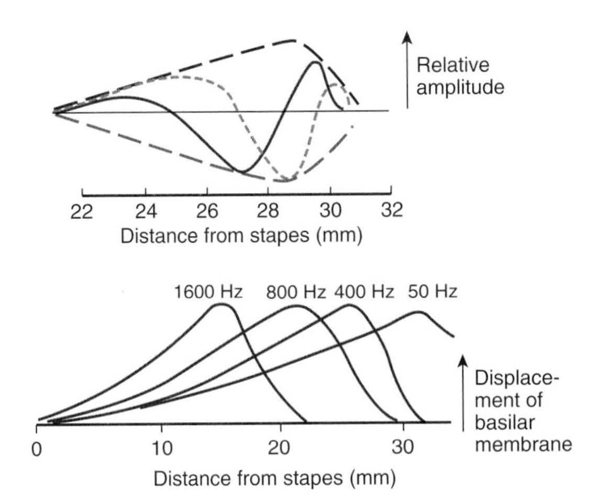

FIGURE 10–11 Traveling waves. Top: The solid and the short-dashed lines represent the wave at two instants of time. The long-dashed line shows the "envelope" of the wave formed by connecting the wave peaks at successive instants. **Bottom:** Displacement of the basilar membrane by the waves generated by stapes vibration of the frequencies shown at the top of each curve.

maximal is determined by the frequency of the sound wave. The tops of the hair cells in the organ of Corti are held rigid by the reticular lamina, and the hairs of the outer hair cells are embedded in the tectorial membrane (Figure 10–4). When the stapes moves, both membranes move in the same direction, but they are hinged on different axes, so a shearing motion bends the hairs. The hairs of the inner hair cells are not attached to the tectorial membrane, but they are apparently bent by fluid moving between the tectorial membrane and the underlying hair cells.

FUNCTIONS OF THE OUTER HAIR CELLS

The inner hair cells are the primary sensory receptors that generate action potentials in the auditory nerves and are stimulated by the fluid movements noted above. The outer hair cells, on the other hand, respond to sound like the inner hair cells, but depolarization makes them shorten and hyperpolarization makes them lengthen. They do this over a very flexible part of the basal membrane, and this action somehow increases the amplitude and clarity of sounds. Thus, outer hair cells amplify sound vibrations entering the inner ear from the middle ear. These changes in outer hair cells occur in parallel with changes in **prestin,** a membrane protein, and this protein may well be the motor protein of outer hair cells.

The **olivocochlear bundle** is a prominent bundle of efferent fibers in each auditory nerve that arises from both ipsilateral and contralateral superior olivary complexes and ends primarily around the bases of the outer hair cells of the organ of Corti. The activity in this nerve bundle modulates the sensitivity of these hair cells via the release of acetylcholine. The effect is inhibitory, and it may function to block background noise while allowing other sounds to be heard.

ACTION POTENTIALS IN AUDITORY NERVE FIBERS

The frequency of the action potentials in single auditory nerve fibers is proportional to the loudness of the sound stimuli. At low sound intensities, each axon discharges to sounds of only one frequency, and this frequency varies from axon to axon depending on the part of the cochlea from which the fiber originates. At higher sound intensities, the individual axons discharge to a wider spectrum of sound frequencies, particularly to frequencies lower than that at which threshold simulation occurs.

The major determinant of the pitch perceived when a sound wave strikes the ear is the place in the organ of Corti that is maximally stimulated. The traveling wave set up by a tone produces peak depression of the basilar membrane, and consequently maximal receptor stimulation, at one point. As noted above, the distance between this point and the stapes is inversely related to the pitch of the sound, with low tones producing maximal stimulation at the apex of the cochlea and high tones producing maximal stimulation at the base. The pathways from the various parts of the cochlea to the brain are distinct. An additional factor involved in pitch perception at sound frequencies of less than 2000 Hz may be the pattern of the action potentials in the auditory nerve. When the frequency is low enough, the nerve fibers begin to respond with an impulse to each cycle of a sound wave. The importance of this **volley effect,** however, is limited; the frequency of the action potentials in a given auditory nerve fiber determines principally the loudness, rather than the pitch, of a sound.

CENTRAL PATHWAY

The afferent fibers in the auditory division of the eighth cranial nerve end in **dorsal** and **ventral cochlear nuclei** (Figure 10–12). From there, auditory impulses pass by various routes to the **inferior colliculi,** the centers for auditory reflexes, and via the **medial geniculate body** in the thalamus to the **auditory cortex** located on the superior temporal gyrus of the temporal lobe. Information from both ears converges on each superior olive, and beyond this, most of the neurons respond to inputs from both sides. In humans, low tones are represented anterolaterally and high tones posteromedially in the auditory cortex.

The responses of individual second-order neurons in the cochlear nuclei to sound stimuli are like those of the individual auditory nerve fibers. The frequency at which sounds of the lowest intensity evoke a response varies from unit to unit; with increased sound intensities, the band of frequencies to which a response occurs becomes wider. The major difference between the responses of the first- and second-order neurons is the presence of a sharper "cutoff" on the low-frequency side in the medullary neurons. This greater specificity of the second-order neurons is probably due to an inhibitory process in the brain stem. In the primary auditory cortex, most neurons

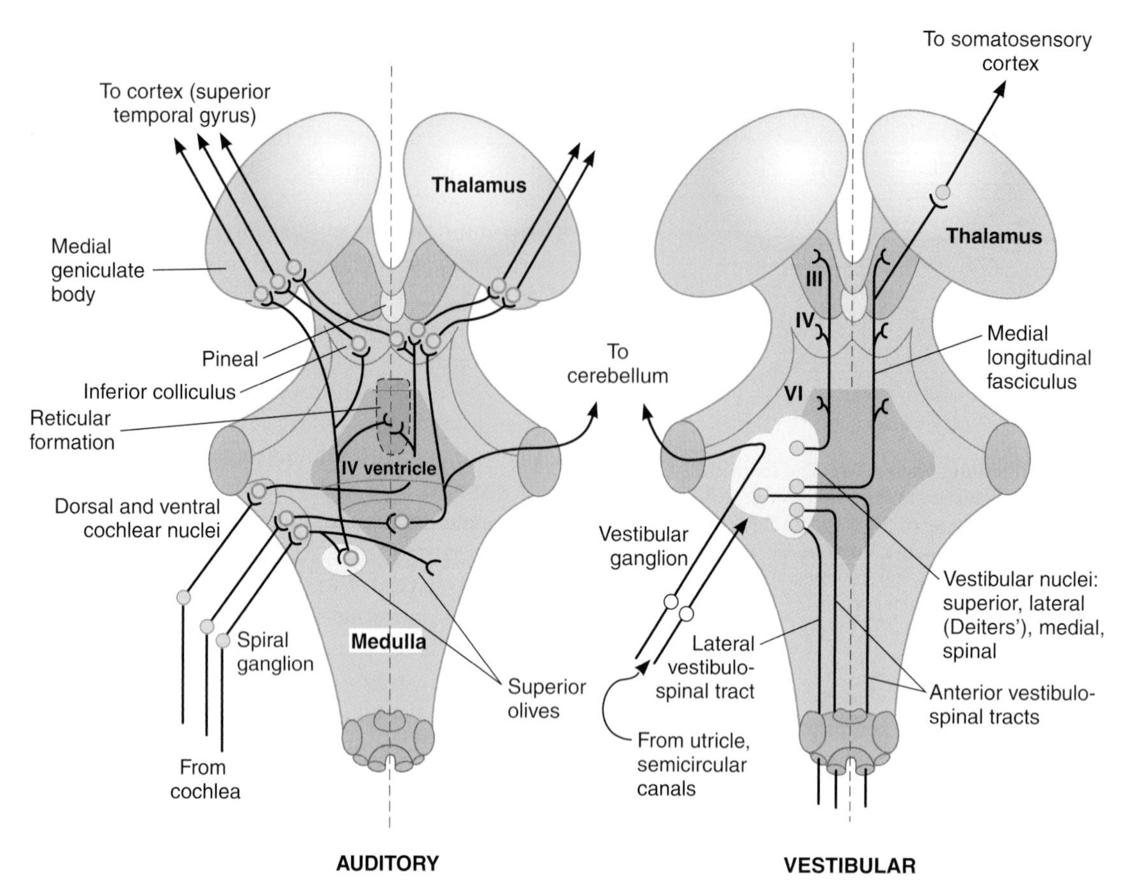

FIGURE 10–12 Simplified diagram of main auditory (left) and vestibular (right) pathways superimposed on a dorsal view of the brain stem. Cerebellum and cerebral cortex have been removed. For the auditory pathway, eighth cranial nerve afferent fibers form the cochlea end in dorsal and ventral cochlear nuclei. From there, most fibers cross the midline and terminate in the contralateral inferior colliculus. From there, fibers project to the medial geniculate body in the thalamus and then to the auditory cortex located on the superior temporal gyrus of the temporal lobe. For the vestibular pathway, the vestibular nerve terminates in the ipsilateral vestibular nucleus. Most fibers from the semicircular canals terminate in the superior and medial divisions of the vestibular nucleus and project to nuclei controlling eye movement. Most fibers from the utricle and saccule terminate in the lateral division, which then projects to the spinal cord. They also terminate on neurons that project to the cerebellum and the reticular formation. The vestibular nuclei also project to the thalamus and from there to the primary somatosensory cortex. The ascending connections to cranial nerve nuclei are concerned with eye movements.

respond to inputs from both ears, but strips of cells are stimulated by input from the contralateral ear and inhibited by input from the ipsilateral ear.

The increasing availability of positron emission tomography (PET) scanning and functional magnetic resonance imaging (fMRI) has greatly improved our level of knowledge about auditory association areas in humans. The auditory pathways in the cortex resemble the visual pathways in that increasingly complex processing of auditory information takes place along them. An interesting observation is that although the auditory areas look very much the same on the two sides of the brain, there is marked hemispheric specialization. For example, **Wernicke's area** (see Figure 8–7) is concerned with the processing of auditory signals related to speech. During language processing, this area is much more active on the left side than on the right side. Wernicke's area on the right side is more concerned with melody, pitch, and sound intensity. The auditory pathways are also very plastic, and, like the visual and somatosensory pathways, they are modified by experience. Examples of auditory plasticity in humans include the observation that in individuals who become deaf before language skills are fully developed, sign language activates auditory association areas. Conversely, individuals who become blind early in life are demonstrably better at localizing sound than individuals with normal eyesight.

Musicians provide additional examples of cortical plasticity. In these individuals, the size of the auditory areas activated by musical tones is increased. In addition, violinists have altered somatosensory representation of the areas to which the fingers they use in playing their instruments project. Musicians also have larger cerebellums than nonmusicians, presumably because of learned precise finger movements.

A portion of the posterior superior temporal gyrus known as the **planum temporale,** which is located between **Heschl's gyrus** (transverse temporal gyrus) and the **sylvian fissure** (Figure 10–13) is regularly larger in the left than in the right

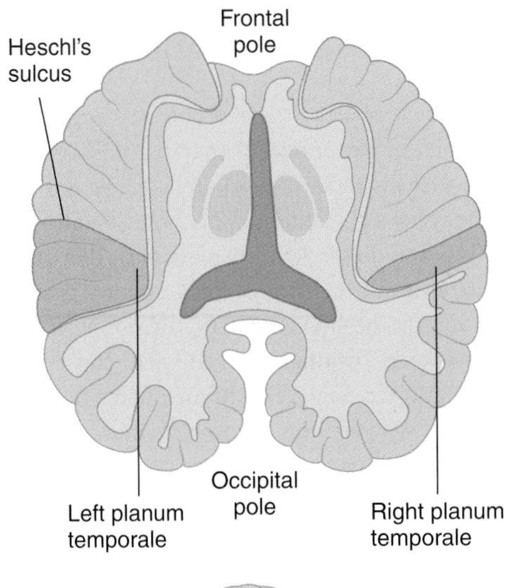

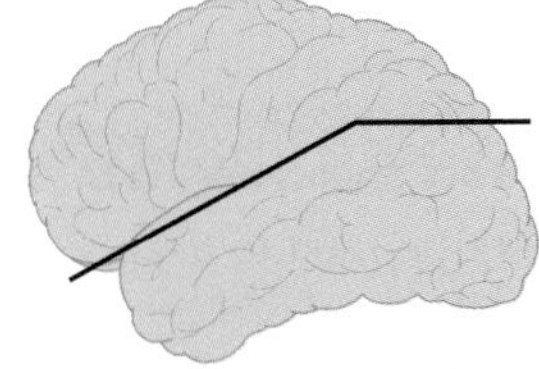

FIGURE 10–13 Left and right planum temporale in a brain sectioned horizontally along the plane of the sylvian fissure. Plane of section shown in the insert at the bottom. (Reproduced with permission from Kandel ER, Schwartz JH, Jessel TM [editors]: *Principles of Neural Science*, 3rd ed. McGraw-Hill, 1991.)]

cerebral hemisphere, particularly in right-handed individuals. This area appears to be involved in language-related auditory processing. A curious observation is that the planum temporale is even larger than normal on the left side in musicians and others who have perfect pitch.

SOUND LOCALIZATION

Determination of the direction from which a sound emanates in the horizontal plane depends on detecting the difference in time between the arrival of the stimulus in the two ears and the consequent difference in phase of the sound waves on the two sides; it also depends on the fact that the sound is louder on the side closest to the source. The detectable time difference, which can be as little as 20 μs, is said to be the most important factor at frequencies below 3000 Hz and the loudness difference the most important at frequencies above 3000 Hz. Neurons in the auditory cortex that receive input from both ears respond maximally or minimally when the time of arrival of a stimulus at one ear is delayed by a fixed period relative to the time of arrival at the other ear. This fixed period varies from neuron to neuron.

Sounds coming from directly in front of the individual differ in quality from those coming from behind because each pinna (the visible portion of the exterior ear) is turned slightly forward. In addition, reflections of the sound waves from the pinnal surface change as sounds move up or down, and the change in the sound waves is the primary factor in locating sounds in the vertical plane. Sound localization is markedly disrupted by lesions of the auditory cortex.

DEAFNESS

Deafness can be divided into two major categories: conductive (or conduction) and sensorineural hearing loss. **Conductive deafness** refers to impaired sound transmission in the external or middle ear and impacts all sound frequencies. Among the causes of conduction deafness are plugging of the external auditory canals with wax (cerumen) or foreign bodies, otitis externa (inflammation of the outer ear, "swimmer's ear") and otitis media (inflammation of the middle ear) causing fluid accumulation, perforation of the eardrum, and osteosclerosis in which bone is resorbed and replaced with sclerotic bone that grows over the oval window.

Sensorineural deafness is most commonly the result of loss of cochlear hair cells but can also be due to problems with the eighth cranial nerve or within central auditory pathways. It often impairs the ability to hear certain pitches while others are unaffected. Aminoglycoside antibiotics such as streptomycin and gentamicin obstruct the mechanosensitive channels in the stereocilia of hair cells (especially outer hair cells) and can cause the cells to degenerate, producing sensorineural hearing loss and abnormal vestibular function. Damage to the hair cells by prolonged exposure to noise is also associated with hearing loss (see Clinical Box 10–1). Other causes include tumors of the eighth cranial nerve and cerebellopontine angle and vascular damage in the medulla.

Auditory acuity is commonly measured with an **audiometer.** This device presents the subject with pure tones of various frequencies through earphones. At each frequency, the threshold intensity is determined and plotted on a graph as a percentage of normal hearing. This provides an objective measurement of the degree of deafness and a picture of the tonal range most affected.

Conduction and sensorineural deafness can be differentiated by simple tests with a tuning fork. Three of these tests, named for the individuals who developed them, are outlined in Table 10–1. The Weber and Schwabach tests demonstrate the important masking effect of environmental noise on the auditory threshold.

VESTIBULAR SYSTEM

The vestibular system can be divided into the **vestibular apparatus** and central **vestibular nuclei.** The vestibular apparatus within the inner ear detects head motion and position and transduces this information to a neural signal (Figure 10–3). The vestibular nuclei are primarily concerned with maintaining the position of the head in space. The tracts that descend from these nuclei mediate head-on-neck and head-on-body adjustments.

CLINICAL BOX 10–1

Hearing Loss

Hearing loss is the most common sensory defect in humans. According to the World Health Organization, over 270 million people worldwide have moderate to profound hearing loss, with one fourth of these cases beginning in childhood. According to the National Institutes of Health, ~15% of Americans between 20 and 69 years of age have high frequency hearing loss due to exposure to loud sounds or noise at work or in leisure activities (**noise-induced hearing loss, NIHL**). Both inner and outer hair cells are damaged by excessive noise, but outer hair cells appear to be more vulnerable. The use of various chemicals also causes hearing loss; such chemicals are called **ototoxins.** These include some antibiotics (streptomycin), loop diuretics (furosemide), and platinum-based chemotherapy agents (cisplatin). These ototoxic agents damage the outer hair cells or the stria vascularis. **Presbycusis,** the gradual hearing loss associated with aging, affects more than one-third of those over 75 and is probably due to gradual cumulative loss of hair cells and neurons. In most cases, hearing loss is a multifactorial disorder caused by both genetic and environmental factors. Single-gene mutations have been shown to cause hearing loss. This type of hearing loss is a monogenic disorder with an autosomal dominant, autosomal recessive, X-linked, or mitochondrial mode of inheritance. Monogenic forms of deafness can be defined as **syndromic** (hearing loss associated with other abnormalities) or **nonsyndromic** (only hearing loss). About 0.1% of newborns have genetic mutations leading to deafness. Nonsyndromic deafness due to genetic mutations can first appear in adults rather than in children and may account for many of the 16% of all adults who have significant hearing impairment. It is now estimated that the products of 100 or more genes are essential for normal hearing, and deafness loci have been described in all but 5 of the 24 human chromosomes. The most common mutation leading to congenital hearing loss is that of the protein connexin 26. This defect prevents the normal recycling of K^+ through the sustenacular cells. Mutations in three nonmuscle myosins also cause deafness. These are myosin-VIIa, associated with the actin in the hair cell processes; myosin-Ib, which is probably part of the "adaptation motor" that adjusts tension on the tip links; and myosin-VI, which is essential in some way for the formation of normal cilia. Deafness is also associated with mutant forms of α-tectin, one of the major proteins in the tectorial membrane. An example of syndromic deafness is **Pendred syndrome,** in which a mutant multifunctional anion exchanger causes deafness and goiter. Another example is one form of the **long QT syndrome** in which one of the K^+ channel proteins, **KVLQT1,** is mutated. In the stria vascularis, the normal form of this protein is essential for maintaining the high K^+ concentration in endolymph, and in the heart it helps maintain a normal QT interval. Individuals who are homozygous for mutant KVLQT1 are deaf and predisposed to the ventricular arrhythmias and sudden death that characterize the long QT syndrome. Mutations of the membrane protein **barttin** can cause deafness as well as the renal manifestations of Bartter syndrome.

THERAPEUTIC HIGHLIGHTS

Cochlear implants are used to treat both children and adults with severe hearing loss. The U.S. Food and Drug Administration has reported that, as of April 2009, approximately 188,000 people worldwide have received cochlear implants. They may be used in children as young as 12 months old. These devices consist of a microphone (picks up environmental sounds), a speech processor (selects and arranges these sounds), a transmitter and receiver/stimulator (converts these sounds into electrical impulses), and an electrode array (sends the impulses to the auditory nerve). Although the implant cannot restore normal hearing, it provides a useful representation of environmental sounds to a deaf person. Those with adult-onset deafness that receive cochlear implants can learn to associate the signals it provides with sounds they remember. Children that receive cochlear implants in conjunction with intensive therapy have been able to acquire speech and language skills. Research is also underway to develop cells that can replace the hair cells in the inner ear. For example, researchers at Stanford University were able to generate cells resembling mechanosensitive hair cells from mouse **embryonic** and **pluripotent stem cells.**

CENTRAL PATHWAY

The cell bodies of the 19,000 neurons supplying the cristae and maculae on each side are located in the vestibular ganglion. Each vestibular nerve terminates in the ipsilateral four-part vestibular nucleus (Figure 10–12) and in the flocculonodular lobe of the cerebellum (not shown in the figure). Fibers from the semicircular canals terminate primarily in the superior and medial divisions of the vestibular nucleus; neurons in this region project mainly to nuclei controlling eye movement. Fibers from the utricle and saccule project predominantly to the lateral division (Deiters nucleus) of the vestibular nucleus which then projects to the spinal cord (lateral vestibulospinal tract). Fibers from the utricle and saccule also terminate on neurons that project to the cerebellum and the reticular formation. The vestibular nuclei also project to the thalamus and from there to two parts of the primary somatosensory cortex. The ascending connections to cranial nerve nuclei are largely concerned with eye movements.

TABLE 10–1 Common tests with a tuning fork to distinguish between sensorineural and conduction deafness.

	Weber	Rinne	Schwabach
Method	Base of vibrating tuning fork placed on vertex of skull	Base of vibrating tuning fork placed on mastoid process until subject no longer hears it, then held in air next to ear	Bone conduction of patient compared with that of normal subject.
Normal	Hears equally on both sides.	Hears vibration in air after bone conduction is over	
Conduction deafness (one ear)	Sound louder in diseased ear because masking effect of environmental noise is absent on diseased side	Vibrations in air not heard after bone conduction is over	Bone conduction better than normal (conduction defect excludes masking noise)
Sensorineural deafness (one ear)	Sound louder in normal ear	Vibration heard in air after bone conduction is over, as long as nerve deafness is partial	Bone conduction worse than normal

RESPONSES TO ROTATIONAL ACCELERATION

Rotational acceleration in the plane of a given semicircular canal stimulates its crista. The endolymph, because of its inertia, is displaced in a direction opposite to the direction of rotation. The fluid pushes on the cupula, deforming it. This bends the processes of the hair cells (Figure 10–3). When a constant speed of rotation is reached, the fluid spins at the same rate as the body and the cupula swings back into the upright position. When rotation is stopped, deceleration produces displacement of the endolymph in the direction of the rotation, and the cupula is deformed in a direction opposite to that during acceleration. It returns to mid position in 25–30 s. Movement of the cupula in one direction commonly causes an increase in the firing rate of single nerve fibers from the crista, whereas movement in the opposite direction commonly inhibits neural activity (Figure 10–14).

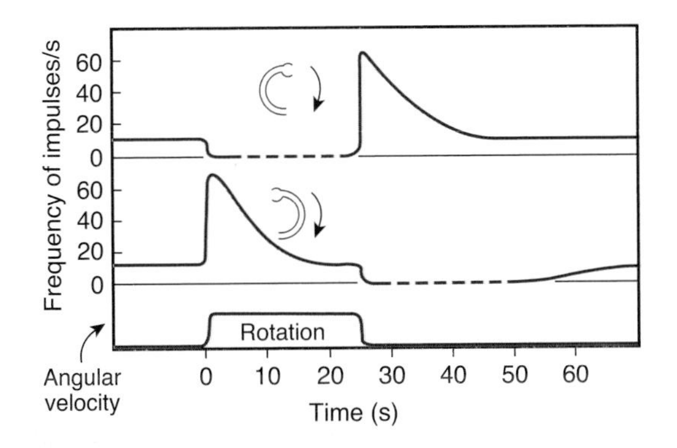

FIGURE 10–14 Ampullary responses to rotation. Average time course of impulse discharge from the ampulla of two semicircular canals during rotational acceleration, steady rotation, and deceleration. Movement of the cupula in one direction increases the firing rate of single nerve fibers from the crista, and movement in the opposite direction inhibits neural activity. (Reproduced with permission from Adrian ED: Discharge from vestibular receptors in the cat. J Physiol [Lond] 1943;101:389.)

Rotation causes maximal stimulation of the semicircular canals most nearly in the plane of rotation. Because the canals on one side of the head are a mirror image of those on the other side, the endolymph is displaced toward the ampulla on one side and away from it on the other. The pattern of stimulation reaching the brain therefore varies with the direction as well as the plane of rotation. Linear acceleration probably fails to displace the cupula and therefore does not stimulate the cristae. However, there is considerable evidence that when one part of the labyrinth is destroyed, other parts take over its functions. Clinical Box 10–2 describes the characteristic eye movements that occur during a period of rotation.

RESPONSES TO LINEAR ACCELERATION

The utriclar macula responds to horizontal acceleration, and the saccular macula responds to vertical acceleration. The otoliths in the surrounding membrane are denser than the endolymph, and acceleration in any direction causes them to be displaced in the opposite direction, distorting the hair cell processes and generating activity in the nerve fibers. The maculae also discharge tonically in the absence of head movement, because of the pull of gravity on the otoliths.

The impulses generated from these receptors are partly responsible for **labyrinth righting reflexes.** These reflexes are a series of responses integrated for the most part in the nuclei of the midbrain. The stimulus for the reflex is tilting of the head, which stimulates the otolithic organs; the response is compensatory contraction of the neck muscles to keep the head level. In cats, dogs, and primates, visual cues can initiate **optical righting reflexes** that right the animal in the absence of labyrinthine or body stimulation. In humans, the operation of these reflexes maintains the head in a stable position and the eyes fixed on visual targets despite movements of the body and the jerks and jolts of everyday life. The responses are initiated by vestibular stimulation, stretching of neck muscles, and movement of visual images on the retina, and the responses

CLINICAL BOX 10–2

Nystagmus

The characteristic jerky movement of the eye observed at the start and end of a period of rotation is called **nystagmus.** It is actually a reflex that maintains visual fixation on stationary points while the body rotates, although it is not initiated by visual impulses and is present in blind individuals. When rotation starts, the eyes move slowly in a direction opposite to the direction of rotation, maintaining visual fixation (**vestibuloocular reflex, VOR**). When the limit of this movement is reached, the eyes quickly snap back to a new fixation point and then again move slowly in the other direction. The slow component is initiated by impulses from the vestibular labyrinths; the quick component is triggered by a center in the brain stem. Nystagmus is frequently horizontal (ie, the eyes move in the horizontal plane), but it can also be vertical (when the head is tipped sideways during rotation) or rotatory (when the head is tipped forward). By convention, the direction of eye movement in nystagmus is identified by the direction of the quick component. The direction of the quick component during rotation is the same as that of the rotation, but the postrotatory nystagmus that occurs owing to displacement of the cupula when rotation is stopped is in the opposite direction. When nystagmus is seen at rest , it is a sign of a pathology. Two examples of this are **congenital nystagmus** that is seen at birth and **acquired nystagmus** that occurs later in life. In these clinical cases, nystagmus can persist for hours at rest. Acquired nystagmus can be seen in patients with acute **temporal bone fracture** affecting **semicircular canals** or after damage to the **flocculonodular lobe** or midline structures such as the **fastigial nucleus.** It can also occur as a result of stroke, multiple sclerosis, head injury, and brain tumors. Some drugs (especially antiseizure drugs), alcohol, and sedatives can cause nystagmus.

Nystagmus can be used as a diagnostic indicator of the integrity of the vestibular system. Caloric stimulation can be used to test the function of the **vestibular labyrinth.** The semicircular canals are stimulated by instilling warm (40°C) or cold (30°C) water into the external auditory meatus. The temperature difference sets up convection currents in the **endolymph,** with consequent motion of the cupula. In healthy subjects, warm water causes nystagmus that bears toward the stimulus, whereas cold water induces nystagmus that bears toward the opposite ear. This test is given the mnemonic **COWS** (Cold water nystagmus is Opposite sides, Warm water nystagmus is Same side). In the case of a unilateral lesion in the vestibular pathway, nystagmus is reduced or absent on the side of the lesion. To avoid nystagmus, vertigo, and nausea when irrigating the ear canals in the treatment of ear infections, it is important to be sure that the fluid used is at body temperature.

THERAPEUTIC HIGHLIGHTS

There is no cure for acquired nystagmus and treatment is dependent upon the cause. Correcting the underlying cause (stopping drug usage, surgical removal of a tumor) is often the treatment of choice. Also, rectus muscle surgery has been used successfully to treat some cases of acquired nystagmus. Short-term correction of nystagmus can result from injections of **botulinum toxin** (Botox) to paralyze the ocular muscles.

are the **vestibulo-ocular reflex** and other remarkably precise reflex contractions of the neck and extraocular muscles.

Although most of the responses to stimulation of the maculae are reflex in nature, vestibular impulses also reach the cerebral cortex. These impulses are presumably responsible for conscious perception of motion and supply part of the information necessary for orientation in space. **Vertigo** is the sensation of rotation in the absence of actual rotation and is a prominent symptom when one labyrinth is inflamed.

SPATIAL ORIENTATION

Orientation in space depends in part on input from the vestibular receptors, but visual cues are also important. Pertinent information is also supplied by impulses from proprioceptors in joint capsules, which supply data about the relative position of the various parts of the body, and impulses from cutaneous exteroceptors, especially touch and pressure receptors. These four inputs are synthesized at a cortical level into a continuous picture of the individual's orientation in space. Clinical Box 10–3 describes some common vestibular disorders.

CHAPTER SUMMARY

- The external ear funnels sound waves to the external auditory meatus and tympanic membrane. From there, sound waves pass through three auditory ossicles (malleus, incus, and stapes) in the middle ear. The inner ear contains the cochlea and organ of Corti.
- The hair cells in the organ of Corti signal hearing. The stereocilia provide a mechanism for generating changes in membrane potential proportional to the direction and distance the hair moves. Sound is the sensation produced when longitudinal vibrations of air molecules strike the tympanic membrane.

CLINICAL BOX 10–3

Vestibular Disorders

Vestibular balance disorders are the ninth most common reason for visits to a primary care physician. It is one of the most common reasons elderly people seek medical advice. Patients often describe balance problems in terms of vertigo, dizziness, lightheadedness, and motion sickness. Neither lightheadedness nor dizziness is necessarily a symptom of vestibular problems, but **vertigo** is a prominent symptom of a disorder of the inner ear or vestibular system, especially when one labyrinth is inflamed. **Benign paroxysmal positional vertigo (BPPV)** is the most common vestibular disorder characterized by episodes of vertigo that occur with particular changes in body position (eg, turning over in bed, bending over). One possible cause is that **otoconia** from the utricle separate from the otolith membrane and become lodged in the canal or cupula of the semicircular canal. This causes abnormal deflections when the head changes position relative to gravity.

Ménière disease is an abnormality of the inner ear causing vertigo or severe dizziness, **tinnitus,** fluctuating hearing loss, and the sensation of pressure or pain in the affected ear lasting several hours. Symptoms can occur suddenly and recur daily or very rarely. The hearing loss is initially transient but can become permanent. The pathophysiology likely involves an immune reaction. An inflammatory response can increase fluid volume within the membranous labyrinth, causing it to rupture and allowing the endolymph and perilymph to mix together. The worldwide prevalence for Ménière's disease is ~12 per 1000 individuals. It is diagnosed most often between the ages of 30 and 60; and it affects both genders similarly.

The nausea, blood pressure changes, sweating, pallor, and vomiting that are the well-known symptoms of **motion sickness** are produced by excessive vestibular stimulation and occurs when conflicting information is fed into the vestibular and other sensory systems. The symptoms are probably due to reflexes mediated via vestibular connections in the brain stem and the flocculonodular lobe of the cerebellum. **Space motion sickness** (ie, the nausea, vomiting, and vertigo experienced by astronauts) develops when they are first exposed to microgravity and often wears off after a few days of space flight. It can then recur with reentry, as the force of gravity increases again. It is believed to be due to mismatches in neural input created by changes in the input from some parts of the vestibular apparatus and other gravity sensors without corresponding changes in the other spatial orientation inputs.

THERAPEUTIC HIGHLIGHTS

Symptoms of BPPV often subside over weeks or months, but if treatment is needed one option is a procedure called **canalith repositioning.** This consists of simple and slow maneuvers to position your head to move the otoconia from the semicircular canals back into the vestibule that houses the utricle. There is no cure for Ménière disease, but the symptoms can be controlled by reducing the fluid retention through dietary changes (low-salt or salt-free diet, no caffeine, no alcohol) or medications such as diuretics (eg, **hydrochlorothiazide**). Individuals with Ménière disease often respond to drugs used to alleviate the symptoms of vertigo. **Vestibulosuppressants** such as **meclizine** (an **antihistamine** drug) decrease the excitability of the middle ear labyrinth and block conduction in middle ear vestibular-cerebellar pathway. Motion sickness commonly can be prevented with the use of antihistamines or **scopolamine,** a **cholinergic muscarinic receptor antagonist.**

- The pressure changes produced by sound waves cause the tympanic membrane to move in and out; thus it functions as a resonator to reproduce the vibrations of the sound source. The auditory ossicles serve as a lever system to convert the vibrations of the tympanic membrane into movements of the stapes against the perilymph-filled scala vestibuli of the cochlea.
- The activity within the auditory pathway passes from the eighth cranial nerve afferent fibers to the dorsal and ventral cochlear nuclei to the inferior colliculi to the thalamic medial geniculate body and then to the auditory cortex.
- Loudness is correlated with the amplitude of a sound wave, pitch with the frequency, and timbre with harmonic vibrations.
- Conductive deafness is due to impaired sound transmission in the external or middle ear and impacts all sound frequencies. Sensorineural deafness is usually due to loss of cochlear hair cells but can result from damage to the eighth cranial nerve or central auditory pathway. Conduction and sensorineural deafness can be differentiated by simple tests with a tuning fork.
- Rotational acceleration stimulates the crista in the semicircular canals, displacing the endolymph in a direction opposite to the direction of rotation, deforming the cupula and bending the hair cell. The utricle responds to horizontal acceleration and the saccule to vertical acceleration. Acceleration in any direction displaces the otoliths, distorting the hair cell processes and generating neural activity.
- Spatial orientation is dependent on input from vestibular receptors, visual cues, proprioceptors in joint capsules, and cutaneous touch and pressure receptors.

CHAPTER 11

Smell & Taste

OBJECTIVES

After studying this chapter, you should be able to:

- Describe the basic features of the neural elements in the olfactory epithelium and olfactory bulb.
- Describe signal transduction in odorant receptors.
- Outline the pathway by which impulses generated in the olfactory epithelium reach the olfactory cortex.
- Describe the location and cellular composition of taste buds.
- Name the five major taste receptors and signal transduction mechanisms in these receptors.
- Outline the pathways by which impulses generated in taste receptors reach the insular cortex.

INTRODUCTION

Smell (**olfaction**) and taste (**gustation**) are generally classified as visceral senses because of their close association with gastrointestinal function. Physiologically, they are related to each other. The flavors of various foods are in large part a combination of their taste and smell. Consequently, food may taste "different" if one has a cold that depresses the sense of smell. Both smell and taste receptors are **chemoreceptors** that are stimulated by molecules in solution in mucus in the nose and saliva in the mouth. Because stimuli arrive from an external source, they are also classified as **exteroceptors.** The sensations of smell and taste allow individuals to distinguish between estimates of up to 30 million compounds that are present in food, predators, and mates and to convert the information received into appropriate behaviors.

SMELL

OLFACTORY EPITHELIUM AND OLFACTORY BULBS

Olfactory sensory neurons are located in a specialized portion of the nasal mucosa, the yellowish pigmented **olfactory epithelium.** In dogs and other animals in which the sense of smell is highly developed (macrosmatic animals), the area covered by this membrane is large; in microsmatic animals, such as humans, it is small. In humans, it covers an area of 10 cm^2 in the roof of the nasal cavity near the septum (Figure 11–1). The olfactory epithelium is said to be the place in the body where the nervous system is closest to the external world.

The human olfactory epithelium contains about 50 million bipolar olfactory sensory neurons interspersed with glia-like **supporting (sustentacular) cells** and **basal stem cells.** New olfactory sensory neurons are generated by basal stem cells as needed to replace those damaged by exposure to the environment. The olfactory epithelium is covered by a thin layer of mucus secreted by the supporting cells and **Bowman glands,** which lie beneath the epithelium.

Each olfactory sensory neuron has a short, thick dendrite that projects into the nasal cavity where it terminates in a knob containing 6–12 **cilia** (Figure 11–1). In humans, the cilia are unmyelinated processes, about 5–10 μm long and 0.1–2 μm in diameter that protrude into the mucus overlying the epithelium. **Odorant** molecules (chemicals) dissolve in the mucus and bind to **odorant receptors** on the cilia of olfactory sensory

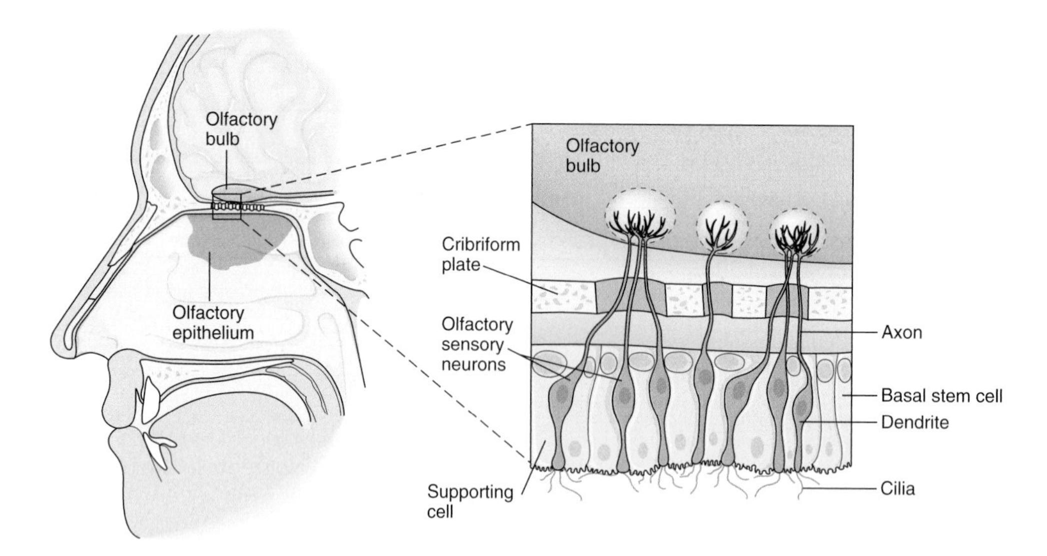

FIGURE 11–1 Structure of the olfactory epithelium. There are three cell types: olfactory sensory neurons, supporting (sustentacular) cells, and basal stem cells at the base of the epithelium. Each olfactory sensory neuron has a dendrite that projects to the epithelial surface. Numerous cilia protrude into the mucus layer lining the nasal lumen. Odorants bind to specific odorant receptors on the cilia and initiate a cascade of events leading to generation of action potentials in the sensory axon. Each olfactory sensory neuron has a single axon that projects to the olfactory bulb, a small ovoid structure that rests on the cribriform plate of the ethmoid bone. (From Kandel ER, Schwartz JH, Jessell TM [editors]: *Principles of Neural Science*, 4th ed. McGraw-Hill, 2000.)

neurons. The mucus provides the appropriate molecular and ionic environment for odor detection.

The axons of the olfactory sensory neurons (first cranial nerve) pass through the **cribriform plate** of the ethmoid bone and enter the **olfactory bulbs** (Figure 11–1). In the olfactory bulbs, the axons of the olfactory sensory neurons contact the primary dendrites of the **mitral cells** and **tufted cells** (Figure 11–2) to form anatomically discrete synaptic units called **olfactory glomeruli.** The olfactory bulbs also contain **periglomerular cells,** which are inhibitory neurons connecting one glomerulus to another, and **granule cells,** which have no axons and make reciprocal synapses with the lateral dendrites of the mitral and tufted cells (Figure 11–2). At these synapses, the mitral or tufted cells excite the granule cell by releasing **glutamate,** and the granule cells in turn inhibit the mitral or tufted cell by releasing **GABA.**

Free endings of many trigeminal pain fibers are found in the olfactory epithelium. They are stimulated by irritating substances, which leads to the characteristic "odor" of such substances as peppermint, menthol, and chlorine. Activation of these endings by nasal irritants also initiates sneezing, lacrimation, respiratory inhibition, and other reflexes.

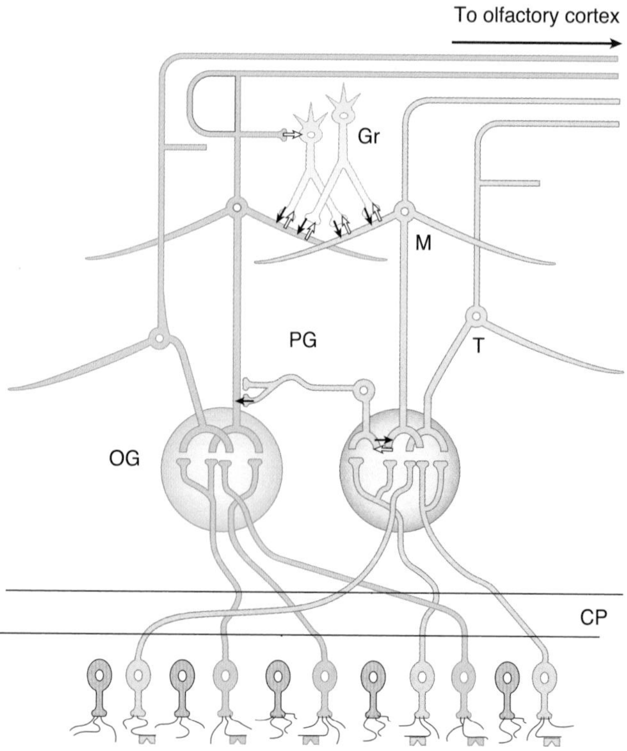

FIGURE 11–2 Basic neural circuits in the olfactory bulb. Note that olfactory receptor cells with one type of odorant receptor project to one olfactory glomerulus (OG) and olfactory receptor cells with another type of receptor project to a different olfactory glomerulus. Solid black arrows signify inhibition via GABA release, and white arrows signify excitatory connections via glutamate release. CP, cribriform plate; Gr, granule cell; M, mitral cell; PG, periglomerular cell; T, tufted cell. (Adapted with permission from Mori K, et al: The olfactory bulb: coding and processing of odor molecular information. Science 1999;286(5440):711–715.)

OLFACTORY CORTEX

The tufted cells are smaller than the mitral cells and they have thinner axons, but they are similar from a functional point of view. The axons of the mitral and tufted cells pass posteriorly through the **lateral olfactory stria** to terminate on apical dendrites of pyramidal cells in five regions of the **olfactory cortex: anterior olfactory nucleus, olfactory tubercle, piriform**

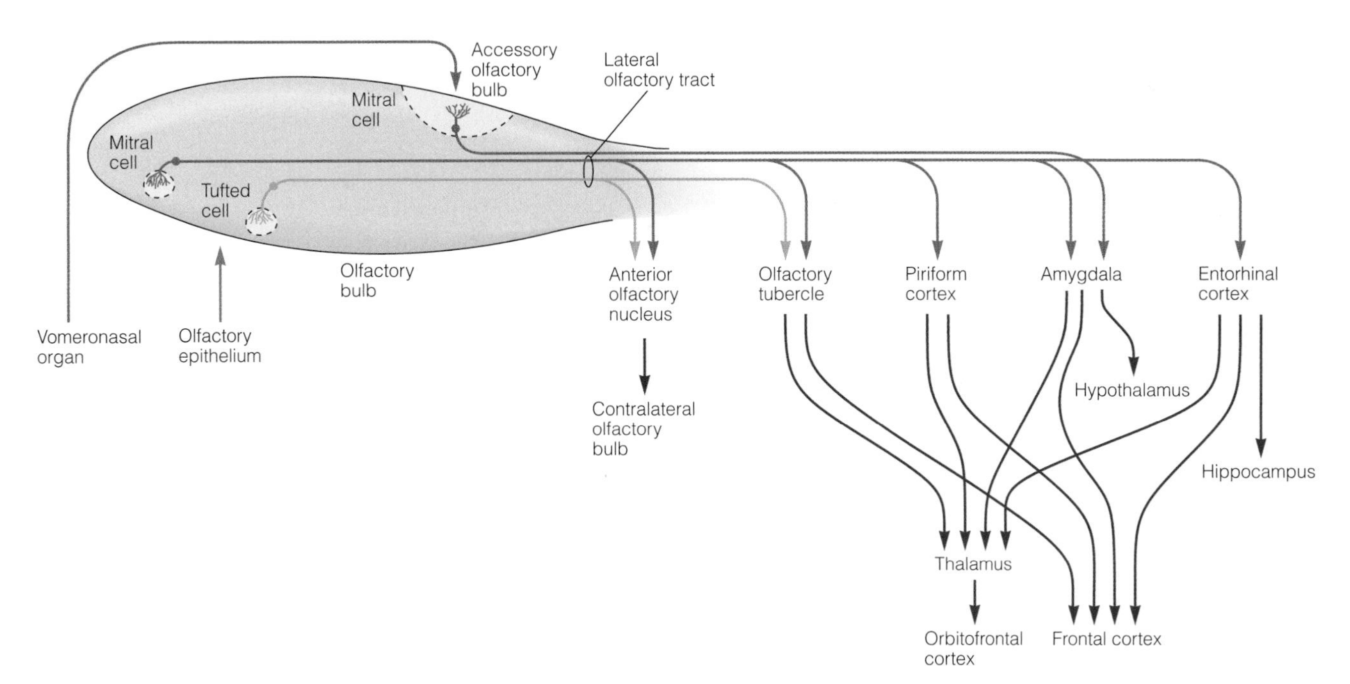

FIGURE 11–3 Diagram of the olfactory pathway. Information is transmitted from the olfactory bulb by axons of mitral and tufted relay neurons in the lateral olfactory tract. Mitral cells project to five regions of the olfactory cortex: anterior olfactory nucleus, olfactory tubercle, piriform cortex, and parts of the amygdala and entorhinal cortex. Tufted cells project to anterior olfactory nucleus and olfactory tubercle; mitral cells in the accessory olfactory bulb project only to the amygdala. Conscious discrimination of odor depends on the neocortex (orbitofrontal and frontal cortices). Emotive aspects of olfaction derive from limbic projections (amygdala and hypothalamus). (From Kandel ER, Schwartz JH, Jessell TM [editors]: *Principles of Neural Science,* 4th ed. McGraw-Hill, 2000.)

cortex, amygdala, and **entorhinal cortex** (Figure 11–3). From these regions, information travels directly to the frontal cortex or via the thalamus to the orbitofrontal cortex. Conscious discrimination of odors is dependent on the pathway to the orbitofrontal cortex. The orbitofrontal activation is generally greater on the right side than the left; thus, cortical representation of olfaction is asymmetric. The pathway to the amygdala is probably involved with the emotional responses to olfactory stimuli, and the pathway to the entorhinal cortex is concerned with olfactory memories.

In rodents and various other mammals, the nasal cavity contains another patch of olfactory epithelium located along the nasal septum in a well-developed **vomeronasal organ.** This structure is concerned with the perception of odors that act as **pheromones.** Vomeronasal sensory neurons project to the **accessory olfactory bulb** (Figure 11–3) and from there to the amygdala and hypothalamus that are concerned with reproduction and ingestive behavior. Vomeronasal input has major effects on these functions. An example is pregnancy block in mice; the pheromones of a male from a different strain prevent pregnancy as a result of mating with that male, but mating with a mouse of the same strain does not produce blockade. The vomeronasal organ has about 100 G protein-coupled odorant receptors that differ in structure from those in the rest of the olfactory epithelium.

The organ is not well developed in humans, but an anatomically separate and biochemically unique area of olfactory epithelium occurs in a pit in the anterior third of the nasal septum, which appears to be a homologous structure. There is evidence for the existence of pheromones in humans, and there is a close relationship between smell and sexual function. Perfume advertisements bear witness to this. The sense of smell is said to be more acute in women than in men, and in women it is most acute at the time of ovulation. Smell, and to a lesser extent, taste, have a unique ability to trigger long-term memories, a fact noted by novelists and documented by experimental psychologists.

ODORANT RECEPTORS AND SIGNAL TRANSDUCTION

The olfactory system has received considerable attention in recent years because of the intriguing biologic question of how a simple sense organ such as the olfactory epithelium and its brain representation, which apparently lacks a high degree of complexity, can mediate discrimination of more than 10,000 different odors. One part of the answer to this question is that there are many different odorant receptors.

There are approximately 500 functional olfactory genes in humans, accounting for about 2% of the human genome. The amino acid sequences of odorant receptors are very diverse, but all the odorant receptors are **G protein coupled receptors** (GPCR). When an odorant molecule binds to its receptor, the G protein subunits (α, β, γ) dissociate (Figure 11–4). The α-subunit activates adenylate cyclase to catalyze the production of cAMP, which acts as a second messenger to open cation channels,

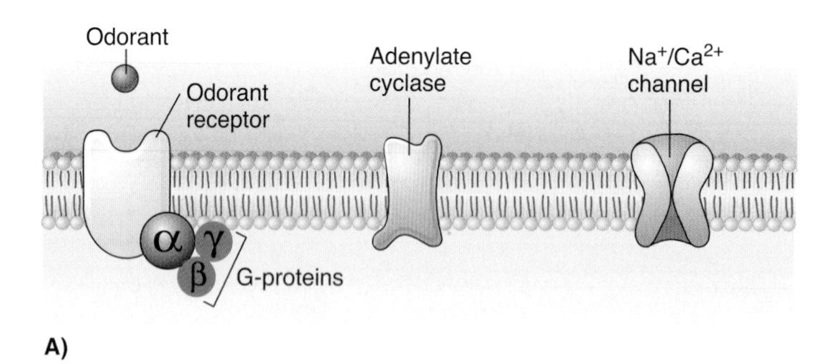

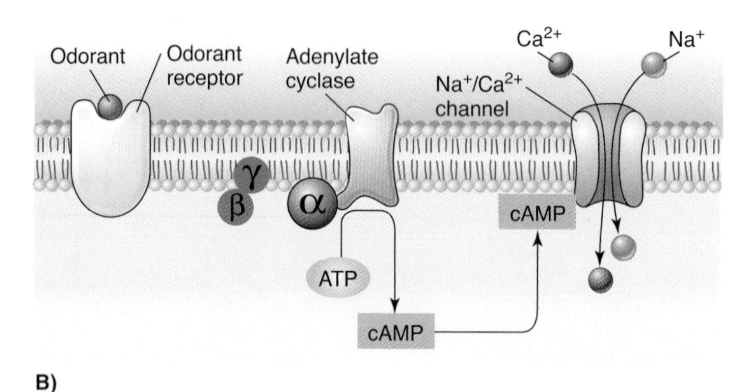

FIGURE 11–4 Signal transduction in an odorant receptor. **A)** Olfactory receptors are examples of G protein-coupled receptors; ie, they are associated with three G protein subunits (α, β, γ). **B)** When an odorant binds to the receptors, the subunits dissociate. The α-subunit of G proteins activates adenylate cyclase to catalyze production of cAMP. cAMP acts as a second messenger to open cation channels. Inward diffusion of Na^+ and Ca^{2+} produces depolarization. (From Fox SI: *Human Physiology*. McGraw-Hill, 2008.)

increasing the permeability to Na^+, K^-, and Ca^{2+}. The net effect is an inward-directed Ca^{2+} current which produces the **graded receptor potential.** This then opens Ca^{2+}-activated Cl^- channels, further depolarizing the cell due to the high intracellular Cl^- levels in olfactory sensory neurons. If the stimulus is sufficient for the receptor potential to exceed its threshold, an action potential in the **olfactory nerve** (first cranial nerve) is triggered.

A second part of the answer to the question of how 10,000 different odors can be detected lies in the neural organization of the olfactory pathway. Although there are millions of olfactory sensory neurons, each expresses only one of the 500 olfactory genes. Each neuron projects to one or two glomeruli (Figure 11–2). This provides a distinct two-dimensional map in the olfactory bulb that is unique to the odorant. The mitral cells with their glomeruli project to different parts of the olfactory cortex.

The olfactory glomeruli demonstrate lateral inhibition mediated by periglomerular cells and granule cells. This sharpens and focuses olfactory signals. In addition, the extracellular field potential in each glomerulus oscillates, and the granule cells appear to regulate the frequency of the oscillation. The exact function of the oscillation is unknown, but it probably also helps to focus the olfactory signals reaching the cortex.

ODOR DETECTION THRESHOLD

Odor-producing molecules **(odorants)** are generally small, containing from 3 to 20 carbon atoms; and molecules with the same number of carbon atoms but different structural configurations have different odors. Relatively high water and lipid solubility are characteristic of substances with strong odors. Some common abnormalities in odor detection are described in Clinical Box 11–1.

The **odor detection thresholds** are the lowest concentration of a chemical that can be detected. The wide range of thresholds illustrates the remarkable sensitivity of the odorant receptors. Some examples of substances detected at very low concentrations include hydrogen sulfide (0.0005 parts per million, ppm), acetic acid (0.016 ppm), kerosene (0.1 ppm), and gasoline (0.3 ppm). On the other end of the spectrum, some toxic substances are essentially odorless; they have odor detection thresholds higher than lethal concentrations. An example is carbon dioxide, which is detected at 74,000 ppm but is lethal at 50,000 ppm. Not all individuals have the same odor detection threshold for a given odorant. While one person may detect and recognize an odorant at a particular concentration, another person may barely notice it.

Olfactory discrimination is remarkable. On the other hand, determination of differences in the intensity of any given odor is poor. The concentration of an odor-producing substance must be changed by about 30% before a difference can be detected. The comparable visual discrimination threshold is a 1% change in light intensity. The direction from which a smell comes may be indicated by the slight difference in the time of arrival of odorant molecules in the two nostrils.

CLINICAL BOX 11–1

Abnormalities in Odor Detection

Anosmia (inability to smell) and **hyposmia** or **hypesthesia** (diminished olfactory sensitivity) can result from simple nasal congestion or nasal polyps. It may also be a sign of a more serious problem such as damage to the olfactory nerves due to fractures of the cribriform plate or head trauma, tumors such as neuroblastomas or meningiomas, and respiratory tract infections (such as abscesses). **Congenital anosmia** is a rare disorder in which an individual is born without the ability to smell. Prolonged use of nasal decongestants can also lead to anosmia, and damage to the olfactory nerves is often seen in patients with Alzheimer disease. According to the National Institutes of Health, 1–2% of the North American population under the age of 65 experiences a significant degree of loss of smell. However, aging is associated with abnormalities in smell sensation; 50% of individuals between the ages of 65 and 80 and >75% of those over the age of 80 have an impaired ability to identify smells. Because of the close relationship between taste and smell, anosmia is associated with a reduction in taste sensitivity (**hypogeusia**). Anosmia is generally permanent in cases in which the olfactory nerve or other neural elements in the olfactory neural pathway are damaged. In addition to not being able to experience the enjoyment of pleasant aromas and a full spectrum of tastes, individuals with anosmia are at risk because they are not able to detect the odor from dangers such as gas leaks, fire, and spoiled food. **Hyperosmia** (enhanced olfactory sensitivity) is less common than loss of smell, but pregnant women commonly become oversensitive to smell. **Dysosmia** (distorted sense of smell) can be caused by several disorders including sinus infections, partial damage to the olfactory nerves, and poor dental hygiene.

THERAPEUTIC HIGHLIGHTS

Quite often anosmia is a temporary condition due to sinus infection or a common cold, but it can be permanent if caused by nasal polyps or trauma. **Antibiotics** can be prescribed to reduce the inflammation caused by polyps and improve the ability to smell. In some cases, surgery is performed to remove the nasal polyps. Topical **corticosteroids** have also been shown to be effective in reversing the loss of smell due to nasal and sinus diseases.

ODORANT-BINDING PROTEINS

The olfactory epithelium contains one or more **odorant-binding proteins (OBP)** that are produced by support cells and released onto the extracellular space. An 18-kDa OBP that is unique to the nasal cavity has been isolated, and other related proteins probably exist. The protein has considerable homology to other proteins in the body that are known to be carriers for small lipophilic molecules. A similar binding protein appears to be associated with taste. These OBP may function in several ways. One, they may concentrate the odorants and transfer them to the receptors. Two, they may partition hydrophobic ligands from the air to an aqueous phase. Three, they may sequester odorants away from the site of odor recognition to allow for odor clearance.

ADAPTATION

It is common knowledge that when one is continuously exposed to even the most disagreeable odor, perception of the odor decreases and eventually ceases. This sometimes beneficent phenomenon is due to the fairly rapid **adaptation,** or **desensitization,** that occurs in the olfactory system. Adaptation in the olfactory system occurs in several stages. The first step may be mediated by a calcium-binding protein (calcium/calmodulin) that binds to the receptor channel protein to lower its affinity for cyclic nucleotides. The next step is called short-term adaptation, which occurs in response to cAMP and implicates a feedback pathway involving calcium/calmodulin-dependent protein kinase II acting on adenylyl cyclase. The next step is called long-term adaptation, which includes activation of guanylate cyclase and cGMP production. A Na^+/Ca^{2+} exchanger to restore ion balance also contributes to long-term adaptation.

TASTE

TASTE BUDS

The specialized sense organ for taste (gustation) consists of approximately 10,000 **taste buds,** which are ovoid bodies measuring 50–70 μm. There are four morphologically distinct types of cells within each taste bud: **basal cells, dark cells, light cells,** and **intermediate cells** (Figure 11–5). The latter three cell types are also referred to as **Type I, II, and III taste cells.** They are the sensory neurons that respond to taste stimuli or **tastants.** Each taste bud has between 50 and 100 taste cells. The three cell types may represent various stages of differentiation of developing taste cells, with the light cells being the most mature. Alternatively, each cell type may represent different cell lineages. The apical ends of taste cells have **microvilli** that project into the taste pore, a small opening on the dorsal surface of the tongue where tastes cells are exposed to the oral contents. Each taste bud is innervated by about 50 nerve fibers, and conversely, each nerve fiber receives input from an average of five taste buds. The basal cells arise from the epithelial

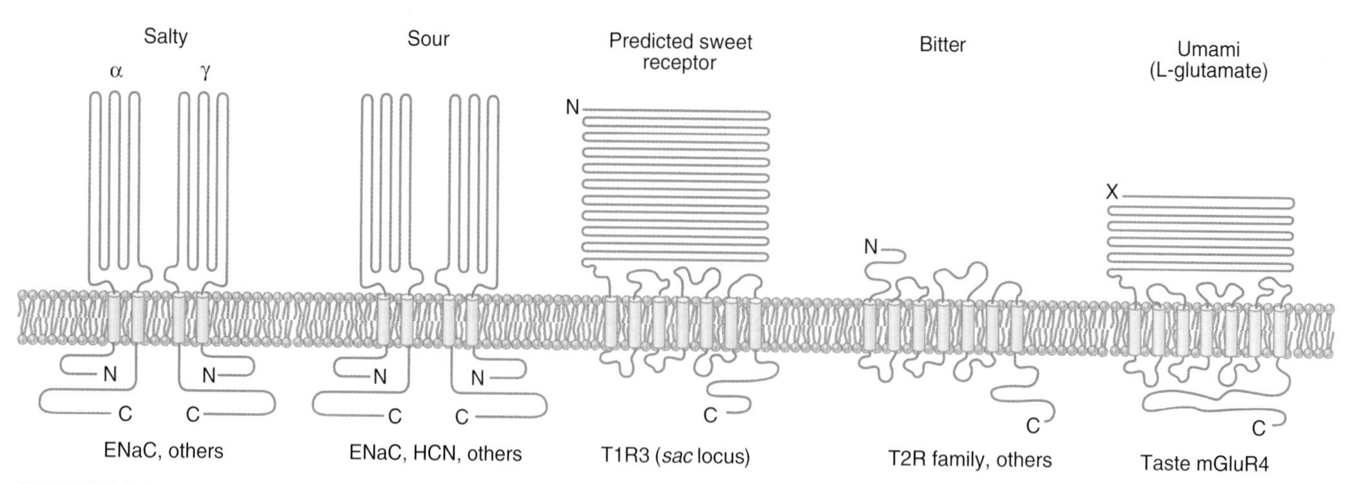

FIGURE 11–7 Signal transduction in taste receptors. Salt-sensitive taste is mediated by a Na^+-selective channel (ENaC); sour taste is mediated by H^+ ions permeable via ENaCs; sweet taste may be dependent on the T1R3 family of G protein-coupled receptors which couple to the G protein gustducin; bitter taste is mediated by the T2R family of G protein-coupled receptors; umami taste is mediated by glutamate acting on a metabotropic receptor, mGluR4. (Adapted with permission from Lindemann B: Receptors and transduction in taste. Nature 2001:413(6852):219–225.)

entirely different structure. Natural sugars such as sucrose and synthetic sweeteners may act on gustducin via different receptors. Sweet-responsive receptors act via cyclic nucleotides and inositol phosphate metabolism.

Bitter taste is produced by a variety of unrelated compounds. Many of these are poisons, and bitter taste serves as a warning to avoid them. Some bitter compounds bind to and block K^+-selective channels. Many GPCR (T2R family) that interact with gustducin are stimulated by bitter substances such as strychnine. Some bitter compounds are membrane permeable and their detection may not involve G proteins; quinine is an example.

Umami taste is due to the activation of a truncated metabotropic glutamate receptor, **mGluR4,** in the taste buds. The way activation of the receptor produces depolarization is unsettled. Glutamate in food may also activate ionotropic glutamate receptors to depolarize umami receptors.

concentrations, preventing accidental ingestion of this chemical, which causes fatal convulsions.

A protein that binds taste-producing molecules has been cloned. It is produced by the von Ebner gland that secretes mucus into the cleft around circumvallate papillae (Figure 11–5) and probably has a concentrating and transport function similar to that of the OBP described for olfaction. Some common abnormalities in taste detection are described in Clinical Box 11–2.

Taste exhibits after reactions and contrast phenomena that are similar in some ways to visual after images and contrasts. Some of these are chemical "tricks," but others may be true central phenomena. A taste modifier protein, **miraculin,** has been discovered in a plant. When applied to the tongue, this protein makes acids taste sweet.

Animals, including humans, form particularly strong aversions to novel foods if eating the food is followed by illness. The survival value of such aversions is apparent in terms of avoiding poisons.

TASTE THRESHOLDS & INTENSITY DISCRIMINATION

The ability of humans to discriminate differences in the intensity of tastes, like intensity discrimination in olfaction, is relatively crude. A 30% change in the concentration of the substance being tasted is necessary before an intensity difference can be detected. **Taste threshold** refers to the minimum concentration at which a substance can be perceived. The threshold concentrations of substances to which the taste buds respond vary with the particular substance (Table 11–1). Bitter substances tend to have the lowest threshold. Some toxic substances such as strychnine have a bitter taste at very low

TABLE 11–1 Some taste thresholds.

Substance	Taste	Threshold Concentration (μmol/L)
Hydrochloric acid	Sour	100
Sodium chloride	Salt	2000
Strychnine hydrochloride	Bitter	1.6
Glucose	Sweet	80,000
Sucrose	Sweet	10,000
Saccharin	Sweet	23

CLINICAL BOX 11–2

Abnormalities in Taste Detection

Ageusia (absence of the sense of taste) and **hypogeusia** (diminished taste sensitivity) can be caused by damage to the lingual or glossopharyngeal nerve. Neurological disorders such as vestibular schwannoma, Bell palsy, familial dysautonomia, multiple sclerosis, certain infections (eg, primary amoeboid meningoencephalopathy), and poor oral hygiene can also cause problems with taste sensitivity. Ageusia can be an adverse side effect of various drugs, including cisplatin and captopril, or vitamin B_3 or zinc deficiencies. Aging and tobacco abuse also contribute to diminished taste. **Dysgeusia** or **parageusia** (unpleasant perception of taste) causes a metallic, salty, foul, or rancid taste. In many cases, dysgeusia is a temporary problem. Factors contributing to ageusia or hypogeusia can also lead to abnormal taste sensitivity. Taste disturbances can also occur under conditions in which **serotonin (5-HT)** and **norepinephrine (NE)** levels are altered (eg, during anxiety or depression). This implies that these neuromodulators contribute to the determination of **taste thresholds**. Administration of a **5-HT reuptake inhibitor** reduces sensitivity to sucrose (sweet taste) and quinine (bitter taste). In contrast, administration of a **NE reuptake inhibitor** reduces bitter taste and sour thresholds. About 25% of the population has a heightened sensitivity to taste, in particular to bitterness. These individuals are called **supertasters**; this may be due to the presence of an increased number of fungiform papillae on their tongue.

THERAPEUTIC HIGHLIGHTS

Improved oral hygiene and adding zinc supplements to one's diet can correct the inability to taste in some individuals.

CHAPTER SUMMARY

- Olfactory sensory neurons, supporting (sustentacular) cells, and basal stem cells are located in the olfactory epithelium within the upper portion of the nasal cavity.
- The cilia located on the dendritic knob of the olfactory sensory neuron contain odorant receptors that are coupled to G proteins. Axons of olfactory sensory neurons contact the dendrites of mitral and tufted cells in the olfactory bulbs to form olfactory glomeruli.
- Information from the olfactory bulb travels via the lateral olfactory stria directly to the olfactory cortex, including the anterior olfactory nucleus, olfactory tubercle, piriform cortex, amygdala, and entorhinal cortex.
- Taste buds are the specialized sense organs for taste and are comprised of basal stem cells and three types of taste cells (dark, light, and intermediate cells). The three types of taste cells may represent various stages of differentiation of developing taste cells, with the light cells being the most mature. Taste buds are located in the mucosa of the epiglottis, palate, and pharynx and in the walls of papillae of the tongue.
- There are taste receptors for sweet, sour, bitter, salt, and umami. Signal transduction mechanisms include passage through ion channels, binding to and blocking ion channels, and GPCR requiring second messenger systems.
- The afferents from taste buds in the tongue travel via the seventh, ninth, and tenth cranial nerves to synapse in the nucleus of the tractus solitarius. From there, axons ascend via the ipsilateral medial lemniscus to the ventral posteromedial nucleus of the thalamus, and on to the anterior insula and frontal operculum in the ipsilateral cerebral cortex.

MULTIPLE-CHOICE QUESTIONS

For all questions, select the single best answer unless otherwise directed.

1. A young boy was diagnosed with congenital anosmia, a rare disorder in which an individual is born without the ability to smell. Odorant receptors are
   A. located in the olfactory bulb.
   B. located on dendrites of mitral and tufted cells.
   C. located on neurons that project directly to the olfactory cortex.
   D. located on neurons in the olfactory epithelium that project to mitral cells and from there directly to the olfactory cortex.
   E. located on sustentacular cells that project to the olfactory bulb.
2. A 37-year-old female was diagnosed with multiple sclerosis. One of the potential consequences of this disorder is diminished taste sensitivity. Taste receptors
   A. for sweet, sour, bitter, salt, and umami are spatially separated on the surface of the tongue.
   B. are synonymous with taste buds.
   C. are a type of chemoreceptor.
   D. are innervated by afferents in the facial, trigeminal, and glossopharyngeal nerves.
   E. all of the above
3. Which of the following does *not* increase the ability to discriminate many different odors?
   A. Many different receptors.
   B. Pattern of olfactory receptors activated by a given odorant.
   C. Projection of different mitral cell axons to different parts of the brain.
   D. High β-arrestin content in olfactory neurons.
   E. Sniffing.
4. As a result of an automobile accident, a 10-year-old boy suffered damage to the brain including the periamygdaloid, piriform, and entorhinal cortices. Which of the following sensory deficits is he most likely to experience?
   A. Visual disturbance
   B. Hyperosmia
   C. Auditory problems
   D. Taste and odor abnormalities
   E. No major sensory deficits

5. Which of the following are *incorrectly* paired?
   A. ENaC : Sour taste
   B. Gustducin : Bitter taste
   C. T1R3 family of GPCR : Sweet taste
   D. Heschel sulcus : Smell
   E. Ebner glands : Taste acuity
6. A 9-year-old boy had frequent episodes of uncontrollable nose bleeds. At the advice of his physician, he underwent surgery to correct a problem in his nasal septum. A few days after the surgery, he told his mother he could not smell the cinnamon rolls she was baking in the oven. Which of the following is true about olfactory transmission?
   A. An olfactory sensory neuron expresses a wide range of odorant receptors.
   B. Lateral inhibition within the olfactory glomeruli reduces the ability to distinguish between different types of odorant receptors.
   C. Conscious discrimination of odors is dependent on the pathway to the orbitofrontal cortex.
   D. Olfaction is closely related to gustation because odorant and gustatory receptors use the same central pathways.
   E. All of the above.
7. A 31-year-old female is a smoker who has had poor oral hygiene for most of her life. In the past few years she has noticed a reduced sensitivity to the flavors in various foods which she used to enjoy eating. Which of the following is *not* true about gustatory sensation?
   A. The sensory nerve fibers from the taste buds on the anterior two-thirds of the tongue travel in the chorda tympani branch of the facial nerve.
   B. The sensory nerve fibers from the taste buds on the posterior third of the tongue travel in the petrosal branch of the glossopharyngeal nerve.
   C. The pathway from taste buds on the left side of the tongue is transmitted ipsilaterally to the cerebral cortex.
   D. Sustentacular cells in the taste buds serve as stem cells to permit growth of new taste buds.
   E. The pathway from taste receptors includes synapses in the nucleus of the tractus solitarius in the brain stem and ventral posterior medial nucleus in the thalamus.
8. A 20-year-old woman was diagnosed with Bell palsy (damage to facial nerve). Which of the following symptoms is she likely to exhibit?
   A. Loss of sense of taste
   B. Facial twitching
   C. Droopy eyelid
   D. Ipsilateral facial paralysis
   E. All of the above

CHAPTER RESOURCES

Adler E, Melichar JK, Nutt DJ, et al: A novel family of mammalian taste receptors. Cell 2000;100:693.

Anholt RRH: Odor recognition and olfactory transduction: The new frontier. Chem Senses 1991;16:421.

Bachmanov AA, Beauchamp GK: Taste receptor genes. Annu Rev Nutrition 2007;27:389.

Gilbertson TA, Damak S, Margolskee RF: The molecular physiology of taste transduction. Curr Opin Neurobiol 2000;10:519.

Gold GH: Controversial issues in vertebrate olfactory transduction. Annu Rev Physiol 1999;61:857.

Heath TP, Melichar JK, Nutt DJ, Donaldson LF. Human taste thresholds are modulated by serotonin and noradrenaline. J Neurosci 2006;26:12664.

Herness HM, Gilbertson TA: Cellular mechanisms of taste transduction. Annu Rev Physiol 1999;61:873.

Kato A, Touhara K. Mammalian olfactory receptors: pharmacology, G protein coupling and desensitization. Cell Mol Life Sci 2009;66:3743.

Lindemann B: Receptors and transduction in taste. Nature 2001;413:219.

Mombaerts P: Genes and ligands for odorant, vomeronasal and taste receptors. Nature Rev Neurosci 2004;5:263.

Reisert J, Restrepo D: Molecular tuning of odorant receptors and its implication for odor signal processing. Chem Senses 2009;34:535.

Ronnett GV, Moon C: G proteins and olfactory signal transduction. Annu Rev Physiol 2002;64:189.

Shepherd GM, Singer MS, Greer CA: Olfactory receptors: A large gene family with broad affinities and multiple functions (Review). Neuroscientist 1996;2:262.

Stern P, Marks J (editors): Making sense of scents. Science 1999;286:703.

CHAPTER 12

Reflex and Voluntary Control of Posture & Movement

OBJECTIVES

After studying this chapter, you should be able to:

- Describe the elements of the stretch reflex and how the activity of γ-motor neurons alters the response to muscle stretch.
- Describe the role of Golgi tendon organs in control of skeletal muscle.
- Describe the elements of the withdrawal reflex.
- Define spinal shock and describe the initial and long-term changes in spinal reflexes that follow transection of the spinal cord.
- Describe how skilled movements are planned and carried out.
- Compare the organization of the central pathways involved in the control of axial (posture) and distal (skilled movement, fine motor movements) muscles.
- Define decerebrate and decorticate rigidity, and comment on the cause and physiologic significance of each.
- Identify the components of the basal ganglia and the pathways that interconnect them, along with the neurotransmitters in each pathway.
- Explain the pathophysiology and symptoms of Parkinson disease and Huntington disease.
- Discuss the functions of the cerebellum and the neurologic abnormalities produced by diseases of this part of the brain.

INTRODUCTION

Somatic motor activity depends ultimately on the pattern and rate of discharge of the spinal motor neurons and homologous neurons in the motor nuclei of the cranial nerves. These neurons, the final common paths to skeletal muscle, are bombarded by impulses from an immense array of descending pathways, other spinal neurons, and peripheral afferents. Some of these inputs end directly on α-motor neurons, but many exert their effects via interneurons or via γ-motor neurons to the muscle spindles and back through the Ia afferent fibers to the spinal cord. It is the integrated activity of these multiple inputs from spinal, medullary, midbrain, and cortical levels that regulates the posture of the body and makes coordinated movement possible.

The inputs converging on motor neurons have three functions: they bring about voluntary activity, they adjust body posture to provide a stable background for movement, and they coordinate the action of the various muscles to make movements smooth and precise. The patterns of voluntary activity are planned within the brain, and the commands are sent to the muscles primarily via the corticospinal and corticobulbar systems. Posture is continually adjusted not only before but also during movement by information carried in descending brain stem pathways and peripheral afferents. Movement is smoothed and coordinated by the medial and intermediate portions of the cerebellum (spinocerebellum) and its connections. The basal ganglia and the lateral portions of the cerebellum (cerebrocerebellum) are part of a feedback circuit to the premotor and motor cortex that is concerned with planning and organizing voluntary movement.

This chapter considers two types of motor output: reflex (involuntary) and voluntary. A subdivision of reflex responses includes some rhythmic movements such as swallowing, chewing, scratching, and walking, which are largely involuntary but subject to voluntary adjustment and control.

GENERAL PROPERTIES OF REFLEXES

The basic unit of integrated reflex activity is the **reflex arc.** This arc consists of a sense organ, an afferent neuron, one or more synapses within a central integrating station, an efferent neuron, and an effector. The afferent neurons enter via the dorsal roots or cranial nerves and have their cell bodies in the dorsal root ganglia or in the homologous ganglia of the cranial nerves. The efferent fibers leave via the ventral roots or corresponding motor cranial nerves.

Activity in the reflex arc starts in a sensory receptor with a **receptor potential** whose magnitude is proportional to the strength of the stimulus (Figure 12–1). This generates all-or-none action potentials in the afferent nerve, the number of action potentials being proportional to the size of the receptor potential. In the central nervous system (CNS), the responses are again graded in terms of excitatory postsynaptic potentials (EPSPs) and inhibitory postsynaptic potentials (IPSPs) at the synaptic junctions. All-or-none responses (action potentials) are generated in the efferent nerve. When these reach the effector, they again set up a graded response. When the effector is smooth muscle, responses summate to produce action potentials in the smooth muscle, but when the effector is skeletal muscle, the graded response is adequate to produce action potentials that bring about muscle contraction. The connection between the afferent and efferent neurons is in the CNS, and activity in the reflex arc is modified by the multiple inputs converging on the efferent neurons or at any synaptic station within the reflex arc.

The stimulus that triggers a reflex is generally very precise. This stimulus is called the **adequate stimulus** for the particular reflex. A dramatic example is the scratch reflex in the dog. This spinal reflex is adequately stimulated by multiple linear touch stimuli such as those produced by an insect crawling across the skin. The response is vigorous scratching of the area stimulated. If the multiple touch stimuli are widely separated or not in a line, the adequate stimulus is not produced and no scratching occurs. Fleas crawl, but they also jump from place to place. This jumping separates the touch stimuli so that an adequate stimulus for the scratch reflex is not produced.

Reflex activity is stereotyped and specific in that a particular stimulus elicits a particular response. The fact that reflex responses are stereotyped does not exclude the possibility of their being modified by experience. Reflexes are adaptable and can be modified to perform motor tasks and maintain balance. Descending inputs from higher brain regions play an important role in modulating and adapting spinal reflexes.

The **α-motor neurons** that supply the extrafusal fibers in skeletal muscles are the efferent side of many reflex arcs. All neural influences affecting muscular contraction ultimately funnel through them to the muscles, and they are therefore called the **final common pathway.** Numerous inputs converge on α-motor neurons. Indeed, the surface of the average motor neuron and its dendrites accommodates about 10,000 synaptic knobs. At least five inputs go from the same spinal segment to a typical spinal motor neuron. In addition to these, there are excitatory and inhibitory inputs, generally relayed via interneurons, from other levels of the spinal cord and multiple long-descending tracts from the brain. All of these pathways converge on and determine the activity in the final common pathways.

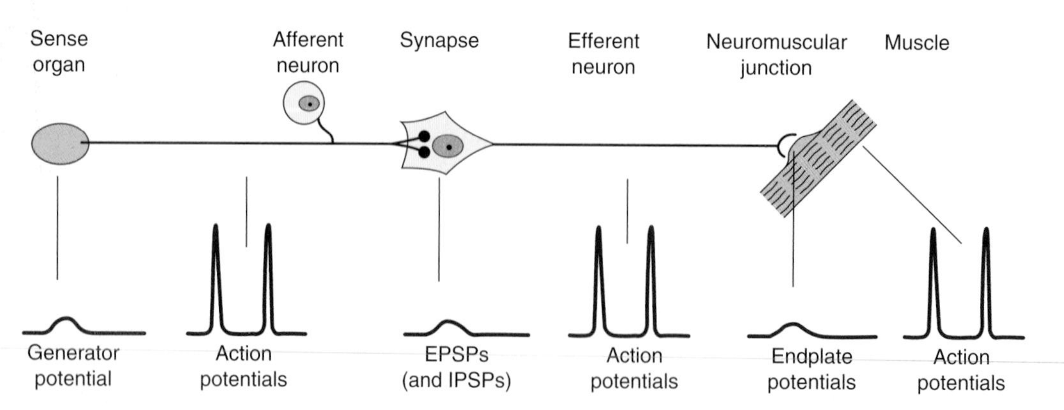

FIGURE 12–1 The reflex arc. Note that at the receptor and in the CNS a nonpropagated graded response occurs that is proportional to the magnitude of the stimulus. The response at the neuromuscular junction is also graded, though under normal conditions it is always large enough to produce a response in skeletal muscle. On the other hand, in the portions of the arc specialized for transmission (afferent and efferent nerve fibers, muscle membrane), the responses are all-or-none action potentials.

MONOSYNAPTIC REFLEXES: THE STRETCH REFLEX

The simplest reflex arc is one with a single synapse between the afferent and efferent neurons, and reflexes occurring in them are called **monosynaptic reflexes.** Reflex arcs in which interneurons are interposed between the afferent and efferent neurons are called **polysynaptic reflexes.** There can be anywhere from two to hundreds of synapses in a polysynaptic reflex arc.

When a skeletal muscle with an intact nerve supply is stretched, it contracts. This response is called the **stretch reflex** or **myotatic reflex.** The stimulus that initiates this reflex is stretch of the muscle, and the response is contraction of the muscle being stretched. The sense organ is a small encapsulated spindlelike or fusiform shaped structure called the **muscle spindle,** located within the fleshy part of the muscle. The impulses originating from the spindle are transmitted to the CNS by fast sensory fibers that pass directly to the motor neurons that supply the same muscle. The neurotransmitter at the central synapse is glutamate. The stretch reflex is the best known and studied monosynaptic reflex and is typified by the **knee jerk reflex** (see Clinical Box 12–1).

CLINICAL BOX 12–1

Knee Jerk Reflex

Tapping the patellar tendon elicits the **knee jerk,** a stretch reflex of the quadriceps femoris muscle, because the tap on the tendon stretches the muscle. A similar contraction is observed if the quadriceps is stretched manually. Stretch reflexes can also be elicited from most of the large muscles of the body. Tapping on the tendon of the triceps brachii, for example, causes an extensor response at the elbow as a result of reflex contraction of the triceps; tapping on the Achilles tendon causes an ankle jerk due to reflex contraction of the gastrocnemius; and tapping on the side of the face causes a stretch reflex in the masseter. The knee jerk reflex is an example of a **deep tendon reflex (DTR)** in a neurological exam and is graded on the following scale: 0 (absent), 1+ (hypoactive), 2+ (brisk, normal), 3+ (hyperactive without clonus), 4+ (hyperactive with mild clonus), and 5+ (hyperactive with sustained clonus). Absence of the knee jerk can signify an abnormality anywhere within the reflex arc, including the muscle spindle, the Ia afferent nerve fibers, or the motor neurons to the quadriceps muscle. The most common cause is a peripheral neuropathy from such things as diabetes, alcoholism, and toxins. A hyperactive reflex can signify an interruption of corticospinal and other descending pathways that influence the reflex arc.

STRUCTURE OF MUSCLE SPINDLES

Each muscle spindle has three essential elements: (1) a group of specialized **intrafusal muscle fibers** with contractile polar ends and a noncontractile center, (2) large diameter myelinated afferent nerves (types Ia and II) originating in the central portion of the intrafusal fibers, and (3) small diameter myelinated efferent nerves supplying the polar contractile regions of the intrafusal fibers (Figure 12–2A). It is important to understand the relationship of these elements to each other and to the muscle itself to appreciate the role of this sense organ in signaling changes in the length of the muscle in which it is located. Changes in muscle length are associated with changes in joint angle; thus muscle spindles provide information on position (ie, **proprioception**).

The **intrafusal fibers** are positioned in parallel to the **extrafusal fibers** (the regular contractile units of the muscle) with the ends of the spindle capsule attached to the tendons at either end of the muscle. Intrafusal fibers do not contribute to the overall contractile force of the muscle, but rather serve a pure sensory function. There are two types of intrafusal fibers in mammalian muscle spindles. The first type contains many nuclei in a dilated central area and is called a **nuclear bag fiber** (Figure 12–2B). There are two subtypes of nuclear bag fibers, **dynamic** and **static.** The second intrafusal fiber type, the **nuclear chain fiber,** is thinner and shorter and lacks a definite bag. Typically, each muscle spindle contains two or three nuclear bag fibers and about five nuclear chain fibers.

There are two kinds of sensory endings in each spindle, a single **primary (group Ia) ending** and up to eight **secondary (group II) endings** (Figure 12–2B). The Ia afferent fiber wraps around the center of the dynamic and static nuclear bag fibers and nuclear chain fibers. Group II sensory fibers are located adjacent to the centers of the static nuclear bag and nuclear chain fibers; these fibers do not innervate the dynamic nuclear bag fibers. Ia afferents are very sensitive to the velocity of the change in muscle length during a stretch **(dynamic response);** thus they provide information about the speed of movements and allow for quick corrective movements. The steady-state (tonic) activity of group Ia and II afferents provide information on steady-state length of the muscle **(static response).** The top trace in Figure 12–2C shows the dynamic and static components of activity in a Ia afferent during muscle stretch. Note that they discharge most rapidly while the muscle is being stretched (shaded area of graphs) and less rapidly during sustained stretch.

The spindles have a motor nerve supply of their own. These nerves are 3–6 μm in diameter, constitute about 30% of the fibers in the ventral roots, and are called **γ-motor neurons.** There are two types of γ-motor neurons: **dynamic,** which supply the dynamic nuclear bag fibers and **static,** which supply the static nuclear bag fibers and the nuclear chain fibers. Activation of dynamic γ-motor neurons increases the dynamic sensitivity of the group Ia endings. Activation of the static

EFFECTS OF γ-MOTOR NEURON DISCHARGE

Stimulation of γ-motor neurons produces a very different picture from that produced by stimulation of the α-motor neurons. Stimulation of γ-motor neurons does not lead directly to detectable contraction of the muscles because the intrafusal fibers are not strong enough or plentiful enough to cause shortening. However, stimulation does cause the contractile ends of the intrafusal fibers to shorten and therefore stretches the nuclear bag portion of the spindles, deforming the endings, and initiating impulses in the Ia fibers (Figure 12–4). This in turn can lead to reflex contraction of the muscle. Thus, muscles can be made to contract via stimulation of the α-motor neurons that innervate the extrafusal fibers or the γ-motor neurons that initiate contraction indirectly via the stretch reflex.

If the whole muscle is stretched during stimulation of the γ-motor neurons, the rate of discharge in the Ia fibers is further increased (Figure 12–4). Increased γ-motor neuron activity thus increases **spindle sensitivity** during stretch.

In response to descending excitatory input to spinal motor circuits, both α- and γ-motor neurons are activated. Because of this **"α–γ coactivation,"** intrafusal and extrafusal fibers shorten together, and spindle afferent activity can occur throughout the period of muscle contraction. In this way, the spindle remains capable of responding to stretch and reflexively adjusting α-motor neuron discharge.

CONTROL OF γ-MOTOR NEURON DISCHARGE

The γ-motor neurons are regulated to a large degree by descending tracts from a number of areas in the brain that also control α-motor neurons (described below). Via these pathways, the sensitivity of the muscle spindles and hence the threshold of the stretch reflexes in various parts of the body can be adjusted and shifted to meet the needs of postural control.

Other factors also influence γ-motor neuron discharge. Anxiety causes an increased discharge, a fact that probably explains the hyperactive tendon reflexes sometimes seen in anxious patients. In addition, unexpected movement is associated with a greater efferent discharge. Stimulation of the skin, especially by noxious agents, increases γ-motor neuron discharge to ipsilateral flexor muscle spindles while decreasing that to extensors and produces the opposite pattern in the opposite limb. It is well known that trying to pull the hands apart when the flexed fingers are hooked together facilitates the knee jerk reflex **(Jendrassik's maneuver),** and this may also be due to increased γ-motor neuron discharge initiated by afferent impulses from the hands.

RECIPROCAL INNERVATION

When a stretch reflex occurs, the muscles that antagonize the action of the muscle involved (antagonists) relax. This phenomenon is said to be due to **reciprocal innervation.** Impulses in the Ia fibers from the muscle spindles of the protagonist muscle cause postsynaptic inhibition of the motor neurons to the antagonists. The pathway mediating this effect is bisynaptic. A collateral from each Ia fiber passes in the spinal cord to an inhibitory interneuron that synapses on a motor neuron supplying the antagonist muscles. This example of postsynaptic inhibition is discussed in Chapter 6, and the pathway is illustrated in Figure 6–6.

INVERSE STRETCH REFLEX

Up to a point, the harder a muscle is stretched, the stronger is the reflex contraction. However, when the tension becomes great enough, contraction suddenly ceases and the muscle relaxes. This relaxation in response to strong stretch is called the **inverse stretch reflex.** The receptor for the inverse stretch reflex is in the **Golgi tendon organ** (Figure 12–5). This organ consists of a netlike collection of knobby nerve endings among the fascicles of a tendon. There are 3–25 muscle fibers per tendon organ. The fibers from the Golgi tendon

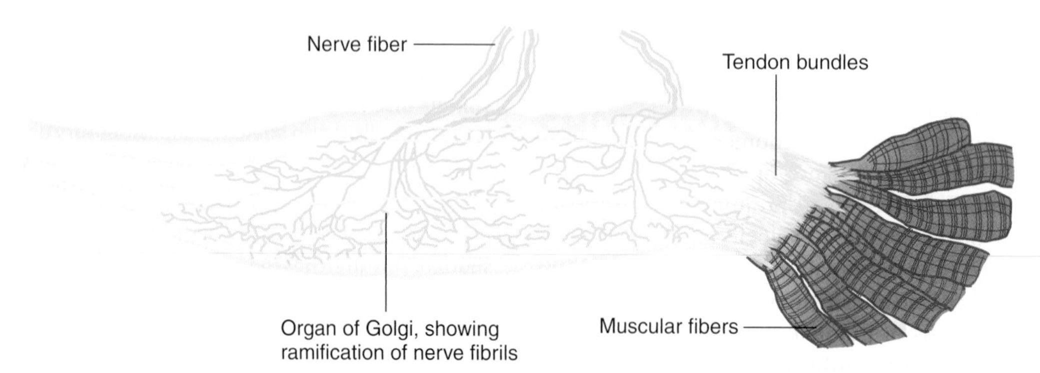

FIGURE 12–5 Golgi tendon organ. This organ is the receptor for the inverse stretch reflex and consists of a netlike collection of knobby nerve endings among the fascicles of a tendon. The innervation is the Ib group of myelinated, rapidly conducting sensory nerve fibers. (Reproduced with permission from Gray H [editor]: *Gray's Anatomy of the Human Body,* 29th ed. Lea & Febiger, 1973.)

CLINICAL BOX 12–2

Clonus

A characteristic of states in which increased γ-motor neuron discharge is present is **clonus.** This neurologic sign is the occurrence of regular, repetitive, rhythmic contractions of a muscle subjected to sudden, maintained stretch. Only sustained clonus with five or more beats is considered abnormal. Ankle clonus is a typical example. This is initiated by brisk, maintained dorsiflexion of the foot, and the response is rhythmic plantar flexion at the ankle. The **stretch reflex–inverse stretch reflex sequence** may contribute to this response. However, it can occur on the basis of synchronized motor neuron discharge without Golgi tendon organ discharge. The spindles of the tested muscle are hyperactive, and the burst of impulses from them discharges all the motor neurons supplying the muscle at once. The consequent muscle contraction stops spindle discharge. However, the stretch has been maintained, and as soon as the muscle relaxes it is again stretched and the spindles stimulated. Clonus may also occur after disruption of descending cortical input to a spinal glycinergic inhibitory interneuron called the **Renshaw cell.** This cell receives excitatory input from α-motor neurons via axon collaterals (and in turn it inhibits the same). In addition, cortical fibers activating ankle flexors contact Renshaw cells (as well as type Ia inhibitory interneurons) that inhibit the antagonistic ankle extensors. This circuitry prevents reflex stimulation of the extensors when flexors are active. Therefore, when the descending cortical fibers are damaged **(upper motor neuron lesion),** the inhibition of antagonists is absent. The result is repetitive, sequential contraction of ankle flexors and extensors (clonus). Clonus may be seen in patients with amyotrophic lateral sclerosis, stroke, multiple sclerosis, spinal cord damage, epilepsy, liver or kidney failure, and hepatic encephalopathy.

THERAPEUTIC HIGHLIGHTS

Since there are numerous causes of clonus, treatment centers on the underlying cause. For some individuals, stretching exercises can reduce episodes of clonus. **Immunosuppressants** (eg, **azathioprine** and **corticosteroids**), **anticonvulsants** (eg, **primidone** and **levetiracetam**), and **tranquilizers** (eg, **clonazepam**) have been shown to be beneficial in the treatment of clonus. **Botulinum toxin** has also been used to block the release of acetylcholine in the muscle, which triggers the rhythmic muscle contractions.

organs make up the Ib group of myelinated, rapidly conducting sensory nerve fibers. Stimulation of these Ib fibers leads to the production of IPSPs on the motor neurons that supply the muscle from which the fibers arise. The Ib fibers end in the spinal cord on inhibitory interneurons that in turn terminate directly on the motor neurons (Figure 12–3). They also make excitatory connections with motor neurons supplying antagonists to the muscle.

Because the Golgi tendon organs, unlike the spindles, are in series with the muscle fibers, they are stimulated by both passive stretch and active contraction of the muscle. The threshold of the Golgi tendon organs is low. The degree of stimulation by passive stretch is not great because the more elastic muscle fibers take up much of the stretch, and this is why it takes a strong stretch to produce relaxation. However, discharge is regularly produced by contraction of the muscle, and the Golgi tendon organ thus functions as a transducer in a feedback circuit that regulates muscle force in a fashion analogous to the spindle feedback circuit that regulates muscle length.

The importance of the primary endings in the spindles and the Golgi tendon organs in regulating the velocity of the muscle contraction, muscle length, and muscle force is illustrated by the fact that section of the afferent nerves to an arm causes the limb to hang loosely in a semiparalyzed state. The interaction of spindle discharge, tendon organ discharge, and reciprocal innervation determines the rate of discharge of α-motor neurons (see Clinical Box 12–2).

MUSCLE TONE

The resistance of a muscle to stretch is often referred to as its **tone** or **tonus.** If the motor nerve to a muscle is severed, the muscle offers very little resistance and is said to be **flaccid.** A **hypertonic (spastic)** muscle is one in which the resistance to stretch is high because of hyperactive stretch reflexes. Somewhere between the states of flaccidity and spasticity is the ill-defined area of normal tone. The muscles are generally **hypotonic** when the rate of γ-motor neuron discharge is low and hypertonic when it is high.

When the muscles are hypertonic, the sequence of moderate stretch → muscle contraction, strong stretch → muscle relaxation is clearly seen. Passive flexion of the elbow, for example, meets immediate resistance as a result of the stretch reflex in the triceps muscle. Further stretch activates the inverse stretch reflex. The resistance to flexion suddenly collapses, and the arm flexes. Continued passive flexion stretches the muscle again, and the sequence may be repeated. This sequence of resistance followed by give when a limb is moved passively is known as the **clasp-knife effect** because of its resemblance to the closing of a pocket knife. It is also known as the **lengthening reaction** because it is the response of a spastic muscle to lengthening.

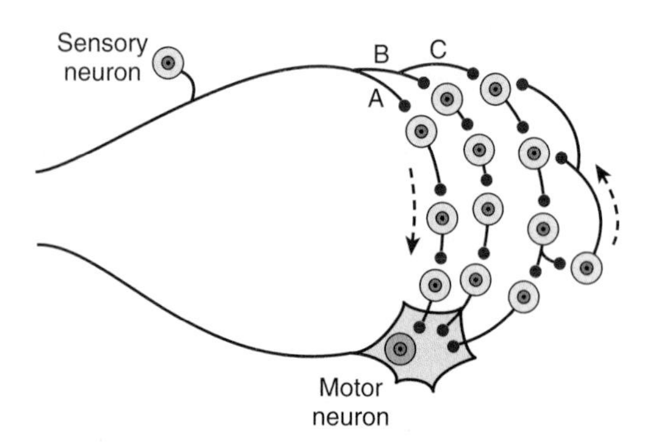

FIGURE 12–6 Diagram of polysynaptic connections between afferent and efferent neurons in the spinal cord. The dorsal root fiber activates pathway A with three interneurons, pathway B with four interneurons, and pathway C with four interneurons. Note that one of the interneurons in pathway C connects to a neuron that doubles back to other interneurons, forming reverberating circuits.

POLYSYNAPTIC REFLEXES: THE WITHDRAWAL REFLEX

Polysynaptic reflex paths branch in a complex fashion (Figure 12–6). The number of synapses in each of their branches varies. Because of the synaptic delay at each synapse, activity in the branches with fewer synapses reaches the motor neurons first, followed by activity in the longer pathways. This causes prolonged bombardment of the motor neurons from a single stimulus and consequently prolonged responses. Furthermore, some of the branch pathways turn back on themselves, permitting activity to reverberate until it becomes unable to cause a propagated transsynaptic response and dies out. Such **reverberating circuits** are common in the brain and spinal cord.

The **withdrawal reflex** is a typical polysynaptic reflex that occurs in response to a noxious stimulus to the skin or subcutaneous tissues and muscle. The response is flexor muscle contraction and inhibition of extensor muscles, so that the body part stimulated is flexed and withdrawn from the stimulus. When a strong stimulus is applied to a limb, the response includes not only flexion and withdrawal of that limb but also extension of the opposite limb. This **crossed extensor response** is properly part of the withdrawal reflex. Strong stimuli can generate activity in the interneuron pool that spreads to all four extremities. This spread of excitatory impulses up and down the spinal cord to more and more motor neurons is called **irradiation of the stimulus,** and the increase in the number of active motor units is called **recruitment of motor units.**

IMPORTANCE OF THE WITHDRAWAL REFLEX

Flexor responses can be produced by innocuous stimulation of the skin or by stretch of the muscle, but strong flexor responses with withdrawal are initiated only by stimuli that are noxious or at least potentially harmful (ie, **nociceptive stimuli**). The withdrawal reflex serves a protective function as flexion of the stimulated limb gets it away from the source of irritation, and extension of the other limb supports the body. The pattern assumed by all four extremities puts one in position to escape from the offending stimulus. Withdrawal reflexes are **prepotent;** that is, they preempt the spinal pathways from any other reflex activity taking place at the moment.

Many of the characteristics of polysynaptic reflexes can be demonstrated by studying the withdrawal reflex. A weak noxious stimulus to one foot evokes a minimal flexion response; stronger stimuli produce greater and greater flexion as the stimulus irradiates to more and more of the motor neuron pool supplying the muscles of the limb. Stronger stimuli also cause a more prolonged response. A weak stimulus causes one quick flexion movement; a strong stimulus causes prolonged flexion and sometimes a series of flexion movements. This prolonged response is due to prolonged, repeated firing of the motor neurons. The repeated firing is called **after-discharge** and is due to continued bombardment of motor neurons by impulses arriving by complicated and circuitous polysynaptic paths.

As the strength of a noxious stimulus is increased, the reaction time is shortened. Spatial and temporal facilitation occurs at synapses in the polysynaptic pathway. Stronger stimuli produce more action potentials per second in the active branches and cause more branches to become active; summation of the EPSPs to the threshold level for action potential generation occurs more rapidly.

FRACTIONATION & OCCLUSION

Another characteristic of the withdrawal response is the fact that supramaximal stimulation of any of the sensory nerves from a limb never produces as strong a contraction of the flexor muscles as that elicited by direct electrical stimulation of the muscles themselves. This indicates that the afferent inputs **fractionate** the motor neuron pool; that is, each input goes to only part of the motor neuron pool for the flexors of that particular extremity. On the other hand, if all the sensory inputs are dissected out and stimulated one after the other, the sum of the tension developed by stimulation of each is greater than that produced by direct electrical stimulation of the muscle or stimulation of all inputs at once. This indicates that the various afferent inputs share some of the motor neurons and that **occlusion** occurs when all inputs are stimulated at once.

SPINAL INTEGRATION OF REFLEXES

The responses of animals and humans to **spinal cord injury (SCI)** illustrate the integration of reflexes at the spinal level. The deficits seen after SCI vary, of course, depending on the level of the injury. Clinical Box 12–3 provides information on long-term problems related to SCI and recent advancements in treatment options.

CLINICAL BOX 12–3

Spinal Cord Injury

It has been estimated that the worldwide annual incidence of sustaining **spinal cord injury (SCI)** is between 10 and 83 per million of the population. Leading causes are vehicle accidents, violence, and sports injuries. The mean age of patients who sustain an SCI is 33 years old, and men outnumber women with a nearly 4:1 ratio. Approximately 52% of SCI cases result in quadriplegia and about 42% lead to paraplegia. In quadriplegic humans, the threshold of the withdrawal reflex is very low; even minor noxious stimuli may cause not only prolonged withdrawal of one extremity but marked flexion–extension patterns in the other three limbs. Stretch reflexes are also hyperactive. Afferent stimuli irradiate from one reflex center to another after SCI. When even a relatively minor noxious stimulus is applied to the skin, it may activate autonomic neurons and produce evacuation of the bladder and rectum, sweating, pallor, and blood pressure swings in addition to the withdrawal response. This distressing **mass reflex** can however sometimes be used to give paraplegic patients a degree of bladder and bowel control. They can be trained to initiate urination and defecation by stroking or pinching their thighs, thus producing an intentional mass reflex. If the cord section is incomplete, the flexor spasms initiated by noxious stimuli can be associated with bursts of pain that are particularly bothersome. They can be treated with considerable success with baclofen, a $GABA_B$ receptor agonist that crosses the blood–brain barrier and facilitates inhibition.

THERAPEUTIC HIGHLIGHTS

Treatment of SCI patients presents complex problems. Administration of large doses of **glucocorticoids** has been shown to foster recovery and minimize loss of function after SCI. They need to be given soon after the injury and then discontinued because of the well-established deleterious effects of long-term steroid treatment. Their immediate value is likely due to reduction of the inflammatory response in the damaged tissue. Due to immobilization, SCI patients develop a negative nitrogen balance and catabolize large amounts of body protein. Their body weight compresses the circulation to the skin over bony prominences, causing **decubitus ulcers** to form. The ulcers heal poorly and are prone to infection because of body protein depletion. The tissues that are broken down include the protein matrix of bone and this, plus the immobilization, cause Ca^{2+} to be released in large amounts, leading to **hypercalcemia, hypercalciuria,** and formation of **calcium stones** in the urinary tract. The combination of stones and bladder paralysis cause urinary stasis, which predisposes to **urinary tract infection,** the most common complication of SCI. The search continues for ways to get axons of neurons in the spinal cord to regenerate. Administration of **neurotrophins** shows some promise in experimental animals, and so does implantation of **embryonic stem cells** at the site of injury. Another possibility being explored is bypassing the site of SCI with **brain-computer interface devices.** However, these novel approaches are a long way from routine clinical use.

In all vertebrates, transection of the spinal cord is followed by a period of **spinal shock** during which all spinal reflex responses are profoundly depressed. Subsequently, reflex responses return and become hyperactive. The duration of spinal shock is proportional to the degree of encephalization of motor function in the various species. In frogs and rats it lasts for minutes; in dogs and cats it lasts for 1–2 h; in monkeys it lasts for days; and in humans it usually lasts for a minimum of 2 weeks.

Cessation of tonic bombardment of spinal neurons by excitatory impulses in descending pathways (see below) undoubtedly plays a role in development of spinal shock. In addition, spinal inhibitory interneurons that normally are themselves inhibited may be released from this descending inhibition to become disinhibited. This, in turn, would inhibit motor neurons. The recovery of reflex excitability may be due to the development of denervation hypersensitivity to the mediators released by the remaining spinal excitatory endings. Another contributing factor may be sprouting of collaterals from existing neurons, with the formation of additional excitatory endings on interneurons and motor neurons.

The first reflex response to appear as spinal shock wears off in humans is often a slight contraction of the leg flexors and adductors in response to a noxious stimulus (ie, the withdrawal reflex). In some patients, the knee jerk reflex recovers first. The interval between cord transection and the return of reflex activity is about 2 weeks in the absence of any complications, but if complications are present it is much longer. Once the spinal reflexes begin to reappear after spinal shock, their threshold steadily drops.

Circuits intrinsic to the spinal cord can produce walking movements when stimulated in a suitable fashion even after spinal cord transection in cats and dogs. There are two **locomotor pattern generators** in the spinal cord: one in the cervical region and one in the lumbar region. However, this does not mean that spinal animals or humans can walk without

CLINICAL BOX 12–4

Cerebral Palsy

Cerebral palsy (CP) is a term used to describe any one of several nonprogressive neurological disorders that occur before or during childbirth or during early childhood. Prenatal factors, including exposure of the developing brain to hypoxia, infections, or toxins, may account for 70–80% of cases of CP. Typical symptoms of the disorder include spasticity, ataxia, deficits in fine motor control, and abnormal gait (crouched or "scissored gait"). Sensory deficits including loss of vision and hearing as well as learning difficulties and seizures often occur in children with CP. In developed countries, the prevalence of CP is 2–2.5 cases per 1000 live births; however, the incidence of CP in children who are born prematurely is much higher compared with children born at term. Based on differences in the resting tone in muscles and the limbs involved, CP is classified into different groups. The most prevalent type is **spastic CP,** which is characterized by **spasticity, hyperreflexia,** clonus, and a **positive Babinski sign.** These are all signs of damage to the corticospinal tract (see Clinical Box 12–5). **Dyskinetic CP** is characterized by abnormal involuntary movements (**chorea** and **athetosis**) and is thought to reflect damage to extrapyramidal motor areas. It is not uncommon to have signs of both types of CP to coexist. The rarest type is **hypotonic CP,** which presents with truncal and extremity hypotonia, hyperreflexia, and persistent primitive reflexes.

THERAPEUTIC HIGHLIGHTS

There is no cure for CP. Treatment often includes physical and occupational therapy. Botulinum toxin injections into affected muscles have been used to reduce muscle spasticity, especially in the gastrocnemius muscle. Other drugs used to treat muscle spasticity in patients with CP include **diazepam** (a benzodiazepine that binds to the $GABA_A$ receptor), **baclofen** (an agonist at presynaptic $GABA_B$ receptors in the spinal cord), and **dantrolene** (a direct muscle relaxant). Various surgeries have been used to treat CP, including **selective dorsal rhizotomy** (section of the dorsal roots) and **tenotomy** (severing the tendon in the gastrocnemius muscles).

stimulation; the pattern generator has to be turned on by tonic discharge of a discrete area in the midbrain, the **mesencephalic locomotor region,** and, of course, this is only possible in patients with incomplete spinal cord transection. Interestingly, the generators can also be turned on in experimental animals by administration of the norepinephrine precursor L-dopa (levodopa) after complete section of the spinal cord. Progress is being made in teaching humans with SCI to take a few steps by placing them, with support, on a treadmill.

GENERAL PRINCIPLES OF CENTRAL ORGANIZATION OF MOTOR PATHWAYS

To voluntarily move a limb, the brain must plan a movement, arrange appropriate motion at many different joints at the same time, and adjust the motion by comparing plan with performance. The motor system "learns by doing" and performance improves with repetition. This involves synaptic plasticity. Damage to the cerebral cortex before or during childbirth or during the first 2–3 years of development can lead to cerebral palsy, a disorder that affects muscle tone, movement, and coordination (Clinical Box 12–4).

There is considerable evidence for the general motor control scheme shown in Figure 12–7. Commands for voluntary movement originate in cortical association areas. The movements are planned in the cortex as well as in the basal ganglia and the lateral portions of the cerebellar hemispheres, as indicated by increased electrical activity before the movement. The basal ganglia and cerebellum funnel information to the premotor and motor cortex by way of the thalamus. Motor commands from the motor cortex are relayed in large part via the corticospinal tracts to the spinal cord and the corresponding corticobulbar tracts to motor neurons in the brain stem. However, collaterals from these pathways and a few direct connections from the motor cortex end on brain stem nuclei, which also project to motor neurons in the brain stem and spinal cord. These pathways can also mediate voluntary movement. Movement sets up alterations in sensory input from the special senses and from muscles, tendons, joints, and the skin. This feedback information, which adjusts and smoothes movement, is relayed directly to the motor cortex and to the spinocerebellum. The spinocerebellum projects in turn to the brain stem. The main brain stem pathways that are concerned with posture and coordination are the rubrospinal, reticulospinal, tectospinal, and vestibulospinal tracts.

MOTOR CORTEX & VOLUNTARY MOVEMENT

PRIMARY MOTOR CORTEX

The reader can refer to Figure 8–8 for the locations of the major cortical regions involved in motor control. The **primary motor cortex (M1)** is in the precentral gyrus of the frontal

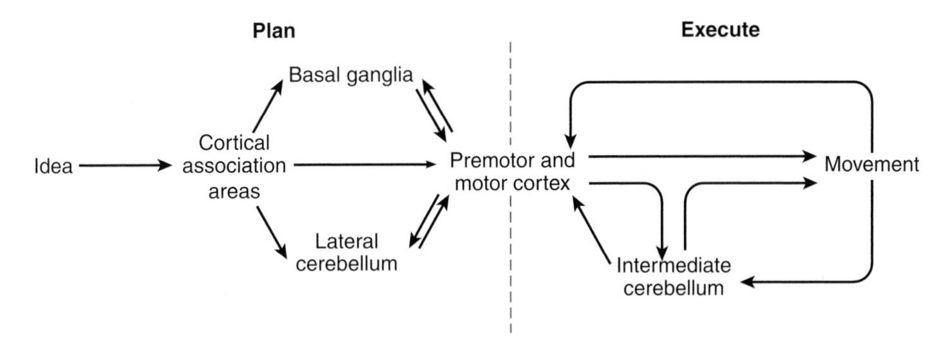

FIGURE 12–7 Control of voluntary movement. Commands for voluntary movement originate in cortical association areas. The cortex, basal ganglia, and cerebellum work cooperatively to plan movements. Movement executed by the cortex is relayed via the corticospinal tracts and corticobulbar tracts to motor neurons. The cerebellum provides feedback to adjust and smooth movement.

lobe, extending into the central sulcus. By means of stimulation experiments in patients undergoing craniotomy under local anesthesia, this region was mapped to show where various parts of the body are represented in the precentral gyrus. Figure 12–8 shows the **motor homunculus** with the feet at the top of the gyrus and the face at the bottom. The facial area is represented bilaterally, but the rest of the representation is generally unilateral, with the cortical motor area controlling the musculature on the opposite side of the body. The cortical representation of each body part is proportional in size to the skill with which the part is used in fine, voluntary movement. The areas involved in speech and hand movements are especially large in the cortex; use of the pharynx, lips, and tongue to form words and of the fingers and opposable thumbs to manipulate the environment are activities in which humans are especially skilled.

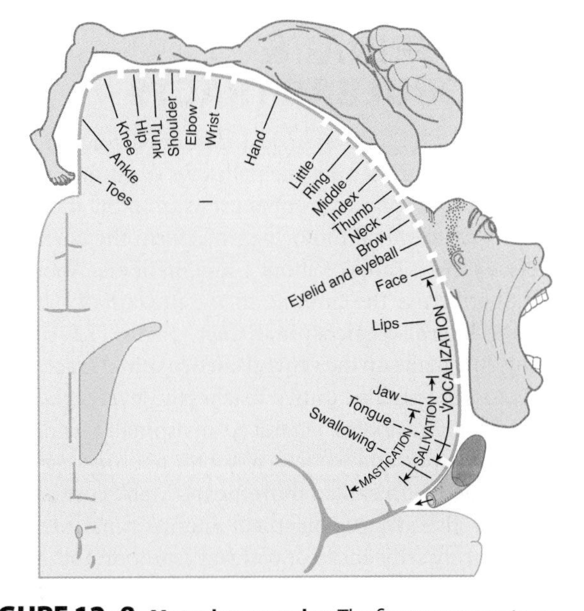

FIGURE 12–8 Motor homunculus. The figure represents, on a coronal section of the precentral gyrus, the location of the cortical representation of the various parts. The size of the various parts is proportional to the cortical area devoted to them. Compare with Figure 8–9 which shows the sensory homunculus. (Reproduced with permission from Penfield W, Rasmussen G: *The Cerebral Cortex of Man*. Macmillan, 1950.)

Modern brain imaging techniques such as **positive emission tomography (PET)** and **functional magnetic resonance imaging (fMRI)** have been used to map the cortex to identify motor areas. Figure 12–9 shows activation of the hand area of the motor cortex while repetitively squeezing a ball with either the right or left hand.

The cells in the cortical motor areas are arranged in a **columnar organization.** The ability to elicit discrete movements of a single muscle by electrical stimulation of a column within M1 led to the view that this area was responsible for control of individual muscles. More recent work has shown that neurons in several cortical columns project to the same muscle; in fact, most stimuli activate more than one muscle. Moreover, the cells in each column receive fairly extensive sensory input from the peripheral area in which they produce movement, providing the basis for feedback control of movement. Some of

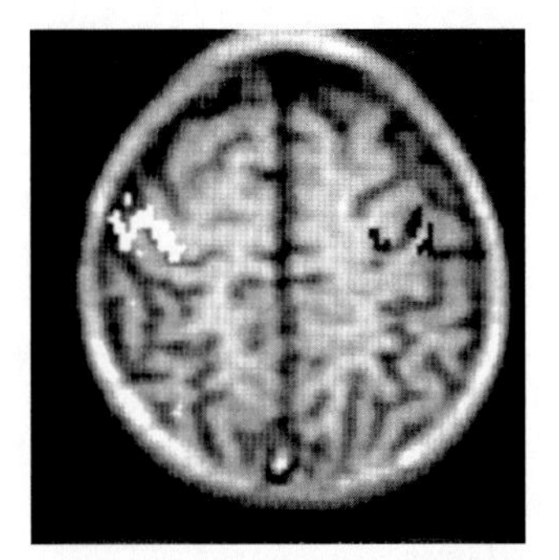

FIGURE 12–9 Hand area of motor cortex demonstrated by functional magnetic resonance imaging (fMRI) in a 7-year-old boy. Changes in signal intensity, measured using a method called echoplanar magnetic resonance imaging, result from changes in the flow, volume, and oxygenation of the blood. The child was instructed to repetitively squeeze a foam-rubber ball at the rate of two to four squeezes per second with the right or left hand. Changes in cortical activity with the ball in the right hand are shown in black. Changes in cortical activity with the ball in the left hand are shown in white. (Data from Novotny EJ, et al: Functional magnetic resonance imaging (fMRI) in pediatric epilepsy. Epilepsia 1994;35(Supp 8):36.)

this input may be direct and some is relayed from other parts of the cortex. The current view is that M1 neurons represent movements of groups of muscles for different tasks.

SUPPLEMENTARY MOTOR AREA

The supplementary motor area is on and above the superior bank of the cingulate sulcus on the medial side of the hemisphere. It projects to the primary motor cortex and also contains a map of the body; but it is less precise than in M1. The supplementary motor area may be involved primarily in organizing or planning motor sequences, while M1 executes the movements. Lesions of this area in monkeys produce awkwardness in performing complex activities and difficulty with bimanual coordination.

When human subjects count to themselves without speaking, the motor cortex is quiescent, but when they speak the numbers aloud as they count, blood flow increases in M1 and the supplementary motor area. Thus, both regions are involved in voluntary movement when the movements being performed are complex and involve planning.

PREMOTOR CORTEX

The premotor cortex is located anterior to the precentral gyrus, on the lateral and medial cortical surface; it also contains a somatotopic map. This region receives input from sensory regions of the parietal cortex and projects to M1, the spinal cord, and the brain stem reticular formation. This region may be concerned with setting posture at the start of a planned movement and with getting the individual prepared to move. It is most involved in control of proximal limb muscles needed to orient the body for movement.

POSTERIOR PARIETAL CORTEX

The somatic sensory area and related portions of the posterior parietal lobe project to the premotor cortex. Lesions of the somatic sensory area cause defects in motor performance that are characterized by inability to execute learned sequences of movements such as eating with a knife and fork. Some of the neurons are concerned with aiming the hands toward an object and manipulating it, whereas other neurons are concerned with hand–eye coordination. As described below, neurons in this posterior parietal cortex contribute to the descending pathways involved in motor control.

PLASTICITY

A striking discovery made possible by PET and fMRI is that the motor cortex shows the same kind of plasticity as already described for the sensory cortex in Chapter 8. For example, the finger areas of the contralateral motor cortex enlarge as a pattern of rapid finger movement is learned with the fingers of one hand; this change is detectable at 1 week and maximal at 4 weeks. Cortical areas of output to other muscles also increase in size when motor learning involves these muscles. When a small focal ischemic lesion is produced in the hand area of the motor cortex of monkeys, the hand area may reappear, with return of motor function, in an adjacent undamaged part of the cortex. Thus, the maps of the motor cortex are not immutable, and they change with experience.

CONTROL OF AXIAL & DISTAL MUSCLES

Within the brain stem and spinal cord, pathways and neurons that are concerned with the control of skeletal muscles of the trunk (axial) and proximal portions of the limbs are located medially or ventrally, whereas pathways and neurons that are concerned with the control of skeletal muscles in the distal portions of the limbs are located laterally. The axial muscles are concerned with postural adjustments and gross movements, whereas the distal limb muscles mediate fine, skilled movements. Thus, for example, neurons in the medial portion of the ventral horn innervate proximal limb muscles, particularly the flexors, whereas lateral ventral horn neurons innervate distal limb muscles. Similarly, the ventral corticospinal tract and medial descending brain stem pathways (tectospinal, reticulospinal, and vestibulospinal tracts) are concerned with adjustments of proximal muscles and posture, whereas the lateral corticospinal and rubrospinal tracts are concerned with distal limb muscles and, particularly in the case of the lateral corticospinal tract, with skilled voluntary movements. Phylogenetically, the lateral pathways are newer.

CORTICOSPINAL & CORTICOBULBAR TRACTS

The somatotopic organization just described for the motor cortex continues throughout the pathways from the cortex to the motor neurons. The axons of neurons from the motor cortex that project to spinal motor neurons form the **corticospinal tracts,** a large bundle of about 1 million fibers. About 80% of these fibers cross the midline in the medullary pyramids to form the **lateral corticospinal tract** (Figure 12–10). The remaining 20% make up the **ventral corticospinal tract,** which does not cross the midline until it reaches the level of the spinal cord at which it terminates. Lateral corticospinal tract neurons make monosynaptic connections to motor neurons, especially those concerned with skilled movements. Many corticospinal tract neurons also synapse on spinal interneurons antecedent to motor neurons; this indirect pathway is important in coordinating groups of muscles.

The trajectory from the cortex to the spinal cord passes through the **corona radiata** to the posterior limb of the **internal capsule.** Within the midbrain they traverse the **cerebral peduncle** and the basilar pons until they reach the **medullary pyramids** on their way to the spinal cord.

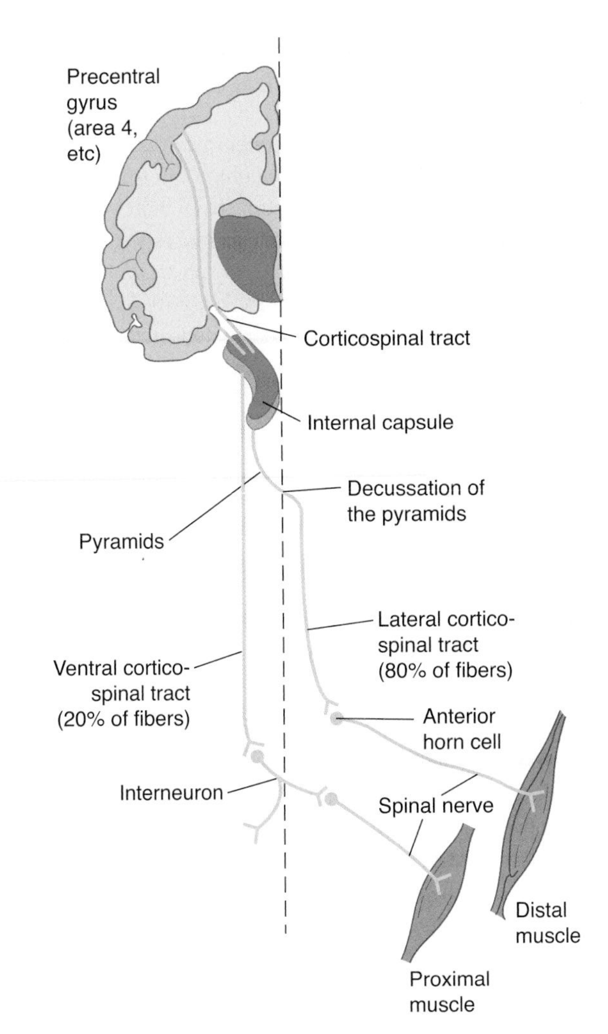

FIGURE 12–10 The corticospinal tracts. This tract originates in the precentral gyrus and passes through the internal capsule. Most fibers decussate in the pyramids and descend in the lateral white matter of the spinal cord to form the lateral division of the tract which can make monosynaptic connections with spinal motor neurons. The ventral division of the tract remains uncrossed until reaching the spinal cord where axons terminate on spinal interneurons antecedent to motor neurons.

The **corticobulbar tract** is composed of the fibers that pass from the motor cortex to motor neurons in the trigeminal, facial, and hypoglossal nuclei. Corticobulbar neurons end either directly on the cranial nerve nuclei or on their antecedent interneurons within the brain stem. Their axons traverse through the genu of the internal capsule, the cerebral peduncle (medial to corticospinal tract neurons), to descend with corticospinal tract fibers in the pons and medulla.

The motor system can be divided into lower and upper motor neurons. **Lower motor neurons** refer to the spinal and cranial motor neurons that directly innervate skeletal muscles. **Upper motor neurons** are those in the cortex and brain stem that activate the lower motor neurons. The pathophysiological responses to damage to lower and upper motor neurons are very distinctive (see Clinical Box 12–5).

ORIGINS OF CORTICOSPINAL & CORTICOBULBAR TRACTS

Corticospinal and corticobulbar tract neurons are pyramidal shaped and located in layer V of the cerebral cortex (see Chapter 14). The cortical areas from which these tracts originate were identified on the basis of electrical stimulation that produced prompt discrete movement. About 31% of the corticospinal tract neurons are from the primary motor cortex. The premotor cortex and supplementary motor cortex account for 29% of the corticospinal tract neurons. The other 40% of corticospinal tract neurons originate in the parietal lobe and primary somatosensory area in the postcentral gyrus.

ROLE IN MOVEMENT

The corticospinal and corticobulbar system is the primary pathway for the initiation of skilled voluntary movement. This does not mean that movement—even skilled movement—is impossible without it. Nonmammalian vertebrates have essentially no corticospinal and corticobulbar system, but they move with great agility. Cats and dogs stand, walk, and run after complete destruction of this system. Only in primates are relatively marked deficits produced.

Careful section of the pyramids producing highly selective destruction of the lateral corticospinal tract in laboratory primates produces prompt and sustained loss of the ability to grasp small objects between two fingers and to make isolated movements of the wrists. However, the animal can still use the hand in a gross fashion and can stand and walk. These deficits are consistent with loss of control of the distal musculature of the limbs, which is concerned with fine-skilled movements. On the other hand, lesions of the ventral corticospinal tract produce axial muscle deficits that cause difficulty with balance, walking, and climbing.

BRAIN STEM PATHWAYS INVOLVED IN POSTURE AND VOLUNTARY MOVEMENT

As mentioned above, spinal motor neurons are organized such that those innervating the most proximal muscles are located most medially and those innervating the more distal muscles are located more laterally. This organization is also reflected in descending brain stem pathways (Figure 12–11).

MEDIAL BRAIN STEM PATHWAYS

The medial brain stem pathways, which work in concert with the ventral corticospinal tract, are the **pontine and medullary reticulospinal, vestibulospinal,** and **tectospinal tracts.** These pathways descend in the ipsilateral ventral columns of the spinal cord and terminate predominantly on interneurons and long propriospinal neurons in the ventromedial part of

CLINICAL BOX 12–5

Lower versus Upper Motor Neuron Damage

Lower motor neurons are those whose axons terminate on skeletal muscles. Damage to these neurons is associated with **flaccid paralysis, muscular atrophy, fasciculations** (visible muscle twitches that appear as flickers under the skin), **hypotonia** (decreased muscle tone), and **hyporeflexia** or **areflexia.** An example of a disease that leads to lower motor neuron damage is **amyotrophic lateral sclerosis (ALS).** "Amyotrophic" means "no muscle nourishment" and describes the atrophy that muscles undergo because of disuse. "Sclerosis" refers to the hardness felt when a pathologist examines the spinal cord on autopsy; the hardness is due to proliferation of astrocytes and scarring of the lateral columns of the spinal cord. ALS is a selective, progressive degeneration of α-motor neurons. This fatal disease is also known as **Lou Gehrig disease** in remembrance of a famous American baseball player who died of ALS. The worldwide annual incidence of ALS has been estimated to be 0.5–3 cases per 100,000 people. The disease has no racial, socioeconomic, or ethnic boundaries. The life expectancy of ALS patients is usually 3–5 years after diagnosis. ALS is most commonly diagnosed in middle age and affects men more often than women. Most cases of ALS are sporadic in origin; but 5–10% of the cases have a familial link. Possible causes include viruses, neurotoxins, heavy metals, DNA defects (especially in familial ALS), immune system abnormalities, and enzyme abnormalities. About 40% of the familial cases have a mutation in the gene for Cu/Zn superoxide dismutase *(SOD-1)* on chromosome 21. SOD is a free radical scavenger that reduces oxidative stress. A defective *SOD-1* gene permits free radicals to accumulate and kill neurons. Some evidence suggests an increase in the excitability of deep cerebellar nuclei due to the inhibition of **small-conductance calcium-activated potassium (SK) channels** contributes to the development of cerebellar ataxia.

Upper motor neurons typically refer to corticospinal tract neurons that innervate spinal motor neurons, but they can also include brain stem neurons that control spinal motor neurons. Damage to these neurons initially causes muscles to become weak and flaccid but eventually leads to spasticity, **hypertonia** (increased resistance to passive movement), hyperactive stretch reflexes, and abnormal plantar extensor reflex (positive Babinski sign). The Babinski sign is dorsiflexion of the great toe and fanning of the other toes when the lateral aspect of the sole of the foot is scratched. In adults, the normal response to this stimulation is plantar flexion in all the toes. The Babinski sign is believed to be a flexor withdrawal reflex that is normally held in check by the lateral corticospinal system. It is of value in the localization of disease processes, but its physiologic significance is unknown. However, in infants whose corticospinal tracts are not well developed, dorsiflexion of the great toe and fanning of the other toes is the natural response to stimuli applied to the sole of the foot.

THERAPEUTIC HIGHLIGHTS

One of the few drugs shown to modestly slow the progression of ALS is **riluzole,** a drug that opens the SK channels. Spasticity associated with motor neuron disease can be reduced by the muscle relaxant **baclofen** (a derivative of GABA); in some cases a subarachnoid infusion of baclofen is given via an implanted lumbar pump. Spasticity can also be treated with **tizanidine,** a centrally acting α_2-adrenoceptor agonist; its effectiveness is thought to be due to increasing presynaptic inhibition of spinal motor neurons. **Botulinum toxin** is also approved for the treatment of spasticity; this toxin acts by binding to receptors on the cholinergic nerve terminals to decrease the release of acetylcholine, causing neuromuscular blockade.

the ventral horn to control axial and proximal muscles. A few medial pathway neurons synapse directly on motor neurons controlling axial muscles.

The medial and lateral vestibulospinal tracts are involved in vestibular function and are briefly described in Chapter 10. The medial tract originates in the medial and inferior vestibular nuclei and projects bilaterally to cervical spinal motor neurons that control neck musculature. The lateral tract originates in the lateral vestibular nuclei and projects ipsilaterally to neurons at all spinal levels. It activates motor neurons to antigravity muscles (eg, proximal limb extensors) to control posture and balance.

The pontine and medullary reticulospinal tracts project to all spinal levels. They are involved in the maintenance of posture and in modulating muscle tone, especially via an input to γ-motor neurons. Pontine reticulospinal neurons are primarily excitatory and medullary reticulospinal neurons are primarily inhibitory.

The tectospinal tract originates in the superior colliculus of the midbrain. It projects to the contralateral cervical spinal cord to control head and eye movements.

LATERAL BRAIN STEM PATHWAY

The main control of distal muscles arise from the lateral corticospinal tract, but neurons within the red nucleus of the midbrain cross the midline and project to interneurons in the dorsolateral part of the spinal ventral horn to also influence motor neurons that control distal limb muscles. This **rubrospinal tract** excites flexor motor neurons and inhibits extensor motor neurons. This

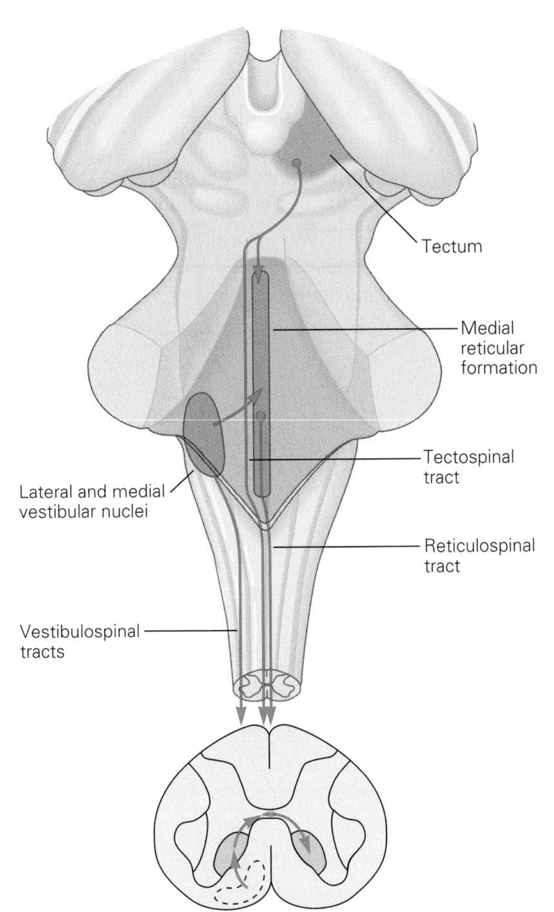

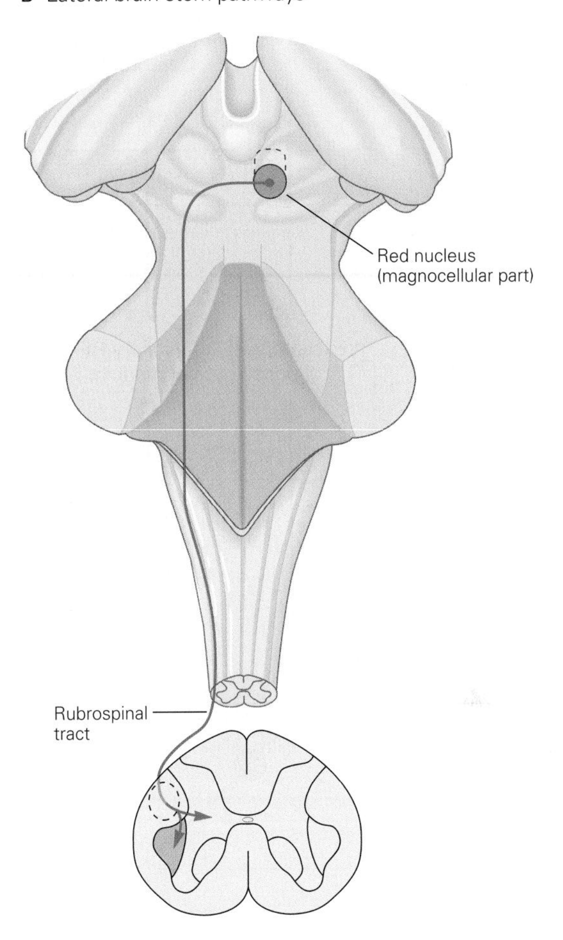

FIGURE 12–11 Medial and lateral descending brain stem pathways involved in motor control. A) Medial pathways (reticulospinal, vestibulospinal, and tectospinal) terminate in ventromedial area of spinal gray matter and control axial and proximal muscles. **B)** Lateral pathway (rubrospinal) terminates in dorsolateral area of spinal gray matter and controls distal muscles. (From Kandel ER, Schwartz JH, Jessell TM [editors]: *Principles of Neural Science,* 4th ed. McGraw-Hill, 2000.)

pathway is not very prominent in humans, but it may play a role in the posture typical of decorticate rigidity (see below).

POSTURE-REGULATING SYSTEMS

In the intact animal, individual motor responses are submerged in the total pattern of motor activity. When the neural axis is transected, the activities integrated below the section are cut off, or released, from the control of higher brain centers and often appear to be accentuated. Release of this type, long a cardinal principle in neurology, may be due in some situations to removal of an inhibitory control by higher neural centers. A more important cause of the apparent hyperactivity is loss of differentiation of the reaction so that it no longer fits into the broader pattern of motor activity. Research using animal models has led to information on the role of cortical and brain stem mechanisms involved in control of voluntary movement and posture. The deficits in motor control seen after various lesions mimic those seen in humans with damage in the same structures.

DECEREBRATION

A complete transection of the brain stem between the superior and inferior colliculi permits the brain stem pathways to function independent of their input from higher brain structures. This is called a **midcollicular decerebration** and is diagramed in Figure 12–12 by the dashed line labeled A. This lesion interrupts all input from the cortex (corticospinal and corticobulbar tracts) and red nucleus (rubrospinal tract), primarily to distal muscles of the extremities. The excitatory and inhibitory reticulospinal pathways (primarily to postural extensor muscles) remain intact. The dominance of drive from ascending sensory pathways to the excitatory reticulospinal pathway leads to hyperactivity in extensor muscles in all four extremities which is called **decerebrate rigidity.** This resembles what ensues after **uncal herniation** due to a supratentorial lesion. Uncal herniation can occur in patients with large tumors or a hemorrhage in the cerebral hemisphere. Figure 12–13A shows the posture typical of such a patient. Clinical Box 12–6 describes complications related to uncal herniation.

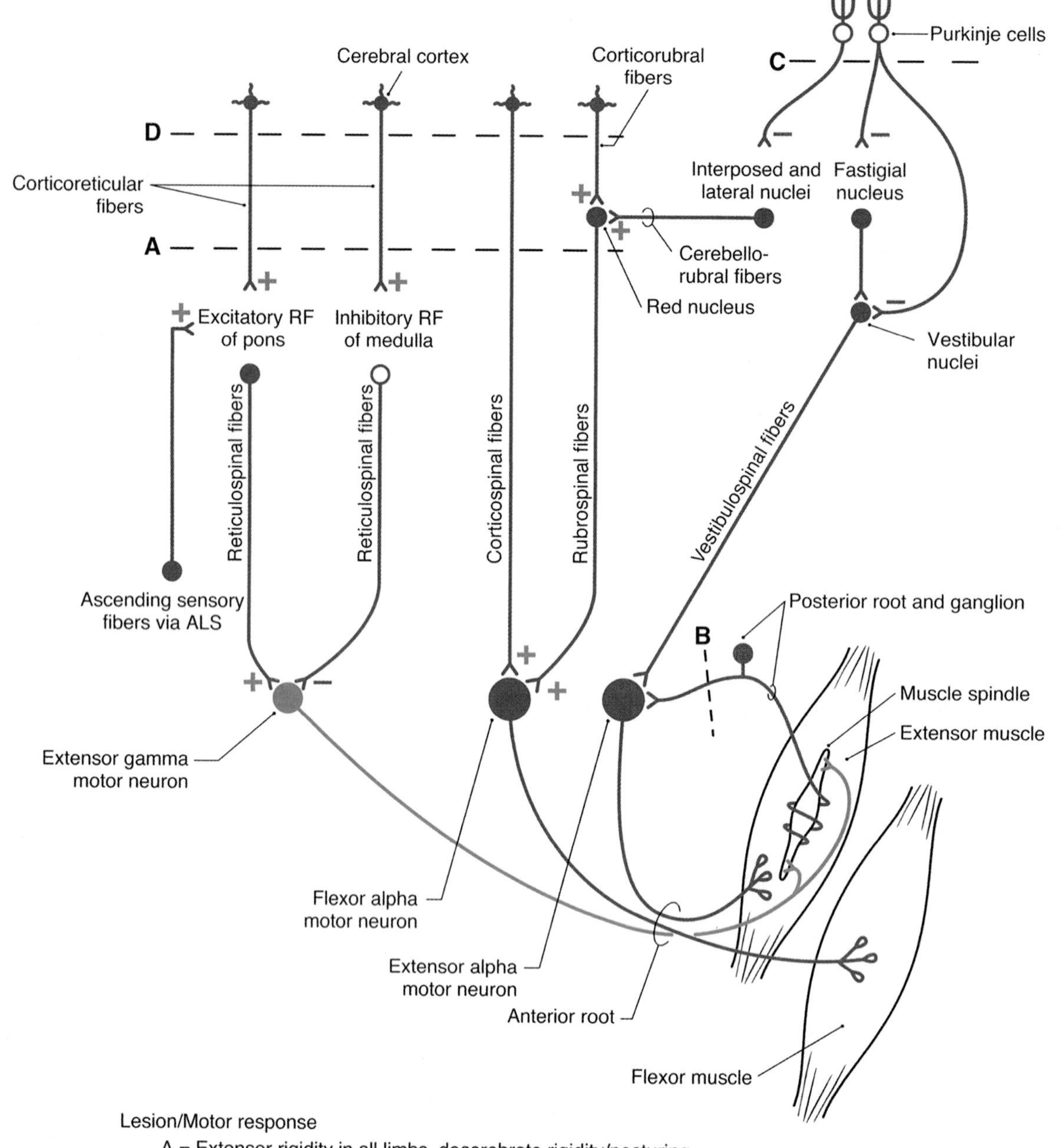

FIGURE 12–12 A circuit drawing representing lesions produced in experimental animals to replicate decerebrate and decorticate deficits seen in humans. Bilateral transections are indicated by dashed lines A, B, C, and D. Decerebration is at a midcollicular level (A), decortication is rostral to the superior colliculus, dorsal roots sectioned for one extremity (B), and removal of anterior lobe of cerebellum (C). The objective was to identify anatomic substrates responsible for decerebrate or decorticate rigidity/posturing seen in humans with lesions that either isolate the forebrain from the brain stem or separate rostral from caudal brain stem and spinal cord. (Reproduced with permission from Haines DE [editor]: *Fundamental Neuroscience for Basic and Clinical Applications,* 3rd ed. Elsevier, 2006.)

In midcollicular decerebrate cats, section of dorsal roots to a limb (dashed line labeled B in Figure 12–12) immediately eliminates the hyperactivity of extensor muscles. This suggests that decerebrate rigidity is spasticity due to facilitation of the myotatic stretch reflex. That is, the excitatory input from the reticulospinal pathway activates γ-motor neurons which indirectly activate α-motor neurons (via Ia spindle afferent activity). This is called the **gamma loop.**

A Upper pontine damage

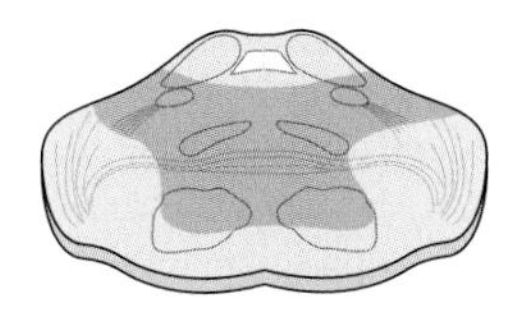

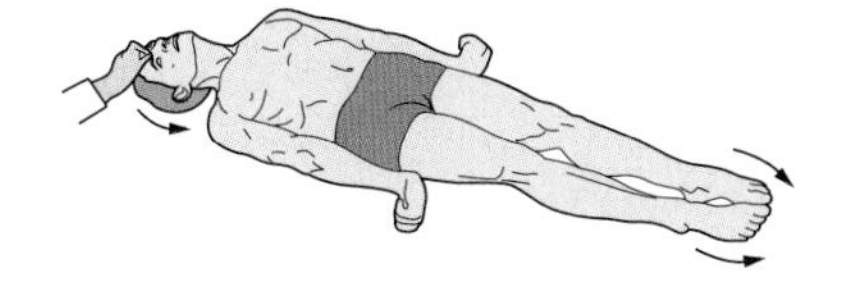

B Upper midbrain damage

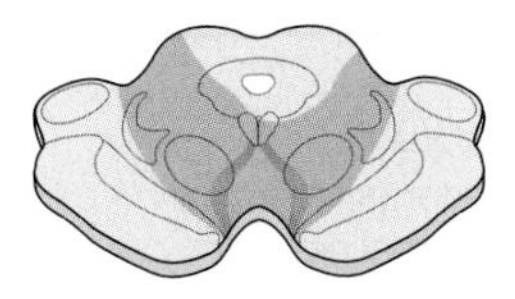

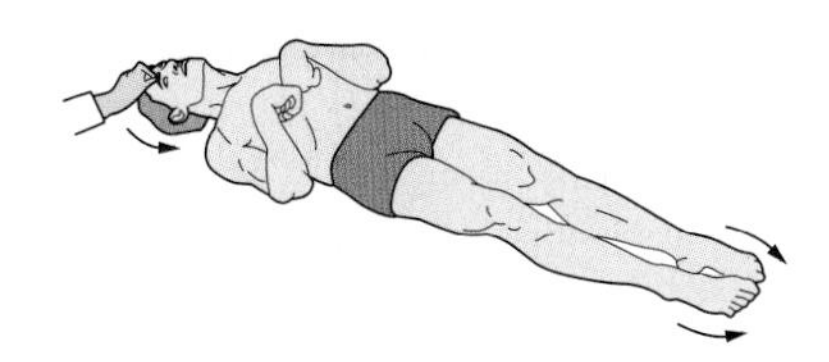

FIGURE 12–13 Decerebrate and decorticate postures. A) Damage to lower midbrain and upper pons causes decerebrate posturing in which lower extremities are extended with toes pointed inward and upper extremities extended with fingers flexed and forearms pronate. Neck and head are extended. **B)** Damage to upper midbrain may cause decorticate posturing in which upper limbs are flexed, lower limbs are extended with toes pointed slightly inward, and head is extended. (Modified from Kandel ER, Schwartz JH, Jessell TM [editors]: *Principles of Neural Science,* 4th ed. McGraw-Hill, 2000.)

The exact site of origin within the cerebral cortex of the fibers that inhibit stretch reflexes is unknown. Under certain conditions, stimulation of the anterior edge of the precentral gyrus can cause inhibition of stretch reflexes and cortically evoked movements. This region, which also projects to the basal ganglia, has been named the **suppressor strip.**

There is also evidence that decerebrate rigidity leads to direct activation of α-motor neurons. If the anterior lobe of the cerebellum is removed in a decerebrate animal (dashed line labeled C in Figure 12–12), extensor muscle hyperactivity is exaggerated **(decerebellate rigidity).** This cut eliminates cortical inhibition of the cerebellar fastigial nucleus and secondarily increases excitation to vestibular nuclei. Subsequent dorsal root section does not reverse the rigidity, thus it was due to activation of α-motor neurons independent of the gamma loop.

CLINICAL BOX 12–6

Uncal Herniation

Space-occupying lesions from large tumors, hemorrhages, strokes, or abscesses in the cerebral hemisphere can drive the uncus of the temporal lobe over the edge of the cerebellar tentorium, compressing the ipsilateral cranial nerve III (**uncal herniation**). Before the herniation these patients experience a decreased level of consciousness, lethargy, poorly reactive pupils, deviation of the eye to a "down and out" position, hyperactive reflexes, and a bilateral Babinski sign (due to compression of the ipsilateral corticospinal tract). After the brain herniates, the patients are decerebrate and comatose, have fixed and dilated pupils, and eye movements are absent. Once damage extends to the midbrain, a **Cheyne–Stokes respiratory pattern** develops, characterized by a pattern of waxing-and-waning depth of respiration with interposed periods of apnea. Eventually, medullary function is lost, breathing ceases, and recovery is unlikely. Hemispheric masses closer to the midline compress the thalamic reticular formation and can cause coma before eye findings develop (**central herniation**). As the mass enlarges, midbrain function is affected, the pupils dilate, and a decerebrate posture ensues. With progressive herniation, pontine vestibular and then medullary respiratory functions are lost.

DECORTICATION

Removal of the cerebral cortex (**decortication;** dashed line labeled D in Figure 12–12) produces **decorticate rigidity** which is characterized by flexion of the upper extremities at the elbow and extensor hyperactivity in the lower extremities (Figure 12–13B). The flexion can be explained by rubrospinal excitation of flexor muscles in the upper extremities; the hyperextension of lower extremities is due to the same changes that occur after midcollicular decerebration.

Decorticate rigidity is seen on the hemiplegic side after hemorrhages or thromboses in the internal capsule. Probably because of their anatomy, the small arteries in the internal capsule are especially prone to rupture or thrombotic obstruction, so this type of decorticate rigidity is fairly common. Sixty percent of intracerebral hemorrhages occur in the internal capsule, as opposed to 10% in the cerebral cortex, 10% in the pons, 10% in the thalamus, and 10% in the cerebellum.

BASAL GANGLIA

ORGANIZATION OF THE BASAL GANGLIA

The term **basal ganglia** (or **basal nuclei**) is applied to five interactive structures on each side of the brain (Figure 12–14). These are the **caudate nucleus, putamen,** and **globus pallidus** (three large nuclear masses underlying the cortical mantle), the **subthalamic nucleus,** and **substantia nigra.** The caudate nucleus

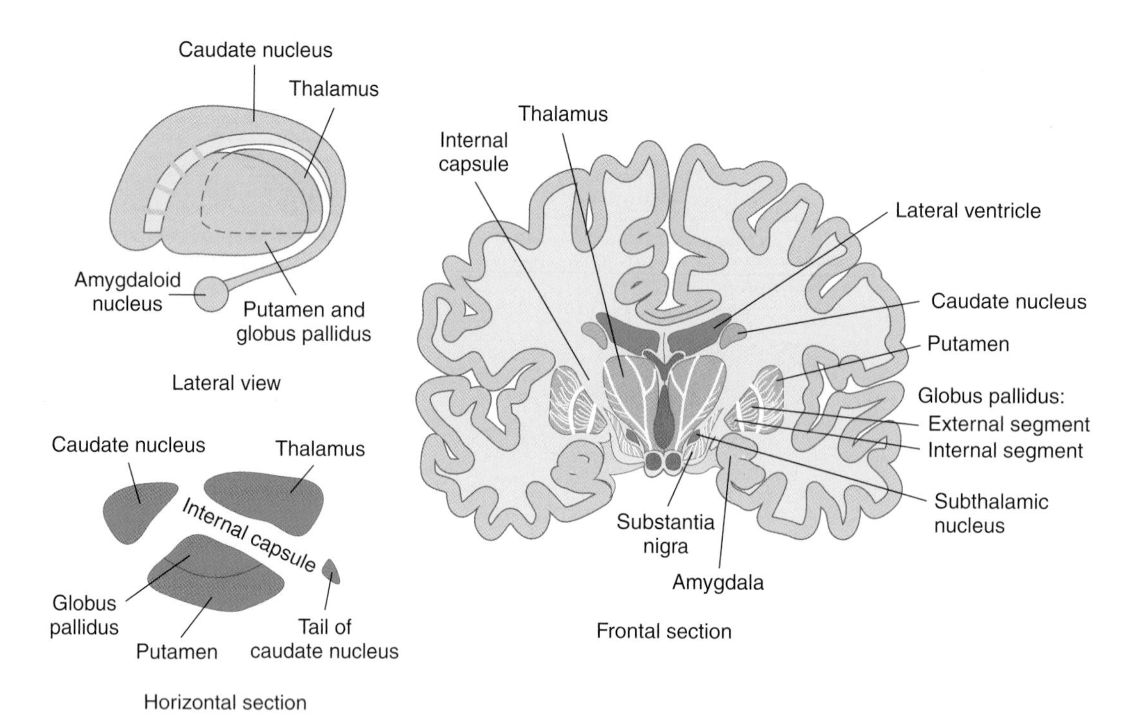

FIGURE 12–14 The basal ganglia. The basal ganglia are composed of the caudate nucleus, putamen, and globus pallidus and the functionally related subthalamic nucleus and substantia nigra. The frontal (coronal) section shows the location of the basal ganglia in relation to surrounding structures.

and putamen collectively form the **striatum;** the putamen and globus pallidus collectively form the **lenticular nucleus.**

The globus pallidus is divided into external and internal segments (GPe and GPi); both regions contain inhibitory GABAergic neurons. The substantia nigra is divided into a **pars compacta** which uses dopamine as a neurotransmitter and a **pars reticulata** which uses GABA as a neurotransmitter. There are at least four neuronal types within the striatum. About 95% of striatal neurons are medium spiny neurons that use GABA as a neurotransmitter. The remaining striatal neurons are all aspiny interneurons that differ in terms of size and neurotransmitters: large (acetylcholine), medium (somatostatin), and small (GABA).

Figure 12–15 shows the major connections to and from and within the basal ganglia along with the neurotransmitters within these pathways. There are two main inputs to the basal ganglia; they are both excitatory (glutamate), and they both terminate in the striatum. They are from a wide region of the cerebral cortex **(corticostriate pathway)** and from intralaminar nuclei of the thalamus **(thalamostriatal pathway).** The two major outputs of the basal ganglia are from GPi and substantia nigra pars reticulata. Both are inhibitory (GABAergic) and both project to the thalamus. From the thalamus, there is an excitatory (presumably glutamate) projection to the prefrontal and premotor cortex. This completes a full cortical-basal ganglia-thalamic-cortical loop.

The connections within the basal ganglia include a dopaminergic **nigrostriatal projection** from the substantia nigra pars compacta to the striatum and a GABAergic projection from the striatum to substantia nigra pars reticulata. There is an inhibitory projection from the striatum to both GPe and GPi. The subthalamic nucleus receives an inhibitory input from GPe, and in turn the subthalamic nucleus has an excitatory (glutamate) projection to both GPe and GPi.

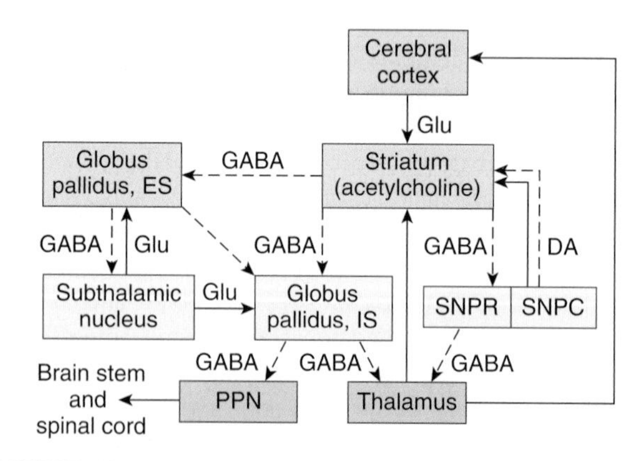

FIGURE 12–15 Diagrammatic representation of the principal connections of the basal ganglia. Solid lines indicate excitatory pathways, dashed lines inhibitory pathways. The transmitters are indicated in the pathways, where they are known. DA, dopamine; Glu, glutamate. Acetylcholine is the transmitter produced by interneurons in the striatum. ES, external segment; IS, internal segment; PPN, pedunculopontine nuclei; SNPC, substantia nigra, pars compacta; SNPR, substantia nigra, pars reticulata. The subthalamic nucleus also projects to the pars compacta of the substantia nigra; this pathway has been omitted for clarity.

FUNCTION

Neurons in the basal ganglia, like those in the lateral portions of the cerebellar hemispheres, discharge before movements begin. This observation, plus careful analysis of the effects of diseases of the basal ganglion in humans and the effects of drugs that destroy dopaminergic neurons in animals, have led to the idea that the basal ganglia are involved in the planning and programming of movement or, more broadly, in the processes by which an abstract thought is converted into voluntary action (Figure 12–7). They influence the motor cortex via the thalamus, and the corticospinal pathways provide the final common pathway to motor neurons. In addition, GPi projects to nuclei in the brain stem, and from there to motor neurons in the brain stem and spinal cord. The basal ganglia, particularly the caudate nuclei, also play a role in some cognitive processes. Possibly because of the interconnections of this nucleus with the frontal portions of the neocortex, lesions of the caudate nuclei disrupt performance on tests involving object reversal and delayed alternation. In addition, lesions of the head of the left but not the right caudate nucleus and nearby white matter in humans are associated with a dysarthric form of aphasia that resembles Wernicke aphasia (see Chapter 15).

DISEASES OF THE BASAL GANGLIA IN HUMANS

Three distinct biochemical pathways in the basal ganglia normally operate in a balanced fashion: (1) the nigrostriatal dopaminergic system, (2) the intrastriatal cholinergic system, and (3) the GABAergic system, which projects from the striatum to the globus pallidus and substantia nigra. When one or more of these pathways become dysfunctional, characteristic motor abnormalities occur. Diseases of the basal ganglia lead to two general types of disorders: **hyperkinetic** and **hypokinetic.** The hyperkinetic conditions are those in which movement is excessive and abnormal, including chorea, athetosis, and ballism. Hypokinetic abnormalities include akinesia and bradykinesia.

Chorea is characterized by rapid, involuntary "dancing" movements. **Athetosis** is characterized by continuous, slow writhing movements. Choreiform and athetotic movements have been likened to the start of voluntary movements occurring in an involuntary, disorganized way. In **ballism,** involuntary flailing, intense, and violent movements occur. **Akinesia** is difficulty in initiating movement and decreased spontaneous movement. **Bradykinesia** is slowness of movement.

In addition to Parkinson disease, which is described below, there are several other disorders known to involve a malfunction within the basal ganglia. A few of these are described in Clinical Box 12–7. **Huntington disease** is one of an increasing number of human genetic diseases affecting the nervous system that are characterized by **trinucleotide repeat** expansion. Most of these involve cytosine-adenine-guanine (CAG) repeats (Table 12–1), but one involves CGG repeats and another involves CTG repeats (T refers to thymine). All of these are in exons; however, a GAA repeat in an intron is associated with Friedreich's ataxia. There is also preliminary evidence that increased numbers of a 12-nucleotide repeat are associated with a rare form of epilepsy.

TABLE 12–1 Examples of trinucleotide repeat diseases.

Disease	Expanded Trinucleotide Repeat	Affected Protein
Huntington disease	CAG	Huntingtin
Spinocerebellar ataxia, types 1, 2, 3, 7	CAG	Ataxin 1, 2, 3, 7
Spinocerebellar ataxia, type 6	CAG	α_{1A} subunit of Ca^{2+} channel
Dentatorubral-pallidoluysian atrophy	CAG	Atrophin
Spinobulbar muscular atrophy	CAG	Androgen receptor
Fragile X syndrome	CGG	FMR-1
Myotonic dystrophy	CTG	DM protein kinase
Friedreich ataxia	GAA	Frataxin

PARKINSON DISEASE

Parkinson disease has both hypokinetic and hyperkinetic features. It was originally described in 1817 by James Parkinson and is named for him. Parkinson disease is the first disease identified as being due to a deficiency in a specific neurotransmitter (see Clinical Box 12–8). In the 1960s, Parkinson disease was shown to result from the degeneration of dopaminergic neurons in the substantia nigra pars compacta. The fibers to the putamen (part of the striatum) are most severely affected.

The hypokinetic features of Parkinson disease are akinesia and bradykinesia, and the hyperkinetic features are **cogwheel rigidity** and **tremor at rest.** The absence of motor activity and the difficulty in initiating voluntary movements are striking. There is a decrease in the normal, unconscious movements such as swinging of the arms during walking, the panorama of facial expressions related to the emotional content of thought and speech, and the multiple "fidgety" actions and gestures that occur in all of us. The rigidity is different from spasticity because motor neuron discharge increases to both the agonist and antagonist muscles. Passive motion

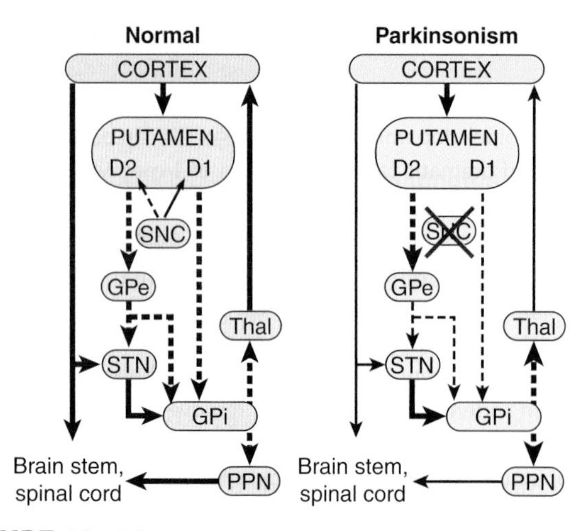

FIGURE 12–16 Probable basal ganglia-thalamocortical circuitry in Parkinson disease. Solid arrows indicate excitatory outputs and dashed arrows inhibitory outputs. The strength of each output is indicated by the width of the arrow. GPe, external segment of the globus pallidus; GPi, internal segment of the globus pallidus; PPN, pedunculopontine nuclei; SNC, pars compacta of the substantia nigra; STN, subthalamic nucleus; Thal, thalamus. See text for details. (Modified from Grafton SC, DeLong M: Tracing the brain circuitry with functional imaging. Nat Med 1997;3:602.)

of an extremity meets with a plastic, dead-feeling resistance that has been likened to bending a lead pipe and is therefore called **lead pipe rigidity.** Sometimes a series of "catches" takes place during passive motion (cogwheel rigidity), but the sudden loss of resistance seen in a spastic extremity is absent. The tremor, which is present at rest and disappears with activity, is due to regular, alternating 8-Hz contractions of antagonistic muscles.

A current view of the pathogenesis of the movement disorders in Parkinson disease is shown in Figure 12–16. In normal individuals, basal ganglia output is inhibitory via GABAergic nerve fibers. The dopaminergic neurons that project from the substantia nigra to the putamen normally have two effects: they stimulate the D_1 dopamine receptors, which inhibit GPi via direct GABAergic receptors, and they inhibit D_2 receptors, which also inhibit the GPi. In addition, the inhibition reduces the excitatory discharge from the subthalamic nucleus to the GPi. This balance between inhibition and excitation somehow maintains normal motor function. In Parkinson disease, the dopaminergic input to the putamen is lost. This results in decreased inhibition and increased excitation from the subthalamic nucleus to the GPi. The overall increase in inhibitory output to the thalamus and brain stem disorganizes movement.

Familial cases of Parkinson disease occur, but these are uncommon. The genes for at least five proteins can be mutated. These proteins appear to be involved in ubiquitination. Two of the proteins, **α-synuclein** and **barkin,** interact and are found in **Lewy bodies.** The Lewy bodies are inclusion bodies in neurons that occur in all forms of Parkinson disease. However, the significance of these findings is still unsettled.

An important consideration in Parkinson disease is the balance between the excitatory discharge of cholinergic interneurons and the inhibitory dopaminergic input in the striatum. Some improvement is produced by decreasing the cholinergic influence with anticholinergic drugs. More dramatic improvement is produced by administration of L-dopa **(levodopa).** Unlike dopamine, this dopamine precursor crosses the blood–brain barrier and helps repair the dopamine deficiency. However, the degeneration of these neurons continues and in 5–7 years the beneficial effects of L-dopa disappear.

CEREBELLUM

The cerebellum sits astride the main sensory and motor systems in the brain stem (Figure 12–17). The medial **vermis** and lateral **cerebellar hemispheres** are more extensively folded and fissured than the cerebral cortex. The cerebellum weighs only 10% as much as the cerebral cortex, but its surface area is about 75% of that of the cerebral cortex. Anatomically, the cerebellum is divided into three parts by two transverse fissures. The posterolateral fissure separates the medial nodulus and the lateral flocculus on either side from the rest of the cerebellum, and the primary fissure divides the remainder into an anterior and a posterior lobe. Lesser fissures divide the vermis into smaller sections, so that it contains 10 primary lobules numbered I–X from superior to inferior.

The cerebellum is connected to the brain stem by three pairs of peduncles that are located above and around the fourth ventricle. The **superior cerebellar peduncle** includes fibers from deep cerebellar nuclei that project to the brain stem, red nucleus, and thalamus. The **middle cerebellar peduncle** contains only afferent fibers from the contralateral pontine nuclei, and the **inferior cerebellar peduncle** a mixture of afferent fibers from the brain stem and spinal cord and efferent fibers to the vestibular nuclei.

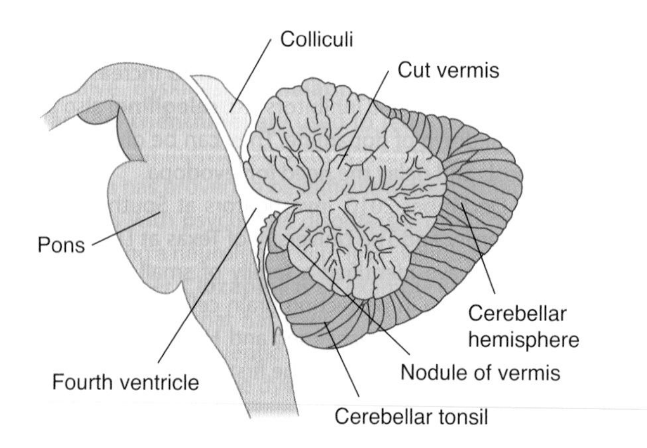

FIGURE 12–17 A midsagittal section through the cerebellum. The medial vermis and lateral cerebellar hemispheres have many narrow, ridge-like folds called folia. Although not shown, the cerebellum is connected to the brain stem by three pairs of peduncles (superior, middle, and inferior). (Reproduced with permission, from Waxman SG: *Clinical Neuroanatomy,* 26th ed. McGraw-Hill, 2010.)

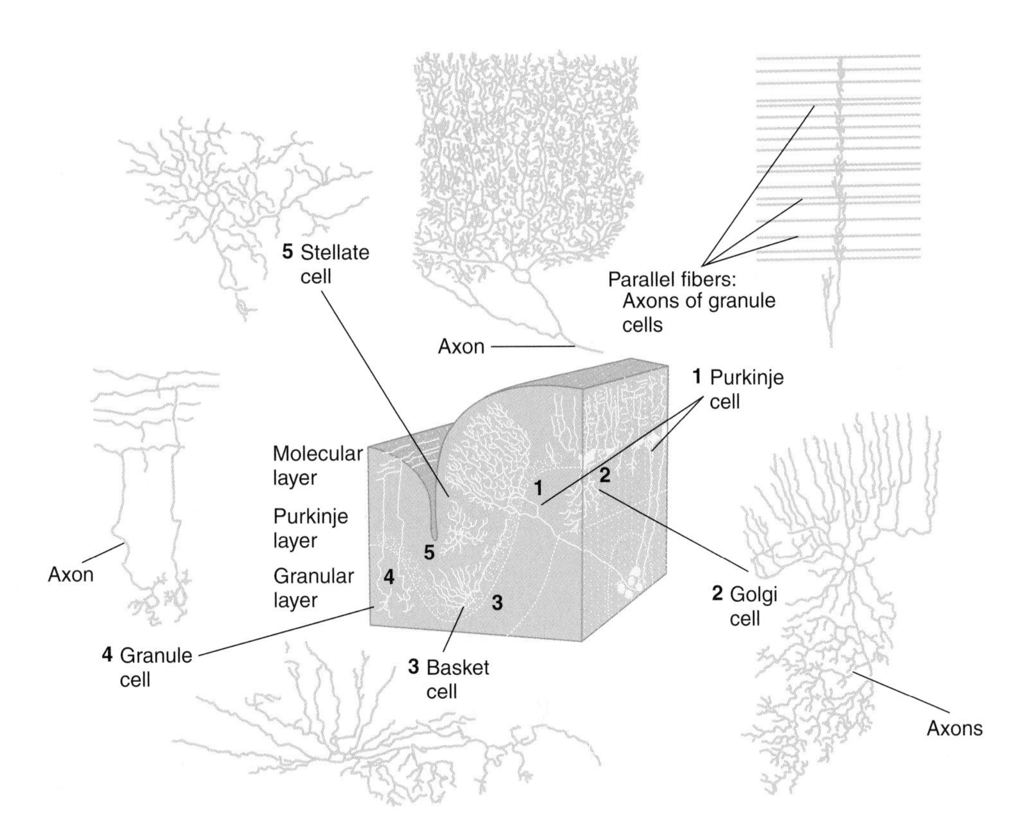

FIGURE 12–18 Location and structure of five neuronal types in the cerebellar cortex. Drawings are based on Golgi-stained preparations. Purkinje cells (1) have processes aligned in one plane; their axons are the only output from the cerebellum. Axons of granule cells (4) traverse and make connections with Purkinje cell processes in molecular layer. Golgi (2), basket (3), and stellate (5) cells have characteristic positions, shapes, branching patterns, and synaptic connections. (For 1 and 2, After Ramon y Cajal S: Histologie du Systeme Nerveux II., C.S.I.C. Madrid; For 3–5, After Palay SL, Chan-Palay V: *Cerebellar Cortex*. Berlin: Springer-Verlag, 1974.)

The cerebellum has an external **cerebellar cortex** separated by white matter from the **deep cerebellar nuclei.** The middle and inferior cerebellar peduncles carry afferent fibers into the cerebellum where they are called **mossy** and **climbing fibers.** These fibers emit collaterals to the deep nuclei and pass to the cortex. There are four deep nuclei: the **dentate,** the **globose,** the **emboliform,** and the **fastigial** nuclei. The globose and the emboliform nuclei are sometimes lumped together as the **interpositus nucleus.**

ORGANIZATION OF THE CEREBELLUM

The cerebellar cortex has three layers: an external molecular layer, a Purkinje cell layer that is only one cell thick, and an internal granular layer. There are five types of neurons in the cortex: Purkinje, granule, basket, stellate, and Golgi cells (Figure 12–18). The **Purkinje cells** are among the biggest neurons in the CNS. They have very extensive dendritic arbors that extend throughout the molecular layer. Their axons, which are the only output from the cerebellar cortex, project to the deep cerebellar nuclei, especially the dentate nucleus, where they form inhibitory synapses. They also make inhibitory connections with neurons in the vestibular nuclei.

The cerebellar **granule cells,** whose cell bodies are in the granular layer, receive excitatory input from the mossy fibers and innervate the Purkinje cells (Figure 12–19). Each sends an axon to the molecular layer, where the axon bifurcates to form a T. The branches of the T are straight and run long distances; thus, they are called **parallel fibers.** The dendritic trees of the Purkinje cells are markedly flattened and oriented at right angles to the parallel fibers. The parallel fibers form excitatory synapses on the dendrites of many Purkinje cells, and the parallel fibers and Purkinje dendritic trees form a grid of remarkably regular proportions.

The other three types of neurons in the cerebellar cortex are inhibitory interneurons. **Basket cells** (Figure 12–18) are located in the molecular layer. They receive excitatory input from the parallel fibers and each projects to many Purkinje cells (Figure 12–19). Their axons form a basket around the cell body and axon hillock of each Purkinje cell they innervate. **Stellate cells** are similar to the basket cells but are located in the more superficial molecular layer. **Golgi cells** are located in the granular layer. Their dendrites, which project into the molecular layer, receive excitatory input from the parallel fibers (Figure 12–19). Their cell bodies receive excitatory input via collaterals from the incoming mossy fibers. Their axons project to the dendrites of the granule cells where they form an inhibitory synapse.

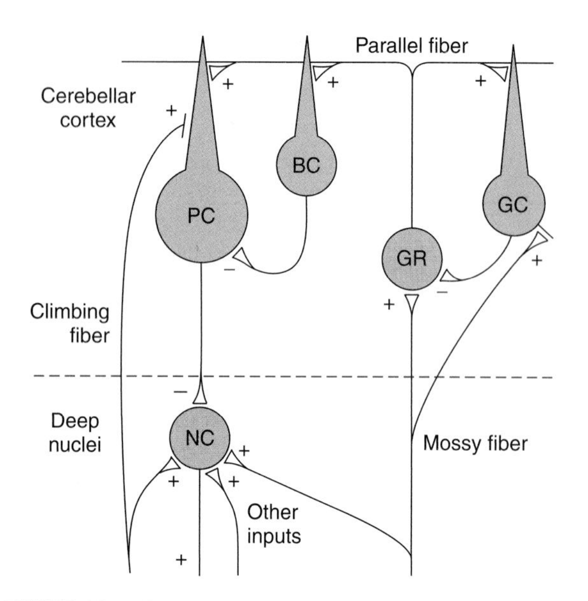

FIGURE 12–19 Diagram of neural connections in the cerebellum. Plus (+) and minus (–) signs indicate whether endings are excitatory or inhibitory. BC, basket cell; GC, Golgi cell; GR, granule cell; NC, cell in deep nucleus; PC, Purkinje cell. Note that PCs and BCs are inhibitory. The connections of the stellate cells, which are not shown, are similar to those of the basket cells, except that they end for the most part on Purkinje cell dendrites.

TABLE 12–2 Function of principal afferent systems to the cerebellum.[a]

Afferent Tracts	Transmits
Vestibulocerebellar	Vestibular impulses from labyrinths, direct and via vestibular nuclei
Dorsal spinocerebellar	Proprioceptive and exteroceptive impulses from body
Ventral spinocerebellar	Proprioceptive and exteroceptive impulses from body
Cuneocerebellar	Proprioceptive impulses, especially from head and neck
Tectocerebellar	Auditory and visual impulses via inferior and superior colliculi
Pontocerebellar	Impulses from motor and other parts of cerebral cortex via pontine nuclei
Olivocerebellar	Proprioceptive input from whole body via relay in inferior olive

[a]The olivocerebellar pathway projects to the cerebellar cortex via climbing fibers; the rest of the listed paths project via mossy fibers. Several other pathways transmit impulses from nuclei in the brain stem to the cerebellar cortex and to the deep nuclei, including a serotonergic input from the raphé nuclei to the granular and molecular layers and a noradrenergic input from the locus coeruleus to all three layers.

As already mentioned, the two main inputs to the cerebellar cortex are climbing fibers and mossy fibers. Both are excitatory (Figure 12–19). The climbing fibers come from a single source, the inferior olivary nuclei. Each projects to the primary dendrites of a Purkinje cell, around which it entwines like a climbing plant. Proprioceptive input to the inferior olivary nuclei comes from all over the body. On the other hand, the mossy fibers provide direct proprioceptive input from all parts of the body plus input from the cerebral cortex via the pontine nuclei to the cerebellar cortex. They end on the dendrites of granule cells in complex synaptic groupings called **glomeruli.** The glomeruli also contain the inhibitory endings of the Golgi cells mentioned above.

The fundamental circuits of the cerebellar cortex are thus relatively simple (Figure 12–19). Climbing fiber inputs exert a strong excitatory effect on single Purkinje cells, whereas mossy fiber inputs exert a weak excitatory effect on many Purkinje cells via the granule cells. The basket and stellate cells are also excited by granule cells via their parallel fibers; and the basket and stellate cells, in turn, inhibit the Purkinje cells **(feed-forward inhibition).** Golgi cells are excited by the mossy fiber collaterals and parallel fibers, and they inhibit transmission from mossy fibers to granule cells. The neurotransmitter released by the stellate, basket, Golgi, and Purkinje cells is GABA, whereas the granule cells release glutamate. GABA acts via $GABA_A$ receptors, but the combinations of subunits in these receptors vary from one cell type to the next. The granule cell is unique in that it appears to be the only type of neuron in the CNS that has a $GABA_A$ receptor containing the α6 subunit.

The output of the Purkinje cells is in turn inhibitory to the deep cerebellar nuclei. As noted above, these nuclei also receive excitatory inputs via collaterals from the mossy and climbing fibers. It is interesting, in view of their inhibitory Purkinje cell input, that the output of the deep cerebellar nuclei to the brain stem and thalamus is always excitatory. Thus, almost all the cerebellar circuitry seems to be concerned solely with modulating or timing the excitatory output of the deep cerebellar nuclei to the brain stem and thalamus. The primary afferent systems that converge to form the mossy fiber or climbing fiber input to the cerebellum are summarized in Table 12–2.

FUNCTIONAL DIVISIONS

From a functional point of view, the cerebellum is divided into three parts (Figure 12–20). The nodulus in the vermis and the flanking flocculus in the hemisphere on each side form the **vestibulocerebellum** (or **flocculonodular lobe**). This lobe, which is phylogenetically the oldest part of the cerebellum, has vestibular connections and is concerned with equilibrium and eye movements. The rest of the vermis and the adjacent medial portions of the hemispheres form the **spinocerebellum,** the region that receives proprioceptive input from the body as well as a copy of the "motor plan" from the motor cortex. By comparing plan with performance, it smoothes and coordinates movements that are ongoing. The vermis projects to the brain stem area concerned with control of axial and proximal limb muscles (medial brain stem pathways), whereas the hemispheres of the spinocerebellum project to the brain stem areas concerned with control of

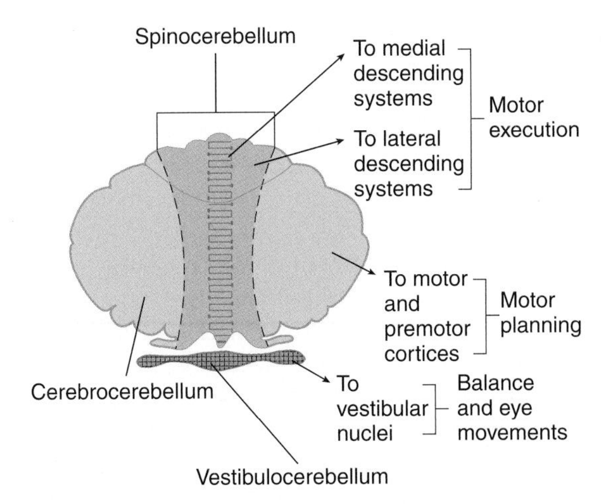

FIGURE 12–20 Three functional divisions of the cerebellum. The nodulus in the vermis and the flanking flocculus in the hemisphere on each side form the vestibulocerebellum which has vestibular connections and is concerned with equilibrium and eye movements. The rest of the vermis and the adjacent medial portions of the hemispheres form the spinocerebellum, the region that receives proprioceptive input from the body as well as a copy of the "motor plan" from the motor cortex. The lateral portions of the cerebellar hemispheres are called the cerebrocerebellum which interacts with the motor cortex in planning and programming movements. (Modified from Kandel ER, Schwartz JH, Jessell TM [editors]: *Principles of Neural Science,* 4th ed. McGraw-Hill, 2000.)

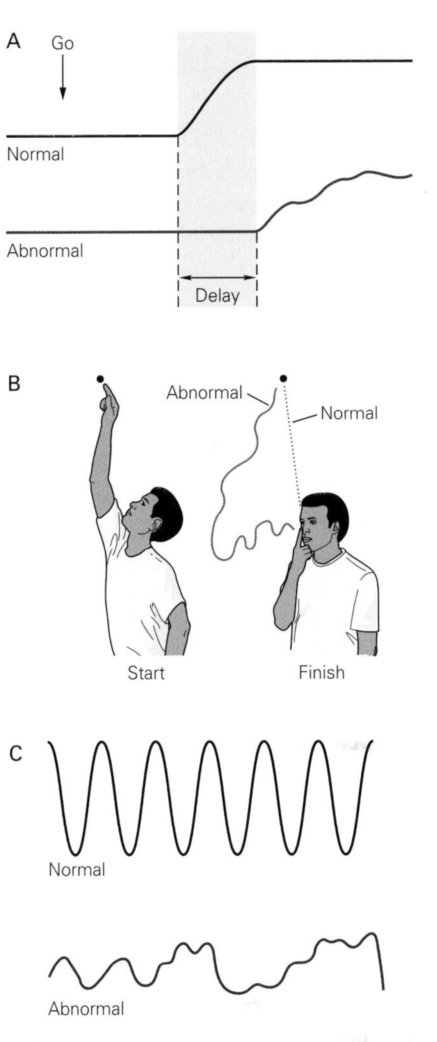

FIGURE 12–21 Typical defects associated with cerebellar disease. A) Lesion of the right cerebellar hemisphere delays initiation of movement. The patient is told to clench both hands simultaneously; right hand clenches later than left (shown by recordings from a pressure bulb transducer squeezed by the patient). B) Dysmetria and decomposition of movement shown by patient moving his arm from a raised position to his nose. Tremor increases on approaching the nose. C) Dysdiadochokinesia occurs in the abnormal position trace of hand and forearm as a cerebellar subject tries alternately to pronate and supinate forearm while flexing and extending elbow as rapidly as possible. (From Kandel ER, Schwartz JH, Jessell TM [editors]: *Principles of Neural Science,* 4th ed. McGraw-Hill, 2000.)

distal limb muscles (lateral brain stem pathways). The lateral portions of the cerebellar hemispheres are called the **cerebrocerebellum.** They are the newest from a phylogenetic point of view, reaching their greatest development in humans. They interact with the motor cortex in planning and programming movements.

Most of the vestibulocerebellar output passes directly to the brain stem, but the rest of the cerebellar cortex projects to the deep nuclei, which in turn project to the brain stem. The deep nuclei provide the only output for the spinocerebellum and the cerebrocerebellum. The medial portion of the spinocerebellum projects to the fastigial nuclei and from there to the brain stem. The adjacent hemispheric portions of the spinocerebellum project to the emboliform and globose nuclei and from there to the brain stem. The cerebrocerebellum projects to the dentate nucleus and from there either directly or indirectly to the ventrolateral nucleus of the thalamus.

CEREBELLAR DISEASE

Damage to the cerebellum leads to several characteristic abnormalities, including **hypotonia, ataxia,** and **intention tremor.** Motor abnormalities associated with cerebellar damage vary depending on the region involved. Figure 12–21 illustrates some of these abnormalities. Additional information is provided in Clinical Box 12–9.

THE CEREBELLUM & LEARNING

The cerebellum is concerned with learned adjustments that make coordination easier when a given task is performed over and over. As a motor task is learned, activity in the brain shifts from the prefrontal areas to the parietal and motor cortex and the cerebellum. The basis of the learning in the cerebellum is probably the input via the olivary nuclei. The mossy fiber–granule cell–Purkinje cell pathway is highly divergent, allowing an individual Purkinje cell to receive input from many mossy fibers arising from different regions. In contrast, a Purkinje cell receives input from a single climbing fiber but it makes

CLINICAL BOX 12–9

Cerebellar Disease

Most abnormalities associated with damage to the cerebellum are apparent during movement. The marked **ataxia** is characterized as incoordination due to errors in the rate, range, force, and direction of movement. Ataxia is manifest not only in the wide-based, unsteady, "drunken" gait of patients, but also in defects of the skilled movements involved in the production of speech, so that slurred, **scanning speech** results. Many types of ataxia are hereditary, including **Friedreich's ataxia** and **Machado-Joseph disease.** There is no cure for hereditary ataxias. Voluntary movements are also highly abnormal when the cerebellum is damaged. For example, attempting to touch an object with a finger results in overshooting. This **dysmetria** or **past-pointing** promptly initiates a gross corrective action, but the correction overshoots to the other side, and the finger oscillates back and forth. This oscillation is called an **intention tremor.** Another characteristic of cerebellar disease is inability to "put on the brakes" to stop movement promptly. Normally, for example, flexion of the forearm against resistance is quickly checked when the resistance force is suddenly broken off. The patient with cerebellar disease cannot stop the movement of the limb, and the forearm flies backward in a wide arc. This abnormal response is known as the **rebound phenomenon.** This is one of the important reasons these patients show **dysdiadochokinesia,** the inability to perform rapidly alternating opposite movements such as repeated pronation and supination of the hands. Finally, patients with cerebellar disease have difficulty performing actions that involve simultaneous motion at more than one joint. They dissect such movements and carry them out one joint at a time, a phenomenon known as **decomposition of movement.** Other signs of cerebellar deficit in humans point to the importance of the cerebellum in the control of movement.

Motor abnormalities associated with cerebellar damage vary depending on the region involved. The major dysfunction seen after damage to the vestibulocerebellum is ataxia, **dysequilibrium,** and **nystagmus.** Damage to the vermis and fastigial nucleus (part of the spinocerebellum) leads to disturbances in control of axial and trunk muscles during attempted antigravity postures and scanning speech. Degeneration of this portion of the cerebellum can result from thiamine deficiency in alcoholics or malnourished individuals. The major dysfunction seen after damage to the cerebrocerebellum is delays in initiating movements and decomposition of movement.

THERAPEUTIC HIGHLIGHTS

Management of ataxia is primarily supportive; it often includes physical, occupational, and speech therapy. Attempts to identify effective drug therapies have met with little success. **Deep brain stimulation** of the ventral intermediate nucleus of the thalamus may reduce cerebellar tremor, but it is less effective in reducing ataxia. There is some evidence that a deficiency in **coenzyme Q10** (CoQ10) contributes to the abnormalities seen in some forms of familial ataxia. If low levels of CoQ10 are detected, treatment to replace the missing CoQ10 has been shown to be beneficial.

2000–3000 synapses on it. Climbing fiber activation produces a large, complex spike in the Purkinje cell, and this spike produces long-term modification of the pattern of mossy fiber input to that particular Purkinje cell. Climbing fiber activity is increased when a new movement is being learned, and selective lesions of the olivary complex abolish the ability to produce long-term adjustments in certain motor responses.

CHAPTER SUMMARY

- A muscle spindle is a group of specialized intrafusal muscle fibers with contractile polar ends and a noncontractile center that is located in parallel to the extrafusal muscle fibers and is innervated by types Ia and II afferent fibers and efferent γ-motor neurons. Muscle stretch activates the muscle spindle to initiate reflex contraction of the extrafusal muscle fibers in the same muscle (stretch reflex).
- A Golgi tendon organ is a netlike collection of knobby nerve endings among the fascicles of a tendon that is located in series with extrafusal muscle fibers and innervated by type Ib afferents. They are stimulated by both passive stretch and active contraction of the muscle to relax the muscle (inverse stretch reflex) and function as a transducer to regulate muscle force.
- The flexor withdrawal reflex is a polysynaptic reflex that is initiated by nociceptive stimuli; it can serve as a protective mechanism to present further injury.
- Spinal cord transection is followed by a period of spinal shock during which all spinal reflex responses are profoundly depressed. In humans, recovery begins about 2 weeks after the injury.
- The supplementary cortex, basal ganglia, and cerebellum participate in the planning of skilled movements. Commands from the primary motor cortex and other cortical regions are relayed via the corticospinal and corticobulbar tracts to spinal and brainstem motor neurons. Cortical areas and motor pathways descending from the cortex are somatotopically organized.
- The ventral corticospinal tract and medial descending brain stem pathways (tectospinal, reticulospinal, and

vestibulospinal tracts) regulate proximal muscles and posture. The lateral corticospinal and rubrospinal tracts control distal limb muscles for fine motor control and skilled voluntary movements.

- Decerebrate rigidity leads to hyperactivity in extensor muscles in all four extremities; it is actually spasticity due to facilitation of the stretch reflex. It resembles what is seen with uncal herniation due to a supratentorial lesion. Decorticate rigidity is flexion of the upper extremities at the elbow and extensor hyperactivity in the lower extremities. It occurs on the hemiplegic side after hemorrhage or thrombosis in the internal capsule.
- The basal ganglia include the caudate nucleus, putamen, globus pallidus, subthalamic nucleus, and substantia nigra. The connections between the parts of the basal ganglia include a dopaminergic nigrostriatal projection from the substantia nigra to the striatum and a GABAergic projection from the striatum to substantia nigra.
- Parkinson disease is due to degeneration of the nigrostriatal dopaminergic neurons and is characterized by akinesia, bradykinesia, cogwheel rigidity, and tremor at rest. Huntington disease is characterized by choreiform movements due to the loss of the GABAergic inhibitory pathway to the globus pallidus.
- The cerebellar cortex contains five types of neurons: Purkinje, granule, basket, stellate, and Golgi cells. The two main inputs to the cerebellar cortex are climbing fibers and mossy fibers. Purkinje cells are the only output from the cerebellar cortex, and they generally project to the deep nuclei. Damage to the cerebellum leads to several characteristic abnormalities, including hypotonia, ataxia, and intention tremor.

MULTIPLE-CHOICE QUESTIONS

For all questions, select the single best answer unless otherwise directed.

1. When dynamic γ-motor neurons are activated at the same time as α-motor neurons to muscle,
   A. prompt inhibition of discharge in spindle Ia afferents takes place.
   B. clonus is likely to occur.
   C. the muscle will not contract.
   D. the number of impulses in spindle Ia afferents is smaller than when α discharge alone is increased.
   E. the number of impulses in spindle Ia afferents is greater than when α discharge alone is increased.
2. The inverse stretch reflex
   A. occurs when Ia spindle afferents are inhibited.
   B. is a monosynaptic reflex initiated by activation of the Golgi tendon organ.
   C. is a disynaptic reflex with a single interneuron inserted between the afferent and efferent limbs.
   D. is a polysynaptic reflex with many interneurons inserted between the afferent and efferent limbs.
   E. uses type II afferent fibers from the Golgi tendon organ.
3. Withdrawal reflexes are *not*
   A. initiated by nociceptive stimuli.
   B. prepotent.
   C. prolonged if the stimulus is strong.
   D. an example of a flexor reflex.
   E. accompanied by the same response on both sides of the body.
4. While exercising, a 42-year-old female developed sudden onset of tingling in her right leg and an inability to control movement in that limb. A neurological exam showed a hyperactive knee jerk reflex and a positive Babinski sign. Which of the following is *not* characteristic of a reflex?
   A. Reflexes can be modified by impulses from various parts of the CNS.
   B. Reflexes may involve simultaneous contraction of some muscles and relaxation of others.
   C. Reflexes are chronically suppressed after spinal cord transection.
   D. Reflexes involve transmission across at least one synapse.
   E. Reflexes often occur without conscious perception.
5. Increased neural activity before a skilled voluntary movement is *first* seen in the
   A. spinal motor neurons.
   B. precentral motor cortex.
   C. midbrain.
   D. cerebellum.
   E. cortical association areas.
6. A 58-year-old woman was brought to the emergency room of her local hospital because of a sudden change of consciousness. All four limbs were extended, suggestive of decerebrate rigidity. A brain CT showed a rostral pontine hemorrhage. Which of the following describes components of the central pathway responsible for control of posture?
   A. The tectospinal pathway terminates on neurons in the dorsolateral area of the spinal ventral horn that innervate limb muscles.
   B. The medullary reticulospinal pathway terminates on neurons in the ventromedial area of the spinal ventral horn that innervate axial and proximal muscles.
   C. The pontine reticulospinal pathway terminates on neurons in the dorsomedial area of the spinal ventral horn that innervate limb muscles.
   D. The medial vestibular pathway terminates on neurons in the dorsomedial area of the spinal ventral horn that innervate axial and proximal muscles.
   E. The lateral vestibular pathway terminates on neurons in the dorsolateral area of the spinal ventral horn that innervate axial and proximal muscles.
7. A 38-year-old female had been diagnosed with a metastatic brain tumor. She was brought to the emergency room of her local hospital because of irregular breathing and progressive loss of consciousness. She also showed signs of decerebrate posturing. Which of the following is *not* true about decerebrate rigidity?
   A. It involves hyperactivity in extensor muscles of all four limbs.
   B. The excitatory input from the reticulospinal pathway activates γ-motor neurons which indirectly activate α-motor neurons.
   C. It is actually a type of spasticity due to inhibition of the stretch reflex.
   D. It resembles what ensues after uncal herniation.
   E. Lower extremities are extended with toes pointed inward.

8. Which of the following describes a connection between components of the basal ganglia?
   A. The subthalamic nucleus releases glutamate to excite the globus pallidus, internal segment.
   B. The substantia nigra pars reticulata releases dopamine to inhibit the striatum.
   C. The substantia nigra pars compacta releases dopamine to excite the globus pallidus, external segment.
   D. The striatum releases acetylcholine to excite the substantia nigra pars reticulata.
   E. The globus pallidus, external segment releases glutamate to excite the striatum.
9. A 60-year old male was diagnosed 15 years ago with Parkinson disease. He has been taking Sinemet and, until recently, has been able to continue to work and help with routine jobs around the house. Now his tremor and rigidity interfere with these activities. His physician has suggested that he undergo deep brain stimulation therapy. The therapeutic effect of L-dopa in patients with Parkinson disease eventually wears off because
   A. antibodies to dopamine receptors develop.
   B. inhibitory pathways grow into the basal ganglia from the frontal lobe.
   C. there is an increase in circulating α-synuclein.
   D. the normal action of nerve growth factor (NGF) is disrupted.
   E. the dopaminergic neurons in the substantia nigra continue to degenerate.
10. An 8-year-old girl was brought to her pediatrician because her parents noted frequent episodes of gait unsteadiness and speech difficulties. Her mother was concerned because of a family history of Friedreich's ataxia. Which of the following is a correct description of connections involving cerebellar neurons?
   A. Basket cells release glutamate to activate Purkinje cells.
   B. Climbing fiber inputs exert a strong excitatory effect on Purkinje cells, and mossy fiber inputs exert a strong inhibitory effect on Purkinje cells.
   C. Granule cells release glutamate to excite basket cells and stellate cells.
   D. The axons of Purkinje cells are the sole output of the cerebellar cortex, and they release glutamate to excite the deep cerebellar nuclei.
   E. Golgi cells are inhibited by mossy fiber collaterals.
11. After falling down a flight of stairs, a young woman is found to have partial loss of voluntary movement on the right side of her body and loss of pain and temperature sensation on the left side below the midthoracic region. It is probable that she has a lesion
   A. transecting the left half of the spinal cord in the lumbar region.
   B. transecting the left half of the spinal cord in the upper thoracic region.
   C. transecting sensory and motor pathways on the right side of the pons.
   D. transecting the right half of the spinal cord in the upper thoracic region.
   E. transecting the dorsal half of the spinal cord in the upper thoracic region.
12. At the age of 30, a male postal worker reported weakness in his right leg. Within a year the weakness had spread to his entire right side. A neurological examination revealed flaccid paralysis, muscular atrophy, fasciculations, hypotonia, and hyporeflexia of muscles in the right arm and leg. Sensory and cognitive function tests were normal. Which of the following diagnosis is likely?
   A. A large tumor in the left primary motor cortex.
   B. A cerebral infarct in the region of the corona radiate.
   C. A vestibulocerebellar tumor.
   D. Damage to the basal ganglia.
   E. Amyotrophic lateral sclerosis.

CHAPTER RESOURCES

Alexi T, Liu X-Z, Qu Y, et al: Neuroprotective strategies for basal ganglia degeneration: Parkinson's and Huntington's diseases. Prog Neurbiol 2000;60:409.

De Zeeuw CI, Strata P, Voogd J: *The Cerebellum: From Structure to Control.* Elsevier, 1997.

Ditunno JF Jr, Formal CF: Chronic spinal cord injury. N Engl J Med 1994; 330:550.

Graybiel AM, Delong MR, Kitai ST: *The Basal Ganglia VI.* Springer, 2003.

Hunt CC: Mammalian muscle spindle: Peripheral mechanisms. Physiol Rev 1990;70: 643.

Jankowska E: Interneuronal relay in spinal pathways from proprioceptors. Prog Neurobiol 1992;38:335.

Jueptner M, Weiller C: A review of differences between basal ganglia and cerebellar control of movements as revealed by functional imaging studies. Brain 1998;121:1437.

Latash ML: *Neurophysiological Basis of Movement,* 2nd ed. Human Kinetics, 2008.

Lemon RN: Descending pathways in motor control. Annu Rev Neurosci 2008;31:195.

Lundberg A: Multisensory control of spinal reflex pathways. Prog Brain Res 1979;50:11.

Manto MU, Pandolfo M: *The Cerebellum and its Disorders.* Cambridge University Press, 2001.

Matyas F, Sreenivasan V, Marbach F, Wacongne C, Barsy B, Mateo C, Aronoff R, Petersen CCH: Motor control of sensory cortex. Science 2010;26:1240.

McDonald JW, Liu X-Z, Qu Y, et al: Transplanted embryonic stem cells survive, differentiate and promote recovery in injured rat spinal cord. Nature Med 1999;5:1410.

Nudo RJ: Postinfarct cortical plasticity and behavioral recovery. Stroke 2007;38:840.

Ramer LM, Ramer MS, Steeves JD: Setting the stage for functional repair of spinal cord injuries: a cast of thousands. Spinal Cord 2005;43:134.

Stein RB, Thompson AK: Muscle reflexes and motion: How, what, and why? Exerc Sport Sci Rev 2006;34:145.

CHAPTER

13

Autonomic Nervous System

OBJECTIVES

After studying this chapter, you should be able to:

- Describe the location of the cell bodies and axonal trajectories of preganglionic and postganglionic sympathetic and parasympathetic neurons.
- Name the neurotransmitters that are released by preganglionic autonomic neurons, postganglionic sympathetic neurons, postganglionic parasympathetic neurons, and adrenal medullary cells.
- Name the types of receptors on autonomic ganglia and on various target organs and list the ways that drugs can act to alter the function of the processes involved in transmission within the autonomic nervous system.
- Describe functions of the sympathetic and parasympathetic nervous systems.
- Describe the location of some forebrain and brainstem neurons that are components of central autonomic pathways.
- Describe the composition and functions of the enteric nervous system.

INTRODUCTION

The autonomic nervous system (ANS) is the part of the nervous system that is responsible for homeostasis. Except for skeletal muscle, which gets its innervation from the somatomotor nervous system, innervation to all other organs is supplied by the ANS. Nerve terminals are located in smooth muscle (eg, blood vessels, the wall of the gastrointestinal tract, urinary bladder), cardiac muscle, and glands (eg, sweat glands, salivary glands). Although survival is possible without an ANS, the ability to adapt to environmental stressors and other challenges is severely compromised (see Clinical Box 13–1). The importance of understanding the functions of the ANS is underscored by the fact that so many drugs used to treat a vast array of diseases exert their actions on elements of the ANS.

The ANS has two major and anatomically distinct divisions: the **sympathetic** and **parasympathetic** nervous systems. As will be described, some target organs are innervated by both divisions and others are controlled by only one. In addition, the ANS includes the **enteric nervous system** within the gastrointestinal tract. The classic definition of the ANS is the preganglionic and postganglionic neurons within the sympathetic and parasympathetic divisions. This would be equivalent to defining the somatomotor nervous system as the cranial and spinal motor neurons. A modern definition of the ANS takes into account the descending pathways from several forebrain and brain stem regions as well as visceral afferent pathways that set the level of activity in sympathetic and parasympathetic nerves. This is analogous to including the many descending and ascending pathways that influence the activity of somatic motor neurons as elements of the somatomotor nervous system.

CLINICAL BOX 13–1

Multiple System Atrophy & Shy–Drager Syndrome

Multiple system atrophy (MSA) is a neurodegenerative disorder associated with autonomic failure due to loss of preganglionic autonomic neurons in the spinal cord and brain stem. In the absence of an autonomic nervous system, it is difficult to regulate body temperature, fluid and electrolyte balance, and blood pressure. In addition to these autonomic abnormalities, MSA presents with cerebellar, basal ganglia, locus coeruleus, inferior olivary nucleus, and pyramidal tract deficits. MSA is defined as "a sporadic, progressive, adult onset disorder characterized by autonomic dysfunction, parkinsonism, and cerebellar ataxia in any combination." **Shy–Drager syndrome** is a subtype of MSA in which autonomic failure dominates. The pathological hallmark of MSA is cytoplasmic and nuclear inclusions in oligodendrocytes and neurons in central motor and autonomic areas. There is also depletion of monoaminergic, cholinergic, and peptidergic markers in several brain regions and in the cerebrospinal fluid. The cause of MSA remains elusive, but there is some evidence that a neuroinflammatory mechanism causing activation of microglia and production of toxic cytokines may occur in brains of MSA patients. Basal levels of sympathetic activity and plasma norepinephrine levels are normal in MSA patients, but they fail to increase in response to standing or other stimuli and leads to severe **orthostatic hypotension.** In addition to the fall in blood pressure, orthostatic hypotension leads to dizziness, dimness of vision, and even fainting. MSA is also accompanied by parasympathetic dysfunction, including urinary and sexual dysfunction. MSA is most often diagnosed in individuals between 50 and 70 years of age; it affects more men than women. Erectile dysfunction is often the first symptom of the disease. There are also abnormalities in baroreceptor reflex and respiratory control mechanisms. Although autonomic abnormalities are often the first symptoms, 75% of patients with MSA also experience motor disturbances.

THERAPEUTIC HIGHLIGHTS

There is no cure for MSA but various therapies are used to treat specific signs and symptoms of the disease. **Corticosteroids** are often prescribed to retain salt and water to increase blood pressure. In some individuals, Parkinsonium-like signs can be alleviated by administration of **levodopa** and **carbidopa.** Various clinical trials are underway to test the effectiveness of using intravenous **immunoglobulins** to counteract the neuroinflammatory process that occurs in MSA; **fluoxetine** (a **serotonin uptake inhibitor**) to prevent orthostatic hypotension, improve mood, and alleviate sleep, pain, and fatigue in MSA patients; and **rasagiline** (a **monoamine oxidase inhibitor**) in MSA patients with parskinsonism.

ANATOMIC ORGANIZATION OF AUTONOMIC OUTFLOW

GENERAL FEATURES

Figure 13–1 compares some fundamental characteristics of the innervation to skeletal muscles with innervation to smooth muscle, cardiac muscle, and glands. As discussed in earlier chapters, the final common pathway linking the central nervous system (CNS) to skeletal muscles is the α-motor neuron. Similarly, sympathetic and parasympathetic neurons serve as the final common pathway from the CNS to visceral targets. However, in marked contrast to the somatomotor nervous system, the peripheral motor portions of the ANS are made up of two neurons: **preganglionic** and **postganglionic neurons.** The cell bodies of the preganglionic neurons are located in the intermediolateral (IML) column of the spinal cord and in motor nuclei of some cranial nerves. In contrast to the large diameter and rapidly conducting α-motor neurons, preganglionic axons are small-diameter, myelinated, relatively slowly conducting B fibers. A preganglionic axon diverges to an average of eight or nine postganglionic neurons. In this way, autonomic output is diffuse. The axons of the postganglionic neurons are mostly unmyelinated C fibers and terminate on the visceral effectors.

One similar feature of autonomic preganglionic neurons and α-motor neurons is that acetylcholine is released at their nerve terminals (Figure 13–1). This is the neurotransmitter released by all neurons whose axons exit the CNS, including cranial motor neurons, α-motor neurons, γ-motor neurons, preganglionic sympathetic neurons, and preganglionic parasympathetic neurons. Postganglionic parasympathetic neurons also release acetylcholine, whereas postganglionic sympathetic neurons release either norepinephrine or acetylcholine.

SYMPATHETIC DIVISION

In contrast to α-motor neurons, which are located at all spinal segments, sympathetic preganglionic neurons are located in the IML of only the first thoracic to the third or fourth lumbar segments. This is why the sympathetic nervous system is sometimes called the thoracolumbar division of the ANS. The axons of the sympathetic preganglionic neurons

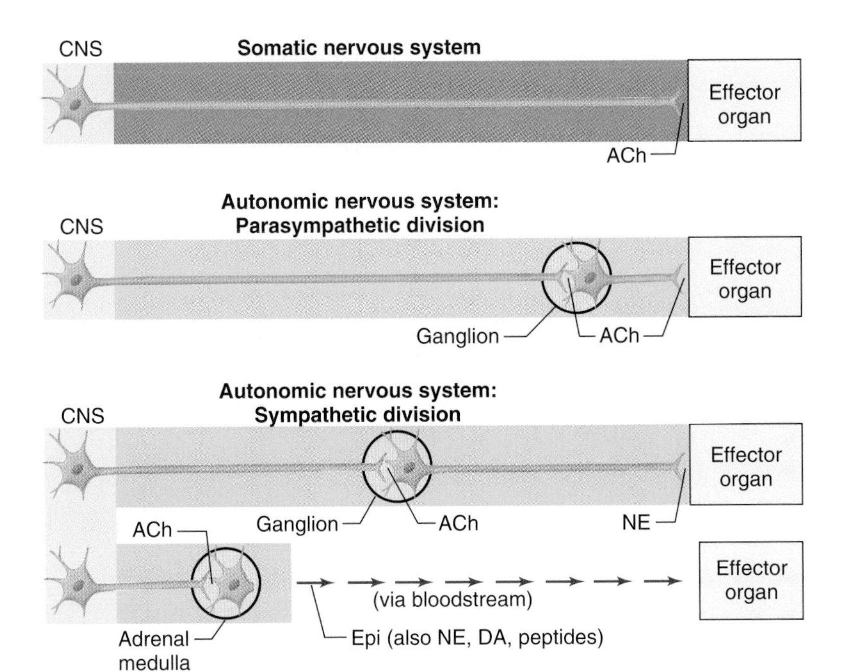

FIGURE 13–1 Comparison of peripheral organization and transmitters released by somatomotor and autonomic nervous systems. In the case of the somatomotor nervous system, the neuron that leaves the spinal cord projects directly to the effector organ. In the case of the autonomic nervous system, there is a synapse between the neuron that leaves the spinal cord and the effector organ (except for neurons that innervate the adrenal medulla). Note that all neurons that leave the central nervous system release acetylcholine (ACh). DA, dopamine; Epi, epinephrine; NE, norepinephrine. (From Widmaier EP, Raff H, Strang KT: *Vander's Human Physiology*. McGraw-Hill, 2008.)

leave the spinal cord at the level at which their cell bodies are located and exit via the ventral root along with axons of α- and γ-motor neurons (Figure 13–2). They then separate from the ventral root via the **white rami communicans** and project to the adjacent **sympathetic paravertebral ganglion,** where some of them end on the cell bodies of the postganglionic neurons. Paravertebral ganglia are located adjacent to each thoracic and upper lumbar spinal segment; in addition, there are a few ganglia adjacent to the cervical and sacral spinal segments. The ganglia are connected to each other via the axons of preganglionic neurons that travel rostrally or caudally to terminate on postganglionic neurons located at some distance. Together these ganglia and axons form the **sympathetic chain** bilaterally. This arrangement is seen in Figure 13–2 and Figure 13–3.

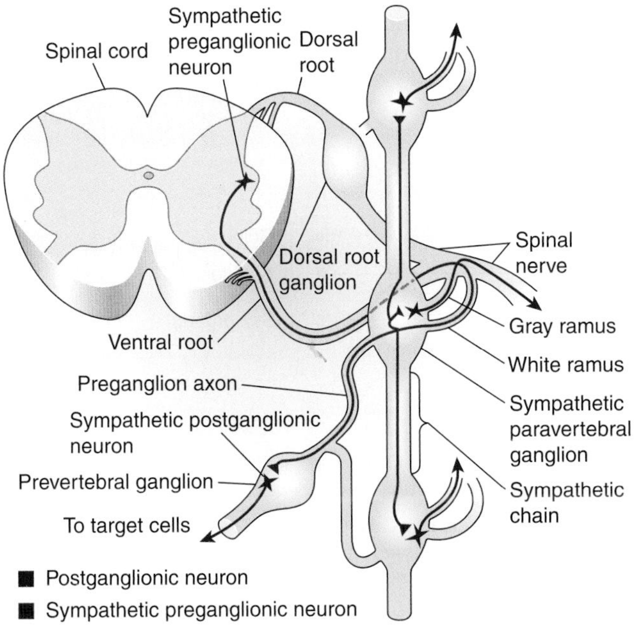

FIGURE 13–2 Projection of sympathetic preganglionic and postganglionic fibers. The drawing shows the thoracic spinal cord, paravertebral, and prevertebral ganglia. Preganglionic neurons are shown in red, and postganglionic neurons in dark blue. (Courtesy of P. Banyas, Michigan State University.)

Some preganglionic neurons pass through the paravertebral ganglion chain and end on postganglionic neurons located in **prevertebral** (or **collateral) ganglia** close to the viscera, including the celiac, superior mesenteric, and inferior mesenteric ganglia (Figure 13–3). There are also preganglionic neurons whose axons terminate directly on the effector organ, the adrenal gland.

The axons of some of the postganglionic neurons leave the chain ganglia and reenter the spinal nerves via the **gray rami communicans** and are distributed to autonomic effectors in the areas supplied by these spinal nerves (Figure 13–2). These postganglionic sympathetic nerves terminate mainly on smooth muscle (eg, blood vessels, hair follicles) and on sweat glands in the limbs. Other postganglionic fibers leave the chain ganglia to enter the thoracic cavity to terminate in visceral organs. Postganglionic fibers from prevertebral ganglia also terminate in visceral targets.

PARASYMPATHETIC DIVISION

The parasympathetic nervous system is sometimes called the **craniosacral division** of the ANS because of the location of its preganglionic neurons; preganglionic neurons are located in

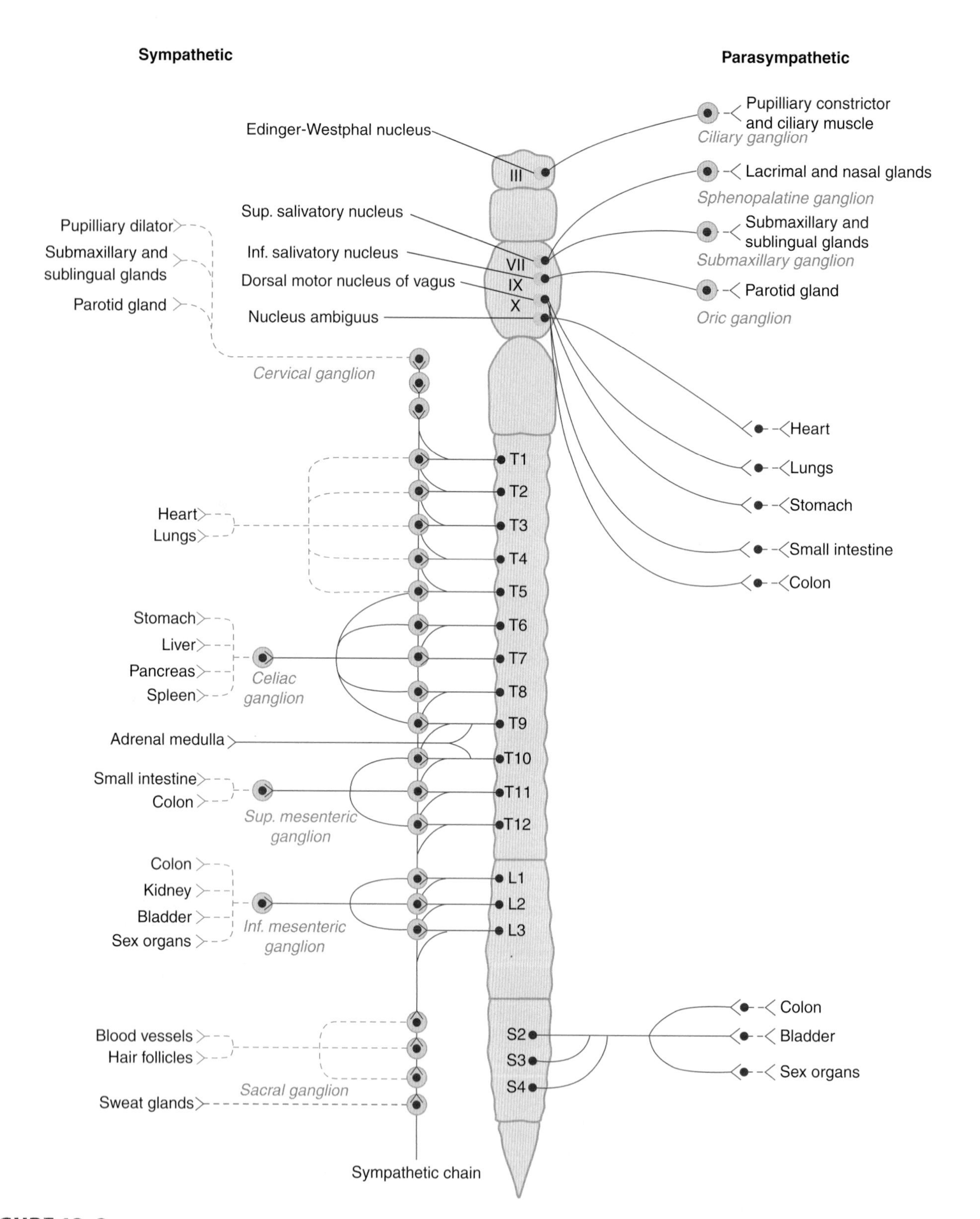

FIGURE 13–3 Organization of sympathetic (left) and parasympathetic (right) nervous systems. Cholinergic nerves are shown in red and noradrenergic nerves are shown in blue. Preganglionic nerves are solid lines; postganglionic nerves are dashed lines. (Courtesy of P. Banyas, Michigan State University.)

several cranial nerve nuclei (III, VII, IX, and X) and in the IML of the sacral spinal cord (Figure 13–3). The cell bodies in the **Edinger-Westphal nucleus** of the oculomotor nerve project to the **ciliary ganglia** to innervate the sphincter (constrictor) muscle of the iris and the ciliary muscle. Neurons in the **superior salivatory nucleus** of the facial nerve project to the **sphenopalatine ganglia** to innervate the lacrimal glands and nasal and palatine mucous membranes and to the **submandibular ganglia** to innervate the submandibular and submaxillary glands. The cell bodies in the **inferior salivatory nucleus** of the glossopharyngeal nerve project to the **otic ganglion** to innervate the parotid salivary gland. Vagal preganglionic fibers synapse on ganglia cells clustered within the walls of visceral organs; thus these parasympathetic postganglionic fibers are very short. Neurons in the **nucleus ambiguus** innervate the sinoatrial (SA) and atrioventricular (AV) nodes in the heart and neurons in the **dorsal motor vagal nucleus** innervate the esophagus, trachea, lungs, and gastrointestinal tract. The parasympathetic sacral outflow **(pelvic nerve)** supplies the pelvic viscera via branches of the second to fourth sacral spinal nerves.

CHEMICAL TRANSMISSION AT AUTONOMIC JUNCTIONS

ACETYLCHOLINE & NOREPINEPHRINE

The first evidence for chemical neurotransmission was provided by a simple yet dramatic study by Otto Loewi in 1920 in which he showed that the slowing of the heart rate produced by stimulation of the vagal parasympathetic nerves was due to the release of acetylcholine (see Chapter 7). Transmission at the synaptic junctions between preganglionic and postganglionic neurons and between the postganglionic neurons and the autonomic effectors are chemically mediated. The principal transmitter agents involved are **acetylcholine** and **norepinephrine.** The autonomic neurons that are **cholinergic** (ie, release acetylcholine) are (1) all preganglionic neurons, (2) all parasympathetic postganglionic neurons, (3) sympathetic postganglionic neurons that innervate sweat glands, and (4) sympathetic postganglionic neurons that end on blood vessels in some skeletal muscles and produce vasodilation when stimulated (sympathetic vasodilator nerves). The remaining sympathetic postganglionic neurons are **noradrenergic** (ie, release norepinephrine). The adrenal medulla is essentially a sympathetic ganglion in which the postganglionic cells have lost their axons and secrete norepinephrine and epinephrine directly into the bloodstream.

Table 13–1 shows the types of cholinergic and adrenergic receptors at various junctions within the ANS. The junctions in the peripheral autonomic motor pathways are a logical site for pharmacologic manipulation of visceral function. The transmitter agents are synthesized, stored in the nerve endings, and released near the neurons, muscle cells, or gland cells where they bind to various ion channel or G protein-coupled receptors (GPCR). They bind to receptors on these cells, thus initiating their characteristic actions, and they are then removed from the area by reuptake or metabolism. Each of these steps can be stimulated or inhibited, with predictable consequences. Table 13–2 lists how various drugs can affect neurotransmission in autonomic neurons and effector sites.

CHOLINERGIC NEUROTRANSMISSION

The processes involved in the synthesis and breakdown of acetylcholine were described in Chapter 7. Acetylcholine does not usually circulate in the blood, and the effects of localized cholinergic discharge are generally discrete and of short duration because of the high concentration of acetylcholinesterase at cholinergic nerve endings. This enzyme rapidly breaks down the acetylcholine, terminating its actions.

Transmission in autonomic ganglia is mediated primarily by the actions of acetylcholine on nicotinic cholinergic receptors that are blocked by hexamethonium (Figure 13–4). These are called N_N receptors to distinguish them from the nicotinic cholinergic receptors (N_M) that are located at the neuromuscular junction and are blocked by D-tubocurare. Nicotinic receptors are examples of ion gated channels; binding of an agonist to nicotinic receptors opens N^+ and K^+ channels to cause depolarization.

The responses produced in postganglionic neurons by stimulation of their preganglionic innervation include both a rapid depolarization called a **fast excitatory postsynaptic potential (EPSP)** that generates action potentials and a prolonged excitatory postsynaptic potential **(slow EPSP).** The slow response may modulate and regulate transmission through the sympathetic ganglia. The initial depolarization is produced by acetylcholine acting on the N_N receptor. The slow EPSP is produced by acetylcholine acting on a muscarinic receptor on the membrane of the postganglionic neuron.

The release of acetylcholine from postganglionic fibers acts on muscarinic cholinergic receptors, which are blocked by atropine. Muscarinic receptors are GPCR and are divided into subtypes M_1–M_5, but M_2 and M_3 are the main subtypes found in autonomic target organs. M_2 receptors are located in the heart; binding of an agonist to these receptors opens K^+ channels and inhibits **adenylyl cyclase.** M_3 receptors are located on smooth muscle and glands; binding of an agonist to these receptors leads to the formation of **inositol 1,4,5-triphosphate (IP_3)** and **diacylglycerol (DAG)** and an increase in intracellular Ca^{2+}.

Compounds with muscarinic actions include congeners of acetylcholine and drugs that inhibit acetylcholinesterase. Clinical Box 13–2 describes some of signs and therapeutic strategies for the treatment of acute intoxication from **organophosphate cholinesterase inhibitors.** Clinical Box 13–3 describes an example of **cholinergic poisoning** resulting from digestion of toxic mushrooms.

TABLE 13–1 Responses of some effector organs to autonomic nerve activity.

Effector Organs	Parasympathetic Nervous System	Sympathetic Nervous System	
		Receptor Type	Response
Eyes			
Radial muscle of iris	—	α_1	Contraction (mydriasis)
Sphincter muscle of iris	Contraction (miosis)		—
Ciliary muscle	Contraction for near vision		—
Heart			
SA node	Decreased heart rate	β_1	Increased heart rate
Atria & ventricle	Decreased atrial contractility	β_1, β_2	Increased contractility
AV node & Purkinje	Decreased conduction velocity	β_1	Increased conduction velocity
Arterioles			
Skin, splanchnic vessels	—	α_1	Constriction
Skeletal muscle	—	α_1/β_2, M	Constriction/Dilation
Systemic veins	—	$\alpha_1, \alpha_2/\beta_2$	Constriction/Dilation
Bronchial smooth muscle	Contraction	β_2	Relaxation
Stomach & Intestine			
Motility and tone	Increased	$\alpha_1, \alpha_2, \beta_2$	Decreased
Sphincters	Relaxation	α_1	Contraction
Secretion	Stimulation	—	
Gall bladder	Contraction	β_2	Relaxation
Urinary bladder			
Detrusor	Contraction	β_2	Relaxation
Sphincter	Relaxation	α_1	Contraction
Uterus (pregnant)	—	α_1/β_2	Contraction/Relaxation
Male sex organs	Erection	α_1	Ejaculation
Skin			
Pilomotor muscles	—	α_1	Contraction
Sweat glands	—	M	Secretion
Liver	—	α_1, β_2	Glycogenolysis
Pancreas			
Acini	Increased secretion	α	Decreased secretion
Islet cells	—	α/β_2	Decreased/Increased secretion
Salivary glands	Profuse, watery secretion	α_1/β	Thick, viscous secretion/ Amylase secretion
Lacrimal glands	Secretion		—
Adipose tissue	—	β_3	Lipolysis

A dash means the target tissue is not innervated by this division of the autonomic nervous system. Modified from Brunton LL, Chabner BA, Knollmann BC (editors): *Goodman and Gilman's The Pharmacological Basis of Therapeutics*, 12th ed. McGraw-Hill, 2011.

NORADRENERGIC NEUROTANSMISSION

The processes involved in the synthesis, reuptake, and breakdown of norepinephrine were described in Chapter 7. Norepinephrine spreads farther and has a more prolonged action than acetylcholine. Norepinephrine, epinephrine, and dopamine are all found in plasma. The epinephrine and some of the dopamine come from the adrenal medulla, but most of the norepinephrine diffuses into the bloodstream from sympathetic nerve endings. Metabolites of norepinephrine and dopamine also enter the circulation.

TABLE 13–2 Examples of drugs that affect processes involved in autonomic neurotransmission.

Transmission Process	Drug	Site of Drug Action	Drug Action
Neurotransmitter synthesis	Hemicholinium	Membrane of cholinergic nerve terminals	Blocks choline uptake; slows synthesis
	Metyrosine	Cytoplasm of noradrenergic nerve terminals	Inhibits tyrosine hydroxylase; blocks synthesis
Neurotransmitter storage mechanism	Vesamicol	Vesicles in cholinergic nerve terminals	Prevents storage of acetylcholine
	Reserpine	Vesicles in noradrenergic nerve terminals	Prevents storage of norepinephrine
Neurotransmitter release mechanism	Norepinephrine, dopamine, acetylcholine, prostaglandins	Receptors on cholinergic and adrenergic nerve terminals	Modulates transmitter release
Neurotransmitter reuptake mechanism	Cocaine, tricyclic antidepressants	Adrenergic nerve terminals	Inhibits uptake; prolongs transmitter's action on postsynaptic receptors
Inactivation of neurotransmitter	Edrophonium, neostigmine, physostigmine,	Acetylcholinesterase in cholinergic synapses	Inhibits enzyme; prolongs and intensifies actions of acetylcholine
Adrenoceptor activation	α_1: Phenylephrine α_2: Clonidine	Sympathetic postganglionic nerve-effector organ junctions (eg, blood vessels, hair follicles, radial muscle)	Binds to and activates α-adrenoceptors; ↑ IP_3/DAG cascade (α_1) or ↓ cAMP (α_2)
	β_1: Dobutamine β_2: Albuterol, ritodrine, salmeterol, terbutaline	Sympathetic postganglionic nerve-effector organ junctions (eg, heart, bronchial smooth muscle, uterine smooth muscle)	Binds to and activates β-adrenoceptors; ↑ cAMP
Adrenoceptor blockade	Nonselective: Phenoxybenzamine α_1: Prazosin, terazosin α_2: Yohimbine	Sympathetic postganglionic nerve-effector organ junctions (eg, blood vessels)	Binds to and blocks α-adrenoceptors
	β_1, β_2: Propranolol $\beta_1 > \beta_2$: Atenolol, esmolol	Sympathetic postganglionic nerve-effector organ junctions (eg, heart, bronchial smooth muscle)	Binds to and blocks β-adrenoceptors
Nicotinic receptor activation	Nicotine	Receptors on autonomic ganglia	Binds to nicotinic receptors; opens Na^+, K^+ channels in postsynaptic membrane
Nicotinic receptor blockade	Hexamethonium, trimethaphan	Receptors on autonomic ganglia	Binds to and blocks nicotinic receptors
Muscarinic receptor activation	Bethanecol	Cholinergic receptors on smooth muscle, cardiac muscle, and glands	Binds to and activates muscarinic receptors; ↑ IP_3/DAG cascade or ↓ cAMP
Muscarinic receptor blockade	Atropine, ipratropium, scopolamine, tropicamide	Cholinergic receptors on smooth muscle, cardiac muscle, and glands	Binds to and blocks muscarinic receptors

Many of these drugs can also act on cholinergic and adrenergic receptors in the central nervous system. For example, the main actions of clonidine and yohimbine to alter blood pressure are by their actions in the brain. cAMP, cyclic adenosine monophosphate; IP_3/DAG, inositol 1,4,5-triphosphate and diacylglycerol.

The norepinephrine released from sympathetic postganglionic fibers binds to adrenoceptors. These are also GPCR and are divided into several subtypes: α_1, α_2, β_1, β_2, and β_3. Table 13–1 shows some of the locations of these receptor subtypes on smooth muscles, cardiac muscle, and glands on autonomic effector targets. Binding of an agonist to α_1-adrenoceptors activates the G_q-coupling protein, which leads to formation of IP_3 and DAG and an increase in intracellular Ca^{2+}. Binding of an agonist to α_2-adrenoceptors causes dissociation of the inhibitory G protein G_i to inhibit adenylyl cyclase and decrease **cyclic adenosine monophosphate (cAMP).** Binding of an agonist to β-adrenoceptors activates the G_s-coupling protein to activate adenylyl cyclase and increase cAMP.

There are several diseases or syndromes that result from dysfunction of sympathetic innervation of specific body regions. Clinical Box 13–4 describes **Horner syndrome,** which is due to interruption of sympathetic nerves to the

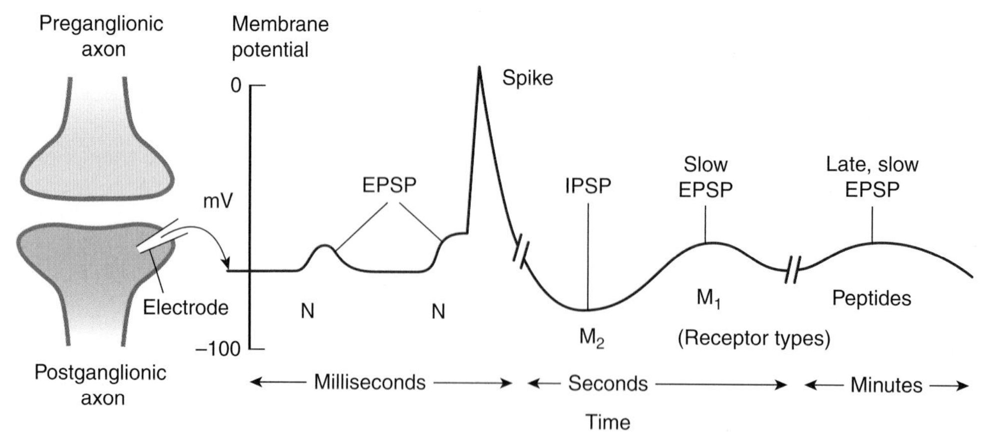

FIGURE 13–4 Schematic of excitatory and inhibitory postsynaptic potentials (EPSP and IPSP) recorded via an electrode in an autonomic ganglion cell. In response acetylcholine release from the preganglionic neuron, two EPSPs were generated in the postganglionic neuron due to nicotinic (N) receptor activation. The first EPSP was below the threshold for eliciting an action potential, but the second EPSP was suprathreshold and evoked an action potential. This was followed by an IPSP, probably evoked by muscarinic (M_2) receptor activation. The IPSP is then followed by a slower, M_1-dependent EPSP, and this can be followed by an even slower peptide-induced EPSP. (From Katzung BG, Maters SB, Trevor AJ: *Basic & Clinical Pharmacology*, 11th ed. McGraw-Hill, 2009.)

CLINICAL BOX 13–2

Organophosphates: Pesticides and Nerve Gases

The World Health Organization estimates that 1–3% of agricultural workers worldwide suffer from **acute pesticide poisoning;** it accounts for significant morbidity and mortality, especially in developing countries. Like **organophosphate** pesticides (eg, **parathion, malathion**), **nerve gases** (eg, **soman, sarin**) used in chemical warfare and terrorism inhibit acetylcholinesterase at peripheral and central cholinergic synapses, prolonging the actions of acetylcholine at these synapses. The organophosphate **cholinesterase inhibitors** are readily absorbed by the skin, lung, gut, and conjunctiva, making them very dangerous. They bind to the enzyme and undergo hydrolysis, resulting in a phosphorylated active site on the enzyme. The covalent phosphorous-enzyme bond is very stable and hydrolyzes at a very slow rate. The phosphorylated enzyme complex may undergo a process called **aging** in which one of the oxygen-phosphorous bonds breaks down, which strengthens the phosphorous-enzyme bond. This process takes only 10 min to occur after exposure to soman. The earliest signs of organophosphate toxicity are usually indicative of excessive activation of autonomic muscarinic receptors; these include miosis, salivation, sweating, bronchial constriction, vomiting, and diarrhea. CNS signs of toxicity include cognitive disturbances, convulsions, seizures, and even coma; these signs are often accompanied by nicotinic effects such as depolarizing neuromuscular blockade.

THERAPEUTIC HIGHLIGHTS

The muscarinic cholinergic receptor antagonist **atropine** is often given parenterally in large doses to control signs of excessive activation of these receptors. When given soon after exposure to the organophosphate and before aging has occurred, nucleophiles such as **pralidoxime** are able to break the bond between the organophosphate and the acetylcholinesterase. Thus this drug is called a "cholinesterase regenerator." If **pyridostigmine** is administered in advance of exposure to a cholinesterase inhibitor, it binds to the enzyme and prevents binding by the toxic organophosphate agent. The protective effects of pyridostigmine dissipate within 3–6 h, but this provides enough time for clearance of the organophosphate from the body. Since the drug cannot cross the blood brain barrier, protection is limited to peripheral cholinergic synapses. A mixture of pyridostigmine, **carbamate,** and atropine can be administered prophylactically to soldiers and civilians who are at risk for exposure to nerve gases. **Benzodiazepines** can be used to abort the seizures caused by exposure to organophosphates.

CLINICAL BOX 13–3

Mushroom Poisoning

Of the more than 5000 species of mushrooms found in the US, approximately 100 are poisonous and ingestion of about 12 of these can result in fatality. Estimates are an annual incidence of five cases per 100,000 individuals in the US; worldwide databases are not available. Mushroom poisoning or **mycetism** is divided into rapid- (15–30 min after ingestion) and delayed-onset (6–12 h after ingestion) types. In rapid-onset cases caused by mushrooms of the *Inocybe* genus, the symptoms are due to excessive activation of muscarinic cholinergic synapses. The major signs of **muscarinic poisoning** include nausea, vomiting, diarrhea, urinary urgency, vasodilation, sweating, and salivation. Ingestion of mushrooms such as the *Amanita muscaria* exhibit signs of the **antimuscarinic syndrome** rather than muscarinic poisoning because they also contain alkaloids that block muscarinic cholinergic receptors. The classic symptoms of this syndrome are being "red as a beet" (flushed skin), "hot as a hare" (hyperthermia), "dry as a bone" (dry mucous membranes, no sweating), "blind as a bat" (blurred vision, cycloplegia), and "mad as a hatter" (confusion, delirium). The delayed-onset type of mushroom poisoning occurs after ingestion of *Amanita phalloides*, *Amanita virosa*, *Galerina autumnalis*, and *Galerina marginata*. These mushrooms cause abdominal cramping, nausea, vomiting, and profuse diarrhea; but the major toxic effects are due to hepatic injury (jaundice, bruising) and associated central effects (confusion, lethargy, coma). These mushrooms contain **amatoxins** that inhibit RNA polymerase. There is a 60% mortality rate associated with ingestion of these mushrooms.

THERAPEUTIC HIGHLIGHTS

The rapid-onset type muscarinic poisoning can be treated effectively with atropine. Individuals who exhibit the antimuscarinic syndrome can be treated with **physostigmine,** which is a cholinesterase inhibitor with a 2–4 h duration of action that acts centrally and peripherally. If agitated, these individuals may require sedation with a benzodiazepine or an **antipsychotic agent.** The delayed onset toxicity due to ingestion of mushrooms containing amatoxins does not respond to cholinergic drugs. Treatment of amatoxin ingestion includes intravenous administration of fluids and electrolytes to maintain adequate hydration. Administering a combination of a high dose of **penicillin G** and **silibinin** (a **flavonolignan** found in certain herbs with **antioxidant** and hepatoprotective properties) has been shown to improve survival. If necessary, vomiting can also be induced by using activated charcoal to reduce the absorption of the toxin.

CLINICAL BOX 13–4

Horner Syndrome

Horner syndrome is a rare disorder resulting from interruption of preganglionic or postganglionic sympathetic innervation to the face. The problem can result from injury to the nerves, injury to the carotid artery, a stroke or lesion in the brainstem, or a tumor in the lung. In most cases the problem is unilateral, with symptoms occurring only on the side of the damage. The hallmark of Horner syndrome is the triad of **anhidrosis** (reduced sweating), **ptosis** (drooping eyelid), and **miosis** (constricted pupil). Symptoms also include **enophthalmos** (sunken eyeball) and vasodilation.

THERAPEUTIC HIGHLIGHTS

There is no specific pharmacological treatment for Horner syndrome, but drugs affecting noradrenergic neurotransmission can be used to determine whether the source of the problem is interruption of the preganglionic or postganglionic innervation to the face. Since the iris of the eye responds to topical **sympathomimetic drugs** (ie, drugs that are direct agonists on adrenoceptors or drugs that increase the release or prevent reuptake of norepinephrine from the nerve terminal), the physician can easily test the viability of the noradrenergic nerves to the eye. If the postganglionic sympathetic fibers are damaged, their terminals would degenerate and there would be a loss of stored catecholamines. If the preganglionic fibers are damaged, the postganglionic noradrenergic nerve would remain intact (but be inactive) and would still have stored catecholamines in its terminal. If one administers a drug that causes release of catecholamine stores (eg, **hydroxyamphetamine**) and the constricted pupil does not dilate, one would conclude that the noradrenergic nerve is damaged. If the eye dilates in response to this drug, the catecholamine stores are still able to be released, so the damage must be preganglionic. Administration of **phenylephrine** (α-adrenoceptor agonist) would dilate the pupil regardless of the site of injury as the drug binds to the receptor on the radial muscle of the iris.

CLINICAL BOX 13–5

Raynaud Phenomenon

Approximately 5% of men and 8% of women experience an episodic reduction in blood flow primarily to the fingers, often during exposure to cold or during a stressful situation. Vasospasms in the toes, tip of nose, ears, and penis can also occur. Smoking is associated with an increase in the incidence and severity of the symptoms of Raynaud phenomenon. The symptoms begin to occur between the age of 15 and 25; it is most common in cold climates. The symptoms often include a triphasic change in color to the skin of the digits. Initially, the skin becomes pale or white **(pallor),** cold, and numb. This can be followed by a **cyanotic** period in which the skin turns blue or even purple, during which time the reduced blood flow can cause intense pain. Once the blood flow recovers, the digits often turn deep red **(rubor)** and there can be swelling and tingling. Primary Raynaud phenomenon or **Raynaud disease** refers to the idiopathic appearance of the symptoms in individuals who do not have another underlying disease to account for the symptoms. In such cases, the vasospastic attacks may merely be an exaggeration of a normal response to cold temperature or stress. Secondary Raynaud phenomenon or **Raynaud syndrome** refers to the presence of these symptoms due to another disorder such as **scleroderma, lupus, rheumatoid arthritis, Sjogren syndrome, carpel tunnel syndrome,** and **anorexia.** Although initially thought to reflect an increase in sympathetic activity to the vasculature of the digits, this is no longer regarded as the mechanism underlying the episodic vasospasms.

THERAPEUTIC HIGHLIGHTS

The first treatment strategy for Raynaud phenomenon is to avoid exposure to the cold, reduce stress, quit smoking, and avoid the use of medications that are vasoconstrictors (eg, β-adrenoceptor antagonists, cold medications, caffeine, and narcotics). If the symptoms are severe, drugs may be needed to prevent tissue damage. These include **calcium channel blockers** (eg, **nifedipine**) and **α-adrenoceptor antagonists** (eg, **prazosin**). In individuals who do not respond to pharmacological treatments, surgical sympathectomy has been done.

face. Clinical Box 13–5 describes a vasospastic condition in which blood flow to the fingers and toes is transiently reduced, typically when a sensitive individual is exposed to stress or cold.

NONADRENERGIC, NONCHOLINERGIC TRANSMITTERS

In addition to the "classical neurotransmitters," some autonomic fibers also release neuropeptides, although their exact functions in autonomic control have not been determined. The small granulated vesicles in postganglionic noradrenergic neurons contain **adenosine triphosphate (ATP)** and norepinephrine, and the large granulated vesicles contain **neuropeptide Y (NPY).** There is some evidence that low-frequency stimulation promotes release of ATP, whereas high-frequency stimulation causes release of NPY. Some visceral organs contain purinergic receptors, and evidence is accumulating that ATP is a mediator in the ANS along with norepinephrine.

Many sympathetic fibers innervating the vasculature of viscera, skin, and skeletal muscles release NPY and **galanin** in addition to norepinephrine. **Vasoactive intestinal polypeptide (VIP), calcitonin gene-related peptide (CGRP),** or **substance P** are co-released with acetylcholine from the sympathetic innervation to sweat glands **(sudomotor fibers).** VIP is co-localized with acetylcholine in many cranial parasympathetic postganglionic neurons supplying glands. Vagal parasympathetic postganglionic neurons in the gastrointestinal tract contain VIP and the enzymatic machinery to synthesize **nitric oxide (NO).**

RESPONSES OF EFFECTOR ORGANS TO AUTONOMIC NERVE IMPULSES

GENERAL PRINCIPLES

The ANS is responsible for regulating and coordinating many physiological functions that include blood flow, blood pressure, heart rate, airflow through the bronchial tree, gastrointestinal motility, urinary bladder contraction, glandular secretions, pupillary diameter, body temperature, and sexual physiology.

The effects of stimulation of the noradrenergic and cholinergic postganglionic nerve fibers are indicated in Figure 13–3 and Table 13–1. These findings point out another difference between the ANS and the somatomotor nervous system. The release of acetylcholine by α-motor neurons only leads to contraction of skeletal muscles. In contrast, release of acetylcholine onto smooth muscle of some organs leads to contraction (eg, walls of the gastrointestinal tract) while release onto other organs promotes relaxation (eg, sphincters in the gastrointestinal tract). The only way to relax a skeletal muscle is to inhibit the discharges of the α-motor neurons; but for some targets innervated by the

ANS, one can shift from contraction to relaxation by switching from activation of the parasympathetic nervous system to activation of the sympathetic nervous system. This is the case for the many organs that receive dual innervation with antagonistic effects, including the digestive tract, airways, and urinary bladder. The heart is another example of an organ with dual antagonistic control. Stimulation of sympathetic nerves increases heart rate; stimulation of parasympathetic nerves decreases heart rate.

In other cases, the effects of sympathetic and parasympathetic activation can be considered complementary. An example is the innervation of salivary glands. Parasympathetic activation causes release of watery saliva, while sympathetic activation causes the production of thick, viscous saliva.

The two divisions of the ANS can also act in a synergistic or cooperative manner in the control of some functions. One example is the control of pupil diameter in the eye. Both sympathetic and parasympathetic innervations are excitatory, but the former contracts the radial muscle to cause mydriasis (widening of the pupil) and the latter contracts the sphincter (or constrictor) muscle to cause miosis (narrowing of the pupil). Another example is the synergistic actions of these nerves on sexual function. Activation of parasympathetic nerves to the penis increases blood flow and leads to erection while activation of sympathetic nerves to the penis causes ejaculation.

There are also several organs that are innervated by only one division of the ANS. In addition to the adrenal gland, most blood vessels, the pilomotor muscles in the skin (hair follicles), and sweat glands are innervated exclusively by sympathetic nerves (sudomotor fibers). The lacrimal muscle (tear gland), ciliary muscle (for accommodation for near vision), and the sublingual salivary gland are innervated exclusively by parasympathetic nerves.

PARASYMPATHETIC CHOLINERGIC & SYMPATHETIC NORADRENERGIC DISCHARGE

In a general way, the functions promoted by activity in the cholinergic division of the ANS are those concerned with the vegetative aspects of day-to-day living. For example, parasympathetic action favors digestion and absorption of food by increasing the activity of the intestinal musculature, increasing gastric secretion, and relaxing the pyloric sphincter. For this reason, the cholinergic division is sometimes called the anabolic nervous system.

The sympathetic (noradrenergic) division discharges as a unit in emergency situations and can be called the catabolic nervous system. The effect of this discharge prepares the individual to cope with an emergency. Sympathetic activity dilates the pupils (letting more light into the eyes), accelerates the heartbeat and raises the blood pressure (providing better perfusion of the vital organs and muscles), and constricts the blood vessels of the skin (which limits bleeding from wounds). Noradrenergic discharge also leads to elevated plasma glucose and free fatty acid levels (supplying more energy). On the basis of effects like these, Walter Cannon called the emergency-induced discharge of the sympathetic nervous system the "preparation for flight or fight."

The emphasis on mass discharge in stressful situations should not obscure the fact that the sympathetic fibers also subserve other functions. For example, tonic sympathetic discharge to the arterioles maintains arterial pressure, and variations in this tonic discharge are the mechanism by which carotid sinus feedback regulation of blood pressure occurs (see Chapter 32). In addition, sympathetic discharge is decreased in fasting animals and increased when fasted animals are again fed. These changes may explain the decrease in blood pressure and metabolic rate produced by fasting and the opposite changes produced by feeding.

DESCENDING INPUTS TO AUTONOMIC PREGANGLIONIC NEURONS

As is the case for α-motor neurons, the activity of autonomic nerves is dependent on both reflexes (eg, baroreceptor and chemoreceptor reflexes) and a balance between descending excitatory and inhibitory inputs from several brain regions. To identify brain regions that provide input to preganglionic sympathetic neurons, neuroanatomical tract tracing chemicals can be injected into the thoracic IML. These chemicals are picked up by axon terminals and transported retrogradely to the cell bodies of origin. Figure 13–5 shows the source of some forebrain and brain stem inputs to sympathetic preganglionic neurons. There are parallel pathways from the hypothalamic paraventricular

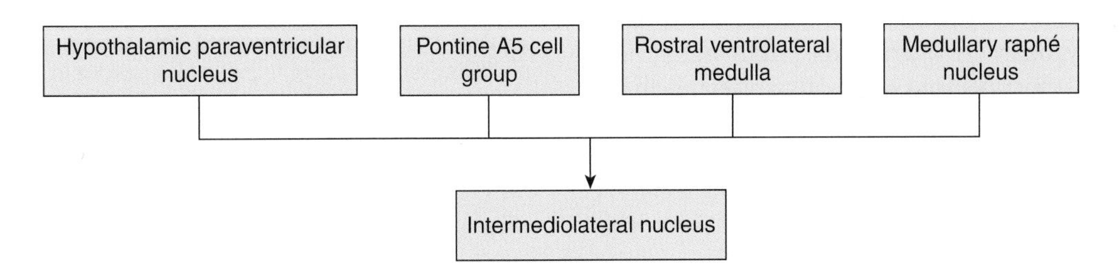

FIGURE 13–5 Pathways that control autonomic responses. Direct projections (solid lines) to autonomic preganglionic neurons include the hypothalamic paraventricular nucleus, pontine A5 cell group, rostral ventrolateral medulla, and medullary raphé.

nucleus, pontine catecholaminergic A5 cell group, rostral ventrolateral medulla, and medullary raphé nuclei. This is analogous to projections from the brainstem and cortex converging on somatomotor neurons in the spinal cord. The rostral ventrolateral medulla is generally considered to major source of excitatory input to sympathetic neurons. In addition to these direct pathways to preganglionic neurons, there are many brain regions that feed into these pathways, including the amygdala, mesencephalic periaqueductal gray, caudal ventrolateral medulla, nucleus of the tractus solitarius, and medullary lateral tegmental field. This is analogous to the control of somatomotor function by areas such as the basal ganglia and cerebellum. Chapter 32 describes the role of some of these brain regions as well as the role of various reflexes in setting the level of activity in autonomic nerves supplying cardiovascular effector organs.

AUTONOMIC DYSFUNCTION

Drugs, neurodegenerative diseases, trauma, inflammatory processes, and neoplasia are a few examples of factors that can lead to dysfunction of the ANS (see Clinical Boxes 13–1 through 13–4). The types of dysfunction can range from complete autonomic failure to autonomic hyperactivity. Among disorders associated with autonomic failure are orthostatic hypotension, neurogenic syncope (vasovagal response), impotence, neurogenic bladder, gastrointestinal dysmotility, sudomotor failure, and Horner's syndrome. Autonomic hyperactivity can be the basic for neurogenic hypertension, cardiac arrhythmias, neurogenic pulmonary edema, myocardial injury, hyperhidrosis, hyperthermia, and hypothermia.

ENTERIC NERVOUS SYSTEM

The enteric nervous system, which can be considered as the third division of the ANS, is located within the wall of the digestive tract, all the way from the esophagus to the anus. It is comprised of two well-organized neural plexuses. The **myenteric plexus** is located between longitudinal and circular layers of muscle; it is involved in control of digestive tract motility. The **submucosal plexus** is located between the circular muscle and the luminal mucosa; it senses the environment of the lumen and regulates gastrointestinal blood flow and epithelial cell function.

The enteric nervous system contains as many neurons as the entire spinal cord. It is sometimes referred to as a "mini-brain" as it contains all the elements of a nervous system including sensory neurons, interneurons, and motor neurons. It contains sensory neurons innervating receptors in the mucosa that respond to mechanical, thermal, osmotic, and chemical stimuli. Motor neurons control motility, secretion, and absorption by acting on smooth muscle and secretory cells. Interneurons integrate information from sensory neurons and feedback to the enteric motor neurons.

Parasympathetic and sympathetic nerves connect the CNS to the enteric nervous system or directly to the digestive tract. Although the enteric nervous system can function autonomously, normal digestive function often requires communication between the CNS and the enteric nervous system (see Chapter 25).

CHAPTER SUMMARY

- Preganglionic sympathetic neurons are located in the IML of the thoracolumbar spinal cord and project to postganglionic neurons in the paravertebral or prevertebral ganglia or the adrenal medulla. Preganglionic parasympathetic neurons are located in motor nuclei of cranial nerves III, VII, IX, and X and the sacral IML. Postganglionic nerve terminals are located in smooth muscle (eg, blood vessels, gut wall, urinary bladder), cardiac muscle, and glands (eg, sweat gland, salivary glands).
- Acetylcholine is released at nerve terminals of all preganglionic neurons, postganglionic parasympathetic neurons, and a few postganglionic sympathetic neurons (sweat glands, sympathetic vasodilator fibers). The remaining sympathetic postganglionic neurons release norepinephrine.
- Ganglionic transmission results from activation of nicotinic receptors. Postganglionic cholinergic transmission is mediated by activation of muscarinic receptors. Postganglionic adrenergic transmission is mediated by activation of α_1, β_1, or β_2 adrenoceptors, depending on the target organ. Many common drugs exert their therapeutic actions by serving as agonists or antagonists at autonomic synapses.
- Sympathetic activity prepares the individual to cope with an emergency by accelerating the heartbeat, raising blood pressure (perfusion of the vital organs), and constricting the blood vessels of the skin (limits bleeding from wounds). Parasympathetic activity is concerned with the vegetative aspects of day-to-day living and favors digestion and absorption of food by increasing the activity of the intestinal musculature, increasing gastric secretion, and relaxing the pyloric sphincter.
- Direct projections to sympathetic preganglionic neurons in the IML originate in the hypothalamic paraventricular nucleus, pontine catecholaminergic A5 cell group, rostral ventrolateral medulla, and medullary raphé nuclei.
- The enteric nervous system is located within the wall of the digestive tract and is composed of the myenteric plexus (control of digestive tract motility) and the submucosal plexus (regulates gastrointestinal blood flow and epithelial cell function).

MULTIPLE-CHOICE QUESTIONS

For all questions, select the single best answer unless otherwise directed.

1. A 26-year-old male developed hypertension after he began taking amphetamine to boost his energy and to suppress his appetite. Which of the following drugs would be expected to mimic the effects of increased sympathetic discharge on blood vessels?
   A. Phenylephrine
   B. Trimethaphan
   C. Atropine
   D. Reserpine
   E. Albuterol

2. A 35-year-old female was diagnosed with multiple system atrophy and had symptoms indicative of failure of sympathetic nerve activity. Which of the following statements about the sympathetic nervous system is correct?
   A. All postganglionic sympathetic nerves release norepinephrine from their terminals.
   B. Cell bodies of preganglionic sympathetic neurons are located in the intermediolateral column of the thoracic and sacral spinal cord.
   C. The sympathetic nervous system is required for survival.
   D. Acetylcholine is released from all sympathetic preganglionic nerve terminals.
   E. The sympathetic nervous system adjusts pupillary diameter by relaxing the pupillary constrictor muscle.
3. A 45-year-old male had a meal containing wild mushrooms that he picked in a field earlier in the day. Within a few hours after eating, he developed nausea, vomiting, diarrhea, urinary urgency, vasodilation, sweating, and salivation. Which of the following statements about the parasympathetic nervous system is correct?
   A. Postganglionic parasympathetic nerves release acetylcholine to activate muscarinic receptors on sweat glands.
   B. Parasympathetic nerve activity affects only smooth muscles and glands.
   C. Parasympathetic nerve activity causes contraction of smooth muscles of the gastrointestinal wall and relaxation of the gastrointestinal sphincter.
   D. Parasympathetic nerve activity causes contraction of the radial muscle of the eye to allow accommodation for near vision.
   E. An increase in parasympathetic activity causes an increase in heart rate.
4. Which of the following is correctly paired?
   A. Sinoatrial node: Nicotinic cholinergic receptors
   B. Autonomic ganglia: Muscarinic cholinergic receptors
   C. Pilomotor smooth muscle: β_2-adrenergic receptors
   D. Vasculature of some skeletal muscles: Muscarinic cholinergic receptors
   E. Sweat glands: α_2-adrenergic receptors
5. A 57-year-old male had severe hypertension that was found to result from a tumor compressing on the surface of the medulla. Which one of the following statements about pathways involved in the control of sympathetic nerve activity is correct?
   A. Preganglionic sympathetic nerves receive inhibitory input from the rostral ventrolateral medulla.
   B. The major source of excitatory input to preganglionic sympathetic nerves is the paraventricular nucleus of the hypothalamus.
   C. The activity of sympathetic preganglionic neurons can be affected by the activity of neurons in the amygdala.
   D. Unlike the activity in δ-motor neurons, sympathetic preganglionic neurons are not under any significant reflex control.
   E. Under resting conditions, the sympathetic nervous system is not active; it is active only during stress giving rise to the term "flight or fight" response.
6. A 53-year-old female with diabetes was diagnosed with diabetic autonomic neuropathy a few years ago. She recently noted abdominal distension and a feeling of being full after eating only a small portion of food, suggesting the neuropathy had extended to her enteric nervous system to cause gastroparesis. Which of the following statements about the enteric nervous system is correct?
   A. The enteric nervous system is a subdivision of the parasympathetic nervous system for control of gastrointestinal function.
   B. The myenteric plexus is a group of motor neurons located within circular layer of muscle in a portion of the gastrointestinal tract.
   C. The submucosal plexus is a group of sensory neurons located between the circular muscle and the luminal mucosa of the gastrointestinal tract.
   D. Neurons comprising the enteric nervous system are located only in the stomach and intestine.
   E. The enteric nervous system can function independent of the autonomic innervation to the gastrointestinal tract.

CHAPTER RESOURCES

Benarroch EE: *Central Autonomic Network. Functional Organization and Clinical Correlations.* Futura Publishing, 1997.

Cheshire WP: Autonomic physiology. In: *Clinical Neurophysiology,* 3rd ed. Oxford University Press, 2009.

Elvin LG, Lindh B, Hokfelt T: The chemical neuroanatomy of sympathetic ganglia. Annu Rev Neurosci 1993;16:471.

Jänig W: *The Integrative Action of the Autonomic Nervous System. Neurobiology of Homeostasis.* Cambridge University Press, 2006.

Loewy AD, Spyer KM (editors): *Central Regulation of Autonomic Function.* Oxford University Press, 1990.

Saper CB: The central autonomic nervous system: Conscious visceral perception and autonomic pattern generation. Annu Rev Neurosci 2002;25:433.

FACULTY OF HEALTH + SOCIAL CARE
Library
10 JUL 2013
Wirral Campus
Clatterbridge
UNIVERSITY OF CHESTER

Electrical Activity of the Brain, Sleep–Wake States, & Circadian Rhythms

CHAPTER 14

OBJECTIVES

After studying this chapter, you should be able to:

- Describe the primary types of rhythms that make up the electroencephalogram (EEG).
- List the main clinical uses of the EEG.
- Summarize the behavioral and EEG characteristics of each of the stages of nonrapid eye movement (NREM) and rapid eye movement (REM) sleep and the mechanisms responsible for their production.
- Describe the pattern of normal nighttime sleep in adults and the variations in this pattern from birth to old age.
- Describe the interplay between brain stem neurons that contain norepinephrine, serotonin, and acetylcholine as well as GABA and histamine in mediating transitions between sleep and wakefulness.
- Discuss the circadian rhythm and the role of the suprachiasmatic nuclei (SCN) in its regulation.
- Describe the diurnal regulation of synthesis of melatonin from serotonin in the pineal gland and its secretion into the bloodstream.

INTRODUCTION

Most of the various sensory pathways described in Chapters 8–11 relay impulses from sense organs via three- and four-neuron chains to particular sites in the cerebral cortex. The impulses are responsible for perception and localization of individual sensations. However, they must be processed in the awake brain to be perceived. There is a spectrum of behavioral states ranging from deep sleep through light sleep, REM sleep, and the two awake states: relaxed awareness and awareness with concentrated attention. Discrete patterns of brain electrical activity correlate with each of these states. Feedback oscillations within the cerebral cortex and between the thalamus and the cortex serve as producers of this activity and possible determinants of the behavioral state. Arousal can be produced by sensory stimulation and by impulses ascending in the reticular core of the midbrain. Many of these activities have rhythmic fluctuations that are approximately 24 h in length; that is, they are **circadian.**

THALAMUS, CEREBRAL CORTEX, & RETICULAR FORMATION

THALAMIC NUCLEI

The **thalamus** is a large collection of neuronal groups within the diencephalon; it participates in sensory, motor, and limbic functions. Virtually all information that reaches the cortex is processed by the thalamus, leading to its being called the "gateway to the cerebral cortex."

The thalamus can be divided into nuclei that project diffusely to wide regions of the neocortex and nuclei that project to specific discrete portions of the neocortex and limbic system. The nuclei that project to wide regions of the neocortex are the **midline and intralaminar nuclei.** The nuclei that project to specific areas include the specific sensory relay

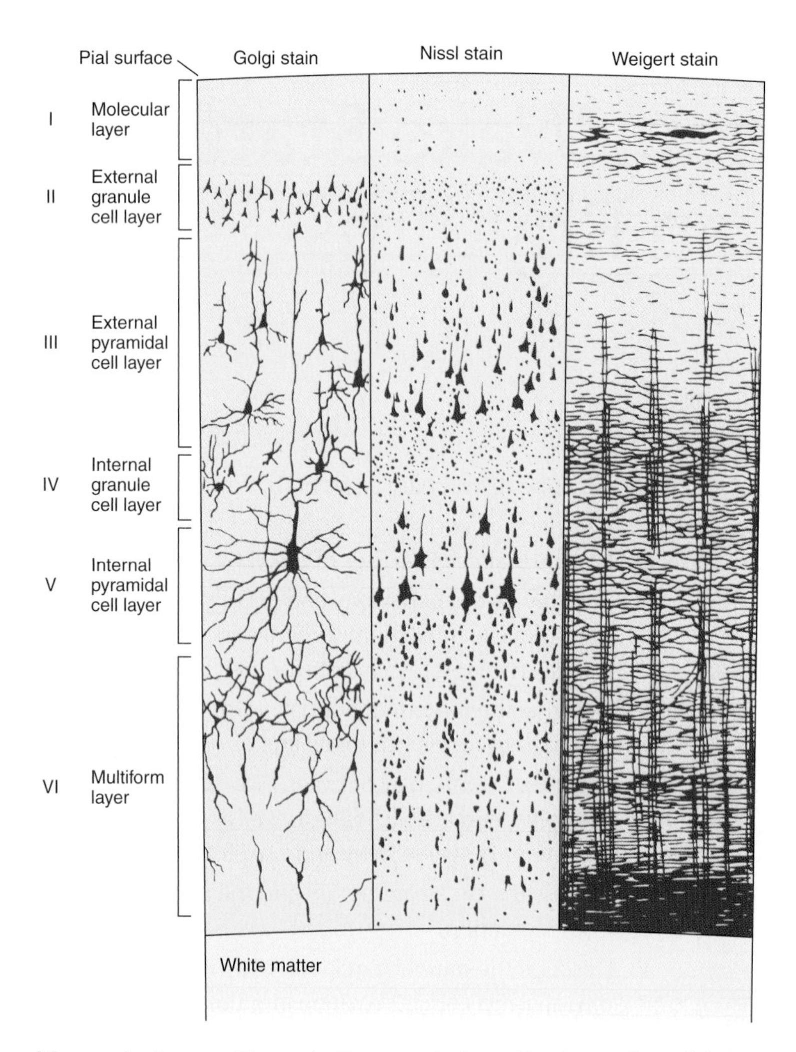

FIGURE 14–1 Structure of the cerebral cortex. The cortical layers are indicated by the numbers. Golgi stain shows neuronal cell bodies and dendrites, Nissl stain shows cell bodies, and Weigert myelin sheath stain shows myelinated nerve fibers. (Modified from Ranson SW, Clark SL: *The Anatomy of the Nervous System*, 10th ed. Saunders, 1959.)

nuclei and the nuclei concerned with efferent control mechanisms. The **specific sensory relay nuclei** include the medial and lateral geniculate bodies, which relay auditory and visual impulses to the auditory and visual cortices; and the ventral posterior lateral (VPL) and ventral posteromedial nuclei, which relay somatosensory information to the postcentral gyrus. The ventral anterior and ventral lateral nuclei are concerned with motor function. They receive input from the basal ganglia and the cerebellum and project to the motor cortex. The anterior nuclei receive afferents from the mamillary bodies and project to the limbic cortex, which may be involved in memory and emotion. Most of the thalamic nuclei described are excitatory neurons that release glutamate. The thalamus also contains inhibitory neurons in the **thalamic reticular nucleus.** These neurons release **GABA,** and unlike the other thalamic neurons just described, their axons do not project to the cortex. Rather, they are thalamic interneurons and modulate the responses of other thalamic neurons to input coming from the cortex.

CORTICAL ORGANIZATION

The neocortex is generally arranged in six layers (Figure 14–1). The most common cell type is the **pyramidal neuron** with an extensive vertical dendritic tree (Figure 14–1 and Figure 14–2) that may reach to the cortical surface. Their cell bodies can be found in all cortical layers except layer I. The axons of these cells usually give off recurrent collaterals that turn back and synapse on the superficial portions of the dendritic trees. Afferents from the specific nuclei of the thalamus terminate primarily in cortical layer IV, whereas the nonspecific afferents are distributed to layers I–IV. Pyramidal neurons are the only projection neurons of the cortex, and they are excitatory neurons that release **glutamate** at their terminals. The other cortical cell types are local circuit neurons (interneurons) that have been classified based on their shape, pattern of projection, and neurotransmitter. Inhibitory interneurons (**basket cells** and **chandelier cells**) release GABA as their neurotransmitter. Basket cells have long axonal endings that surround

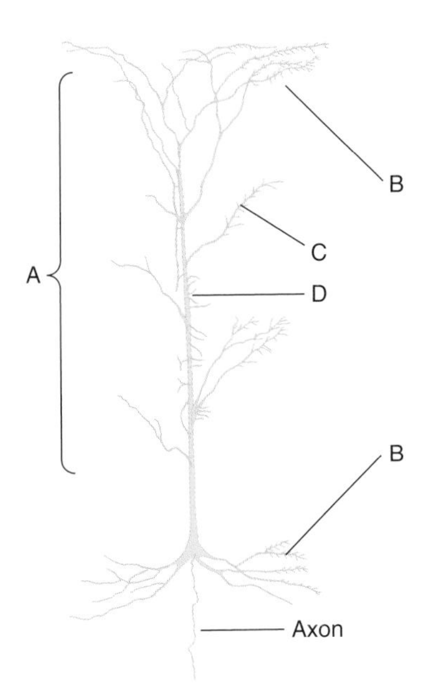

FIGURE 14–2 Neocortical pyramidal cell, showing the distribution of neurons that terminate on it. A denotes nonspecific afferents from the reticular formation and the thalamus; **B** denotes recurrent collaterals of pyramidal cell axons; **C** denotes commissural fibers from mirror image sites in the contralateral hemisphere; **D** denotes specific afferents from thalamic sensory relay nuclei. (Based on Scheibel ME, Scheibel AB: Structural organization of nonspecific thalamic nuclei and their projection toward cortex. Brain Res 1967 Sep;6(1):60–94.)

the soma of pyramidal neurons; they account for most inhibitory synapses on the pyramidal soma and dendrites. Chandelier cells are a powerful source of inhibition of pyramidal neurons because they have axonal endings that terminate exclusively on the initial segment of the pyramidal cell axon. Their terminal boutons form short vertical rows that resemble candlesticks, thus accounting for their name. **Spiny stellate cells** are excitatory interneurons that release glutamate as a neurotransmitter. These cells are located primarily in layer IV and are a major recipient of sensory information arising from the thalamus; they are an example of a multipolar neuron (Chapter 4) with local dendritic and axonal arborizations.

In addition to being organized into layers, the cerebral cortex is also organized into columns. Neurons within a column have similar response properties, suggesting they comprise a local processing network (eg, orientation and ocular dominance columns in the visual cortex).

RETICULAR ACTIVATING SYSTEM

The **reticular formation,** the phylogenetically old reticular core of the brain, occupies the central portion of the medulla and midbrain, surrounding the fourth ventricle and cerebral aqueduct. The reticular formation contains the cell bodies and fibers of many of the serotonergic, noradrenergic, and cholinergic systems. These pathways were shown in Figure 7–2. The reticular formation also contains many of the areas concerned with regulation of heart rate, blood pressure, and respiration. The reticular formation plays an important role in determining the level of arousal, thus it is called the ascending **reticular activating system (RAS).**

The RAS is a complex polysynaptic pathway arising from the brain stem reticular formation and hypothalamus with projections to the intralaminar and reticular nuclei of the thalamus which, in turn, project diffusely and nonspecifically to wide regions of the cortex including the frontal, parietal, temporal, and occipital cortices (Figure 14–3). Collaterals funnel into it not only from the long ascending sensory tracts but also from the trigeminal, auditory, visual, and olfactory systems. The complexity of the neuron net and the degree of convergence in it abolish modality specificity, and most reticular neurons are activated with equal facility by different sensory stimuli. The system is therefore **nonspecific,** whereas the classic sensory pathways are **specific** in that the fibers in them are activated by only one type of sensory stimulation.

EVOKED CORTICAL POTENTIALS

The electrical events that occur in the cortex after stimulation of a sense organ can be monitored with a recording electrode. If the electrode is over the primary receiving area for a particular sense, a surface-positive wave appears with a latency of 5–12 ms. This is followed by a small negative wave, and then a larger, more prolonged positive deflection frequently occurs with a latency of 20–80 ms. The first positive–negative wave sequence is the **primary evoked potential;** the second is the **diffuse secondary response.**

The primary evoked potential is highly specific in its location and can be observed only where the pathways from a particular sense organ end. The positive–negative wave sequence recorded from the surface of the cortex occurs because the superficial cortical layers are positive relative to the initial negativity, then negative relative to the deep hyperpolarization. The surface-positive diffuse secondary response, unlike the primary response, is not highly localized. It appears at the same time over most of the cortex and is due to activity in projections from the midline and related thalamic nuclei.

PHYSIOLOGIC BASIS OF THE ELECTROENCEPHALOGRAM

The background electrical activity of the brain in unanesthetized animals was first described in the 19th century. Subsequently, it was analyzed in systematic fashion by the German psychiatrist Hans Berger, who introduced the term **electroencephalogram (EEG)** to denote the recording of the variations in brain potential. The EEG can be recorded with scalp electrodes through the unopened skull or with electrodes on or in the brain. The term **electrocorticogram (ECoG)** is used for the recording obtained with electrodes on the pial surface of the cortex.

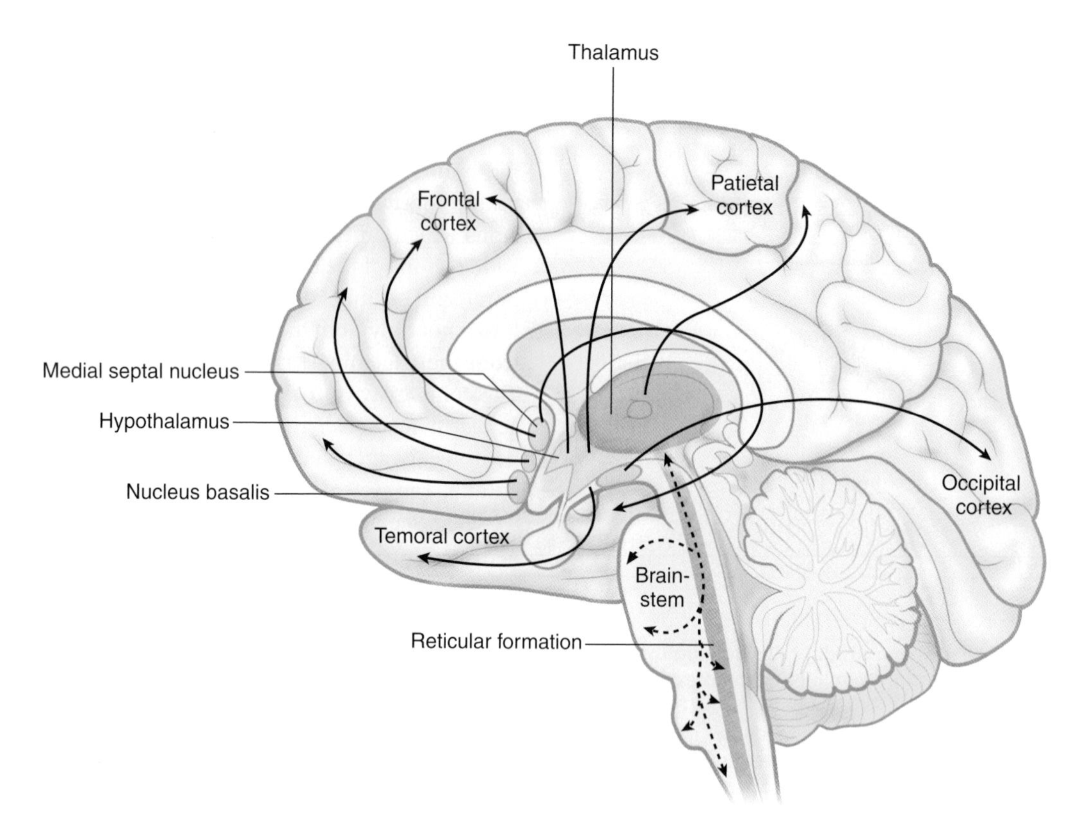

FIGURE 14–3 Cross-section through the midline of the human brain showing the ascending reticular activating system in the brainstem with projections to the intralaminar nuclei of the thalamus and the output from the intralaminar nuclei to many parts of the cerebral cortex. Activation of these areas can be shown by positive emission tomography scans when subjects shift from a relaxed awake state to an attention-demanding task.

The EEG recorded from the scalp is a measure of the summation of dendritic postsynaptic potentials rather than action potentials (Figure 14–4). The dendrites of the cortical neurons are a forest of similarly oriented, densely packed units in the superficial layers of the cerebral cortex (Figure 14–1). Propagated potentials can be generated in dendrites. In addition, recurrent axon collaterals end on dendrites in the superficial layers. As excitatory and inhibitory endings on the dendrites of each cell become active, current flows into and out of these current sinks and sources from the rest of the dendritic processes and the cell body. The cell body–dendrite relationship is therefore that of a constantly shifting dipole. Current flow in this dipole produces wave-like potential fluctuations in a volume conductor (Figure 14–4). When the sum of the dendritic activity is negative relative to the cell body, the neuron is depolarized and hyperexcitable; when it is positive, the neuron is hyperpolarized and less excitable.

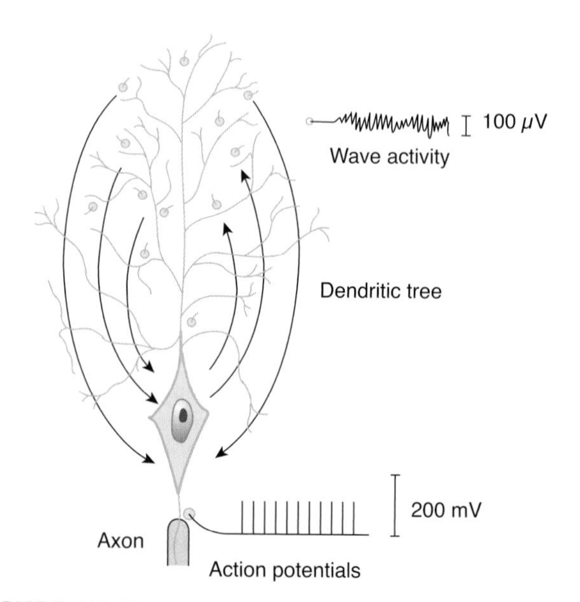

FIGURE 14–4 Diagrammatic comparison of the electrical responses of the axon and the dendrites of a large cortical neuron. Current flow to and from active synaptic knobs on the dendrites produces wave activity, while all-or-none action potentials are transmitted along the axon. When the sum of the dendritic activity is negative relative to the cell body, the neuron is depolarized; when it is positive, the neuron is hyperpolarized. The electroencephalogram recorded from the scalp is a measure of the summation of dendritic postsynaptic potentials rather than action potentials.

SLEEP–WAKE CYCLE: ALPHA, BETA, & GAMMA RHYTHMS

In adult humans who are awake but at rest with the mind wandering and the eyes closed, the most prominent component of the EEG is a fairly regular pattern of waves at a frequency

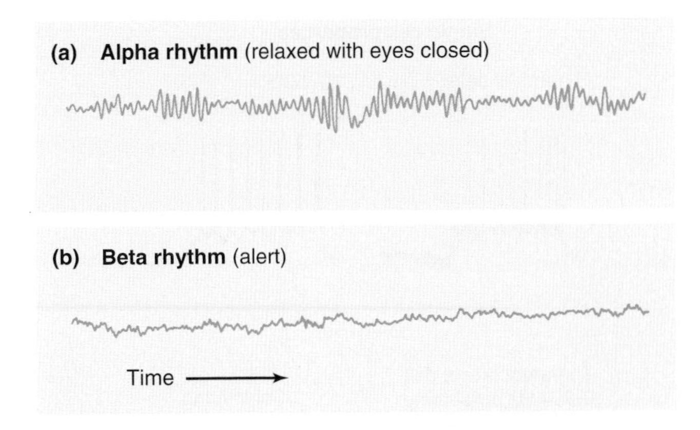

FIGURE 14–5 EEG records showing the alpha and beta rhythms. When attention is focused on something, the 8–13 Hz alpha rhythm is replaced by an irregular 13–30 Hz low-voltage activity, the beta rhythm. This phenomenon is referred to as alpha block, arousal, or the alerting response. (From Widmaier EP, Raff H, Strang KT: *Vander's Human Physiology,* 11th ed. McGraw-Hill, 2008.)

of 8–13 Hz and amplitude of 50–100 μV when recorded from the scalp. This pattern is the **alpha rhythm** (Figure 14–5). It is most marked in the parietal and occipital lobes and is associated with decreased levels of attention. A similar rhythm has been observed in a wide variety of mammalian species. There are some minor variations from species to species, but in all mammals the pattern is remarkably similar (see Clinical Box 14–1).

When attention is focused on something, the alpha rhythm is replaced by an irregular 13–30 Hz low-voltage activity, the **beta rhythm** (Figure 14–5). This phenomenon is called alpha block and can be produced by any form of sensory stimulation or mental concentration, such as solving arithmetic problems. Another term for this phenomenon is the **arousal** or **alerting response,** because it is correlated with the aroused, alert state. It has also been called **desynchronization,** because it represents breaking up of the obviously synchronized neural activity necessary to produce regular waves. However, the rapid EEG activity seen in the alert state is also synchronized, but at a higher rate. Therefore, the term *desynchronization* is misleading. **Gamma oscillations** at 30–80 Hz are often seen when an individual is aroused and focuses attention on something. This is often replaced by irregular fast activity as the individual initiates motor activity in response to the stimulus.

CLINICAL BOX 14–1

Variations in the Alpha Rhythm

In humans, the frequency of the dominant EEG rhythm at rest varies with age. In infants, there is fast, beta-like activity, but the occipital rhythm is a slow 0.5–2-Hz pattern. During childhood this latter rhythm speeds up, and the adult alpha pattern gradually appears during adolescence. The frequency of the alpha rhythm is decreased by low blood glucose levels, low body temperature, low levels of adrenal glucocorticoid hormones, and high arterial partial pressure of CO_2 ($PaCO_2$). It is increased by the reverse conditions. Forced over-breathing to lower the $PaCO_2$ is sometimes used clinically to bring out latent EEG abnormalities. The frequency and magnitude of the alpha rhythm is also decreased by metabolic and toxic encephalopathies including those due to hyponatremia and vitamin B12 deficiency. The frequency of the alpha rhythm is reduced during acute intoxication with **alcohol, amphetamines, barbiturates, phenytoin,** and **antipsychotics. Propofol,** a hypnotic/sedative drug, can induce a rhythm in the EEG that is analogous to the classic alpha rhythm.

THALAMOCORTICAL LOOP

A circuit linking the cortex and thalamus is thought to be important in generating patterns of brain activity in sleep–wake states. Figure 14–6 shows properties of activity in such a thalamocortical circuit hypothesized to be involved in generating rhythmic activity. Augmented activation of low threshold **T-type Ca^{2+} channels** in thalamic neurons likely contributes to both physiological and pathophysiological synchrony in thalamocortical circuits. Although not shown, inhibitory thalamic reticular neurons are elements of this network. The EEG shows the characteristic awake, light sleep, and deep sleep patterns of activity. Likewise, recordings from individual thalamic and cortical neurons show different patterns of rhythmic activity. In the waking state, corticocortical and thalamocortical networks generate higher-frequency rhythmic activity (30–80 Hz; gamma rhythm). This rhythm may be generated within the cells and networks of the cerebral cortex or within thalamocortical loops. The gamma rhythm has been suggested as a mechanism to "bind" together diverse sensory information into a single percept and action, but this theory is still controversial. In fact, disturbances in the integrity of this thalamocortical loop and its interaction with other brain structures may underlie some neurological disorders, including seizure activity.

SLEEP STAGES

There are two kinds of sleep: **rapid eye movement (REM) sleep** and **nonREM (NREM),** or **slow-wave sleep.** REM sleep is so named because of the characteristic eye movements that occur during this stage of sleep. NREM sleep is divided into four stages (Figure 14–7). As a person begins to fall asleep and enters stage 1, the EEG shows a low-voltage, mixed frequency pattern. A **theta rhythm** (4–7 Hz) can be seen at this early stage of slow-wave sleep. Throughout NREM sleep, there is some activity of skeletal muscle but no eye movements occur. Stage 2 of NREM sleep is marked by the appearance of sinusoidal waves called **sleep spindles** (12–14 Hz) and occasional high voltage biphasic waves called **K com-**

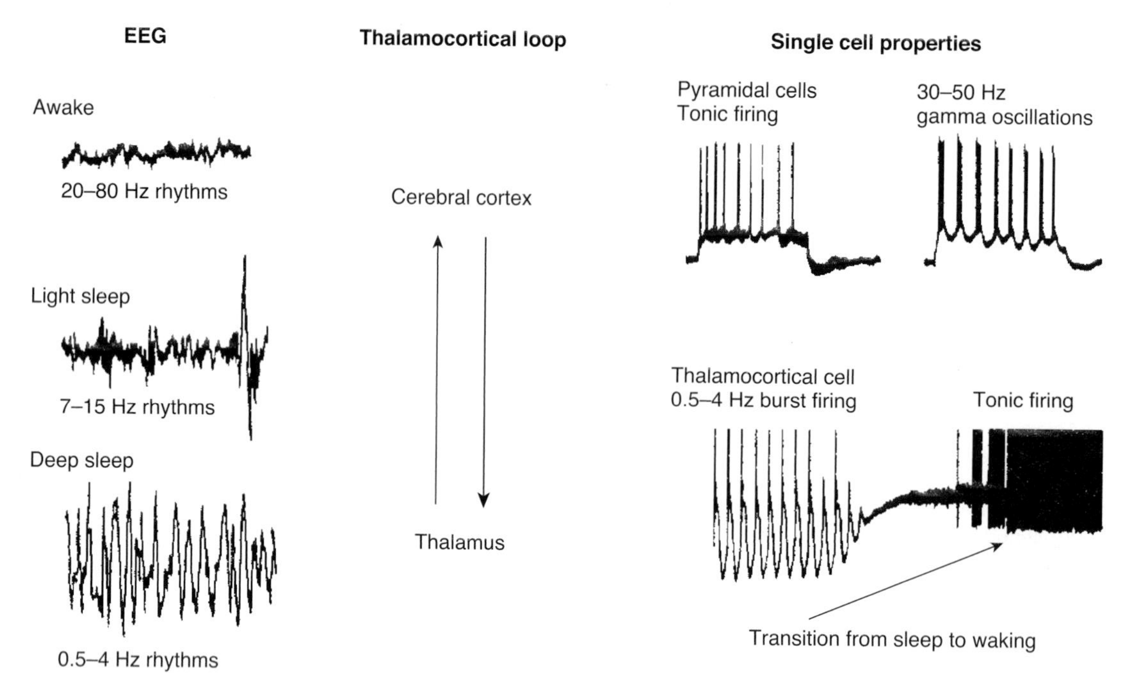

FIGURE 14–6 Correlation between behavioral states, EEG, and single-cell responses in the cerebral cortex and thalamus. The EEG is characterized by high-frequency oscillations in the awake state and low-frequency rhythms during sleep. Thalamic and cortical neurons can also show different patterns of rhythmic activity. Thalamocortical neurons show slow rhythmic oscillations during deep sleep, and fire tonic trains of action potentials in the awake state. Most pyramidal neurons in the cortex generate only tonic trains of action potentials, although others may participate in the generation of high frequency rhythms through activation of rhythmic bursts of spikes. The thalamus and cerebral cortex are connected together in a loop. (Modified from McCormick DA: Are thalamocortical rhythms the Rosetta stone of a subset of neurological disorders? Nat Med 1999;5:1349.)

plexes. In stage 3 of NREM sleep, a high-amplitude **delta rhythm** (0.5–4 Hz) dominates the EEG waves. Maximum slowing with large waves is seen in stage 4 of NREM sleep. Thus, the characteristic of deep sleep is a pattern of rhythmic slow waves, indicating marked **synchronization;** it is sometimes referred to as slow-wave sleep. While the occurrence of theta and delta rhythms is normal during sleep, their appearance during wakefulness is a sign of brain dysfunction.

REM SLEEP

The high-amplitude slow waves seen in the EEG during sleep are periodically replaced by rapid, low-voltage EEG activity, which resembles that seen in the awake, aroused state and in stage 1 sleep (Figure 14–7). For this reason, REM sleep is also called **paradoxical sleep.** However, sleep is not interrupted; indeed, the threshold for arousal by sensory stimuli and by

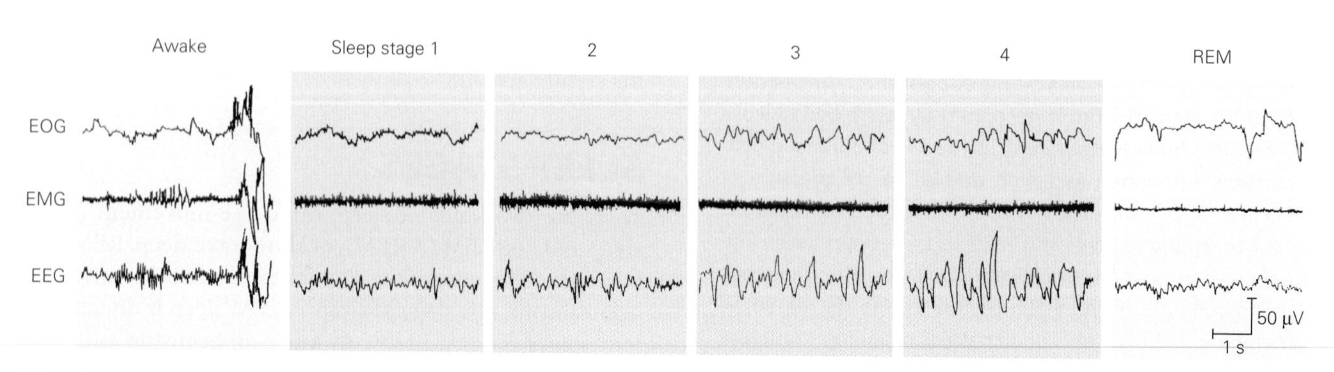

FIGURE 14–7 EEG and muscle activity during various stages of the sleep–wake cycle. NREM sleep has four stages. Stage 1 is characterized by a slight slowing of the EEG. Stage 2 has high-amplitude K complexes and spindles. Stages 3 and 4 have slow, high-amplitude delta waves. REM sleep is characterized by eye movements, loss of muscle tone, and a low-amplitude, high-frequency activity pattern. The higher voltage activity in the EOG tracings during stages 2 and 3 reflect high amplitude EEG activity in the prefrontal areas rather than eye movements. EOG, electro-oculogram registering eye movements; EMG, electromyogram registering skeletal muscle activity. (Reproduced with permission from Rechtschaffen A, Kales A: *A Manual of Standardized Terminology, Techniques and Scoring System and Sleep Stages of Human Subjects.* Los Angeles: University of California Brain Information Service, 1968.)

stimulation of the reticular formation is elevated. Rapid, roving movements of the eyes occur during paradoxical sleep, and it is for this reason that it is also called REM sleep. Another characteristic of REM sleep is the occurrence of large phasic potentials that originate in the cholinergic neurons in the pons and pass rapidly to the lateral geniculate body and from there to the occipital cortex. They are called **pontogeniculo-occipital (PGO) spikes.** The tone of the skeletal muscles in the neck is markedly reduced during REM sleep.

Humans aroused at a time when they show the EEG characteristics of REM sleep generally report that they were dreaming, whereas individuals awakened from slow-wave sleep do not. This observation and other evidence indicate that REM sleep and dreaming are closely associated.

Positron emission tomography (PET) scans of humans in REM sleep show increased activity in the pontine area, amygdala, and anterior cingulate gyrus, but decreased activity in the prefrontal and parietal cortex. Activity in visual association areas is increased, but there is a decrease in the primary visual cortex. This is consistent with increased emotion and operation of a closed neural system cut off from the areas that relate brain activity to the external world.

DISTRIBUTION OF SLEEP STAGES

In a typical night of sleep, a young adult first enters NREM sleep, passes through stages 1 and 2, and spends 70–100 min in stages 3 and 4. Sleep then lightens, and a REM period follows. This cycle is repeated at intervals of about 90 min throughout the night (Figure 14–8). The cycles are similar, though there is less stage 3 and 4 sleep and more REM sleep toward morning. Thus, 4–6 REM periods occur per night. REM sleep occupies 80% of total sleep time in premature infants and 50% in full-term neonates. Thereafter, the proportion of REM sleep falls rapidly and plateaus at about 25% until it falls to about 20% in the elderly. Children have more total sleep time (8–10 h) compared to most adults (about 6 h).

IMPORTANCE OF SLEEP

Sleep has persisted throughout evolution of mammals and birds, so it is likely that it is functionally important. Indeed, if humans are awakened every time they show REM sleep, then permitted to sleep without interruption, they show a great deal more than the normal amount of REM sleep for a few nights. Relatively prolonged REM deprivation does not seem to have adverse psychological effects. However, rats deprived of all sleep for long periods lose weight in spite of increased caloric intake and eventually die. Various studies imply that sleep is needed to maintain metabolic-caloric balance, thermal equilibrium, and immune competence.

In experimental animals, sleep is necessary for learning and memory consolidation. Learning sessions do not improve performance until a period of slow-wave or slow-wave plus REM sleep has occurred. Clinical Box 14–2 describes several common sleep disorders.

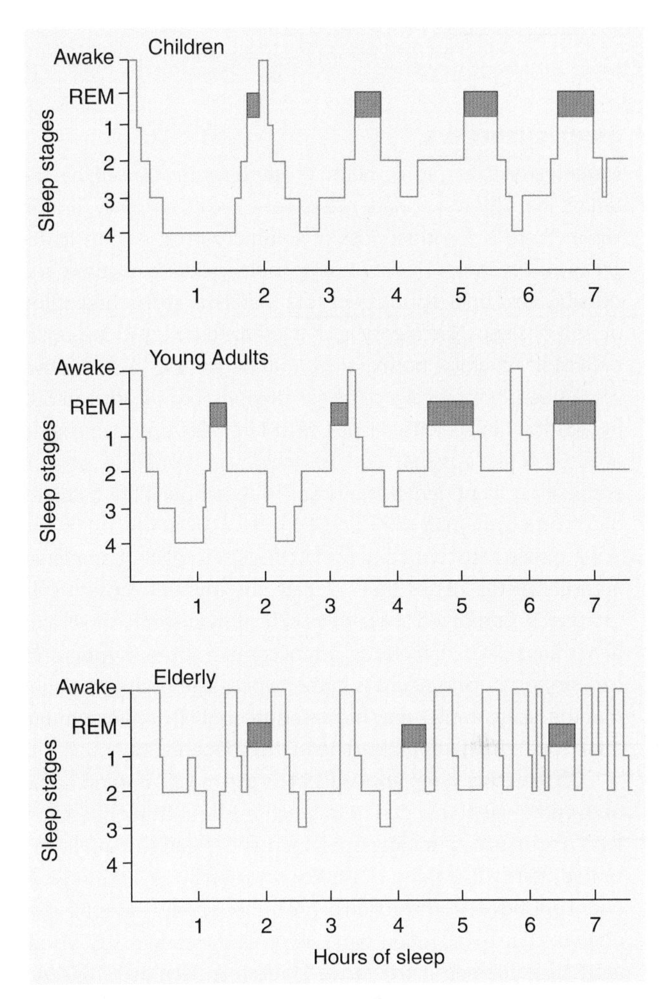

FIGURE 14–8 Normal sleep cycles at various ages. REM sleep is indicated by the darker colored areas. In a typical night of sleep, a young adult first enters NREM sleep, passes through stages 1 and 2, and spends 70–100 min in stages 3 and 4. Sleep then lightens, and a REM period follows. This cycle is repeated at intervals of about 90 min throughout the night. The cycles are similar, though there is less stage 3 and 4 sleep and more REM sleep toward morning. REM sleep occupies 50% of total sleep time in neonates; this proportion declines rapidly and plateaus at ~25% until it falls further in the elderly. (Reproduced with permission from Kales AM, Kales JD: Sleep disorders. N Engl J Med 1974;290:487.)

CLINICAL USES OF THE EEG

The EEG is sometimes of value in localizing pathologic processes. When a collection of fluid overlies a portion of the cortex, activity over this area may be damped. This fact may aid in diagnosing and localizing conditions such as subdural hematomas. Lesions in the cerebral cortex cause local formation of transient disturbances in brain activity, marked by high-voltage abnormal waves that can be recorded with an EEG. Seizure activity can occur because of increased firing of neurons that are excitatory (eg, release of glutamate) or decreased firing of neurons that are inhibitory (eg, release GABA).

CLINICAL BOX 14–2

Sleep Disorders

Narcolepsy is a chronic neurological disorder caused by the brain's inability to regulate sleep-wake cycles normally, and in which there is a sudden loss of voluntary muscle tone **(cataplexy),** an eventual irresistible urge to sleep during daytime, and possibly also brief episodes of total paralysis at the beginning or end of sleep. Narcolepsy is characterized by a sudden onset of REM sleep, unlike normal sleep that begins with NREM, slow-wave sleep. The prevalence of narcolepsy ranges from 1 in 600 in Japan to 1 in 500,000 in Israel, with 1 in 1000 Americans being affected. Narcolepsy has a familial incidence strongly associated with a class II antigen of the major histocompatibility complex on chromosome 6 at the HLA-DR2 or HLA-DQW1 locus, implying a genetic susceptibility to narcolepsy. The HLA complexes are interrelated genes that regulate the immune system (see Chapter 3). Compared to brains from healthy subjects, the brains of humans with narcolepsy often contain fewer **hypocretin (orexin)**-producing neurons in the hypothalamus. It is thought that the HLA complex may increase susceptibility to an immune attack on these neurons, leading to their degeneration.

Obstructive sleep apnea (OSA) is the most common cause of daytime sleepiness due to fragmented sleep at night and affects about 24% of middle-aged men and 9% of women in the United States. Breathing ceases for more than 10 s during frequent episodes of obstruction of the upper airway (especially the pharynx) due to reduction in muscle tone. The apnea causes brief arousals from sleep in order to reestablish upper airway tone. An individual with OSA typically begins to snore soon after falling asleep. The snoring gets progressively louder until it interrupted by an episode of apnea, which is then followed by a loud snort and gasp, as the individual tries to breathe. OSA is not associated with a reduction in total sleep time, but individuals with OSA experience a much greater time in stage 1 NREM sleep (from an average of 10% of total sleep to 30–50%) and a marked reduction in slow-wave sleep (stages 3 and 4 NREM sleep). The pathophysiology of OSA includes both a reduction in neuromuscular tone at the onset of sleep and a change in the central respiratory drive.

Periodic limb movement disorder (PLMD) is a stereotypical rhythmic extension of the big toe and dorsiflexion of the ankle and knee during sleep lasting for about 0.5–10 s and recurring at intervals of 20–90 s. Movements can actually range from shallow, continual movement of the ankle or toes, to wild and strenuous kicking and flailing of the legs and arms. Electromyograph (EMG) recordings show bursts of activity during the first hours of NREM sleep associated with brief EEG signs of arousal. The duration of stage 1 NREM sleep may be increased and that of stages 3 and 4 may be decreased compared to age-matched controls. PLMD is reported to occur in 5% of individuals between the ages of 30 and 50 years and increases to 44% of those over the age of 65. PLMD is similar to **restless leg syndrome** in which individuals have an irresistible urge to move their legs while at rest all day long.

Sleepwalking **(somnambulism),** bed-wetting **(nocturnal enuresis),** and **night terrors** are referred to as **parasomnias,** which are sleep disorders associated with arousal from NREM and REM sleep. Episodes of sleepwalking are more common in children than in adults and occur predominantly in males. They may last several minutes. Somnambulists walk with their eyes open and avoid obstacles, but when awakened they cannot recall the episodes.

THERAPEUTIC HIGHLIGHTS

Excessive daytime sleepiness in patients with narcolepsy can be treated with amphetamine-like stimulants, including **modafinil, methylphenidate (Ritalin),** and **methamphetamine. Gamma hydroxybutyrate (GHB)** is used to reduce the frequency of cataplexy attacks and the incidences of daytime sleepiness. Cataplexy is often treated with antidepressants such as **imipramine** and **desipramine,** but these drugs are not officially approved by the US Federal Drug Administration for such use. The most common treatment for OSA is **continuous positive airflow pressure (CPAP),** a machine that increases airway pressure to prevent airway collapse. Drugs have generally proven to have little or no benefit in treating OSA. Drugs used to treat Parkinson disease, **dopamine agonists,** can be used to treat PLMD.

TYPES OF SEIZURES

Epilepsy is a condition in which there are recurring, unprovoked seizures that may result from damage to the brain. The seizures represent abnormal, highly synchronous neuronal activity. Epilepsy is a syndrome with multiple causes. In some forms, characteristic EEG patterns occur during seizures; between attacks; however, abnormalities are often difficult to demonstrate. Seizures are divided into **partial (focal) seizures** and **generalized seizures.**

Partial seizures originate in a small group of neurons and can result from head injury, brain infection, stroke, or tumor, but often the cause is unknown. Symptoms depend on the seizure focus. They are further subdivided into **simple partial seizures** (without loss of consciousness) and **complex partial seizures** (with altered consciousness). An example of a simple partial seizure is localized jerking movements in one hand progressing to clonic movements of the entire arm lasting about 60–90 s. **Auras** typically precede the onset of a partial seizure and include abnormal sensations. The time after the seizure until normal neurological function returns is called the **postictal period.**

Generalized seizures are associated with widespread electrical activity and involve both hemispheres simultaneously.

They are further subdivided into **convulsive** and **nonconvulsive** categories depending on whether tonic or clonic movements occur. **Absence seizures** (formerly called petit mal seizures) are one of the forms of nonconvulsive generalized seizures characterized by a momentary loss of consciousness. They are associated with 3/s doublets, each consisting of a typical **spike-and-wave** pattern of activity that lasts for about 10 s (Figure 14–9). They are not accompanied by auras or postictal periods. These spike and waves are likely generated by low threshold T-type Ca^{2+} channels in thalamic neurons.

The most common convulsive generalized seizure is **tonic-clonic seizure** (formerly called grand mal seizure). This is associated with sudden onset of contraction of limb muscles **(tonic phase)** lasting about 30 s, followed by a clonic phase with symmetric jerking of the limbs as a result of alternating contraction and relaxation **(clonic phase)** lasting 1–2 min. There is fast EEG activity during the tonic phase. Slow waves, each preceded by a spike, occur at the time of each clonic jerk. For a while after the attack, slow waves are present.

Recent research provides insight into a possible role of release of glutamate from astrocytes in the pathophysiology of epilepsy. Also, there is evidence to support the view that reorganization of astrocytes along with dendritic sprouting and new synapse formation form the structural basis for recurrent excitation in the epileptic brain. Clinical Box 14–3 describes information regarding the role of genetic mutations in some forms of epilepsy.

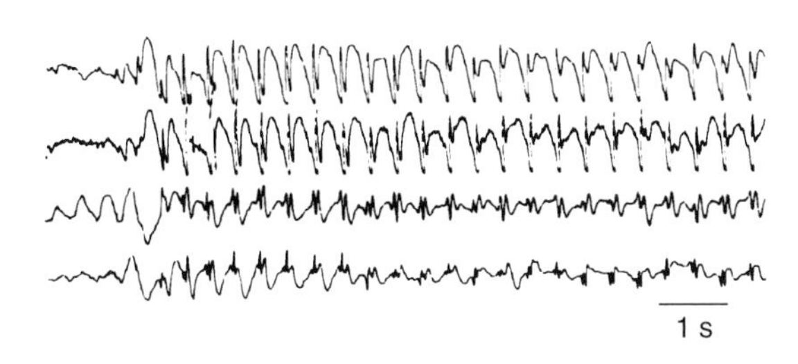

FIGURE 14–9 Absence seizures. This is a recording of four cortical EEG leads from a 6-year-old boy who, during the recording, had one of his "blank spells" in which he was transiently unaware of his surroundings and blinked his eyelids. Absence seizures are associated with 3/s doublets, each consisting of a typical spike-and-wave pattern of activity that lasts for about 10 s. Time is indicated by the horizontal calibration line. (Reproduced with permission from Waxman SG: *Neuroanatomy with Clinical Correlations,* 25th ed. McGraw-Hill, 2003.)

TREATMENT OF SEIZURES

Only about 2/3 of those suffering from seizure activity respond to drug therapies. Some respond to surgical interventions (eg, those with temporal lobe seizures), whereas others respond to vagal nerve stimulation (eg, those with partial seizures). Prior

CLINICAL BOX 14–3

Genetic Mutations & Epilepsy

Epilepsy has no geographical, racial, gender, or social bias. It can occur at any age, but is most often diagnosed in infancy, childhood, adolescence, and old age. It is the second most common neurological disorder after stroke. According to the World Health Organization, it is estimated that 50 million people worldwide (8.2 per 1000 individuals) experience epileptic seizures. The prevalence in developing countries (such as Colombia, Ecuador, India, Liberia, Nigeria, Panama, United Republic of Tanzania, and Venezuela) is more than 10 per 1000. Many affected individuals experience unprovoked seizures, for no apparent reason, and without any other neurological abnormalities. These are called **idiopathic epilepsies** and are assumed to be genetic in origin. Mutations in voltage-gated potassium, sodium, and chloride channels have been linked to some forms of idiopathic epilepsy. Mutated ion channels can lead to neuronal hyperexcitability via various pathogenic mechanisms. Scientists have recently identified the mutated gene responsible for development of **childhood absence epilepsy (CAE).** Several patients with CAE were found to have mutations in a subunit gene of the GABA receptor called **GABRB3.** Also, **SCN1A and SCN1B mutations** have been identified in an inherited form of epilepsy called **generalized epilepsy with febrile seizures.** SCN1A and SCN1B are sodium channel subunit genes that are widely expressed within the nervous system. SCN1A mutations are suspected in several other forms of epilepsy.

THERAPEUTIC HIGHLIGHTS

There are three broad mechanisms of action of new and old anticonvulsant drugs: enhancing inhibitory neurotransmission (increase GABA release), reducing excitatory neurotransmission (decrease glutamate release), or altering ionic conductance. **Gabapentin** is a GABA analog that acts by decreasing Ca^{2+} entry into cells and reducing glutamate release; it is used to treat generalized seizures. **Topiramate** blocks voltage-gated Na^{+} channels associated with glutamate receptors and potentiates the inhibitory effect of GABA; it is also used to treat generalized seizures. **Ethosuximide** reduces the low threshold T-type Ca^{2+} currents in thalamic neurons, and thus is particularly effective in treatment of absence seizures. **Valproate** and **phenytoin** block high frequency firing of neurons by acting on voltage-gated Na^{+} channels to reduce glutamate release.

reticular formation. This pattern of activity contributes to the appearance of the awake state. The reverse of this pattern leads to REM sleep. When there is a more even balance in the activity of the aminergic and cholinergic neurons, NREM sleep occurs. The orexin released from hypothalamic neurons may regulate the changes in activity in these brainstem neurons.

In addition, an increased release of GABA and reduced release of histamine increase the likelihood of NREM sleep via deactivation of the thalamus and cortex. Wakefulness occurs when GABA release is reduced and histamine release is increased.

MELATONIN AND THE SLEEP–WAKE STATE

In addition to the previously described neurochemical mechanisms promoting changes in the sleep–wake state, **melatonin** release from the richly vascularized **pineal gland** plays a role in sleep mechanisms (Figure 14–12). The pineal arises from the roof of the third ventricle in the diencephalon and is encapsulated by the meninges. The pineal stroma contains glial cells and pinealocytes with features suggesting that they have a secretory function. Like other endocrine glands, it has highly permeable fenestrated capillaries. In infants, the pineal is large and the cells tend to be arranged in alveoli. It begins to involute before puberty and small concretions of calcium phosphate and carbonate **(pineal sand)** appear in the tissue. Because the concretions are radiopaque, the pineal is often visible on x-ray films of the skull in adults. Displacement of a calcified pineal from its normal position indicates the presence of a space-occupying lesion such as a tumor in the brain.

Melatonin and the enzymes responsible for its synthesis from serotonin by N-acetylation and O-methylation are present in pineal pinealocytes, and the hormone is secreted by them into the blood and the cerebrospinal fluid (Figure 14–12). Two melatonin receptors (MT_1 and MT_2) have been found on neurons in the SCN. Both are G protein coupled receptors, with MT_1 receptors inhibiting adenylyl cyclase and resulting in sleepiness. MT_2 receptors stimulate phosphoinositide hydrolysis and may function in synchronization of the light–dark cycle.

The diurnal change in melatonin secretion may function as a timing signal to coordinate events with the light–dark cycle in the environment. Melatonin synthesis and secretion are increased during the dark period of the day and maintained at a low level during daylight hours (Figure 14–12). This diurnal variation in secretion is brought about by norepinephrine secreted by the postganglionic sympathetic nerves that innervate the pineal gland (Figure 14–10). Norepinephrine acts via β-adrenergic receptors to increase intracellular cAMP, and the cAMP in turn produces a marked increase in N-acetyltransferase activity. This results in increased melatonin synthesis and secretion. Circulating melatonin is rapidly metabolized in the liver by 6-hydroxylation followed by conjugation, and over 90% of the melatonin that appears in the urine is in the form of 6-hydroxy conjugates and 6-sulfatoxymelatonin. The pathway by which the brain metabolizes melatonin is unsettled but may involve cleavage of the indole nucleus.

The discharge of the sympathetic nerves to the pineal is entrained to the light–dark cycle in the environment via the retinohypothalamic nerve fibers to the SCN. From the hypothalamus, descending pathways converge onto preganglionic sympathetic neurons that in turn innervate the superior cervical ganglion, the site of origin of the postganglionic neurons to the pineal gland.

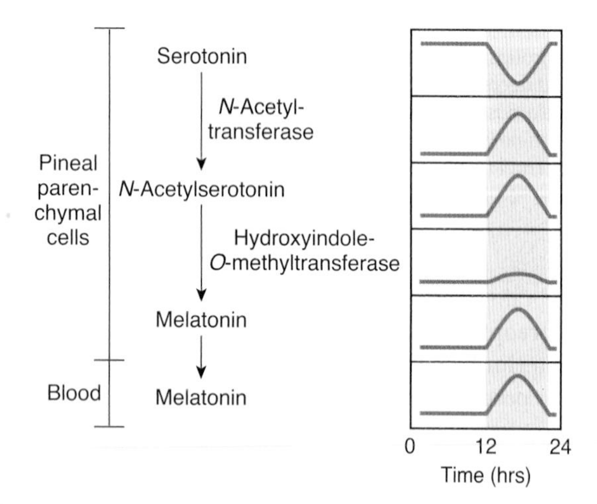

FIGURE 14–12 Diurnal rhythms of compounds involved in melatonin synthesis in the pineal. Melatonin and the enzymes responsible for its synthesis from serotonin are found in pineal pinealocytes; melatonin is secreted into the bloodstream. Melatonin synthesis and secretion are increased during the dark period (shaded area) and maintained at a low level during the light period.

CHAPTER SUMMARY

- The major rhythms in the EEG are alpha (8–13 Hz), beta (13–30 Hz), theta (4–7 Hz), delta (0.5–4 Hz), and gamma (30–80 Hz) oscillations.
- The EEG is of some value in localizing pathologic processes, and it is useful in characterizing different types of seizures.
- Throughout NREM sleep, there is some activity of skeletal muscle. A theta rhythm can be seen during stage 1 of sleep. Stage 2 is marked by the appearance of sleep spindles and occasional K complexes. In stage 3, a delta rhythm is dominant. Maximum slowing with slow waves is seen in stage 4. REM sleep is characterized by low-voltage, high-frequency EEG activity and rapid, roving movements of the eyes.
- A young adult typically passes through stages 1 and 2, and spends 70–100 min in stages 3 and 4. Sleep then lightens, and a REM period follows. This cycle repeats at 90-min intervals throughout the night. REM sleep occupies 50% of total sleep time in full-term neonates; this proportion declines rapidly and plateaus at about 25% until it falls further in old age.
- Transitions from sleep to wakefulness may involve alternating reciprocal activity of different groups of RAS neurons. When the activity of norepinephrine- and serotonin-containing neurons is dominant, the activity in acetylcholine-containing neurons is reduced, leading to the appearance of wakefulness.

The reverse of this pattern leads to REM sleep. Also, wakefulness occurs when GABA release is reduced and histamine release is increased.

- The entrainment of biological processes to the light–dark cycle is regulated by the SCN.
- The diurnal change in melatonin secretion from serotonin in the pineal gland may function as a timing signal to coordinate events with the light–dark cycle, including the sleep–wake cycle.

MULTIPLE-CHOICE QUESTIONS

For all questions, select the single best answer unless otherwise directed.

1. In a healthy, alert adult sitting with their eyes closed, the dominant EEG rhythm observed with electrodes over the occipital lobes is
   A. delta (0.5–4 Hz).
   B. theta (4–7 Hz).
   C. alpha (8–13 Hz).
   D. beta (18–30 Hz).
   E. fast, irregular low-voltage activity.
2. A 35-year-old male spent the evening in a sleep clinic to determine whether he had obstructive sleep apnea. The tests showed that NREM sleep accounted for over 30% of his total sleep time. Which of the following pattern of changes in central neurotransmitters or neuromodulators are associated with the transition from NREM to wakefulness?
   A. Decrease in norepinephrine, increase in serotonin, increase in acetylcholine, decrease in histamine, and decrease in GABA.
   B. Decrease in norepinephrine, increase in serotonin, increase in acetylcholine, decrease in histamine, and increase in GABA.
   C. Decrease in norepinephrine, decrease in serotonin, increase in acetylcholine, increase in histamine, and increase in GABA.
   D. Increase in norepinephrine, increase in serotonin, decrease in acetylcholine, increase in histamine, and decrease in GABA.
   E. Increase in norepinephrine, decrease in serotonin, decrease in acetylcholine, increase in histamine, and decrease in GABA.
3. A gamma rhythm (30–80 Hz)
   A. is characteristic of seizure activity.
   B. is seen in an individual who is awake but not focused.
   C. may be a mechanism to bind together sensory information into a single percept and action.
   D. is independent of thalamocortical loops.
   E. is generated in the hippocampus.
4. For the past several months, a 67-year-old female experienced difficulty initiating and/or maintaining sleep several times a week. A friend suggested that she take melatonin to regulate her sleep–wake cycle. Melatonin secretion would probably not be increased by
   A. stimulation of the superior cervical ganglia.
   B. intravenous infusion of tryptophan.
   C. intravenous infusion of epinephrine.
   D. stimulation of the optic nerve.
   E. induction of pineal hydroxyindole-O-methyltransferase.
5. A 10-year-old boy was diagnosed with childhood absence epilepsy. His EEG showed a bilateral synchronous, symmetrical 3-Hz spike-and-wave discharge. Absence seizures
   A. are a form of nonconvulsive generalized seizures accompanied by momentary loss of consciousness.
   B. are a form of complex partial seizures accompanied by momentary loss of consciousness.
   C. are a form of nonconvulsive generalized seizures without a loss of consciousness.
   D. are a form of simple partial seizures without a loss of consciousness.
   E. are a form of convulsive generalized seizures accompanied by momentary loss of consciousness.
6. A 57-year-old professor at a medical school experienced numerous episodes of a sudden loss of muscle tone and an irresistible urge to sleep in the middle of the afternoon. He was diagnosed with narcolepsy which
   A. is characterized by a sudden onset of NREM sleep.
   B. has a familial incidence associated with a class II antigen of the major histocompatibility complex.
   C. may be due to the presence of an excessive number of orexin-producing neurons in the hypothalamus.
   D. is often effectively treated with dopamine receptor agonists.
   E. is the most common cause of daytime sleepiness.

CHAPTER RESOURCES

Blackman S: *Consciousness: An Introduction.* Oxford University Press, 2004.

Feely M: Drug treatment of epilepsy. British Med J 1999;318:106.

McCormick DA, Contreras D: Of the cellular and network bases of epileptic seizures. Annu Rev Physiol 2001;63:815.

Merica H, Fortune RD: State transitions between wake and sleep, and within the ultradian cycle, with focus on the link to neuronal activity. Sleep Med Rev 2004;8:473.

Oberheim NA, Tian GF, Han X, et al: Loss of astrocytic domain organization in the epileptic brain. J Neurosci 2008;28:3264.

Sakurai T: The neural circuit of orexin (hypocretin): maintaining sleep and wakefulness. Nature Rev Neurosci 2007;8:171.

Saper CB, Fuller PM, Pedersen NP, Lu J, Scrammell TE: Sleep state switching. Neuron 2010;68:1023.

Shaw JC (editor): *The Brain's Alpha Rhythms and the Mind.* Elsevier, 2003.

Siegel JM: Narcolepsy. Sci Am 2000;282:76.

Stafstrom CE: Epilepsy: A review of selected clinical syndromes and advances in basic science. J Cereb Blood Flow Metab 2006;26:983.

Steinlein O: Genetic mechanisms that underlie epilepsy. Nat Rev Neurosci 2004;5:400.

Steriade M, McCarley RW: *Brain Stem Control of Wakefulness and Sleep.* Plenum, 1990.

Steriade M, Paré D: *Gating in Cerebral Networks.* Cambridge University Press, 2007.

Thorpy M (editor): *Handbook of Sleep Disorders.* Marcel Dekker, 1990.

CHAPTER

15

Learning, Memory, Language, & Speech

OBJECTIVES

After studying this chapter, you should be able to:

- Describe the various forms of memory.
- Identify the parts of the brain involved in memory processing and storage.
- Define synaptic plasticity, long-term potentiation (LTP), long-term depression (LTD), habituation, and sensitization, and their roles in learning and memory.
- Describe the abnormalities of brain structure and function found in Alzheimer disease.
- Define the terms categorical hemisphere and representational hemisphere and summarize the difference between these hemispheres.
- Summarize the differences between fluent and nonfluent aphasia, and explain each type on the basis of its pathophysiology.

INTRODUCTION

A revolution in our understanding of brain function in humans has been brought about by the development and widespread availability of **positron emission tomographic (PET), functional magnetic resonance imaging (fMRI), computed tomography (CT) scanning,** and other imaging and diagnostic techniques. PET is often used to measure local glucose metabolism, which is proportional to neural activity, and fMRI is used to measure local amounts of oxygenated blood. These techniques provide an index of the level of the activity in various parts of the brain in completely intact healthy humans and in those with different diseases or brain injuries (see Clinical Box 15–1). They have been used to study not only simple responses but complex aspects of learning, memory, and perception. Different portions of the cortex are activated when hearing, seeing, speaking, or generating words. Figure 15–1 shows examples of the use of imaging to compare the functions of the cerebral cortex in processing words in a male versus a female subject.

Other techniques that have provided information on cortical function include stimulation of the exposed cerebral cortex in conscious humans undergoing neurosurgical procedures and, in a few instances, studies with chronically implanted electrodes. Valuable information has also been obtained from investigations in laboratory primates. However, in addition to the difficulties in communicating with them, the brain of the rhesus monkey is only one-fourth the size of the brain of the chimpanzee, our nearest primate relative, and the chimpanzee brain is in turn one-fourth the size of the human brain.

LEARNING & MEMORY

A characteristic of animals and particularly of humans is their ability to alter behavior on the basis of experience. **Learning** is acquisition of the information that makes this possible and **memory** is the retention and storage of that information. The two are obviously closely related and are considered together in this Chapter.

FORMS OF MEMORY

From a physiologic point of view, memory is divided into explicit and implicit forms (Figure 15–2). **Explicit** or **declarative memory** is associated with consciousness, or at least awareness, and is dependent on the **hippocampus** and other parts of the **medial temporal lobes** of the brain for its retention. Clinical Box 15–2 describes how tracking a patient

CLINICAL BOX 15–1

Traumatic Brain Injury

Traumatic brain injury (TBI) is defined as a nondegenerative, noncongenital insult to the brain due to an excessive mechanical force or penetrating injury to the head. It can lead to a permanent or temporary impairment of cognitive, physical, emotional, and behavioral functions, and it can be associated with a diminished or altered state of consciousness. TBI is one of the leading causes of death or disability worldwide. According to the Center for Disease Control, each year at least 1.5 million individuals in the United States sustain a TBI. It is most common in children under age 4, in adolescents aged 15–19 years of age, and in adults over the age of 65. In all age groups, the incidence of TBI occurrence is about twice as high in males compared to females. In about 75% of the cases, the TBI is considered mild and manifests as a concussion. Adults with severe TBI who are treated have a mortality rate of about 30%, but about 50% regain most if not all of their functions with therapy. The leading causes of TBI include falls, motor vehicle accidents, being struck by an object, and assaults. In some cases, areas remote from the actual injury also begin to malfunction, a process called **diaschisis.** TBI is often divided into primary and secondary stages. Primary injury is that caused by the mechanical force (eg, skull fracture and surface contusions) or acceleration–deceleration due to unrestricted movement of the head leading to shear, tensile, and compressive strains. These injuries can cause **intracranial hematoma** (epidural, subdural, or subarachnoid) and **diffuse axonal injury.** Secondary injury is often a delayed response and may be due to impaired cerebral blood flow that can eventually lead to cell death. A **Glasgow Coma Scale** is the most common system used to define the severity of TBI and evaluates motor responses, verbal responses, and eye opening to assess the levels of consciousness and neurologic functioning after an injury. Symptoms of mild TBI include headache, confusion, dizziness, blurred vision, ringing in the ears, a bad taste in the mouth, fatigue, disturbances in sleep, mood changes, and problems with memory, concentration, or thinking. Individuals with moderate or severe TBI show these symptoms as well as vomiting or nausea, convulsions or seizures, an inability to be roused, fixed and dilated pupils, slurred speech, limb weakness, loss of coordination, and increased confusion, restlessness, or agitation. In the most severe cases of TBI, the affected individual may go into a **permanent vegetative state.**

THERAPEUTIC HIGHLIGHTS

The advancements in brain imaging technology have improved the ability of medical personnel to diagnose and evaluate the extent of brain damage. Since little can be done to reverse the brain damage, therapy is initially directed at stabilizing the patient and trying to prevent further (secondary) injury. This is followed by rehabilitation that includes physical, occupational, and speech/language therapies. Recovery of brain function can be due to several factors: brain regions that were suppressed but not damaged can regain their function, axonal sprouting and redundancy allows other areas of the brain to take over the functions that were lost due to the injury, and behavioral substitution, by learning new strategies to compensate for the deficits.

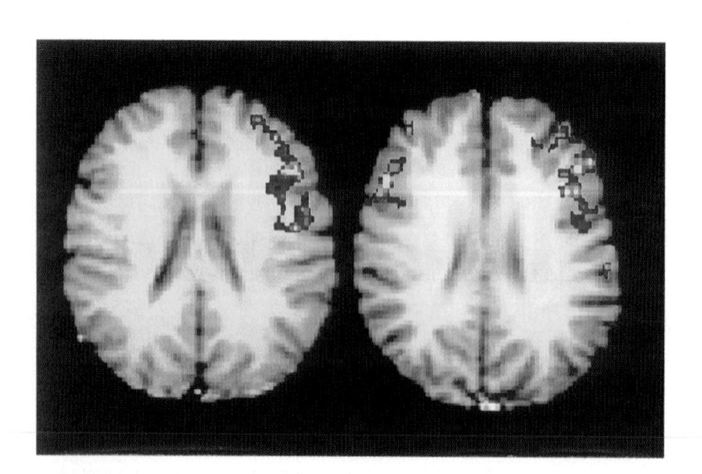

FIGURE 15–1 Comparison of the images of the active areas of the brain in a man (left) and a woman (right) during a language-based activity. Women use both sides of their brain whereas men use only a single side. This difference may reflect different strategies used for language processing. (From Shaywitz et al, 1995. NMR Research/Yale Medical School.)

with brain damage has led to an awareness of the role of the temporal lobe in declarative memory. **Implicit** or **nondeclarative memory** does not involve awareness, and its retention does not usually involve processing in the hippocampus.

Explicit memory is for factual knowledge about people, places, and things. It is divided into **semantic memory** for facts (eg, words, rules, and language) and **episodic memory** for events. Explicit memories that are initially required for activities such as riding a bicycle can become implicit once the task is thoroughly learned.

Implicit memory is important for training reflexive motor or perceptual skills and is subdivided into four types. **Priming** is the facilitation of the recognition of words or objects by prior exposure to them and is dependent on the **neocortex.** An example of priming is the improved recall of a word when presented with the first few letters of it. **Procedural memory** includes skills and habits, which, once acquired, become unconscious and automatic. This type of memory is processed in the **striatum.** **Associative learning** relates to **classical** and

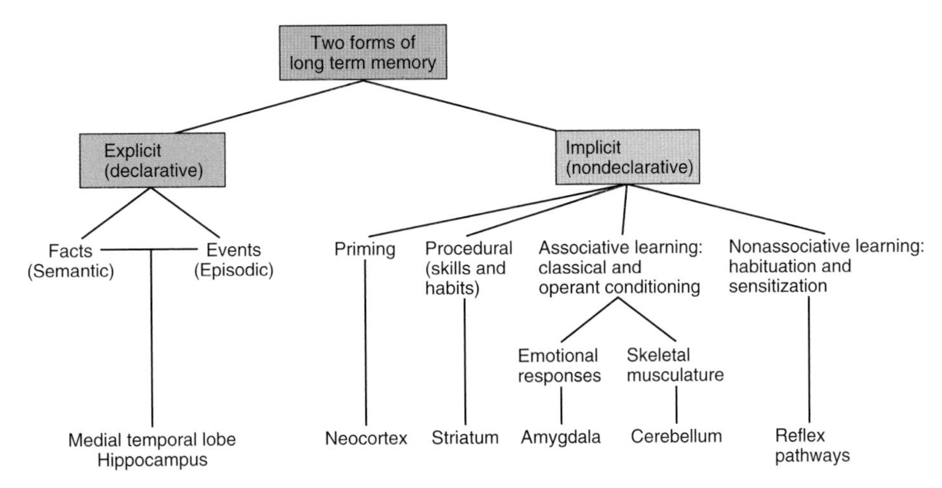

FIGURE 15–2 Forms of memory. Explicit (declarative) memory is associated with consciousness and is dependent on the hippocampus and other parts of the medial temporal lobes of the brain for its retention. It is for factual knowledge about people, places, and things. Implicit (nondeclarative) memory does not involve awareness, and it does not involve processing in the hippocampus. It is important for training reflexive motor or perceptual skills. (Modified from Kandel ER, Schwartz JH, Jessell TM [editors]: *Principles of Neural Science,* 4th ed. McGraw-Hill, 2000.)

operant conditioning in which one learns about the relationship between one stimulus and another. This type of memory is dependent on the **amygdala** for its emotional responses and the **cerebellum** for the motor responses. **Nonassociative** learning includes **habituation** and **sensitization** and is dependent on various reflex pathways.

Explicit memory and many forms of implicit memory involve (1) **short-term memory,** which lasts seconds to hours, during which processing in the hippocampus and elsewhere lays down long-term changes in synaptic strength; and (2) **long-term memory,** which stores memories for years and sometimes for life. During short-term memory, the memory traces are subject to disruption by trauma and various drugs, whereas long-term memory traces are remarkably resistant to disruption. **Working memory** is a form of short-term memory that keeps information available, usually for very short periods, while the individual plans action based on it.

CLINICAL BOX 15–2

The Case of HM: Defining a Link between Brain Function & Memory

HM was a patient who suffered from bilateral temporal lobe seizures that began following a bicycle accident at age 9. His case has been studied by many scientists and has led to a greater understanding of the link between the **temporal lobe** and **declarative memory.** HM had partial seizures for many years, and then several tonic-clonic seizures by age 16. In 1953, at the age of 27, HM underwent bilateral surgical removal of the amygdala, large portions of the hippocampal formation, and portions of the association area of the temporal cortex. HM's seizures were better controlled after surgery, but removal of the temporal lobes led to devastating memory deficits. He maintained **long-term memory** for events that occurred prior to surgery, but he suffered from **anterograde amnesia.** His **short-term memory** was intact, but he could not commit new events to long-term memory. He had normal procedural memory, and he could learn new puzzles and motor tasks. His case was the first to bring attention to the critical role of temporal lobes in formation of long-term declarative memories and to implicate this region in the conversion of short-term to long-term memories. Later work showed that the **hippocampus** is the primary structure within the temporal lobe involved in this conversion. Because HM retained memories from before surgery, his case also shows that the hippocampus is not involved in the storage of declarative memory. HM died in 2008 and only at that time was his identity released. An audio-recording by National Public Radio from the 1990s of HM talking to scientists was released in 2007 and is available at http://www.npr.org/templates/story/story.php?storyId=7584970.

NEURAL BASIS OF MEMORY

The key to memory is alteration in the strength of selected synaptic connections. Second messenger systems contribute to the changes in neural circuitry required for learning and memory. Alterations in cellular membrane channels are often correlated to learning and memory. In all but the simplest of cases, the alteration involves the synthesis of proteins and the activation of genes. This occurs during the change from short-term working memory to long-term memory.

In animals, acquisition of long-term learned responses is prevented if, within 5 min after each training session, the animals are anesthetized, given electroshock, subjected to hypothermia, or given drugs, antibodies, or oligonucleotides that block the synthesis of proteins. If these interventions are performed 4 h after the training sessions, there is no effect on acquisition. The human counterpart of this phenomenon is the loss of memory for the events immediately preceding a brain concussion or electroshock therapy **(retrograde amnesia).** This amnesia encompasses longer periods than it does in experimental animals (sometimes many days) but remote memories remain intact.

SYNAPTIC PLASTICITY & LEARNING

Short- and long-term changes in synaptic function can occur as a result of the history of discharge at a synapse; that is, synaptic conduction can be strengthened or weakened on the basis of past experience. These changes are of great interest because they represent forms of learning and memory. They can be presynaptic or postsynaptic in location.

One form of plastic change is **posttetanic potentiation,** the production of enhanced postsynaptic potentials in response to stimulation. This enhancement lasts up to 60 s and occurs after a brief tetanizing train of stimuli in the presynaptic neuron. The tetanizing stimulation causes Ca^{2+} to accumulate in the presynaptic neuron to such a degree that the intracellular binding sites that keep cytoplasmic Ca^{2+} low are overwhelmed.

Habituation is a simple form of learning in which a neutral stimulus is repeated many times. The first time it is applied it is novel and evokes a reaction (the orienting reflex or "what is it?" response). However, it evokes less and less electrical response as it is repeated. Eventually, the subject becomes habituated to the stimulus and ignores it. This is associated with decreased release of neurotransmitter from the presynaptic terminal because of decreased intracellular Ca^{2+}. The decrease in intracellular Ca^{2+} is due to a gradual inactivation of Ca^{2+} channels. It can be short term, or it can be prolonged if exposure to the benign stimulus is repeated many times. Habituation is a classic example of nonassociative learning.

Sensitization is in a sense the opposite of habituation. Sensitization is the prolonged occurrence of augmented postsynaptic responses after a stimulus to which one has become habituated is paired once or several times with a noxious stimulus. At least in the sea snail *Aplysia,* the noxious stimulus causes discharge of serotonergic neurons that end on the presynaptic endings of sensory neurons. Thus, sensitization is due to presynaptic facilitation. Sensitization may occur as a transient response, or if it is reinforced by additional pairings of the noxious stimulus and the initial stimulus, it can exhibit features of short-term or long-term memory. The short-term prolongation of sensitization is due to a Ca^{2+}-mediated change in adenylyl cyclase that leads to a greater production of cAMP. The **long-term potentiation (LTP)** also involves protein synthesis and growth of the presynaptic and postsynaptic neurons and their connections.

LTP is a rapidly developing persistent enhancement of the postsynaptic potential response to presynaptic stimulation after a brief period of rapidly repeated stimulation of the presynaptic neuron. It resembles posttetanic potentiation but is much more prolonged and can last for days. There are multiple mechanisms by which LTP can occur, some are dependent on changes in the **N-methyl-D-aspartate (NMDA) receptor** and some are independent of the NMDA receptor. LTP is initiated by an increase in intracellular Ca^{2+} in either the presynaptic or postsynaptic neuron.

LTP occurs in many parts of the nervous system but has been studied in greatest detail in a synapse within the hippocampus, specifically the connection of a pyramidal cell in the CA3 region and a pyramidal cell in the CA1 region via the **Schaffer collateral.** This is an example of an NMDA receptor-dependent form of LTP involving an increase in Ca^{2+} in the postsynaptic neuron. Recall that NMDA receptors are permeable to Ca^{2+} as well as to Na^+ and K^+. The hypothetical basis of the Schaffer collateral LTP is summarized in Figure 15–3. At the resting membrane potential, glutamate release from a presynaptic neuron binds to both NMDA and non-NMDA receptors on the postsynaptic neuron. In the case of the Schaffer collateral the non-NMDA receptor of interest is the **α-amino-3-hydroxy-5-methylisoxazole-4 propionic acid (AMPA) receptor.** Na^+ and K^- can flow only through the AMPA receptor because the presence of Mg^{2+} on the NMDA receptor blocks it. However, the membrane depolarization that occurs in response to high frequency tetanic stimulation of the presynaptic neuron is sufficient to expel the Mg^{2+} from the NMDA receptor, allowing the influx of Ca^{2+} into the postsynaptic neuron. This leads to activation of Ca^{2+}/calmodulin kinase, protein kinase C, and tyrosine kinase which together induce LTP. The Ca^{2+}/calmodulin kinase phosphorylates the AMPA receptors, increasing their conductance, and moves more of these receptors into the synaptic cell membrane from cytoplasmic storage sites. In addition, once LTP is induced, a chemical signal (possibly nitric oxide, NO) is released by the postsynaptic neuron and passes retrogradely to the presynaptic neuron, producing a long-term increase in the quantal release of glutamate.

LTP identified in the mossy fibers of the hippocampus (connecting granule cells in the dentate cortex) is due to an increase in Ca^{2+} in the presynaptic rather than the postsynaptic neuron in response to tetanic stimulation and is independent of NMDA receptors. The influx of Ca^{2+} in the presynaptic neuron is thought to activate Ca^{2+}/calmodulin-dependent adenylyl cyclase to increase cAMP.

Long-term depression (LTD) was first noted in the hippocampus but was subsequently shown to be present throughout the brain in the same fibers as LTP. LTD is the opposite of LTP. It resembles LTP in many ways, but it is characterized by a decrease in synaptic strength. It is produced by slower stimulation of presynaptic neurons and is associated with a smaller rise in intracellular Ca^{2+} than occurs in LTP. In the cerebellum, its occurrence appears to require the phosphorylation of the GluR2 subunit of the AMPA receptors. It may be involved in the mechanism by which learning occurs in the cerebellum.

INTERCORTICAL TRANSFER OF MEMORY

If a cat or monkey is conditioned to respond to a visual stimulus with one eye covered and then tested with the blindfold transferred to the other eye, it performs the conditioned response. This is true even if the optic chiasm has been cut, making the

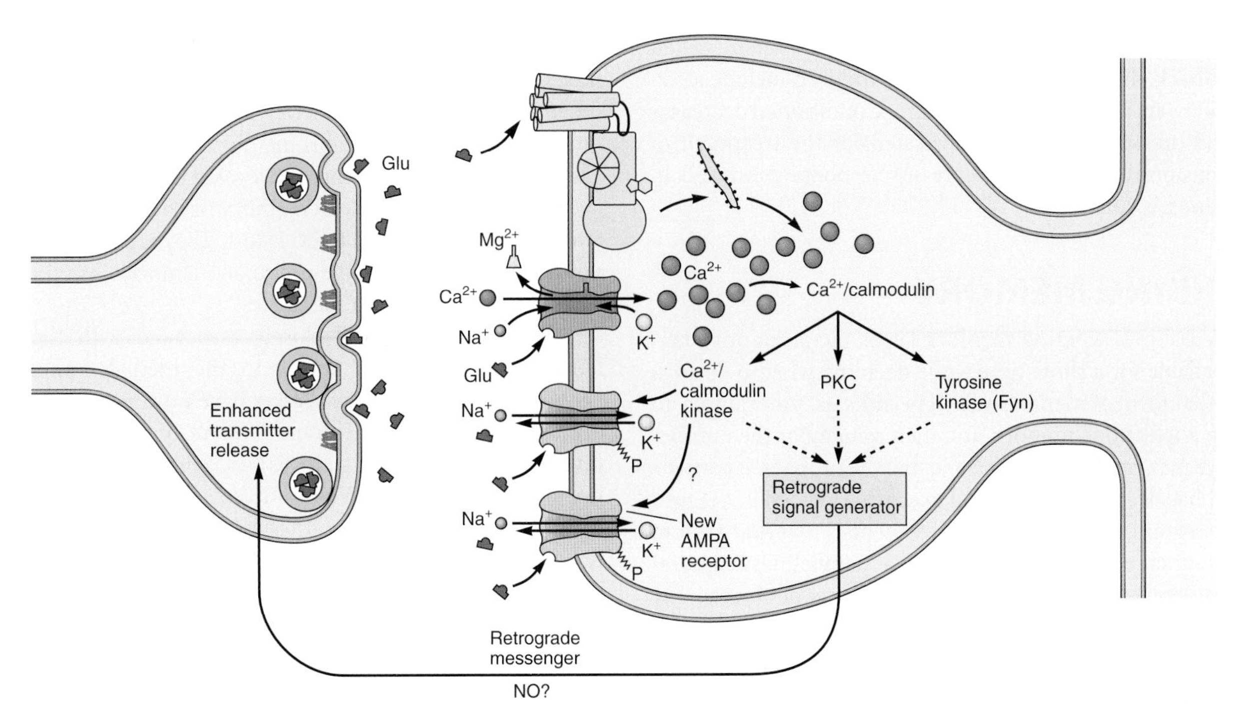

FIGURE 15–3 Production of LTP in Schaffer collaterals in the hippocampus. Glutamate (Glu) released from the presynaptic neuron binds to AMPA and NMDA receptors in the membrane of the postsynaptic neuron. The depolarization triggered by activation of the AMPA receptors relieves the Mg^{2+} block in the NMDA receptor channel, and Ca^{2+} enters the neuron with Na^{+}. The increase in cytoplasmic Ca^{2+} activates Ca^{2+}/calmodulin kinase, protein kinase C, and tyrosine kinase which together induce LTP. The Ca^{2+}/calmodulin kinase II phosphorylates the AMPA receptors, increasing their conductance, and moves more AMPA receptors into the synaptic cell membrane from cytoplasmic storage sites. In addition, once LTF is induced, a chemical signal (possibly nitric oxide, NO) is released by the postsynaptic neuron and passes retrogradely to the presynaptic neuron, producing a long-term increase in the quantal release of glutamate. (Modified from Kandel ER, Schwartz JH, Jessell TM [editors]: *Principles of Neural Science*, 4th ed. McGraw-Hill, 2000.)

visual input from each eye go only to the ipsilateral cortex. If, in addition to the optic chiasm, the anterior and posterior commissures and the corpus callosum are sectioned ("split-brain animal"), no memory transfer occurs. Experiments in which the corpus callosum was partially sectioned indicate that the memory transfer occurs in the anterior portion of the corpus callosum. Similar results have been obtained in humans in whom the corpus callosum is congenitally absent or in whom it has been sectioned surgically in an effort to control epileptic seizures. This demonstrates that the neural coding necessary for "remembering with one eye what has been learned with the other" has been transferred to the opposite cortex via the commissures. Evidence suggests that similar transfer of information is acquired through other sensory pathways.

NEUROGENESIS

It is now established that the traditional view that brain cells are not added after birth is wrong; new neurons form from stem cells throughout life in at least two areas: the olfactory bulb and the hippocampus. This is a process called **neurogenesis.** There is evidence implicating that experience-dependent growth of new granule cells in the dentate gyrus of the hippocampus may contribute to learning and memory. A reduction in the number of new neurons formed reduces at least one form of hippocampal memory production. However, a great deal more work is needed before the relation of new cells to memory processing can be considered established.

ASSIOCIATIVE LEARNING: CONDITIONED REFLEXES

A classic example of associative learning is a **conditioned reflex.** A conditioned reflex is a reflex response to a stimulus that previously elicited little or no response, acquired by repeatedly pairing the stimulus with another stimulus that normally does produce the response. In Pavlov's classic experiments, the salivation normally induced by placing meat in the mouth of a dog was studied. A bell was rung just before the meat was placed in the dog's mouth, and this was repeated a number of times until the animal would salivate when the bell was rung even though no meat was placed in its mouth. In this experiment, the meat placed in the mouth was the **unconditioned stimulus (US),** the stimulus that normally produces a particular innate response. The **conditioned stimulus (CS)** was the bell ringing. After the CS and US had been paired a sufficient number of times, the CS produced the response originally evoked only by the US. The CS had to precede the US. An immense number of somatic, visceral, and other neural changes can be made to occur as conditioned reflex responses.

Conditioning of visceral responses is often called **biofeedback.** The changes that can be produced include alterations in heart rate and blood pressure. Conditioned decreases in blood pressure have been advocated for the treatment of hypertension; however, the depressor response produced in this fashion is small.

WORKING MEMORY

As noted above, working memory keeps incoming information available for a short time while deciding what to do with it. It is that form of memory which permits us, for example, to look up a telephone number, and then remember the number while we pick up the telephone and dial the number. It consists of what has been called a **central executive** located in the prefrontal cortex, and two "rehearsal systems:" a **verbal system** for retaining verbal memories and a parallel **visuospatial system** for retaining visual and spatial aspects of objects. The executive steers information into these rehearsal systems.

HIPPOCAMPUS & MEDIAL TEMPORAL LOBE

Working memory areas are connected to the hippocampus and the adjacent parahippocampal portions of the medial temporal cortex (Figure 15–4). Output from the hippocampus leaves via the subiculum and the entorhinal cortex and somehow binds together and strengthens circuits in many different neocortical areas, forming over time the stable remote memories that can now be triggered by many different cues.

In humans, bilateral destruction of the ventral hippocampus, or Alzheimer disease and similar disease processes that destroy its CA1 neurons, can cause striking defects in short-term memory. Humans with such destruction have intact working memory and remote memory. Their implicit memory processes are generally intact. They perform adequately in terms of conscious memory as long as they concentrate on what they are doing. However, if they are distracted for even a very short period, all memory of what they were doing and what they proposed to do is lost. They are thus capable of new learning and retain old prelesion memories, but they cannot form new long-term memories.

The hippocampus is closely associated with the overlying parahippocampal cortex in the medial frontal lobe (Figure 15–4). Memory processes have now been studied not only with fMRI but with measurement of evoked potentials (event-related potentials; ERPs) in epileptic patients with implanted electrodes. When subjects recall words, activity in their left frontal lobe and their left parahippocampal cortex increases, but when they recall pictures or scenes, activity takes place in their right frontal lobe and the parahippocampal cortex on both sides.

The connections of the hippocampus to the diencephalon are also involved in memory. Some people with alcoholism-related brain damage develop impairment of recent memory, and the memory loss correlates well with the presence of pathologic changes in the mamillary bodies, which have extensive efferent connections to the hippocampus via the fornix. The mamillary bodies project to the anterior thalamus via the mamillothalamic tract, and in monkeys, lesions of the thalamus cause loss of recent memory. From the thalamus, the fibers concerned with memory project to the prefrontal cortex and from there to the basal forebrain. From the **nucleus basalis of Meynert** in the basal forebrain, a diffuse cholinergic projection goes to all of the neocortex, the amygdala, and the hippocampus. Severe loss of these fibers occurs in Alzheimer disease.

The amygdala is closely associated with the hippocampus and is concerned with encoding and recalling emotionally charged memories. During retrieval of fearful memories, the theta rhythms of the amygdala and the hippocampus become synchronized. In normal humans, events associated with strong emotions are remembered better than events without an emotional charge, but in patients with bilateral lesions of the amygdala, this difference is absent.

Confabulation is an interesting though poorly understood condition that sometimes occurs in individuals with lesions of the ventromedial portions of the frontal lobes. These individuals perform poorly on memory tests, but they spontaneously describe events that never occurred. This has been called "honest lying."

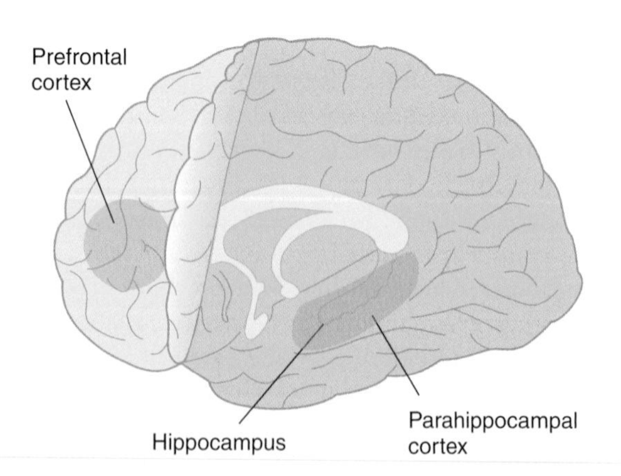

FIGURE 15–4 Areas concerned with encoding explicit memories. The prefrontal cortex and the parahippocampal cortex of the brain are active during the encoding of memories. Output from the hippocampus leaves via the subiculum and the entorhinal cortex and strengthens circuits in many neocortical areas, forming stable remote memories that can be triggered by various cues. (Modified from Rugg MD: Memories are made of this. Science 1998;281:1151.)

LONG-TERM MEMORY

While the encoding process for short-term explicit memory involves the hippocampus, long-term memories are stored in various parts of the neocortex. Apparently, the various parts of the memories—visual, olfactory, auditory, etc—are located in the cortical regions concerned with these functions, and the pieces are tied together by long-term changes in the strength of transmission at relevant synaptic junctions so that all the components are brought to consciousness when the memory is recalled.

Once long-term memories have been established, they can be recalled or accessed by a large number of different associations. For example, the memory of a vivid scene can be evoked not only by a similar scene but also by a sound or smell associated with the scene and by words such as "scene," "vivid," and "view." Thus, each stored memory must have multiple routes or keys. Furthermore, many memories have an emotional component or "color," that is, in simplest terms, memories can be pleasant or unpleasant.

STRANGENESS & FAMILIARITY

It is interesting that stimulation of some parts of the temporal lobes in humans causes a change in interpretation of one's surroundings. For example, when the stimulus is applied, the subject may feel strange in a familiar place or may feel that what is happening now has happened before. The occurrence of a sense of familiarity or a sense of strangeness in appropriate situations probably helps the normal individual adjust to the environment. In strange surroundings, one is alert and on guard, whereas in familiar surroundings, vigilance is relaxed. An inappropriate feeling of familiarity with new events or in new surroundings is known clinically as the **déjà vu phenomenon,** from the French words meaning "already seen." The phenomenon occurs from time to time in normal individuals, but it also may occur as an aura (a sensation immediately preceding a seizure) in patients with temporal lobe epilepsy.

ALZHEIMER DISEASE & SENILE DEMENTIA

Alzheimer disease is the most common age-related neurodegenerative disorder. Memory decline initially manifests as a loss of episodic memory, which impedes recollection of recent events. Loss of short-term memory is followed by general loss of cognitive and other brain functions, agitation, depression, the need for constant care, and, eventually, death. Clinical Box 15–3 describes the etiology and therapeutic strategies for the treatment of Alzheimer disease.

Figure 15–5 summarizes some of the risk factors, pathogenic processes, and clinical signs linked to cellular

CLINICAL BOX 15–3

Alzheimer Disease

Alzheimer disease was originally characterized in middle-aged people, and similar deterioration in elderly individuals is technically **senile dementia** of the alzheimer type, though it is frequently just called Alzheimer disease. Both genetic and environmental factors are thought to contribute to the etiology of the disease. Most cases are sporadic, but a familial form of the disease (accounting for about 5% of the cases) is seen in an early-onset form of the disease. In these cases, the disease is caused by mutations in genes for the amyloid precursor protein on chromosome 21, presenilin I on chromosome 14, or presenilin II on chromosome 1. It is transmitted in an autosomal dominant mode, so offspring in the same generation have a 50/50 chance of developing familial Alzheimer disease if one of their parents is affected. Each mutation leads to an overproduction of the β-amyloid protein found in neuritic plaques. Senile dementia can be caused by vascular disease and other disorders, but Alzheimer disease is the most common cause, accounting for 50–60% of the cases. Alzheimer disease is present in 8–17% of the population over the age of 65, with the incidence increasing steadily with age (nearly doubling every 5 years after reaching the age of 60). In those who are 95 years of age and older, the incidence is 40–50%. It is estimated that by the year 2050, up to 16 million people age 65 and older in the US alone will have Alzheimer disease. Although the prevalence of the disease appears to be higher in women, this may be due to their longer life span as the incidence rates are similar for men and women. Alzheimer disease plus the other forms of senile dementia are a major medical problem.

THERAPEUTIC HIGHLIGHTS

Research is aimed at identifying strategies to prevent the occurrence, delay the onset, slow the progression, or alleviate the symptoms of Alzheimer disease. The use of **acetylcholinesterase inhibitors** (eg, **rivastigmine, donepezil,** or **galantamine**) in early stages of the disease increases the availability of acetylcholine in the synaptic cleft. It has shown some promise in ameliorating global cognitive dysfunction, but not learning and memory impairments in these patients. These drugs also delay the worsening of symptoms for up to 12 months in about 50% of the cases studied. **Antidepressants** (eg, **paroxetine, imipramine**) have been useful for treating depression in Alzheimer patients. **Memantine** (an NMDA receptor antagonist) prevents glutamate-induced excitotoxicity in the brain and is used to treat moderate to severe Alzheimer disease. It has been shown to delay the worsening of symptoms in some patients. Drugs used to block the production of β-amyloid proteins are under development. An example is **R-flurbiprofen.** Also attempts are underway to develop vaccines that would allow the body's immune system to produce antibodies to attack these proteins.

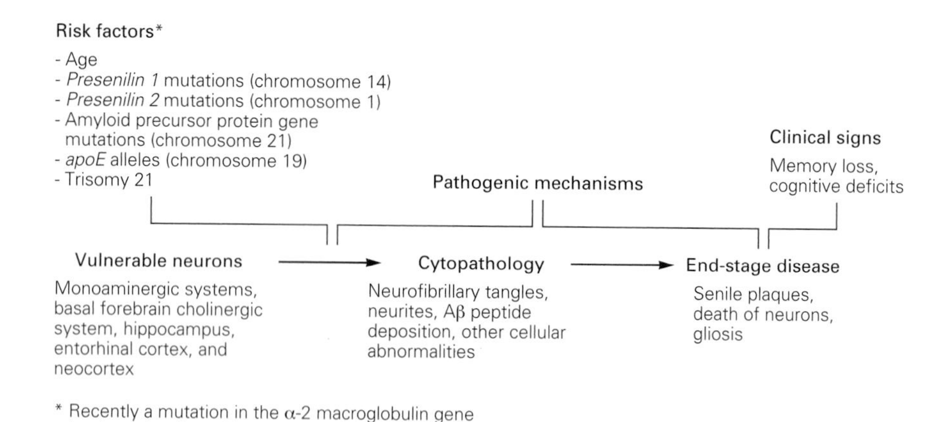

FIGURE 15–5 Relationships of risk factors, pathogenic processes, and clinical signs to cellular abnormalities in the brain during Alzheimer disease. (From Kandel ER, Schwartz JH, Jessell TM [editors]: *Principles of Neural Science,* 4th ed. McGraw-Hill, 2000.)

abnormalities that occur in Alzheimer disease. The cytopathologic hallmarks of Alzheimer disease are intracellular **neurofibrillary tangles,** made up in part of hyperphosphorylated forms of the **tau protein** that normally binds to microtubules, and extracellular **senile plaques,** which have a core of **β-amyloid peptides** surrounded by altered nerve fibers and reactive glial cells. Figure 15–6 compares a normal nerve cell to one showing abnormalities associated with Alzheimer disease.

The β-amyloid peptides are products of a normal protein, **amyloid precursor protein (APP),** a transmembrane protein that projects into the extracellular fluid (ECF) from all nerve cells. This protein is hydrolyzed at three different sites by α-secretase, β-secretase, and γ-secretase, respectively. When APP is hydrolyzed by α-secretase, nontoxic peptide products are produced. However, when it is hydrolyzed by β-secretase and γ-secretase, polypeptides with 40–42 amino acids are produced; the actual length varies because of variation in the site at which γ-secretase cuts the protein chain. These polypeptides are toxic, the most toxic being $A\beta\sigma^{1-42}$. The polypeptides form extracellular aggregates, which can stick to AMPA receptors and Ca^{2+} ion channels, increasing Ca^{2+} influx. The polypeptides also initiate an inflammatory response, with production of intracellular tangles. The damaged cells eventually die.

An interesting finding that may well have broad physiologic implications is the observation—now confirmed in a rigorous prospective study—that frequent effortful mental activities, such as doing difficult crossword puzzles and playing board games, slow the onset of cognitive dementia due to Alzheimer disease and vascular disease. The explanation for this "use it or lose it" phenomenon is as yet unknown, but it certainly suggests that the hippocampus and its connections have plasticity like other parts of the brain and skeletal and cardiac muscles.

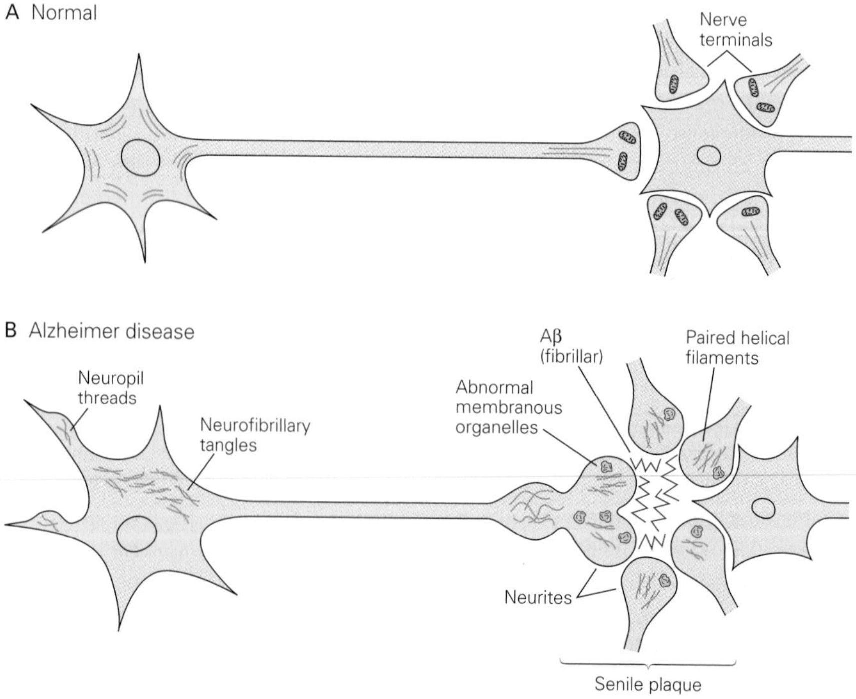

FIGURE 15–6 Comparison of a normal neuron and one with abnormalities associated with Alzheimer disease. The cytopathologic hallmarks are intracellular neurofibrillary tangles and extracellular senile plaques that have a core of β-amyloid peptides surrounded by altered nerve fibers and reactive glial cells. (From Kandel ER, Schwartz JH, Jessell TM [editors]: *Principles of Neural Science,* 4th ed. McGraw-Hill, 2000.)

LANGUAGE & SPEECH

Memory and learning are functions of large parts of the brain, but the centers controlling some of the other "higher functions of the nervous system," particularly the mechanisms related to language, are more or less localized to the neocortex. Speech and other intellectual functions are especially well developed in humans—the animal species in which the neocortical mantle is most highly developed.

COMPLEMENTARY SPECIALIZATION OF THE HEMISPHERES VERSUS "CEREBRAL DOMINANCE"

One group of functions localized to the neocortex in humans consists of those related to language; that is, the understanding of the spoken and printed word and expressing ideas in speech and writing. It is a well-established fact that human language functions depend more on one cerebral hemisphere than on the other. This hemisphere is concerned with categorization and symbolization and has often been called the **dominant hemisphere.** However, the other hemisphere is not simply less developed or "nondominant;" instead, it is specialized in the area of spatiotemporal relations. It is this hemisphere that is concerned, for example, with the identification of objects by their form and the recognition of musical themes. It also plays a primary role in the recognition of faces. Consequently, the concept of "cerebral dominance" and a dominant and nondominant hemisphere has been replaced by a concept of complementary specialization of the hemispheres, one for sequential-analytic processes (the **categorical hemisphere**) and one for visuospatial relations (the **representational hemisphere**). The categorical hemisphere is concerned with language functions, but hemispheric specialization is also present in monkeys, so it predates the evolution of language. Clinical Box 15–4 describes deficits that occur in subjects with representational or categorical hemisphere lesions.

Hemispheric specialization is related to handedness. Handedness appears to be genetically determined. In 96% of

CLINICAL BOX 15–4

Lesions of Representational & Categorical Hemispheres

Lesions in the categorical hemisphere produce language disorders, whereas extensive lesions in the representational hemisphere do not. Instead, lesions in the representational hemisphere produce **astereognosis—**the inability to identify objects by feeling them—and other agnosias. **Agnosia** is the general term used for the inability to recognize objects by a particular sensory modality even though the sensory modality itself is intact. Lesions producing these defects are generally in the parietal lobe. Especially when they are in the representational hemisphere, lesions of the inferior parietal lobule, a region in the posterior part of the parietal lobe that is close to the occipital lobe, cause **unilateral inattention** and **neglect.** Individuals with such lesions do not have any apparent primary visual, auditory, or somatesthetic defects, but they ignore stimuli from the contralateral portion of their bodies or the space around these portions. This leads to failure to care for half their bodies and, in extreme cases, to situations in which individuals shave half their faces, dress half their bodies, or read half of each page. This inability to put together a picture of visual space on one side is due to a shift in visual attention to the side of the brain lesion and can be improved, if not totally corrected, by wearing eyeglasses that contain prisms. Hemispheric specialization extends to other parts of the cortex as well. Patients with lesions in the categorical hemisphere are disturbed about their disability and often depressed, whereas patients with lesions in the representational hemisphere are sometimes unconcerned and even euphoric. Lesions of different parts of the categorical hemisphere produce **fluent, nonfluent,** and **anomic aphasias.** Although aphasias are produced by lesions of the categorical hemisphere, lesions in the representational hemisphere also have effects. For example, they may impair the ability to tell a story or make a joke. They may also impair a subject's ability to get the point of a joke and, more broadly, to comprehend the meaning of differences in inflection and the "color" of speech. This is one more example of the way the hemispheres are specialized rather than simply being dominant and nondominant.

THERAPEUTIC HIGHLIGHTS

Treatments for agnosia and aphasia are symptomatic and supportive. Individuals with agnosia can be taught exercises to help them identify objects that are a necessity for independence. Therapy for individuals with aphasia helps them to use remaining language abilities, compensate for language problems, and learn other methods of communicating. Some individuals with aphasia experience recovery but often some disabilities remain. Factors that influence the degree of improvement include the cause and extent of the brain damage, the area of the brain that was damaged, and the age and health of the individual. Computer assisted therapies have been shown to improve retrieval of certain parts of speech as well as allowing an alternative way to communicate.

CLINICAL BOX 15–5

Dyslexia

Dyslexia, which is a broad term applied to impaired ability to read, is characterized by difficulties with learning how to decode at the word level, to spell, and to read accurately and fluently despite having a normal or even higher than normal level of intelligence. It is frequently due to an inherited abnormality that affects 5% of the population with a similar incidence in boys and girls. Dyslexia is the most common and prevalent of all known learning disabilities. It often coexists with attention deficit disorder. Many individuals with dyslexic symptoms also have problems with short-term memory skills and problems processing spoken language. Although its precise cause is unknown, dyslexia is of neurological origin. Acquired dyslexias often occur due to brain damage in the left hemisphere's key language areas. Also, in many cases, there is a decreased blood flow in the angular gyrus in the categorical hemisphere. There are numerous theories to explain the causes of dyslexia. The **phonological hypothesis** is that dyslexics have a specific impairment in the representation, storage, and/or retrieval of speech sounds. The **rapid auditory processing theory** proposes that the primary deficit is the perception of short or rapidly varying sounds. The **visual theory** is that a defect in the magnocellular portion of the visual system slows processing and also leads to phonemic deficit. More selective speech defects have also been described. For example, lesions limited to the left temporal pole cause inability to retrieve names of places and persons but preserves the ability to retrieve common nouns, that is, the names of nonunique objects. The ability to retrieve verbs and adjectives is also intact.

THERAPEUTIC HIGHLIGHTS

Treatments for children with dyslexia frequently rely on modified teaching strategies that include the involvement of various senses (hearing, vision, and touch) to improve reading skills. The sooner the diagnosis is made and interventions are applied, the better the prognosis.

right-handed individuals, who constitute 91% of the human population, the left hemisphere is the dominant or categorical hemisphere, and in the remaining 4%, the right hemisphere is dominant. In approximately 15% of left-handed individuals, the right hemisphere is the categorical hemisphere and in 15%, there is no clear lateralization. However, in the remaining 70% of left-handers, the left hemisphere is the categorical hemisphere. It is interesting that learning disabilities such as **dyslexia** (see Clinical Box 15–5), an impaired ability to learn to read, are 12 times as common in left-handers as they are in right-handers, possibly because some fundamental abnormality in the left hemisphere led to a switch in handedness early in development. However, the spatial talents of left-handers may be well above average; a disproportionately large number of artists, musicians, and mathematicians are left-handed. For unknown reasons, left-handers have slightly but significantly shorter life spans than right-handers.

Some anatomic differences between the two hemispheres may correlate with the functional differences. The **planum temporale,** an area of the superior temporal gyrus that is involved in language-related auditory processing, is regularly larger on the left side than the right (see Figure 10-13). It is also larger on the left in the brain of chimpanzees, even though language is almost exclusively a human trait. Imaging studies show that other portions of the upper surface of the left temporal lobe are larger in right-handed individuals, the right frontal lobe is normally thicker than the left, and the left occipital lobe is wider and protrudes across the midline. Chemical differences also exist between the two sides of the brain. For example, the concentration of dopamine is higher in the nigrostriatal pathway on the left side in right-handed humans but higher on the right in left-handers. The physiologic significance of these differences is unknown.

In patients with schizophrenia, MRI studies have demonstrated reduced volumes of gray matter on the left side in the anterior hippocampus, amygdala, parahippocampal gyrus, and posterior superior temporal gyrus. The degree of reduction in the left superior temporal gyrus correlates with the degree of disordered thinking in the disease. There are also apparent abnormalities of dopaminergic systems and cerebral blood flow in this disease.

PHYSIOLOGY OF LANGUAGE

Language is one of the fundamental bases of human intelligence and a key part of human culture. The primary brain areas concerned with language are arrayed along and near the sylvian fissure (lateral cerebral sulcus) of the categorical hemisphere. A region at the posterior end of the superior temporal gyrus called **Wernicke's area** (Figure 15–7) is concerned with comprehension of auditory and visual information. It projects via the **arcuate fasciculus** to **Broca's area** in the frontal lobe immediately in front of the inferior end of the motor cortex. Broca's area processes the information received from Wernicke's area into a detailed and coordinated pattern for vocalization and then projects the pattern via a speech articulation area in the insula to the motor cortex, which initiates the appropriate movements of the lips, tongue, and larynx to produce speech. The probable sequence of events that occurs when a subject names a visual object is shown in Figure 15–8. The angular

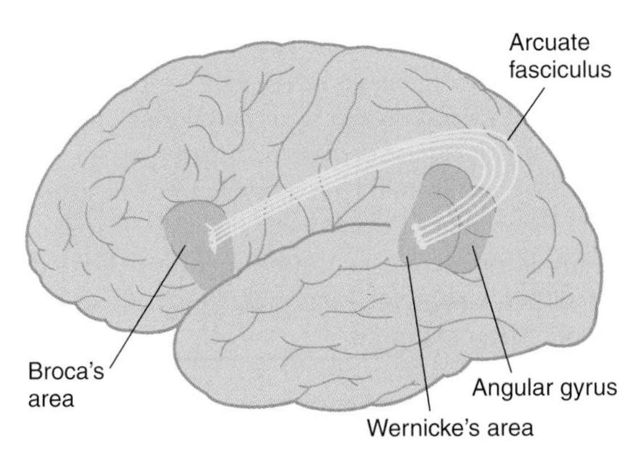

FIGURE 15–7 Location of some of the areas in the categorical hemisphere that are concerned with language functions. Wernicke's area is in the posterior end of the superior temporal gyrus and is concerned with comprehension of auditory and visual information. It projects via the arcuate fasciculus to Broca's area in the frontal lobe. Broca's area processes information received from Wernicke's area into a detailed and coordinated pattern for vocalization and then projects the pattern via a speech articulation area in the insula to the motor cortex, which initiates the appropriate movements of the lips, tongue, and larynx to produce speech.

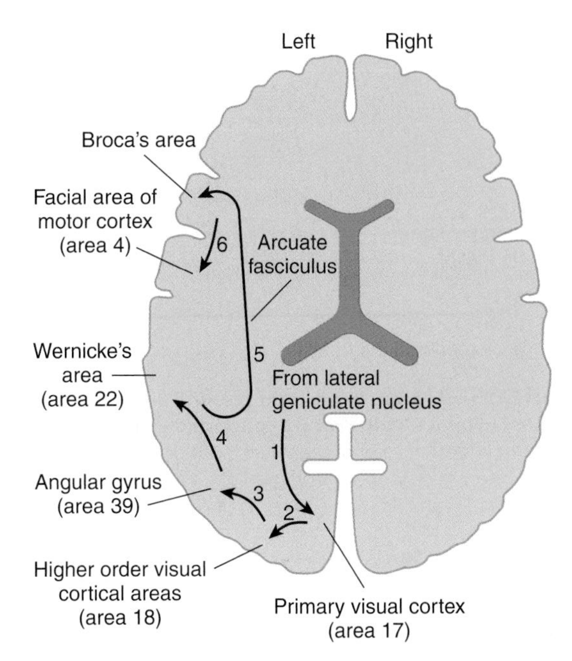

FIGURE 15–8 Path taken by impulses when a subject identifies a visual object, projected on a horizontal section of the human brain. Information travels from the lateral geniculate nucleus in the thalamus to the primary visual cortex, to higher order visual critical areas, and to the angular gyrus. Information then travels from Wernicke's area to Brocas's area via the arcuate fasciculus. Broca's area processes the information into a detailed and coordinated pattern for vocalization and then projects the pattern via a speech articulation area in the insula to the motor cortex, which initiates the appropriate movements of the lips, tongue, and larynx to produce speech.

gyrus behind Wernicke's area appears to process information from words that are read in such a way that they can be converted into the auditory forms of the words in Wernicke's area.

It is interesting that in individuals who learn a second language in adulthood, fMRI reveals that the portion of Broca's area concerned with it is adjacent to but separate from the area concerned with the native language. However, in children who learn two languages early in life, only a single area is involved with both. It is well known, of course, that children acquire fluency in a second language more easily than adults.

LANGUAGE DISORDERS

Aphasias are abnormalities of language functions that are not due to defects of vision or hearing or to motor paralysis. They are caused by lesions in the categorical hemisphere (see Clinical Box 15–4). The most common cause is embolism or thrombosis of a cerebral blood vessel. Many different classifications of the aphasias have been published, but a convenient classification divides them into **nonfluent, fluent,** and **anomic aphasias.** In nonfluent aphasia, the lesion is in Broca's area. Speech is slow, and words are hard to come by. Patients with severe damage to this area are limited to two or three words with which to express the whole range of meaning and emotion. Sometimes the words retained are those that were being spoken at the time of the injury or vascular accident that caused the aphasia.

In one form of fluent aphasia, the lesion is in Wernicke's area. In this condition, speech itself is normal and sometimes the patients talk excessively. However, what they say is full of jargon and neologisms that make little sense. The patient also fails to comprehend the meaning of spoken or written words, so other aspects of the use of language are compromised.

Another form of fluent aphasia is a condition in which patients can speak relatively well and have good auditory comprehension but cannot put parts of words together or conjure up words. This is called **conduction aphasia** because it was thought to be due to lesions of the arcuate fasciculus connecting Wernicke's and Broca's areas. However, it now appears that it is due to lesions near the auditory cortex in the posterior perisylvian gyrus.

When a lesion damages the angular gyrus in the categorical hemisphere without affecting Wernicke's or Broca's areas, there is no difficulty with speech or the understanding of auditory information; instead there is trouble understanding written language or pictures, because visual information is not processed and transmitted to Wernicke's area. The result is a condition called **anomic aphasia.**

The isolated lesions that cause the selective defects described above occur in some patients, but brain destruction is often more general. Consequently, more than one form of aphasia is often present. Frequently, the aphasia is general **(global),** involving both receptive and expressive functions. In this situation, speech is scant as well as nonfluent. Writing is abnormal in all aphasias in which speech is abnormal, but the neural circuits involved are unknown. In addition, deaf subjects who develop a lesion in the categorical hemisphere lose their ability to communicate in sign language.

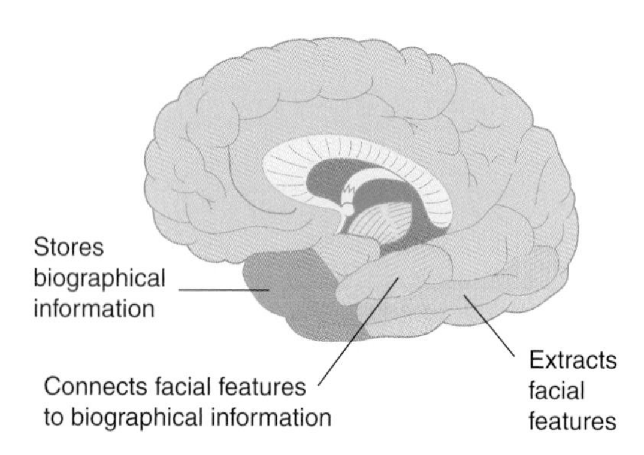

FIGURE 15–9 Areas in the right cerebral hemisphere, in right-handed individuals, that are concerned with recognition of faces. An important part of the visual input goes to the inferior temporal lobe, where representations of objects, particularly faces, are stored. In humans, storage and recognition of faces is more strongly represented in the right inferior temporal lobe in right-handed individuals, though the left lobe is also active. (Modified from Szpir M: Accustomed to your face. Am Sci 1992;80:539.)

Stuttering has been found to be associated with right cerebral dominance and widespread overactivity in the cerebral cortex and cerebellum. This includes increased activity of the supplementary motor area. Stimulation of part of this area has been reported to produce **laughter,** with the duration and intensity of the laughter proportional to the intensity of the stimulus.

RECOGNITION OF FACES

An important part of the visual input goes to the inferior temporal lobe, where representations of objects, particularly faces, are stored (Figure 15–9). Faces are particularly important in distinguishing friends from foes and the emotional state of those seen. In humans, storage and recognition of faces is more strongly represented in the right inferior temporal lobe in right-handed individuals, though the left lobe is also active. Damage to this area can cause **prosopagnosia,** the inability to recognize faces. Patients with this abnormality can recognize forms and reproduce them. They can recognize people by their voices, and many of them show autonomic responses when they see familiar as opposed to unfamiliar faces. However, they cannot identify the familiar faces they see. The left hemisphere is also involved, but the role of the right hemisphere is primary. The presence of an autonomic response to a familiar face in the absence of recognition has been explained by postulating the existence of a separate dorsal pathway for processing information about faces that leads to recognition at only a subconscious level.

LOCALIZATION OF OTHER FUNCTIONS

Use of fMRI and PET scanning combined with study of patients with strokes and head injuries has provided further insight into the ways serial processing of sensory information produce cognition, reasoning, comprehension, and language. Analysis of the brain regions involved in arithmetic calculations has highlighted two areas. In the inferior portion of the left frontal lobe is an area concerned with number facts and exact calculations. Frontal lobe lesions can cause **acalculia,** a selective impairment of mathematical ability. There are areas around the intraparietal sulci of both parietal lobes that are concerned with visuospatial representations of numbers and, presumably, finger counting.

Two right-sided subcortical structures play a role in accurate navigation in humans. One is the right hippocampus, which is concerned with learning where places are located, and the other is the right caudate nucleus, which facilitates movement to the places. Men have larger brains than women and are said to have superior spatial skills and ability to navigate.

Other defects seen in patients with localized cortical lesions include, for example, the inability to name animals, though the ability to name other living things and objects is intact. One patient with a left parietal lesion had difficulty with the second half but not the first half of words. Some patients with parietooccipital lesions write only with consonants and omit vowels. The pattern that emerges from studies of this type is one of precise sequential processing of information in localized brain areas. Additional research of this type should greatly expand our understanding of the functions of the neocortex.

CHAPTER SUMMARY

- Memory is divided into explicit (declarative) and implicit (nondeclarative). Explicit is further subdivided into semantic and episodic. Implicit is further subdivided into priming, procedural, associative learning, and nonassociative learning.
- Declarative memory involves the hippocampus and the medial temporal lobe for retention. Priming is dependent on the neocortex. Procedural memory is processed in the striatum. Associative learning is dependent on the amygdala for its emotional responses and the cerebellum for the motor responses. Nonassociative learning is dependent on various reflex pathways.
- Synaptic plasticity is the ability of neural tissue to change as reflected by LTP (an increased effectiveness of synaptic activity) or LTD (a reduced effectiveness of synaptic activity) after continued use. Habituation is a simple form of learning in which a neutral stimulus is repeated many times. Sensitization is the prolonged occurrence of augmented postsynaptic responses after a stimulus to which one has become habituated is paired once or several times with a noxious stimulus.
- Alzheimer disease is characterized by progressive loss of short-term memory followed by general loss of cognitive function. The cytopathologic hallmarks of Alzheimer disease are intracellular neurofibrillary tangles and extracellular senile plaques.
- Categorical and representational hemispheres are for sequential-analytic processes and visuospatial relations, respectively. Lesions in the categorical hemisphere produce

language disorders, whereas lesions in the representational hemisphere produce astereognosis.

- Aphasias are abnormalities of language functions and are caused by lesions in the categorical hemisphere. They are classified as fluent (Wernicke's area), nonfluent (Broca's area), and anomic (angular gyrus) based on the location of brain lesions.

MULTIPLE-CHOICE QUESTIONS

For all questions, select the single best answer unless otherwise directed.

1. A 17-year-old male suffered a traumatic brain injury as a result of a motor cycle accident. He was unconscious and was rushed to the emergency room of the local hospital. A CT scan was performed and appropriate interventions were taken. About 6 months later he still had memory deficits. Which of the following is correctly paired to show the relationship between a brain area and a type of memory?
   A. Hippocampus and implicit memory
   B. Neocortex and associative learning
   C. Medial temporal lobe and declarative memory
   D. Angular gyrus and procedural memory
   E. Striatum and priming
2. The optic chiasm and corpus callosum are sectioned in a dog, and with the right eye covered, the animal is trained to bark when it sees a red square. The right eye is then uncovered and the left eye covered. The animal will now
   A. fail to respond to the red square because the square does not produce impulses that reach the right occipital cortex.
   B. fail to respond to the red square because the animal has bitemporal hemianopia.
   C. fail to respond to the red square if the posterior commissure is also sectioned.
   D. respond to the red square only after retraining.
   E. respond promptly to the red square in spite of the lack of input to the left occipital cortex.
3. A 32-year-old male had medial temporal lobe epilepsy for over 10 years. This caused bilateral loss of hippocampal function. As a result, this individual might be expected to experience a
   A. disappearance of remote memories.
   B. loss of working memory.
   C. loss of the ability to encode events of the recent past into long-term memory.
   D. loss of the ability to recall faces and forms but not the ability to recall printed or spoken words.
   E. production of inappropriate emotional responses when recalling events of the recent past.
4. A 70-year-old woman fell down a flight of stairs, hitting her head on the concrete sidewalk. The trauma caused a severe intracranial hemorrhage. The symptoms she might experience are dependent on the area of the brain most affected. Which of the following is *incorrectly* paired?
   A. Damage to the parietal lobe of the representational hemisphere : Unilateral inattention and neglect
   B. Loss of cholinergic neurons in the nucleus basalis of Meynert and related areas of the forebrain : Loss of recent memory
   C. Damage to the mammillary bodies : Loss of recent memory
   D. Damage to the angular gyrus in the categorical hemisphere : Nonfluent aphasia
   E. Damage to Broca's area in the categorical hemisphere : Slow speech
5. The representational hemisphere is better than the categorical hemisphere at
   A. language functions.
   B. recognition of objects by their form.
   C. understanding printed words.
   D. understanding spoken words.
   E. mathematical calculations.
6. A 67-year-old female suffered a stroke that damaged the posterior end of the superior temporal gyrus . A lesion of Wernicke's area in the categorical hemisphere causes her to
   A. lose her short-term memory.
   B. experience nonfluent aphasia in which she speaks in a slow, halting voice.
   C. experience déjà vu.
   D. talk rapidly but make little sense, which is characteristic of fluent aphasia.
   E. lose the ability to recognize faces, which is called prosopagnosia.
7. Which of the following is most likely *not* involved in production of LTP?
   A. NO
   B. Ca^{2+}
   C. NMDA receptors
   D. Membrane hyperpolarization
   E. Membrane depolarization
8. An 79-year-old woman has been experiencing difficulty finding her way back home after her morning walks. Her husband has also noted that she takes much longer to do routine chores around the home and often appears to be confused. He is hoping that this is just due to "old age" but fears it may be a sign of Alzheimer disease. Which of the following is the definitive sign of this disease?
   A. Loss of short-term memory.
   B. The presence of intracellular neurofibrillary tangles and extracellular neuritic plaques with a core of β-amyloid peptides.
   C. A mutation in genes for amyloid precursor protein (APP) on chromosome 21.
   D. Rapid reversal of symptoms with the use of acetylcholinesterase inhibitors.
   E. A loss of cholinergic neurons in the nucleus basalis of Meynert.

CHAPTER RESOURCES

Aimone JB, Wiles J, Gage FH: Computational influence of adult neurogenesis on memory encoding. Neuron 2009;61:187.

Andersen P, Morris R, Amaral D, Bliss T, O'Keefe J: *The Hippocampus Book*. Oxford University Press, 2007.

Bird CM, Burgess N: The hippocampus and memory: Insights from spatial processing. Nature Rev Neurosci 2008;9:182.

Eichenbaum H: A cortical-hippocampal system for declarative memory. Nat Neurosci Rev 2000;1:41.

Goodglass H: *Understanding Aphasia.* Academic Press, 1993.
Ingram VM: Alzheimer's disease. Am Scientist 2003;91:312.
Kandel ER: The molecular biology of memory: A dialogue between genes and synapses. Science 2001;294:1028.
LaFerla FM, Green KN, Oddo S: Intracellular amyloid-β in Alzheimer's disease. Nature Rev Neurosci 2007;8:499.
Ramus F: Developmental dyslexia: Specific phonological defect or general sensorimotor dysfunction. Curr Opin Neurobiol 2003;13:212.
Russ MD: Memories are made of this. Science 1998;281:1151.
Selkoe DJ: Translating cell biology into therapeutic advances in Alzheimer's disease. Nature 1999;399 (Suppl): A23.
Shaywitz S: Dyslexia. N Engl J Med 1998;338:307.
Squire LR, Stark CE, Clark RE: The medial temporal lobe. Annu Rev Neurosci 2004;27:279.
Squire LR, Zola SM: Structure and function of declarative and nondeclarative memory systems. Proc Natl Acad Sci 1996;93:13515.

SECTION III Endocrine and Reproductive Physiology

The role of the endocrine system is to maintain whole body homeostasis. This is accomplished via the coordination of hormonal signaling pathways that regulate cellular activity in target organs throughout the body. Endocrine mechanisms are also concerned with the ability of humans to reproduce, and the sexual maturation required for this function. Classic **endocrine glands** are scattered throughout the body and secrete **hormones** into the circulatory system, usually via ductless secretion into the interstitial fluid. **Target organs** express receptors that bind the specific hormone to initiate a cellular response. The endocrine system can be contrasted with the neural regulation of physiological function that was the focus of the previous section. Endocrine effectors typically provide "broadcast" regulation of multiple tissues and organs simultaneously, with specificity provided for by the expression of relevant receptors. A change in environmental conditions, for example, often calls for an integrated response across many organ systems. Neural regulation, on the other hand, is often exquisitely spatially delimited, such as the ability to contract just a single muscle. Nevertheless, both systems must work collaboratively to allow for minute-to-minute as well as longer term stability of the body's interior milieu.

Hormones are the soluble messengers of the endocrine system and are classified into steroids, peptides, and amines (see Chapters 1 and 2). Steroid hormones can cross the lipid-containing plasma membrane of cells and usually bind to intracellular receptors. Peptide and amine hormones bind to cell surface receptors. Steroid hormones are produced by the adrenal cortex (chapter 20), the gonads, testes (chapter 23), and ovaries (chapter 22) in addition to steroid hormones that are made by the placenta during pregnancy (chapter 22). Amine hormones are derivatives of the amino acid tyrosine and are made by the thyroid (chapter 19) and the adrenal medulla (chapter 20). Interestingly, the tyrosine-derived thyroid hormone behaves more like a steroid than a peptide hormone by binding to an intracellular receptor. The majority of hormones, however, are peptides and they are usually synthesized as preprohormones before being cleaved first to prohormones in the endoplasmic reticulum and then to the active hormone in secretory vesicles.

Diseases of the endocrine system are numerous. Indeed, endocrine and metabolic disorders are among the most common afflictions in developed countries, particularly when nutrition and access to health care provisions are generous and high risk individuals are identified by regular screening. At least 11 endocrine and metabolic disorders are present in 5% or more of the adult US population, including diabetes mellitus, osteopenia, dyslipidemia, metabolic syndrome, and thyroiditis. For example, type 2 diabetes mellitus is one of the most prevalent endocrine disorders of the 21st century and involves an inability of the body to respond to insulin. The resulting high blood glucose damages many tissues leading to secondary complications (see Chapter 24). In large part, the high and increasing prevalence of diabetes and other metabolic disorders rests on the substantial prevalence of obesity in developed countries, with as many as a third of the US adult population now considered to be obese, and two thirds overweight. Indeed, based on a 2009 report, obesity also affects 28% of US children aged 12–17, and while the current prevalence of type 2 diabetes in children is quite low, this prevalence is accordingly expected to rise. Further, a number of endocrine disorders are more prevalent in specific ethnic groups, or in a particular gender. Overall, the burden of endocrine and metabolic disorders, with their protean manifestations and complications, is a serious public health crisis and even highlights an apparent national shortage of trained endocrinologists. Many endocrine disorders must be managed by primary care physicians as a result.

CHAPTER

Basic Concepts of Endocrine Regulation

16

OBJECTIVES

After studying this chapter, you should be able to:

- Describe hormones and their contribution to whole body homeostatic mechanisms.
- Understand the chemical nature of different classes of hormones and how this determines their mechanism of action on target cells.
- Define how hormones are synthesized and secreted by cells of endocrine glands, including how peptide hormones are cleaved from longer precursors.
- Explain the relevance of protein carriers in the blood for hydrophobic hormones, and the mechanisms that determine the level of free circulating hormones.
- Understand the principles of feedback control for hormone release and its relevance for homeostasis.
- Understand the principles governing disease states that result from over- or under-production of key hormones.

INTRODUCTION

This section of the text deals with the various endocrine glands that control the function of multiple organ systems of the body. In general, endocrine physiology is concerned with the maintenance of various aspects of **homeostasis.** The mediators of such control mechanisms are soluble factors known as **hormones.** The word hormone was derived from the Greek *horman,* meaning to set in motion. In preparation for specific discussions of the various endocrine systems and their hormones, this chapter will address some concepts of endocrine regulation that are common among all systems.

Another feature of endocrine physiology to keep in mind is that, unlike other physiological systems that are considered in this text, the endocrine system cannot be cleanly defined along anatomic lines. Rather, the endocrine system is a distributed system of glands and circulating messengers that is often stimulated by the central nervous system and/or autonomic nervous system.

EVOLUTION OF HORMONES AND THEIR ACTIONS ON TARGET CELLS

As noted in the introduction to this section, hormones comprise steroids, amines, and peptides. Peptide hormones are by far the most numerous. Many hormones can be grouped into families reflecting their structural similarities as well as the similarities of the receptors they activate. However, the number of hormones and their diversity increases as one moves from simple to higher life forms, reflecting the added challenges in providing for homeostasis in more complex organisms. For example, among the peptide hormones, several are heterodimers that share a common α chain, with specificity being conferred by the β-chain. In the specific case of thyroid-stimulating hormone (TSH), follicle-stimulating hormone (FSH), and luteinizing hormone (LH), there is evidence that the distinctive β-chains arose from a series of duplications of a common ancestral gene. For these and other hormones, moreover, this molecular evolution implies that hormone receptors also needed to evolve to allow for spreading of hormone actions/specificity. This was accomplished by co-evolution of the basic G-protein coupled receptors (GPCR) and receptor tyrosine kinases that mediate the effects of peptide and amine hormones that act at the cell surface (see Chapter 2). The underlying ancestral relationships sometimes re-emerge, however, in the cross-reactivity that may be seen when hormones rise to unusually high levels (eg, endocrine tumors).

Steroids and thyroid hormones are distinguished by their predominantly intracellular sites of action, since they can diffuse freely through the cell membrane. They bind to a family of largely cytoplasmic proteins known as nuclear receptors. Upon ligand binding, the receptor–ligand complex translocates to the nucleus where it either homodimerizes, or associates with a distinct liganded nuclear receptor to form a heterodimer. In either case, the dimer binds to DNA to either increase or decrease gene transcription in the target tissue. Individual members of the nuclear receptor family have a considerable degree of homology, perhaps implying a common ancestral gene, and share many functional domains, such as the zinc fingers that permit DNA binding. However, sequence variations allow for ligand specificity as well as binding to specific DNA motifs. In this way, the transcription of distinct genes is regulated by individual hormones.

HORMONE SECRETION

SYNTHESIS AND PROCESSING

The regulation of hormone synthesis, of course, depends on their chemical nature. For peptide hormones as well as hormone receptors, synthesis is controlled predominantly at the level of transcription. For amine and steroid hormones, synthesis is controlled indirectly by regulating the production of key synthetic enzymes, as well as by substrate availability.

Interestingly, the majority of peptide hormones are synthesized initially as much larger polypeptide chains, and then processed intracellularly by specific proteases to yield the final hormone molecule. In some cases, multiple hormones may be derived from the same initial precursor, depending on the specific processing steps present in a given cell type. Presumably this provides for a level of genetic "economy." It is also notable that the hormone precursors themselves are typically inactive. This may be a mechanism that provides for an additional measure of regulatory control, or, in the case of thyroid hormones, may dictate the site of highest hormone availability.

The synthesis of all of the proteins/peptides discussed above is subject to the normal mechanisms of transcriptional control in the cell (see Chapter 2). In addition, there is provision for exquisitely specific regulation by other hormones, since the regulatory regions of many peptide hormone genes contain binding motifs for the nuclear receptors discussed above. For example, thyroid hormone directly suppresses TSH expression via the thyroid hormone receptor. These specific mechanisms to regulate hormone transcription are essential to the function of feedback loops, as will be addressed in greater detail below. In some cases, the abundance of selected hormones may also be regulated via effects on translation. For example, elevated levels of circulating glucose stimulate the translation of insulin mRNA. These effects are mediated by the ability of glucose to increase the interaction of the insulin mRNA with specific RNA binding proteins, which increase its stability and enhance its translation. The net effect is a more precise and timely regulation of insulin levels, and thus energy metabolism, than could be accomplished with transcriptional regulation alone.

The precursors for peptide hormones are processed through the cellular machinery that handles proteins destined for export, including trafficking through specific vesicles where the propeptide form can be cleaved to the final active hormones. Mature hormones are also subjected to a variety of posttranslational processing steps, such as glycosylation, which can influence their ultimate biological activity and/or stability in the circulation. Ultimately, all hormones enter either the constitutive or regulated secretory pathway (see Chapter 2).

SECRETION

The secretion of many hormones is via a process of exocytosis of stored granules, as discussed in Chapter 2. The exocytotic machinery is activated when the cell type that synthesizes and stores the hormone in question is activated by a specific signal, such as a neurotransmitter or peptide releasing factor. One should, however, contrast the secretion of stored hormones with that of those that are continually released by diffusion (eg, steroids). Control of the secretion of the latter molecules occurs via kinetic influences on the synthetic enzymes or carrier proteins involved in hormone production. For example, the steroidogenic acute regulatory protein (StAR) is a labile protein whose expression, activation, and deactivation are regulated by intracellular signaling cascades and their effectors, including a variety of protein kinases and phosphatases. StAR traffics cholesterol from the outer to the inner membrane leaflet of the mitochondrion. Because this is a rate-limiting first step in the synthesis of the steroid precursor, pregnenolone, this arrangement permits changes in the rate of steroid synthesis, and thus secretion, in response to homeostatic cues such as trophic hormones, cytokines and stress (Figure 16–1).

An additional complexity related to hormone secretion relates to the fact that some hormones are secreted in a pulsatile fashion. Secretion rates may peak and ebb relative to circadian rhythms, in response to the timing of meals, or as regulated by other pattern generators whose periodicity may range from milliseconds to years. Pulsatile secretion is often related to the activity of oscillators in the hypothalamus that regulate the membrane potential of neurons, in turn secreting bursts of hormone releasing factors into the hypophysial blood flow that then cause the release of pituitary and other downstream hormones in a similar pulsatile fashion (see Chapters 17 and 18). There is evidence that these hormone pulses convey different information to the target tissues that they act upon than steady exposure to a single concentration of the hormone. Therapeutically, pulsatile secretion may pose challenges if, due to deficiency, it proves necessary to replace a particular hormone that is normally secreted in this way.

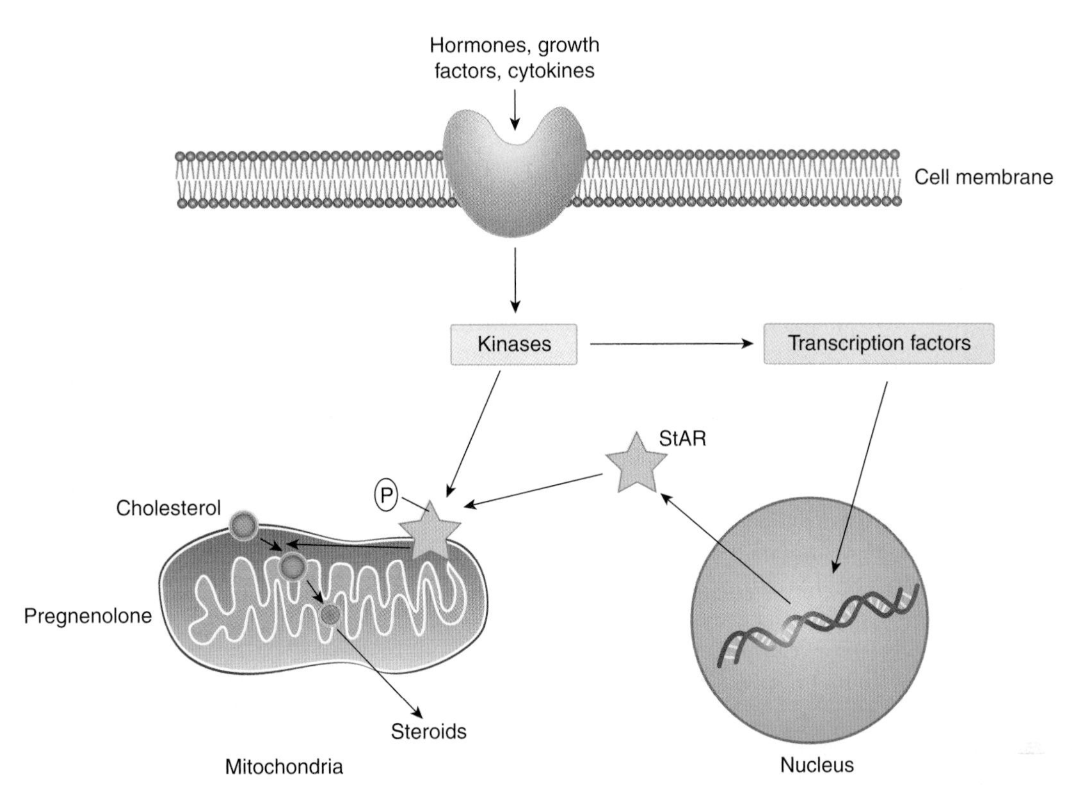

FIGURE 16–1 Regulation of steroid biosynthesis by the steroidogenic acute regulatory protein (StAR). Extracellular signals activate intracellular kinases that, in turn, phosphorylate transcription factors that upregulate StAR expression. StAR is activated by phosphorylation, and facilitates transfer of cholesterol from the outer to inner mitochondrial membrane leaflet. This then allows entry of cholesterol into the steroid biosynthetic pathway, beginning with pregnenolone.

HORMONE TRANSPORT IN THE BLOOD

In addition to the rate of secretion and its nature (steady vs. pulsatile), a number of factors influence the circulating levels of hormones. These include the rates of hormone degradation and/or uptake, receptor binding and availability of receptors, and the affinity of a given hormone for plasma carriers (Figure 16–2). Stability influences the circulating half-life of a given hormone and has therapeutic implications for hormone replacement therapy, in addition to those posed by pulsatile secretion as discussed above.

Plasma carriers for specific hormones have a number of important physiological functions. First, they serve as a reservoir of inactive hormone and thus provide a hormonal reserve. Bound hormones are typically prevented from degradation or uptake. Thus, the bound hormone reservoir can allow fluctuations in hormonal levels to be smoothed over time. Plasma carriers also restrict the access of the hormone to some sites. Ultimately, plasma carriers may be vital in modulating levels of the free hormone in question. Typically, it is only the free hormone that is biologically active in target tissues or can mediate feedback regulation (see below) since it is the only form able to access the extravascular compartment.

Catecholamine and most peptide hormones are soluble in plasma and are transported as such. In contrast steroid

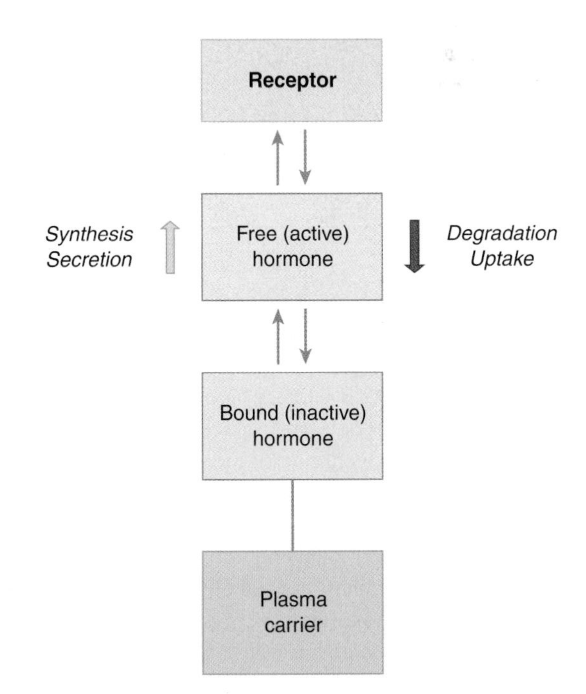

FIGURE 16–2 Summary of factors that determine the level of free hormones circulating in the bloodstream. Factors that increase (green upward arrow) or decrease (red downward arrow) hormone levels are shown. Free hormones also equilibrate with the forms bound to either receptors or plasma carrier proteins.

hormones are hydrophobic and are mostly bound to large proteins called **steroid binding proteins** (SBP), which are synthesized in the liver. As a result, only small amounts of the free hormone are dissolved in the plasma. Specifically, **sex hormone-binding globulin** (SHBG) is a glycoprotein that binds to the sex hormones, testosterone and 17β-estradiol. Progesterone, cortisol, and other corticosteroids are bound by transcortin.

The SBP-hormone complex and the free hormone are in equilibrium in the plasma, and only the free hormone is able to diffuse across cell membranes. SBP have three main functions: they increase the solubility of lipid based hormones in the blood, they reduce the rate of hormone loss in the urine by preventing the hormones from being filtered in the kidney, and as mentioned above, they provide a source of hormone in the bloodstream that can release free hormone as the equilibrium changes. It follows that an additional way to regulate the availability of hormones that bind to carrier proteins, such as steroids, is to regulate the expression and secretion of the carrier proteins themselves. This is a critical mechanism that regulates the bioavailability of thyroid hormones, for example (see Chapter 19).

In a pathophysiological setting, some medications can alter levels of binding proteins or displace hormones that are bound to them. In addition, some binding proteins are promiscuous and bind multiple hormones (eg, SHBG). These observations may have clinical implications for endocrine homeostasis, since free hormones are needed to feedback and control their rates of synthesis and secretion (see below).

Finally, the anatomic relationship of sites of release and action of hormones may play a key role in their regulation. For example, a number of hormones are destroyed by passage through the pulmonary circulation or the liver. This may markedly curtail the temporal window within which a given hormone can act.

HORMONE ACTION

As we will see in later chapters, hormones exert a wide range of distinctive actions on a huge number of target cells to effect changes in metabolism, release of other hormones and regulatory substances, changes in ion channel activity, and cell growth, among others (Clinical Box 16–1). Ultimately, the concerted action of the hormones of the body ensures the maintenance of homeostasis. Indeed, all hormones affect homeostasis to some degree. However, a subset of the hormones, as detailed in Table 16–1, are the key contributors to homeostasis. These include thyroid hormone, cortisol, parathyroid hormone, vasopressin, the mineralocorticoids, and insulin. Detailed information on the precise biological effects of these molecules can be found in subsequent chapters.

Hydrophilic hormones, including peptides and catecholamines, exert their acute effects by binding to cell surface receptors. Most of these are from the GPCR family. Hydrophobic hormones, in the other hand, predominantly exert their actions via nuclear receptors. There are two classes of nuclear receptors that are important in endocrine physiology. The first of these provide for direct stimulation of transcription via induction of the binding of a transcriptional co-activator when the hormonal ligand is bound. In the second class, hormone binding triggers simultaneous dislodging of a transcriptional co-repressor and recruitment of a co-activator. The latter class of receptor allows for a wider dynamic range of regulation of the genes targeted by the hormone in question.

CLINICAL BOX 16–1

Breast Cancer

Breast cancer is the most common malignancy of women, with about 1 million new cases diagnosed each year worldwide. The proliferation of more than two-thirds of breast tumors are driven by the ovarian hormone, estrogen, by virtue of the fact that the tumor cells express high levels of posttranslationally modified estrogen receptors (ER). The clinical significance of these molecular findings has been known for more than 100 years, since the Scottish surgeon, Sir Thomas Beatson, reported delayed disease progression in patients with advanced breast cancer following removal of their ovaries. In modern times, determination of whether a given breast cancer is, or is not, **ER-positive** is a critical diagnostic test that guides treatment decisions, as well as an important prognosticator. ER-positive tumors are typically of lower grade, and patients with such tumors have improved survival (although the latter is likely due, at least in part, to the availability of excellent treatment options for ER-positive tumors compared with those that are ER-negative—see below).

THERAPEUTIC HIGHLIGHTS

Estrogen-responsive breast tumors are dependent on the presence of the hormone for growth. In modern times, cells can be deprived of the effects of estrogen pharmacologically, rather than resorting to oophorectomy. **Tamoxifen** and related agents specifically inhibit the receptor and may also hasten its degradation. In postmenopausal women, where estrogen is derived from the metabolism of testosterone in extragonadal tissues rather than from the ovaries, **aromatase inhibitors** inhibit the conversion of androgens to estrogen, and thereby deprive tumor cells of their critical signal for continued proliferation.

TABLE 16–1 Major hormonal contributors to homeostasis.

Hormone	Source	Action
Thyroid hormone	Thyroid	Controls basal metabolism in most tissues
Cortisol	Adrenal cortex	Energy metabolism; permissive action for other hormones
Mineralocorticoids	Adrenal cortex	Regulate plasma volume via effects on serum electrolytes
Vasopressin	Posterior pituitary	Regulates plasma osmolality via effects on water excretion
Parathyroid hormone	Parathyroids	Regulates calcium and phosphorus levels
Insulin	Pancreas	Regulates plasma glucose concentration

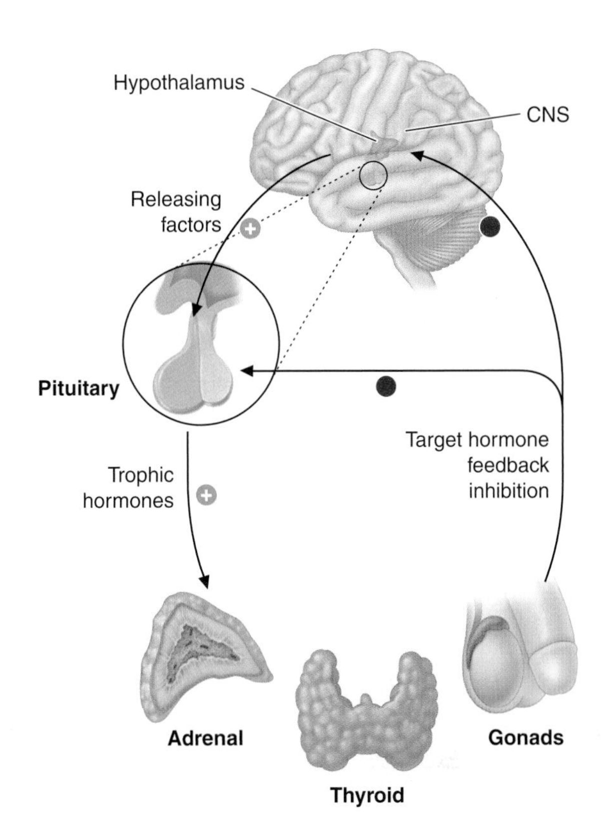

FIGURE 16–3 Summary of feedback loops regulating endocrine axes. CNS, central nervous system. (Reproduced with permission from Jameson JL (editor): *Harrison's Endocrinology* 2nd ed. McGraw Hill, 2010.)

In recent years, it has become apparent that a number of receptors for steroid and other hydrophobic hormones are extranuclear, and some may even be present on the cell surface. The characterization of such receptors at a molecular level, their associated signaling pathways, and indeed proof of their existence has been complicated by the ability of hydrophobic hormones to diffuse relatively freely into all cellular compartments. These **extranuclear receptors,** some of which may be structurally related or even identical to the more classical nuclear receptors, are proposed to mediate rapid responses to steroids and other hormones that do not require alterations in gene transcription. The physiological effects at these receptors may therefore be distinct from those classically associated with a given hormone. Evidence is accumulating, for example, that plasma membrane receptors for estrogen can mediate acute arterial vasodilation as well as reducing cardiac hypertrophy in pathophysiological settings. Functions such as these may account for differences in the prevalence of cardiovascular disease in pre and postmenopausal women. In any event, this active area of biomedical investigation is likely to broaden our horizons of the full spectrum of action of steroid hormones.

PRINCIPLES OF FEEDBACK CONTROL

A final general principle that is critical for endocrine physiology is that of **feedback regulation.** This holds that the responsiveness of target cells to hormonal action subsequently "feeds back" to control the inciting endocrine organ. Feedback can regulate the further release of the hormone in either a negative feedback or (more rarely) a positive feedback loop. Positive feedback relates to the enhancement or continued stimulation of the original release mechanism/stimulus. Such mechanisms are really only seen in settings that need to gather momentum to an eventual outcome, such as parturition. Negative feedback is a far more common control mechanism and involves the inhibition or dampening of the initial hormone release mechanism/stimulus. A general scheme for feedback inhibition of endocrine axes is depicted in Figure 16–3.

In general, the endocrine system uses a network of feedback responses to maintain a steady state. Steady state can be explained using blood osmolality as an example (Figure 16–4). Blood osmolality in humans must be maintained within a physiological range of 275–299 mOsm, and to maintain homeostasis this variable should not exceed that range. To ensure that osmolality does not change in the context of an open system, processes are in place that will add or remove water from the system to ensure a constant osmolality. The osmolality of blood will increase with dehydration and decrease with overhydration. If blood osmolality increases outside the ideal range (by 10 mOsm or more), osmoreceptors are activated. These signal release of the peptide hormone, vasopressin, into the circulation (from the pituitary). Vasopressin acts on the renal collecting duct, and increases the permeability of the plasma membrane to water via the insertion of a protein called an aquaporin. Water is then moved from the urine into the circulation via transcellular transport. The reabsorption of water from the

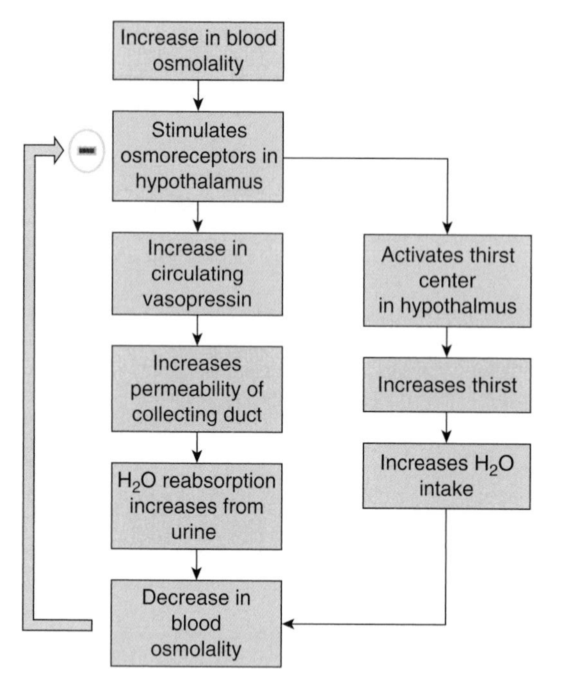

FIGURE 16–4 Feedback loop that ensures homeostasis of blood osmolality. An increase in blood osmolality triggers the thirst mechanism as well as renal conservation of water via the release of vasopressin from the hypothalamus. Both outcomes decrease blood osmolality back towards the normal range, which feeds back to terminate hypothalamic signaling.

urine to the blood resets the osmolality of the blood to within the physiological range. The decrease in blood osmolality then exerts a negative feedback on the cells of the hypothalamus and the pituitary and vasopressin release is inhibited, meaning that water reabsorption from the urine is reduced. Further details of this collaboration between the kidneys, hypothalamus and pituitary are found in Chapter 38.

Negative feedback control systems such as those described are the most common feedback/homeostatic systems in the body. Other examples include temperature regulation (see Chapter 17) and the regulation of blood glucose concentrations (see Chapter 24). Feedback control loops also provide for diagnostic strategies in evaluating patients with suspected endocrine disorders. For example, in a patient being evaluated for hypothyroidism, normal levels of TSH (see Chapter 19) tend to rule out a primary defect at the level of the thyroid gland itself, and rather suggest that a defect at the level of the anterior pituitary should be sought. Conversely, if TSH is elevated, it suggests that the normal ability of circulating thyroid hormone to suppress TSH synthesis has been lost, likely due to a reduction in the ability of the thyroid gland to synthesize the hormone (Clinical Box 16–2).

TYPES OF ENDOCRINE DISORDERS

It is pertinent also to discuss briefly the types of disease states where endocrine physiology can become deranged. Additional details of these disease states can be found in ensuing chapters.

HORMONE DEFICIENCY

Deficiencies of particular hormones are most commonly seen in the setting where there is destruction of the glandular structure responsible for their production, often as a result of

CLINICAL BOX 16–2

Approach to the Patient with Suspected Endocrine Disease

Unlike many of the disorders of individual organ systems considered elsewhere in this volume, the symptoms of endocrine disease may be protean because of the number of body systems that are impacted by hormonal action. Further, many endocrine glands are relatively inaccessible to direct physical examination. Endocrine disorders must therefore be diagnosed on the basis of the symptoms they produce in concert with appropriate biochemical testing. Radioimmunoassays for specific hormones remain the mainstay of diagnostic endocrinolology and can be used to establish steady state concentrations as well as dynamic changes of the hormone in question (the latter requiring repeated blood sampling over time). In addition, the principles of feedback regulation of hormone synthesis and release may allow the physician to pinpoint the likely locus of any defect by comparing the levels of hormones in the same axis. For example, if testosterone levels are low but those of luteinizing hormone (LH) are high, this suggests that the testes are unable to respond to LH. Conversely, if both testosterone and LH are low, the problem is more likely to be at the level of the pituitary. Synthetic hormones can also be administered exogenously to test whether increased basal levels of a given hormone can be suppressed, or abnormally low levels can be stimulated by a relevant upstream agent. An example of applying this type of reasoning to the evaluation of suspected hypothyroidism is provided in Figure 16–5.

THERAPEUTIC HIGHLIGHTS

The appropriate treatment of endocrine disorders depends on their underlying basis. For example, if a particular hormone or its releasing factor is deficient, hormone replacement therapy is often indicated to ameliorate symptoms as well as long-term negative outcomes (Figure 16–5).

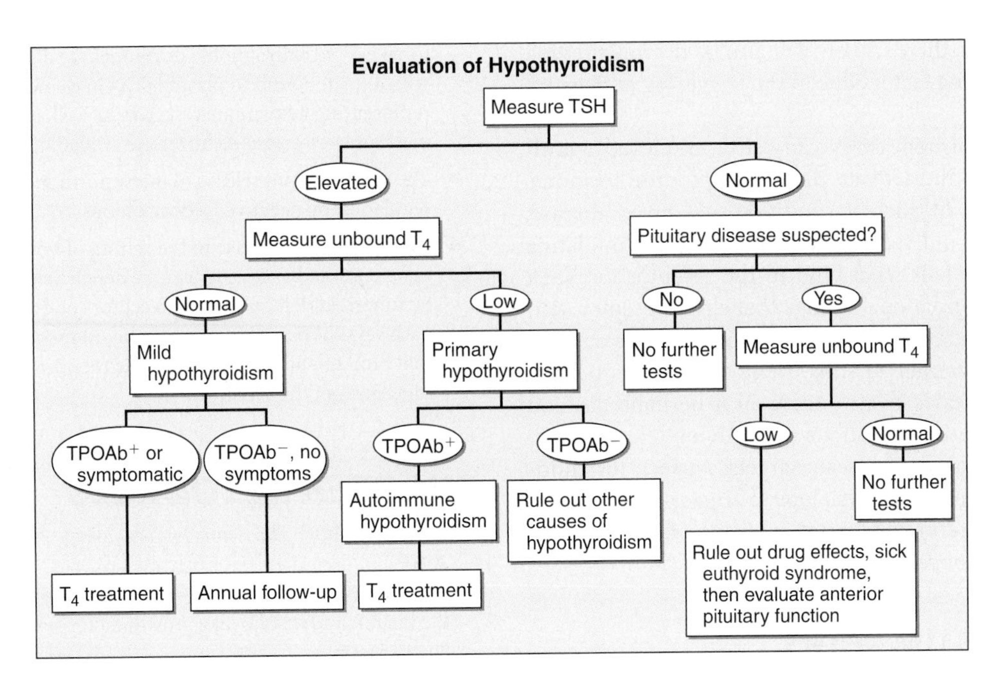

FIGURE 16–5 Summary of a strategy for the laboratory evaluation of hypothyroidism. TSH, thyroid stimulating hormone; T_4, thyroid hormone; TPOAb$^+$, positive for autoantibodies to thyroid peroxidase; TPOAb$^-$, antiperoxidase antibodies not present. (Reproduced with permission from Jameson JL (editor): *Harrison's Endocrinology* 2nd ed. McGraw Hill, 2010.)

inappropriate autoimmune attack. For example, in type 1 diabetes mellitus, pancreatic β cells are destroyed leading to an inability to synthesize insulin, often from a very young age. Similarly, hormonal deficiencies arise when there are inherited mutations in the factors responsible for their release or in the receptors for these releasing factors. Defects in the enzymatic machinery needed for hormone production, or a lack of appropriate precursors (eg, iodine deficiency leads to hypothyroidism) will also reduce the amount of the relevant hormone available for bodily requirements.

HORMONE RESISTANCE

Many of the consequences of hormone deficiency can by reproduced in disease states where adequate levels of a given hormone are synthesized and released, but the target tissues become resistant to the hormone's effects. Indeed, there is often overproduction of the implicated hormone in these conditions because the feedback loops that normally serve to shut off hormone synthesis and/or secretion are similarly desensitized. Mutations in hormone receptors (especially nuclear receptors) may result in heritable syndromes of hormone resistance. These syndromes, while relatively rare, usually exhibit severe outcomes, and have provided insights into the basic cell biology of hormone signaling. Functional hormone resistance that develops over time is also seen. Resistance arises from a relative failure of receptor signaling to couple efficiently to downstream intracellular effector pathways that normally mediate the effects of the hormone. The most common example of this is seen in type 2 diabetes mellitus. Target tissues for insulin gradually become more and more resistant to its actions, secondary to reduced activation of phosphatidylinositol 3-kinase and other intracellular signaling pathways. A key factor precipitating this outcome is obesity. In addition, because of excessive insulin secretion, pancreatic β cells become "exhausted" and may eventually fail, necessitating treatment with exogenous insulin. An important therapeutic goal, therefore, is to minimize progression to β cell exhaustion before irreversible insulin resistance sets in, with diet, exercise, and treatment with so-called **insulin sensitizers** (such as metformin and rosiglitazone).

HORMONE EXCESS

The converse of disorders of hormone deficiency or resistance is seen in diseases associated with hormone excess and/or over-stimulation of hormone receptors. A wide variety of endocrine tumors may produce hormones in an excessive and uncontrolled fashion. Note that the secretion of hormones from tumor cells may not be subject to the same types of feedback regulation that are seen for the normal source of that hormone. In the setting of an endocrine tumor, exaggerated effects of the hormone are seen. For example, acromegaly, or gigantism, occurs in patients afflicted with an adenoma derived from pituitary somatotropes that secretes excessive quantities of growth hormone (see Chapter 18). In addition, other endocrine tumors may secrete hormones other than those characteristic of the cell type or tissue from which they are originally derived. When hormone production is increased in all of these cases, there usually will also be downregulation of upstream releasing factors due to the triggering of negative feedback loops.

Disorders of hormone excess can also be mimicked by antibodies that bind to, and activate, the receptor for the hormone. A classic example of such a condition is Graves' disease, where susceptible individuals generate thyroid-stimulating immunoglobulins (TSIs) that bind to the receptor for TSH. This causes a conformational change that elicits receptor activation, and thus secretion of thyroid hormone in the absence of a physiological trigger for this event. Diseases associated with hormone excess can also occur in a heritable fashion secondary to activating mutations of hormone releasing factor receptors or their downstream targets. As seen for endocrine tumors, these pathophysiological triggers of excessive hormone release are of course not subject to dampening by negative feedback loops.

CHAPTER SUMMARY

- The endocrine system consists of a distributed set of glands and the chemical messengers that they produce, referred to as hormones. Hormones play a critical role in ensuring the relative stability of body systems, that is homeostasis.
- Hormones can be grouped into peptide/protein, amine, and steroid categories. Water-soluble hormones (peptides and catecholamines) bind to cell surface receptors; hydrophobic hormones diffuse into the cell and activate nuclear receptors to regulate gene transcription. The receptors and hormones appear to have evolved in parallel.
- Hormone availability is dictated by the rate of synthesis, the presence of releasing factors, and rates of degradation or uptake. Free hydrophobic hormones are also in equilibrium with a form bound to plasma protein carriers, the latter representing a hormone reservoir as well as an additional mechanism to regulate hormone availability.
- The synthesis and release of many hormones is subject to regulation by negative feedback loops.
- Disease states can arise in the setting of both hormone deficiency and excess. Hormone deficiencies may be mimicked by inherited defects in their receptors or downstream signaling pathways; hormone excess may be mimicked by autoantibodies that bind to and activate hormone receptors, or by activating mutations of these receptors.

CHAPTER RESOURCES

Jameson JL (editor): *Harrison's Endocrinology*, 2nd ed. McGraw Hill, 2010.

Lee EK, Gorospe M: Minireview: Posttranslational regulation of the insulin and insulin-like growth factor systems. Endocrinol 2010;151:1403.

Levin ER: Minireview: Extranuclear steroid receptors: Roles in modulation of cell functions. Mol Endocrinol 2011;25:377.

Manna PR, Stocco DM: The role of specific mitogen-activated protein kinase signaling cascades in the regulation of steroidogenesis. J Signal Transduct 2011. Article ID 821615; 13 pp.

Musso C, Cochran E, Moran SA, Skarulis MC, Oral EA, Taylor S, Gorden P: Clinical course of genetic diseases of the insulin receptor (Type A and Rabson-Mendenhall syndromes). A 30-year perspective. Medicine 2004;83:209.

Walker JJ, Terry JR, Tsaneva-Atanasova K, Armstrong SP, McArdle CA, Lightman SL: Encoding and decoding mechanisms of pulsatile hormone secretion. J Neuroendocrinol 2010;22:1226.

CHAPTER 17

Hypothalamic Regulation of Hormonal Functions

OBJECTIVES

After reading this chapter you should be able to:

- Describe the anatomic connections between the hypothalamus and the pituitary gland and the functional significance of each connection.
- List the factors that control water intake, and outline the way in which they exert their effects.
- Describe the synthesis, processing, storage, and secretion of the hormones of the posterior pituitary.
- Discuss the effects of vasopressin, the receptors on which it acts, and how its secretion is regulated.
- Discuss the effects of oxytocin, the receptors on which it acts, and how its secretion is regulated.
- Name the hypophysiotropic hormones, and outline the effects that each has an anterior pituitary function.
- List the mechanisms by which heat is produced in and lost from the body, and comment on the differences in temperature in the hypothalamus, rectum, oral cavity, and skin.
- List the temperature-regulating mechanisms, and describe the way in which they are integrated under hypothalamic control to maintain normal body temperature.
- Discuss the pathophysiology of fever.

INTRODUCTION

Many of the complex autonomic mechanisms that maintain the chemical constancy and temperature of the internal environment are integrated in the hypothalamus. The hypothalamus also functions with the limbic system as a unit that regulates emotional and instinctual behavior.

HYPOTHALAMUS: ANATOMIC CONSIDERATIONS

The hypothalamus (Figure 17–1) is the portion of the anterior end of the diencephalon that lies below the hypothalamic sulcus and in front of the interpeduncular nuclei. It is divided into a variety of nuclei and nuclear areas.

AFFERENT & EFFERENT CONNECTIONS OF THE HYPOTHALAMUS

The principal afferent and efferent neural pathways to and from the hypothalamus are mostly unmyelinated. Many connect the hypothalamus to the limbic system. Important connections

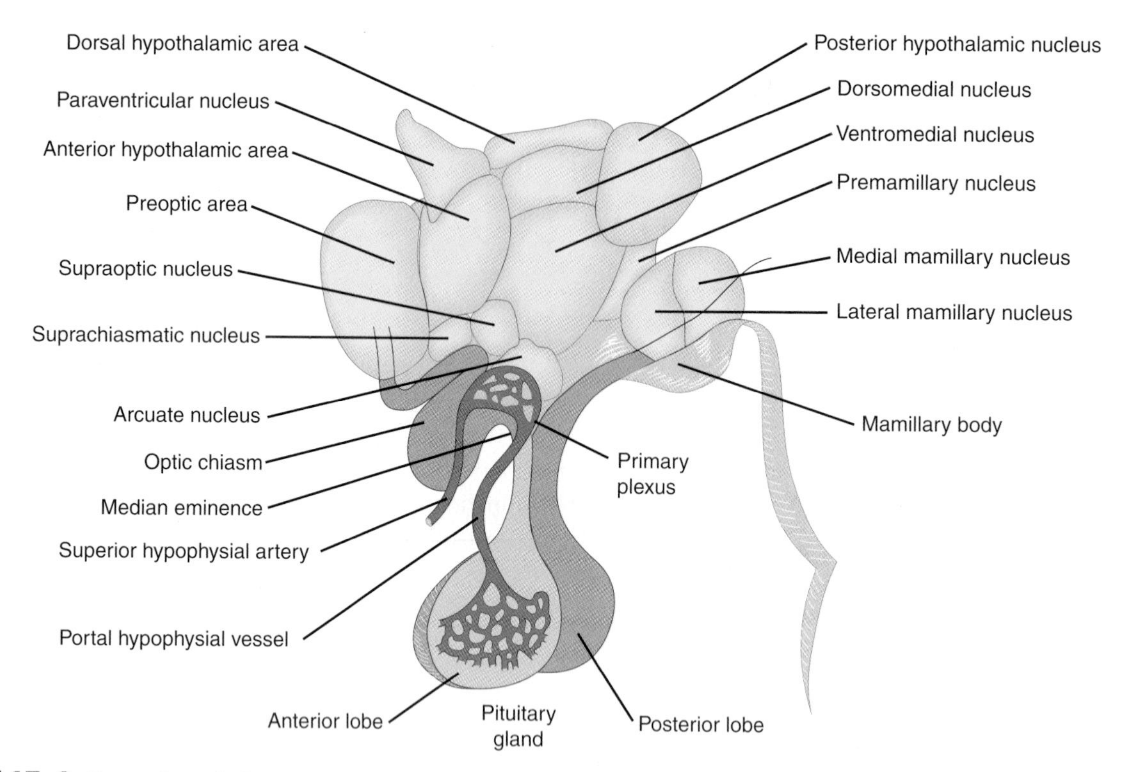

FIGURE 17–1 Human hypothalamus, with a superimposed diagrammatic representation of the portal hypophysial vessels.

also exist between the hypothalamus and nuclei in the midbrain tegmentum, pons, and hindbrain.

Norepinephrine-secreting neurons with their cell bodies in the hindbrain end in many different parts of the hypothalamus (see Figure 7–2). Paraventricular neurons that secrete oxytocin and vasopressin project in turn to the hindbrain and the spinal cord. Neurons that secrete epinephrine have their cell bodies in the hindbrain and end in the ventral hypothalamus.

An intrahypothalamic system is comprised of dopamine-secreting neurons that have their cell bodies in the arcuate nucleus and end on or near the capillaries that form the portal vessels in the median eminence. Serotonin-secreting neurons project to the hypothalamus from the raphe nuclei.

RELATION TO THE PITUITARY GLAND

There are neural connections between the hypothalamus and the posterior lobe of the pituitary gland and vascular connections between the hypothalamus and the anterior lobe. Embryologically, the posterior pituitary arises as an evagination of the floor of the third ventricle. It is made up in large part of the endings of axons that arise from cell bodies in the supraoptic and paraventricular nuclei and pass to the posterior pituitary (Figure 17–2) via the **hypothalamohypophysial tract.** Most of the supraoptic fibers end in the posterior lobe itself, whereas some of the paraventricular fibers end in the median eminence. The anterior and intermediate lobes of the pituitary arise in the embryo from the Rathke pouch, an evagination from the roof of the pharynx (see Figure 18–1). Sympathetic nerve fibers

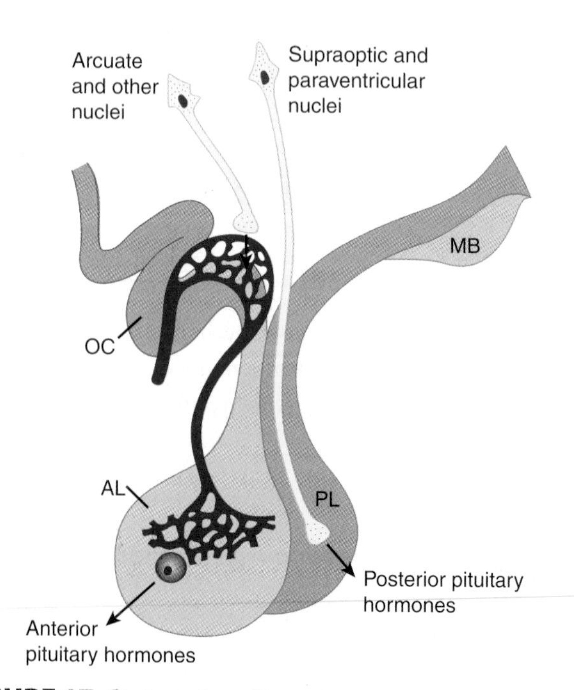

FIGURE 17–2 Secretion of hypothalamic hormones. The hormones of the posterior lobe (PL) are released into the general circulation from the endings of supraoptic and paraventricular neurons, whereas hypophysiotropic hormones are secreted into the portal hypophysial circulation from the endings of arcuate and other hypothalamic neurons. AL, anterior lobe; MB, mamillary bodies; OC, optic chiasm.

TABLE 17–1 Summary of principal hypothalamic regulatory mechanisms.

Function	Afferents from	Integrating Areas
Temperature regulation	Temperature receptors in the skin, deep tissues, spinal cord, hypothalamus, and other parts of the brain	Anterior hypothalamus, response to heat; posterior hypothalamus, response to cold
Neuroendocrine control of:		
Catecholamines	Limbic areas concerned with emotion	Dorsal and posterior hypothalamus
Vasopressin	Osmoreceptors, "volume receptors," others	Supraoptic and paraventricular nuclei
Oxytocin	Touch receptors in breast, uterus, genitalia	Supraoptic and paraventricular nuclei
Thyroid-stimulating hormone (thyrotropin, TSH) via TRH	Temperature receptors in infants, perhaps others	Paraventricular nuclei and neighboring areas
Adrenocorticotropic hormone (ACTH) and β-lipotropin (β-LPH) via CRH	Limbic system (emotional stimuli); reticular formation ("systemic" stimuli); hypothalamic and anterior pituitary cells sensitive to circulating blood cortisol level; suprachiasmatic nuclei (diurnal rhythm)	Paraventricular nuclei
Follicle-stimulating hormone (FSH) and luteinizing hormone (LH) via GnRH	Hypothalamic cells sensitive to estrogens, eyes, touch receptors in skin and genitalia of reflex ovulating species	Preoptic area; other areas
Prolactin via PIH and PRH	Touch receptors in breasts, other unknown receptors	Arcuate nucleus; other areas (hypothalamus inhibits secretion)
Growth hormone via somatostatin and GRH	Unknown receptors	Periventricular nucleus, arcuate nucleus
"Appetitive" behavior:		
Thirst	Osmoreceptors, probably located in the organum vasculosum of the lamina terminalis; angiotensin II uptake in the subfornical organ	Lateral superior hypothalamus
Hunger	Glucostat cells sensitive to rate of glucose utilization; leptin receptors; receptors for other polypeptides	Ventromedial, arcuate, and paraventricular nuclei; lateral hypothalamus
Sexual behavior	Cells sensitive to circulating estrogen and androgen, others	Anterior ventral hypothalamus plus, in the male, piriform cortex
Defensive reactions (fear, rage)	Sense organs and neocortex, paths unknown	Diffuse, in limbic system and hypothalamus
Control of body rhythms	Retina via retinohypothalamic fibers	Suprachiasmatic nuclei

reach the anterior lobe from its capsule, and parasympathetic fibers reach it from the petrosal nerves, but few if any nerve fibers pass to it from the hypothalamus. However, the **portal hypophysial vessels** form a direct vascular link between the hypothalamus and the anterior pituitary. Arterial twigs from the carotid arteries and circle of Willis form a network of fenestrated capillaries called the **primary plexus** on the ventral surface of the hypothalamus (Figure 17–1). Capillary loops also penetrate the median eminence. The capillaries drain into the sinusoidal portal hypophysial vessels that carry blood down the pituitary stalk to the capillaries of the anterior pituitary. This system begins and ends in capillaries without going through the heart and is therefore a true portal system. In birds and some mammals, including humans, there is no other anterior hypophysial arterial supply other than capsular vessels and anastomotic connections from the capillaries of the posterior pituitary. The **median eminence** is generally defined as the portion of the ventral hypothalamus from which the portal vessels arise. This region is outside the blood–brain barrier (see Chapter 33).

HYPOTHALAMIC FUNCTION

The major functions of the hypothalamus are summarized in Table 17–1. Some are fairly clear-cut visceral reflexes, and others include complex behavioral and emotional reactions; however, all involve a particular response to a particular stimulus. It is important to keep this in mind in considering hypothalamic function.

RELATION TO AUTONOMIC FUNCTION

Many years ago, Sherrington called the hypothalamus "the head ganglion of the autonomic system." Stimulation of the hypothalamus produces autonomic responses, but the hypothalamus does not seem to be concerned with the regulation of visceral function per se. Rather, the autonomic responses triggered in the hypothalamus are part of more complex phenomena such as eating, and emotions such as rage. For example, stimulation of various parts of the hypothalamus, especially the lateral areas, produces diffuse sympathetic discharge and increased adrenal medullary secretion—the mass sympathetic discharge seen in animals exposed to stress (the flight or fight reaction; see Chapter 13).

It has been claimed that separate hypothalamic areas control epinephrine and norepinephrine secretion. Differential secretion of one or the other of these adrenal medullary catecholamines does occur in certain situations (see Chapter 20), but the selective increases are small.

Body weight depends on the balance between caloric intake and utilization of calories. Obesity results when the former exceeds the latter. The hypothalamus and related parts of the brain play a key role in the regulation of food intake. Obesity is considered in detail in Chapter 26, and the relation of obesity to diabetes mellitus is discussed in Chapter 24.

Hypothalamic regulation of sleep and circadian rhythms are discussed in Chapter 14.

THIRST

Another appetitive mechanism under hypothalamic control is thirst. Drinking is regulated by plasma osmolality and extracellular fluid (ECF) volume in much the same fashion as vasopressin secretion (see Chapter 38). Water intake is increased by increased effective osmotic pressure of the plasma (Figure 17–3), by decreases in ECF volume, and by psychologic and other factors. Osmolality acts via **osmoreceptors,** receptors that sense the osmolality of the body fluids. These osmoreceptors are located in the anterior hypothalamus.

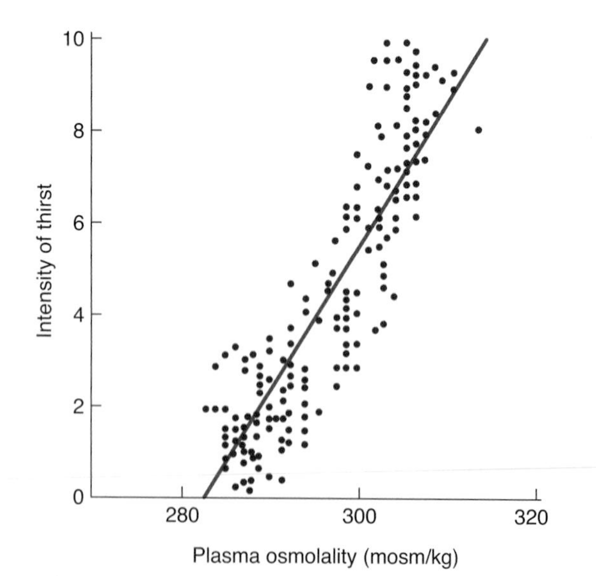

FIGURE 17–3 Relation of plasma osmolality to thirst in healthy adult humans during infusion of hypertonic saline. The intensity of thirst is measured on a special analog scale. (Reproduced with permission from Thompson CJ et al: The osmotic thresholds for thirst and vasopressin release are similar in healthy humans. Clin Sci Lond 1986;71:651.)

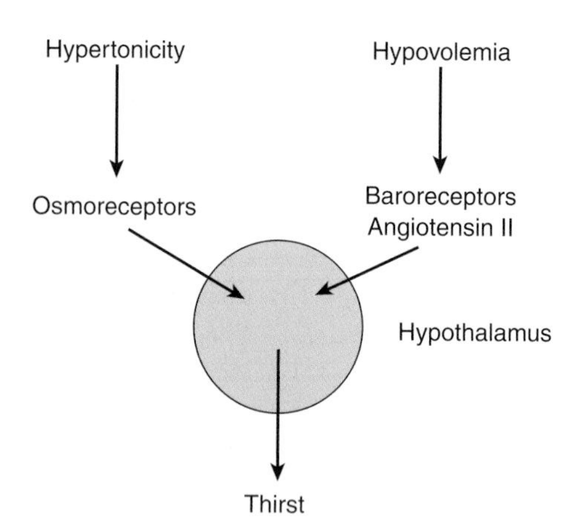

FIGURE 17–4 Diagrammatic representation of the way in which changes in plasma osmolality and changes in ECF volume affect thirst by separate pathways.

Decreases in ECF volume also stimulate thirst by a pathway independent of that mediating thirst in response to increased plasma osmolality (Figure 17–4). Thus, hemorrhage causes increased drinking even if there is no change in the osmolality of the plasma. The effect of ECF volume depletion on thirst is mediated in part via the renin–angiotensin system (see Chapter 38). Renin secretion is increased by hypovolemia and results in an increase in circulating angiotensin II. The angiotensin II acts on the **subfornical organ,** a specialized receptor area in the diencephalon (see Figure 33–7), to stimulate the neural areas concerned with thirst. Some evidence suggests that it acts on the **organum vasculosum of the lamina terminalis (OVLT)** as well. These areas are highly permeable and are two of the circumventricular organs located outside the blood–brain barrier (see Chapter 33). However, drugs that block the action of angiotensin II do not completely block the thirst response to hypovolemia, and it appears that the baroreceptors in the heart and blood vessels are also involved.

The intake of liquids is increased during eating **(prandial drinking).** The increase has been called a learned or habit response, but it has not been investigated in detail. One factor is an increase in plasma osmolality that occurs as food is absorbed. Another may be an action of one or more gastrointestinal hormones on the hypothalamus.

When the sensation of thirst is obtunded, either by direct damage to the diencephalon or by depressed or altered states of consciousness, patients stop drinking adequate amounts of fluid. Dehydration results if appropriate measures are not instituted to maintain water balance. If the protein intake is high, the products of protein metabolism cause an osmotic diuresis (see Chapter 38), and the amounts of water required to maintain hydration are large. Most cases of **hypernatremia** are actually due to simple dehydration in patients with psychoses or hypothalamic disease who do not or cannot increase their water intake when their thirst mechanism is stimulated. Lesions of the anterior communicating artery can also obtund thirst because branches of this artery supply the hypothalamic areas concerned with thirst.

OTHER FACTORS REGULATING WATER INTAKE

A number of other well-established factors contribute to the regulation of water intake. Psychologic and social factors are important. Dryness of the pharyngeal mucous membrane causes a sensation of thirst. Patients in whom fluid intake must be restricted sometimes get appreciable relief of thirst by sucking ice chips or a wet cloth.

Dehydrated dogs, cats, camels, and some other animals rapidly drink just enough water to make up their water deficit. They stop drinking before the water is absorbed (while their plasma is still hypertonic), so some kind of pharyngeal gastrointestinal "metering" must be involved. Some evidence suggests that humans have a similar metering ability, though it is not well developed.

CONTROL OF POSTERIOR PITUITARY SECRETION

VASOPRESSIN & OXYTOCIN

In most mammals, the hormones secreted by the posterior pituitary gland are **arginine vasopressin (AVP)** and **oxytocin.** In hippopotami and most pigs, arginine in the vasopressin molecule is replaced by lysine to form **lysine vasopressin.** The posterior pituitaries of some species of pigs and marsupials contain a mixture of arginine and lysine vasopressin. The posterior lobe hormones are nanopeptides with a disulfide ring at one end (Figure 17–5).

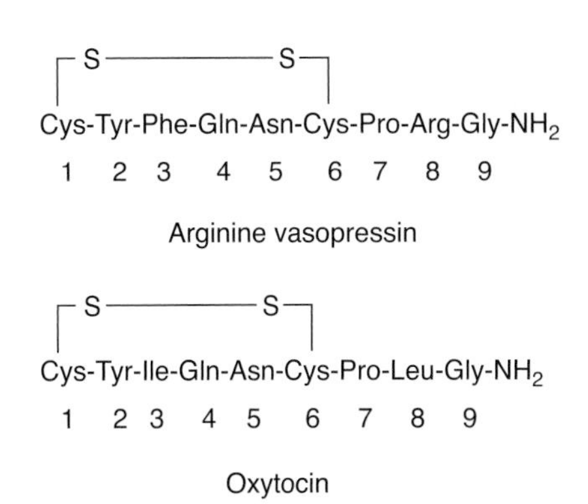

FIGURE 17–5 Arginine vasopressin and oxytocin.

BIOSYNTHESIS, INTRANEURONAL TRANSPORT, & SECRETION

The hormones of the posterior pituitary gland are synthesized in the cell bodies of the magnocellular neurons in the supraoptic and paraventricular nuclei and transported down the axons of these neurons to their endings in the posterior lobe, where they are secreted in response to electrical activity in the endings. Some of the neurons make oxytocin and others make vasopressin, and oxytocin-containing and vasopressin-containing cells are found in both nuclei.

Oxytocin and vasopressin are typical **neural hormones,** that is, hormones secreted into the circulation by nerve cells. This type of neural regulation is compared with other types in Figure 17–6. The term **neurosecretion** was originally coined to describe the secretion of hormones by neurons, but the term is somewhat misleading because it appears that all neurons secrete chemical messengers (see Chapter 7).

Like other peptide hormones, the posterior lobe hormones are synthesized as part of larger precursor molecules. Vasopressin and oxytocin each have a characteristic **neurophysin** associated with them in the granules in the neurons that secrete them—neurophysin I in the case of oxytocin and neurophysin II in the case of vasopressin. The neurophysins were originally thought to be binding polypeptides, but it now appears that they are simply parts of the precursor molecules. The precursor for AVP, **prepropressophysin,** contains a 19-amino-acid residue leader sequence followed by AVP, neurophysin II, and a glycopeptide (Figure 17–7). **Prepro-oxyphysin,** the precursor for oxytocin, is a similar but smaller molecule that lacks the glycopeptide.

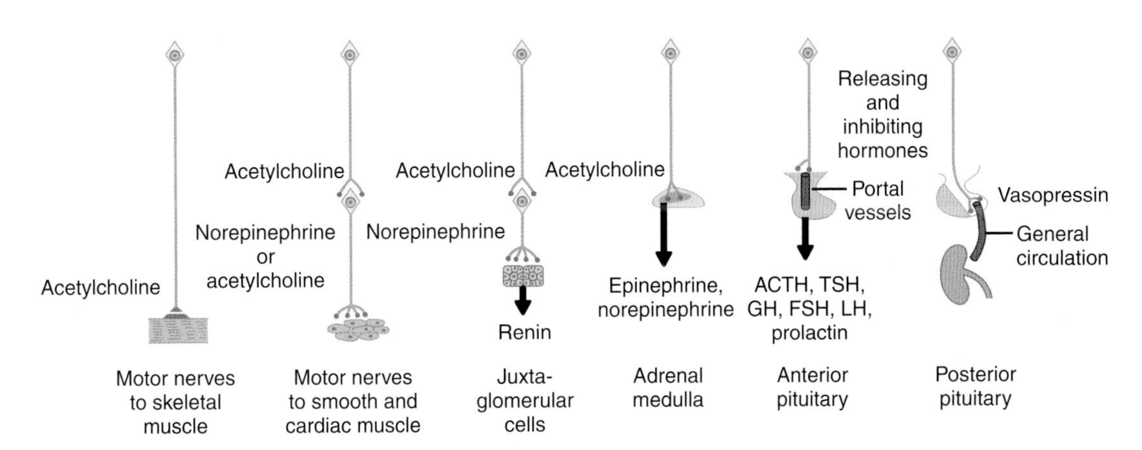

FIGURE 17–6 Neural control mechanisms. In the two situations on the left, neurotransmitters act at nerve endings on muscle; in the two in the middle, neurotransmitters regulate the secretion of endocrine glands; and in the two on the right, neurons secrete hormones into the hypophysial portal or general circulation.

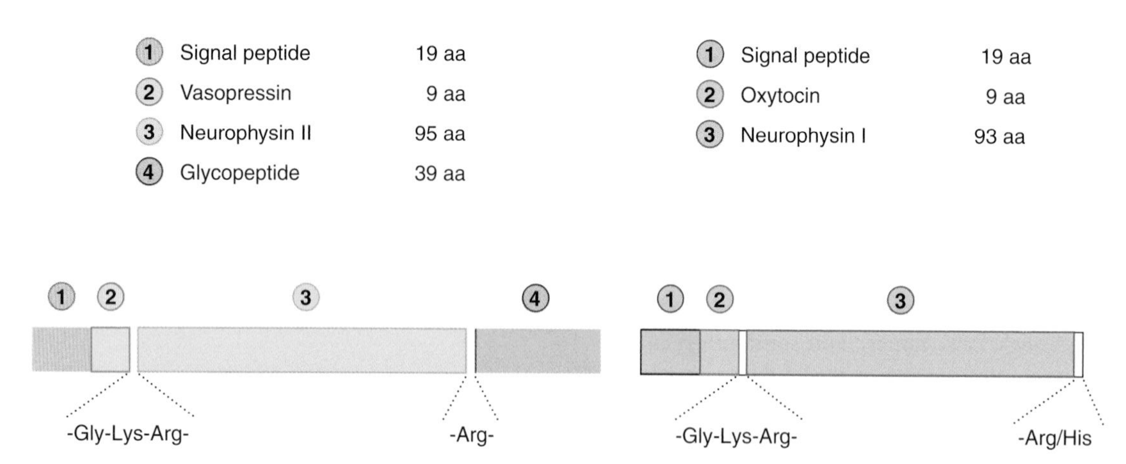

FIGURE 17–7 Structure of bovine prepropressophysin (left) and prepro-oxyphysin (right). Gly in the 10 position of both peptides is necessary for amidation of the Gly residue in position 9. aa, amino acid residues. (Reproduced with permission from Richter D: Molecular events in expression of vasopressin and oxytocin and their cognate receptors. Am J Physiol 1988;255:F207.)

The precursor molecules are synthesized in the ribosomes of the cell bodies of the neurons. They have their leader sequences removed in the endoplasmic reticulum, are packaged into secretory granules in the Golgi apparatus, and are transported down the axons by axoplasmic flow to the endings in the posterior pituitary. The secretory granules, called **Herring bodies,** are easy to stain in tissue sections, and they have been extensively studied. Cleavage of the precursor molecules occurs as they are being transported, and the storage granules in the endings contain free vasopressin or oxytocin and the corresponding neurophysin. In the case of vasopressin, the glycopeptide is also present. All these products are secreted, but the functions of the components other than the established posterior pituitary hormones are unknown. Physiological control of vasopressin secretion is described in detail in Chapter 38.

ELECTRICAL ACTIVITY OF MAGNOCELLULAR NEURONS

The oxytocin-secreting and vasopressin-secreting neurons also generate and conduct action potentials, and action potentials reaching their endings trigger the release of hormones by Ca^{2+}-dependent exocytosis. At least in anesthetized rats, these neurons are silent at rest or discharge at low, irregular rates (0.1–3 spikes/s). However, their response to stimulation varies (Figure 17–8). Stimulation of the nipples causes a synchronous, high-frequency discharge of the oxytocin neurons after an appreciable latency. This discharge causes release of a pulse of oxytocin and consequent milk ejection in postpartum females. On the other hand, stimulation of vasopressin-secreting neurons by a stimulus such as an increase in blood osmolality during dehydration, or loss of blood volume due to hemorrhage, causes an initial steady increase in firing rate followed by a prolonged pattern of phasic discharge in which periods of high-frequency discharge alternate with periods of electrical quiescence **(phasic bursting).** These phasic bursts are generally not synchronous in different vasopressin-secreting neurons. They are well suited to maintain a prolonged increase in the output of vasopressin, as opposed to the synchronous, relatively short, high-frequency discharge of oxytocin-secreting neurons in response to stimulation of the nipples.

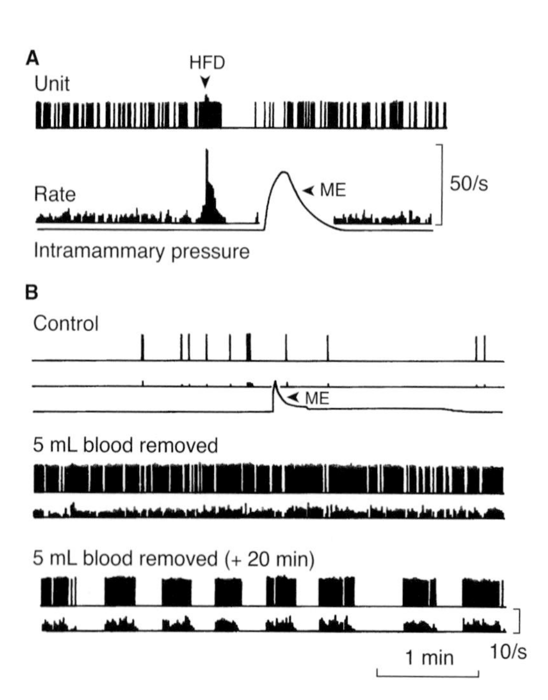

FIGURE 17–8 Responses of magnocellular neurons to stimulation. The tracings show individual extracellularly recorded action potentials, discharge rates, and intramammary duct pressure. **A)** Response of an oxytocin-secreting neuron. HFD, high-frequency discharge; ME, milk ejection. Stimulation of nipples started before the onset of recording. **B)** Responses of a vasopressin-secreting neuron, showing no change in the slow firing rate in response to stimulation of nipples and a prompt increase in the firing rate when 5 mL of blood was drawn, followed by typical phasic discharge. (Modified from Wakerly JB: Hypothalamic neurosecretory function: Insights from electrophysiological studies of the magno-cellular nuclei. IBRO News 1985;4:15.)

VASOPRESSIN & OXYTOCIN IN OTHER LOCATIONS

Vasopressin-secreting neurons are found in the suprachiasmatic nuclei, and vasopressin and oxytocin are also found in the endings of neurons that project from the paraventricular nuclei to the brain stem and spinal cord. These neurons appear to be involved in cardiovascular control. In addition, vasopressin and oxytocin are synthesized in the gonads and the adrenal cortex, and oxytocin is present in the thymus. The functions of the peptides in these organs are unsettled.

Vasopressin Receptors

There are at least three kinds of vasopressin receptors: V_{1A}, V_{1B}, and V_2. All are G protein-coupled. The V_{1A} and V_{1B} receptors act through phosphatidylinositol hydrolysis to increase intracellular Ca^{2+} concentrations. The V_2 receptors act through G_s to increase cAMP levels.

Effects of Vasopressin

Because one of its principal physiologic effects is the retention of water by the kidney, vasopressin is often called the **antidiuretic hormone (ADH).** It increases the permeability of the collecting ducts of the kidney so that water enters the hypertonic interstitium of the renal pyramids (see Chapter 37). The urine becomes concentrated and its volume decreases. The overall effect is therefore retention of water in excess of solute; consequently, the effective osmotic pressure of the body fluids is decreased. In the absence of vasopressin, the urine is hypotonic to plasma, urine volume is increased, and there is a net water loss. Consequently, the osmolality of the body fluid rises.

Effects of Oxytocin

In humans, oxytocin acts primarily on the breasts and uterus, though it appears to be involved in luteolysis as well (see Chapter 22). A G protein-coupled oxytocin receptor has been identified in human myometrium, and a similar or identical receptor is found in mammary tissue and the ovary. It triggers increases in intracellular Ca^{2+} levels.

The Milk Ejection Reflex

Oxytocin causes contraction of the **myoepithelial cells** that line the ducts of the breast. This squeezes the milk out of the alveoli of the lactating breast into the large ducts (sinuses) and thence out of the nipple **(milk ejection).** Many hormones acting in concert are responsible for breast growth and the secretion of milk into the ducts (see Chapter 22), but milk ejection in most species requires oxytocin.

Milk ejection is normally initiated by a neuroendocrine reflex. The receptors involved are touch receptors, which are plentiful in the breast—especially around the nipple. Impulses generated in these receptors are relayed from the somatic touch pathways to the supraoptic and paraventricular nuclei. Discharge of the oxytocin-containing neurons causes secretion of oxytocin from the posterior pituitary (Figure 17–8). The suckling of an infant at the breast stimulates the touch receptors, the nuclei are stimulated, oxytocin is released, and the milk is expressed into the sinuses, ready to flow into the mouth of the waiting infant. In lactating women, genital stimulation and emotional stimuli also produce oxytocin secretion, sometimes causing milk to spurt from the breasts.

Other Actions of Oxytocin

Oxytocin causes contraction of the smooth muscle of the uterus. The sensitivity of the uterine musculature to oxytocin is enhanced by estrogen and inhibited by progesterone. The inhibitory effect of progesterone is due to a direct action of the steroid on uterine oxytocin receptors. In late pregnancy, the uterus becomes very sensitive to oxytocin coincident with a marked increase in the number of oxytocin receptors and oxytocin receptor mRNA (see Chapter 22). Oxytocin secretion is then increased during labor. After dilation of the cervix, descent of the fetus down the birth canal initiates impulses in the afferent nerves that are relayed to the supraoptic and paraventricular nuclei, causing secretion of sufficient oxytocin to enhance labor (Figure 22–24). The amount of oxytocin in plasma is normal at the onset of labor. It is possible that the marked increase in oxytocin receptors at this time allows normal oxytocin levels to initiate contractions, setting up a positive feedback. However, the amount of oxytocin in the uterus is also increased, and locally produced oxytocin may also play a role.

Oxytocin may also act on the nonpregnant uterus to facilitate sperm transport. The passage of sperm up the female genital tract to the uterine tubes, where fertilization normally takes place, depends not only on the motile powers of the sperm but also, at least in some species, on uterine contractions. The genital stimulation involved in coitus releases oxytocin, but whether oxytocin initiates the rather specialized uterine contractions that transport the sperm is as yet unproven. The secretion of oxytocin is also increased by stressful stimuli and, like that of vasopressin, is inhibited by alcohol.

Circulating oxytocin increases at the time of ejaculation in males, and it is possible that this causes increased contraction of the smooth muscle of the vas deferens, propelling sperm toward the urethra.

CONTROL OF ANTERIOR PITUITARY SECRETION

ANTERIOR PITUITARY HORMONES

The anterior pituitary secretes six hormones: **adrenocorticotropic hormone (corticotropin, ACTH), thyroid-stimulating hormone (thyrotropin, TSH), growth hormone, follicle-stimulating hormone (FSH), luteinizing hormone (LH),**

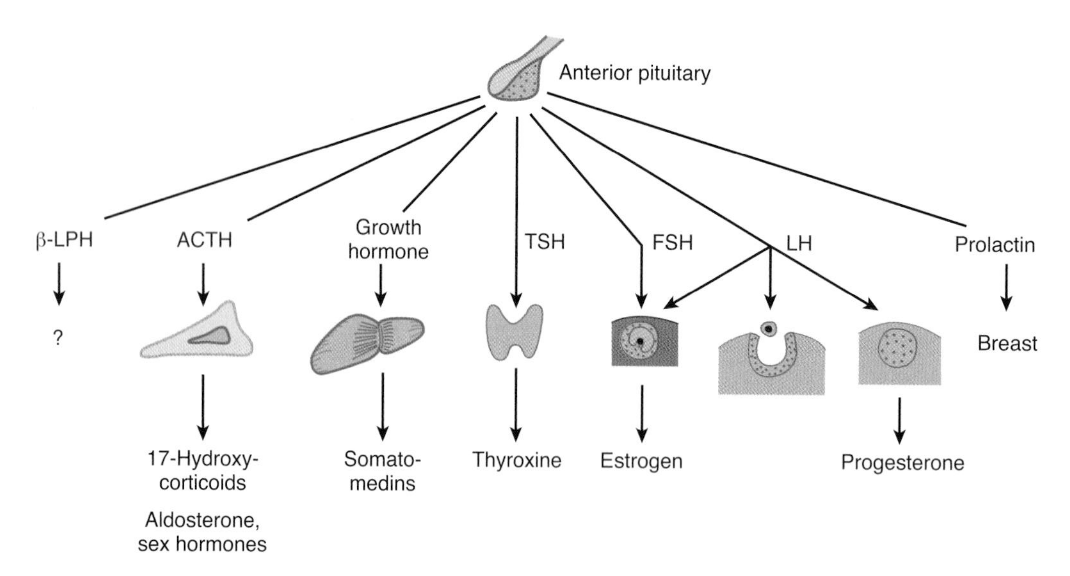

FIGURE 17–9 Anterior pituitary hormones. In women, FSH and LH act in sequence on the ovary to produce growth of the ovarian follicle, ovulation, and formation and maintenance of the corpus luteum. Prolactin stimulates lactation. In men, FSH and LH control the functions of the testes.

and **prolactin (PRL).** An additional polypeptide, β-lipotropin (β-LPH), is secreted with ACTH, but its physiologic role is unknown. The actions of the anterior pituitary hormones are summarized in Figure 17–9. The hormones are discussed in detail in subsequent chapters. The hypothalamus plays an important stimulatory role in regulating the secretion of ACTH, β-LPH, TSH, growth hormone, FSH, and LH. It also regulates prolactin secretion, but its effect is predominantly inhibitory rather than stimulatory.

NATURE OF HYPOTHALAMIC CONTROL

Anterior pituitary secretion is controlled by chemical agents carried in the portal hypophysial vessels from the hypothalamus to the pituitary. These substances used to be called releasing and inhibiting factors, but now they are commonly called **hypophysiotropic hormones.** The latter term seems appropriate since they are secreted into the bloodstream and act at a distance from their site of origin. Small amounts escape into the general circulation, but they are at their highest concentration in portal hypophysial blood.

HYPOPHYSIOTROPIC HORMONES

There are six established hypothalamic releasing and inhibiting hormones (Figure 17–10): **corticotropin-releasing hormone (CRH); thyrotropin-releasing hormone (TRH); growth hormone-releasing hormone (GRH); growth hormone-inhibiting hormone (GIH,** now generally called **somatostatin); luteinizing hormone-releasing hormone (LHRH,** now generally known as **gonadotropin-releasing hormone (GnRH));** and **prolactin-inhibiting hormone (PIH).** In addition, hypothalamic extracts contain prolactin-releasing activity, and a **prolactin-releasing hormone (PRH)** has been postulated to exist. TRH, VIP, and several other polypeptides found in the hypothalamus stimulate prolactin secretion, but it is uncertain whether one or more of these peptides is the physiologic PRH. Recently, an orphan receptor was isolated from the anterior pituitary, and the search for its ligand led to the isolation of a 31-amino-acid polypeptide from the human

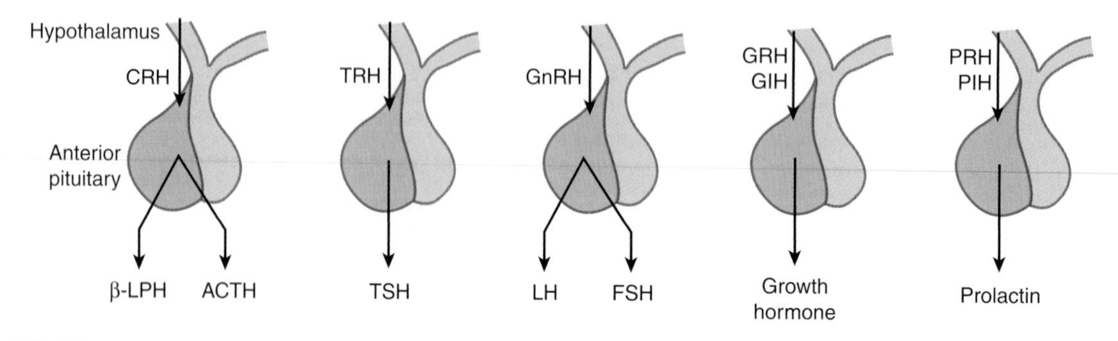

FIGURE 17–10 Effects of hypophysiotropic hormones on the secretion of anterior pituitary hormones.

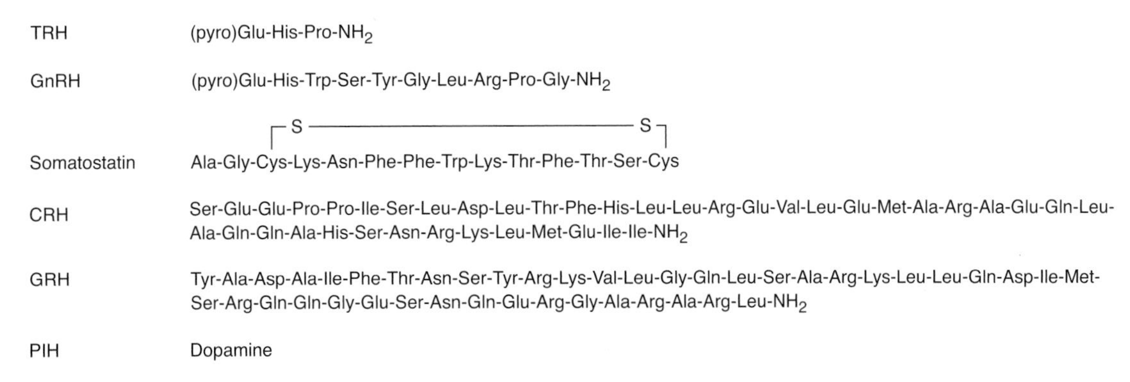

FIGURE 17–11 Structure of hypophysiotropic hormones in humans. Preprosomatostatin is processed to a tetradecapeptide (somatostatin 14, [SS14], shown above) and also to a polypeptide containing 28 amino acid residues (SS28).

hypothalamus. This polypeptide stimulated prolactin secretion by an action on the anterior pituitary receptor, but additional research is needed to determine if it is the physiologic PRH. GnRH stimulates the secretion of FSH as well as that of LH, and it seems unlikely that a separate FSH-releasing hormone exists.

The structures of the six established hypophysiotropic hormones are shown in Figure 17–11. The structures of the genes and preprohormones for TRH, GnRH, somatostatin, CRH, and GRH are known. PreproTRH contains six copies of TRH. Several other preprohormones may contain other hormonally active peptides in addition to the hypophysiotropic hormones.

The area from which the hypothalamic releasing and inhibiting hormones are secreted is the median eminence of the hypothalamus. This region contains few nerve cell bodies, but many nerve endings are in close proximity to the capillary loops from which the portal vessels originate.

The locations of the cell bodies of the neurons that project to the external layer of the median eminence and secrete the hypophysiotropic hormones are shown in Figure 17–12, which also shows the location of the neurons secreting oxytocin and vasopressin. The GnRH-secreting neurons are primarily in the medial preoptic area, the somatostatin-secreting neurons are in the periventricular nuclei, the TRH-secreting and CRH-secreting neurons are in the medial parts of the paraventricular nuclei, and the GRH-secreting (and dopamine-secreting) neurons are in the arcuate nuclei.

Most, if not all, of the hypophysiotropic hormones affect the secretion of more than one anterior pituitary hormone (Figure 17–10). The FSH-stimulating activity of GnRH has been mentioned previously. TRH stimulates the secretion of prolactin as well as TSH. Somatostatin inhibits the secretion of TSH as well as growth hormone. It does not normally inhibit the secretion of the other anterior pituitary hormones, but it inhibits the abnormally elevated secretion of ACTH in patients with Nelson's syndrome. CRH stimulates the secretion of ACTH and β-LPH.

Hypophysiotropic hormones function as neurotransmitters in other parts of the brain, the retina, and the autonomic nervous system (see Chapter 7). In addition, somatostatin is found in the pancreatic islets (see Chapter 24), GRH is secreted by pancreatic tumors, and somatostatin and TRH are found in the gastrointestinal tract (see Chapter 25).

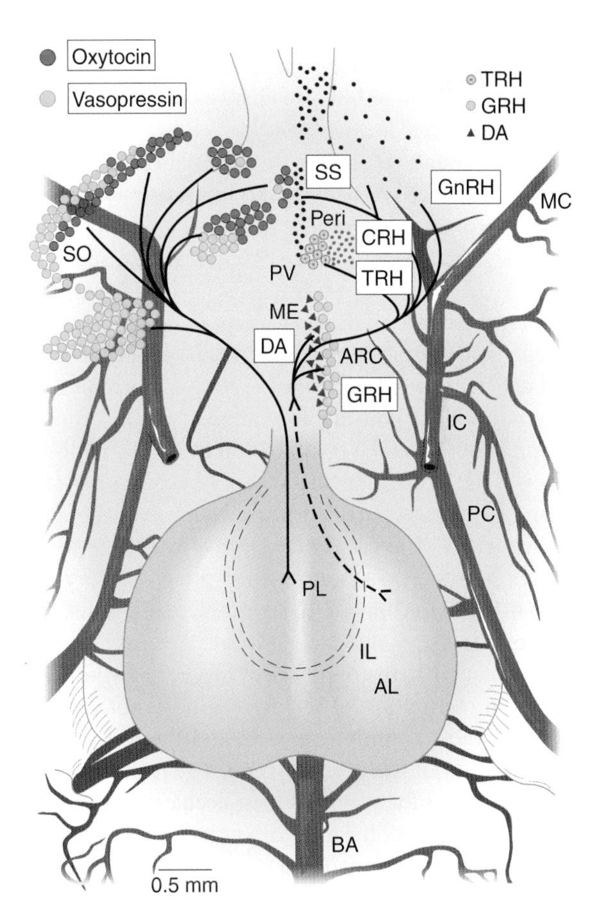

FIGURE 17–12 Location of cell bodies of hypophysiotropic hormone-secreting neurons projected on a ventral view of the hypothalamus and pituitary of the rat. AL, anterior lobe; ARC, arcuate nucleus; BA, basilar artery; DA, dopamine; IC, internal carotid artery; IL, intermediate lobe; MC, middle cerebral artery; ME, median eminence; PC, posterior cerebral artery; Peri, periventricular nucleus; PL, posterior lobe; PV, paraventricular nucleus; SO, supraoptic nucleus. The names of the hormones are enclosed in boxes. (Courtesy of LW Swanson and ET Cunningham Jr)

Receptors for most of the hypophysiotropic hormones are coupled to G proteins. There are two human CRH receptors: hCRH-RI and hCRH-RII. The physiologic role of hCRH-RII is unsettled, though it is found in many parts of the brain. In addition, a **CRH-binding protein** in the peripheral circulation inactivates CRH. It is also found in the cytoplasm of corticotropes in the anterior pituitary, and in this location it might play a role in receptor internalization. However, the exact physiologic role of this protein is unknown. Other hypophysiotropic hormones do not have known binding proteins.

SIGNIFICANCE & CLINICAL IMPLICATIONS

Research delineating the multiple neuroendocrine regulatory functions of the hypothalamus is important because it helps explain how endocrine secretion is matched to the demands of a changing environment. The nervous system receives information about changes in the internal and external environment from the sense organs. It brings about adjustments to these changes through effector mechanisms that include not only somatic movement but also changes in the rate at which hormones are secreted.

The manifestations of hypothalamic disease are neurologic defects, endocrine changes, and metabolic abnormalities such as hyperphagia and hyperthermia. The relative frequencies of the signs and symptoms of hypothalamic disease in one large series of cases are shown in Table 17–2. The possibility of hypothalamic pathology should be kept in mind in evaluating all patients with pituitary dysfunction, especially those with isolated deficiencies of single pituitary tropic hormones.

TABLE 17–2 Symptoms and signs in 60 patients with hypothalamic disease.

Symptoms and Signs	Percentage of Cases
Endocrine and metabolic findings	
Precocious puberty	40
Hypogonadism	32
Diabetes insipidus	35
Obesity	25
Abnormalities of temperature regulation	22
Emaciation	18
Bulimia	8
Anorexia	7
Neurologic findings	
Eye signs	78
Pyramidal and sensory deficits	75
Headache	65
Extrapyramidal signs	62
Vomiting	40
Psychic disturbances, rage attacks, etc	35
Somnolence	30
Convulsions	15

Data from Bauer HG: Endocrine and other clinical manifestations of hypothalamic disease. J Clin Endocrinol 1954;14:13. See also Kahana L, et al: Endocrine manifestations of intracranial extrasellar lesions. J Clin Endocrinol 1962;22:304.

A condition of considerable interest in this context is **Kallmann syndrome,** the combination of hypogonadism due to low levels of circulating gonadotropins (**hypogonadotropic hypogonadism**) with partial or complete loss of the sense of smell (**hyposmia** or **anosmia**). Embryologically, GnRH neurons develop in the nose and migrate up the olfactory nerves and then through the brain to the hypothalamus. If this migration is prevented by congenital abnormalities in the olfactory pathways, the GnRH neurons do not reach the hypothalamus and pubertal maturation of the gonads does not occur. The syndrome is most common in men, and the cause in many cases is mutation of the *KALIG1* gene, a gene on the X chromosome that codes for an adhesion molecule necessary for the normal development of the olfactory nerve. However, the condition also occurs in women and can be due to other genetic abnormalities.

TEMPERATURE REGULATION

In the body, heat is produced by muscular exercise, assimilation of food, and all the vital processes that contribute to the basal metabolic rate. It is lost from the body by radiation, conduction, and vaporization of water in the respiratory passages and on the skin. Small amounts of heat are also removed in the urine and feces. The balance between heat production and heat loss determines the body temperature. Because the speed of chemical reactions varies with temperature and because the enzyme systems of the body have narrow temperature ranges in which their function is optimal, normal body function depends on a relatively constant body temperature.

Invertebrates generally cannot adjust their body temperatures and so are at the mercy of the environment. In vertebrates, mechanisms for maintaining body temperature by adjusting heat production and heat loss have evolved. In reptiles, amphibians, and fish, the adjusting mechanisms are relatively rudimentary, and these species are called "cold-blooded" (**poikilothermic**) because their body temperature fluctuates over a considerable range. In birds and mammals, the "warm-blooded" (**homeothermic**) animals, a group of reflex responses that are primarily integrated in the hypothalamus operate to maintain body temperature within a narrow range in spite of wide fluctuations in environmental temperature. The hibernating mammals are a partial exception. While awake they are homeothermic, but during hibernation their body temperature falls.

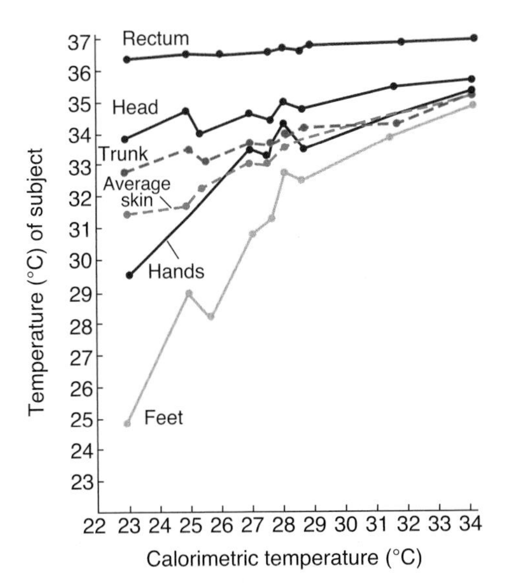

FIGURE 17–13 Temperatures of various parts of the body of a naked subject at various ambient temperatures in a calorimeter. (Redrawn and reproduced, with permission, from Hardy JD, DuBois EF: Basal heat production and elimination of thirteen normal women at temperatures from 22 degrees C. to 35 degrees C. J Nutr 1938 Oct; 48(2):257–293.)

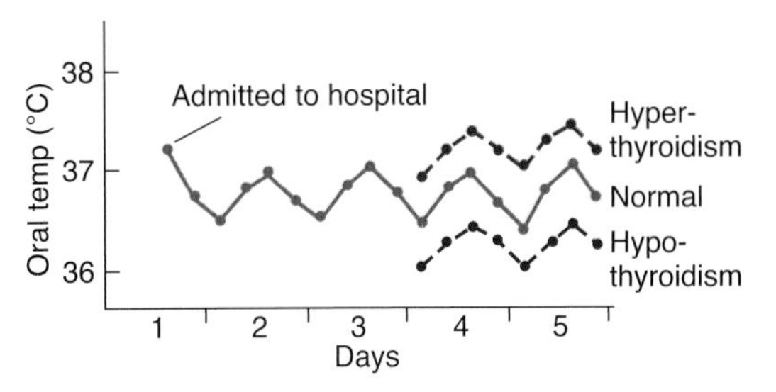

FIGURE 17–14 Typical temperature chart of a hospitalized patient who does not have a febrile disease. Note the slight rise in temperature, due to excitement and apprehension, at the time of admission to the hospital, and the regular circadian temperature cycle.

NORMAL BODY TEMPERATURE

In homeothermic animals, the actual temperature at which the body is maintained varies from species to species and, to a lesser degree, from individual to individual. In humans, the traditional normal value for the oral temperature is 37°C (98.6°F), but in one large series of normal young adults, the morning oral temperature averaged 36.7°C, with a standard deviation of 0.2°C. Therefore, 95% of all young adults would be expected to have a morning oral temperature of 36.3–37.1°C (97.3–98.8°F; mean ± 1.96 standard deviations). Various parts of the body are at different temperatures, and the magnitude of the temperature difference between the parts varies with the environmental temperature (Figure 17–13). The extremities are generally cooler than the rest of the body. The temperature of the scrotum is carefully regulated at 32°C. The rectal temperature is representative of the temperature at the core of the body and varies least with changes in environmental temperature. The oral temperature is normally 0.5°C lower than the rectal temperature, but it is affected by many factors, including ingestion of hot or cold fluids, gum chewing, smoking, and mouth breathing.

The normal human core temperature undergoes a regular circadian fluctuation of 0.5–0.7°C. In individuals who sleep at night and are awake during the day (even when hospitalized at bed rest), it is lowest at about 6:00 AM and highest in the evenings (Figure 17–14). It is lowest during sleep, is slightly higher in the awake but relaxed state, and rises with activity. In women, an additional monthly cycle of temperature variation is characterized by a rise in basal temperature at the time of ovulation (Figure 22–15). Temperature regulation is less precise in young children and they may normally have a temperature that is 0.5°C or so above the established norm for adults.

During exercise, the heat produced by muscular contraction accumulates in the body and the rectal temperature normally rises as high as 40°C (104 °F). This rise is due in part to the inability of the heat-dissipating mechanisms to handle the greatly increased amount of heat produced, but evidence suggests that during exercise in addition there is an elevation of the body temperature at which the heat-dissipating mechanisms are activated. Body temperature also rises slightly during emotional excitement, probably owing to unconscious tensing of the muscles. It is chronically elevated by as much as 0.5°C when the metabolic rate is high, as in hyperthyroidism, and lowered when the metabolic rate is low, as in hypothyroidism (Figure 17–14). Some apparently normal adults chronically have a temperature above the normal range (constitutional hyperthermia).

HEAT PRODUCTION

A variety of basic chemical reactions contribute to body heat production at all times. Ingestion of food increases heat production, but the major source of heat is the contraction of skeletal muscle (Table 17–3). Heat production can be varied by endocrine mechanisms in the absence of food intake or muscular exertion. Epinephrine and norepinephrine produce a rapid but short-lived increase in heat production; thyroid hormones produce a slowly developing but prolonged increase.

TABLE 17–3 Body heat production and heat loss.

Body heat is produced by:	
Basic metabolic processes	
Food intake (specific dynamic action)	
Muscular activity	
Body heat is lost by:	**Percentage of heat lost at 21°C**
Radiation and conduction	70
Vaporization of sweat	27
Respiration	2
Urination and defecation	1

Furthermore, sympathetic discharge decreases during fasting and is increased by feeding.

A source of considerable heat, particularly in infants, is **brown fat.** This fat has a high rate of metabolism and its thermogenic function has been likened to that of an electric blanket.

HEAT LOSS

The processes by which heat is lost from the body when the environmental temperature is below body temperature are listed in Table 17–3. **Conduction** is heat exchange between objects or substances at different temperatures that are in contact with one another. A basic characteristic of matter is that its molecules are in motion, with the amount of motion proportional to the temperature. These molecules collide with the molecules in cooler objects, transferring thermal energy to them. The amount of heat transferred is proportional to the temperature difference between the objects in contact **(thermal gradient).** Conduction is aided by **convection,** the movement of molecules away from the area of contact. Thus, for example, an object in contact with air at a different temperature changes the specific gravity of the air, and because warm air rises and cool air falls, a new supply of air is brought into contact with the object. Of course, convection is greatly aided if the object moves about in the medium or the medium moves past the object, for example, if a subject swims through water or a fan blows air through a room. **Radiation** is the transfer of heat by infrared electromagnetic radiation from one object to another at a different temperature with which it is not in contact. When an individual is in a cold environment, heat is lost by conduction to the surrounding air and by radiation to cool objects in the vicinity. Conversely, of course, heat is transferred to an individual and the heat load is increased by these processes when the environmental temperature is above body temperature. Note that because of radiation, an individual can feel chilly in a room with cold walls even though the room is relatively warm. On a cold but sunny day, the heat of the sun reflected off bright objects exerts an appreciable warming effect. It is the heat reflected from the snow, for example, that in part makes it possible to ski in fairly light clothes even though the air temperature is below freezing.

Because conduction occurs from the surface of one object to the surface of another, the temperature of the skin determines to a large extent the degree to which body heat is lost or gained. The amount of heat reaching the skin from the deep tissues can be varied by changing the blood flow to the skin. When the cutaneous vessels are dilated, warm blood wells into the skin, whereas in the maximally vasoconstricted state, heat is held centrally in the body. The rate at which heat is transferred from the deep tissues to the skin is called the **tissue conductance.** Further, birds have a layer of feathers next to the skin, and most mammals have a significant layer of hair or fur. Heat is conducted from the skin to the air trapped in this layer and from the trapped air to the exterior. When the thickness of the trapped layer is increased by fluffing the feathers or erection of the hairs **(horripilation),** heat transfer across the layer is reduced and heat losses (or, in a hot environment, heat gains) are decreased. "Goose pimples" are the result of horripilation in humans; they are the visible manifestation of cold-induced contraction of the piloerector muscles attached to the rather meager hair supply. Humans usually supplement this layer of hair with one or more layers of clothes. Heat is conducted from the skin to the layer of air trapped by the clothes, from the inside of the clothes to the outside, and from the outside of the clothes to the exterior. The magnitude of the heat transfer across the clothing, a function of its texture and thickness, is the most important determinant of how warm or cool the clothes feel, but other factors, especially the size of the trapped layer of warm air, are also important. Dark clothes absorb radiated heat and light-colored clothes reflect it back to the exterior.

The other major process transferring heat from the body in humans and other animals that sweat is vaporization of water on the skin and mucous membranes of the mouth and respiratory passages. Vaporization of 1 g of water removes about 0.6 kcal of heat. A certain amount of water is vaporized at all times. This **insensible water loss** amounts to 50 mL/h in humans. When sweat secretion is increased, the degree to which the sweat vaporizes depends on the humidity of the environment. It is common knowledge that one feels hotter on a humid day. This is due in part to the decreased vaporization of sweat, but even under conditions in which vaporization of sweat is complete, an individual in a humid environment feels warmer than an individual in a dry environment. The reason for this difference is unknown, but it seems related to the fact that in the humid environment sweat spreads over a greater area of skin before it evaporates. During muscular exertion in a hot environment, sweat secretion reaches values as high as 1600 mL/h, and in a dry atmosphere, most of this sweat is vaporized. Heat loss by vaporization of water therefore varies from 30 to over 900 kcal/h.

Some mammals lose heat by **panting.** This rapid, shallow breathing greatly increases the amount of water vaporization in the mouth and respiratory passages and therefore the amount of heat lost. Because the breathing is shallow, it produces relatively little change in the composition of alveolar air (see Chapter 34).

The relative contribution of each of the processes that transfer heat away from the body (Table 17–3) varies with the environmental temperature. At 21°C, vaporization is a minor component in humans at rest. As the environmental temperature approaches body temperature, radiation losses decline and vaporization losses increase.

TEMPERATURE-REGULATING MECHANISMS

The reflex and semireflex thermoregulatory responses in humans are listed in Table 17–4. They include autonomic, somatic, endocrine, and behavioral changes. One group of

TABLE 17–4 Temperature-regulating mechanisms.

Mechanisms activated by cold
Shivering
Hunger
Increased voluntary activity
Increased secretion of norepinephrine and epinephrine
Decreased heat loss
Cutaneous vasoconstriction
Curling up
Horripilation
Mechanisms activated by heat
Increased heat loss
Cutaneous vasodilation
Sweating
Increased respiration
Decreased heat production
Anorexia
Apathy and inertia

responses increases heat loss and decreases heat production; the other decreases heat loss and increases heat production. In general, exposure to heat stimulates the former group of responses and inhibits the latter, whereas exposure to cold does the opposite.

Curling up "in a ball" is a common reaction to cold in animals and has a counterpart in the position some people assume on climbing into a cold bed. Curling up decreases the body surface exposed to the environment. Shivering is an involuntary response of the skeletal muscles, but cold also causes a semiconscious general increase in motor activity. Examples include foot stamping and dancing up and down on a cold day. Increased catecholamine secretion is an important endocrine response to cold. Mice unable to make norepinephrine and epinephrine because their dopamine β-hydroxylase gene is knocked out do not tolerate cold; they have deficient vasoconstriction and are unable to increase thermogenesis in brown adipose tissue through UCP 1. TSH secretion is increased by cold and decreased by heat in laboratory animals, but the change in TSH secretion produced by cold in adult humans is small and of questionable significance. It is common knowledge that activity is decreased in hot weather—the "it's too hot to move" reaction.

Thermoregulatory adjustments involve local responses as well as more general reflex responses. When cutaneous blood vessels are cooled they become more sensitive to catecholamines and the arterioles and venules constrict. This local effect of cold directs blood away from the skin. Another heat-conserving mechanism that is important in animals living in cold water is heat transfer from arterial to venous blood in the limbs. The deep veins **(venae comitantes)** run alongside the arteries supplying the limbs and heat is transferred from the warm arterial blood going to the limbs to the cold venous blood coming from the extremities (**countercurrent exchange;** see Chapter 37). This limits the ability to maintain heat in the tips of the extremities but conserves body heat.

The reflex responses activated by cold are controlled from the posterior hypothalamus. Those activated by warmth are controlled primarily from the anterior hypothalamus, although some thermoregulation against heat still occurs after decerebration at the level of the rostral midbrain. Stimulation of the anterior hypothalamus causes cutaneous vasodilation and sweating, and lesions in this region cause hyperthermia, with rectal temperatures sometimes reaching 43°C (109.4°F). Posterior hypothalamic stimulation causes shivering, and the body temperature of animals with posterior hypothalamic lesions falls toward that of the environment.

AFFERENTS

The hypothalamus is said to integrate body temperature information from sensory receptors (primarily cold receptors) in the skin, deep tissues, spinal cord, extrahypothalamic portions of the brain, and the hypothalamus itself. Each of these five inputs contributes about 20% of the information that is integrated. There are threshold core temperatures for each of the main temperature-regulating responses and when the threshold is reached the response begins. The threshold is 37°C for sweating and vasodilation, 36.8°C for vasoconstriction, 36°C for nonshivering thermogenesis, and 35.5°C for shivering.

FEVER

Fever is perhaps the oldest and most universally known hallmark of disease. It occurs not only in mammals but also in birds, reptiles, amphibia, and fish. When it occurs in homeothermic animals, the thermoregulatory mechanisms behave as if they were adjusted to maintain body temperature at a higher than normal level, that is, "as if the thermostat had been reset" to a new point above 37°C. The temperature receptors then signal that the actual temperature is below the new set point, and the temperature-raising mechanisms are activated. This usually produces chilly sensations due to cutaneous vasoconstriction and occasionally enough shivering to produce a shaking chill. However, the nature of the response depends on the ambient temperature. The temperature rise in experimental animals injected with a pyrogen is due mostly to increased heat production if they are in a cold environment and mostly to decreased heat loss if they are in a warm environment.

The pathogenesis of fever is summarized in Figure 17–15. Toxins from bacteria, such as endotoxin, act on monocytes, macrophages, and Kupffer cells to produce cytokines that act as **endogenous pyrogens (EPs).** There is good evidence that

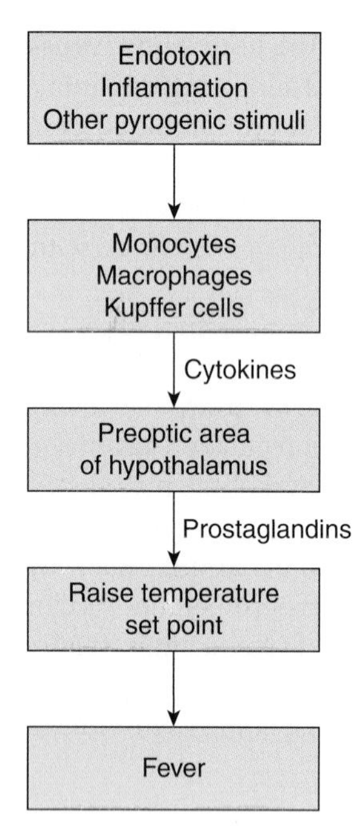

FIGURE 17–15 Pathogenesis of fever.

IL-1β, IL-6, IFN-β, IFN-γ, and TNF-α (see Chapter 3) can act independently to produce fever. These circulating cytokines are polypeptides and it is unlikely that they penetrate the brain. Instead, evidence suggests that they act on the OVLT, one of the circumventricular organs (see Chapter 33). This in turn activates the preoptic area of the hypothalamus. Cytokines are also produced by cells in the central nervous system (CNS) when these are stimulated by infection, and these may act directly on the thermoregulatory centers.

The fever produced by cytokines is probably due to local release of prostaglandins in the hypothalamus. Intrahypothalamic injection of prostaglandins produces fever. In addition, the antipyretic effect of aspirin is exerted directly on the hypothalamus, and aspirin inhibits prostaglandin synthesis. PGE_2 is one of the prostaglandins that causes fever. It acts on four sub-types of prostaglandin receptors—EP_1, EP_2, EP_3, and EP_4—and knockout of the EP_3 receptor impairs the febrile response to PGE_2, IL-1β, and endotoxin, or bacterial lipopolysaccharide (LPS).

The benefit of fever to the organism is uncertain. A beneficial effect is assumed because fever has evolved and persisted as a response to infections and other diseases. Many microorganisms grow best within a relatively narrow temperature range and a rise in temperature inhibits their growth. In addition, antibody production is increased when body temperature is elevated. Before the advent of antibiotics, fevers were artificially induced for the treatment of neurosyphilis and proved to be beneficial. Hyperthermia also benefits individuals infected with anthrax, pneumococcal pneumonia, leprosy, and various fungal, rickettsial, and viral diseases. Hyperthermia also slows the growth of some tumors. However, very high temperatures are harmful. A rectal temperature over 41°C (106°F) for prolonged periods results in some permanent brain damage. When the temperature is over 43°C, heat stroke develops and death is common.

In **malignant hyperthermia,** various mutations of the gene coding for the ryanodine receptor (see Chapter 5) lead to excess Ca^{2+} release during muscle contraction triggered by stress. This in turn leads to contractures of the muscles, increased muscle metabolism, and a great increase in heat production in muscle. The increased heat production causes a marked rise in body temperature that is fatal if not treated.

Periodic fevers also occur in humans with mutations in the gene for **pyrin,** a protein found in neutrophils; the gene for mevalonate kinase, an enzyme involved in cholesterol synthesis; and the gene for the type 1 TNF receptor, which is involved in inflammatory responses. However, how any of these three mutant gene products cause fever is unknown.

HYPOTHERMIA

In hibernating mammals, body temperature drops to low levels without causing any demonstrable ill effects on subsequent arousal. This observation led to experiments on induced hypothermia. When the skin or the blood is cooled enough to lower the body temperature in nonhibernating animals or in humans, metabolic and physiologic processes slow down. Respiration and heart rate are very slow, blood pressure is low, and consciousness is lost. At rectal temperatures of about 28°C, the ability to spontaneously return the temperature to normal is lost, but the individual continues to survive and, if rewarmed with external heat, returns to a normal state. If care is taken to prevent the formation of ice crystals in the tissues, the body temperature of experimental animals can be lowered to subfreezing levels without producing any detectable damage after subsequent rewarming.

Humans tolerate body temperatures of 21–24°C (70–75°F) without permanent ill effects, and induced hypothermia has been used in surgery. On the other hand, accidental hypothermia due to prolonged exposure to cold air or cold water is a serious condition and requires careful monitoring and prompt rewarming.

CHAPTER SUMMARY

- Neural connections run between the hypothalamus and the posterior lobe of the pituitary gland, and vascular connections between the hypothalamus and the anterior lobe of the pituitary.
- In most mammals, the hormones secreted by the posterior pituitary gland are vasopressin and oxytocin. Vasopressin increases the permeability of the collecting ducts of the kidney to water, thus concentrating the urine. Oxytocin acts on the breasts (lactation) and the uterus (contraction).
- The anterior pituitary secretes six hormones: adrenocorticotropic hormone (corticotropin, ACTH), thyroid-stimulating hormone (thyrotropin, TSH), growth hormone, follicle-stimulating hormone (FSH), luteinizing hormone (LH), and prolactin (PRL).
- Other complex autonomic mechanisms that maintain the chemical constancy and temperature of the internal environment are integrated in the hypothalamus.

MULTIPLE-CHOICE QUESTIONS

For all questions, select the single best answer unless otherwise directed.

1. Thirst is stimulated by
   A. increases in plasma osmolality and volume.
   B. an increase in plasma osmolality and a decrease in plasma volume.
   C. a decrease in plasma osmolality and an increase in plasma volume.
   D. decreases in plasma osmolality and volume.
   E. injection of vasopressin into the hypothalamus.
2. When an individual is naked in a room in which the air temperature is 21°C (69.8°F) and the humidity 80%, the greatest amount of heat is lost from the body by
   A. elevated metabolism.
   B. respiration.
   C. urination.
   D. vaporization of sweat.
   E. radiation and conduction.

In questions 3–8, select A if the item is associated with (a) below, B if the item is associated with (b) below, C if the item is associated with both (a) and (b), and D if the item is associated with neither (a) nor (b).

(a) V_{1A} vasopressin receptors
(b) V_2 vasopressin receptors

3. Activation of G_s
4. Vasoconstriction
5. Increase in intracellular inositol triphosphate
6. Movement of aquaporin
7. Proteinuria
8. Milk ejection

CHAPTER RESOURCES

Brunton PJ, Russell JA, Douglas AJ: Adaptive responses of the maternal hypothalamic-pituitary-adrenal axis during pregnancy and lactation. J Neuroendocrinol 2008;20:764.
Lamberts SWJ, Hofland LJ, Nobels FRE: Neuroendocrine tumor markers. Front Neuroendocrinol 2001;22:309.
Loh JA, Verbalis JG: Disorders of water and salt metabolism associated with pituitary disease. Endocrinol Metab Clin 2008;37:213.
McKinley MS, Johnson AK: The physiologic regulation of thirst and fluid intake. News Physiol Sci 2004;19:1.

CHAPTER

18

The Pituitary Gland

OBJECTIVES

After studying this chapter, you should be able to:

- Describe the structure of the pituitary gland and how it relates to its function.
- Define the cell types present in the anterior pituitary and understand how their numbers are controlled in response to physiologic demands.
- Understand the function of hormones derived from proopiomelanocortin in humans, and how they are involved in regulating pigmentation in humans, other mammals, and lower vertebrates.
- Define the effects of the growth hormone in growth and metabolic function, and how insulin-like growth factor I (IGF-I) may mediate some of its actions in the periphery.
- List the stimuli that regulate growth hormone secretion and define their underlying mechanisms.
- Understand the relevance of pituitary secretion of gonadotropins and prolactin, and how these are regulated.
- Understand the basis of conditions where pituitary function and growth hormone secretion and function are abnormal, and how they can be treated.

INTRODUCTION

The pituitary gland, or hypophysis, lies in a pocket of the sphenoid bone at the base of the brain. It is a coordinating center for control of many downstream endocrine glands, some of which are discussed in subsequent chapters. In many ways, it can be considered to consist of at least two (and in some species, three) separate endocrine organs that contain a plethora of hormonally active substances. The anterior pituitary secretes **thyroid-stimulating hormone (TSH, thyrotropin), adrenocorticotropic hormone (ACTH), luteinizing hormone (LH), follicle-stimulating hormone (FSH), prolactin,** and **growth hormone** (see Figure 17–9), and receives almost all of its blood supply from the portal hypophysial vessels that pass initially through the median eminence, a structure immediately below the hypothalamus. This vascular arrangement positions the cells of the anterior pituitary to respond efficiently to regulatory factors released from the hypothalamus. Of the listed hormones, prolactin acts on the breast. The remaining five are, at least in part, **tropic hormones;** that is, they stimulate secretion of hormonally active substances by other endocrine glands or, in the case of growth hormone, the liver and other tissues (see below). The tropic hormones for some endocrine glands are discussed in the chapter on that gland: TSH in Chapter 19; and ACTH in Chapter 20. However, the gonadotropins FSH and LH, along with prolactin, are covered here.

The posterior pituitary in mammals consists predominantly of nerves that have their cell bodies in the hypothalamus, and stores **oxytocin** and **vasopressin** in the termini of these neurons, to be released into the bloodstream. The secretion of these hormones, as well as a discussion of the overall role of the hypothalamus and median eminence in regulating both the anterior and posterior pituitary, was covered in Chapter 17. In some species, there is also a well-developed intermediate lobe of the pituitary, whereas in humans it is rudimentary. Nevertheless, the intermediate lobe, as well as the anterior pituitary, contains hormonally active derivatives of the proopiomelanocortin (POMC) molecule that regulate skin pigmentation, among other functions (see below). To avoid redundancy, this chapter will focus predominantly on growth hormone and its role in growth and facilitating the activity of other hormones, along with a number of general considerations about the pituitary. The melanocyte-stimulating hormones (MSHs) of the intermediate lobe of the pituitary, α-MSH and β-MSH, will also be touched upon.

MORPHOLOGY

GROSS ANATOMY

The anatomy of the pituitary gland is summarized in Figure 18–1 and discussed in detail in Chapter 17. The posterior pituitary is made up largely of the endings of axons from the supraoptic and paraventricular nuclei of the hypothalamus and arises initially as an extension of this structure. The anterior pituitary, on the other hand, contains endocrine cells that store its characteristic hormones and arises embryologically as an invagination of the pharynx **(Rathke's pouch).** In species where it is well developed, the intermediate lobe is formed in the embryo from the dorsal half of Rathke's pouch, but is closely adherent to the posterior lobe in the adult. It is separated from the anterior lobe by the remains of the cavity in Rathke's pouch, the **residual cleft.**

HISTOLOGY

In the posterior lobe, the endings of the supraoptic and paraventricular axons can be observed in close relation to blood vessels. **Pituicytes,** stellate cells that are modified astrocytes, are also present.

As noted above, the intermediate lobe is rudimentary in humans and a few other mammalian species. In these species, most of its cells are incorporated in the anterior lobe. Along the residual cleft are small thyroid-like follicles, some containing a little colloid (see Chapter 19). The function of the colloid, if any, is unknown.

The anterior pituitary is made up of interlacing cell cords and an extensive network of sinusoidal capillaries. The endothelium of the capillaries is fenestrated, like that in other endocrine organs. The cells contain granules of stored hormone that are extruded from the cells by exocytosis. Their constituents then enter the capillaries to be conveyed to target tissues.

CELL TYPES IN THE ANTERIOR PITUITARY

Five types of secretory cells have been identified in the anterior pituitary by immunocytochemistry and electron microscopy. The cell types are the somatotropes, which secrete growth hormone; the lactotropes (also called mammotropes), which secrete prolactin; the corticotropes, which secrete ACTH; the thyrotropes, which secrete TSH; and the gonadotropes, which secrete FSH and LH. The characteristics of these cells are summarized in Table 18–1. Some cells may contain two or more hormones. It is also notable that the three pituitary glycoprotein hormones, FSH, LH, and TSH, while being made up of two subunits, all share a common α subunit that is the product of a single gene and has the same amino acid composition in each hormone, although their carbohydrate residues vary. The α subunit must be combined with a β subunit characteristic of each hormone for maximal physiologic activity. The β subunits, which are produced by separate genes and differ in structure, confer hormonal specificity (see Chapter 16). The α subunits are remarkably interchangeable and hybrid molecules can be created. In addition, the placental glycoprotein gonadotropin human chorionic gonadotropin (hCG) has α and β subunits (see Chapter 22).

The anterior pituitary also contains folliculostellate cells that send processes between the granulated secretory

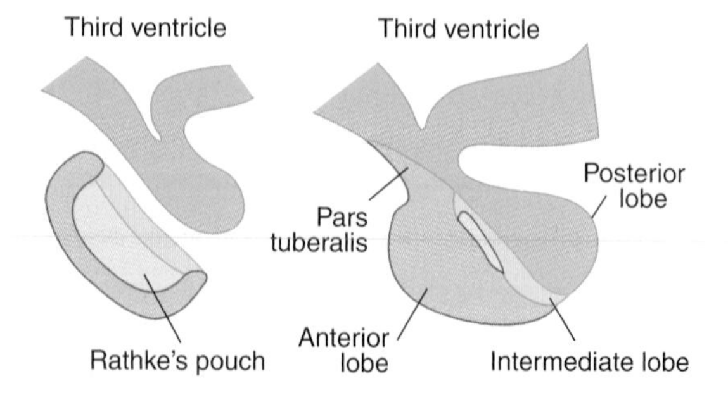

FIGURE 18–1 Diagrammatic outline of the formation of the pituitary (left) and the various parts of the organ in the adult (right).

TABLE 18–1 Hormone-secreting cells of the human anterior pituitary gland.

Cell Type	Hormones Secreted	Percentage of Total Secretory Cells
Somatotrope	Growth hormone	50
Lactotrope	Prolactin	10–30
Corticotrope	ACTH	10
Thyrotrope	TSH	5
Gonadotrope	FSH, LH	20

cells. These cells produce paracrine factors that regulate the growth and function of the secretory cells discussed above. Indeed, the anterior pituitary can adjust the relative proportion of secretory cell types to meet varying requirements for different hormones at different life stages. This plasticity has recently been ascribed to the presence of a small number of pluripotent stem cells that persist in the adult gland.

PROOPIOMELANOCORTIN & DERIVATIVES

BIOSYNTHESIS

Intermediate-lobe cells, if present, and corticotropes of the anterior lobe both synthesize a large precursor protein that is cleaved to form a family of hormones. After removal of the signal peptide, this prohormone is known as **proopiomelanocortin (POMC).** This molecule is also synthesized in the hypothalamus, the lungs, the gastrointestinal tract, and the placenta. The structure of POMC, as well as its derivatives, is shown in Figure 18–2. In corticotropes, it is hydrolyzed to ACTH and β-lipotropin (LPH), plus a small amount of β-endorphin, and these substances are secreted. In the intermediate lobe cells, POMC is hydrolyzed to corticotropin-like intermediate-lobe peptide (CLIP), γ-LPH, and appreciable quantities of β-endorphin. The functions, if any, of CLIP and γ-LPH are unknown, whereas β-endorphin is an opioid peptide (see Chapter 7) that has the five amino acid residues of met-enkephalin at its amino terminal end. The **melanotropins** α- and β-MSH are also formed. However, the intermediate lobe in humans is rudimentary, and it appears that neither α-MSH nor β-MSH is secreted in adults. In some species, however, the melanotropins have important physiological functions, as discussed below.

CONTROL OF SKIN COLORATION & PIGMENT ABNORMALITIES

Fish, reptiles, and amphibia change the color of their skin for thermoregulation, camouflage, and behavioral displays. They do this in part by moving black or brown granules into or out of the periphery of pigment cells called **melanophores.** The granules are made up of **melanins,** which are synthesized from dopamine (see Chapter 7) and dopaquinone. The movement of these granules is controlled by a variety of hormones and neurotransmitters, including α- and β-MSH, melanin-concentrating hormone, melatonin, and catecholamines.

Mammals have no melanophores containing pigment granules that disperse and aggregate, but they do have **melanocytes,** which have multiple processes containing melanin granules. Melanocytes express **melanotropin-1** receptors. Treatment with MSHs accelerates melanin synthesis, causing readily detectable darkening of the skin in humans in 24 h. As noted above, α- and β-MSH do not circulate in adult humans, and their function is unknown. However, ACTH binds to melanotropin-1 receptors. Indeed, the pigmentary changes in several human endocrine diseases are due to changes in circulating ACTH. For example, abnormal pallor is a hallmark of hypopituitarism. Hyperpigmentation occurs in patients with adrenal insufficiency due to primary adrenal disease. Indeed, the presence of hyperpigmentation in association with adrenal insufficiency rules out the possibility that the insufficiency is secondary to pituitary or hypothalamic disease because in these conditions, plasma ACTH is not increased (see Chapter 20). Other disorders of pigmentation result from peripheral mechanisms. Thus, **albinos** have a congenital inability to synthesize melanin. This can result from a variety of different genetic defects in the pathways for melanin synthesis. **Piebaldism** is characterized by patches of skin that lack melanin as a result of congenital defects in the migration of pigment cell precursors from the

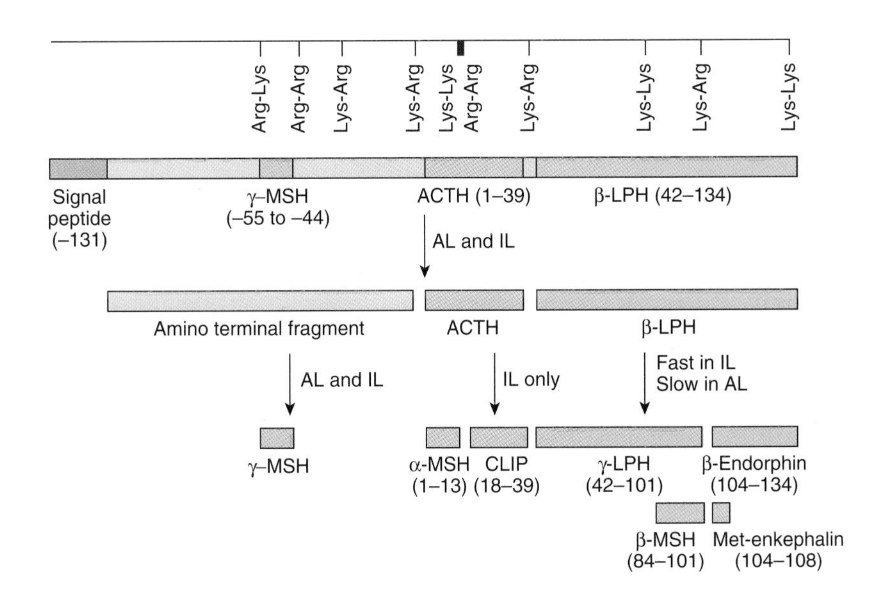

FIGURE 18–2 Schematic representation of the preproopiomelanocortin molecule formed in pituitary cells, neurons, and other tissues. The numbers in parentheses identify the amino acid sequences in each of the polypeptide fragments. For convenience, the amino acid sequences are numbered from the amino terminal of ACTH and read toward the carboxyl terminal portion of the parent molecule, whereas the amino acid sequences in the other portion of the molecule read to the left to - 131, the amino terminal of the parent molecule. The locations of Lys–Arg and other pairs of basic amino acids residues are also indicated; these are the sites of proteolytic cleavage in the formation of the smaller fragments of the parent molecule. AL, anterior lobe; IL, intermediate lobe.

neural crest during embryonic development. Not only the condition but also the precise pattern of the loss is passed from one generation to the next. **Vitiligo** involves a similar patchy loss of melanin, but the loss develops progressively after birth secondary to an autoimmune process that targets melanocytes.

GROWTH HORMONE

BIOSYNTHESIS & CHEMISTRY

The long arm of human chromosome 17 contains the growth hormone-hCS cluster that contains five genes: one, *hGH-N*, codes for the most abundant ("normal") form of growth hormone; a second, *hGH-V*, codes for the variant form of growth hormone (see below); two code for human chorionic somatomammotropin (hCS) (see Chapter 22); and the fifth is probably an hCS pseudogene.

Growth hormone that is secreted into the circulation by the pituitary gland consists of a complex mixture of hGH-N, peptides derived from this molecule with varying degrees of posttranslational modifications, such as glycosylation, and a splice variant of hGH-N that lacks amino acids 32–46. The physiologic significance of this complex array of hormones has yet to be fully understood, particularly since their structural similarities make it difficult to assay for each species separately. Nevertheless, there is emerging evidence that, while the various peptides share a broad range of functions, they may occasionally exert actions in opposition to one another. hGH-V and hCS, on the other hand, are primarily products of the placenta, and as a consequence are only found in appreciable quantities in the circulation during pregnancy (see Chapter 22).

SPECIES SPECIFICITY

The structure of growth hormone varies considerably from one species to another. Porcine and simian growth hormones have only a transient effect in the guinea pig. In monkeys and humans, bovine and porcine growth hormones do not even have a transient effect on growth, although monkey and human growth hormones are fully active in both monkeys and humans. These facts are relevant to public health discussions surrounding the presence of bovine growth hormones (used to increase milk production) in dairy products, as well as the popularity of growth hormone supplements, marketed via the Internet, with body builders. Controversially, recombinant human growth hormone has also been given to children who are short in stature, but otherwise healthy (ie, without growth hormone deficiency), with apparently limited results.

PLASMA LEVELS, BINDING, & METABOLISM

A portion of circulating growth hormone is bound to a plasma protein that is a large fragment of the extracellular domain of the growth hormone receptor (see below). It appears to be produced by cleavage of receptors in humans, and its concentration is an index of the number of growth hormone receptors in the tissues. Approximately 50% of the circulating pool of growth hormone activity is in the bound form, providing a reservoir of the hormone to compensate for the wide fluctuations that occur in secretion (see below).

The basal plasma growth hormone level measured by radioimmunoassay in adult humans is normally less than 3 ng/mL. This represents both the protein-bound and free forms. Growth hormone is metabolized rapidly, at least in part in the liver. The half-life of circulating growth hormone in humans is 6–20 min, and the daily growth hormone output has been calculated to be 0.2–1.0 mg/d in adults.

GROWTH HORMONE RECEPTORS

The growth hormone receptor is a 620-amino-acid protein with a large extracellular portion, a transmembrane domain, and a large cytoplasmic portion. It is a member of the cytokine receptor superfamily, which is discussed in Chapter 3. Growth hormone has two domains that can bind to its receptor, and when it binds to one receptor, the second binding site attracts another, producing a homodimer (Figure 18–3). Dimerization is essential for receptor activation.

Growth hormone has widespread effects in the body (see below), so even though it is not yet possible precisely to correlate intracellular and whole body effects, it is not surprising that, like insulin, growth hormone activates many different intracellular signaling cascades (Figure 18–3). Of particular note is its activation of the JAK2–STAT pathway. JAK2 is a member of the Janus family of cytoplasmic tyrosine kinases. STATs (for signal transducers and activators of transcription) are a family of cytoplasmic transcription factors that, upon phosphorylation by JAK kinases, migrate to the nucleus where they activate various genes. JAK–STAT pathways are known also to mediate the effects of prolactin and various other growth factors.

EFFECTS ON GROWTH

In young animals in which the epiphyses have not yet fused to the long bones (see Chapter 21), growth is inhibited by hypophysectomy and stimulated by growth hormone. Chondrogenesis is accelerated, and as the cartilaginous epiphysial plates widen, they lay down more bone matrix at the ends of long bones. In this way, stature is increased. Prolonged treatment of animals with growth hormone leads to gigantism.

When the epiphyses are closed, linear growth is no longer possible. In this case, an overabundance of growth hormone produces the pattern of bone and soft tissue deformities known in humans as **acromegaly.** The sizes of most of the viscera are increased. The protein content of the body is increased, and the fat content is decreased (see Clinical Box 18–1).

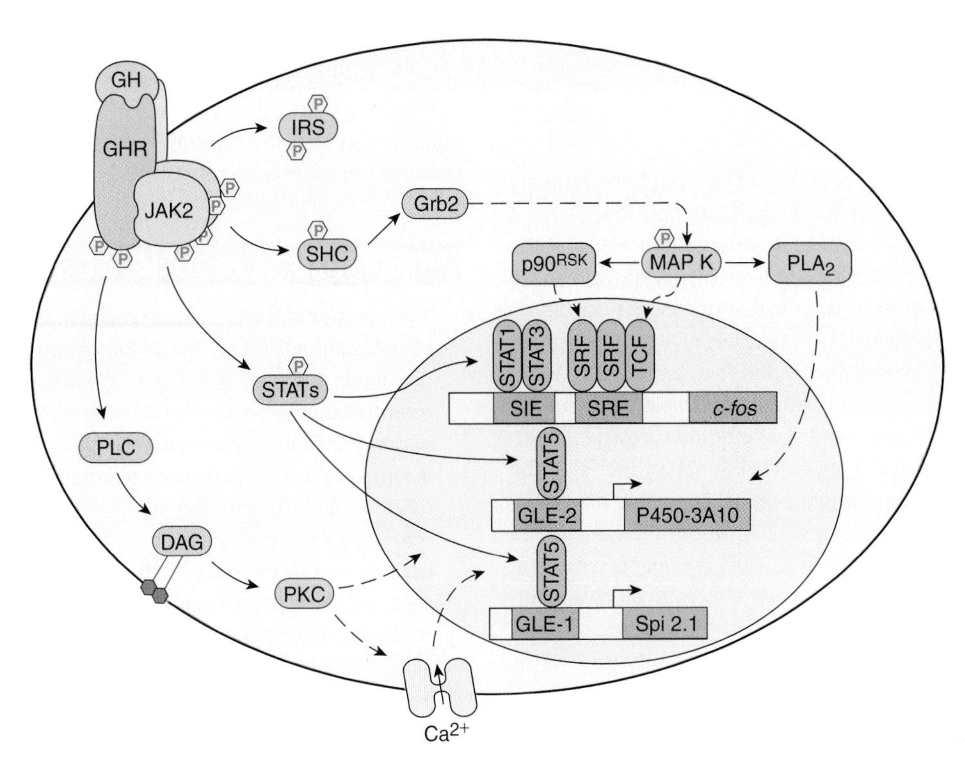

FIGURE 18–3 Some of the principal signaling pathways activated by the dimerized growth hormone receptor (GHR). Solid arrows indicate established pathways; dashed arrows indicate probable pathways. The details of the PLC pathway and the pathway from Grb2 to MAP K are discussed in Chapter 2. The small uppercase letter P's in yellow hexagons represent phosphorylation of the factor indicated. GLE-1 and GLE-2, interferon γ-activated response elements; IRS, insulin receptor substrate; $p90^{RSK}$, an S6 kinase; PLA_2, phospholipase A_2; SIE, Sis-induced element; SRE, serum response element; SRF, serum response factor; TCF, ternary complex factor.

EFFECTS ON PROTEIN & ELECTROLYTE HOMEOSTASIS

Growth hormone is a protein anabolic hormone and produces a positive nitrogen and phosphorus balance, a rise in plasma phosphorus, and a fall in blood urea nitrogen and amino acid levels. In adults with growth hormone deficiency, recombinant human growth hormone produces an increase in lean body mass and a decrease in body fat, along with an increase in metabolic rate and a fall in plasma cholesterol. Gastrointestinal absorption of Ca^{2+} is increased. Na^+ and K^+ excretion is reduced by an action independent of the adrenal glands, probably because these electrolytes are diverted from the kidneys to the growing tissues. On the other hand, excretion of the amino acid 4-hydroxyproline is increased during this growth, reflective of the ability of growth hormone to stimulate the synthesis of soluble collagen.

EFFECTS ON CARBOHYDRATE & FAT METABOLISM

The actions of growth hormone on carbohydrate metabolism are discussed in Chapter 24. At least some forms of growth hormone are diabetogenic because they increase hepatic glucose output and exert an anti-insulin effect in muscle. Growth hormone is also ketogenic and increases circulating free fatty acid (FFA) levels. The increase in plasma FFA, which takes several hours to develop, provides a ready source of energy for the tissues during hypoglycemia, fasting, and stressful stimuli. Growth hormone does not stimulate β cells of the pancreas directly, but it increases the ability of the pancreas to respond to insulinogenic stimuli such as arginine and glucose. This is an additional way growth hormone promotes growth, since insulin has a protein anabolic effect (see Chapter 24).

SOMATOMEDINS

The effects of growth hormone on growth, cartilage, and protein metabolism depend on an interaction between growth hormone and **somatomedins,** which are polypeptide growth factors secreted by the liver and other tissues. The first of these factors isolated was called sulfation factor because it stimulated the incorporation of sulfate into cartilage. However, it also stimulated collagen formation, and its name was changed to somatomedin. It then became clear that there are a variety of different somatomedins and that they are members of an increasingly large family of **growth factors** that affect many different tissues and organs.

The principal (and in humans probably the only) circulating somatomedins are **insulin-like growth factor I (IGF-I, somatomedin C)** and **IGF-II.** These factors are closely related to insulin, except that their C chains are not

CLINICAL BOX 18–1

Gigantism & Acromegaly

Tumors of the somatotropes of the anterior pituitary (pituitary adenoma) secrete large amounts of growth hormone, leading to **gigantism** in children and to **acromegaly** in adults. If the tumor arises before puberty, the individual may grow to an extraordinary height. After linear growth is no longer possible, on the other hand, the characteristic features of acromegaly arise, including greatly enlarged hands and feet, vertebral changes attributable to osteoarthritis, soft tissue swelling, hirsutism, and protrusion of the brow and jaw. Abnormal growth of internal organs may eventually impair their function such that the condition, which has an insidious onset, can prove fatal if left untreated. Hypersecretion of growth hormone is accompanied by hypersecretion of prolactin in 20–40% of patients with acromegaly. About 25% of patients have abnormal glucose tolerance tests, and 4% develop lactation in the absence of pregnancy. Acromegaly can be caused by extra-pituitary as well as intrapituitary growth hormone-secreting tumors and by hypothalamic tumors that secrete GHRH, but the latter are rare.

THERAPEUTIC HIGHLIGHTS

The mainstay of therapy for acromegaly remains the use of somatostatin analogues that inhibit the secretion of growth hormone. A growth hormone receptor antagonist has recently become available and has been found to reduce plasma IGF-I and produce clinical improvement in cases of acromegaly that fail to respond to other treatments. Surgical removal of the pituitary tumor is also helpful in both acromegaly and gigantism, but sometimes challenging to perform due to the tumor's often invasive nature. In any case, adjuvant pharmacological therapy must often be continued after surgery to control ongoing symptoms.

separated (Figure 18–4) and they have an extension of the A chain called the D domain. The hormone relaxin (see Chapter 22) is also a member of this family. Humans have two related relaxin isoforms, and both resemble IGF-II. In humans a variant form of IGF-I lacking three amino terminal amino acid residues has been found in the brain, and there are several variant forms of human IGF-II (Figure 18–4). The mRNAs for IGF-I and IGF-II are found in the liver, in

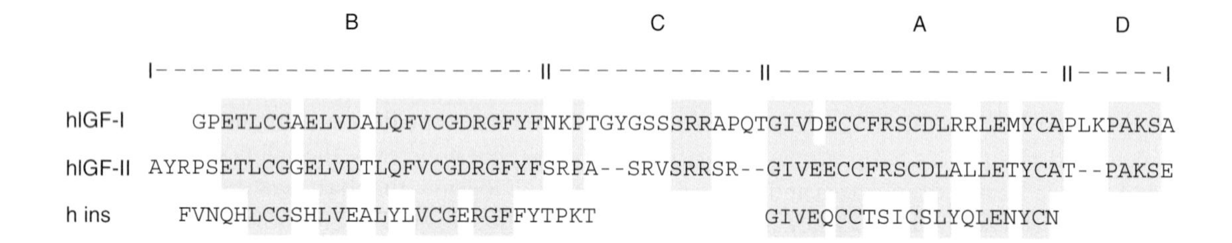

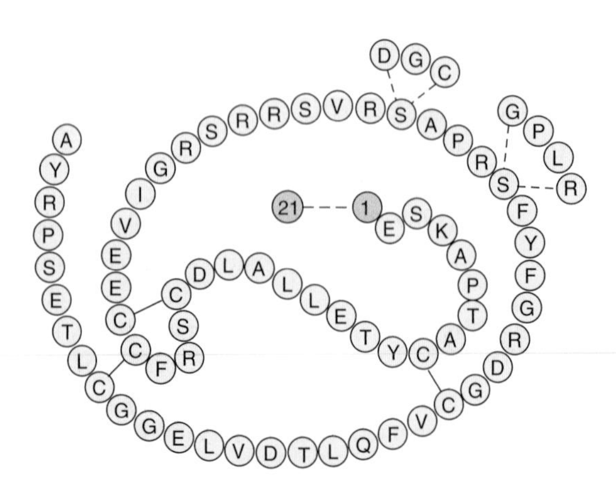

FIGURE 18–4 Structure of human IGF-I, IGF-II, and insulin (ins) (top). The lower panel shows the structure of human IGF-II with its disulfide bonds, as well as three variant structures shown: a 21-aa extension of the C-terminus, a tetrapeptide substitution at Ser-29, and a tripeptide substitution of Ser-33.

TABLE 18–2 Comparison of insulin and the insulin-like growth factors.

	Insulin	IGF-I	IGF-II
Other names	...	Somatomedin C	Multiplication-stimulating activity (MSA)
Number of amino acids	51	70	67
Source	Pancreatic B cells	Liver and other tissues	Diverse tissues
Level regulated by	Glucose	Growth hormone after birth, nutritional status	Unknown
Plasma levels	0.3–2 ng/mL	10–700 ng/mL; peaks at puberty	300–800 ng/mL
Plasma-binding proteins	No	Yes	Yes
Major physiologic role	Control of metabolism	Skeletal and cartilage growth	Growth during fetal development

cartilage, and in many other tissues, indicating that they are likely synthesized in these tissues.

The properties of IGF-I, IGF-II, and insulin are compared in Table 18–2. Both IGF-I and IGF-II are tightly bound to proteins in the plasma, and, at least for IGF-I, this prolongs their half-life in the circulation. Six different IGF-binding proteins, with different patterns of distribution in various tissues, have been identified. All are present in plasma, with IGF-binding protein-3 (IGFBP-3) accounting for 95% of the binding in the circulation. The contribution of the IGFs to the insulin-like activity in blood is discussed in Chapter 24. The IGF-I receptor is very similar to the insulin receptor and probably uses similar or identical intracellular signaling pathways. The IGF-II receptor has a distinct structure (see Figure 24–5) and is involved in the intracellular targeting of acid hydrolases and other proteins to intracellular organelles. Secretion of IGF-I is independent of growth hormone before birth but is stimulated by growth hormone after birth, and it has pronounced growth-stimulating activity. Its concentration in plasma rises during childhood and peaks at the time of puberty, then declines to low levels in old age. IGF-II is largely independent of growth hormone and plays a role in the growth of the fetus before birth. In human fetuses in which it is overexpressed, several organs, especially the tongue, other muscles, kidneys, heart, and liver, develop out of proportion to the rest of the body. In adults, the gene for IGF-II is expressed only in the choroid plexus and meninges.

DIRECT & INDIRECT ACTIONS OF GROWTH HORMONE

Our understanding of the mechanism of action of growth hormone has evolved. It was originally thought to produce growth by a direct action on tissues, and then later was believed to act solely through its ability to induce somatomedins. However, if growth hormone is injected into one proximal tibial epiphysis, a unilateral increase in cartilage width is produced, and cartilage, like other tissues, makes IGF-I. A current hypothesis to explain these results holds that growth hormone acts on cartilage to convert stem cells into cells that respond to IGF-I. Locally produced as well as circulating IGF-I then makes the cartilage grow. However, the independent role of circulating IGF-I remains important, since infusion of IGF-I in hypophysectomized rats restores bone and body growth. Overall, it seems that growth hormone and somatomedins can act both in cooperation and independently to stimulate pathways that lead to growth. The situation is almost certainly complicated further by the existence of multiple forms of growth hormone in the circulation that can, in some situations, have opposing actions.

Figure 18–5 is a summary of current views of the other actions of growth hormone and IGF-I. However, growth hormone probably combines with circulating and locally produced IGF-I in various proportions to produce at least some of the latter effects.

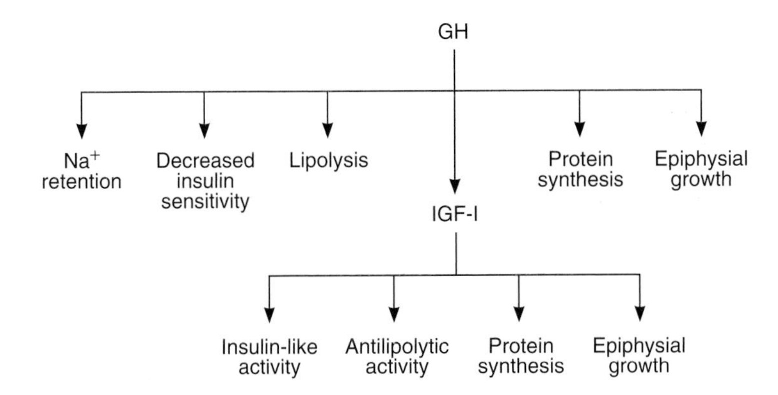

FIGURE 18–5 Direct and indirect actions of growth hormone (GH). The latter are mediated by the ability of growth hormone to induce production of IGF-I. (Courtesy of R Clark and N Gesundheit.)

HYPOTHALAMIC & PERIPHERAL CONTROL OF GROWTH HORMONE SECRETION

The secretion of growth hormone is not stable over time. Adolescents have the highest circulating levels of growth hormone, followed by children and finally adults. Levels decline in old age, and there has been considerable interest in injecting growth hormone to counterbalance the effects of aging. The hormone increases lean body mass and decreases body fat, but it does not produce statistically significant increases in muscle strength or mental status. There are also diurnal variations in growth hormone secretion superimposed on these developmental stages. Growth hormone is found at relatively low levels during the day, unless specific triggers for its release are present (see below). During sleep, on the other hand, large pulsatile bursts of growth hormone secretion occur. Therefore, it is not surprising that the secretion of growth hormone is under hypothalamic control. The hypothalamus controls growth hormone production by secreting growth hormone-releasing hormone (GHRH) as well as somatostatin, which inhibits growth hormone release (see Chapter 17). Thus, the balance between the effects of these hypothalamic factors on the pituitary will determine the level of growth hormone release. The stimuli of growth hormone secretion can therefore act by increasing hypothalamic secretion of GHRH, decreasing secretion of somatostatin, or both. A third regulator of growth hormone secretion is **ghrelin.** The main site of ghrelin synthesis and secretion is the stomach, but it is also produced in the hypothalamus and has marked growth hormone-stimulating activity. In addition, it appears to be involved in the regulation of food intake (see Chapter 26).

Growth hormone secretion is under feedback control (see Chapter 16), like the secretion of other anterior pituitary hormones. It acts on the hypothalamus to antagonize GHRH release. Growth hormone also increases circulating IGF-I, and IGF-I in turn exerts a direct inhibitory action on growth hormone secretion from the pituitary. It also stimulates somatostatin secretion (Figure 18–6).

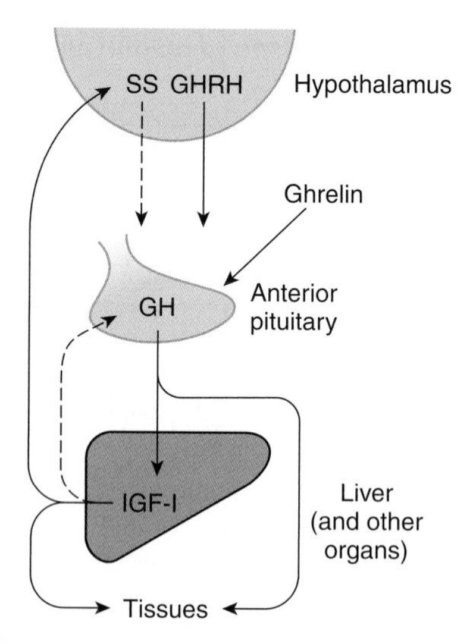

FIGURE 18–6 Feedback control of growth hormone secretion. Solid arrows represent positive effects and dashed arrows represent inhibition. GH, growth hormone; GHRH, growth hormone releasing hormone; IGF-I, insulin-like growth factor-I; SS, somatostatin.

Stimuli Affecting Growth Hormone Secretion

The basal plasma growth hormone concentration ranges from 0 to 3 ng/mL in normal adults. However, secretory rates cannot be estimated from single values because of their irregular nature. Thus, average values over 24 h (see below) and peak values may be more meaningful, albeit difficult to assess in the clinical setting. The stimuli that increase growth hormone secretion are summarized in Table 18–3. Most of them fall into three general categories: (1) conditions such as hypoglycemia and/or fasting in which there is an actual or threatened decrease in the substrate for energy production in cells, (2) conditions in which the amounts of certain amino acids are increased in the plasma, and (3) stressful stimuli. The response to glucagon has been used as a test of growth hormone reserve. Growth hormone secretion is also increased in subjects deprived of rapid eye movement

TABLE 18–3 Stimuli that affect growth hormone secretion in humans.

Stimuli that increase secretion
Hypoglycemia
2-Deoxyglucose
Exercise
Fasting
Increase in circulating levels of certain amino acids
Protein meal
Infusion of arginine and some other amino acids
Glucagon
Stressful stimuli
Pyrogen
Lysine vasopressin
Various psychologic stresses
Going to sleep
L-Dopa and α-adrenergic agonists that penetrate the brain
Apomorphine and other dopamine receptor agonists
Estrogens and androgens
Stimuli that decrease secretion
REM sleep
Glucose
Cortisol
FFA
Medroxyprogesterone
Growth hormone and IGF-I

(REM) sleep (see Chapter 14) and inhibited during normal REM sleep.

Glucose infusions lower plasma growth hormone levels and inhibit the response to exercise. The increase produced by 2-deoxyglucose is presumably due to intracellular glucose deficiency, since this compound blocks the catabolism of glucose 6-phosphate. Sex hormones induce growth hormone secretion, increase growth hormone responses to provocative stimuli such as arginine and insulin, and also serve as permissive factors for the action of growth hormone in the periphery. This likely contributes to the relatively high levels of circulating growth hormone and associated growth spurt in puberty. Growth hormone secretion is also induced by thyroid hormones. Growth hormone secretion is inhibited, on the other hand, by cortisol, FFA, and medroxyprogesterone.

Growth hormone secretion is increased by L-dopa, which increases the release of dopamine and norepinephrine in the brain, and by the dopamine receptor agonist apomorphine.

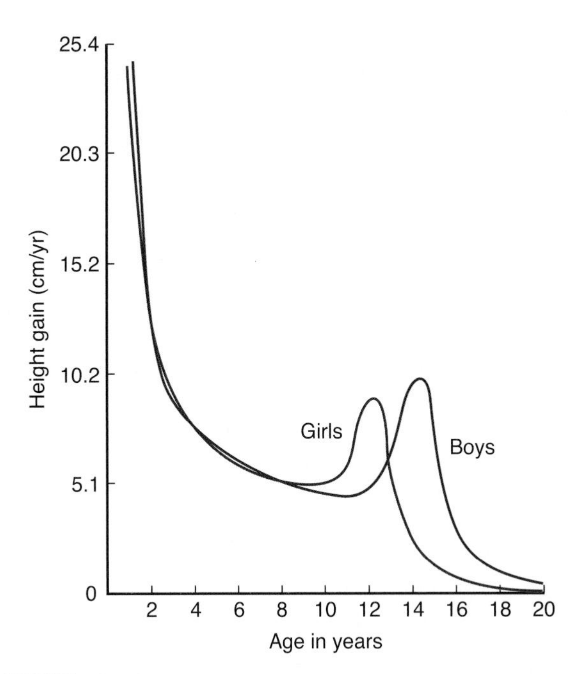

FIGURE 18–7 Rate of growth in boys and girls from birth to age 20.

PHYSIOLOGY OF GROWTH

Growth hormone, while being essentially unimportant for fetal development, is the most important hormone for postnatal growth. However, growth overall is a complex phenomenon that is affected not only by growth hormone and somatomedins, but also, as would be predicted by the previous discussion, by thyroid hormones, androgens, estrogens, glucocorticoids, and insulin. It is also affected, of course, by genetic factors, and it depends on adequate nutrition. It is normally accompanied by an orderly sequence of maturational changes, and it involves accretion of protein and an increase in length and size, not just an increase in weight (which could reflect the formation of fat or retention of salt and water rather than growth per se).

ROLE OF NUTRITION

The food supply is the most important extrinsic factor affecting growth. The diet must be adequate not only in protein content but also in essential vitamins and minerals (see Chapter 26) and in calories, so that ingested protein is not burned for energy. However, the age at which a dietary deficiency occurs appears to be an important consideration. For example, once the pubertal growth spurt has commenced, considerable linear growth continues even if caloric intake is reduced. Injury and disease, on the other hand, stunt growth because they increase protein catabolism.

GROWTH PERIODS

Patterns of growth vary somewhat from species to species. Rats continue to grow, although at a declining rate, throughout life. In humans, two periods of rapid growth occur (Figure 18–7): the first in infancy and the second in late puberty just before growth stops. The first period of accelerated growth is partly a continuation of the fetal growth period. The second growth spurt, at the time of puberty, is due to growth hormone, androgens, and estrogens, and the subsequent cessation of growth is due in large part to closure of the epiphyses in the long bones by estrogens (see Chapter 21). After this time, further increases in height are not possible. Because girls mature earlier than boys, this growth spurt appears earlier in girls. Of course, in both sexes the rate of growth of individual tissues varies (Figure 18–8).

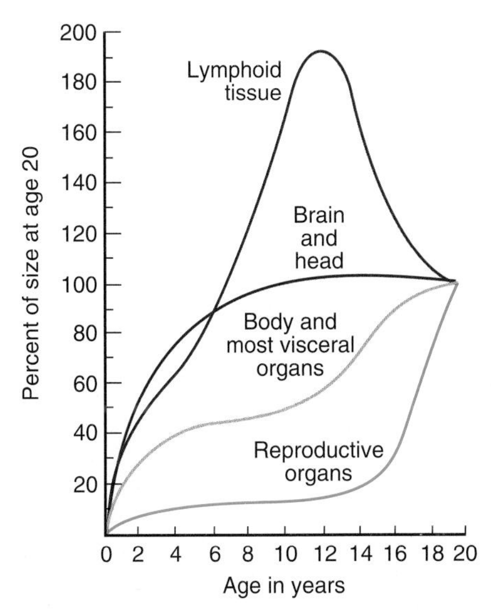

FIGURE 18–8 Growth of different tissues at various ages as a percentage of size at age 20. The curves are composites that include data for both boys and girls.

It is interesting that at least during infancy, growth is not a continuous process but is episodic or saltatory. Increases in length of human infants of 0.5–2.5 cm in a few days are separated by periods of 2–63 days during which no measurable growth can be detected. The cause of the episodic growth is unknown.

HORMONAL EFFECTS

The contributions of hormones to growth after birth are shown diagrammatically in Figure 18–9. Plasma growth hormone is elevated in newborns. Subsequently, average resting levels fall but the spikes of growth hormone secretion are larger, especially during puberty, so the mean plasma level over 24 h is increased; it is 2–4 ng/mL in normal adults, but 5–8 ng/mL in children. One of the factors stimulating IGF-I secretion is growth hormone, and plasma IGF-I levels rise during childhood, reaching a peak at 13–17 years of age. In contrast, IGF-II levels are constant throughout postnatal growth.

The growth spurt that occurs at the time of puberty (Figure 18–7) is due in part to the protein anabolic effect of androgens, and the secretion of adrenal androgens increases at this time in both sexes; however, it is also due to an interaction among sex steroids, growth hormone, and IGF-I. Treatment with estrogens and androgens increases the secretion of growth hormone in response to various stimuli and increases plasma IGF-I secondary to this increase in circulating growth hormone. This, in turn, causes growth.

Although androgens and estrogens initially stimulate growth, estrogens ultimately terminate growth by causing the epiphyses to fuse to the long bones (epiphysial closure). Once the epiphyses have closed, linear growth ceases (see Chapter 21). This is why patients with sexual precocity are apt to be dwarfed. On the other hand, men who were castrated before puberty tend to be tall because their estrogen production is decreased and their epiphyses remain open, allowing some growth to continue past the normal age of puberty.

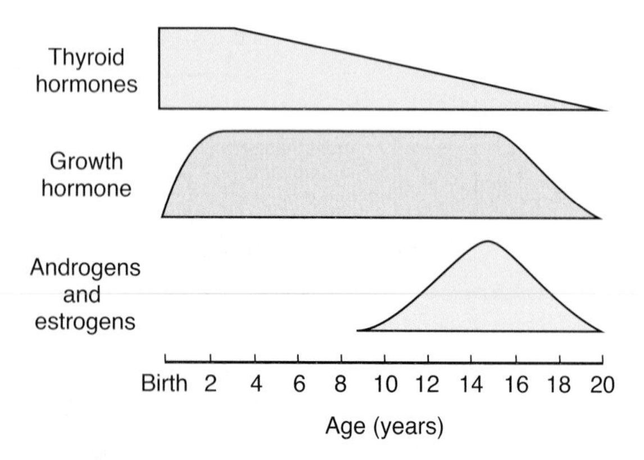

FIGURE 18–9 Relative importance of hormones in human growth at various ages. (Courtesy of Fisher DA.)

In hypophysectomized animals, growth hormone increases growth but this effect is potentiated by thyroid hormones, which by themselves have no effect on growth. The action of thyroid hormones in this situation is therefore permissive to that of growth hormone, possibly via potentiation of the actions of somatomedins. Thyroid hormones also often appear to be necessary for the normal rate of growth hormone secretion; basal growth hormone levels are normal in hypothyroidism, but the response to hypoglycemia is frequently blunted. Thyroid hormones have widespread effects on the ossification of cartilage, the growth of teeth, the contours of the face, and the proportions of the body. Hypothyroid dwarfs (also known as **cretins**) therefore have infantile features (Figure 18–10). Patients who are dwarfed because of panhypopituitarism have features consistent with their chronologic age until puberty, but since they do not mature sexually, they have juvenile features in adulthood (Clinical Box 18–2).

The effect of insulin on growth is discussed in Chapter 24. Diabetic animals fail to grow, and insulin causes growth in hypophysectomized animals. However, the growth is appreciable only when large amounts of carbohydrate and protein are supplied with the insulin.

Adrenocortical hormones other than androgens exert a permissive action on growth in the sense that adrenalectomized animals fail to grow unless their blood pressures and circulations are maintained by replacement therapy. On the other hand, glucocorticoids are potent inhibitors of growth because of their direct action on cells, and treatment of children with pharmacologic doses of steroids slows or stops growth for as long as the treatment is continued.

CATCH-UP GROWTH

Following illness or starvation in children, a period of **catch-up growth** (Figure 18–11) takes place during which the growth rate is greater than normal. The accelerated growth usually continues until the previous growth curve is reached, then slows to normal. The mechanisms that bring about and control catch-up growth are unknown.

PITUITARY GONADOTROPINS & PROLACTIN

CHEMISTRY

FSH and LH are each made up of an α and a β subunit. They are glycoproteins that contain the hexoses mannose and galactose, the hexosamines N-acetylgalactosamine and N-acetylglycosamine, and the methylpentose fucose. They also contain sialic acid. The carbohydrate in the gonadotropin molecules increases their potency by markedly slowing their metabolism. The half-life of human FSH is about 170 min; the half-life of LH is about 60 min. Loss-of-function

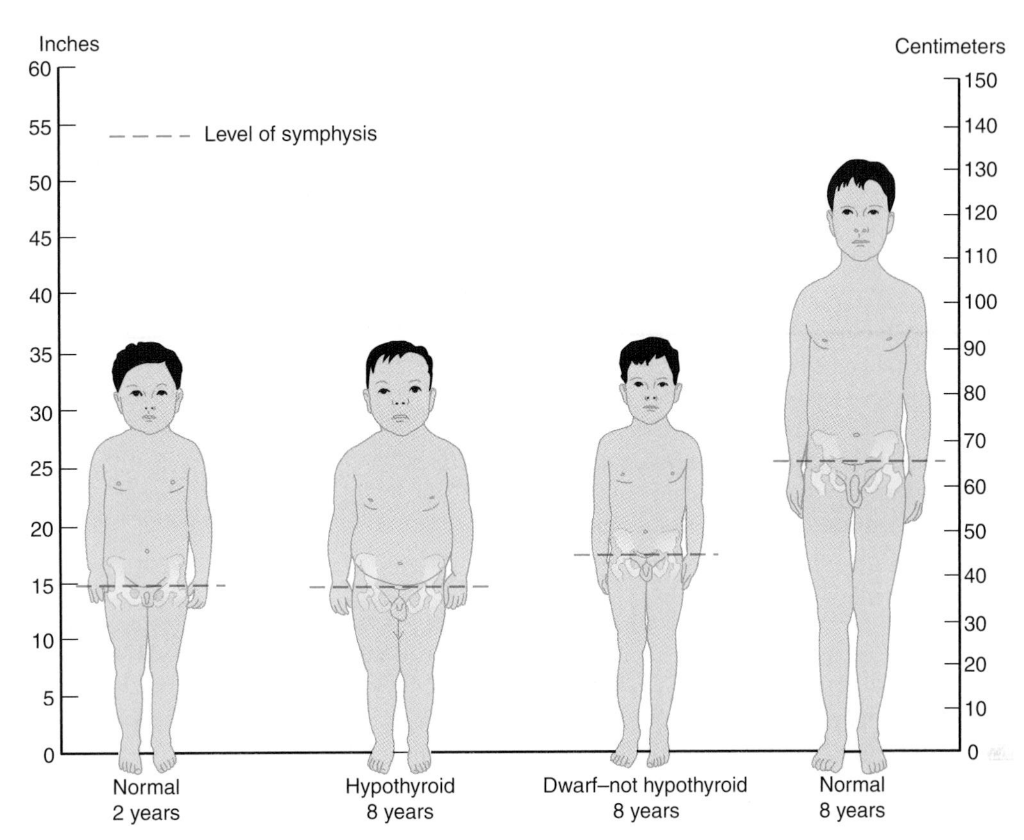

FIGURE 18–10 Normal and abnormal growth. Hypothyroid dwarfs (cretins) retain their infantile proportions, whereas dwarfs of the constitutional type and, to a lesser extent, of the hypopituitary type have proportions characteristic of their chronologic age. See also Clinical Box 18–2. (Reproduced, with permission, from Wilkins L: *The Diagnosis and Treatment of Endocrine Disorders in Childhood and Adolescence*, 3rd ed. Thomas, 1966.)

mutations in the FSH receptor cause hypogonadism. Gain-of-function mutations cause a spontaneous form of **ovarian hyperstimulation syndrome,** a condition in which many follicles are stimulated and cytokines are released from the ovary, causing increased vascular permeability and shock.

Human pituitary prolactin contains 199 amino acid residues and three disulfide bridges and has considerable structural similarity to human growth hormone and human chorionic somatomammotropin (hCS). The half-life of prolactin, like that of growth hormone, is about 20 min. Structurally similar prolactins are secreted by the endometrium and by the placenta.

RECEPTORS

The receptors for FSH and LH are G-protein coupled receptors coupled to adenylyl cyclase through a stimulatory G protein (G_s; see Chapter 2). In addition, each has an extended, glycosylated extracellular domain.

The human prolactin receptor resembles the growth hormone receptor and is one of the superfamily of receptors that includes the growth hormone receptor and receptors for many cytokines and hematopoietic growth factors (see Chapters 2 and 3). It dimerizes and activates the Janus kinase/signal transducers and activators of transcription (JAK–STAT) pathway and other intracellular enzyme cascades (Figure 18–3).

ACTIONS

The testes and ovaries become atrophic when the pituitary is removed or destroyed. The actions of prolactin and the gonadotropins FSH and LH, as well as those of the gonadotropin secreted by the placenta, are described in detail in Chapters 22 and 23. In brief, FSH helps maintain the spermatogenic epithelium by stimulating Sertoli cells in the male and is responsible for the early growth of ovarian follicles in the female. LH is tropic for the Leydig cells and, in females, is responsible for the final maturation of the ovarian follicles and estrogen secretion from them. It is also responsible for ovulation, the initial formation of the corpus luteum, and secretion of progesterone.

Prolactin causes milk secretion from the breast after estrogen and progesterone priming. Its effect on the breast involves increasing mRNA levels and subsequent production of casein and lactalbumin. However, the action of the hormone is not exerted on the cell nucleus and is prevented by inhibitors of microtubules. Prolactin also inhibits the effects of

CLINICAL BOX 18–2

Dwarfism

The accompanying discussion of growth control should suggest several possible etiologies of short stature. It can be due to GHRH deficiency, growth hormone deficiency, or deficient secretion of IGF-I. Isolated growth hormone deficiency is often due to GHRH deficiency, and in these instances, the growth hormone response to GHRH is normal. However, some patients with isolated growth hormone deficiency have abnormalities of their growth hormone secreting cells. In another group of dwarfed children, the plasma growth hormone concentration is normal or elevated but their growth hormone receptors are unresponsive as a result of loss-of-function mutations. The resulting condition is known as **growth hormone insensitivity or Laron dwarfism.** Plasma IGF-I is markedly reduced, along with IGFBP 3, which is also growth hormone-dependent. African pygmies have normal plasma growth hormone levels and a modest reduction in the plasma level of growth hormone-binding protein. However, their plasma IGF-I concentration fails to increase at the time of puberty and they experience less growth than nonpygmy controls throughout the prepubertal period.

Short stature may also be caused by mechanisms independent of specific defects in the growth hormone axis. It is characteristic of childhood hypothyroidism (cretinism) and occurs in patients with precocious puberty. It is also part of the syndrome of **gonadal dysgenesis** seen in patients who have an XO chromosomal pattern instead of an XX or XY pattern (see Chapter 22). Various bone and metabolic diseases also cause stunted growth, and in many cases there is no known cause ("constitutional delayed growth"). Chronic abuse and neglect can also cause dwarfism in children, independent of malnutrition. This condition is known as **psychosocial dwarfism** or the **Kaspar Hauser syndrome,** named for the patient with the first reported case. Finally, **achondroplasia,** the most common form of dwarfism in humans, is characterized by short limbs with a normal trunk. It is an autosomal dominant condition caused by a mutation in the gene that codes for **fibroblast growth factor receptor 3 (FGFR3).** This member of the fibroblast growth receptor family is normally expressed in cartilage and the brain.

THERAPEUTIC HIGHLIGHTS

The treatment of dwarfism is dictated by its underlying cause. If treatment to replace the relevant hormone is commenced promptly in appropriate childhood cases, almost normal stature can often be attained. Thus, the availability of recombinant forms of growth hormone and IGF-I has greatly improved treatment in cases where these hormones are deficient.

gonadotropins, possibly by an action at the level of the ovary. It also prevents ovulation in lactating women. The function of prolactin in normal males is unsettled, but excess prolactin secreted by tumors causes impotence.

REGULATION OF PROLACTIN SECRETION

The regulatory factors for prolactin secretion by the pituitary overlap, in part, with those causing secretion of growth hormone, but there are important differences, and some stimuli increase prolactin secretion while decreasing that of growth hormone (and vice versa) (Table 18–4). The normal plasma prolactin concentration is approximately 5 ng/mL in men and 8 ng/mL in women. Secretion is tonically inhibited by the hypothalamus, and section of the pituitary stalk leads to an increase in circulating prolactin. Thus, the effect of the hypothalamic prolactin-inhibiting hormone, dopamine, must normally be greater than the effects of the various hypothalamic peptides with prolactin-releasing activity. In humans, prolactin secretion is increased by exercise, surgical and psychologic stresses, and stimulation of the nipple (Table 18–4). The plasma prolactin level rises during sleep, the rise starting after the onset of sleep and persisting throughout the sleep period. Secretion is increased during pregnancy, reaching a peak at the time of parturition. After delivery, the plasma concentration falls to nonpregnant levels in about 8 days. Suckling produces a prompt increase in secretion, but the magnitude of this rise gradually declines after a woman has been nursing for more than 3 months. With prolonged lactation, milk secretion occurs with prolactin levels that are in the normal range.

L-Dopa decreases prolactin secretion by increasing the formation of dopamine; bromocriptine, and other dopamine agonists inhibit secretion because they stimulate dopamine receptors. Chlorpromazine and related drugs that block dopamine receptors increase prolactin secretion. Thyrotropin-releasing hormone (TRH) stimulates the secretion of prolactin in addition to TSH, and additional polypeptides with prolactin-releasing activity are present in hypothalamic tissue. Estrogens produce a slowly developing increase in prolactin secretion as a result of a direct action on the lactotropes.

It has now been established that prolactin facilitates the secretion of dopamine in the median eminence. Thus,

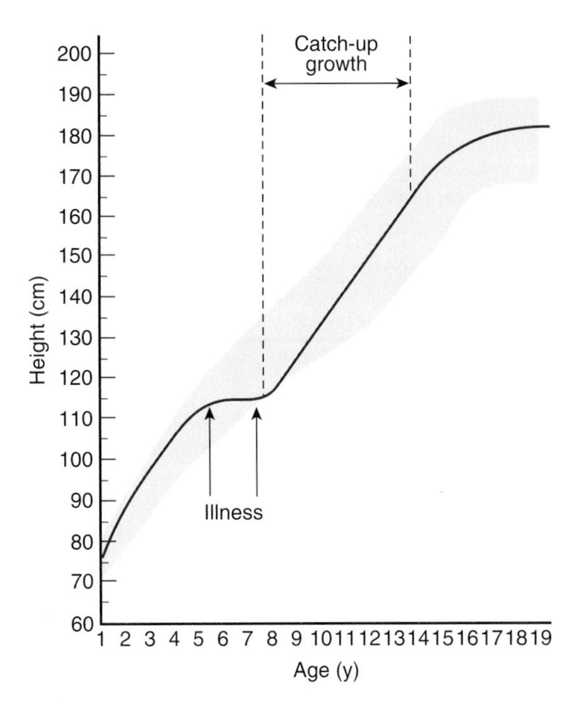

FIGURE 18–11 Growth curve for a normal boy who had an illness beginning at age 5 and ending at age 7. The shaded area shows the range of normal heights for a given age. The red line shows actual growth of the boy studied. Catch-up growth eventually returned his height to his previous normal growth curve. (Modified from Boersma B, Wit JM: Catch-up growth. Endocr Rev 1997;18:646.)

prolactin acts in the hypothalamus in a negative feedback fashion to inhibit its own secretion.

TABLE 18–4 Comparison of factors affecting the secretion of human prolactin and growth hormone.

Factor	Prolactin	Growth Hormone
Sleep	I+	I+
Nursing	I++	N
Breast stimulation in nonlactating women	I	N
Stress	I+	I+
Hypoglycemia	I	I+
Strenuous exercise	I	I
Sexual intercourse in women	I	N
Pregnancy	I++	N
Estrogens	I	I
Hypothyroidism	I	N
TRH	I+	N
Phenothiazines, butyrophenones	I+	N
Opioids	I	I
Glucose	N	D
Somatostatin	N	D+
L-Dopa	D+	I+
Apomorphine	D+	I+
Bromocriptine and related ergot derivatives	D+	I

I, moderate increase; I+, marked increase; I++, very marked increase; N, no change; D, moderate decrease; D+, marked decrease; TRH, thyrotropin-releasing hormone.

EFFECTS OF PITUITARY INSUFFICIENCY

CHANGES IN OTHER ENDOCRINE GLANDS

The widespread changes that develop when the pituitary is removed surgically or destroyed by disease in humans or animals are predictable in terms of the known hormonal functions of the gland. In hypopituitarism, the adrenal cortex atrophies, and the secretion of adrenal glucocorticoids and sex hormones falls to low levels. Stress induced increases in aldosterone secretion are absent, but basal aldosterone secretion and increases induced by salt depletion are normal, at least for some time. Since no mineralocorticoid deficiency is present, salt loss and hypovolemic shock do not develop, but the inability to increase glucocorticoid secretion makes patients with pituitary insufficiency sensitive to stress. The development of salt loss in long-standing hypopituitarism is discussed in Chapter 20. Growth is inhibited (see above). Thyroid function is depressed to low levels, and cold is tolerated poorly. The gonads atrophy, sexual cycles stop, and some of the secondary sex characteristics disappear.

INSULIN SENSITIVITY

Hypophysectomized animals have a tendency to become hypoglycemic, especially when fasted. Hypophysectomy ameliorates diabetes mellitus (see Chapter 24) and markedly increases the hypoglycemic effect of insulin. This is due in part to the deficiency of adrenocortical hormones, but hypophysectomized animals are more sensitive to insulin than adrenalectomized animals because they also lack the anti-insulin effect of growth hormone.

WATER METABOLISM

Although selective destruction of the supraoptic–posterior pituitary causes diabetes insipidus (see Chapter 17), removal of both the anterior and posterior pituitary usually causes no more than a transient polyuria. In the past, there was speculation that the anterior pituitary secreted a "diuretic hormone," but the amelioration of the diabetes insipidus is actually explained by a decrease in the osmotic load presented for excretion. Osmotically active particles hold water in the renal tubules (see Chapter 38). Because of the ACTH deficiency, the rate of protein catabolism is decreased in hypophysectomized animals. Because of the TSH deficiency, the metabolic rate is low. Consequently, fewer osmotically active products of catabolism are filtered and urine volume declines, even in the absence of vasopressin. Growth hormone deficiency contributes to the depression of the glomerular filtration rate in hypophysectomized animals, and growth hormone increases the glomerular filtration rate and renal plasma flow in humans. Finally, because of the glucocorticoid deficiency, there is the same defective excretion of a water load that is seen in adrenalectomized animals. The "diuretic" activity of the anterior pituitary can thus be explained in terms of the actions of ACTH, TSH, and growth hormone.

OTHER DEFECTS

When growth hormone deficiency develops in adulthood, it is usually accompanied by deficiencies in other anterior pituitary hormones. The deficiency of ACTH and other pituitary hormones with MSH activity may be responsible for the pallor of the skin in patients with hypopituitarism. There may be some loss of protein in adults, but wasting is not a feature of hypopituitarism in humans, and most patients with pituitary insufficiency are well nourished.

CAUSES OF PITUITARY INSUFFICIENCY IN HUMANS

Tumors of the anterior pituitary cause pituitary insufficiency. Suprasellar cysts, remnants of Rathke's pouch that enlarge and compress the pituitary, are another cause of hypopituitarism. In women who have an episode of shock due to postpartum hemorrhage, the pituitary may become infarcted, with the subsequent development of postpartum necrosis **(Sheehan syndrome)**. The blood supply to the anterior lobe is vulnerable because it descends on the pituitary stalk through the rigid diaphragma sellae, and during pregnancy the pituitary is enlarged. Pituitary infarction is usually extremely rare in men.

CHAPTER SUMMARY

- The pituitary gland plays a critical role in regulating the function of downstream glands, and also exerts independent endocrine actions on a wide variety of peripheral organs and tissues. It consists of two functional sections in humans: the anterior pituitary, which secretes mainly tropic hormones; and the posterior pituitary, which contains nerve endings that release oxytocin and vasopressin. The intermediate lobe is prominent in lower vertebrates but not in humans or other mammals.
- Corticotropes of the anterior lobe synthesize proopiomelanocortin, which is the precursor of ACTH, endorphins, and melanotropins. The latter have a critical role in the control of skin coloration in fish, amphibians, and reptiles, whereas ACTH is a primary regulator of skin pigmentation in mammals.
- Growth hormone is synthesized by somatotropes. It is secreted in an episodic fashion in response to hypothalamic factors, and secretion is subject to feedback inhibition. A portion of the circulating pool is protein-bound.
- Growth hormone activates growth and influences protein, carbohydrate, and fat metabolism to react to stressful conditions. Many, but not all, of the peripheral actions of growth hormone can be attributed to its ability to stimulate production of IGF-I.
- Growth reflects a complex interplay of growth hormone, IGF-I, and many other hormones as well as extrinsic influences and genetic factors. The consequences of over- or underproduction of such influences depends on whether this occurs before or after puberty. Deficiencies in components of the growth hormone pathway in childhood lead to dwarfism; overproduction results in gigantism, acromegaly, or both.
- The pituitary also supplies hormones that regulate reproductive tissues and lactation—follicle-stimulating hormone, luteinizing hormone, and prolactin. Prolactin, in particular, is regulated by many of the factors that also regulate growth hormone secretion, although specific regulators may have opposing effects.

MULTIPLE-CHOICE QUESTIONS

For all questions, select the single best answer unless otherwise directed.

1. A neuroscientist is studying communication between the hypothalamus and pituitary in a rat model. She interrupts blood flow through the median eminence and then measures circulating levels of pituitary hormones following appropriate physiological stimulation. Secretion of which of the following hormones will be unaffected by the experimental manipulation?
   A. Growth hormone
   B. Prolactin
   C. Thyroid stimulating hormone
   D. Follicle-stimulating hormone
   E. Vasopressin
2. Which of the following pituitary hormones is an opioid peptide?
   A. α-melanocyte-stimulating hormone (α-MSH)
   B. β-MSH
   C. ACTH
   D. Growth hormone
   E. β-endorphin
3. During childbirth, a woman suffers a serious hemorrhage and goes into shock. After she recovers, she displays symptoms of hypopituitarism. Which of the following will not be expected in this patient?
   A. Cachexia
   B. Infertility
   C. Pallor
   D. Low basal metabolic rate
   E. Intolerance to stress

4. A scientist finds that infusion of growth hormone into the median eminence of the hypothalamus in experimental animals inhibits the secretion of growth hormone and concludes that this proves that growth hormone feeds back to inhibit GHRH secretion. Do you accept this conclusion?
   A. No, because growth hormone does not cross the blood–brain barrier.
   B. No, because the infused growth hormone could be stimulating dopamine secretion.
   C. No, because substances placed in the median eminence could be transported to the anterior pituitary.
   D. Yes, because systemically administered growth hormone inhibits growth hormone secretion.
   E. Yes, because growth hormone binds GHRH, inactivating it.
5. The growth hormone receptor
   A. activates G_s.
   B. requires dimerization to exert its effects.
   C. must be internalized to exert its effects.
   D. resembles the IGF-I receptor.
   E. resembles the ACTH receptor.

CHAPTER RESOURCES

Ayuk J, Sheppard MC: Growth hormone and its disorders. Postgrad Med J 2006;82:24.

Boissy RE, Nordlund JJ: Molecular basis of congenital hypopigmentary disorders in humans: A review. Pigment Cell Res 1997;10:12.

Brooks AJ, Waters MJ: The growth hormone receptor: mechanism of activation and clinical implications. Nat Rev Endocrinol 2010;6:515.

Buzi F, Mella P, Pilotta A, Prandi E, Lanfranchi F, Carapella T: Growth hormone receptor polymorphisms. Endocr Dev 2007;11:28.

Fauquier T, Rizzoti K, Dattani M, Lovell-Badge R, Robinson ICAF: SOX2-expressing progenitor cells generate all of the major cell types in the adult mouse pituitary gland. Proc Natl Acad Sci USA 2008;105:2907.

Hindmarsh PC, Dattani MT: Use of growth hormone in children. Nat Clin Pract Endocrinol Metab 2006;2:260.

CHAPTER 19

The Thyroid Gland

OBJECTIVES

After studying this chapter, you should be able to:

- Describe the structure of the thyroid gland and how it relates to its function.
- Define the chemical nature of the thyroid hormones and how they are synthesized.
- Understand the critical role of iodine in the thyroid gland and how its transport is controlled.
- Describe the role of protein binding in the transport of thyroid hormones and peripheral metabolism.
- Identify the role of the hypothalamus and pituitary in regulating thyroid function.
- Define the effects of the thyroid hormones in homeostasis and development.
- Understand the basis of conditions where thyroid function is abnormal and how they can be treated.

INTRODUCTION

The thyroid gland is one of the larger endocrine glands of the body. The gland has two primary functions. The first is to secrete the thyroid hormones, which maintain the level of metabolism in the tissues that is optimal for their normal function. Thyroid hormones stimulate O_2 consumption by most of the cells in the body, help to regulate lipid and carbohydrate metabolism, and thereby influence body mass and mentation. Consequences of thyroid gland dysfunction depend on the life stage at which they occur. The thyroid is not essential for life, but its absence or hypofunction during fetal and neonatal life results in severe mental retardation and dwarfism. In adults, hypothyroidism is accompanied by mental and physical slowing and poor resistance to cold. Conversely, excess thyroid secretion leads to body wasting, nervousness, tachycardia, tremor, and excess heat production. Thyroid function is controlled by the thyroid-stimulating hormone (TSH, thyrotropin) of the anterior pituitary. The secretion of this hormone is in turn increased by thyrotropin-releasing hormone (TRH) from the hypothalamus and is also subject to negative feedback control by high circulating levels of thyroid hormones acting on the anterior pituitary and the hypothalamus.

The second function of the thyroid gland is to secrete calcitonin, a hormone that regulates circulating levels of calcium. This function of the thyroid gland is discussed in Chapter 21 in the broader context of whole body calcium homeostasis.

ANATOMIC CONSIDERATIONS

The thyroid is a butterfly-shaped gland that straddles the trachea in the front of the neck. It develops from an evagination of the floor of the pharynx, and a **thyroglossal duct** marking the path of the thyroid from the tongue to the neck sometimes persists in the adult. The two lobes of the human thyroid are connected by a bridge of tissue, the **thyroid isthmus,** and there is sometimes a **pyramidal lobe** arising from the isthmus in front of the larynx (Figure 19–1). The gland is well vascularized, and the thyroid has one of the highest rates of blood flow per gram of tissue of any organ in the body.

The portion of the thyroid concerned with the production of thyroid hormone consists of multiple **acini**

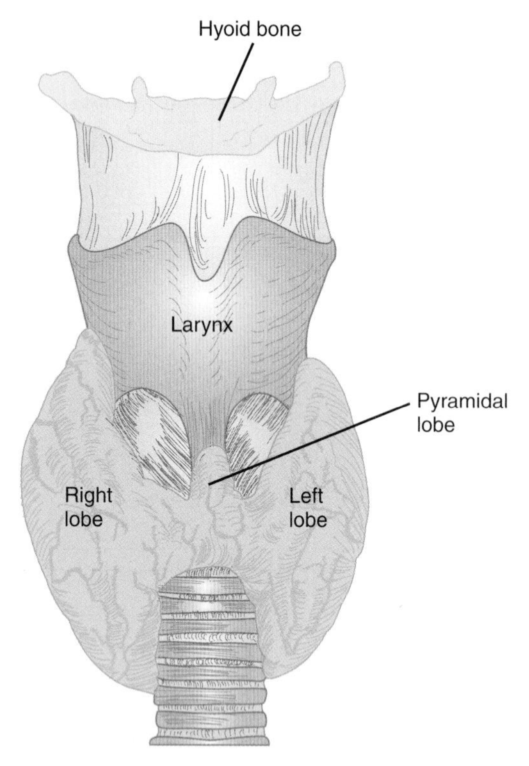

FIGURE 19–1 The human thyroid.

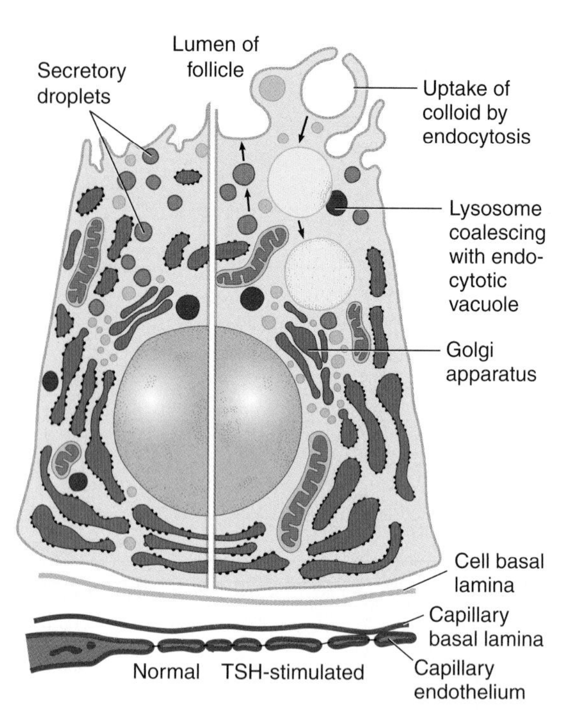

FIGURE 19–3 Thyroid cell. Left: Normal pattern. Right: After TSH stimulation. The arrows on the right show the secretion of thyroglobulin into the colloid. On the right, endocytosis of the colloid and merging of a colloid-containing vacuole with a lysosome are also shown. The cell rests on a capillary with gaps (fenestrations) in the endothelial wall.

(follicles). Each spherical follicle is surrounded by a single layer of polarized epithelial cells and filled with pink-staining proteinaceous material called **colloid.** Colloid consists predominantly of the glycoprotein, thyroglobulin. When the gland is inactive, the colloid is abundant, the follicles are large, and the cells lining them are flat. When the gland is active, the follicles are small, the cells are cuboid or columnar, and areas where the colloid is being actively reabsorbed into the thyrocytes are visible as "reabsorption lacunae" (Figure 19–2).

Microvilli project into the colloid from the apexes of the thyroid cells and canaliculi extend into them. The endoplasmic reticulum is prominent, a feature common to most glandular cells, and secretory granules containing thyroglobulin are seen (Figure 19–3). The individual thyroid cells rest on a basal lamina that separates them from the adjacent capillaries. The capillaries are fenestrated, like those of other endocrine glands (see Chapter 31).

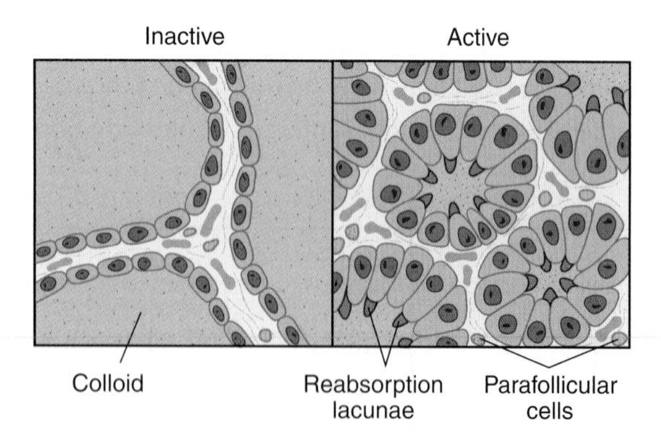

FIGURE 19–2 Thyroid histology. The appearance of the gland when it is inactive (left) and actively secreting (right) is shown. Note the small, punched-out "reabsorption lacunae" in the colloid next to the cells in the active gland.

FORMATION & SECRETION OF THYROID HORMONES

CHEMISTRY

The primary hormone secreted by the thyroid is **thyroxine (T_4)**, along with much lesser amounts of **triiodothyronine (T_3)**. T_3 has much greater biological activity than T_4 and is specifically generated at its site of action in peripheral tissues by deiodination of T_4 (see below). Both hormones are iodine-containing amino acids (Figure 19–4). Small amounts of reverse triiodothyronine (3,3′,5′-triiodothyronine, RT_3) and other compounds are also found in thyroid venous blood. RT_3 is not biologically active.

IODINE HOMEOSTASIS

Iodine is an essential raw material for thyroid hormone synthesis. Dietary iodide is absorbed by the intestine and enters the circulation; its subsequent fate is summarized in Figure 19–5.

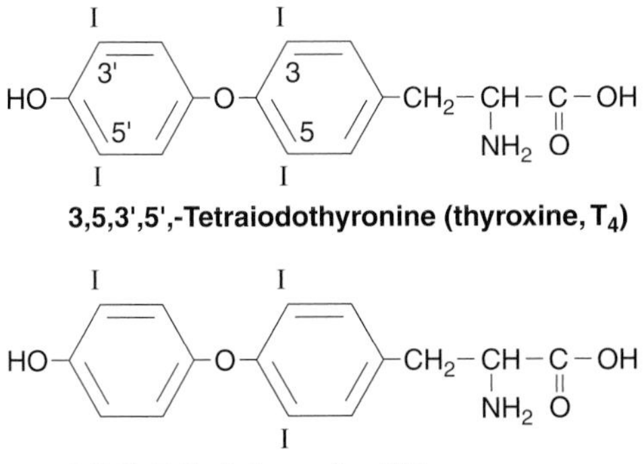

3,5,3',5',-Tetraiodothyronine (thyroxine, T_4)

3,5,3',-Triiodothyronine (T_3)

FIGURE 19–4 Thyroid hormones. The numbers in the rings in the T_4 formula indicate the number of positions in the molecule. RT_3 is 3,3′,5′-triiodothyronine.

The minimum daily iodine intake that will maintain normal thyroid function is 150 μg in adults. In most developed countries, supplementation of table salt means that the average dietary intake is approximately 500 μg/d. The principal organs that take up circulating I^- are the thyroid, which uses it to make thyroid hormones, and the kidneys, which excrete it in the urine. About 120 μg/d enter the thyroid at normal rates of thyroid hormone synthesis and secretion. The thyroid secretes 80 μg/d in the form of T_3 and T_4, while 40 μg/d diffuses back into the extracellular fluid (ECF). Circulating T_3 and T_4 are metabolized in the liver and other tissues, with the release of a further 60 μg of I^- per day into the ECF. Some thyroid hormone derivatives are excreted in the bile, and some of the iodine in them is reabsorbed (enterohepatic circulation), but there is a net loss of I^- in the stool of approximately 20 μg/d. The total amount of I^- entering the ECF is thus 500 + 40 + 60, or 600 μg/d; 20% of this I^- enters the thyroid, whereas 80% is excreted in the urine.

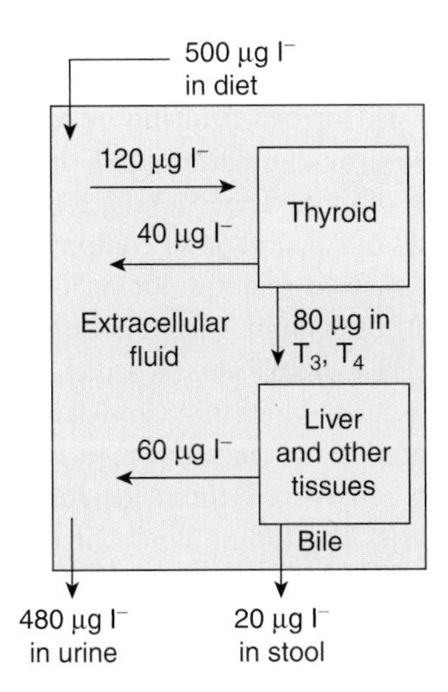

FIGURE 19–5 Iodine metabolism. The figure shows the movement of iodide amongst various body compartments on a daily basis.

IODIDE TRANSPORT ACROSS THYROCYTES

The basolateral membranes of thyrocytes facing the capillaries contain a **symporter** that transports two Na^+ ions and one I^- ion into the cell with each cycle, against the electrochemical gradient for I^-. This Na^+/I^- symporter **(NIS)** is capable of producing intracellular I^- concentrations that are 20–40 times as great as the concentration in plasma. The process involved is secondary active transport (see Chapter 2), with the energy provided by active transport of Na^+ out of thyroid cells by Na, K ATPase. NIS is regulated both by transcriptional means and by active trafficking into and out of the thyrocyte basolateral membrane; in particular, thyroid stimulating hormone (TSH; see below) induces both NIS expression and the retention of NIS in the basolateral membrane, where it can mediate sustained iodide uptake.

Iodide must also exit the thyrocyte across the apical membrane to access the colloid, where the initial steps of thyroid hormone synthesis occur. This transport step is believed to be mediated, at least in part, by a Cl^-/I^- exchanger known as **pendrin.** This protein was first identified as the product of the gene responsible for the Pendred syndrome, whose patients suffer from thyroid dysfunction and deafness. Pendrin (SLC26A4) is one member of the larger family of SLC26 anion exchangers.

The relation of thyroid function to iodide is unique. As discussed in more detail below, iodide is essential for normal thyroid function, but iodide deficiency and iodide excess both inhibit thyroid function.

The salivary glands, the gastric mucosa, the placenta, the ciliary body of the eye, the choroid plexus, the mammary glands, and certain cancers derived from these tissues also express NIS and can transport iodide against a concentration gradient, but the transporter in these tissues is not affected by TSH. The physiologic significance of all these extrathyroidal iodide-concentrating mechanisms is obscure, but they may provide pathways for radioablation of NIS-expressing cancer cells using iodide radioisotopes. This approach is also useful for the ablation of thyroid cancers.

THYROID HORMONE SYNTHESIS & SECRETION

At the interface between the thyrocyte and the colloid, iodide undergoes a process referred to as organification. First, it is oxidized to iodine, and then incorporated into the carbon 3 position of tyrosine residues that are part of the thyroglobulin molecule in the colloid (Figure 19–6). **Thyroglobulin** is a glycoprotein made up of two subunits and has a molecular weight of 660 kDa. It contains 10% carbohydrate by weight. It also contains 123 tyrosine residues, but only 4–8 of these are normally incorporated into thyroid hormones. Thyroglobulin is synthesized in the thyroid cells and secreted into the colloid by exocytosis of granules. The oxidation and reaction of iodide with the secreted thyroglobulin is mediated by **thyroid**

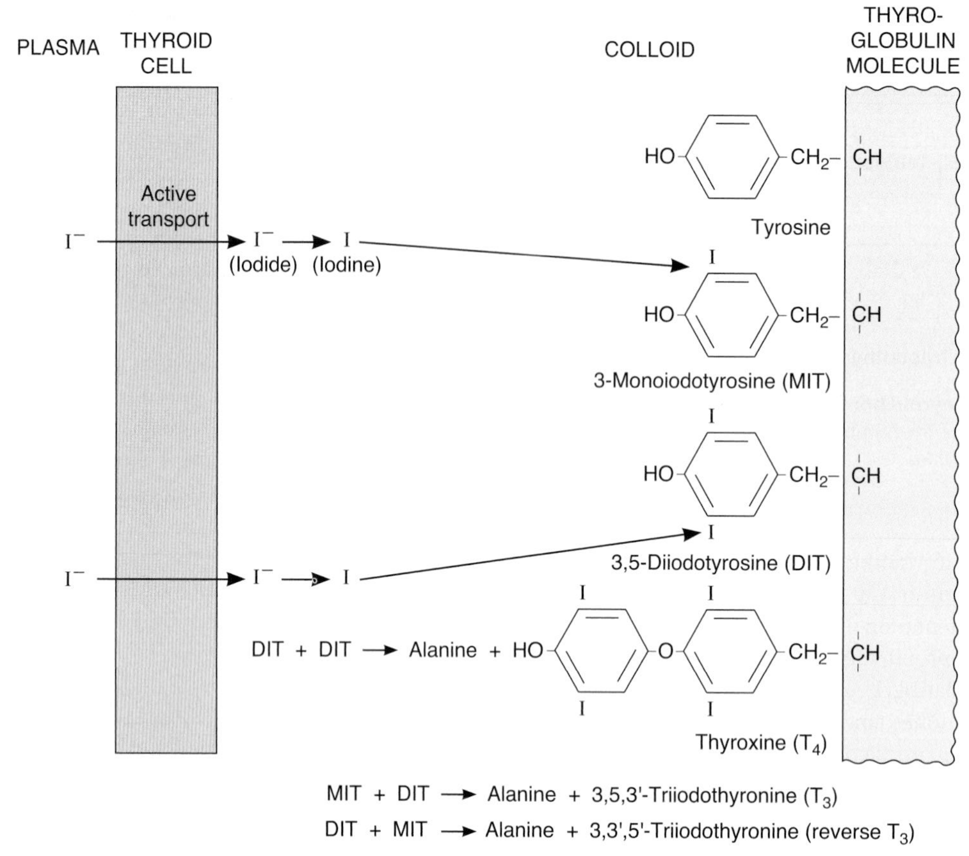

FIGURE 19–6 Outline of thyroid hormone biosynthesis. Iodide is transported from the plasma across the cells of the thyroid gland by both secondary active and passive transport. The iodide is converted to iodine, which reacts with tyrosine residues exposed on the surface of thyroglobulin molecules resident in the colloid. Iodination of tyrosine takes place at the apical border of the thyroid cells while the molecules are bound in peptide linkage in thyroglobulin.

peroxidase, a membrane-bound enzyme found in the thyrocyte apical membrane. The thyroid hormones so produced remain part of the thyroglobulin molecule until needed. As such, colloid represents a reservoir of thyroid hormones, and humans can ingest a diet completely devoid of iodide for up to 2 months before a decline in circulating thyroid hormone levels is seen. When there is a need for thyroid hormone secretion, colloid is internalized by the thyrocytes by endocytosis, and directed toward lysosomal degradation. Thus, the peptide bonds of thyroglobulin are hydrolyzed, and free T_4 and T_3 are discharged into cytosol and thence to the capillaries (see below). Thyrocytes thus have four functions: They collect and transport iodine, they synthesize thyroglobulin and secrete it into the colloid, they fix iodine to the thyroglobulin to generate thyroid hormones, and they remove the thyroid hormones from thyroglobulin and secrete them into the circulation.

Thyroid hormone synthesis is a multistep process. Thyroid peroxidase generates reactive iodine species that can attack thyroglobulin. The first product is monoiodotyrosine (MIT). MIT is next iodinated on the carbon 5 position to form diiodotyrosine (DIT). Two DIT molecules then undergo an oxidative condensation to form T_4 with the elimination of the alanine side chain from the molecule that forms the outer ring. There are two theories of how this **coupling reaction** occurs. One holds that the coupling occurs with both DIT molecules attached to thyroglobulin (intramolecular coupling). The other holds that the DIT that forms the outer ring is first detached from thyroglobulin (intermolecular coupling). In either case, thyroid peroxidase is involved in coupling as well as iodination. T_3 is formed by condensation of MIT with DIT. A small amount of RT_3 is also formed, probably by condensation of DIT with MIT. In the normal human thyroid, the average distribution of iodinated compounds is 3% MIT, 33% DIT, 35% T_4, and 7% T_3. Only traces of RT_3 and other components are present.

The human thyroid secretes about 80 μg (103 nmol) of T_4, 4 μg (7 nmol) of T_3, and 2 μg (3.5 nmol) of RT_3 per day (Figure 19–7). MIT and DIT are not secreted. These iodinated tyrosines are deiodinated by a microsomal **iodotyrosine deiodinase.** This represents a mechanism to recover iodine and bound tyrosines and recycle them for additional rounds of hormone synthesis. The iodine liberated by deiodination of MIT and DIT is reutilized in the gland and normally provides about twice as much iodide for hormone synthesis as NIS does. In patients with congenital absence of the iodotyrosine deiodinase, MIT and DIT appear in the urine and there are symptoms of iodine deficiency (see below). Iodinated thyronines are resistant to the activity of iodotyrosine deiodinase, thus allowing T_4 and T_3 to pass into the circulation.

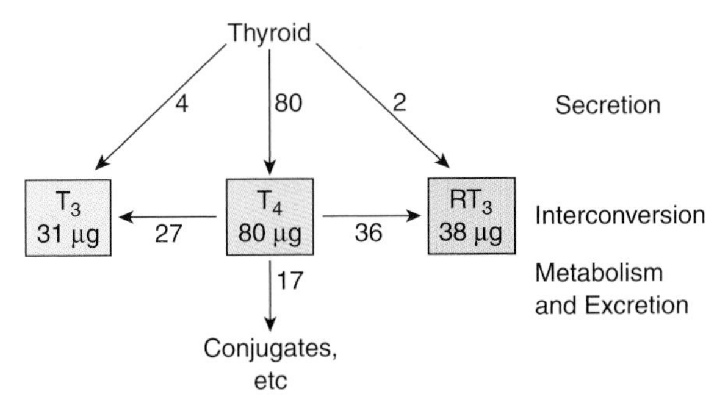

FIGURE 19–7 Secretion and interconversion of thyroid hormones in normal adult humans. Figures are in micrograms per day. Note that most of the T_3 and RT_3 are formed from T_4 deiodination in the tissues and only small amounts are secreted by the thyroid. T_4 is also conjugated for subsequent excretion from the body.

TRANSPORT & METABOLISM OF THYROID HORMONES

PROTEIN BINDING

The normal total **plasma T_4** level in adults is approximately 8 µg/dL (103 nmol/L), and the **plasma T_3** level is approximately 0.15 µg/dL (2.3 nmol/L). T_4 and T_3 are relatively lipophilic; thus, their free forms in plasma are in equilibrium with a much larger pool of protein-bound thyroid hormones in plasma and in tissues. Free thyroid hormones are added to the circulating pool by the thyroid. It is the free thyroid hormones in plasma that are physiologically active and that feed back to inhibit pituitary secretion of TSH (Figure 19–8). The function of protein-binding appears to be maintenance of a large pool of hormone that can readily be mobilized as needed. In addition, at least for T_3, hormone binding prevents excess uptake by the first cells encountered and promotes uniform tissue distribution. Total

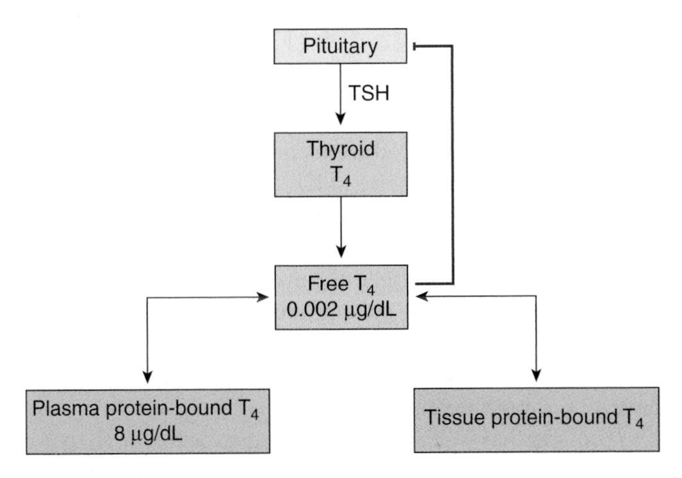

FIGURE 19–8 Regulation of thyroid hormone synthesis. T4 is secreted by the thyroid in response to TSH. Free T_4 secreted by the thyroid into the circulation is in equilibrium with T_4 bound to both plasma and tissue proteins. Free T_4 also feeds back to inhibit TSH secretion by the pituitary.

TABLE 19–1 Binding of thyroid hormones to plasma proteins in normal adult humans.

Protein	Plasma Concentration (mg/dL)	Amount of Circulating Hormone Bound (%)	
		T_4	T_3
Thyroxine-binding globulin (TBG)	2	67	46
Transthyretin (thyroxine-binding prealbumin, TBPA)	15	20	1
Albumin	3500	13	53

T_4 and T_3 can both be measured by radioimmunoassay. There are also direct assays that specifically measure only the free forms of the hormones. The latter are the more clinically relevant measures given that these are the active forms, and also due to both acquired and congenital variations in the concentrations of binding proteins between individuals.

The plasma proteins that bind thyroid hormones are **albumin,** a prealbumin called **transthyretin** (formerly called **thyroxine-binding prealbumin),** and a globulin known as **thyroxine-binding globulin (TBG).** Of the three proteins, albumin has the largest **capacity** to bind T_4 (ie, it can bind the most T_4 before becoming saturated) and TBG has the smallest capacity. However, the **affinities** of the proteins for T_4 (ie, the avidity with which they bind T_4 under physiologic conditions) are such that most of the circulating T_4 is bound to TBG (Table 19–1), with over a third of the binding sites on the protein occupied. Smaller amounts of T_4 are bound to transthyretin and albumin. The half-life of transthyretin is 2 days, that of TBG is 5 days, and that of albumin is 13 days.

Normally, 99.98% of the T_4 in plasma is bound; the free T_4 level is only about 2 ng/dL. There is very little T_4 in the urine. Its biologic half-life is long (about 6–7 days), and its volume of distribution is less than that of ECF (10 L, or about 15% of body weight). All of these properties are characteristic of a substance that is strongly bound to protein.

T_3 is not bound to quite as great an extent; of the 0.15 µg/dL normally found in plasma, 0.2% (0.3 ng/dL) is free. The remaining 99.8% is protein-bound, 46% to TBG and most of the remainder to albumin, with very little binding to transthyretin (Table 19–1). The lesser binding of T_3 correlates with the facts that T_3 has a shorter half-life than T_4 and that its action on the tissues is much more rapid. RT_3 also binds to TBG.

FLUCTUATIONS IN BINDING

When a sudden, sustained increase in the concentration of thyroid-binding proteins in the plasma takes place, the concentration of free thyroid hormones falls. This change is temporary, however, because the decrease in the concentration of free thyroid hormones in the circulation stimulates TSH secretion, which in turn causes an increase in the production of free thyroid hormones. A new equilibrium is eventually reached at

TABLE 19–2 Effect of variations in the concentrations of thyroid hormone-binding proteins in the plasma on various parameters of thyroid function after equilibrium has been reached.

Condition	Concentrations of Binding Proteins	Total Plasma T_4, T_3, RT_3	Free Plasma T_4, T_3, RT_3	Plasma TSH	Clinical State
Hyperthyroidism	Normal	High	High	Low	Hyperthyroid
Hypothyroidism	Normal	Low	Low	High	Hypothyroid
Estrogens, methadone, heroin, major tranquilizers, clofibrate	High	High	Normal	Normal	Euthyroid
Glucocorticoids, androgens, danazol, asparaginase	Low	Low	Normal	Normal	Euthyroid

which the total quantity of thyroid hormones in the blood is elevated but the concentration of free hormones, the rate of their metabolism, and the rate of TSH secretion are normal. Corresponding changes in the opposite direction occur when the concentration of thyroid-binding protein is reduced. Consequently, patients with elevated or decreased concentrations of binding proteins, particularly TBG, are typically neither hyper- nor hypothyroid; that is, they are **euthyroid.**

TBG levels are elevated in estrogen-treated patients and during pregnancy, as well as after treatment with various drugs (Table 19–2). They are depressed by glucocorticoids, androgens, the weak androgen danazol, and the cancer chemotherapeutic agent L-asparaginase. A number of other drugs, including salicylates, the anti-convulsant phenytoin, and the cancer chemotherapeutic agents mitotane (o, p′-DDD) and 5-fluorouracil inhibit binding of T_4 and T_3 to TBG and consequently produce changes similar to those produced by a decrease in TBG concentration. Changes in total plasma T_4 and T_3 can also be produced by changes in plasma concentrations of albumin and prealbumin.

METABOLISM OF THYROID HORMONES

T_4 and T_3 are deiodinated in the liver, the kidneys, and many other tissues. These deiodination reactions serve not only to catabolize the hormones, but also to provide a local supply specifically of T_3, which is believed to be the primary mediator of the physiological effects of thyroid secretion. One third of the circulating T_4 is normally converted to T_3 in adult humans, and 45% is converted to RT_3. As shown in Figure 19–7, only about 13% of the circulating T_3 is secreted by the thyroid while 87% is formed by deiodination of T_4; similarly, only 5% of the circulating RT_3 is secreted by the thyroid and 95% is formed by deiodination of T_4. It should be noted as well that marked differences in the ratio of T_3 to T_4 occur in various tissues. Two tissues that have very high T_3/T_4 ratios are the pituitary and the cerebral cortex, due to the expression of specific deiodinases, as discussed below.

Three different deiodinases act on thyroid hormones: D_1, D_2, and D_3. All are unique in that they contain the rare amino acid selenocysteine, with selenium in place of sulfur, which is essential for their enzymatic activity. D_1 is present in high concentrations in the liver, kidneys, thyroid, and pituitary. It appears primarily to be responsible for maintaining the formation of T_3 from T_4 in the periphery. D_2 is present in the brain, pituitary, and brown fat. It also contributes to the formation of T_3. In the brain, it is located in astroglia and produces a supply of T_3 to neurons. D_3 is also present in the brain and in reproductive tissues. It acts only on the 5 position of T_4 and T_3 and is probably the main source of RT_3 in the blood and tissues. Overall, the deiodinases appear to be responsible for maintaining differences in T_3/T_4 ratios in the various tissues in the body. In the brain, in particular, high levels of deiodinase activity ensure an ample supply of active T_3.

Some of the T_4 and T_3 is further converted to deiodotyrosines by deiodinases. T_4 and T_3 are also conjugated in the liver to form sulfates and glucuronides. These conjugates enter the bile and pass into the intestine. The thyroid conjugates are hydrolyzed, and some are thereafter reabsorbed (enterohepatic circulation), but others are excreted in the stool. In addition, some T_4 and T_3 passes directly from the circulation to the intestinal lumen. The iodide lost by these routes amounts to about 4% of the total daily iodide loss.

FLUCTUATIONS IN DEIODINATION

Much more RT_3 and much less T_3 are formed during fetal life, and the ratio shifts to that of adults about 6 weeks after birth. Various drugs inhibit deiodinases, producing a fall in plasma T_3 levels and a reciprocal rise in RT_3. Selenium deficiency has the same effect. A wide variety of nonthyroidal illnesses also suppress deiodinases. These include burns, trauma, advanced cancer, cirrhosis, renal failure, myocardial infarction, and febrile states. The low-T_3 state produced by these conditions disappears with recovery. It is difficult to decide whether individuals with the low-T_3 state produced by drugs and illness have mild hypothyroidism.

Diet also has a clear-cut effect on conversion of T_4 to T_3. In fasted individuals, plasma T_3 is reduced by 10–20% within 24 h and by about 50% in 3–7 days, with a corresponding rise in RT_3 (Figure 19–9). Free and bound T_4 levels remain essentially normal. During more prolonged starvation, RT_3 returns

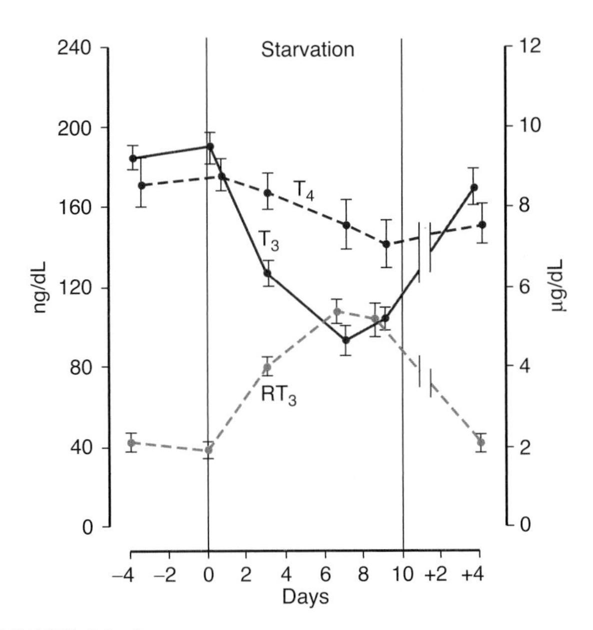

FIGURE 19–9 Effect of starvation on plasma levels of T_4, T_3, and RT_3 in humans. The scale for T_3 and RT_3 is on the left and the scale for T_4 is on the right. The most pronounced effect is a reduction in T_3 levels with a reciprocal rise in RT_3. The changes, which conserve calories by reducing tissue metabolism, are reversed promptly by re-feeding. Similar changes occur in wasting diseases. (Reproduced with permission from Burger AG: New aspects of the peripheral action of thyroid hormones. Triangle 1983;22:175. Copyright © 1983 Sandoz Ltd., Basel, Switzerland.)

to normal but T_3 remains depressed. At the same time, the basal metabolic rate (BMR) falls and urinary nitrogen excretion, an index of protein breakdown, is decreased. Thus, the decline in T_3 conserves calories and protein. Conversely, overfeeding increases T_3 and reduces RT_3.

REGULATION OF THYROID SECRETION

Thyroid function is regulated primarily by variations in the circulating level of pituitary TSH (Figure 19–8). TSH secretion is increased by the hypothalamic hormone TRH (see Chapter 17) and inhibited in a negative feedback fashion by circulating free T_4 and T_3. The effect of T_4 is enhanced by production of T_3 in the cytoplasm of the pituitary cells by the 5′-D_2 they contain. TSH secretion is also inhibited by stress, and in experimental animals it is increased by cold and decreased by warmth.

CHEMISTRY & METABOLISM OF TSH

Human TSH is a glycoprotein that contains 211 amino acid residues. It is made up of two subunits, designated α and β. The α subunit is encoded by a gene on chromosome 6 and the β subunit by a gene on chromosome 1. The α and β subunits become noncovalently linked in the pituitary thyrotropes. TSH-α is identical to the α subunit of LH, FSH, and hCG-α (see Chapters 18 and 22). The functional specificity of TSH is conferred by the β subunit. The structure of TSH varies from species to species, but other mammalian TSHs are biologically active in humans.

The biologic half-life of human TSH is about 60 min. TSH is degraded for the most part in the kidneys and to a lesser extent in the liver. Secretion is pulsatile, and mean output starts to rise at about 9:00 PM, peaks at midnight, and then declines during the day. The normal secretion rate is about 110 μg/d. The average plasma level is about 2 μg/mL.

Because the α subunit in hCG is the same as that in TSH, large amounts of hCG can activate thyroid receptors (TR) nonspecifically. In some patients with benign or malignant tumors of placental origin, plasma hCG levels can rise so high that they produce mild hyperthyroidism.

EFFECTS OF TSH ON THE THYROID

When the pituitary is removed, thyroid function is depressed and the gland atrophies; when TSH is administered, thyroid function is stimulated. Within a few minutes after the injection of TSH, there are increases in iodide binding; synthesis of T_3, T_4, and iodotyrosines; secretion of thyroglobulin into the colloid; and endocytosis of colloid. Iodide trapping is increased in a few hours; blood flow increases; and, with chronic TSH treatment, the cells hypertrophy and the weight of the gland increases.

Whenever TSH stimulation is prolonged, the thyroid becomes detectably enlarged. Enlargement of the thyroid is called a **goiter.**

TSH RECEPTORS

The TSH receptor is a typical G protein-coupled, seven-transmembrane receptor that activates adenylyl cyclase through G_s. It also activates phospholipase C (PLC). Like other glycoprotein hormone receptors, it has an extended, glycosylated extracellular domain.

OTHER FACTORS AFFECTING THYROID GROWTH

In addition to TSH receptors, thyrocytes express receptors for insulin-like growth factor I (IGF-I), EGF, and other growth factors. IGF-I and EGF promote growth, whereas interferon γ and tumor necrosis factor α inhibit growth. The exact physiologic role of these factors in the thyroid has not been established, but the effect of the cytokines implies that thyroid function might be inhibited in the setting of chronic inflammation, which could contribute to cachexia, or weight loss.

CONTROL MECHANISMS

The mechanisms regulating thyroid secretion are summarized in Figure 19–8. The negative feedback effect of thyroid hormones on TSH secretion is exerted in part at the hypothalamic level, but it is also due in large part to an action on the pituitary, since T_4 and T_3 block the increase in TSH secretion produced

CLINICAL BOX 19–1

Reduced Thyroid Function

The syndrome of adult **hypothyroidism** is generally called **myxedema,** although this term is also used to refer specifically to the skin changes in the syndrome. Hypothyroidism may be the end result of a number of diseases of the thyroid gland, or it may be secondary to pituitary or hypothalamic failure. In the latter two conditions, the thyroid remains able to respond to TSH. Thyroid function may be reduced by a number of conditions (Table 19–3). For example, when the dietary iodine intake falls below 50 μg/d, thyroid hormone synthesis is inadequate and secretion declines. As a result of increased TSH secretion, the thyroid hypertrophies, producing an **iodine deficiency goiter** that may become very large. Such "endemic goiters" have been substantially reduced by the practice of adding iodide to table salt. Drugs may also inhibit thyroid function. Most do so either by interfering with the iodide-trapping mechanism or by blocking the organic binding of iodine. In either case, TSH secretion is stimulated by the decline in circulating thyroid hormones, and a goiter is produced. Paradoxically, another substance that inhibits thyroid function under certain conditions is iodide itself. In normal individuals, large doses of iodide act directly on the thyroid to produce a mild and transient inhibition of organic binding of iodide and hence of hormone synthesis. This inhibition is known as the **Wolff–Chaikoff effect.**

In completely athyreotic adults, the BMR falls to about 40%. The hair is coarse and sparse, the skin is dry and yellowish (carotenemia), and cold is poorly tolerated. Mentation is slow, memory is poor, and in some patients there are severe mental symptoms ("myxedema madness"). Plasma cholesterol is elevated. Children who are hypothyroid from birth or before are called **cretins.** They are dwarfed and mentally retarded. Worldwide, congenital hypothyroidism is one of the most common causes of preventable mental retardation. The main causes are included in Table 19–3. They include not only maternal iodine deficiency and various congenital abnormalities of the fetal hypothalamo–pituitary–thyroid axis, but also maternal antithyroid antibodies that cross the placenta and damage the fetal thyroid. T_4 crosses the placenta, and unless the mother is hypothyroid, growth and development are normal until birth. If treatment is started at birth, the prognosis for normal growth and development is good, and mental retardation can generally be avoided; for this reason, screening tests for congenital hypothyroidism are becoming routine. When the mother is hypothyroid as well, as in the case of iodine deficiency, the mental deficiency is more severe and less responsive to treatment after birth. It has been estimated that 20 million people in the world now have various degrees of brain damage caused by iodine deficiency in utero.

Uptake of tracer doses of radioactive iodine can be used to assess thyroid function (contrast this with the use of large doses to ablate thyroid tissue in cases of hyperthyroidism (Clinical Box 19–2).

THERAPEUTIC HIGHLIGHTS

The treatment of hypothyroidism depends on the underlying mechanisms. Iodide deficiency can be addressed by adding it to the diet, as is done routinely in developed countries with the use of iodized salt. In congenital hypothyroidism, levothyroxine—a synthetic form of the thyroid hormone T_4—can be given. It is important that this take place as soon as possible after birth, with levels regularly monitored, to minimize long-term adverse effects.

by TRH. Infusion of either T_4 or T_3 reduces the circulating level of TSH, which declines measurably within 1 h. In experimental animals, there is an initial rise in pituitary TSH content before the decline, indicating that thyroid hormones inhibit secretion before they inhibit synthesis. The day-to-day maintenance of thyroid secretion depends on the feedback interplay of thyroid hormones with TSH and TRH (Figure 19–8). The adjustments that appear to be mediated via TRH include the increased secretion of thyroid hormones produced by cold and, presumably, the decrease produced by heat. It is worth noting that although cold produces clear-cut increases in circulating TSH in experimental animals and human infants, the rise produced by cold in adult humans is negligible. Consequently, in adults, increased heat production due to increased thyroid hormone secretion **(thyroid hormone thermogenesis)** plays little if any role in the response to cold. Stress has an inhibitory effect on TRH secretion. Dopamine and somatostatin act at the pituitary level to inhibit TSH secretion, but it is not known whether they play a physiologic role in the regulation of TSH secretion. Glucocorticoids also inhibit TSH secretion.

TABLE 19–3 Causes of congenital hypothyroidism.

Maternal iodine deficiency
Fetal thyroid dysgenesis
Inborn errors of thyroid hormone synthesis
Maternal antithyroid antibodies that cross the placenta
Fetal hypopituitary hypothyroidism

The amount of thyroid hormone necessary to maintain normal cellular function in thyroidectomized individuals used to be defined as the amount necessary to normalize the BMR, but it is now defined as the amount necessary to return plasma TSH to normal. Indeed, with the accuracy and sensitivity of modern assays for TSH and the marked inverse correlation

CLINICAL BOX 19–2

Hyperthyroidism

The symptoms of an overactive thyroid gland follow logically from the actions of thyroid hormone discussed in this chapter. Thus, hyperthyroidism is characterized by nervousness; weight loss; hyperphagia; heat intolerance; increased pulse pressure; a fine tremor of the outstretched fingers; warm, soft skin; sweating; and a BMR from +10 to as high as +100. It has various causes (Table 19–4); however, the most common cause is **Graves disease (Graves hyperthyroidism),** which accounts for 60–80% of the cases. This is an autoimmune disease, more common in women, in which antibodies to the TSH receptor stimulate the receptor. This produces marked T_4 and T_3 secretion and enlargement of the thyroid gland (goiter). However, due to the feedback effects of T_4 and T_3, plasma TSH is low, not high. Another hallmark of Graves disease is the occurrence of swelling of tissues in the orbits, producing protrusion of the eyeballs **(exophthalmos).** This occurs in 50% of patients and often precedes the development of obvious hyperthyroidism. Other antithyroid antibodies are present in Graves disease, including antibodies to thyroglobulin and thyroid peroxidase. In Hashimoto thyroiditis, autoimmune antibodies and infiltrating cytotoxic T cells ultimately destroy the thyroid, but during the early stage the inflammation of the gland causes excess thyroid hormone secretion and thyrotoxicosis similar to that seen in Graves disease.

THERAPEUTIC HIGHLIGHTS

Some of the symptoms of hyperthyroidism can be controlled by the **thioureylenes.** These are a group of compounds related to thiourea, which inhibit the iodination of monoiodotyrosine and block the coupling reaction. The two used clinically are propylthiouracil and methimazole. Iodination of tyrosine is inhibited because propylthiouracil and methimazole compete with tyrosine residues for iodine and become iodinated. In addition, propylthiouracil but not methimazole inhibits D_2 deiodinase, reducing the conversion of T_4 to T_3 in many extrathyroidal tissues. In severe cases, hyperthyroidism can also be treated by the infusion of radioactive iodine, which accumulates in the gland and then partially destroys it. Surgery is also considered if the thyroid becomes so large that it affects swallowing and/or breathing.

between plasma free thyroid hormone levels and plasma TSH, measurement of TSH is now widely regarded as one of the best tests of thyroid function. The amount of T_4 that normalizes plasma TSH in athyreotic individuals averages 112 μg of T_4 by mouth per day in adults. About 80% of this dose is absorbed from the gastrointestinal tract. It produces a slightly greater than normal FT_4I but a normal FT_3I, indicating that in humans, unlike some experimental animals, it is circulating T_3 rather than T_4 that is the principal feedback regulator of TSH secretion (see Clinical Boxes 19–1 and 19–2).

TABLE 19–4 Causes of hyperthyroidism.

Thyroid overactivity
Graves disease
Solitary toxic adenoma
Toxic multinodular goiter
Early stages of Hashimoto thyroiditis[a]
TSH-secreting pituitary tumor
Mutations causing constitutive activation of TSH receptor
Other rare causes
Extrathyroidal
Administration of T_3 or T_4 (factitious or iatrogenic hyperthyroidism)
Ectopic thyroid tissue

[a]Note that ultimately the thyroid will be destroyed in Hashimoto disease, resulting in hypothyroidism. Many patients only present after they become hypothyroid, and do not recall a transient phase of hyperthyroidism.

EFFECTS OF THYROID HORMONES

Some of the widespread effects of thyroid hormones in the body are secondary to stimulation of O_2 consumption **(calorigenic action),** although the hormones also affect growth and development in mammals, help regulate lipid metabolism, and increase the absorption of carbohydrates from the intestine (Table 19–5). They also increase the dissociation of oxygen from hemoglobin by increasing red cell 2,3-diphosphoglycerate (DPG) (see Chapter 35).

MECHANISM OF ACTION

Thyroid hormones enter cells and T_3 binds to TR in the nuclei. T_4 can also bind, but not as avidly. The hormone–receptor complex then binds to DNA via zinc fingers and increases (or in some cases, decreases) the expression of a variety of different genes that code for proteins that regulate cell function (see Chapters 1 and 16). Thus, the nuclear receptors for thyroid hormones are members of the superfamily of hormone-sensitive nuclear transcription factors.

There are two human TR genes: an α receptor gene on chromosome 17 and a β receptor gene on chromosome 3. By

TABLE 19–5 Physiologic effects of thyroid hormones.

Target Tissue	Effect	Mechanism
Heart	Chronotropic and Inotropic	Increased number of β-adrenergic receptors Enhanced responses to circulating catecholamines Increased proportion of α-myosin heavy chain (with higher ATPase activity)
Adipose tissue	Catabolic	Stimulated lipolysis
Muscle	Catabolic	Increased protein breakdown
Bone	Developmental	Promote normal growth and skeletal development
Nervous system	Developmental	Promote normal brain development
Gut	Metabolic	Increased rate of carbohydrate absorption
Lipoprotein	Metabolic	Formation of LDL receptors
Other	Calorigenic	Stimulated oxygen consumption by metabolically active tissues (exceptions: testes, uterus, lymph nodes, spleen, anterior pituitary) Increased metabolic rate

Modified and reproduced with permission from McPhee SJ, Lingarra VR, Ganong WF (editors): *Pathophysiology of Disease*, 6th ed. McGraw-Hill, 2010.

alternative splicing, each forms at least two different mRNAs and therefore two different receptor proteins. TRβ2 is found only in the brain, but TRα1, TRα2, and TRβ1 are widely distributed. TRα2 differs from the other three in that it does not bind T3 and its function is not yet fully established. TRs bind to DNA as monomers, homodimers, and heterodimers with other nuclear receptors, particularly the retinoid X receptor **(RXR).** The TR/RXR heterodimer does not bind to 9-cis retinoic acid, the usual ligand for RXR, but TR binding to DNA is greatly enhanced in response to thyroid hormones when the receptor is in the form of this heterodimer. There are also coactivator and corepressor proteins that affect the actions of TRs. Presumably, this complexity underlies the ability of thyroid hormones to produce many different effects in the body.

In most of its actions, T_3 acts more rapidly and is three to five times more potent than T_4 (Figure 19–10). This is because T_3 is less tightly bound to plasma proteins than is T_4, but binds more avidly to thyroid hormone receptors. As previously noted, RT_3 is inert (see Clinical Box 19–3).

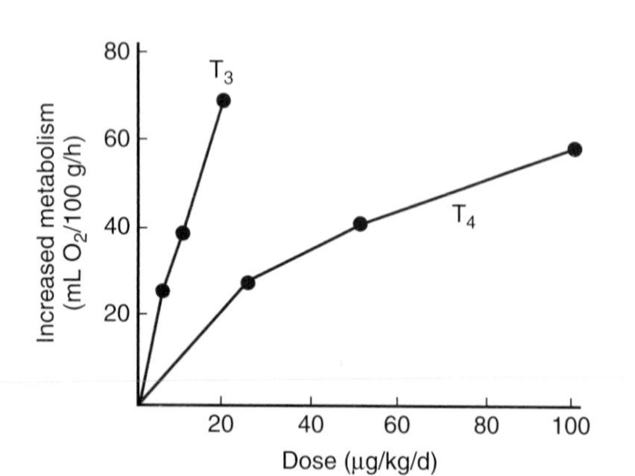

FIGURE 19–10 Calorigenic responses of thyroidectomized rats to subcutaneous injections of T_4 and T_3. Note the substantially greater potency of T_3. (Redrawn and reproduced with permission from Barker SB: Peripheral actions of thyroid hormones. Fed Proc 1962;21:635.)

CALORIGENIC ACTION

T_4 and T_3 increase the O_2 consumption of almost all metabolically active tissues. The exceptions are the adult brain, testes, uterus, lymph nodes, spleen, and anterior pituitary. T_4 actually depresses the O_2 consumption of the anterior pituitary, presumably because it inhibits TSH secretion. The increase in metabolic rate produced by a single dose of T_4 becomes measurable after a latent period of several hours and lasts 6 days or more.

Some of the calorigenic effect of thyroid hormones is due to metabolism of the fatty acids they mobilize. In addition, thyroid hormones increase the activity of the membrane-bound Na, K ATPase in many tissues.

Effects Secondary to Calorigenesis

When the metabolic rate is increased by T_4 and T_3 in adults, nitrogen excretion is increased; if food intake is not increased, endogenous protein and fat stores are catabolized and weight is lost. In hypothyroid children, small doses of thyroid hormones cause a positive nitrogen balance because they stimulate growth, but large doses cause protein catabolism similar to that produced in the adult. The potassium liberated during protein catabolism appears in the urine, and there is also an increase in urinary hexosamine and uric acid excretion.

When the metabolic rate is increased, the need for all vitamins is increased and vitamin deficiency syndromes may be precipitated. Thyroid hormones are necessary for hepatic conversion of carotene to vitamin A, and the accumulation of carotene in the bloodstream **(carotenemia)** in hypothyroidism is responsible for the yellowish tint of the skin. Carotenemia

CLINICAL BOX 19–3

Thyroid Hormone Resistance

Some mutations in the gene that codes for TRβ are associated with resistance to the effects of T_3 and T_4. Most commonly, there is resistance to thyroid hormones in the peripheral tissues and the anterior pituitary gland. Patients with this abnormality are usually not clinically hypothyroid, because they maintain plasma levels of T_3 and T_4 that are high enough to overcome the resistance, and hTRα is unaffected. However, plasma TSH is inappropriately high relative to the high circulating T_3 and T_4 levels and is difficult to suppress with exogenous thyroid hormone. Some patients have thyroid hormone resistance only in the pituitary. They have hypermetabolism and elevated plasma T_3 and T_4 levels with normal, nonsuppressible levels of TSH. A few patients apparently have peripheral resistance with normal pituitary sensitivity. They have hypometabolism despite normal plasma levels of T_3, T_4, and TSH An interesting finding is that **attention deficit hyperactivity disorder,** a condition frequently diagnosed in children who are overactive and impulsive, is much more common in individuals with thyroid hormone resistance than in the general population. This suggests that hTRβ may play a special role in brain development.

THERAPEUTIC HIGHLIGHTS

Most patients remain euthyroid in this condition, even in the face of a goiter. It is important to consider thyroid hormone resistance in the differential diagnosis of Graves disease to avoid the inappropriate use of antithyroid medications or even thyroid ablation. Isolated peripheral resistance to thyroid hormones can be treated by supplying large doses of synthetic T_4 exogenously. These are sufficient to overcome the resistance and increase the metabolic rate.

can be distinguished from jaundice because in the former condition the sclera are not yellow.

The skin normally contains a variety of proteins combined with polysaccharides, hyaluronic acid, and chondroitin sulfuric acid. In hypothyroidism, these complexes accumulate, promoting water retention and the characteristic puffiness of the skin (myxedema). When thyroid hormones are administered, the proteins are metabolized, and diuresis continues until the myxedema is cleared.

Milk secretion is decreased in hypothyroidism and stimulated by thyroid hormones, a fact sometimes put to practical use in the dairy industry. Thyroid hormones do not stimulate the metabolism of the uterus but are essential for normal menstrual cycles and fertility.

EFFECTS ON THE CARDIOVASCULAR SYSTEM

Large doses of thyroid hormones cause enough extra heat production to lead to a slight rise in body temperatures (Chapter 17), which in turn activates heat-dissipating mechanisms. Peripheral resistance decreases because of cutaneous vasodilation, and this increases levels of renal Na^+ and water absorption, expanding blood volume. Cardiac output is increased by the direct action of thyroid hormones, as well as that of catecholamines, on the heart, so that pulse pressure and cardiac rate are increased and circulation time is shortened.

T_3 is not formed from T_4 in cardiac myocytes to any degree, but circulatory T_3 enters the myocytes, combines with its receptors, and enters the nucleus, where it promotes the expression of some genes and inhibits the expression of others. Those that are enhanced include the genes for α-myosin heavy chain, sarcoplasmic reticulum Ca^{2+} ATPase, β-adrenergic receptors, G proteins, Na, K ATPase, and certain K^+ channels. Those that are inhibited include the genes for β-myosin heavy chain, phospholamban, two types of adenylyl cyclase, T_3 nuclear receptors, and NCX, the Na^+–Ca^{2+} exchanger. The net result is increased heart rate and force of contraction.

The two myosin heavy chain (MHC) isoforms, α-MHC and β-MHC, produced by the heart are encoded by two highly homologous genes located on the short arm of chromosome 17. Each myosin molecule consists of two heavy chains and two pairs of light chains (see Chapter 5). The myosin containing β-MHC has less ATPase activity than the myosin containing α-MHC. α-MHC predominates in the atria in adults, and its level is increased by treatment with thyroid hormone. This increases the speed of cardiac contraction. Conversely, expression of the α-MHC gene is depressed and that of the β-MHC gene is enhanced in hypothyroidism.

EFFECTS ON THE NERVOUS SYSTEM

In hypothyroidism, mentation is slow and the cerebrospinal fluid (CSF) protein level is elevated. Thyroid hormones reverse these changes, and large doses cause rapid mentation, irritability, and restlessness. Overall, cerebral blood flow and glucose and O_2 consumption by the brain are normal in adult hypo- and hyperthyroidism. However, thyroid hormones enter the brain in adults and are found in gray matter in numerous different locations. In addition, astrocytes in the brain convert T_4 to T_3, and there is a sharp increase in brain D_2 activity after thyroidectomy that is reversed within 4 h by a single intravenous dose of T_3. Some of the effects of thyroid hormones on the brain are probably secondary to increased responsiveness to catecholamines, with consequent increased activation of the reticular activating system (see Chapter 14). In addition, thyroid hormones have marked effects on brain development. The parts of the central nervous system (CNS) most affected are the cerebral cortex and the basal ganglia. In addition, the cochlea is also affected. Consequently, thyroid hormone deficiency during development causes mental retardation, motor rigidity, and deaf–mutism. Deficiencies in thyroid hormone synthesis secondary to a failure of thyrocytes to transport iodide presumably also contribute to deafness in Pendred syndrome, discussed above.

Thyroid hormones also exert effects on reflexes. The reaction time of stretch reflexes (see Chapter 12) is shortened in hyperthyroidism and prolonged in hypothyroidism. Measurement of the reaction time of the ankle jerk (Achilles reflex) has attracted attention as a clinical test for evaluating thyroid function, but this reaction time is also affected by other diseases and thus is not a specific assessment of thyroid activity.

RELATION TO CATECHOLAMINES

The actions of thyroid hormones and the catecholamines norepinephrine and epinephrine are intimately interrelated. Epinephrine increases the metabolic rate, stimulates the nervous system, and produces cardiovascular effects similar to those of thyroid hormones, although the duration of these actions is brief. Norepinephrine has generally similar actions. The toxicity of the catecholamines is markedly increased in rats treated with T_4. Although plasma catecholamine levels are normal in hyperthyroidism, the cardiovascular effects, tremulousness, and sweating that are seen in the setting of excess thyroid hormones can be reduced or abolished by sympathectomy. They can also be reduced by drugs such as propranolol that block β-adrenergic receptors. Indeed, propranolol and other β blockers are used extensively in the treatment of thyrotoxicosis and in the treatment of the severe exacerbations of hyperthyroidism called **thyroid storms.** However, even though β blockers are weak inhibitors of extrathyroidal conversion of T_4 to T_3, and consequently may produce a small fall in plasma T_3, they have little effect on the other actions of thyroid hormones. Presumably, the functional synergism observed between catecholamines and thyroid hormones, particularly in pathological settings, arises from their overlapping biological functions as well as the ability of thyroid hormones to increase expression of catecholamine receptors and the signaling effectors to which they are linked.

EFFECTS ON SKELETAL MUSCLE

Muscle weakness occurs in most patients with hyperthyroidism **(thyrotoxic myopathy),** and when the hyperthyroidism is severe and prolonged, the myopathy may be severe. The muscle weakness may be due in part to increased protein catabolism. Thyroid hormones affect the expression of the MHC genes in skeletal as well as cardiac muscle (see Chapter 5). However, the effects produced are complex and their relation to the myopathy is not established. Hypothyroidism is also associated with muscle weakness, cramps, and stiffness.

EFFECTS ON CARBOHYDRATE METABOLISM

Thyroid hormones increase the rate of absorption of carbohydrates from the gastrointestinal tract, an action that is probably independent of their calorigenic action. In hyperthyroidism, therefore, the plasma glucose level rises rapidly after a carbohydrate meal, sometimes exceeding the renal threshold. However, it falls again at a rapid rate.

EFFECTS ON CHOLESTEROL METABOLISM

Thyroid hormones lower circulating cholesterol levels. The plasma cholesterol level drops before the metabolic rate rises, which indicates that this action is independent of the stimulation of O_2 consumption. The decrease in plasma cholesterol concentration is due to increased formation of low-density lipoprotein (LDL) receptors in the liver, resulting in increased hepatic removal of cholesterol from the circulation. Despite considerable effort, however, it has not been possible to produce a clinically useful thyroid hormone analog that lowers plasma cholesterol without increasing metabolism.

EFFECTS ON GROWTH

Thyroid hormones are essential for normal growth and skeletal maturation (see Chapter 21). In hypothyroid children, bone growth is slowed and epiphysial closure delayed. In the absence of thyroid hormones, growth hormone secretion is also depressed. This further impairs growth and development, since thyroid hormones normally potentiate the effect of growth hormone on tissues.

CHAPTER SUMMARY

- The thyroid gland transports and fixes iodide to amino acids present in thyroglobulin to generate the thyroid hormones thyroxine (T_4) and triiodothyronine (T_3).
- Synthesis and secretion of thyroid hormones is stimulated by thyroid-stimulating hormone (TSH) from the pituitary, which in turn is released in response to thyrotropin-releasing hormone (TRH) from the hypothalamus. These releasing factors are controlled by changes in whole body status (eg, exposure to cold or stress).
- Thyroid hormones circulate in the plasma predominantly in protein-bound forms. Only the free hormones are biologically active, and both feed back to reduce secretion of TSH.
- Thyroid hormones exert their effects by entering cells and binding to thyroid receptors. The liganded forms of thyroid receptors are nuclear transcription factors that alter gene expression.
- Thyroid hormones stimulate metabolic rate, calorigenesis, cardiac function, and normal mentation, and interact synergistically with catecholamines. Thyroid hormones

also play critical roles in development, particularly of the nervous system, and growth.

- Disease results with both under- and overactivity of the thyroid gland. Hypothyroidism is accompanied by mental and physical slowing in adults, and by mental retardation and dwarfism if it occurs in neonatal life. Overactivity of the thyroid gland, which most commonly is caused by autoantibodies that trigger secretion (Graves disease) results in body wasting, nervousness, and tachycardia.

MULTIPLE-CHOICE QUESTIONS

For all questions, select the single best answer unless otherwise directed.

1. A 40-year-old woman comes to her primary care physician complaining of nervousness and an unexplained weight loss of 20 pounds over the past 3 months despite her impression that she is eating all the time. On physical examination, her eyes are found to be protruding, her skin is moist and warm, and her fingers have a slight tremor. Compared to a normal individual, a biopsy of her thyroid gland would most likely reveal which of the following:
   A. Decreased numbers of reabsorption lacunae
   B. Decreased evidence of endocytosis
   C. A decrease in the cross-sectional area occupied by colloid
   D. Increased levels of NIS in the basolateral membrane of thyrocytes
   E. Decreased evidence of lysosomal activity
2. Which of the following is *not* essential for normal biosynthesis of thyroid hormones?
   A. Iodine
   B. Ferritin
   C. Thyroglobulin
   D. Protein synthesis
   E. TSH
3. Increasing intracellular I^- due to the action of NIS is an example of
   A. Endocytosis
   B. Passive diffusion
   C. Na^+ and K^+ cotransport
   D. Primary active transport
   E. Secondary active transport
4. The metabolic rate is *least* affected by an increase in the plasma level of
   A. TSH
   B. TRH
   C. TBG
   D. Free T_4
   E. Free T_3
5. In which of the following conditions is it *most* likely that the TSH response to TRH will be reduced?
   A. Hypothyroidism due to tissue resistance to thyroid hormone
   B. Hypothyroidism due to disease destroying the thyroid gland
   C. Hyperthyroidism due to circulating antithyroid antibodies with TSH activity
   D. Hyperthyroidism due to diffuse hyperplasia of thyrotropes of the anterior pituitary
   E. Iodine deficiency
6. Hypothyroidism due to disease of the thyroid gland is associated with increased plasma levels of
   A. Cholesterol
   B. Albumin
   C. RT_3
   D. Iodide
   E. TBG
7. A young woman has puffy skin and a hoarse voice. Her plasma TSH concentration is low but increases markedly when she is given TRH. She probably has
   A. hyperthyroidism due to a thyroid tumor.
   B. hypothyroidism due to a primary abnormality in the thyroid gland.
   C. hypothyroidism due to a primary abnormality in the pituitary gland.
   D. hypothyroidism due to a primary abnormality in the hypothalamus.
   E. hyperthyroidism due to a primary abnormality in the hypothalamus.
8. The enzyme primarily responsible for the conversion of T_4 to T_3 in the periphery is
   A. D_1 thyroid deiodinase
   B. D_2 thyroid deiodinase
   C. D_3 thyroid deiodinase
   D. Thyroid peroxidase
   E. None of the above
9. Which of the following would be *least* affected by injections of TSH?
   A. Thyroidal uptake of iodine
   B. Synthesis of thyroglobulin
   C. Cyclic adenosine monophosphate (AMP) in thyroid cells
   D. Cyclic guanosine monophosphate (GMP) in thyroid cells
   E. Size of the thyroid
10. Thyroid hormone receptors bind to DNA in which of the following forms?
    A. A heterodimer with the prolactin receptor
    B. A heterodimer with the growth hormone receptor
    C. A heterodimer with the retinoid X receptor
    D. A heterodimer with the insulin receptor
    E. A heterodimer with the progesterone receptor

CHAPTER RESOURCES

Brent GA: Graves' disease. N Engl J Med 2008;358:2594.
Dohan O, Carrasco N: Advances in Na^+/I^- symporter (NIS) research in the thyroid and beyond. Mol Cell Endocrinol 2003;213:59.
Glaser B: Pendred syndrome. Pediatr Endocrinol Rev 2003;1(Suppl 2):199.
Peeters RP, van der Deure WM, Visser TJ: Genetic variation in thyroid hormone pathway genes: Polymorphisms in the TSH receptor and the iodothyronine deiodinases. Eur J Endocrinol 2006;155:655.

CHAPTER

20

The Adrenal Medulla & Adrenal Cortex

OBJECTIVES

After reading this chapter, you should be able to:

- Name the three catecholamines secreted by the adrenal medulla and summarize their biosynthesis, metabolism, and function.
- List the stimuli that increase adrenal medullary secretion.
- Differentiate between C_{18}, C_{19}, and C_{21} steroids and give examples of each.
- Outline the steps involved in steroid biosynthesis in the adrenal cortex.
- Name the plasma proteins that bind adrenocortical steroids and discuss their physiologic role.
- Name the major site of adrenocortical hormone metabolism and the principal metabolites produced from glucocorticoids, adrenal androgens, and aldosterone.
- Describe the mechanisms by which glucocorticoids and aldosterone produce changes in cellular function.
- List and briefly describe the physiologic and pharmacologic effects of glucocorticoids.
- Contrast the physiologic and pathologic effects of adrenal androgens.
- Describe the mechanisms that regulate secretion of glucocorticoids and adrenal sex hormones.
- List the actions of aldosterone and describe the mechanisms that regulate aldosterone secretion.
- Describe the main features of the diseases caused by excess or deficiency of each of the hormones of the adrenal gland.

INTRODUCTION

There are two endocrine organs in the adrenal gland, one surrounding the other. The main secretions of the inner **adrenal medulla** (Figure 20–1) are the catecholamines **epinephrine, norepinephrine,** and **dopamine;** the outer **adrenal cortex** secretes steroid hormones.

The adrenal medulla is in effect a sympathetic ganglion in which the postganglionic neurons have lost their axons and become secretory cells. The cells secrete when stimulated by the preganglionic nerve fibers that reach the gland via the splanchnic nerves. Adrenal medullary hormones work mostly to prepare the body for emergencies, the so-called "fight-or-flight" responses.

The adrenal cortex secretes **glucocorticoids,** steroids with widespread effects on the metabolism of carbohydrate and protein; and a **mineralocorticoid** essential to the maintenance of Na^+ balance and extracellular fluid (ECF) volume. It is also a secondary site of **androgen** synthesis, secreting sex hormones such as testosterone, which can exert effects on reproductive function. Mineralocorticoids and the glucocorticoids are necessary for survival. Adrenocortical secretion is controlled primarily by adrenocorticotropic hormone (ACTH) from the anterior pituitary, but mineralocorticoid secretion is also subject to independent control by circulating factors, of which the most important is **angiotensin II,** a peptide formed in the bloodstream by the action of **renin.**

ADRENAL MORPHOLOGY

The adrenal medulla, which constitutes 28% of the mass of the adrenal gland, is made up of interlacing cords of densely innervated granule-containing cells that abut on venous sinuses. Two cell types can be distinguished morphologically: an epinephrine-secreting type that has larger, less dense granules; and a norepinephrine-secreting type in which smaller, very dense granules fail to fill the vesicles in which they are contained. In humans, 90% of the cells are the epinephrine-secreting type and 10% are the norepinephrine-secreting type. The type of cell that secretes dopamine is unknown. **Paraganglia,** small groups of cells resembling those in the adrenal medulla, are found near the thoracic and abdominal sympathetic ganglia (Figure 20–1).

In adult mammals, the adrenal cortex is divided into three zones (Figure 20–2). The outer **zona glomerulosa** is made up of whorls of cells that are continuous with the columns of cells that form the **zona fasciculata.** These columns are separated by venous sinuses. The inner portion of the zona fasciculata merges into the **zona reticularis,** where the cell columns become interlaced in a network. The zona glomerulosa makes up 15% of the mass of the adrenal gland; the zona fasciculata, 50%; and the zona reticularis, 7%. The adrenocortical cells contain abundant lipid, especially in the outer portion of the zona fasciculata. All three cortical zones secrete **corticosterone,** but the active enzymatic mechanism for aldosterone biosynthesis is limited to the zona glomerulosa, whereas the enzymatic mechanisms for forming cortisol and sex hormones are found in the two inner zones. Furthermore, subspecialization occurs within the inner two zones, with the zona fasciculata secreting mostly glucocorticoids and the zona reticularis secreting mainly sex hormones.

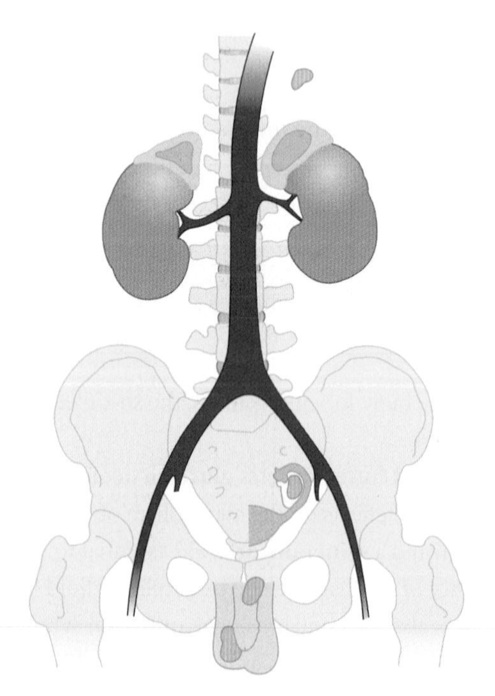

FIGURE 20–1 Human adrenal glands. Adrenocortical tissue is yellow; adrenal medullary tissue is orange. Note the location of the adrenals at the superior pole of each kidney. Also shown are extra-adrenal sites (gray) at which cortical and medullary tissue is sometimes found. (Reproduced with permission from Williams RH: *Textbook of Endocrinology,* 4th ed. Williams RH [editor]: Saunders, 1968.)

Arterial blood reaches the adrenal from many small branches of the phrenic and renal arteries and the aorta. From a plexus in the capsule, blood flows through the cortex to the sinusoids of the medulla. The medulla is also supplied by a few arterioles that pass directly to it from the capsule. In most species, including humans, blood from the medulla flows into a central adrenal vein. The blood flow through the adrenal is large, as it is in most endocrine glands.

During fetal life, the human adrenal is large and under pituitary control, but the three zones of the permanent cortex represent only 20% of the gland. The remaining 80% is the large **fetal adrenal cortex,** which undergoes rapid degeneration at the time of birth. A major function of this fetal adrenal is synthesis and secretion of sulfate conjugates of androgens that are converted in the placenta to estrogens (see Chapter 22). No structure is comparable to the human fetal adrenal in laboratory animals.

An important function of the zona glomerulosa, in addition to aldosterone synthesis, is the formation of new cortical cells. The adrenal medulla does not regenerate, but when the inner two zones of the cortex are removed, a new zona fasciculata and zona reticularis regenerate from glomerular cells attached to the capsule. Small capsular remnants regrow large pieces of adreno-cortical tissue. Immediately after hypophysectomy, the zona fasciculata and zona reticularis begin to atrophy, whereas the zona glomerulosa is unchanged because of the action of angiotensin II on this zone. The ability to secrete aldosterone and conserve Na^+ is normal for some time after hypophysectomy, but in long-standing hypopituitarism, aldosterone deficiency may develop, apparently because of the absence of a pituitary factor that maintains the responsiveness of the zona glomerulosa. Injections of ACTH and stimuli that cause endogenous ACTH secretion produce hypertrophy of the zona fasciculata and zona reticularis but actually decrease, rather than increase, the size of the zona glomerulosa.

The cells of the adrenal cortex contain large amounts of smooth endoplasmic reticulum, which is involved in the steroid-forming process. Other steps in steroid biosynthesis occur in the mitochondria. The structure of steroid-secreting cells is very similar throughout the body. The typical features of such cells are shown in Figure 20–3.

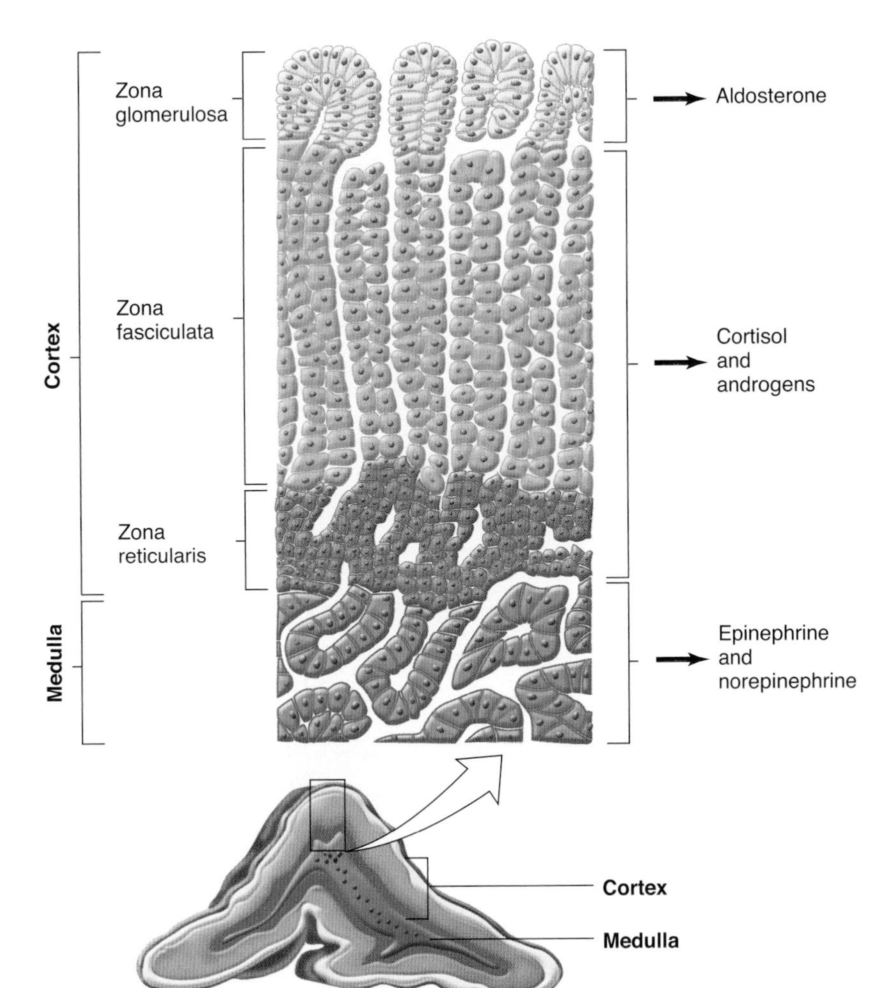

FIGURE 20–2 Section through an adrenal gland showing both the medulla and the zones of the cortex, as well as the hormones they secrete. (Reproduced with permission from Widmaier EP, Raff H, Strang KT: *Vander's Human Physiology: The Mechanisms of Body Function,* 11th ed. McGraw-Hill, 2008.)

ADRENAL MEDULLA: STRUCTURE & FUNCTION OF MEDULLARY HORMONES

CATECHOLAMINES

Norepinephrine, epinephrine, and small amounts of dopamine are synthesized by the adrenal medulla. Cats and some other species secrete mainly norepinephrine, but in dogs and humans, most of the catecholamine output in the adrenal vein is epinephrine. Norepinephrine also enters the circulation from noradrenergic nerve endings.

The structures of norepinephrine, epinephrine, and dopamine and the pathways for their biosynthesis and metabolism are discussed in Chapter 7. Norepinephrine is formed by hydroxylation and decarboxylation of tyrosine, and epinephrine by methylation of norepinephrine. Phenylethanolamine-N-methyltransferase (PNMT), the enzyme that catalyzes the formation of epinephrine from norepinephrine, is found in appreciable quantities only in the brain and the adrenal medulla. Adrenal medullary PNMT is induced by glucocorticoids. Although relatively large amounts are required, the glucocorticoid concentration is high in the blood draining from the cortex to the medulla. After hypophysectomy, the glucocorticoid concentration of this blood falls and epinephrine synthesis is decreased. In addition, glucocorticoids are apparently necessary for the normal development of the adrenal medulla; in 21β-hydroxylase deficiency, glucocorticoid secretion is reduced during fetal life and the adrenal medulla is dysplastic. In untreated 21β-hydroxylase deficiency, circulating catecholamines are low after birth.

In plasma, about 95% of the dopamine and 70% of the norepinephrine and epinephrine are conjugated to sulfate. Sulfate conjugates are inactive and their function is unsettled. In recumbent humans, the normal plasma level of free norepinephrine is about 300 pg/mL (1.8 nmol/L). On standing, the level increases 50–100% (Figure 20–4). The plasma norepinephrine level is generally unchanged after adrenalectomy, but the free epinephrine level, which is normally about 30 pg/mL (0.16 nmol/L), falls to essentially zero. The epinephrine found in tissues other than the adrenal medulla and the brain is for the most part absorbed from the bloodstream rather than synthesized in situ. Interestingly, low levels of epinephrine reappear in the blood some time after bilateral adrenalectomy, and these levels are regulated like those secreted by the adrenal medulla. They may come from cells such as the intrinsic cardiac adrenergic (ICA) cells (see Chapter 13), but their exact source is unknown.

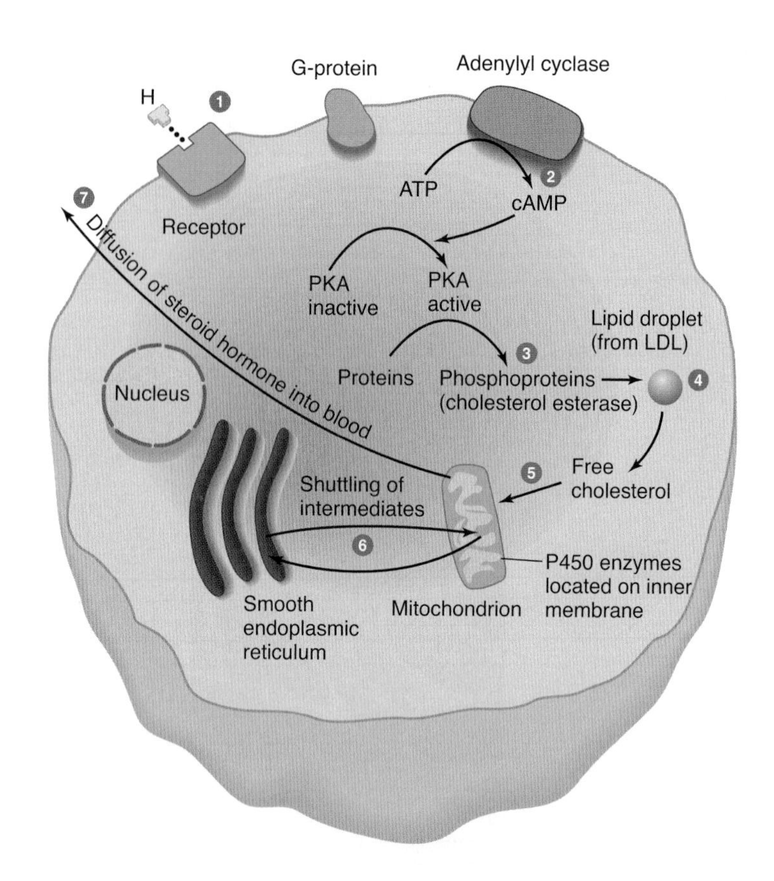

FIGURE 20–3 Schematic overview of the structures of steroid-secreting cells and the intracellular pathway of steroid synthesis. LDL, low-density lipoprotein; PKA, protein kinase A. (Reproduced with permission from Widmaier EP, Raff H, Strang KT: *Vander's Human Physiology: The Mechanisms of Body Function,* 11th ed. McGraw-Hill, 2008.)

Plasma dopamine levels are normally very low, about 0.13 nmol/L. Most plasma dopamine is thought to be derived from sympathetic noradrenergic ganglia.

The catecholamines have a half-life of about 2 min in the circulation. For the most part, they are methoxylated and then oxidized to 3-methoxy-4-hydroxymandelic acid (vanillylmandelic acid [VMA]; see Chapter 7). About 50% of the secreted catecholamines appear in the urine as free or conjugated metanephrine and normetanephrine, and 35% as VMA. Only small amounts of free norepinephrine and epinephrine are excreted. In normal humans, about 30 μg of norepinephrine, 6 μg of epinephrine, and 700 μg of VMA are excreted per day.

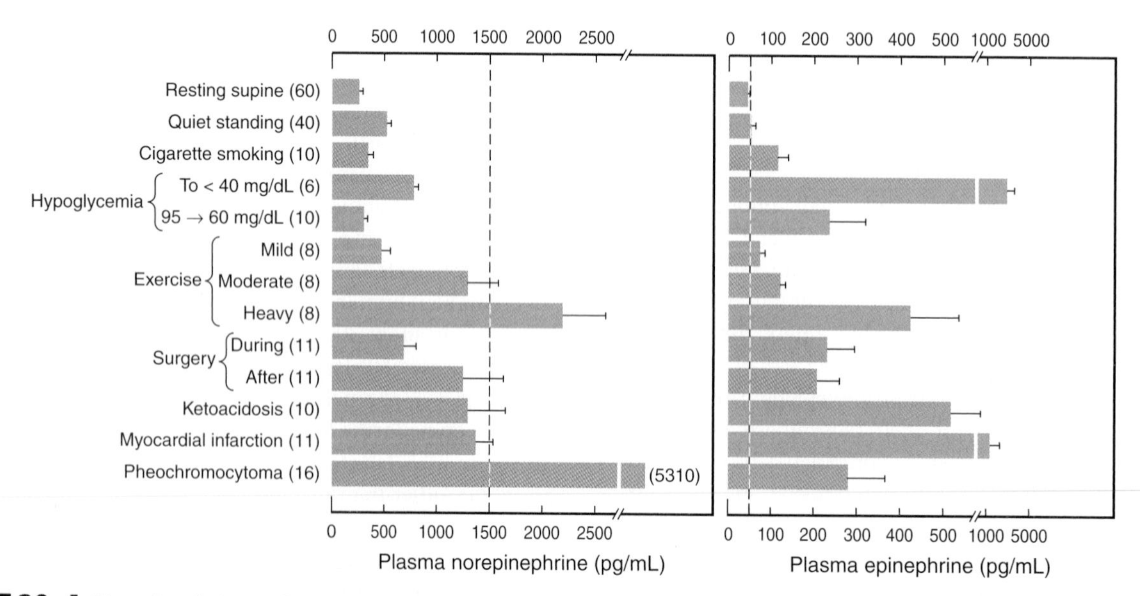

FIGURE 20–4 Norepinephrine and epinephrine levels in human venous blood in various physiologic and pathologic states. Note that the horizontal scales are different. The numbers to the left in parentheses are the numbers of subjects tested. In each case, the vertical dashed line identifies the threshold plasma concentration at which detectable physiologic changes are observed. (Modified and reproduced with permission from Cryer PE: Physiology and pathophysiology of the human sympathoadrenal neuroendocrine system. N Engl J Med 1980;303:436.)

OTHER SUBSTANCES SECRETED BY THE ADRENAL MEDULLA

In the medulla, norepinephrine and epinephrine are stored in granules with ATP. The granules also contain chromogranin A (see Chapter 7). Secretion is initiated by acetylcholine released from the preganglionic neurons that innervate the secretory cells. Acetylcholine activates cation channels allowing Ca^{2+} to enter the cells from the ECF and trigger the exocytosis of the granules. In this fashion, catecholamines, ATP, and proteins from the granules are all released into the blood together.

Epinephrine-containing cells of the medulla also contain and secrete opioid peptides (see Chapter 7). The precursor molecule is preproenkephalin. Most of the circulating metenkephalin comes from the adrenal medulla. The circulating opioid peptides do not cross the blood–brain barrier.

Adrenomedullin, a vasodepressor polypeptide found in the adrenal medulla, is discussed in Chapter 32.

EFFECTS OF EPINEPHRINE & NOREPINEPHRINE

In addition to mimicking the effects of noradrenergic nervous discharge, norepinephrine and epinephrine exert metabolic effects that include glycogenolysis in liver and skeletal muscle, mobilization of free fatty acids (FFA), increased plasma lactate, and stimulation of the metabolic rate. The effects of norepinephrine and epinephrine are brought about by actions on two classes of receptors: α- and β-adrenergic receptors. α receptors are subdivided into two groups, α_1 and α_2 receptors, and β receptors into β_1, β_2, and β_3 receptors, as outlined in Chapter 7. There are three subtypes of α_1 receptors and three subtypes of α_2 receptors (see Table 7–2).

Norepinephrine and epinephrine both increase the force and rate of contraction of the isolated heart. These responses are mediated by β_1 receptors. The catecholamines also increase myocardial excitability, causing extrasystoles and, occasionally, more serious cardiac arrhythmias. Norepinephrine produces vasoconstriction in most if not all organs via α_1 receptors, but epinephrine dilates the blood vessels in skeletal muscle and the liver via β_2 receptors. This usually overbalances the vasoconstriction produced by epinephrine elsewhere, and the total peripheral resistance drops. When norepinephrine is infused slowly in normal animals or humans, the systolic and diastolic blood pressures rise. The **hypertension** stimulates the carotid and aortic baroreceptors, producing reflex bradycardia that overrides the direct cardioacceleratory effect of norepinephrine. Consequently, cardiac output per minute falls. Epinephrine causes a widening of the pulse pressure, but because baroreceptor stimulation is insufficient to obscure the direct effect of the hormone on the heart, cardiac rate, and output increase. These changes are summarized in Figure 20–5.

Catecholamines increase alertness (see Chapter 14). Epinephrine and norepinephrine are equally potent in this regard, although in humans epinephrine usually evokes more anxiety and fear.

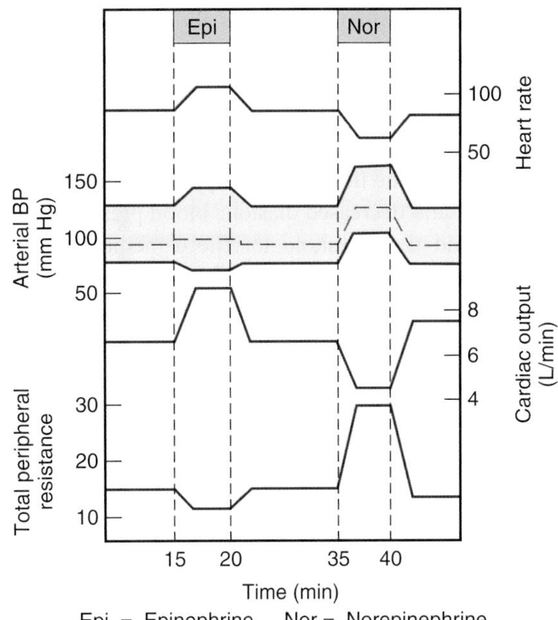

FIGURE 20–5 Circulatory changes produced in humans by the slow intravenous infusion of epinephrine and norepinephrine.

The catecholamines have several different actions that affect blood glucose. Epinephrine and norepinephrine both cause glycogenolysis. They produce this effect via β-adrenergic receptors that increase cyclic adenosine monophosphate (cAMP), with activation of phosphorylase, and via α-adrenergic receptors that increase intracellular Ca^{2+} (see Chapter 7). In addition, the catecholamines increase the secretion of insulin and glucagon via β-adrenergic mechanisms and inhibit the secretion of these hormones via α-adrenergic mechanisms.

Norepinephrine and epinephrine also produce a prompt rise in the metabolic rate that is independent of the liver and a smaller, delayed rise that is abolished by hepatectomy and coincides with the rise in blood lactate concentration. The initial rise in metabolic rate may be due to cutaneous vasoconstriction, which decreases heat loss and leads to a rise in body temperature, or to increased muscular activity, or both. The second rise is probably due to oxidation of lactate in the liver. Mice unable to make norepinephrine or epinephrine because their dopamine β-hydroxylase gene is knocked out are intolerant of cold, but surprisingly, their basal metabolic rate is elevated. The cause of this elevation is unknown.

When injected, epinephrine and norepinephrine cause an initial rise in plasma K^+ because of release of K^+ from the liver and then a prolonged fall in plasma K^+ because of an increased entry of K^+ into skeletal muscle that is mediated by β_2-adrenergic receptors. Some evidence suggests that activation of α receptors opposes this effect.

The increases in plasma norepinephrine and epinephrine that are needed to produce the various effects listed above have been determined by infusion of catecholamines in resting humans. In general, the threshold for the cardiovascular and the

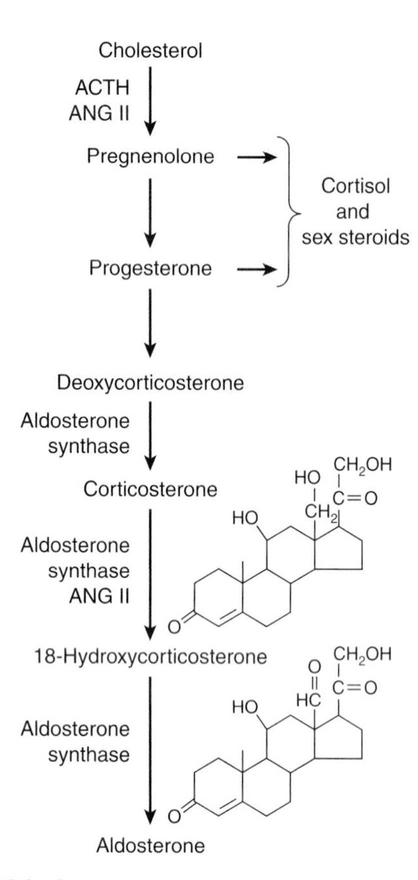

FIGURE 20–8 Hormone synthesis in the zona glomerulosa. The zona glomerulosa lacks 17α-hydroxylase activity, and only the zona glomerulosa can convert corticosterone to aldosterone because it is the only zone that normally contains aldosterone synthase. ANG II, angiotensin II.

sheep, monkeys, and humans secrete predominantly cortisol. In humans, the ratio of secreted cortisol to corticosterone is approximately 7:1.

STEROID BIOSYNTHESIS

The major paths by which the naturally occurring adrenocortical hormones are synthesized in the body are summarized in Figures 20–7 and 20–8. The precursor of all steroids is cholesterol. Some of the cholesterol is synthesized from acetate, but most of it is taken up from LDL in the circulation. LDL receptors are especially abundant in adrenocortical cells. The cholesterol is esterified and stored in lipid droplets. **Cholesterol ester hydrolase** catalyzes the formation of free cholesterol in the lipid droplets (Figure 20–9). The cholesterol is transported to mitochondria by a sterol carrier protein. In the mitochondria, it is converted to pregnenolone in a reaction catalyzed by an enzyme known as **cholesterol desmolase** or **side-chain cleavage enzyme.** This enzyme, like most of the enzymes involved in steroid biosynthesis, is a member of the cytochrome P450 superfamily and is also known as **P450scc** or **CYP11A1.** For convenience, the various names of the enzymes involved in adrenocortical steroid biosynthesis are summarized in Table 20–3.

Pregnenolone moves to the smooth endoplasmic reticulum, where some of it is dehydrogenated to form progesterone in a reaction catalyzed by **3β-hydroxysteroid dehydrogenase.** This enzyme has a molecular weight of 46,000 and is not a cytochrome P450. It also catalyzes the conversion of 17α-hydroxypregnenolone to 17α-hydroxyprogesterone, and dehydroepiandrosterone to androstenedione (Figure 22–7) in the smooth endoplasmic reticulum. The 17α-hydroxypregnenolone and the 17α-hydroxyprogesterone are formed from pregnenolone and progesterone, respectively (Figure 20–7) by the action of **17α-hydroxylase.** This is another mitochondrial P450, and it is also known as **P450c17**

CLINICAL BOX 20–1

Synthetic Steroids

As with many other naturally occurring substances, the activity of adrenocortical steroids can be increased by altering their structure. A number of synthetic steroids are available that have many times the activity of cortisol. The relative glucocorticoid and mineralocorticoid potencies of the natural steroids are compared with those of the synthetic steroids 9α-fluorocortisol, prednisolone, and dexamethasone in Table 20–2. The potency of dexamethasone is due to its high affinity for glucocorticoid receptors and its long half-life. Prednisolone also has a long half-life.

TABLE 20–1 Principal adrenocortical hormones in adult humans.[a]

Name	Synonyms	Average Plasma Concentration (Free and Bound)[a] μg/dL)	Average Amount Secreted (mg/24 h)
Cortisol	Compound F, hydrocortisone	13.9	10
Corticosterone	Compound B	0.4	3
Aldosterone		0.0006	0.15
Deoxycorticosterone	DOC	0.0006	0.20
Dehydroepiandrosterone sulfate	DHEAS	175.0	20

[a]All plasma concentration values except DHEAS are fasting morning values after overnight recumbency.

TABLE 20–2 Relative potencies of corticosteroids compared with cortisol.[a]

Steroid	Glucocorticoid Activity	Mineralocorticoid Activity
Cortisol	1.0	1.0
Corticosterone	0.3	15
Aldosterone	0.3	3000
Deoxycorticosterone	0.2	100
Costisone	0.7	0.8
Prednisolone	4	0.8
9α-Fluorocortisol	10	125
Dexamethasone	25	~0

[a]Values are approximations based on liver glycogen deposition or anti-inflammatory assays for glucocorticoid activity, and effect on urinary Na^+/K^+ or maintenance of adrenalectomized animals for mineralocorticoid activity. The last three steroids listed are synthetic compounds that do not occur naturally.

or **CYP17.** Located in another part of the same enzyme is **17,20-lyase** activity that breaks the 17,20 bond, converting 17α-pregnenolone and 17α-progesterone to the C_{19} steroids dehydroepiandrosterone and androstenedione.

Hydroxylation of progesterone to 11-deoxycorticosterone and of 17α-hydroxyprogesterone to 11-deoxycortisol occurs in the smooth endoplasmic reticulum. These reactions are catalyzed by 21β-hydroxylase, a cytochrome P450 that is also known as **P450c21** or **CYP21A2.**

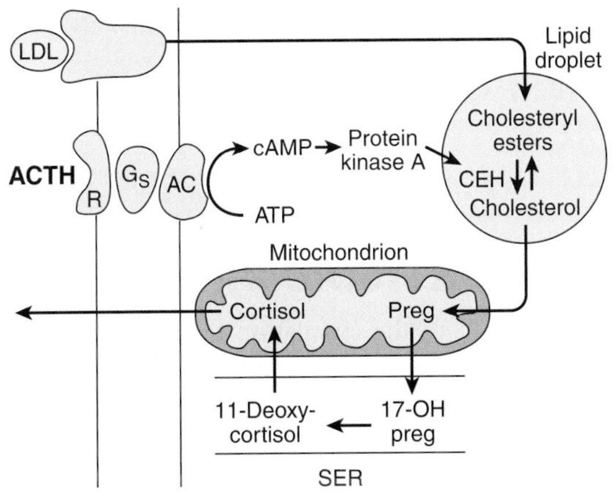

FIGURE 20–9 Mechanism of action of ACTH on cortisol-secreting cells in the inner two zones of the adrenal cortex. When ACTH binds to its receptor (R), adenylyl cyclase (AC) is activated via Gs. The resulting increase in cAMP activates protein kinase A, and the kinase phosphorylates cholesteryl ester hydrolase (CEH), increasing its activity. Consequently, more free cholesterol is formed and converted to pregnenolone. Note that in the subsequent steps in steroid biosynthesis, products are shuttled between the mitochondria and the smooth endoplasmic reticulum (SER). Corticosterone is also synthesized and secreted.

11-deoxycorticosterone and the 11-deoxycortisol move back to the mitochondria, where they are 11-hydroxylated to form corticosterone and cortisol. These reactions occur in the zona fasciculata and zona reticularis and are catalyzed by 11β-hydroxylase, a cytochrome P450 also known as **P450c11** or **CYP11B1.**

In the zona glomerulosa there is no 11β-hydroxylase but a closely related enzyme called **aldosterone synthase** is present. This cytochrome P450 is 95% identical to 11β-hydroxylase and is also known as **P450c11AS** or **CYP11B2.** The genes that code CYP11B1 and CYP11B2 are both located on chromosome 8. However, aldosterone synthase is normally found only in the zona glomerulosa. The zona glomerulosa also lacks 17α-hydroxylase. This is why the zona glomerulosa makes aldosterone but fails to make cortisol or sex hormones.

Furthermore, subspecialization occurs within the inner two zones. The zona fasciculata has more 3β-hydroxysteroid dehydrogenase activity than the zona reticularis, and the zona reticularis has more of the cofactors required for the 17,20-lyase activity of 17α-hydroxylase. Therefore, the zona fasciculata makes more cortisol and corticosterone, and the zona reticularis makes more androgens. Most of the dehydroepiandrosterone that is formed is converted to dehydroepiandrosterone sulfate (DHEAS) by **adrenal sulfokinase,** and this enzyme is localized in the zona reticularis as well.

ACTION OF ACTH

ACTH binds to high-affinity receptors on the plasma membrane of adrenocortical cells. This activates adenylyl cyclase via G_s. The resulting reactions (Figure 20–9) lead to a prompt increase in the formation of pregnenolone and its derivatives, with secretion of the latter. Over longer periods, ACTH also increases the synthesis of the P450s involved in the synthesis of glucocorticoids.

ACTIONS OF ANGIOTENSIN II

Angiotensin II binds to AT_1 receptors (see Chapter 38) in the zona glomerulosa that act via a G protein to activate phospholipase C. The resulting increase in protein kinase C fosters the conversion of cholesterol to pregnenolone (Figure 20–8) and facilitates the action of aldosterone synthase, resulting in increased secretion of aldosterone.

ENZYME DEFICIENCIES

The consequences of inhibiting any of the enzyme systems involved in steroid biosynthesis can be predicted from Figures 20–7 and 20–8. Congenital defects in the enzymes lead to deficient cortisol secretion and the syndrome of **congenital adrenal hyperplasia.** The hyperplasia is due to increased ACTH secretion. Cholesterol desmolase deficiency is fatal in utero because it prevents the placenta from making the progesterone necessary for pregnancy to continue. A cause of severe congenital adrenal hyperplasia in newborns is a loss of function mutation

TABLE 20–3 Nomenclature for adrenal steroidogenic enzymes and their location in adrenal cells.

Trivial Name	P450	CYP	Location
Cholesterol desmolase; side-chain cleavage enzyme	P450scc	CYP11A1	Mitochondria
3β-Hydroxysteroid dehydrogenase	. . .	. . .	SER
17α-Hydroxylase, 17,20-lyase	P450c17	CYP17	Mitochondria
21β-Hydroxylase	P450c21	CYP21A2	SER
11β-Hydroxylase	P450c11	CYP11B1	Mitochondria
Aldosterone synthase	P450c11AS	CYP11B2	Mitochondria

SER, smooth endoplasmic reticulum.

of the gene for the **steroidogenic acute regulatory (StAR) protein.** This protein is essential in the adrenals and gonads but not in the placenta for the normal movement of cholesterol into the mitochondria to reach cholesterol desmolase, which is located on the matrix space side of the internal mitochondrial membrane (see Chapter 16). In its absence, only small amounts of steroids are formed. The degree of ACTH stimulation is marked, resulting eventually in accumulation of large numbers of lipoid droplets in the adrenal. For this reason, the condition is called **congenital lipoid adrenal hyperplasia.** Because androgens are not formed, female genitalia develop regardless of genetic sex (see Chapter 22). In 3β hydroxysteroid dehydrogenase deficiency, another rare condition, DHEA secretion is increased. This steroid is a weak androgen that can cause some masculinization in females with the disease, but it is not adequate to produce full masculinization of the genitalia in genetic males. Consequently, **hypospadias,** a condition where the opening of the urethra is on the underside of the penis rather than its tip, is common. In fully developed 17α-hydroxylase deficiency, a third rare condition due to a mutated gene for **CYP17,** no sex hormones are produced, so female external genitalia are present. However, the pathway leading to corticosterone and aldosterone is intact, and elevated levels of 11-deoxycorticosterone and other mineralocorticoids produce hypertension and hypokalemia. Cortisol is deficient, but this is partially compensated by the glucocorticoid activity of corticosterone.

Unlike the defects discussed in the preceding paragraph, 21β-hydroxylase deficiency is common, accounting for 90% or more of the enzyme deficiency cases. The 21β-hydroxylase gene, which is in the human leukocyte antigen (HLA) complex of genes on the short arm of chromosome 6 (see Chapter 3) is one of the most polymorphic in the human genome. Mutations occur at many different sites in the gene, and the abnormalities that are produced therefore range from mild to severe. Production of cortisol and aldosterone are generally reduced, so ACTH secretion and consequently production of precursor steroids are increased. These steroids are converted to androgens, producing **virilization.** The characteristic pattern that develops in females in the absence of treatment is the **adrenogenital syndrome.** Masculization may not be marked until later in life and mild cases can be detected only by laboratory tests. In 75% of the cases, aldosterone deficiency causes appreciable loss of Na^+ (**salt-losing form** of adrenal hyperplasia). The resulting hypovolemia can be severe.

In 11β-hydroxylase deficiency, virilization plus excess secretion of 11-deoxycortisol and 11-deoxycorticosterone take place. Because the former is an active mineralocorticoid, patients with this condition also have salt and water retention and, in two-thirds of the cases, hypertension (**hypertensive form** of congenital adrenal hyperplasia).

Glucocorticoid treatment is indicated in all of the virilizing forms of congenital adrenal hyperplasia because it repairs the glucocorticoid deficit and inhibits ACTH secretion, reducing the abnormal secretion of androgens and other steroids.

Expression of the cytochrome P450 enzymes responsible for steroid hormone biosynthesis depends on **steroid factor-1 (SF-1),** an orphan nuclear receptor. If *Ft2-F1,* the gene for SF-1, is knocked out, the gonads as well as adrenals fail to develop and additional abnormalities are present at the pituitary and hypothalamic level.

TRANSPORT, METABOLISM, & EXCRETION OF ADRENOCORTICAL HORMONES

GLUCOCORTICOID BINDING

Cortisol is bound in the circulation to an α globulin called **transcortin** or **corticosteroid-binding globulin (CBG).** A minor degree of binding to albumin also takes place. Corticosterone is similarly bound, but to a lesser degree. The half-life of cortisol in the circulation is therefore longer (about 60–90 min) than that of corticosterone (50 min). Bound steroids are physiologically inactive (see Chapter 16). In addition, relatively little free cortisol and corticosterone are found in the urine because of protein binding.

The equilibrium between cortisol and its binding protein and the implications of binding in terms of tissue supplies and ACTH secretion are summarized in Figure 20–10. The bound cortisol functions as a circulating reservoir of hormone that keeps a supply of free cortisol available to the tissues.

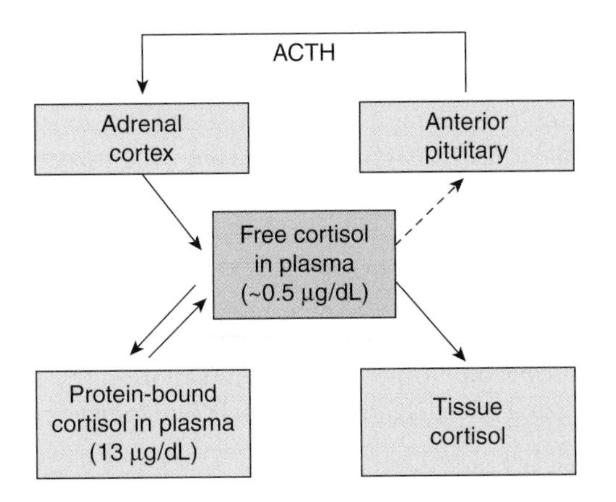

FIGURE 20–10 The interrelationships of free and bound cortisol. The dashed arrow indicates that cortisol inhibits ACTH secretion. The value for free cortisol is an approximation; in most studies, it is calculated by subtracting the protein-bound cortisol from the total plasma cortisol.

The relationship is similar to that of T_4 and its binding protein (see Chapter 19). At normal levels of total plasma cortisol (13.5 µg/dL or 375 nmol/L), very little free cortisol is present in the plasma, but the binding sites on CBG become saturated when the total plasma cortisol exceeds 20 µg/dL. At higher plasma levels, binding to albumin increases, but the main increase is in the unbound fraction.

CBG is synthesized in the liver and its production is increased by estrogen. CBG levels are elevated during pregnancy and depressed in cirrhosis, nephrosis, and multiple myeloma. When the CBG level rises, more cortisol is bound, and initially the free cortisol level drops. This stimulates ACTH secretion, and more cortisol is secreted until a new equilibrium is reached at which the bound cortisol is elevated but the free cortisol is normal. Changes in the opposite direction occur when the CBG level falls. This explains why pregnant women have high total plasma cortisol levels without symptoms of glucocorticoid excess and, conversely, why some patients with nephrosis have low total plasma cortisol without symptoms of glucocorticoid deficiency.

METABOLISM & EXCRETION OF GLUCOCORTICOIDS

Cortisol is metabolized in the liver, which is the principal site of glucocorticoid catabolism. Most of the cortisol is reduced to dihydrocortisol and then to tetrahydrocortisol, which is conjugated to glucuronic acid (Figure 20–11). The glucuronyl transferase system responsible for this conversion also catalyzes the formation of the glucuronides of bilirubin (see Chapter 28) and a number of hormones and drugs. Competitive inhibition takes place between these substrates for the enzyme system.

The liver and other tissues contain the enzyme 11β hydroxysteroid dehydrogenase. There are at least two forms of this enzyme. Type 1 catalyzes the conversion of cortisol to cortisone and the reverse reaction, though it functions primarily as a reductase, forming cortisol from corticosterone. Type 2 catalyzes almost exclusively the one-way conversion of cortisol to cortisone. Cortisone is an active glucocorticoid because it is

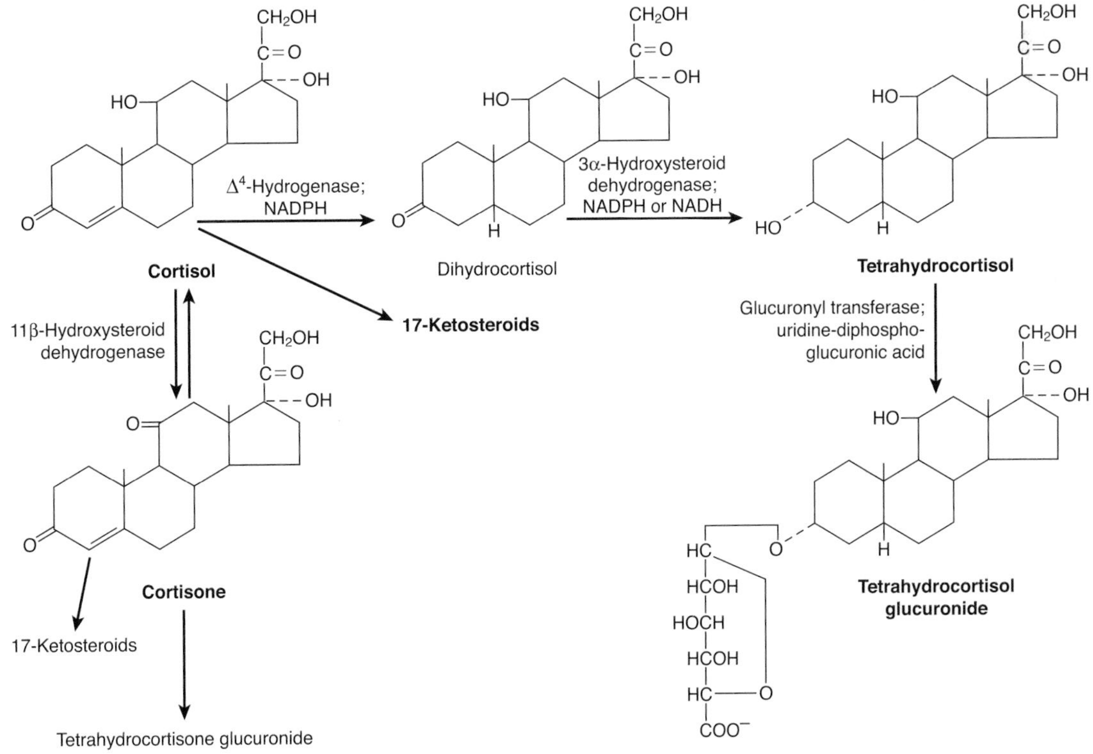

Cortisol
Δ^4-Hydrogenase; NADPH
Dihydrocortisol
3α-Hydroxysteroid dehydrogenase; NADPH or NADH
Tetrahydrocortisol
17-Ketosteroids
11β-Hydroxysteroid dehydrogenase
Glucuronyl transferase; uridine-diphospho-glucuronic acid
Cortisone
17-Ketosteroids
Tetrahydrocortisone glucuronide
Tetrahydrocortisol glucuronide

FIGURE 20–11 Outline of hepatic metabolism of cortisol.

CLINICAL BOX 20–2

Variations in the Rate of Hepatic Metabolism

The rate of hepatic inactivation of glucocorticoids is depressed in liver disease and, interestingly, during surgery and other stresses. Thus, in stressed humans, the plasma-free cortisol level rises higher than it does with maximal ACTH stimulation in the absence of stress.

converted to cortisol, and it is well known because of its extensive use in medicine. It is not secreted in appreciable quantities by the adrenal glands. Little, if any, of the cortisone formed in the liver enters the circulation, because it is promptly reduced and conjugated to form tetrahydrocortisone glucuronide. The tetrahydroglucuronide derivatives ("conjugates") of cortisol and corticosterone are freely soluble. They enter the circulation, where they do not become bound to protein. They are rapidly excreted in the urine.

About 10% of the secreted cortisol is converted in the liver to the 17-ketosteroid derivatives of cortisol and cortisone. The ketosteroids are conjugated for the most part to sulfate and then excreted in the urine. Other metabolites, including 20-hydroxy derivatives, are formed. There is an enterohepatic circulation of glucocorticoids and about 15% of the secreted cortisol is excreted in the stool. The metabolism of corticosterone is similar to that of cortisol, except that it does not form a 17-ketosteroid derivative (see Clinical Box 20–2).

ALDOSTERONE

Aldosterone is bound to protein to only a slight extent, and its half-life is short (about 20 min). The amount secreted is small (Table 20–1), and the total plasma aldosterone level in humans is normally about 0.006 μg/dL (0.17 nmol/L), compared with a cortisol level (bound and free) of about 13.5 μg/dL (375 nmol/L). Much of the aldosterone is converted in the liver to the tetrahydroglucuronide derivative, but some is changed in the liver and in the kidneys to an 18-glucuronide. This glucuronide, which is unlike the breakdown products of other steroids, is converted to free aldosterone by hydrolysis at pH 1.0, and it is therefore often referred to as the "acid-labile conjugate." Less than 1% of the secreted aldosterone appears in the urine in the free form. Another 5% is in the form of the acid-labile conjugate, and up to 40% is in the form of the tetrahydroglucuronide.

17-KETOSTEROIDS

The major adrenal androgen is the 17-ketosteroid dehydroepiandrosterone, although androstenedione is also secreted. The 11-hydroxy derivative of androstenedione and the 17-ketosteroids formed from cortisol and cortisone by side chain cleavage in the liver are the only 17-ketosteroids that have an =O or an —OH group in the 11 position ("11-oxy-17-ketosteroids"). Testosterone is also converted to a 17-ketosteroid. Because the daily 17-ketosteroid excretion in normal adults is 15 mg in men and 10 mg in women, about two thirds of the urinary ketosteroids in men are secreted by the adrenal or formed from cortisol in the liver and about one third are of testicular origin.

Etiocholanolone, one of the metabolites of the adrenal androgens and testosterone, can cause fever when it is unconjugated (see Chapter 17). Certain individuals have episodic bouts of fever due to periodic accumulation in the blood of unconjugated etiocholanolone ("etiocholanolone fever").

EFFECTS OF ADRENAL ANDROGENS & ESTROGENS

ANDROGENS

Androgens are the hormones that exert masculinizing effects and they promote protein anabolism and growth (see Chapter 23). Testosterone from the testes is the most active androgen and the adrenal androgens have less than 20% of its activity. Secretion of the adrenal androgens is controlled acutely by ACTH and not by gonadotropins. However, the concentration of DHEAS increases until it peaks at about 225 mg/dL in the early 20s, and then falls to very low values in old age (Figure 20–12). These long-term changes are not due to changes in ACTH secretion and appear to be due instead to a rise and then a gradual fall in the lyase activity of 17α-hydroxylase.

All but about 0.3% of the circulating DHEA is conjugated to sulfate (DHEAS). The secretion of adrenal androgens is nearly as great in castrated males and females as it is in normal males, so it is clear that these hormones exert very little masculinizing effect when secreted in normal amounts. However, they can produce appreciable masculinization when secreted

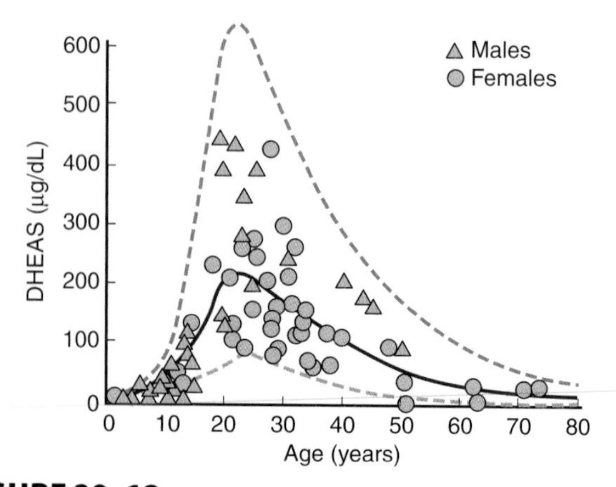

FIGURE 20–12 Change in serum dehydroepiandrosterone sulfate (DHEAS) with age. The middle line is the mean, and the dashed lines identify ±1.96 standard deviations. (Reproduced, with permission, from Smith MR, et al: A radioimmunoassay for the estimation of serum dehydroepiandrosterone sulfate in normal and pathological sera. Clin Chim Acta 1975;65:5.)

in excessive amounts. In adult males, excess adrenal androgens merely accentuate existing characteristics, but in prepubertal boys they can cause precocious development of the secondary sex characteristics without testicular growth **(precocious pseudopuberty).** In females they cause female pseudo-hermaphroditism and the adrenogenital syndrome. Some health practitioners recommend injections of dehydroepiandrosterone to combat the effects of aging (see Chapter 1), but results to date are controversial at best.

ESTROGENS

The adrenal androgen androstenedione is converted to testosterone and to estrogens (aromatized) in fat and other peripheral tissues. This is an important source of estrogens in men and postmenopausal women (see Chapters 22 and 23).

PHYSIOLOGIC EFFECTS OF GLUCOCORTICOIDS

ADRENAL INSUFFICIENCY

In untreated adrenal insufficiency, Na^+ loss and shock occurs due to the lack of mineralocorticoid activity, as well as abnormalities of water, carbohydrate, protein, and fat metabolism due to the lack of glucocorticoids. These metabolic abnormalities are eventually fatal despite mineralocorticoid treatment. Small amounts of glucocorticoids correct the metabolic abnormalities, in part directly and in part by permitting other reactions to occur. It is important to separate these physiologic actions of glucocorticoids from the quite different effects produced by large amounts of the hormones.

MECHANISM OF ACTION

The multiple effects of glucocorticoids are triggered by binding to glucocorticoid receptors, and the steroid–receptor complexes act as transcription factors that promote the transcription of certain segments of DNA (see Chapter 1). This, in turn, leads via the appropriate mRNAs to synthesis of enzymes that alter cell function. In addition, it seems likely that glucocorticoids have nongenomic actions.

EFFECTS ON INTERMEDIARY METABOLISM

The actions of glucocorticoids on the intermediary metabolism of carbohydrate, protein, and fat are discussed in Chapter 24. They include increased protein catabolism and increased hepatic glycogenesis and gluconeogenesis. Glucose 6-phosphatase activity is increased, and the plasma glucose level rises. Glucocorticoids exert an anti-insulin action in peripheral tissues and make diabetes worse. However, the brain and the heart are spared, so the increase in plasma glucose provides extra glucose to these vital organs. In diabetics, glucocorticoids raise plasma lipid levels and increase ketone body formation, but in normal individuals, the increase in insulin secretion provoked by the rise in plasma glucose obscures these actions. In adrenal insufficiency, the plasma glucose level is normal as long as an adequate caloric intake is maintained, but fasting causes hypoglycemia that can be fatal. The adrenal cortex is not essential for the ketogenic response to fasting.

PERMISSIVE ACTION

Small amounts of glucocorticoids must be present for a number of metabolic reactions to occur, although the glucocorticoids do not produce the reactions by themselves. This effect is called their **permissive action.** Permissive effects include the requirement for glucocorticoids to be present for glucagon and catecholamines to exert their calorigenic effects (see above and Chapter 24), for catecholamines to exert their lipolytic effects, and for catecholamines to produce pressor responses and bronchodilation.

EFFECTS ON ACTH SECRETION

Glucocorticoids inhibit ACTH secretion, which represents a negative feedback response on the pituitary. ACTH secretion is increased in adrenalectomized animals. The consequences of the negative feedback action of cortisol on ACTH secretion are discussed below in the section on regulation of glucocorticoid secretion.

VASCULAR REACTIVITY

In adrenally insufficient animals, vascular smooth muscle becomes unresponsive to norepinephrine and epinephrine. The capillaries dilate and, terminally, become permeable to colloidal dyes. Failure to respond to the norepinephrine liberated at noradrenergic nerve endings probably impairs vascular compensation for the hypovolemia of adrenal insufficiency and promotes vascular collapse. Glucocorticoids restore vascular reactivity.

EFFECTS ON THE NERVOUS SYSTEM

Changes in the nervous system in adrenal insufficiency that are reversed only by glucocorticoids include the appearance of electroencephalographic waves slower than the normal β rhythm, and personality changes. The latter, which are mild, include irritability, apprehension, and inability to concentrate.

EFFECTS ON WATER METABOLISM

Adrenal insufficiency is characterized by an inability to excrete a water load, causing the possibility of water intoxication. Only glucocorticoids repair this deficit. In patients with adrenal insufficiency who have not received glucocorticoids, glucose infusion may cause high fever ("glucose fever") followed by collapse and death. Presumably, the glucose is metabolized, the water dilutes the plasma, and the resultant osmotic gradient between the plasma and the cells causes the cells of the thermoregulatory centers in the hypothalamus to swell to such an extent that their function is disrupted.

The cause of defective water excretion in adrenal insufficiency is unsettled. Plasma vasopressin levels are elevated in adrenal insufficiency and reduced by glucocorticoid treatment. The glomerular filtration rate is low, and this probably contributes to the reduction in water excretion. The selective effect of glucocorticoids on the abnormal water excretion is consistent with this possibility, because even though the mineralocorticoids improve filtration by restoring plasma volume, the glucocorticoids raise the glomerular filtration rate to a much greater degree.

EFFECTS ON THE BLOOD CELLS & LYMPHATIC ORGANS

Glucocorticoids decrease the number of circulating eosinophils by increasing their sequestration in the spleen and lungs. Glucocorticoids also lower the number of basophils in the circulation and increase the number of neutrophils, platelets, and red blood cells (Table 20–4).

Glucocorticoids decrease the circulating lymphocyte count and the size of the lymph nodes and thymus by inhibiting lymphocyte mitotic activity. They reduce secretion of cytokines by inhibiting the effect of NF-κB on the nucleus. The reduced secretion of the cytokine IL-2 leads to reduced proliferation of lymphocytes (see Chapter 3), and these cells undergo apoptosis.

RESISTANCE TO STRESS

The term **stress** as used in biology has been defined as any change in the environment that changes or threatens to change an existing optimal steady state. Most, if not all, of these stresses activate processes at the molecular, cellular, or systemic level that tend to restore the previous state, that is, they are homeostatic reactions. Some, but not all, of the stresses stimulate ACTH secretion. The increase in ACTH secretion is essential for survival when the stress is severe. If animals are then hypophysectomized, or adrenalectomized but treated with maintenance doses of glucocorticoids, they die when exposed to the same stress.

TABLE 20–4 Typical effects of cortisol on the white and red blood cell counts in humans (cells/ μL).

Cell	Normal	Cortisol-Treated
White blood cells		
Total	9000	10,000
PMNs	5760	8330
Lymphocytes	2370	1080
Eosinophils	270	20
Basophils	60	30
Monocytes	450	540
Red blood cells	5 million	5.2 million

The reason an elevated circulating ACTH, and hence glucocorticoid level, is essential for resisting stress remains for the most part unknown. Most of the stressful stimuli that increase ACTH secretion also activate the sympathetic nervous system, and part of the function of circulating glucocorticoids may be maintenance of vascular reactivity to catecholamines. Glucocorticoids are also necessary for the catecholamines to exert their full FFA-mobilizing action, and the FFAs are an important emergency energy supply. However, sympathectomized animals tolerate a variety of stresses with relative impunity. Another theory holds that glucocorticoids prevent other stress-induced changes from becoming excessive. At present, all that can be said is that stress causes increases in plasma glucocorticoids to high "pharmacologic" levels that in the short run are life-saving.

It should also be noted that the increase in ACTH, which is beneficial in the short term, becomes harmful and disruptive in the long term, causing among other things, the abnormalities of Cushing syndrome.

PHARMACOLOGIC & PATHOLOGIC EFFECTS OF GLUCOCORTICOIDS

CUSHING SYNDROME

The clinical picture produced by prolonged increases in plasma glucocorticoids was described by Harvey Cushing and is called **Cushing syndrome** (Figure 20–13). It may be **ACTH-independent** or **ACTH-dependent.** The causes of ACTH-independent Cushing syndrome include glucocorticoid-secreting adrenal tumors, adrenal hyperplasia, and prolonged administration of exogenous glucocorticoids for diseases such as rheumatoid arthritis. Rare but interesting ACTH-independent cases have been reported in which adrenocortical cells abnormally express receptors for gastric inhibitory polypeptide (GIP) (see Chapter 25), vasopressin (see Chapter 38), β-adrenergic agonists, IL-1, or gonadotropin-releasing hormone (GnRH; see Chapter 22), causing these peptides to increase glucocorticoid secretion. The causes of ACTH-dependent Cushing syndrome include ACTH-secreting tumors of the anterior pituitary gland and tumors of other organs, usually the lungs, that secrete ACTH (ectopic ACTH syndrome) or corticotropin releasing hormone (CRH). Cushing syndrome due to anterior pituitary tumors is often called **Cushing disease** because these tumors were the cause of the cases described by Cushing. However, it is confusing to speak of Cushing disease as a subtype of

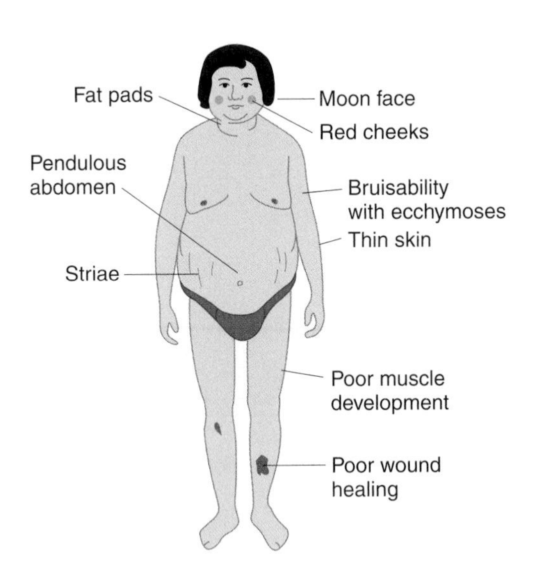

FIGURE 20–13 Typical findings in Cushing syndrome. (Reproduced with permission from Forsham PH, Di Raimondo VC: *Traumatic Medicine and Surgery for the Attorney.* Butterworth, 1960.)

Cushing syndrome, and the distinction seems to be of little more than historical value.

Patients with Cushing syndrome are protein-depleted as a result of excess protein catabolism. The skin and subcutaneous tissues are therefore thin and the muscles are poorly developed. Wounds heal poorly, and minor injuries cause bruises and ecchymoses. The hair is thin and scraggly. Many patients with the disease have some increase in facial hair and acne, but this is caused by the increased secretion of adrenal androgens and often accompanies the increase in glucocorticoid secretion.

Body fat is redistributed in a characteristic way. The extremities are thin, but fat collects in the abdominal wall, face, and upper back, where it produces a "buffalo hump." As the thin skin of the abdomen is stretched by the increased subcutaneous fat depots, the subdermal tissues rupture to form prominent reddish purple **striae.** These scars are seen normally whenever a rapid stretching of skin occurs, but in normal individuals the striae are usually inconspicuous and lack the intense purplish color.

Many of the amino acids liberated from catabolized proteins are converted into glucose in the liver and the resultant hyperglycemia and decreased peripheral utilization of glucose may be sufficient to precipitate insulin-resistant diabetes mellitus, especially in patients genetically predisposed to diabetes. Hyperlipidemia and ketosis are associated with the diabetes, but acidosis is usually not severe.

The glucocorticoids are present in such large amounts in Cushing syndrome that they may exert a significant mineralocorticoid action. Deoxycorticosterone secretion is also elevated in cases due to ACTH hypersecretion. The salt and water retention plus the facial obesity cause the characteristic plethoric, rounded "moon-faced" appearance, and there may be significant K^+ depletion and weakness. About 85% of patients with Cushing syndrome are hypertensive. The hypertension may be due to increased deoxycorticosterone secretion, increased angiotensinogen secretion, or a direct glucocorticoid effect on blood vessels (see Chapter 32).

Glucocorticoid excess leads to bone dissolution by decreasing bone formation and increasing bone resorption. This leads to **osteoporosis,** a loss of bone mass that leads eventually to collapse of vertebral bodies and other fractures. The mechanisms by which glucocorticoids produce their effects on bone are discussed in Chapter 21.

Glucocorticoids in excess accelerate the basic electroencephalographic rhythms and produce mental aberrations ranging from increased appetite, insomnia, and euphoria to frank toxic psychoses. As noted above, glucocorticoid deficiency is also associated with mental symptoms, but the symptoms produced by glucocorticoid excess are more severe.

ANTI-INFLAMMATORY & ANTI-ALLERGIC EFFECTS OF GLUCOCORTICOIDS

Glucocorticoids inhibit the inflammatory response to tissue injury. The glucocorticoids also suppress manifestations of allergic disease that are due to the release of histamine from mast cells and basophils. Both of these effects require high levels of circulating glucocorticoids and cannot be produced by administering steroids without producing the other manifestations of glucocorticoid excess. Furthermore, large doses of exogenous glucocorticoids inhibit ACTH secretion to the point that severe adrenal insufficiency can be a dangerous problem when therapy is stopped. However, local administration of glucocorticoids, for example, by injection into an inflamed joint or near an irritated nerve, produces a high local concentration of the steroid, often without enough systemic absorption to cause serious side effects.

The actions of glucocorticoids in patients with bacterial infections are dramatic but dangerous. For example, in pneumococcal pneumonia or active tuberculosis, the febrile reaction, the toxicity, and the lung symptoms disappear, but unless antibiotics are given at the same time, the bacteria spread throughout the body. It is important to remember that the symptoms are the warning that disease is present; when these symptoms are masked by treatment with glucocorticoids, there may be serious and even fatal delays in diagnosis and the institution of treatment with antimicrobial drugs.

The role of NF-κB in the anti-inflammatory and anti-allergic effects of glucocorticoids has been mentioned above and is discussed in Chapter 3. An additional action that combats local inflammation is inhibition of phospholipase A_2. This reduces the release of arachidonic acid from tissue phospholipids and consequently reduces the formation of leukotrienes, thromboxanes, prostaglandins, and prostacyclin (see Chapter 32).

OTHER EFFECTS

Large doses of glucocorticoids inhibit growth, decrease growth hormone secretion (see Chapter 18), induce PNMT, and decrease thyroid-stimulating hormone (TSH) secretion. During fetal life, glucocorticoids accelerate the maturation of surfactant in the lungs (see Chapter 34).

REGULATION OF GLUCOCORTICOID SECRETION

ROLE OF ACTH

Both basal secretion of glucocorticoids and the increased secretion provoked by stress are dependent upon ACTH from the anterior pituitary. Angiotensin II also stimulates the adrenal cortex, but its effect is mainly on aldosterone secretion. Large doses of a number of other naturally occurring substances, including vasopressin, serotonin, and vasoactive intestinal polypeptide (VIP), are capable of stimulating the adrenal directly, but there is no evidence that these agents play any role in the physiologic regulation of glucocorticoid secretion.

CHEMISTRY & METABOLISM OF ACTH

ACTH is a single-chain polypeptide containing 39 amino acids. Its origin from proopiomelanocortin (POMC) in the pituitary is discussed in Chapter 18. The first 23 amino acids in the chain generally constitute the active "core" of the molecule. Amino acids 24–39 constitute a "tail" that stabilizes the molecule and varies slightly in composition from species to species The ACTHs that have been isolated are generally active in all species but antigenic in heterologous species.

ACTH is inactivated in blood in vitro more slowly than in vivo; its half-life in the circulation in humans is about 10 min. A large part of an injected dose of ACTH is found in the kidneys, but neither nephrectomy nor evisceration appreciably enhances its in vivo activity, and the site of its inactivation is not known.

EFFECT OF ACTH ON THE ADRENAL

After hypophysectomy, glucocorticoid synthesis and output decline within 1 h to very low levels, although some hormone is still secreted. Within a short time after an injection of ACTH (in dogs, less than 2 min), glucocorticoid output is increased. With low doses of ACTH, the relationship between the log of the dose and the increase in glucocorticoid secretion is linear. However, the maximal rate at which glucocorticoids can be secreted is rapidly reached, and this "ceiling on output" also exists in humans. The effects of ACTH on adrenal morphology and the mechanism by which it increases steroid secretion have been discussed above.

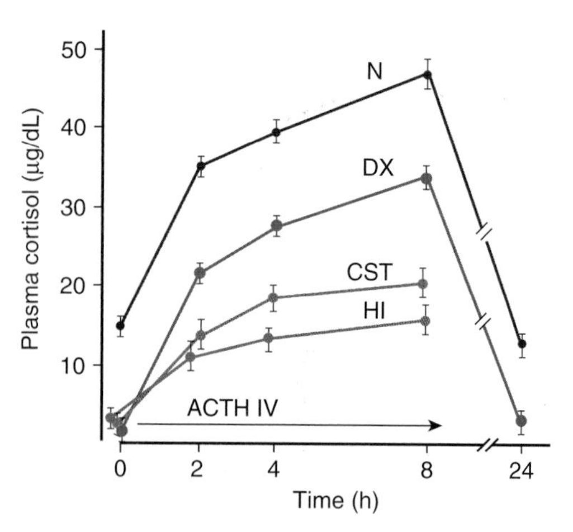

FIGURE 20–14 Loss of ACTH responsiveness when ACTH secretion is decreased in humans. The 1- to 24-amino-acid sequence of ACTH was infused intravenously (IV) in a dose of 250 μg over 8 h. CST, long-term corticosteroid therapy; DX, dexamethasone 0.75 mg every 8 h for 3 days; HI, anterior pituitary insufficiency; N, normal subjects. (Reproduced with permission from Kolanowski J, et al: Adrenocortical response upon repeated stimulation with corticotropin in patients lacking endogenous corticotropin secretion. Acta Endocrinol [Kbh] 1977;85:595.)

ADRENAL RESPONSIVENESS

ACTH not only produces prompt increases in glucocorticoid secretion but also increases the sensitivity of the adrenal to subsequent doses of ACTH. Conversely, single doses of ACTH do not increase glucocorticoid secretion in chronically hypophysectomized animals and patients with hypopituitarism, and repeated injections or prolonged infusions of ACTH are necessary to restore normal adrenal responses to ACTH. Decreased responsiveness is also produced by doses of glucocorticoids that inhibit ACTH secretion. The decreased adrenal responsiveness to ACTH is detectable within 24 h after hypophysectomy and increases progressively with time (Figure 20–14). It is marked when the adrenal is atrophic but develops before visible changes occur in adrenal size or morphology.

CIRCADIAN RHYTHM

ACTH is secreted in irregular bursts throughout the day and plasma cortisol tends to rise and fall in response to these bursts (Figure 20–15). In humans, the bursts are most frequent in the early morning, and about 75% of the daily production of cortisol occurs between 4:00 AM and 10:00 AM. The bursts are least frequent in the evening. This **diurnal (circadian) rhythm** in ACTH secretion is present in patients with adrenal insufficiency receiving constant doses of glucocorticoids. It is not due to the stress of getting up in the morning, traumatic as that may be, because the increased ACTH secretion occurs before waking up. If the "day" is lengthened experimentally to more than 24 h, that is, if the individual is isolated and the day's activities are spread over more than 24 h, the adrenal cycle also lengthens, but the increase in ACTH secretion still occurs during the period of sleep. The

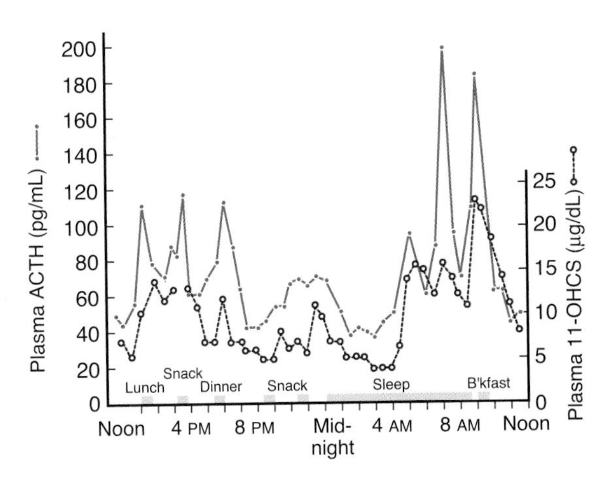

FIGURE 20–15 Fluctuations in plasma ACTH and glucocorticoids throughout the day in a normal girl (age 16). The ACTH was measured by immunoassay and the glucocorticoids as 11-oxysteroids (11-OHCS). Note the greater ACTH and glucocorticoid rises in the morning, before awakening. (Reproduced, with permission, from Krieger DT, et al: Characterization of the normal temporal pattern of plasma corticosteroid levels. J Clin Endocrinol Metab 1971;32:266.)

biologic clock responsible for the diurnal ACTH rhythm is located in the suprachiasmatic nuclei of the hypothalamus (see Chapter 14).

THE RESPONSE TO STRESS

The morning plasma ACTH concentration in a healthy resting human is about 25 pg/mL (5.5 pmol/L). ACTH and cortisol values in various abnormal conditions are summarized in Figure 20–16. During severe stress, the amount of ACTH secreted exceeds the amount necessary to produce maximal glucocorticoid output. However, prolonged exposure to ACTH in conditions such as the ectopic ACTH syndrome increases the adrenal maximum.

Increases in ACTH secretion to meet emergency situations are mediated almost exclusively through the hypothalamus via release of CRH. This polypeptide is produced by neurons in the paraventricular nuclei. It is secreted in the median eminence and transported in the portal-hypophysial vessels to the anterior pituitary, where it stimulates ACTH secretion (see Chapter 18). If the median eminence is destroyed, increased secretion in response to many different stresses is blocked. Afferent nerve pathways from many parts of the brain converge on the paraventricular nuclei. Fibers from the amygdaloid nuclei mediate responses to emotional stresses, and fear, anxiety, and apprehension cause marked increases in ACTH secretion. Input from the suprachiasmatic nuclei provides the drive for the diurnal rhythm. Impulses ascending to the hypothalamus via the nociceptive pathways and the reticular formation trigger increased ACTH secretion in response to injury (Figure 20–16). The baroreceptors exert an inhibitory input via the nucleus of the tractus solitarius.

GLUCOCORTICOID FEEDBACK

Free glucocorticoids inhibit ACTH secretion, and the degree of pituitary inhibition is proportional to the circulating glucocorticoid level. The inhibitory effect is exerted at both the pituitary and the hypothalamic levels. The inhibition is due primarily to an action on DNA, and maximal inhibition takes several hours to develop, although more rapid "fast feedback" also occurs. The ACTH-inhibiting activity of the various steroids parallels their glucocorticoid potency. A drop in resting corticoid levels stimulates ACTH secretion, and in chronic adrenal insufficiency the rate of ACTH synthesis and secretion is markedly increased.

Thus, the rate of ACTH secretion is determined by two opposing forces: the sum of the neural and possibly other stimuli converging through the hypothalamus to increase ACTH secretion, and the magnitude of the braking action of glucocorticoids on ACTH secretion, which is proportional to their level in the circulating blood (Figure 20–17).

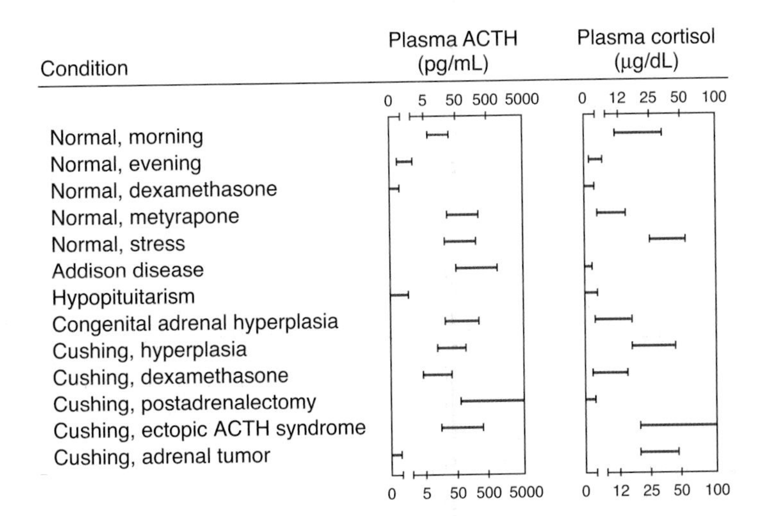

FIGURE 20–16 Plasma concentrations of ACTH and cortisol in various clinical states. (Reproduced with permission from Williams RH [editor]: *Textbook of Endocrinology*, 5th ed. Saunders, 1974.)

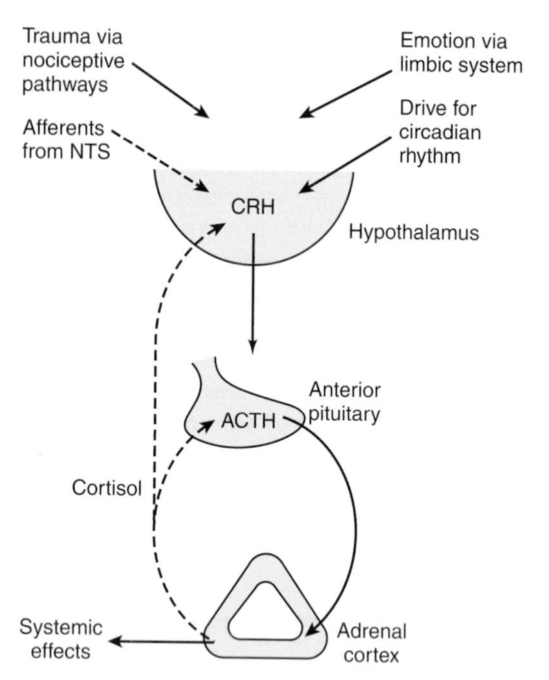

FIGURE 20–17 Feedback control of the secretion of cortisol and other glucocorticoids via the hypothalamic-pituitary-adrenal axis. The dashed arrows indicate inhibitory effects and the solid arrows indicate stimulating effects. NTS, nucleus tractus solitarius.

The dangers involved when prolonged treatment with anti-inflammatory doses of glucocorticoids is stopped deserve emphasis. Not only is the adrenal atrophic and unresponsive after such treatment, but even if its responsiveness is restored by injecting ACTH, the pituitary may be unable to secrete normal amounts of ACTH for as long as a month. The cause of the deficiency is presumably diminished ACTH synthesis. Thereafter, ACTH secretion slowly increases to supranormal levels. These in turn stimulate the adrenal, and glucocorticoid output rises, with feedback inhibition gradually reducing the elevated ACTH levels to normal (Figure 20–18). The complications of sudden cessation of steroid therapy can usually be avoided by slowly decreasing the steroid dose over a long period of time.

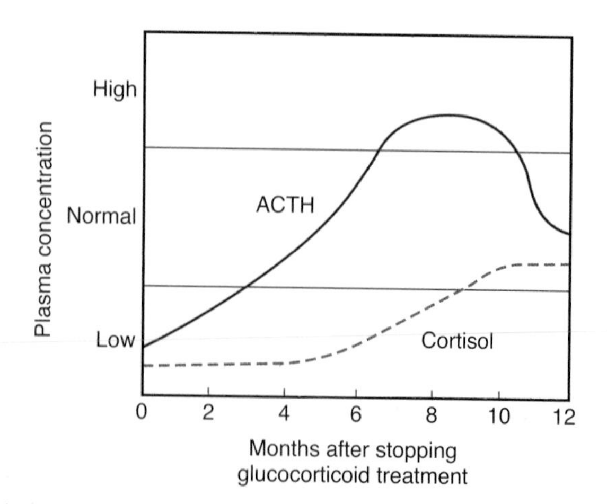

FIGURE 20–18 Pattern of plasma ACTH and cortisol values in patients recovering from prior long-term daily treatment with large doses of glucocorticoids. (Courtesy of R Ney.)

EFFECTS OF MINERALOCORTICOIDS

ACTIONS

Aldosterone and other steroids with mineralocorticoid activity increase the reabsorption of Na^+ from the urine, sweat, saliva, and the contents of the colon. Thus, mineralocorticoids cause retention of Na^+ in the ECF. This expands ECF volume. In the kidneys, they act primarily on the **principal cells (P cells)** of the collecting ducts (see Chapter 37). Under the influence of aldosterone, increased amounts of Na^+ are in effect exchanged for K^+ and H^+ in the renal tubules, producing a K^+ diuresis (Figure 20–19) and an increase in urine acidity.

MECHANISM OF ACTION

Like many other steroids, aldosterone binds to a cytoplasmic receptor, and the receptor–hormone complex moves to the nucleus where it alters the transcription of mRNAs. This in turn increases the production of proteins that alter cell function. The aldosterone-stimulated proteins have two effects—a rapid effect, to increase the activity of epithelial sodium channels (ENaCs) by increasing the insertion of these channels into the cell membrane from a cytoplasmic pool; and a slower effect to increase the synthesis of ENaCs. Among the genes activated by aldosterone is the gene for **serum- and glucocorticoid regulated kinase (sgk)**, a serine-threonine protein kinase. The gene for sgk is an early response gene, and sgk increases ENaC activity. Aldosterone also increases the mRNAs for the three subunits

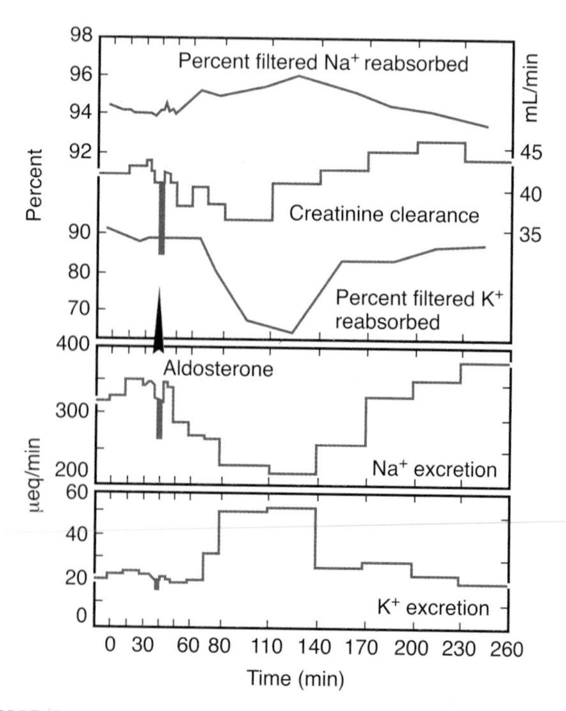

FIGURE 20–19 Effect of aldosterone (5 μg as a single dose injected into the aorta) on electrolyte excretion in an adrenalectomized dog. The scale for creatinine clearance is on the right.

that make up ENaCs. The fact that sgk is activated by glucocorticoids as well as aldosterone is not a problem because glucocorticoids are inactivated at mineralocorticoid receptor sites. However, aldosterone activates the genes for other proteins in addition to sgk and ENaCs and inhibits others. Therefore, the exact mechanism by which aldosterone-induced proteins increase Na^+ reabsorption is still unsettled.

Evidence is accumulating that aldosterone also binds to the cell membrane and by a rapid, nongenomic action increases the activity of membrane Na^+–K^+ exchangers. This produces an increase in intracellular Na^+, and the second messenger involved is probably IP_3. In any case, the principal effect of aldosterone on Na^+ transport takes 10–30 min to develop and peaks even later (Figure 20–19), indicating that it depends on the synthesis of new proteins by a genomic mechanism.

RELATION OF MINERALOCORTICOID TO GLUCOCORTICOID RECEPTORS

It is intriguing that in vitro, the mineralocorticoid receptor has an appreciably higher affinity for glucocorticoids than the glucocorticoid receptor does, and glucocorticoids are present in large amounts in vivo. This raises the question of why glucocorticoids do not bind to the mineralocorticoid receptors in the kidneys and other locations and produce mineralocorticoid effects. At least in part, the answer is that the kidneys and other mineralocorticoid-sensitive tissues also contain the enzyme **11β-hydroxysteroid dehydrogenase type 2.** This enzyme leaves aldosterone untouched, but it converts cortisol to cortisone (Figure 20–11) and corticosterone to its 11-oxy derivative. These 11-oxy derivatives do not bind to the receptor (Clinical Box 20–3).

CLINICAL BOX 20–3

Apparent Mineralocorticoid Excess

If 11β-hydroxysteroid dehydrogenase type 2 is inhibited or absent, cortisol has marked mineralocorticoid effects. The resulting syndrome is called **apparent mineralocorticoid excess (AME).** Patients with this condition have the clinical picture of hyperaldosteronism because cortisol is acting on their mineralocorticoid receptors, and their plasma aldosterone level as well as their plasma renin activity is low. The condition can be due to congenital absence of the enzyme.

THERAPEUTIC HIGHLIGHTS

Prolonged ingestion of licorice can also cause an increase in blood pressure. Outside of the United States, licorice contains glycyrrhetinic acid, which inhibits 11β-hydroxysteroid dehydrogenase type 2. Individuals who eat large amounts of licorice have an increase in MR-activated sodium absorption via the epithelial sodium channel ENaC in the renal collecting duct, and blood pressure can rise.

OTHER STEROIDS THAT AFFECT NA⁺ EXCRETION

Aldosterone is the principal mineralocorticoid secreted by the adrenal, although corticosterone is secreted in sufficient amounts to exert a minor mineralocorticoid effect (Tables 20–1 and 20–2). Deoxycorticosterone, which is secreted in appreciable amounts only in abnormal situations, has about 3% of the activity of aldosterone. Large amounts of progesterone and some other steroids cause natriuresis, but there is little evidence that they play any normal role in the control of Na^+ excretion.

EFFECT OF ADRENALECTOMY

In adrenal insufficiency, Na^+ is lost in the urine; K^+ is retained, and the plasma K^+ rises. When adrenal insufficiency develops rapidly, the amount of Na^+ lost from the ECF exceeds the amount excreted in the urine, indicating that Na^+ also must be entering cells. When the posterior pituitary is intact, salt loss exceeds water loss, and the plasma Na^+ falls (Table 20–5). However, the plasma volume is also reduced, resulting in hypotension, circulatory insufficiency, and, eventually, fatal shock. These changes can be prevented to a degree by increasing dietary NaCl intake. Rats survive indefinitely on extra salt alone, but in dogs and most humans, the amount of supplementary salt needed is so large that it is almost impossible to prevent eventual collapse and death unless mineralocorticoid treatment is also instituted (see Clinical Box 20–4).

REGULATION OF ALDOSTERONE SECRETION

STIMULI

The principal conditions that increase aldosterone secretion are summarized in Table 20–6. Some of them also increase glucocorticoid secretion; others selectively affect the output

TABLE 20–5 Typical plasma electrolyte levels in normal humans and patients with adrenocortical diseases.

	Plasma Electolytes (mEq/L)			
State	Na^+	K^+	Cl^-	HCO_3^-
Normal	142	4.5	105	25
Adrenal insufficiency	120	6.7	85	25
Primary hyperaldosteronism	145	2.4	96	41

CLINICAL BOX 20–4

Secondary Effects of Excess Mineralocorticoids

A prominent feature of prolonged mineralocorticoid excess (Table 20–5) is K^+ depletion due to prolonged K^+ diuresis. H^+ is also lost in the urine. Na^+ is retained initially, but the plasma Na^+ is elevated only slightly if at all, because water is retained with the osmotically active sodium ions. Consequently, ECF volume is expanded and the blood pressure rises. When the ECF expansion passes a certain point, Na^+ excretion is usually increased in spite of the continued action of mineralocorticoids on the renal tubules. This **escape phenomenon** (Figure 20–20) is probably due to increased secretion of ANP (see Chapter 38). Because of increased excretion of Na^+ when the ECF volume is expanded, mineralocorticoids do not produce edema in normal individuals and patients with hyperaldosteronism. However, escape may not occur in certain disease states, and in these situations, continued expansion of ECF volume leads to edema (see Chapters 37 and 38).

of aldosterone. The primary regulatory factors involved are ACTH from the pituitary, renin from the kidney via angiotensin II, and a direct stimulatory effect on the adrenal cortex of a rise in plasma K^+ concentration.

EFFECT OF ACTH

When first administered, ACTH stimulates the output of aldosterone as well as that of glucocorticoids and sex hormones. Although the amount of ACTH required to increase aldosterone output is somewhat greater than the amount that stimulates maximal glucocorticoid secretion (Figure 20–21), it is well within the range of endogenous ACTH secretion. The effect is transient, and even if ACTH secretion remains elevated, aldosterone output declines in 1 or 2 days. On the other hand, the output of the mineralocorticoid deoxycorticosterone remains elevated. The decline in aldosterone output is partly due to decreased renin secretion secondary to hypervolemia, but it is possible that some other factor also decreases the conversion of corticosterone to aldosterone. After hypophysectomy, the basal rate of aldosterone secretion is normal. The increase normally produced by surgical and other stresses is absent, but the increase produced by dietary salt restriction is unaffected for some time. Later on, atrophy of the zona glomerulosa complicates the picture in long-standing hypopituitarism, and this may lead to salt loss and hypoaldosteronism.

Normally, glucocorticoid treatment does not suppress aldosterone secretion. However, an interesting recently described syndrome is **glucocorticoid-remediable aldosteronism (GRA).** This is an autosomal dominant disorder in which the increase in aldosterone secretion produced by ACTH is no longer transient. The hypersecretion of aldosterone and the accompanying hypertension are remedied when ACTH secretion is suppressed by administering glucocorticoids. The genes encoding aldosterone synthase and 11β-hydroxylase are 95% identical and are close together on chromosome 8. In individuals with GRA, there is unequal crossing over so that the 5′-regulatory region of the 11βhydroxylase gene is fused to the coding region of the aldosterone synthase gene. The product of this hybrid gene is an ACTH-sensitive aldosterone synthase.

TABLE 20–6 Conditions that increase aldosterone secretion.

Glucocorticoid secretion also increased
Surgery
Anxiety
Physical trauma
Hemorrhage
Glucocorticoid secretion unaffected
High potassium intake
Low sodium intake
Constriction of inferior vena cava in thorax
Standing
Secondary hyperaldosteronism (in some cases of congestive heart failure, cirrhosis, and nephrosis)

EFFECTS OF ANGIOTENSIN II & RENIN

The octapeptide angiotensin II is formed in the body from angiotensin I, which is liberated by the action of renin on circulating angiotensinogen (see Chapter 38). Injections of angiotensin II stimulate adrenocortical secretion and, in small doses, affect primarily the secretion of aldosterone (Figure 20–22). The sites of action of angiotensin II are both early and late in the steroid biosynthetic pathway. The early action is on the conversion of cholesterol to pregnenolone, and the late action is on the conversion of corticosterone to aldosterone (Figure 20–8). Angiotensin II does not increase the secretion of deoxycorticosterone, which is controlled by ACTH.

Renin is secreted from the juxtaglomerular cells that surround the renal afferent arterioles as they enter the glomeruli (see Chapter 38). Aldosterone secretion is regulated via the renin–angiotensin system in a feedback fashion (Figure 20–23). A drop in ECF volume or intra-arterial vascular volume leads to a reflex increase in renal nerve discharge and decreases renal arterial pressure. Both changes increase renin secretion, and the angiotensin II formed by the action of renin increases the rate of secretion of aldosterone. The aldosterone causes Na^+ and, secondarily, water retention, expanding ECF volume, and shutting off the stimulus that initiated increased renin secretion.

Hemorrhage stimulates ACTH and renin secretion. Like hemorrhage, standing and constriction of the thoracic inferior

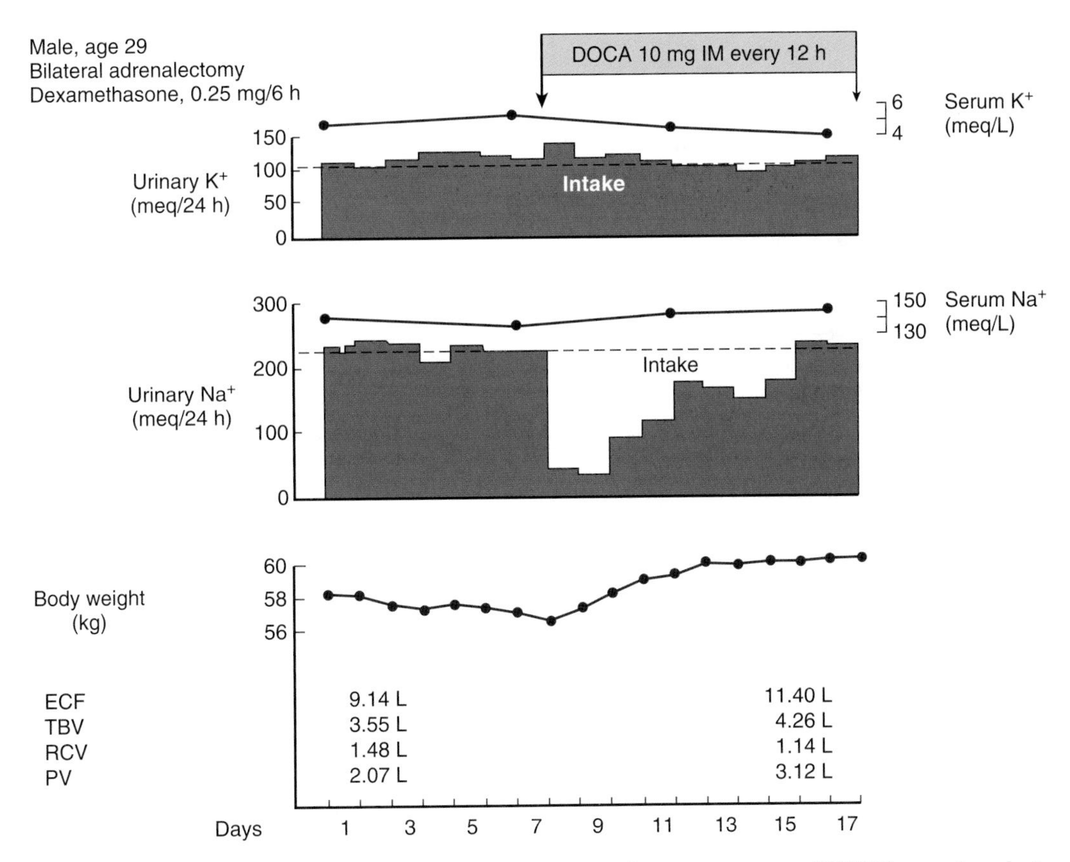

FIGURE 20–20 **"Escape" from the sodium-retaining effect of desoxycorticosterone acetate (DOCA) in an adrenalectomized patient.** ECF, extracellular fluid volume; PV, plasma volume; RCV, red cell volume; TBV, total blood volume. (Courtesy of EG Biglieri.)

vena cava decrease intrarenal arterial pressure. Dietary sodium restriction also increases aldosterone secretion via the renin–angiotensin system (Figure 20–24). Such restriction reduces ECF volume, but aldosterone and renin secretion are increased before any consistent decrease in blood pressure takes place. Consequently, the initial increase in renin secretion produced by dietary sodium restriction is probably due to a reflex increase in the activity of the renal nerves. The increase in circulating angiotensin II produced by salt depletion upregulates the angiotensin II receptors in the adrenal cortex and hence increases the response to angiotensin II, whereas it down-regulates the angiotensin II receptors in the blood vessels.

ELECTROLYTES & OTHER FACTORS

An acute decline in plasma Na^+ of about 20 mEq/L stimulates aldosterone secretion, but changes of this magnitude are rare. However, the plasma K^+ level need increase only 1 mEq/L to stimulate aldosterone secretion, and transient increases of this

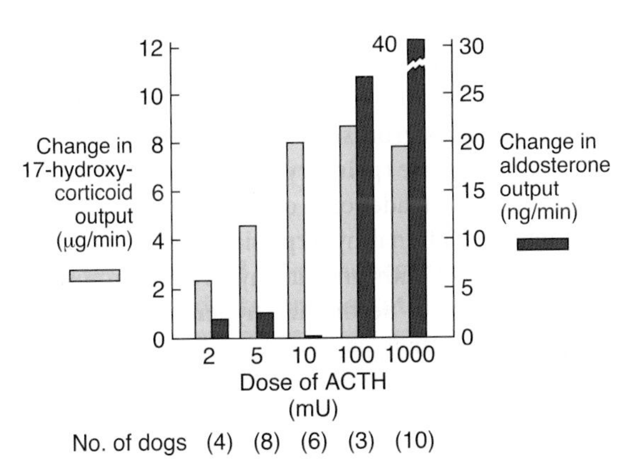

FIGURE 20–21 **Changes in adrenal venous output of steroids produced by ACTH in nephrectomized hypophysectomized dogs.**

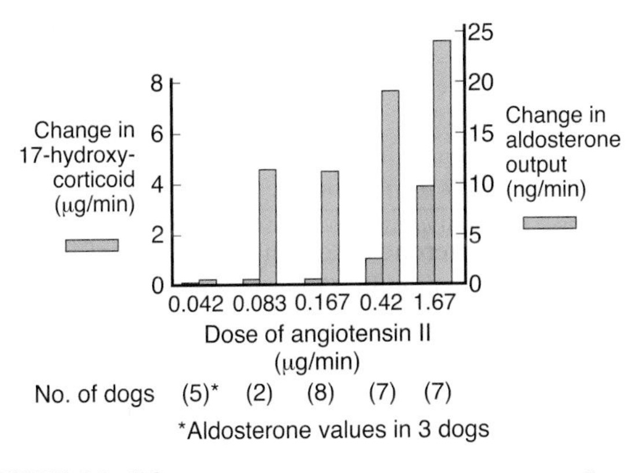

FIGURE 20–22 **Changes in adrenal venous output of steroids produced by angiotensin II in nephrectomized hypophysectomized dogs.**

tissues. This is an important source of estrogens in men and postmenopausal women.

- The mineralocorticoid aldosterone has effects on Na^+ and K^+ excretion and glucocorticoids affect glucose and protein metabolism.
- Glucocorticoid secretion is dependent on ACTH from the anterior pituitary and is increased by stress. Angiotensin II increases the secretion of aldosterone.

MULTIPLE-CHOICE QUESTIONS

For all questions, select the single best answer unless otherwise directed.

1. Which of the following is produced only by *large amounts* of glucocorticoids?
   A. Normal responsiveness of fat depots to norepinephrine
   B. Maintenance of normal vascular reactivity
   C. Increased excretion of a water load
   D. Inhibition of the inflammatory response
   E. Inhibition of ACTH secretion
2. Which of the following are *incorrectly* paired?
   A. Gluconeogenesis : Cortisol
   B. Free fatty acid mobilization : Dehydroepiandrosterone
   C. Muscle glycogenolysis : Epinephrine
   D. Kaliuresis : Aldosterone
   E. Hepatic glycogenesis : Insulin
3. Which of the following hormones has the shortest plasma half-life?
   A. Corticosterone
   B. Renin
   C. Dehydroepiandrosterone
   D. Aldosterone
   E. Norepinephrine
4. Mole for mole, which of the following has the greatest effect on Na^+ excretion?
   A. Progesterone
   B. Cortisol
   C. Vasopressin
   D. Aldosterone
   E. Dehydroepiandrosterone
5. Mole for mole, which of the following has the greatest effect on plasma osmolality?
   A. Progesterone
   B. Cortisol
   C. Vasopressin
   D. Aldosterone
   E. Dehydroepiandrosterone
6. The secretion of which of the following would be *least* affected by a decrease in extracellular fluid volume?
   A. CRH
   B. Arginine vasopressin
   C. Dehydroepiandrosterone
   D. Estrogens
   E. Aldosterone
7. A young man presents with a blood pressure of 175/110 mm Hg. He is found to have a high circulating aldosterone but a low circulating cortisol. Glucocorticoid treatment lowers his circulating aldosterone and lowers his blood pressure to 140/85 mm Hg. He probably has an abnormal
   A. 17α-hydroxylase.
   B. 21β-hydroxylase.
   C. 3β-hydroxysteroid dehydrogenase.
   D. aldosterone synthase.
   E. cholesterol desmolase.
8. A 32-year-old woman presents with a blood pressure of 155/96 mm Hg. In response to questioning, she admits that she loves licorice and eats some at least three times a week. She probably has a low level of
   A. type 2 11β-hydroxysteroid dehydrogenase activity.
   B. ACTH.
   C. 11β-hydroxylase activity.
   D. glucuronyl transferase.
   E. norepinephrine.
9. In its action in cells, aldosterone
   A. increases transport of ENaCs from the cytoplasm to the cell membrane.
   B. does not act on the cell membrane.
   C. binds to a receptor excluded from the nucleus.
   D. may activate a heat shock protein.
   E. also binds to glucocorticoid receptors.

CHAPTER RESOURCES

Goldstein JL, Brown MS: The cholesterol quartet. Science 2001;292:1510.

Goodman HM (editor): *Handbook of Physiology,* Section 7: *The Endocrine System.* Oxford University Press, 2000.

Larsen PR, Kronenberg HM, Melmed S, et al. (editors). *Williams Textbook of Endocrinology,* 9th ed. Saunders, 2003.

Stocco DM: A review of the characteristics of the protein required for the acute regulation of steroid hormone biosynthesis: The case for the steroidogenic acute regulatory (StAR) protein. Proc Soc Exp Biol Med 1998;217:123.

White PC: Disorders of aldosterone biosynthesis and action. N Engl J Med 1994;331:250.

Hormonal Control of Calcium & Phosphate Metabolism & the Physiology of Bone

CHAPTER

21

OBJECTIVES

After studying this chapter, you should be able to:

- Understand the importance of maintaining homeostasis of body calcium and phosphate concentrations, and how this is accomplished.
- Describe the body pools of calcium, their rates of turnover, and the organs that play central roles in regulating movement of calcium between stores.
- Delineate the mechanisms of calcium and phosphate absorption and excretion.
- Identify the major hormones and other factors that regulate calcium and phosphate homeostasis and their sites of synthesis as well as targets of their action.
- Define the basic anatomy of bone.
- Delineate cells and their functions in bone formation and resorption.

INTRODUCTION

Calcium is an essential intracellular signaling molecule and also plays a variety of extracellular functions, thus the control of body calcium concentrations is vitally important. The components of the system that maintains calcium homeostasis include cell types that sense changes in extracellular calcium and release calcium-regulating hormones, and the targets of these hormones, including the kidneys, bones, and intestine, that respond with changes in calcium mobilization, excretion, or uptake. Three hormones are primarily concerned with the regulation of calcium homeostasis. **Parathyroid hormone (PTH)** is secreted by the parathyroid glands. Its main action is to mobilize calcium from bone and increase urinary phosphate excretion. **1,25-Dihydroxycholecalciferol** is a steroid hormone formed from vitamin D by successive hydroxylations in the liver and kidneys. Its primary action is to increase calcium absorption from the intestine. **Calcitonin,** a calcium-lowering hormone that in mammals is secreted primarily by cells in the thyroid gland, inhibits bone resorption. Although the role of calcitonin seems to be relatively minor, all three hormones probably operate in concert to maintain the constancy of the calcium level in the body fluids. Phosphate homeostasis is likewise critical to normal body function, particularly given its inclusion in adenosine triphosphate (ATP), its role as a biological buffer, and its role as a modifier of proteins, thereby altering their functions. Many of the systems that regulate calcium homeostasis also contribute to that of phosphate, albeit sometimes in a reciprocal fashion, and thus will also be discussed in this chapter.

CALCIUM & PHOSPHORUS METABOLISM

CALCIUM

The body of a young adult human contains about 1100 g (27.5 moles) of calcium. Ninety-nine per cent of the calcium is in the skeleton. Plasma calcium, normally at a concentration of around 10 mg/dL (5 mEq/L, 2.5 mmol/L), is partly bound to protein and partly diffusible (Table 21–1). The distribution of calcium inside cells and the role of Ca^{2+} as a second messenger molecule is discussed in Chapter 2.

It is the free, ionized calcium (Ca^{2+}) in the body fluids that is a vital second messenger and is necessary for blood coagulation, muscle contraction, and nerve function. A decrease in extracellular Ca^{2+} exerts a net excitatory effect on nerve

TABLE 21–1 Distribution (mg/dL) of calcium in normal human plasma.

Total diffusible		**5.36**
Ionized (Ca^{2+})	4.72	
Complexed to HCO_3^-, citrate, etc	0.64	
Total nondiffusible (protein-bound)		**4.64**
Bound to albumin	3.68	
Bound to globulin	0.96	
Total plasma calcium		**10.00**

and muscle cells in vivo (see Chapters 4 and 5). The result is **hypocalcemic tetany,** which is characterized by extensive spasms of skeletal muscle, involving especially the muscles of the extremities and the larynx. Laryngospasm can become so severe that the airway is obstructed and fatal asphyxia is produced. Ca^{2+} also plays an important role in blood clotting (see Chapter 31), but in vivo, fatal tetany would occur before compromising the clotting reaction.

Because the extent of Ca^{2+} binding by plasma proteins is proportional to the plasma protein level, it is important to know the plasma protein level when evaluating the total plasma calcium. Other electrolytes and pH also affect the free Ca^{2+} level. Thus, for example, symptoms of tetany appear at higher total calcium levels if the patient hyperventilates, thereby increasing plasma pH. Plasma proteins are more ionized when the pH is high, providing more protein anions to bind with Ca^{2+}.

The calcium in bone is of two types: a readily exchangeable reservoir and a much larger pool of stable calcium that is only slowly exchangeable. Two independent but interacting homeostatic systems affect the calcium in bone. One is the system that regulates plasma Ca^{2+}, providing for the movement of about 500 mmol of Ca^{2+} per day into and out of the readily exchangeable pool in the bone (Figure 21–1). The other system involves bone remodeling by the constant interplay of bone resorption and deposition (see following text). However, the Ca^{2+} interchange between plasma and this stable pool of bone calcium is only about 7.5 mmol/d.

Ca^{2+} is transported across the brush border of intestinal epithelial cells via channels known as transient receptor potential vanilloid type 6 (TRPV6) and binds to an intracellular protein known as calbindin-D_{9k}. Calbindin-D_{9k} sequesters the absorbed calcium so that it does not disturb epithelial signaling processes that involve calcium. The absorbed Ca^{2+} is thereby delivered to the basolateral membrane of the epithelial cell, from where it can be transported into the bloodstream by either a Na^+/Ca^{2+} exchanger (NCX1) or a Ca^{2+}-dependent ATPase. Nevertheless, it should be noted that recent studies indicate that some intestinal Ca^{2+} uptake persists even in the absence of TRPV6 and calbindin-D_{9k}, suggesting that additional pathways are likely also involved in this critical process. The overall transport process is regulated by 1,25-dihydroxycholecalciferol (see below). As Ca^{2+} uptake rises, moreover, 1,25-dihydroxycholecalciferol levels fall in response to increased plasma Ca^{2+}.

Plasma Ca^{2+} is filtered in the kidneys, but 98–99% of the filtered Ca^{2+} is reabsorbed. About 60% of the reabsorption occurs in the proximal tubules and the remainder in the ascending limb of the loop of Henle and the distal tubule. Distal tubular reabsorption depends on the TRPV5 channel, which is related to TRPV6 discussed previously, and whose expression is regulated by PTH.

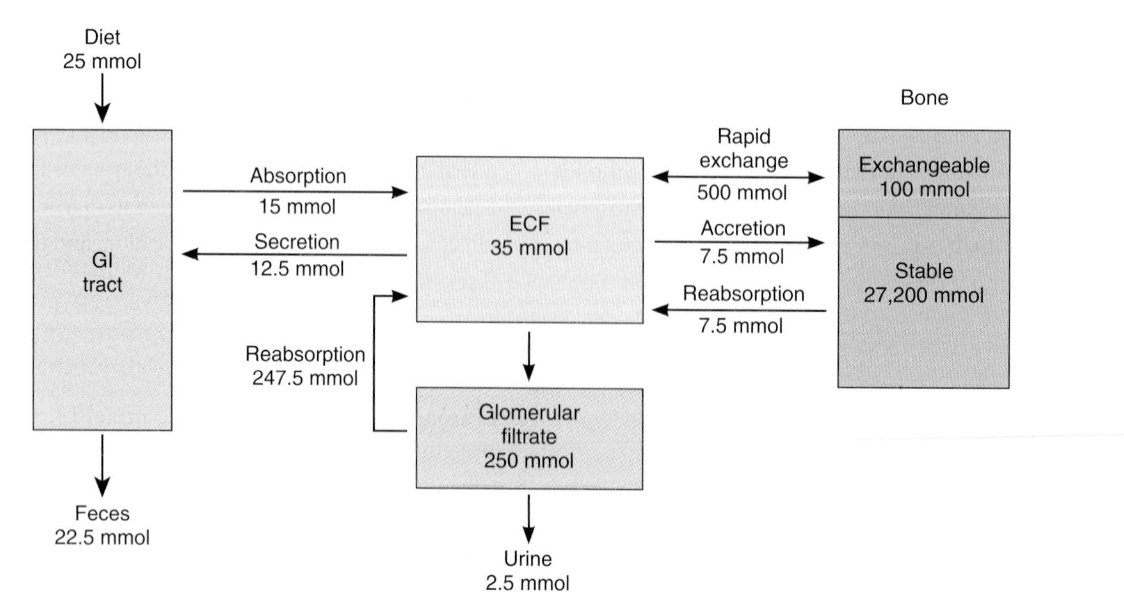

FIGURE 21–1 Calcium metabolism in an adult human. A typical daily dietary intake of 25 mmol Ca^{2+} (1000 mg) moves through many body compartments. Note that the majority of body calcium is in bones, in a pool that is only slowly exchangeable with the extracellular fluid (ECF).

PHOSPHORUS

Phosphate is found in ATP, cyclic adenosine monophosphate (cAMP), 2,3-diphosphoglycerate, many proteins, and other vital compounds in the body. Phosphorylation and dephosphorylation of proteins are involved in the regulation of cell function (see Chapter 2). Therefore, it is not surprising that, like calcium, phosphate metabolism is closely regulated. Total body phosphorus is 500–800 g (16.1–25.8 moles), 85–90% of which is in the skeleton. Total plasma phosphorus is about 12 mg/dL, with two-thirds of this total in organic compounds and the remaining inorganic phosphorus (P_i) mostly in PO_4^{3-}, HPO_4^{2-}, and $H_2PO_4^-$. The amount of phosphorus normally entering bone is about 3 mg (97 μmol)/kg/d, with an equal amount leaving via reabsorption.

P_i in the plasma is filtered in the glomeruli, and 85–90% of the filtered P_i is reabsorbed. Active transport in the proximal tubule accounts for most of the reabsorption and involves two related sodium-dependent P_i cotransporters, NaP_i-IIa and NaP_i-IIc. NaP_i-IIa is powerfully inhibited by PTH, which causes its internalization and degradation and thus a reduction in renal P_i reabsorption (see below).

P_i is absorbed in the duodenum and small intestine. Uptake occurs by a transporter related to those in the kidney, NaP_i-IIb, that takes advantage of the low intracellular Na^+ concentration established by the Na, K ATPase on the basolateral membrane of intestinal epithelial cells to load P_i against its concentration gradient. However, the pathway by which P_i exits into the bloodstream is not known. Many stimuli that increase Ca^{2+} absorption, including 1,25-dihydroxycholecalciferol, also increase P_i absorption via increased NaP_i-IIb expression and/or its insertion into the enterocyte apical membrane.

VITAMIN D & THE HYDROXYCHOLECALCIFEROLS

CHEMISTRY

The active transport of Ca^{2+} and PO_4^{3-} from the intestine is increased by a metabolite of **vitamin D.** The term "vitamin D" is used to refer to a group of closely related sterols produced by the action of ultraviolet light on certain provitamins (Figure 21–2). Vitamin D_3, which is also called cholecalciferol, is produced in the skin of mammals from 7-dehydrocholesterol by the action of sunlight. The reaction involves the rapid formation of previtamin D_3, which is then converted more slowly to vitamin D_3. Vitamin D_3 and its hydroxylated derivatives are transported in the plasma bound to a globulin vitamin D-binding protein (DBP). Vitamin D_3 is also ingested in the diet.

Vitamin D_3 is metabolized by enzymes that are members of the cytochrome P450 (CYP) superfamily (see Chapters 1 and 28). In the liver, vitamin D_3 is converted to **25-hydroxycholecalciferol** (calcidiol, 25-OHD$_3$). The 25-hydroxycholecalciferol is converted in the cells of the proximal tubules of the kidneys to the more active

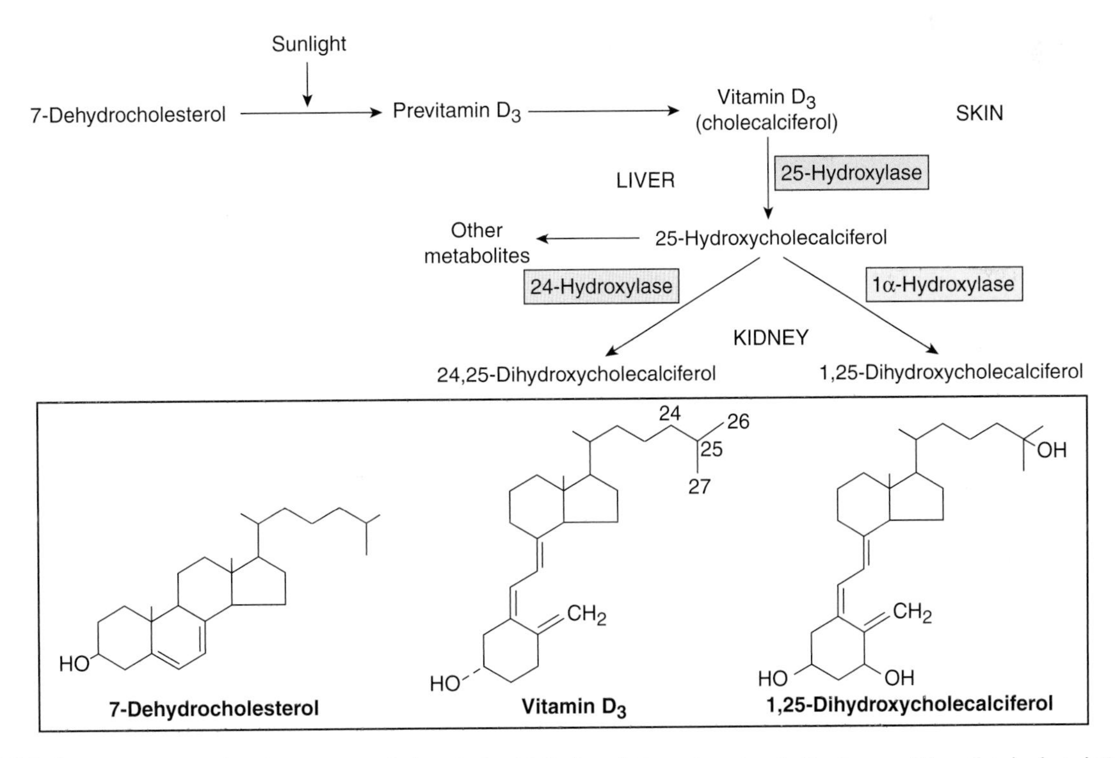

FIGURE 21–2 Formation and hydroxylation of vitamin D_3. 25-Hydroxylation takes place in the liver, and the other hydroxylations occur primarily in the kidneys. The structures of 7-dehydrocholesterol, vitamin D_3, and 1,25-dihydroxycholecalciferol are also shown in the boxed area.

metabolite **1,25-dihydroxycholecalciferol,** which is also called calcitriol or 1,25-$(OH)_2D_3$. 1,25-Dihydroxycholecalciferol is also made in the placenta, in keratinocytes in the skin, and in macrophages. The normal plasma level of 25-hydroxycholecalciferol is about 30 ng/mL, and that of 1,25-dihydroxycholecalciferol is about 0.03 ng/mL (approximately 100 pmol/L). The less active metabolite 24,25-dihydroxycholecalciferol is also formed in the kidneys (Figure 21–2).

MECHANISM OF ACTION

1,25 Dihydroxycholecalciferol stimulates the expression of a number of gene products involved in Ca^{2+} transport and handling via its receptor, which acts as a transcriptional regulator in its ligand-bound form. One group is the family of **calbindin-D** proteins. These are members of the troponin C superfamily of Ca^{2+}-binding proteins that also includes calmodulin (see Chapter 2). Calbindin-Ds are found in human intestine, brain, and kidneys. In the intestinal epithelium and many other tissues, two calbindins are induced: calbindinD_{9K} and calbindin-D_{28K}, with molecular weights of 9000 and 28,000, respectively. 1,25-Dihydroxycholecalciferol also increases the number of Ca^{2+}–ATPase and TRPV6 molecules in the intestinal cells, and thus, the overall capacity for absorption of dietary calcium.

In addition to increasing Ca^{2+} absorption from the intestine, 1,25-dihydroxycholecalciferol facilitates Ca^{2+} reabsorption in the kidneys via increased TRPV5 expression in the proximal tubules, increases the synthetic activity of osteoblasts, and is necessary for normal calcification of matrix (see Clinical Box 21–1). The stimulation of osteoblasts brings about a secondary increase in the activity of osteoclasts (see below).

REGULATION OF SYNTHESIS

The formation of 25-hydroxycholecalciferol does not appear to be stringently regulated. However, the formation of 1,25-dihydroxycholecalciferol in the kidneys, which is catalyzed by the renal 1α-hydroxylase, is regulated in a feedback fashion by plasma Ca^{2+} and PO_4^{3+} (Figure 21–3). When the plasma Ca^{2+} level is high, little 1,25-dihydroxycholecalciferol is produced, and the kidneys produce the relatively inactive metabolite 24,25-dihydroxycholecalciferol instead. This effect of Ca^{2+} on production of 1,25-dihydroxycholecalciferol is the mechanism that brings about adaptation of Ca^{2+} absorption from the intestine (see previous text). Conversely, expression of 1α-hydroxylase is stimulated by PTH, and when the plasma Ca^{2+} level is low, PTH secretion is increased. The production of 1,25-dihydroxycholecalciferol is also increased by low and inhibited by high plasma PO_4^{3-} levels, by a direct inhibitory effect of PO_4^{3-} on the 1α-hydroxylase. Additional control of 1,25-dihydroxycholecalciferol formation results from a direct negative feedback effect of the metabolite on 1α-hydroxylase, a positive feedback action on the formation of 24,25-dihydroxycholecalciferol, and a direct action on the parathyroid gland to inhibit PTH expression.

An "anti-aging" protein called α-Klotho (named after Klotho, a daughter of Zeus in Greek mythology who spins the thread of life) has also recently been discovered to play important roles in calcium and phosphate homeostasis, in part by reciprocal effects on 1,25-dihydroxycholecalciferol levels. Mice deficient in α-Klotho displayed accelerated aging, decreased bone mineral density, calcifications, and hypercalcemia and hyperphosphatemia. α-Klotho plays an important role in stabilizing the membrane localization of proteins important in calcium and phosphate (re)absorption,

CLINICAL BOX 21–1

Rickets & Osteomalacia

Vitamin D deficiency causes defective calcification of bone matrix and the disease called **rickets** in children and **osteomalacia** in adults. Even though 1,25-dihydroxycholecalciferol is necessary for normal mineralization of bone matrix, the main defect in this condition is failure to deliver adequate amounts of Ca^{2+} and PO_4^{3-} to the sites of mineralization. The full-blown condition in children is characterized by weakness and bowing of weight-bearing bones, dental defects, and hypocalcemia. In adults, the condition is less obvious. It used to be most commonly due to inadequate exposure to the sun in smoggy cities, but now it is more commonly due to inadequate intake of the provitamins on which the sun acts in the skin. These cases respond to administration of vitamin D. The condition can also be caused by inactivating mutations of the gene for renal 1α-hydroxylase, or in severe renal or liver diseases, in which case there is no response to vitamin D but a normal response to 1,25-dihydroxycholecalciferol **(type I vitamin D-resistant rickets).** In rare instances, it can be due to inactivating mutations of the gene for the 1,25-dihydroxycholecalciferol receptor **(type II vitamin D-resistant rickets),** in which case there is a deficient response to both vitamin D and 1,25-dihydroxycholecalciferol.

THERAPEUTIC HIGHLIGHTS

Treatment of these conditions depends on the underlying biochemical basis, as indicated above. Routine supplementation of milk with vitamin D has greatly reduced the occurrence of rickets in Western countries, but the condition remains among the most common childhood diseases in developing countries. Orthopedic surgery may be necessary in severely affected children.

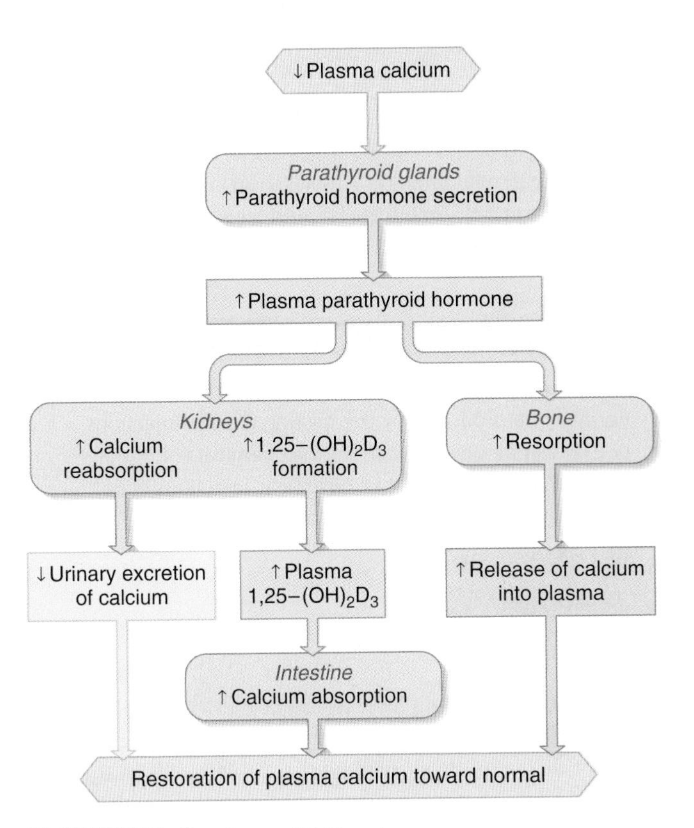

FIGURE 21–3 Effects of PTH and 1,25-dihydroxycholecalciferol on whole body calcium homeostasis. A reduction in plasma calcium stimulates parathyroid hormone secretion. PTH in turn causes calcium conservation and production of 1,25-dihydroxycholecalciferol in the kidneys, the latter of which increases calcium uptake in the intestine. PTH also releases calcium from the readily exchangeable pool in the bone. All of these actions act to restore normal plasma calcium. (Reproduced with permission from Widmaier EP, Raff H, Strang KT: *Vander's Human Physiology,* 10th ed. McGraw-Hill, 2006.)

such as TRPV5 and Na, K ATPase. Likewise, it enhances the activity of another factor, fibroblast growth factor 23 (FGF23), at its receptor. FGF23 thereby decreases renal NaP_i-IIa and NaP_i-IIc expression and inhibits the production of 1α-hydroxylase, reducing levels of 1,25-dihydroxycholecalciferol (Clinical Box 21–1).

THE PARATHYROID GLANDS

ANATOMY

Humans usually have four parathyroid glands: two embedded in the superior poles of the thyroid and two in its inferior poles (Figure 21–4). Each parathyroid gland is a richly vascularized disk, about 3 × 6 × 2 mm, containing two distinct types of cells (Figure 21–5). The abundant **chief cells,** which contain a prominent Golgi apparatus plus endoplasmic reticulum and secretory granules, synthesize and secrete **PTH.** The less abundant and larger **oxyphil cells** contain oxyphil granules and

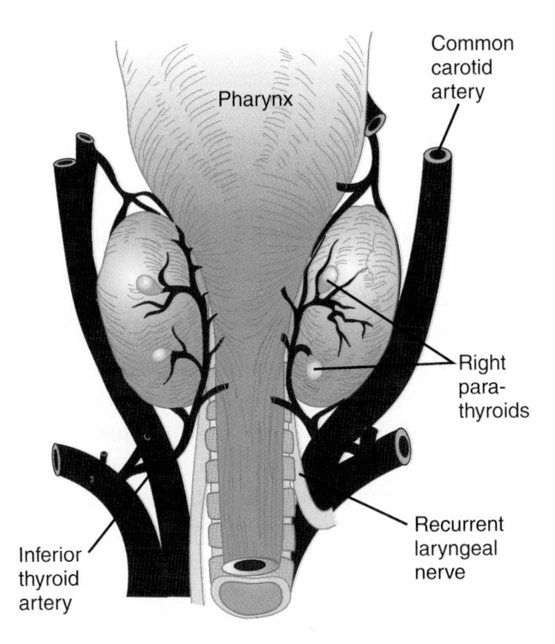

FIGURE 21–4 The human parathyroid glands, viewed from behind. The glands are small structures adherent to the posterior surface of the thyroid gland.

large numbers of mitochondria in their cytoplasm. In humans, few oxyphil cells are seen before puberty, and thereafter they increase in number with age. Their function is unknown. Consequences of loss of the parathyroid glands are discussed in Clinical Box 21–2.

SYNTHESIS & METABOLISM OF PTH

Human PTH is a linear polypeptide with a molecular weight of 9500 that contains 84 amino acid residues (Figure 21–6). It is synthesized as part of a larger molecule containing 115 amino acid residues **(preproPTH).** On entry of preproPTH

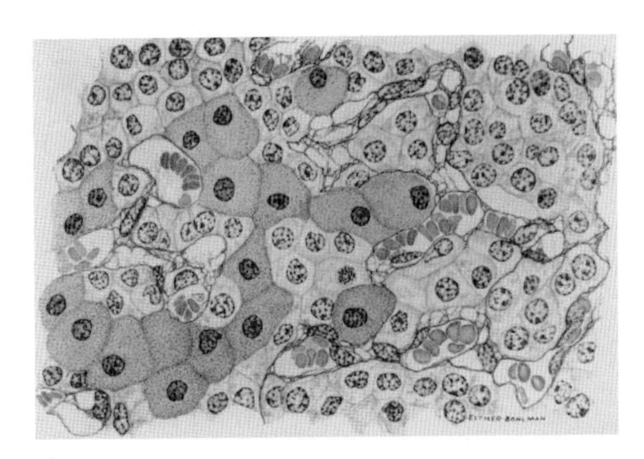

FIGURE 21–5 Section of human parathyroid (Reduced 50% from × 960). Small cells are chief cells; large stippled cells (especially prominent in the lower left of picture) are oxyphil cells. (Reproduced with permission from Fawcett DW: *Bloom and Fawcett, A Textbook of Histology,* 11th ed. Saunders, 1986.)

CLINICAL BOX 21–2

Effects of Parathyroidectomy

Occasionally, inadvertent parathyroidectomy occurs in humans during thyroid surgery. This can have serious consequences as PTH is essential for life. After parathyroidectomy, there is a steady decline in the plasma Ca^{2+} level. Signs of neuromuscular hyperexcitability appear, followed by full-blown hypocalcemic tetany (see text). Plasma phosphate levels usually rise as the plasma Ca^{2+} level falls. Symptoms usually develop 2–3 days postoperatively but may not appear for several weeks or more. The signs of tetany in humans include **Chvostek's sign,** a quick contraction of the ipsilateral facial muscles elicited by tapping over the facial nerve at the angle of the jaw, and **Trousseau's sign,** a spasm of the muscles of the upper extremity that causes flexion of the wrist and thumb with extension of the fingers. In individuals with mild tetany in whom spasm is not yet evident, Trousseau's sign can sometimes be produced by occluding the circulation for a few minutes with a blood pressure cuff.

THERAPEUTIC HIGHLIGHTS

Treatment centers around replacing the PTH that would normally be produced by the missing glands. Injections of PTH can be given to correct the chemical abnormalities, and the symptoms then disappear. Injections of Ca^{2+} salts can also give temporary relief.

into the endoplasmic reticulum, a leader sequence is removed from the amino terminal to form the 90-amino-acid polypeptide **proPTH.** Six additional amino acid residues are removed from the amino terminal of proPTH in the Golgi apparatus, and the 84-amino-acid polypeptide PTH is packaged in secretory granules and released as the main secretory product of the chief cells.

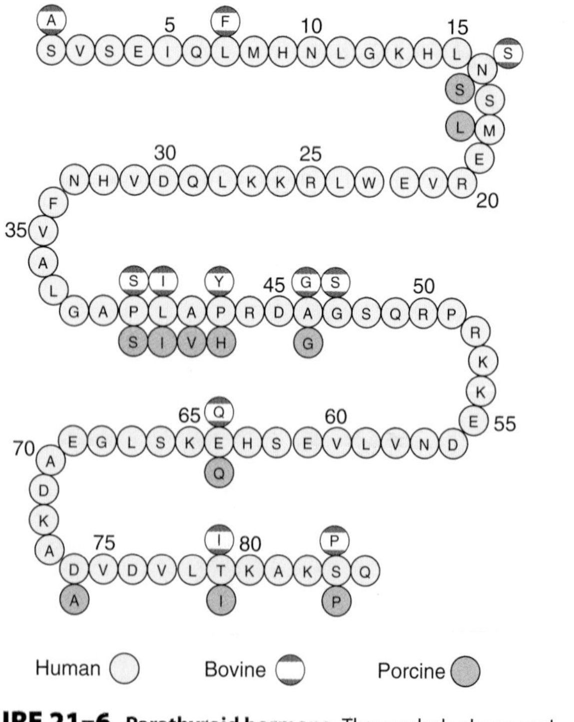

Human Bovine Porcine

FIGURE 21–6 Parathyroid hormone. The symbols above and below the human structure show where amino acid residues are different in bovine and porcine PTH. (Reproduced with permission from Keutmann HT, et al: Complete amino acid sequence of human parathyroid hormone. Biochemistry 1978;17:5723. Copyright © 1978 by the American Chemical Society.)

The normal plasma level of intact PTH is 10–55 pg/mL. The half-life of PTH is approximately 10 min, and the secreted polypeptide is rapidly cleaved by the Kupffer cells in the liver into fragments that are probably biologically inactive. PTH and these fragments are then cleared by the kidneys. Currently used immunoassays for PTH are designed only to measure mature PTH (ie, 84 amino acids) and not these fragments to obtain an accurate measure of "active" PTH.

ACTIONS

PTH acts directly on bone to increase bone resorption and mobilize Ca^{2+}. In addition to increasing plasma Ca^{2+}, PTH increases phosphate excretion in the urine and thereby depresses plasma phosphate levels. This **phosphaturic action** is due to a decrease in reabsorption of phosphate via effects on NaP_i-IIa in the proximal tubules, as discussed previously. PTH also increases reabsorption of Ca^{2+} in the distal tubules, although Ca^{2+} excretion in the urine is often increased in hyperparathyroidism because the increase in the load of filtered calcium overwhelms the effect on reabsorption (Clinical Box 21–3). PTH also increases the formation of 1,25-dihydroxycholecalciferol, and this increases Ca^{2+} absorption from the intestine. On a longer time scale, PTH stimulates both osteoblasts and osteoclasts.

MECHANISM OF ACTION

It now appears that there are at least three different PTH receptors. One also binds parathyroid hormone-related protein (PTHrP; see below) and is known as the hPTH/PTHrP receptor. A second receptor, PTH2 (hPTH2-R), does not bind PTHrP and is found in the brain, placenta, and pancreas. In addition, there is evidence for a third receptor, CPTH, which reacts with the carboxyl terminal rather than the amino

CLINICAL BOX 21–3

Diseases of Parathyroid Excess

Hyperparathyroidism due to hypersecretion from a functioning parathyroid tumor in humans is characterized by hypercalcemia and hypophosphatemia. Humans with PTH-secreting adenomas are usually asymptomatic, with the condition detected when plasma Ca^{2+} is measured in conjunction with a routine physical examination. However, there may be minor changes in personality, and calcium-containing kidney stones occasionally form. In conditions such as chronic renal disease and rickets, in which the plasma Ca^{2+} level is chronically low, stimulation of the parathyroid glands causes compensatory parathyroid hypertrophy and secondary hyperparathyroidism. The plasma Ca^{2+} level is low in chronic renal disease primarily because the diseased kidneys lose the ability to form 1,25-dihydroxycholecalciferol. Finally, mutations in the Ca^{2+} receptor (CaR) gene cause predictable long-term changes in plasma Ca^{2+}. Individuals heterozygous for inactivating mutations have familial benign hypocalciuric hypercalcemia, a condition in which there is a chronic moderate elevation in plasma Ca^{2+} because the feedback inhibition of PTH secretion by Ca^{2+} is reduced. Plasma PTH levels are normal or even elevated. However, children who are homozygous for inactivating mutations develop neonatal severe primary hyperparathyroidism. Conversely, individuals with gain-of-function mutations of the CaR gene develop familial hypercalciuric hypocalcemia due to increased sensitivity of the parathyroid glands to plasma Ca^{2+}.

THERAPEUTIC HIGHLIGHTS

Subtotal parathyroidectomy is sometimes necessary in patients who develop parathyroid adenoma or hyperplasia with associated hypercalcemia and resulting symptoms. However, because parathyroid disease is often benign or only slowly progressing, surgery remains controversial in most patients and is typically reserved for those who have experienced life-threatening complications of hypercalcemia.

terminal of PTH. The first two receptors are coupled to G_s, and via this heterotrimeric G protein they activate adenylyl cyclase, increasing intracellular cAMP. The hPTH/PTHrP receptor also activates PLC via G_q, increasing intracellular Ca^{2+} concentrations and activating protein kinase C (Figure 21–7). However, the way these second messengers affect Ca^{2+} in bone is unsettled.

In the disease called **pseudohypoparathyroidism,** the signs and symptoms of hypoparathyroidism develop but the circulating level of PTH is normal or even elevated. Because tissues fail to respond to the hormone, this is a receptor disease. There are two forms. In the more common form, a congenital 50% reduction of the activity of G_s occurs and PTH fails to produce a normal increase in cAMP concentration. In a different, less common form, the cAMP response is normal but the phosphaturic action of the hormone is defective.

REGULATION OF SECRETION

Circulating Ca^{2+} acts directly on the parathyroid glands in a negative feedback fashion to regulate the secretion of PTH. The key to this regulation is a cell membrane Ca^{2+} receptor, CaR. Activation of this G-protein coupled receptor leads to phosphoinositide turnover in many tissues. In the parathyroid, its activation inhibits PTH secretion. In this way, when the plasma Ca^{2+} level is high, PTH secretion is inhibited and Ca^{2+} is deposited in the bones. When it is low, secretion is increased and Ca^{2+} is mobilized from the bones.

1,25-Dihydroxycholecalciferol acts directly on the parathyroid glands to decrease preproPTH mRNA. Increased plasma phosphate stimulates PTH secretion by lowering plasma levels of free Ca^{2+} and inhibiting the formation of 1,25-dihydroxycholecalciferol. Magnesium is required to maintain normal parathyroid secretory responses. Impaired PTH release along with diminished target organ responses to PTH account for the hypocalcemia that occasionally occurs in magnesium deficiency (Clinical Box 21–2 and Clinical Box 21–3).

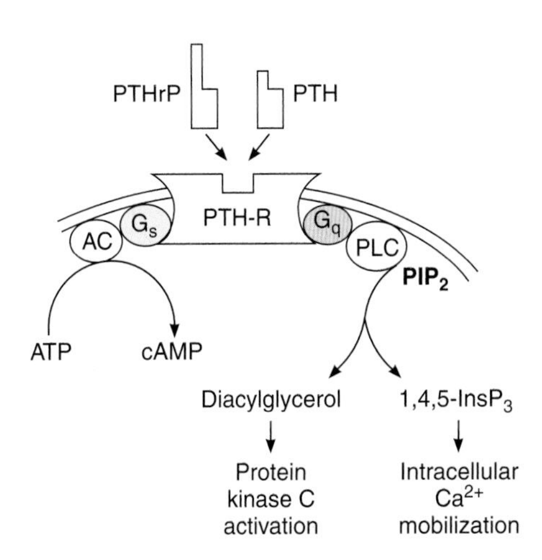

FIGURE 21–7 Signal transduction pathways activated by PTH or PTHrP binding to the hPTH/hPTHrP receptor. Intracellular cAMP is increased via G_s and adenylyl cyclase (AC). Diacylglycerol and IP_3 (1,4,5-InsP_3) are increased via G_q and phospholipase C (PLC). (Modified and reproduced with permission from McPhee SJ, Lingappa VR, Ganong WF [editors]: *Pathophysiology of Disease*, 6th ed. McGraw-Hill, 2010.)

PTHrP

Another protein with PTH activity, **parathyroid hormone-related protein (PTHrP),** is produced by many different tissues in the body. It has 140 amino acid residues, compared with 84 in PTH, and is encoded by a gene on human chromosome 12, whereas PTH is encoded by a gene on chromosome 11. PTHrP and PTH have marked homology at their amino terminal ends and they both bind to the hPTH/PTHrP receptor, yet their physiologic effects are very different. How is this possible when they bind to the same receptor? For one thing, PTHrP is primarily a paracrine factor, acting close to where it is produced. It may be that circulating PTH cannot reach at least some of these sites. Second, subtle conformational differences may be produced by binding of PTH versus PTHrP to their receptor, despite their structural similarities. Another possibility is action of one or the other hormone on additional, more selective receptors.

PTHrP has a marked effect on the growth and development of cartilage in utero. Mice in which both alleles of the PTHrP gene are knocked out have severe skeletal deformities and die soon after birth. In normal animals, on the other hand, PTHrP-stimulated cartilage cells proliferate and their terminal differentiation is inhibited. PTHrP is also expressed in the brain, where evidence indicates that it inhibits excitotoxic damage to developing neurons. In addition, there is evidence that it is involved in Ca^{2+} transport in the placenta. PTHrP is also found in keratinocytes in the skin, in smooth muscle, and in the teeth, where it is present in the enamel epithelium that caps each tooth. In the absence of PTHrP, teeth cannot erupt.

HYPERCALCEMIA OF MALIGNANCY

Hypercalcemia is a common metabolic complication of cancer. About 20% of hypercalcemic patients have bone metastases that produce the hypercalcemia by eroding bone **(local osteolytic hypercalcemia).** Evidence suggests that this erosion is produced by prostaglandins such as prostaglandin E_2 arising from the tumor. The hypercalcemia in the remaining 80% of the patients is due to elevated circulating levels of PTHrP **(humoral hypercalcemia of malignancy).** The tumors responsible for this hypersecretion include cancers of the breast, kidney, ovary, and skin.

CALCITONIN

ORIGIN

In dogs, perfusion of the thyroparathyroid region with solutions containing high concentrations of Ca^{2+} leads to a fall in peripheral plasma Ca^{2+}, and after damage to this region, Ca^{2+} infusions cause a greater increase in plasma Ca^{2+} than they do in control animals. These and other observations led to the discovery that a Ca^{2+}-lowering as well as a Ca^{2+}-elevating hormone was secreted by structures in the neck. The Ca^{2+}-lowering hormone has been named **calcitonin.** In mammals, calcitonin is produced by the **parafollicular cells** of the thyroid gland, which are also known as the clear or C cells.

SECRETION & METABOLISM

Human calcitonin has a molecular weight of 3500 and contains 32 amino acid residues. Its secretion is increased when the thyroid gland is exposed to a plasma calcium level of approximately 9.5 mg/dL. Above this level, plasma calcitonin is directly proportional to plasma calcium. β-Adrenergic agonists, dopamine, and estrogens also stimulate calcitonin secretion. Gastrin, cholecystokinin (CCK), glucagon, and secretin have also been reported to stimulate calcitonin secretion, with gastrin being the most potent stimulus (see Chapter 25). Thus, the plasma calcitonin level is elevated in Zollinger–Ellison syndrome and in pernicious anemia (see Chapter 25). However, the dose of gastrin needed to stimulate calcitonin secretion is supraphysiological and not seen after eating in normal individuals, so dietary calcium in the intestine probably does not induce secretion of a calcium-lowering hormone prior to the calcium being absorbed. In any event, the actions of calcitonin are short-lived because it has a half-life of less than 10 min in humans.

ACTIONS

Receptors for calcitonin are found in bones and the kidneys. Calcitonin lowers circulating calcium and phosphate levels. It exerts its calcium-lowering effect by inhibiting bone resorption. This action is direct, and calcitonin inhibits the activity of osteoclasts in vitro. It also increases Ca^{2+} excretion in the urine.

The exact physiologic role of calcitonin is uncertain. The calcitonin content of the human thyroid is low, and after thyroidectomy, bone density and plasma Ca^{2+} level are normal as long as the parathyroid glands are intact. In addition, after thyroidectomy, there are only transient abnormalities of Ca^{2+} homeostasis when a Ca^{2+} load is injected. This may be explained in part by secretion of calcitonin from tissues other than the thyroid. However, there is general agreement that the hormone has little long-term effect on the plasma Ca^{2+} level in adult animals and humans. Further, unlike PTH and 1,25-dihydroxycholecalciferol, calcitonin does not appear to be involved in phosphate homeostasis. Moreover, patients with medullary carcinoma of the thyroid have a very high circulating calcitonin level but no symptoms directly attributable to the hormone, and their bones are essentially normal. No syndrome due to calcitonin deficiency has been described. More hormone is secreted in young individuals, and it may play a role in skeletal development. In addition, it may protect the bones of the mother from excess calcium loss during pregnancy. Bone formation in the infant and lactation are major drains on Ca^{2+} stores, and plasma concentrations of

1,25-dihydroxycholecalciferol are elevated in pregnancy. They would cause bone loss in the mother if bone resorption were not simultaneously inhibited by an increase in the plasma calcitonin level.

SUMMARY OF CALCIUM HOMEOSTATIC MECHANISMS

The actions of the three principal hormones that regulate the plasma concentration of Ca^{2+} can now be summarized. PTH increases plasma Ca^{2+} by mobilizing this ion from bone. It increases Ca^{2+} reabsorption in the kidney, but this may be offset by the increase in filtered Ca^{2+}. It also increases the formation of 1,25-dihydroxycholecalciferol. 1,25-Dihydroxycholecalciferol increases Ca^{2+} absorption from the intestine and increases Ca^{2+} reabsorption in the kidneys. Calcitonin inhibits bone resorption and increases the amount of Ca^{2+} in the urine.

EFFECTS OF OTHER HORMONES & HUMORAL AGENTS ON CALCIUM METABOLISM

Calcium metabolism is affected by various hormones in addition to 1,25-dihydroxycholecalciferol, PTH, and calcitonin. **Glucocorticoids** lower plasma Ca^{2+} levels by inhibiting osteoclast formation and activity, but over long periods they cause osteoporosis by decreasing bone formation and increasing bone resorption. They decrease bone formation by inhibiting protein synthesis in osteoblasts. They also decrease the absorption of Ca^{2+} and PO_4^{3-} from the intestine and increase the renal excretion of these ions. The decrease in plasma Ca^{2+} concentration also increases the secretion of PTH, and bone resorption is facilitated. **Growth hormone** increases Ca^{2+} excretion in the urine, but it also increases intestinal absorption of Ca^{2+}, and this effect may be greater than the effect on excretion, with a resultant positive calcium balance. Insulin-like growth factor I (IGF-I) generated by the action of growth hormone stimulates protein synthesis in bone. As noted previously, **thyroid hormones** may cause hypercalcemia, hypercalciuria, and, in some instances, osteoporosis. **Estrogens** prevent osteoporosis by inhibiting the stimulatory effects of certain cytokines on osteoclasts. **Insulin** increases bone formation, and there is significant bone loss in untreated diabetes.

BONE PHYSIOLOGY

Bone is a special form of connective tissue with a collagen framework impregnated with Ca^{2+} and PO_4^{3-} salts, particularly **hydroxyapatites,** which have the general formula $Ca_{10}(PO_4)_6(OH)_2$. Bone is also involved in overall Ca^{2+} and PO_4^{3-} homeostasis. It protects vital organs, and the rigidity it provides permits locomotion and the support of loads against gravity. Old bone is constantly being resorbed and new bone formed, permitting remodeling that allows it to respond to the stresses and strains that are put upon it. It is a living tissue that is well vascularized and has a total blood flow of 200–400 mL/min in adult humans.

STRUCTURE

Bone in children and adults is of two types: **compact** or **cortical bone,** which makes up the outer layer of most bones (Figure 21–8) and accounts for 80% of the bone in the body; and **trabecular** or **spongy bone** inside the cortical bone, which makes up the remaining 20% of bone in the body. In compact bone, the surface-to-volume ratio is low, and bone cells lie in lacunae. They receive nutrients by way of canaliculi that ramify throughout the compact bone (Figure 21–8). Trabecular bone is made up of spicules or plates, with a high surface to volume ratio and many cells sitting on the surface of the plates. Nutrients diffuse from bone extracellular fluid (ECF) into the trabeculae, but in compact bone, nutrients are provided via **haversian canals** (Figure 21–8), which contain blood vessels. Around each Haversian canal, collagen is arranged in concentric layers, forming cylinders called **osteons** or **haversian systems.**

The protein in bone matrix is over 90% type I collagen, which is also the major structural protein in tendons and skin. This collagen, which weight for weight is as strong as steel, is made up of a triple helix of three polypeptides bound tightly together. Two of these are identical α_1 polypeptides encoded by one gene, and one is an α_2 polypeptide encoded by a different gene. Collagens make up a family of structurally related proteins that maintain the integrity of many different organs. Fifteen different types of collagens encoded by more than 20 different genes have so far been identified.

BONE GROWTH

During fetal development, most bones are modeled in cartilage and then transformed into bone by ossification **(enchondral bone formation).** The exceptions are the clavicles, the mandibles, and certain bones of the skull in which mesenchymal cells form bone directly **(intramembranous bone formation).**

During growth, specialized areas at the ends of each long bone **(epiphyses)** are separated from the shaft of the bone by a plate of actively proliferating cartilage, the **epiphysial plate** (Figure 21–9). The bone increases in length as this plate lays down new bone on the end of the shaft. The width of the epiphysial plate is proportional to the rate of growth. The width is affected by a number of hormones, but most markedly by the pituitary growth hormone and IGF-I (see Chapter 18).

Linear bone growth can occur as long as the epiphyses are separated from the shaft of the bone, but such growth ceases after the epiphyses unite with the shaft **(epiphysial closure).** The cartilage cells stop proliferating, become

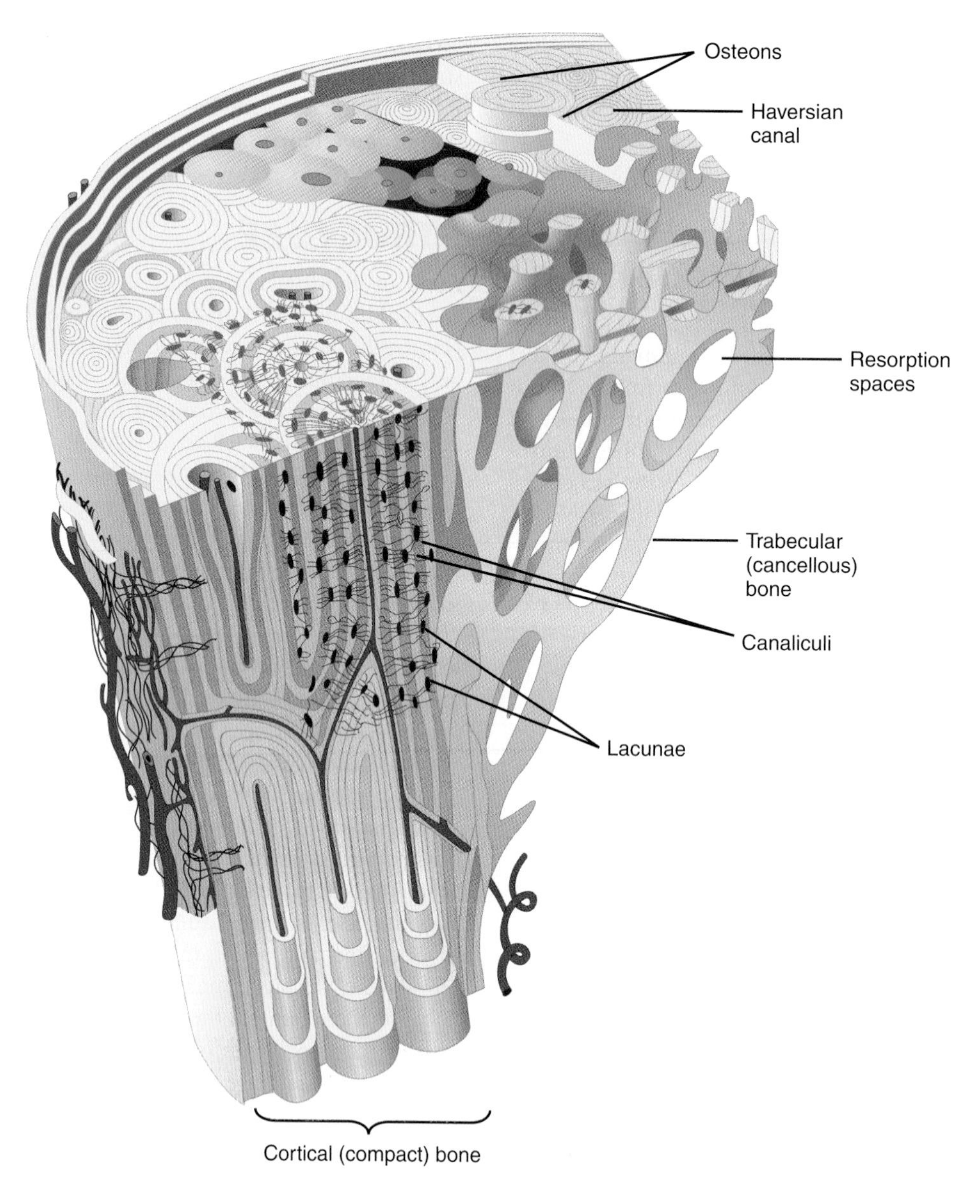

FIGURE 21–8 Structure of compact and trabecular bone. The compact bone is shown in horizontal section (top) and vertical section (bottom). (Reproduced with permission from Williams PL et al (editors): *Gray's Anatomy,* 37th ed. Churchill Livingstone, 1989.)

hypertrophic, and secrete vascular endothelial growth factor (VEGF), leading to vascularization and ossification. The epiphyses of the various bones close in an orderly temporal sequence, the last epiphyses closing after puberty. The normal age at which each of the epiphyses closes is known, and the "bone age" of a young individual can be determined by X-raying the skeleton and noting which epiphyses are open and which are closed.

The **periosteum** is a dense fibrous, vascular, and innervated membrane that covers the surface of bones. This layer consists of an outer layer of collagenous tissue and an inner layer of fine elastic fibers that can include cells that have the potential to contribute to bone growth. The periosteum covers all surfaces of the bone except for those capped with cartilage (eg, at the joints) and serves as a site of attachment of ligaments and tendons. As one ages, the periosteum becomes thinner and loses some of its vasculature. This renders bones more susceptible to injury and disease.

BONE FORMATION & RESORPTION

The cells responsible for bone formation are **osteoblasts** and the cells responsible for bone resorption are **osteoclasts.**

Osteoblasts are modified fibroblasts. Their early development from the mesenchyme is the same as that of fibroblasts, with extensive growth factor regulation. Later, ossification-specific transcription factors, such as Cbfa1/Runx2, contribute to their differentiation. The importance of this transcription

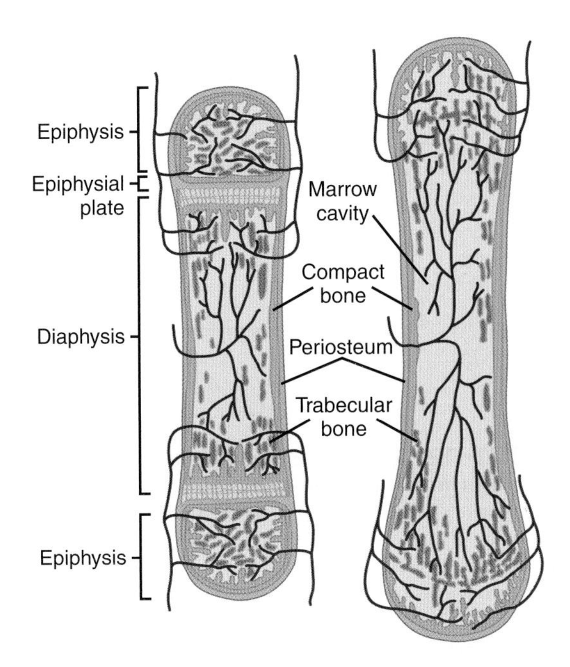

FIGURE 21–9 Structure of a typical long bone before (left) and after (right) epiphysial closure. Note the rearrangement of cells and growth of the bone as the epiphysial plate closes (see text for details).

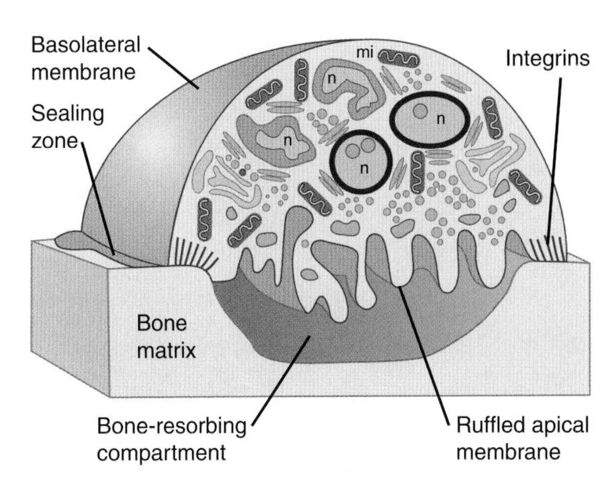

FIGURE 21–10 Osteoclast resorbing bone. The edges of the cell are tightly sealed to bone, permitting secretion of acid from the ruffled apical membrane and consequent erosion of the bone underneath the cell. Note the multiple nuclei (n) and mitochondria (mi). (Courtesy of R Baron.)

factor in bone development is underscored in knockout mice deficient for the Cbfa1/Runx gene. These mice develop to term with their skeletons made exclusively of cartilage; no ossification occurs. Normal osteoblasts are able to lay down type 1 collagen and form new bone.

Osteoclasts, on the other hand, are members of the monocyte family. Stromal cells in the bone marrow, osteoblasts, and T lymphocytes all express receptor activator for nuclear factor kappa beta ligand (RANKL) on their surface. When these cells come in contact with appropriate monocytes expressing RANK (ie, the RANKL receptor) two distinct signaling pathways are initiated: (1) there is a RANKL/RANK interaction between the cell pairs, (2) mononuclear phagocyte colony stimulating factor (M-CSF) is secreted by the nonmonocytic cells and it binds to its corresponding receptor on the monocytes (c-fin). The combination of these two signaling events leads to differentiation of the monocytes into osteoclasts. The precursor cells also secrete **osteoprotegerin (OPG),** which controls for differentiation of the monocytes by competing with RANK for binding of RANKL.

Osteoclasts erode and absorb previously formed bone. They become attached to bone via integrins in a membrane extension called the **sealing zone.** This creates an isolated area between the bone and a portion of the osteoclast. Proton pumps (ie, H^+-dependent ATPases) then move from endosomes into the cell membrane apposed to the isolated area, and they acidify the area to approximately pH 4.0. Similar proton pumps are found in the endosomes and lysosomes of all eukaryotic cells, but in only a few other instances do they move into the cell membrane. Note in this regard that the sealed-off space formed by the osteoclast resembles a large lysosome. The acidic pH dissolves hydroxyapatite, and acid proteases secreted by the cell break down collagen, forming a shallow depression in the bone (Figure 21–10). The products of digestion are then endocytosed and move across the osteoclast by transcytosis (see Chapter 2), with release into the interstitial fluid. The collagen breakdown products have pyridinoline structures, and pyridinolines can be measured in the urine as an index of the rate of bone resorption.

Throughout life, bone is being constantly resorbed and new bone is being formed. The calcium in bone turns over at a rate of 100% per year in infants and 18% per year in adults. Bone remodeling is mainly a local process carried out in small areas by populations of cells called bone-remodeling units. First, osteoclasts resorb bone, and then osteoblasts lay down new bone in the same general area. This cycle takes about 100 days. Modeling drifts also occur in which the shapes of bones change as bone is resorbed in one location and added in another. Osteoclasts tunnel into cortical bone followed by osteoblasts, whereas trabecular bone remodeling occurs on the surface of the trabeculae. About 5% of the bone mass is being remodeled by about 2 million bone-remodeling units in the human skeleton at any one time. The renewal rate for bone is about 4% per year for compact bone and 20% per year for trabecular bone. The remodeling is related in part to the stresses and strains imposed on the skeleton by gravity.

At the cellular level, there is some regulation of osteoclast formation by osteoblasts via the RANKL–RANK and the M-CSF–OPG mechanism; however, specific feedback mechanisms of osteoclasts on osteoblasts are not well defined. In a broader sense, the bone remodeling process is primarily under endocrine control. PTH accelerates bone resorption, and estrogens slow bone resorption by inhibiting the production of bone-eroding cytokines. An interesting new observation

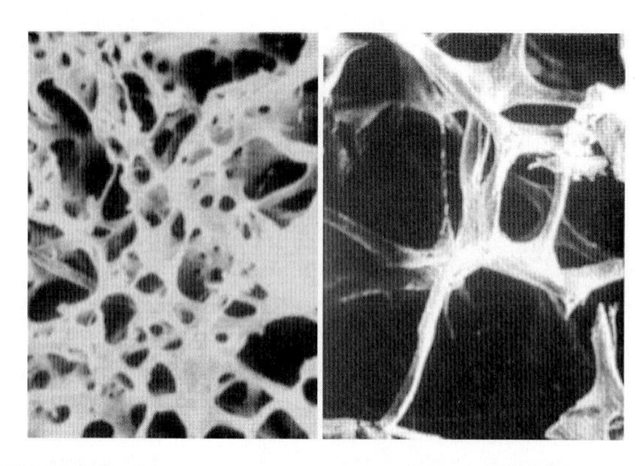

FIGURE 21–11 Normal trabecular bone (left) compared with trabecular bone from a patient with osteoporosis (right). The loss of mass in osteoporosis leaves bones more susceptible to breakage.

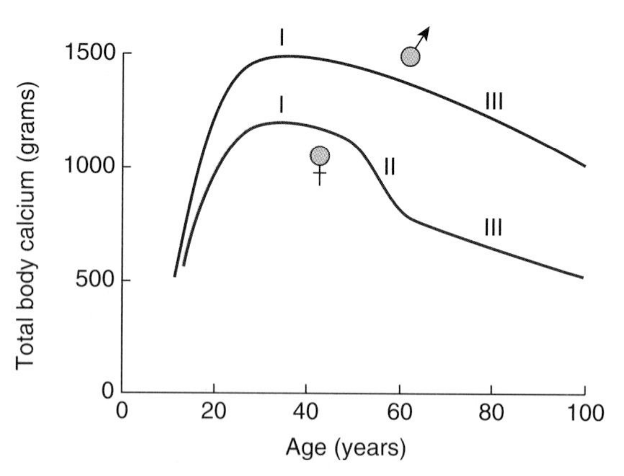

FIGURE 21–12 Total body calcium, an index of bone mass, at various ages in men and women. Note the rapid increase to young adult levels (phase I) followed by the steady loss of bone with advancing age in both sexes (phase III) and the superimposed rapid loss in women after menopause (phase II). (Reproduced by permission of Oxford University Press from Riggs BL, Melton LJ III: Involutional osteoporosis. In Evans TG, Williams TF (editors): *Oxford Textbook of Geriatric Medicine.* Oxford University Press, 1992.)

is that intracerebroventricular, but not intravenous, leptin decreases bone formation. This finding is consistent with the observations that obesity protects against bone loss and that most obese humans are resistant to the effects of leptin on appetite. Thus, there may be neuroendocrine regulation of bone mass via leptin.

BONE DISEASE

The diseases produced by selective abnormalities of the cells and processes discussed above illustrate the interplay of factors that maintain normal bone function.

In **osteopetrosis,** a rare and often severe disease, the osteoclasts are defective and are unable to resorb bone in their usual fashion so the osteoblasts operate unopposed. The result is a steady increase in bone density, neurologic defects due to narrowing and distortion of foramina through which nerves normally pass, and hematologic abnormalities due to crowding out of the marrow cavities. Mice lacking the protein encoded by the immediate-early gene c-*fos* develop osteopetrosis; osteopetrosis also occurs in mice lacking the PU.1 transcription factor. This suggests that all these factors are involved in normal osteoclast development and function.

On the other hand, **osteoporosis** is caused by a relative excess of osteoclastic function. Loss of bone matrix in this condition (Figure 21–11) is marked, and the incidence of fractures is increased. Fractures are particularly common in the distal forearm (Colles fracture), vertebral body, and hip. All of these areas have a high content of trabecular bone, and because trabecular bone is more active metabolically, it is lost more rapidly. Fractures of the vertebrae with compression cause kyphosis, with the production of a typical “widow’s hump” that is common in elderly women with osteoporosis. Fractures of the hip in elderly individuals are associated with a mortality rate of 12–20%, and half of those who survive require prolonged expensive care.

Osteoporosis has multiple causes, but by far the most common form is **involutional osteoporosis.** All normal humans gain bone early in life, during growth. After a plateau, they begin to lose bone as they grow older (Figure 21–12). When this loss is accelerated or exaggerated, it leads to osteoporosis (see Clinical Box 21–4). Increased intake of calcium, particularly from natural sources such as milk, and moderate exercise may help prevent or slow the progress of osteoporosis, although their effects are not great. Bisphosphonates such as etidronate, which inhibit osteoclastic activity, increase the mineral content of bone and decrease the rate of new vertebral fractures when administered in a cyclical fashion. Fluoride stimulates osteoblasts, making bone more dense, but it has proven to be of little value in the treatment of the disease.

CHAPTER SUMMARY

- Circulating levels of calcium and phosphate ions are controlled by cells that sense the levels of these electrolytes in the blood and release hormones, and effects of these hormones are evident in mobilization of the minerals from the bones, intestinal absorption, and/or renal wasting.
- The majority of the calcium in the body is stored in the bones but it is the free, ionized calcium in the cells and extracellular fluids that fulfills physiological roles in cell signaling, nerve function, muscle contraction, and blood coagulation, among others.
- Phosphate is likewise predominantly stored in the bones and regulated by many of the same factors that influence calcium levels, sometimes reciprocally.
- The two major hormones regulating calcium and phosphate homeostasis are 1,25-dihydroxycholecalciferol (a derivative of vitamin D) and parathyroid hormone; calcitonin is also capable of regulating levels of these ions, but its full physiologic contribution is unclear.

CLINICAL BOX 21–4

Osteoporosis

Adult women have less bone mass than adult men, and after menopause they initially lose it more rapidly than men of comparable age. Consequently, they are more prone to development of serious osteoporosis. The cause of bone loss after menopause is primarily estrogen deficiency, and estrogen treatment arrests the progress of the disease. Estrogens inhibit secretion of cytokines such as interleukin-1 (IL-1), IL-6, and tumor necrosis factor (TNF-α), cytokines that otherwise foster the development of osteoclasts. Estrogen also stimulates production of transforming growth factor (TGF-β), and this cytokine increases apoptosis of osteoclasts.

Bone loss can also occur in both men and women as a result of inactivity. In patients who are immobilized for any reason, and during space flight, bone resorption exceeds bone formation and **disuse osteoporosis** develops. The plasma calcium level is not markedly elevated, but plasma concentrations of parathyroid hormone and 1,25-dihydroxycholecalciferol fall and large amounts of calcium are lost in the urine.

THERAPEUTIC HIGHLIGHTS

Hormone therapy has traditionally been used to offset osteoporosis. **Estrogen replacement therapy** begun shortly after menopause can help to maintain bone density. However, it now appears that even small doses of estrogens may increase the incidence of uterine and breast cancer, and in carefully controlled studies, estrogens do not protect against cardiovascular disease. Therefore, treatment of a postmenopausal woman with estrogens is no longer used as a primary option. **Raloxifene** is a selective estrogen receptor modulator that can mimic the beneficial effects of estrogen on bone density in postmenopausal women without some of the risks associated with estrogen. However, this too carries risk of side effects (eg, blood clots). Other hormone treatments include the use of **calcitonin** and the parathyroid hormone analogue **Teriparatide.** An alternative to hormone treatments is the **bisphosphonates.** These drugs can inhibit bone breakdown, preserve bone mass, and even increase bone density in the spine and hip to reduce the risk of fractures. Unfortunately these drugs also can cause mild to serious side effects and require monitoring for patient suitability. In addition to hormones and medications listed above, **physical therapy** to increase appropriate mechanical load and improve balance and muscle strength can significantly improve quality of life.

- 1,25-Dihydroxycholecalciferol acts to elevate plasma calcium and phosphate by predominantly transcriptional mechanisms, whereas parathyroid hormone elevates calcium but decreases phosphate by increasing the latter's renal excretion. Calcitonin lowers both calcium and phosphate levels.
- Deficiencies of 1,25-dihydroxycholecalciferol, or mutations in its receptor, lead to decreases in circulating calcium, defective calcification of the bones, and bone weakness. Disease states also result from either deficiencies or overproduction of parathyroid hormone, with reciprocal effects on calcium and phosphate.
- Bone is a highly structured mass with outer cortical and inner trabecular layers. The larger cortical layer has a high surface to volume layer with haversian canals that provide nutrients and gaps (lacunae) inhabited by bone cells that are connected by a canaliculi network. The smaller trabecular layer has a much higher surface to volume layer that relies on diffusion for nutrients supply.
- Regulated bone growth through puberty occurs through epiphysial plates. These plates are located near the end of the bone shaft and fuse with the shaft of the bone to cease linear bone growth.
- Bone is constantly remodeled by osteoclasts, which erode and absorb bone, and osteoblasts, which lay down new bone.

MULTIPLE-CHOICE QUESTIONS

For all questions, select the single best answer unless otherwise directed.

1. A patient with parathyroid deficiency 10 days after inadvertent damage to the parathyroid glands during thyroid surgery would probably have
   A. low plasma phosphate and Ca^{2+} levels and tetany.
   B. low plasma phosphate and Ca^{2+} levels and tetanus.
   C. a low plasma Ca^{2+} level, increased muscular excitability, and spasm of the muscles of the upper extremity (Trousseau sign).
   D. high plasma phosphate and Ca^{2+} levels and bone demineralization.
   E. increased muscular excitability, a high plasma Ca^{2+} level, and bone demineralization.
2. In an experiment, a rat is infused with a small volume of a calcium chloride solution, or sodium chloride as a control. Compared to the control condition, which of the following would result from the calcium load?
   A. Bone demineralization.
   B. Increased formation of 1,25-dihydroxycholecalciferol.
   C. Decreased secretion of calcitonin.
   D. Decreased blood coagulability.
   E. Increased formation of 24,25-dihydroxycholecalciferol.

the pubic symphysis and softens the cervix, facilitating delivery of the fetus. In both sexes, the gonads secrete other polypeptides, including **inhibin B,** a polypeptide that inhibits follicle-stimulating hormone (FSH) secretion.

The secretory and gametogenic functions of the gonads are both dependent on the secretion of the anterior pituitary gonadotropins, FSH, and luteinizing hormone (LH). The sex hormones and inhibin B feed back to inhibit gonadotropin secretion. In males, gonadotropin secretion is noncyclic; but in postpubertal females an orderly, sequential secretion of gonadotropins is necessary for the occurrence of menstruation, pregnancy, and lactation.

SEX DIFFERENTIATION & DEVELOPMENT

CHROMOSOMAL SEX

The Sex Chromosomes

Sex is determined genetically by two chromosomes, called the **sex chromosomes,** to distinguish them from the **somatic chromosomes (autosomes).** In humans and many other mammals, the sex chromosomes are called X and Y. The Y chromosome is necessary and sufficient for the production of testes, and the testis-determining gene product is called SRY (for sex-determining region of the Y chromosome). SRY is a DNA-binding regulatory protein. It bends the DNA and acts as a transcription factor that initiates transcription of a cascade of genes necessary for testicular differentiation, including the gene for **müllerian inhibiting substance** (**MIS;** see below). The gene for SRY is located near the tip of the short arm of the human Y chromosome. Diploid male cells contain an X and a Y chromosome (XY pattern), whereas female cells contain two X chromosomes (XX pattern). As a consequence of meiosis during gametogenesis, each normal ovum contains a single X chromosome, but half of the normal sperm contain an X chromosome and half contain a Y chromosome (Figure 22–1). When a sperm containing a Y chromosome fertilizes an ovum, an XY pattern results and the zygote develops into a **genetic male.** When fertilization occurs with an X-containing sperm, an XX pattern and a **genetic female** results. Cell division and the chemical nature of chromosomes are discussed in Chapter 1.

Human Chromosomes

Human chromosomes can be studied in detail. Human cells are grown in tissue culture; treated with the drug colchicine, which arrests mitosis at the metaphase; exposed to a hypotonic solution that makes the chromosomes swell and disperse; and then "squashed" onto slides. Staining techniques make it possible to identify the individual chromosomes (Figure 22–2). There are 46 chromosomes: in males, 22 pairs of autosomes plus an X chromosome and a Y chromosome; in females, 22 pairs of autosomes plus two X chromosomes. The individual chromosomes are usually arranged in an arbitrary pattern (**karyotype**). The individual autosome pairs are identified by the numbers 1–22 on the basis of their morphologic characteristics.

Sex Chromatin

Soon after cell division has started during embryonic development, one of the two X chromosomes of the somatic cells in normal females becomes functionally inactive. In abnormal individuals with more than two X chromosomes, only

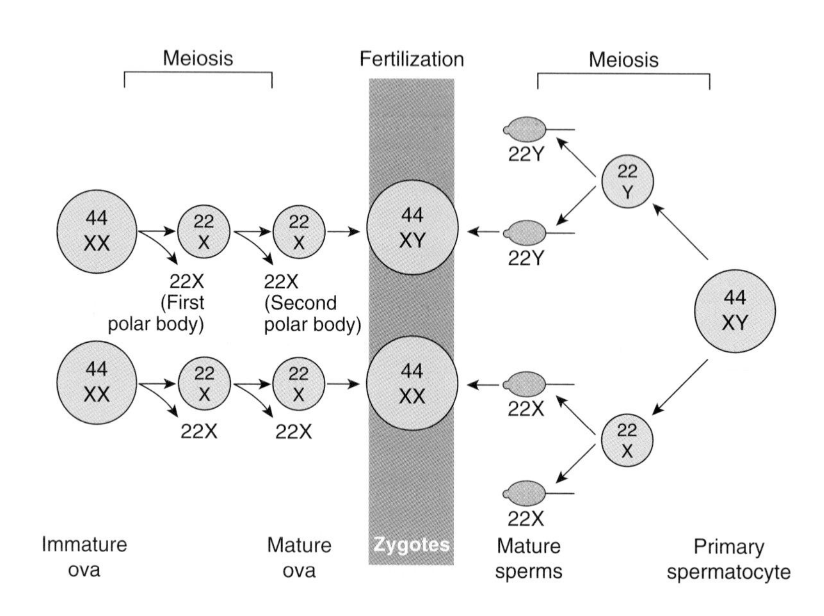

FIGURE 22–1 Basis of genetic sex determination. In the two-stage meiotic division in the female, only one cell survives as the mature ovum. In the male, the meiotic division results in the formation of four sperms, two containing the X and two the Y chromosome. Fertilization thus produces a male zygote with 22 pairs of autosomes plus an X and a Y or a female zygote with 22 pairs of autosomes and two X chromosomes. Note that for clarity, this figure and Figures 25–6 and 25–7 differ from the current international nomenclature for karyotypes, which lists the total number of chromosomes followed by the sex chromosome pattern. Thus, XO is 45, X; XY is 46, XY; XXY is 47, XXY, and so on.

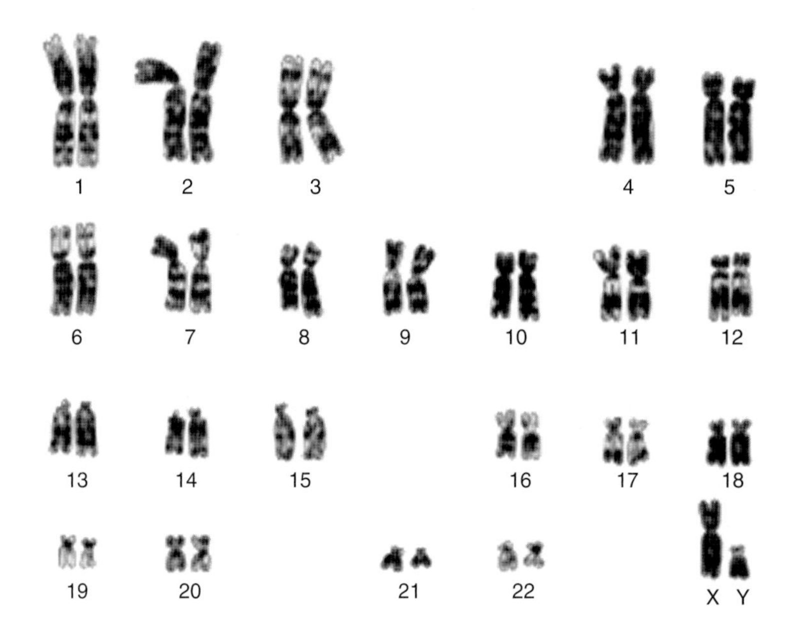

FIGURE 22–2 Karyotype of chromosomes from a normal male. The chromosomes have been stained with Giemsa's stain, which produces a characteristic banding pattern. (Reproduced with permission, from Lingappa VJ, Farey K: *Physiological Medicine.* McGraw-Hill, 2000.)

one remains active. The process that is normally responsible for inactivation is initiated in an X-inactivation center in the chromosome, probably via the transactivating factor CTCF (for CCCTC-binding factor), which is also induced during gene imprinting. However, the details of the inactivation process are still incompletely understood. The choice of which X chromosome remains active is random, so normally one X chromosome remains active in approximately half of the cells and the other X chromosome is active in the other half. The selection persists through subsequent divisions of these cells, and consequently some of the somatic cells in adult females contain an active X chromosome of paternal origin and some contain an active X chromosome of maternal origin.

In normal cells, the inactive X chromosome condenses and can be seen in various types of cells, usually near the nuclear membrane, as the **Barr body,** also called sex chromatin (Figure 22–3). Thus, there is a Barr body for each X chromosome in excess of one in the cell. The inactive X chromosome is also visible as a small "drumstick" of chromatin projecting from the nuclei of 1–15% of the polymorphonuclear leukocytes in females but not in males (Figure 22–3).

EMBRYOLOGY OF THE HUMAN REPRODUCTIVE SYSTEM

Development of the Gonads

On each side of the embryo, a primitive gonad arises from the genital ridge, a condensation of tissue near the adrenal gland. The gonad develops a **cortex** and a **medulla.** Until the 6th week of development, these structures are identical in both sexes. In genetic males, the medulla develops during the 7th and 8th weeks into a testis, and the cortex regresses. Leydig and Sertoli cells appear, and testosterone and MIS are secreted. In genetic females, the cortex develops into an ovary and the medulla regresses. The embryonic ovary does not secrete hormones. Hormonal treatment of the mother has no effect on gonadal (as opposed to ductal and genital) differentiation in humans, although it does in some experimental animals.

Embryology of the Genitalia

The embryology of the gonads is summarized in Figures 22–4 and 22–5. In the 7th week of gestation, the embryo has both

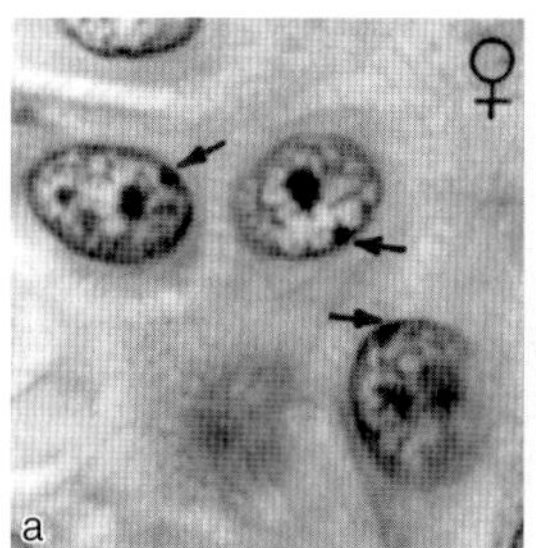

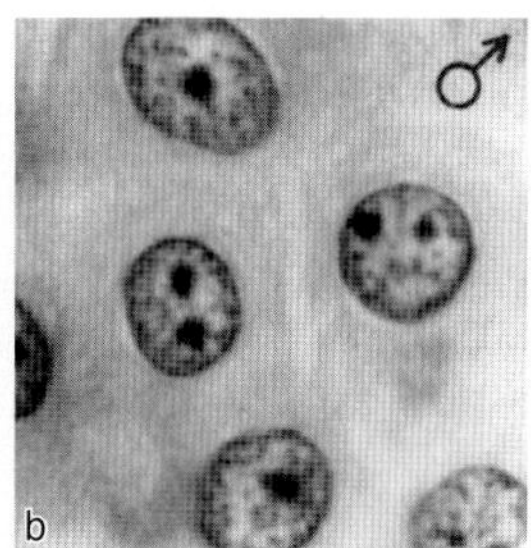

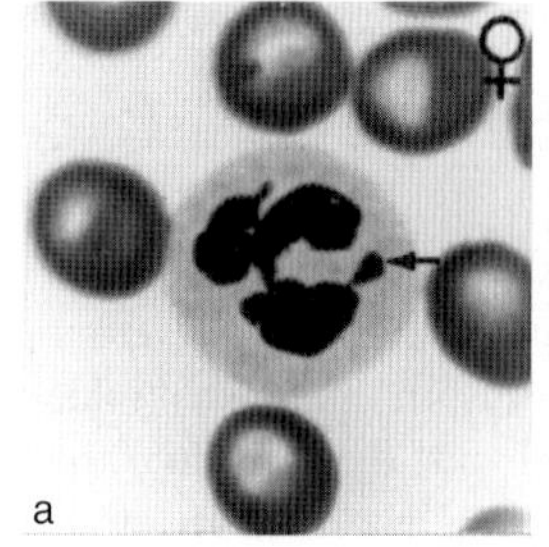

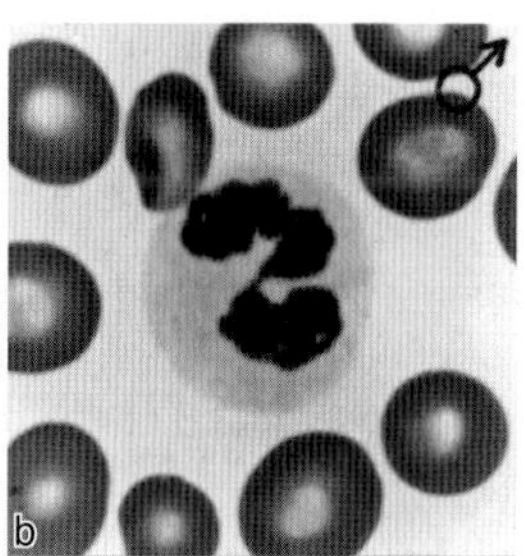

FIGURE 22–3 Left: Barr body (arrows) in the epidermal spinous cell layer. **Right:** Nuclear appendage ("drumstick") identified by arrow in white blood cells. (Reproduced with permission from Grumbach MM, Barr ML: Cytologic tests of chromosomal sex in relation to sex anomalies in man. Recent Prog Horm Res 1958;14:255.)

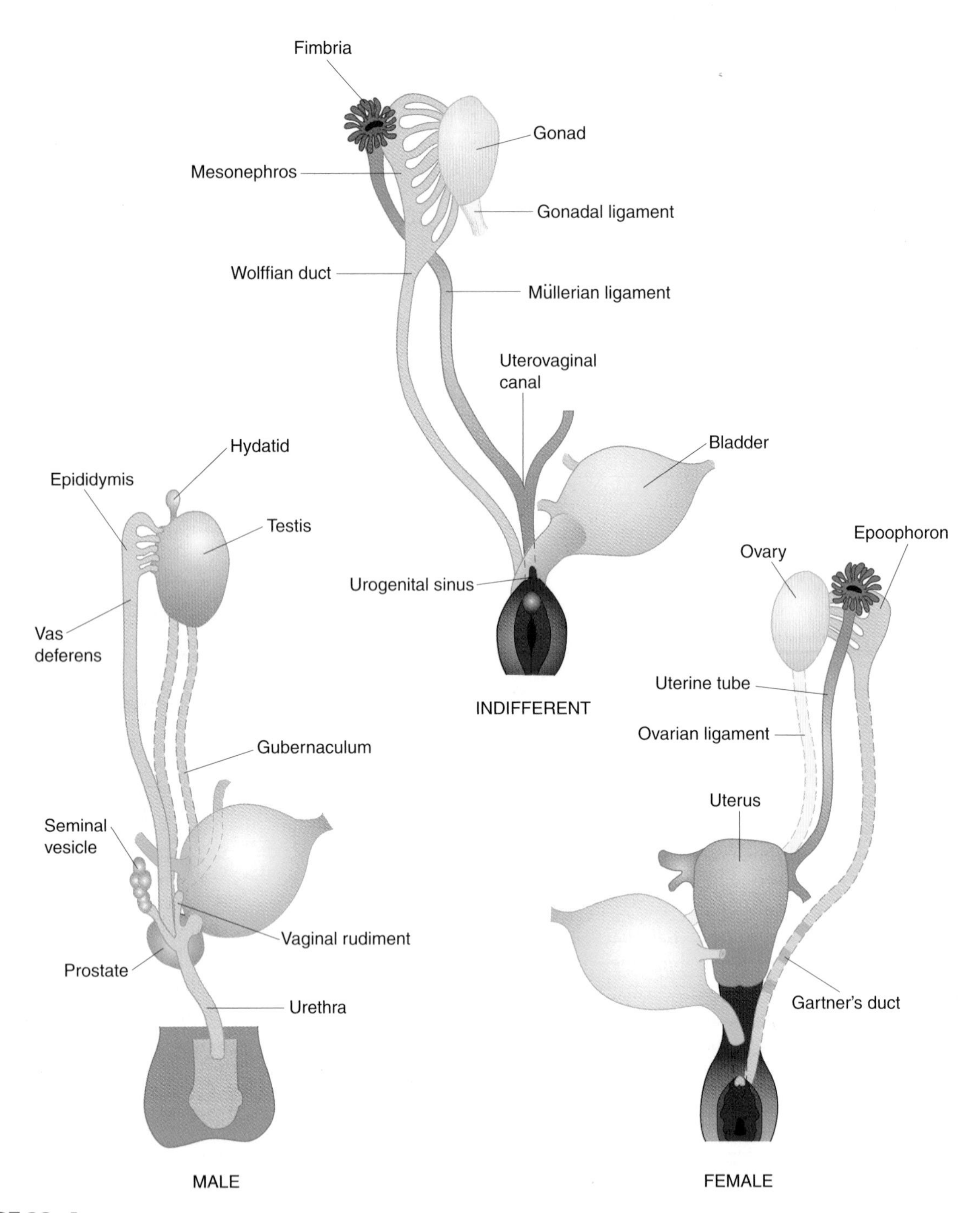

FIGURE 22–4 Embryonic differentiation of male and female internal genitalia (genital ducts) from wolffian (male) and müllerian (female) primordia. (After Corning HK, Wilkins L. Redrawn and reproduced with permission from *Williams Textbook of Endocrinology,* 7th ed. Wilson JD, Foster DW [editors]. Saunders, 1985.)

male and female primordial genital ducts (Figure 22–4). In a normal female fetus, the müllerian duct system then develops into uterine tubes (oviducts) and a uterus. In the normal male fetus, the wolffian duct system on each side develops into the epididymis and vas deferens. The external genitalia are similarly bipotential until the 8th week (Figure 22–5). Thereafter, the urogenital slit disappears and male genitalia form, or, alternatively, it remains open and female genitalia form.

When the embryo has functional testes, male internal and external genitalia develop. The Leydig cells of the fetal testis secrete testosterone, and the Sertoli cells secrete MIS (also called müllerian regression factor, or MRF). MIS is a 536-amino-acid homodimer that is a member of the transforming growth factor β (TGF-β) super-family of growth factors, which includes inhibins and activins.

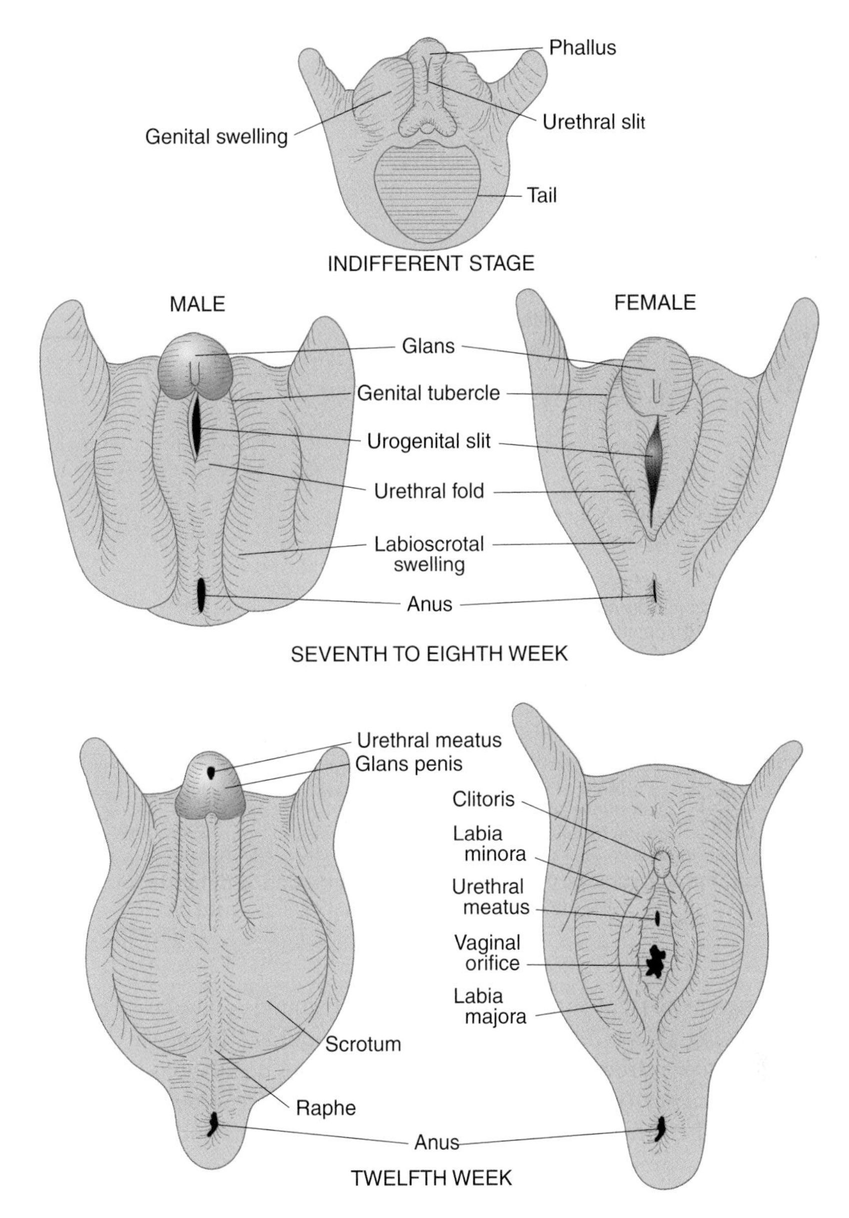

FIGURE 22–5 **Differentiation of male and female external genitalia from indifferent primordial structures in the embryo.**

In their effects on the internal as opposed to the external genitalia, MIS and testosterone act unilaterally. MIS causes regression of the müllerian ducts by apoptosis on the side on which it is secreted, and testosterone fosters the development of the vas deferens and related structures from the wolffian ducts. The testosterone metabolite dihydrotestosterone induces the formation of male external genitalia and male secondary sex characteristics (Figure 22–6).

MIS continues to be secreted by the Sertoli cells, and it reaches mean values of 48 ng/mL in plasma in 1- to 2-year-old boys. Thereafter, it declines to low levels by the time of puberty and persists at low but detectable levels throughout life. In girls, MIS is produced by granulosa cells in small follicles in the ovaries, but plasma levels are very low or undetectable until puberty. Thereafter, plasma MIS is about the same as in adult men, that is, about 2 ng/mL. The functions of MIS after early embryonic life are unsettled, but it is probably involved in germ cell maturation in both sexes and in control of testicular descent in boys.

Development of the Brain

At least in some species, the development of the brain as well as the external genitalia is affected by androgens early in life. In rats, a brief exposure to androgens during the first few days of life causes the male pattern of sexual behavior and the male pattern of hypothalamic control of gonadotropin secretion to develop after puberty. In the absence of androgens, female patterns develop (see Chapter 17). In monkeys, similar effects on sexual behavior are produced by exposure to androgens in utero, but the pattern of gonadotropin secretion remains cyclical. Early exposure

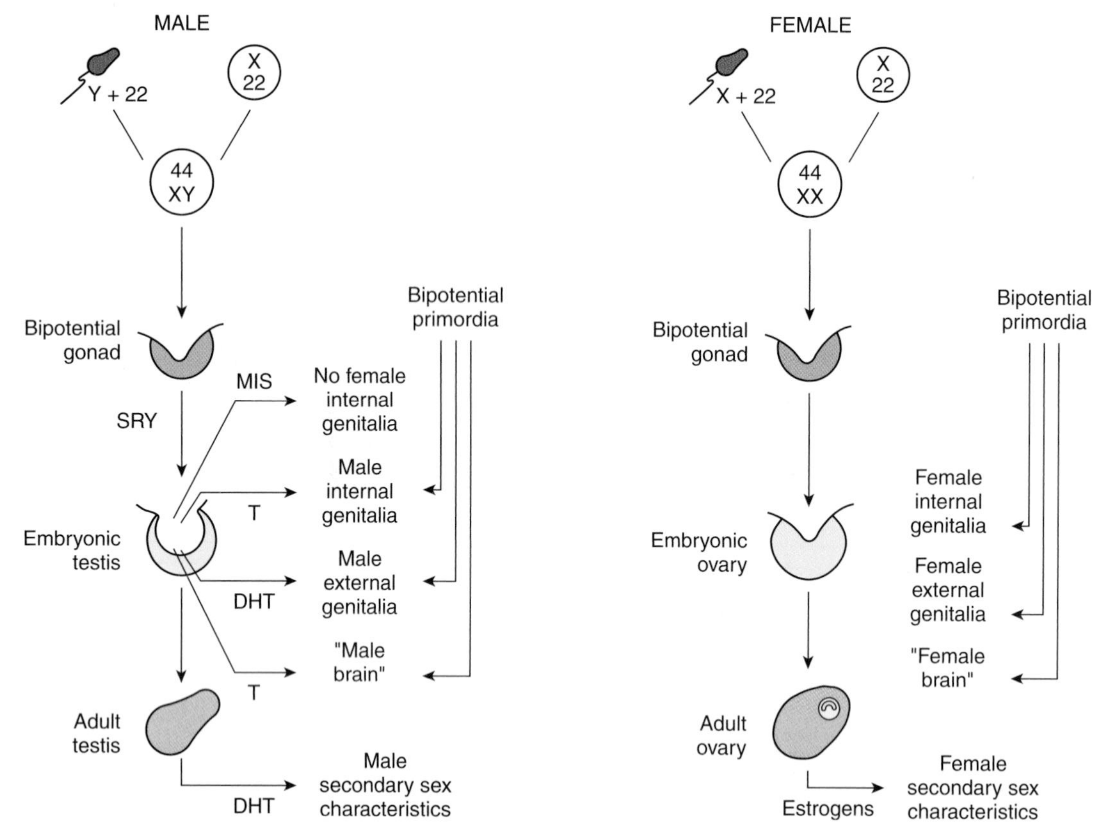

FIGURE 22–6 Diagrammatic summary of normal sex determination, differentiation, and development in humans. DHT, dihydrotestosterone, MIS, müllerian inhibiting substance; T, testosterone.

of female human fetuses to androgens also appears to cause subtle but significant masculinizing effects on behavior. However, women with adrenogenital syndrome due to congenital adrenocortical enzyme deficiency (see Chapter 20) develop normal menstrual cycles when treated with cortisol. Thus, the human, like the monkey, appears to retain the cyclical pattern of gonadotropin secretion despite exposure to androgens in utero.

ABERRANT SEXUAL DIFFERENTIATION

Chromosomal Abnormalities

From the preceding discussion, it might be expected that abnormalities of sexual development could be caused by genetic or hormonal abnormalities as well as by other nonspecific teratogenic influences, and this is indeed the case. The major classes of abnormalities are listed in Table 22–1.

Nondisjunction of sex chromosomes during the first division in meiosis results in distinct defects (see Clinical Box 22–1; Figure 22–7). Meiosis is a two-stage process, and although nondisjunction usually occurs during the first meiotic division, it can occur in the second, producing more complex chromosomal abnormalities. In addition, nondisjunction or simple loss of a sex chromosome can occur during the early mitotic divisions after fertilization. The consequence of faulty mitoses in the early zygote is **mosaicism,** in which two or more populations of cells have different chromosome complements. **True hermaphroditism,** the condition in which the individual has both ovaries and testes, is probably due to XX/XY mosaicism and related mosaic patterns, although other genetic aberrations are possible.

Chromosomal abnormalities also include transposition of parts of chromosomes to other chromosomes. Rarely, genetic males are found to have the XX karyotype because the short arm of their father's Y chromosome was transposed to their father's X chromosome during meiosis and they received that X chromosome along with their mother's. Similarly, deletion of the small portion of the Y chromosome containing SRY produces females with the XY karyotype.

Hormonal Abnormalities

Development of the male external genitalia occurs normally in genetic males in response to androgen secreted by the embryonic testes, but male genital development may also occur in genetic females exposed to androgens from some other source during the 8th to the 13th weeks of gestation. The syndrome that results is **female pseudohermaphroditism.** A pseudohermaphrodite is an individual with the genetic constitution and gonads of one sex and the genitalia of the other. After the

TABLE 22–1 Classification of the major disorders of sex differentiation in humans.[a]

Chromosomal disorders
Gonadal dysgenesis (XO and variants)
"Superfemales" (XXX)
Seminiferous tubule dysgenesis (XXY and variants)
True hermaphroditism
Developmental disorders
Female pseudohermaphroditism
Congenital virilizing adrenal hyperplasia of fetus
Maternal androgen excess
Virilizing ovarian tumor
Iatrogenic: Treatment with androgens or certain synthetic progestational drugs
Male pseudohermaphroditism
Androgen resistance
Defective testicular development
Congenital 17α-hydroxylase deficiency
Congenital adrenal hyperplasia due to blockade of pregnenolone formation
Various nonhormonal anomalies

[a]Many of these syndromes can have great variation in degree and, consequently, in manifestations.

13th week, the genitalia are fully formed, but exposure to androgens can cause hypertrophy of the clitoris. Female pseudohermaphroditism may be due to congenital virilizing adrenal hyperplasia (see Chapter 20), or it may be caused by androgens administered to the mother. Conversely, one cause of the development of female external genitalia in genetic males **(male pseudohermaphroditism)** is defective testicular development. Because the testes also secrete MIS, genetic males with defective testes have female internal genitalia.

Another cause of male pseudohermaphroditism is **androgen resistance,** in which, as a result of various congenital abnormalities, male hormones cannot exert their full effects on the tissues. One form of androgen resistance is a **5α-reductase deficiency,** in which the enzyme responsible for the formation of dihydrotestosterone, the active form of testosterone, is decreased (Figure 22-8). The consequences of this deficiency are discussed in Chapter 23. Other forms of androgen resistance are due to various mutations in the androgen receptor gene, and the resulting defects in receptor function range from minor to severe. Mild defects cause infertility with or without gynecomastia. When the loss of receptor function is complete, the **testicular feminizing syndrome,** now known as **complete androgen resistance syndrome,** results. In this condition, MIS is present and testosterone is secreted at normal or even elevated rates. The external genitalia are female,

CLINICAL BOX 22–1

Chromosomal Abnormalities

An established defect in gametogenesis is **nondisjunction,** a phenomenon in which a pair of chromosomes fail to separate, so that both go to one of the daughter cells during meiosis. Four of the abnormal zygotes that can form as a result of nondisjunction of one of the X chromosomes during oogenesis are shown in Figure 25–7. In individuals with the XO chromosomal pattern, the gonads are rudimentary or absent, so that female external genitalia develop, stature is short, other congenital abnormalities are often present, and no sexual maturation occurs at puberty. This syndrome is called **gonadal dysgenesis** or, alternatively, **ovarian agenesis** or **Turner syndrome.** Individuals with the XXY pattern, the most common sex chromosome disorder, have the genitalia of a normal male. Testosterone secretion at puberty is often great enough for the development of male characteristics; however, the seminiferous tubules are abnormal, and the incidence of mental retardation is higher than normal. This syndrome is known as **seminiferous tubule dysgenesis** or **Klinefelter syndrome.** The XXX ("superfemale") pattern is second in frequency only to the XXY pattern and may be even more common in the general population, since it does not seem to be associated with any characteristic abnormalities. The YO combination is probably lethal.

Nondisjunction of chromosome 21 produces **trisomy 21,** the chromosomal abnormality associated with **Down syndrome** (mongolism). The additional chromosome 21 is normal, so Down syndrome is a pure case of gene excess causing abnormalities.

Many other chromosomal abnormalities occur as well as numerous diseases due to defects in single genes. These conditions are generally diagnosed in utero by analysis of fetal cells in a sample of amniotic fluid collected by inserting a needle through the abdominal wall **(amniocentesis)** or, earlier in pregnancy, by examining fetal cells obtained by a needle biopsy of chorionic villi **(chorionic villus sampling).**

THERAPEUTIC HIGHLIGHTS

Many of the syndromes mentioned have effects on multiple organ systems, and patients must be carefully followed with a multidisciplinary approach to avert the consequences of cardiovascular defects, infections secondary to urinary tract and renal malformations, and the psychological impact of reproductive implications. Girls with Turner syndrome and evidence of gonadal failure are also treated with low dose estrogen to evoke puberty, followed by gradual replacement of mature estrogen levels to permit feminization. Conversely, patients with Klinefelter syndrome are often supplemented with androgens to improve virilization and libido.

CLINICAL BOX 22–3

Hyperprolactinemia

Up to 70% of the patients with chromophobe adenomas of the anterior pituitary have elevated plasma prolactin levels. In some instances, the elevation may be due to damage to the pituitary stalk, but in most cases, the tumor cells are actually secreting the hormone. The hyperprolactinemia may cause galactorrhea, but in many individuals no demonstrable endocrine abnormalities are present. Conversely, most women with galactorrhea have normal prolactin levels; definite elevations are found in less than a third of patients with this condition.

Another interesting observation is that 15–20% of women with secondary amenorrhea have elevated prolactin levels, and when prolactin secretion is reduced, normal menstrual cycles and fertility return. Prolactin may produce amenorrhea by blocking the action of gonadotropins on the ovaries. The hypogonadism produced by prolactinomas is associated with osteoporosis due to estrogen deficiency.

As noted previously, hyperprolactinemia in men is associated with impotence and hypogonadism that disappear when prolactin secretion is reduced.

THERAPEUTIC HIGHLIGHTS

Prescription drug use is a common cause of hyperprolactinemia. Prolactin secretion in the pituitary is suppressed by the brain chemical dopamine. Use of drugs that block the effects of dopamine can cause the pituitary to secrete prolactin. Examples of some prescription drugs that can cause hyperprolactinemia include the major tranquilizers haloperidol (Haldol) and phenothiazines, most antipsychotic medications, and cisapride, which is used to treat nausea and gastro-oesophageal reflux in cancer patients. If possible, the drug suspected as causing hyperprolactinemia should be withdrawn, or the dose titrated. Regardless of the etiology, treatment should endeavor to restore normal prolactin levels to avoid suppressive effects on the ovaries and preserve bone density. Dopamine agonists also provide benefit in many cases, including prolactinoma, and can be used in patients for whom an inciting pharmaceutical agent cannot be withdrawn.

increased because of an activating mutation in the G protein that couples the receptors to adenylyl cyclase.

Delayed or Absent Puberty

The normal variation in the age at which adolescent changes occur is so wide that puberty cannot be considered to be pathologically delayed until the menarche has failed to occur by the age of 17 or testicular development by the age of 20. Failure of maturation due to panhypopituitarism is associated with dwarfing and evidence of other endocrine abnormalities. Patients with the XO chromosomal pattern and gonadal dysgenesis are also dwarfed. In some individuals, puberty is delayed even though the gonads are present and other endocrine functions are normal. In males, this clinical picture is called **eunuchoidism.** In females, it is called **primary amenorrhea** (see Clinical Box 22–3).

MENOPAUSE

The human ovaries become unresponsive to gonadotropins with advancing age, and their function declines, so that sexual cycles disappear **(menopause).** This unresponsiveness is associated with and probably caused by a decline in the number of primordial follicles, which becomes precipitous at the time of menopause (Figure 22–10). The ovaries no longer secrete progesterone and 17β-estradiol in appreciable quantities, and estrogen is formed only in small amounts by aromatization of androstenedione in peripheral tissues (see Chapter 20). The uterus and the vagina gradually become atrophic. As the negative feedback effect of estrogens and progesterone is reduced, secretion of FSH is increased, and

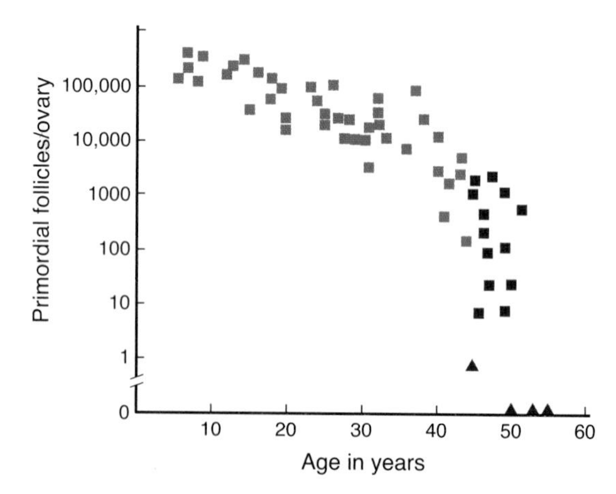

FIGURE 22–10 Number of primordial follicles per ovary in women at various ages. Blue squares, premenopausal women (regular menses); red squares, perimenopausal women (irregular menses for at least 1 year); red triangles, postmenopausal women (no menses for at least 1 year). Note that the vertical scale is a log scale and that the values are from one rather than two ovaries. (Redrawn by PM Wise and reproduced with permission from Richardson SJ, Senikas V, Nelson JF: Follicular depletion during the menopausal transition: Evidence for accelerated loss and ultimate exhaustion. J Clin Endocrinol Metab 1987;65:1231.)

plasma FSH increases to high levels, LH levels are moderately high. Old female mice and rats have long periods of diestrus and increased levels of gonadotropin secretion. In women, a period called perimenopause precedes menopause, and can last up to 10 years. During perimenopause FSH levels will increase before an increase in LH is observed due to a decrease in estrogen, progesterone, and inhibins and the menses become irregular. This usually occurs between the ages of 45 and 55. The average age at onset of the menopause is 52 years.

The loss of ovarian function causes many symptoms such as sensations of warmth spreading from the trunk to the face (hot flushes; also called hot flashes) and night sweats. In addition, the onset of menopause increases the risk of many diseases such as osteoporosis, ischemic heart disease, and renal disease.

Hot flushes are said to occur in 75% of menopausal women and may continue intermittently for as long as 40 years. They also occur when early menopause is produced by bilateral ovariectomy, and they are prevented by estrogen treatment. In addition, they occur after castration in men. Their cause is unknown. However, they coincide with surges of LH secretion. LH is secreted in episodic bursts at intervals of 30–60 min or more **(circhoral secretion),** and in the absence of gonadal hormones these bursts are large. Each hot flush begins with the start of a burst. However, LH itself is not responsible for the symptoms, because they can continue after removal of the pituitary. Instead, it appears that some estrogen-sensitive event in the hypothalamus initiates both the release of LH and the episode of flushing.

Although the function of the testes tends to decline slowly with advancing age, the evidence is unclear whether there is a "male menopause" **(andropause)** similar to that occurring in women.

THE FEMALE REPRODUCTIVE SYSTEM

THE MENSTRUAL CYCLE

The reproductive system of women (Figure 22–11), unlike that of men, shows regular cyclic changes that teleologically may be regarded as periodic preparations for fertilization and pregnancy. In humans and other primates, the cycle is a **menstrual** cycle, and its most conspicuous feature is the periodic vaginal bleeding that occurs with the shedding of the uterine mucosa **(menstruation).** The length of the cycle is notoriously variable in women, but an average figure is 28 days from the start of one menstrual period to the start of the next. By common usage, the days of the cycle are identified by number, starting with the first day of menstruation.

Ovarian Cycle

From the time of birth, there are many **primordial follicles** under the ovarian capsule. Each contains an immature ovum (Figure 22–11). At the start of each cycle, several of these follicles enlarge, and a cavity forms around the ovum **(antrum formation).** This cavity is filled with follicular fluid. In humans, usually one of the follicles in one ovary starts to grow rapidly on about the 6th day and becomes the **dominant follicle,** while the others regress, forming **atretic follicles.** The atretic process involves apoptosis. It is uncertain how one follicle is selected to be the dominant follicle in this **follicular phase** of the menstrual cycle, but it seems to be related to the ability of the follicle to secrete the estrogen inside it that is needed for final maturation. When women are given human pituitary gonadotropin preparations by injection, many follicles develop simultaneously.

The structure of a maturing ovarian **(graafian)** follicle is shown in Figure 22–11. The primary source of circulating estrogen is the granulosa cells of the ovaries; however, the cells of the **theca interna** of the follicle are necessary for the production of estrogen as they secrete androgens that are aromatized to estrogen by the granulosa cells.

At about the 14th day of the cycle, the distended follicle ruptures, and the ovum is extruded into the abdominal cavity. This is the process of **ovulation.** The ovum is picked up by the fimbriated ends of the uterine tubes (oviducts). It is transported to the uterus and, unless fertilization occurs, out through the vagina.

The follicle that ruptures at the time of ovulation promptly fills with blood, forming what is sometimes called a **corpus hemorrhagicum.** Minor bleeding from the follicle into the abdominal cavity may cause peritoneal irritation and fleeting lower abdominal pain ("mittelschmerz"). The granulosa and theca cells of the follicle lining promptly begin to proliferate, and the clotted blood is rapidly replaced with yellowish, lipid-rich **luteal cells,** forming the **corpus luteum.** This initiates the **luteal phase** of the menstrual cycle, during which the luteal cells secrete estrogen and progesterone. Growth of the corpus luteum depends on its developing an adequate blood supply, and there is evidence that vascular endothelial growth factor (VEGF) (see Chapter 31) is essential for this process.

If pregnancy occurs, the corpus luteum persists and usually there are no more periods until after delivery. If pregnancy does not occur, the corpus luteum begins to degenerate about 4 days before the next menses (24th day of the cycle) and is eventually replaced by scar tissue, forming a **corpus albicans.**

The ovarian cycle in other mammals is similar, except that in many species more than one follicle ovulates and multiple births are the rule. Corpora lutea form in some submammalian species but not in others.

In humans, no new ova are formed after birth. During fetal development, the ovaries contain over 7 million primordial follicles. However, many undergo atresia (involution) before birth and others are lost after birth. At the time of birth, there are 2 million ova, but 50% of these are atretic. The million that are normal undergo the first part of the first meiotic division at about this time and enter a stage of arrest in prophase in which those that survive persist until adulthood. Atresia continues during development, and the number of ova in both of the ovaries at the time of puberty is less than 300,000

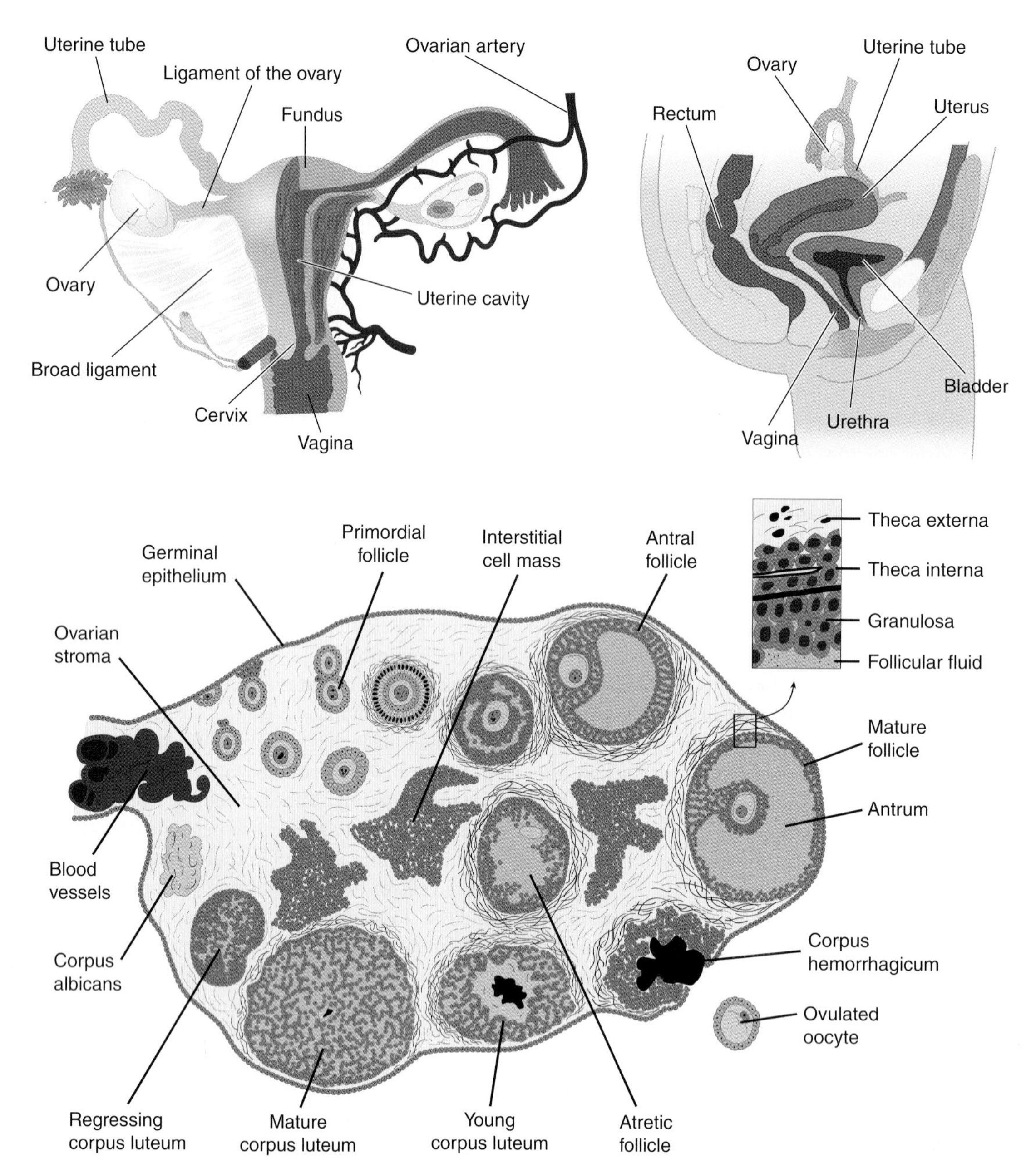

FIGURE 22–11 Functional anatomy of the female reproductive tract. The female reproductive organs include the ovaries, the uterus and the fallopian tubes, and the breast/mammary glands. The sequential development of a follicle, the formation of a corpus luteum and follicular atresia are shown.

(Figure 22–10). Only one of these ova per cycle (or about 500 in the course of a normal reproductive life) normally reaches maturity; the remainder degenerate. Just before ovulation, the first meiotic division is completed. One of the daughter cells, the **secondary oocyte,** receives most of the cytoplasm, while the other, the **first polar body,** fragments and disappears. The secondary oocyte immediately begins the second meiotic division, but this division stops at metaphase and is completed only when a sperm penetrates the oocyte. At that time, the **second polar body** is cast off and the fertilized ovum proceeds to form a new individual. The arrest in metaphase is due, at least in some species, to formation in the ovum of the protein **pp39**^{mos}, which is encoded by the **c-*mos*** protooncogene. When fertilization occurs, the pp39^{mos} is destroyed within 30 min by **calpain,** a calcium-dependent cysteine protease.

Uterine Cycle

At the end of menstruation, all but the deep layers of the endometrium have sloughed. A new endometrium then regrows under the influence of estrogens from the developing follicle. The endometrium increases rapidly in thickness from the 5th to the 14th days of the menstrual cycle. As the thickness increases, the uterine glands are drawn out so that

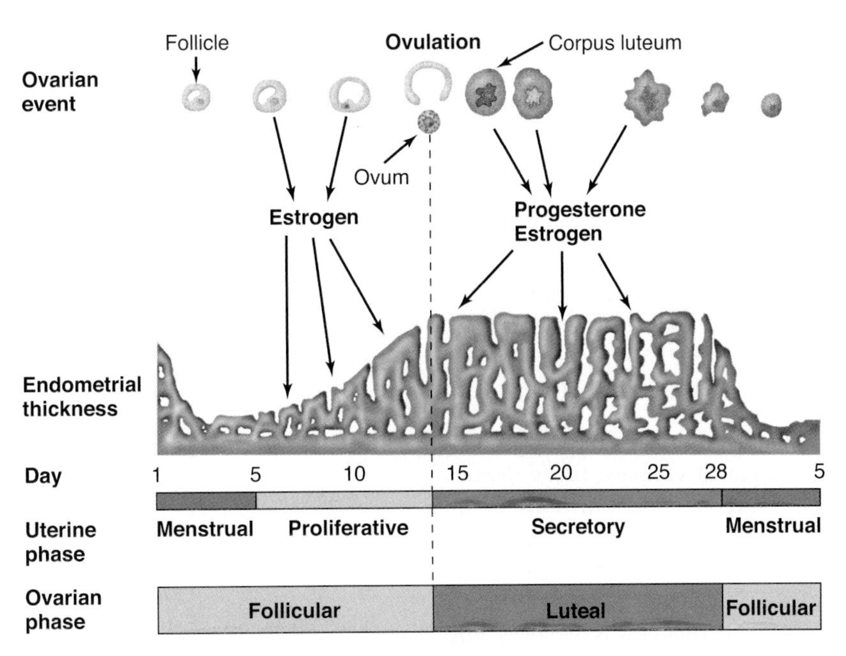

FIGURE 22–12 Relationship between ovarian and uterine changes during the menstrual cycle. (Reproduced with permission from Windmaier EP, Raff H, Strang KT: *Vander's Human Physiology: The Mechanisms of Body Function*, 11th ed. McGraw-Hill, 2008.)

they lengthen (Figure 22–12), but they do not become convoluted or secrete to any degree. These endometrial changes are called proliferative, and this part of the menstrual cycle is sometimes called the **proliferative phase.** It is also called the preovulatory or follicular phase of the cycle. After ovulation, the endometrium becomes more highly vascularized and slightly edematous under the influence of estrogen and progesterone from the corpus luteum. The glands become coiled and tortuous and they begin to secrete a clear fluid. Consequently, this phase of the cycle is called the **secretory** or **luteal phase.** Late in the luteal phase, the endometrium, like the anterior pituitary, produces prolactin, but the function of this endometrial prolactin is unknown.

The endometrium is supplied by two types of arteries. The superficial two thirds of the endometrium that is shed during menstruation, the **stratum functionale,** is supplied by long, coiled **spiral arteries** (Figure 22–13), whereas the deep layer that is not shed, the **stratum basale,** is supplied by short, straight **basilar arteries.**

When the corpus luteum regresses, hormonal support for the endometrium is withdrawn. The endometrium becomes thinner, which adds to the coiling of the spiral arteries. Foci of necrosis appear in the endometrium, and these coalesce. In addition, spasm and degeneration of the walls of the spiral arteries take place, leading to spotty hemorrhages that become confluent and produce the menstrual flow.

The vasospasm is probably produced by locally released prostaglandins. Large quantities of prostaglandins are present in the secretory endometrium and in menstrual blood, and infusions of prostagladin $F_{2\alpha}$ ($PGF_{2\alpha}$) produce endometrial necrosis and bleeding.

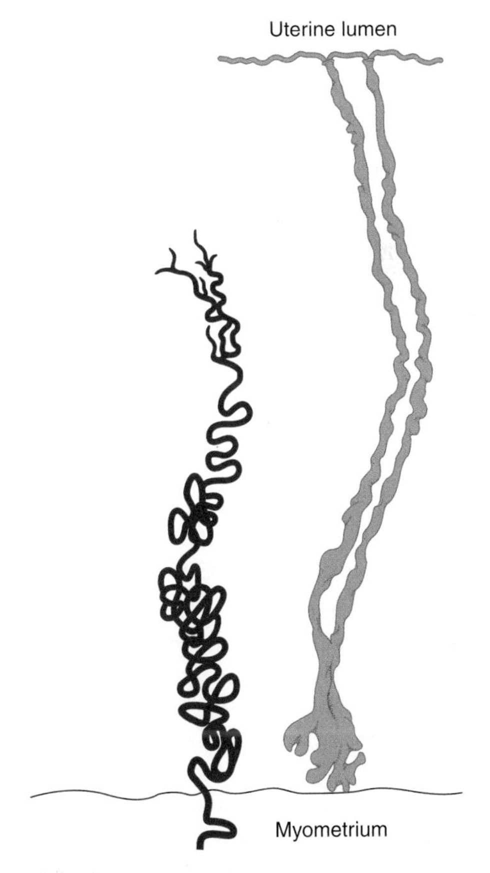

FIGURE 22–13 Spiral artery of endometrium. Drawing of a spiral artery (left) and two uterine glands (right) from the endometrium of a rhesus monkey; early secretory phase. (Reproduced with permission from Daron GH: The arterial pattern of the tunica mucosa of the uterus in the *Macacus rhesus.* Am J Anat 1936;58:349.)

From the point of view of endometrial function, the proliferative phase of the menstrual cycle represents restoration of the epithelium from the preceding menstruation, and the secretory phase represents preparation of the uterus for implantation of the fertilized ovum. The length of the secretory phase is remarkably constant at about 14 days, and the variations seen in the length of the menstrual cycle are due for the most part to variations in the length of the proliferative phase. When fertilization fails to occur during the secretory phase, the endometrium is shed and a new cycle starts.

Normal Menstruation

Menstrual blood is predominantly arterial, with only 25% of the blood being of venous origin. It contains tissue debris, prostaglandins, and relatively large amounts of fibrinolysin from endometrial tissue. The fibrinolysin lyses clots, so that menstrual blood does not normally contain clots unless the flow is excessive.

The usual duration of the menstrual flow is 3–5 days, but flows as short as 1 day and as long as 8 days can occur in normal women. The amount of blood lost may range normally from slight spotting to 80 mL; the average amount lost is 30 mL. Loss of more than 80 mL is abnormal. Obviously, the amount of flow can be affected by various factors, including the thickness of the endometrium, medication, and diseases that affect the clotting mechanism.

Anovulatory Cycles

In some instances, ovulation fails to occur during the menstrual cycle. Such anovulatory cycles are common for the first 12–18 months after menarche and again before the onset of the menopause. When ovulation does not occur, no corpus luteum is formed and the effects of progesterone on the endometrium are absent. Estrogens continue to cause growth, however, and the proliferative endometrium becomes thick enough to break down and begins to slough. The time it takes for bleeding to occur is variable, but it usually occurs in less than 28 days from the last menstrual period. The flow is also variable and ranges from scanty to relatively profuse.

Cyclical Changes in the Uterine Cervix

Although it is continuous with the body of the uterus, the cervix of the uterus is different in a number of ways. The mucosa of the uterine cervix does not undergo cyclical desquamation, but there are regular changes in the cervical mucus. Estrogen makes the mucus thinner and more alkaline, changes that promote the survival and transport of sperm. Progesterone makes it thick, tenacious, and cellular. The mucus is thinnest at the time of ovulation, and its elasticity, or **spinnbarkeit,** increases so that by midcycle, a drop can be stretched into a long, thin thread that may be 8–12 cm or more in length. In addition, it dries in an arborizing, fern-like pattern (Figure 22–14) when a thin layer is spread on a slide. After ovulation and during pregnancy, it becomes thick and fails to form the fern pattern.

Normal cycle, 14th day

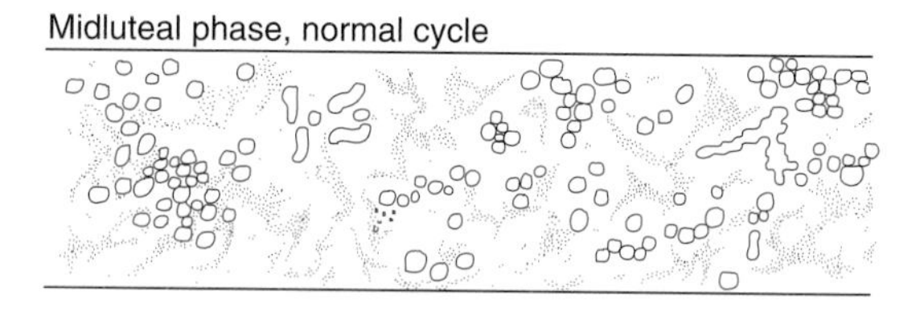

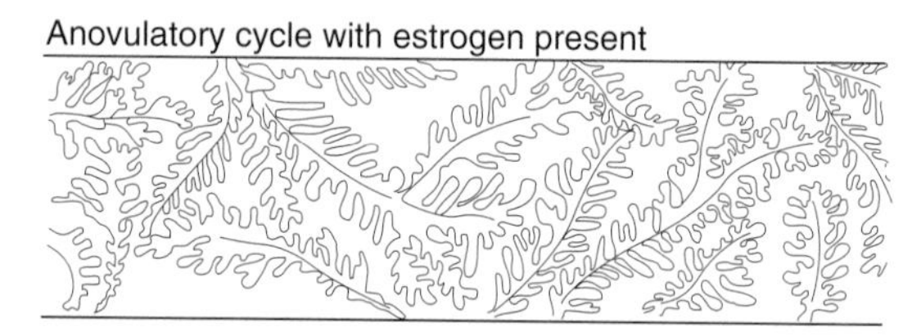

FIGURE 22–14 Patterns formed when cervical mucus is smeared on a slide, permitted to dry, and examined under the microscope. Progesterone makes the mucus thick and cellular. In the smear from a patient who failed to ovulate **(bottom),** no progesterone is present to inhibit the estrogen-induced fern pattern.

Vaginal Cycle

Under the influence of estrogens, the vaginal epithelium becomes cornified, and cornified epithelial cells can be identified in the vaginal smear. Under the influence of progesterone, a thick mucus is secreted, and the epithelium proliferates and becomes infiltrated with leukocytes. The cyclical changes in the vaginal smear in rats are relatively marked. The changes in humans and other species are similar but not so clear-cut.

Cyclical Changes in the Breasts

Although lactation normally does not occur until the end of pregnancy, cyclical changes take place in the breasts during the menstrual cycle. Estrogens cause proliferation of mammary ducts, whereas progesterone causes growth of lobules and alveoli. The breast swelling, tenderness, and pain experienced by many women during the 10 days preceding menstruation are probably due to distention of the ducts, hyperemia, and edema of the interstitial tissue of the breast. All these changes regress, along with the symptoms, during menstruation.

Changes During Intercourse

During sexual excitement in women, fluid is secreted onto the vaginal walls, probably because of release of VIP from vaginal nerves. A lubricating mucus is also secreted by the vestibular glands. The upper part of the vagina is sensitive to stretch, while tactile stimulation from the labia minora and clitoris adds to the sexual excitement. These stimuli are reinforced by tactile stimuli from the breasts and, as in men, by visual, auditory, and olfactory stimuli, which may build to the crescendo known as orgasm. During orgasm, autonomically mediated rhythmic contractions occur in the vaginal walls. Impulses also travel via the pudendal nerves and produce rhythmic contraction of the bulbocavernosus and ischiocavernosus muscles. The vaginal contractions may aid sperm transport but are not essential for it, since fertilization of the ovum is not dependent on female orgasm.

Indicators of Ovulation

Knowing when during the menstrual cycle ovulation occurs is important in increasing fertility or, conversely, in family planning. A convenient and reasonably reliable indicator of the time of ovulation is a change—usually a rise—in the basal body temperature (Figure 22–15). The rise starts 1–2 days after ovulation. Women interested in obtaining an accurate temperature chart should use a digital thermometer and take their temperatures (oral or rectal) in the morning before getting out

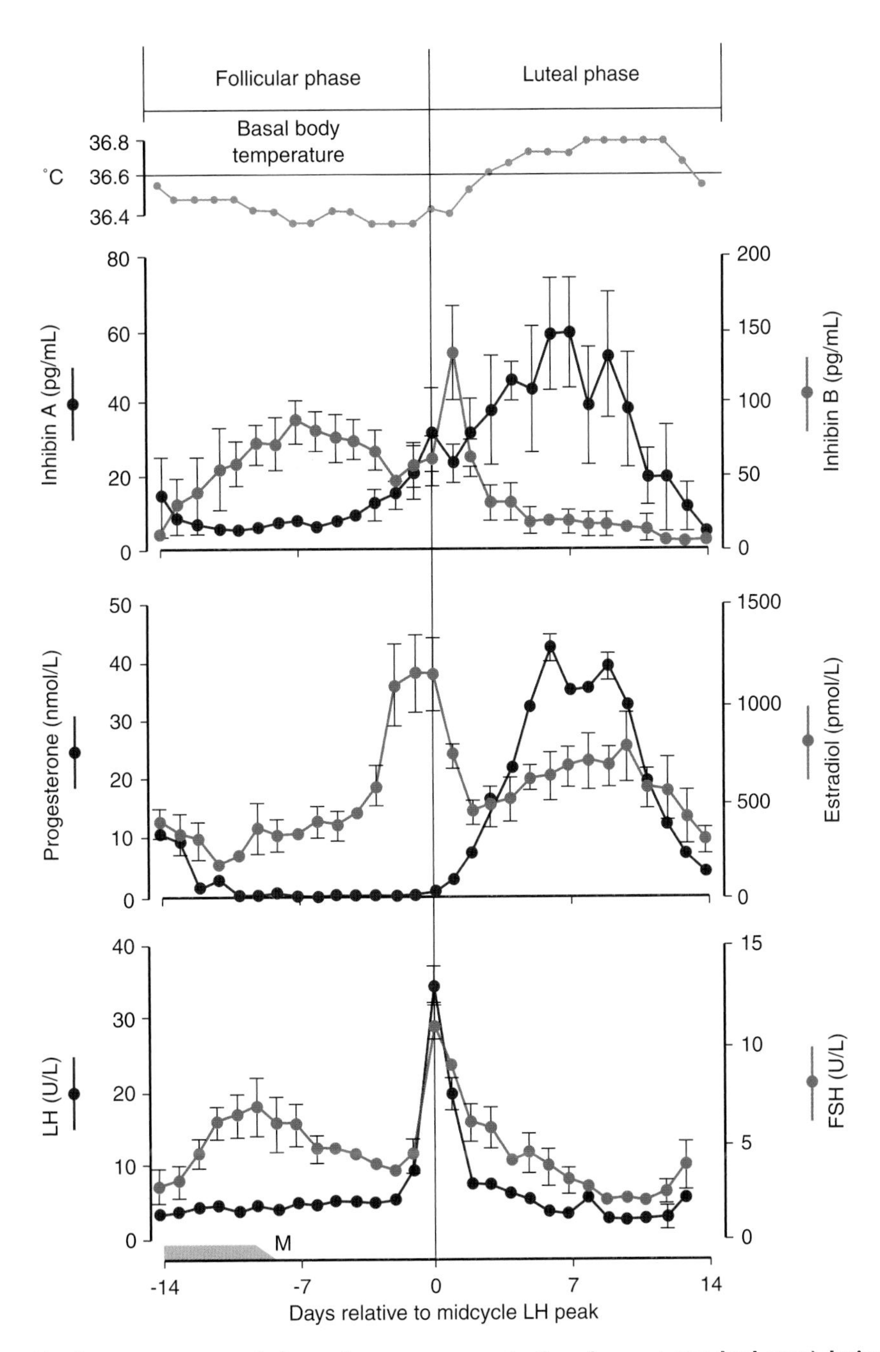

FIGURE 22–15 Basal body temperature and plasma hormone concentrations (mean ± standard error) during the normal human menstrual cycle. Values are aligned with respect to the day of the midcycle LH peak. FSH, follicle-stimulating hormone; LH, luteinizing hormone; M, menses.

treatment has been used commercially to increase the weight of domestic animals. They cause epiphysial closure in humans (see Chapter 21).

Effects on the Central Nervous System

The estrogens are responsible for estrous behavior in animals, and they increase libido in humans. They apparently exert this action by a direct effect on certain neurons in the hypothalamus Estrogens also increase the proliferation of dendrites on neurons and the number of synaptic knobs in rats.

Effects on the Breasts

Estrogens produce duct growth in the breasts and are largely responsible for breast enlargement at puberty in girls; they have been called the growth hormones of the breast. They are responsible for the pigmentation of the areolas, although pigmentation usually becomes more intense during the first pregnancy than it does at puberty. The role of estrogens in the overall control of breast growth and lactation is discussed below.

Female Secondary Sex Characteristics

The body changes that develop in girls at puberty—in addition to enlargement of breasts, uterus, and vagina—are due in part to estrogens, which are the "feminizing hormones," and in part simply to the absence of testicular androgens. Women have narrow shoulders and broad hips, thighs that converge, and arms that diverge (wide **carrying angle**). This body configuration, plus the female distribution of fat in the breasts and buttocks, is seen also in castrate males. In women, the larynx retains its prepubertal proportions and the voice stays high-pitched. Women have less body hair and more scalp hair, and the pubic hair generally has a characteristic flat-topped pattern (female escutcheon). However, growth of pubic and axillary hair in both sexes is due primarily to androgens rather than estrogens.

Other Actions

Normal women retain salt and water and gain weight just before menstruation. Estrogens cause some degree of salt and water retention. However, aldosterone secretion is slightly elevated in the luteal phase, and this also contributes to the premenstrual fluid retention.

Estrogens are said to make sebaceous gland secretions more fluid and thus to counter the effect of testosterone and inhibit formation of **comedones** ("black-heads") and acne. The liver palms, spider angiomas, and slight breast enlargement seen in advanced liver disease are due to increased circulating estrogens. The increase appears to be due to decreased hepatic metabolism of androstenedione, making more of this androgen available for conversion to estrogens.

Estrogens have a significant plasma cholesterol-lowering action, and they rapidly produce vasodilation by increasing the local production of NO. Their action on bone is discussed in Chapter 21.

Mechanism of Action

There are two principal types of nuclear estrogen receptors: estrogen receptor α (ERα) encoded by a gene on chromosome 6; and estrogen receptor β (ERβ), encoded by a gene on chromosome 14. Both are members of the nuclear receptor super-family (see Chapter 2). After binding estrogen, they form homodimers and bind to DNA, altering its transcription. Some tissues contain one type or the other, but overlap also occurs, with some tissues containing both ERα and ERβ. ERα is found primarily in the uterus, kidneys, liver, and heart, whereas ERβ is found primarily in the ovaries, prostate, lungs, gastrointestinal tract, hemopoietic system, and central nervous system (CNS). ERα and ERβ can also form heterodimers. Male and female mice in which the gene for ERα has been knocked out are sterile, develop osteoporosis, and continue to grow because their epiphyses do not close. ERβ female knockouts are infertile, but ERβ male knockouts are fertile even though they have hyperplastic prostates and loss of fat. Both receptors exist in isoforms and, like thyroid receptors, can bind to various activating and stimulating factors. In some situations, ERβ can inhibit ERα transcription. Thus, their actions are complex, multiple, and varied.

Most of the effects of estrogens are genomic, that is, due to actions on the nucleus, but some are so rapid that it is difficult to believe they are mediated via production of mRNAs. These include effects on neuronal discharge in the brain and, possibly, feedback effects on gonadotropin secretion. Evidence is accumulating that these effects are mediated by cell membrane receptors that appear to be structurally related to the nuclear receptors and produce their effects by intracellular mitogen-activated protein kinase pathways. Similar rapid effects of progesterone, testosterone, glucocorticoids, aldosterone, and 1,25-dihydroxycholecalciferol may also be produced by membrane receptors (see Chapter 16).

Synthetic and Environmental Estrogens

The ethinyl derivative of estradiol is a potent estrogen and, unlike the naturally occurring estrogens, is relatively active when given by mouth because it is resistant to hepatic metabolism. The activity of the naturally occurring hormones is low when they are administered by mouth because the portal venous drainage of the intestine carries them to the liver, where they are inactivated before they can reach the general circulation. Some nonsteroidal substances and a few compounds found in plants also have estrogenic activity. The plant estrogens are rarely a problem in human nutrition, but they may cause undesirable effects in farm animals. **Dioxins,** which are found in the environment and are produced by a variety of industrial processes, can activate estrogen response elements on genes. However, they have been reported to have antiestrogenic as well as estrogenic effects, and their role, if any, in the production of human disease remains a matter of disagreement and debate.

Because natural estrogens have undesirable as well as desirable effects (for example, they preserve bone in osteoporosis but can cause uterine and breast cancer), there has been an active search for "tailor-made" estrogens that have selective effects in humans. Two compounds, **tamoxifen** and **raloxifene,** show promise in this regard. Neither combats the symptoms of menopause, but both have the bone-preserving effects of estradiol. In addition, tamoxifen does not stimulate the breast, and raloxifene does not stimulate the breast or uterus. The way the effects of these selective estrogen receptor modulators **(SERMs)** are brought about is related to the complexity of the estrogen receptors and hence to differences in the way that the receptor–ligand complexes they form bind to DNA.

Chemistry, Biosynthesis, & Metabolism of Progesterone

Progesterone is a C_{21} steroid (Figure 22–18) secreted by the corpus luteum, the placenta, and (in small amounts) the follicle. It is an important intermediate in steroid biosynthesis in all tissues that secrete steroid hormones, and small amounts apparently enter the circulation from the testes and adrenal cortex. About 2% of the circulating progesterone is free whereas 80% is bound to albumin and 18% is bound to corticosteroid-binding globulin. Progesterone has a short half-life and is converted in the liver to pregnanediol, which is conjugated to glucuronic acid and excreted in the urine.

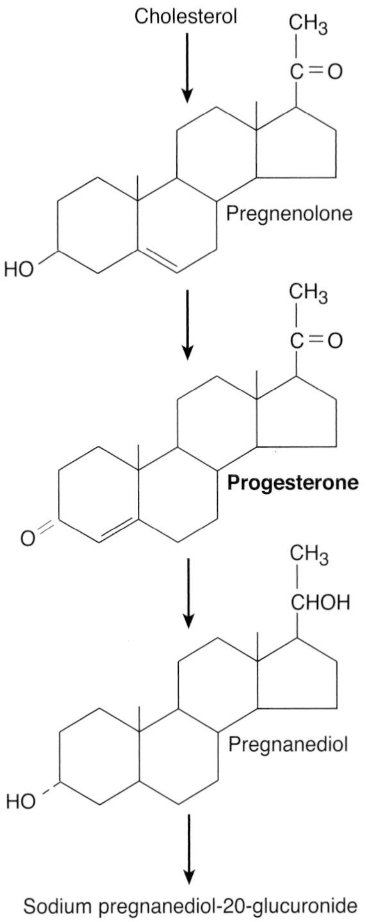

Cholesterol
Pregnenolone
Progesterone
Pregnanediol
Sodium pregnanediol-20-glucuronide

FIGURE 22–18 Biosynthesis of progesterone and major pathway for its metabolism. Other metabolites are also formed.

Secretion

In men, the plasma progesterone level is approximately 0.3 ng/mL (1 nmol/L). In women, the level is approximately 0.9 ng/mL (3 nmol/L) during the follicular phase of the menstrual cycle (Figure 22–15). The difference is due to secretion of small amounts of progesterone by cells in the ovarian follicles; theca cells provide pregnenolone to the granulosa cells, which convert it to progesterone. Late in the follicular phase, progesterone secretion begins to increase. During the luteal phase, the corpus luteum produces large quantities of progesterone (Table 22–3) and plasma progesterone is markedly increased to a peak value of approximately 18 ng/mL (60 nmol/L).

The stimulating effect of LH on progesterone secretion by the corpus luteum is due to activation of adenylyl cyclase and involves a subsequent step that is dependent on protein synthesis.

Actions

The principal target organs of progesterone are the uterus, the breasts, and the brain. Progesterone is responsible for the progestational changes in the endometrium and the cyclical changes in the cervix and vagina described above. It has an antiestrogenic effect on the myometrial cells, decreasing their excitability, their sensitivity to oxytocin, and their spontaneous electrical activity while increasing their membrane potential. It also decreases the number of estrogen receptors in the endometrium and increases the rate of conversion of 17β-estradiol to less active estrogens.

In the breast, progesterone stimulates the development of lobules and alveoli. It induces differentiation of estrogen-prepared ductal tissue and supports the secretory function of the breast during lactation.

The feedback effects of progesterone are complex and are exerted at both the hypothalamic and pituitary levels. Large doses of progesterone inhibit LH secretion and potentiate the inhibitory effect of estrogens, preventing ovulation.

Progesterone is thermogenic and is probably responsible for the rise in basal body temperature at the time of ovulation. It stimulates respiration, and the alveolar PCO_2 (see Chapter 34) in women during the luteal phase of the menstrual cycle is lower than that in men. In pregnancy, the PCO_2 falls as progesterone secretion rises. However, the physiologic significance of this respiratory response is unknown.

Large doses of progesterone produce natriuresis, probably by blocking the action of aldosterone on the kidney. The hormone does not have a significant anabolic effect.

Mechanism of Action

The effects of progesterone, like those of other steroids, are brought about by an action on DNA to initiate synthesis of new mRNA. The progesterone receptor is bound to a heat shock protein in the absence of the steroid, and progesterone binding releases the heat shock protein, exposing the DNA-binding domain of the receptor. The synthetic steroid **mifepristone (RU 486)** binds to the receptor but does not release the heat shock protein, and it blocks the binding of progesterone. Because the maintenance of early pregnancy depends on the stimulatory effect of progesterone on endometrial growth and its inhibition of uterine contractility, mifepristone combined with a prostaglandin can be used to produce elective abortions.

There are two isoforms of the progesterone receptor—PR_A and PR_B—that are produced by differential processing from a single gene. PR_A is a truncated form, but it is likely that both isoforms mediate unique subsets of progesterone action.

Substances that mimic the action of progesterone are sometimes called **progestational agents, gestagens,** or **progestins.** They are used along with synthetic estrogens as oral contraceptive agents.

Relaxin

Relaxin is a polypeptide hormone that is produced in the corpus luteum, uterus, placenta, and mammary glands in women and in the prostate gland in men. During pregnancy, it relaxes the pubic symphysis and other pelvic joints and softens and dilates the uterine cervix. Thus, it facilitates delivery. It also inhibits uterine contractions and may play a role in the development of the mammary glands. In nonpregnant women, relaxin is found in the corpus luteum and the endometrium during the secretory but not the proliferative phase of the menstrual cycle. Its function in nonpregnant women is unknown. In men, it is found in semen, where it may help maintain sperm motility and aid in sperm penetration of the ovum.

In most species there is only one relaxin gene, but in humans there are two genes on chromosome 9 that code for two structurally different polypeptides that both have relaxin activity. However, only one of these genes is active in the ovary and the prostate. The structure of the polypeptide produced in these two tissues is shown in Figure 22–19.

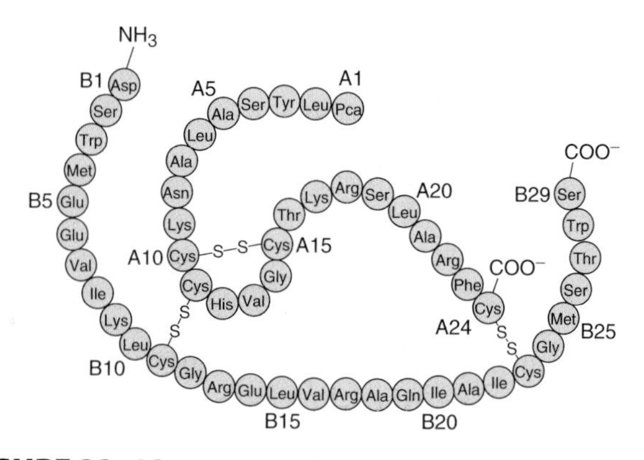

FIGURE 22–19 Structure of human luteal and seminal relaxin. Pca, pyroglutamic acid. (Modified and reproduced with permission from Winslow JW, et al: Human seminal relaxin is a product of the same gene as human luteal relaxin. Endocrinology 1992;130:2660. Copyright © 1992 by The Endocrine Society.)

CONTROL OF OVARIAN FUNCTION

FSH from the pituitary is responsible for the early maturation of the ovarian follicles, and FSH and LH together are responsible for their final maturation. A burst of LH secretion (Figure 22–15) is responsible for ovulation and the initial formation of the corpus luteum. A smaller midcycle burst of FSH secretion also occurs, the significance of which is uncertain. LH stimulates the secretion of estrogen and progesterone from the corpus luteum.

Hypothalamic Components

The hypothalamus occupies a key position in the control of gonadotropin secretion. Hypothalamic control is exerted by GnRH secreted into the portal hypophysial vessels. GnRH stimulates the secretion of FSH as well as LH.

GnRH is normally secreted in episodic bursts, and these bursts produce the circhoral peaks of LH secretion. They are essential for normal secretion of gonadotropins. If GnRH is administered by constant infusion, the GnRH receptors in the anterior pituitary down-regulate and LH secretion declines to zero. However, if GnRH is administered episodically at a rate of one pulse per hour, LH secretion is stimulated. This is true even when endogenous GnRH secretion has been prevented by a lesion of the ventral hypothalamus.

It is now clear not only that episodic secretion of GnRH is a general phenomenon but also that fluctuations in the frequency and amplitude of the GnRH bursts are important in generating the other hormonal changes that are responsible for the menstrual cycle. Frequency is increased by estrogens and decreased by progesterone and testosterone. The frequency increases late in the follicular phase of the cycle, culminating in the LH surge. During the secretory phase, the frequency decreases as a result of the action of progesterone (Figure 22–20), but when estrogen and progesterone secretion decrease at the end of the cycle, the frequency once again increases.

At the time of the midcycle LH surge, the sensitivity of the gonadotropes to GnRH is greatly increased because of their exposure to GnRH pulses at a specific frequency. This self-priming effect of GnRH is important in producing a maximum LH response.

The nature and the exact location of the GnRH pulse generator in the hypothalamus are still unsettled. However, it is known in a general way that norepinephrine and

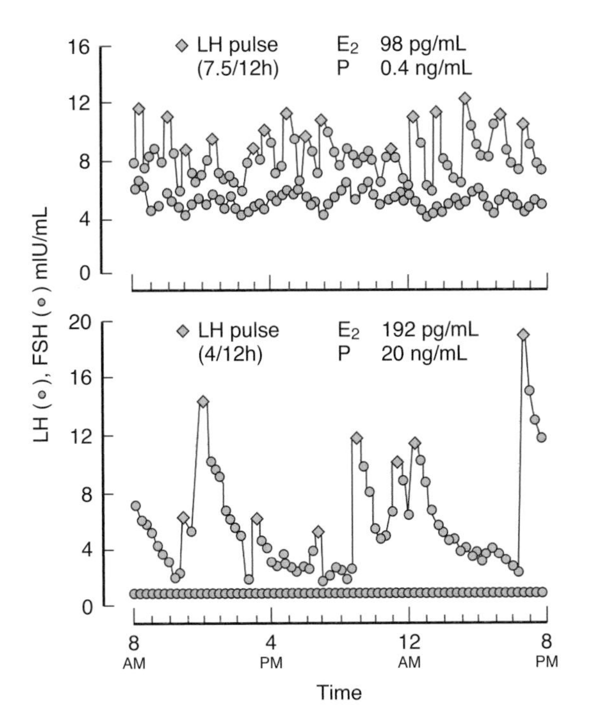

FIGURE 22–20 Episodic secretion of LH (s) and FSH (d) during the follicular stage (top) and the luteal stage (bottom) of the menstrual cycle. The numbers above each graph indicate the numbers of LH pulses per 12 hours and the plasma estradiol (E_2) and progesterone (P) concentrations at these two times of the cycle. (Reproduced with permission from Marshall JC, Kelch RO: Gonadotropin-releasing hormone: Role of pulsatile secretion in the regulation of reproduction. N Engl J Med 1986;315:1459.)

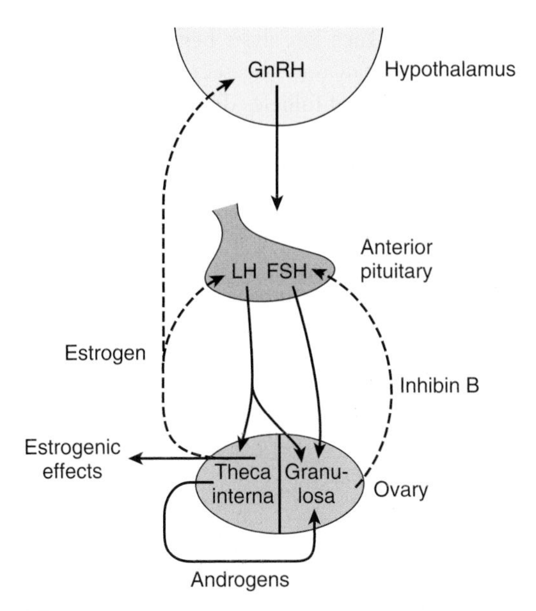

FIGURE 22–21 Feedback regulation of ovarian function. The cells of the theca interna provide androgens to the granulosa cells, and theca cells also produce the circulating estrogens that inhibit the secretion of GnRH, LH, and FSH. Inhibin from the granulosa cells inhibits FSH secretion. LH regulates the thecal cells, whereas the granulosa cells are regulated by both LH and FSH. The dashed arrows indicate inhibitory effects and the solid arrows stimulatory effects.

possibly epinephrine in the hypothalamus increase GnRH pulse frequencies. Conversely, opioid peptides such as the enkephalins and β-endorphin reduce the frequency of GnRH pulses.

The down-regulation of pituitary receptors and the consequent decrease in LH secretion produced by constantly elevated levels of GnRH has led to the use of long-acting GnRH analogs to inhibit LH secretion in precocious puberty and in cancer of the prostate.

Feedback Effects

Changes in plasma LH, FSH, sex steroids, and inhibin during the menstrual cycle are shown in Figure 22–15, and their feedback relations are diagrammed in Figure 22–21. During the early part of the follicular phase, inhibin B is low and FSH is modestly elevated, fostering follicular growth. LH secretion is held in check by the negative feedback effect of the rising plasma estrogen level. At 36–48 h before ovulation, the estrogen feedback effect becomes positive, and this initiates the burst of LH secretion (LH surge) that produces ovulation. Ovulation occurs about 9 h after the LH peak. FSH secretion also peaks, despite a small rise in inhibin, probably because of the strong stimulation of gonadotropes by GnRH. During the luteal phase, the secretion of LH and FSH is low because of the elevated levels of estrogen, progesterone, and inhibin.

It should be emphasized that a moderate, constant level of circulating estrogen exerts a negative feedback effect on LH secretion, whereas during the cycle, an elevated estrogen level exerts a positive feedback effect and stimulates LH secretion. In monkeys, it has been demonstrated that estrogens must also be elevated for a minimum time to produce positive feedback. When circulating estrogen was increased about 300% for 24 h, only negative feedback was seen; but when it was increased about 300% for 36 h or more, a brief decline in secretion was followed by a burst of LH secretion that resembled the midcycle surge. When circulating levels of progesterone were high, the positive feedback effect of estrogen was inhibited. In primates there is evidence that both the negative and the positive feedback effects of estrogen are exerted in the mediobasal hypothalamus, but exactly how negative feedback is switched to positive feedback and then back to negative feedback in the luteal phase remains unknown.

Control of the Cycle

In an important sense, regression of the corpus luteum (**luteolysis**) starting 3–4 days before menses is the key to the menstrual cycle. $PGF_{2\alpha}$ appears to be a physiologic luteolysin, but this prostaglandin is only active when endothelial cells producing ET-1 (see Chapter 32) are present. Therefore, it appears that at least in some species luteolysis is produced by the combined action of $PGF_{2\alpha}$ and ET-1. In some domestic animals, oxytocin secreted by the corpus luteum appears to exert a local luteolytic effect, possibly by causing the release of prostaglandins. Once luteolysis begins, the estrogen and progesterone levels fall and the secretion of FSH and LH increases. A new crop of follicles develops, and then a single dominant follicle matures as a result of the action of FSH and LH. Near midcycle, estrogen secretion from the follicle rises. This rise augments the responsiveness of the pituitary to GnRH and triggers a burst of LH secretion. The resulting ovulation is followed by formation of a corpus luteum. Estrogen secretion drops, but progesterone and estrogen levels then rise together, along with inhibin B. The elevated levels inhibit FSH and LH secretion for a while, but luteolysis again occurs and a new cycle starts.

Reflex Ovulation

Female cats, rabbits, mink, and some other animals have long periods of estrus, during which they ovulate only after copulation. Such **reflex ovulation** is brought about by afferent impulses from the genitalia and the eyes, ears, and nose that converge on the ventral hypothalamus and provoke an ovulation-inducing release of LH from the pituitary. In species such as rats, monkeys, and humans, ovulation is a spontaneous periodic phenomenon, but neural mechanisms are also involved. Ovulation can be delayed 24 h in rats by administering pentobarbital or various other neurally active drugs 12 h before the expected time of follicle rupture.

Contraception

Methods commonly used to prevent conception are listed in Table 22–4, along with their failure rates. Once conception has occurred, abortion can be produced by progesterone antagonists such as mifepristone.

Implantation of foreign bodies in the uterus causes changes in the duration of the sexual cycle in a number of mammalian species. In humans, such foreign bodies do not alter the menstrual cycle, but they act as effective contraceptive devices. Intrauterine implantation of pieces of metal or plastic **(intrauterine devices, IUDs)** has been used in programs aimed at controlling population growth. Although the mechanism of action of IUDs is still unsettled, they seem in general to prevent sperm from fertilizing ova. Those containing copper appear to exert a spermatocidal effect. IUDs that slowly release progesterone or synthetic progestins have the additional effect of thickening cervical mucus so that entry of sperm into the uterus is impeded. IUDs can cause intrauterine infections, but these usually occur in the first month after insertion and in women exposed to sexually transmitted diseases.

Women undergoing long-term treatment with relatively large doses of estrogen do not ovulate, probably because they have depressed FSH levels and multiple irregular bursts of LH secretion rather than a single midcycle peak. Women treated with similar doses of estrogen plus a progestational agent do not ovulate because the secretion of both gonadotropins is suppressed. In addition, the progestin makes the cervical mucus thick and unfavorable to sperm migration, and it may also interfere with implantation. For contraception, an orally active estrogen such as ethinyl estradiol is often combined with a synthetic progestin such as norethindrone. The pills are administered for 21 days, then withdrawn for 5–7 days to permit menstrual flow, and started again. Like ethinyl estradiol, norethindrone has an ethinyl group on position 17 of the steroid nucleus, so it is resistant to hepatic metabolism and consequently is effective by mouth. In addition to being a progestin, it is partly metabolized to ethinyl estradiol, and for this reason it also has estrogenic activity. Small as well as large doses of estrogen are effective (Table 22–4).

Implants made up primarily of progestins such as levonorgestrel are now seeing increased use in some parts of the world. These are inserted under the skin and can prevent pregnancy for up to 5 years. They often produce amenorrhea, but otherwise they appear to be effective and well tolerated.

TABLE 22–4 Relative effectiveness of frequently used contraceptive methods.

Method	Failures per 100 Woman-Years
Vasectomy	0.02
Tubal ligation and similar procedures	0.13
Oral contraceptives	
>50 mg estrogen and progestin	0.32
<50 mg estrogen and progestin	0.27
Progestin only	1.2
IUD	
Copper 7	1.5
Loop D	1.3
Diaphragm	1.9
Condom	3.6
Withdrawal	6.7
Spermicide	11.9
Rhythm	15.5

Data from Vessey M, Lawless M, Yeates D: Efficacy of different contraceptive methods. Lancet 1982;1:841. Reproduced with permission.

ABNORMALITIES OF OVARIAN FUNCTION

Menstrual Abnormalities

Some women who are infertile have **anovulatory cycles;** they fail to ovulate but have menstrual periods at fairly regular intervals. As noted above, anovulatory cycles are the rule for the first 1–2 years after menarche and again before the menopause. **Amenorrhea** is the absence of menstrual periods. If menstrual bleeding has never occurred, the condition is called **primary amenorrhea.** Some women with primary amenorrhea have small breasts and other signs of failure to mature sexually. Cessation of cycles in a woman with previously normal periods is called **secondary amenorrhea.** The most common cause of secondary amenorrhea is pregnancy, and the old clinical maxim that "secondary amenorrhea should be considered to be due to pregnancy until proved otherwise" has considerable merit. Other causes of amenorrhea include emotional stimuli and changes in the environment, hypothalamic diseases, pituitary disorders, primary ovarian disorders, and various systemic diseases. Evidence suggests that in some women with hypothalamic amenorrhea, the frequency of GnRH pulses is slowed as a result of excess opioid activity in the hypothalamus. In encouraging preliminary studies, the frequency of GnRH pulses was increased by administration of the orally active opioid blocker naltrexone.

The terms **hypomenorrhea** and **menorrhagia** refer to scanty and abnormally profuse flow, respectively, during regular periods. **Metrorrhagia** is bleeding from the uterus between periods, and **oligomenorrhea** is reduced frequency of periods. **Dysmenorrhea** is painful menstruation. The severe menstrual cramps that are common in young women quite often disappear after the first pregnancy. Most of the symptoms of dysmenorrhea are due to accumulation of prostaglandins in the uterus, and symptomatic relief has been obtained by treatment with inhibitors of prostaglandin synthesis.

Some women develop symptoms such as irritability, bloating, edema, decreased ability to concentrate, depression, headache, and constipation during the last 7–10 days of their menstrual cycles. These symptoms of the **premenstrual syndrome (PMS)** have been attributed to salt and water retention. However, it seems unlikely that this or any of the other hormonal alterations that occur in the late luteal phase are responsible because the time course and severity of the symptoms are not modified if the luteal phase is terminated early and menstruation produced by administration of mifepristone. The antidepressant fluoxetine (Prozac), which is a serotonin reuptake inhibitor, and the benzodiazepine alprazolam (Xanax) produce symptomatic relief, and so do GnRH-releasing agonists in doses that suppress the pituitary–ovarian axis. How these diverse clinical observations fit together to produce a picture of the pathophysiology of PMS is still unknown (see Clinical Box 22–4).

CLINICAL BOX 22–4

Genetic Defects Causing Reproductive Abnormalities

A number of single-gene mutations cause reproductive abnormalities when they occur in women. Examples include (1) Kallmann syndrome, which causes hypogonadotropic hypogonadism; (2) GnRH resistance, FSH resistance, and LH resistance, which are due to defects in the GnRH, FSH, or LH receptors, respectively; and (3) aromatase deficiency, which prevents the formation of estrogens. These are all caused by loss-of-function mutations. An interesting gain-of-function mutation causes the **McCune–Albright syndrome,** in which Gsα becomes constitutively active in certain cells but not others (mosaicism) because a somatic mutation after initial cell division has occurred in the embryo. It is associated with multiple endocrine abnormalities, including precocious puberty and amenorrhea with galactorrhea.

PREGNANCY

Fertilization & Implantation

In humans, **fertilization** of the ovum by the sperm (see Chapter 23) usually occurs in the ampulla of the uterine tube. Fertilization involves (1) chemoattraction of the sperm to the ovum by substances produced by the ovum; (2) adherence to the **zona pellucida,** the membranous structure surrounding the ovum; (3) penetration of the zona pellucida and the acrosome reaction; and (4) adherence of the sperm head to the cell membrane of the ovum, with breakdown of the area of fusion and release of the sperm nucleus into the cytoplasm of the ovum (Figure 22–22). Millions of sperm are deposited in the vagina during intercourse. Eventually, 50–100 sperm reach the ovum, and many of them contact the zona pellucida. Sperm bind to a receptor in the zona, and this is followed by the **acrosomal reaction,** that is, the breakdown of the acrosome, the lysosome-like organelle on the head of the sperm (Figure 23–4). Various enzymes are released, including the trypsin-like protease **acrosin.** Acrosin facilitates but is not

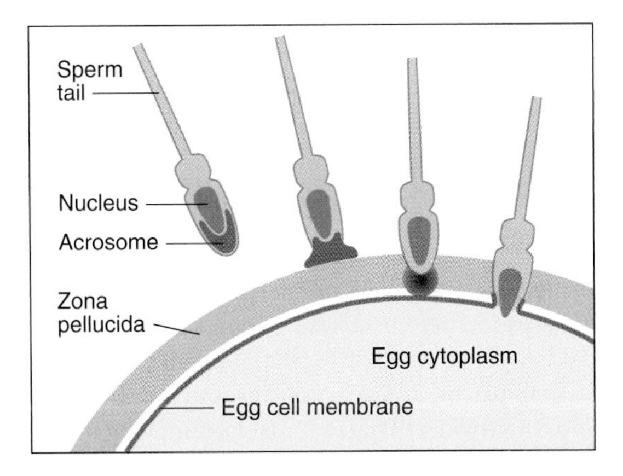

FIGURE 22–22 Sequential events in fertilization in mammals. Sperm are attracted to the ovum, bind to the zona pellucida, release acrosomal enzymes, penetrate the zona pellucida, and fuse with the membrane of the ovum, releasing the sperm nucleus into its cytoplasm. Current evidence indicates that the side, rather than the tip, of the sperm head fuses with the egg cell membrane. (Modified from Vacquier VD: Evolution of gamete recognition proteins. Science 1998;281:1995.)

required for the penetration of the sperm through the zona pellucida. When one sperm reaches the membrane of the ovum, fusion to the ovum membrane is mediated by **fertilin,** a protein on the surface of the sperm head that resembles the viral fusion proteins that permit some viruses to attack cells. The fusion provides the signal that initiates development. In addition, the fusion sets off a reduction in the membrane potential of the ovum that prevents polyspermy, the fertilization of the ovum by more than one sperm. This transient potential change is followed by a structural change in the zona pellucida that provides protection against polyspermy on a more long-term basis.

The developing embryo, now called a **blastocyst**, moves down the tube into the uterus. This journey takes about 3 days, during which the blastocyst reaches the 8- or 16-cell stage. Once in contact with the endometrium, the blastocyst becomes surrounded by an outer layer of **syncytiotrophoblast,** a multi-nucleate mass with no discernible cell boundaries, and an inner layer of **cytotrophoblast** made up of individual cells. The syncytiotrophoblast erodes the endometrium, and the blastocyst burrows into it **(implantation).** The implantation site is usually on the dorsal wall of the uterus. A placenta then develops, and the trophoblast remains associated with it.

Failure to Reject the "Fetal Graft"

It should be noted that the fetus and the mother are two genetically distinct individuals, and the fetus is in effect a transplant of foreign tissue in the mother. However, the transplant is tolerated, and the rejection reaction that is characteristically produced when other foreign tissues are transplanted (see Chapter 3) fails to occur. The way the "fetal graft" is protected is unknown. However, one explanation may be that the placental trophoblast, which separates maternal and fetal tissues, does not express the polymorphic class I and class II MHC genes and instead expresses *HLA-G,* a nonpolymorphic gene. Therefore, antibodies against the fetal proteins do not develop. In addition, there is Fas ligand on the surface of the placenta, and this binds to T cells, causing them to undergo apoptosis (see Chapter 3).

Infertility

The vexing clinical problem of infertility often requires extensive investigation before a cause is found. In 30% of cases the problem is in the man; in 45%, the problem is in the woman; in 20%, both partners have a problem; and in 5% no cause can be found. **In vitro fertilization,** that is, removing mature ova, fertilizing them with sperm, and implanting one or more of them in the uterus at the four-cell stage is of some value in these cases. It has a 5–10% chance of producing a live birth.

Endocrine Changes

In all mammals, the corpus luteum in the ovary at the time of fertilization fails to regress and instead enlarges in response to stimulation by gonadotropic hormones secreted by the placenta. The placental gonadotropin in humans is called **human chorionic gonadotropin (hCG).** The enlarged **corpus luteum of pregnancy** secretes estrogens, progesterone, and relaxin. Progesterone and relaxin help maintain pregnancy by inhibiting myometrial contractions; progesterone prevents prostaglandin production by the uterus, which stops contractions from occurring. In humans, the placenta produces sufficient estrogen and progesterone from maternal and fetal precursors to take over the function of the corpus luteum after the 6th week of pregnancy. Ovariectomy before the 6th week thus leads to abortion, but ovariectomy thereafter has no effect on the pregnancy. The function of the corpus luteum begins to decline after 8 weeks of pregnancy, but it persists throughout pregnancy. hCG secretion decreases after an initial marked rise, but estrogen and progesterone secretion increase until just before parturition (Table 22–5).

Human Chorionic Gonadotropin

hCG is a glycoprotein that contains galactose and hexosamine. It is produced by the syncytiotrophoblast. Like the pituitary glycoprotein hormones, it is made up of α and β subunits. hCG-α is identical to the α subunit of LH, FSH, and TSH. The molecular weight of hCG-α is 18,000, and that of hCG-β is 28,000. hCG is primarily luteinizing and luteotropic and has little FSH activity. It can be measured by radioimmunoassay and detected in the blood as early as 6 days after conception. Its presence in the urine in early pregnancy is the basis of the various laboratory tests for pregnancy, and it can sometimes be detected in the urine as early as 14 days after conception. It appears to act on the same receptor as LH. hCG is not absolutely specific for pregnancy. Small amounts are secreted by a variety of gastrointestinal and other tumors in both sexes, and hCG has been measured in individuals with suspected tumors as a "tumor marker." It also appears that the fetal liver and kidney normally produce small amounts of hCG.

TABLE 22–5 Hormone levels in human maternal blood during normal pregnancy.

Hormone	Approximate Peak Value	Time of Peak Secretion
hCG	5 mg/mL	First trimester
Relaxin	1 ng/mL	First trimester
hCS	15 mg/mL	Term
Estradiol	16 ng/mL	Term
Estriol	14 ng/mL	Term
Progesterone	190 ng/mL	Term
Prolactin	200 ng/mL	Term

Human Chorionic Somatomammotropin

The syncytiotrophoblast also secretes large amounts of a protein hormone that is lactogenic and has a small amount of growth-stimulating activity. This hormone has been called **chorionic growth hormone-prolactin (CGP)** and **human placental lactogen (hPL),** but it is now generally called **human chorionic somatomammotropin (hCS).** The structure of hCS is very similar to that of human growth hormone (see Chapter 18), and it appears that these two hormones and prolactin evolved from a common progenitor hormone. Large quantities of hCS are found in maternal blood, but very little reaches the fetus. Secretion of growth hormone from the maternal pituitary is not increased during pregnancy and may actually be decreased by hCS. However, hCS has most of the actions of growth hormone and apparently functions as a "maternal growth hormone of pregnancy" to bring about the nitrogen, potassium, and calcium retention, lipolysis, and decreased glucose utilization seen in this state. These latter two actions divert glucose to the fetus. The amount of hCS secreted is proportional to the size of the placenta, which normally weighs about one-sixth as much as the fetus. Low hCS levels are a sign of placental insufficiency.

Other Placental Hormones

In addition to hCG, hCS, progesterone, and estrogens, the placenta secretes other hormones. Human placental fragments probably produce proopiomelanocortin (POMC). In culture, they release corticotropin-releasing hormone (CRH), β-endorphin, α-melanocyte-stimulating hormone (MSH), and dynorphin A, all of which appear to be identical to their hypothalamic counterparts. They also secrete GnRH and inhibin, and since GnRH stimulates and inhibin inhibits hCG secretion, locally produced GnRH and inhibin may act in a paracrine fashion to regulate hCG secretion. The trophoblast cells and amnion cells also secrete leptin, and moderate amounts of this satiety hormone enter the maternal circulation. Some also enters the amniotic fluid. Its function in pregnancy is unknown. The placenta also secretes prolactin in a number of forms.

Finally, the placenta secretes the α subunits of hCG, and the plasma concentration of free α subunits rises throughout pregnancy. These α subunits acquire a carbohydrate composition that makes them unable to combine with β subunits, and their prominence suggests that they have a function of their own. It is interesting in this regard that the secretion of the prolactin produced by the endometrium also appears to increase throughout pregnancy, and it may be that the circulating α subunits stimulate endometrial prolactin secretion.

The cytotrophoblast of the human chorion contains prorenin (see Chapter 38). A large amount of prorenin is also present in amniotic fluid, but its function in this location is unknown.

Fetoplacental Unit

The fetus and the placenta interact in the formation of steroid hormones. The placenta synthesizes pregnenolone and progesterone from cholesterol. Some of the progesterone enters the fetal circulation and provides the substrate for the formation of cortisol and corticosterone in the fetal adrenal glands (Figure 22–23). Some of the pregnenolone enters the fetus and, along with pregnenolone synthesized in the fetal liver, is the substrate for the formation of dehydroepiandrosterone sulfate (DHEAS) and 16-hydroxydehydroepiandrosterone sulfate (16-OHDHEAS) in the fetal adrenal. Some 16-hydroxylation also occurs in the fetal liver. DHEAS and 16-OHDHEAS are transported back to the placenta, where DHEAS forms estradiol and 16-OHDHEAS forms estriol. The principal estrogen formed is estriol, and since fetal 16-OHDHEAS is the principal substrate for the estrogens, the urinary estriol excretion of the mother can be monitored as an index of the state of the fetus.

Parturition

The duration of pregnancy in humans averages 270 days from fertilization (284 days from the first day of the menstrual period preceding conception). Irregular uterine contractions increase in frequency in the last month of pregnancy.

The difference between the body of the uterus and the cervix becomes evident at the time of delivery. The cervix, which is firm in the nonpregnant state and throughout pregnancy until near the time of delivery, softens and dilates, while the body of the uterus contracts and expels the fetus.

There is still considerable uncertainty about the mechanisms responsible for the onset of labor. One factor is the increase in circulating estrogens produced by increased circulating DHEAS. This makes the uterus more excitable, increases the number of gap junctions between myometrial cells, and causes production of more prostaglandins, which

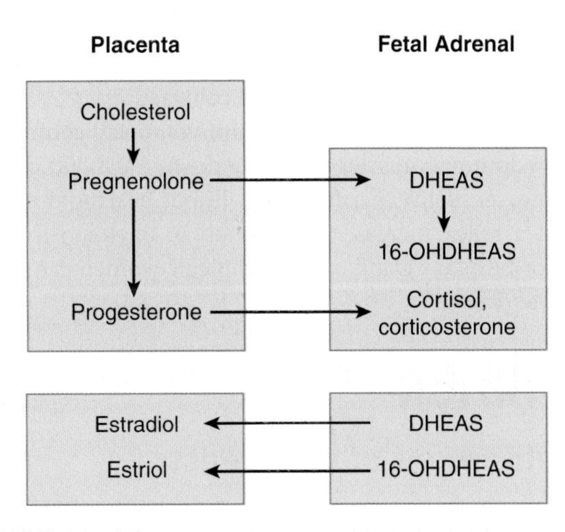

FIGURE 22–23 Interactions between the placenta and the fetal adrenal cortex in the production of steroids.

in turn cause uterine contractions. In humans, CRH secretion by the fetal hypothalamus increases and is supplemented by increased placental production of CRH. This increases circulating adrenocorticotropic hormone (ACTH) in the fetus, and the resulting increase in cortisol hastens the maturation of the respiratory system. Thus, in a sense, the fetus picks the time to be born by increasing CRH secretion.

The number of oxytocin receptors in the myometrium and the decidua (the endometrium of pregnancy) increases more than 100-fold during pregnancy and reaches a peak during early labor. Estrogens increase the number of oxytocin receptors, and uterine distention late in pregnancy may also increase their formation. In early labor, the oxytocin concentration in maternal plasma is not elevated from the prelabor value of about 25 pg/mL. It is possible that the marked increase in oxytocin receptors causes the uterus to respond to normal plasma oxytocin concentrations. However, at least in rats, the amount of oxytocin mRNA in the uterus increases, reaching a peak at term; this suggests that locally produced oxytocin also participates in the process.

Premature onset of labor is a problem because premature infants have a high mortality rate and often require intensive, expensive care. Intramuscular 17α-hydroxyprogesterone causes a significant decrease in the incidence of premature labor. The mechanism by which it exerts its effect is uncertain, but it may be that the steroid provides a stable level of circulating progesterone. Progesterone relaxes uterine smooth muscle, inhibits the action of oxytocin on the muscle, and reduces the formation of gap junctions between the muscle fibers. All these actions would be expected to inhibit the onset of labor.

Once labor is started, the uterine contractions dilate the cervix, and this dilation in turn sets up signals in afferent nerves that increase oxytocin secretion (Figure 22–24). The plasma oxytocin level rises and more oxytocin becomes available to act on the uterus. Thus, a positive feedback loop is established that aids delivery and terminates on expulsion of the products of conception. Oxytocin increases uterine contractions in two ways: (1) It acts directly on uterine smooth muscle cells to make them contract and (2) it stimulates the formation of prostaglandins in the decidua. The prostaglandins enhance the oxytocin-induced contractions.

During labor, spinal reflexes and voluntary contractions of the abdominal muscles ("bearing down") also aid in delivery. However, delivery can occur without bearing down and without a reflex increase in secretion of oxytocin from the posterior pituitary gland, since paraplegic women can go into labor and deliver.

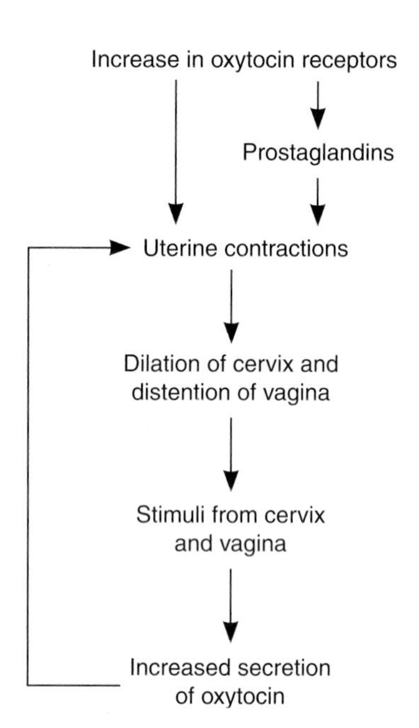

FIGURE 22–24 **Role of oxytocin in parturition.**

LACTATION

Development of the Breasts

Many hormones are necessary for full mammary development. In general, estrogens are primarily responsible for proliferation of the mammary ducts and progesterone for the development of the lobules. In rats, some prolactin is also needed for development of the glands at puberty, but it is not known if prolactin is necessary in humans. During pregnancy, prolactin levels increase steadily until term, and levels of estrogens and progesterone are elevated as well, producing full lobuloalveolar development.

Secretion & Ejection of Milk

The composition of human and cows' milk is shown in Table 22–6. In estrogen- and progesterone-primed rodents, injections of prolactin cause the formation of milk droplets and their secretion into the ducts. Oxytocin causes contraction of the myoepithelial cells lining the duct walls, with consequent ejection of the milk through the nipple.

Initiation of Lactation after Delivery

The breasts enlarge during pregnancy in response to high circulating levels of estrogens, progesterone, prolactin, and possibly hCG. Some milk is secreted into the ducts as early as the 5th month, but the amounts are small compared with the surge of milk secretion that follows delivery. In most animals, milk is secreted within an hour after delivery, but in women it takes 1–3 days for the milk to "come in."

After expulsion of the placenta at parturition, the levels of circulating estrogens and progesterone abruptly decline. The drop in circulating estrogen initiates lactation. Prolactin and estrogen synergize in producing breast growth, but estrogen antagonizes the milk-producing effect of prolactin on the breast. Indeed, in women who do not wish to nurse their babies, estrogens may be administered to stop lactation.

TABLE 22–6 Composition of colostrum and milk.[a]

Component	Human Colostrum	Human Milk	Cows' Milk
Water (g)	...	88	88
Lactose (g)	5.3	6.8	5.0
Protein (g)	2.7	1.2	3.3
Casein: lactalbumin ratio	...	1:2	3:1
Fat (g)	2.9	3.8	3.7
Linoleic acid	...	8.3% of fat	1.6% of fat
Sodium (mg)	92	15	58
Potassium (mg)	55	55	138
Chloride (mg)	117	43	103
Calcium (mg)	31	33	125
Magnesium (mg)	4	4	12
Phosphorus (mg)	14	15	100
Iron (mg)	0.09[2]	0.15[b]	0.10[b]
Vitamin A (μg)	89	53	34
Vitamin D (μg)	...	0.03[b]	0.06[b]
Thiamine (μg)	15	16	42
Riboflavin (μg)	30	43	157
Nicotinic acid (μg)	75	172	85
Ascorbic acid (mg)	4.4[b]	4.3[b]	1.6[b]

[a]Weights per deciliter.

[b]Poor source.

Reproduced with permission from Findlay ALR: Lactation. Res Reprod (Nov) 1974;6(6).

Suckling not only evokes reflex oxytocin release and milk ejection, it also maintains and augments the secretion of milk because of the stimulation of prolactin secretion it produces.

Effect of Lactation on Menstrual Cycles

Women who do not nurse their infants usually have their first menstrual period 6 weeks after delivery. However, women who nurse regularly have amenorrhea for 25–30 weeks. Nursing stimulates prolactin secretion, and evidence suggests that prolactin inhibits GnRH secretion, inhibits the action of GnRH on the pituitary, and antagonizes the action of gonadotropins on the ovaries. Ovulation is inhibited, and the ovaries are inactive, so estrogen and progesterone output falls to low levels. Consequently, only 5–10% of women become pregnant again during the suckling period, and nursing has long been known to be an important, if only partly effective, method of birth control. Furthermore, almost 50% of the cycles in the first 6 months after resumption of menses are anovulatory (see Clinical Box 22–5).

CLINICAL BOX 22–5

Chiari–Frommel Syndrome

An interesting, although rare, condition is persistence of lactation **(galactorrhea)** and amenorrhea in women who do not nurse after delivery. This condition, called the **Chiari–Frommel syndrome**, may be associated with some genital atrophy and is due to persistent prolactin secretion without the secretion of the FSH and LH necessary to produce maturation of new follicles and ovulation. A similar pattern of galactorrhea and amenorrhea with high circulating prolactin levels is seen in nonpregnant women with chromophobe pituitary tumors and in women in whom the pituitary stalk has been sectioned during treatment of cancer.

Gynecomastia

Breast development in the male is called **gynecomastia.** It may be unilateral but is more commonly bilateral. It is common in newborns because of transplacental passage of maternal estrogens (occurring in about 75%). It also occurs in a mild, transient form in 70% of normal boys at the time of puberty and in many men over the age of 50. It occurs in androgen resistance. It is a complication of estrogen therapy and is seen in patients with estrogen-secreting tumors. It is found in a wide variety of seemingly unrelated conditions, including eunuchoidism, hyperthyroidism, and cirrhosis of the liver. Digitalis can produce it, apparently because cardiac glycosides are weakly estrogenic. It can also be caused by many other drugs. It has been seen in malnourished prisoners of war, but only after they were liberated and eating an adequate diet. A feature common to many and perhaps all cases of gynecomastia is an increase in the plasma estrogen:androgen ratio due to either increased circulating estrogens or decreased circulating androgens.

HORMONES & CANCER

About 35% of carcinomas of the breast in women of childbearing age are **estrogen-dependent;** their continued growth depends on the presence of estrogens in the circulation. The tumors are not cured by decreasing estrogen secretion, but symptoms are dramatically relieved, and the tumor regresses for months or years before recurring (see Chapter 16). Women with estrogen-dependent tumors often have a remission when their ovaries are removed. Inhibition of the action of estrogens with **tamoxifen** also produces remissions, and inhibition of estrogen formation with drugs that inhibit **aromatase** (Figure 22–16) is even more effective.

CHAPTER SUMMARY

- Differences between males and females depend primarily on a single chromosome (the Y chromosome) and a single pair of endocrine structures (the gonads); testes in the male and ovaries in the female.
- The gonads have a dual function: the production of germ cells (gametogenesis) and the secretion of sex hormones. The testes secrete large amounts of androgens, principally testosterone, but they also secrete small amounts of estrogens. The ovaries secrete large amounts of estrogens and small amounts of androgens.
- The reproductive system of women has regular cyclical changes that can be thought of as periodic preparations for fertilization and pregnancy. In humans and other primates, the cycle is a **menstrual** cycle, and features the periodic vaginal bleeding that occurs with the shedding of the uterine mucosa **(menstruation).**
- Ovaries also secrete progesterone, a steroid that has special functions in preparing the uterus for pregnancy. During pregnancy the ovaries secrete relaxin, which facilitates the delivery of the fetus. In both sexes, the gonads secrete other polypeptides, including inhibin B, a polypeptide that inhibits FSH secretion.
- In women, a period called perimenopause precedes menopause, and can last up to 10 years; during this time the menstrual cycles become irregular and the level of inhibins decrease.
- Once in menopause, the ovaries no longer secrete progesterone and 17β-estradiol and estrogen is formed only in small amounts by aromatization of androstenedione in peripheral tissues.
- The naturally occurring estrogens are **17β estradiol, estrone,** and **estriol.** They are secreted primarily by the granulosa cells of the ovarian follicles, the corpus luteum, and the placenta. Their biosynthesis depends on the enzyme **aromatase** (CYP19), which converts testosterone to estradiol and androstenedione to estrone. The latter reaction also occurs in fat, liver, muscle, and the brain.

MULTIPLE-CHOICE QUESTIONS

For all questions, select the single best answer unless otherwise directed.

1. If a young woman has high plasma levels of T_3, cortisol, and renin activity but her blood pressure is only slightly elevated and she has no symptoms or signs of thyrotoxicosis or Cushing syndrome, the most likely explanation is that
   A. she has been treated with TSH and ACTH.
   B. she has been treated with T_3 and cortisol.
   C. she is in the third trimester of pregnancy.
   D. she has an adrenocortical tumor.
   E. she has been subjected to chronic stress.
2. In humans, fertilization usually occurs in the
   A. vagina.
   B. cervix.
   C. uterine cavity.
   D. uterine tubes.
   E. abdominal cavity.
3. Which of the following is *not* a steroid?
   A. 17α-hydroxyprogesterone
   B. Estrone
   C. Relaxin
   D. Pregnenolone
   E. Etiocholanolone
4. Which of the following probably triggers the onset of labor?
   A. ACTH in the fetus
   B. ACTH in the mother
   C. Prostaglandins
   D. Oxytocin
   E. Placental renin

CHAPTER RESOURCES

Bole-Feysot C, Goffin V, Edery M et al: Prolactin (PRL) and its receptor: Actions, signal transduction pathways, and phenotypes observed in PRL receptor knockout mice. Endocrinol Rev 1998;19:225.

Mather JP, Moore A, Li R-H: Activins, inhibins, and follistatins: Further thoughts on a growing family of regulators. Proc Soc Exper Biol Med 1997;215:209.

Matthews J, Gustafson J-A: Estrogen signaling: A subtle balance between ER_α and ER_β. Mol Interv 2003;3:281.

McLaughlin DT, Donahoe PR: Sex determination and differentiation. N Engl J Med 2004;350:367.

Naz RK (editor): *Endocrine Disruptors.* CRC Press, 1998.

Norwitz ER, Robinson JN, Challis JRG: The control of labor. N Engl J Med 1999;341:660.

Primakoff P, Nyles DG: Penetration, adhesion, and fusion in mammalian sperm–egg interaction. Science 2002;296:2183.

Simpson ER, Clyne C, Rubin G, et al. Aromatase—A brief overview. Annu Rev Physiol. 2002;64:93-127.

Yen SSC, Jaffe RB, Barbieri RL: *Reproductive Endocrinology: Physiology, Pathophysiology, and Clinical Management*, 4th ed. Saunders, 1999.

CHAPTER

Function of the Male Reproductive System

23

OBJECTIVES

After studying this chapter, you should be able to:

- Name the key hormones secreted by Leydig cells and Sertoli cells of the testes.
- Outline the steps involved in spermatogenesis.
- Outline the mechanisms that produce erection and ejaculation.
- Know the general structure of testosterone, and describe its biosynthesis, transport, metabolism, and actions.
- Describe the processes involved in regulation of testosterone secretion.

INTRODUCTION

The role for a functional, secreting testis in the formation of male genitalia, the action of male hormones on the brain in early development, and development of the male reproductive system through adolescence and into adulthood were discussed in the previous chapter. As observed in the female, male gonads have a dual function: the production of germ cells **(gametogenesis)** and the secretion of **sex hormones.** The **androgens** are the steroid sex hormones that are masculinizing in their action. The testes secrete large amounts of androgens, principally **testosterone,** but they also secrete small amounts of estrogens. Unlike that observed in females, male gonadotropin secretion is noncyclical, and once mature, male gonadal function slowly declines with advancing age, but the ability to produce viable gametes persists. In this chapter we will focus discussion on the structure and physiology of the mature male reproductive system.

THE MALE REPRODUCTIVE SYSTEM

STRUCTURE

The testes are made up of loops of convoluted **seminiferous tubules,** in the walls of which the spermatozoa are formed from the primitive germ cells **(spermatogenesis).** Both ends of each loop drain into a network of ducts in the head of the **epididymis.** From there, spermatozoa pass through the tail of the epididymis into the **vas deferens.** They enter through the **ejaculatory ducts** into the urethra in the body of the **prostate** at the time of ejaculation (Figure 23–1). Between the tubules in the testes are nests of cells containing lipid granules, the **interstitial cells of Leydig** (Figures 23–2 and 23–3), which secrete testosterone into the bloodstream. The spermatic arteries to the testes are tortuous, and blood in them runs parallel but in the opposite direction to blood in the pampiniform plexus of spermatic veins. This anatomic arrangement may permit countercurrent exchange of heat and testosterone. The principles of countercurrent exchange are considered in detail in relation to the kidney in Chapter 37.

GAMETOGENESIS & EJACULATION

Blood–Testis Barrier

The walls of the seminiferous tubules are lined by primitive germ cells and **Sertoli cells,** large, complex glycogen-containing cells that stretch from the basal lamina of the tubule to the lumen (Figure 23–3). Germ cells must stay in contact with Sertoli cells to survive; this contact is maintained by cytoplasmic bridges. Tight junctions between adjacent Sertoli cells near the basal lamina form a **blood–testis barrier** that prevents many large molecules from passing from the interstitial tissue and the part of the tubule near the basal lamina (basal compartment) to the region near the tubular lumen (adluminal compartment) and

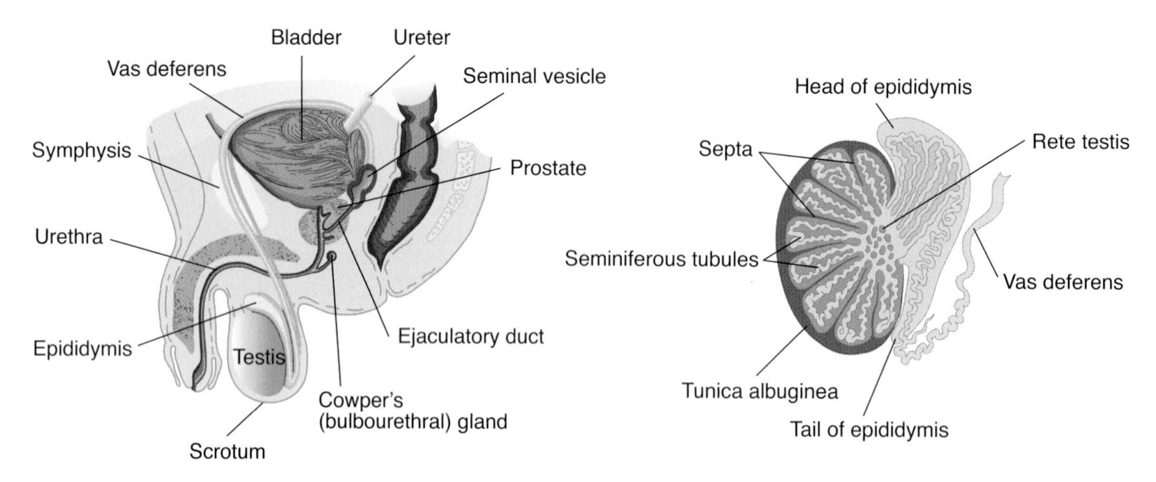

FIGURE 23–1 Anatomical features of the male reproductive system. Left: Male reproductive system. **Right:** Duct system of the testis.

the lumen. However, steroids penetrate this barrier with ease, and evidence suggests that some proteins also pass from the Sertoli cells to the Leydig cells, and vice versa, to function in a paracrine fashion. In addition, maturing germ cells must pass through the barrier as they move to the lumen. This appears to occur without disruption of the barrier by coordinated breakdown of the tight junctions above the germ cells and formation of new tight junctions below them.

The fluid in the lumen of the seminiferous tubules is quite different from plasma; it contains very little protein and glucose but is rich in androgens, estrogens, K^+, inositol, and glutamic and aspartic acids. Maintenance of its composition depends on the blood–testis barrier. The barrier also protects the germ cells from bloodborne noxious agents, prevents antigenic products of germ cell division and maturation from entering the circulation and generating an autoimmune response, and may help establish an osmotic gradient that facilitates movement of fluid into the tubular lumen.

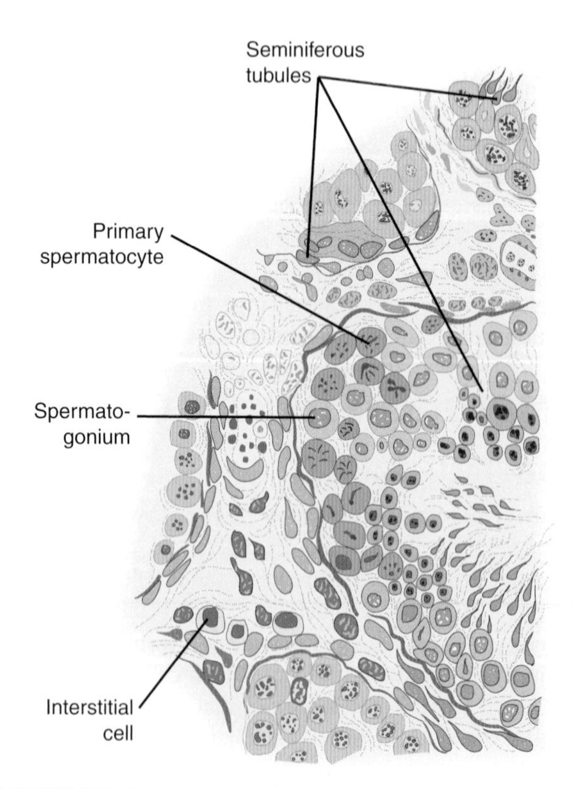

FIGURE 23–2 Section of human testis.

Spermatogenesis

Spermatogonia, the primitive germ cells next to the basal lamina of the seminiferous tubules, mature into **primary spermatocytes** (Figure 23–3). This process begins during adolescence. The primary spermatocytes undergo meiotic division, reducing the number of chromosomes. In this two-stage process, they divide into **secondary spermatocytes** and then into **spermatids,** which contain the haploid number of 23 chromosomes. The spermatids mature into **spermatozoa (sperm).** As a single spermatogonium divides and matures, its descendants remain tied together by cytoplasmic bridges until the late spermatid stage. This arrangement helps to ensure synchrony of the differentiation of each clone of germ cells. The estimated number of spermatids formed from a single spermatogonium is 512. The formation of a mature sperm from a primitive germ cell by spermatogenesis in humans spans approximately 74 days.

Each sperm is an intricate motile cell, rich in DNA, with a head that is made up mostly of chromosomal material (Figure 23–4). Covering the head like a cap is the **acrosome,** a lysosome-like organelle rich in enzymes involved in sperm penetration of the ovum and other events associated with fertilization. The motile tail of the sperm is wrapped in its proximal portion by a sheath holding numerous mitochondria. The membranes of late spermatids and spermatozoa contain a special small form of angiotensin converting enzyme called **germinal angiotensin converting enzyme (gACE).** Germinal ACE is transcribed from the same gene as the somatic ACE (sACE); however, gACE displays tissue specific expression based on alternative transcription initiation sites and alternate splicing patterns. The full function of gACE has yet to be elucidated, although gACE-specific knockout mouse models are sterile.

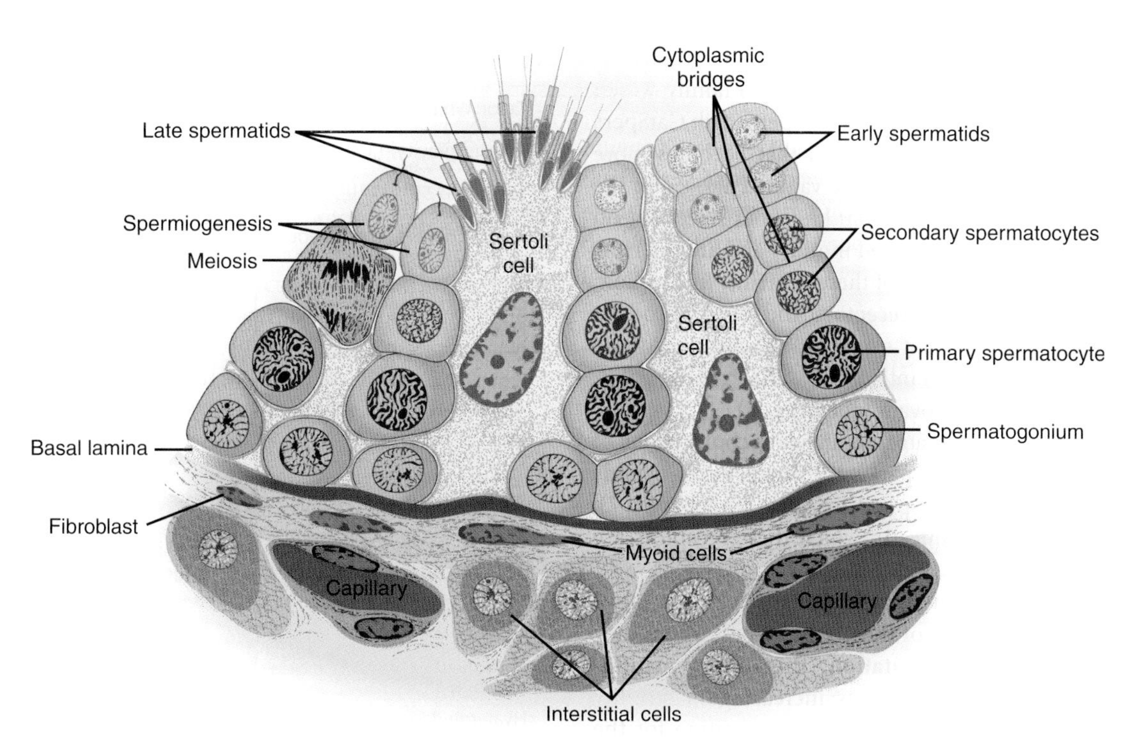

FIGURE 23–3 Seminiferous epithelium. Note that maturing germ cells remain connected by cytoplasmic bridges through the early spermatid stage and that these cells are closely invested by Sertoli cell cytoplasm as they move from the basal lamina to the lumen. (Reproduced with permission from Junqueira LC, Carneiro J: *Basic Histology: Text & Atlas,* 10th ed. McGraw-Hill, 2003.)

Spermatids mature into spermatozoa in deep folds of the cytoplasm of the Sertoli cells (Figure 23–3). Mature spermatozoa are released from the Sertoli cells and become free in the lumens of the tubules. The Sertoli cells secrete **androgen-binding protein (ABP), inhibin,** and **MIS.** They do not synthesize androgens, but they contain **aromatase (CYP19),** the enzyme responsible for conversion of androgens to estrogens, and they can produce estrogens. ABP probably functions to maintain a high, stable supply of androgen in the tubular fluid. Inhibin inhibits follicle-stimulating hormone (FSH) secretion.

FSH and androgens maintain the gametogenic function of the testis. After hypophysectomy, injection of luteinizing hormone (LH) produces a high local concentration of androgen in the testes, and this maintains spermatogenesis. The stages from spermatogonia to spermatids appear to be androgen-independent. However, the maturation from spermatids to spermatozoa depends on androgen acting on the Sertoli cells in which the developing spermatozoa are embedded. FSH acts on the Sertoli cells to facilitate the last stages of spermatid maturation. In addition, it promotes the production of ABP.

An interesting observation is that the estrogen content of the fluid in the rete testis (Figure 23–1) is high, and the walls of the rete testis contain numerous α estrogen receptors (ERα). In this region, fluid is reabsorbed and the spermatozoa are concentrated. If this does not occur, the sperm entering the epididymis are diluted in a large volume of fluid, resulting in reduced fertility.

Further Development of Spermatozoa

Spermatozoa leaving the testes are not fully mobile. They continue their maturation and acquire motility during their passage through the epididymis. Motility is obviously important in vivo, but fertilization occurs in vitro if an immotile spermatozoon from the head of the epididymis is microinjected directly into an ovum. The ability to move forward **(progressive**

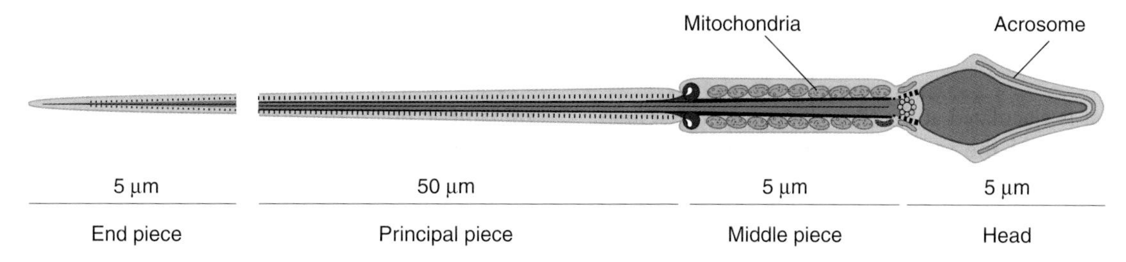

FIGURE 23–4 Human spermatozoon, profile view. Note the acrosome, an organelle that covers half the sperm head inside the plasma membrane of the sperm. (Reproduced with permission from Junqueira LC, Carneiro J: *Basic Histology: Text & Atlas,* 11th ed. McGraw-Hill, 2005.)

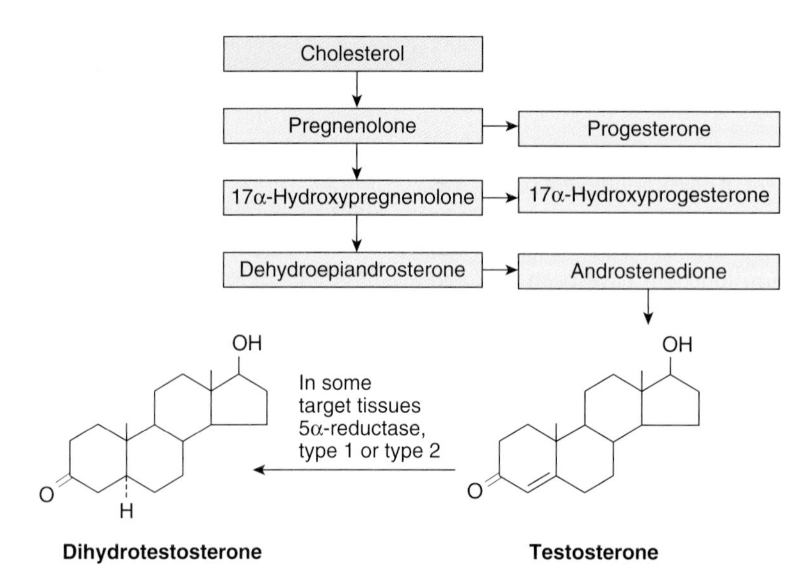

FIGURE 23–5 Biosynthesis of testosterone. The formulas of the precursor steroids are shown in Figure 22–7. Although the main secretory product of the Leydig cells is testosterone, some of the precursors also enter the circulation.

that form steroid hormones are similar, the organs differing only in the enzyme systems they contain. In the Leydig cells, the 11- and 21-hydroxylases found in the adrenal cortex (see Figure 20–7) are absent, but 17(α-hydroxylase is present. Pregnenolone is therefore hydroxylated in the 17 position and then subjected to side chain cleavage to form dehydroepiandrosterone. Androstenedione is also formed via progesterone and 17-hydroxyprogesterone, but this pathway is less prominent in humans. Dehydroepiandrosterone and androstenedione are then converted to testosterone.

The secretion of testosterone is under the control of LH, and the mechanism by which LH stimulates Leydig cells involves increased formation of cAMP via the G protein-coupled LH receptor and G_s. Cyclic AMP increases the formation of cholesterol from cholesterol esters and the conversion of cholesterol to pregnenolone via the activation of protein kinase A.

Secretion

The testosterone secretion rate is 4–9 mg/d (13.9–31.33 μmol/d) in normal adult males. Small amounts of testosterone are also secreted in females, with the major source being the ovary, but possibly from the adrenal as well.

Transport & Metabolism

Ninety-eight per cent of the testosterone in plasma is bound to protein: 65% is bound to a β-globulin called **gonadal steroid-binding globulin (GBG)** or **sex steroid-binding globulin,** and 33% to albumin (Table 23–2). GBG also binds estradiol. The plasma testosterone level (free and bound) is 300–1000 ng/dL (10.4–34.7 nmol/L) in adult men (Figure 22–8), compared with 30–70 ng/dL (1.04–2.43 nmol/L) in adult women. It declines somewhat with age in males.

A small amount of circulating testosterone is converted to estradiol, but most of the testosterone is converted to 17-ketosteroids, principally androsterone and its isomer etiocholanolone (Figure 23–6), and excreted in the urine. About two thirds of the urinary 17-ketosteroids are of adrenal origin, and one third are of testicular origin. Although most of the 17-ketosteroids are weak androgens (they have 20% or less the potency of testosterone), it is worth emphasizing that not all 17-ketosteroids are androgens and not all androgens are 17-ketosteroids. Etiocholanolone, for example, has no androgenic activity, and testosterone itself is not a 17-ketosteroid.

Actions

In addition to their actions during development, testosterone and other androgens exert an inhibitory feedback effect on pituitary LH secretion; develop and maintain the male secondary sex characteristics; exert an important protein-anabolic, growth-promoting effect; and, along with FSH, maintain spermatogenesis.

TABLE 23–2 Distribution of gonadal steroids and cortisol in plasma.

		% Bound to		
Steroid	**% Free**	**CBG**	**GBG**	**Albumin**
Testosterone	2	0	65	33
Androstenedione	7	0	8	85
Estradiol	2	0	38	60
Progesterone	2	18	0	80
Cortisol	4	90	0	6

CBG, corticosteroid-binding globulin; GBG, gonadal steroid-binding globulin. (Courtesy of S Munroe.)

HO H O

Androsterone

HO H O

Etiocholanolone

FIGURE 23–6 **Two 17-ketosteroid metabolites of testosterone.**

TABLE 23–3 **Changes at puberty in boys (male secondary sex characteristics).**

External genitalia: Penis increases in length and width. Scrotum becomes pigmented and rugose.
Internal genitalia: Seminal vesicles enlarge and secrete and begin to form fructose. Prostate and bulbourethral glands enlarge and secrete.
Voice: Larynx enlarges, vocal cords increase in length and thickness, and voice becomes deeper.
Hair growth: Beard appears. Hairline on scalp recedes anterolaterally. Pubic hair grows with male (triangle with apex up) pattern. Hair appears in axillas, on chest, and around anus; general body hair increases.
Mental: More aggressive, active attitude. Interest in opposite sex develops.
Body conformation: Shoulders broaden; muscles enlarge.
Skin: Sebaceous gland secretion thickens and increases (predisposing to acne).

Secondary Sex Characteristics

The widespread changes in hair distribution, body configuration, and genital size that develop in boys at puberty—the male **secondary sex characteristics**—are summarized in Table 23–3. The prostate and seminal vesicles enlarge, and the seminal vesicles begin to secrete fructose. This sugar appears to function as the main nutritional supply for the spermatozoa. The psychic effects of testosterone are difficult to define in humans, but in experimental animals, androgens provoke boisterous and aggressive play. The effects of androgens and estrogens on sexual behavior are considered in detail in Chapter 15. Although body hair is increased by androgens, scalp hair is decreased (Figure 23–7). Hereditary baldness often fails to develop unless dihydrotestosterone (DHT) is present.

Anabolic Effects

Androgens increase the synthesis and decrease the breakdown of protein, leading to an increase in the rate of growth. It used to be argued that they cause the epiphyses to fuse to the long bones, thus eventually stopping growth, but it now appears that epiphysial closure is due to estrogens (see Chapter 21). Secondary to their anabolic effects, androgens cause moderate Na^+, K^+, H_2O, Ca^{2+}, SO_4^-, and PO_4^- retention; and they also increase the size of the kidneys. Doses of exogenous testosterone that exert significant anabolic effects are also masculinizing and increase libido, which limits the usefulness of the hormone as an anabolic agent in patients with wasting diseases. Attempts to develop synthetic steroids in which the anabolic action is separated from the androgenic action have not been successful.

Mechanism of Action

Like other steroids, testosterone binds to an intracellular receptor, and the receptor/steroid complex then binds to DNA in the nucleus, facilitating transcription of various genes. In addition, testosterone is converted to **DHT** by 5α-reductase in some target cells (Figure 23–5 and Figure 23–8), and DHT binds to the same intracellular receptor as testosterone. DHT also circulates, with a plasma level that is about 10% of the testosterone level. Testosterone–receptor complexes are less stable than DHT–receptor complexes in target cells, and

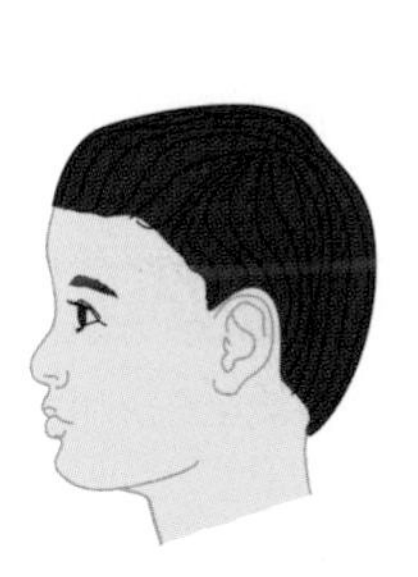
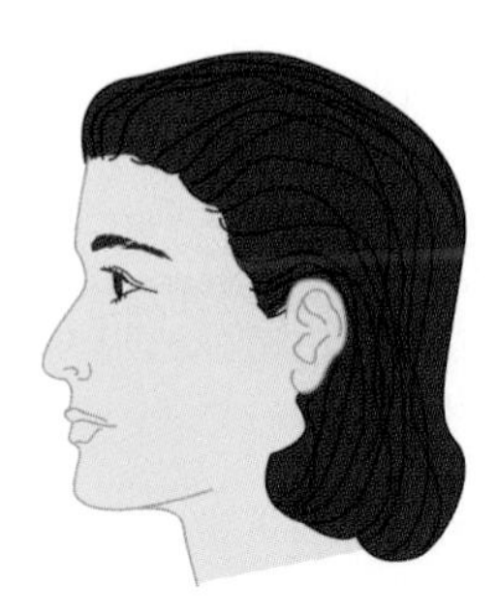
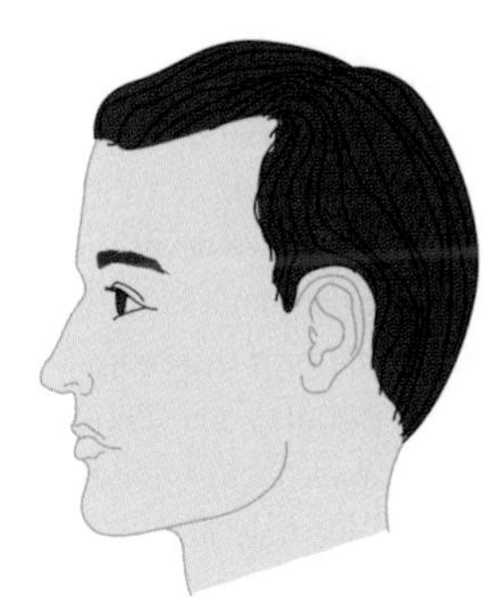

FIGURE 23–7 **Hairline in children and adults.** The hairline of the woman is like that of the child, whereas that of the man is indented in the lateral frontal region.

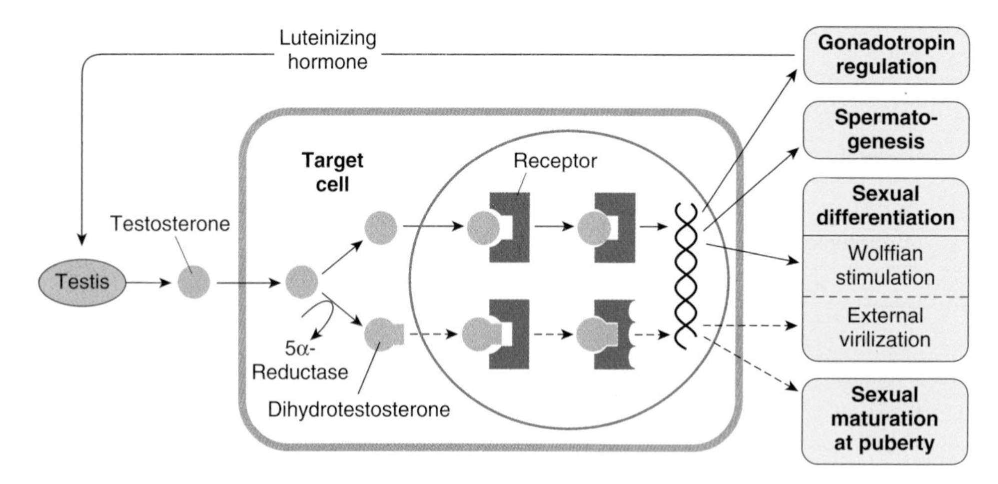

FIGURE 23–8 Schematic diagram of the actions of testosterone (solid arrows) and dihydrotestosterone (dashed arrows). Note that they both bind to the same receptor, but DHT binds more effectively. (Reproduced with permission from Wilson JD, Griffin JE, Russell W: Steroid 5α-reductase 2 deficiency. Endocr Rev 1993;14:577. Copyright © 1993 by The Endocrine Society.)

they conform less well to the DNA-binding state. Thus, DHT formation is a way of amplifying the action of testosterone in target tissues. Humans have two 5α-reductases that are encoded by different genes. Type 1 5α-reductase is present in skin throughout the body and is the dominant enzyme in the scalp. Type 2 5α-reductase is present in genital skin, the prostate, and other genital tissues.

Testosterone–receptor complexes are responsible for the maturation of Wolffian duct structures and consequently for the formation of male internal genitalia during development, but DHT–receptor complexes are needed to form male external genitalia (Figure 23–8). DHT–receptor complexes are also primarily responsible for enlargement of the prostate and probably of the penis at the time of puberty, as well as for the facial hair, the acne, and the temporal recession of the hairline. On the other hand, the increase in muscle mass and the development of male sex drive and libido depend primarily on testosterone rather than DHT (see Clinical Box 23–2).

Testicular Production of Estrogens

Over 80% of the estradiol and 95% of the estrone in the plasma of adult men is formed by extragonadal and extraadrenal aromatization of circulating testosterone and androstenedione. The remainder comes from the testes. Some of the estradiol in testicular venous blood comes from the Leydig cells, but some is also produced by aromatization of androgens in Sertoli cells. In men, the plasma estradiol level is 20–50 pg/mL (73–184 pmol/L) and the total production rate is approximately 50 μg/ d (184 nmol/d). In contrast to the situation in women, estrogen production moderately increases with advancing age in men.

CONTROL OF TESTICULAR FUNCTION

FSH is tropic for Sertoli cells, and FSH and androgens maintain the gametogenic function of the testes. FSH also stimulates the secretion of ABP and inhibin. Inhibin feeds back to

CLINICAL BOX 23–2

Congenital 5α-Reductase Deficiency

Congenital 5α-reductase deficiency, in which the gene for type 2 5α-reductase is mutated, is common in certain parts of the Dominican Republic. It produces an interesting form of male pseudohermaphroditism. Individuals with this syndrome are born with male internal genitalia including testes, but they have female external genitalia and are usually raised as girls. However, when they reach puberty, LH secretion and circulating testosterone levels are increased. Consequently, they develop male body contours and male libido. At this point, they usually change their gender identities and "become boys." The clitoris enlarges ("penis-at-12 syndrome") to the point that some of the individuals can have intercourse with women. This enlargement probably occurs because with the high LH, enough testosterone is produced to overcome the need for DHT amplification in the genitalia.

THERAPEUTIC HIGHLIGHTS

5α-Reductase-inhibiting drugs are now being used clinically to treat benign prostatic hyperplasia, and **finasteride,** the most extensively used drug, has its greatest effect on type 2 5α-reductase.

inhibit FSH secretion. LH is tropic for Leydig cells and stimulates the secretion of testosterone, which in turn feeds back to inhibit LH secretion. Hypothalamic lesions in animals and hypothalamic disease in humans lead to atrophy of the testes and loss of their function.

Inhibins

Testosterone reduces plasma LH but, except in large doses, it has no effect on plasma FSH. Plasma FSH is elevated in patients who have atrophy of the seminiferous tubules but normal levels of testosterone and LH secretion. These observations led to the search for **inhibin,** a factor of testicular origin that inhibits FSH secretion. There are two inhibins in extracts of testes in men and in antral fluid from ovarian follicles in women. They are formed from three polypeptide subunits: a glycosylated α subunit with a molecular weight of 18,000; and two nonglycosylated β subunits, β_A and β_B, each with a molecular weight of 14,000. The subunits are formed from precursor proteins (Figure 23–9). The α subunit combines with β_A to form a heterodimer and with β_B to form another heterodimer, with the subunits linked by disulfide bonds. Both $\alpha\beta_A$ (inhibin A) and $\alpha\beta_B$ (inhibin B) inhibit FSH secretion by a direct action on the pituitary, though it now appears that it is inhibin B that is the FSH-regulating inhibin in adult men and women. Inhibins are produced by Sertoli cells in males and granulosa cells in females.

The heterodimer $\beta_A\beta_B$ and the homodimers $\beta_A\beta_A$ and $\beta_B\beta_B$ are also formed. They stimulate rather than inhibit FSH secretion and consequently are called **activins.** Their function in reproduction is unsettled. However, the inhibins and activins are members of the TGFβ superfamily of dimeric growth factors that also includes MIS. **Activin receptors** have been identified and belong to the serine/threonine kinase receptor family. Inhibins and activins are found not only in the gonads but also in the brain and many other tissues. In the bone marrow, activins are involved in the development of white blood cells. In embryonic life, activins are involved in the formation of mesoderm. All mice with a targeted deletion of the α-inhibin subunit gene initially exhibit normal growth but then develop gonadal stromal tumors, so the gene is a tumor suppressor gene.

In plasma, α_2-macroglobulin binds activins and inhibins. In tissues, activins bind to a family of four glycoproteins called **follistatins.** Binding of the activins inactivates their biologic activity, but the relation of follistatins to inhibin and their physiologic function remain unsettled.

Steroid Feedback

A current "working hypothesis" of the way the functions of the testes are regulated by steroids is shown in Figure 23–10. Castration is followed by a rise in the pituitary content and secretion of FSH and LH, and hypothalamic lesions prevent this rise. Testosterone inhibits LH secretion by acting directly on the anterior pituitary and by inhibiting the secretion of GnRH from the hypothalamus. Inhibin acts directly on the anterior pituitary to inhibit FSH secretion.

In response to LH, some of the testosterone secreted from the Leydig cells bathes the seminiferous epithelium and provides the high local concentration of androgen to the Sertoli cells that is necessary for normal spermatogenesis. Systemically administered testosterone does not raise the androgen level in the testes to as great a degree, and it inhibits LH secretion. Consequently, the net effect of systemically administered testosterone is generally a decrease in sperm count. Testosterone therapy has been suggested as a means of male contraception. However, the dose of testosterone needed to suppress spermatogenesis causes sodium and water retention. The possible use of inhibins as male contraceptives is now being explored.

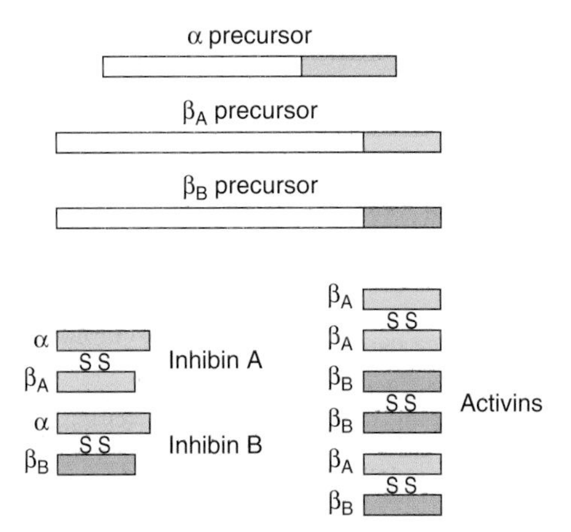

FIGURE 23–9 Inhibin precursor proteins and the various inhibins and activins that are formed from the carboxyl terminal regions of these precursors. SS, disulfide bonds.

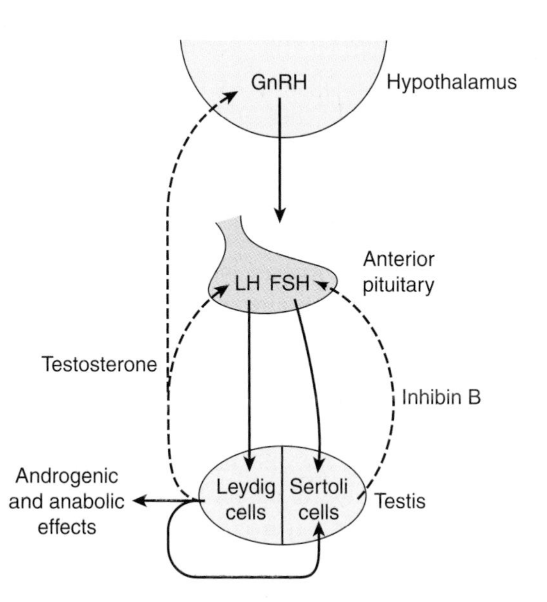

FIGURE 23–10 Postulated interrelationships between the hypothalamus, anterior pituitary, and testes. Solid arrows indicate excitatory effects; dashed arrows indicate inhibitory effects.

ABNORMALITIES OF TESTICULAR FUNCTION

Cryptorchidism

The testes develop in the abdominal cavity and normally migrate to the scrotum during fetal development. **Testicular descent** to the inguinal region depends on MIS, and descent from the inguinal region to the scrotum depends on other factors. Descent is incomplete on one or, less commonly, both sides in 10% of newborn males, with the testes remaining in the abdominal cavity or inguinal canal. Gonadotropic hormone treatment speeds descent in some cases, or the defect can be corrected surgically. Spontaneous descent of the testes is the rule, and the proportion of boys with undescended testes **(cryptorchidism)** falls to 2% at age 1 year and 0.3% after puberty. However, early treatment is now recommended despite these figures because the incidence of malignant tumors is higher in undescended than in scrotal testes and because after puberty the higher temperature in the abdomen eventually causes irreversible damage to the spermatogenic epithelium.

Male Hypogonadism

The clinical picture of male hypogonadism depends on whether testicular deficiency develops before or after puberty. In adults, if it is due to testicular disease, circulating gonadotropin levels are elevated **(hypergonadotropic hypogonadism);** if it is secondary to disorders of the pituitary or the hypothalamus (eg, Kallmann syndrome), circulating gonadotropin levels are depressed **(hypogonadotropic hypogonadism).** If the endocrine function of the testes is lost in adulthood, the secondary sex characteristics regress slowly because it takes very little androgen to maintain them once they are established. The growth of the larynx during adolescence is permanent, and the voice remains deep. Men castrated in adulthood suffer some loss of libido, although the ability to copulate persists for some time. They occasionally have hot flushes and are generally more irritable, passive, and depressed than men with intact testes. When the Leydig cell deficiency dates from childhood, the clinical picture is that of **eunuchoidism.** Eunuchoid individuals over the age of 20 are characteristically tall, although not as tall as hyperpituitary giants, because their epiphyses remain open and some growth continues past the normal age of puberty. They have narrow shoulders and small muscles, a body configuration resembling that of the adult female. The genitalia are small and the voice high-pitched. Pubic hair and axillary hair are present because of adrenocortical androgen secretion. However, the hair is sparse, and the pubic hair has the female "triangle with the base up" distribution rather than the "triangle with the base down" pattern (male escutcheon) seen in normal males.

Androgen-Secreting Tumors

"Hyperfunction" of the testes in the absence of tumor formation is not a recognized entity. Androgen-secreting Leydig cell tumors are rare and cause detectable endocrine symptoms only in prepubertal boys, who develop precocious pseudopuberty (Table 22–2).

Hormones and Cancer

Some carcinomas of the prostate are **androgen-dependent** and regress temporarily after the removal of the testes or treatment with GnRH agonists in doses that are sufficient to produce down-regulation of the GnRH receptors on gonadotropes and decrease LH secretion.

CHAPTER SUMMARY

- The gonads have a dual function: the production of germ cells (gametogenesis) and the secretion of sex hormones. The testes secrete large amounts of androgens, principally testosterone, but they also secrete small amounts of estrogens.
- Spermatogonia develop into mature spermatozoa in the seminiferous tubules via a process called spermatogenesis. This is a multistep process that includes maturation of spermatogonia into primary spermatocytes, which undergo meiotic division, resulting in haploid secondary spermatocytes. Several further divisions result in spermatids. Each cell division from a spermatogonium to a spermatid is incomplete, with cells remaining connected via cytoplasmic bridges. Spermatids eventually mature into motile spermatozoa to complete spermatogenesis; this last part of maturation is called spermiogenesis.
- Testosterone is the principal hormone of the testis. It is synthesized from cholesterol in Leydig cells. The secretion of testosterone from Leydig cells is under control of luteinizing hormone at a rate of 4–9 mg/day in adult males. Most testosterone is bound to albumin or to gonadal steroid-binding globulin in the plasma. Testosterone plays an important role in the development and maintenance of male secondary sex characteristics, as well as other defined functions.

MULTIPLE-CHOICE QUESTIONS

For all questions, select the single best answer unless otherwise directed.

1. Full development and function of the seminiferous tubules require
   A. somatostatin.
   B. LH.
   C. oxytocin.
   D. FSH.
   E. androgens and FSH.
2. In human males, testosterone is produced mainly by the
   A. Leydig cells.
   B. Sertoli cells.
   C. seminiferous tubules.
   D. epididymis.
   E. vas deferens.

3. Nitric oxide synthase contributes to erection by:
   A. raising cAMP levels that relax smooth muscles and increase blood flow.
   B. blocking phosphodiesterases to increase cGMP levels that release smooth muscle and increase blood flow.
   C. activating soluble guanylate cyclases to increase cGMP levels that relax smooth muscle and increase blood flow.
   D. raising inctracellular Ca^{2+} concentrations that relax smooth muscles and increase blood flow.
4. Testosterone is produced
   A. in the testes after reduction of dihydrotestosterone.
   B. in Leydig cells from cholesterol and pregnenolone precusors.
   C. by leutinizing hormone in Leydig cells.
   D. as a precursor for several membrane lipids.

CHAPTER RESOURCES

Mather JP, Moore A, Li R-H: Activins, inhibins, and follistatins: Further thoughts on a growing family of regulators. Proc Soc Exper Biol Med 1997;215:209.

Primakoff P, Nyles DG: Penetration, adhesion, and fusion in mammalian sperm–egg interaction. Science 2002;296:2183.

Yen SSC, Jaffe RB, Barbieri RL: *Reproductive Endocrinology: Physiology, Pathophysiology, and Clinical Management,* 4th ed. Saunders, 1999.

Yu H-S: *Human Reproductive Biology,* CRC Press, 1994.

Endocrine Functions of the Pancreas & Regulation of Carbohydrate Metabolism

CHAPTER

24

OBJECTIVES

After reading this chapter, you should be able to:

- List the hormones that affect the plasma glucose concentration and briefly describe the action of each.
- Describe the structure of the pancreatic islets and name the hormones secreted by each of the cell types in the islets.
- Describe the structure of insulin and outline the steps involved in its biosynthesis and release into the bloodstream.
- List the consequences of insulin deficiency and explain how each of these abnormalities is produced.
- Describe insulin receptors, the way they mediate the effects of insulin, and the way they are regulated.
- Describe the types of glucose transporters found in the body and the function of each.
- List the major factors that affect the secretion of insulin.
- Describe the structure of glucagon and other physiologically active peptides produced from its precursor.
- List the physiologically significant effects of glucagon and the factors that regulate glucagon secretion.
- Describe the physiologic effects of somatostatin in the pancreas.
- Outline the mechanisms by which thyroid hormones, adrenal glucocorticoids, catecholamines, and growth hormone affect carbohydrate metabolism.
- Understand the major differences between type 1 and type 2 diabetes.

INTRODUCTION

At least four polypeptides with regulatory activity are secreted by the islets of Langerhans in the pancreas. Two of these, **insulin** and **glucagon,** are hormones and have important functions in the regulation of the intermediary metabolism of carbohydrates, proteins, and fats. The third polypeptide, **somatostatin,** plays a role in the regulation of islet cell secretion, and the fourth, **pancreatic polypeptide,** is probably concerned primarily with the regulation of ion transport in the intestine. Glucagon, somatostatin, and possibly pancreatic polypeptide are also secreted by cells in the mucosa of the gastrointestinal tract.

Insulin is anabolic, increasing the storage of glucose, fatty acids, and amino acids. Glucagon is catabolic, mobilizing glucose, fatty acids, and the amino acids from stores into the bloodstream. The two hormones are thus reciprocal in their overall action and are reciprocally secreted in most circumstances. Insulin excess causes hypoglycemia, which leads to convulsions and coma. Insulin deficiency, either absolute or relative, causes **diabetes mellitus** (chronic elevated blood glucose), a complex and debilitating disease that if untreated is eventually fatal. Glucagon deficiency can cause hypoglycemia, and glucagon excess makes diabetes worse. Excess pancreatic production of somatostatin causes hyperglycemia and other manifestations of diabetes.

A variety of other hormones also have important roles in the regulation of carbohydrate metabolism.

ISLET CELL STRUCTURE

The islets of Langerhans (Figure 24–1) are ovoid, 76- × 175-μm collections of cells. The islets are scattered throughout the pancreas, although they are more plentiful in the tail than in the body and head. β-islets make up about 2% of the volume of the gland, whereas the exocrine portion of the pancreas (see Chapter 25) makes up 80%, and ducts and blood vessels make up the remainder. Humans have 1 to 2 million islets. Each has a copious blood supply; blood from the islets, like that from the gastrointestinal tract (but unlike that from any other endocrine organs) drains into the hepatic portal vein.

The cells in the islets can be divided into types on the basis of their staining properties and morphology. Humans have at least four distinct cell types: A, B, D, and F cells. A, B, and D cells are also called α, β, and δ cells. However, this leads to confusion in view of the use of Greek letters to refer to other structures in the body, particularly adrenergic receptors (see Chapter 7). The A cells secrete glucagon, the B cells secrete insulin, the D cells secrete somatostatin, and the F cells secrete pancreatic polypeptide. The B cells, which are the most common and account for 60–75% of the cells in the islets, are generally located in the center of each islet. They tend to be surrounded by the A cells, which make up 20% of the total, and the less common D and F cells. The islets in the tail, the body, and the anterior and superior part of the head of the human pancreas have many A cells and few if any F cells in the outer rim, whereas in rats and probably in humans, the islets in the posterior part of the head of the pancreas have a relatively large number of F cells and few A cells. The A-cell-rich (glucagon-rich) islets arise embryologically from the dorsal pancreatic bud, and the F-cell-rich (pancreatic polypeptide-rich) islets arise from the ventral pancreatic bud. These buds arise separately from the duodenum.

The B cell granules are packets of insulin in the cell cytoplasm. The shape of the packets varies from species to species; in humans, some are round whereas others are rectangular (Figure 24–2). In the B cells, the insulin molecule forms polymers and also complexes with zinc. The differences in the shape of the packets are probably due to differences in the size of polymers or zinc aggregates of insulin. The A granules, which contain glucagon, are relatively uniform from species to species (Figure 24–3). The D cells also contain large numbers of relatively homogeneous granules.

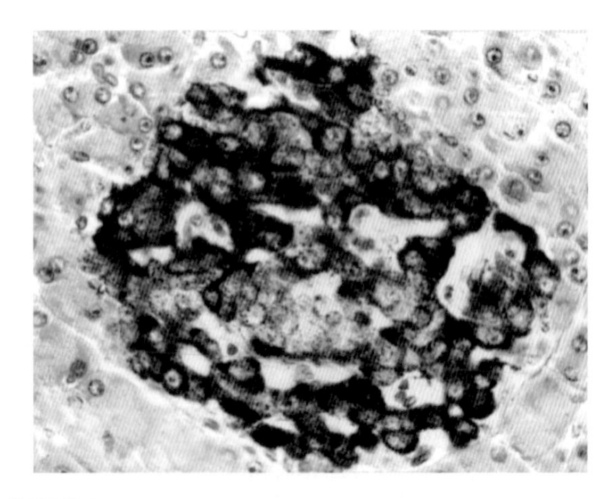

FIGURE 24–1 Islet of Langerhans in the rat pancreas. Darkly stained cells are B cells. Surrounding pancreatic acinar tissue is light-colored (× 400). (Courtesy of LL Bennett.)

STRUCTURE, BIOSYNTHESIS, & SECRETION OF INSULIN

STRUCTURE & SPECIES SPECIFICITY

Insulin is a polypeptide containing two chains of amino acids linked by disulfide bridges Minor differences occur in the amino acid composition of the molecule from species to species. The differences are generally not sufficient to affect the

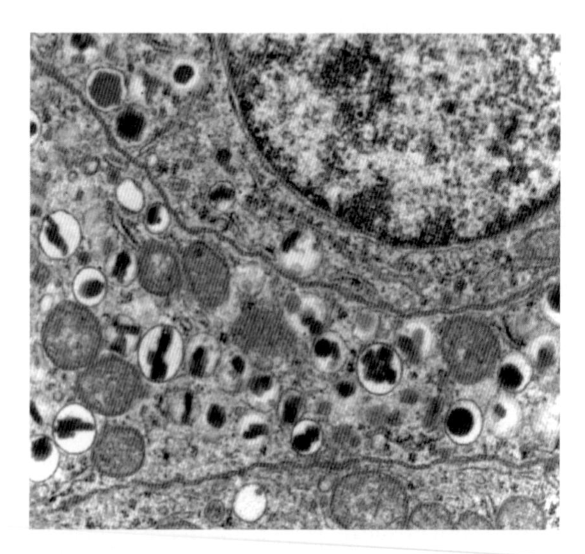

FIGURE 24–2 Electronmicrograph of two adjoining B cells in a human pancreatic islet. The B granules are the crystals in the membrane-lined vesicles. They vary in shape from rhombic to round (× 26,000). (Courtesy of A Like. Reproduced, with permission, from Fawcett DW: *Bloom and Fawcett, A Textbook of Histology,* 11th ed. Saunders, 1986.)

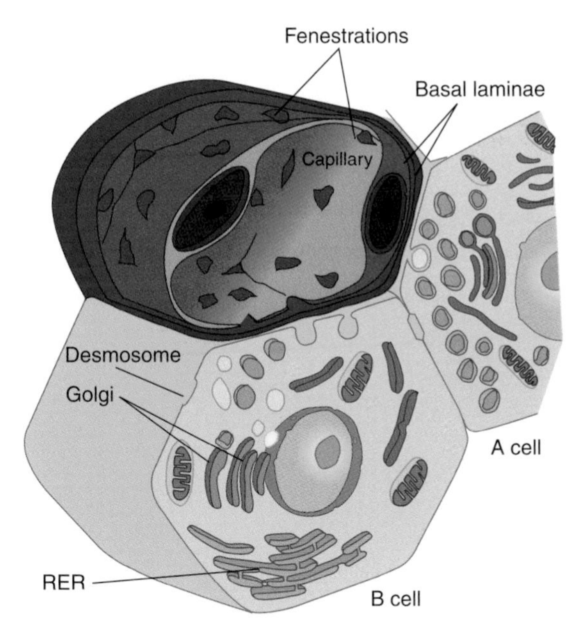

FIGURE 24–3 A and B cells, showing their relation to a blood vessel. RER, rough endoplasmic reticulum. Insulin from the B cell and glucagon from the A cell are secreted by exocytosis and cross the basal lamina of the cell and the basal lamina of the capillary before entering the lumen of the fenestrated capillary. (Reproduced with permission from Junqueira IC, Carneiro J: *Basic Histology: Text and Atlas,* 10th ed. McGraw-Hill, 2003.)

biologic activity of a particular insulin in heterologous species but are sufficient to make the insulin antigenic. If insulin of one species is injected for a prolonged period into another species, the anti-insulin antibodies formed inhibit the injected insulin. Almost all humans who have received commercial bovine insulin for more than 2 months have antibodies against bovine insulin, but the titer is usually low. Porcine insulin differs from human insulin by only one amino acid residue and has low antigenicity. Human insulin produced in bacteria by recombinant DNA technology is now widely used to avoid antibody formation.

BIOSYNTHESIS & SECRETION

Insulin is synthesized in the rough endoplasmic reticulum of the B cells (Figure 24–3). It is then transported to the Golgi apparatus, where it is packaged into membrane-bound granules. These granules move to the plasma membrane by a process involving microtubules, and their contents are expelled by exocytosis (see Chapters 2 and 16). The insulin then crosses the basal lamina of the B cell and a neighboring capillary and the fenestrated endothelium of the capillary to reach the bloodstream. The fenestrations are discussed in detail in Chapter 31.

Like other polypeptide hormones and related proteins that enter the endoplasmic reticulum, insulin is synthesized as part of a larger preprohormone (see Chapter 1). The gene for insulin is located on the short arm of chromosome 11 in humans. It has two introns and three exons. **Preproinsulin** originates from the endoplasmic reticulum. The remainder of the molecule is then folded, and the disulfide bonds are formed to make **proinsulin.** The peptide segment connecting the A and B chains, the **connecting peptide (C peptide),** facilitates the folding and then is detached in the granules before secretion. Two proteases are involved in processing the proinsulin. Normally, 90–97% of the product released from the B cells is insulin along with equimolar amounts of C peptide. The rest is mostly proinsulin. C peptide can be measured by radioimmunoassay, and its level in blood provides an index of B cell function in patients receiving exogenous insulin.

FATE OF SECRETED INSULIN

INSULIN & INSULIN-LIKE ACTIVITY IN BLOOD

Plasma contains a number of substances with insulin-like activity in addition to insulin. The activity that is not suppressed by anti-insulin antibodies has been called **nonsuppressible insulin-like activity (NSILA).** Most, if not all, of this activity persists after pancreatectomy and is due to the insulin-like growth factors **IGF-I** and **IGF-II** (see Chapter 18). These IGFs are polypeptides. Small amounts are free in the plasma (low-molecular-weight fraction), but large amounts are bound to proteins (high-molecular-weight fraction).

One may well ask why pancreatectomy causes diabetes mellitus when NSILA persists in the plasma. However, the insulin-like activities of IGF-I and IGF-II are weak compared to that of insulin and likely subserve other specific functions.

METABOLISM

The half-life of insulin in the circulation in humans is about 5 min. Insulin binds to insulin receptors, and some is internalized. It is destroyed by proteases in the endosomes formed by the endocytotic process.

EFFECTS OF INSULIN

The physiologic effects of insulin are far-reaching and complex. They are conveniently divided into rapid, intermediate, and delayed actions, as listed in Table 24–1. The best known is the hypoglycemic effect, but there are additional effects on amino acid and electrolyte transport, many enzymes, and growth. The net effect of the hormone is storage of carbohydrate, protein, and fat. Therefore, insulin is appropriately called the "hormone of abundance."

The actions of insulin on adipose tissue; skeletal, cardiac, and smooth muscle; and the liver are summarized in Table 24–2.

TABLE 24–1 Principal actions of insulin.

Rapid (seconds)
Increased transport of glucose, amino acids, and K^+ into insulin-sensitive cells
Intermediate (minutes)
Stimulation of protein synthesis
Inhibition of protein degradation
Activation of glycolytic enzymes and glycogen synthase
Inhibition of phosphorylase and gluconeogenic enzymes
Delayed (hours)
Increase in mRNAs for lipogenic and other enzymes

Courtesy of ID Goldfine.

TABLE 24–2 Effects of insulin on various tissues.

Adipose tissue
Increased glucose entry
Increased fatty acid synthesis
Increased glycerol phosphate synthesis
Increased triglyceride deposition
Activation of lipoprotein lipase
Inhibition of hormone-sensitive lipase
Increased K^+ uptake
Muscle
Increased glucose entry
Increased glycogen synthesis
Increased amino acid uptake
Increased protein synthesis in ribosomes
Decreased protein catabolism
Decreased release of gluconeogenic amino acids
Increased ketone uptake
Increased K^+ uptake
Liver
Decreased ketogenesis
Increased protein synthesis
Increased lipid synthesis
Decreased glucose output due to decreased gluconeogenesis, increased glycogen synthesis, and increased glycolysis
General
Increased cell growth

GLUCOSE TRANSPORTERS

Glucose enters cells by **facilitated diffusion** (see Chapter 1) or, in the intestine and kidneys, by secondary active transport with Na^+. In muscle, adipose, and some other tissues, insulin stimulates glucose entry into cells by increasing the number of glucose transporters (GLUTs) in the cell membranes.

The GLUTs that are responsible for facilitated diffusion of glucose across cell membranes are a family of closely related proteins that span the cell membrane 12 times and have their amino and carboxyl terminals inside the cell. They differ from and have no homology with the sodium-dependent glucose transporters, SGLT 1 and SGLT 2, responsible for the secondary active transport of glucose in the intestine (see Chapter 26) and renal tubules (see Chapter 38), although the SGLTs also have 12 transmembrane domains.

Seven different GLUTs, named GLUT 1–7 in order of discovery, have been characterized (Table 24–3). They contain 492–524 amino acid residues and their affinity for glucose varies. Each transporter appears to have evolved for special tasks. GLUT 4 is the transporter in muscle and adipose tissue that is stimulated by insulin. A pool of GLUT 4 molecules is maintained within vesicles in the cytoplasm of insulin-sensitive cells. When the insulin receptors of these cells are activated, the vesicles move rapidly to the cell membrane and fuse with it, inserting the transporters into the cell membrane (Figure 24–4). When insulin action ceases, the transporter-containing patches of membrane are endocytosed and the vesicles are ready for the next exposure to insulin. Activation of the insulin receptor brings about the movement of the vesicles to the cell membrane by activating phosphatidylinositol 3-kinase (Figure 24–4). Most of the other GLUT transporters that are not insulin-sensitive appear to be constitutively expressed in the cell membrane.

In the tissues in which insulin increases the number of GLUTs in cell membranes, the rate of phosphorylation of the glucose, once it has entered the cells, is regulated by other hormones. Growth hormone and cortisol both inhibit phosphorylation in certain tissues. Transport is normally so rapid that it is not a rate-limiting step in glucose metabolism. However, it is rate-limiting in B cells.

Insulin also increases the entry of glucose into liver cells, but it does not exert this effect by increasing the number of GLUT 4 transporters in the cell membranes. Instead, it induces glucokinase, and this increases the phosphorylation of glucose, so that the intracellular free glucose concentration stays low, facilitating the entry of glucose into the cell.

Insulin-sensitive tissues also contain a population of GLUT 4 vesicles that move into the cell membrane in response to exercise, a process that occurs independent of the action of insulin. This is why exercise lowers blood sugar. A 5'-AMP-activated kinase may tigger the insertion of these vesicles into the cell membrane.

INSULIN PREPARATIONS

The maximal decline in plasma glucose occurs 30 min after intravenous injection of insulin. After subcutaneous administration, the maximal fall occurs in 2–3 h. A wide

TABLE 24–3 Glucose transporters in mammals.

	Function	K_m (mM)[a]	Major Sites of Expression
Secondary active transport (Na^{+}-glucose cotransport)			
SGLT 1	Absorption of glucose	0.1–1.0	Small intestine, renal tubules
SGLT 2	Absorption of glucose	1.6	Renal tubules
Facilitated diffusion			
GLUT 1	Basal glucose uptake	1–2	Placenta, blood-brain barrier, brain, red cells, kidneys, colon, many other organs
GLUT 2	B-cell glucose sensor; transport out of intestinal and renal epithelial cells	12–20	B cells of islets, liver, epithelial cells of small intestine, kidneys
GLUT 3	Basal glucose uptake	<1	Brain, placenta, kidneys, many other organs
GLUT 4	Insulin-stimulated glucose uptake	5	Skeletal and cardiac muscle, adipose tissue, other tissues
GLUT 5	Fructose transport	1–2	Jejunum, sperm
GLUT 6	Unknown	—	Brain, spleen and leukocytes
GLUT 7	Glucose 6-phosphate transporter in endoplasmic reticulum	—	Liver

[a]The K_m is the glucose concentration at which transport is half-maximal.

Data from Stephens JM, Pilch PF: The metabolic regulation and vesicular transport of GLUT 4, the major insulin-responsive glucose transporter. Endocr Rev 1995;16:529.

variety of insulin preparations are now available commercially. These include insulins that have been complexed with protamine and other polypeptides to delay absorption and degradation, and synthetic insulins in which there have been changes in amino acid residues. In general, they fall into three categories: rapid, intermediate-acting, and long-acting (24–36 h).

RELATION TO POTASSIUM

Insulin causes K^+ to enter cells, with a resultant lowering of the extracellular K^+ concentration. Infusions of insulin and glucose significantly lower the plasma K^+ level in normal individuals and are very effective for the temporary relief of hyperkalemia in patients with renal failure. **Hypokalemia**

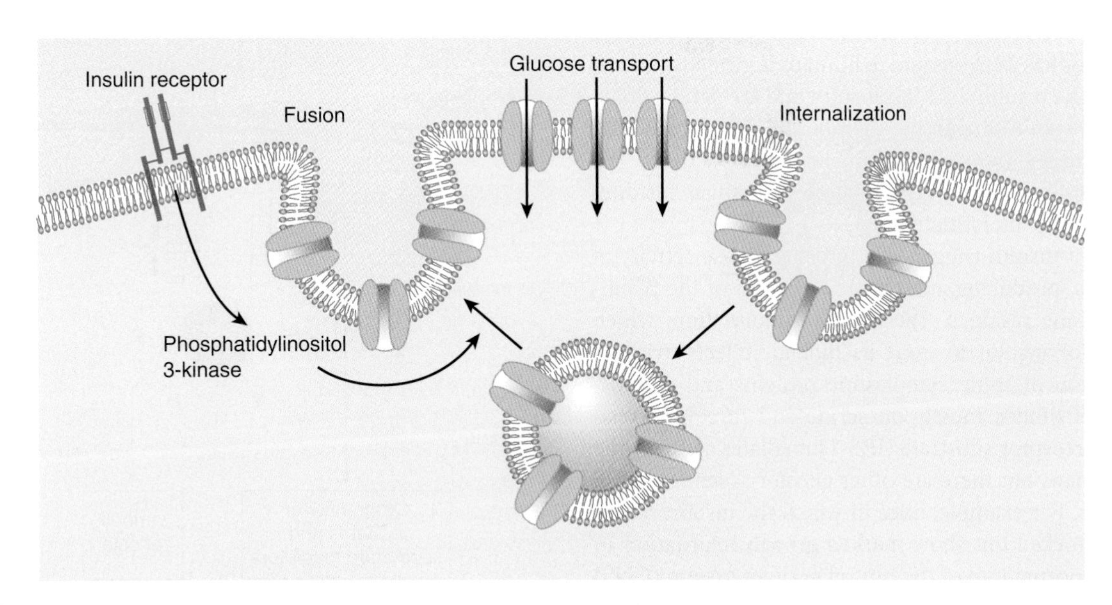

FIGURE 24–4 Cycling of GLUT 4 transporters through endosomes in insulin-sensitive tissues. Activation of the insulin receptor causes activation of phosphatidylinositol 3-kinase, which speeds translocation of the GLUT 4-containing endosomes into the cell membrane. The GLUT 4 transporters then mediate glucose transport into the cell.

often develops when patients with diabetic acidosis are treated with insulin. The reason for the intracellular migration of K^+ is still uncertain. However, insulin increases the activity of Na, K ATPase in cell membranes, so that more K^+ is pumped into cells.

OTHER ACTIONS

The hypoglycemic and other effects of insulin are summarized in temporal terms in Table 24–1, and the net effects on various tissues are summarized in Table 24–2. The action on glycogen synthase fosters glycogen storage, and the actions on glycolytic enzymes favor glucose metabolism to two carbon fragments (see Chapter 1), with resulting promotion of lipogenesis. Stimulation of protein synthesis from amino acids entering the cells and inhibition of protein degradation foster growth.

The anabolic effect of insulin is aided by the protein-sparing action of adequate intracellular glucose supplies. Failure to grow is a symptom of diabetes in children, and insulin stimulates the growth of immature hypophysectomized rats to almost the same degree as growth hormone.

MECHANISM OF ACTION

INSULIN RECEPTORS

Insulin receptors are found on many different cells in the body, including cells in which insulin does not increase glucose uptake.

The insulin receptor, which has a molecular weight of approximately 340,000, is a tetramer made up of two α and two β glycoprotein subunits (Figure 24–5). All these are synthesized on a single mRNA and then proteolytically separated and bound to each other by disulfide bonds. The gene for the insulin receptor has 22 exons and in humans is located on chromosome 19. The α subunits bind insulin and are extracellular, whereas the β subunits span the membrane. The intracellular portions of the β subunits have tyrosine kinase activity. The α and β subunits are both glycosylated, with sugar residues extending into the interstitial fluid.

Binding of insulin triggers the tyrosine kinase activity of the β subunits, producing autophosphorylation of the β subunits on tyrosine residues. The autophosphorylation, which is necessary for insulin to exert its biologic effects, triggers phosphorylation of some cytoplasmic proteins and dephosphorylation of others, mostly on serine and threonine residues. Insulin receptor substrate (IRS-1) mediates some of the effects in humans but there are other effector systems as well (Figure 24–6). For example, mice in which the insulin receptor gene is knocked out show marked growth retardation in utero, have abnormalities of the central nervous system (CNS) and skin, and die at birth of respiratory failure, whereas IRS-1 knockouts show only moderate growth retardation in utero, survive, and are insulin-resistant but otherwise nearly normal.

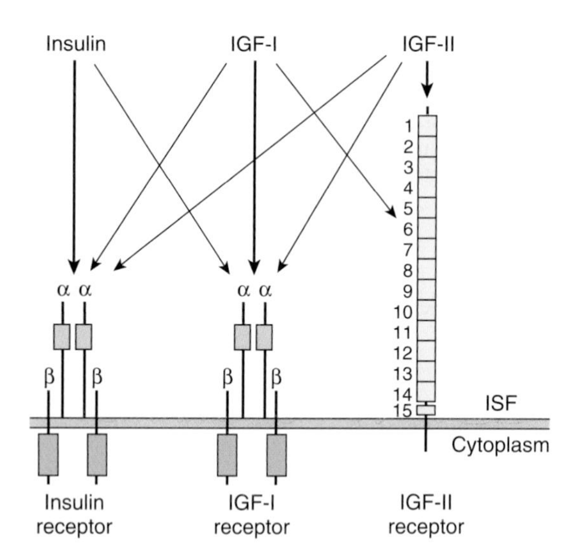

FIGURE 24–5 Insulin, IGF-I, and IGF-II receptors. Each hormone binds primarily to its own receptor, but insulin also binds to the IGF-I receptor, and IGF-I and IGF-II bind to all three. The purple boxes are intracellular tyrosine kinase domains. Note the marked similarity between the insulin receptor and the IGF-I receptor; also note the 15 repeat sequences in the extracellular portion of the IGF-II receptor. ISF, interstitial fluid.

The growth-promoting protein anabolic effects of insulin are mediated via **phosphatidylinositol 3-kinase (PI3K),** and evidence indicates that in invertebrates, this pathway is involved in the growth of nerve cells and axon guidance in the visual system.

It is interesting to compare the insulin receptor with other related receptors. The insulin receptor is very similar to the receptor for IGF-I but different from the receptor for IGF-II

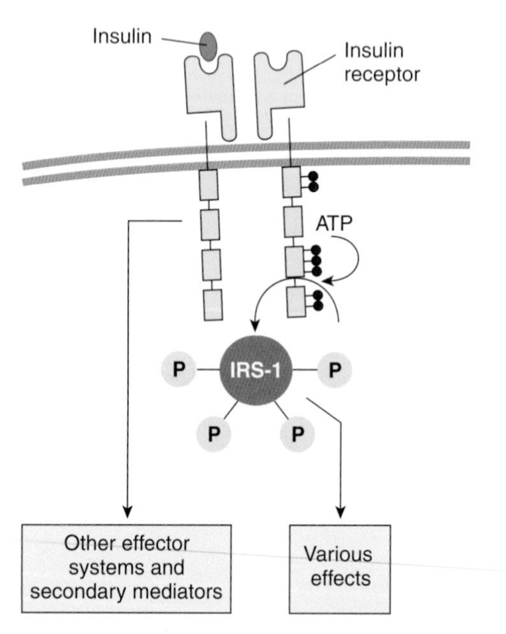

FIGURE 24–6 Intracellular responses triggered by insulin binding to the insulin receptor. Red circles and circles labeled P represent phosphate groups. IRS-1, insulin receptor substrate-1.

(Figure 24–5). Other receptors for growth factors and receptors for various oncogenes also are tyrosine kinases. However, the amino acid composition of these receptors is quite different.

When insulin binds to its receptors, they aggregate in patches and are taken up into the cell by receptor-mediated endocytosis (see Chapter 2). Eventually, the insulin–receptor complexes enter lysosomes, where the receptors are broken down or recycled. The half-life of the insulin receptor is about 7 h.

CONSEQUENCES OF INSULIN DEFICIENCY

The far-reaching physiologic effects of insulin are highlighted by a consideration of the extensive and serious consequences of insulin deficiency (Clinical Box 24–1).

In humans, insulin deficiency is a common pathologic condition. In animals, it can be produced by pancreatectomy; by administration of alloxan, streptozocin, or other toxins that in appropriate doses cause selective destruction of the B cells of the pancreatic islets; by administration of drugs that inhibit insulin secretion; and by administration of anti-insulin antibodies. Strains of mice, rats, hamsters, guinea pigs, miniature swine, and monkeys that have a high incidence of spontaneous diabetes mellitus have also been described.

GLUCOSE TOLERANCE

In diabetes, glucose piles up in the bloodstream, especially after meals. If a glucose load is given to a diabetic, the plasma glucose rises higher and returns to the baseline more slowly than it does in normal individuals. The response to a standard oral test dose of glucose, the **oral glucose tolerance test,** is used in the clinical diagnosis of diabetes (Figure 24–7).

Impaired glucose tolerance in diabetes is due in part to reduced entry of glucose into cells **(decreased peripheral utilization).** In the absence of insulin, the entry of glucose into skeletal, cardiac, and smooth muscle and other tissues is decreased (Figure 24–8). Glucose uptake by the liver is also reduced, but the effect is indirect. Intestinal absorption of glucose is unaffected, as is its reabsorption from the urine by the cells of the proximal tubules of the kidneys. Glucose uptake by most of the brain and the red blood cells is also normal.

The second and the major cause of hyperglycemia in diabetes is derangement of the glucostatic function of the liver (see Chapter 28). The liver takes up glucose from the bloodstream and stores it as glycogen, but because the liver contains glucose 6-phosphatase it also discharges glucose into the bloodstream. Insulin facilitates glycogen synthesis and inhibits hepatic glucose output. When the plasma glucose is high, insulin secretion is normally increased and hepatic glucogenesis is decreased. This response does not occur in type 1 diabetes (as insulin is absent) and in type 2 diabetes (as tissues are insulin-resistant). Glucagon can contribute to hyperglycemia

CLINICAL BOX 24–1

Diabetes Mellitus

The constellation of abnormalities caused by insulin deficiency is called **diabetes mellitus.** Greek and Roman physicians used the term "diabetes" to refer to conditions in which the cardinal finding was a large urine volume, and two types were distinguished: "diabetes mellitus," in which the urine tasted sweet; and "diabetes insipidus," in which the urine had little taste. Today, the term "diabetes insipidus" is reserved for conditions in which there is a deficiency of the production or action of vasopressin (see Chapter 38), and the unmodified word "diabetes" is generally used as a synonym for diabetes mellitus.

The cause of clinical diabetes is always a deficiency of the effects of insulin at the tissue level. **Type 1 diabetes,** or **insulin-dependent diabetes mellitus (IDDM),** is due to insulin deficiency caused by autoimmune destruction of the B cells in the pancreatic islets, and it accounts for 3–5% of cases and usually presents in children. **Type 2 diabetes, or non-insulin-dependent diabetes mellitus (NIDDM),** is characterized by the dysregulation of insulin release from the B cells, along with insulin resistance in peripheral tissues such as skeletal muscle, brain, and liver. Type 2 diabetes historically presented in overweight or obese adults, although it is increasingly being diagnosed in children as childhood obesity increases.

Diabetes is characterized by polyuria (passage of large volumes of urine), polydipsia (excessive drinking), weight loss in spite of polyphagia (increased appetite), hyperglycemia, glycosuria, ketosis, acidosis, and coma. Widespread biochemical abnormalities are present, but the fundamental defects to which most of the abnormalities can be traced are (1) reduced entry of glucose into various "peripheral" tissues and (2) increased liberation of glucose into the circulation from the liver. Therefore there is an extracellular glucose excess and, in many cells, an intracellular glucose deficiency—a situation that has been called "starvation in the midst of plenty." Also, the entry of amino acids into muscle is decreased and lipolysis is increased.

THERAPEUTIC HIGHLIGHTS

In type 1 diabetes, the mainstay of therapy is provision of exogenous insulin, carefully titrated to dietary intake of glucose. In type 2 diabetes, lifestyle changes such as alterations in the diet or increased exercise can often delay symptoms in early disease, but these are difficult to secure. Insulin-sensitizing drugs represent second-line agents (see Chapter 16).

Accumulation of lactate in the blood **(lactic acidosis)** may also complicate diabetic ketoacidosis if the tissues become hypoxic, and lactic acidosis may itself cause coma. Brain edema occurs in about 1% of children with ketoacidosis, and it can cause coma. Its cause is unsettled, but it is a serious complication, with a mortality rate of about 25%.

CHOLESTEROL METABOLISM

In diabetes, the plasma cholesterol level is usually elevated and this plays a role in the accelerated development of the atherosclerotic vascular disease that is a major long-term complication of diabetes in humans. The rise in plasma cholesterol level is due to an increase in the plasma concentration of very low-density lipoprotein (VLDL) and low-density lipoprotein (LDL) (see Chapter 1). These in turn may be due to increased hepatic production of VLDL or decreased removal of VLDL and LDL from the circulation.

SUMMARY

Because of the complexities of the metabolic abnormalities in diabetes, a summary is in order. One of the key features of insulin deficiency (Figure 24–9) is decreased entry of glucose into many tissues (decreased peripheral utilization). Also, the net release of glucose from the liver is increased (increased production), due in part to glucagon excess. The resultant hyperglycemia leads to glycosuria and a dehydrating osmotic diuresis. Dehydration leads to polydipsia. In the face of intracellular glucose deficiency, appetite is stimulated, glucose is formed from protein (gluconeogenesis), and energy supplies are maintained by metabolism of proteins and fats. Weight loss, debilitating protein deficiency, and inanition are the result.

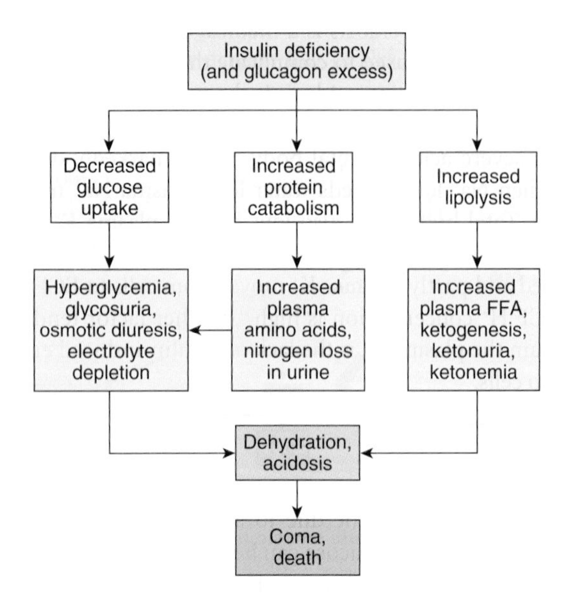

FIGURE 24–9 Effects of insulin deficiency. (Courtesy of RJ Havel.)

Fat catabolism is increased and the system is flooded with triglycerides and FFA. Fat synthesis is inhibited and the overloaded catabolic pathways cannot handle the excess acetyl-CoA that is formed. In the liver, the acetyl-CoA is converted to ketone bodies. Two of these are organic acids, and metabolic acidosis develops as ketones accumulate. Na^+ and K^+ depletion is added to the acidosis because these plasma cations are excreted with the organic anions not covered by the H^+ and NH_4^+ secreted by the kidneys. Finally, the acidotic, hypovolemic, hypotensive, depleted animal or patient becomes comatose because of the toxic effects of acidosis, dehydration, and hyperosmolarity on the nervous system and dies if treatment is not instituted.

All of these abnormalities are corrected by administration of insulin. Although emergency treatment of acidosis also includes administration of alkali to combat the acidosis and parenteral water, Na^+, and K^+ to replenish body stores, only insulin repairs the fundamental defects in a way that permits a return to normal.

INSULIN EXCESS

SYMPTOMS

All the known consequences of insulin excess are manifestations, directly or indirectly, of the effects of hypoglycemia on the nervous system. Except in individuals who have been fasting for some time, glucose is the only fuel used in appreciable quantities by the brain. The carbohydrate reserves in neural tissue are very limited and normal function depends on a continuous glucose supply. As the plasma glucose level falls, the first symptoms are palpitations, sweating, and nervousness due to autonomic discharge. These appear at plasma glucose values slightly lower than the value at which autonomic activation first begins, because the threshold for symptoms is slightly above the threshold for initial activation. At lower plasma glucose levels, so-called **neuroglycopenic symptoms** begin to appear. These include hunger as well as confusion and the other cognitive abnormalities. At even lower plasma glucose levels, lethargy, coma, convulsions, and eventually death occur. Obviously, the onset of hypoglycemic symptoms calls for prompt treatment with glucose or glucose-containing drinks such as orange juice. Although a dramatic disappearance of symptoms is the usual response, abnormalities ranging from intellectual dulling to coma may persist if the hypoglycemia was severe or prolonged.

COMPENSATORY MECHANISMS

One important compensation for hypoglycemia is cessation of the secretion of endogenous insulin. Inhibition of insulin secretion is complete at a plasma glucose level of about 80 mg/dL (Figure 24–10). In addition, hypoglycemia triggers increased secretion of at least four counter-regulatory

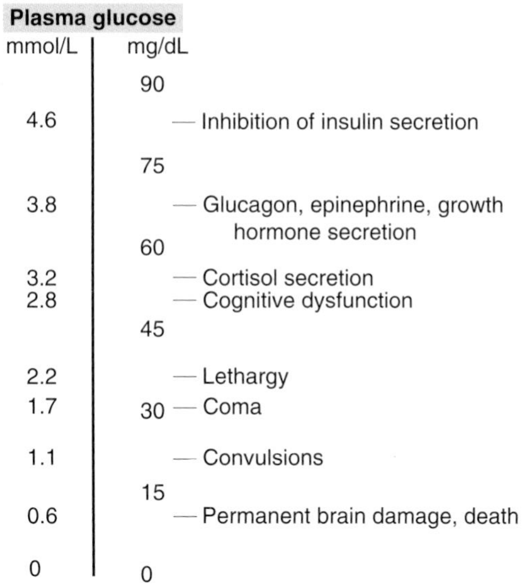

Plasma glucose mmol/L	mg/dL	
	90	
4.6		Inhibition of insulin secretion
	75	
3.8		Glucagon, epinephrine, growth hormone secretion
	60	
3.2		Cortisol secretion
2.8		Cognitive dysfunction
	45	
2.2		Lethargy
1.7	30	Coma
1.1		Convulsions
	15	
0.6		Permanent brain damage, death
0	0	

FIGURE 24–10 Plasma glucose levels at which various effects of hypoglycemia appear.

hormones: glucagon, epinephrine, growth hormone, and cortisol. The epinephrine response is reduced during sleep. Glucagon and epinephrine increase the hepatic output of glucose by increasing glycogenolysis. Growth hormone decreases the utilization of glucose in various peripheral tissues, and cortisol has a similar action. The keys to counter-regulation appear to be epinephrine and glucagon: if the plasma concentration of either increases, the decline in the plasma glucose level is reversed; but if both fail to increase, there is little if any compensatory rise in the plasma glucose level. The actions of the other hormones are supplementary.

Note that the autonomic discharge and release of counter-regulatory hormones normally occurs at a higher plasma glucose level than the cognitive deficits and other more serious CNS changes (Figure 24–10). For diabetics treated with insulin, the symptoms caused by the autonomic discharge serve as a warning to seek glucose replacement. However, particularly in long-term diabetics who have been tightly regulated, the autonomic symptoms may not occur, and the resulting **hypoglycemia unawareness** can be a clinical problem of some magnitude.

REGULATION OF INSULIN SECRETION

The normal concentration of insulin measured by radioimmunoassay in the peripheral venous plasma of fasting normal humans is 0–70 μU/mL (0–502 pmol/L). The amount of insulin secreted in the basal state is about 1 U/h, with a fivefold to 10-fold increase following ingestion of food. Therefore, the average amount secreted per day in a normal human is about 40 U (287 nmol).

TABLE 24–4 Factors affecting insulin secretion.

Stimulators	Inhibitors
Glucose	Somatostatin
Mannose	2-Deoxyglucose
Amino acids (leucine, arginine, others)	Mannoheptulose
Intestinal hormones (GIP, GLP-1 [7–36], gastrin, secretin, CCK; others?)	α-Adrenergic stimulators (norepinephrine, epinephrine)
β-Keto acids	β-Adrenergic blockers (propranolol)
Acetylcholine	
Glucagon	Galanin
Cyclic AMP and various cAMP-generating substances	Diazoxide
	Thiazide diuretics
β-Adrenergic stimulators	K^+ depletion
Theophylline	Phenytoin
Sulfonylureas	Alloxan
	Microtubule inhibitors
	Insulin

Factors that stimulate and inhibit insulin secretion are summarized in Table 24–4.

EFFECTS OF THE PLASMA GLUCOSE LEVEL

It has been known for many years that glucose acts directly on pancreatic B cells to increase insulin secretion. The response to glucose is biphasic; there is a rapid but short-lived increase in secretion followed by a more slowly developing prolonged increase.

Glucose enters the B cells via GLUT 2 transporters and is phosphorylated by glucokinase then metabolized to pyruvate in the cytoplasm (Figure 24–11). The pyruvate enters the mitochondria and is metabolized to CO_2 and H_2O via the citric acid cycle with the formation of ATP by oxidative phosphorylation. The ATP enters the cytoplasm, where it inhibits ATP-sensitive K^+ channels, reducing K^+ efflux. This depolarizes the B cell, and Ca^{2+} enters the cell via voltage-gated Ca^{2+} channels. The Ca^{2+} influx causes exocytosis of a readily releasable pool of insulin-containing secretory granules, producing the initial spike of insulin secretion.

Metabolism of pyruvate via the citric acid cycle also causes an increase in intracellular glutamate. The glutamate appears to act on a second pool of secretory granules, committing them to the releasable form. The action of glutamate may be to decrease the pH in the secretory granules, a necessary step in their maturation. The release of these granules then produces the prolonged second phase of the insulin response to glucose. Thus, glutamate appears to act as an intracellular second messenger that primes secretory granules for secretion.

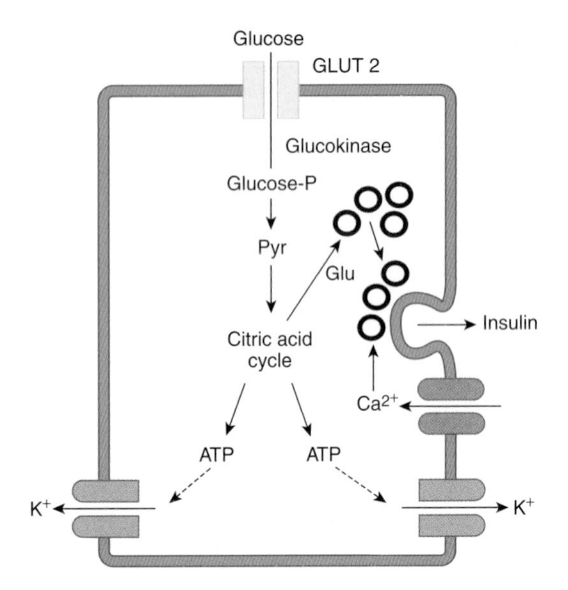

FIGURE 24–11 Insulin secretion. Glucose enters B cells by GLUT 2 transporters. It is phosphorylated and metabolized to pyruvate (Pyr) in the cytoplasm. The Pyr enters the mitochondria and is metabolized via the citric acid cycle. The ATP formed by oxidative phosphorylation inhibits ATP-sensitive K^+ channels, reducing K^+ efflux. This depolarizes the B cell, and Ca^{2+} influx is increased. The Ca^{2+} stimulates release of insulin by exocytosis. Glutamate (Glu) is also formed, and this primes secretory granules, preparing them for exocytosis.

The feedback control of plasma glucose on insulin secretion normally operates with great precision so that plasma glucose and insulin levels parallel each other with remarkable consistency.

PROTEIN & FAT DERIVATIVES

Insulin stimulates the incorporation of amino acids into proteins and combats the fat catabolism that produces the β-keto acids. Therefore, it is not surprising that arginine, leucine, and certain other amino acids stimulate insulin secretion, as do β–keto acids such as acetoacetate. Like glucose, these compounds generate ATP when metabolized, and this closes ATP-sensitive K^+ channels in the B cells. In addition, L-arginine is the precursor of NO, and NO stimulates insulin secretion.

ORAL HYPOGLYCEMIC AGENTS

Tolbutamide and other sulfonylurea derivatives such as acetohexamide, tolazamide, glipizide, and glyburide are orally active hypoglycemic agents that lower blood glucose by increasing the secretion of insulin. They only work in patients with some remaining B cells and are ineffective after pancreatectomy or in type 1 diabetes. They bind to the ATP-inhibited K^+ channels in the B cell membranes and inhibit channel activity, depolarizing the B cell membrane and increasing Ca^{2+} influx and hence insulin release, independent of increases in plasma glucose.

Persistent hyperinsulinemic hypoglycemia of infancy is a condition in which plasma insulin is elevated despite the hypoglycemia. The condition is caused by mutations in the genes for various enzymes in B cells that decrease K^+ efflux via the ATP-sensitive K^+ channels. Treatment consists of administration of diazoxide, a drug that increases the activity of the K^+ channels or, in more severe cases, subtotal pancreatectomy.

The biguanide metformin is an oral hypoglycemic agent that acts in the absence of insulin. Metformin acts primarily by reducing gluconeogenesis and therefore decreasing hepatic glucose output. It is sometimes combined with a sulfonylurea in the treatment of type 2 diabetes. Metformin can cause lactic acidosis, but the incidence is usually low.

Troglitazone (Rezulin) and related **thiazolidinediones** are also used in the treatment of diabetes because they increase insulin-mediated peripheral glucose disposal, thus reducing insulin resistance. They bind to and activate peroxisome proliferator-activated receptor γ (PPARγ) in the nucleus of cells. Activation of this receptor, which is a member of the superfamily of hormone-sensitive nuclear transcription factors, has a unique ability to normalize a variety of metabolic functions.

CYCLIC AMP & INSULIN SECRETION

Stimuli that increase cAMP levels in B cells increase insulin secretion, including β-adrenergic agonists, glucagon, and phosphodiesterase inhibitors such as theophylline.

Catecholamines have a dual effect on insulin secretion; they inhibit insulin secretion via α_2-adrenergic receptors and stimulate insulin secretion via β-adrenergic receptors. The net effect of epinephrine and norepinephrine is usually inhibition. However, if catecholamines are infused after administration of α-adrenergic blocking drugs, the inhibition is converted to stimulation.

EFFECT OF AUTONOMIC NERVES

Branches of the right vagus nerve innervate the pancreatic islets, and stimulation of this parasympathetic pathway causes increased insulin secretion via M_4 receptors (see Table 7–2). Atropine blocks the response and acetylcholine stimulates insulin secretion. The effect of acetylcholine, like that of glucose, is due to increased cytoplasmic Ca^{2+}, but acetylcholine activates phospholipase C, with the released IP_3 releasing the Ca^{2+} from the endoplasmic reticulum.

Stimulation of the sympathetic nerves to the pancreas inhibits insulin secretion. The inhibition is produced by released norepinephrine acting on α_2-adrenergic receptors. However, if α-adrenergic receptors are blocked, stimulation of the sympathetic nerves causes increased insulin secretion mediated by β_2-adrenergic receptors. The polypeptide galanin is found in some of the autonomic nerves innervating the islets, and galanin inhibits insulin secretion by activating the K^+ channels that are inhibited by ATP. Thus, although the denervated pancreas responds to glucose, the autonomic innervation of the pancreas is involved in the overall regulation of insulin secretion (Clinical Box 24–3).

INTESTINAL HORMONES

Orally administered glucose exerts a greater insulin-stimulating effect than intravenously administered glucose, and orally administered amino acids also produce a greater insulin response than intravenous amino acids. These observations led to exploration of the possibility that a substance secreted by the gastrointestinal mucosa stimulated insulin secretion. Glucagon, glucagon derivatives, secretin, cholecystokinin (CCK), gastrin, and gastric inhibitory peptide (GIP) all have such an action (see Chapter 25), and CCK potentiates the insulin-stimulating effects of amino acids. However, GIP is the only one of these peptides that produces stimulation when administered in doses that reflect blood GIP levels produced by an oral glucose load.

Recently, attention has focused on glucagon-like polypeptide 1 (7–36) (GLP-1 [7–36]) as an additional gut factor that stimulates insulin secretion. This polypeptide is a product of preproglucagon. B cells have GLP-1 (7–36) receptors as well as GIP receptors, and GLP-1 (7–36) is a more potent insulinotropic hormone than GIP. GIP and GLP-1 (7–36) both appear to act by increasing Ca^{2+} influx through voltage-gated Ca^{2+} channels.

The possible roles of pancreatic somatostatin and glucagon in the regulation of insulin secretion are discussed below.

CLINICAL BOX 24–3

Effects of K^+ Depletion

K^+ depletion decreases insulin secretion, and K^+-depleted patients, for example, patients with primary hyperaldosteronism (see Chapter 20), develop diabetic glucose tolerance curves. These curves are restored to normal by K^+ repletion.

THERAPEUTIC HIGHLIGHTS

The thiazide diuretics, which cause loss of K^+ as well as Na^+ in the urine (see Chapter 37), decrease glucose tolerance and make diabetes worse. They apparently exert this effect primarily because of their K^+-depleting effects, although some of them also cause pancreatic islet cell damage. Potassium-sparing diuretics, such as amiloride, should be substituted in the diabetic patient who needs such treatment.

LONG-TERM CHANGES IN B CELL RESPONSES

The magnitude of the insulin response to a given stimulus is determined in part by the secretory history of the B cells. Individuals fed a high-carbohydrate diet for several weeks not only have higher fasting plasma insulin levels but also show a greater secretory response to a glucose load than individuals fed an isocaloric low-carbohydrate diet.

Although B cells respond to stimulation with hypertrophy like other endocrine cells, they become exhausted and stop secreting **(B cell exhaustion)** when the stimulation is marked or prolonged. The pancreatic reserve is large and it is difficult to produce B cell exhaustion in normal animals, but if the pancreatic reserve is reduced by partial pancreatectomy, exhaustion of the remaining B cells can be initiated by any procedure that chronically raises the plasma glucose level. For example, diabetes can be produced in animals with limited pancreatic reserves by anterior pituitary extracts, growth hormone, thyroid hormones, or the prolonged continuous infusion of glucose alone. The diabetes precipitated by hormones in animals is at first reversible, but with prolonged treatment it becomes permanent. The transient diabetes is usually named for the agent producing it, for example, "hypophysial diabetes" or "thyroid diabetes." Permanent diabetes persisting after treatment has been discontinued is indicated by the prefix meta-, for example, **"metahypophysial diabetes"** or **"metathyroid diabetes."** When insulin is administered along with the diabetogenic hormones, the B cells are protected, probably because the plasma glucose is lowered, and diabetes does not develop.

It is interesting in this regard that genetic factors may be involved in the control of B cell reserve. In mice in which the gene for IRS-1 has been knocked out (see above), a robust compensatory B cell response occurs. However, in IRS-2 knockouts, the compensation is reduced and a more severe diabetic phenotype is produced.

GLUCAGON

CHEMISTRY

Human glucagon, a linear polypeptide with a molecular weight of 3485, is produced by the A cells of the pancreatic islets and the upper gastrointestinal tract. It contains 29 amino acid residues. All mammalian glucagons appear to have the same structure. Human preproglucagon (Figure 24–12) is a 179-amino-acid protein that is found in pancreatic A cells, in L cells in the lower gastrointestinal tract, and in the brain. It is the product of a single mRNA, but it is processed differently in different tissues. In A cells, it is processed primarily to glucagon and **major proglucagon fragment (MPGF).** In L cells, it is processed primarily to **glicentin,** a polypeptide that consists of glucagon extended by additional amino acid residues at either end, plus **glucagon-like polypeptides 1 and 2 (GLP-1 and GLP-2).** Some **oxyntomodulin** is also formed, and in

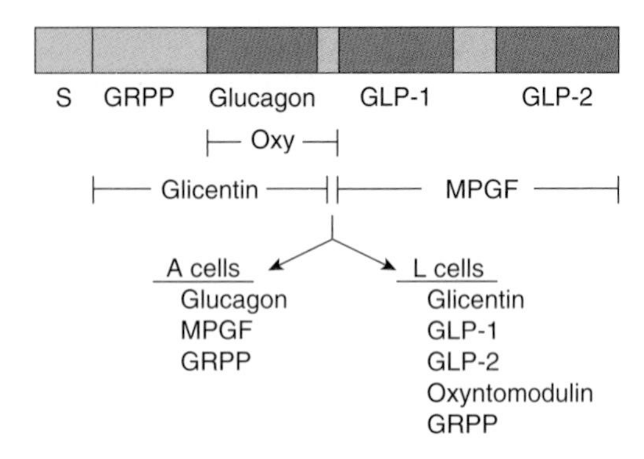

FIGURE 24–12 Posttranslational processing of preproglucagon in A and L cells. S, signal peptide; GRPP, glicentin-related polypeptide; GLP, glucagon-like polypeptide; Oxy, oxyntomodulin; MPGF, major proglucagon fragment. (Modified from Drucker DJ: Glucagon and glucagon-like peptides. Pancreas 1990;5:484.)

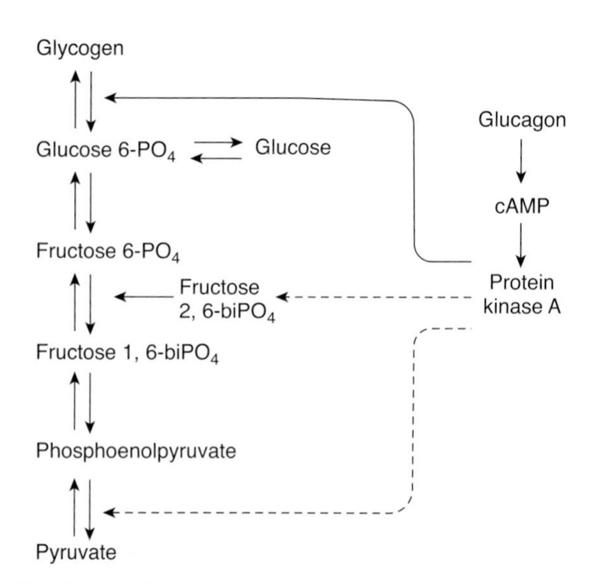

FIGURE 24–13 Mechanisms by which glucagon increases glucose output from the liver. Solid arrows indicate facilitation; dashed arrows indicate inhibition.

both A and L cells, residual **glicentin-related polypeptide (GRPP)** is left. Glicentin has some glucagon activity. GLP-1 and GLP-2 have no definite biologic activity by themselves. However, GLP-1 is processed further by removal of its amino-terminal amino acid residues and the product, **GLP-1 (7–36),** is a potent stimulator of insulin secretion that also increases glucose utilization (see above). GLP-1 and GLP-2 are also produced in the brain. The function of GLP-1 in this location is uncertain, but GLP-2 appears to be the mediator in a pathway from the nucleus tractus solitarius (NTS) to the dorsomedial nuclei of the hypothalamus, and injection of GLP-2 lowers food intake. Oxyntomodulin inhibits gastric acid secretion, though its physiologic role is unsettled, and GRPP does not have any established physiologic effects.

ACTION

Glucagon is glycogenolytic, gluconeogenic, lipolytic, and ketogenic. It acts on G-protein coupled receptors with a molecular weight of about 190,000. In the liver, it acts via G_s to activate adenylyl cyclase and increase intracellular cAMP. This leads via protein kinase A to activation of phosphorylase and therefore to increased breakdown of glycogen and an increase in plasma glucose. However, glucagon acts on different glucagon receptors located on the same hepatic cells to activate phospholipase C, and the resulting increase in cytoplasmic Ca^{2+} also stimulates glycogenolysis. Protein kinase A also decreases the metabolism of glucose 6-phosphate (Figure 24–13) by inhibiting the conversion of phosphoenolpyruvate to pyruvate. It also decreases the concentration of fructose 2,6-diphosphate and this in turn inhibits the conversion of fructose 6-phosphate to fructose 1,6-diphosphate. The resultant buildup of glucose 6-phosphate leads to increased glucose synthesis and release.

Glucagon does not cause glycogenolysis in muscle. It increases gluconeogenesis from available amino acids in the liver and elevates the metabolic rate. It increases ketone body formation by decreasing malonyl-CoA levels in the liver. Its lipolytic activity, which leads in turn to increased ketogenesis, is discussed in Chapter 1. The calorigenic action of glucagon is not due to the hyperglycemia per se but probably to the increased hepatic deamination of amino acids.

Large doses of exogenous glucagon exert a positive inotropic effect on the heart (see Chapter 30) without producing increased myocardial excitability, presumably because they increase myocardial cAMP. Use of this hormone in the treatment of heart disease has been advocated, but there is no evidence for a physiologic role of glucagon in the regulation of cardiac function. Glucagon also stimulates the secretion of growth hormone, insulin, and pancreatic somatostatin.

METABOLISM

Glucagon has a half-life in the circulation of 5–10 min. It is degraded by many tissues but particularly by the liver. Because glucagon is secreted into the portal vein and reaches the liver before it reaches the peripheral circulation, peripheral blood levels are relatively low. The rise in peripheral blood glucagon levels produced by excitatory stimuli is exaggerated in patients with cirrhosis, presumably because of decreased hepatic degradation of the hormone.

REGULATION OF SECRETION

The principal factors known to affect glucagon secretion are summarized in Table 24–5. Secretion is increased by hypoglycemia and decreased by a rise in plasma glucose. Pancreatic B cells contain GABA, and evidence suggests that coincident with the increased insulin secretion produced by hyperglycemia, GABA is released and acts on the A cells to inhibit glucagon secretion by activating $GABA_A$ receptors. The $GABA_A$

TABLE 24–5 Factors affecting glucagon secretion.

Stimulators	Inhibitors
Amino acids (particularly the glucogenic amino acids: alanine, serine, glycine, cysteine, and threonine)	Glucose
CCK, gastrin	Somatostatin
Cortisol	Secretin
Exercise	FFA
Infections	Ketones
Other stresses	Insulin
β-Adrenergic stimulators	Phenytoin
Theophylline	α-Adrenergic stimulators
Acetylcholine	GABA

receptors are Cl^- channels, and the resulting Cl^- influx hyperpolarizes the A cells.

Secretion is also increased by stimulation of the sympathetic nerves to the pancreas, and this sympathetic effect is mediated via β-adrenergic receptors and cAMP. It appears that the A cells are like the B cells in that stimulation of β-adrenergic receptors increases secretion and stimulation of α-adrenergic receptors inhibits secretion. However, the pancreatic response to sympathetic stimulation in the absence of blocking drugs is increased secretion of glucagon, so the effect of β-receptors predominates in the glucagon-secreting cells. The stimulatory effects of various stresses and possibly of exercise and infection are mediated at least in part via the sympathetic nervous system. Vagal stimulation also increases glucagon secretion.

A protein meal and infusion of various amino acids increase glucagon secretion. It seems appropriate that the glucogenic amino acids are particularly potent in this regard, since these are the amino acids that are converted to glucose in the liver under the influence of glucagon. The increase in glucagon secretion following a protein meal is also valuable, since the amino acids stimulate insulin secretion and the secreted glucagon prevents the development of hypoglycemia while the insulin promotes storage of the absorbed carbohydrates and lipids. Glucagon secretion increases during starvation. It reaches a peak on the third day of a fast, at the time of maximal gluconeogenesis. Thereafter, the plasma glucagon level declines as fatty acids and ketones become the major sources of energy.

During exercise, there is an increase in glucose utilization that is balanced by an increase in glucose production caused by an increase in circulating glucagon levels.

The glucagon response to oral administration of amino acids is greater than the response to intravenous infusion of amino acids, suggesting that a glucagon-stimulating factor is secreted from the gastrointestinal mucosa. CCK and gastrin increase glucagon secretion, whereas secretin inhibits it. Because CCK and gastrin secretion are both increased by a protein meal, either hormone could be the gastrointestinal mediator of the glucagon response. The inhibition produced by somatostatin is discussed below.

Glucagon secretion is also inhibited by FFA and ketones. However, this inhibition can be overridden, since plasma glucagon levels are high in diabetic ketoacidosis.

INSULIN–GLUCAGON MOLAR RATIOS

As noted previously, insulin is glycogenic, antigluconeogenetic, antilipolytic, and antiketotic in its actions. It thus favors storage of absorbed nutrients and is a "hormone of energy storage." Glucagon, on the other hand, is glycogenolytic, gluconeogenetic, lipolytic, and ketogenic. It mobilizes energy stores and is a "hormone of energy release." Because of their opposite effects, the blood levels of both hormones must be considered in any given situation. It is convenient to think in terms of the molar ratios of these hormones.

The insulin–glucagon molar ratios fluctuate markedly because the secretion of glucagon and insulin are both modified by the conditions that preceded the application of any given stimulus (Table 24–6). Thus, for example, the insulin–glucagon molar ratio on a balanced diet is approximately 2.3. An infusion of arginine increases the secretion of both hormones and raises the ratio to 3.0. After 3 days of starvation, the ratio falls to 0.4, and an infusion of arginine in this state lowers the ratio to 0.3. Conversely, the ratio is 25 in individuals receiving a constant infusion of glucose and rises to 170 on ingestion of a protein meal during the infusion (Table 24–6). The rise occurs because insulin secretion rises sharply, while

TABLE 24–6 Insulin-glucagon molar ratios (I/G) in blood in various conditions.

Condition	Hepatic Glucose Storage (S) or Production (P)[a]	I/G
Glucose availability		
Large carbohydrate meal	4+ (S)	70
Intravenous glucose	2+ (S)	25
Small meal	1+ (S)	7
Glucose need		
Overnight fast	1+ (P)	2.3
Low-carbohydrate diet	2+ (P)	1.8
Starvation	4+ (P)	0.4

[a]1+ to 4+ indicate relative magnitude.

Courtesy of RH Unger.

CLINICAL BOX 24–4

Macrosomia & GLUT 1 Deficiency

Infants born to diabetic mothers often have high birth weights and large organs (**macrosomia**). This condition is caused by excess circulating insulin in the fetus, which in turn is caused in part by stimulation of the fetal pancreas by high blood glucose and amino acids from the diabetic mother.

Free insulin in maternal blood is destroyed by proteases in the placenta, but antibody-bound insulin is protected, so it reaches the fetus. Therefore, fetal macrosomia also occurs in women who develop antibodies against various animal insulins and then continue to receive the animal insulin during pregnancy.

Infants with **GLUT 1 deficiency** have defective transport of glucose across the blood–brain barrier. They have low cerebrospinal fluid glucose in the presence of normal plasma glucose, seizures, and developmental delay.

the usual glucagon response to a protein meal is abolished. Thus, when energy is needed during starvation, the insulin–glucagon molar ratio is low, favoring glycogen breakdown and gluconeogenesis; conversely, when the need for energy mobilization is low, the ratio is high, favoring the deposition of glycogen, protein, and fat (Clinical Box 24–4).

OTHER ISLET CELL HORMONES

In addition to insulin and glucagon, the pancreatic islets secrete somatostatin and pancreatic polypeptide into the bloodstream. In addition, somatostatin may be involved in regulatory processes within the islets that adjust the pattern of hormones secreted in response to various stimuli.

SOMATOSTATIN

Somatostatin and its receptors are discussed in Chapter 7. Somatostatin 14 (SS 14) and its amino terminal-extended form somatostatin 28 (SS 28) are found in the D cells of pancreatic islets. Both forms inhibit the secretion of insulin, glucagon, and pancreatic polypeptide and act locally within the pancreatic islets in a paracrine fashion. SS 28 is more active than SS 14 in inhibiting insulin secretion, and it apparently acts via the SSTR5 receptor (see Chapter 7). Patients with somatostatin-secreting pancreatic tumors (**somatostatinomas**) develop hyperglycemia and other manifestations of diabetes that disappear when the tumor is removed. They also develop dyspepsia due to slow gastric emptying and decreased gastric acid secretion, and gallstones, which are precipitated by decreased gallbladder contraction due to inhibition of CCK secretion. The secretion of pancreatic somatostatin is increased by several of the same stimuli that increase insulin secretion, that is, glucose and amino acids, particularly arginine and leucine. It is also increased by CCK. Somatostatin is released from the pancreas and the gastrointestinal tract into the peripheral blood.

PANCREATIC POLYPEPTIDE

Human pancreatic polypeptide is a linear polypeptide that contains 36 amino acid residues and is produced by F cells in the islets. It is closely related to two other 36-amino acid polypeptides, **polypeptide YY,** a gastrointestinal peptide (see Chapter 25), and **neuropeptide Y,** which is found in the brain and the autonomic nervous system (see Chapter 7). All end in tyrosine and are amidated at their carboxyl terminal. At least in part, pancreatic polypeptide secretion is under cholinergic control; plasma levels fall after administration of atropine. Its secretion is increased by a meal containing protein and by fasting, exercise, and acute hypoglycemia. Secretion is decreased by somatostatin and intravenous glucose. Infusions of leucine, arginine, and alanine do not affect it, so the stimulatory effect of a protein meal may be mediated indirectly. Pancreatic polypeptide slows the absorption of food in humans, and it may smooth out the peaks and valleys of absorption. However, its exact physiologic function is still uncertain.

ORGANIZATION OF THE PANCREATIC ISLETS

The presence in the pancreatic islets of hormones that affect the secretion of other islet hormones suggests that the islets function as secretory units in the regulation of nutrient homeostasis. Somatostatin inhibits the secretion of insulin, glucagon, and pancreatic polypeptide (Figure 24–14); insulin inhibits the secretion of glucagon; and glucagon stimulates the secretion of insulin and somatostatin. As noted above, A and D cells and pancreatic polypeptide-secreting cells are generally located around the periphery of the islets, with the B cells in the center. There are clearly two types of

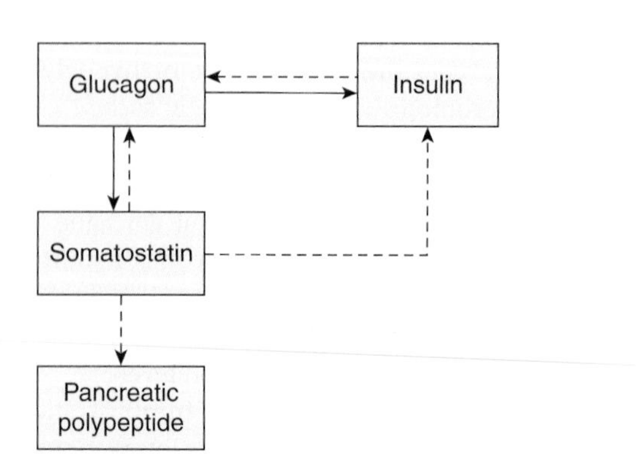

FIGURE 24–14 Effects of islet cell hormones on the secretion of other islet cell hormones. Solid arrows indicate stimulation; dashed arrows indicate inhibition.

islets, glucagon-rich islets and pancreatic polypeptide-rich islets, but the functional significance of this separation is not known. The islet cell hormones released into the ECF probably diffuse to other islet cells and influence their function (paracrine communication; see Chapter 25). It has been demonstrated that gap junctions are present between A, B, and D cells and that these permit the passage of ions and other small molecules from one cell to another, which could coordinate their secretory functions.

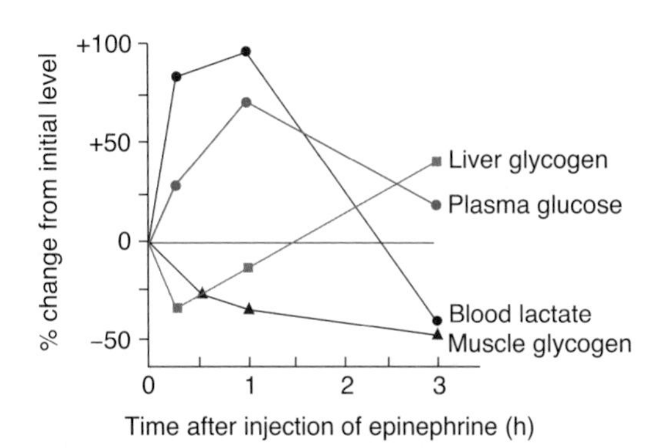

FIGURE 24–15 Effect of epinephrine on tissue glycogen, plasma glucose, and blood lactate levels in fed rats. (Reproduced with permission from Ruch TC, Patton HD [editors]: *Physiology and Biophysics,* 20th ed, Vol 3. Saunders, 1973.)

EFFECTS OF OTHER HORMONES & EXERCISE ON CARBOHYDRATE METABOLISM

Exercise has direct effects on carbohydrate metabolism. Many hormones in addition to insulin, IGF-I, IGF-II, glucagon, and somatostatin also have important roles in the regulation of carbohydrate metabolism. They include epinephrine, thyroid hormones, glucocorticoids, and growth hormone. The other functions of these hormones are considered elsewhere, but it seems wise to summarize their effects on carbohydrate metabolism in the context of the present chapter.

EXERCISE

The entry of glucose into skeletal muscle is increased during exercise in the absence of insulin by causing an insulin-independent increase in the number of GLUT 4 transporters in muscle cell membranes (see above). This increase in glucose entry persists for several hours after exercise, and regular exercise training can also produce prolonged increases in insulin sensitivity. Exercise can precipitate hypoglycemia in diabetics not only because of the increase in muscle uptake of glucose but also because absorption of injected insulin is more rapid during exercise. Patients with diabetes should take in extra calories or reduce their insulin dosage when they exercise.

CATECHOLAMINES

The activation of phosphorylase in liver by catecholamines is discussed in Chapter 1. Activation occurs via β-adrenergic receptors, which increase intracellular cAMP, and α-adrenergic receptors, which increase intracellular Ca^{2+}. Hepatic glucose output is increased, producing hyperglycemia. In muscle, the phosphorylase is also activated via cAMP and presumably via Ca^{2+}, but the glucose 6-phosphate formed can be catabolized only to pyruvate because of the absence of glucose 6-phosphatase. For reasons that are not entirely clear, large amounts of pyruvate are converted to lactate, which diffuses from the muscle into the circulation (Figure 24–15). The lactate is oxidized in the liver to pyruvate and converted to glycogen. Therefore, the response to an injection of epinephrine is an initial glycogenolysis followed by a rise in hepatic glycogen content. Lactate oxidation may be responsible for the calorigenic effect of epinephrine (see Chapter 20). Epinephrine and norepinephrine also liberate FFA into the circulation, and epinephrine decreases peripheral utilization of glucose.

THYROID HORMONES

Thyroid hormones make experimental diabetes worse; thyrotoxicosis aggravates clinical diabetes; and metathyroid diabetes can be produced in animals with decreased pancreatic reserve. The principal diabetogenic effect of thyroid hormones is to increase absorption of glucose from the intestine, but the hormones also cause (probably by potentiating the effects of catecholamines) some degree of hepatic glycogen depletion. Glycogen-depleted liver cells are easily damaged. When the liver is damaged, the glucose tolerance curve is diabetic because the liver takes up less of the absorbed glucose. Thyroid hormones may also accelerate the degradation of insulin. All these actions have a hyperglycemic effect and, if the pancreatic reserve is low, may lead to B cell exhaustion.

ADRENAL GLUCOCORTICOIDS

Glucocorticoids from the adrenal cortex (see Chapter 20) elevate blood glucose and produce a diabetic type of glucose tolerance curve. In humans, this effect may occur only in individuals with a genetic predisposition to diabetes. Glucose tolerance is reduced in 80% of patients with Cushing syndrome (see Chapter 20), and 20% of these patients have frank diabetes. The glucocorticoids are necessary for glucagon to exert its gluconeogenic action during fasting. They are gluconeogenic themselves, but their role is mainly permissive. In adrenal insufficiency, the blood glucose is normal as long as food intake is maintained, but fasting precipitates hypoglycemia and collapse. The plasma-glucose-lowering effect of insulin is greatly enhanced in patients with adrenal insufficiency. In animals with experimental diabetes, adrenalectomy markedly ameliorates the diabetes. The major diabetogenic effects are an increase in protein catabolism with increased gluconeogenesis in the liver; increased hepatic glycogenesis and ketogenesis; and a decrease in peripheral glucose utilization relative to the blood insulin level that may be due to inhibition of glucose phosphorylation.

GROWTH HORMONE

Human growth hormone makes clinical diabetes worse, and 25% of patients with growth hormone-secreting tumors of the anterior pituitary have diabetes. Hypophysectomy ameliorates diabetes and decreases insulin resistance even more than adrenalectomy, whereas growth hormone treatment increases insulin resistance.

The effects of growth hormone are partly direct and partly mediated via IGF-I (see Chapter 18). Growth hormone mobilizes FFA from adipose tissue, thus favoring ketogenesis. It decreases glucose uptake into some tissues ("anti-insulin action"), increases hepatic glucose output, and may decrease tissue binding of insulin. Indeed, it has been suggested that the ketosis and decreased glucose tolerance produced by starvation are due to hypersecretion of growth hormone. Growth hormone does not stimulate insulin secretion directly, but the hyperglycemia it produces secondarily stimulates the pancreas and may eventually exhaust the B cells.

HYPOGLYCEMIA & DIABETES MELLITUS IN HUMANS

HYPOGLYCEMIA

"Insulin reactions" are common in type 1 diabetics and occasional hypoglycemic episodes are the price of good diabetic control in most diabetics. Glucose uptake by skeletal muscle and absorption of injected insulin both increase during exercise (see above).

Symptomatic hypoglycemia also occurs in nondiabetics, and a review of some of the more important causes serves to emphasize the variables affecting plasma glucose homeostasis. Chronic mild hypoglycemia can cause incoordination and slurred speech, and the condition can be mistaken for drunkenness. Mental aberrations and convulsions in the absence of frank coma also occur. When the level of insulin secretion is chronically elevated by an **insulinoma,** a rare, insulin-secreting tumor of the pancreas, symptoms are most common in the morning. This is because a night of fasting has depleted hepatic glycogen reserves. However, symptoms can develop at any time, and in such patients, the diagnosis may be missed. Some cases of insulinoma have been erroneously diagnosed as epilepsy or psychosis. Hypoglycemia also occurs in some patients with large malignant tumors that do not involve the pancreatic islets, and the hypoglycemia in these cases is apparently due to excess secretion of IGF-II.

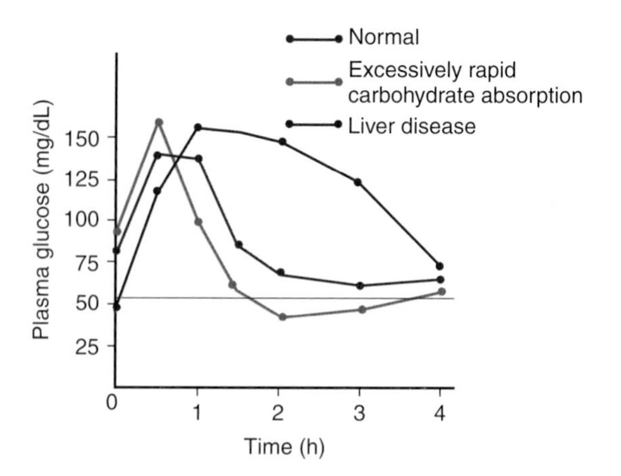

FIGURE 24–16 Typical glucose tolerance curves after an oral glucose load in liver disease and in conditions causing excessively rapid absorption of glucose from the intestine. The horizontal line is the approximate plasma glucose level at which hypoglycemic symptoms may appear.

As noted above, the autonomic discharge caused by lowered blood glucose that produces shakiness, sweating, anxiety, and hunger normally occurs at plasma glucose levels that are higher than the glucose levels that cause cognitive dysfunction, thereby serving as a warning to ingest sugar. However, in some individuals, these warning symptoms fail to occur before the cognitive symptoms, due to cerebral dysfunction (desensitization), and this **hypoglycemia unawareness** is potentially dangerous. The condition is prone to develop in patients with insulinomas and in diabetics receiving intensive insulin therapy, so it appears that repeated bouts of hypoglycemia cause the eventual development of hypoglycemia unawareness. If blood sugar rises again for some time, the warning symptoms again appear at a higher plasma glucose level than cognitive abnormalities and coma. The reason why prolonged hypoglycemia causes loss of the warning symptoms is unsettled.

In liver disease, the glucose tolerance curve is diabetic but the fasting plasma glucose level is low (Figure 24–16). In **functional hypoglycemia,** the plasma glucose rise is normal after a test dose of glucose, but the subsequent fall overshoots to hypoglycemic levels, producing symptoms 3–4 h after meals. This pattern is sometimes seen in individuals who later develop diabetes. Patients with this syndrome should be distinguished from the more numerous patients with similar symptoms due to psychologic or other problems who do not have hypoglycemia when blood is drawn during the symptomatic episode. It has been postulated that the overshoot of the plasma glucose is due to insulin secretion stimulated by impulses in the right vagus, but cholinergic blocking agents do not routinely correct the abnormality. In some thyrotoxic patients and in patients who have had gastrectomies or other operations that speed the passage of food into the intestine, glucose absorption is abnormally rapid. The plasma glucose rises to a high, early peak, but it then falls rapidly to hypoglycemic levels because the wave of hyperglycemia evokes a greater than normal rise in insulin secretion. Symptoms characteristically occur about 2 h after meals.

DIABETES MELLITUS

The incidence of diabetes mellitus in the human population has reached epidemic proportions worldwide and it is increasing at a rapid rate. In 2010 an estimated 285 million people worldwide had diabetes, according to the *International Diabetes Federation.* The federation predicts as many as 438 million will have diabetes by 2030. Ninety per cent of the present cases are type 2 diabetes, and most of the increase will be in type 2, paralleling the increase in the incidence of obesity.

Diabetes is sometimes complicated by acidosis and coma, and in long-standing diabetes additional complications occur. These include microvascular, macrovascular, and neuropathic disease. The microvascular abnormalities are proliferative scarring of the retina **(diabetic retinopathy)** leading to blindness; and renal disease **(diabetic nephropathy)** leading to renal failure. The macrovascular abnormalities are due to accelerated atherosclerosis, which is secondary to increased plasma LDL. The result is an increased incidence of stroke and myocardial infarction. The neuropathic abnormalities **(diabetic neuropathy)** involve the autonomic nervous system and peripheral nerves. The neuropathy plus the atherosclerotic circulatory insufficiency in the extremities and reduced resistance to infection can lead to chronic ulceration and gangrene, particularly in the feet.

The ultimate cause of the microvascular and neuropathic complications is chronic hyperglycemia, and tight control of the diabetes reduces their incidence. Intracellular hyperglycemia activates the enzyme aldose reductase. This increases the formation of sorbitol in cells, which in turn reduces cellular Na, K ATPase. In addition, intracellular glucose can be converted to so-called Amadori products, and these in turn can form **advanced glycosylation end products (AGEs),** which cross-link matrix proteins. This damages blood vessels. The AGEs also interfere with leukocyte responses to infection.

TYPES OF DIABETES

The cause of clinical diabetes is always a deficiency of the effects of insulin at the tissue level, but the deficiency may be relative. One of the common forms, **type 1,** or **insulin-dependent diabetes mellitus (IDDM),** is due to insulin deficiency caused by autoimmune destruction of the B cells in the pancreatic islets; the A, D, and F cells remain intact. The second common form, **type 2,** or **noninsulin-dependent diabetes mellitus (NIDDM),** is characterized by insulin resistance.

In addition, some cases of diabetes are due to other diseases or conditions such as chronic pancreatitis, total pancreatectomy, Cushing syndrome (see Chapter 20), and acromegaly (see Chapter 18). These make up 5% of the total cases and are sometimes classified as **secondary diabetes.**

Type 1 diabetes usually develops before the age of 40 and hence is called **juvenile diabetes.** Patients with this disease are not obese and they have a high incidence of ketosis and acidosis. Various anti-B cell antibodies are present in plasma, but the current thinking is that type 1 diabetes is primarily a T lymphocyte-mediated disease. Definite genetic susceptibility is present as well; if one identical twin develops the disease, the chances are 1 in 3 that the other twin will also do so. In other words, the **concordance rate** is about 33%. The main genetic abnormality is in the major histocompatibility complex on chromosome 6, making individuals with certain types of histocompatibility antigens (see Chapter 3) much more prone to the disease. Other genes are also involved.

Immunosuppression with drugs such as cyclosporine ameliorate type 1 diabetes if given early in the disease before all islet B cells are lost. Attempts have been made to treat type 1 diabetes by transplanting pancreatic tissue or isolated islet cells, but results to date have been poor, largely because B cells are easily damaged and it is difficult to transplant enough of them to normalize glucose responses.

As mentioned above, type 2 is the most common type of diabetes and is usually associated with obesity. It usually develops after age 40 and is not associated with total loss of the ability to secrete insulin. It has an insidious onset, is rarely associated with ketosis, and is usually associated with normal B cell morphology and insulin content if the B cells have not become exhausted. The genetic component in type 2 diabetes is actually stronger than the genetic component in type 1 diabetes; in identical twins, the concordance rate is higher, ranging in some studies to nearly 100%.

In some patients, type 2 diabetes is due to defects in identified genes. Over 60 of these defects have been described. They include defects in glucokinase (about 1% of the cases), the insulin molecule itself (about 0.5% of the cases), the insulin receptor (about 1% of the cases), GLUT 4 (about 1% of the cases), or IRS-1 (about 15% of the cases). In maturity-onset diabetes occurring in young individuals (MODY), which accounts for about 1% of the cases of type 2 diabetes, loss-of-function mutations have been described in six different genes. Five code for transcription factors affecting the production of enzymes involved in glucose metabolism. The sixth is the gene for glucokinase (Figure 24–11), the enzyme that controls the rate of glucose phosphorylation and hence its metabolism in the B cells. However, the vast majority of cases of type 2 diabetes are almost certainly polygenic in origin, and the actual genes involved are still unknown.

OBESITY, THE METABOLIC SYNDROME, & TYPE 2 DIABETES

Obesity is increasing in incidence, and relates to the regulation of food intake and energy balance and overall nutrition. It deserves additional consideration in this chapter because of its special relation to disordered carbohydrate metabolism and diabetes. As body weight increases, insulin resistance increases, that is, there is a decreased ability of insulin to move glucose into fat and muscle and to shut off glucose release from the liver. Weight reduction decreases insulin resistance. Associated with obesity there is hyperinsulinemia, dyslipidemia (characterized by high circulating triglycerides and low high-density lipoprotein [HDL]), and accelerated development of atherosclerosis. This combination of findings is commonly called the **metabolic syndrome,** or **syndrome X.** Some of the patients with the syndrome are prediabetic, whereas others have frank type 2 diabetes. It has not been proved but it is logical to assume that the hyperinsulinemia is a compensatory response to the increased insulin resistance and that frank diabetes develops in individuals with reduced B cell reserves.

These observations and other data strongly suggest that fat produces a chemical signal or signals that act on muscles and the liver to increase insulin resistance. Evidence for this includes the recent observation that when GLUTs are selectively knocked out in adipose tissue, there is an associated decrease in glucose transport in muscle in vivo, but when the muscles of those animals are tested in vitro their transport is normal.

One possible signal is the circulating level of free fatty acids, which is elevated in many insulin-resistant states. Other possibilities are peptides and proteins secreted by fat cells. It is now clear that white fat depots are not inert lumps but are actually endocrine tissues that secrete not only leptin but also other hormones that affect fat metabolism. These adipose derived hormones are commonly termed **adipokines** as they are *cytokines* secreted by *adipose tissue.* Known adipokines are leptin, adiponectin, and resistin.

Some adipokines decrease, rather than increase, insulin resistance. Leptin and adiponectin, for example, decrease insulin resistance, whereas resistin increases insulin resistance. Further complicating the situation, marked insulin resistance is present in the rare metabolic disease **congenital lipodystrophy,** in which fat depots fail to develop. This resistance is reduced by leptin and adiponectin. Finally, a variety of knockouts of intracellular second messengers have been reported to increase insulin resistance. It is unclear how, or indeed if, these findings fit together to provide an explanation of the relation of obesity to insulin tolerance, but the topic is obviously an important one and it is under intensive investigation.

CHAPTER SUMMARY

- Four polypeptides with hormonal activity are secreted by the pancreas: insulin, glucagon, somatostatin, and pancreatic polypeptide.
- Insulin increases the entry of glucose into cells. In skeletal muscle cell it increases the number of GLUT 4 transporters in the cell membranes. In liver it induces glucokinase, which increases the phosphorylation of glucose, facilitating the entry of glucose into the cell.
- Insulin causes K^+ to enter cells, with a resultant lowering of the extracellular K^+ concentration. Insulin increases the activity of Na, K ATPase in cell membranes, so that more K^+ is pumped into cells. Hypokalemia often develops when patients with diabetic acidosis are treated with insulin.
- Insulin receptors are found on many different cells in the body and have two subunits, α and β. Binding of insulin to its receptor triggers a signaling pathway that involves autophosphorylation of the β subunits on tyrosine residues. This triggers phosphorylation of some cytoplasmic proteins and dephosphorylation of others, mostly on serine and threonine residues.
- The constellation of abnormalities caused by insulin deficiency is called diabetes mellitus. Type 1 diabetes is due to insulin deficiency caused by autoimmune destruction of the B cells in the pancreatic islets; Type 2 diabetes is characterized by the dysregulation of insulin release from the B cells, along with insulin resistance in peripheral tissues such as skeletal muscle, brain, and liver.

MULTIPLE-CHOICE QUESTIONS

For all questions, select the single best answer unless otherwise directed.

1. Which of the following are incorrectly paired?
   A. B cells: insulin
   B. D cells: somatostatin
   C. A cells: glucagons
   D. Pancreatic exocrine cells: chymotrypsinogen
   E. F cells: gastrin
2. Which of the following are *incorrectly* paired?
   A. Epinephrine: increased glycogenolysis in skeletal muscle
   B. Insulin: increased protein synthesis
   C. Glucagon: increased gluconeogenesis
   D. Progesterone: increased plasma glucose level
   E. Growth hormone: increased plasma glucose level
3. Which of the following would be *least* likely to be seen 14 days after a rat is injected with a drug that kills all of its pancreatic B cells?
   A. A rise in the plasma H^+ concentration
   B. A rise in the plasma glucagon concentration
   C. A fall in the plasma HCO_3^- concentration
   D. A fall in the plasma amino acid concentration
   E. A rise in plasma osmolality
4. When the plasma glucose concentration falls to low levels, a number of different hormones help combat the hypoglycemia. After intravenous administration of a large dose of insulin, the return of a low blood sugar level to normal is delayed in
   A. adrenal medullary insufficiency.
   B. glucagon deficiency.
   C. combined adrenal medullary insufficiency and glucagon deficiency.
   D. thyrotoxicosis.
   E. acromegaly.

5. Insulin increases the entry of glucose into
   A. all tissues.
   B. renal tubular cells.
   C. the mucosa of the small intestine.
   D. most neurons in the cerebral cortex.
   E. skeletal muscle.
6. Glucagon increases glycogenolysis in liver cells but ACTH does not because
   A. cortisol increases the plasma glucose level.
   B. liver cells have an adenylyl cyclase different from that in adrenocortical cells.
   C. ACTH cannot enter the nucleus of liver cells.
   D. the membranes of liver cells contain receptors different from those in adrenocortical cells.
   E. liver cells contain a protein that inhibits the action of ACTH.
7. A meal rich in proteins containing the amino acids that stimulate insulin secretion but low in carbohydrates does not cause hypoglycemia because
   A. the meal causes a compensatory increase in T_4 secretion.
   B. cortisol in the circulation prevents glucose from entering muscle.
   C. glucagon secretion is also stimulated by the meal.`
   D. the amino acids in the meal are promptly converted to glucose.
   E. insulin does not bind to insulin receptors if the plasma concentration of amino acids is elevated.

CHAPTER RESOURCES

Banerjee RR, Rangwala SM, Shapiro JS et al: Regulation of fasted blood glucose by resistin. Science 2004;303:1195.

Gehlert DR: Multiple receptors for the pancreatic polypeptide (PP-fold) family: Physiological implications. Proc Soc Exper Biol Med 1998;218:7.

Harmel AP, Mothur R: *Davidson's Diabetes Mellitus,* 5th ed. Elsvier, 2004.

Kjos SL, Buchanan TA: Gestational diabetes mellitus. N Engl J Med 1999;341:1749.

Kulkarni RN, Kahn CR: HNFs-linking the liver and pancreatic islets in diabetes. Science 2004;303:1311.

Larsen PR, et al (editors): *Williams Textbook of Endocrinology,* 9th ed. Saunders, 2003.

Lechner D, Habner JF: Stem cells for the treatment of diabetes mellitus. Endocrinol Rounds 2003;2(2).

LeRoith D: Insulin-like growth factors. N Engl J Med 1997;336:633.

Meigs JB, Avruch J: The metabolic syndrome. Endocrinol Rounds 2003;2(5).

Sealey RJ (basic research), Rolls BJ (clinical research), Hensrud DD (clinical practice): Three perspectives on obesity. Endocrine News 2004;29:7.

SECTION IV Gastrointestinal Physiology

For unicellular organisms that exist in a sea of nutrients, it is possible to satisfy nutritional requirements simply with the activity of membrane transport proteins that permit the uptake of specific molecules into the cytosol. However, for multicellular organisms, including humans, the challenges of delivering nutrients to appropriate sites in the body are significantly greater, particularly if the organisms are terrestrial. Further, most of the food we eat is in the form of macromolecules, and even when these are digested to their component monomers, most of the end products are water-soluble and do not readily cross cell membranes (a notable exception are the constituents of dietary lipids). Thus, the gastrointestinal system has evolved to permit nutrient acquisition and assimilation into the body, while prohibiting the uptake of undesirable substances (toxins and microbial products, as well as microbes themselves). The latter situation is complicated by the fact that the intestine maintains a lifelong relationship with a rich microbial ecosystem residing in its lumen, a relationship that is largely mutually beneficial if the microbes are excluded from the systemic compartment.

The intestine is a continuous tube that extends from mouth to anus and is formally contiguous with the external environment. A single cell layer of columnar epithelial cells comprises the semipermeable barrier across which controlled uptake of nutrients takes place. Various glandular structures empty into the intestinal lumen at points along its length, providing for digestion of food components, signaling to distal segments, and regulation of the microbiota. There are also important motility functions that move the intestinal contents and resulting waste products along the length of the gut, and a rich innervation that regulates motility, secretion and nutrient uptake, in many cases in a manner that is independent of the central nervous system. There is also a large number of endocrine cells that release hormones that work together with neurotransmitters to coordinate overall regulation of the GI system. In general, there is considerable redundancy of control systems as well as excess capacity for nutrient digestion and uptake. This served us well in ancient times when food sources were scarce, but may now contribute to the modern epidemic of obesity.

The liver, while playing important roles in whole body metabolism, is usually considered a part of the gastrointestinal system for two main reasons. First, it provides for excretion from the body of lipid-soluble waste products that cannot enter the urine. These are secreted into the bile and thence into the intestine to be excreted with the feces. Second, the blood flow draining the intestine is arranged such that substances that are absorbed pass first through the liver, allowing for the removal and metabolism of any toxins that have inadvertently been taken up, as well as clearance of particulates, such as small numbers of enteric bacteria.

In this section, the function of the gastrointestinal system and liver will be considered, and the ways in which the various segments communicate to provide an integrated response to a mixed meal (proteins, carbohydrates, and lipids). The relevance of gastrointestinal physiology for the development of digestive diseases will also be considered. While many are rarely life-threatening (with some notable exceptions, such as specific cancers) digestive diseases represent a substantial burden in terms of morbidity and lost productivity. A 2009 report of the U.S. National Institutes of Diabetes, Digestive and Kidney Diseases found that on an annual basis, for every 100 U.S. residents, there were 35 ambulatory care visits and nearly five overnight hospital stays that involved a gastrointestinal diagnosis. Digestive diseases also appear to be increasing in this population (although mortality, principally from cancers, is thankfully in decline). On the other hand, digestive diseases, and in particular infectious diarrhea, remain important causes of mortality in developing countries where clean sources of food and water cannot be assured. In any event, the burden of digestive diseases provides an important impetus for gaining a full understanding of gastrointestinal physiology, since it is a failure of such physiology that most often leads to disease. Conversely, an understanding of specific digestive conditions can often illuminate physiological principles, as will be stressed in this section.

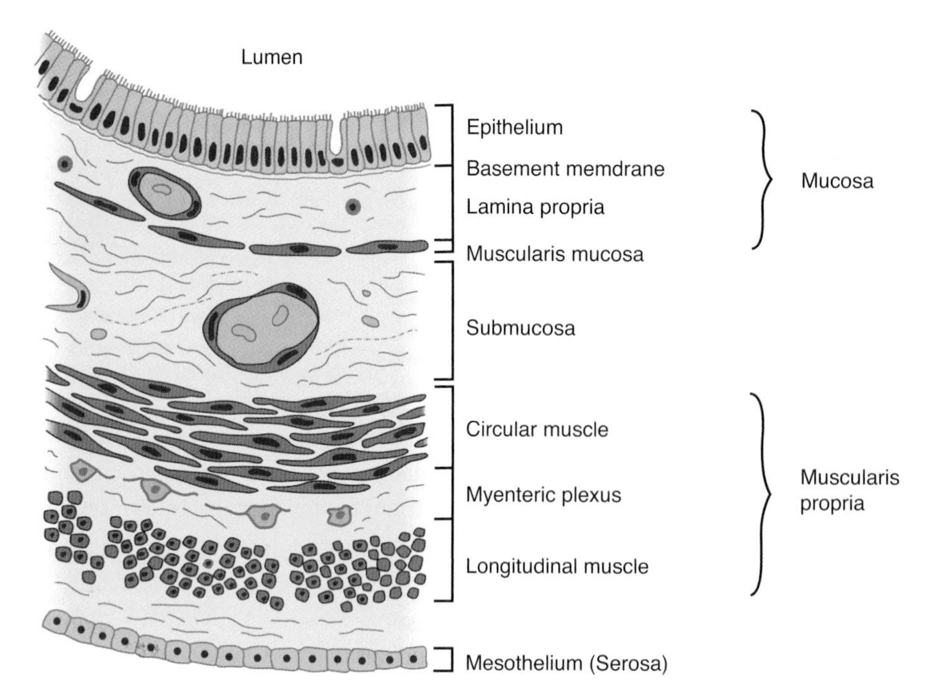

FIGURE 25–1 Organization of the wall of the intestine into functional layers. (Adapted from Yamada: *Textbook of Gasteronenterology,* 4th ed, pp 151–165. Copyright LWW, 2003.)

represents the barrier that nutrients must traverse to enter the body. Below the epithelium is a layer of loose connective tissue known as the lamina propria, which in turn is surrounded by concentric layers of smooth muscle, oriented circumferentially and then longitudinally to the axis of the gut (the circular and longitudinal muscle layers, respectively). The intestine is also amply supplied with blood vessels, nerve endings, and lymphatics, which are all important in its function.

The epithelium of the intestine is also further specialized in a way that maximizes the surface area available for nutrient absorption. Throughout the small intestine, it is folded up into fingerlike projections called villi (Figure 25–2). Between the villi are infoldings known as crypts. Stem cells that give rise to both crypt and villus epithelial cells reside toward the base of the crypts and are responsible for completely renewing the epithelium every few days or so. Indeed, the gastrointestinal epithelium is one of the most rapidly dividing tissues in the body. Daughter cells undergo several rounds of cell division in the crypts then migrate out onto the villi, where they are eventually shed and lost in the stool. The villus epithelial cells are also notable for the extensive microvilli that characterize their apical membranes. These microvilli are endowed with a dense glycocalyx (the brush border) that probably protects the cells to some extent from the effects of digestive enzymes. Some digestive enzymes are also actually part of the brush border, being membrane-bound proteins. These so-called "brush border hydrolases" perform the final steps of digestion for specific nutrients.

GASTROINTESTINAL SECRETIONS

SALIVARY SECRETION

The first secretion encountered when food is ingested is saliva. Saliva is produced by three pairs of salivary glands (the **parotid, submandibular,** and **sublingual glands**) that drain into the oral cavity. It has a number of organic constituents that serve to initiate digestion (particularly of starch, mediated by amylase) and which also protect the oral cavity from bacteria (such as immunoglobulin A and lysozyme). Saliva also serves to lubricate the food bolus (aided by mucins). Secretions of the three glands differ in their relative proportion of proteinaceous and mucinous components, which results from the relative number of serous and mucous salivary acinar cells, respectively. Saliva is also hypotonic compared with plasma and alkaline; the latter feature is important to neutralize any gastric secretions that reflux into the esophagus.

The salivary glands consist of blind end pieces (acini) that produce the primary secretion containing the organic constituents dissolved in a fluid that is essentially identical in its composition to plasma. The salivary glands are actually extremely active when maximally stimulated, secreting their own weight in saliva every minute. To accomplish this, they are richly endowed with surrounding blood vessels that dilate when salivary secretion is initiated. The composition of the saliva is then modified as it flows from the acini out into ducts that eventually coalesce and deliver the saliva into the mouth.

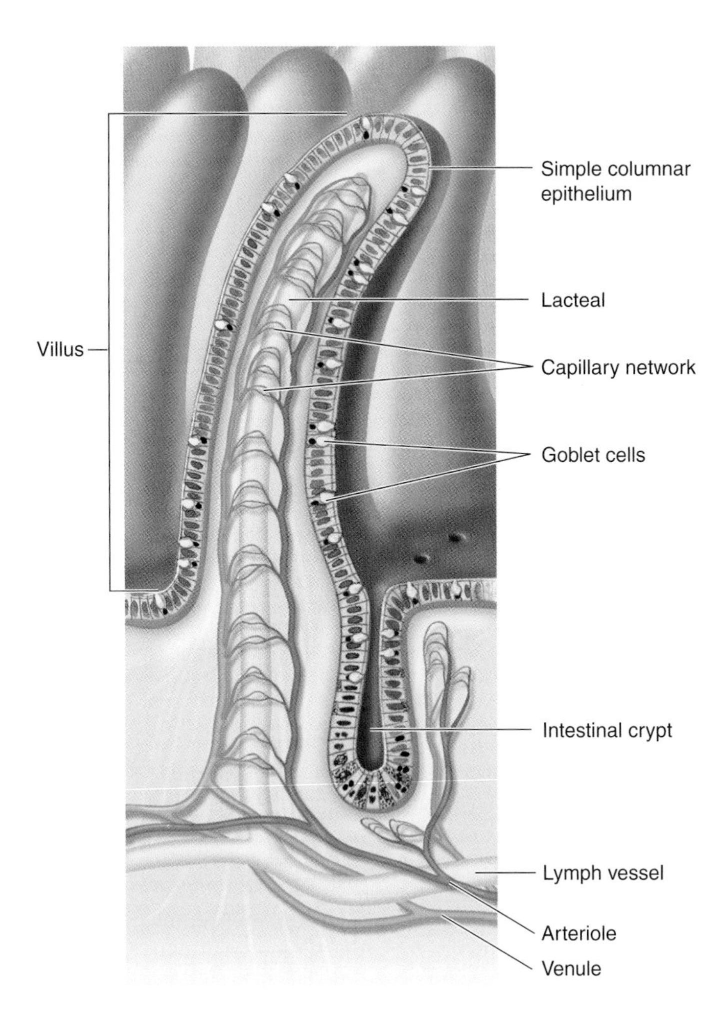

FIGURE 25–2 The structure of intestinal villi and crypts. The epithelial layer also contains scattered endocrine cells and intraepithelial lymphocytes. The crypt base contains Paneth cells, which secrete antimicrobial peptides, as well as the stem cells that provide for continual turnover of the crypt and villus epithelium. The epithelium turns over every 3–5 days in healthy adult humans. (Reproduced with permission from Fox SI: *Human Physiology,* 10th ed. McGraw-Hill, 2008.)

Na^+ and Cl^- are extracted and K^+ and bicarbonate are added. Because the ducts are relatively impermeable to water, the loss of NaCl renders the saliva hypotonic, particularly at low secretion rates. As the rate of secretion increases, there is less time for NaCl to be extracted and the tonicity of the saliva rises, but it always stays somewhat hypotonic with respect to plasma. Overall, the three pairs of salivary glands that drain into the mouth supply 1000–1500 mL of saliva per day.

Salivary secretion is almost entirely controlled by neural influences, with the parasympathetic branch of the autonomic nervous system playing the most prominent role (Figure 25–3). Sympathetic input slightly modifies the composition of saliva (particularly by increasing proteinaceous content), but has little influence on volume. Secretion is triggered by reflexes that are stimulated by the physical act of chewing, but is actually initiated even before the meal is taken into the mouth as a result of central triggers that are prompted by thinking about, seeing, or smelling food. Indeed, salivary secretion can readily be conditioned, as in the classical experiments of Pavlov where dogs were conditioned to salivate in response to a ringing bell by associating this stimulus with a meal. Salivary secretion is also prompted by nausea, but inhibited by fear or during sleep.

Saliva performs a number of important functions: it facilitates swallowing, keeps the mouth moist, serves as a solvent for the molecules that stimulate the taste buds, aids speech by facilitating movements of the lips and tongue, and keeps the mouth and teeth clean. The saliva also has some antibacterial action, and patients with deficient salivation **(xerostomia)** have a higher than normal incidence of dental caries. The buffers in saliva help maintain the oral pH at about 7.0.

GASTRIC SECRETION

Food is stored in the stomach; mixed with acid, mucus, and pepsin; and released at a controlled, steady rate into the duodenum (see Clinical Box 25–1).

ANATOMIC CONSIDERATIONS

The gross anatomy of the stomach is shown in Figure 25–4. The gastric mucosa contains many deep glands. In the cardia and the pyloric region, the glands secrete mucus. In the body of the stomach, including the fundus, the glands also contain **parietal (oxyntic) cells,** which secrete hydrochloric acid and intrinsic factor, and **chief (zymogen, peptic) cells,** which secrete pepsinogens (Figure 25–5). These secretions mix with mucus secreted by the cells in the necks of the glands. Several of the glands open on a common chamber **(gastric pit)** that opens in turn on the surface of the mucosa. Mucus is also secreted along with HCO_3^- by mucus cells on the surface of the epithelium between glands.

The stomach has a very rich blood and lymphatic supply. Its parasympathetic nerve supply comes from the vagi and its sympathetic supply from the celiac plexus.

ORIGIN & REGULATION OF GASTRIC SECRETION

The stomach also adds a significant volume of digestive juices to the meal. Like salivary secretion, the stomach actually readies itself to receive the meal before it is actually taken in, during the so-called cephalic phase that can be influenced by food preferences. Subsequently, there is a gastric phase of secretion that is quantitatively the most significant, and finally an intestinal phase once the meal has left the stomach. Each phase is closely regulated by both local and distant triggers.

The gastric secretions (Table 25–1) arise from glands in the wall of the stomach that drain into its lumen, and also from the surface cells that secrete primarily mucus and bicarbonate to protect the stomach from digesting itself, as well as

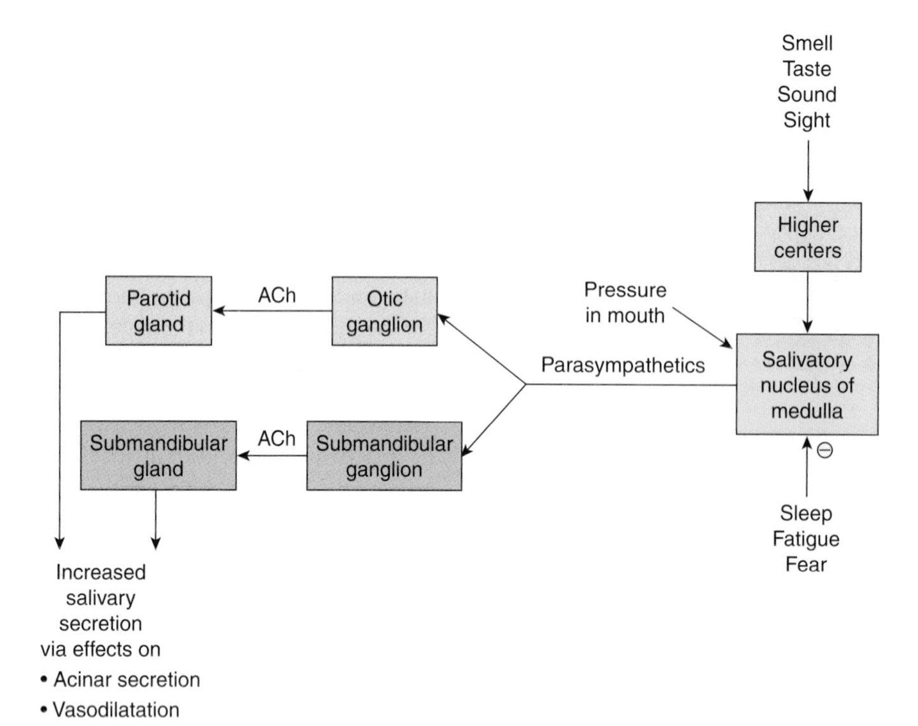

FIGURE 25–3 Regulation of salivary secretion by the parasympathetic nervous system. ACh, acetylcholine. Saliva is also produced by the sublingual glands (not depicted), but these are a minor contributor to both resting and stimulated salivary flows. (Adapted from Barrett KE: *Gastrointestinal Physiology*. McGraw-Hill, 2006.)

substances known as trefoil peptides that stabilize the mucus-bicarbonate layer. The glandular secretions of the stomach differ in different regions of the organ. The most characteristic secretions derive from the glands in the fundus or body of the stomach. These contain the distinctive parietal cells, which secrete hydrochloric acid and intrinsic factor; and chief cells, which produce pepsinogens and gastric lipase (Figure 25–5). The acid secreted by parietal cells serves to sterilize the meal and also to begin the hydrolysis of dietary macromolecules. Intrinsic factor is important for the later absorption of vitamin B_{12}, or cobalamin. Pepsinogen is the precursor of pepsin, which initiates protein digestion. Lipase similarly begins the digestion of dietary fats.

There are three primary stimuli of gastric secretion, each with a specific role to play in matching the rate of secretion to functional requirements (Figure 25–6). Gastrin is a hormone

CLINICAL BOX 25–1

Peptic Ulcer Disease

Gastric and duodenal ulceration in humans is related primarily to a breakdown of the barrier that normally prevents irritation and autodigestion of the mucosa by the gastric secretions. Infection with the bacterium *Helicobacter pylori* disrupts this barrier, as do aspirin and other nonsteroidal anti-inflammatory drugs (NSAIDs), which inhibit the production of prostaglandins and consequently decrease mucus and HCO_3^- secretion. The NSAIDs are widely used to combat pain and treat arthritis. An additional cause of ulceration is prolonged excess secretion of acid. An example of this is the ulcers that occur in the **Zollinger–Ellison syndrome.** This syndrome is seen in patients with gastrinomas. These tumors can occur in the stomach and duodenum, but most of them are found in the pancreas. The gastrin causes prolonged hypersecretion of acid, and severe ulcers are produced.

THERAPEUTIC HIGHLIGHTS

Gastric and duodenal ulcers can be given a chance to heal by inhibition of acid secretion with drugs such as omeprazole and related drugs that inhibit H^+–K^+ ATPase ("proton pump inhibitors"). If present, *H. pylori* can be eradicated with antibiotics, and NSAID-induced ulcers can be treated by stopping the NSAID or, when this is not advisable, by treatment with the prostaglandin agonist misoprostol. Gastrinomas can sometimes be removed surgically.

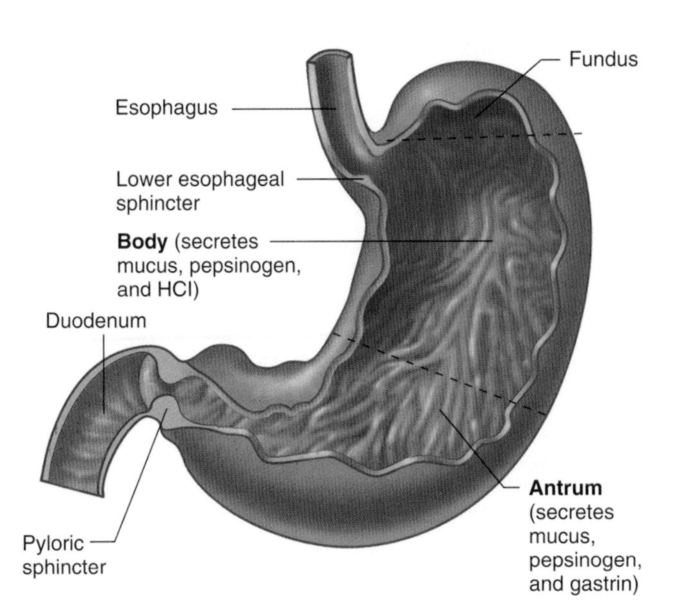

FIGURE 25–4 Anatomy of the stomach. The principal secretions of the body and antrum are listed in parentheses. (Reproduced with permission from Widmaier EP, Raff H, Strang KT: *Vander's Human Physiology: The Mechanisms of Body Function,* 11th ed. McGraw-Hill, 2008.)

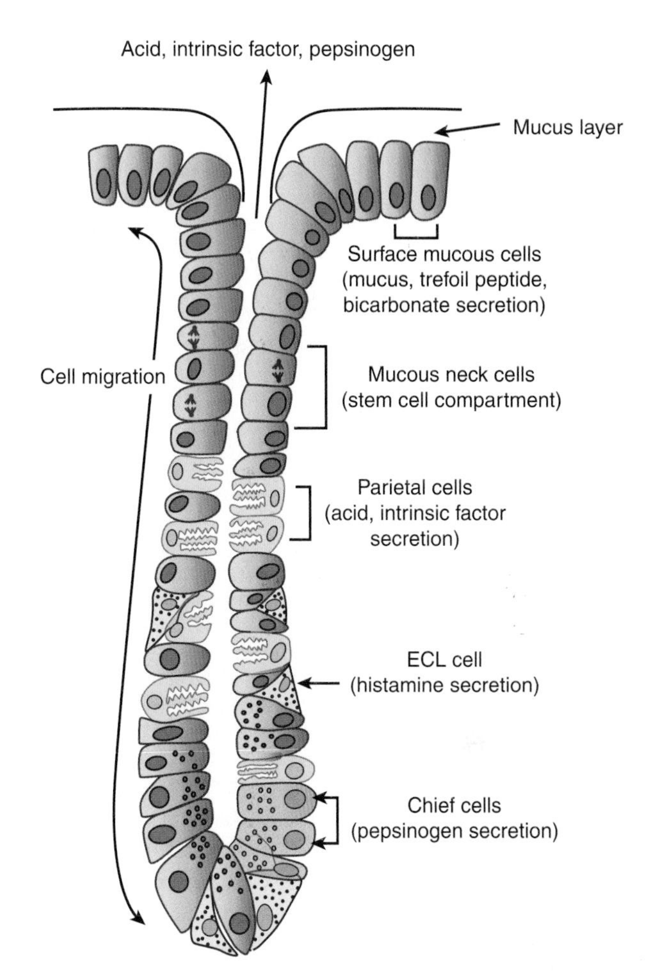

FIGURE 25–5 Structure of a gastric gland from the fundus or body of the stomach. These acid- and pepsinogen-producing glands are referred to as "oxyntic" glands in some sources. Similarly, some sources refer to parietal cells as oxyntic cells. (Adapted from Barrett KE: *Gastrointestinal Physiology.* McGraw-Hill, 2006.)

that is released by G cells in the antrum of the stomach both in response to a specific neurotransmitter released from enteric nerve endings, known as gastrin releasing peptide (GRP) or bombesin, and also in response to the presence of oligopeptides in the gastric lumen. Gastrin is then carried through the bloodstream to the fundic glands, where it binds to receptors not only on parietal (and likely, chief cells) to activate secretion, but also on so-called enterochromaffin-like cells (ECL cells) that are located in the gland, and release histamine. Histamine is also a trigger of parietal cell secretion, via binding to H_2 histamine receptors. Finally, parietal and chief cells can also be stimulated by acetylcholine, released from enteric nerve endings in the fundus.

During the cephalic phase of gastric secretion, secretion is predominantly activated by vagal input that originates from the brain region known as the dorsal vagal complex, which coordinates input from higher centers. Vagal outflow to the stomach then releases GRP and acetylcholine, thereby initiating secretory function. However, before the meal enters the stomach, there are few additional triggers and thus the amount of secretion is limited. Once the meal is swallowed, on the other hand, meal constituents trigger substantial release of gastrin and the physical presence of the meal also distends the stomach and activates stretch receptors, which provoke a "vago-vagal" as well as local reflexes that further amplify secretion. The presence of the meal also buffers gastric acidity that would otherwise serve as a feedback inhibitory signal to shut off secretion secondary to the release of somatostatin, which inhibits both G and ECL cells as well as secretion by parietal cells themselves (Figure 25–6). This probably represents a key mechanism whereby gastric secretion is terminated after the meal moves from the stomach into the small intestine.

Gastric parietal cells are highly specialized for their unusual task of secreting concentrated acid (Figure 25–7). The cells are packed with mitochondria that supply energy to drive the apical H,K-ATPase, or proton pump, that moves H^+ ions out of the parietal cell against a concentration gradient of more than a million-fold. At rest, the proton pumps are

TABLE 25–1 Contents of normal gastric juice (fasting state).

Cations: Na^+, K^+, Mg^{2+}, H^+ (pH approximately 3.0)
Anions: Cl^-, HPO_4^{2-}, SO_4^{2-}
Pepsins
Lipase
Mucus
Intrinsic factor

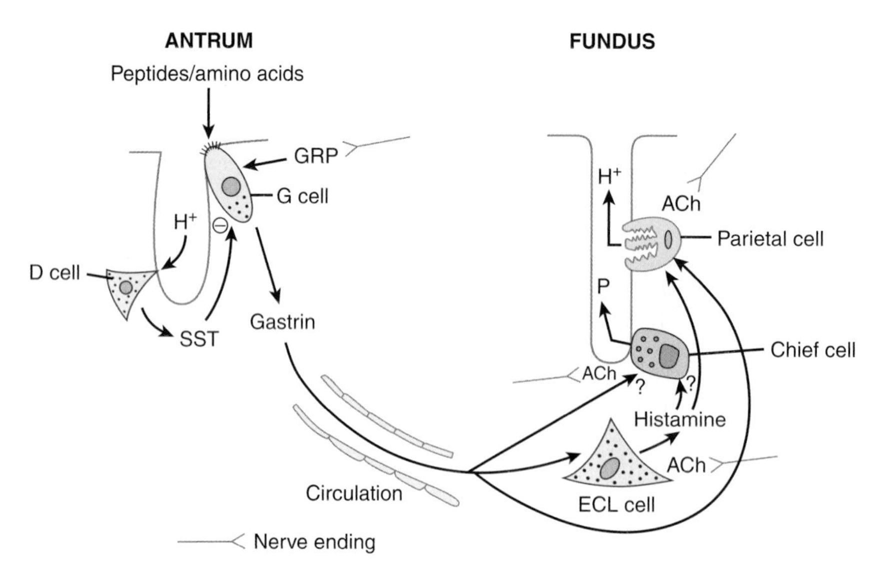

FIGURE 25–6 Regulation of gastric acid and pepsin secretion by soluble mediators and neural input. Gastrin is released from G cells in the antrum in response to gastrin releasing peptide (GRP) and travels through the circulation to influence the activity of ECL cells and parietal cells. ECL cells release histamine, which also acts on parietal cells. Acetylcholine (ACh), released from nerves, is an agonist for ECL cells, chief cells, and parietal cells. Other specific agonists of the chief cell are not well understood. Gastrin release is negatively regulated by luminal acidity via the release of somatostatin from antral D cells. P, pepsinogen. (Adapted from Barrett KE: *Gastrointestinal Physiology.* McGraw-Hill, 2006.)

sequestered within the parietal cell in a series of membrane compartments known as tubulovesicles. When the parietal cell begins to secrete, on the other hand, these vesicles fuse with invaginations of the apical membrane known as canaliculi, thereby substantially amplifying the apical membrane area and positioning the proton pumps to begin acid secretion (Figure 25–8). The apical membrane also contains potassium channels, which supply the K^+ ions to be exchanged for H^+, and Cl^- channels that supply the counterion for HCl secretion (Figure 25–9). The secretion of protons is also accompanied by the release of equivalent numbers of bicarbonate ions into the bloodstream, which as we will see are later used to neutralize gastric acidity once its function is complete (Figure 25–9).

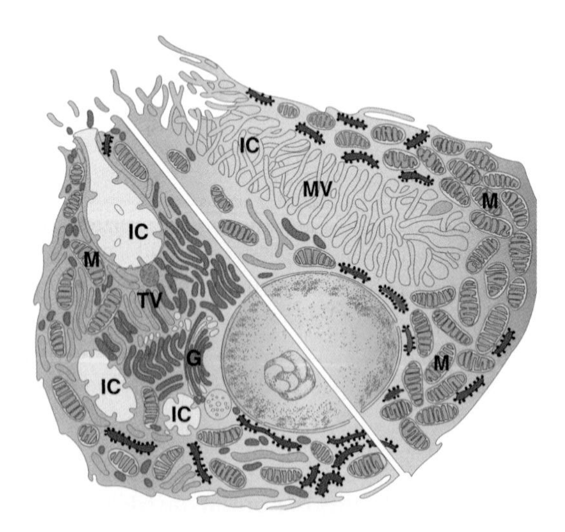

FIGURE 25–7 Composite diagram of a parietal cell, showing the resting state (lower left) and the active state (upper right). The resting cell has intracellular canaliculi (IC), which open on the apical membrane of the cell, and many tubulovesicular structures (TV) in the cytoplasm. When the cell is activated, the TVs fuse with the cell membrane and microvilli (MV) project into the canaliculi, so the area of cell membrane in contact with gastric lumen is greatly increased. M, mitochondrion; G, Golgi apparatus. (Based on the work of Ito S, Schofield GC: Studies on the depletion and accumulation of microvilli and changes in the tubulovesicular compartment of mouse parietal cells in relation to gastric acid secretion. J Cell Biol 1974; Nov;63(2 Pt 1):364–382.)

The three agonists of the parietal cell—gastrin, histamine, and acetylcholine—each bind to distinct receptors on the basolateral membrane (Figure 25–8). Gastrin and acetylcholine promote secretion by elevating cytosolic free calcium concentrations, whereas histamine increases intracellular cyclic adenosine 3′,5′-monophosphate (cAMP). The net effects of these second messengers are the transport and morphological changes described above. However, it is important to be aware that the two distinct pathways for activation are synergistic, with a greater than additive effect on secretion rates when histamine plus gastrin or acetylcholine, or all three, are present simultaneously. The physiologic significance of this synergism is that high rates of secretion can be stimulated with relatively small changes in availability of each of the stimuli. Synergism is also therapeutically significant because secretion can be markedly inhibited by blocking the action of only one of the triggers (most commonly that of histamine, via H_2 histamine antagonists that are widely used therapies for adverse effects of excessive gastric secretion, such as reflux).

Gastric secretion adds about 2.5 L per day to the intestinal contents. However, despite their substantial volume and fine control, gastric secretions are dispensable for the full

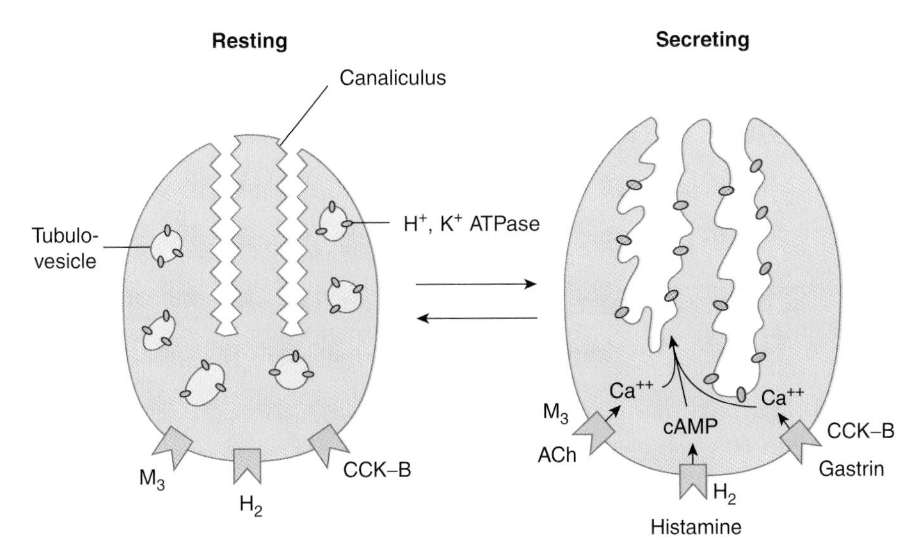

FIGURE 25–8 Parietal cell receptors and schematic representation of the morphological changes depicted in Figure 25–7. Amplification of the apical surface area is accompanied by an increased density of H^+, K^+–ATPase molecules at this site. Note that acetylcholine (ACh) and gastrin signal via calcium, whereas histamine signals via cAMP. (Adapted from Barrett KE: *Gastrointestinal Physiology.* McGraw-Hill, 2006.)

digestion and absorption of a meal, with the exception of cobalamin absorption. This illustrates an important facet of gastrointestinal physiology, namely that digestive and absorptive capacities are markedly in excess of normal requirements. On the other hand, if gastric secretion is chronically reduced, individuals may display increased susceptibility to infections acquired via the oral route.

PANCREATIC SECRETION

The pancreatic juice contains enzymes that are of major importance in digestion (see Table 25–2). Its secretion is controlled in part by a reflex mechanism and in part by the gastrointestinal hormones secretin and cholecystokinin (CCK).

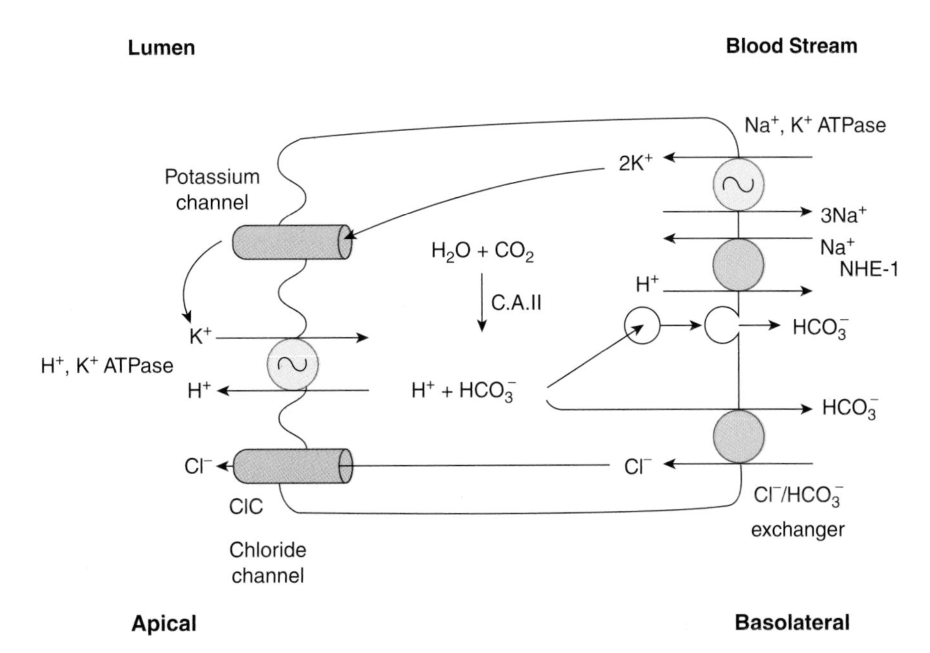

FIGURE 25–9 Ion transport proteins of parietal cells. Protons are generated in the cytoplasm via the action of carbonic anhydrase II (C.A. II). Bicarbonate ions are exported from the basolateral pole of the cell either by vesicular fusion or via a chloride/bicarbonate exchanger. The sodium/hydrogen exchanger, NHE1, on the basolateral membrane is considered a "housekeeping" transporter that maintains intracellular pH in the face of cellular metabolism during the unstimulated state. (Adapted from Barrett KE: *Gastrointestinal Physiology.* McGraw-Hill, 2006.)

TABLE 25–2 Principal digestive enzymes.[a]

Source	Enzyme	Activator	Substrate	Catalytic Function or Products
Salivary glands	Salivary α-amylase	Cl^-	Starch	Hydrolyzes 1:4α linkages, producing α-limit dextrins, maltotriose, and maltose
Stomach	Pepsins (pepsinogens)	HCl	Proteins and polypeptides	Cleave peptide bonds adjacent to aromatic amino acids
	Gastric lipase		Triglycerides	Fatty acids and glycerol
Exocrine pancreas	Trypsin (trypsinogen)	Enteropeptidase	Proteins and polypeptides	Cleave peptide bonds on carboxyl side of basic amino acids (arginine or lysine)
	Chymotrypsins (chymotrypsinogens)	Trypsin	Proteins and polypeptides	Cleave peptide bonds on carboxyl side of aromatic amino acids
	Elastase (proelastase)	Trypsin	Elastin, some other proteins	Cleaves bonds on carboxyl side of aliphatic amino acids
	Carboxypeptidase A (procarboxypeptidase A)	Trypsin	Proteins and polypeptides	Cleave carboxyl terminal amino acids that have aromatic or branched aliphatic side chains
	Carboxypeptidase B (procarboxypeptidase B)	Trypsin	Proteins and polypeptides	Cleave carboxyl terminal amino acids that have basic side chains
	Colipase (procolipase)	Trypsin	Fat droplets	Binds pancreatic lipase to oil droplet in the presence of bile acids
	Pancreatic lipase	...	Triglycerides	Monoglycerides and fatty acids
	Cholesteryl ester hydrolase	...	Cholesteryl esters	Cholesterol
	Pancreatic α-amylase	Cl^-	Starch	Same as salivary α-amylase
	Ribonuclease	...	RNA	Nucleotides
	Deoxyribonuclease	...	DNA	Nucleotides
	Phospholipase A_2 (pro-phospholipase A^2)	Trypsin	Phospholipids	Fatty acids, lysophospholipids
Intestinal mucosa	Enteropeptidase	...	Trypsinogen	Trypsin
	Aminopeptidases	...	Polypeptides	Cleave amino terminal amino acid from peptide
	Carboxypeptidases	...	Polypeptides	Cleave carboxyl terminal amino acid from peptide
	Endopeptidases	...	Polypeptides	Cleave between residues in midportion of peptide
	Dipeptidases	...	Dipeptides	Two amino acids
	Maltase	...	Maltose, maltotriose	Glucose
	Lactase	...	Lactose	Galactose and glucose
	Sucrase[b]	...	Sucrose; also maltotriose and maltose	Fructose and glucose
	Isomaltase[b]	...	α-limit dextrins, maltose maltotriose	Glucose
	Nuclease and related enzymes	...	Nucleic acids	Pentoses and purine and pyrimidine bases
Cytoplasm of mucosal cells	Various peptidases	...	Di-, tri-, and tetrapeptides	Amino acids

[a]Corresponding proenzymes, where relevant, are shown in parentheses.

[b]Sucrase and isomaltase are separate subunits of a single protein.

ANATOMIC CONSIDERATIONS

The portion of the pancreas that secretes pancreatic juice is a compound alveolar gland resembling the salivary glands. Granules containing the digestive enzymes **(zymogen granules)** are formed in the cell and discharged by exocytosis (see Chapter 2) from the apexes of the cells into the lumens of the pancreatic ducts (Figure 25–10). The small duct radicles coalesce into a single duct (pancreatic duct of Wirsung), which usually joins the common bile duct to form the ampulla of Vater (Figure 25–11). The ampulla opens through the duodenal papilla, and its orifice is encircled by the sphincter of Oddi. Some individuals have an accessory pancreatic duct (duct of Santorini) that enters the duodenum more proximally.

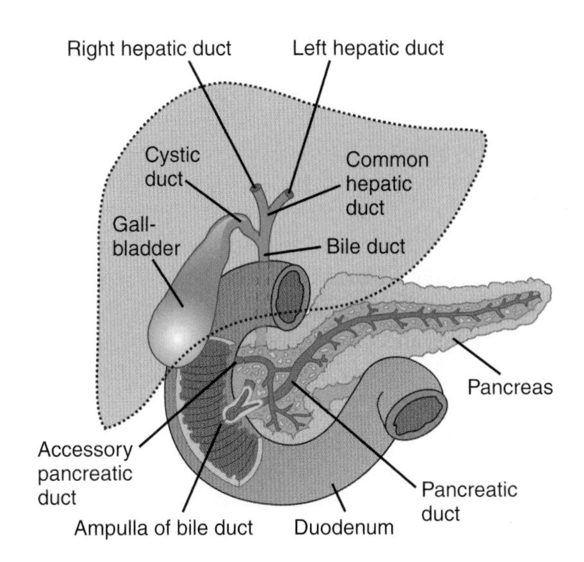

FIGURE 25–11 Connections of the ducts of the gallbladder, liver, and pancreas. (Adapted from Bell GH, Emslie-Smith D, Paterson CR: *Textbook of Physiology and Biochemistry,* 9th ed. Churchill Livingstone, 1976.)

COMPOSITION OF PANCREATIC JUICE

The pancreatic juice is alkaline (Table 25–3) and has a high HCO_3^- content (approximately 113 mEq/L vs 24 mEq/L in plasma). About 1500 mL of pancreatic juice is secreted per day. Bile and intestinal juices are also neutral or alkaline, and these three secretions neutralize the gastric acid, raising the pH of the duodenal contents to 6.0–7.0. By the time the chyme reaches the jejunum, its pH is nearly neutral, but the intestinal contents are rarely alkaline.

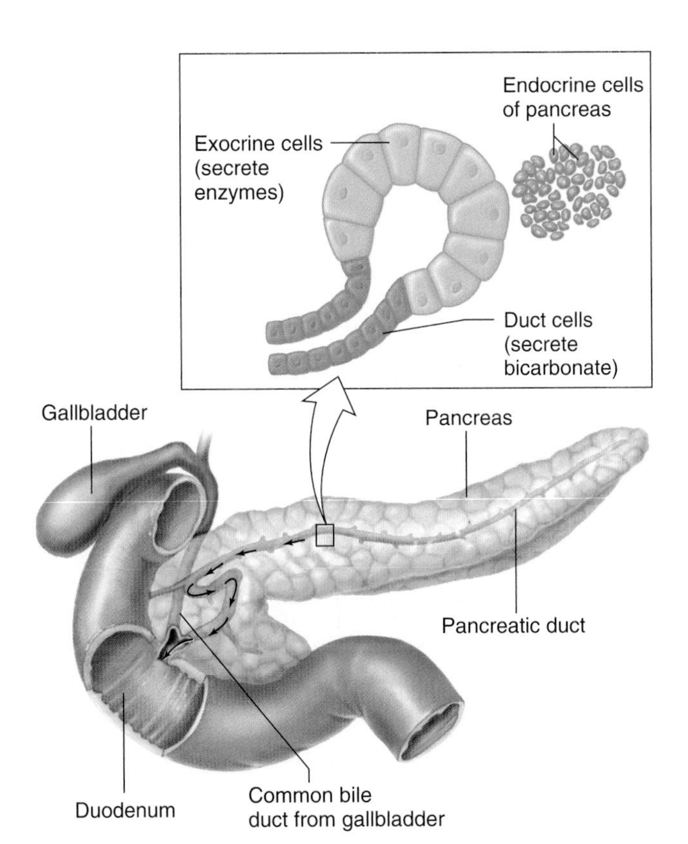

FIGURE 25–10 Structure of the pancreas. (Reproduced with permission from Widmaier EP, Raff H, Strang KT: *Vander's Human Physiology: The Mechanisms of Body Function,* 11th ed. McGraw-Hill, 2008.)

The pancreatic juice contains also contains a range of digestive enzymes, but most of these are released in inactive forms and only activated when they reach the intestinal lumen (see Chapter 26). The enzymes are activated following proteolytic cleavage by trypsin, itself a pancreatic protease that is released as an inactive precursor (trypsinogen). The potential danger of the release into the pancreas of a small amount of trypsin is apparent; the resulting chain reaction would produce active enzymes that could digest the pancreas. It is therefore not surprising that the pancreas also normally secretes a trypsin inhibitor.

Another enzyme activated by trypsin is phospholipase A_2. This enzyme splits a fatty acid off phosphatidylcholine (PC), forming lyso-PC. Lyso-PC damages cell membranes. It has been hypothesized that in **acute pancreatitis,** a severe and sometimes fatal disease, phospholipase A_2 is activated prematurely in the pancreatic ducts, with the formation of lyso-PC from the PC that is a normal constituent of bile. This causes disruption of pancreatic tissue and necrosis of surrounding fat.

Small amounts of pancreatic digestive enzymes normally leak into the circulation, but in acute pancreatitis, the circulating levels of the digestive enzymes rise markedly. Measurement of the plasma amylase or lipase concentration is therefore of value in diagnosing the disease.

TABLE 25–3 Composition of normal human pancreatic juice.

Cations: Na^+, K^+, Ca^{2+}, Mg^{2+} (pH approximately 8.0)
Anions: HCO_3^-, Cl^-, SO_4^{2-}, HPO_4^{2-}
Digestive enzymes (see Table 25–1; 95% of protein in juice)
Other proteins

REGULATION OF THE SECRETION OF PANCREATIC JUICE

Secretion of pancreatic juice is primarily under hormonal control. Secretin acts on the pancreatic ducts to cause copious secretion of a very alkaline pancreatic juice that is rich in HCO_3^- and poor in enzymes. The effect on duct cells is due to an increase in intracellular cAMP. Secretin also stimulates bile secretion. CCK acts on the acinar cells to cause the release of zymogen granules and production of pancreatic juice rich in enzymes but low in volume. Its effect is mediated by phospholipase C (see Chapter 2).

The response to intravenous secretin is shown in Figure 25–12. Note that as the volume of pancreatic secretion increases, its Cl^- concentration falls and its HCO_3^- concentration increases. Although HCO_3^- is secreted in the small ducts, it is reabsorbed in the large ducts in exchange for Cl^- (Figure 25–13). The magnitude of the exchange is inversely proportionate to the rate of flow.

Like CCK, acetylcholine acts on acinar cells via phospholipase C to cause discharge of zymogen granules, and stimulation of the vagi causes secretion of a small amount of pancreatic juice rich in enzymes. There is evidence for vagally mediated conditioned reflex secretion of pancreatic juice in response to the sight or smell of food.

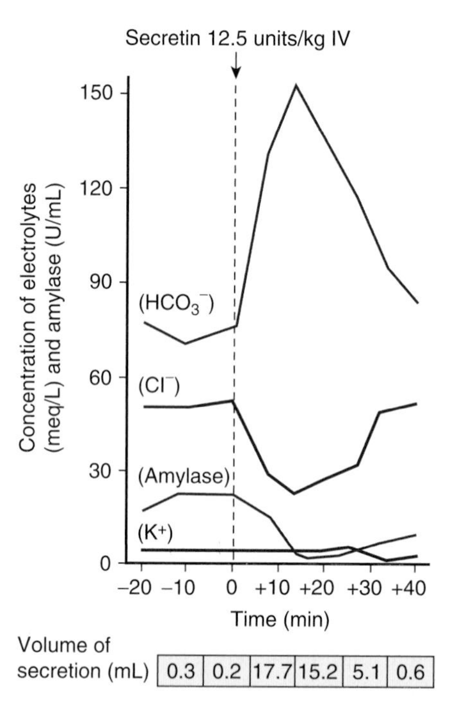

FIGURE 25–12 Effect of a single dose of secretin on the composition and volume of the pancreatic juice in humans. Note the reciprocal changes in the concentrations of chloride and bicarbonate after secretin is infused. The fall in amylase concentration reflects dilution as the volume of pancreatic juice increases.

BILIARY SECRETION

An additional secretion important for gastrointestinal function, bile, arises from the liver. The bile acids contained therein are important in the digestion and absorption of fats. In addition, bile serves as a critical excretory fluid by which the body disposes of lipid soluble end products of metabolism as well as lipid soluble xenobiotics. Bile is also the only route by which the body can dispose of cholesterol—either in its native form, or following conversion to bile acids. In this chapter and the next, we will be concerned with the role of bile as a digestive fluid. In Chapter 28, a more general consideration of the transport and metabolic functions of the liver will be presented.

Bile

Bile is made up of the bile acids, bile pigments, and other substances dissolved in an alkaline electrolyte solution that resembles pancreatic juice. About 500 mL is secreted per day.

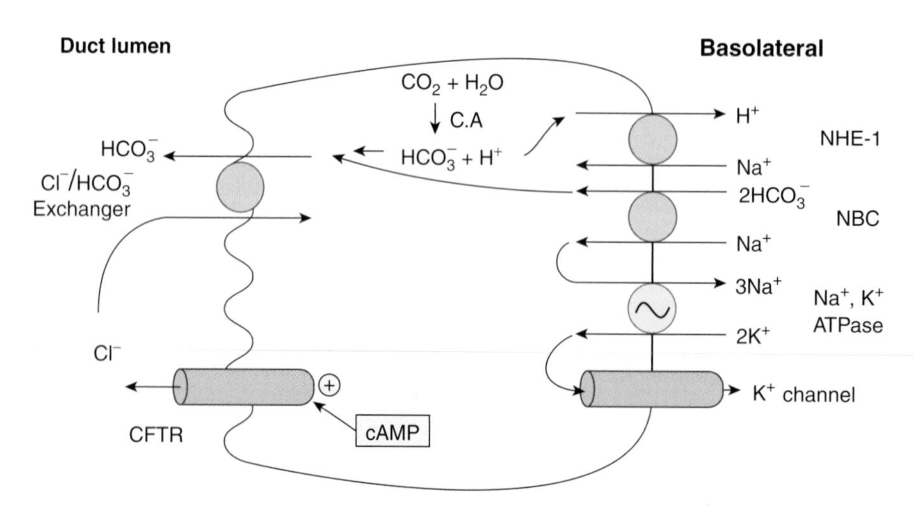

FIGURE 25–13 Ion transport pathways present in pancreatic duct cells. CA, carbonic anhydrase; NHE-1, sodium/hydrogen exchanger-1; NBC, sodium-bicarbonate cotransporter. (Adapted from Barrett KE: *Gastrointestinal Physiology*. McGraw-Hill, 2006.)

Some of the components of the bile are reabsorbed in the intestine and then excreted again by the liver **(enterohepatic circulation).**

The glucuronides of the **bile pigments,** bilirubin and biliverdin, are responsible for the golden yellow color of bile. The formation of these breakdown products of hemoglobin is discussed in detail in Chapter 28.

When considering bile as a digestive secretion, it is the **bile acids** that represent the most important components. They are synthesized from cholesterol and secreted into the bile conjugated to glycine or taurine, a derivative of cysteine. The four major bile acids found in humans are listed in Figure 25–14. In common with vitamin D, cholesterol, a variety of steroid hormones, and the digitalis glycosides, the bile acids contain the steroid nucleus (see Chapter 20). The two principal (primary) bile acids formed in the liver are cholic acid and chenodeoxycholic acid. In the colon, bacteria convert cholic acid to deoxycholic acid and chenodeoxycholic acid to lithocholic acid. In addition, small quantities of ursodeoxycholic acid are formed from chenodeoxycholic acid. Ursodeoxycholic acid is a tautomer of chenodeoxycholic acid at the 7-position. Because they are formed by bacterial action, deoxycholic, lithocholic, and ursodeoxycholic acids are called secondary bile acids.

The bile acids have a number of important actions: they reduce surface tension and, in conjunction with phospholipids and monoglycerides, are responsible for the emulsification of fat preparatory to its digestion and absorption in the small intestine (see Chapter 26). They are **amphipathic,** that is, they have both hydrophilic and hydrophobic domains; one surface of the molecule is hydrophilic because the polar peptide bond and the carboxyl and hydroxyl groups are on that surface, whereas the other surface is hydrophobic. Therefore, the bile acids tend to form cylindrical disks called **micelles.** (Figure 25–15). Their hydrophilic portions face out and their hydrophobic portions face in. Above a certain concentration, called the **critical micelle concentration,** all bile salts added to a solution form micelles. Ninety to 95% of the bile acids are absorbed from the small intestine. Once they are deconjugated, they can be absorbed by nonionic diffusion, but most are absorbed in their conjugated forms from the terminal ileum (Figure 25–16) by an extremely efficient Na^+–bile salt cotransport system (ABST) whose activity is secondarily driven by the low intracellular sodium concentration established by the basolateral Na, K ATPase. The remaining 5–10% of the bile salts enter the colon and are converted to the salts of deoxycholic acid and lithocholic acid. Lithocholate is relatively insoluble and is mostly excreted in the stools; only 1% is absorbed. However, deoxycholate is absorbed.

The absorbed bile acids are transported back to the liver in the portal vein and reexcreted in the bile (enterohepatic circulation) (Figure 25–16). Those lost in the stool are replaced by synthesis in the liver; the normal rate of bile acid synthesis is 0.2–0.4 g/d. The total bile acid pool of approximately 3.5 g recycles repeatedly via the enterohepatic circulation; it has been calculated that the entire pool recycles twice per meal and 6–8 times per day.

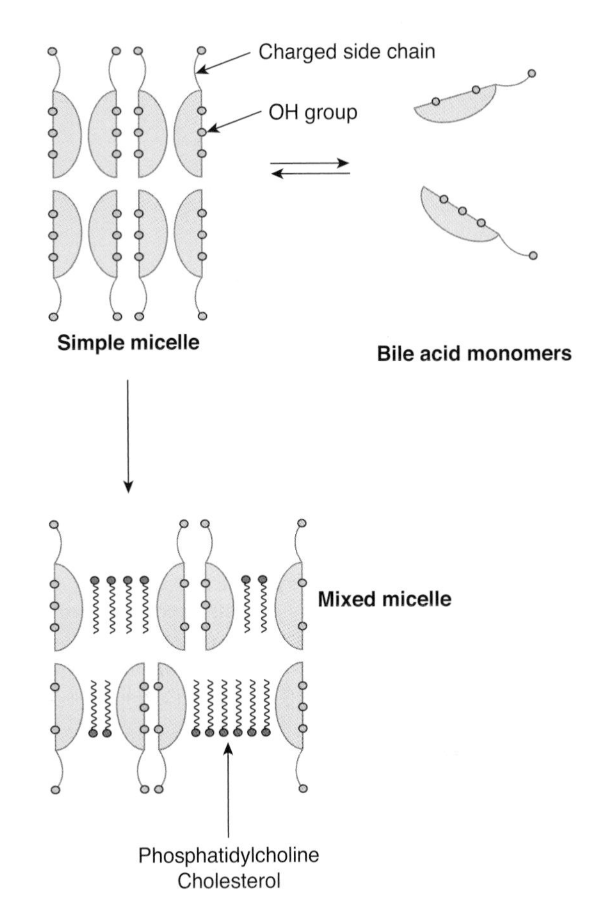

FIGURE 25–15 Physical forms adopted by bile acids in solution. Micelles are shown in cross-section, and are actually thought to be cylindrical in shape. Mixed micelles of bile acids present in hepatic bile also incorporate cholesterol and phosphatidylcholine. (Adapted from Barrett KE: *Gastrointestinal Physiology*. McGraw-Hill, 2006.)

Cholic acid

	Group at position 3	7	12	Percent in human bile
Cholic acid	OH	OH	OH	50
Chenodeoxycholic acid	OH	OH	H	30
Deoxycholic acid	OH	H	OH	15
Lithocholic acid	OH	H	H	5

FIGURE 25–14 Human bile acids. The numbers in the formula for cholic acid refer to the positions in the steroid ring.

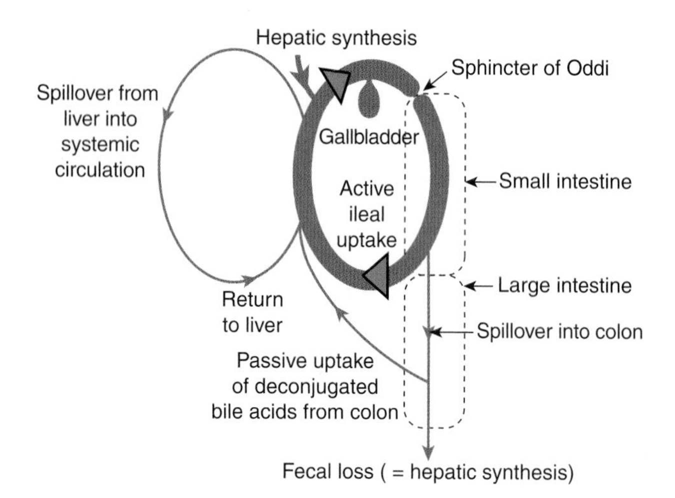

FIGURE 25–16 Quantitative aspects of the circulation of bile acids. The majority of the bile acid pool circulates between the small intestine and liver. A minority of the bile acid pool is in the systemic circulation (due to incomplete hepatocyte uptake from the portal blood) or spills over into the colon and is lost to the stool. Fecal loss must be equivalent to hepatic synthesis of bile acids at steady state. (Adapted from Barrett KE: *Gastrointestinal Physiology.* McGraw-Hill, 2006.)

TABLE 25–4 Daily water turnover (mL) in the gastrointestinal tract.

Ingested		2000
Endogenous secretions		7000
Salivary glands	1500	
Stomach	2500	
Bile	500	
Pancreas	1500	
Intestine	+1000	
	7000	
Total input		9000
Reabsorbed		8800
Jejunum	5500	
Ileum	2000	
Colon	+1300	
	8800	
Balance in stool		200

Data from Moore EW: *Physiology of Intestinal Water and Electrolyte Absorption.* American Gastroenterological Society, 1976.

INTESTINAL FLUID & ELECTROLYTE TRANSPORT

The intestine itself also supplies a fluid environment in which the processes of digestion and absorption can occur. Then, when the meal has been assimilated, fluid used during digestion and absorption is reclaimed by transport back across the epithelium to avoid dehydration. Water moves passively into and out of the gastrointestinal lumen, driven by electrochemical gradients established by the active transport of ions and other solutes. In the period after a meal, much of the fluid reuptake is driven by the coupled transport of nutrients, such as glucose, with sodium ions. In the period between meals, absorptive mechanisms center exclusively around electrolytes. In both cases, secretory fluxes of fluid are largely driven by the active transport of chloride ions into the lumen, although absorption still predominates overall.

Overall water balance in the gastrointestinal tract is summarized in Table 25–4. The intestines are presented each day with about 2000 mL of ingested fluid plus 7000 mL of secretions from the mucosa of the gastrointestinal tract and associated glands. Ninety-eight per cent of this fluid is reabsorbed, with a daily fluid loss of only 200 mL in the stools.

In the small intestine, secondary active transport of Na^+ is important in bringing about absorption of glucose, some amino acids, and other substances such as bile acids (see above). Conversely, the presence of glucose in the intestinal lumen facilitates the reabsorption of Na^+. In the period between meals, when nutrients are not present, sodium and chloride are absorbed together from the lumen by the coupled activity of a sodium/hydrogen exchanger (NHE) and chloride/bicarbonate exchanger in the apical membrane, in a so-called electroneutral mechanism (Figure 25–17). Water then follows to maintain an osmotic balance. In the colon, moreover, an additional electrogenic mechanism for sodium absorption is expressed, particularly in the distal colon. In this mechanism, sodium enters across the apical membrane via an ENaC (epithelial sodium) channel that is identical to that expressed in the distal tubule of the kidney (Figure 25–18). This underpins the ability of the colon to desiccate the stool and ensure that only a small portion of the fluid load used daily in the digestion and absorption of meals is lost from the body. Following a low-salt diet, increased expression of ENaC in response to aldosterone increases the ability to reclaim sodium from the stool.

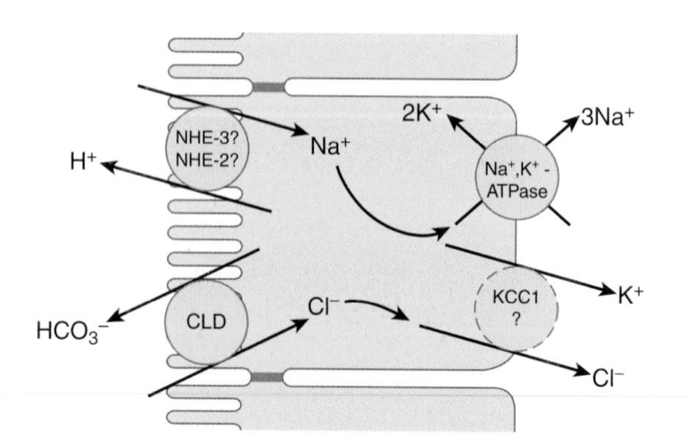

FIGURE 25–17 Electroneutral NaCl absorption in the small intestine and colon. NaCl enters across the apical membrane via the coupled activity of a sodium/hydrogen exchanger (NHE) and a chloride/bicarbonate exchanger (CLD). A putative potassium/chloride cotransporter (KCC1) in the basolateral membrane provides for chloride exit, whereas sodium is extruded by the Na, K ATPase.

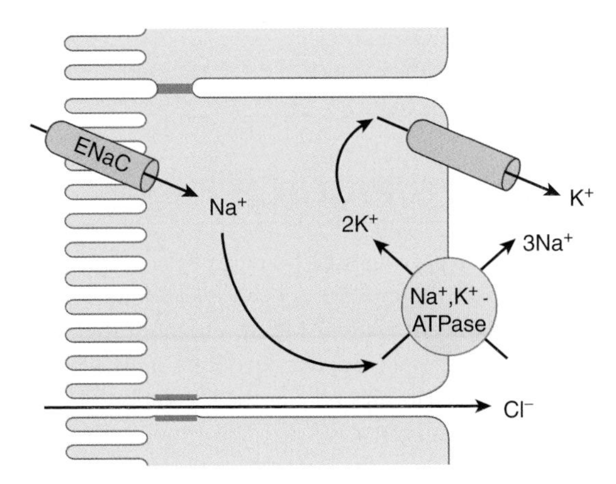

FIGURE 25–18 Electrogenic sodium absorption in the colon. Sodium enters the epithelial cell via apical epithelial sodium channels (ENaC), and exits via the Na, K ATPase.

Despite the predominance of absorptive mechanisms, secretion also takes place continuously throughout the small intestine and colon to adjust the local fluidity of the intestinal contents as needed for mixing, diffusion, and movement of the meal and its residues along the length of the gastrointestinal tract. Cl^- normally enters enterocytes from the interstitial fluid via Na^+–K^+–$2Cl^-$ cotransporters in their basolateral membranes (Figure 25–19), and the Cl^- is then secreted into the intestinal lumen via channels that are regulated by various protein kinases. The cystic fibrosis transmembrane conductance regulator (CFTR) channel that is defective in the disease of cystic fibrosis is quantitatively most important, and is activated by protein kinase A and hence by cAMP (see Clinical Box 25–2).

Water moves into or out of the intestine until the osmotic pressure of the intestinal contents equals that of the plasma.

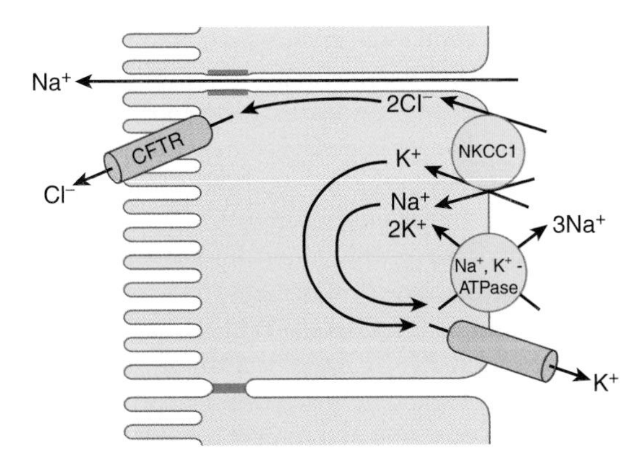

FIGURE 25–19 Chloride secretion in the small intestine and colon. Chloride uptake occurs via the sodium/potassium/2 chloride cotransporter, NKCC1. Chloride exit is via the cystic fibrosis transmembrane conductance regulator (CFTR) as well as perhaps via other chloride channels, not shown.

CLINICAL BOX 25–2

Cholera

Cholera is a severe secretory diarrheal disease that often occurs in epidemics associated with natural disasters where normal sanitary practices break down. Along with other secretory diarrheal illnesses produced by bacteria and viruses, cholera causes a significant amount of morbidity and mortality, particularly among the young and in developing countries. The cAMP concentration in intestinal epithelial cells is increased in cholera. The cholera bacillus stays in the intestinal lumen, but it produces a toxin that binds to GM-1 ganglioside receptors on the apical membrane of intestinal epithelial cells, and this permits part of the A subunit (A^1 peptide) of the toxin to enter the cell. The A^1 peptide binds adenosine diphosphate ribose to the α subunit of G^s, inhibiting its GTPase activity (see Chapter 2). Therefore, the constitutively activated G protein produces prolonged stimulation of adenylyl cyclase and a marked increase in the intracellular cAMP concentration. In addition to increased Cl^- secretion, the function of the mucosal NHE transporter for Na^+ is reduced, thus reducing NaCl absorption. The resultant increase in electrolyte and water content of the intestinal contents causes the diarrhea. However, Na, K ATPase and the Na^+/glucose cotransporter are unaffected, so coupled reabsorption of glucose and Na^+ bypasses the defect.

THERAPEUTIC HIGHLIGHTS

Treatment for cholera is mostly supportive, since the infection will eventually clear, although antibiotics are sometimes used. The most important therapeutic approach is to ensure that the large volumes of fluid, along with electrolytes, lost to the stool are replaced to avoid dehydration. Stool volumes can approach 20 L per day. When sterile supplies are available, fluids and electrolytes can most conveniently be replaced intravenously. However, this is often not possible in the setting of an epidemic. Instead, the persistent activity of the Na^+/glucose cotransporter provides a physiologic basis for the treatment of Na^+ and water loss by oral administration of solutions containing NaCl and glucose. Cereals containing carbohydrates to which salt has been added are also useful in the treatment of diarrhea. Oral rehydration solution, a prepackaged mixture of sugar and salt to be dissolved in water, is a simple remedy that has dramatically reduced mortality in epidemics of cholera and other diarrheal diseases in developing countries.

The osmolality of the duodenal contents may be hypertonic or hypotonic, depending on the meal ingested, but by the time the meal enters the jejunum, its osmolality is close to that of plasma. This osmolality is maintained throughout the rest of the small intestine; the osmotically active particles produced by digestion are removed by absorption, and water moves passively out of the gut along the osmotic gradient thus generated. In the colon, Na^+ is pumped out and water moves passively with it, again along the osmotic gradient. **Saline cathartics** such as magnesium sulfate are poorly absorbed salts that retain their osmotic equivalent of water in the intestine, thus increasing intestinal volume and consequently exerting a laxative effect.

Some K^+ is secreted into the intestinal lumen, especially as a component of mucus. K^+ channels are present in the luminal as well as the basolateral membrane of the enterocytes of the colon, so K^+ is secreted into the colon. In addition, K^+ moves passively down its electrochemical gradient. The accumulation of K^+ in the colon is partially offset by H^+–K^+ ATPase in the luminal membrane of cells in the distal colon, with resulting active transport of K^+ into the cells. Nevertheless, loss of ileal or colonic fluids in chronic diarrhea can lead to severe hypokalemia. When the dietary intake of K^+ is high for a prolonged period, aldosterone secretion is increased and more K^+ enters the colonic lumen. This is due in part to the appearance of more Na, K ATPase pumps in the basolateral membranes of the cells, with a consequent increase in intracellular K^+ and K^+ diffusion across the luminal membranes of the cells.

GASTROINTESTINAL REGULATION

The various functions of the gastrointestinal tract, including secretion, digestion, and absorption (Chapter 26) and motility (Chapter 27) must be regulated in an integrated way to ensure efficient assimilation of nutrients after a meal. There are three main modalities for gastrointestinal regulation that operate in a complementary fashion to ensure that function is appropriate. First, **endocrine** regulation is mediated by the release of hormones by triggers associated with the meal. These hormones travel through the bloodstream to change the activity of a distant segment of the gastrointestinal tract, an organ draining into it (eg, the pancreas), or both. Second, some similar mediators are not sufficiently stable to persist in the bloodstream, but instead alter the function of cells in the local area where they are released, in a **paracrine** fashion. Finally, the intestinal system is endowed with extensive neural connections. These include connections to the central nervous system **(extrinsic innervation),** but also the activity of a largely autonomous **enteric nervous system** that comprises both sensory and secreto-motor neurons. The enteric nervous system integrates central input to the gut, but can also regulate gut function independently in response to changes in the luminal environment. In some cases, the same substance can mediate regulation by endocrine, paracrine, and neurocrine pathways (eg, CCK, see below).

HORMONES/PARACRINES

Biologically active polypeptides that are secreted by nerve cells and gland cells in the mucosa act in a paracrine fashion, but they also enter the circulation. Measurement of their concentrations in blood after a meal has shed light on the roles these **gastrointestinal hormones** play in the regulation of gastrointestinal secretion and motility.

When large doses of the hormones are given, their actions overlap. However, their physiologic effects appear to be relatively discrete. On the basis of structural similarity and, to a degree, similarity of function, the key hormones fall into one of two families: the gastrin family, the primary members of which are gastrin and CCK; and the secretin family, the primary members of which are secretin, glucagon, vasoactive intestinal peptide (VIP; actually a neurotransmitter, or neurocrine), and gastric inhibitory polypeptide (also known as glucose-dependent insulinotropic peptide, or GIP). There are also other hormones that do not fall readily into these families.

ENTEROENDOCRINE CELLS

More than 15 types of hormone-secreting **enteroendocrine cells** have been identified in the mucosa of the stomach, small intestine, and colon. Many of these secrete only one hormone and are identified by letters (G cells, S cells, etc). Others manufacture serotonin or histamine and are called **enterochromaffin** or **ECL cells,** respectively.

GASTRIN

Gastrin is produced by cells called G cells in the antral portion of the gastric mucosa (Figure 25–20). G cells are flask-shaped, with a broad base containing many gastrin granules and a narrow apex that reaches the mucosal surface. Microvilli project from the apical end into the lumen. Receptors mediating gastrin responses to changes in gastric contents are present on the microvilli. Other cells in the gastrointestinal tract that secrete hormones have a similar morphology.

The precursor for gastrin, preprogastrin is processed into fragments of various sizes. Three main fragments contain 34, 17, and 14 amino acid residues. All have the same carboxyl terminal configuration (Table 25–5). These forms are also known as G 34, G 17, and G 14 gastrins, respectively. Another form is the carboxyl terminal tetrapeptide, and there is also a large form that is extended at the amino terminal and contains more than 45 amino acid residues. One form of derivatization is sulfation of the tyrosine that is the sixth amino acid residue from the carboxyl terminal. Approximately equal amounts of nonsulfated and sulfated forms are present in blood and tissues, and they are equally active. Another derivatization

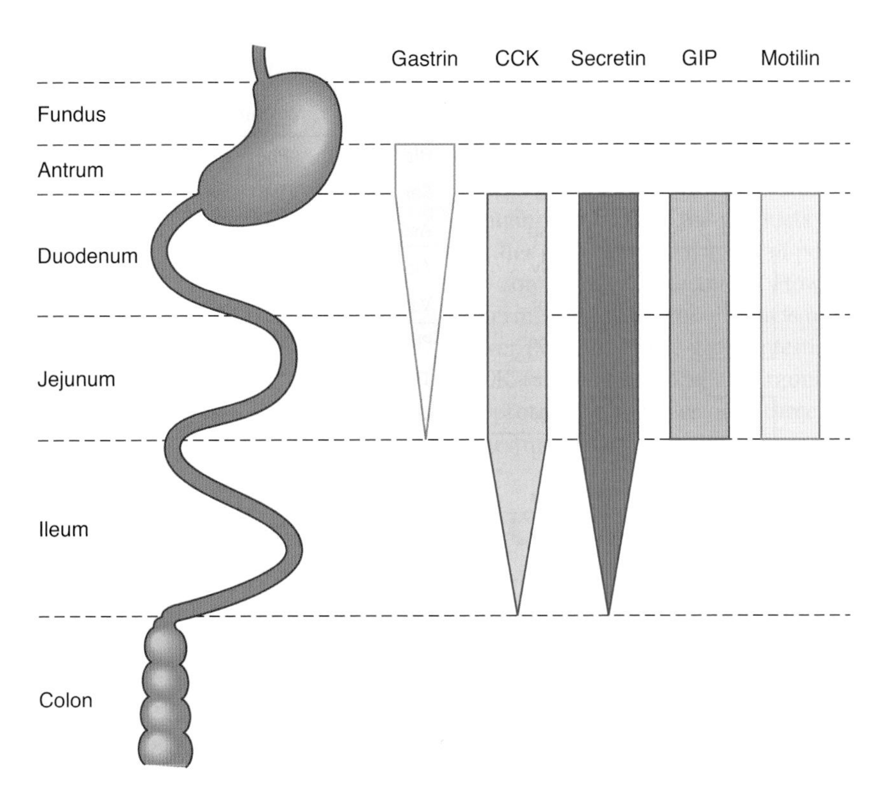

FIGURE 25–20 Sites of production of the five gastrointestinal hormones along the length of the gastrointestinal tract. The width of the bars reflects the relative abundance at each location.

is amidation of the carboxyl terminal phenylalanine, which likely enhances the peptide's stability in the plasma by rendering it resistant to carboxypeptidases.

Some differences in activity exist between the various gastrin peptides, and the proportions of the components also differ in the various tissues in which gastrin is found. This suggests that different forms are tailored for different actions. However, all that can be concluded at present is that G 17 is the principal form with respect to gastric acid secretion. The carboxyl terminal tetrapeptide has all the activities of gastrin but only 10% of the potency of G 17.

G 14 and G 17 have half-lives of 2–3 min in the circulation, whereas G 34 has a half-life of 15 min. Gastrins are inactivated primarily in the kidney and small intestine.

In large doses, gastrin has a variety of actions, but its principal physiologic actions are stimulation of gastric acid and pepsin secretion and stimulation of the growth of the mucosa of the stomach and small and large intestines **(trophic action).** Gastrin secretion is affected by the contents of the stomach, the rate of discharge of the vagus nerves, and bloodborne factors (Table 25–6). Atropine does not inhibit the gastrin response to a test meal in humans, because the transmitter secreted by the postganglionic vagal fibers that innervate the G cells is gastrin-releasing polypeptide (GRP; see below) rather than acetylcholine. Gastrin secretion is also increased by the presence of the products of protein digestion in the stomach, particularly amino acids, which act directly on the G cells. Phenylalanine and tryptophan are particularly effective. Gastrin acts via a receptor (CCK-B) that is related to the primary receptor (CCK-A) for cholecystokinin (see below). This likely reflects the structural similarity of the two hormones, and may result in some overlapping actions if excessive quantities of either hormone are present (eg, in the case of a gastrin-secreting tumor, or gastrinoma).

Acid in the antrum inhibits gastrin secretion, partly by a direct action on G cells and partly by release of somatostatin, a relatively potent inhibitor of gastrin secretion. The effect of acid is the basis of a negative feedback loop regulating gastrin secretion. Increased secretion of the hormone increases acid secretion, but the acid then feeds back to inhibit further gastrin secretion. In conditions such as pernicious anemia in which the acid-secreting cells of the stomach are damaged, gastrin secretion is chronically elevated.

CHOLECYSTOKININ

CCK is secreted by endocrine cells known as I cells in the mucosa of the upper small intestine. It has a plethora of actions in the gastrointestinal system, but the most important appear to be the stimulation of pancreatic enzyme secretion, the contraction of the gallbladder (the action for which it was named), and relaxation of the sphincter of Oddi, which allows both bile and pancreatic juice to flow into the intestinal lumen.

Like gastrin, CCK is produced from a larger precursor. Prepro-CCK is also processed into many fragments. A large CCK contains 58 amino acid residues (CCK 58). In addition, there are CCK peptides that contain 39 amino acid residues (CCK 39) and 33 amino acid residues (CCK 33), several forms

and because in large doses it inhibits gastric secretion and motility, it was named gastric inhibitory peptide. However, it now appears that it does not have significant gastric inhibiting activity when administered in smaller amounts comparable to those seen after a meal. In the meantime, it was found that GIP stimulates insulin secretion. Gastrin, CCK, secretin, and glucagon also have this effect, but GIP is the only one of these that stimulates insulin secretion when administered at blood levels comparable to those produced by oral glucose. For this reason, it is often called **glucose-dependent insulinotropic peptide.** The glucagon derivative GLP-1 (7–36) (see Chapter 24) also stimulates insulin secretion and is said to be more potent in this regard than GIP. Therefore, it may also be a physiologic B cell-stimulating hormone of the gastrointestinal tract.

The integrated action of gastrin, CCK, secretin, and GIP in facilitating digestion and utilization of absorbed nutrients is summarized in Figure 25–21.

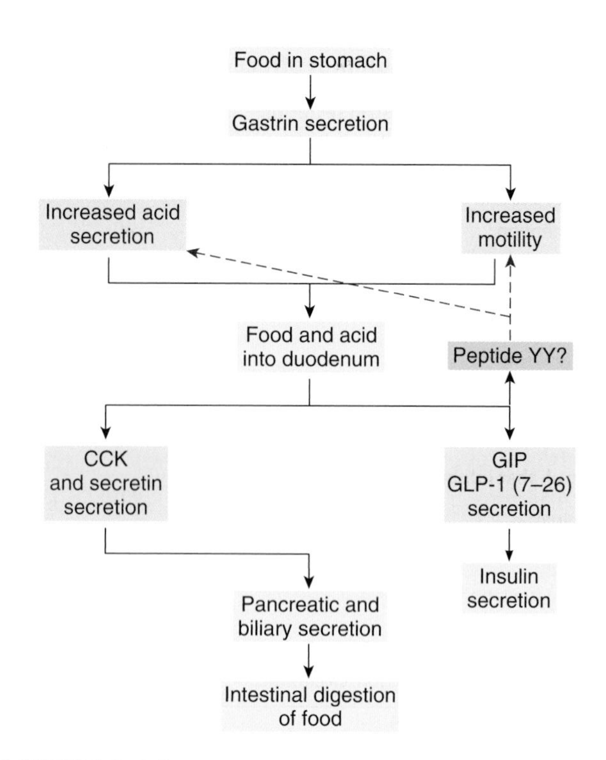

FIGURE 25–21 Integrated action of gastrointestinal hormones in regulating digestion and utilization of absorbed nutrients. The dashed arrows indicate inhibition. The exact identity of the hormonal factor or factors from the intestine that inhibit(s) gastric acid secretion and motility is unsettled, but it may be peptide YY.

VIP

VIP contains 28 amino acid residues (Table 25–5). It is found in nerves in the gastrointestinal tract and thus is not itself a hormone, despite its similarities to secretin. VIP is, however, found in blood, in which it has a half-life of about 2 min. In the intestine, it markedly stimulates intestinal secretion of electrolytes and hence of water. Its other actions include relaxation of intestinal smooth muscle, including sphincters; dilation of peripheral blood vessels; and inhibition of gastric acid secretion. It is also found in the brain and many autonomic nerves (see Chapter 7), where it often occurs in the same neurons as acetylcholine. It potentiates the action of acetylcholine in salivary glands. However, VIP and acetylcholine do not coexist in neurons that innervate other parts of the gastrointestinal tract. VIP-secreting tumors (VIPomas) have been described in patients with severe diarrhea.

MOTILIN

Motilin is a polypeptide containing 22 amino acid residues that is secreted by enterochromaffin cells and Mo cells in the stomach, small intestine, and colon. It acts on G protein-coupled receptors on enteric neurons in the duodenum and colon and produces contraction of smooth muscle in the stomach and intestines in the period between meals (see Chapter 27).

SOMATOSTATIN

Somatostatin, the growth-hormone-inhibiting hormone originally isolated from the hypothalamus, is secreted as a paracrine by D cells in the pancreatic islets (see Chapter 24) and by similar D cells in the gastrointestinal mucosa. It exists in tissues in two forms, somatostatin 14 and somatostatin 28, and both are secreted. Somatostatin inhibits the secretion of gastrin, VIP, GIP, secretin, and motilin. Its secretion is stimulated by acid in the lumen, and it probably acts in a paracrine fashion to mediate the inhibition of gastrin secretion produced by acid. It also inhibits pancreatic exocrine secretion; gastric acid secretion and motility; gallbladder contraction; and the absorption of glucose, amino acids, and triglycerides.

OTHER GASTROINTESTINAL PEPTIDES

Peptide YY

The structure of peptide YY is discussed in Chapter 24. It also inhibits gastric acid secretion and motility and is a good candidate to be the gastric inhibitory peptide (Figure 25–21). Its release from the jejunum is stimulated by fat.

Others

Ghrelin is secreted primarily by the stomach and appears to play an important role in the central control of food intake (see Chapter 26). It also stimulates growth hormone secretion by acting directly on receptors in the pituitary (see Chapter 18).

Substance P (Table 25–5) is found in endocrine and nerve cells in the gastrointestinal tract and may enter the circulation. It increases the motility of the small intestine. The neurotransmitter **GRP** contains 27 amino acid residues, and the 10 amino acid residues at its carboxyl terminal are almost identical to those of amphibian **bombesin.** It is present in the vagal nerve endings that terminate on G cells and is the neurotransmitter producing vagally mediated increases in gastrin secretion. **Glucagon** from the gastrointestinal tract may be responsible (at least in part) for the hyperglycemia seen after pancreatectomy.

Guanylin is a gastrointestinal polypeptide that binds to guanylyl cyclase. It is made up of 15 amino acid residues (Table 25–5) and is secreted by cells of the intestinal mucosa. Stimulation of guanylyl cyclase increases the concentration of intracellular cyclic 3′,5′-guanosine monophosphate (cGMP), and this in turn causes increased secretion of Cl^- into the intestinal lumen. Guanylin appears to act predominantly in a paracrine fashion, and it is produced in cells from the pylorus to the rectum. In an interesting example of molecular mimicry, the heat-stable enterotoxin of certain diarrhea-producing strains of *E. coli* has a structure very similar to guanylin and activates guanylin receptors in the intestine. Guanylin receptors are also found in the kidneys, the liver, and the female reproductive tract, and guanylin may act in an endocrine fashion to regulate fluid movement in these tissues as well, and particularly to integrate the actions of the intestine and kidneys.

THE ENTERIC NERVOUS SYSTEM

Two major networks of nerve fibers are intrinsic to the gastrointestinal tract: the **myenteric plexus** (Auerbach's plexus), between the outer longitudinal and middle circular muscle layers, and the **submucous plexus** (Meissner's plexus), between the middle circular layer and the mucosa (Figure 25–1). Collectively, these neurons constitute the **enteric nervous system.** The system contains about 100 million sensory neurons, inter-neurons, and motor neurons in humans—as many as are found in the whole spinal cord—and the system is probably best viewed as a displaced part of the central nervous system (CNS) that is concerned with the regulation of gastrointestinal function. It is sometimes referred to as the "little brain" for this reason. It is connected to the CNS by parasympathetic and sympathetic fibers but can function autonomously without these connections (see below). The myenteric plexus innervates the longitudinal and circular smooth muscle layers and is concerned primarily with motor control, whereas the submucous plexus innervates the glandular epithelium, intestinal endocrine cells, and submucosal blood vessels and is primarily involved in the control of intestinal secretion. The neurotransmitters in the system include acetylcholine, the amines norepinephrine and serotonin, the amino acid γ-aminobutyrate (GABA), the purine adenosine triphosphate (ATP), the gases NO and CO, and many different peptides and polypeptides. Some of these peptides also act in a paracrine fashion, and some enter the bloodstream, becoming hormones. Not surprisingly, most of them are also found in the brain.

EXTRINSIC INNERVATION

The intestine receives a dual extrinsic innervation from the autonomic nervous system, with parasympathetic cholinergic activity generally increasing the activity of intestinal smooth muscle and sympathetic noradrenergic activity generally decreasing it while causing sphincters to contract. The preganglionic parasympathetic fibers consist of about 2000 vagal efferents and other efferents in the sacral nerves. They generally end on cholinergic nerve cells of the myenteric and submucous plexuses. The sympathetic fibers are postganglionic, but many of them end on postganglionic cholinergic neurons, where the norepinephrine they secrete inhibits acetylcholine secretion by activating α_2 presynaptic receptors. Other sympathetic fibers appear to end directly on intestinal smooth muscle cells. The electrical properties of intestinal smooth muscle are discussed in Chapter 5. Still other fibers innervate blood vessels, where they produce vasoconstriction. It appears that the intestinal blood vessels have a dual innervation: they have an extrinsic noradrenergic innervation and an intrinsic innervation by fibers of the enteric nervous system. VIP and NO are among the mediators in the intrinsic innervation, which seems, among other things, to be responsible for the increase in local blood flow **(hyperemia)** that accompanies digestion of food. It is unsettled whether the blood vessels have an additional cholinergic innervation.

GASTROINTESTINAL (MUCOSAL) IMMUNE SYSTEM

The mucosal immune system was mentioned in Chapter 3, but it bears repeating here that the continuity of the intestinal lumen with the outside world also makes the gastrointestinal system an important portal for infection. Similarly, the intestine benefits from interactions with a complex community of commensal (ie, nonpathogenic) bacteria that provide beneficial metabolic functions as well as likely increasing resistance to pathogens. In the face of this constant microbial stimulation, it is not surprising that the intestine of mammals has developed a sophisticated set of both innate and adaptive immune mechanisms to distinguish friend from foe. Indeed, the intestinal mucosa contains more lymphocytes than are found in the circulation, as well as large numbers of inflammatory cells that are placed to rapidly defend the mucosa if epithelial defenses are breached. It is likely that immune cells, and their products, also impact the physiological function of the epithelium, endocrine cells, nerves and smooth muscle, particularly at times of infection and if inappropriate immune responses are perpetuated, such as in inflammatory bowel diseases (see Chapter 3).

GASTROINTESTINAL (SPLANCHNIC) CIRCULATION

A final general point that should be made about the gastrointestinal tract relates to its unusual circulatory features. The blood flow to the stomach, intestines, pancreas, and liver is arranged in a series of parallel circuits, with all the blood from the intestines and pancreas draining via the portal vein to the liver (Figure 25–22). The blood from the intestines, pancreas, and spleen drains via the hepatic portal vein to the liver and from the liver via the hepatic veins to the inferior vena cava. The viscera and the liver receive about 30% of the cardiac output via the celiac, superior mesenteric, and inferior mesenteric arteries. The liver receives about 1300 mL/min from the portal vein and 500 mL/min from the hepatic artery during fasting, and the portal supply increases still further after meals.

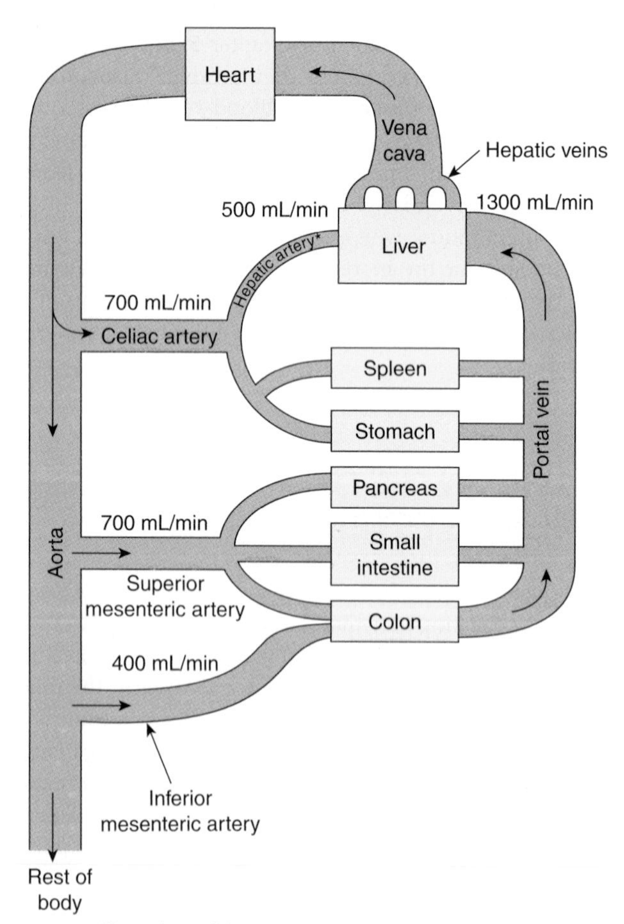

FIGURE 25–22 Schematic of the splanchnic circulation under fasting conditions. Note that even during fasting, the liver receives the majority of its blood supply via the portal vein.

CHAPTER SUMMARY

- The gastrointestinal system evolved as a portal to permit controlled nutrient uptake in multicellular organisms. It is functionally continuous with the outside environment.
- Digestive secretions serve to chemically alter the components of meals (particularly macromolecules) such that their constituents can be absorbed across the epithelium. Meal components are acted on sequentially by saliva, gastric juice, pancreatic juice, and bile, which contain enzymes, ions, water, and other specialized components.
- The intestine and the organs that drain into it secrete about 8 L of fluid per day, which are added to water consumed in food and beverages. Most of this fluid is reabsorbed, leaving only approximately 200 mL to be lost to the stool. Fluid secretion and absorption are both dependent on the active epithelial transport of ions, nutrients, or both.
- Gastrointestinal functions are regulated in an integrated fashion by endocrine, paracrine, and neurocrine mechanisms. Hormones and paracrine factors are released from enteroendocrine cells in response to signals coincident with the intake of meals.
- The enteric nervous system conveys information from the central nervous system to the gastrointestinal tract, but also often can activate programmed responses of secretion and motility in an autonomous fashion.
- The intestine harbors an extensive mucosal immune system that regulates responses to the complex microbiota normally resident in the lumen, as well as defending the body against invasion by pathogens.
- The intestine has an unusual circulation, in that the majority of its venous outflow does not return directly to the heart, but rather is directed initially to the liver via the portal vein.

MULTIPLE-CHOICE QUESTIONS

For all questions, select the single best answer unless otherwise directed.

1. Water is absorbed in the jejunum, ileum, and colon and excreted in the feces. Arrange these in order of the amount of water absorbed or excreted from greatest to smallest.
   A. Colon, jejunum, ileum, feces
   B. Feces, colon, ileum, jejunum
   C. Jejunum, ileum, colon, feces
   D. Colon, ileum, jejunum, feces
   E. Feces, jejunum, ileum, colon
2. Following a natural disaster in Haiti, there is an outbreak of cholera among displaced persons living in a tent encampment.

The affected individuals display severe diarrheal symptoms because of which of the following changes in intestinal transport?

A. Increased Na^+–K^+ cotransport in the small intestine.
B. Increased K^+ secretion into the colon.
C. Reduced K^+ absorption in the crypts of Lieberkühn.
D. Increased Na^+ absorption in the small intestine.
E. Increased Cl^- secretion into the intestinal lumen.

3. A 50-year-old man presents to his physician complaining of severe epigastric pain, frequent heartburn, and unexplained weight loss of 20 pounds over a 6-month period. He claims to have obtained no relief from over-the-counter H_2 antihistamine drugs. He is referred to a gastroenterologist, and upper endoscopy reveals erosions and ulcerations in the proximal duodenum and an increased output of gastric acid in the fasting state. The patient is most likely to have a tumor secreting which of the following hormones?

A. Secretin
B. Somatostatin
C. Motilin
D. Gastrin
E. Cholecystokinin

4. Which of the following has the highest pH?

A. Gastric juice
B. Colonic luminal contents
C. Pancreatic juice
D. Saliva
E. Contents of the intestinal crypts

5. A 60-year-old woman undergoes total pancreatectomy because of the presence of a tumor. Which of the following outcomes would *not* be expected after she recovers from the operation?

A. Steatorrhea
B. Hyperglycemia
C. Metabolic acidosis
D. Weight gain
E. Decreased absorption of amino acids

CHAPTER RESOURCES

Baron TH, Morgan DE: Current concepts: Acute necrotizing pancreatitis. N Engl J Med 1999;340:1412.

Barrett KE: *Gastrointestinal Physiology.* McGraw-Hill, 2006.

Bengmark S: Econutrition and health maintenance—A new concept to prevent GI inflammation, ulceration, and sepsis. Clin Nutr 1996;15:1.

Chong L, Marx J (editors): Lipids in the limelight. Science 2001; 294:1861.

Go VLW, et al: *The Pancreas: Biology, Pathobiology and Disease*, 2nd ed. Raven Press, 1993.

Hersey SJ, Sachs G: Gastric acid secretion. Physiol Rev 1995; 75:155.

Hofmann AF: Bile acids: The good, the bad, and the ugly. News Physiol Sci 1999;14:24.

Hunt RH, Tytgat GN (editors): *Helicobacter pylori: Basic Mechanisms to Clinical Cure.* Kluwer Academic, 2000.

Itoh Z: Motilin and clinical application. Peptides 1997;18:593.

Johnston DE, Kaplan MM: Pathogenesis and treatment of gallstones. N Engl J Med 1993;328:412.

Kunzelmann K, Mall M: Electrolyte transport in the mammalian colon: Mechanisms and implications for disease. Physiol Rev 2002;82:245.

Lamberts SWJ, et al: Octreotide. N Engl J Med 1996;334:246.

Lewis JH (editor): *A Pharmacological Approach to Gastrointestinal Disorders.* Williams & Wilkins, 1994.

Meier PJ, Stieger B: Molecular mechanisms of bile formation. News Physiol Sci 2000;15:89.

Montecucco C, Rappuoli R: Living dangerously: How *Helicobacter pylori* survives in the human stomach. Nat Rev Mol Cell Biol 2001;2:457.

Nakazato M: Guanylin family: New intestinal peptides regulating electrolyte and water homeostasis. J Gastroenterol 2001; 36:219.

Rabon EC, Reuben MA: The mechanism and structure of the gastric H^+, K^+–ATPase. Annu Rev Physiol 1990;52:321.

Sachs G, Zeng N, Prinz C: Pathophysiology of isolated gastric endocrine cells. Annu Rev Physiol 1997;59:234.

Sellin JH: SCFAs: The enigma of weak electrolyte transport in the colon. News Physiol Sci 1999;14:58.

Specian RD, Oliver MG: Functional biology of intestinal goblet cells. Am J Med 1991;260:C183.

Topping DL, Clifton PM: Short-chain fatty acids and human colonic function: Select resistant starch and nonstarch polysaccharides. Physiol Rev 2001;81:1031.

Trauner M, Meier PJ, Boyer JL: Molecular mechanisms of cholestasis. N Engl J Med 1998;339:1217.

Walsh JH (editor): *Gastrin.* Raven Press, 1993.

Williams JA, Blevins GT Jr: Cholecystokinin and regulation of pancreatic acinar cell function. Physiol Rev 1993; 73:701.

Wolfe MM, Lichtenstein DR, Singh G: Gastrointestinal toxicity of nonsteroidal anti-inflammatory drugs. N Engl J Med 1999;340:1888.

Wright EM: The intestinal Na^+/glucose cotransporter. Annu Rev Physiol 1993;55:575.

Young JA, van Lennep EW: *The Morphology of Salivary Glands.* Academic Press, 1978.

Zoetendal EG, et al: Molecular ecological analysis of the gastrointestinal microbiota: A review. J Nutr 2004;134:465.

CHAPTER 26

Digestion, Absorption, & Nutritional Principles

OBJECTIVES

After studying this chapter, you should be able to:

- Understand how nutrients are delivered to the body and the chemical processes needed to convert them to a form suitable for absorption.
- List the major dietary carbohydrates and define the luminal and brush border processes that produce absorbable monosaccharides as well as the transport mechanisms that provide for the uptake of these hydrophilic molecules.
- Understand the process of protein assimilation, and the ways in which it is comparable to, or converges from, that used for carbohydrates.
- Define the stepwise processes of lipid digestion and absorption, the role of bile acids in solubilizing the products of lipolysis, and the consequences of fat malabsorption.
- Identify the source and functions of short-chain fatty acids in the colon.
- Delineate the mechanisms of uptake for vitamins and minerals.
- Understand basic principles of energy metabolism and nutrition.

INTRODUCTION

The gastrointestinal system is the portal through which nutritive substances, vitamins, minerals, and fluids enter the body. Proteins, fats, and complex carbohydrates are broken down into absorbable units **(digested),** principally, although not exclusively, in the small intestine. The products of digestion and the vitamins, minerals, and water cross the mucosa and enter the lymph or the blood **(absorption).** The digestive and absorptive processes are the subject of this chapter.

Digestion of the major foodstuffs is an orderly process involving the action of a large number of **digestive enzymes** discussed in the previous chapter. Enzymes from the salivary glands attack carbohydrates (and fats in some species); enzymes from the stomach attack proteins and fats; and enzymes from the exocrine portion of the pancreas attack carbohydrates, proteins, lipids, DNA, and RNA. Other enzymes that complete the digestive process are found in the luminal membranes and the cytoplasm of the cells that line the small intestine. The action of the enzymes is aided by the hydrochloric acid secreted by the stomach and the bile secreted by the liver.

Most substances pass from the intestinal lumen into the enterocytes and then out of the enterocytes to the interstitial fluid. The processes responsible for movement across the luminal cell membrane are often quite different from those responsible for movement across the basal and lateral cell membranes to the interstitial fluid.

DIGESTION & ABSORPTION: CARBOHYDRATES

DIGESTION

The principal dietary carbohydrates are polysaccharides, disaccharides, and monosaccharides. Starches (glucose polymers) and their derivatives are the only polysaccharides that are digested to any degree in the human gastrointestinal tract. Amylopectin, which typically constitutes around 75% of dietary starch, is a branched molecule, whereas amylose is a straight chain with only 1:4α linkages (Figure 26–1). The disaccharides **lactose** (milk sugar) and **sucrose** (table sugar) are also ingested, along with the monosaccharides fructose and glucose.

In the mouth, starch is attacked by salivary α-amylase. The optimal pH for this enzyme is 6.7. However, it remains partially active even once it moves into the stomach, despite the acidic gastric juice, because the active site is protected in the presence of substrate to some degree. In the small intestine, both the salivary and the pancreatic α-amylase also act on the ingested polysaccharides. Both the salivary and the pancreatic α-amylases hydrolyze 1:4α linkages but spare 1:6α linkages and terminal 1:4α linkages. Consequently, the end products of α-amylase digestion are oligosaccharides: the disaccharide **maltose;** the trisaccharide **maltotriose;** and **α-limit dextrins,** polymers of glucose containing an average of about eight glucose molecules with 1:6α linkages (Figure 26–1).

The oligosaccharidases responsible for the further digestion of the starch derivatives are located in the brush border of small intestinal epithelial cells (Figure 26–1). Some of these enzymes have more than one substrate. **Isomaltase** is mainly responsible for hydrolysis of 1:6α linkages. Along with **maltase and sucrase,** it also breaks down maltotriose and maltose. Sucrase and isomaltase are initially synthesized as a single glycoprotein chain that is inserted into the brush border membrane. It is then hydrolyzed by pancreatic proteases into sucrase and isomaltase subunits.

Sucrase hydrolyzes sucrose into a molecule of glucose and a molecule of fructose. In addition, **lactase** hydrolyzes lactose to glucose and galactose.

Deficiency of one or more of the brush border oligosaccharidases may cause diarrhea, bloating, and flatulence after ingestion of sugar (Clinical Box 26–1). The diarrhea is due to the increased number of osmotically active oligosaccharide molecules that remain in the intestinal lumen, causing

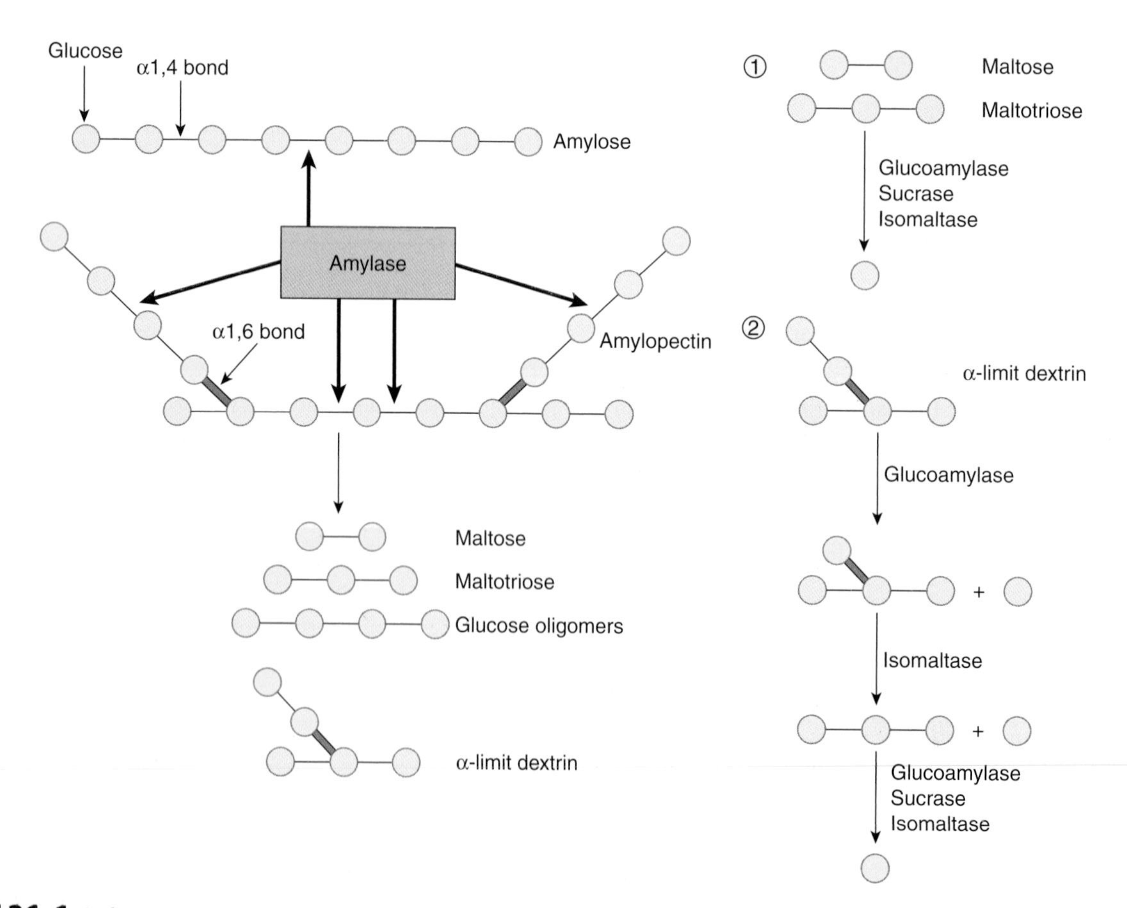

FIGURE 26–1 **Left:** Structure of amylose and amylopectin, which are polymers of glucose (indicated by circles). These molecules are partially digested by the enzyme amylase, yielding the products shown at the bottom of the figure. **Right:** Brush border hydrolases responsible for the sequential digestion of the products of luminal starch digestion (1, linear oligomers; 2, alpha-limit dextrins).

CLINICAL BOX 26–1

Lactose Intolerance

In most mammals and in many races of humans, intestinal lactase activity is high at birth, then declines to low levels during childhood and adulthood. The low lactase levels are associated with intolerance to milk **(lactose intolerance).** Most Europeans and their American descendants retain sufficient intestinal lactase activity in adulthood; the incidence of lactase deficiency in northern and western Europeans is only about 15%. However, the incidence in blacks, American Indians, Asians, and Mediterranean populations is 70–100%. When such individuals ingest dairy products, they are unable to digest lactose sufficiently, and so symptoms such as bloating, pain, gas, and diarrhea are produced by the unabsorbed osmoles that are subsequently digested by colonic bacteria.

THERAPEUTIC HIGHLIGHTS

The simplest treatment for lactose intolerance is to avoid dairy products in the diet, but this can sometimes be challenging (or undesirable for the individual who loves ice cream). Symptoms can be ameliorated by administration of commercial lactase preparations, but this is expensive. Yogurt is better tolerated than milk in intolerant individuals because it contains its own bacterial lactase.

the volume of the intestinal contents to increase. In the colon, bacteria break down some of the oligosaccharides, further increasing the number of osmotically active particles. The bloating and flatulence are due to the production of gas (CO_2 and H_2) from disaccharide residues in the lower small intestine and colon.

ABSORPTION

Hexoses are rapidly absorbed across the wall of the small intestine (Table 26–1). Essentially all the hexoses are removed before the remains of a meal reach the terminal part of the ileum. The sugar molecules pass from the

TABLE 26–1 Normal transport of substances by the intestine and location of maximum absorption or secretion.[a]

	Small Intestine			
Absorption of:	**Upper[b]**	**Mid**	**Lower**	**Colon**
Sugars (glucose, galactose, etc)	++	+++	++	0
Amino acids	++	++	++	0
Water-soluble and fat-soluble vitamins except vitamin B_{12}	+++	++	0	0
Betaine, dimethylglycine, sarcosine	+	++	++	?
Antibodies in newborns	+	++	+++	?
Pyrimidines (thymine and uracil)	+	+	?	?
Long-chain fatty acid absorption and conversion to triglyceride	+++	++	+	0
Bile acids	+	+	+++	
Vitamin B_{12}	0	+	+++	0
Na^+	+++	++	+++	+++
K^+	+	+	+	Sec
Ca^{2+}	+++	++	+	?
Fe^{2+}	+++	+	+	?
Cl^-	+++	++	+	+
SO_4^{2-}	++	+	0	?

[a] Amount of absorption is graded + to +++. Sec, secreted when luminal K^+ is low.

[b] Upper small intestine refers primarily to jejunum, although the duodenum is similar in most cases studied (with the notable exception that the duodenum secretes HCO_3^- and shows little net absorption or secretion of NaCl).

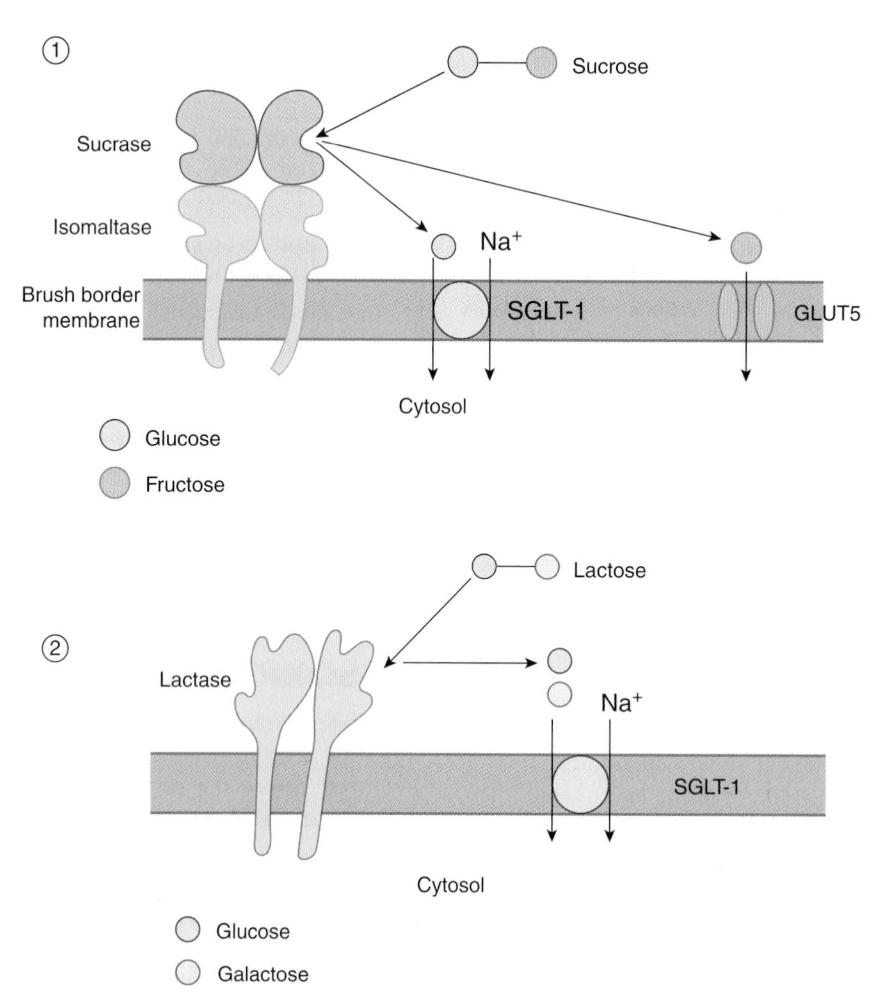

FIGURE 26–2 Brush border digestion and assimilation of the disaccharides sucrose (panel 1) and lactose (panel 2). Uptake of glucose and galactose is driven secondarily by the low intracellular sodium concentration established by the basolateral Na, K ATPase (not shown). SGLT-1, sodium-glucose cotransporter-1.

mucosal cells to the blood in the capillaries draining into the portal vein.

The transport of glucose and galactose is dependent on Na^+ in the intestinal lumen; a high concentration of Na^+ on the mucosal surface of the cells facilitates and a low concentration inhibits sugar influx into the epithelial cells. This is because these sugars and Na^+ share the same **cotransporter,** or **symport,** the **sodium-dependent glucose transporter** (SGLT, Na^+ glucose cotransporter) (Figure 26–2). The members of this family of transporters, SGLT 1 and SGLT 2, resemble the glucose transporters (GLUTs) responsible for facilitated diffusion (see Chapter 24) in that they cross the cell membrane 12 times and have their –COOH and $–NH_2$ terminals on the cytoplasmic side of the membrane. However, there is no homology to the GLUT series of transporters. SGLT-1 is responsible for uptake of dietary glucose from the gut. The related transporter, SGLT-2, is responsible for glucose transport out of the renal tubules (see Chapter 37).

Because the intracellular Na^+ concentration is low in intestinal cells (as it is in other cells), Na^+ moves into the cell along its concentration gradient. Glucose moves with the Na^+ and is released in the cell (Figure 26–2). The Na^+ is transported into the lateral intercellular spaces, and the glucose is transported by GLUT 2 into the interstitium and thence to the capillaries. Thus, glucose transport is an example of secondary active transport (see Chapter 2); the energy for glucose transport is provided indirectly, by the active transport of Na^+ out of the cell. This maintains the concentration gradient across the luminal border of the cell, so that more Na^+ and consequently more glucose enter. When the Na^+/glucose cotransporter is congenitally defective, the resulting **glucose/galactose malabsorption** causes severe diarrhea that is often fatal if glucose and galactose are not promptly removed from the diet. Glucose and its polymers can also be used to retain Na^+ in diarrheal disease, as was discussed in Chapter 25.

As indicated, SGLT-1 also transports galactose, but fructose utilizes a different mechanism. Its absorption is independent of Na^+ or the transport of glucose and galactose; it is transported instead by facilitated diffusion from the intestinal lumen into the enterocytes by GLUT 5 and out of the enterocytes into the interstitium by GLUT 2. Some fructose is converted to glucose in the mucosal cells.

Insulin has little effect on intestinal transport of sugars. In this respect, intestinal absorption resembles glucose reabsorption in the proximal convoluted tubules of the kidneys (see Chapter 37); neither process requires phosphorylation, and both are essentially normal in diabetes but are depressed by the drug phlorizin. The maximal rate of glucose absorption from the intestine is about 120 g/h.

PROTEINS & NUCLEIC ACIDS

PROTEIN DIGESTION

Protein digestion begins in the stomach, where pepsins cleave some of the peptide linkages. Like many of the other enzymes concerned with protein digestion, pepsins are secreted in the form of inactive precursors **(proenzymes)** and activated in the gastrointestinal tract. The pepsin precursors are called pepsinogens and are activated by gastric acid. Human gastric mucosa contains a number of related pepsinogens, which can be divided into two immunohistochemically distinct groups, pepsinogen I and pepsinogen II. Pepsinogen I is found only in acid-secreting regions, whereas pepsinogen II is also found in the pyloric region. Maximal acid secretion correlates with pepsinogen I levels.

Pepsins hydrolyze the bonds between aromatic amino acids such as phenylalanine or tyrosine and a second amino acid, so the products of peptic digestion are polypeptides of very diverse sizes. Because pepsins have a pH optimum of 1.6–3.2, their action is terminated when the gastric contents are mixed with the alkaline pancreatic juice in the duodenum and jejunum. The pH of the intestinal contents in the duodenal bulb is 3.0–4.0, but rapidly rises; in the rest of the duodenum it is about 6.5.

In the small intestine, the polypeptides formed by digestion in the stomach are further digested by the powerful proteolytic enzymes of the pancreas and intestinal mucosa. Trypsin, the chymotrypsins, and elastase act at interior peptide bonds in the peptide molecules and are called **endopeptidases.** The formation of the active endopeptidases from their inactive precursors occurs only when they have reached their site of action, secondary to the action of the brush border hydrolase, **enterokinase** (Figure 26–3). The powerful protein-splitting enzymes of the pancreatic juice are secreted as inactive proenzymes. Trypsinogen is converted to the active enzyme trypsin by **enterokinase** when the pancreatic juice enters the duodenum. Enterokinase contains 41% polysaccharide, and this high polysaccharide content apparently prevents it from being digested itself before it can exert its effect. Trypsin converts

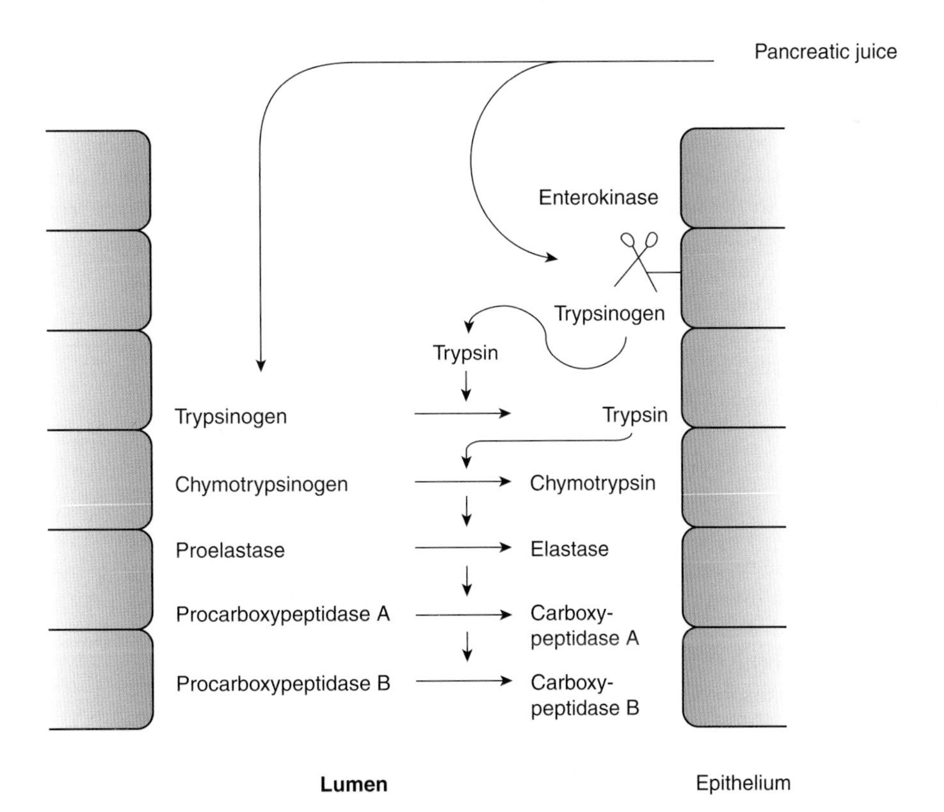

FIGURE 26–3 Mechanism to avoid activation of pancreatic proteases until they are in the duodenal lumen. Pancreatic juice contains proteolytic enzymes in their inactive, precursor forms. When the juice enters the duodenal lumen, trypsinogen contacts enterokinase expressed on the apical surface of enterocytes, Trypsinogen is thereby cleaved to trypsin, which in turn can activate additional trypsin molecules as well as the remaining proteolytic enzymes.

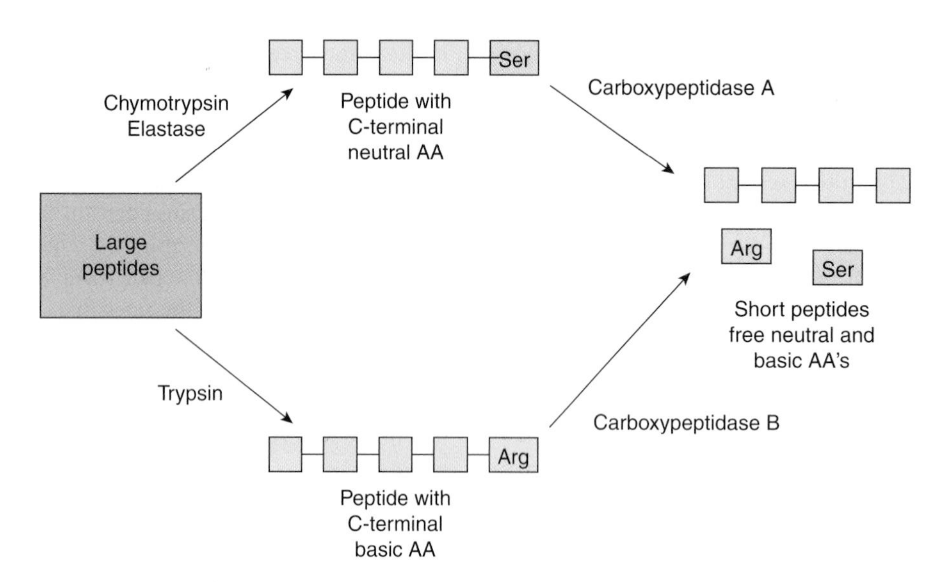

FIGURE 26–4 Luminal digestion of peptides by pancreatic endopeptidases and exopeptidases. Individual amino acids are shown as squares.

chymotrypsinogens into chymotrypsins and other proenzymes into active enzymes (Figure 26–3). Trypsin can also activate trypsinogen; therefore, once some trypsin is formed, there is an auto-catalytic chain reaction. Enterokinase deficiency occurs as a congenital abnormality and leads to protein malnutrition.

The carboxypeptidases of the pancreas are **exopeptidases** that hydrolyze the amino acids at the carboxyl ends of the polypeptides (Figure 26–4). Some free amino acids are liberated in the intestinal lumen, but others are liberated at the cell surface by the aminopeptidases, carboxypeptidases, endopeptidases, and dipeptidases in the brush border of the mucosal cells. Some di- and tripeptides are actively transported into the intestinal cells and hydrolyzed by intracellular peptidases, with the amino acids entering the bloodstream. Thus, the final digestion to amino acids occurs in three locations: the intestinal lumen, the brush border, and the cytoplasm of the mucosal cells.

ABSORPTION

At least seven different transport systems transport amino acids into enterocytes. Five of these require Na^+ and cotransport amino acids and Na^+ in a fashion similar to the cotransport of Na^+ and glucose (Figure 26–3). Two of these five also require Cl^-. In two systems, transport is independent of Na^+.

The di- and tripeptides are transported into enterocytes by a system known as PepT1 (or peptide transporter 1) that requires H^+ instead of Na^+ (Figure 26–5). There is very little absorption of larger peptides. In the enterocytes, amino acids released from the peptides by intracellular hydrolysis plus the amino acids absorbed from the intestinal lumen and brush border are transported out of the enterocytes along their basolateral borders by at least five transport systems. From there, they enter the hepatic portal blood.

Absorption of amino acids is rapid in the duodenum and jejunum. There is little absorption in the ileum in health, because the majority of the free amino acids have already been assimilated at that point. Approximately 50% of the digested protein comes from ingested food, 25% from proteins in digestive juices, and 25% from desquamated mucosal cells. Only 2–5% of the protein in the small intestine escapes digestion and absorption. Some of this is eventually digested by bacterial action in the colon. Almost all of the protein in the stools is not of dietary origin but comes from bacteria and cellular debris. Evidence suggests that the peptidase activities of the

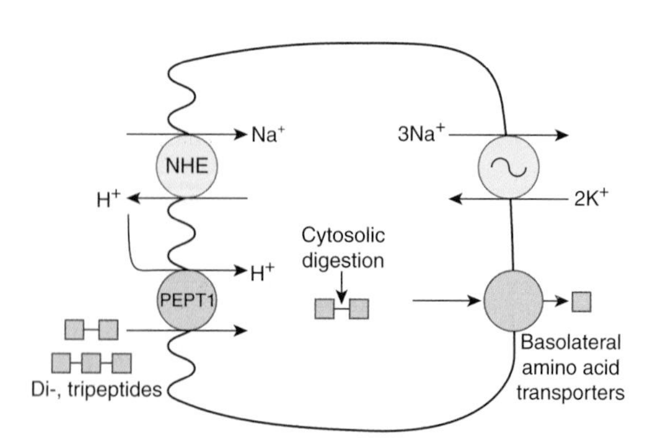

FIGURE 26–5 Disposition of short peptides in intestinal epithelial cells. Peptides are absorbed together with a proton supplied by an apical sodium/hydrogen exchanger (NHE) by the peptide transporter 1 (PepT1). Absorbed peptides are digested by cytosolic proteases, and any amino acids that are surplus to the needs of the epithelial cell are transported into the bloodstream by a series of basolateral transport proteins.

brush border and the mucosal cell cytoplasm are increased by resection of part of the ileum and that they are independently altered in starvation. Thus, these enzymes appear to be subject to homeostatic regulation. In humans, a congenital defect in the mechanism that transports neutral amino acids in the intestine and renal tubules causes **Hartnup disease.** A congenital defect in the transport of basic amino acids causes **cystinuria.** However, most patients do not experience nutritional deficiencies of these amino acids because peptide transport compensates.

In infants, moderate amounts of undigested proteins are also absorbed. The protein antibodies in maternal colostrum are largely secretory immunoglobulins (IgAs), the production of which is increased in the breast in late pregnancy. They cross the mammary epithelium by transcytosis and enter the circulation of the infant from the intestine, providing passive immunity against infections. Absorption is by endocytosis and subsequent exocytosis.

Absorption of intact proteins declines sharply after weaning, but adults still absorb small quantities. Foreign proteins that enter the circulation provoke the formation of antibodies, and the antigen–antibody reaction occurring on subsequent entry of more of the same protein may cause allergic symptoms. Thus, absorption of proteins from the intestine may explain the occurrence of allergic symptoms after eating certain foods. The incidence of food allergy in children is said to be as high as 8%. Certain foods are more allergenic than others. Crustaceans, mollusks, and fish are common offenders, and allergic responses to legumes, cows' milk, and egg white are also relatively frequent. However, in most individuals food allergies do not occur, and there is evidence for a genetic component in susceptibility.

Absorption of protein antigens, particularly bacterial and viral proteins, takes place in large **microfold cells** or **M cells,** specialized intestinal epithelial cells that overlie aggregates of lymphoid tissue (Peyer's patches). These cells pass the antigens to the lymphoid cells, and lymphocytes are activated. The activated lymphoblasts enter the circulation, but they later return to the intestinal mucosa and other epithelia, where they secrete IgA in response to subsequent exposures to the same antigen. This **secretory immunity** is an important defense mechanism (see Chapter 3).

NUCLEIC ACIDS

Nucleic acids are split into nucleotides in the intestine by the pancreatic nucleases, and the nucleotides are split into the nucleosides and phosphoric acid by enzymes that appear to be located on the luminal surfaces of the mucosal cells. The nucleosides are then split into their constituent sugars and purine and pyrimidine bases. The bases are absorbed by active transport. Families of equilibrative (ie, passive) and concentrative (ie, secondary active) nucleoside transporters have recently been identified and are expressed on the apical membrane of enterocytes.

LIPIDS

FAT DIGESTION

A lingual lipase is secreted by Ebner's glands on the dorsal surface of the tongue in some species, and the stomach also secretes a lipase (Table 26–1). They are of little quantitative significance for lipid digestion other than in the setting of pancreatic insufficiency, but they may generate free fatty acids that signal to most distal parts of the GI tract (eg, causing the release of CCK; see Chapter 25).

Most fat digestion therefore begins in the duodenum, pancreatic lipase being one of the most important enzymes involved. This enzyme hydrolyzes the 1- and 3-bonds of the triglycerides (triacylglycerols) with relative ease but acts on the 2-bonds at a very low rate, so the principal products of its action are free fatty acids and 2-monoglycerides (2-monoacylglycerols). It acts on fats that have been emulsified (see below). Its activity is facilitated when an amphipathic helix that covers the active site like a lid is bent back. **Colipase,** a protein with a molecular weight of about 11,000, is also secreted in the pancreatic juice, and when this molecule binds to the –COOH-terminal domain of the pancreatic lipase, opening of the lid is facilitated. Colipase is secreted in an inactive proform (Table 26–1) and is activated in the intestinal lumen by trypsin. Colipase is also critical for the action of lipase because it allows lipase to remain associated with droplets of dietary lipid even in the presence of bile acids.

Another pancreatic lipase that is activated by bile acids has been characterized. This 100,000-kDa **cholesterol esterase** represents about 4% of the total protein in pancreatic juice. In adults, pancreatic lipase is 10–60 times more active, but unlike pancreatic lipase, cholesterol esterase catalyzes the hydrolysis of cholesterol esters, esters of fat-soluble vitamins, and phospholipids, as well as triglycerides. A very similar enzyme is found in human milk.

Fats are relatively insoluble, which limits their ability to cross the unstirred layer and reach the surface of the mucosal cells. However, they are finely emulsified in the small intestine by the detergent action of bile acids, phosphatidylcholine, and monoglycerides. When the concentration of bile acids in the intestine is high, as it is after contraction of the gallbladder, lipids and bile salts interact spontaneously to form **micelles** (Figure 26–6). These cylindrical aggregates take up lipids, and although their lipid concentration varies, they generally contain fatty acids, monoglycerides, and cholesterol in their hydrophobic centers. Micellar formation further solubilizes the lipids and provides a mechanism for their transport to the enterocytes. Thus, the micelles move down their concentration gradient through the unstirred layer to the brush border of the mucosal cells. The lipids diffuse out of the micelles, and a saturated aqueous solution of the lipids is maintained in contact with the brush border of the mucosal cells (Figure 26–6).

Lipids collect in the micelles, with cholesterol in the hydrophobic center and amphipathic phospholipids and

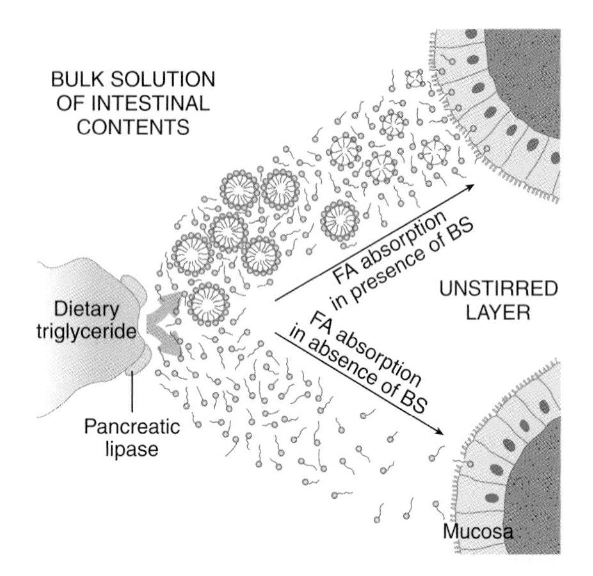

FIGURE 26–6 Lipid digestion and passage to intestinal mucosa. Fatty acids (FA) are liberated by the action of pancreatic lipase on dietary triglycerides and, in the presence of bile acids (BS), form micelles (the circular structures), which diffuse through the unstirred layer to the mucosal surface. Not shown, colipase binds to bile acids on the surface of the triglyceride droplet to anchor lipase to the surface and allow for its lipolytic activity. (Modified from Westergaard H, Dietschy JM: Normal mechanisms of fat absorption and derangements induced by various gastrointestinal diseases. Med Clin North Am 1974 Nov;58(6):1413–1427.)

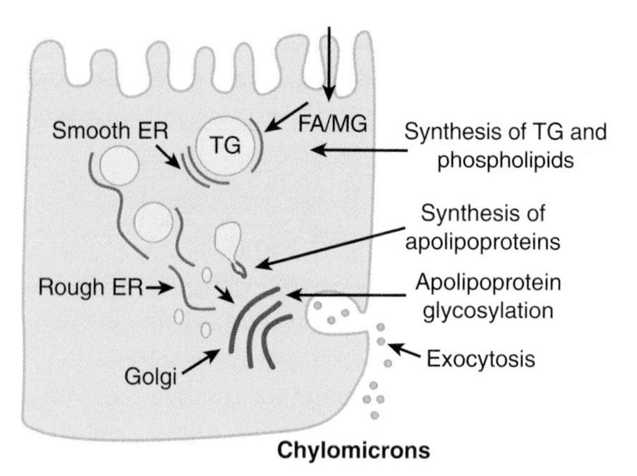

FIGURE 26–7 Intracellular handling of the products of lipid digestion. Absorbed fatty acids (FA) and monoglycerides (MG) are reesterified to form triglyceride (TG) in the smooth endoplasmic reticulum. Apoproteins synthesized in the rough endoplasmic reticulum are coated around lipid cores, and the resulting chylomicrons are secreted from the basolateral pole of epithelial cells by exocytosis.

monoglycerides lined up with their hydrophilic heads on the outside and their hydrophobic tails in the center. The micelles play an important role in keeping lipids in solution and transporting them to the brush border of the intestinal epithelial cells, where they are absorbed.

STEATORRHEA

Pancreatectomized animals and patients with diseases that destroy the exocrine portion of the pancreas have fatty, bulky, clay-colored stools **(steatorrhea)** because of the impaired digestion and absorption of fat. The steatorrhea is mostly due to lipase deficiency. However, acid inhibits the lipase, and the lack of alkaline secretion from the pancreas also contributes by lowering the pH of the intestine contents. In some cases, hypersecretion of gastric acid can cause steatorrhea. Another cause of steatorrhea is defective reabsorption of bile salts in the distal ileum (see Chapter 29).

When bile is excluded from the intestine, up to 50% of ingested fat appears in the feces. A severe malabsorption of fat-soluble vitamins also results. When bile salt reabsorption is prevented by resection of the terminal ileum or by disease in this portion of the small intestine, the amount of fat in the stools is also increased because when the enterohepatic circulation is interrupted, the liver cannot increase the rate of bile salt production to a sufficient degree to compensate for the loss.

FAT ABSORPTION

Traditionally, lipids were thought to enter the enterocytes by passive diffusion, but some evidence now suggests that carriers are involved. Inside the cells, the lipids are rapidly esterified, maintaining a favorable concentration gradient from the lumen into the cells (Figure 26–7). There are also carriers that export certain lipids back into the lumen, thereby limiting their oral availability. This is the case for plant sterols as well as cholesterol.

The fate of the fatty acids in enterocytes depends on their size. Fatty acids containing less than 10–12 carbon atoms are water-soluble enough that they pass through the enterocyte unmodified and are actively transported into the portal blood. They circulate as free (unesterified) fatty acids. The fatty acids containing more than 10–12 carbon atoms are too insoluble for this. They are reesterified to triglycerides in the enterocytes. In addition, some of the absorbed cholesterol is esterified. The triglycerides and cholesterol esters are then coated with a layer of protein, cholesterol, and phospholipid to form chylomicrons. These leave the cell and enter the lymphatics, because they are too large to pass through the junctions between capillary endothelial cells (Figure 26–7).

In mucosal cells, most of the triglyceride is formed by the acylation of the absorbed 2-monoglycerides, primarily in the smooth endoplasmic reticulum. However, some of the triglyceride is formed from glycerophosphate, which in turn is a product of glucose catabolism. Glycerophosphate is also converted into glycerophospholipids that participate in chylomicron formation. The acylation of glycerophosphate and the formation of lipoproteins occur in the rough endoplasmic reticulum. Carbohydrate moieties are added to the proteins in the Golgi apparatus, and the finished chylomicrons are extruded by exocytosis from the basolateral aspect of the cell.

Absorption of long-chain fatty acids is greatest in the upper parts of the small intestine, but appreciable amounts are also absorbed in the ileum. On a moderate fat intake, 95% or more of the ingested fat is absorbed. The processes involved in fat absorption are not fully mature at birth, and infants fail to absorb 10–15% of ingested fat. Thus, they are more susceptible to the ill effects of disease processes that reduce fat absorption.

SHORT-CHAIN FATTY ACIDS IN THE COLON

Increasing attention is being focused on short-chain fatty acids (SCFAs) that are produced in the colon and absorbed from it. SCFAs are 2-5-carbon weak acids that have an average normal concentration of about 80 mmol/L in the lumen. About 60% of this total is acetate, 25% propionate, and 15% butyrate. They are formed by the action of colonic bacteria on complex carbohydrates, resistant starches, and other components of the dietary fiber, that is, the material that escapes digestion in the upper gastrointestinal tract and enters the colon.

Absorbed SCFAs are metabolized and make a significant contribution to the total caloric intake. In addition, they exert a trophic effect on the colonic epithelial cells, combat inflammation, and are absorbed in part by exchange for H^+, helping to maintain acid–base equilibrium. SCFAs are absorbed by specific transporters present in colonic epithelial cells. SCFAs also promote the absorption of Na^+, although the exact mechanism for coupled Na^+–SCFA absorption is unsettled.

ABSORPTION OF VITAMINS & MINERALS

VITAMINS

Vitamins are defined as small molecules that play vital roles in bodily biochemical reactions, and which must be obtained from the diet because they cannot be synthesized endogenously. A discussion of the vitamins that are critical for human nutrition is provided towards the end of this chapter, but here we are concerned with general principles of their digestion and absorption. The fat-soluble vitamins A, D, E, and K are ingested as esters and must be digested by cholesterol esterase prior to absorption. These vitamins are also highly insoluble in the gut, and their absorption is therefore entirely dependent on their incorporation into micelles. Their absorption is deficient if fat absorption is depressed because of lack of pancreatic enzymes or if bile is excluded from the intestine by obstruction of the bile duct.

Most vitamins are absorbed in the upper small intestine, but vitamin B_{12} is absorbed in the ileum. This vitamin binds to intrinsic factor, a protein secreted by the parietal cells of the stomach, and the complex is absorbed across the ileal mucosa.

Vitamin B_{12} absorption and folate absorption are Na^+-independent, but all seven of the remaining water-soluble vitamins—thiamin, riboflavin, niacin, pyridoxine, pantothenate, biotin, and ascorbic acid—are absorbed by carriers that are Na^+ cotransporters.

CALCIUM

A total of 30–80% of ingested calcium is absorbed. The absorptive process and its relation to 1,25-dihydroxycholecalciferol are discussed in Chapter 21. Through this vitamin D derivative, Ca^{2+} absorption is adjusted to body needs; absorption is increased in the presence of Ca^{2+} deficiency and decreased in the presence of Ca^{2+} excess. Ca^{2+} absorption is also facilitated by protein. It is inhibited by phosphates and oxalates because these anions form insoluble salts with Ca^{2+} in the intestine. Magnesium absorption is also facilitated by protein.

IRON

In adults, the amount of iron lost from the body is relatively small. The losses are generally unregulated, and total body stores of iron are regulated by changes in the rate at which it is absorbed from the intestine. Men lose about 0.6 mg/d, largely in the stools. Premenopausal women have a variable, larger loss averaging about twice this value because of the additional iron lost during menstruation. The average daily iron intake in the United States and Europe is about 20 mg, but the amount absorbed is equal only to the losses. Thus, the amount of iron absorbed is normally about 3–6% of the amount ingested. Various dietary factors affect the availability of iron for absorption; for example, the phytic acid found in cereals reacts with iron to form insoluble compounds in the intestine, as do phosphates and oxalates.

Most of the iron in the diet is in the ferric (Fe^{3+}) form, whereas it is the ferrous (Fe^{2+}) form that is absorbed. Fe^{3+} reductase activity is associated with the iron transporter in the brush borders of the enterocytes (Figure 26–8). Gastric secretions dissolve the iron and permit it to form soluble complexes with ascorbic acid and other substances that aid its reduction to the Fe^{2+} form. The importance of this function in humans is indicated by the fact that iron deficiency anemia is a troublesome and relatively frequent complication of partial gastrectomy.

Almost all iron absorption occurs in the duodenum. Transport of Fe^{2+} into the enterocytes occurs via divalent metal transporter 1 **(DMT1)** (Figure 26–8). Some is stored in ferritin, and the remainder is transported out of the enterocytes by a basolateral transporter named **ferroportin 1.** A protein called **hephaestin (Hp)** is associated with ferroportin 1. It is not a transporter itself, but it facilitates basolateral transport. In the plasma, Fe^{2+} is converted to Fe^{3+} and bound to the iron transport protein **transferrin.** This protein has two iron-binding sites. Normally, transferrin is about 35% saturated with iron, and the normal plasma iron level is about 130 μg/dL (23 μmol/L) in men and 110 μg/dL (19 μmol/L) in women.

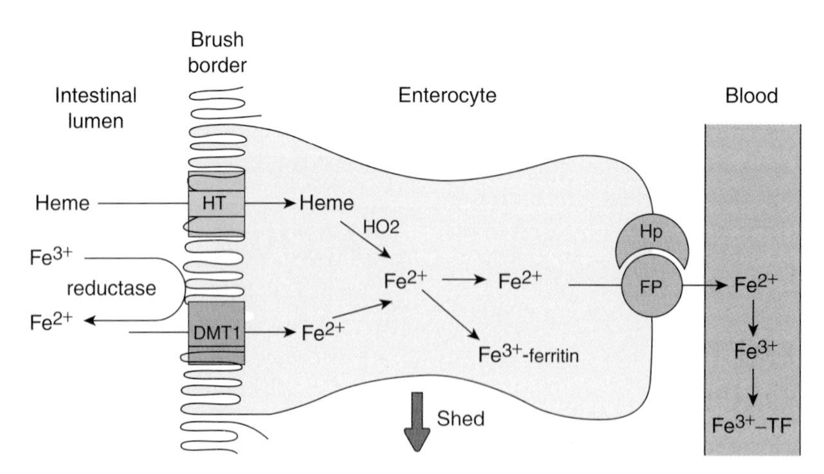

FIGURE 26–8 Absorption of iron. Fe^{3+} is converted to Fe^{2+} by ferric reductase, and Fe^{2+} is transported into the enterocyte by the apical membrane iron transporter DMT1. Heme is transported into the enterocyte by a separate heme transporter (HT), and HO2 releases Fe^{2+} from the heme. Some of the intracellular Fe^{2+} is converted to Fe^{3+} and bound to ferritin. The rest binds to the basolateral Fe^{2+} transporter ferroportin (FP) and is transported to the interstitial fluid. The transport is aided by hephaestin (Hp). In plasma, Fe^{2+} is converted to Fe^{3+} and bound to the iron transport protein transferrin (TF).

Heme (see Chapter 31) binds to an apical transport protein in enterocytes and is carried into the cytoplasm. In the cytoplasm, HO2, a subtype of heme oxygenase, removes Fe^{2+} from the porphyrin and adds it to the intracellular Fe^{2+} pool.

Seventy per cent of the iron in the body is in hemoglobin, 3% in myoglobin, and the rest in ferritin, which is present not only in enterocytes, but also in many other cells. Apoferritin is a globular protein made up of 24 subunits. Ferritin is readily visible under the electron microscope and has been used as a tracer in studies of phagocytosis and related phenomena. Ferritin molecules in lysosomal membranes may aggregate in deposits that contain as much as 50% iron. These deposits are called **hemosiderin.**

Intestinal absorption of iron is regulated by three factors: recent dietary intake of iron, the state of the iron stores in the body, and the state of erythropoiesis in the bone marrow. The normal operation of the factors that maintain iron balance is essential for health (Clinical Box 26–2).

CLINICAL BOX 26–2

Disorders of Iron Uptake

Iron deficiency causes anemia. Conversely, iron overload causes hemosiderin to accumulate in the tissues, producing **hemosiderosis.** Large amounts of **hemosiderin** can damage tissues, such as is seen in the common genetic disorder of hemochromatosis. This syndrome is characterized by pigmentation of the skin, pancreatic damage with diabetes ("bronze diabetes"), cirrhosis of the liver, a high incidence of hepatic carcinoma, and gonadal atrophy. Hemochromatosis may be hereditary or acquired. The most common cause of the hereditary forms is a mutated *HFE* gene that is common in the Caucasian population. It is located on the short arm of chromosome 6 and is closely linked to the human leukocyte antigen-A (HLA-A) locus. It is still unknown precisely how mutations in *HFE* cause hemochromatosis, but individuals who are homogenous for *HFE* mutations absorb excess amounts of iron because *HFE* normally inhibits expression of the duodenal transporters that participate in iron uptake. Acquired hemochromatosis occurs when the iron-regulating system is overwhelmed by excess iron loads due to chronic destruction of red blood cells, liver disease, or repeated transfusions in diseases such as intractable anemia.

THERAPEUTIC HIGHLIGHTS

If hereditary hemochromatosis is diagnosed before excessive amounts of iron accumulate in the tissues, life expectancy can be prolonged substantially by repeated withdrawal of blood.

CONTROL OF FOOD INTAKE

The intake of nutrients is under complex control involving signals from both the periphery and the central nervous system. Complicating the picture, higher functions also modulate the response to both central and peripheral cues that either trigger or inhibit food intake. Thus, food preferences, emotions, environment, lifestyle, and circadian rhythms may all have profound effects on whether food is or is not sought, and the type of food that is ingested.

Many of the hormones and other factors that are released coincident with a meal, and may play other important roles in digestion and absorption (see Chapter 25) are also involved in the regulation of feeding behavior (Figure 26–9). For example, CCK either produced by I cells in the intestine, or released

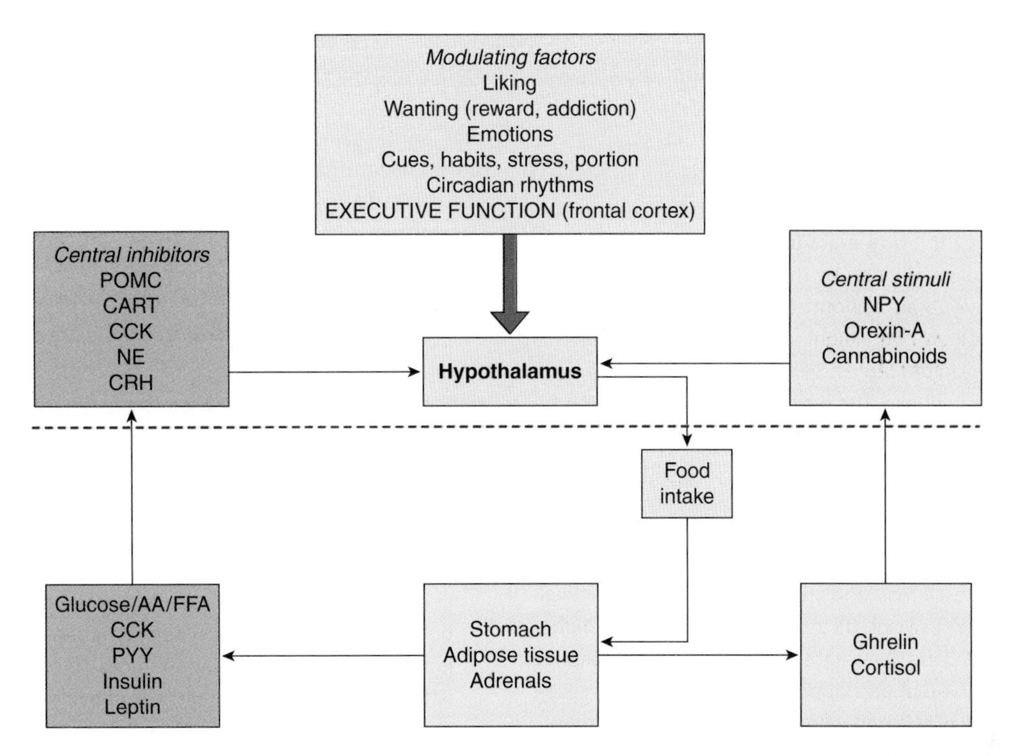

FIGURE 26–9 Summary of mechanisms controlling food intake. Peripheral stimuli and inhibitors, release in anticipation of or in response to food intake, cross the blood-brain barrier (indicated by the broken red line) and activate the release and/or synthesis of central factors in the hypothalamus that either increase or decrease subsequent food intake. Food intake can also be modulated by signals from higher centers, as shown. Not shown, peripheral orexins can reduce production of central inhibitors, and vice versa. (Based on a figure kindly provided by Dr Samuel Klein, Washington University.)

by nerve endings in the brain, inhibits further food intake and thus is defined as a **satiety factor** or **anorexin.** CCK and other similar factors have attracted great interest from the pharmaceutical industry in the hopes that derivatives might be useful as aids to dieting, an objective that is lent greater urgency given the current epidemic of obesity in Western countries (Clinical Box 26–3).

Leptin and ghrelin are peripheral factors that act reciprocally on food intake, and have emerged as critical regulators in this regard. Both activate their receptors in the hypothalamus that initate signaling cascades leading to changes in food intake. Leptin is produced by adipose tissue, and signals the status of the fat stores therein. As adipocytes increase in size, they release greater quantities of leptin and this tends to decrease food intake, in part by increasing the expression of other anorexigenic factors in the hypothalamus such as pro-opiomelanocortin (POMC), cocaine- and amphetamine-regulated transcript (CART), neurotensin, and corticotropin-releasing hormone (CRH). Leptin also stimulates the metabolic rate (see Chapter 18). Animal studies have shown that it is possible to become resistant to the effects of leptin, however, and in this setting, food intake persists despite adequate (or even growing) adipose stores—obesity therefore results.

Ghrelin, on the other hand, is a predominantly fast-acting **orexin** that stimulates food intake. It is produced mainly by the stomach, as well as other tissues such as the pancreas and adrenal glands in responses to changes in nutritional status—circulating ghrelin levels increase preprandially, then decrease after a meal. It is believed to be involved primarily in meal initiation, unlike the longer-term effects of leptin. Like leptin, however, the effects of ghrelin are produced mostly via actions in the hypothalamus. It increases synthesis and/or release of central orexins, including neuropeptide Y and cannabinoids, and suppresses the ability of leptin to stimulate the anorexigenic factors discussed above. Loss of the activity of ghrelin may account in part for the effectiveness of gastric bypass procedures for obesity. Its secretion may also be inhibited by leptin, underscoring the reciprocity of these hormones. There is some evidence to suggest, however, that the ability of leptin to reduce ghrelin secretion is lost in the setting of obesity.

NUTRITIONAL PRINCIPLES & ENERGY METABOLISM

Humans oxidize carbohydrates, proteins, and fats, producing principally CO_2, H_2O, and the energy necessary for life processes (Clinical Box 26–3). CO_2, H_2O, and energy are also produced when food is burned outside the body. However, in the

CLINICAL BOX 26–3

Obesity

Obesity is the most common and most expensive nutritional problem in the United States. A convenient and reliable indicator of body fat is the **body mass index (BMI),** which is body weight (in kilograms) divided by the square of height (in meters). Values above 25 are abnormal. Individuals with values of 25–30 are considered overweight, and those with values >30 are obese. In the United States, 34% of the population is overweight and 34% is obese. The incidence of obesity is also increasing in other countries. Indeed, the Worldwatch Institute has estimated that although starvation continues to be a problem in many parts of the world, the number of over-weight people in the world is now as great as the number of underfed. Obesity is a problem because of its complications. It is associated with accelerated atherosclerosis and an increased incidence of gallbladder and other diseases. Its association with type 2 diabetes is especially striking. As weight increases, insulin resistance increases and frank diabetes appears. At least in some cases, glucose tolerance is restored when weight is lost. In addition, the mortality rates from many kinds of cancer are increased in obese individuals. The causes of the high incidence of obesity in the general population are probably multiple. Studies of twins raised apart show a definite genetic component. It has been pointed out that through much of human evolution, famines were common, and mechanisms that permitted increased energy storage as fat had survival value. Now, however, food is plentiful in many countries, and the ability to gain and retain fat has become a liability. As noted above, the fundamental cause of obesity is still an excess of energy intake in food over energy expenditure. If human volunteers are fed a fixed high-calorie diet, some gain weight more rapidly than others, but the slower weight gain is due to increased energy expenditure in the form of small, fidgety movements **(nonexercise activity thermogenesis; NEAT).** Body weight generally increases at a slow but steady rate throughout adult life. Decreased physical activity is undoubtedly a factor in this increase, but decreased sensitivity to leptin may also play a role.

THERAPEUTIC HIGHLIGHTS

Obesity is such a vexing medical and public health problem because its effective treatment depends so dramatically on lifestyle changes. Long-term weight loss can only be achieved with decreased food intake, increased energy expenditure, or, ideally, some combination of both. Exercise alone is rarely sufficient because it typically induces the patient to ingest more calories. For those who are seriously obese and who have developed serious health complications as a result, a variety of surgical approaches have been developed that reduce the size of the stomach reservoir and/or bypass it altogether. These surgical maneuvers are intended to reduce the size of meals that can be tolerated, but also have dramatic metabolic effects even before significant weight loss occurs, perhaps as a result of reduced production of peripheral orexins by the gut. Pharmaceutical companies are also actively exploring the science of orexins and anorexins to develop drugs that might act centrally to modify food intake (Figure 26–9).

body, oxidation is not a one-step, semiexplosive reaction but a complex, slow, stepwise process called **catabolism,** which liberates energy in small, usable amounts. Energy can be stored in the body in the form of special energy-rich phosphate compounds and in the form of proteins, fats, and complex carbohydrates synthesized from simpler molecules. Formation of these substances by processes that take up rather than liberate energy is called **anabolism.** This chapter consolidates consideration of endocrine function by providing a brief summary of the production and utilization of energy and the metabolism of carbohydrates, proteins, and fats.

METABOLIC RATE

The amount of energy liberated by the catabolism of food in the body is the same as the amount liberated when food is burned outside the body. The energy liberated by catabolic processes in the body is used for maintaining body functions, digesting and metabolizing food, thermoregulation, and physical activity. It appears as external work, heat, and energy storage:

$$\text{Energy output} = \text{External work} + \text{Energy storage} + \text{Heat}$$

The amount of energy liberated per unit of time is the **metabolic rate.** Isotonic muscle contractions perform work at a peak efficiency approximating 50%:

$$\text{Efficiency} = \frac{\text{Work done}}{\text{Total energy expended}}$$

Essentially all of the energy of isometric contractions appears as heat, because little or no external work (force multiplied by the distance that the force moves a mass) is done (see Chapter 5). Energy is stored by forming energy-rich compounds. The amount of energy storage varies, but in fasting individuals it is zero or negative. Therefore, in an adult individual who has not eaten recently and who is not moving (or growing, reproducing, or lactating), all of the energy output appears as heat.

CALORIES

The standard unit of heat energy is the **calorie (cal)**, defined as the amount of heat energy necessary to raise the temperature of 1 g of water 1°, from 15 to 16°C. This unit is also called the gram calorie, small calorie, or standard calorie. The unit commonly used in physiology and medicine is the **Calorie (kilocalorie; kcal)**, which equals 1000 cal.

The caloric values of the common foodstuffs, as measured in a bomb calorimeter, are found to be 4.1 kcal/g of carbohydrate, 9.3 kcal/g of fat, and 5.3 kcal/g of protein. In the body, similar values are obtained for carbohydrate and fat, but the oxidation of protein is incomplete, the end products of protein catabolism being urea and related nitrogenous compounds in addition to CO_2 and H_2O (see below). Therefore, the caloric value of protein in the body is only 4.1 kcal/g.

RESPIRATORY QUOTIENT

The **respiratory quotient (RQ)** is the ratio in the steady state of the volume of CO_2 produced to the volume of O_2 consumed per unit of time. It should be distinguished from the **respiratory exchange ratio (R)**, which is the ratio of CO_2 to O_2 at any given time whether or not equilibrium has been reached. R is affected by factors other than metabolism. RQ and R can be calculated for reactions outside the body, for individual organs and tissues, and for the whole body. The RQ of carbohydrate is 1.00, and that of fat is about 0.70. This is because H and O are present in carbohydrate in the same proportions as in water, whereas in the various fats, extra O_2 is necessary for the formation of H_2O.

Carbohydrate:

$$C_6H_{12}O_6 + 6O_2 \rightarrow 6CO_2 + 6H_2O$$
(glucose)
$$RQ = 6/6 = 1.00$$

Fat:

$$2C_{51}H_{98}O_6 + 145O_2 \rightarrow 102CO_2 + 98H_2O$$
(tripalmitin)
$$RQ = 102/145 = 0.703$$

Determining the RQ of protein in the body is a complex process, but an average value of 0.82 has been calculated. The approximate amounts of carbohydrate, protein, and fat being oxidized in the body at any given time can be calculated from the RQ and the urinary nitrogen excretion. RQ and R for the whole body differ in various conditions. For example, during hyperventilation, R rises because CO_2 is being blown off. During strenuous exercise, R may reach 2.00 because CO_2 is being blown off and lactic acid from anaerobic glycolysis is being converted to CO_2 (see below). After exercise, R may fall for a while to 0.50 or less. In metabolic acidosis, R rises because respiratory compensation for the acidosis causes the amount of CO_2 expired to rise (see Chapter 35). In severe acidosis, R may be greater than 1.00. In metabolic alkalosis, R falls.

TABLE 26–2 Factors affecting the metabolic rate.

Factors affecting the metabolic rate
Muscular exertion during or just before measurement
Recent ingestion of food
High or low environmental temperature
Height, weight, and surface area
Sex
Age
Growth
Reproduction
Lactation
Emotional state
Body temperature
Circulating levels of thyroid hormones
Circulating epinephrine and norepinephrine levels

The O_2 consumption and CO_2 production of an organ can be calculated at equilibrium by multiplying its blood flow per unit of time by the arteriovenous differences for O_2 and CO_2 across the organ, and the RQ can then be calculated. Data on the RQ of individual organs are of considerable interest in drawing inferences about the metabolic processes occurring in them. For example, the RQ of the brain is regularly 0.97–0.99, indicating that its principal but not its only fuel is carbohydrate. During secretion of gastric juice, the stomach has a negative R because it takes up more CO_2 from the arterial blood than it puts into the venous blood (see Chapter 26).

FACTORS AFFECTING THE METABOLIC RATE

The metabolic rate is affected by many factors (Table 26–2). The most important is muscular exertion. O_2 consumption is elevated not only during exertion but also for as long afterward as is necessary to repay the O_2 debt (see Chapter 5). Recently ingested foods also increase the metabolic rate because of their **specific dynamic action (SDA).** The SDA of a food is the obligatory energy expenditure that occurs during its assimilation into the body. It takes 30 kcal to assimilate the amount of protein sufficient to raise the metabolic rate 100 kcal; 6 kcal to assimilate a similar amount of carbohydrate; and 5 kcal to assimilate a similar amount of fat. The cause of the SDA, which may last up to 6 h, is uncertain.

Another factor that stimulates metabolism is the environmental temperature. The curve relating the metabolic rate to the environmental temperature is U-shaped. When the environmental temperature is lower than body temperature, heat-producing mechanisms such as shivering are activated and the metabolic rate rises. When the temperature is high enough to raise the body temperature, metabolic processes generally accelerate, and the metabolic rate rises about 14% for each degree Celsius of elevation.

The metabolic rate determined at rest in a room at a comfortable temperature in the thermoneutral zone 12–14 h after the last meal is called the **basal metabolic rate (BMR).** This value falls about 10% during sleep and up to 40% during prolonged starvation. The rate during normal daytime activities is, of course, higher than the BMR because of muscular activity and food intake. The **maximum metabolic rate** reached during exercise is often said to be 10 times the BMR, but trained athletes can increase their metabolic rate as much as 20-fold.

The BMR of a man of average size is about 2000 kcal/d. Large animals have higher absolute BMRs, but the ratio of BMR to body weight in small animals is much greater. One variable that correlates well with the metabolic rate in different species is the body surface area. This would be expected, since heat exchange occurs at the body surface. The actual relation to body weight (W) would be

$$BMR = 3.52W^{0.67}$$

However, repeated measurements by numerous investigators have come up with a higher exponent, averaging 0.75:

$$BMR = 3.52W^{0.75}$$

Thus, the slope of the line relating metabolic rate to body weight is steeper than it would be if the relation were due solely to body area (Figure 26–10). The cause of the greater slope has been much debated but remains unsettled.

For clinical use, the BMR is usually expressed as a percentage increase or decrease above or below a set of generally used standard normal values. Thus, a value of +65 means that the individual's BMR is 65% above the standard for that age and sex.

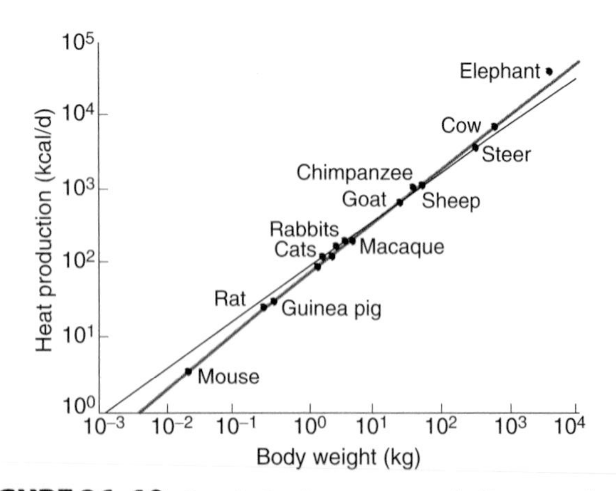

FIGURE 26–10 Correlation between metabolic rate and body weight, plotted on logarithmic scales. The slope of the colored line is 0.75. The black line represents the way surface area increases with weight for geometrically similar shapes and has a slope of 0.67. (Reproduced with permission from McMahon TA: Size and shape in biology. Science 1973;179:1201. Copyright © 1973 by the American Association for the Advancement of Science.)

The decrease in metabolic rate related to a decrease in body weight is part of the explanation of why, when an individual is trying to lose weight, weight loss is initially rapid and then slows down.

ENERGY BALANCE

The first law of thermodynamics, the principle that states that energy is neither created nor destroyed when it is converted from one form to another, applies to living organisms as well as inanimate systems. One may therefore speak of an **energy balance** between caloric intake and energy output. If the caloric content of the food ingested is less than the energy output, that is, if the balance is negative, endogenous stores are utilized. Glycogen, body protein, and fat are catabolized, and the individual loses weight. If the caloric value of the food intake exceeds energy loss due to heat and work and the food is properly digested and absorbed, that is, if the balance is positive, energy is stored, and the individual gains weight.

To balance basal output so that the energy-consuming tasks essential for life can be performed, the average adult must take in about 2000 kcal/d. Caloric requirements above the basal level depend on the individual's activity. The average sedentary student (or professor) needs another 500 kcal, whereas a lumber-jack needs up to 3000 additional kcal per day.

NUTRITION

The aim of the science of nutrition is the determination of the kinds and amounts of foods that promote health and well-being. This includes not only the problems of undernutrition but those of overnutrition, taste, and availability (Clinical Box 26–4). However, certain substances are essential constituents of any human diet. Many of these compounds have been mentioned in previous sections of this chapter, and a brief summary of the essential and desirable dietary components is presented below.

ESSENTIAL DIETARY COMPONENTS

An optimal diet includes, in addition to sufficient water (see Chapter 37), adequate calories, protein, fat, minerals, and vitamins.

CALORIC INTAKE & DISTRIBUTION

As noted above, the caloric value of the dietary intake must be approximately equal to the energy expended if body weight is to be maintained. In addition to the 2000 kcal/d necessary to meet basal needs, 500–2500 kcal/d (or more) are required to meet the energy demands of daily activities.

CLINICAL BOX 26–4

The Malabsorption Syndrome

The digestive and absorptive functions of the small intestine are essential for life. However, the digestive and absorptive capacity of the intestine is larger than needed for normal function (the **anatomic reserve).** Removal of short segments of the jejunum or ileum generally does not cause severe symptoms, and compensatory hypertrophy and hyperplasia of the remaining mucosa occur. However, when more than 50% of the small intestine is resected or bypassed **(short gut syndrome),** the absorption of nutrients and vitamins is so compromised that it is very difficult to prevent malnutrition and wasting **(malabsorption).** Resection of the terminal ileum also prevents the absorption of bile acids, and this leads in turn to deficient fat absorption. It also causes diarrhea because the unabsorbed bile salts enter the colon, where they activate chloride secretion (see Chapter 25). Other complications of intestinal resection or bypass include hypocalcemia, arthritis, and possibly fatty infiltration of the liver, followed by cirrhosis. Various disease processes can also impair absorption without a loss of intestinal length. The pattern of deficiencies that results is sometimes called the **malabsorption syndrome.** This pattern varies somewhat with the cause, but it can include deficient absorption of amino acids, with marked body wasting and, eventually, hypoproteinemia and edema. Carbohydrate and fat absorption are also depressed. Because of defective fat absorption, the fat-soluble vitamins (vitamins A, D, E, and K) are not absorbed in adequate amounts. One of the most interesting conditions causing the malabsorption syndrome is the autoimmune disease **celiac disease.** This disease occurs in genetically predisposed individuals who have the major histocompatibility complex (MHC) class II antigen HLA-DQ2 or DQ8 (see Chapter 3). In these individuals gluten and closely related proteins cause intestinal T cells to mount an inappropriate immune response that damages the intestinal epithelial cells and results in a loss of villi and a flattening of the mucosa. The proteins are found in wheat, rye, barley, and to a lesser extent in oats—but not in rice or corn. When grains containing gluten are omitted from the diet, bowel function is generally restored to normal.

THERAPEUTIC HIGHLIGHTS

Treatment of malabsorption depends on the underlying cause. In celiac disease, the mucosa returns to normal if foods containing gluten are strictly excluded from the diet, although this may be difficult to achieve. The diarrhea that accompanies bile acid malabsorption can be treated with a resin (cholestyramine) that binds the bile acids in the lumen and prevents their secretory action on colonocytes. Patients who become deficient in fat soluble-vitamins may be given these compounds as water soluble derivatives. For serious cases of short bowel syndrome, it may be necessary to supply nutrients parenterally. There is hope that small bowel transplantation may eventually become routine, but of course transplantation carries its own long-term disadvantages and also requires a reliable supply of donor tissues.

The distribution of the calories among carbohydrate, protein, and fat is determined partly by physiologic factors and partly by taste and economic considerations. A daily protein intake of 1 g/kg body weight to supply the eight nutritionally essential amino acids and other amino acids is desirable. The source of the protein is also important. **Grade I proteins,** the animal proteins of meat, fish, dairy products, and eggs, contain amino acids in approximately the proportions required for protein synthesis and other uses. Some of the plant proteins are also grade I, but most are **grade II** because they supply different proportions of amino acid and some lack one or more of the essential amino acids. Protein needs can be met with a mixture of grade II proteins, but the intake must be large because of the amino acid wastage.

Fat is the most compact form of food, since it supplies 9.3 kcal/g. However, often it is also the most expensive. Indeed, internationally there is a reasonably good positive correlation between fat intake and standard of living. In the past, Western diets have contained large amounts (100 g/d or more). The evidence indicating that a high unsaturated/saturated fat ratio in the diet is of value in the prevention of atherosclerosis and the current interest in preventing obesity may change this. In Central and South American Indian communities where corn (carbohydrate) is the dietary staple, adults live without ill effects for years on a very low fat intake. Therefore, provided that the needs for essential fatty acids are met, a low-fat intake does not seem to be harmful, and a diet low in saturated fats is desirable.

Carbohydrate is the cheapest source of calories and provides 50% or more of the calories in most diets. In the average middle-class American diet, approximately 50% of the calories come from carbohydrate, 15% from protein, and 35% from fat. When calculating dietary needs, it is usual to meet the protein requirement first and then split the remaining calories between fat and carbohydrate, depending on taste, income, and other

TABLE 26–3 Trace elements believed essential for life.

Arsenic	Manganese
Chromium	Molybdenum
Cobalt	Nickel
Copper	Selenium
Fluorine	Silicon
Iodine	Vanadium
Iron	Zinc

factors. For example, a 65-kg man who is moderately active needs about 2800 kcal/d. He should eat at least 65 g of protein daily, supplying 267 (65 × 4.1) kcal. Some of this should be grade I protein. A reasonable figure for fat intake is 50–60 g. The rest of the caloric requirement can be met by supplying carbohydrate.

MINERAL REQUIREMENTS

A number of minerals must be ingested daily for the maintenance of health. Besides those for which recommended daily dietary allowances have been set, a variety of different trace elements should be included. Trace elements are defined as elements found in tissues in minute amounts. Those believed to be essential for life, at least in experimental animals, are listed in Table 26–3. In humans, iron deficiency causes anemia. Cobalt is part of the vitamin B_{12} molecule, and vitamin B_{12} deficiency leads to megaloblastic anemia (see Chapter 31). Iodine deficiency causes thyroid disorders (see Chapter 19). Zinc deficiency causes skin ulcers, depressed immune responses, and hypogonadal dwarfism. Copper deficiency causes anemia and changes in ossification. Chromium deficiency causes insulin resistance. Fluorine deficiency increases the incidence of dental caries.

Conversely, some minerals can be toxic when present in the body in excess. For example, severe iron overload with toxic effects is seen hemochromatosis, a disease where the normal homeostatic mechanisms that regulate uptake of iron from the diet (Figure 26–8) are genetically deranged. Similarly, copper excess causes brain damage (Wilson disease), and aluminum poisoning in patients with renal failure who are receiving dialysis treatment causes a rapidly progressive dementia that resembles Alzheimer disease (see Chapter 15).

Sodium and potassium are also essential minerals, but listing them is academic, because it is very difficult to prepare a sodium-free or potassium-free diet. A low-salt diet is, however, well tolerated for prolonged periods because of the compensatory mechanisms that conserve Na^+.

VITAMINS

Vitamins were discovered when it was observed that certain diets otherwise adequate in calories, essential amino acids, fats, and minerals failed to maintain health (for example, in sailors engaged in long voyages without access to fresh fruits and vegetables). The term **vitamin** has now come to refer to any organic dietary constituent necessary for life, health, and growth that does not function by supplying energy.

Because there are minor differences in metabolism between mammalian species, some substances are vitamins in one species and not in another. The sources and functions of the major vitamins in humans are listed in Table 26–4. Most vitamins have important functions in intermediary metabolism or the special metabolism of the various organ systems. Those that are water-soluble (vitamin B complex, vitamin C) are easily absorbed, but the fat-soluble vitamins (vitamins A, D, E, and K) are poorly absorbed in the absence of bile and/or pancreatic enzymes. Some dietary fat intake is necessary for their absorption, and in obstructive jaundice or disease of the exocrine pancreas, deficiencies of the fat-soluble vitamins can develop even if their intake is adequate. Vitamin A and vitamin D are bound to transfer proteins in the circulation. The α-tocopherol form of vitamin E is normally bound to chylomicrons. In the liver, it is transferred to very low density lipoprotein (VLDL) and distributed to tissues by an α-tocopherol transfer protein. When this protein is abnormal due to mutation of its gene in humans, there is cellular deficiency of vitamin E and the development of a condition resembling Friedreich ataxia. Two Na^+-dependent L-ascorbic acid transporters have recently been isolated. One is found in the kidneys, intestines, and liver, and the other in the brain and eyes.

The diseases caused by deficiency of each of the vitamins are also listed in Table 26–4. It is worth remembering, however, particularly in view of the advertising campaigns for vitamin pills and supplements, that very large doses of the fat-soluble vitamins are definitely toxic. **Hypervitaminosis A** is characterized by anorexia, headache, hepatosplenomegaly, irritability, scaly dermatitis, patchy loss of hair, bone pain, and hyperostosis. Acute vitamin A intoxication was first described by Arctic explorers, who developed headache, diarrhea, and dizziness after eating polar bear liver. The liver of this animal is particularly rich in vitamin A. **Hypervitaminosis D** is associated with weight loss, calcification of many soft tissues, and eventual renal failure. **Hypervitaminosis K** is characterized by gastrointestinal disturbances and anemia. Large doses of water-soluble vitamins have been thought to be less likely to cause problems because they can be rapidly cleared from the body. However, it has been demonstrated that ingestion of megadoses of pyridoxine (vitamin B_6) can produce peripheral neuropathy.

TABLE 26–4 Vitamins essential or probably essential to human nutrition.[a]

Vitamin	Action	Deficiency Symptoms	Sources	Chemistry
A (A_1, A_2)	Constituents of visual pigments (see Chapter 12: Vision); necessary for fetal development and for cell development throughout life	Night blindness, dryskin	Yellow vegetables and fruit	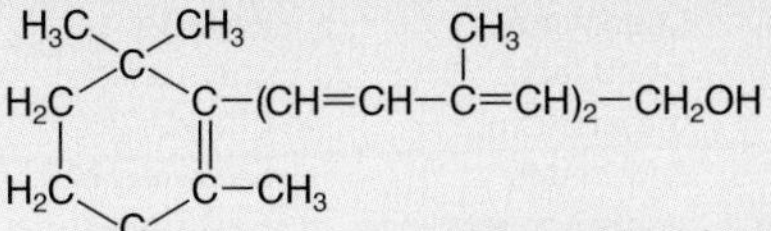 Vitamin A_1 alcohol (retinol).
B complex				
Thiamin (vitamin B_1)	Cofactor in decarboxylations	Beriberi, neuritis	Liver, unrefined cereal grains	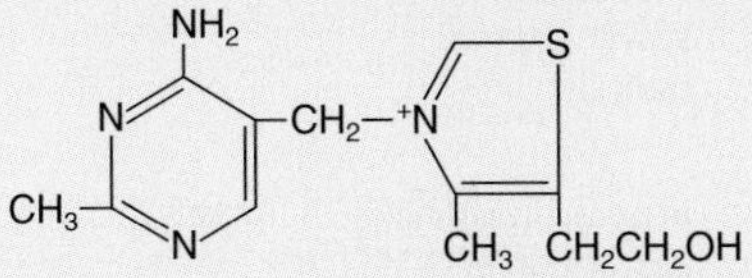
Riboflavin (vitamin B_2)	Constituent of flavoproteins	Glossitis, cheilosis	Liver, milk	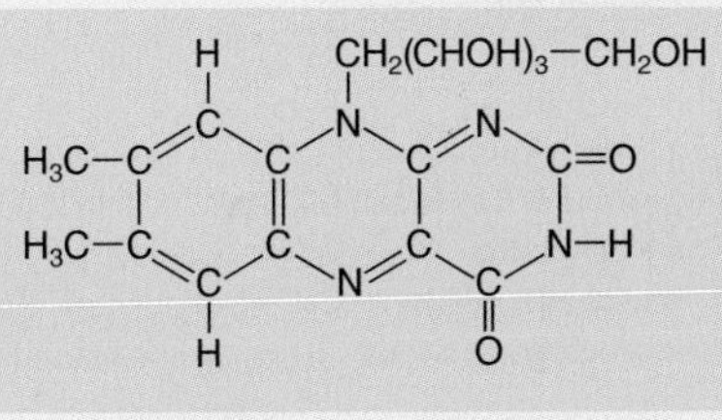
Niacin	Constituent of NAD^+ and $NADP^+$	Pellagra	Yeast, lean meat, liver	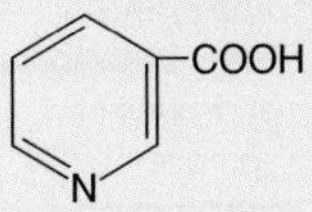 Can be synthesized in body from tryptophan.
Pyridoxine (vitamin B_6)	Forms prosthetic group of certain decarboxylases and transaminases. Converted in body into pyridoxal phosphate and pyridoxamine phosphate	Convulsions, hyperirritability	Yeast, wheat, corn, liver	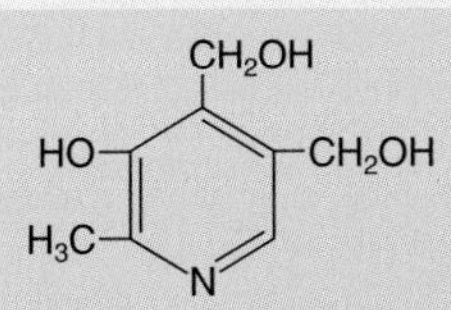
Pantothenic acid	Constituent of CoA	Dermatitis, enteritis, alopecia, adrenal insufficiency	Eggs, liver, yeast	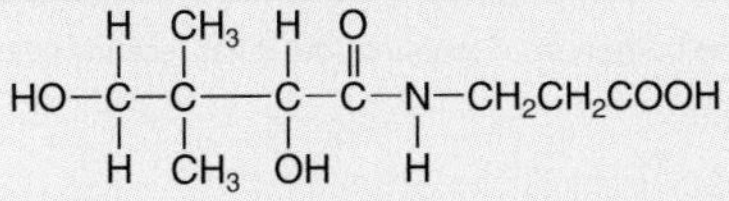
Biotin	Catalyzes CO_2 "fixation" (in fatty acid synthesis, etc)	Dermatitis, enteritis	Egg yolk, liver, tomatoes	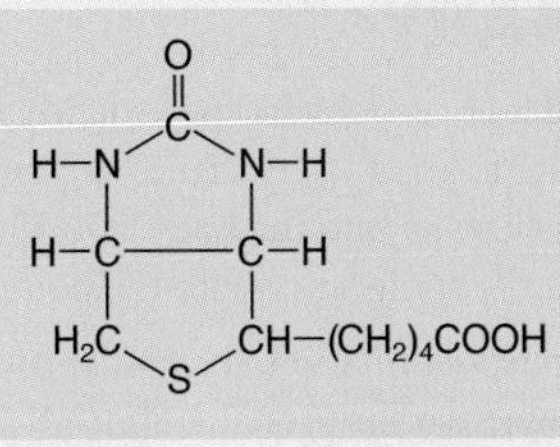
Folates (folic acid) and related compounds	Coenzymes for "1-carbon" transfer; involved in methylating reactions	Sprue, anemia. Neural tube defects in children born to folate-deficient women	Leafy green vegetables	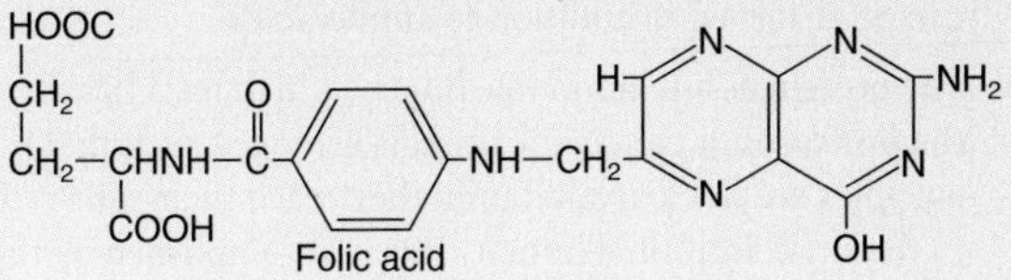 Folic acid

(continued)

TABLE 26–4 Vitamins essential or probably essential to human nutrition.[a] *(continued)*

Vitamin	Action	Deficiency Symptoms	Sources	Chemistry
Cyanocobalamin (vitamin B_{12})	Coenzyme in amino acid metabolism. Stimulates erythropoiesis	Pernicious anemia (see Chapter 26: Overview of Gastrointestinal Function & Regulation)	Liver, meat, eggs, milk	Complex of four substituted pyrrole rings around a cobalt atom (see Chapter 26: Overview of Gastrointestinal Function & Regulation)
C	Maintains prosthetic metal ions in their reduced form; scavenges free radicals	Scurvy	Citrus fruits, leafy green vegetables	CH_2OH, HO—C—H, O, CO, C, H, C=C, OH OH. Ascorbic acid (synthesized in most mammals except guinea pigs and primates, including humans).
D group	Increase intestinal absorption of calcium and phosphate (see Chapter 21: Hormonal Control of Calcium & Phosphate Metabolism & the Physiology of Bone)	Rickets	Fish liver	Family of sterols (see Chapter 21: Hormonal Control of Calcium & Phosphate Metabolism & the Physiology of Bone)
E group	Antioxidants; cofactors in electron transport in cytochrome chain?	Ataxia and other symptoms and signs of spinocerebellar dysfunction	Milk, eggs, meat, leafy vegetables	CH_3, H_2, HO, H_2, H_3C, O, CH_3, $(CH_2)_3$—CH—$(CH_2)_3$—CH—$(CH_2)_3$—CH, CH_3, CH_3, CH_3, CH_3. α-Tocopherol; β- and γ-tocopherol also active.
K group	Catalyze γ carboxylation of glutamic acid residues on various proteins concerned with blood clotting	Hemorrhagic phenomena	Leafy green vegetables	O, CH_3, O. Vitamin K_3; a large number of similar compounds have biological activity.

[a]Choline is synthesized in the body in small amounts, but it has recently been added to the list of essential nutrients.

CHAPTER SUMMARY

- A typical mixed meal consists of carbohydrates, proteins, and lipids (the latter largely in the form of triglycerides). Each must be digested to allow its uptake into the body. Specific transporters carry the products of digestion into the body.
- In the process of carbohydrate assimilation, the epithelium can only transport monomers, whereas for proteins, short peptides can be absorbed in addition to amino acids.
- The protein assimilation machinery, which rests heavily on the proteases in pancreatic juice, is arranged such that these enzymes are not activated until they reach their substrates in the small intestinal lumen. This is accomplished by the restricted localization of an activating enzyme, enterokinase.
- Lipids face special challenges to assimilation given their hydrophobicity. Bile acids solubilize the products of lipolysis in micelles and accelerate their ability to diffuse to the epithelial surface. The assimilation of triglycerides is enhanced by this mechanism, whereas that of cholesterol and fat-soluble vitamins absolutely requires it.
- The catabolism of nutrients provides energy to the body in a controlled fashion, via stepwise oxidations and other reactions.
- A balanced diet is important for health, and certain substances obtained from the diet are essential to life. The caloric value of dietary intake must be approximately equal to energy expenditure for homeostasis.

MULTIPLE-CHOICE QUESTIONS

For all questions, select the single best answer unless otherwise directed.

1. Maximum absorption of short-chain fatty acids produced by bacteria occurs in the
   A. stomach.
   B. duodenum.
   C. jejunum.
   D. ileum.
   E. colon.
2. A premenopausal woman who is physically active seeks advice from her primary care physician regarding measures she can take to ensure adequate availability of dietary calcium to ensure bone health later in life. Which of the following dietary components should enhance calcium uptake?
   A. Protein
   B. Oxalates
   C. Iron
   D. Vitamin D
   E. Sodium
3. A decrease in which of the following would be expected in a child exhibiting a congenital absence of enterokinase?
   A. Incidence of pancreatitis
   B. Glucose absorption
   C. Bile acid reabsorption
   D. Gastric pH
   E. Protein assimilation
4. In Hartnup disease (a defect in the transport of neutral amino acids), patients do not become deficient in these amino acids due to the activity of
   A. PepT1.
   B. brush border peptidases.
   C. Na, K ATPase.
   D. cystic fibrosis transmembrane conductance regulator (CFTR).
   E. trypsin.
5. A newborn baby is brought to the pediatrician suffering from severe diarrhea that worsens with meals. The symptoms diminish when nutrients are delivered intravenously. The child most likely has a mutation in which of the following intestinal transporters?
   A. Na, K ATPase
   B. NHE3
   C. SGLT1
   D. H^+,K^+ ATPase
   E. NKCC1

CHAPTER RESOURCES

Andrews NC: Disorders of iron metabolism. N Engl J Med 1999;341:1986.

Chong L, Marx J (editors): Lipids in the limelight. Science 2001;294:1861.

Farrell RJ, Kelly CP: Celiac sprue. N Engl J Med 2002;346:180.

Hofmann AF: Bile acids: The good, the bad, and the ugly. News Physiol Sci 1999;14:24.

Klok MD, Jakobsdottir S, Drent ML: The role of leptin and ghrelin in the regulation of food intake and body weight in humans: a review. Obesity Rev 2007;8:21.

Levitt MD, Bond JH: Volume, composition and source of intestinal gas. Gastroenterology 1970;59:921.

Mann NS, Mann SK: Enterokinase. Proc Soc Exp Biol Med 1994;206:114.

Meier PJ, Stieger B: Molecular mechanisms of bile formation. News Physiol Sci 2000;15:89.

Topping DL, Clifton PM: Short-chain fatty acids and human colonic function: Select resistant starch and nonstarch polysaccharides. Physiol Rev 2001;81:1031.

Wright EM: The intestinal Na^+/glucose cotransporter. Annu Rev Physiol 1993;55:575.

CHAPTER

27 Gastrointestinal Motility

OBJECTIVES

After studying this chapter, you should be able to:

- List the major forms of motility in the gastrointestinal tract and their roles in digestion and excretion.
- Distinguish between peristalsis and segmentation.
- Explain the electrical basis of gastrointestinal contractions and the role of basic electrical activity in governing motility patterns.
- Describe how gastrointestinal motility changes during fasting.
- Understand how food is swallowed and transferred to the stomach.
- Define the factors that govern gastric emptying and the abnormal response of vomiting.
- Define how the motility patterns of the colon subserve its function to desiccate and evacuate the stool.

INTRODUCTION

The digestive and absorptive functions of the gastrointestinal system outlined in the previous chapter depend on a variety of mechanisms that soften the food, propel it through the length of the gastrointestinal tract (Table 27–1), and mix it with bile from the gallbladder and digestive enzymes secreted by the salivary glands and pancreas. Some of these mechanisms depend on intrinsic properties of the intestinal smooth muscle. Others involve the operation of reflexes involving the neurons intrinsic to the gut, reflexes involving the central nervous system (CNS), paracrine effects of chemical messengers, and gastrointestinal hormones.

GENERAL PATTERNS OF MOTILITY

PERISTALSIS

Peristalsis is a reflex response that is initiated when the gut wall is stretched by the contents of the lumen, and it occurs in all parts of the gastrointestinal tract from the esophagus to the rectum. The stretch initiates a circular contraction behind the stimulus and an area of relaxation in front of it (Figure 27–1). The wave of contraction then moves in an oral-to-caudal direction, propelling the contents of the lumen forward at rates that vary from 2 to 25 cm/s. Peristaltic activity can be increased or decreased by the autonomic input to the gut, but its occurrence is independent of extrinsic innervation. Indeed, progression of the contents is not blocked by removal and resuture of a segment of intestine in its original position and is blocked only if the segment is reversed before it is sewn back into place. Peristalsis is an excellent example of the integrated activity of the enteric nervous system. It appears that local stretch releases serotonin, which activates sensory neurons that activate the myenteric plexus. Cholinergic neurons passing in a retrograde direction in this plexus activate neurons that release substance P and acetylcholine, causing smooth muscle contraction behind the bolus. At the same time, cholinergic neurons passing in an anterograde direction activate neurons that secrete NO and vasoactive intestinal polypeptide (VIP), producing the relaxation ahead of the stimulus.

TABLE 27–1 Mean lengths of various segments of the gastrointestinal tract as measured by intubation in living humans.

Segment	Length (cm)
Pharynx, esophagus, and stomach	65
Duodenum	25
Jejunum and ileum	260
Colon	110

Data from Hirsch JE, Ahrens EH Jr, Blankenhorn DH: Measurement of human intestinal length in vivo and some causes of variation. Gastroenterology 1956;31:274.

SEGMENTATION & MIXING

When the meal is present, the enteric nervous system promotes a motility pattern that is related to peristalsis, but is designed to retard the movement of the intestinal contents along the length of the intestinal tract to provide time for digestion and absorption (Figure 27–1). This motility pattern is known as segmentation, and it provides for ample mixing of the intestinal contents (known as chyme) with the digestive juices. A segment of bowel contracts at both ends, and then a second contraction occurs in the center of the segment to force the chyme both backward and forward. Unlike peristalsis, therefore, retrograde movement of the chyme occurs routinely in the setting of segmentation. This mixing pattern persists for as long as nutrients remain in the lumen to be absorbed. It presumably reflects programmed activity of the bowel dictated by the enteric nervous system, and can occur independent of central input, although the latter can modulate it.

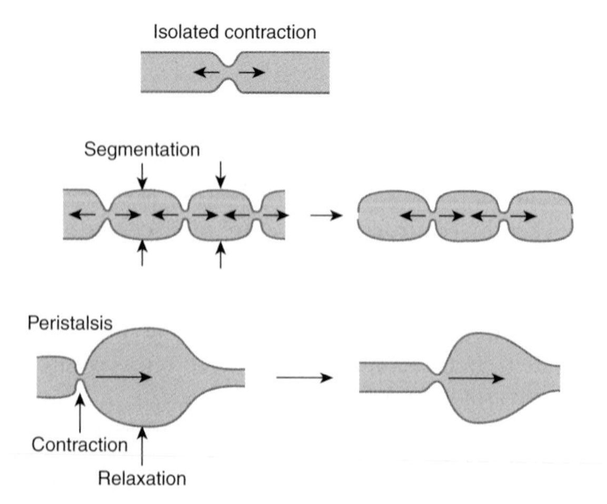

FIGURE 27–1 Patterns of gastrointestinal motility and propulsion. An isolated contraction moves contents orally and aborally. Segmentation mixes contents over a short stretch of intestine, as indicated by the time sequence from left to right. In the diagram on the left, the vertical arrows indicate the sites of subsequent contraction. Peristalsis involves both contraction and relaxation, and moves contents aborally.

BASIC ELECTRICAL ACTIVITY & REGULATION OF MOTILITY

Except in the esophagus and the proximal portion of the stomach, the smooth muscle of the gastrointestinal tract has spontaneous rhythmic fluctuations in membrane potential between about –65 and –45 mV. This **basic electrical rhythm (BER)** is initiated by the **interstitial cells of Cajal,** stellate mesenchymal pacemaker cells with smooth muscle-like features that send long multiply branched processes into the intestinal smooth muscle. In the stomach and the small intestine, these cells are located in the outer circular muscle layer near the myenteric plexus; in the colon, they are at the submucosal border of the circular muscle layer. In the stomach and small intestine, there is a descending gradient in pacemaker frequency, and as in the heart, the pacemaker with the highest frequency usually dominates.

The BER itself rarely causes muscle contraction, but **spike potentials** superimposed on the most depolarizing portions of the BER waves do increase muscle tension (Figure 27–2). The depolarizing portion of each spike is due to Ca^{2+} influx, and the repolarizing portion is due to K^+ efflux. Many polypeptides and neurotransmitters affect the BER. For example, acetylcholine increases the number of spikes and the tension of the smooth muscle, whereas epinephrine decreases the number of spikes and the tension. The rate of the BER is about 4/min in the stomach. It is about 12/min in the duodenum and falls to about 8/min in the distal ileum. In the colon, the BER rate rises from about 2/min at the cecum to about 6/min at the sigmoid. The function of the BER is to coordinate peristaltic and other motor activity, such as setting the rhythm of segmentation; contractions can occur only during the depolarizing part of the waves. After vagotomy or transection of the stomach wall, for example, peristalsis in the stomach becomes irregular and chaotic.

MIGRATING MOTOR COMPLEX

During fasting between periods of digestion, the pattern of electrical and motor activity in gastrointestinal smooth muscle becomes modified so that cycles of motor activity migrate from the stomach to the distal ileum. Each cycle, or **migrating motor complex (MMC),** starts with a quiescent period (phase I), continues with a period of irregular electrical and mechanical activity (phase II), and ends with a burst of regular activity (phase III) (Figure 27–3). The MMCs are initiated by motilin. The circulating level of this hormone increases at intervals of approximately 100 min in the interdigestive state, coordinated with the contractile phases of the MMC. The contractions migrate aborally at a rate of about 5 cm/min, and also occur at intervals of approximately 100 min. Gastric secretion, bile flow, and pancreatic secretion increase during each MMC. They likely serve to clear the stomach and small intestine of luminal contents in preparation for the next meal.

Conversely, when a meal is ingested, secretion of motilin is suppressed (ingestion of food suppresses motilin release via

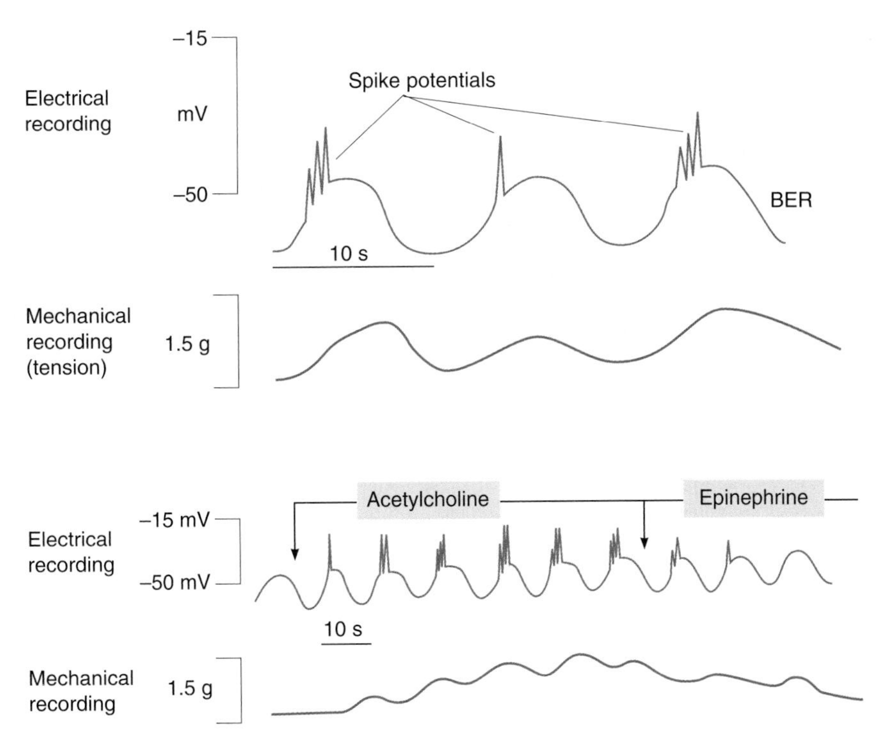

FIGURE 27–2 Basic electrical rhythm (BER) of gastrointestinal smooth muscle. Top: Morphology, and relation to muscle contraction. **Bottom:** Stimulatory effect of acetylcholine and inhibitory effect of epinephrine. (Modified and reproduced with permission from Chang EB, Sitrin MD, Black DD: *Gastrointestinal, Hepatobiliary, and Nutritional Physiology*. Lippincott-Raven, 1996.)

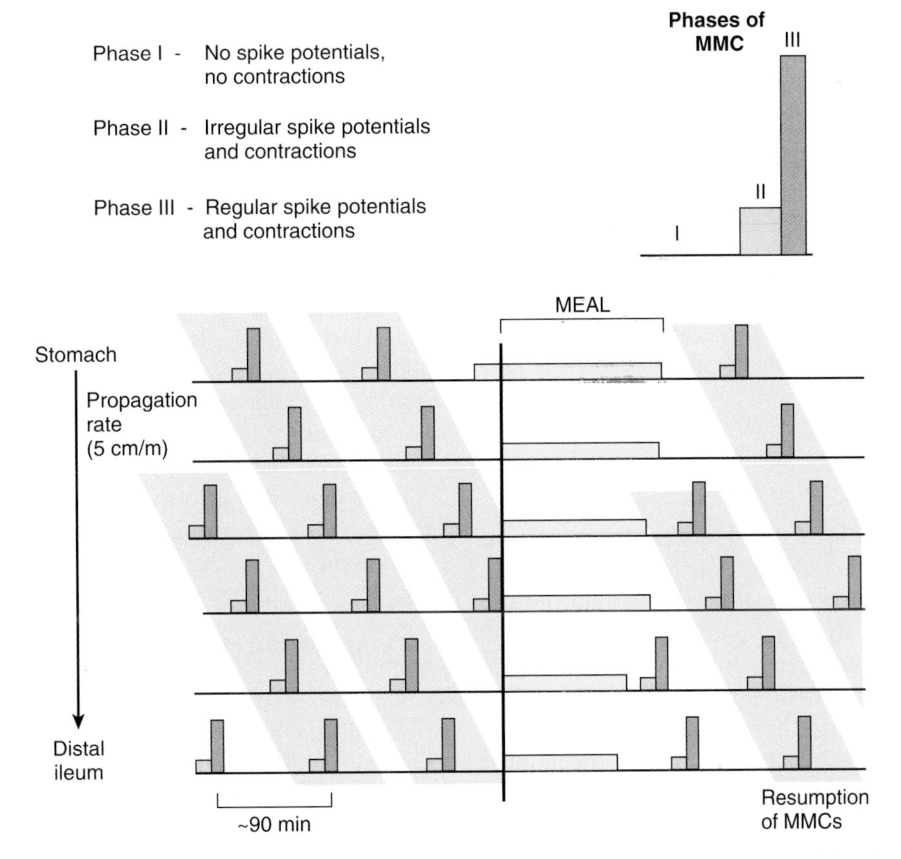

FIGURE 27–3 Migrating motor complexes (MMCs). Note that the complexes move down the gastrointestinal tract at a regular rate during fasting, that they are completely inhibited by a meal, and that they resume 90–120 min after the meal. (Reproduced with permission from Chang EB, Sitrin MD, Black DD: *Gastrointestinal, Hepatobiliary, and Nutritional Physiology*. Lippincott-Raven, 1996.)

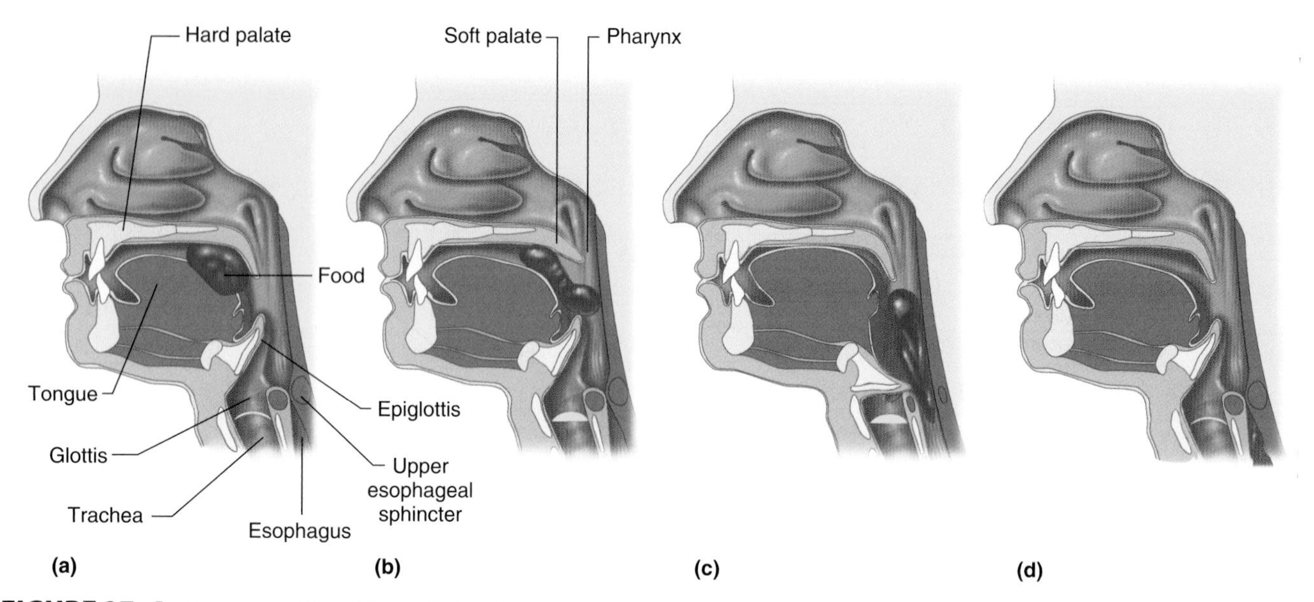

FIGURE 27–4 Movement of food through the pharynx and upper esophagus during swallowing. (a) The tongue pushes the food bolus to the back of the mouth. **(b)** The soft palate elevates to prevent food from entering the nasal passages. **(c)** The epiglottis covers the glottis to prevent food from entering the trachea and the upper esophageal sphincter relaxes. **(d)** Food descends into the esophagus.

mechanisms that have not yet been elucidated), and the MMC is abolished, until digestion and absorption are complete. Instead, there is a return to peristalsis and the other forms of BER and spike potentials during this time. The antibiotic erythromycin binds to motilin receptors, and derivatives of this compound may be of value in treating patients in whom gastrointestinal motility is decreased.

SEGMENT-SPECIFIC PATTERNS OF MOTILITY

MOUTH & ESOPHAGUS

In the mouth, food is mixed with saliva and propelled into the esophagus. Peristaltic waves in the esophagus move the food into the stomach.

MASTICATION

Chewing **(mastication)** breaks up large food particles and mixes the food with the secretions of the salivary glands. This wetting and homogenizing action aids swallowing and subsequent digestion. Large food particles can be digested, but they cause strong and often painful contractions of the esophageal musculature. Particles that are small tend to disperse in the absence of saliva and also make swallowing difficult because they do not form a bolus. The number of chews that is optimal depends on the food, but usually ranges from 20 to 25.

Edentulous patients are generally restricted to a soft diet and have considerable difficulty eating dry food.

SWALLOWING

Swallowing (deglutition) is a reflex response that is triggered by afferent impulses in the trigeminal, glossopharyngeal, and vagus nerves (Figure 27–4). These impulses are integrated in the nucleus of the tractus solitarius and the nucleus ambiguus. The efferent fibers pass to the pharyngeal musculature and the tongue via the trigeminal, facial, and hypoglossal nerves. Swallowing is initiated by the voluntary action of collecting the oral contents on the tongue and propelling them backward into the pharynx. This starts a wave of involuntary contraction in the pharyngeal muscles that pushes the material into the esophagus. Inhibition of respiration and glottic closure are part of the reflex response. A peristaltic ring contraction of the esophageal muscle forms behind the material, which is then swept down the esophagus at a speed of approximately 4 cm/s. When humans are in an upright position, liquids and semisolid foods generally fall by gravity to the lower esophagus ahead of the peristaltic wave. However, if any food remains in the esophagus, it is cleared by a second wave of peristalsis that occurs by the mechanisms discussed above. It is therefore possible to swallow food while standing on one's head.

LOWER ESOPHAGEAL SPHINCTER

Unlike the rest of the esophagus, the musculature of the gastroesophageal junction **(lower esophageal sphincter; LES)** is tonically active but relaxes on swallowing. The tonic activity of the LES between meals prevents reflux of gastric contents into the esophagus. The LES is made up of three components (Figure 27–5). The esophageal smooth muscle is more prominent at the junction with the stomach (intrinsic sphincter).

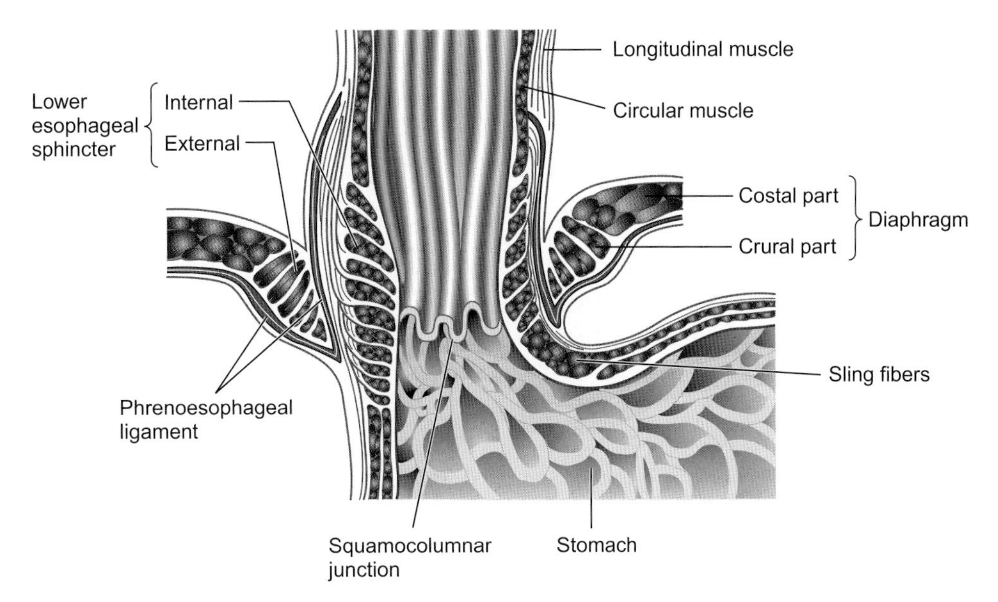

FIGURE 27–5 Esophagogastric junction. Note that the lower esophageal sphincter (intrinsic sphincter) is supplemented by the crural portion of the diaphragm (extrinsic sphincter), and that the two are anchored to each other by the phrenoesophageal ligament.

(Reproduced with permission, from Mittal RK, Balaban DH: The esophagogastric junction. N Engl J Med 1997;336:924.)

Fibers of the crural portion of the diaphragm, a skeletal muscle, surround the esophagus at this point (extrinsic sphincter) and exert a pinchcock-like action on the esophagus. In addition, the oblique or sling fibers of the stomach wall create a flap valve that helps close off the esophagogastric junction and prevent regurgitation when intragastric pressure rises.

The tone of the LES is under neural control. Release of acetylcholine from vagal endings causes the intrinsic sphincter to contract, and release of NO and VIP from interneurons innervated by other vagal fibers causes it to relax. Contraction of the crural portion of the diaphragm, which is innervated by the phrenic nerves, is coordinated with respiration and contractions of chest and abdominal muscles. Thus, the intrinsic and extrinsic sphincters operate together to permit orderly flow of food into the stomach and to prevent reflux of gastric contents into the esophagus (Clinical Box 27–1).

CLINICAL BOX 27–1

Motor Disorders of the Esophagus

Achalasia (literally, failure to relax) is a condition in which food accumulates in the esophagus and the organ can become massively dilated. It is due to increased resting LES tone and incomplete relaxation on swallowing. The myenteric plexus of the esophagus is deficient at the LES in this condition and the release of NO and VIP is defective. The opposite condition is LES incompetence, which permits reflux of acid gastric contents into the esophagus **(gastroesophageal reflux disease).** This common condition is the most frequent digestive disorder causing patients to seek care from a physician. It causes heartburn and esophagitis and can lead to ulceration and stricture of the esophagus due to scarring. In severe cases, the intrinsic sphincter, the extrinsic sphincter, and sometimes both are weak, but less severe cases are caused by intermittent periods of poorly understood decreases in the neural drive to both sphincters.

THERAPEUTIC HIGHLIGHTS

Achalasia can be treated by pneumatic dilation of the sphincter or incision of the esophageal muscle (myotomy). Inhibition of acetylcholine release by injection of botulinum toxin into the LES is also effective and produces relief that lasts for several months. Gastroesophageal reflux disease can be treated by inhibition of acid secretion with H_2 receptor blockers or proton pump inhibitors (see Chapter 25). Surgical treatment in which a portion of the fundus of the stomach is wrapped around the lower esophagus so that the LES is inside a short tunnel of stomach **(fundoplication)** can also be tried, although in many patients who undergo this procedure the symptoms eventually return.

AEROPHAGIA & INTESTINAL GAS

Some air is unavoidably swallowed in the process of eating and drinking **(aerophagia).** Some of the swallowed air is regurgitated (belching), and some of the gases it contains are absorbed, but much of it passes on to the colon. Here, some of the oxygen is absorbed, and hydrogen, hydrogen sulfide, carbon dioxide, and methane formed by the colonic bacteria from carbohydrates and other substances are added to it. It is then expelled as **flatus.** The smell is largely due to sulfides. The volume of gas normally found in the human gastrointestinal tract is about 200 mL, and the daily production is 500–1500 mL. In some individuals, gas in the intestines causes cramps, **borborygmi** (rumbling noises), and abdominal discomfort.

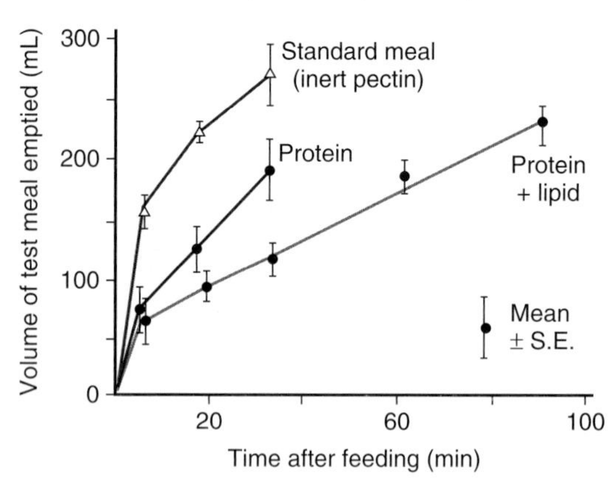

FIGURE 27–6 Effect of protein and fat on the rate of emptying of the human stomach. Subjects were fed 300-mL liquid meals. (Reproduced with permission from Brooks FP: Integrative lecture. Response of the GI tract to a meal. *Undergraduate Teaching Project.* American Gastroenterological Association, 1974.)

STOMACH

Food is stored in the stomach; mixed with acid, mucus, and pepsin; and released at a controlled, steady rate into the duodenum.

GASTRIC MOTILITY & EMPTYING

When food enters the stomach, the fundus and upper portion of the body relax and accommodate the food with little if any increase in pressure **(receptive relaxation).** Peristalsis then begins in the lower portion of the body, mixing and grinding the food and permitting small, semiliquid portions of it to pass through the pylorus and enter the duodenum.

Receptive relaxation is, in part, vagally mediated and triggered by movement of the pharynx and esophagus. Intrinsic reflexes also lead to relaxation as the stomach wall is stretched. Peristaltic waves controlled by the gastric BER begin soon thereafter and sweep toward the pylorus. The contraction of the distal stomach caused by each wave is sometimes called **antral systole** and can last up to 10 s. Waves occur 3–4 times per minute.

In the regulation of gastric emptying, the antrum, pylorus, and upper duodenum apparently function as a unit. Contraction of the antrum is followed by sequential contraction of the pyloric region and the duodenum. In the antrum, partial contraction ahead of the advancing gastric contents prevents solid masses from entering the duodenum, and they are mixed and crushed instead. The more liquid gastric contents are squirted a bit at a time into the small intestine. Normally, regurgitation from the duodenum does not occur, because the contraction of the pyloric segment tends to persist slightly longer than that of the duodenum. The prevention of regurgitation may also be due to the stimulating action of cholecystokinin (CCK) and secretin on the pyloric sphincter.

REGULATION OF GASTRIC MOTILITY & EMPTYING

The rate at which the stomach empties into the duodenum depends on the type of food ingested. Food rich in carbohydrate leaves the stomach in a few hours. Protein-rich food leaves more slowly, and emptying is slowest after a meal containing fat (Figure 27–6). The rate of emptying also depends on the osmotic pressure of the material entering the duodenum. Hyperosmolality of the duodenal contents is sensed by "duodenal osmoreceptors" that initiate a decrease in gastric emptying, which is probably neural in origin.

Fats, carbohydrates, and acid in the duodenum inhibit gastric acid and pepsin secretion and gastric motility via neural and hormonal mechanisms. The messenger involved is probably peptide YY. CCK has also been implicated as an inhibitor of gastric emptying (Clinical Box 27–2).

VOMITING

Vomiting is an example of central regulation of gut motility functions. Vomiting starts with salivation and the sensation of nausea. Reverse peristalsis empties material from the upper part of the small intestine into the stomach. The glottis closes, preventing aspiration of vomitus into the trachea. The breath is held in mid inspiration. The muscles of the abdominal wall contract, and because the chest is held in a fixed position, the contraction increases intra-abdominal pressure. The lower esophageal sphincter and the esophagus relax, and the gastric contents are ejected. The "vomiting center" in the reticular formation of the medulla (Figure 27–7) consists of various scattered groups of neurons in this region that control the different components of the vomiting act.

Irritation of the mucosa of the upper gastrointestinal tract is one trigger for vomiting. Impulses are relayed from the mucosa to the medulla over visceral afferent pathways in the sympathetic nerves and vagi. Other causes of vomiting can arise centrally. For example, afferents from the vestibular nuclei mediate the nausea and vomiting of motion sickness. Other afferents presumably reach the vomiting control areas from the diencephalon and limbic system, because emetic

CLINICAL BOX 27–2

Consequences of Gastric Bypass Surgery

Patients who are morbidly obese often undergo a surgical procedure in which the stomach is stapled so that most of it is bypassed, and thus the reservoir function of the stomach is lost. As a result, such patients must eat frequent small meals. If larger meals are taken, because of rapid absorption of glucose from the intestine and the resultant hyperglycemia and abrupt rise in insulin secretion, gastrectomized patients sometimes develop hypoglycemic symptoms about 2 h after meals. Weakness, dizziness, and sweating after meals, due in part to hypoglycemia, are part of the picture of the **"dumping syndrome,"** a distressing syndrome that develops in patients in whom portions of the stomach have been removed or the jejunum has been anastomosed to the stomach. Another cause of the symptoms is rapid entry of hypertonic meals into the intestine; this provokes the movement of so much water into the gut that significant hypovolemia and hypotension are produced.

THERAPEUTIC HIGHLIGHTS

There are no treatments, per se, for the dumping syndrome, other than avoiding large meals, and particularly those with high concentrations of simple sugars. Indeed, its occurrence may account for the overall success of bypass surgery in reducing food intake, and thus obesity, in many patients who undergo this operation.

responses to emotionally charged stimuli also occur. Thus, we speak of "nauseating smells" and "sickening sights."

Chemoreceptor cells in the medulla can also initiate vomiting when they are stimulated by certain circulating chemical agents. The **chemoreceptor trigger zone** in which these cells are located (Figure 27–7) is in the **area postrema,** a V-shaped band of tissue on the lateral walls of the fourth ventricle near the obex. This structure is one of the circumventricular organs (see Chapter 33) and is not protected by the blood–brain barrier. Lesions of the area postrema have little effect on the vomiting response to gastrointestinal irritation or motion sickness, but abolish the vomiting that follows injection of apomorphine and a number of other emetic drugs. Such lesions also decrease vomiting in uremia and radiation sickness, both of which may be associated with endogenous production of circulating emetic substances.

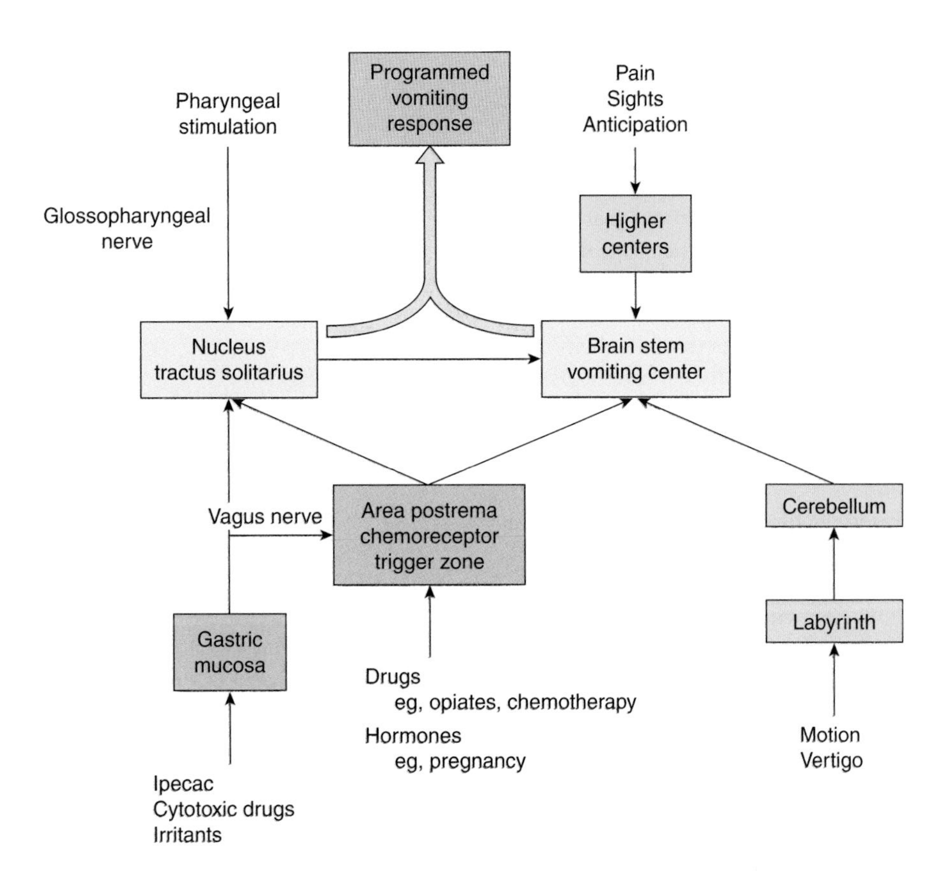

FIGURE 27–7 Neural pathways leading to the initiation of vomiting in response to various stimuli.

Serotonin (5-HT) released from enterochromaffin cells in the small intestine appears to initiate impulses via 5-HT_3 receptors that trigger vomiting. In addition, there are dopamine D_2 receptors and 5-HT_3 receptors in the area postrema and adjacent nucleus of the solitary tract. 5-HT_3 antagonists such as ondansetron and D_2 antagonists such as chlorpromazine and haloperidol are effective antiemetic agents. Corticosteroids, cannabinoids, and benzodiazepines, alone or in combination with 5-HT_3 and D_2 antagonists, are also useful in treatment of the vomiting produced by chemotherapy. The mechanisms of action of corticosteroids and cannabinoids are unknown, whereas the benzodiazepines probably reduce the anxiety associated with chemotherapy.

SMALL INTESTINE

In the small intestine, the intestinal contents are mixed with the secretions of the mucosal cells and with pancreatic juice and bile.

INTESTINAL MOTILITY

The MMCs that pass along the intestine at regular intervals in the fasting state and their replacement by peristaltic and other contractions controlled by the BER are described above. In the small intestine, there are an average of 12 BER cycles/min in the proximal jejunum, declining to 8/min in the distal ileum. There are three types of smooth muscle contractions: peristaltic waves, segmentation contractions, and tonic contractions. **Peristalsis** is described above. It propels the intestinal contents **(chyme)** toward the large intestines. **Segmentation contractions** (Figure 27–1), also described above, move the chyme to and fro and increase its exposure to the mucosal surface. These contractions are initiated by focal increases in Ca^{2+} influx with waves of increased Ca^{2+} concentration spreading from each focus. **Tonic contractions** are relatively prolonged contractions that in effect isolate one segment of the intestine from another. Note that these last two types of contractions slow transit in the small intestine to the point that the transit time is actually longer in the fed than in the fasted state. This permits longer contact of the chyme with the enterocytes and fosters absorption (Clinical Box 27–3).

CLINICAL BOX 27–3

Ileus

When the intestines are traumatized, there is a direct inhibition of smooth muscle, which causes a decrease in intestinal motility. It is due in part to activation of opioid receptors. When the peritoneum is irritated, reflex inhibition occurs due to increased discharge of noradrenergic fibers in the splanchnic nerves. Both types of inhibition operate to cause **paralytic (adynamic) ileus** after abdominal operations. Because of the diffuse decrease in peristaltic activity in the small intestine, its contents are not propelled into the colon, and it becomes irregularly distended by pockets of gas and fluid. Intestinal peristalsis returns in 6–8 h, followed by gastric peristalsis, but colonic activity takes 2–3 days to return.

THERAPEUTIC HIGHLIGHTS

Adynamic ileus can be relieved by passing a tube through the nose down to the small intestine and aspirating the fluid and gas for a few days until peristalsis returns. The occurrence of ileus has been reduced by more widespread use of minimally invasive (eg, laparoscopic) surgery. Post-surgical regimens also now encourage early ambulation, where possible, which tends to enhance intestinal motility. There are also ongoing trials of specific opioid antagonists in this condition.

COLON

The colon serves as a reservoir for the residues of meals that cannot be digested or absorbed (Figure 27–8). Motility in this segment is likewise slowed to allow the colon to absorb water, Na^+, and other minerals. By removing about 90% of the fluid, it converts the 1000–2000 mL of isotonic chyme that enters it each day from the ileum to about 200–250 mL of semisolid feces.

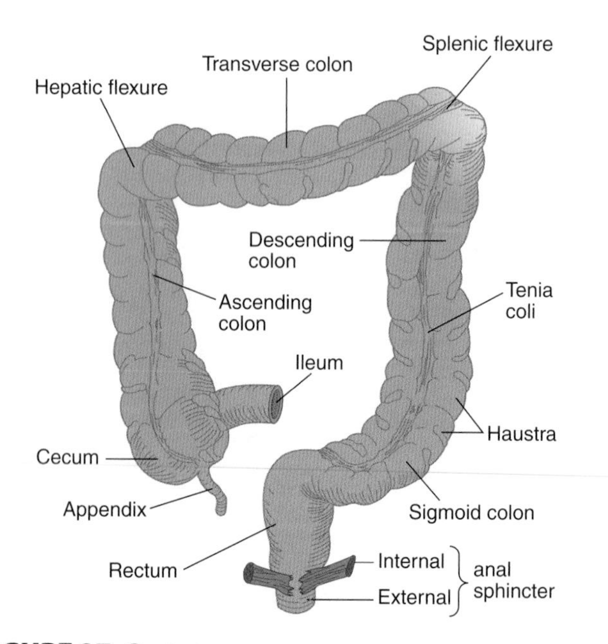

FIGURE 27–8 The human colon.

MOTILITY OF THE COLON

The ileum is linked to the colon by a structure known as the ileocecal valve, which restricts reflux of colonic contents, and particularly the large numbers of commensal bacteria, into the relatively sterile ileum. The portion of the ileum containing the ileocecal valve projects slightly into the cecum, so that increases in colonic pressure squeeze it shut, whereas increases in ileal pressure open it. It is normally closed. Each time a peristaltic wave reaches it, it opens briefly, permitting some of the ileal chyme to squirt into the cecum. When food leaves the stomach, the cecum relaxes and the passage of chyme through the ileocecal valve increases **(gastroileal reflex).** This is presumably a vago-vagal reflex.

The movements of the colon include segmentation contractions and peristaltic waves like those occurring in the small intestine. Segmentation contractions mix the contents of the colon and, by exposing more of the contents to the mucosa, facilitate absorption. Peristaltic waves propel the contents toward the rectum, although weak antiperistalsis is sometimes seen. A third type of contraction that occurs only in the colon is the **mass action contraction,** occurring about 10 times per day, in which there is simultaneous contraction of the smooth muscle over large confluent areas. These contractions move material from one portion of the colon to another (Clinical Box 27–4). They also move material into the rectum, and rectal distention initiates the defecation reflex (see below).

CLINICAL BOX 27–4

Hirschsprung Disease

Some children present with a genetically determined condition of abnormal colonic motility known as Hirschsprung disease or **aganglionic megacolon,** which is characterized by abdominal distention, anorexia, and lassitude. The disease is typically diagnosed in infancy, and affects as many as 1 in 5000 live births. It is due to a congenital absence of the ganglion cells in both the myenteric and submucous plexuses of a segment of the distal colon, as a result of failure of the normal cranial-to-caudal migration of neural crest cells during development. The action of endothelins on the endothelin B receptor (see Chapter 7) are necessary for normal migration of certain neural crest cells, and knockout mice lacking endothelin B receptors developed megacolon. In addition, one cause of congenital aganglionic megacolon in humans appears to be a mutation in the endothelin B receptor gene. The absence of peristalsis in patients with this disorder causes feces to pass the aganglionic region with difficulty, and children with the disease may defecate as infrequently as once every 3 weeks.

THERAPEUTIC HIGHLIGHTS

The symptoms of Hirschsprung disease can be relieved completely if the aganglionic portion of the colon is resected and the portion of the colon above it anastomosed to the rectum. However, this is not possible if an extensive segment is involved. In this case, patients may require a colectomy.

The movements of the colon are coordinated by the BER of the colon. The frequency of this wave, unlike the wave in the small intestine, increases along the colon, from about 2/min at the ileocecal valve to 6/min at the sigmoid.

TRANSIT TIME IN THE SMALL INTESTINE & COLON

The first part of a test meal reaches the cecum in about 4 h in most individuals, and all the undigested portions have entered the colon in 8 or 9 h. On average, the first remnants of the meal traverse the first third of the colon in 6 h, the second third in 9 h, and reach the terminal part of the colon (the sigmoid colon) in 12 h. From the sigmoid colon to the anus, transport is much slower (Clinical Box 27–5). When small colored beads

CLINICAL BOX 27–5

Constipation

Constipation refers to a pathological decrease in bowel movements. It was previously considered to reflect changes in motility, but the recent success of a drug designed to enhance chloride secretion for the treatment of chronic constipation suggests alterations in the balance between secretion and absorption in the colon could also contribute to symptom generation. Patients with persistent constipation, and particularly those with a recent change in bowel habits, should be examined carefully to rule out underlying organic disease. However, many normal humans defecate only once every 2–3 days, even though others defecate once a day and some as often as three times a day. Furthermore, the only symptoms caused by constipation are slight anorexia and mild abdominal discomfort and distention. These symptoms are not due to absorption of "toxic substances," because they are promptly relieved by evacuating the rectum and can be reproduced by distending the rectum with inert material. In western societies, the amount of misinformation and undue apprehension about constipation probably exceeds that about any other health topic. Symptoms other than those described above that are attributed by the lay public to constipation are due to anxiety or other causes.

THERAPEUTIC HIGHLIGHTS

Most cases of constipation are relieved by a change in the diet to include more fiber, or the use of laxatives that retain fluid in the colon, thereby increasing the bulk of the stool and promoting reflexes that lead to evacuation. As noted above, lubiprostone has recently joined the armamentarium for the treatment of constipation, and is assumed to act by enhancing chloride, and thus water, secretion into the colon thereby increasing the fluidity of the colonic contents.

are fed with a meal, an average of 70% of them are recovered in the stool in 72 h, but total recovery requires more than a week. Transit time, pressure fluctuations, and changes in pH in the gastrointestinal tract can be observed by monitoring the progress of a small pill that contains sensors and a miniature radio transmitter.

DEFECATION

Distention of the rectum with feces initiates reflex contractions of its musculature and the desire to defecate. In humans, the sympathetic nerve supply to the internal (involuntary) anal sphincter is excitatory, whereas the parasympathetic supply is inhibitory. This sphincter relaxes when the rectum is distended. The nerve supply to the external anal sphincter, a skeletal muscle, comes from the pudendal nerve. The sphincter is maintained in a state of tonic contraction, and moderate distention of the rectum increases the force of its contraction (Figure 27–9). The urge to defecate first occurs when rectal pressure increases to about 18 mm Hg. When this pressure reaches 55 mm Hg, the external as well as the internal sphincter relaxes and there is reflex expulsion of the contents of the rectum. This is why reflex evacuation of the rectum can occur even in the setting of spinal injury.

Before the pressure that relaxes the external anal sphincter is reached, voluntary defecation can be initiated by straining. Normally, the angle between the anus and the rectum is approximately 90° (Figure 27–10), and this plus contraction of the puborectalis muscle inhibits defecation. With straining, the abdominal muscles contract, the pelvic floor is lowered 1–3 cm, and the puborectalis muscle relaxes. The anorectal angle is reduced to 15° or less. This is combined with relaxation of the external anal sphincter and defecation occurs.

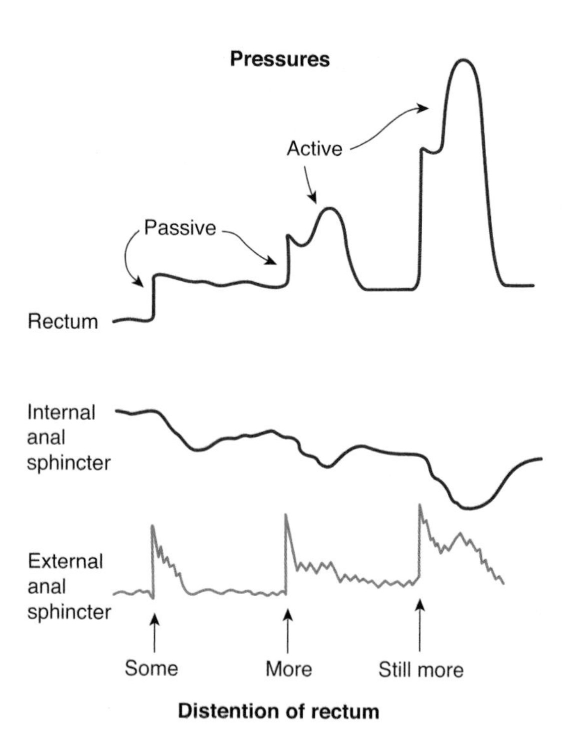

FIGURE 27–9 Responses to distention of the rectum by pressures less than 55 mm Hg. Distention produces passive tension due to stretching of the wall of the rectum, and additional active tension when the smooth muscle in the wall contracts. (For Internal, Adapted from Denny-Brown D, Robertson EG: An investigation of the nervous control of defaecation. Brain 1935; 58:256–310; For external, Adapted from Schuster MM et al: Simultaneous manometric recording of internal and external anal sphincteric reflexes. Bull Johns Hopkins Hosp 1965 Feb;116:79-88.)

Defecation is therefore a spinal reflex that can be voluntarily inhibited by keeping the external sphincter contracted or facilitated by relaxing the sphincter and contracting the abdominal muscles.

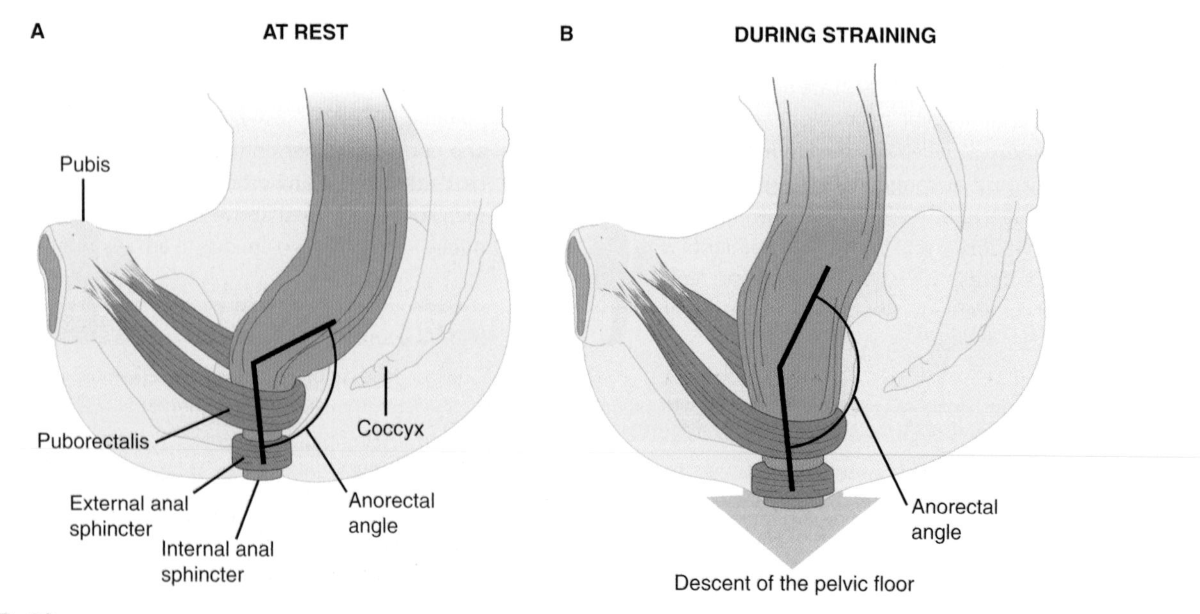

FIGURE 27–10 Sagittal view of the anorectal area at rest (above) and during straining (below). Note the reduction of the anorectal angle and lowering of the pelvic floor during straining. (Modified and reproduced with permission from Lembo A, Camilleri, M: Chronic constipation. N Engl J Med 2003;349:1360.)

Distention of the stomach by food initiates contractions of the rectum and, frequently, a desire to defecate. The response is called the **gastrocolic reflex,** and may be amplified by an action of gastrin on the colon. Because of the response, defecation after meals is the rule in children. In adults, habit and cultural factors play a large role in determining when defecation occurs.

CHAPTER SUMMARY

- The regulatory factors that govern gastrointestinal secretion also regulate its motility to soften the food, mix it with secretions, and propel it along the length of the tract.
- Two major patterns of motility are peristalsis and segmentation, which serve to propel or retard/mix the luminal contents, respectively. Peristalsis involves coordinated contractions and relaxations above and below the food bolus.
- The membrane potential of the majority of gastrointestinal smooth muscle undergoes rhythmic fluctuations that sweep along the length of the gut. The rhythm varies in different gut segments and is established by pacemaker cells known as interstitial cells of Cajal. This basic electrical rhythm provides for sites of muscle contraction when stimuli superimpose spike potentials on the depolarizing portion of the BER waves.
- In the period between meals, the intestine is relatively quiescent, but every 90 min or so it is swept through by a large peristaltic wave triggered by the hormone motilin. This migrating motor complex presumably serves a "housekeeping" function.
- Swallowing is triggered centrally and is coordinated with a peristaltic wave along the length of the esophagus that drives the food bolus to the stomach, even against gravity. Relaxation of the lower esophageal sphincter is timed to just precede the arrival of the bolus, thereby limiting reflux of the gastric contents. Nevertheless, gastroesophageal reflux disease is one of the most common gastrointestinal complaints
- The stomach accommodates the meal by a process of receptive relaxation. This permits an increase in volume without a significant increase in pressure. The stomach then serves to mix the meal and to control its delivery to downstream segments.
- Luminal contents move slowly through the colon, which enhances water recovery. Distension of the rectum causes reflex contraction of the internal anal sphincter and the desire to defecate. After toilet training, defecation can be delayed till a convenient time via voluntary contraction of the external anal sphincter.

MULTIPLE-CHOICE QUESTIONS

For all questions, select the single best answer unless otherwise directed.

1. In infants, defecation often follows a meal. The cause of colonic contractions in this situation is
   A. histamine.
   B. increased circulating levels of CCK.
   C. the gastrocolic reflex.
   D. increased circulating levels of somatostatin.
   E. the enterogastric reflex.
2. The symptoms of the dumping syndrome (discomfort after meals in patients with intestinal short circuits such as anastomosis of the jejunum to the stomach) are caused in part by
   A. increased blood pressure.
   B. increased secretion of glucagon.
   C. increased secretion of CCK.
   D. hypoglycemia.
   E. hyperglycemia.
3. Gastric pressures seldom rise above the levels that breach the lower esophageal sphincter, even when the stomach is filled with a meal, due to which of the following processes?
   A. Peristalsis
   B. Gastroileal reflex
   C. Segmentation
   D. Stimulation of the vomiting center
   E. Receptive relaxation
4. The migrating motor complex is triggered by which of the following?
   A. Motilin
   B. NO
   C. CCK
   D. Somatostatin
   E. Secretin
5. A patient is referred to a gastroenterologist because of persistent difficulties with swallowing. Endoscopic examination reveals that the lower esophageal sphincter fails to fully open as the bolus reaches it, and a diagnosis of achalasia is made. During the examination, or in biopsies taken from the sphincter region, a decrease would be expected in which of the following?
   A. Esophageal peristalsis
   B. Expression of neuronal NO synthase
   C. Acetylcholine receptors
   D. Substance P release
   E. Contraction of the crural diaphragm

CHAPTER RESOURCES

Barrett KE: *Gastrointestinal Physiology.* McGraw-Hill, 2006.
Cohen S, Parkman HP: Heartburn—A serious symptom. N Engl J Med 1999;340:878.
Itoh Z: Motilin and clinical application. Peptides 1997;18:593.
Lembo A, Camilleri M: Chronic constipation. N Engl J Med 2003;349:1360.
Levitt MD, Bond JH: Volume, composition and source of intestinal gas. Gastroenterology 1970;59:921.
Mayer EA, Sun XP, Willenbucher RF: Contraction coupling in colonic smooth muscle. Annu Rev Physiol 1992;54:395.
Mittal RK, Balaban DH: The esophagogastric junction. N Engl J Med 1997;336:924.
Sanders KM, Ward SM: Nitric oxide as a mediator of noncholinergic neurotransmission. Am J Physiol 1992;262:G379.
Ward SM, Sanders KM: Involvement of intramuscular interstitial cells of Cajal in neuroeffector transmission in the gastrointestinal tract. J Physiol 2006;576:675.

CHAPTER 28

Transport & Metabolic Functions of the Liver

OBJECTIVES

After studying this chapter, you should be able to:

- Describe the major functions of the liver with respect to metabolism, detoxification, and excretion of hydrophobic substances.
- Understand the functional anatomy of the liver and the relative arrangements of hepatocytes, cholangiocytes, endothelial cells, and Kupffer cells.
- Define the characteristics of the hepatic circulation and its role in subserving the liver's functions.
- Identify the plasma proteins that are synthesized by the liver.
- Describe the formation of bile, its constituents, and its role in the excretion of cholesterol and bilirubin.
- Outline the mechanisms by which the liver contributes to whole body ammonia homeostasis and the consequences of the failure of these mechanisms, particularly for brain function.
- Identify the mechanisms that permit normal functioning of the gallbladder and the basis of gallstone disease.

INTRODUCTION

The liver is the largest gland in the body. It is essential for life because it conducts a vast array of biochemical and metabolic functions, including ridding the body of substances that would otherwise be injurious if allowed to accumulate, and excreting drug metabolites. It is also the first port of call for most nutrients absorbed across the gut wall, supplies most of the plasma proteins, and synthesizes the bile that optimizes the absorption of fats as well as serving as an excretory fluid. The liver and associated biliary system have therefore evolved an array of structural and physiologic features that underpin this broad range of critical functions.

THE LIVER

FUNCTIONAL ANATOMY

An important function of the liver is to serve as a filter between the blood coming from the gastrointestinal tract and the blood in the rest of the body. Blood from the intestines and other viscera reach the liver via the portal vein. This blood percolates in sinusoids between plates of hepatic cells and eventually drains into the hepatic veins, which enter the inferior vena cava. During its passage through the hepatic plates, it is extensively modified chemically. Bile is formed on the other side at each plate. The bile passes to the intestine via the hepatic duct (Figure 28–1).

In each hepatic lobule, the plates of hepatic cells are usually only one cell thick. Large gaps occur between the endothelial cells, and plasma is in intimate contact with the cells (Figure 28–2). Hepatic artery blood also enters the sinusoids. The central veins coalesce to form the hepatic veins, which drain into the inferior vena cava. The average transit time for blood across the liver lobule from the portal venule to the central hepatic vein is about 8.4 s. Additional details of the features of the hepatic micro- and macrocirculation, which are critical to organ function, are provided below. Numerous

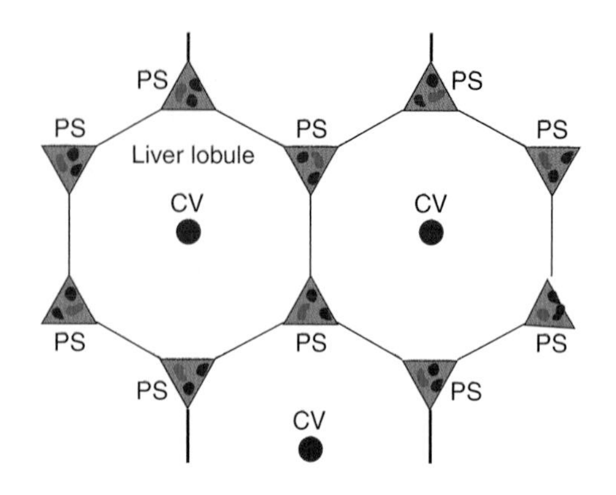

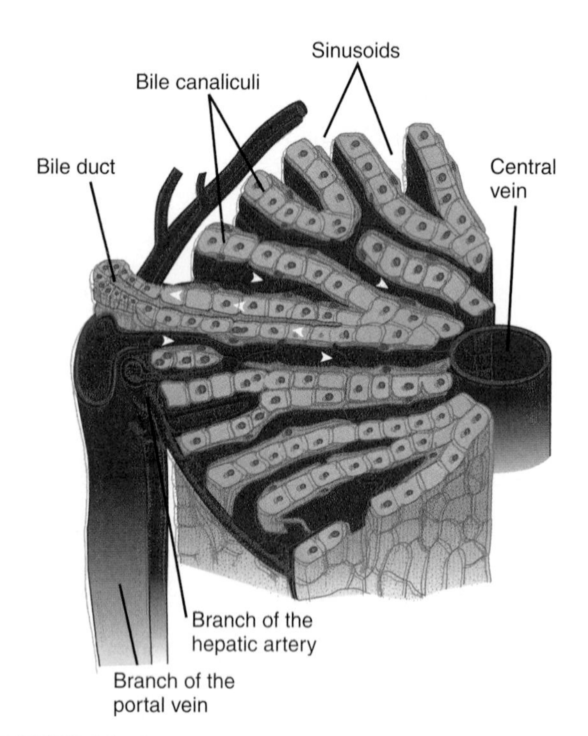

FIGURE 28–1 Schematic anatomy of the liver. Top: Organization of the liver. CV, central vein. PS, portal space containing branches of bile duct (green), portal vein (blue), and hepatic artery (red). **Bottom:** Arrangement of plates of liver cells, sinusoids, and bile ducts in a liver lobule, showing centripetal flow of blood in sinusoids to central vein and centrifugal flow of bile in bile canaliculi to bile ducts. (Redrawn and modified from Ham AW: *Textbook of Histology,* 5th ed. Philadelphia: JB Lippincott Co., 1965.)

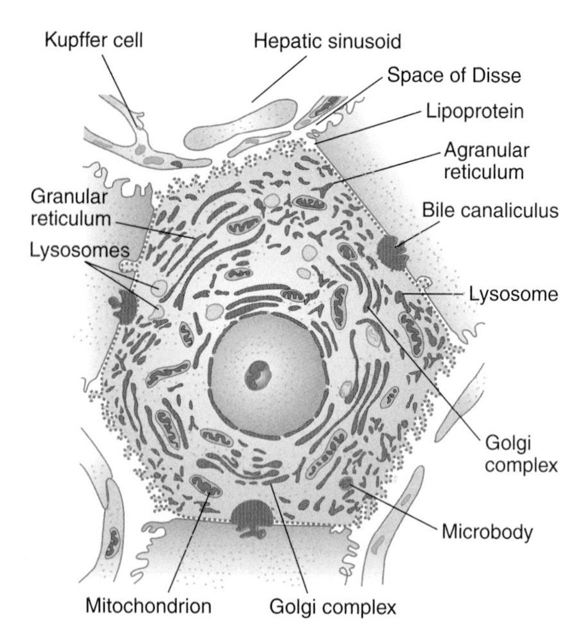

FIGURE 28–2 Hepatocyte. Note the relation of the cell to bile canaliculi and sinusoids. Note also the wide openings (fenestrations) between the endothelial cells next to the hepatocyte. (Drawing by Sylvia Colard Keene.)

macrophages **(Kupffer cells)** are anchored to the endothelium of the sinusoids and project into the lumen. The functions of these phagocytic cells are discussed in Chapter 3.

Each liver cell is also apposed to several **bile canaliculi** (Figure 28–2). The canaliculi drain into intralobular bile ducts, and these coalesce via interlobular bile ducts to form the right and left hepatic ducts. These ducts join outside the liver to form the common hepatic duct. The cystic duct drains the gallbladder. The hepatic duct unites with the cystic duct to form the common bile duct (Figure 28–1). The common bile duct enters the duodenum at the duodenal papilla. Its orifice is surrounded by the **sphincter of Oddi,** and it usually unites with the main pancreatic duct just before entering the duodenum. The sphincter is usually closed, but when the gastric contents enter the duodenum, cholecystokinin (CCK) is released and the gastrointestinal hormone relaxes the sphincter and makes the gallbladder contract.

The walls of the extrahepatic biliary ducts and the gallbladder contain fibrous tissue and smooth muscle. They are lined by a layer of columnar cells with scattered mucous glands. In the gallbladder, the surface is extensively folded; this increases its surface area and gives the interior of the gallbladder a honeycombed appearance. The cystic duct is also folded to form the so-called spiral valves. This arrangement is believed to increase the turbulence of bile as it flows out of the gallbladder, thereby reducing the risk that it will precipitate and form gallstones.

HEPATIC CIRCULATION

Large gaps occur between endothelial cells in the walls of hepatic sinusoids, and the sinusoids are highly permeable. The way the intrahepatic branches of the hepatic artery and portal vein converge on the sinusoids and drain into the central lobular veins of the liver is shown in Figure 28–1. The functional unit of the liver is the acinus. Each acinus is at the end of a vascular stalk containing terminal branches of portal veins, hepatic arteries, and bile ducts. Blood flows from the center of this functional unit to the terminal branches of the hepatic veins at the periphery (Figure 28–3). This is why the central portion of the acinus, sometimes called zone 1, is well

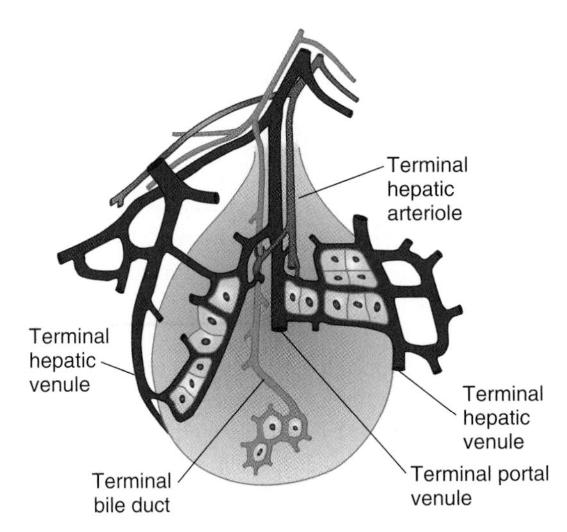

FIGURE 28–3 Concept of the acinus as the functional unit of the liver. In each acinus, blood in the portal venule and hepatic arteriole enters the center of the acinus and flows outward to the hepatic venule. (Based on the acinar concept of Rappaport AM: The microcirculatory hepatic unit. Microvasc Res 1973 Sep;6(2):212–228.)

oxygenated, the intermediate zone (zone 2) is moderately well oxygenated, and the peripheral zone (zone 3) is least well oxygenated and most susceptible to anoxic injury. The hepatic veins drain into the inferior vena cava. The acini have been likened to grapes or berries, each on a vascular stem. The human liver contains about 100,000 acini.

Portal venous pressure is normally about 10 mm Hg in humans, and hepatic venous pressure is approximately 5 mm Hg. The mean pressure in the hepatic artery branches that converge on the sinusoids is about 90 mm Hg, but the pressure in the sinusoids is lower than the portal venous pressure, so a marked pressure drop occurs along the hepatic arterioles. This pressure drop is adjusted so that there is an inverse relationship between hepatic arterial and portal venous blood flow. This inverse relationship may be maintained in part by the rate at which adenosine is removed from the region around the arterioles. According to this hypothesis, adenosine is produced by metabolism at a constant rate. When portal flow is reduced, it is washed away more slowly, and the local accumulation of adenosine dilates the terminal arterioles. In the period between meals, moreover, many of the sinusoids are collapsed. Following a meal, on the other hand, when portal flow to the liver from the intestine increases considerably, these "reserve" sinusoids are recruited. This arrangement means that portal pressures do not increase linearly with portal flow until all sinusoids have been recruited. This may be important to prevent fluid loss from the highly permeable liver under normal conditions. Indeed, if hepatic pressures are increased in disease states (such as the hardening of the liver that is seen in cirrhosis), many liters of fluid can accumulate in the peritoneal cavity as **ascites.**

The intrahepatic portal vein radicles have smooth muscle in their walls that is innervated by noradrenergic vasoconstrictor nerve fibers reaching the liver via the 3rd to 11th thoracic ventral roots and the splanchnic nerves. The vasoconstrictor innervation of the hepatic artery comes from the hepatic sympathetic plexus. No known vasodilator fibers reach the liver. When systemic venous pressure rises, the portal vein radicles are dilated passively and the amount of blood in the liver increases. In congestive heart failure, this hepatic venous congestion may be extreme. Conversely, when diffuse noradrenergic discharge occurs in response to a drop in systemic blood pressure, the intrahepatic portal radicles constrict, portal pressure rises, and blood flow through the liver is brisk, bypassing most of the organ. Most of the blood in the liver enters the systemic circulation. Constriction of the hepatic arterioles diverts blood from the liver, and constriction of the mesenteric arterioles reduces portal inflow. In severe shock, hepatic blood flow may be reduced to such a degree that patchy necrosis of the liver takes place.

FUNCTIONS OF THE LIVER

The liver has many complex functions that are summarized in Table 28–1. Several will be touched upon briefly here.

TABLE 28–1 Principal functions of the liver.

Formation and secretion of bile
Nutrient and vitamin metabolism
Glucose and other sugars
Amino acids
Lipids
Fatty acids
Cholesterol
Lipoproteins
Fat-soluble vitamins
Water-soluble vitamins
Inactivation of various substances
Toxins
Steroids
Other hormones
Synthesis of plasma proteins
Acute-phase proteins
Albumin
Clotting factors
Steroid-binding and other hormone-binding proteins
Immunity
Kupffer cells

METABOLISM & DETOXIFICATION

It is beyond the scope of this volume to touch upon all of the metabolic functions of the liver. Instead, we will describe here those aspects most closely aligned to gastrointestinal physiology. First, the liver plays key roles in carbohydrate metabolism, including glycogen storage, conversion of galactose and fructose to glucose, and gluconeogenesis, as well as many of the reactions covered in Chapter 1. The substrates for these reactions derive from the products of carbohydrate digestion and absorption that are transported from the intestine to the liver in the portal blood. The liver also plays a major role in maintaining the stability of blood glucose levels in the postprandial period, removing excess glucose from the blood and returning it as needed—the so-called **glucose buffer function** of the liver. In liver failure, hypoglycemia is commonly seen. Similarly, the liver contributes to fat metabolism. It supports a high rate of fatty acid oxidation for energy supply to the liver itself and other organs. Amino acids and two carbon fragments derived from carbohydrates are also converted in the liver to fats for storage. The liver also synthesizes most of the lipoproteins required by the body and preserves cholesterol homeostasis by synthesizing this molecule and also converting excess cholesterol to bile acids.

The liver also detoxifies the blood of substances originating from the gut or elsewhere in the body (Clinical Box 28–1). Part of this function is physical in nature—bacteria and other particulates are trapped in and broken down by the strategically located Kupffer cells. The remaining reactions are biochemical, and mediated in their first stages by the large number of cytochrome P450 enzymes expressed in hepatocytes. These convert xenobiotics and other toxins to inactive, less lipophilic metabolites. Detoxification reactions are divided into phase I (oxidation, hydroxylation, and other reactions mediated by cytochrome P450s) and phase II (esterification). Ultimately, metabolites are secreted into the bile for elimination via the gastrointestinal tract. In this regard, in addition to disposing of drugs, the liver is responsible for metabolism of essentially all steroid hormones. Liver disease can therefore result in the apparent overactivity of the relevant hormone systems.

SYNTHESIS OF PLASMA PROTEINS

The principal proteins synthesized by the liver are listed in Table 28–1. Albumin is quantitatively the most significant, and accounts for the majority of plasma oncotic pressure. Many of the products are **acute-phase proteins,** proteins synthesized and secreted into the plasma on exposure to stressful stimuli (see Chapter 3). Others are proteins that transport steroids and other hormones in the plasma, and still others are clotting factors. Following blood loss, the liver replaces the plasma proteins in days to weeks. The only major class of plasma proteins not synthesized by the liver is the immunoglobulins.

CLINICAL BOX 28–1

Hepatic Encephalopathy

The clinical importance of hepatic ammonia metabolism is seen in liver failure, when increased levels of circulating ammonia cause the condition of hepatic encephalopathy. Initially, patients may seem merely confused, but if untreated, the condition can progress to coma and irreversible changes in cognition. The disease results not only from the loss of functional hepatocytes, but also shunting of portal blood around the hardened liver, meaning that less ammonia is removed from the blood by the remaining hepatic mass. Additional substances that are normally detoxified by the liver likely also contribute to the mental status changes.

THERAPEUTIC HIGHLIGHTS

The congnitive symptoms of advanced liver disease can be minimized by reducing the load of ammonia coming to the liver from the colon (eg, by feeding the nonabsorbable carbohydrate, lactulose, which is converted into short-chain fatty acids in the colonic lumen and thereby traps luminal ammonia in its ionized form). However, in severe disease, the only truly effective treatment is to perform a liver transplant, although the paucity of available organs means that there is great interest in artificial liver assist devices that could clean the blood.

BILE

Bile is made up of the bile acids, bile pigments, and other substances dissolved in an alkaline electrolyte solution that resembles pancreatic juice (Table 28–2). About 500 mL is secreted per day. Some of the components of the bile are

TABLE 28–2 Composition of human hepatic duct bile.

Water	97.0%
Bile salts	0.7%
Bile pigments	0.2%
Cholesterol	0.06%
Inorganic salts	0.7%
Fatty acids	0.15%
Phosphatidylcholine	0.2%
Fat	0.1%
Alkaline phosphatase	...

reabsorbed in the intestine and then excreted again by the liver **(enterohepatic circulation).** In addition to its role in digestion and absorption of fats (Chapter 26), bile (and subsequently the feces) is the major excretory route for lipid-soluble waste products.

The glucuronides of the **bile pigments,** bilirubin and biliverdin, are responsible for the golden yellow color of bile. The formation of these breakdown products of hemoglobin is discussed in detail in Chapter 31, and their excretion is discussed in the following text.

BILIRUBIN METABOLISM & EXCRETION

Most of the bilirubin in the body is formed in the tissues by the break down of hemoglobin (see Chapter 31 and Figure 28–4). The bilirubin is bound to albumin in the circulation. Some of it is tightly bound, but most of it can dissociate in the liver, and free bilirubin enters liver cells via a member of the organic anion transporting polypeptide (OATP) family, and then becomes bound to cytoplasmic proteins (Figure 28–5). It is next conjugated to glucuronic acid in a reaction catalyzed by the enzyme **glucuronyl transferase** (UDP-glucuronosyltransferase). This enzyme is located primarily in the smooth endoplasmic reticulum. Each bilirubin molecule reacts with two uridine diphosphoglucuronic acid (UDPG) molecules to form bilirubin diglucuronide. This glucuronide, which is more water-soluble than the free bilirubin, is then transported against a concentration gradient most likely by an active transporter known as multidrug resistance protein-2 (MRP-2) into the bile canaliculi. A small amount of the bilirubin glucuronide escapes into the blood, where it is bound less tightly to albumin than is free bilirubin, and is excreted in the urine. Thus, the total plasma bilirubin normally includes free bilirubin plus a small amount of conjugated bilirubin. Most of the bilirubin glucuronide passes via the bile ducts to the intestine.

The intestinal mucosa is relatively impermeable to conjugated bilirubin but is permeable to unconjugated bilirubin and to urobilinogens, a series of colorless derivatives of bilirubin formed by the action of bacteria in the intestine. Consequently, some of the bile pigments and urobilinogens are reabsorbed in the portal circulation. Some of the reabsorbed substances are again excreted by the liver (enterohepatic circulation), but small amounts of urobilinogens enter the general circulation and are excreted in the urine.

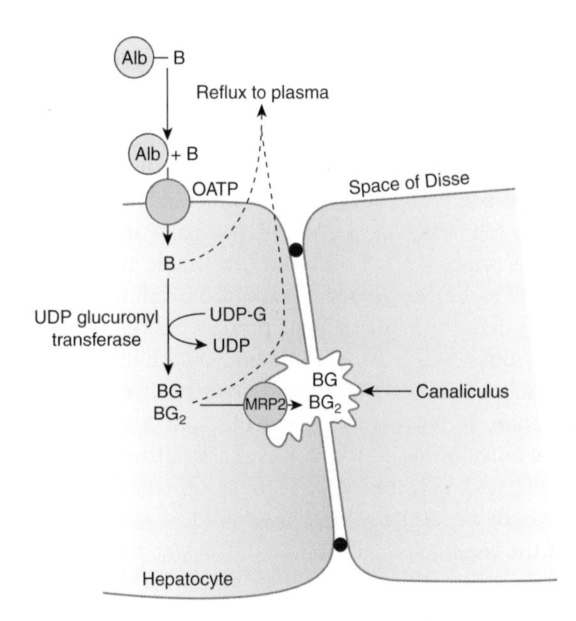

FIGURE 28–5 Handling of bilirubin by hepatocytes. Albumin (Alb)-bound bilirubin (B) enters the space of Disse adjacent to the basolateral membrane of hepatocytes, and bilirubin is selectively transported into the hepatocyte. Here, it is conjugated with glucuronic acid (G). The conjugates are secreted into bile via the multidrug resistance protein 2 (MRP-2). Some unconjugated and conjugated bilirubin also refluxes into the plasma. OATP, organic anion transporting polypeptide.

Heme

Heme oxygenase: $NADPH + O_2$ → $CO + Fe^{3+} + NADP^+$

Biliverdin

Biliverdin reductase: NADPH → $NADP^+$

Bilirubin

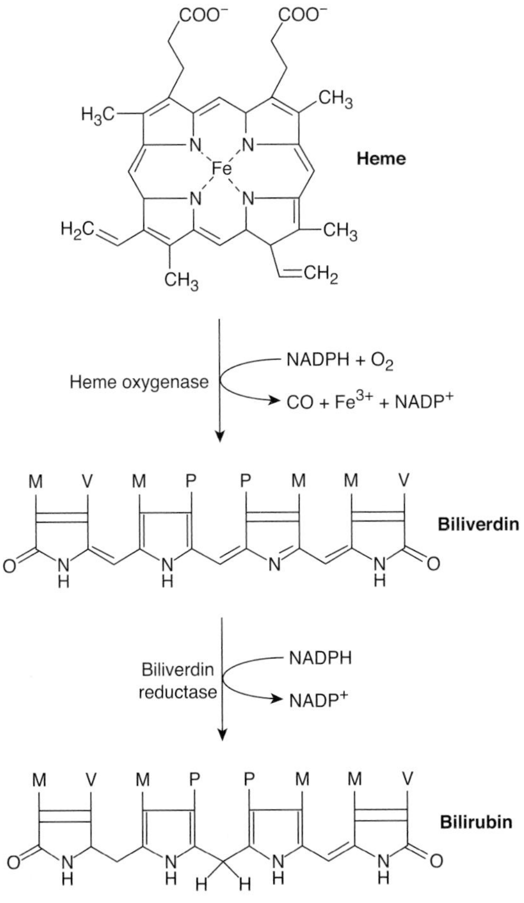

FIGURE 28–4 Conversion of heme to bilirubin is a two-step reaction catalyzed by heme oxygenase and biliverdin reductase. M, methyl; P, propionate; V, vinyl.

JAUNDICE

When free or conjugated bilirubin accumulates in the blood, the skin, scleras, and mucous membranes turn yellow. This yellowness is known as **jaundice** (icterus) and is usually detectable when the total plasma bilirubin is greater than 2 mg/dL (34 μmol/L). Hyperbilirubinemia may be due to (1) excess production of bilirubin (hemolytic anemia, etc; see Chapter 31), (2) decreased uptake of bilirubin into hepatic cells, (3) disturbed intracellular protein binding or conjugation, (4) disturbed secretion of conjugated bilirubin into the bile canaliculi, or (5) intrahepatic or extrahepatic bile duct obstruction. When it is due to one of the first three processes, the free bilirubin rises. When it is due to disturbed secretion of conjugated bilirubin or bile duct obstruction, bilirubin glucuronide regurgitates into the blood, and it is predominantly the conjugated bilirubin in the plasma that is elevated.

OTHER SUBSTANCES CONJUGATED BY GLUCURONYL TRANSFERASE

The glucuronyl transferase system in the smooth endoplasmic reticulum catalyzes the formation of the glucuronides of a variety of substances in addition to bilirubin. As discussed above, the list includes steroids (see Chapter 20) and various drugs. These other compounds can compete with bilirubin for the enzyme system when they are present in appreciable amounts. In addition, several barbiturates, antihistamines, anticonvulsants, and other compounds cause marked proliferation of the smooth endoplasmic reticulum in the hepatic cells, with a concurrent increase in hepatic glucuronyl transferase activity. Phenobarbital has been used successfully for the treatment of a congenital disease in which there is a relative deficiency of glucuronyl transferase (type 2 UDP-glucuronosyltransferase deficiency).

OTHER SUBSTANCES EXCRETED IN THE BILE

Cholesterol and alkaline phosphatase are excreted in the bile. In patients with jaundice due to intra- or extrahepatic obstruction of the bile duct, the blood levels of these two substances usually rise. A much smaller rise is generally seen when the jaundice is due to nonobstructive hepatocellular disease. Adrenocortical and other steroid hormones and a number of drugs are excreted in the bile and subsequently reabsorbed (enterohepatic circulation).

AMMONIA METABOLISM & EXCRETION

The liver is critical for ammonia handling in the body. Ammonia levels must be carefully controlled because it is toxic to the central nervous system (CNS), and freely permeable across the blood–brain barrier. The liver is the only organ in which the complete urea cycle (also known as the Krebs–Henseleit cycle) is expressed (Figure 28–6). This converts circulating ammonia to urea, which can then be excreted in the urine (Figure 28–7).

Ammonia in the circulation comes primarily from the colon and kidneys with lesser amounts deriving from the breakdown of red blood cells and from metabolism in the muscles. As it passes through the liver, the vast majority of ammonia in the circulation is cleared into the hepatocytes. There, it is converted in the mitochondria to carbamoyl phosphate, which in turn reacts with ornithine to generate citrulline. A series of subsequent cytoplasmic reactions eventually produce arginine, and this can be dehydrated to urea and ornithine. The latter returns to the mitochondria to begin another cycle, and urea, as a small molecule, diffuses readily back out into the sinusoidal blood. It is then filtered in the kidneys and lost from the body in the urine.

THE BILIARY SYSTEM

BILE FORMATION

Bile contains substances that are actively secreted into it across the canalicular membrane, such as bile acids, phosphatidylcholine, conjugated bilirubin, cholesterol, and xenobiotics. Each of these enters the bile by means of a specific canalicular transporter. It is the active secretion of bile acids, however, that is believed to be the primary driving force for the initial formation of canalicular bile. Because they are osmotically active, the canalicular bile is transiently hypertonic. However, the tight junctions that join adjacent hepatocytes are relatively permeable and thus a number of additional substances passively enter the bile from the plasma by diffusion. These substances include water, glucose, calcium, glutathione, amino acids, and urea.

Phosphatidylcholine that enters the bile forms mixed micelles with the bile acids and cholesterol. The ratio of bile acids:phosphatidylcholine:cholesterol in canalicular bile is approximately 10:3:1. Deviations from this ratio may cause cholesterol to precipitate, leading to one type of gallstones (Figure 28–8).

The bile is transferred to progressively larger bile ductules and ducts, where it undergoes modification of its composition. The bile ductules are lined by cholangiocytes, specialized columnar epithelial cells. Their tight junctions are less permeable than those of the hepatocytes, although they remain freely permeable to water and thus bile remains isotonic. The ductules scavenge plasma constituents, such as glucose and amino acids, and return them to the circulation by active transport. Glutathione is also hydrolyzed to its constituent amino acids by an enzyme, gamma glutamyltranspeptidase (GGT), expressed on the apical membrane of the cholangiocytes. Removal of glucose and amino acids is likely important to prevent bacterial overgrowth of the bile, particularly during

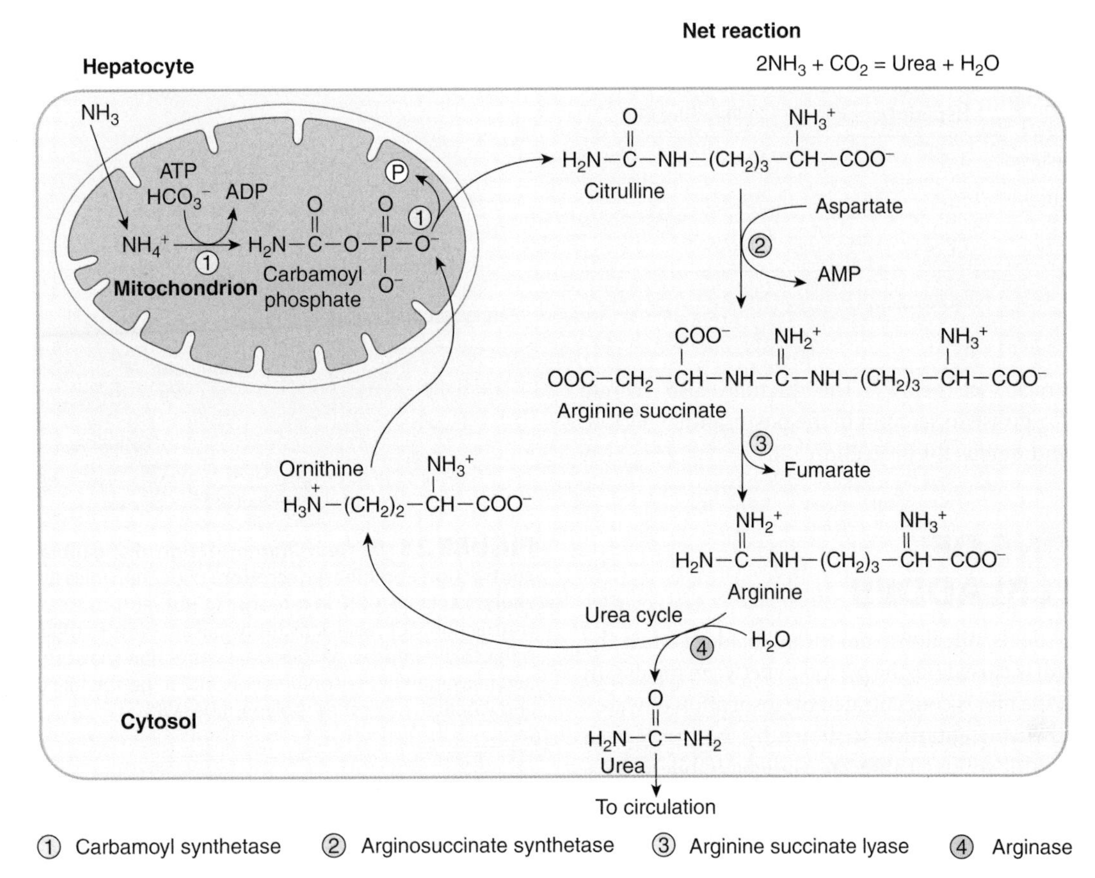

FIGURE 28–6 The urea cycle, which converts ammonia to urea, takes place in the mitochondria and cytosol of hepatocytes.

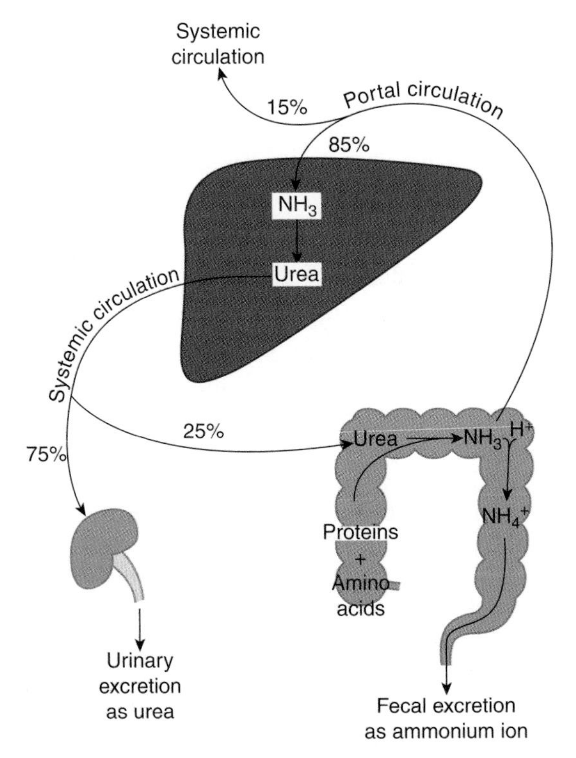

FIGURE 28–7 Whole body ammonia homeostasis in health. The majority of ammonia produced by the body is excreted by the kidneys in the form of urea.

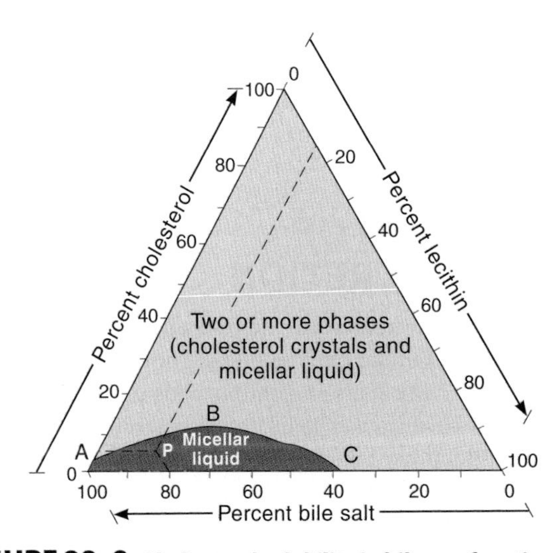

FIGURE 28–8 Cholesterol solubility in bile as a function of the proportions of lecithin, bile salts, and cholesterol. In bile that has a composition described by any point below line ABC (eg, point P), cholesterol is solely in micellar solution; points above line ABC describe bile in which there are cholesterol crystals as well. (Reproduced with permission from Small DM: Gallstones. N Engl J Med 1968;279:588.)

TABLE 28–3 Comparison of human hepatic duct bile and gallbladder bile.

	Hepatic Duct Bile	Gallbladder Bile
Percentage of solids	2–4	10–12
Bile acids (mmol/L)	10–20	50–200
pH	7.8–8.6	7.0–7.4

gallbladder storage (see below). The ductules also secrete bicarbonate in response to secretin in the postprandial period, as well as IgA and mucus for protection.

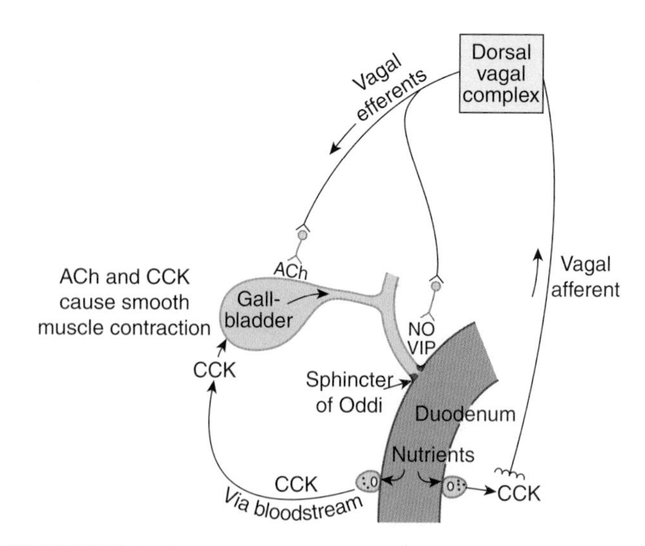

FIGURE 28–9 Neurohumoral control of gallbladder contraction and biliary secretion. Endocrine release of cholecystokinin (CCK) in response to nutrients causes gallbladder contraction. CCK, also activates vagal afferents to trigger a vago-vagal reflex that reinforces gallbladder contraction (via acetylcholine (ACh)) and relaxation of the sphincter of Oddi to permit bile outflow (via NO and vasoactive intestinal polypeptide (VIP)).

FUNCTIONS OF THE GALLBLADDER

In normal individuals, bile flows into the gallbladder when the sphincter of Oddi is closed (ie, the period in between meals). In the gallbladder, the bile is concentrated by absorption of water. The degree of this concentration is shown by the increase in the concentration of solids (Table 28–3); hepatic bile is 97% water, whereas the average water content of gallbladder bile is 89%. However, because the bile acids are a micellar solution, the micelles simply become larger, and since osmolarity is a colligative property, bile remains isotonic. However, bile becomes slightly acidic as sodium ions are exchanged for protons (although the overall concentration of sodium ions rises with a concomitant loss of chloride and bicarbonate as the bile is concentrated).

When the bile duct and cystic duct are clamped, the intrabiliary pressure rises to about 320 mm of bile in 30 min, and bile secretion stops. However, when the bile duct is clamped and the cystic duct is left open, water is reabsorbed in the gallbladder, and the intrabiliary pressure rises only to about 100 mm of bile in several hours.

REGULATION OF BILIARY SECRETION

When food enters the mouth, the resistance of the sphincter of Oddi decreases under both neural and hormonal influences (Figure 28–9). Fatty acids and amino acids in the duodenum release CCK, which causes gallbladder contraction.

The production of bile is increased by stimulation of the vagus nerves and by the hormone secretin, which increases the water and HCO_3^- content of bile. Substances that increase the secretion of bile are known as **choleretics.** Bile acids themselves are among the most important physiologic choleretics.

EFFECTS OF CHOLECYSTECTOMY

The periodic discharge of bile from the gallbladder aids digestion but is not essential for it. Cholecystectomized patients maintain good health and nutrition with a constant slow discharge of bile into the duodenum, although eventually the bile duct becomes somewhat dilated, and more bile tends to enter the duodenum after meals than at other times. Cholecystectomized patients can even tolerate fried foods, although they generally must avoid foods that are particularly high in fat content.

VISUALIZING THE GALLBLADDER

Exploration of the right upper quadrant with an ultrasonic beam **(ultrasonography)** and computed tomography (CT) have become the most widely used methods for visualizing the gallbladder and detecting gallstones. A third method of diagnosing gallbladder disease is **nuclear cholescintigraphy.** When administered intravenously, technetium-99m-labeled derivatives of iminodiacetic acid are excreted in the bile and provide excellent gamma camera images of the gallbladder and bile ducts. The response of the gallbladder to CCK can then be observed following intravenous administration of the hormone. The biliary tree can also be visualized by injecting contrast media from an endoscope channel maneuvered into the sphincter of Oddi, in a procedure known as endoscopic retrograde cholangiopancreatography

CLINICAL BOX 28–2

Gallstones

Cholelithiasis, that is, the presence of gallstones, is a common condition. Its incidence increases with age, so that in the United States, for example, 20% of the women and 5% of the men between the ages of 50 and 65 have gallstones. The stones are of two types: calcium bilirubinate stones and cholesterol stones. In the United States and Europe, 85% of the stones are cholesterol stones. Three factors appear to be involved in the formation of cholesterol stones. One is bile stasis; stones form in the bile that is sequestrated in the gallbladder rather than the bile that is flowing in the bile ducts. A second is supersaturation of the bile with cholesterol. Cholesterol is very insoluble in bile, and it is maintained in solution in micelles only at certain concentrations of bile salts and lecithin. At concentrations above line ABC in Figure 28–8, the bile is supersaturated and contains small crystals of cholesterol in addition to micelles. However, many normal individuals who do not develop gallstones also have supersaturated bile. The third factor is a mix of nucleation factors that favors formation of stones from the supersaturated bile. Outside the body, bile from patients with cholelithiasis forms stones in 2–3 days, whereas it takes more than 2 weeks for stones to form in bile from normal individuals. The exact nature of the nucleation factors is unsettled, although glycoproteins in gallbladder mucus have been implicated. In addition, it is unsettled whether stones form as a result of excess production of components that favor nucleation or decreased production of antinucleation components that prevent stones from forming in normal individuals.

Gallstones that obstruct bile outflow from the liver can result in **obstructive jaundice.** If the flow of bile out of the liver is completely blocked, substances normally excreted in the bile, such as cholesterol, accumulate in the bloodstream. The interruption of the enterohepatic circulation of bile acids also induces the liver to synthesize bile acids at a greater rate. Some of these bile acids can be excreted by the kidney, and thus represent a mechanism for indirect excretion of at least a portion of cholesterol. However, retained biliary constituents may also cause liver toxicity.

THERAPEUTIC HIGHLIGHTS

The treatment of gallstones depends on their nature, and the severity of any symptoms. Many, particularly if small and retained in the gallbladder, may be asymptomatic. Larger stones that cause obstruction may need to be removed surgically, or via ERCP. Oral dissolution agents may dissolve small stones composed of cholesterol, but the effect is slow and stones often return once therapy is stopped. A definitive cure for patients who suffer from recurrent attacks of symptomatic cholelithiasis is to have the gallbladder removed, which is usually now performed laparoscopically.

(ERCP). It is even possible to insert small instruments with which to remove gallstone fragments that may be obstructing the flow of bile, the flow of pancreatic juice, or both (Clinical Box 28–2).

CHAPTER SUMMARY

- The liver conducts a huge number of metabolic reactions and serves to detoxify and dispose of many exogenous substances, as well as metabolites endogenous to the body that would be harmful if allowed to accumulate.
- The structure of the liver is such that it can filter large volumes of blood and remove even hydrophobic substances that are protein-bound. This function is provided for by a fenestrated endothelium. The liver also receives essentially all venous blood from the intestine prior to its delivery to the remainder of the body.
- The liver serves to buffer blood glucose, synthesize the majority of plasma proteins, contribute to lipid metabolism, and preserve cholesterol homeostasis.
- Bilirubin is an end product of heme metabolism that is glucuronidated by the hepatocyte to permit its excretion in bile. Bilirubin and its metabolites impart color to the bile and stools.
- The liver removes ammonia from the blood and converts it to urea for excretion by the kidneys. An accumulation of ammonia as well as other toxins causes hepatic encephalopathy in the setting of liver failure.
- Bile contains substances actively secreted across the canalicular membrane by hepatocytes, and notably bile acids, phosphatidylcholine, and cholesterol. The composition of bile is modified as it passes through the bile ducts and is stored in the gallbladder. Gallbladder contraction is regulated to coordinate bile availability with the timing of meals.

Origin of the Heartbeat & the Electrical Activity of the Heart

CHAPTER 29

OBJECTIVES

After studying this chapter, you should be able to:

- Describe the structure and function of the conduction system of the heart and compare the action potentials in each part.
- Describe the way the electrocardiogram (ECG) is recorded, the waves of the ECG, and the relationship of the ECG to the electrical axis of the heart.
- Name the common cardiac arrhythmias and describe the processes that produce them.
- List the principal early and late ECG manifestations of myocardial infarction and explain the early changes in terms of the underlying ionic events that produce them.
- Describe the ECG changes and the changes in cardiac function produced by alterations in the ionic composition of the body fluids.

INTRODUCTION

The parts of the heart normally beat in orderly sequence: Contraction of the atria **(atrial systole)** is followed by contraction of the ventricles **(ventricular systole)**, and during **diastole** all four chambers are relaxed. The heartbeat originates in a specialized **cardiac conduction system** and spreads via this system to all parts of the myocardium. The structures that make up the conduction system are the **sinoatrial node (SA node)**, the **internodal atrial pathways**, the **atrioventricular node (AV node)**, the **bundle of His** and its branches, and the **Purkinje system.** The various parts of the conduction system and, under abnormal conditions, parts of the myocardium, are capable of spontaneous discharge. However, the SA node normally discharges most rapidly, with depolarization spreading from it to the other regions before they discharge spontaneously. The SA node is therefore the normal **cardiac pacemaker,** with its rate of discharge determining the rate at which the heart beats. Impulses generated in the SA node pass through the atrial pathways to the AV node, through this node to the bundle of His, and through the branches of the bundle of His via the Purkinje system to the ventricular muscle. Each of the cell types in the heart contains a unique electrical discharge pattern; the sum of these electrical discharges can be recorded as the electrocardiogram (ECG).

ORIGIN & SPREAD OF CARDIAC EXCITATION

ANATOMIC CONSIDERATIONS

In the human heart, the SA node is located at the junction of the superior vena cava with the right atrium. The AV node is located in the right posterior portion of the interatrial septum (Figure 29–1). There are three bundles of atrial fibers that contain Purkinje-type fibers and connect the SA node to the AV node: the anterior, middle (tract of Wenckebach), and posterior (tract of Thorel) tracts. Bachmann's bundle is sometimes used to identify a branch of the anterior intermodal tract that connects the right and left atria. Conduction also occurs through atrial myocytes, but it is more rapid in these bundles. The AV node is continuous with the bundle of His, which gives off a left bundle branch at the top of the interventricular

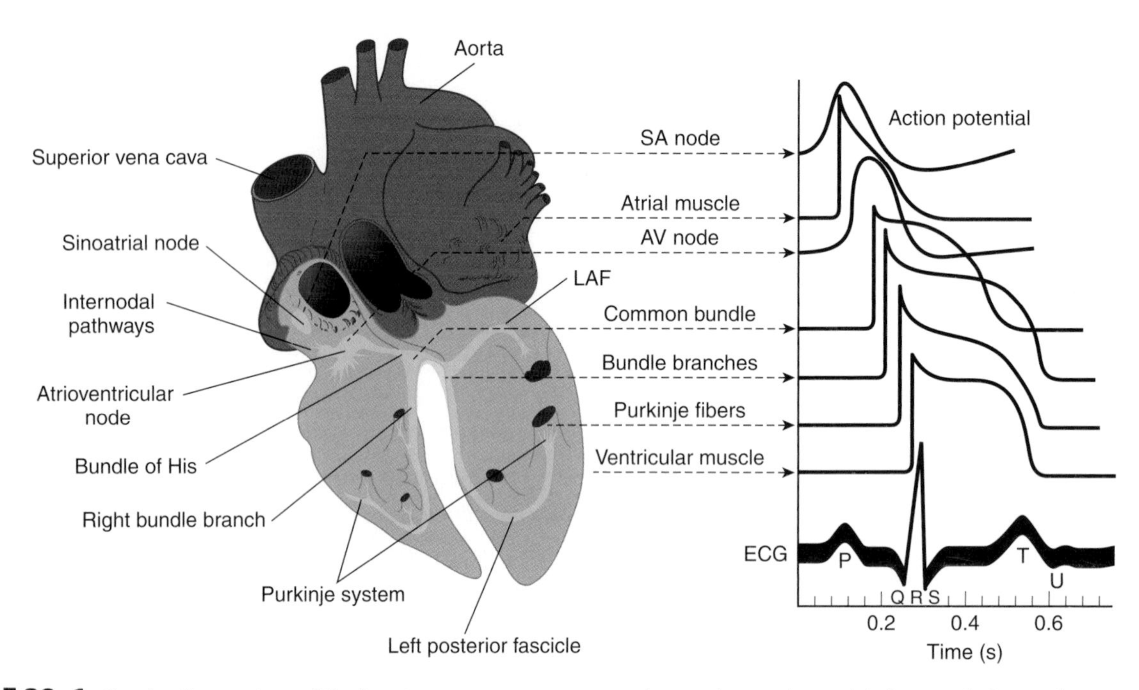

FIGURE 29–1 Conducting system of the heart. **Left:** Anatomical depiction of the human heart with additional focus on areas of the conduction system. **Right:** Typical transmembrane action potentials for the SA and AV nodes, other parts of the conduction system, and the atrial and ventricular muscles are shown along with the correlation to the extracellularly recorded electrical activity, that is, the electrocardiogram (ECG). The action potentials and ECG are plotted on the same time axis but with different zero points on the vertical scale for comparison. LAF, left anterior fascicle.

septum and continues as the right bundle branch. The left bundle branch divides into an anterior fascicle and a posterior fascicle. The branches and fascicles run subendocardially down either side of the septum and come into contact with the Purkinje system, whose fibers spread to all parts of the ventricular myocardium.

The histology of a typical cardiac muscle cell (eg, a ventricular myocyte) is described in Chapter 5. The conduction system is composed, for the most part, of modified cardiac muscle that has fewer striations and indistinct boundaries. Individual cells within regions of the heart have unique histological features. Purkinje fibers, specialized conducting cells, are large with fewer mitochondria and striations and distinctly different from a myocyte specialized for contraction. Cells within the SA node and, to a lesser extent the AV node are smaller and sparsely striated, but unlike Purkinje fibers, are less conductive due to their higher internal resistance. The atrial muscle fibers are separated from those of the ventricles by a fibrous tissue ring, and normally the only conducting tissue between the atria and ventricles is the bundle of His.

The SA node develops from structures on the right side of the embryo and the AV node from structures on the left. This is why in the adult the right vagus is distributed mainly to the SA node and the left vagus mainly to the AV node. Similarly, the sympathetic innervation on the right side is distributed primarily to the SA node and the sympathetic innervation on the left side primarily to the AV node. On each side, most sympathetic fibers come from the stellate ganglion. Noradrenergic fibers are epicardial, whereas the vagal fibers are endocardial. However, connections exist for reciprocal inhibitory effects of the sympathetic and parasympathetic innervation of the heart on each other. Thus, acetylcholine acts presynaptically to reduce norepinephrine release from the sympathetic nerves, and conversely, neuropeptide Y released from noradrenergic endings may inhibit the release of acetylcholine.

PROPERTIES OF CARDIAC MUSCLE

The electrical responses of cardiac muscle and nodal tissue and the ionic fluxes that underlie them are discussed in detail in Chapter 5 and are briefly reviewed here for comparison with the pacemaker cells below. Myocardial fibers have a resting membrane potential of approximately –90 mV (Figure 29–2A). The individual fibers are separated by membranes, but depolarization spreads radially through them as if they were a syncytium because of the presence of gap junctions. The transmembrane action potential of single cardiac muscle cells is characterized by rapid depolarization (phase 0), an initial rapid repolarization (phase 1), a plateau (phase 2), and a slow repolarization process (phase 3) that allows return to the resting membrane potential (phase 4). The initial depolarization is due to Na^+ influx through rapidly opening Na^+ channels (the Na^+ current, I_{Na}). The inactivation of Na^+ channels contributes to the rapid repolarization phase. Ca^{2+} influx through more slowly opening Ca^{2+} channels (the Ca^{2+} current, I_{Ca}) produces the plateau phase, and repolarization is due to net K^+ efflux through multiple types

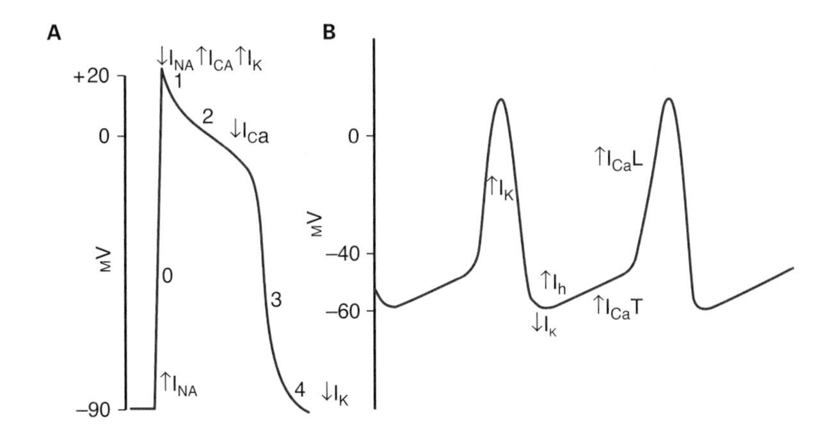

FIGURE 29–2 Comparison of action potentials in ventricular muscle and diagram of the membrane potential of pacemaker tissue. A) Phases of action potential in ventricular myocyte (0–4, see text for details) are superimposed with principal changes in current that contribute to changes in membrane potential. **B)** The principal current responsible for each part of the potential of pacemaker tissue is shown under or beside the component. L, long-lasting; T, transient. Other ion channels contribute to the electrical response. Note that the resting membrane potential of pacemaker tissue is somewhat lower than that of atrial and ventricular muscle.

of K^+ channels. Recorded extracellularly, the summed electrical activity of all the cardiac muscle fibers is the ECG (discussed below). The timing of the discharge of the individual units relative to the ECG is shown in Figure 29–1. Note that the ECG is a combined electrical record and thus the overall shape reflects electrical activity from cells from different regions of the heart.

PACEMAKER POTENTIALS

Rhythmically discharging cells have a membrane potential that, after each impulse, declines to the firing level. Thus, this **prepotential** or **pacemaker potential** (Figure 29–2B) triggers the next impulse. At the peak of each impulse, I_K begins and brings about repolarization. I_K then declines, and a channel permeable to both Na^+ and K^+ is activated. Because this channel is activated following hyperpolarization, it is referred to as an "h" channel; however, because of its unusual (funny) activation it has also been dubbed an "f" channel and the current produced as "funny current." As I_h increases, the membrane begins to depolarize, forming the first part of the prepotential. Ca^{2+} channels then open. These are of two types in the heart, the **T** (for transient) **channels** and the **L** (for long-lasting) **channels.** The calcium current (I_{Ca}) due to opening of T channels completes the prepotential, and I_{Ca} due to opening of L channels produces the impulse. Other ion channels are also involved, and there is evidence that local Ca^{2+} release from the sarcoplasmic reticulum (**Ca^{2+} sparks**) occurs during the prepotential.

The action potentials in the SA and AV nodes are largely due to Ca^{2+}, with no contribution by Na^+ influx. Consequently, there is no sharp, rapid depolarizing spike before the plateau, as there is in other parts of the conduction system and in the atrial and ventricular fibers. In addition, prepotentials are normally prominent only in the SA and AV nodes. However, "latent pacemakers" are present in other portions of the conduction system that can take over when the SA and AV nodes are depressed or conduction from them is blocked. Atrial and ventricular muscle fibers do not have prepotentials, and they discharge spontaneously only when injured or abnormal.

When the cholinergic vagal fibers to nodal tissue are stimulated, the membrane becomes hyperpolarized and the slope of the prepotentials is decreased (Figure 29–3) because the acetylcholine released at the nerve endings increases the K^+ conductance of nodal tissue. This action is mediated by M_2 muscarinic receptors, which, via the $\beta\gamma$ subunit of a G protein, open a special set of K^+ channels. The resulting I_{KAch} slows the depolarizing effect of I_h. In addition, activation of the M_2 receptors decreases cyclic adenosine 3′,5′-monophosphate (cAMP) in the cells, and this slows the opening of Ca^{2+} channels. The result is a decrease in firing rate. Strong vagal stimulation may abolish spontaneous discharge for some time.

Conversely, stimulation of the sympathetic cardiac nerves speeds the depolarizing effect of I_h, and the rate of spontaneous discharge increases (Figure 29–3). Norepinephrine secreted by the sympathetic endings binds to β_1 receptors, and the resulting increase in intracellular cAMP facilitates the opening of L channels, increasing I_{Ca} and the rapidity of the depolarization phase of the impulse.

The rate of discharge of the SA node and other nodal tissue is influenced by temperature and by drugs. The discharge frequency is increased when the temperature rises, and this may contribute to the tachycardia associated with fever. Digitalis depresses nodal tissue and exerts an effect like that of vagal stimulation, particularly on the AV node (Clinical Box 29–1; also see Clinical Box 5–6).

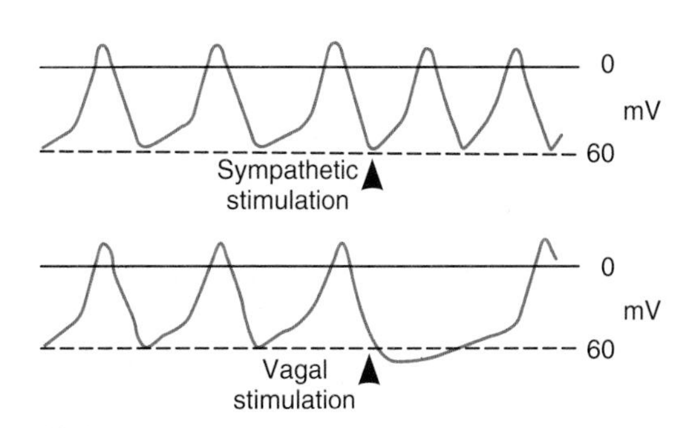

FIGURE 29–3 Effect of sympathetic (noradrenergic) and vagal (cholinergic) and sympathetic (noradrenergic) stimulation on the membrane potential of the SA node. Note the reduced slope of the prepotential after vagal stimulation and the increased spontaneous discharge after sympathetic stimulation.

CLINICAL BOX 29–1

Use of Digitalis

Digitalis, or its clinically useful preparations (digoxin and digitoxin) has been described in medical literature for over 200 years. It was originally derived from the foxglove plant (*Digitalis purpurea* is the name of the common foxglove). Correct administration can strengthen contractions through digitalis inhibitory effects on the Na, K ATPase, resulting in greater amounts of Ca^{2+} release and subsequent changes in contraction forces. Digitalis can also have an electrical effect in decreasing AV nodal conduction velocity and thus altering AV transmission to the ventricles.

THERAPEUTIC HIGHLIGHTS

Digitalis has been used for treatment of systolic heart failure. It augments contractility, thereby improving cardiac output, improving left ventricle emptying, and decreasing ventricular filling pressures. Digitalis has also been used to treat atrial fibrillation and atrial flutter. In this scenario, digitalis reduces the number of impulses transmitted through the AV node and thus, provides effective rate control.

In both these instances alternative treatments developed over the past 20 years and the need to tightly regulate dose due to significant potential for side effects have reduced the use of digitalis. However, with better understanding of mechanism and toxicity, digitalis and its clinically prepared derivatives remain important drugs in modern medicine.

SPREAD OF CARDIAC EXCITATION

Depolarization initiated in the SA node spreads radially through the atria, then converges on the AV node. Atrial depolarization is complete in about 0.1 s. Because conduction in the AV node is slow (Table 29–1), a delay of about 0.1 s **(AV nodal delay)** occurs before excitation spreads to the ventricles. It is interesting to note here that when there is a lack of contribution of I_{Na} in the depolarization (phase 0) of the action potential, a marked loss of conduction is observed. This delay is shortened by stimulation of the sympathetic nerves to the heart and lengthened by stimulation of the vagi. From the top of the septum, the wave of depolarization spreads in the rapidly conducting Purkinje fibers to all parts of the ventricles in 0.08–0.1 s. In humans, depolarization of the ventricular muscle starts at the left side of the interventricular septum and moves first to the right across the mid portion of the septum. The wave of depolarization then spreads down the septum to the apex of the heart. It returns along the ventricular walls to the AV groove, proceeding from the endocardial to the epicardial surface (Figure 29–4). The last parts of the heart to be depolarized are the posterobasal portion of the left ventricle, the pulmonary conus, and the uppermost portion of the septum.

TABLE 29–1 Conduction speeds in cardiac tissue.

Tissue	Conduction Rate (m/s)
SA node	0.05
Atrial pathways	1
AV node	0.05
Bundle of His	1
Purkinje system	4
Ventricular muscle	1

THE ELECTROCARDIOGRAM

Because the body fluids are good conductors (ie, because the body is a **volume conductor**), fluctuations in potential, representing the algebraic sum of the action potentials of myocardial fibers, can be recorded extracellularly. The record of these fluctuations in potential during the cardiac cycle is the **ECG.**

The ECG may be recorded by using an **active or exploring electrode** connected to an indifferent electrode at zero potential **(unipolar recording)** or by using two active electrodes **(bipolar recording).** In a volume conductor, the sum of the potentials at the points of an equilateral triangle with a current source in the center is zero at all times. A triangle with the heart at its center **(Einthoven's triangle,** see below) can be approximated by placing electrodes on both arms and on the left leg. These are the three **standard limb leads** used in electrocardiography. If these electrodes are connected to a common terminal, an indifferent electrode that stays near zero potential is obtained. Depolarization moving toward an active electrode in a volume conductor produces a positive deflection, whereas depolarization moving in the opposite direction produces a negative deflection.

The names of the various waves and segments of the ECG in humans are shown in Figure 29–5. By convention, an upward deflection is written when the active electrode becomes positive relative to the indifferent electrode, and a downward deflection is written when the active electrode becomes negative. As can be seen in Figure 29–1, the P wave is primarily produced by atrial depolarization, the QRS complex is dominated by ventricular depolarization, and the T wave by ventricular repolarization. The U wave is an inconstant finding that may be due to ventricular myocytes with long action potentials. However, the contributions to this segment are still undetermined. The intervals between the various waves of the ECG and the events in the heart that occur during these intervals are shown in Table 29–2.

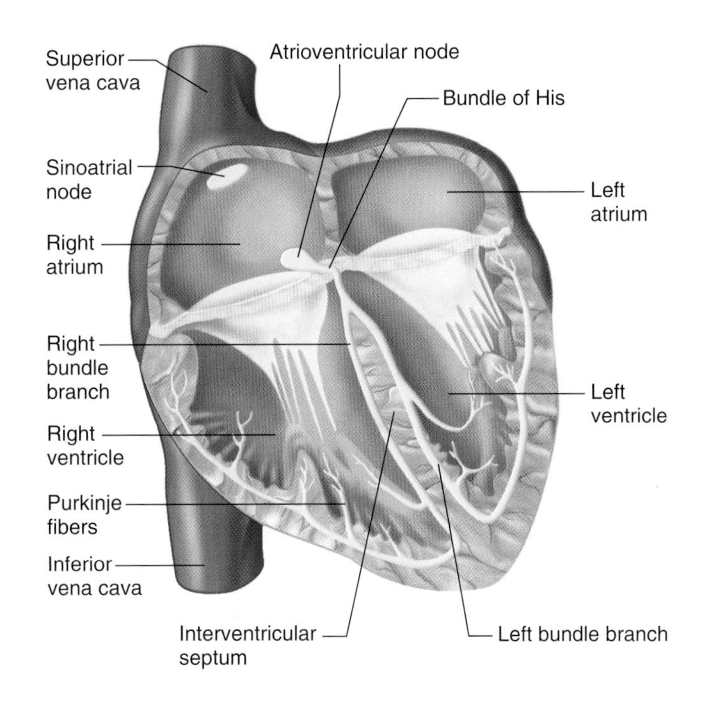

FIGURE 29–4 Normal spread of electrical activity in the heart. A) Conducting system of the heart. **B)** Sequence of cardiac excitation. **Top:** Anatomical position of electrical activity. **Bottom:** Corresponding electrocardiogram. The yellow color denotes areas that are depolarized. (Reproduced with permission from Goldman MJ: *Principles of Clinical Electrocardiography*, 12th ed. Originally published by Appleton & Lange. Copyright © 1986 by McGraw-Hill.)

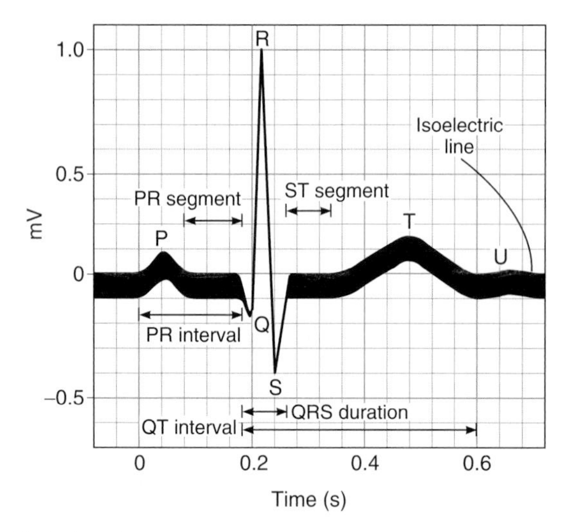

FIGURE 29–5 Waves of the ECG. Standard names for individual waves and segments that make up the ECG are shown. Electrical activity that contributes the observed deflections are discussed in the text and in Table 29–2.

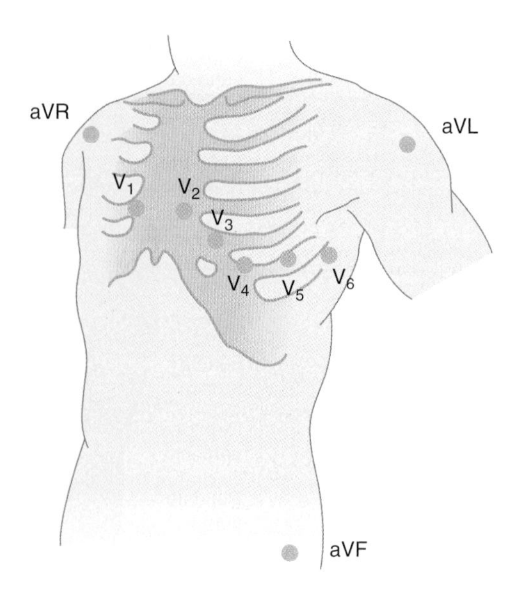

FIGURE 29–6 Unipolar electrocardiographic leads. Positional for standard unipolar leads are shown. The augmented extremity leads (aVR, aVL, and aVF) are shown on the right arm, left arm, and left leg, respectively. The six chest leads (V_1–V_6) are shown in their proper placement.

BIPOLAR LEADS

Bipolar leads were used before unipolar leads were developed. The **standard limb leads**—leads I, II, and III (Figure 29–6)—each record the differences in potential between two limbs. Because current flows only in the body fluids, the records obtained are those that would be obtained if the electrodes were at the points of attachment of the limbs, no matter where on the limbs the electrodes are placed. In lead I, the electrodes are connected so that an upward deflection is inscribed when the left arm becomes positive relative to the right (left arm positive). In lead II, the electrodes are on the right arm and left leg, with the leg positive; and in lead III, the electrodes are on the left arm and left leg, with the leg positive.

TABLE 29–2 ECG intervals.

	Normal Durations		Events in the Heart during Interval
Intervals	Average	Range	
PR interval[a]	0.18[b]	0.12–0.20	Atrioventricular conduction
QRS duration	0.08	to 0.10	Ventricular depolarization
QT interval	0.40[c]	to 0.43	Ventricular action potential
ST interval (QT minus QRS)	0.32	. . .	Plateau portion of the ventricular action potential

[a]Measured from the beginning of the P wave to the beginning of the QRS complex.

[b]Shortens as heart rate increases from average of 0.18 s at a rate of 70 beats/min to 0.14 s at a rate of 130 beats/min.

[c]Can be lower (0.35) depending on the heart rate

UNIPOLAR (V) LEADS

An additional nine unipolar leads, that is, leads that record the potential difference between an **exploring electrode** and an **indifferent electrode,** are commonly used in clinical electrocardiography. There are six unipolar chest leads (precordial leads) designated V_1–V_6 (Figure 29–6) and three unipolar limb leads: VR (right arm), VL (left arm), and VF (left foot). The indifferent electrode is constructed by connecting electrodes placed on the two arms and the left leg to a central terminal. This "V" lead effectively records a "zero" potential because they are situated such that the electrical activity should be cancelled out. **Augmented limb leads,** designated by the letter a (aVR, aVL, aVF), are generally used. The augmented limb leads do not use the "V" electrode as the zero, rather, they are recordings between the one, augmented limb and the other two limbs. This increases the size of the potentials by 50% without any change in configuration from the nonaugmented record.

Unipolar leads can also be placed at the tips of catheters and inserted into the esophagus or heart. Although sensitivity can be increased, this is obviously more invasive and thus, not a first step in obtaining electrical readings.

NORMAL ECG

The ECG tracings of a normal individual are shown in Figure 29–4b and Figure 29–7. The sequence in which the parts of the heart are depolarized (Figure 29–4) and the position of the heart relative to the electrodes are the important considerations (Figure 29–7) in interpreting the configurations of the waves in each lead. The atria are located

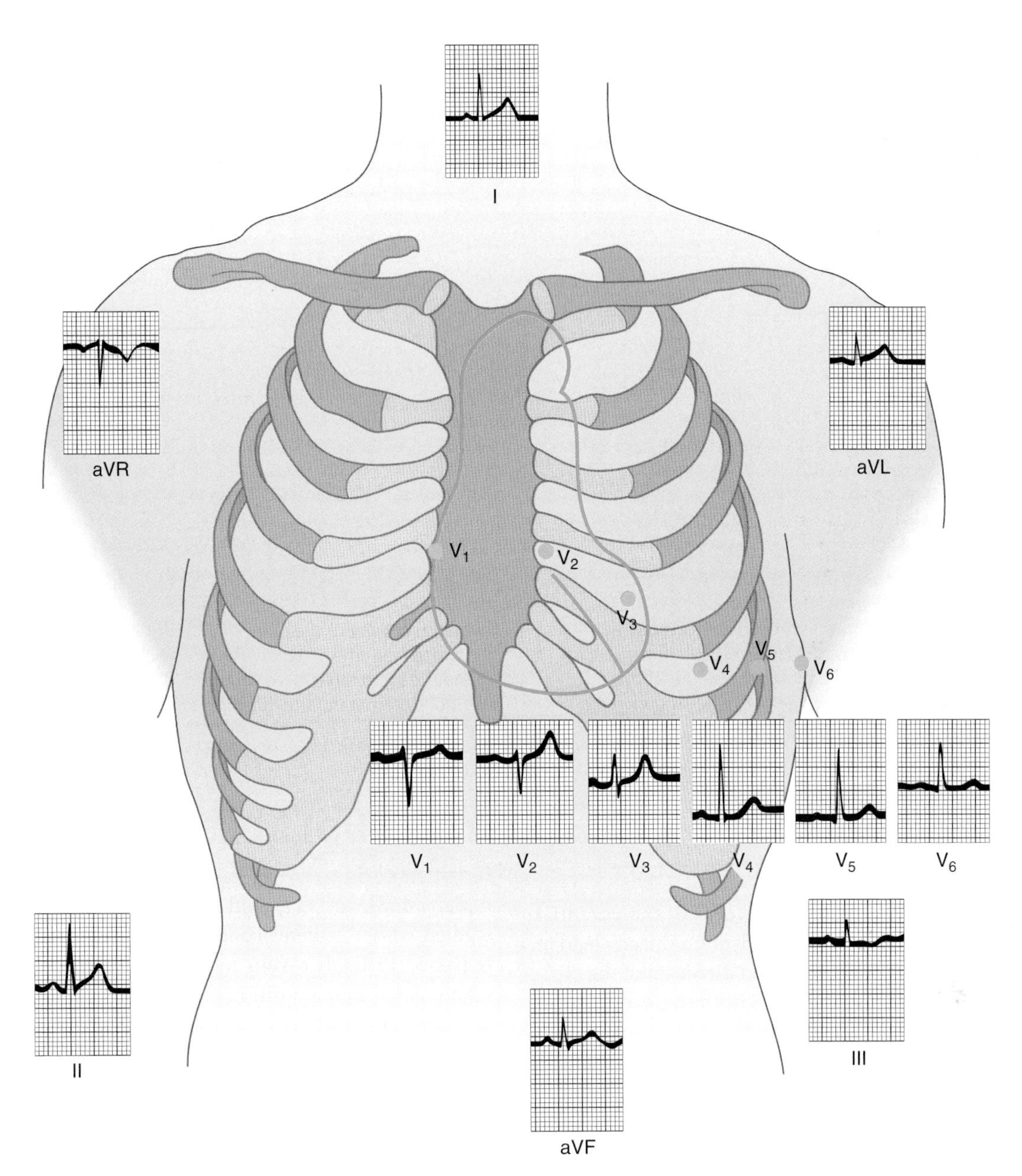

FIGURE 29–7 Normal ECG. Tracings from individual electrodes (positions marked in figure) are shown for a normal ECG. See text for additional details. (Reproduced with permission from Goldman MJ: *Principles of Clinical Electrocardiography,* 12th ed. Originally published by Appleton & Lange. Copyright © 1986 by McGraw-Hill.)

posteriorly in the chest. The ventricles form the base and anterior surface of the heart, and the right ventricle is anterolateral to the left. Thus, aVR "looks at" the cavities of the ventricles. Atrial depolarization, ventricular depolarization, and ventricular repolarization move away from the exploring electrode, and the P wave, QRS complex, and T wave are therefore all negative (downward) deflections; aVL and aVF look at the ventricles, and the deflections are therefore predominantly positive or biphasic. There is no Q wave in V_1 and V_2, and the initial portion of the QRS complex is a small upward deflection because ventricular depolarization first moves across the midportion of the septum from left to right toward the exploring electrode. The wave of excitation then moves down the septum and into the left ventricle away from the electrode, producing a large S wave. Finally, it moves back along the ventricular wall toward the electrode, producing the return to the isoelectric line. Conversely, in the left ventricular leads (V_4–V_6) there may be an initial small Q wave (left to right septal depolarization), and there is a large R wave (septal and left ventricular depolarization) followed in V_4 and V_5 by a moderate S wave (late depolarization of the ventricular walls moving back toward the AV junction). It should be noted that there is considerable variation in the position of the normal heart, and the position affects the configuration of the electrocardiographic complexes in the various leads.

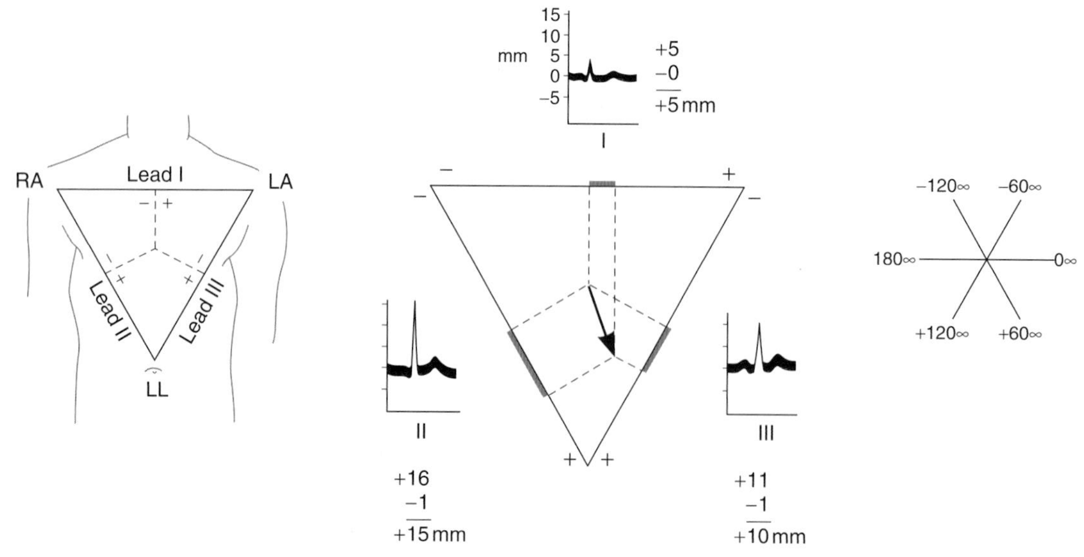

FIGURE 29–8 Cardiac vector. Left: Einthoven's triangle. Perpendiculars dropped from the midpoints of the sides of the equilateral triangle intersect at the center of electrical activity. RA, right arm; LA, left arm; LL, left leg. **Center:** Calculation of mean QRS vector. In each lead, distances equal to the height of the R wave minus the height of the largest negative deflection in the QRS complex are measured off from the midpoint of the side of the triangle representing that lead. An arrow drawn from the center of electrical activity to the point of intersection of perpendiculars extended from the distances measured off on the sides represents the magnitude and direction of the mean QRS vector. **Right:** Reference axes for determining the direction of the vector.

BIPOLAR LIMB LEADS & THE CARDIAC VECTOR

Because the standard limb leads are records of the potential differences between two points, the deflection in each lead at any instant indicates the magnitude and direction of the electromotive force generated in the heart in the axis of the lead **(cardiac vector** or **axis)**. The vector at any given moment in the two dimensions of the frontal plane can be calculated from any two standard limb leads (Figure 29–8) if it is assumed that the three electrode locations form the points of an equilateral triangle (Einthoven's triangle) and that the heart lies in the center of the triangle. These assumptions are not completely warranted, but calculated vectors are useful approximations. An approximate **mean QRS vector** ("electrical axis of the heart") is often plotted by using the average QRS deflection in each lead, as shown in Figure 29–8. This is a **mean** vector as opposed to an **instantaneous** vector, and the average QRS deflections should be measured by integrating the QRS complexes. However, they can be approximated by measuring the net differences between the positive and negative peaks of the QRS. The normal direction of the mean QRS vector is generally said to be –30 to +110° on the coordinate system shown in Figure 29–8. **Left** or **right axis deviation** is said to be present if the calculated axis falls to the left of –30° or to the right of +110°, respectively. Right axis deviation suggests right ventricular hypertrophy. Left axis deviation may be due to left ventricular hypertrophy, but there are better and more reliable electrocardiographic criteria for this condition.

HIS BUNDLE ELECTROGRAM

In patients with heart block, the electrical events in the AV node, bundle of His, and Purkinje system are frequently studied with a catheter containing an electrode at its tip that is passed through a vein to the right side of the heart and manipulated into a position close to the tricuspid valve. Three or more standard electrocardiographic leads are recorded simultaneously. The record of the electrical activity obtained with the catheter (Figure 29–9) is the **His bundle electrogram (HBE).** It normally shows an A deflection when the AV node is activated, an H spike during transmission through the His bundle, and a V deflection during ventricular depolarization. With the HBE and the standard electrocardiographic leads, it is possible to time three intervals accurately: (1) the PA interval, the time from the first appearance of atrial depolarization to the A wave in the HBE, which represents conduction time

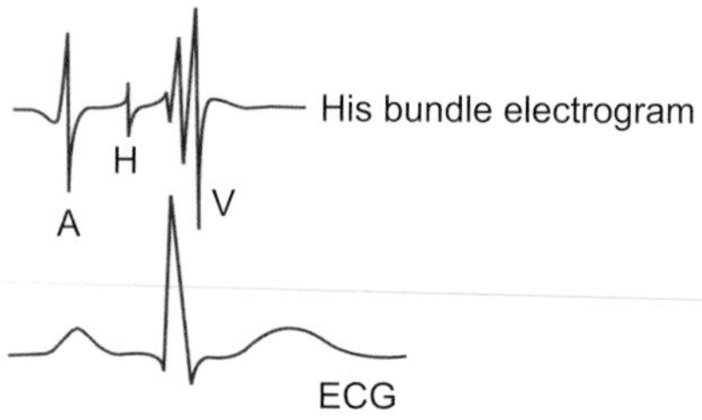

FIGURE 29–9 Normal His bundle electrogram (HBE) with simultaneously recorded ECG. An HBE recorded with an invasive electrode is superimposed on a standard ECG reading. Timing of depolarizations of the HBE are described in the text.

from the SA node to the AV node; (2) the AH interval, from the A wave to the start of the H spike, which represents the AV nodal conduction time; and (3) the HV interval, the time from the start of the H spike to the start of the QRS deflection in the ECG, which represents conduction in the bundle of His and the bundle branches. The approximate normal values for these intervals in adults are PA, 27 ms; AH, 92 ms; and HV, 43 ms. These values illustrate the relative slowness of conduction in the AV node.

MONITORING

The ECG has long been used in normal patient care. In the past, it was often recorded continuously in hospital coronary care units, with alarms arranged to sound at the onset of life-threatening arrhythmias. Using a small portable tape recorder **(Holter monitor),** it is also possible to record the ECG in ambulatory individuals as they go about their normal activities. The recording is later played back at high speed and analyzed. Recordings obtained with monitors have proved valuable in the diagnosis of arrhythmias and in planning the treatment of patients recovering from myocardial infarctions. Currently, modern systems can be hooked up to individuals and obtain and store heart rhythm data over days to better evaluate long-term electrical activity.

CLINICAL APPLICATIONS: CARDIAC ARRHYTHMIAS

NORMAL CARDIAC RATE

In the normal human heart, each beat originates in the SA node **(normal sinus rhythm, NSR).** The heart beats about 70 times a minute at rest. The rate is slowed **(bradycardia)** during sleep and accelerated **(tachycardia)** by emotion, exercise, fever, and many other stimuli. In healthy young individuals breathing at a normal rate, the heart rate varies with the phases of respiration: It accelerates during inspiration and decelerates during expiration, especially if the depth of breathing is increased. This **sinus arrhythmia** (Figure 29–10) is a normal phenomenon and is primarily due to fluctuations in parasympathetic output to the heart. During inspiration, impulses in the vagi from the stretch receptors in the lungs inhibit the cardio-inhibitory area in the medulla oblongata. The tonic vagal discharge that keeps the heart rate slow decreases, and the heart rate rises. Disease processes affecting the sinus node lead to marked bradycardia accompanied by dizziness and syncope (Clinical Box 29–2).

ABNORMAL PACEMAKERS

The AV node and other portions of the conduction system can, in abnormal situations, become the cardiac pacemaker. In addition, diseased atrial and ventricular muscle fibers can have their membrane potentials reduced and discharge repetitively.

As noted above, the discharge rate of the SA node is more rapid than that of the other parts of the conduction system,

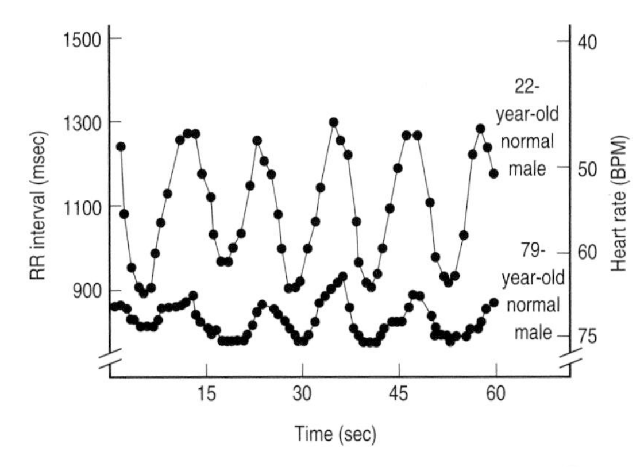

FIGURE 29–10 Sinus arrhythmia in a young man and an old man. Each subject breathed five times per minute. With each inspiration the RR interval (the interval between R waves) declined, indicating an increase in heart rate. Note the marked reduction in the magnitude of the arrhythmia in the older man. These records were obtained after β-adrenergic blockade, but would have been generally similar in its absence. (Reproduced with permission from Pfeifer MA et al: Differential changes of autonomic nervous system function with age in man. Am J Med 1983;75:249.)

CLINICAL BOX 29–2

Sick Sinus Syndrome

Sick sinus syndrome (bradycardia-tachycardia syndrome; sinus node dysfunction) is a collection of heart rhythm disorders that include **sinus bradycardia** (slow heart rates from the natural pacemaker of the heart), **tachycardias** (fast heart rates), and **bradycardia-tachycardia** (alternating slow and fast heart rhythms). Sick sinus syndrome is relatively uncommon and is usually found in people older than 50, in whom the cause is often a nonspecific, scar-like degeneration of the heart's conduction system. When found in younger people, especially in children, a common cause of sick sinus syndrome is heart surgery, especially on the upper chambers. Holter monitoring is an effective tool for diagnosing sick sinus syndrome because of the episodic nature of the disorder. Extremely slow heart rate and prolonged pauses may be seen during Holter monitoring, along with episodes of atrial tachycardias.

THERAPEUTIC HIGHLIGHTS

Treatment is dependent on the severity and type of disease. Tachycardias are frequently treated with medication. When there is marked bradycardia in patients with sick sinus syndrome or third-degree heart block, an electronic pacemaker is frequently implanted. These devices, which have become sophisticated and reliable, are useful in patients with sinus node dysfunction, AV block, and bifascicular or trifascicular block. They are useful also in patients with severe neurogenic syncope, in whom carotid sinus stimulation produces pauses of more than 3 s between heartbeats.

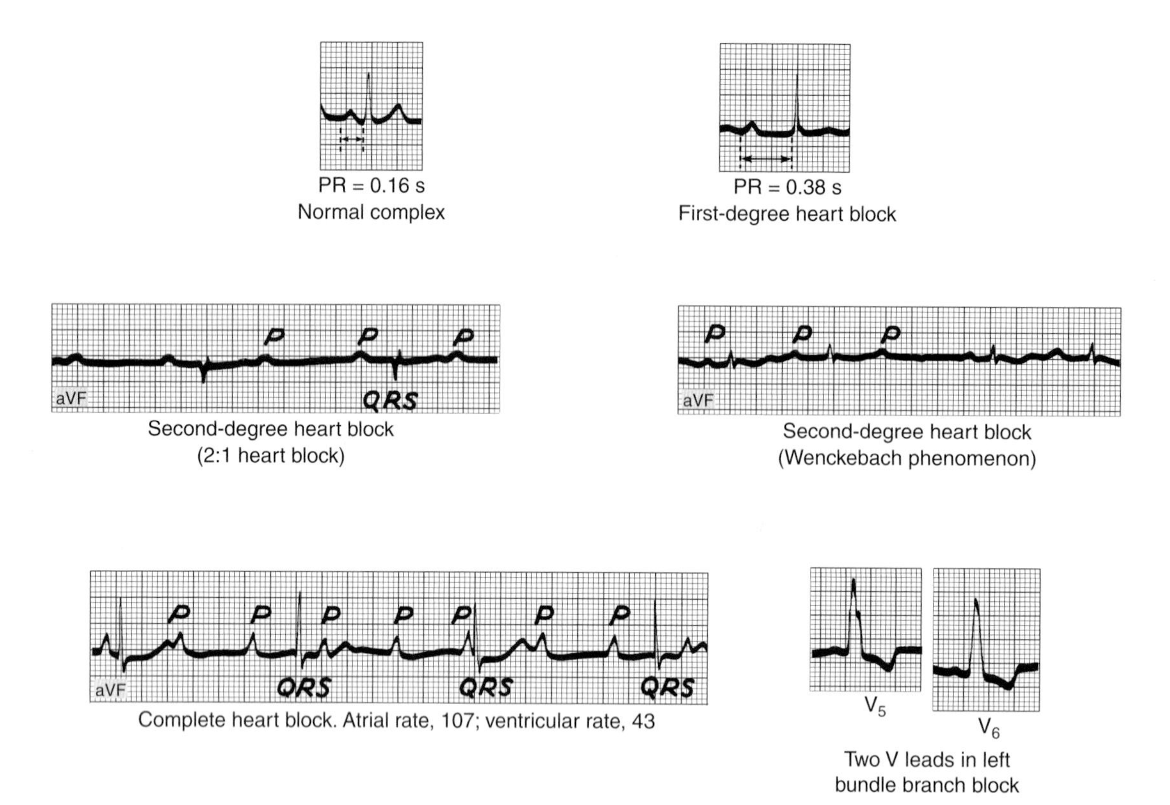

FIGURE 29–11 ECG with heart block. Individual traces that depict various forms of heart block are shown. When appropriate, unipolar leads are noted. See text for further details.

and this is why the SA node normally controls the heart rate. When conduction from the atria to the ventricles is completely interrupted, **complete (third-degree) heart block** results, and the ventricles beat at a low rate **(idioventricular rhythm)** independently of the atria (Figure 29–11). The block may be due to disease in the AV node **(AV nodal block)** or in the conducting system below the node **(infranodal block).** In patients with AV nodal block, the remaining nodal tissue becomes the pacemaker and the rate of the idioventricular rhythm is approximately 45 beats/min. In patients with infranodal block due to disease in the bundle of His, the ventricular pacemaker is located more peripherally in the conduction system and the ventricular rate is lower; it averages 35 beats/min, but in individual cases it can be as low as 15 beats/min. In such individuals, there may also be periods of asystole lasting a minute or more. The resultant cerebral ischemia causes dizziness and fainting **(Stokes–Adams syndrome).** Causes of third-degree heart block include septal myocardial infarction and damage to the bundle of His during surgical correction of congenital interventricular septal defects.

When conduction between the atria and ventricles is slowed but not completely interrupted, **incomplete heart block** is present. In the form called **first-degree heart block,** all the atrial impulses reach the ventricles but the PR interval is abnormally long. In the form called **second-degree heart block,** not all atrial impulses are conducted to the ventricles. For example, a ventricular beat may follow every second or every third atrial beat (2:1 block, 3:1 block, etc). In another form of incomplete heart block, there are repeated sequences of beats in which the PR interval lengthens progressively until a ventricular beat is dropped **(Wenckebach phenomenon).** The PR interval of the cardiac cycle that follows each dropped beat is usually normal or only slightly prolonged (Figure 29–11).

Sometimes one branch of the bundle of His is interrupted, causing **right** or **left bundle branch block.** In bundle branch block, excitation passes normally down the bundle on the intact side and then sweeps back through the muscle to activate the ventricle on the blocked side. The ventricular rate is therefore normal, but the QRS complexes are prolonged and deformed (Figure 29–11). Block can also occur in the anterior or posterior fascicle of the left bundle branch, producing the condition called **hemiblock** or **fascicular block.** Left anterior hemiblock produces abnormal left axis deviation in the ECG, whereas left posterior hemiblock produces abnormal right axis deviation. It is not uncommon to find combinations of fascicular and branch blocks **(bifascicular** or **trifascicular block).** The HBE permits detailed analysis of the site of block when there is a defect in the conduction system.

ECTOPIC FOCI OF EXCITATION

Normally, myocardial cells do not discharge spontaneously, and the possibility of spontaneous discharge of the His bundle and Purkinje system is low because the normal pacemaker discharge of the SA node is more rapid than their rate of spontaneous discharge. However, in abnormal conditions, the His–Purkinje fibers or the myocardial fibers may discharge spontaneously. In these conditions, **increased automaticity** of the heart is said to be present. If an irritable **ectopic focus** discharges once, the result is a beat that occurs before the expected next normal beat and transiently interrupts the cardiac rhythm (atrial, nodal, or ventricular **extrasystole** or **premature beat**). If the focus discharges repetitively at a rate higher than that of the SA node, it produces rapid, regular tachycardia (atrial, ventricular, or nodal **paroxysmal tachycardia** or **atrial flutter**).

REENTRY

A more common cause of paroxysmal arrhythmias is a defect in conduction that permits a wave of excitation to propagate continuously within a closed circuit **(circus movement).** For example, if a transient block is present on one side of a portion of the conducting system, the impulse can go down the other side. If the block then wears off, the impulse may conduct in a retrograde direction in the previously blocked side back to the origin and then descend again, establishing a circus movement. An example of this in a ring of tissue is shown in Figure 29–12. If the reentry is in the AV node, the reentrant activity depolarizes the atrium, and the resulting atrial beat is called an echo beat. In addition, the reentrant activity in the node propagates back down to the ventricle, producing paroxysmal nodal tachycardia. Circus movements can also become established in the atrial or ventricular muscle fibers. In individuals with an abnormal extra bundle of conducting tissue connecting the atria to the ventricles (bundle of Kent), the circus activity can pass in one direction through the AV node and in the other direction through the bundle, thus involving both the atria and the ventricles.

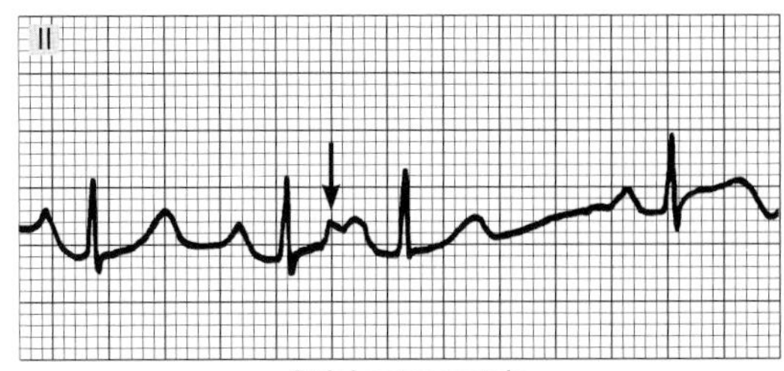

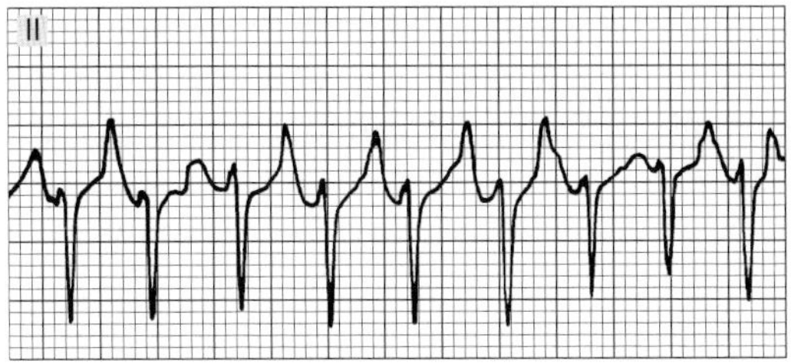

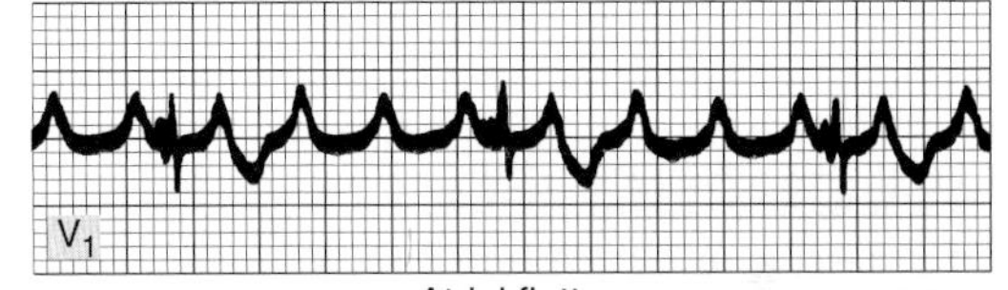

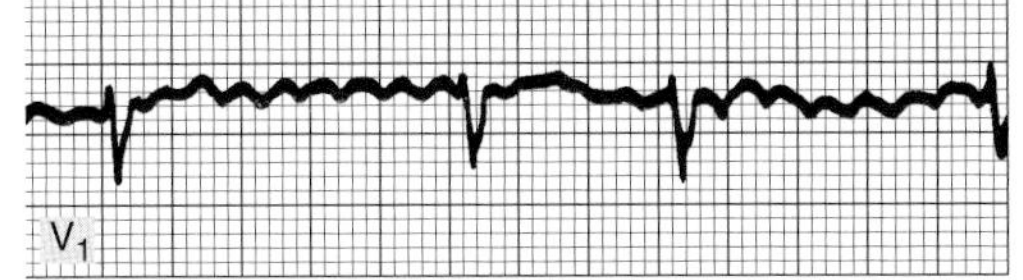

FIGURE 29–13 Atrial arrhythmias. The illustration shows an atrial premature beat with its P wave superimposed on the T wave of the preceding beat *(arrow)*; atrial tachycardia; atrial flutter with 4:1 AV block; and atrial fibrillation with a totally irregular ventricular rate. Leads used to capture electrical activity are marked in each trace. (Tracings reproduced with permission from Goldschlager N, Goldman MJ: *Principles of Clinical Electrocardiography,* 13th ed. Originally published by Appleton & Lange. Copyright © 1989 by McGraw-Hill.)

ATRIAL ARRHYTHMIAS

Excitation spreading from an independently discharging focus in the atria stimulates the AV node prematurely and is conducted to the ventricles. The P waves of atrial extrasystoles are abnormal, but the QRST configurations are usually normal (Figure 29–13). The excitation may depolarize the SA node, which must repolarize and then depolarize to the firing level before it can initiate the next normal beat. Consequently, a pause occurs between the extrasystole and the next normal beat that is usually equal in length to the interval between the normal beats preceding the extrasystole, and the rhythm is "reset" (see below).

Atrial tachycardia occurs when an atrial focus discharges regularly or there is reentrant activity producing atrial rates up to 220/min. Sometimes, especially in digitalized patients, some degree of atrioventricular block is associated with the tachycardia **(paroxysmal atrial tachycardia with block).**

In atrial flutter, the atrial rate is 200–350/min (Figure 29–13). In the most common form of this arrhythmia, there is large counterclockwise circus movement in the right atrium. This produces a characteristic sawtooth pattern of flutter waves due

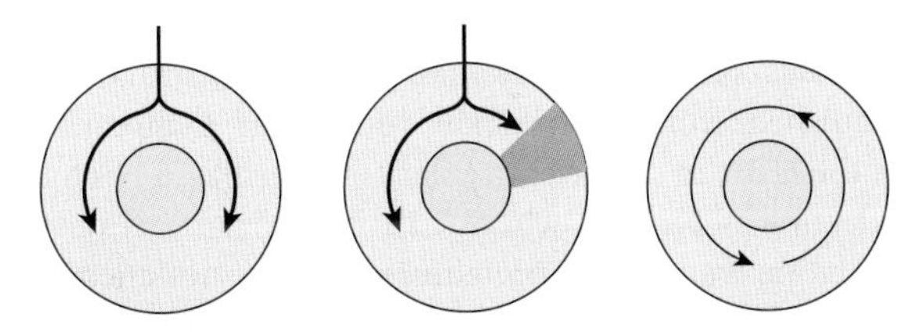

FIGURE 29–12 Depolarization of a ring of cardiac tissue. Normally, the impulse spreads in both directions in the ring **(left)** and the tissue immediately behind each branch of the impulse is refractory. When a transient block occurs on one side **(center),** the impulse on the other side goes around the ring, and if the transient block has now worn off **(right),** the impulse passes this area and continues to circle indefinitely (circus movement).

to atrial contractions. It is almost always associated with 2:1 or greater AV block, because in adults the AV node cannot conduct more than about 230 impulses per minute.

In **atrial fibrillation,** the atria beat very rapidly (300–500/min) in a completely irregular and disorganized fashion. Because the AV node discharges at irregular intervals, the ventricles also beat at a completely irregular rate, usually 80–160/min (Figure 29–13). The condition can be paroxysmal or chronic, and in some cases there appears to be a genetic predisposition. The cause of atrial fibrillation is still a matter of debate, but in most cases it appears to be due to multiple concurrently circulating reentrant excitation waves in both atria. However, some cases of paroxysmal atrial fibrillation seem to be produced by discharge of one or more ectopic foci. Many of these foci appear to be located in the pulmonary veins as much as 4 cm from the heart. Atrial muscle fibers extend along the pulmonary veins and are the origin of these discharges.

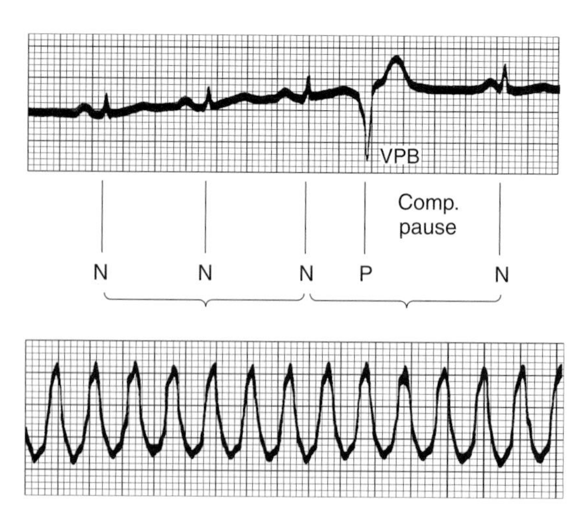

FIGURE 29–14 **Top:** Ventricular premature beats (VPB). The lines under the tracing illustrate the compensatory pause and show that the duration of the premature beat plus the preceding normal beat is equal to the duration of two normal beats. **Bottom:** Ventricular tachycardia.

CONSEQUENCES OF ATRIAL ARRHYTHMIAS

Occasional atrial extrasystoles occur from time to time in most normal humans and have no pathologic significance. In paroxysmal atrial tachycardia and flutter, the ventricular rate may be so high that diastole is too short for adequate filling of the ventricles with blood between contractions. Consequently, cardiac output is reduced and symptoms of heart failure appear. Heart failure may also complicate atrial fibrillation when the ventricular rate is high. Acetylcholine liberated at vagal endings depresses conduction in the atrial musculature and AV node. This is why stimulating reflex vagal discharge by pressing on the eyeball **(oculocardiac reflex)** or massaging the carotid sinus often converts tachycardia and sometimes converts atrial flutter to normal sinus rhythm. Alternatively, vagal stimulation increases the degree of AV block, abruptly lowering the ventricular rate. Digitalis also depresses AV conduction and is used to lower a rapid ventricular rate in atrial fibrillation.

VENTRICULAR ARRHYTHMIAS

Premature beats that originate in an ectopic ventricular focus usually have bizarrely shaped prolonged QRS complexes (Figure 29–14) because of the slow spread of the impulse from the focus through the ventricular muscle to the rest of the ventricle. They are usually incapable of exciting the bundle of His, and retrograde conduction to the atria therefore does not occur. In the meantime, the next succeeding normal SA nodal impulse depolarizes the atria. The P wave is usually buried in the QRS of the extrasystole. If the normal impulse reaches the ventricles, they are still in the refractory period following depolarization from the ectopic focus.

However, the second succeeding impulse from the SA node produces a normal beat. Thus, ventricular premature beats are followed by a **compensatory pause** that is often longer than the pause after an atrial extrasystole. Furthermore, ventricular premature beats do not interrupt the regular discharge of the SA node, whereas atrial premature beats often interrupt and "reset" the normal rhythm.

Atrial and ventricular premature beats are not strong enough to produce a pulse at the wrist if they occur early in diastole, when the ventricles have not had time to fill with blood and the ventricular musculature is still in its relatively refractory period. They may not even open the aortic and pulmonary valves, in which case there is, in addition, no second heart sound.

Paroxysmal ventricular tachycardia (Figure 29–14) is in effect a series of rapid, regular ventricular depolarizations usually due to a circus movement involving the ventricles. **Torsade de pointes** is a form of ventricular tachycardia in which the QRS morphology varies (Figure 29–15). Tachycardias originating above the ventricles (supraventricular tachycardias such as paroxysmal nodal tachycardia) can be distinguished from paroxysmal ventricular tachycardia by use of the HBE; in supraventricular tachycardias, a His bundle H deflection is present, whereas in ventricular tachycardias, there is none. Ventricular premature beats are not uncommon and, in the absence of ischemic heart disease, usually benign. Ventricular tachycardia is more serious because cardiac output is decreased, and ventricular fibrillation is an occasional complication of ventricular tachycardia.

In **ventricular fibrillation** (Figure 29–15), the ventricular muscle fibers contract in a totally irregular and ineffective

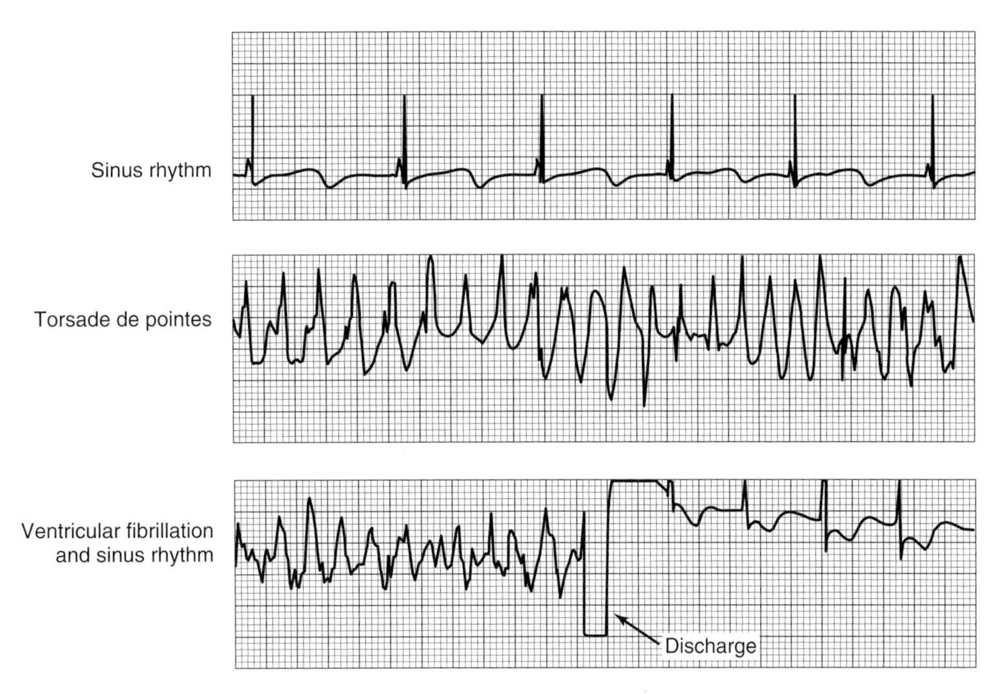

FIGURE 29–15 Record obtained from an implanted cardioverter–defibrillator in a 12-year-old boy with congenital long QT syndrome who collapsed while answering a question in school. Top: Normal sinus rhythm with long QT interval. Middle: Torsade de pointes. **Bottom:** Ventricular fibrillation with discharge of defibrillator, as programmed 7.5 s after the start of ventricular tachycardia, converting the heart to normal sinus rhythm. The boy recovered consciousness in 2 min and had no neurologic sequelae. (Reproduced with permission from Moss AJ, Daubert JP: Images in clinical medicine. Internal ventricular fibrillation. N Engl J Med 2000;342:398.)

way because of the very rapid discharge of multiple ventricular ectopic foci or a circus movement. The fibrillating ventricles, like the fibrillating atria, look like a quivering "bag of worms." Ventricular fibrillation can be produced by an electric shock or an extrasystole during a critical interval, the **vulnerable period.** The vulnerable period coincides in time with the mid-portion of the T wave; that is, it occurs at a time when some of the ventricular myocardium is depolarized, some is incompletely repolarized, and some is completely repolarized. These are excellent conditions in which to establish reentry and a circus movement. The fibrillating ventricles cannot pump blood effectively, and circulation of the blood stops. Therefore, in the absence of emergency treatment, ventricular fibrillation that lasts more than a few minutes is fatal. The most frequent cause of sudden death in patients with myocardial infarcts is ventricular fibrillation.

LONG QT SYNDROME

An indication of vulnerability of the heart during repolarization is the fact that in patients in whom the QT interval is prolonged, cardiac repolarization is irregular and the incidence of ventricular arrhythmias and sudden death increases. The syndrome can be caused by a number of different drugs, by electrolyte abnormalities, and by myocardial ischemia. It can also be congenital. Mutations of eight different genes have been reported to cause the syndrome. Six cause reduced function of various K^+ channels by alterations in their structure; one inhibits a K^+ channel by reducing the amount of the ankyrin isoform that links it to the cytoskeleton; and one increases the function of the cardiac Na^+ channel. Long QT Syndrome is discussed in Clinical Box 5–5.

ACCELERATED AV CONDUCTION

An interesting condition seen in some otherwise normal individuals who are prone to attacks of paroxysmal atrial arrhythmias is **accelerated AV conduction (Wolff–Parkinson–White syndrome).** Normally, the only conducting pathway between the atria and the ventricles is the AV node. Individuals with Wolff–Parkinson–White syndrome have an additional aberrant muscular or nodal tissue connection **(bundle of Kent)** between the atria and ventricles. This conducts more rapidly than the slowly conducting AV node, and one ventricle is excited early. The manifestations of its activation merge with the normal QRS pattern, producing a short PR interval and a prolonged QRS deflection slurred on the upstroke (Figure 29–16), with a normal interval between the start of the P wave and the end of the QRS complex ("PJ interval"). The paroxysmal atrial tachycardias seen in this syndrome often follow an atrial premature beat. This beat conducts normally down the AV node but spreads to the ventricular end of the

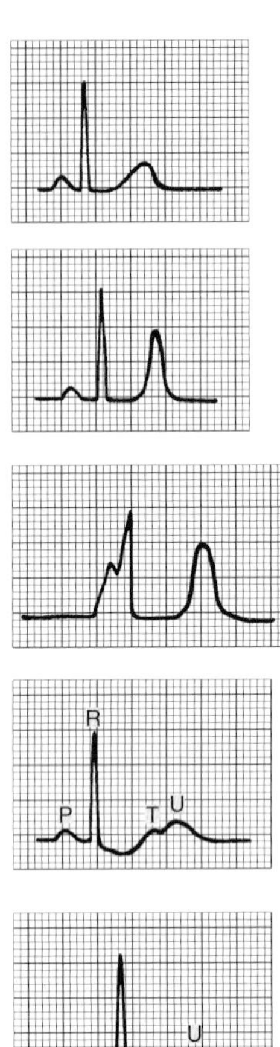

Normal tracing (plasma K^+ 4–5.5 meq/L). PR interval = 0.16 s; QRS interval = 0.06 s; QT interval = 0.4 s (normal for an assumed heart rate of 60).

Hyperkalemia (plasma K^+ ±7.0 meq/L). The PR and QRS intervals are within normal limits. Very tall, slender peaked T waves are now present.

Hyperkalemia (plasma K^+ ±8.5 meq/L). There is no evidence of atrial activity; the QRS complex is broad and slurred and the QRS interval has widened to 0.2 s. The T waves remain tall and slender. Further elevation of the plasma K^+ level may result in ventricular tachycardia and ventricular fibrillation.

Hypokalemia (plasma K^+ ±3.5 meq/L). PR interval = 0.2 s; QRS interval = 0.06 s; ST segment depression. A prominent U wave is now present immediately following the T. The actual QT interval remains 0.4 s. If the U wave is erroneously considered a part of the T, a falsely prolonged QT interval of 0.6 s will be measured.

Hypokalemia (plasma K^+ ±2.5 meq/L). The PR interval is lengthened to 0.32 s; the ST segment is depressed; the T wave is inverted; a prominent U wave is seen. The true QT interval remains normal.

FIGURE 29–18 Correlation of plasma K^+ level and the ECG, assuming that the plasma Ca^{2+} level is normal. The diagrammed complexes are left ventricular epicardial leads. (Reproduced with permission from Goldman MJ: *Principles of Clinical Electrocardiography,* 12th ed. Originally published by Appleton & Lange. Copyright © 1986 by McGraw-Hill.)

infarct, the membrane potential of the area is greater than it is in the normal area during the latter part of repolarization, making the normal region negative relative to the infarct. Extracellularly, current therefore flows out of the infarct into the normal area (since, by convention, current flow is from positive to negative). This current flows toward electrodes over the injured area, causing increased positivity between the S and T waves of the ECG. Similarly, the delayed depolarization of the infarcted cells causes the infarcted area to be positive relative to the healthy tissue (Table 29–3) during the early part of repolarization, and the result is also ST segment elevation. The remaining change—the decline in resting membrane potential during diastole—causes a current flow into the infarct during ventricular diastole. The result of this current flow is a depression of the TQ segment of the ECG. However, the electronic arrangement in electrocardiographic recorders is such that a TQ segment depression is recorded as an ST segment elevation. Thus, the hallmark of acute myocardial infarction is elevation of the ST segments in the leads overlying the area of infarction (Figure 29–17). Leads on the opposite side of the heart show ST segment depression.

After some days or weeks, the ST segment abnormalities subside. The dead muscle and scar tissue become electrically silent. The infarcted area is therefore negative relative to the normal myocardium during systole, and it fails to contribute its share of positivity to the electrocardiographic complexes. The manifestations of this negativity are multiple and subtle. Common changes include the appearance of a Q wave in some of the leads in which it was not previously present and an increase in the size of the normal Q wave in some of the other leads, although so-called non-Q-wave infarcts are also seen. These latter infarcts tend to be less severe, but there is a high incidence of subsequent reinfarction. Another finding in infarction of the anterior left ventricle is "failure of progression of the R wave;" that is, the R wave fails to become successively larger in the precordial leads as the electrode is moved from right to left over the left ventricle. If the septum is infarcted, the conduction system may be damaged, causing bundle branch block or other forms of heart block.

Myocardial infarctions are often complicated by serious ventricular arrhythmias, with the threat of ventricular fibrillation and death. In experimental animals, and presumably in humans, ventricular arrhythmias occur during three periods. During the first 30 min of an infarction, arrhythmias due to reentry are common. There follows a period relatively free from arrhythmias, but, starting 12 h after infarction, arrhythmias occur as a result of increased automaticity. Arrhythmias occurring 3 days to several weeks after infarction are once again usually due to reentry. It is worth noting in this regard that infarcts that damage the epicardial portions of the myocardium interrupt sympathetic nerve fibers, producing denervation supersensitivity to catecholamines in the area beyond the infarct. Alternatively, endocardial lesions can selectively interrupt vagal fibers, leaving the actions of sympathetic fibers unopposed.

EFFECTS OF CHANGES IN THE IONIC COMPOSITION OF THE BLOOD

Changes in the Na^+ and K^+ concentrations of the extracellular fluids would be expected to affect the potentials of the myocardial fibers because the electrical activity of the heart depends upon the distribution of these ions across the muscle cell membranes. Clinically, a fall in the plasma level of Na^+ may be associated with low-voltage electrocardiographic complexes, but changes in the plasma K^+ level produce severe cardiac abnormalities. Hyperkalemia is a very dangerous and potentially lethal condition because of its effects on the heart. As the plasma K^+ level rises, the first change in the ECG is the appearance of tall peaked T waves, a manifestation of altered repolarization (Figure 29–18). At higher K^+ levels, paralysis of the atria and prolongation of the QRS complexes occur. Ventricular arrhythmias may develop. The resting membrane potential of the muscle fibers decreases as the extracellular K^+ concentration increases. The fibers eventually become unexcitable, and the heart stops in diastole. Conversely, a decrease in the plasma K^+ level causes prolongation of the PR interval, prominent U waves, and, occasionally, late T wave inversion in the precardial leads. If the T and U waves merge, the apparent QT interval is often prolonged; if the T and U waves are separated, the true QT interval is seen to be of normal duration. Hypokalemia is a serious condition, but it is not as rapidly fatal as hyperkalemia.

Increases in extracellular Ca^{2+} concentration enhance myocardial contractility. When large amounts of Ca^{2+} are infused into experimental animals, the heart relaxes less during diastole and eventually stops in systole **(calcium rigor).** However, in clinical conditions associated with hypercalcemia, the plasma calcium level is rarely if ever high enough to affect the heart. Hypocalcemia causes prolongation of the ST segment and consequently of the QT interval, a change that is also produced by phenothiazines and tricyclic antidepressant drugs and by various diseases of the central nervous system.

CHAPTER SUMMARY

- Contractions in the heart are controlled via a well-regulated electrical signaling cascade that originates in pacemaker cells in the sinoatrial (SA) node and is passed via internodal atrial pathways to the atrioventrical (AV) node, the bundle of His, the Purkinje system, and to all parts of the ventricle.
- Most cardiac cells have an action potential that includes a rapid depolarization, an initial rapid repolarization, a plateau, and a slow repolarization process to return to resting potential. These changes are defined by sequential activation and inactivation of Na^+, Ca^{2+}, and K^+ channels.
- Compared to typical myocytes, pacemaker cells have a slightly different sequence of events. After repolarization to the resting potential, there is a slow depolarization that occurs due to a channel that can pass both Na^+ and K^+. As this "funny" current continues to depolarize the cell, Ca^{2+} channels are activated to rapidly depolarize the cell. The hyperpolarization phase is again dominated by K^+ current.
- Spread of the electrical signal from cell to cell is via gap junctions. The rate of spread is dependent on anatomical features, but also can be altered (to a certain extent) via neural input.
- The electrocardiogram (ECG) is an algebraic sum of the electrical activity in the heart. The normal ECG includes well-defined waves and segments, including the P wave (atrial depolarization), the QRS complex (ventricular depolarization), and the T wave (ventricular repolarization). Various arrhythmias can be detected in irregular ECG recordings.
- Because of the contribution of ionic movement to cardiac muscle contraction, heart tissue is sensitive to ionic composition of the blood. Most serious are increases in $[K^+]$ that can produce severe cardiac abnormalities, including paralysis of the atria and ventricular arrhythmias.

MULTIPLE-CHOICE QUESTIONS

For all questions, select the single best answer unless otherwise directed.

1. Which part of the ECG (eg, Figure 29–5) corresponds to ventricular repolarization?
   A. The P wave
   B. The QRS duration
   C. The T wave
   D. The U wave
   E. The PR interval
2. Which of the following normally has a slowly depolarizing "prepotential"?
   A. Sinoatrial node
   B. Atrial muscle cells
   C. Bundle of His
   D. Purkinje fibers
   E. Ventricular muscle cells

3. In second-degree heart block
   A. the ventricular rate is lower than the atrial rate.
   B. the ventricular ECG complexes are distorted.
   C. there is a high incidence of ventricular tachycardia.
   D. stroke volume is decreased.
   E. cardiac output is increased.
4. Currents caused by opening of which of the following channels contribute to the repolarization phase of the action potential of ventricular muscle fibers?
   A. Na^+ channels
   B. Cl^- channels
   C. Ca^{2+} channels
   D. K^+ channels
   E. HCO_3^- channels
5. In complete heart block
   A. fainting may occur because the atria are unable to pump blood into the ventricles.
   B. ventricular fibrillation is common.
   C. the atrial rate is lower than the ventricular rate.
   D. fainting may occur because of prolonged periods during which the ventricles fail to contract.

CHAPTER RESOURCES

Hile B: *Ionic Channels of Excitable Membranes*, 3rd ed. Sinauer Associates, Inc., 2001.

Jackson WF: Ion channels and vascular tone. Hypertension 2000;35:173.

Jessup M, Brozena S: Heart failure. N Engl J Med 2003;348:2007.

Katz, AM: *Physiology of the Heart*, 4th ed. Lippincott Williams and Wilkins, 2006.

Morady F: Radiofrequency ablation as treatment for cardiac arrhythmias. N Engl J Med 1999;340:534.

Nabel EG: Genomic medicine: cardiovascular disease. N Engl J Med 2003;349:60.

Opie, LH: *Heart Physiology from Cell to Circulation*. Lipincott Williams and Wilkins, 2004.

Roder DM: Drug-induced prolongation of the Q-T interval. N Engl J Med 2004;350:1013.

Rowell LB: *Human Cardiovascular Control*. Oxford University Press, 1993.

Wagner GS: *Marriott's Practical Electrocardiography*, 10th ed. Lippincott Williams and Wilkins, 2000.

CHAPTER

30

The Heart as a Pump

OBJECTIVES

After studying this chapter, you should be able to:

- Describe how the sequential pattern of contraction and relaxation in the heart results in a normal pattern of blood flow.
- Understand the pressure, volume, and flow changes that occur during the cardiac cycle.
- Explain the basis of the arterial pulse, heart sounds, and murmurs.
- Delineate the ways by which cardiac output can be up-regulated in the setting of specific physiologic demands for increased oxygen supply to the tissues, such as exercise.
- Describe how the pumping action of the heart can be compromised in the setting of specific disease states.

INTRODUCTION

Of course, the electrical activity of the heart discussed in the previous chapter is designed to subserve the heart's primary physiological role—to pump blood through the lungs, where gas exchange can occur, and thence to the remainder of the body (Clinical Box 30–1). This is accomplished when the orderly depolarization process described in the previous chapter triggers a wave of contraction that spreads through the myocardium. In single muscle fibers, contraction starts just after depolarization and lasts until about 50 ms after repolarization is completed (see Figure 5–15). Atrial systole starts after the P wave of the electrocardiogram (ECG); ventricular systole starts near the end of the R wave and ends just after the T wave. In this chapter, we will consider how these changes in contraction produce sequential changes in pressures and flows in the heart chambers and blood vessels, and thereby propel blood appropriately as needed by whole body demands for oxygen and nutrients. As an aside, it should be noted that the term **systolic pressure** in the vascular system refers to the peak pressure reached during systole, not the mean pressure; similarly, the **diastolic pressure** refers to the lowest pressure during diastole.

MECHANICAL EVENTS OF THE CARDIAC CYCLE

EVENTS IN LATE DIASTOLE

Late in diastole, the mitral (bicuspid) and tricuspid valves between the atria and ventricles (atrioventricular [AV] valves) are open and the aortic and pulmonary valves are closed. Blood flows into the heart throughout diastole, filling the atria and ventricles. The rate of filling declines as the ventricles become distended, and, especially when the heart rate is low, the cusps of the AV valves drift toward the closed position (Figure 30–1). The pressure in the ventricles remains low. About 70% of the ventricular filling occurs passively during diastole.

ATRIAL SYSTOLE

Contraction of the atria propels some additional blood into the ventricles. Contraction of the atrial muscle narrows the orifices of the superior and inferior vena cava and pulmonary

CLINICAL BOX 30–1

Heart Failure

Heart failure occurs when the heart is unable to put out an amount of blood that is adequate for the needs of the tissues. It can be acute and associated with sudden death, or chronic. The failure may involve primarily the right ventricle (cor pulmonale), but much more commonly it involves the larger, thicker left ventricle or both ventricles. Heart failure may also be systolic or diastolic. In **systolic failure,** stroke volume is reduced because ventricular contraction is weak. This causes an increase in the end-systolic ventricular volume, so that the **ejection fraction** falls from 65% to as low as 20%. The initial response to failure is activation of the genes that cause cardiac myocytes to hypertrophy, and thickening of the ventricular wall **(cardiac remodeling).** The incomplete filling of the arterial system leads to increased discharge of the sympathetic nervous system and increased secretion of renin and aldosterone, so Na^+ and water are retained. These responses are initially compensatory, but eventually the failure worsens and the ventricles dilate.

In **diastolic failure,** the ejection fraction is initially maintained, but the elasticity of the myocardium is reduced so filling during diastole is reduced. This leads to inadequate stroke volume and the same cardiac remodeling and Na^+ and water retention that occur in systolic failure. It should be noted that the inadequate cardiac output in failure may be relative rather than absolute. When a large arteriovenous fistula is present, in thyrotoxicosis and in thiamine deficiency, cardiac output may be elevated in absolute terms but still be inadequate to meet the needs of the tissues **(high-output failure).**

THERAPEUTIC HIGHLIGHTS

Treatment of congestive heart failure is aimed at improving cardiac contractility, treating the symptoms, and decreasing the load on the heart. Currently, the most effective treatment in general use is inhibition of the production of angiotensin II with angiotensin-converting enzyme (ACE) inhibitors. Blockade of the effects of angiotensin II on AT_1 receptors with nonpeptide antagonists is also of value. Blocking the production of angiotensin II or its effects also reduces the circulating aldosterone level and decreases blood pressure, reducing the afterload against which the heart pumps. The effects of aldosterone can be further reduced by administering aldosterone receptor blockers. Reducing venous tone with nitrates or hydralazine increases venous capacity so that the amount of blood returned to the heart is reduced, lowering the preload. Diuretics reduce the fluid overload. Drugs that block β-adrenergic receptors have been shown to decrease mortality and morbidity. Digitalis derivatives such as digoxin have classically been used to treat congestive heart failure because of their ability to increase stores of intracellular Ca^{2+} and hence exert a positive inotropic effect, but they are now used in a secondary role to treat systolic dysfunction and slow the ventricular rate in patients with atrial fibrillation.

veins, and the inertia of the blood moving toward the heart tends to keep blood in it. However, despite these inhibitory influences, there is some regurgitation of blood into the veins.

VENTRICULAR SYSTOLE

At the start of ventricular systole, the AV valves close. Ventricular muscle initially shortens relatively little, but intraventricular pressure rises sharply as the myocardium presses on the blood in the ventricle (Figure 30–2). This period of **isovolumetric (isovolumic, isometric) ventricular contraction** lasts about 0.05 s, until the pressures in the left and right ventricles exceed the pressures in the aorta (80 mm Hg; 10.6 kPa) and pulmonary artery (10 mm Hg) and the aortic and pulmonary valves open. During isovolumetric contraction, the AV valves bulge into the atria, causing a small but sharp rise in atrial pressure (Figure 30–3).

When the aortic and pulmonary valves open, the phase of **ventricular ejection** begins. Ejection is rapid at first, slowing down as systole progresses. The intraventricular pressure rises to a maximum and then declines somewhat before ventricular systole ends. Peak pressures in the left and right ventricles are about 120 and 25 mm Hg, respectively. Late in systole, pressure in the aorta actually exceeds that in the left ventricle, but for a short period momentum keeps the blood moving forward. The AV valves are pulled down by the contractions of the ventricular muscle, and atrial pressure drops. The amount of blood ejected by each ventricle per stroke at rest is 70–90 mL. The **end-diastolic ventricular volume** is about 130 mL. Thus, about 50 mL of blood remains in each ventricle at the end of systole **(end-systolic ventricular volume),** and the **ejection fraction,** the percentage of the end-diastolic ventricular volume that is ejected with each stroke, is about 65%. The ejection fraction is a valuable index of ventricular function. It can be measured by injecting radionuclide-labeled red blood cells and imaging the cardiac blood pool at the end of diastole and the end of systole (equilibrium radionuclide angiocardiography), or by computed tomography.

EARLY DIASTOLE

Once the ventricular muscle is fully contracted, the already falling ventricular pressures drop more rapidly. This is the period of **protodiastole,** which lasts about 0.04 s. It ends when

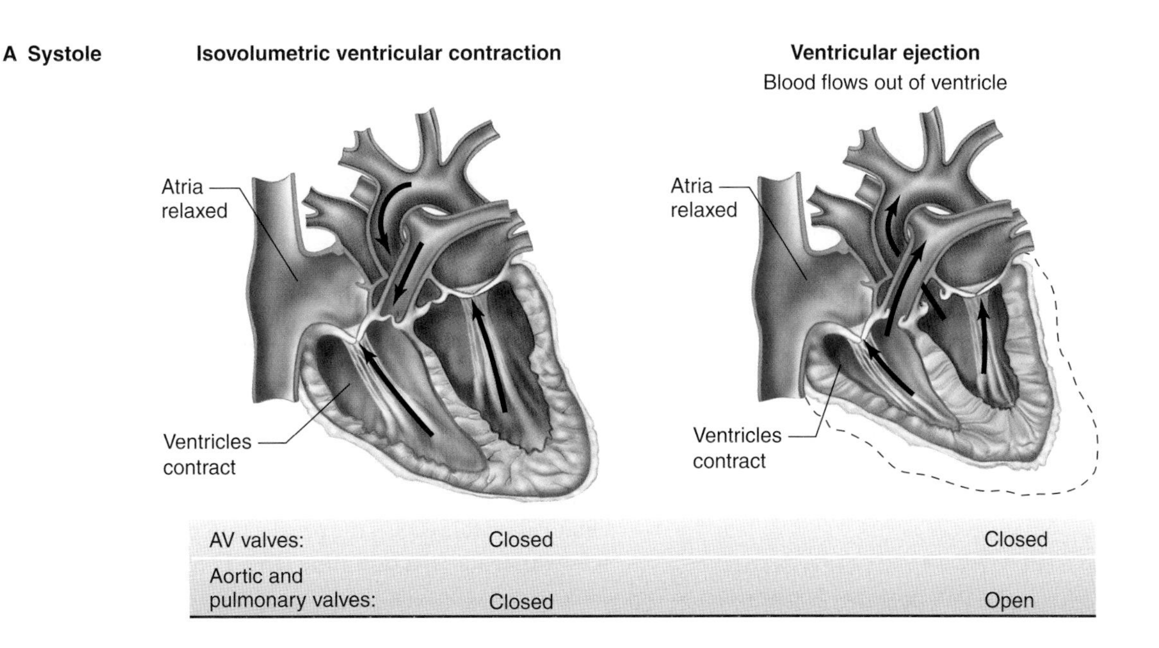

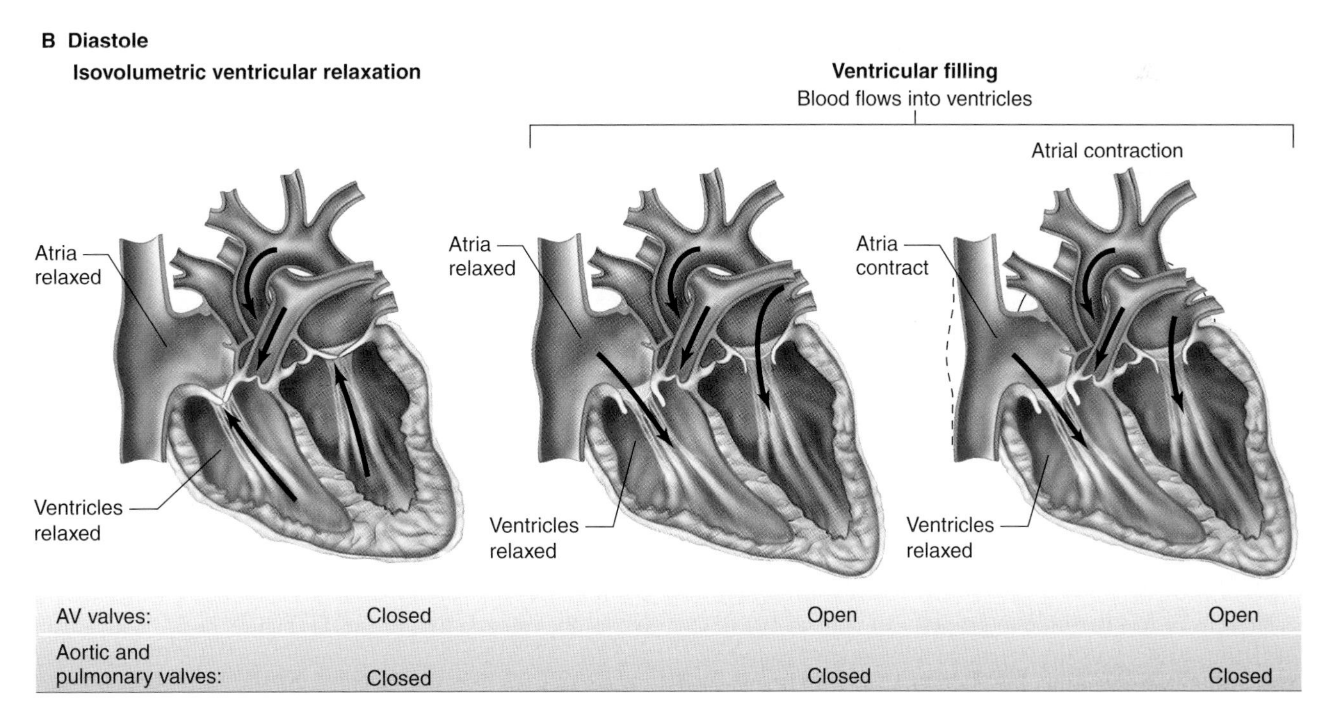

FIGURE 30–1 Divisions of the cardiac cycle: A) systole and B) diastole. The phases of the cycle are identical in both halves of the heart. The direction in which the pressure difference favors flow is denoted by an arrow; note, however, that flow will not actually occur if a valve prevents it. AV, atrioventricular.

the momentum of the ejected blood is overcome and the aortic and pulmonary valves close, setting up transient vibrations in the blood and blood vessel walls. After the valves are closed, pressure continues to drop rapidly during the period of **isovolumetric ventricular relaxation.** Isovolumetric relaxation ends when the ventricular pressure falls below the atrial pressure and the AV valves open, permitting the ventricles to fill. Filling is rapid at first, then slows as the next cardiac contraction approaches. Atrial pressure continues to rise after the end of ventricular systole until the AV valves open, then drops and slowly rises again until the next atrial systole.

TIMING

Although events on the two sides of the heart are similar, they are somewhat asynchronous. Right atrial systole precedes left atrial systole, and contraction of the right ventricle starts after

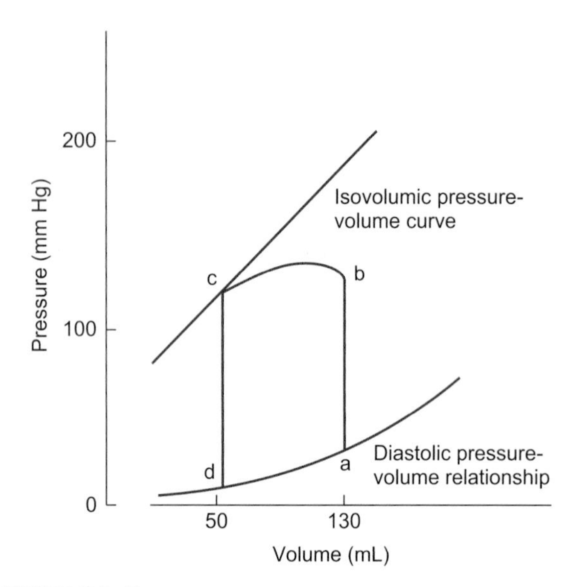

FIGURE 30–2 Normal pressure–volume loop of the left ventricle. During diastole, the ventricle fills and pressure increases from d to a. Pressure then rises sharply from a to b during isovolumetric contraction and from b to c during ventricular ejection. At c, the aortic valves close and pressure falls during isovolumetric relaxation from c back to d. (Reproduced with permission from McPhee SJ, Lingappa VR, Ganong WF [editors]: *Pathophysiology of Disease*, 6th ed. McGraw-Hill, 2010.)

that of the left (see Chapter 29). However, since pulmonary arterial pressure is lower than aortic pressure, right ventricular ejection begins before that of the left. During expiration, the pulmonary and aortic valves close at the same time; but during inspiration, the aortic valve closes slightly before the pulmonary. The slower closure of the pulmonary valve is due to lower impedance of the pulmonary vascular tree. When measured over a period of minutes, the outputs of the two ventricles are, of course, equal, but transient differences in output during the respiratory cycle occur in normal individuals.

LENGTH OF SYSTOLE & DIASTOLE

Cardiac muscle has the unique property of contracting and repolarizing faster when the heart rate is high (see Chapter 5), and the duration of systole decreases from 0.27 s at a heart rate of 65 to 0.16 s at a rate of 200 beats/min (Table 30–1). The reduced time interval is mainly due to a decrease in the duration of systolic ejection. However, the duration of systole is much more fixed than that of diastole, and when the heart rate is increased, diastole is shortened to a much greater degree. For example, at a heart rate of 65, the duration of diastole is 0.62 s, whereas at a heart rate of 200, it is only 0.14 s. This fact has important physiologic and clinical implications. It is during diastole that the heart muscle rests, and coronary blood flow to the subendocardial portions of the left ventricle occurs only during diastole (see Chapter 33). Furthermore, most of the ventricular filling occurs in diastole. At heart rates up to about 180, filling is adequate as long as there is ample venous return, and cardiac output per minute is increased by an increase in rate. However, at very high heart rates, filling may be compromised to such a degree that cardiac output per minute falls.

Because it has a prolonged action potential, cardiac muscle cannot contract in response to a second stimulus until near the end of the initial contraction (see Figure 5–15). Therefore, cardiac muscle cannot be tetanized like skeletal muscle. The highest rate at which the ventricles can contract is theoretically about 400/min, but in adults the AV node will not conduct more than about 230 impulses/min because of its long refractory period. A ventricular rate of more than 230 is seen only in paroxysmal ventricular tachycardia (see Chapter 29).

Exact measurement of the duration of isovolumetric ventricular contraction is difficult in clinical situations, but it is relatively easy to measure the duration of **total electromechanical systole (QS_2)**, the **preejection period (PEP),** and the **left ventricular ejection time (LVET)** by recording the ECG, phonocardiogram, and carotid pulse simultaneously. QS_2 is the period from the onset of the QRS complex to the closure of the aortic valves, as determined by the onset of the second heart sound. LVET is the period from the beginning of the carotid pressure rise to the dicrotic notch (see below). PEP is the difference between QS_2 and LVET and represents the time for the electrical as well as the mechanical events that precede systolic ejection. The ratio PEP/LVET is normally about 0.35, and it increases without a change in QS_2 when left ventricular performance is compromised in a variety of cardiac diseases.

ARTERIAL PULSE

The blood forced into the aorta during systole not only moves the blood in the vessels forward but also sets up a pressure wave that travels along the arteries. The pressure wave expands the arterial walls as it travels, and the expansion is palpable as the **pulse.** The rate at which the wave travels, which is independent of and much higher than the velocity of blood flow, is about 4 m/s in the aorta, 8 m/s in the large arteries, and 16 m/s in the small arteries of young adults. Consequently, the pulse is felt in the radial artery at the wrist about 0.1 s after the peak of systolic ejection into the aorta (Figure 30–3). With advancing age, the arteries become more rigid, and the pulse wave moves faster.

The strength of the pulse is determined by the pulse pressure and bears little relation to the mean pressure. The pulse is weak ("thready") in shock. It is strong when stroke volume is large; for example, during exercise or after the administration of histamine. When the pulse pressure is high, the pulse waves may be large enough to be felt or even heard by the individual (palpitation, "pounding heart"). When the aortic valve is incompetent (aortic insufficiency), the pulse is particularly strong, and the force of systolic ejection may be sufficient to make the head nod with each heartbeat. The pulse in aortic insufficiency is called a **collapsing, Corrigan,** or **water-hammer pulse.**

The **dicrotic notch,** a small oscillation on the falling phase of the pulse wave caused by vibrations set up when the aortic

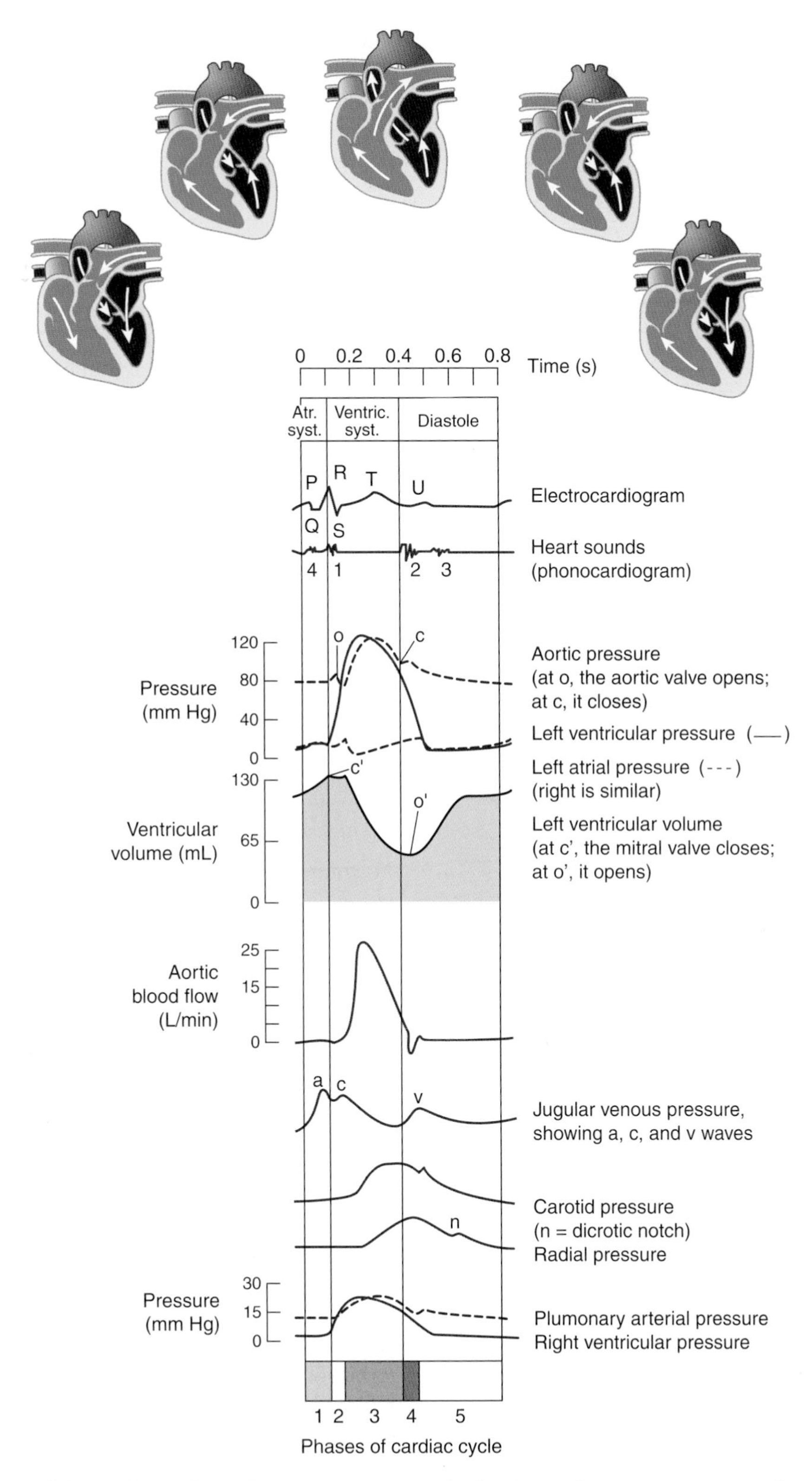

FIGURE 30–3 Events of the cardiac cycle at a heart rate of 75 beats/min. The phases of the cardiac cycle identified by the numbers at the bottom are as follows: 1, atrial systole; 2, isovolumetric ventricular contraction; 3, ventricular ejection; 4, isovolumetric ventricular relaxation; 5, ventricular filling. Note that late in systole, aortic pressure actually exceeds left ventricular pressure. However, the momentum of the blood keeps it flowing out of the ventricle for a short period. The pressure relationships in the right ventricle and pulmonary artery are similar. Atr. syst., atrial systole; Ventric. syst., ventricular systole.

TABLE 30–1 Variation in length of action potential and associated phenomena with cardiac rate.[a]

	Heart Rate 75/min	Heart Rate 200/min	Skeletal Muscle
Duration, each cardiac cycle	0.80	0.30	...
Duration of systole	0.27	0.16	...
Duration of action potential	0.25	0.15	0.007
Duration of absolute refractory period	0.20	0.13	0.004
Duration of relative refractory period	0.05	0.02	0.003
Duration of diastole	0.53	0.14	...

[a]All values are in seconds.
Courtesy of AC Barger and GS Richardson.

valve snaps shut (Figure 30–3), is visible if the pressure wave is recorded but is not palpable at the wrist. The pulmonary artery pressure curve also has a dicrotic notch produced by the closure of the pulmonary valves.

ATRIAL PRESSURE CHANGES & THE JUGULAR PULSE

Atrial pressure rises during atrial systole and continues to rise during isovolumetric ventricular contraction when the AV valves bulge into the atria. When the AV valves are pulled down by the contracting ventricular muscle, pressure falls rapidly and then rises as blood flows into the atria until the AV valves open early in diastole. The return of the AV valves to their relaxed position also contributes to this pressure rise by reducing atrial capacity. The atrial pressure changes are transmitted to the great veins, producing three characteristic waves in the record of jugular pressure (Figure 30–3). The **a wave** is due to atrial systole. As noted above, some blood regurgitates into the great veins when the atria contract. In addition, venous inflow stops, and the resultant rise in venous pressure contributes to the a wave. The **c wave** is the transmitted manifestation of the rise in atrial pressure produced by the bulging of the tricuspid valve into the atria during isovolumetric ventricular contraction. The **v wave** mirrors the rise in atrial pressure before the tricuspid valve opens during diastole. The jugular pulse waves are superimposed on the respiratory fluctuations in venous pressure. Venous pressure falls during inspiration as a result of the increased negative intrathoracic pressure and rises again during expiration.

HEART SOUNDS

Two sounds are normally heard through a stethoscope during each cardiac cycle. The first is a low, slightly prolonged "lub" **(first sound),** caused by vibrations set up by the sudden closure of the AV valves at the start of ventricular systole (Figure 30–3). The second is a shorter, high-pitched "dup" **(second sound),** caused by vibrations associated with closure of the aortic and pulmonary valves just after the end of ventricular systole. A soft, low-pitched **third sound** is heard about one third of the way through diastole in many normal young individuals. It coincides with the period of rapid ventricular filling and is probably due to vibrations set up by the inrush of blood. A **fourth sound** can sometimes be heard immediately before the first sound when atrial pressure is high or the ventricle is stiff in conditions such as ventricular hypertrophy. It is due to ventricular filling and is rarely heard in normal adults.

The first sound has a duration of about 0.15 s and a frequency of 25–45 Hz. It is soft when the heart rate is low, because the ventricles are well filled with blood and the leaflets of the AV valves float together before systole. The second sound lasts about 0.12 s, with a frequency of 50 Hz. It is loud and sharp when the diastolic pressure in the aorta or pulmonary artery is elevated, causing the respective valves to shut briskly at the end of systole. The interval between aortic and pulmonary valve closure during inspiration is frequently long enough for the second sound to be reduplicated (physiologic splitting of the second sound). Splitting also occurs in various diseases. The third sound, when present, has a duration of 0.1 s.

MURMURS

Murmurs, or **bruits,** are abnormal sounds heard in various parts of the vascular system. The two terms are used interchangeably, though "murmur" is more commonly used to denote noise heard over the heart than over blood vessels. As discussed in detail in Chapter 31, blood flow is laminar, nonturbulent, and silent up to a critical velocity; above this velocity (such as beyond an obstruction), blood flow is turbulent and creates sounds. Blood flow speeds up when an artery or a heart valve is narrowed.

Examples of vascular sounds outside the heart are the bruit heard over a large, highly vascular goiter, the bruit heard over a carotid artery when its lumen is narrowed and distorted by atherosclerosis, and the murmurs heard over an aneurysmal dilation of one of the large arteries, an arteriovenous (A-V) fistula, or a patent ductus arteriosus.

The major—but certainly not the only—cause of cardiac murmurs is disease of the heart valves. When the orifice of a valve is narrowed **(stenosis),** blood flow through it is accelerated and turbulent. When a valve is incompetent, blood flows through it backward (**regurgitation** or **insufficiency**), again through a narrow orifice that accelerates flow. The timing (systolic or diastolic) of a murmur due to any particular valve (Table 30–2) can be predicted from a knowledge of the mechanical events of the cardiac cycle. Murmurs due to disease of a particular valve can generally be heard best when the stethoscope is directly over the valve. There are also other aspects of the duration, character, accentuation, and transmission of the sound that help to locate its origin in one valve or another. One of the loudest murmurs is that produced when blood flows backward in diastole through

TABLE 30–2 Heart murmurs.

Valve	Abnormality	Timing of Murmur
Aortic or pulmonary	Stenosis	Systolic
	Insufficiency	Diastolic
Mitral or tricuspid	Stenosis	Diastolic
	Insufficiency	Systolic

a hole in a cusp of the aortic valve. Most murmurs can be heard only with the aid of the stethoscope, but this high-pitched musical diastolic murmur is sometimes audible to the unaided ear several feet from the patient.

In patients with congenital interventricular septal defects, flow from the left to the right ventricle causes a systolic murmur. Soft murmurs may also be heard in patients with interatrial septal defects, although they are not a constant finding.

Soft systolic murmurs are also common in individuals, especially children, who have no cardiac disease. Systolic murmurs are also heard in anemic patients as a result of the low viscosity of the blood and associated rapid flow (see Chapter 31).

ECHOCARDIOGRAPHY

Wall movement and other aspects of cardiac function can be evaluated by the noninvasive technique of **echocardiography.** Pulses of ultrasonic waves are emitted from a transducer that also functions as a receiver to detect waves reflected back from various parts of the heart. Reflections occur wherever acoustic impedance changes, and a recording of the echoes displayed against time on an oscilloscope provides a record of the movements of the ventricular wall, septum, and valves during the cardiac cycle. When combined with Doppler techniques, echocardiography can be used to measure velocity and volume of flow through valves. It has considerable clinical usefulness, particularly in evaluating and planning therapy in patients with valvular lesions.

CARDIAC OUTPUT

METHODS OF MEASUREMENT

In experimental animals, cardiac output can be measured with an electromagnetic flow meter placed on the ascending aorta. Two methods of measuring output that are applicable to humans, in addition to Doppler combined with echocardiography, are the **direct Fick method** and the **indicator dilution method.**

The **Fick principle** states that the amount of a substance taken up by an organ (or by the whole body) per unit of time is equal to the arterial level of the substance minus the venous level **(A-V difference)** times the blood flow. This principle can be applied, of course, only in situations in which the arterial blood is the sole source of the substance taken up. The principle can be used to determine cardiac output by measuring the amount of O_2 consumed by the body in a given period and dividing this value by the A-V difference across the lungs. Because systemic arterial blood has effectively the same O_2 content in all parts of the body, the arterial O_2 content can be measured in a sample obtained from any convenient artery. A sample of venous blood in the pulmonary artery is obtained by means of a cardiac catheter. It has now become commonplace to insert a long catheter through a forearm vein and to guide its tip into the heart with the aid of a fluoroscope. The procedure is generally benign. Catheters can be inserted through the right atrium and ventricle into the small branches of the pulmonary artery. An example of the calculation of cardiac output using a typical set of values is as follows:

Output of left ventricle

$$= \frac{O_2 \text{ consumption (mL/min)}}{[A_{O_2}] - [V_{O_2}]}$$

$$= \frac{250 \text{ mL/min}}{190 \text{ mL/L arterial blood} - 140 \text{ mL/L venous blood in pulmonary artery}}$$

$$= \frac{250 \text{ mL/min}}{50 \text{ mL/L}}$$

$$= 5 \text{ L/min}$$

In the indicator dilution technique, a known amount of a substance such as a dye or, more commonly, a radioactive isotope is injected into an arm vein and the concentration of the indicator in serial samples of arterial blood is determined. The output of the heart is equal to the amount of indicator injected divided by its average concentration in arterial blood after a single circulation through the heart (Figure 30–4). The indicator must, of course, be a substance that stays in the bloodstream during the test and has no harmful or hemodynamic effects. In practice, the log of the indicator concentration in the serial arterial samples is plotted against time as the concentration rises, falls, and then rises again as the indicator recirculates. The initial decline in concentration, linear on a semilog plot, is extrapolated to the abscissa, giving the time for first passage of the indicator through the circulation. The cardiac output for that period is calculated (Figure 30–4) and then converted to output per minute.

A popular indicator dilution technique is **thermodilution,** in which the indicator used is cold saline. The saline is injected into the right atrium through one channel of a double-lumen catheter, and the temperature change in the blood is recorded in the pulmonary artery, using a thermistor in the other, longer side of the catheter. The temperature change is inversely proportional to the amount of blood flowing through the pulmonary artery; that is, to the extent that the cold saline is diluted by blood. This technique has two important advantages: (1) the saline is completely innocuous; and (2) the cold is dissipated in the tissues so recirculation is not a problem, and it is easy to make repeated determinations.

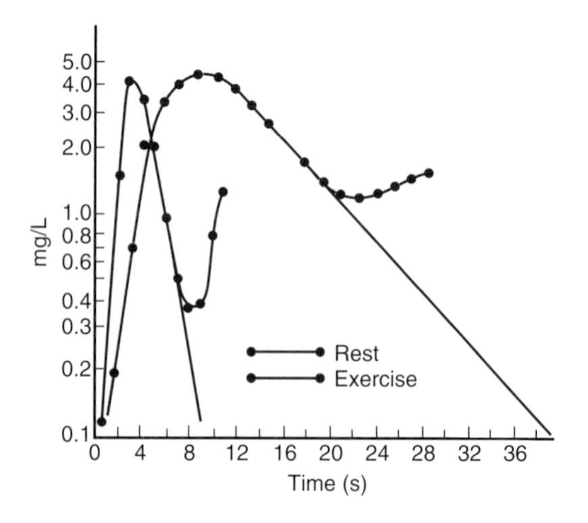

$$F = \frac{E}{\int_{0}^{\alpha} C\,dt}$$

F = flow
E = amount of indicator injected
C = instantaneous concentration of indicator in arterial blood

In the **rest** example above,

$$\text{Flow in 39 s (time of first passage)} = \frac{\text{5 mg injection}}{\text{1.6 mg/L (avg concentration)}}$$

Flow = 3.1 L in 39 s

$$\text{Flow (cardiac output)/min} = 3.1 \times \frac{60}{39} = 4.7\text{ L}$$

For the **exercise** example,

$$\text{Flow in 9 s} = \frac{\text{5 mg}}{\text{1.51 mg/L}} = 3.3\text{ L}$$

$$\text{Flow/min} = 3.3 \times \frac{60}{9} = 22.0\text{ L}$$

FIGURE 30–4 Determination of cardiac output by indicator (dye) dilution. Two examples are shown—at rest and during exercise.

CARDIAC OUTPUT IN VARIOUS CONDITIONS

The amount of blood pumped out of the heart per beat, the **stroke volume,** is about 70 mL from each ventricle in a resting man of average size in the supine position. The output of the heart per unit of time is the **cardiac output.** In a resting, supine man, it averages about 5.0 L/min (70 mL × 72 beats/min). There is a correlation between resting cardiac output and body surface area. The output per minute per square meter of body surface (the **cardiac index**) averages 3.2 L. The effects of various conditions on cardiac output are summarized in Table 30–3.

TABLE 30–3 Effect of various conditions on cardiac output.

	Condition or Factor[a]
No change	Sleep
	Moderate changes in environmental temperature
Increase	Anxiety and excitement (50–100%)
	Eating (30%)
	Exercise (up to 700%)
	High environmental temperature
	Pregnancy
	Epinephrine
Decrease	Sitting or standing from lying position (20–30%)
	Rapid arrhythmias
	Heart disease

[a]Approximate percentage changes are shown in parentheses.

FACTORS CONTROLLING CARDIAC OUTPUT

Predictably, changes in cardiac output that are called for by physiologic conditions can be produced by changes in cardiac rate, or stroke volume, or both (Figure 30–5). The cardiac rate is controlled primarily by the autonomic nerves, with sympathetic stimulation increasing the rate and parasympathetic stimulation decreasing it (see Chapter 29). Stroke volume is also determined in part by neural input, with sympathetic stimuli making the myocardial muscle fibers contract with greater strength at any given length and parasympathetic

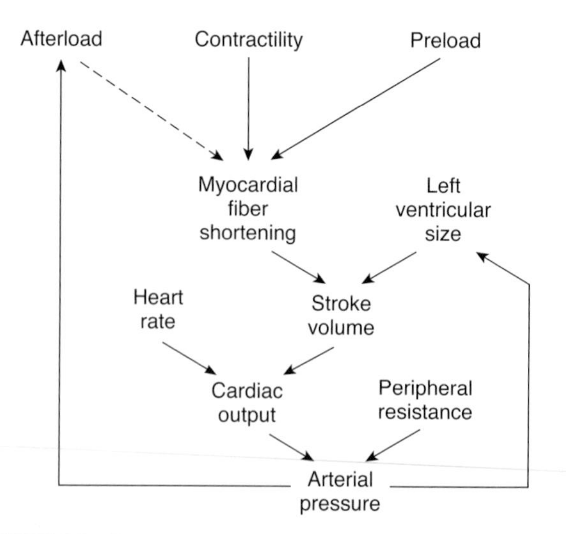

FIGURE 30–5 Interactions between the components that regulate cardiac output and arterial pressure. Solid arrows indicate increases, and the dashed arrow indicates a decrease.

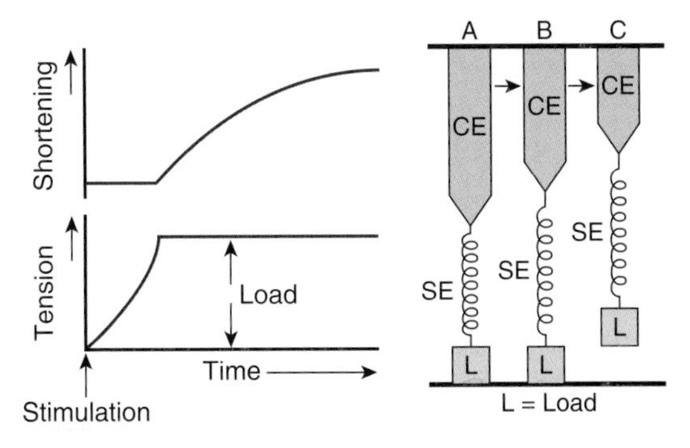

FIGURE 30–6 Model for contraction of afterloaded muscles. A) Rest. B) Partial contraction of the contractile element of the muscle (CE), with stretching of the series elastic element (SE) but no shortening. C) Complete contraction, with shortening. (Reproduced with permission from Sonnenblick EH: *The Myocardial Cell: Structure, Function and Modification*. Briller SA, Conn HL [editors]. University Pennsylvania Press, 1966.)

stimuli having the opposite effect. When the strength of contraction increases without an increase in fiber length, more of the blood that normally remains in the ventricles is expelled; that is, the ejection fraction increases. The cardiac accelerator action of the catecholamines liberated by sympathetic stimulation is referred to as their **chronotropic action,** whereas their effect on the strength of cardiac contraction is called their **inotropic action.**

The force of contraction of cardiac muscle depends on its preloading and its afterloading. These factors are illustrated in Figure 30–6, in which a muscle strip is stretched by a load (the **preload**) that rests on a platform. The initial phase of the contraction is isometric; the elastic component in series with the contractile element is stretched, and tension increases until it is sufficient to lift the load. The tension at which the load is lifted is the **afterload.** The muscle then contracts isotonically without developing further tension. In vivo, the preload is the degree to which the myocardium is stretched before it contracts and the afterload is the resistance against which blood is expelled.

RELATION OF TENSION TO LENGTH IN CARDIAC MUSCLE

The length–tension relationship in cardiac muscle (see Figure 5–17) is similar to that in skeletal muscle (see Figure 5–11); when the muscle is stretched, the developed tension increases to a maximum and then declines as stretch becomes more extreme. Starling pointed this out when he stated that the "energy of contraction is proportional to the initial length of the cardiac muscle fiber" (**Starling's law of the heart** or the **Frank–Starling law**). For the heart, the length of the muscle fibers (ie, the extent of the preload) is proportional to the end-diastolic volume. The relation between ventricular stroke volume and end-diastolic volume is called the Frank–Starling curve.

When cardiac output is regulated by changes in cardiac muscle fiber length, this is referred to as **heterometric regulation.** Conversely, regulation due to changes in contractility independent of length is sometimes called **homometric regulation.**

FACTORS AFFECTING END-DIASTOLIC VOLUME

Alterations in systolic and diastolic function have different effects on the heart. When systolic contractions are reduced, there is a primary reduction in stroke volume. Diastolic function also affects stroke volume, but in a different way.

The myocardium is covered by a fibrous layer known as the epicardium. This, in turn, is surrounded by the pericardium, which separates the heart from the rest of the thoracic viscera. The space between the epicardium and pericardium (the pericardial sac) normally contains 5–30 mL of clear fluid, which lubricates the heart and permits it to contract with minimal friction.

An increase in intrapericardial pressure (eg, as a result of infection or pressure from a tumor) limits the extent to which the ventricle can fill, as does a decrease in ventricular compliance; that is, an increase in ventricular stiffness produced by myocardial infarction, infiltrative disease, and other abnormalities. Atrial contractions aid ventricular filling. Factors affecting the amount of blood returning to the heart likewise influence the degree of cardiac filling during diastole. An increase in total blood volume increases venous return (Clinical Box 30–2). Constriction of the veins reduces the size of the venous reservoirs, decreasing venous pooling and thus increasing venous return. An increase in the normal negative intrathoracic pressure increases the pressure gradient along which blood flows to the heart, whereas a decrease impedes venous return. Standing decreases venous return, and muscular activity increases it as a result of the pumping action of skeletal muscle.

The effects of systolic and diastolic dysfunction on the pressure–volume loop of the left ventricle are summarized in Figure 30–7.

MYOCARDIAL CONTRACTILITY

The contractility of the myocardium exerts a major influence on stroke volume. When the sympathetic nerves to the heart are stimulated, the whole length–tension curve shifts upward and to the left (Figure 30–8). The positive inotropic effect of norepinephrine liberated at the nerve endings is augmented by circulating norepinephrine, and epinephrine has a similar effect. Conversely, there is a negative inotropic effect of vagal stimulation on both atrial and (to a lesser extent) ventricular muscle.

Changes in cardiac rate and rhythm also affect myocardial contractility (known as the force–frequency relation, Figure 30–8). Ventricular extrasystoles condition the myocardium in such a way that the next succeeding contraction

CLINICAL BOX 30–2

Shock

Circulatory shock comprises a collection of different entities that share certain common features; however, the feature that is common to all the entities is inadequate tissue perfusion with a relatively or absolutely inadequate cardiac output. The cardiac output may be inadequate because the amount of fluid in the vascular system is inadequate to fill it **(hypovolemic shock).** Alternatively, it may be inadequate in the relative sense because the size of the vascular system is increased by vasodilation even though the blood volume is normal (**distributive, vasogenic,** or **low-resistance shock).** Shock may also be caused by inadequate pumping action of the heart as a result of myocardial abnormalities **(cardiogenic shock),** and by inadequate cardiac output as a result of obstruction of blood flow in the lungs or heart **(obstructive shock).**

Hypovolemic shock is also called "cold shock." It is characterized by hypotension; a rapid, thready pulse; cold, pale, clammy skin; intense thirst; rapid respiration; and restlessness or, alternatively, torpor. None of these findings, however, are invariably present. Hypovolemic shock is commonly subdivided into categories on the basis of cause. Of these, it is useful to consider the effects of hemorrhage in some detail because of the multiple compensatory reactions that come into play to defend extracellular fluid (ECF) volume. Thus, the decline in blood volume produced by bleeding decreases venous return, and cardiac output falls. The heart rate is increased, and with severe hemorrhage, a fall in blood pressure always occurs. With moderate hemorrhage (5–15 mL/kg body weight), pulse pressure is reduced but mean arterial pressure may be normal. The blood pressure changes vary from individual to individual, even when exactly the same amount of blood is lost. The skin is cool and pale and may have a grayish tinge because of stasis in the capillaries and a small amount of cyanosis. Inadequate perfusion of the tissues leads to increased anaerobic glycolysis, with the production of large amounts of lactic acid. In severe cases, the blood lactate level rises from the normal value of about 1 mmol/L to 9 mmol/ L or more. The resulting **lactic acidosis** depresses the myocardium, decreases peripheral vascular responsiveness to catecholamines, and may be severe enough to cause coma. When blood volume is reduced and venous return is decreased, moreover, stimulation of arterial baroreceptors is reduced, increasing sympathetic output. Even if there is no drop in mean arterial pressure, the decrease in pulse pressure decreases the rate of discharge in the arterial baroreceptors, and reflex tachycardia and vasoconstriction result.

With more severe blood loss, tachycardia is replaced by bradycardia; this occurs while shock is still reversible. The bradycardia is presumably due to unmasking a vagally mediated depressor reflex, and the response may have evolved as a mechanism for stopping further blood loss. With even greater hemorrhage, the heart rate rises again. Vasoconstriction is generalized, sparing only the vessels of the brain and heart. A widespread reflex venoconstriction also helps maintain the filling pressure of the heart. In the kidneys, both afferent and efferent arterioles are constricted, but the efferent vessels are constricted to a greater degree. The glomerular filtration rate is depressed, but renal plasma flow is decreased to a greater extent, so that the filtration fraction increases. Na^+ retention is marked, and the nitrogenous products of metabolism are retained in the blood **(azotemia** or **uremia).** If the hypotension is prolonged, renal tubular damage may be severe **(acute renal failure).** After a moderate hemorrhage, the circulating plasma volume is restored in 12–72 h. Preformed albumin also enters rapidly from extravascular stores, but most of the tissue fluids that are mobilized are protein-free. After the initial influx of preformed albumin, the rest of the plasma protein losses are replaced, presumably by hepatic synthesis, over a period of 3–4 days. Erythropoietin appears in the circulation, and the reticulocyte count increases, reaching a peak in 10 days. The red cell mass is restored to normal in 4–8 weeks.

THERAPUETIC HIGHLIGHTS

The treatment of shock is aimed at correcting the cause and helping the physiologic compensatory mechanisms to restore an adequate level of tissue perfusion. If the primary cause of the shock is blood loss, the treatment should include early and rapid transfusion of adequate amounts of compatible whole blood. In shock due to burns and other conditions in which there is hemoconcentration, plasma is the treatment of choice to restore the fundamental defect, the loss of plasma. Concentrated human serum albumin and other hypertonic solutions expand the blood volume by drawing fluid out of the interstitial spaces. They are valuable in emergency treatment but have the disadvantage of further dehydrating the tissues of an already dehydrated patient.

is stronger than the preceding normal contraction. This **postextrasystolic potentiation** is independent of ventricular filling, since it occurs in isolated cardiac muscle and is due to increased availability of intracellular Ca^{2+}. A sustained increment in contractility can be produced therapeutically by delivering paired electrical stimuli to the heart in such a way that the second stimulus is delivered shortly after the refractory period of the first. It has also been shown that

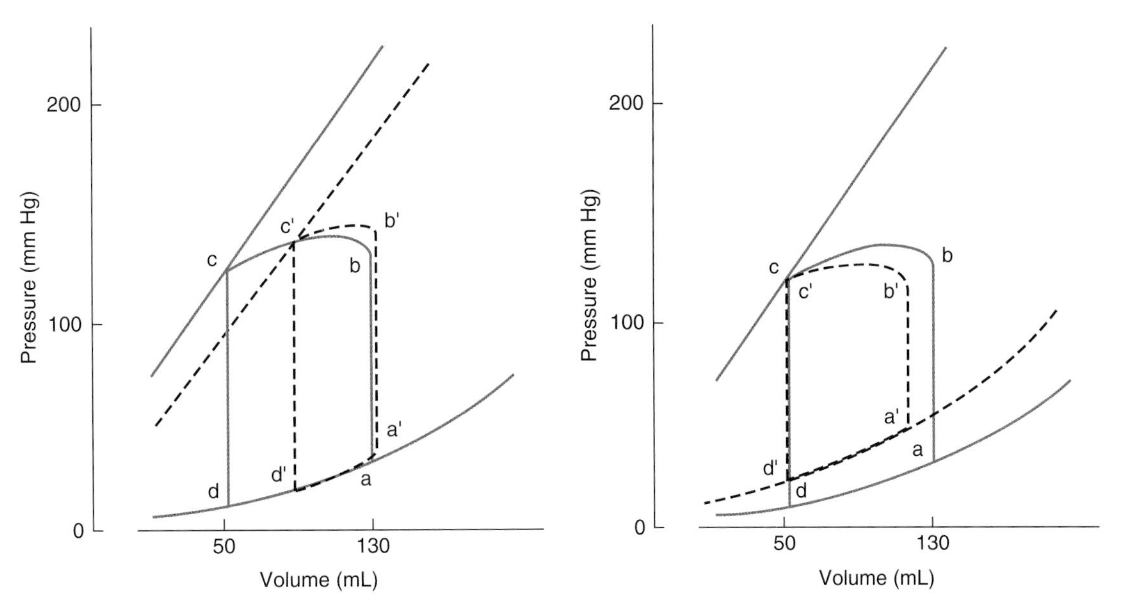

FIGURE 30–7 Effect of systolic and diastolic dysfunction on the pressure–volume loop of the left ventricle. In both panels, the solid lines represents the normal pressure–volume loop (equivalent to that shown in Figure 30–2 and the dashed lines show how the loop is shifted by the disease process represented. Left: Systolic dysfunction shifts the isovolumic pressure–volume curve to the right, decreasing the stroke volume from b–c to b′–c′. Right: Diastolic dysfunction increases end-diastolic volume and shifts the diastolic pressure–volume relationship upward and to the left. This reduces the stroke volume from b–c to b′–c′. (Reproduced with permission from McPhee SJ, Lingappa VR, Ganong WF [editors]: *Pathophysiology of Disease,* 6th ed. McGraw-Hill, 2010.)

myocardial contractility increases as the heart rate increases, although this effect is relatively small.

Catecholamines exert their inotropic effect via an action on cardiac β_1-adrenergic receptors and Gs, with resultant activation of adenylyl cyclase and increased intracellular cyclic adenosine 3′,5′-monophosphate (cAMP). Xanthines such as caffeine and theophylline that inhibit the breakdown of cAMP are predictably positively inotropic. The positively inotropic effect of digitalis and related drugs (Figure 30–8), on the other hand, is due to their inhibitory effect on the

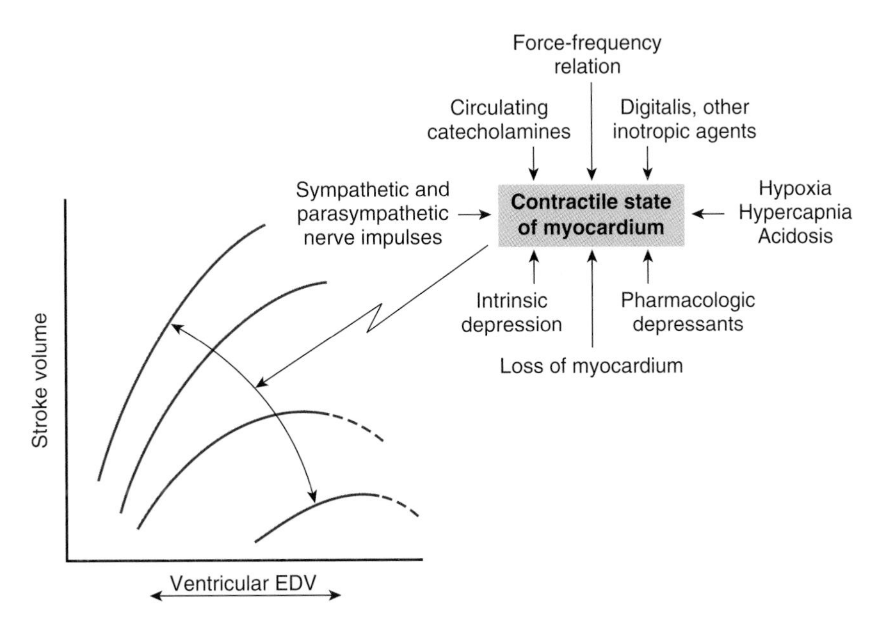

FIGURE 30–8 Effect of changes in myocardial contractility on the Frank–Starling curve. The curve shifts downward and to the right as contractility is decreased. The major factors influencing contractility are summarized on the right. The dashed lines indicate portions of the ventricular function curves where maximum contractility has been exceeded; that is, they identify points on the "descending limb" of the Frank–Starling curve. EDV, end-diastolic volume. (Reproduced with permission from Braunwald E, Ross J, Sonnenblick EH: Mechanisms of contraction of the normal and failing heart. N Engl J Med 1967;277:794.)

Na, K ATPase in the myocardium, and a subsequent decrease in calcium removal from the cytosol by Na^+/Ca^{2+} exchange (see Chapter 5). Hypercapnia, hypoxia, acidosis, and drugs such as quinidine, procainamide, and barbiturates depress myocardial contractility. The contractility of the myocardium is also reduced in heart failure (intrinsic depression). The causes of this depression are not fully understood, but may reflect down-regulation of β-adrenergic receptors and associated signaling pathways and impaired calcium liberation from the sarcoplasmic reticulum. In acute heart failure, such as that associated with sepsis, this response could be considered an appropriate adaptation (so-called "myocardial hibernation") to a situation where energy supply to the heart is limited, thereby reducing energy expenditure and avoiding cell death.

INTEGRATED CONTROL OF CARDIAC OUTPUT

The mechanisms listed above operate in an integrated way to maintain cardiac output. For example, during muscular exercise, there is increased sympathetic discharge, so that myocardial contractility is increased and the heart rate rises. The increase in heart rate is particularly prominent in normal individuals, and there is only a modest increase in stroke volume (see Table 30–4 and Clinical Box 30–3). However, patients with transplanted hearts are able to increase their cardiac output during exercise in the absence of cardiac innervation through the operation of the Frank–Starling mechanism (Figure 30–9). Circulating catecholamines also contribute. If venous return increases and there is no change in sympathetic tone, venous pressure rises, diastolic inflow is greater, ventricular end-diastolic pressure increases, and the heart muscle contracts more forcefully. During muscular exercise, venous return is increased by the pumping action of the muscles and the increase in respiration (see Chapter 32). In addition, because of vasodilation in the contracting muscles, peripheral resistance and, consequently, afterload are decreased. The end result in both normal and transplanted hearts is thus a prompt and marked increase in cardiac output.

One of the differences between untrained individuals and trained athletes is that the athletes have lower heart rates, greater end-systolic ventricular volumes, and greater stroke volumes at rest. Therefore, they can potentially achieve a given increase in cardiac output by further increases in stroke volume without increasing their heart rate to as great a degree as an untrained individual.

OXYGEN CONSUMPTION BY THE HEART

Basal O_2 consumption by the myocardium is about 2 mL/100 g/min. This value is considerably higher than that of resting skeletal muscle. O_2 consumption by the beating heart is about 9 mL/100 g/min at rest. Increases occur during exercise and in a number of different states. Cardiac venous O_2 tension is low, and little additional O_2 can be extracted from the blood in the coronaries, so increases in O_2 consumption require increases in coronary blood flow. The regulation of coronary flow is discussed in Chapter 33.

O_2 consumption by the heart is determined primarily by the intramyocardial tension, the contractile state of the myocardium, and the heart rate. Ventricular work per beat correlates with O_2 consumption. The work is the product of stroke volume and mean arterial pressure in the pulmonary artery or the aorta (for the right and left ventricle, respectively). Because aortic pressure is seven times greater than pulmonary artery pressure, the stroke work of the left ventricle is approximately seven times the stroke work of the right. In theory, a 25% increase in stroke volume without a change in arterial pressure should produce the same increase in O_2 consumption as a 25% increase in arterial pressure without a change

TABLE 30–4 Changes in cardiac function with exercise. Note that stroke volume levels off, then falls somewhat (as a result of the shortening of diastole) when the heart rate rises to high values.

Work (kg-m/min)	O_2 Usage (mL/min)	Pulse Rate (per min)	Cardiac Output (L/min)	Stroke Volume (mL)	A-V O_2 Difference (mL/dL)
Rest	267	64	6.4	100	4.3
288	910	104	13.1	126	7.0
540	1430	122	15.2	125	9.4
900	2143	161	17.8	110	12.3
1260	3007	173	20.9	120	14.5

Reproduced with permission from Asmussen E, Nielsen M: The cardiac output in rest and work determined by the acetylene and the dye injection methods. Acta Physiol Scand 1952;27:217.

CLINICAL BOX 30–3

Circulatory Changes during Exercise

The blood flow of resting skeletal muscle is low (2–4 mL/100 g/min). When a muscle contracts, it compresses the vessels in it if it develops more than 10% of its maximal tension; when it develops more than 70% of its maximal tension, blood flow is completely stopped. Between contractions, however, flow is so greatly increased that blood flow per unit of time in a rhythmically contracting muscle is increased as much as 30-fold. Local mechanisms maintaining a high blood flow in exercising muscle include a fall in tissue PO_2, a rise in tissue PCO_2, and accumulation of K^+ and other vasodilator metabolites. The temperature rises in active muscle, and this further dilates the vessels. Dilation of the arterioles and precapillary sphincters causes a 10- to 100-fold increase in the number of open capillaries. The average distance between the blood and the active cells—and the distance O_2 and metabolic products must diffuse—is thus greatly decreased. The dilation increases the cross-sectional area of the vascular bed, and the velocity of flow therefore decreases.

The systemic cardiovascular response to exercise that provides for the additional blood flow to contracting muscle depends on whether the muscle contractions are primarily isometric or primarily isotonic with the performance of external work. With the start of an isometric muscle contraction, the heart rate rises, probably as a result of psychic stimuli acting on the medulla oblongata. The increase is largely due to decreased vagal tone, although increased discharge of the cardiac sympathetic nerves plays some role. Within a few seconds of the onset of an isometric muscle contraction, systolic and diastolic blood pressures rise sharply. Stroke volume changes relatively little, and blood flow to the steadily contracting muscles is reduced as a result of compression of their blood vessels. The response to exercise involving isotonic muscle contraction is similar in that there is a prompt increase in heart rate, but different in that a marked increase in stroke volume occurs. In addition, there is a net fall in total peripheral resistance due to vasodilation in exercising muscles. Consequently, systolic blood pressure rises only moderately, whereas diastolic pressure usually remains unchanged or falls.

The difference in response to isometric and isotonic exercise is explained in part by the fact that the active muscles are tonically contracted during isometric exercise and consequently contribute to increased total peripheral resistance. Cardiac output is increased during isotonic exercise to values that may exceed 35 L/min, the amount being proportional to the increase in O_2 consumption. The maximal heart rate achieved during exercise decreases with age. In children, it rises to 200 or more beats/min; in adults it rarely exceeds 195 beats/min, and in elderly individuals the rise is even smaller. Both at rest and at any given level of exercise, trained athletes have a larger stroke volume and lower heart rate than untrained individuals and they tend to have larger hearts. Training increases the maximal oxygen consumption (VO_{2max}) that can be produced by exercise in an individual. VO_{2max} averages about 38 mL/kg/min in active healthy men and about 29 mL/kg/min in active healthy women. It is lower in sedentary individuals. VO_{2max} is the product of maximal cardiac output and maximal O_2 extraction by the tissues, and both increase with training.

A great increase in venous return also takes place with exercise, although the increase in venous return is not the primary cause of the increase in cardiac output. Venous return is increased by the activity of the muscle and thoracic pumps; by mobilization of blood from the viscera; by increased pressure transmitted through the dilated arterioles to the veins; and by noradrenergically mediated venoconstriction, which decreases the volume of blood in the veins. Blood mobilized from the splanchnic area and other reservoirs may increase the amount of blood in the arterial portion of the circulation by as much as 30% during strenuous exercise. After exercise, the blood pressure may transiently drop to subnormal levels, presumably because accumulated metabolites keep the muscle vessels dilated for a short period. However, the blood pressure soon returns to the preexercise level. The heart rate returns to normal more slowly.

in stroke volume. However, for reasons that are incompletely understood, pressure work produces a greater increase in O_2 consumption than volume work. In other words, an increase in afterload causes a greater increase in cardiac O_2 consumption than does an increase in preload. This is why angina pectoris due to deficient delivery of O_2 to the myocardium is more common in aortic stenosis than in aortic insufficiency. In aortic stenosis, intraventricular pressure must be increased to force blood through the stenotic valve, whereas in aortic insufficiency, regurgitation of blood produces an increase in stroke volume with little change in aortic impedance.

It is worth noting that the increase in O_2 consumption produced by increased stroke volume when the myocardial fibers are stretched is an example of the operation of the law of Laplace. This law, which is discussed in detail in Chapter 31, states that the tension developed in the wall of a hollow viscus is proportional to the radius of the viscus. When the heart is dilated, its radius is increased. O_2 consumption per unit time

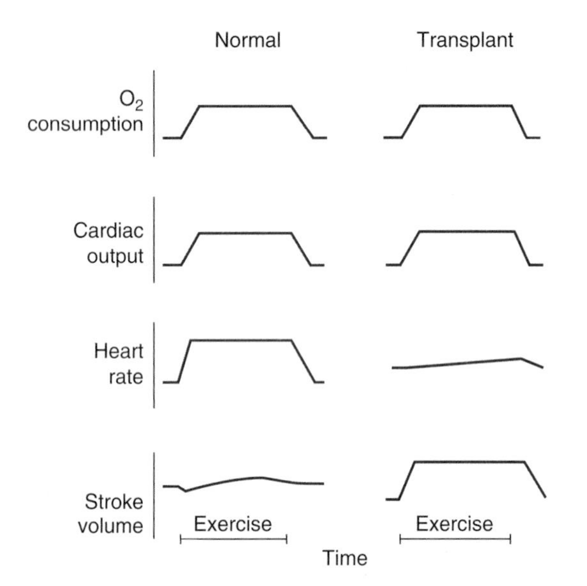

FIGURE 30–9 Cardiac responses to moderate supine exercise in normal humans and patients with transplanted and hence denervated hearts. Note that the transplanted heart, without the benefit of neural input, relies primarily on an increase in stroke volume rather than heart rate to raise cardiac output in the setting of exercise. (Reproduced with permission from Kent KM, Cooper T: The denervated heart. N Engl J Med 1974;291:1017.)

increases when the heart rate is increased by sympathetic stimulation because of the increased number of beats and the increased velocity and strength of each contraction. However, this is somewhat offset by the decrease in end-systolic volume and hence in the radius of the heart.

CHAPTER SUMMARY

- Blood flows into the atria and then the ventricles of the heart during diastole and atrial systole, and is ejected during systole when the ventricles contract and pressure exceeds the pressures in the pulmonary artery and aorta.
- Careful timing of the opening and closing of the atrioventricular (AV), pulmonary, and aortic valves allows blood to move in an appropriate direction through the heart with minimal regurgitation.
- The proportion of blood leaving the ventricles in each cardiac cycle is called the ejection fraction and is a sensitive indicator of cardiac health.
- The arterial pulse represents a pressure wave set up when blood is forced into the aorta; it travels much faster than the blood itself.
- Heart sounds reflect the normal vibrations set up by abrupt valve closures; heart murmurs can arise from abnormal flow often (although not exclusively) caused by diseased valves.
- Changes in cardiac output reflect variations in heart rate, stroke volume, or both; these are controlled, in turn, by neural and hormonal input to cardiac myocytes.
- Cardiac output is strikingly increased during exercise.
- In heart failure, the ejection fraction of the heart is reduced due to impaired contractility in systole or reduced filling during diastole; this results in inadequate blood supplies to meet the body's needs. Initially, this is manifested only during exercise, but eventually the heart will not be able to supply sufficient blood flow even at rest.

MULTIPLE-CHOICE QUESTIONS

For all questions, select the single best answer unless otherwise directed.

1. The second heart sound is caused by
   A. closure of the aortic and pulmonary valves.
   B. vibrations in the ventricular wall during systole.
   C. ventricular filling.
   D. closure of the mitral and tricuspid valves.
   E. retrograde flow in the vena cava.
2. The fourth heart sound is caused by
   A. closure of the aortic and pulmonary valves.
   B. vibrations in the ventricular wall during systole.
   C. ventricular filling.
   D. closure of the mitral and tricuspid valves.
   E. retrograde flow in the vena cava.
3. The dicrotic notch on the aortic pressure curve is caused by
   A. closure of the mitral valve.
   B. closure of the tricuspid valve.
   C. closure of the aortic valve.
   D. closure of the pulmonary valve.
   E. rapid filling of the left ventricle.
4. During exercise, a man consumes 1.8 L of oxygen per minute. His arterial O_2 content is 190 mL/L, and the O_2 content of his mixed venous blood is 134 mL/L. His cardiac output is approximately
   A. 3.2 L/min.
   B. 16 L/min.
   C. 32 L/min.
   D. 54 L/min.
   E. 160 mL/min.
5. The work performed by the left ventricle is substantially greater than that performed by the right ventricle, because in the left ventricle
   A. the contraction is slower.
   B. the wall is thicker.
   C. the stroke volume is greater.
   D. the preload is greater.
   E. the afterload is greater.
6. Starling's law of the heart
   A. does not operate in the failing heart.
   B. does not operate during exercise.
   C. explains the increase in heart rate produced by exercise.
   D. explains the increase in cardiac output that occurs when venous return is increased.
   E. explains the increase in cardiac output when the sympathetic nerves supplying the heart are stimulated.

CHAPTER RESOURCES

Hunter JD, Doddi M: Sepsis and the heart. Br J Anaesth 2010;104:3.

Leach JK, Priola DV, Grimes LA, Skipper BJ: Shortening deactivation of cardiac muscle: Physiological mechanisms and clinical implications. J Investig Med 1999;47:369.

Overgaard CB, Dzavik V: Inotropes and vasopressors: Review of physiology and clinical use in cardiovascular disease. Circulation 2008;118:1047.

Wang J, Nagueh SF: Current perspectives on cardiac function in patients with diastolic heart failure. Circulation 2009;119:1146.

Blood as a Circulatory Fluid & the Dynamics of Blood & Lymph Flow

CHAPTER

31

OBJECTIVES

After studying this chapter, you should be able to:

- Describe the components of blood and lymph, their origins, and the role of hemoglobin in transporting oxygen in red blood cells.
- Understand the molecular basis of blood groups and the reasons for transfusion reactions.
- Delineate the process of hemostasis that restricts blood loss when vessels are damaged, and the adverse consequences of intravascular thrombosis.
- Identify the types of blood and lymphatic vessels that make up the circulatory system and the regulation and function of their primary constituent cell types.
- Describe how physical principles dictate the flow of blood and lymph around the body.
- Understand the basis of methods used to measure blood flow and blood pressure in various vascular segments.
- Understand the basis of disease states where components of the blood and vasculature are abnormal, dysregulated, or both.

INTRODUCTION

The **circulatory system** supplies inspired O_2 as well as substances absorbed from the gastrointestinal tract to the tissues, returns CO_2 to the lungs and other products of metabolism to the kidneys, functions in the regulation of body temperature, and distributes hormones and other agents that regulate cell function. The blood, the carrier of these substances, is pumped through a closed system of blood vessels by the heart. From the left ventricle, blood is pumped through the arteries and arterioles to the capillaries, where it equilibrates with the interstitial fluid. The capillaries drain through venules into the veins and back to the right atrium. Some tissue fluids enter another system of closed vessels, the lymphatics, which drain lymph via the thoracic duct and the right lymphatic duct into the venous system. The circulation is controlled by multiple regulatory systems that function in general to maintain adequate capillary blood flow when possible in all organs, but particularly in the heart and brain.

Blood flows through the circulation primarily because of the forward motion imparted to it by the pumping of the heart, although in the case of the systemic circulation, diastolic recoil of the walls of the arteries, compression of the veins by skeletal muscles during exercise, and the negative pressure in the thorax during inspiration also move the blood forward. The resistance to flow depends to a minor degree on the viscosity of the blood but mostly on the diameter of the vessels, and principally that of the arterioles. The blood flow to each tissue is regulated by local chemical and general neural and humoral mechanisms that dilate or constrict its vessels. All of the blood flows through the lungs, but the systemic circulation is made up of numerous different circuits in parallel (Figure 31–1). The arrangement permits wide variations in regional blood flow without changing total systemic flow.

This chapter is concerned with blood and lymph and with the multiple functions of the cells they contain. It will also address general principles that apply to all parts of the circulation and pressure and flow in the systemic circulation. The homeostatic mechanisms operating to adjust flow are the subject of Chapter 32. The special characteristics of pulmonary and renal circulation are discussed in Chapters 34 and 37. Likewise, the role of blood as the carrier of many immune effector cells will not be discussed here, but rather was covered in Chapter 3.

BLOOD AS A CIRCULATORY FLUID

Blood consists of a protein-rich fluid known as plasma, in which are suspended cellular elements: white blood cells, red blood cells, and platelets. The normal total circulating blood volume is about 8% of the body weight (5600 mL in a 70-kg man). About 55% of this volume is plasma.

BONE MARROW

In the adult, red blood cells, many white blood cells, and platelets are formed in the bone marrow. In the fetus, blood cells are also formed in the liver and spleen, and in adults such **extramedullary hematopoiesis** may occur in diseases in which the bone marrow becomes destroyed or fibrosed. In children, blood cells are actively produced in the marrow cavities of all the bones. By age 20, the marrow in the cavities of the long bones, except for the upper humerus and femur, has become inactive (Figure 31–2). Active cellular marrow is called **red marrow;** inactive marrow that is infiltrated with fat is called **yellow marrow.**

The bone marrow is actually one of the largest organs in the body, approaching the size and weight of the liver. It is also one of the most active. Normally, 75% of the cells in the marrow belong to the white blood cell-producing myeloid series and only 25% are maturing red cells, even though there are over 500 times as many red cells in the circulation as there are white cells. This difference in the marrow reflects the fact that the average life span of white cells is short, whereas that of red cells is long.

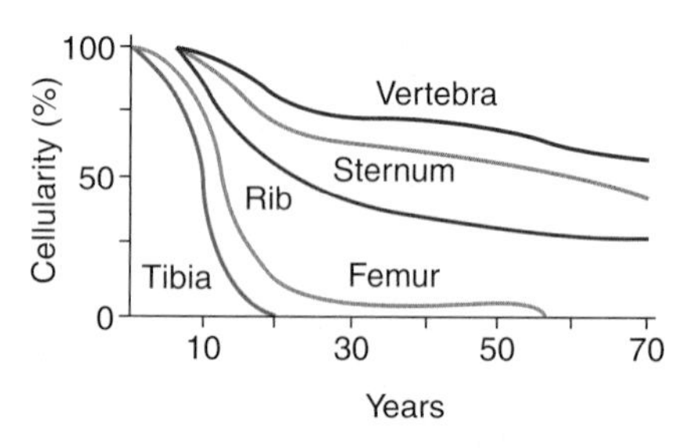

FIGURE 31–2 Changes in red bone marrow cellularity in various bones with age. Hundred per cent equals the degree of cellularity at birth. (Reproduced with permission from Whitby LEH, Britton CJC: *Disorders of the Blood*, 10th ed. Churchill Livingstone, 1969.)

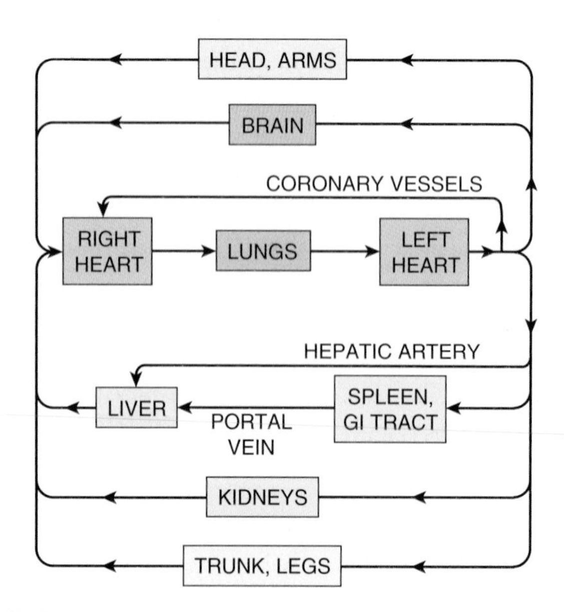

FIGURE 31–1 Diagram of the circulation in the adult.

Hematopoietic stem cells (HSCs) are bone marrow cells that are capable of producing all types of blood cells. They differentiate into one or another type of committed stem cells **(progenitor cells).** These in turn form the various differentiated types of blood cells. There are separate pools of progenitor cells for megakaryocytes, lymphocytes, erythrocytes, eosinophils, and basophils; neutrophils and monocytes arise from a common precursor. The bone marrow stem cells are also the source of osteoclasts (see Chapter 21), Kupffer cells (see Chapter 28), mast cells, dendritic cells, and Langerhans cells. The HSCs are few in number but are capable of completely replacing the bone marrow when injected into a host whose own bone marrow has been entirely destroyed.

The HSCs are derived from uncommitted, totipotent stem cells that can be stimulated to form any cell in the body. Adults have a few of these, but they are more readily obtained from the blastocysts of embryos. There is not surprisingly immense interest in stem cell research due to its potential to regenerate diseased tissues, but ethical issues are involved, and debate on these issues will undoubtedly continue.

WHITE BLOOD CELLS

Normally, human blood contains 4000–11,000 white blood cells per microliter (Table 31–1). Of these, the **granulocytes (polymorphonuclear leukocytes, PMNs)** are the most numerous. Young granulocytes have horseshoe-shaped nuclei that become multilobed as the cells grow older (Figure 31–3).

TABLE 31–1 Normal values for the cellular elements in human blood.

Cell	Cells/μL (average)	Approximate Normal Range	Percentage of Total White Cells
Total white blood cells	9000	4000–11,000	...
Granulocytes			
Neutrophils	5400	3000–6000	50–70
Eosinophils	275	150–300	1–4
Basophils	35	0–100	0.4
Lymphocytes	2750	1500–4000	20–40
Monocytes	540	300–600	2–8
Erythrocytes			
Females	4.8×10^6	...	...
Males	5.4×10^6	...	...
Platelets	300,000	200,000–500,000	...

Most of them contain neutrophilic granules **(neutrophils)**, but a few contain granules that stain with acidic dyes **(eosinophils)**, and some have basophilic granules **(basophils)**. The other two cell types found normally in peripheral blood are **lymphocytes,** which have large round nuclei and scanty cytoplasm, and **monocytes,** which have abundant agranular cytoplasm and kidney-shaped nuclei (Figure 31–3). Acting together, these cells provide the body with the powerful defenses against tumors and viral, bacterial, and parasitic infections that were discussed in Chapter 3.

PLATELETS

Platelets are small, granulated bodies that aggregate at sites of vascular injury. They lack nuclei and are 2–4 μm in diameter (Figure 31–3). There are about 300,000/μL of circulating blood, and they normally have a half-life of about 4 days. The **megakaryocytes,** giant cells in the bone marrow, form platelets by pinching off bits of cytoplasm and extruding them into the circulation. Between 60 and 75% of the platelets that have been extruded from the bone marrow are in the circulating blood, and the remainder are mostly in the spleen. Splenectomy causes an increase in the platelet count **(thrombocytosis).**

RED BLOOD CELLS

The red blood cells **(erythrocytes)** carry hemoglobin in the circulation. They are biconcave disks (Figure 31–4) that are manufactured in the bone marrow. In mammals, they lose their nuclei before entering the circulation. In humans, they survive in the circulation for an average of 120 days. The average normal red blood cell count is 5.4 million/μL in men and 4.8 million/μL in women. The number of red cells is also conveniently expressed as the **hematocrit,** or the percentage of the blood, by volume, that is occupied by erythrocytes. Each human red blood cell is about 7.5 μm in diameter and 2 μm thick, and each contains approximately 29 pg of hemoglobin (Table 31–2). There are thus about 3×10^{13} red blood cells and about 900 g of hemoglobin in the circulating blood of an adult man (Figure 31–5).

The feedback control of erythropoiesis by erythropoietin is discussed in Chapter 38, and the role of IL-1, IL-3, IL-6 (interleukin), and GM-CSF (granulocyte-macrophage colony-stimulating factor) in development of the relevant erythroid stem cells is shown in Figure 31–3.

ROLE OF THE SPLEEN

The spleen is an important blood filter that removes aged or abnormal red cells. It also contains many platelets and plays a significant role in the immune system. Abnormal red cells are removed if they are not as flexible as normal red cells and consequently are unable to squeeze through the slits between the endothelial cells that line the splenic sinuses (see Clinical Box 31–1).

HEMOGLOBIN

The red, oxygen-carrying pigment in the red blood cells of vertebrates is **hemoglobin,** a protein with a molecular weight of 64,450. Hemoglobin is a globular molecule made up of four subunits (Figure 31–6). Each subunit contains a **heme** moiety conjugated to a polypeptide. Heme is an iron-containing porphyrin derivative (Figure 31–7). The polypeptides are referred to collectively as the **globin** portion of the hemoglobin molecule. There are two pairs of polypeptides in each hemoglobin molecule. In normal adult human hemoglobin **(hemoglobin A)**, the two polypeptides are called α chains, and β chains. Thus, hemoglobin A is designated $\alpha_2\beta_2$. Not all the hemoglobin in the blood of normal adults is hemoglobin A. About 2.5% of the hemoglobin is hemoglobin A_2, in which β chains are replaced by δ chains ($\alpha_2\delta_2$). The δ chains contain 10 individual amino acid residues that differ from those in β chains.

There are small amounts of hemoglobin A derivatives closely associated with hemoglobin A that represent glycated hemoglobins. One of these, hemoglobin A_{1c} (HbA_{1c}), has a glucose attached to the terminal valine in each β chain and is of special interest because it increases in the blood of patients with poorly controlled diabetes mellitus (see Chapter 24), and is measured clinically as a marker of the progression of that disease and/or the effectiveness of treatment.

REACTIONS OF HEMOGLOBIN

O_2 binds to the Fe^{2+} in the heme moiety of hemoglobin to form **oxyhemoglobin.** The affinity of hemoglobin for O_2 is affected by pH, temperature, and the concentration in the red cells of 2,3-bisphosphoglycerate (2,3-BPG). 2,3-BPG and H^+

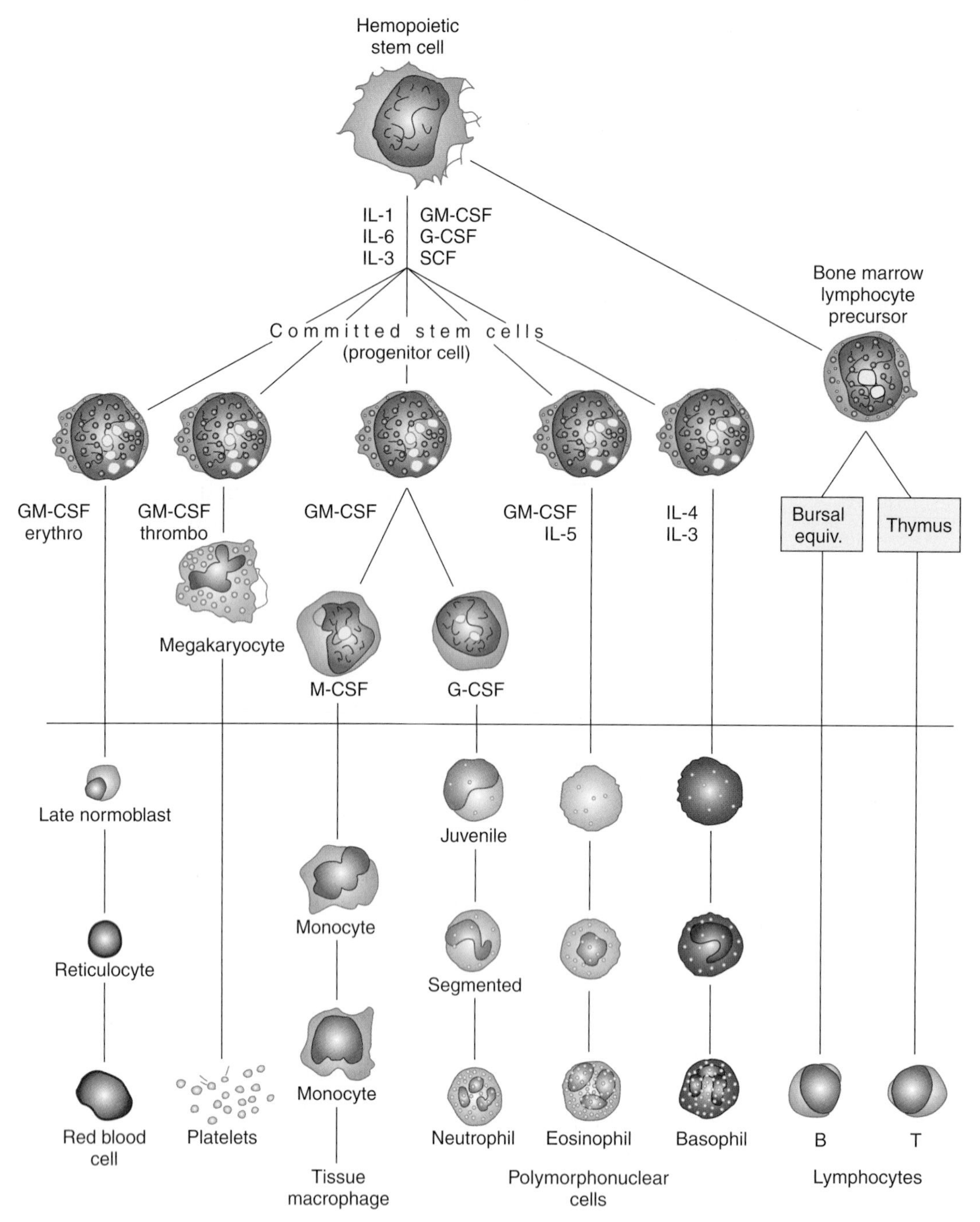

FIGURE 31–3 Development of various formed elements of the blood from bone marrow cells. Cells below the horizontal line are found in normal peripheral blood. The principal sites of action of erythropoietin (erythro) and the various colony-stimulating factors (CSF) that stimulate the differentiation of the components are indicated. G, granulocyte; M, macrophage; IL, interleukin; thrombo, thrombopoietin; erythro, erythropoietin; SCF, stem cell factor.

compete with O_2 for binding to deoxygenated hemoglobin, decreasing the affinity of hemoglobin for O_2 by shifting the positions of the four peptide chains (quaternary structure). The details of the oxygenation and deoxygenation of hemoglobin and the physiologic role of these reactions in O_2 transport are discussed in Chapter 35.

When blood is exposed to various drugs and other oxidizing agents in vitro or in vivo, the ferrous iron (Fe^{2+}) that is normally present in hemoglobin is converted to ferric iron (Fe^{3+}), forming **methemoglobin.** Methemoglobin is dark-colored, and when it is present in large quantities in the circulation, it causes a dusky discoloration of the skin resembling cyanosis (see Chapter 35). Some oxidation of hemoglobin to methemoglobin occurs normally, but an enzyme system in the red cells, the dihydronicotinamide adenine dinucleotide (NADH)-methemoglobin reductase system, converts

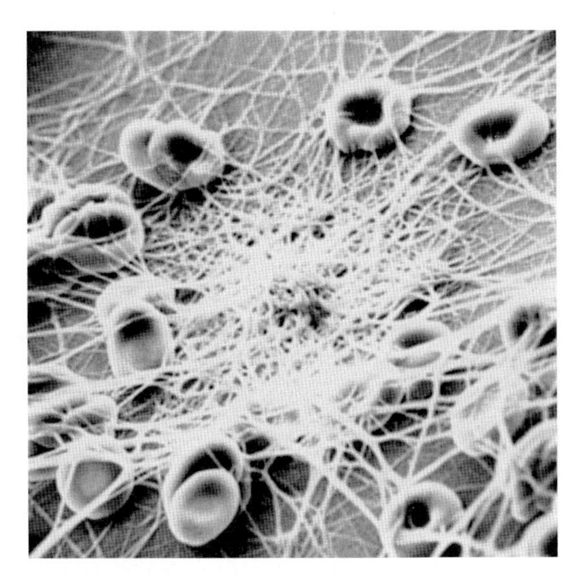

FIGURE 31–4 Human red blood cells and fibrin fibrils. Blood was placed on a polyvinyl chloride surface, fixed, and photographed with a scanning electron microscope. Reduced from ×2590. (Courtesy of NF Rodman.)

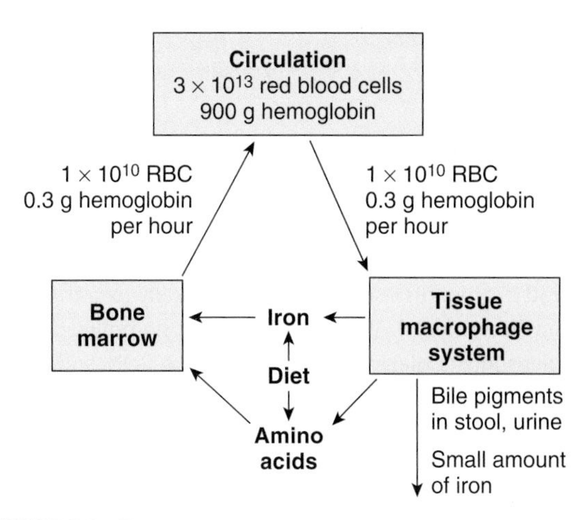

FIGURE 31–5 Red cell formation and destruction. RBC, red blood cells.

methemoglobin back to hemoglobin. Congenital absence of this system is one cause of hereditary methemoglobinemia.

Carbon monoxide reacts with hemoglobin to form **carbon monoxyhemoglobin (carboxyhemoglobin).** The affinity of hemoglobin for O_2 is much lower than its affinity for carbon monoxide, which consequently displaces O_2 on hemoglobin, reducing the oxygen-carrying capacity of blood (see Chapter 35).

TABLE 31–2 Characteristics of human red cells.[a]

		Male	Female
Hematocrit (Hct) (%)		47	42
Red blood cells (RBC) ($10^6/\mu L$)		5.4	4.8
Hemoglobin (Hb) (g/dL)		16	14
Mean corpuscular volume (MCV) (fL)	$= \frac{\text{Hct} \times 10}{\text{RBC } (10^6/\mu L)}$	87	87
Mean corpuscular hemoglobin (MCH) (pg)	$= \frac{\text{Hb} \times 10}{\text{RBC } (10^6/\mu L)}$	29	29
Mean corpuscular hemoglobin concentration (MCHC) (g/dL)	$= \frac{\text{Hb} \times 100}{\text{Hct}}$	34	34
Mean cell diameter (MCD) μm)	= Mean diameter of 500 cells in smear	7.5	7.5

[a]Cells with MCVs > 95 fL are called macrocytes; cells with MCVs < 80 fL are called microcytes; cells with MCHCs < 25 g/dL are called hypochromic.

HEMOGLOBIN IN THE FETUS

The blood of the human fetus normally contains **fetal hemoglobin (hemoglobin F).** Its structure is similar to that of hemoglobin A except that the β chains are replaced by γ chains; that is, hemoglobin F is $\alpha_2\gamma_2$. The γ chains have 37 amino acid residues that differ from those in the β chain. Fetal hemoglobin is normally replaced by adult hemoglobin soon after birth (Figure 31–8). In certain individuals, it fails to disappear and persists throughout life. In the body, its O_2 content at a given PO_2 is greater than that of adult hemoglobin because it binds 2,3-BPG less avidly. Hemoglobin F is therefore critical to facilitate movement of O_2 from the maternal to the fetal circulation, particularly at later stages of gestation where oxygen demand increases (see Chapter 33). In young embryos there are, in addition, ζ and ε chains, forming Gower 1 hemoglobin ($\zeta_2\varepsilon_2$) and Gower 2 hemoglobin ($\alpha_2\varepsilon_2$). Switching from one form of hemoglobin to another during development seems to be regulated largely by oxygen availability, with relative hypoxia favoring the production of hemoglobin F both via direct effects on globin gene expression, as well as up-regulated production of erythropoietin.

SYNTHESIS OF HEMOGLOBIN

The average normal hemoglobin content of blood is 16 g/dL in men and 14 g/dL in women, all of it in red cells. In the body of a 70-kg man, there are about 900 g of hemoglobin, and 0.3 g of hemoglobin is destroyed and 0.3 g synthesized every hour (Figure 31–5). The heme portion of the hemoglobin molecule is synthesized from glycine and succinyl-CoA (see Clinical Box 31–2).

CATABOLISM OF HEMOGLOBIN

When old red blood cells are destroyed by tissue macrophages, the globin portion of the hemoglobin molecule is split off, and the heme is converted to **biliverdin.** The enzyme involved is a

CLINICAL BOX 31–1

Red Cell Fragility

Red blood cells, like other cells, shrink in solutions with an osmotic pressure greater than that of normal plasma. In solutions with a lower osmotic pressure they swell, become spherical rather than disk-shaped, and eventually lose their hemoglobin **(hemolysis).** The hemoglobin of hemolyzed red cells dissolves in the plasma, coloring it red. A 0.9% sodium chloride solution is isotonic with plasma. When **osmotic fragility** is normal, red cells begin to hemolyze when suspended in 0.5% saline; 50% lysis occurs in 0.40–0.42% saline, and lysis is complete in 0.35% saline. In **hereditary spherocytosis** (congenital hemolytic icterus), the cells are spherocytic in normal plasma and hemolyze more readily than normal cells in hypotonic sodium chloride solutions. Abnormal spherocytes are also trapped and destroyed in the spleen, meaning that hereditary spherocytosis is one of the most common causes of **hereditary hemolytic anemia.** The spherocytosis is caused by mutations in proteins that make up the membrane skeleton of the erythrocyte, which normally maintain the shape and flexibility of the red cell membrane, including **spectrin,** the transmembrane protein band 3, and the linker protein, **ankyrin.** Red cells can also be lysed by drugs (especially penicillin and sulfa drugs) and infections. The susceptibility of red cells to hemolysis by these agents is increased by deficiency of the enzyme glucose 6-phosphate dehydrogenase (G6PD), which catalyzes the initial step in the oxidation of glucose via the hexose mono-phosphate pathway (see Chapter 1). This pathway generates dihydronicotinamide adenine dinucleotide phosphate (NADPH), which is needed for the maintenance of normal red cell fragility. Severe G6PD deficiency also inhibits the killing of bacteria by granulocytes and predisposes to severe infections.

THERAPEUTIC HIGHLIGHTS

Severe cases of hereditary spherocytosis can be treated by splenectomy, but this is not without other risks, such as sepsis. Milder cases can be treated with dietary folate supplementation and/or blood transfusions. Treatment of other forms of hemolytic anemia depends on the underlying cause. Some forms are autoimmune in nature, and have been shown to benefit from treatment with corticosteroids.

subtype of heme oxygenase (see Figure 28–4), and CO is formed in the process. CO is an intercellular messenger, like NO (see Chapters 2 and 3). In humans, most of the biliverdin is converted to **bilirubin** and excreted in the bile (see Chapter 28). The iron from the heme is reused for hemoglobin synthesis.

β β α α $^{\oplus}NH_3$ $COO^{\ominus}$ 1 nm

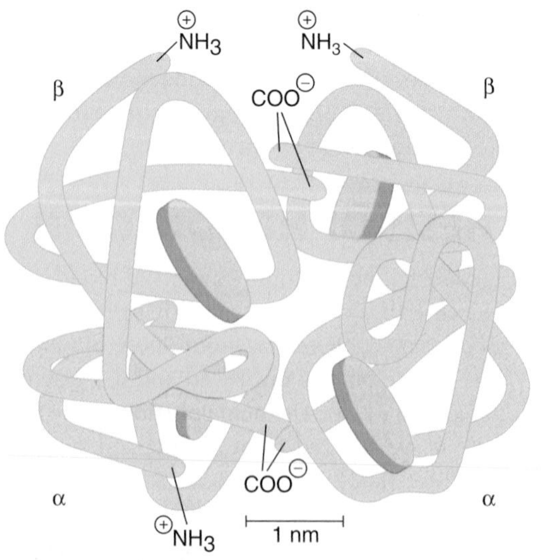

FIGURE 31–6 Diagrammatic representation of a molecule of hemoglobin A, showing the four subunits. There are two α and two β polypeptide chains, each containing a heme moiety. These moieties are represented by the blue disks. (Reproduced with permission from Harper HA, et al: *Physiologische Chemie*. Springer, 1975.)

Exposure of the skin to white light converts bilirubin to lumirubin, which has a shorter half-life than bilirubin. **Phototherapy** (exposure to light) is of value in treating infants with jaundice due to hemolysis. Iron is essential for hemoglobin synthesis; if blood is lost from the body and the iron deficiency is not corrected, **iron deficiency anemia** results.

BLOOD TYPES

The membranes of human red cells contain a variety of **blood group antigens**, which are also called **agglutinogens.** The most important and best known of these are the A and B antigens, but there are many more.

THE ABO SYSTEM

The A and B antigens are inherited as mendelian dominants, and individuals are divided into four major **blood types** on this basis. Type A individuals have the A antigen, type B have the B, type AB have both, and type O have neither. The A and B antigens are complex oligosaccharides that differ in their terminal sugar. An *H* gene codes for a fucose transferase that adds a terminal fucose, forming the H antigen that is usually present in individuals of all blood types (Figure 31–9). Individuals who are type A also express a second transferase that catalyzes placement of a terminal N-acetylgalactosamine on the H antigen, whereas individuals who are type B express a transferase that places a terminal galactose. Individuals who

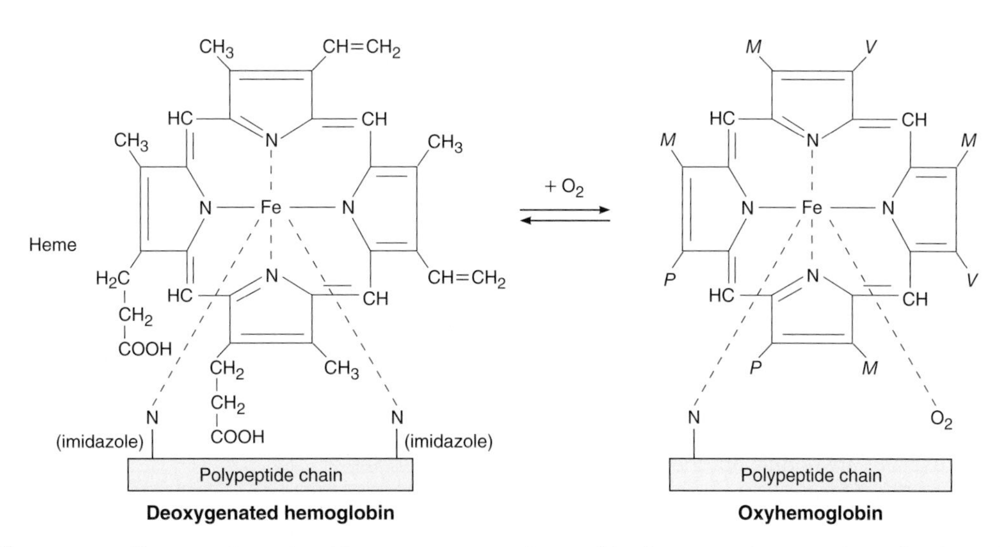

FIGURE 31–7 Reaction of heme with O_2. The abbreviations M, V, and P stand for the groups shown on the molecule on the left.

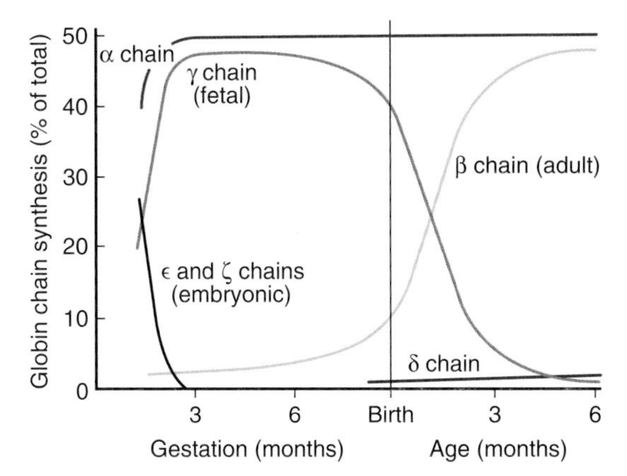

FIGURE 31–8 Development of human hemoglobin chains. The graph shows the normal rate of synthesis of the various hemoglobin chains in utero, and how these change after birth.

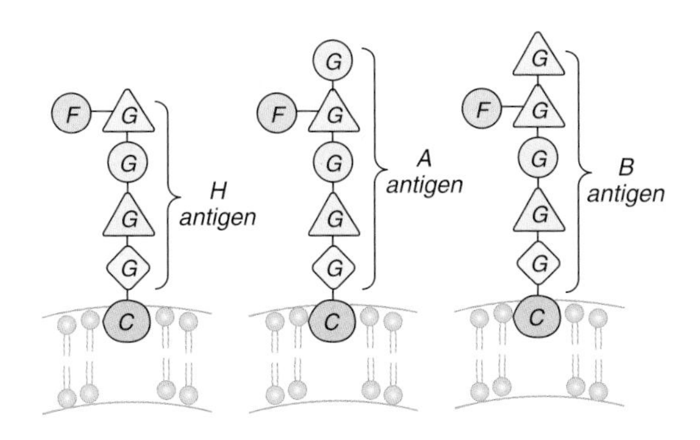

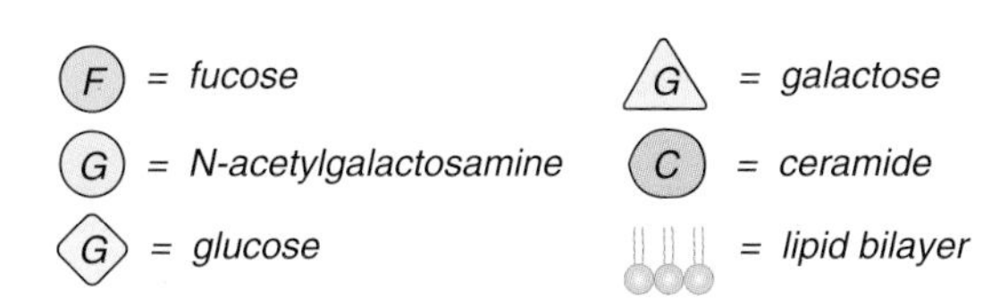

FIGURE 31–9 Antigens of the ABO system on the surface of red blood cells.

are type AB have both transferases. Individuals who are type O have neither, so the H antigen persists.

Antibodies against red cell agglutinogens are called **agglutinins.** Antigens very similar to A and B are common in intestinal bacteria and possibly in foods to which newborn individuals are exposed. Therefore, infants rapidly develop antibodies against the antigens not present in their own cells. Thus, type A individuals develop anti-B antibodies, type B individuals develop anti-A antibodies, type O individuals develop both, and type AB individuals develop neither (Table 31–3). When the plasma of a type A individual is mixed with type B red cells, the anti-B antibodies cause the type B red cells to clump (agglutinate), as shown in Figure 31–10. The other agglutination reactions produced by mismatched plasma and red cells are summarized in Table 31–3. ABO **blood typing** is performed by mixing an individual's red blood cells with antisera containing the various agglutinins on a slide and seeing whether agglutination occurs.

TRANSFUSION REACTIONS

Dangerous **hemolytic transfusion reactions** occur when blood is transfused into an individual with an incompatible blood type; that is, an individual who has agglutinins against

TABLE 31–3 Summary of ABO system.

Blood Type	Agglutinins in Plasma	Frequency in United States %	Plasma Agglutinates Red Cells of Type:
O	Anti-A, anti-B	45	A, B, AB
A	Anti-B	41	B, AB
B	Anti-A	10	A, AB
AB	None	4	None

CLINICAL BOX 31–2

Abnormalities of Hemoglobin Production

There are two major types of inherited disorders of hemoglobin in humans: the **hemoglobinopathies,** in which abnormal globin polypeptide chains are produced, and the **thalassemias** and related disorders, in which the chains are normal in structure but produced in decreased amounts or absent because of defects in the regulatory portion of the globin genes. Mutant genes that cause the production of abnormal hemoglobins are widespread, and over 1000 abnormal hemoglobins have been described in humans. In one of the most common examples, hemoglobin S, the α chains are normal but the β chains have a single substitution of a valine residue for one glutamic acid, leading to **sickle cell anemia** (Table 31–4). When an abnormal gene inherited from one parent dictates formation of an abnormal hemoglobin (ie, when the individual is heterozygous), half the circulating hemoglobin is abnormal and half is normal. When identical abnormal genes are inherited from both parents, the individual is homozygous and all the hemoglobin is abnormal. It is theoretically possible to inherit two different abnormal hemoglobins, one from the father and one from the mother. Studies of the inheritance and geographic distribution of abnormal hemoglobins have made it possible in some cases to decide where the mutant gene originated and approximately how long ago the mutation occurred. In general, harmful mutations tend to die out, but mutant genes that confer traits with survival value persist and spread in the population. Many of the abnormal hemoglobins are harmless; however, some have abnormal O_2 equilibriums, while others cause anemia. For example, hemoglobin S polymerizes at low O_2 tensions, and this causes the red cells to become sickle-shaped, hemolyze, and form aggregates that block blood vessels. The sickle cell gene is an example of a gene that has persisted and spread in the population due to its beneficial effect when present in heterozygous form. It originated in Africa, and confers resistance to one type of malaria. In some parts of Africa, 40% of the population is heterozygous for hemoglobin S. There is a corresponding prevalence of 10% among African Americans in the United States.

THERAPEUTIC HIGHLIGHTS

Hemoglobin F decreases the polymerization of deoxygenated hemoglobin S, and hydroxyurea stimulates production of hemoglobin F in children and adults. Hydroxyurea has therefore proven to be a very valuable agent for the treatment of sickle cell disease. In patients with severe sickle cell disease, bone marrow transplantation has also been shown to have some benefit, and prophylactic treatment with antibiotics has also been shown to be helpful. The clinically important thalassemias result in severe anemia, often requiring repeated blood transfusions. However, these run the risk of iron overload, and often must be accompanied by treatment with drugs that chelate iron. Bone marrow transplantation is also being explored for treatment of the thalassemias.

the red cells in the transfusion. The plasma in the transfusion is usually so diluted in the recipient that it rarely causes agglutination even when the titer of agglutinins against the recipient's cells is high. However, when the recipient's plasma has agglutinins against the donor's red cells, the cells agglutinate and hemolyze. Free hemoglobin is liberated into the plasma. The severity of the resulting transfusion reaction may vary from an asymptomatic minor rise in the plasma bilirubin level to severe jaundice and renal tubular damage leading to anuria and death.

Incompatibilities in the ABO blood group system are summarized in Table 31–3. Persons with type AB blood are "universal recipients" because they have no circulating agglutinins and can be given blood of any type without developing a transfusion reaction due to ABO incompatibility. Type O individuals are "universal donors" because they lack A and B antigens, and type O blood can be given to anyone without producing a transfusion reaction due to ABO incompatibility. This does not mean, however, that blood should ever be transfused without being cross-matched except in the most extreme emergencies, since the possibility of reactions or sensitization due to incompatibilities in systems other than ABO systems always exists. In cross-matching, donor red cells are mixed with recipient plasma on a slide and checked for agglutination. It is advisable to check the action of the donor's plasma on the recipient cells in addition, even though, as noted above, this is rarely a source of trouble.

A procedure that has recently become popular is to withdraw the patient's own blood in advance of elective surgery and then infuse this blood back **(autologous transfusion)** if a transfusion is needed during the surgery. With iron treatment, 1000–1500 mL can be withdrawn over a 3-weeks period. The popularity of banking one's own blood is primarily due to fear of transmission of infectious diseases by heterologous transfusions, but of course another advantage is elimination of the risk of transfusion reactions.

INHERITANCE OF A & B ANTIGENS

The A and B antigens are inherited as mendelian allelomorphs, A and B being dominants. For example, an individual with type B blood may have inherited a B antigen from each parent

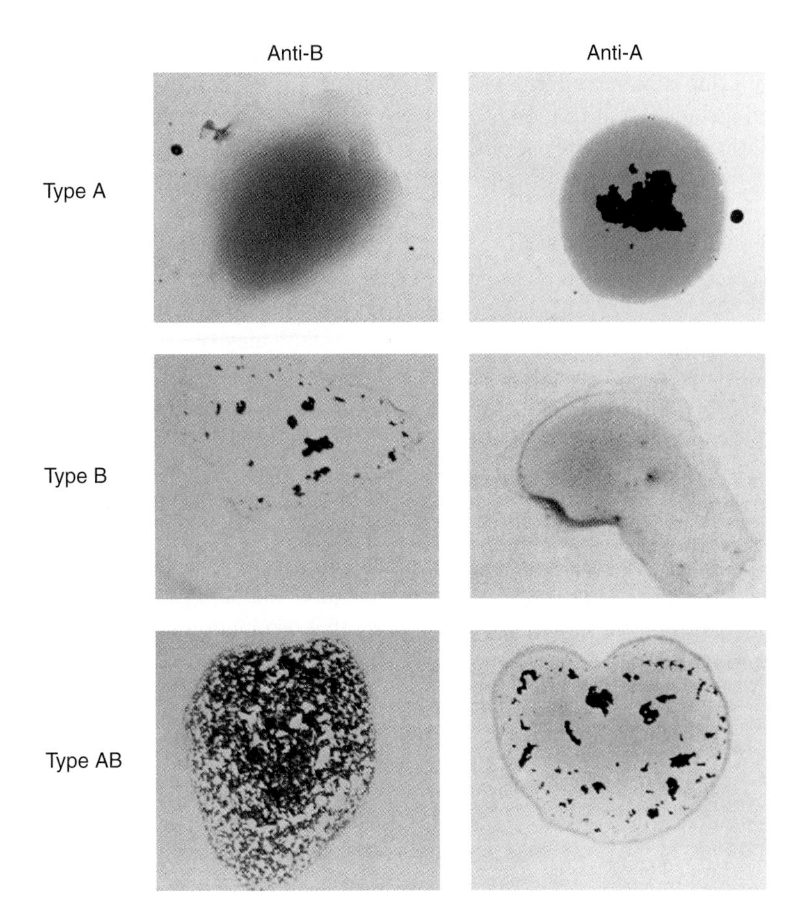

FIGURE 31–10 Red cell agglutination in incompatible plasma.

or a B antigen from one parent and an O from the other; thus, an individual whose **phenotype** is B may have the **genotype** BB **(homozygous)** or BO **(heterozygous).**

When the blood types of the parents are known, the possible genotypes of their children can be stated. When both parents are type B, they could have children with genotype BB (B antigen from both parents), BO (B antigen from one parent, O from the other heterozygous parent), or OO (O antigen from both parents, both being heterozygous). When the blood types of a mother and her child are known, typing can prove

TABLE 30–4 Partial amino acid composition of normal human β chain, and some hemoglobins with abnormal β chains.[a]

Hemoglobin	Positions on Polypeptide Chain of Hemoglobin						
	1 2 3	6 7	26	63	67	121	146
A (normal)	Val-His-Leu	Glu-Glu	Glu	His	Val	Glu	His
S (sickle cell)		Val					
C		Lys					
$G_{San\ Jose}$		Gly					
E			Lys				
$M_{Saskatoon}$				Tyr			
$M_{Milwaukee}$					Glu		
O_{Arabia}						Lys	

[a]Other hemoglobins have abnormal α chains. Abnormal hemoglobins that are very similar electrophoretically but differ slightly in composition are indicated by the same letter and a subscript indicating the geographic location where they were first discovered; hence, $M_{Saskatoon}$ and $M_{Milwaukee}$.

that a man cannot be the father, although it cannot prove that he is the father. The predictive value is increased if the blood typing of the parties concerned includes identification of antigens other than the ABO agglutinogens. With the use of DNA fingerprinting (see Chapter 1), the exclusion rate for paternity rises to close to 100%.

OTHER AGGLUTINOGENS

In addition to the ABO system of antigens in human red cells, there are systems such as the Rh, MNSs, Lutheran, Kell, Kidd, and many others. There are over 500 billion possible known blood group phenotypes, and because undiscovered antigens undoubtedly exist, it has been calculated that the number of phenotypes is actually in the trillions.

The number of blood groups in animals is as large as it is in humans. An interesting question is why this degree of polymorphism developed and has persisted through evolution. Certain diseases are more common in individuals with one blood type or another, but the differences are not great. The significance of a recognition code of this complexity is therefore unknown.

THE RH GROUP

Aside from the antigens of the ABO system, those of the Rh system are of the greatest clinical importance. The Rh factor, named for the rhesus monkey because it was first studied using the blood of this animal, is a system composed primarily of the C, D, and E antigens, although it actually contains many more. Unlike the ABO antigens, the system has not been detected in tissues other than red cells. D is by far the most antigenic component, and the term Rh-positive as it is generally used means that the individual has agglutinogen D. The D protein is not glycosylated, and its function is unknown. The Rh-negative individual has no D antigen and forms the anti-D agglutinin when injected with D-positive cells. The Rh typing serum used in routine blood typing is anti-D serum. Eighty-five per cent of Caucasians are D-positive and 15% are D-negative; over 99% of Asians are D-positive. Unlike the antibodies of the ABO system, anti-D antibodies do not develop without exposure of a D-negative individual to D-positive red cells by transfusion or entrance of fetal blood into the maternal circulation. However, D-negative individuals who have received a transfusion of D-positive blood (even years previously) can have appreciable anti-D titers and thus may develop transfusion reactions when transfused again with D-positive blood.

HEMOLYTIC DISEASE OF THE NEWBORN

Another complication due to Rh incompatibility arises when an Rh-negative mother carries an Rh-positive fetus. Small amounts of fetal blood leak into the maternal circulation at the time of delivery, and some mothers develop significant titers of anti-Rh agglutinins during the postpartum period. During the next pregnancy, the mother's agglutinins cross the placenta to the fetus. In addition, there are some cases of fetal–maternal hemorrhage during pregnancy, and sensitization can occur during pregnancy. In any case, when anti-Rh agglutinins cross the placenta to an Rh-positive fetus, they can cause hemolysis and various forms of **hemolytic disease of the newborn (erythroblastosis fetalis).** If hemolysis in the fetus is severe, the infant may die in utero or may develop anemia, severe jaundice, and edema **(hydrops fetalis). Kernicterus,** a neurologic syndrome in which unconjugated bilirubin is deposited in the basal ganglia, may also develop, especially if birth is complicated by a period of hypoxia. Bilirubin rarely penetrates the brain in adults, but it does in infants with erythroblastosis, possibly in part because the blood–brain barrier is more permeable in infancy. However, the main reasons that the concentration of unconjugated bilirubin is very high in this condition are that production is increased and the bilirubin-conjugating system is not yet mature.

About 50% of Rh-negative individuals are sensitized (develop an anti-Rh titer) by transfusion of Rh-positive blood. Because sensitization of Rh-negative mothers by carrying an Rh-positive fetus generally occurs at birth, the first child is usually normal. However, hemolytic disease occurs in about 17% of the Rh-positive fetuses born to Rh-negative mothers who have previously been pregnant one or more times with Rh-positive fetuses. Fortunately, it is usually possible to prevent sensitization from occurring the first time by administering a single dose of anti-Rh antibodies in the form of Rh immune globulin during the postpartum period. Such passive immunization does not harm the mother and has been demonstrated to prevent active antibody formation by the mother. In obstetric clinics, the institution of such treatment on a routine basis to unsensitized Rh-negative women who have delivered an Rh-positive baby has reduced the overall incidence of hemolytic disease by more than 90%. In addition, fetal Rh typing with material obtained by amniocentesis or chorionic villus sampling is now possible, and treatment with a small dose of Rh immune serum will prevent sensitization during pregnancy.

PLASMA

The fluid portion of the blood, the **plasma,** is a remarkable solution containing an immense number of ions, inorganic molecules, and organic molecules that are in transit to various parts of the body or aid in the transport of other substances. Normal plasma volume is about 5% of body weight, or roughly 3500 mL in a 70-kg man. Plasma clots on standing, remaining fluid only if an anticoagulant is added. If whole blood is allowed to clot and the clot is removed, the remaining fluid is called **serum.** Serum has essentially the same composition as plasma, except that its fibrinogen and clotting factors II, V, and VIII (Table 31–5) have been removed and it has a higher

TABLE 31–5 System for naming blood-clotting factors.

Factor[a]	Names
I	Fibrinogen
II	Prothrombin
III	Thromboplastin
IV	Calcium
V	Proaccelerin, labile factor, accelerator globulin
VII	Proconvertin, SPCA, stable factor
VIII	Antihemophilic factor (AHF), antihemophilic factor A, antihemophilic globulin (AHG)
IX	Plasma thromboplastic component (PTC), Christmas factor, antihemophilic factor B
X	Stuart–Prower factor
XI	Plasma thromboplastin antecedent (PTA), antihemophilic factor C
XII	Hageman factor, glass factor
XIII	Fibrin-stabilizing factor, Laki–Lorand factor
HMW-K	High-molecular-weight kininogen, Fitzgerald factor
Pre-Ka	Prekallikrein, Fletcher factor
Ka	Kallikrein
PL	Platelet phospholipid

[a]Factor VI is not a separate entity and has been dropped.

serotonin content because of the breakdown of platelets during clotting.

PLASMA PROTEINS

The plasma proteins consist of **albumin, globulin,** and **fibrinogen** fractions. Most capillary walls are relatively impermeable to the proteins in plasma, and the proteins therefore exert an osmotic force of about 25 mm Hg across the capillary wall (**oncotic pressure;** see Chapter 1) that pulls water into the blood. The plasma proteins are also responsible for 15% of the buffering capacity of the blood (see Chapter 39) because of the weak ionization of their substituent COOH and NH_2 groups. At the normal plasma pH of 7.40, the proteins are mostly in the anionic form (see Chapter 1). Plasma proteins may have specific functions (eg, antibodies and the proteins concerned with blood clotting), whereas others function as nonspecific carriers for various hormones, other solutes, and drugs.

ORIGIN OF PLASMA PROTEINS

Circulating antibodies are manufactured by lymphocytes. Most of the other plasma proteins are synthesized in the liver. These proteins and their principal functions are listed in Table 31–6.

Data on the turnover of albumin show that synthesis plays an important role in the maintenance of normal levels. In normal adult humans, the plasma albumin level is 3.5–5.0 g/dL, and the total exchangeable albumin pool is 4.0–5.0 g/kg body weight; 38–45% of this albumin is intravascular, and much of the rest of it is in the skin. Between 6 and 10% of the exchangeable pool is degraded per day, and the degraded albumin is replaced by hepatic synthesis of 200–400 mg/kg/d. The albumin is probably transported to the extravascular areas by vesicular transport across the walls of the capillaries (see Chapter 2). Albumin synthesis is carefully regulated. It is decreased during fasting and increased in conditions such as nephrosis in which there is excessive albumin loss.

HYPOPROTEINEMIA

Plasma protein levels are maintained during starvation until body protein stores are markedly depleted. However, in prolonged starvation and in malabsorption syndromes due to intestinal diseases, plasma protein levels are low **(hypoproteinemia).** They are also low in liver disease, because hepatic protein synthesis is depressed, and in nephrosis, because large amounts of albumin are lost in the urine. Because of the decrease in the plasma oncotic pressure, edema tends to develop. Rarely, there is congenital absence of one or another plasma protein. An example of congenital protein deficiency is the congenital form of **afibrinogenemia**, characterized by defective blood clotting.

HEMOSTASIS

Hemostasis is the process of forming clots in the walls of damaged blood vessels and preventing blood loss while maintaining blood in a fluid state within the vascular system. A collection of complex interrelated systemic mechanisms operates to maintain a balance between coagulation and anticoagulation.

RESPONSE TO INJURY

When a small blood vessel is transected or damaged, the injury initiates a series of events (Figure 31–11) that lead to the formation of a clot. This seals off the damaged region and prevents further blood loss. The initial event is constriction of the vessel and formation of a temporary **hemostatic plug** of platelets that is triggered when platelets bind to collagen and aggregate. This is followed by conversion of the plug into the definitive clot. The constriction of an injured arteriole or small artery may be so marked that its lumen is obliterated, at least temporarily. The vasoconstriction is due to serotonin and other vasoconstrictors liberated from platelets that adhere to the walls of the damaged vessels.

TABLE 31–6 Some of the proteins synthesized by the liver: Physiologic functions and properties.

Name	Principal Function	Binding Characteristics	Serum or Plasma Concentration
Albumin	Binding and carrier protein; osmotic regulator	Hormones, amino acids, steroids, vitamins, fatty acids	4500–5000 mg/dL
Orosomucoid	Uncertain; may have a role in inflammation		Trace; rises in inflammation
α_1-Antiprotease	Trypsin and general protease inhibitor	Proteases in serum and tissue secretions	1.3–1.4 mg/dL
α-Fetoprotein	Osmotic regulation; binding and carrier protein[a]	Hormones, amino acids	Found normally in fetal blood
α_2-Macroglobulin	Inhibitor of serum endoproteases	Proteases	150–420 mg/dL
Antithrombin-III	Protease inhibitor of intrinsic coagulation system	1:1 binding to proteases	17–30 mg/dL
Ceruloplasmin	Transport of copper	Six atoms copper/molecule	15–60 mg/dL
C-reactive protein	Uncertain; has role in tissue inflammation	Complement C1q	< 1 mg/dL; rises in inflammation
Fibrinogen	Precursor to fibrin in hemostasis		200–450 mg/dL
Haptoglobin	Binding, transport of cell-free hemoglobin	Hemoglobin 1:1 binding	40–180 mg/dL
Hemopexin	Binds to porphyrins, particularly heme for heme recycling	1:1 with heme	50–100 mg/dL
Transferrin	Transport of iron	Two atoms iron/molecule	3.0–6.5 mg/dL
Apolipoprotein B	Assembly of lipoprotein particles	Lipid carrier	
Angiotensinogen	Precursor to pressor peptide angiotensin II		
Proteins, coagulation factors II, VII, IX, X	Blood clotting		20 mg/dL
Antithrombin C, protein C	Inhibition of blood clotting		
Insulinlike growth factor I	Mediator of anabolic effects of growth hormone	IGF-I receptor	
Steroid hormone-binding globulin	Carrier protein for steroids in bloodstream	Steroid hormones	3.3 mg/dL
Thyroxine-binding globulin	Carrier protein for thyroid hormone in bloodstream	Thyroid hormones	1.5 mg/dL
Transthyretin (thyroid-binding prealbumin)	Carrier protein for thyroid hormone in bloodstream	Thyroid hormones	25 mg/dL

[a]The function of α-fetoprotein is uncertain, but because of its structural homology to albumin it is often assigned these functions.

THE CLOTTING MECHANISM

The loose aggregation of platelets in the temporary plug is bound together and converted into the definitive clot by **fibrin.** Fibrin formation involves a cascade of enzymatic reactions and a series of numbered clotting factors (Table 31–5). The fundamental reaction is conversion of the soluble plasma protein fibrinogen to insoluble fibrin (Figure 31–12). The process involves the release of two pairs of polypeptides from each fibrinogen molecule. The remaining portion, **fibrin monomer,** then polymerizes with other monomer molecules to form **fibrin.** The fibrin is initially a loose mesh of interlacing strands. It is converted by the formation of covalent cross-linkages to a dense, tight aggregate (stabilization). This latter reaction is catalyzed by activated factor XIII and requires Ca^{2+}.

The conversion of fibrinogen to fibrin is catalyzed by thrombin. Thrombin is a serine protease that is formed from its circulating precursor, prothrombin, by the action of activated factor X. It has additional actions, including activation of platelets, endothelial cells, and leukocytes via so-called proteinase activated receptors, which are G protein-coupled.

Factor X can be activated by either of two systems, known as intrinsic and extrinsic (Figure 31–12). The initial reaction in the **intrinsic system** is conversion of inactive

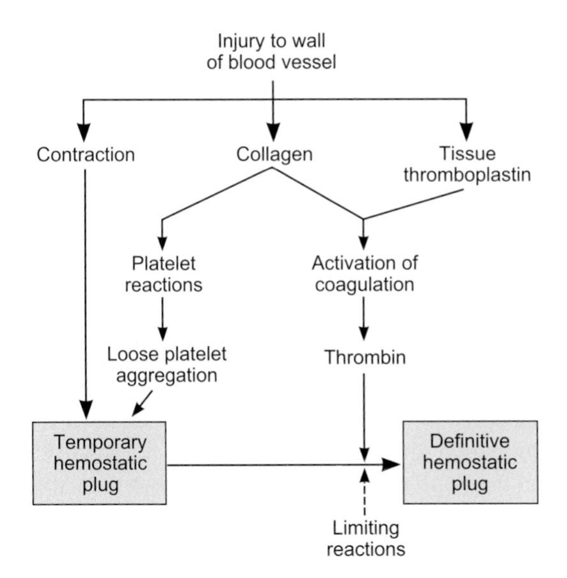

FIGURE 31–11 Summary of reactions involved in hemostasis. The dashed arrow indicates inhibition. (Modified from Deykin D: Thrombogenesis. N Engl J Med 1967;276:622.)

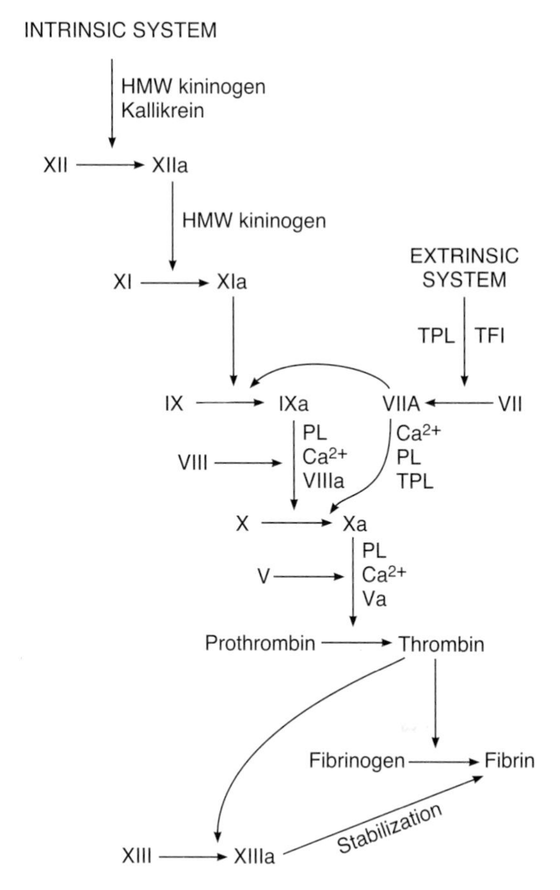

FIGURE 31–12 The clotting mechanism. a, active form of clotting factor. TFI, tissue factor pathway inhibitor; TPL, tissue thromboplastin. For other abbreviations, see Table 31–5.

factor XII to active factor XII (XIIa). This activation, which is catalyzed by high-molecular-weight kininogen and kallikrein (see Chapter 32), can be brought about in vitro by exposing the blood to glass, or in vivo by collagen fibers underlying the endothelium. Active factor XII then activates factor XI, and active factor XI activates factor IX. Activated factor IX forms a complex with active factor VIII, which is activated when it is separated from von Willebrand factor. The complex of IXa and VIIIa activate factor X. Phospholipids from aggregated platelets (PL) and Ca^{2+} are necessary for full activation of factor X. The **extrinsic system** is triggered by the release of tissue thromboplastin, a protein–phospholipid mixture that activates factor VII. Tissue thromboplastin and factor VII activate factors IX and X. In the presence of PL, Ca^{2+}, and factor V, activated factor X catalyzes the conversion of prothrombin to thrombin. The extrinsic pathway is inhibited by a **tissue factor pathway inhibitor** that forms a quaternary structure with tissue thromboplastin (TPL), factor VIIa, and factor Xa.

ANTICLOTTING MECHANISMS

The tendency of blood to clot is balanced in vivo by reactions that prevent clotting inside the blood vessels, break down any clots that do form, or both. These reactions include the interaction between the platelet-aggregating effect of thromboxane A_2 and the antiaggregating effect of prostacyclin, which causes clots to form at the site when a blood vessel is injured but keeps the vessel lumen free of clot (see Chapter 32 and Clinical Box 31–3).

Antithrombin III is a circulating protease inhibitor that binds to serine proteases in the coagulation system, blocking their activity as clotting factors. This binding is facilitated by **heparin,** a naturally occurring anticoagulant that is a mixture of sulfated polysaccharides. The clotting factors that are inhibited are the active forms of factors IX, X, XI, and XII.

The endothelium of the blood vessels also plays an active role in preventing the extension of clots. All endothelial cells except those in the cerebral microcirculation produce **thrombomodulin,** a thrombin-binding protein, on their surfaces. In circulating blood, thrombin is a procoagulant that activates factors V and VIII, but when it binds to thrombomodulin, it becomes an anticoagulant in that the thrombomodulin–thrombin complex activates protein C (Figure 33–13). Activated protein C (APC), along with its cofactor protein S, inactivates factors V and VIII and inactivates an inhibitor of tissue plasminogen activator, increasing the formation of plasmin.

Plasmin (fibrinolysin) is the active component of the **plasminogen (fibrinolytic) system** (Figure 31–13). This enzyme lyses fibrin and fibrinogen, with the production of fibrinogen degradation products (FDP) that inhibit thrombin. Plasmin is formed from its inactive precursor, plasminogen, by the action of thrombin and **tissue-type plasminogen activator (t-PA).** Plasminogen is converted to active plasmin when t-PA hydrolyzes the bond between Arg 560 and Val 561.It is also activated by **urokinase-type plasminogen activator (u-PA).** If the t-PA gene or the u-PA gene is knocked out in mice, some fibrin

CLINICAL BOX 31–3

Abnormalities of Hemostasis

In addition to clotting abnormalities due to platelet disorders, hemorrhagic diseases can be produced by selective deficiencies of most of the clotting factors (Table 31–7). Hemophilia A, which is caused by factor VIII deficiency, is relatively common. von Willebrand factor deficiency likewise causes a bleeding disorder (von Willebrand disease) by reducing platelet adhesion and by lowering plasma factor VIII. The condition can be congenital or acquired. The large von Willebrand molecule is subject to cleavage and resulting inactivation by the plasma metalloprotease ADAM 13 in vascular areas where fluid shear stress is elevated. Finally, when absorption of vitamin K is depressed along with absorption of other fat-soluble vitamins (see Chapter 26), the resulting clotting factor deficiencies may cause the development of a significant bleeding tendency.

Formation of clots inside blood vessels is called **thrombosis** to distinguish it from the normal extravascular clotting of blood. Thromboses are a major medical problem. They are particularly prone to occur where blood flow is sluggish because the slow flow permits activated clotting factors to accumulate instead of being washed away. They also occur in vessels when the intima is damaged by atherosclerotic plaques, and over areas of damage to the endocardium. They frequently occlude the arterial supply to the organs in which they form, and bits of thrombus **(emboli)** sometimes break off and travel in the bloodstream to distant sites, damaging other organs. An example is obstruction of the pulmonary artery or its branches by thrombi from the leg veins **(pulmonary embolism).** Congenital absence of protein C leads to uncontrolled intravascular coagulation and, in general, death in infancy. If this condition is diagnosed and treatment is instituted, the coagulation defect disappears. Resistance to activated protein C is another cause of thrombosis, and this condition is common. It is due to a point mutation in the gene for factor V, which prevents activated protein C from inactivating the factor. Mutations in protein S and antithrombin III may less commonly increase the incidence of thrombosis.

Disseminated intravascular coagulation is another serious complication of septicemia, extensive tissue injury, and other diseases in which fibrin is deposited in the vascular system and many small- and medium-sized vessels are thrombosed. The increased consumption of platelets and coagulation factors causes bleeding to occur at the same time. The cause of the condition appears to be increased generation of thrombin due to increased TPL activity without adequate tissue factor inhibitory pathway activity.

THERAPEUTIC HIGHLIGHTS

Hemophilia has been treated with factor VIII-rich preparations made from plasma, or, more recently, factor VIII produced by recombinant DNA techniques. Some patients with von Willebrand's disease are treated with desmopressin, which stimulates production of Factor VIII, particularly prior to dental procedures or surgery. Thrombotic disorders, on the other hand, are treated with anticoagulants such as heparin.

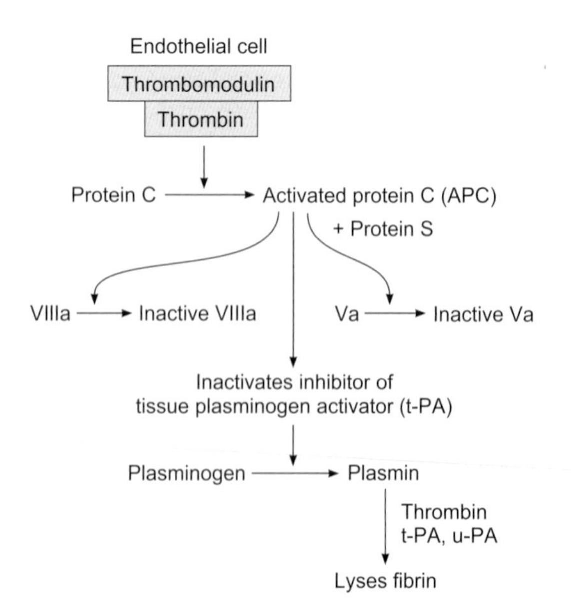

FIGURE 31–13 The fibrinolytic system and its regulation by protein C.

deposition occurs and clot lysis is slowed. However, when both are knocked out, spontaneous fibrin deposition is extensive.

Plasminogen receptors are located on the surfaces of many different types of cells and are plentiful on endothelial cells. When plasminogen binds to its receptors, it becomes activated, so intact blood vessel walls are provided with a mechanism that discourages clot formation.

Human t-PA is now produced by recombinant DNA techniques for clinical use in myocardial infarction and stroke. Streptokinase, a bacterial enzyme, is also fibrinolytic and is also used in the treatment of early myocardial infarction (see Chapter 33).

ANTICOAGULANTS

As noted above, heparin is a naturally occurring anticoagulant that facilitates the action of antithrombin III. Low-molecular-weight fragments have been produced from unfractionated heparin, and these are seeing increased clinical use because they have a longer half-life and produce a more predictable anticoagulant response than unfractionated heparin. The

TABLE 31–7 Examples of diseases due to deficiency of clotting factors.

Deficiency of Factor:	Clinical Syndrome	Cause
I	Afibrinogenemia	Depletion during pregnancy with premature separation of placenta; also congenital (rare)
II	Hypoprothrombinemia (hemorrhagic tendency in liver disease)	Decreased hepatic synthesis, usually secondary to vitamin K deficiency
V	Parahemophilia	Congenital
VII	Hypoconvertinemia	Congenital
VIII	Hemophilia A (classic hemophilia)	Congenital defect due to various abnormalities of the gene on X chromosome that codes for factor VIII; disease is therefore inherited as a sex-linked characteristic
IX	Hemophilia B (Christmas disease)	Congenital
X	Stuart–Prower factor deficiency	Congenital
XI	PTA deficiency	Congenital
XII	Hageman trait	Congenital

highly basic protein protamine forms an irreversible complex with heparin and is used clinically to neutralize heparin.

In vivo, a plasma Ca^{2+} level low enough to interfere with blood clotting is incompatible with life, but clotting can be prevented in vitro if Ca^{2+} is removed from the blood by the addition of substances such as oxalates, which form insoluble salts with Ca^{2+}, or **chelating agents**, which bind Ca^{2+}. Coumarin derivatives such as **dicumarol** and **warfarin** are also effective anticoagulants. They inhibit the action of vitamin K, which is a necessary cofactor for the enzyme that catalyzes the conversion of glutamic acid residues to γ-carboxyglutamic acid residues. Six of the proteins involved in clotting require conversion of a number of glutamic acid residues to γ-carboxyglutamic acid residues before being released into the circulation, and hence all six are vitamin K-dependent. These proteins are factors II (prothrombin), VII, IX, and X, protein C, and protein S (see above).

LYMPH

Lymph is tissue fluid that enters the lymphatic vessels. It drains into the venous blood via the thoracic and right lymphatic ducts. It contains clotting factors and clots on standing in vitro. In most locations, it also contains proteins that have traversed capillary walls, and can then return to the blood via the lymph. Nevertheless, its protein content is generally lower than that of plasma, which contains about 7 g/dL, but lymph protein content varies with the region from which the lymph drains (Table 31–8). Water-insoluble fats are absorbed from the intestine into the lymphatics, and the lymph in the thoracic duct after a meal is milky because of its high fat content (see Chapter 26). Lymphocytes also enter the circulation principally through the lymphatics, and there are appreciable numbers of lymphocytes in thoracic duct lymph.

STRUCTURAL FEATURES OF THE CIRCULATION

Here, we will first describe the two major cell types that make up the blood vessels and then how they are arranged into the various vessel types that subserve the needs of the circulation.

ENDOTHELIUM

Located between the circulating blood and the media and adventitia of the blood vessels, the endothelial cells constitute a large and important organ. They respond to flow changes, stretch, a variety of circulating substances, and inflammatory mediators. They secrete growth regulators and vasoactive substances (see below and Chapter 32).

VASCULAR SMOOTH MUSCLE

The smooth muscle in blood vessel walls has been one of the most-studied forms of visceral smooth muscle because of its importance in the regulation of blood pressure and hypertension. The membranes of the muscle cells contain various types of K^+, Ca^{2+}, and Cl^- channels. Contraction is produced primarily by the myosin light chain mechanism described in Chapter 5. However, vascular smooth muscle also undergoes prolonged contractions that determine vascular tone. These may be due in part to the latch-bridge mechanism (see Chapter 5), but other factors also play a role. Some of the molecular mechanisms that appear to be involved in contraction and relaxation are shown in Figure 31–14.

Vascular smooth muscle cells provide an interesting example of the way high and low cytosolic Ca^{2+} can have different and even opposite effects (see Chapter 2). In these cells, influx of Ca^{2+} via voltage-gated Ca^{2+} channels produces a diffuse increase in cytosolic Ca^{2+} that initiates contraction. However, the Ca^{2+} influx also initiates Ca^{2+} release from the sarcoplasmic reticulum via ryanodine receptors (see Chapter 5), and the high local Ca^{2+} concentration produced by these Ca^{2+} sparks increases the activity of **Ca^{2+}-activated K^+ channels** in the cell membrane. These are also known as big K or **BK channels** because K^+ flows through them at a high rate. The increased K^+ efflux increases the membrane potential, shutting off voltage-gated Ca^{2+} channels and producing relaxation. The site of action of the Ca^{2+} sparks is the

TABLE 31–8 Approximate protein content of lymph in humans.

Source of Lymph	Protein Content (g/dL)
Choroid plexus	0
Ciliary body	0
Skeletal muscle	2
Skin	2
Lung	4
Gastrointestinal tract	4.1
Heart	4.4
Liver	6.2

Data largely from JN Diana.

β_1-subunit of the BK channel, and mice in which this subunit is knocked out develop increased vascular tone and blood pressure. Obviously, therefore, the sensitivity of the β_1 subunit to Ca^{2+} sparks plays an important role in the control of vascular tone.

ARTERIES & ARTERIOLES

The characteristics of the various types of blood vessels are listed in Table 31–9. The walls of all arteries are made up of an outer layer of connective tissue, the adventitia; a middle layer of smooth muscle, the media; and an inner layer, the intima, made up of the endothelium and underlying connective tissue (Figure 31–15). The walls of the aorta and other large diameter arteries contain a relatively large amount of elastic tissue, primarily located in the inner and external elastic laminas. They are stretched during systole and recoil on the blood during diastole. The walls of the arterioles contain less elastic tissue but much more smooth muscle. The muscle is innervated by noradrenergic nerve fibers, which function as constrictors, and in some instances by cholinergic fibers, which dilate the vessels. The arterioles are the major site of the resistance to blood flow, and small changes in their caliber cause large changes in the total peripheral resistance.

CAPILLARIES

The arterioles divide into smaller muscle-walled vessels, sometimes called **metarterioles,** and these in turn feed into capillaries (Figure 31–16). The openings of the capillaries are

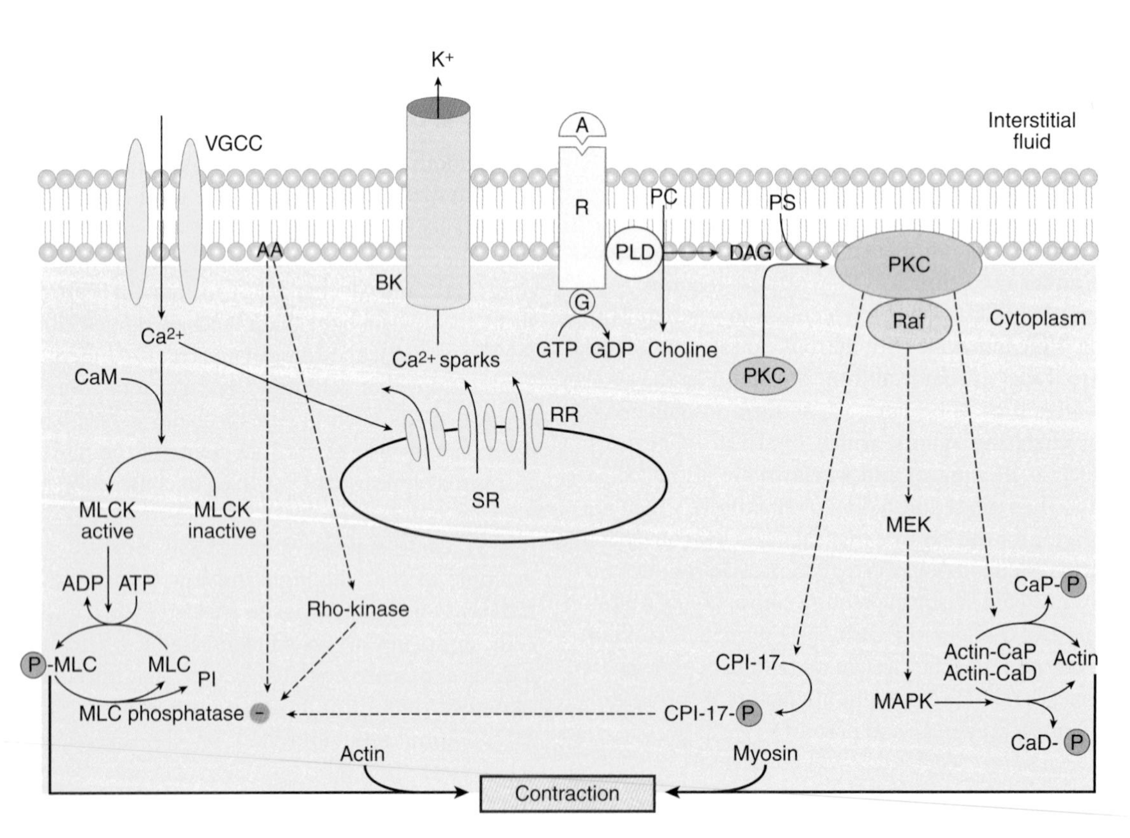

FIGURE 31–14 Some of the established and postulated mechanisms involved in the contraction and relaxation of vascular smooth muscle. A, agonist; AA, arachidonic acid; BK, Ca^{+}-activated K^{+} channel; G, heterotrimeric G protein; MLC, myosin light chain; MLCK, myosin light chain kinase; PLD, phospholipase D; R, receptor; RR, ryanodine receptors; SF, sarcoplasmic reticulum; VGCC, voltage-gated Ca^{2+} channel. For other abbreviations, see Chapter 2 and Appendix. (Modified from Khahl R: Mechanisms of vascular smooth muscle contraction. Council for High Blood Pressure Newsletter, Spring 2001.)

TABLE 31–9 Characteristics of various types of blood vessels in humans.

Vessel	Lumen Diameter	Wall Thickness	All Vessels of Each Type: Approximate Total Cross-Sectional Area (cm²)	All Vessels of Each Type: Percentage of Blood Volume Contained[a]
Aorta	2.5 cm	2 mm	4.5	2
Artery	0.4 cm	1 mm	20	8
Arteriole	30 μm	20 μm	400	1
Capillary	5 μm	1 μm	4500	5
Venule	20 μm	2 μm	4000	
Vein	0.5 cm	0.5 mm	40	54
Vena cava	3 cm	1.5 mm	18	

[a]In systemic vessels; there is an additional 12% in the heart and 18% in the pulmonary circulation.

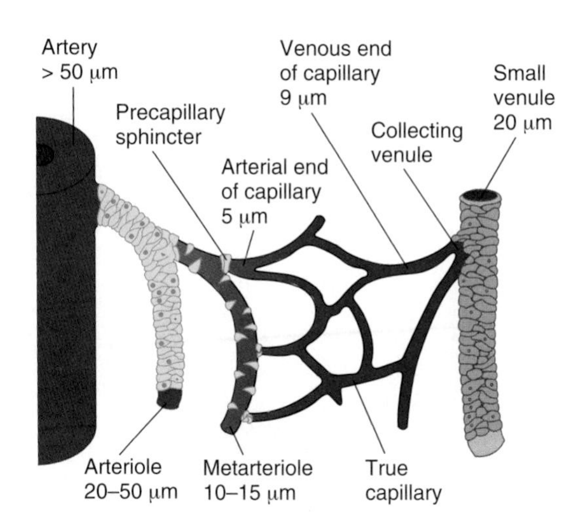

FIGURE 31–16 The microcirculation. Arterioles give rise to metarterioles, which give rise to capillaries. The capillaries drain via short collecting venules to the venules. The walls of the arteries, arterioles, and small venules contain relatively large amounts of smooth muscle. There are scattered smooth muscle cells in the walls of the metarterioles, and the openings of the capillaries are guarded by muscular precapillary sphincters. The diameters of the various vessels are also shown. (Courtesy of JN Diana.)

surrounded on the upstream side by minute smooth muscle **precapillary sphincters.** It is unsettled whether the metarterioles are innervated, and it appears that the precapillary sphincters are not. However, they can of course respond to local or circulating vasoconstrictor substances. The capillaries are about 5 μm in diameter at the arterial end and 9 μm in diameter at the venous end. When the sphincters are dilated, the diameter of the capillaries is just sufficient to permit red blood cells to squeeze through in "single file." As they pass through the capillaries, the red cells become thimble- or parachute-shaped, with the flow pushing the center ahead of the edges. This configuration appears to be simply due to the pressure in the center of the vessel, whether or not the edges of the red blood cell are in contact with the capillary walls.

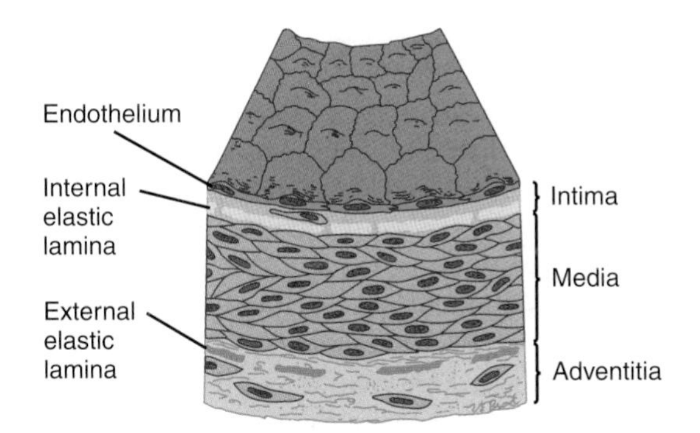

FIGURE 31–15 Structure of a normal muscle artery. (Reproduced with permission from Ross R, Glomset JA: The pathogenesis of atherosclerosis. N Engl J Med 1976;295:369.)

The total area of all the capillary walls in the body exceeds 6300 m² in the adult. The walls, which are about 1 μm thick, are made up of a single layer of endothelial cells. The structure of the walls varies from organ to organ. In many beds, including those in skeletal, cardiac, and smooth muscle, the junctions between the endothelial cells (Figure 31–17) permit the passage of molecules up to 10 nm in diameter. It also appears that plasma and its dissolved proteins can be taken up by endocytosis, transported across the endothelial cells, and discharged by exocytosis (**vesicular transport**; see Chapter 2). However, this process can account for only a small portion of the transport across the endothelium. In the brain, the capillaries resemble the capillaries in muscle, but the junctions between endothelial cells are tighter, and transport across them is largely limited to small molecules (although pathological conditions may open these junctions). In most endocrine glands, the intestinal villi, and parts of the kidneys, on the other hand, the cytoplasm of the endothelial cells is attenuated to form gaps called **fenestrations.** These fenestrations are 20–100 nm in diameter and may be opened to permit the passage of larger molecules, although their permeability is likely significantly reduced under normal circumstances by a thick layer of endothelial glycocalyx. An exception to this, however, is found in the liver, where the sinusoidal capillaries are extremely porous, the endothelium is discontinuous, and gaps occur between endothelial cells that are not closed by membranes (see Figure 28–2). Some of the gaps are 600 nm in diameter, and others may be as large as 3000 nm. They therefore permit the passage of large molecules, including plasma proteins, which is important for hepatic function (see Chapter 28). The permeability of capillaries in various parts of the

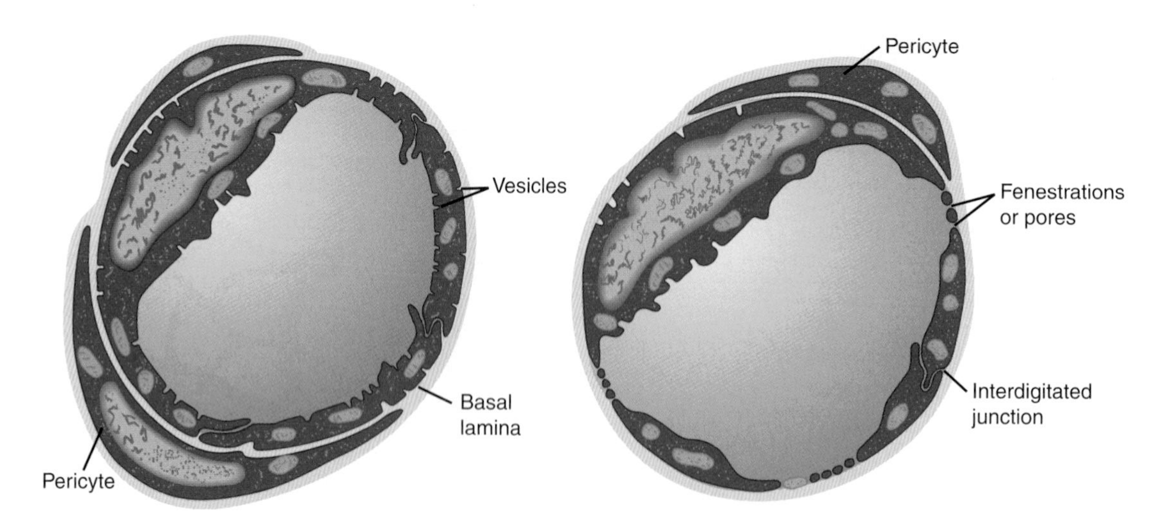

FIGURE 31–17 Cross-sections of capillaries. Left: Type of capillary found in muscle. **Right:** Fenestrated type of capillary. (Reproduced with permission from Orbison JL and D Smith, editors: *Peripheral Blood Vessels.* Baltimore: Williams & Wilkins, 1962.)

body, expressed in terms of their hydraulic conductivity, are summarized in Table 31–10.

Capillaries and postcapillary venules have **pericytes** around their endothelial cells (Figure 31–17). These cells have long processes that wrap around the vessels. They are contractile and release a wide variety of vasoactive agents. They also synthesize and release constituents of the basement membrane and extracellular matrix. One of their physiologic functions appears to be regulation of flow through the junctions between endothelial cells, particularly in the presence of inflammation. They are closely related to the mesangial cells in the renal glomeruli (see Chapter 37).

TABLE 31–10 Hydraulic conductivity of capillaries in various parts of the body.

Organ	Conductivity[a]	Type of Endothelium
Brain (excluding circumventricular organs)	3	
Skin	100	Continuous
Skeletal muscle	250	
Lung	340	
Heart	860	
Gastrointestinal tract (intestinal mucosa)	13,000	
		Fenestrated
Glomerulus in kidney	15,000	

[a]Units of conductivity are 10^{-13} cm^3 s^{-1} $dyne^{-1}$.
Data courtesy of JN Diana.

LYMPHATICS

The lymphatics serve to collect plasma and its constituents that have exuded from the capillaries into the interstitial space (ie, the lymph). They drain from the body tissues via a system of vessels that coalesce and eventually enter the right and left subclavian veins at their junctions with the respective internal jugular veins. The lymph vessels contain valves and regularly traverse lymph nodes along their course. The ultrastructure of the small lymph vessels differs from that of the capillaries in several details: No fenestrations are visible in the lymphatic endothelium; very little if any basal lamina is present under the endothelium; and the junctions between endothelial cells are open, with no tight intercellular connections.

ARTERIOVENOUS ANASTOMOSES

In the fingers, palms, and ear lobes, short channels connect arterioles to venules, bypassing the capillaries. These **arteriovenous (A-V) anastomoses,** or **shunts,** have thick, muscular walls and are abundantly innervated, presumably by vasoconstrictor nerve fibers.

VENULES & VEINS

The walls of the venules are only slightly thicker than those of the capillaries. The walls of the veins are also thin and easily distended. They contain relatively little smooth muscle, but considerable venoconstriction is produced by activity in the noradrenergic nerves to the veins and by circulating vasoconstrictors such as endothelins. Variations in venous tone are important in circulatory adjustments.

The intima of the limb veins is folded at intervals to form **venous valves** that prevent retrograde flow. The way these valves function was first demonstrated by William Harvey in the 17th century. No valves are present in the very small veins, the great veins, or the veins from the brain and viscera.

ANGIOGENESIS

When tissues grow, blood vessels must proliferate if the tissue is to maintain a normal blood supply. Therefore, angiogenesis, the formation of new blood vessels, is important during fetal life and growth to adulthood. It is also important in adulthood for processes such as wound healing, formation of the corpus luteum after ovulation, and formation of new endometrium after menstruation. Abnormally, it is important in tumor growth; if tumors do not develop a blood supply, they do not grow.

During embryonic development, a network of leaky capillaries is formed in tissues from angioblasts: this process is sometimes called **vasculogenesis.** Vessels then branch off from nearby vessels, hook up with the capillaries, and provide them with smooth muscle, which brings about their maturation. Angiogenesis in adults is presumably similar, but consists of new vessel formation by branching from preexisting vessels rather than from angioblasts.

Many factors are involved in angiogenesis. A key compound is **vascular endothelial growth factor (VEGF).** This factor exists in multiple isoforms, and there are three VEGF receptors that are tyrosine kinases, which also cooperate with nonkinase co-receptors known as neuropilins in some cell types. VEGF appears to be primarily responsible for vasculogenesis, whereas the budding of vessels that connect to the immature capillary network is regulated by other as yet unidentified factors. Some of the VEGF isoforms and receptors may play a more prominent role in the formation of lymphatic vessels **(lymphangiogenesis)** than that of blood vessels.

The actions of VEGF and related factors have received considerable attention in recent years because of the requirement for angiogenesis in the development of tumors. VEGF antagonists and other angiogenesis inhibitors have now entered clinical practice as adjunctive therapies for many malignancies and are being tested as first line therapies as well.

BIOPHYSICAL CONSIDERATIONS FOR CIRCULATORY PHYSIOLOGY

FLOW, PRESSURE, & RESISTANCE

Blood always flows, of course, from areas of high pressure to areas of low pressure, except in certain situations when momentum transiently sustains flow (see Figure 30–3). The relationship between mean flow, mean pressure, and resistance in the blood vessels is analogous in a general way to the relationship between the current, electromotive force (voltage), and resistance in an electrical circuit as expressed in Ohm's law:

$$\text{Current (I)} = \frac{\text{Electromotive force (E)}}{\text{Resistance (R)}}$$

$$\text{Flow (F)} = \frac{\text{Pressure (P)}}{\text{Resistance (R)}}$$

Flow in any portion of the vascular system is therefore equal to the **effective perfusion pressure** in that portion divided by the **resistance.** The effective perfusion pressure is the mean intraluminal pressure at the arterial end minus the mean pressure at the venous end. The units of resistance (pressure divided by flow) are dyne·s/cm^5. To avoid dealing with such complex units, resistance in the cardiovascular system is sometimes expressed in **R units,** which are obtained by dividing pressure in mm Hg by flow in mL/s (see also Table 33–1). Thus, for example, when the mean aortic pressure is 90 mm Hg and the left ventricular output is 90 mL/s, the total peripheral resistance is

$$\frac{90 \text{ mm Hg}}{90 \text{ mL/s}} = 1 \text{ R unit}$$

METHODS FOR MEASURING BLOOD FLOW

Blood flow can be measured by cannulating a blood vessel, but this has obvious limitations. Various noninvasive devices have therefore been developed to measure flow. Most commonly, blood velocity can be measured with **Doppler flow meters.** Ultrasonic waves are sent into a vessel diagonally, and the waves reflected from the red and white blood cells are picked up by a downstream sensor. The frequency of the reflected waves is higher by an amount that is proportional to the rate of flow toward the sensor because of the Doppler effect.

Indirect methods for measuring the blood flow of various organs in humans include adaptations of the Fick and indicator dilution techniques described in Chapter 30. One example is the use of the Kety N_2O method for measuring cerebral blood flow (see Chapter 33). Another is determination of the renal blood flow by measuring the clearance of para-aminohippuric acid (see Chapter 37). A considerable amount of data on blood flow in the extremities has been obtained by **plethysmography**. The forearm, for example, is sealed in a watertight chamber **(plethysmograph).** Changes in the volume of the forearm, reflecting changes in the amount of blood and interstitial fluid it contains, displace the water, and this displacement is measured with a volume recorder. When the venous drainage of the forearm is occluded, the rate of increase in the volume of the forearm is a function of the arterial blood flow **(venous occlusion plethysmography).**

APPLICABILITY OF PHYSICAL PRINCIPLES TO FLOW IN BLOOD VESSELS

Physical principles and equations that describe the behavior of perfect fluids in rigid tubes have often been used indiscriminately to explain the behavior of blood in blood vessels. Blood vessels are not rigid tubes, and the blood is not a perfect fluid but a two-phase system of liquid and cells. Therefore, the behavior of the circulation deviates, sometimes markedly, from that predicted by these principles. However, the physical principles are of value when used as an aid to understanding what goes on in the body.

LAMINAR FLOW

The flow of blood in straight blood vessels, like the flow of liquids in narrow rigid tubes, is normally **laminar.** Within the blood vessels, an infinitely thin layer of blood in contact with the wall of the vessel does not move. The next layer within the vessel has a low velocity, the next a higher velocity, and so forth, velocity being greatest in the center of the stream (Figure 31–18). Laminar flow occurs at velocities up to a certain **critical velocity.** At or above this velocity, flow is turbulent. The probability of turbulence is also related to the diameter of the vessel and the viscosity of the blood. This probability can be expressed by the ratio of inertial to viscous forces as follows:

$$Re = \frac{\rho D\dot{V}}{\eta}$$

where Re is the Reynolds number, named for the man who described the relationship; ρ is the density of the fluid; D is the diameter of the tube under consideration; V is the velocity of the flow; and η is the viscosity of the fluid. The higher the value of Re, the greater the probability of turbulence. When D is in cm, V is in cm/s^{-1}, and η is in poise; flow is usually not turbulent if Re is less than 2000. When Re is more than 3000, turbulence is almost always present. Laminar flow can be disturbed at the branching points of arteries, and the resulting turbulence may increase the likelihood that atherosclerotic plaques will be deposited. Constriction of an artery likewise increases the velocity of blood flow through the constriction, producing turbulence and sound beyond the constriction (Figure 31–19). Examples are bruits heard over arteries constricted by atherosclerotic plaques and the sounds of Korotkoff heard when measuring blood pressure (see below). In healthy humans, the critical velocity is sometimes exceeded in the ascending aorta at the peak of systolic ejection, but it is usually exceeded only when an artery is constricted.

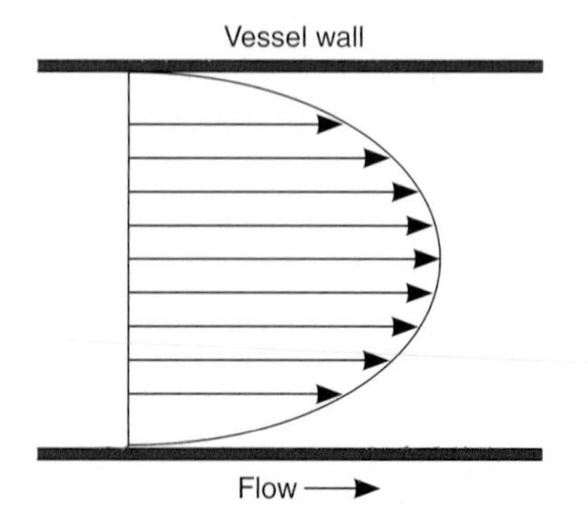

FIGURE 31–18 Diagram of the velocities of concentric laminas of a viscous fluid flowing in a tube, illustrating the parabolic distribution of velocities.

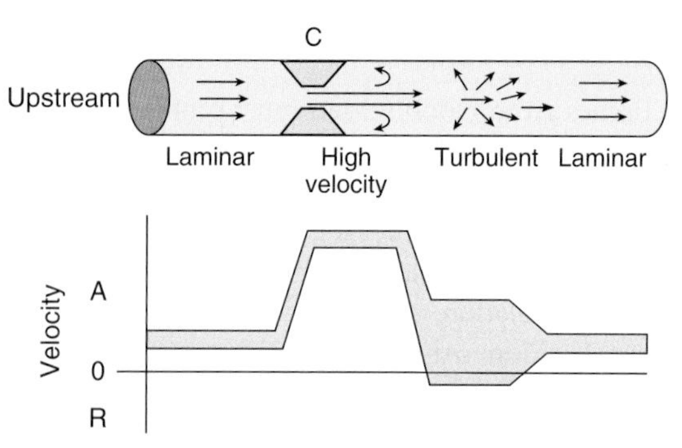

FIGURE 31–19 Top: Effect of constriction (C) on the profile of velocities in a blood vessel. The arrows indicate direction of velocity components, and their length is proportional to their magnitude. **Bottom:** Range of velocities at each point along the vessel. In the area of turbulence, there are many different anterograde (A) and some retrograde (R) velocities. (Modified and reproduced with permission from Richards KE: Doppler echocardiographic diagnosis and quantification of vascular heart disease. Curr Probl Cardiol 1985 Feb;10(2):1–49.)

SHEAR STRESS & GENE ACTIVATION

Flowing blood creates a force on the endothelium that is parallel to the long axis of the vessel. This **shear stress** (γ) is proportional to viscosity (η) times the shear rate (dy/dr), which is the rate at which the axial velocity increases from the vessel wall toward the lumen.

$$\gamma = \eta \,(dy/dr)$$

Change in shear stress and other physical variables, such as cyclic strain and stretch, produce marked changes in the expression of genes by endothelial cells. The genes that are activated include those that produce growth factors, integrins, and related molecules (Table 31–11).

AVERAGE VELOCITY

When considering flow in a system of tubes, it is important to distinguish between velocity, which is displacement per unit time (eg, cm/s), and flow, which is volume per unit time (eg, cm^3/s). Velocity (V) is proportional to flow (Q) divided by the area of the conduit (A):

$$\dot{V} = \frac{Q}{A}$$

Therefore, Q = A × V, and if flow stays constant, velocity increases in direct proportion to any decrease in A (Figure 31–19).

The average velocity of fluid movement at any point in a system of tubes in parallel is inversely proportional to the *total* cross-sectional area at that point. Therefore, the average velocity of the blood is high in the aorta, declines steadily in the smaller vessels, and is lowest in the capillaries,

TABLE 31–11 Genes in human, bovine, and rabbit endothelial cells that are affected by shear stress, and transcription factors involved.

Gene	Transcription Factors
Endothelin-1	AP-1
VCAM-1	AP-1, NF-κB
ACE	SSRE, AP-1, Egr-1
Tissue factor	SP1, Egr-1
TM	AP-1
PDGF-α	SSRE, Egr-1
PDGF-β	SSRE
ICAM-1	SSRE, AP-1, NF-κB
TGF-β	SSRE, AP-1, NF-κB
Egr-1	SREs
c-fos	SSRE
c-jun	SSRE, AP-1
NOS 3	SSRE, AP-1, NF-κB
MCP-1	SSRE, AP-1, NF-κB

ACE, angiotensin converting enzyme; AP-1, activator protein-1; Egr-1, early growth response protein-1; ICAM-1; intercellular adhesion molecule-1; MCP-1, monocyte chemoattractant protein-1; NF-κB, nuclear factor-κB; NOS 3, nitric oxide synthase 3; PDGF, platelet-derived growth factor; SP1, specificity protein 1; SSRE, shear stress responsive element; TGF-β, transforming growth factor-β; TM, thrombomodulin; VCAM-1, vascular cell adhesion molecule-1.

Modified from Braddock M, et al: Fluid shear stress modulation of gene expression in endothelial cells. News Physiol Sci 1998;13:241.

which have 1000 times the *total* cross-sectional area of the aorta (Table 31–9). The average velocity of blood flow increases again as the blood enters the veins and is relatively high in the vena cava, although not so high as in the aorta. Clinically, the velocity of the circulation can be measured by injecting a bile salt preparation into an arm vein and timing the first appearance of the bitter taste it produces (Figure 31–20). The average normal arm-to-tongue **circulation time** is 15 s.

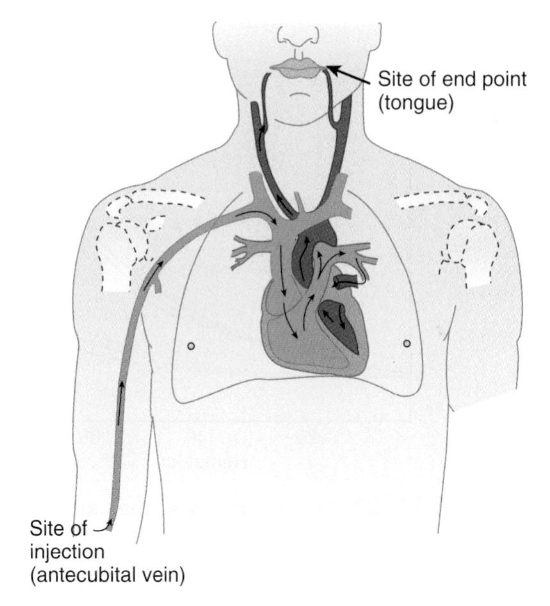

FIGURE 31–20 Pathway traversed by the injected material when the arm-to-tongue circulation time is measured.

POISEUILLE–HAGEN FORMULA

The relationship between the flow in a long narrow tube, the viscosity of the fluid, and the radius of the tube is expressed mathematically in the **Poiseuille–Hagen formula:**

$$F = (P_A - P_B) \times \left(\frac{\pi}{8}\right) \times \left(\frac{1}{\eta}\right) \times \left(\frac{r^4}{L}\right)$$

where

F = flow

$P_A - P_B$ = pressure difference between two ends of the tube

η = viscosity

r = radius of tube

L = length of tube

Because flow is equal to pressure difference divided by resistance (R),

$$R = \frac{8\eta L}{\pi r^4}$$

Because flow varies directly and resistance inversely with the fourth power of the radius, blood flow, and resistance in vivo are markedly affected by small changes in the caliber of the vessels. Thus, for example, flow through a vessel is doubled by an increase of only 19% in its radius; and when the radius is doubled, resistance is reduced to 6% of its previous value. This is why organ blood flow is so effectively regulated by small changes in the caliber of the arterioles and why variations in arteriolar diameter have such a pronounced effect on systemic arterial pressure.

VISCOSITY & RESISTANCE

The resistance to blood flow is determined not only by the radius of the blood vessels (**vascular hindrance**) but also by the viscosity of the blood. Plasma is about 1.8 times as viscous as water, whereas whole blood is 3–4 times as viscous as water. Thus, viscosity depends for the most part on the **hematocrit.** The effect of viscosity in vivo deviates from that predicted by the Poiseuille–Hagen formula. In large vessels, increases in hematocrit cause appreciable increases in viscosity. However, in vessels smaller than 100 μm in diameter—that is, in arterioles, capillaries, and venules—the viscosity change per unit change in hematocrit is much less than it is in large-bore vessels. This is due to a difference in the nature of flow through the small vessels. Therefore, the net change in viscosity per unit change in hematocrit is considerably smaller in the body than

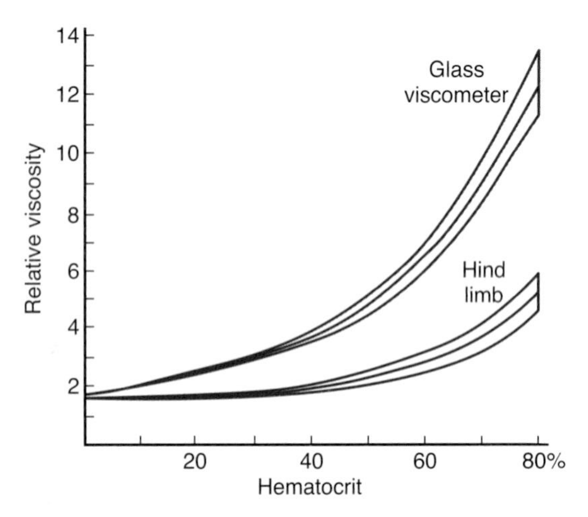

FIGURE 31–21 Effect of changes in hematocrit on the relative viscosity of blood measured in a glass viscometer and in the hind leg of a dog. In each case, the middle line represents the mean and the upper and lower lines the standard deviation. (Reproduced with permission from Whittaker SRF, Winton FR: The apparent viscosity of blood flowing in the isolated hind limb of the dog, and its variation with corpuscular concentration. J Physiol [Lond] 1933;78:338.)

it is in vitro (Figure 31–21). This is why hematocrit changes have relatively little effect on the peripheral resistance except when the changes are large. In severe polycythemia (excessive production of red blood cells), the increase in resistance does increase the work of the heart. Conversely, in marked anemia, peripheral resistance is decreased, in part because of the decline in viscosity. Of course, the decrease in hemoglobin decreases the O_2-carrying ability of the blood, but the improved blood flow due to the decrease in viscosity partially compensates for this.

Viscosity is also affected by the composition of the plasma and the resistance of the cells to deformation. Clinically significant increases in viscosity are seen in diseases in which plasma proteins such as immunoglobulins are markedly elevated as well as when red blood cells are abnormally rigid (hereditary spherocytosis).

CRITICAL CLOSING PRESSURE

In rigid tubes, the relationship between pressure and flow of homogeneous fluids is linear, but in thin-walled blood vessels in vivo it is not. When the pressure in a small blood vessel is reduced, a point is reached at which no blood flows, even though the pressure is not zero (Figure 31–22). This is because the vessels are surrounded by tissues that exert a small but definite pressure on them, and when the intraluminal pressure falls below the tissue pressure, they collapse. In inactive tissues, for example, the pressure in many capillaries is low because the precapillary sphincters and metarterioles are constricted, and many of these capillaries are collapsed. The pressure at which flow ceases is called the **critical closing pressure.**

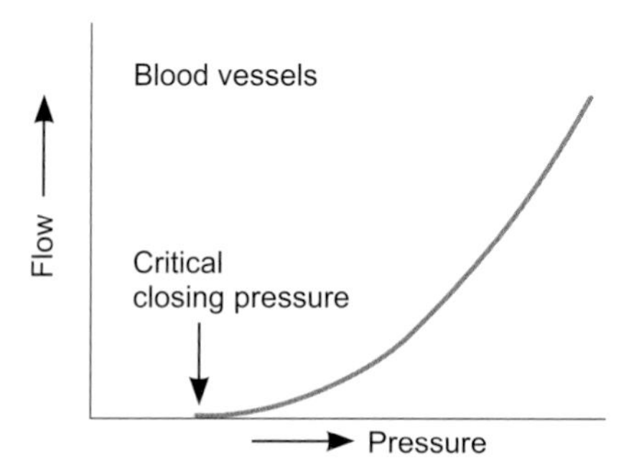

FIGURE 31–22 Relation of pressure to flow in thin-walled blood vessels.

LAW OF LAPLACE

It is perhaps surprising that structures as thin-walled and delicate as the capillaries are not more prone to rupture. The principal reason for their relative invulnerability is their small diameter. The protective effect of small size in this case is an example of the operation of the **law of Laplace,** an important physical principle with several other applications in physiology. This law states that tension in the wall of a cylinder (T) is equal to the product of the transmural pressure (P) and the radius (r) divided by the wall thickness (w):

$$T = Pr/w$$

The **transmural pressure** is the pressure inside the cylinder minus the pressure outside the cylinder, but because tissue pressure in the body is low, it can generally be ignored and P equated to the pressure inside the viscus. In a thin-walled viscus, w is very small and it too can be ignored, but it becomes a significant factor in vessels such as arteries. Therefore, in a thin-walled viscus, P = T divided by the two principal radii of curvature of the viscus:

$$P = T\left(\frac{1}{r_1} + \frac{1}{r_2}\right)$$

In a sphere, $r_1 = r_2$, so

$$P = \frac{2T}{r}$$

In a cylinder such as a blood vessel, one radius is infinite, so

$$P = \frac{T}{r}$$

Consequently, the smaller the radius of a blood vessel, the lower the tension in the wall necessary to balance the distending pressure. In the human aorta, for example, the tension at normal pressures is about 170,000 dynes/cm, and in the vena cava it is about 21,000 dynes/cm; but in the capillaries, it is approximately 16 dynes/cm.

The law of Laplace also makes clear a disadvantage faced by dilated hearts. When the radius of a cardiac chamber is increased, a greater tension must be developed in the myocardium to produce any given pressure; consequently, a dilated heart must do more work than a nondilated heart. In the lungs, the radii of curvature of the alveoli become smaller during expiration, and these structures would tend to collapse because of the pull of surface tension if the tension were not reduced by the surface-tension-lowering agent, surfactant (see Chapter 34). Another example of the operation of this law is seen in the urinary bladder (see Chapter 37).

RESISTANCE & CAPACITANCE VESSELS

In vivo, the veins are an important blood reservoir. Normally, they are partially collapsed and oval in cross-section. A large amount of blood can be added to the venous system before the veins become distended to the point where further increments in volume produce a large rise in venous pressure. The veins are therefore called **capacitance vessels.** The small arteries and arterioles are referred to as **resistance vessels** because they are the principal site of the peripheral resistance (see below).

At rest, at least 50% of the circulating blood volume is in the systemic veins, 12% is in the heart cavities, and 18% is in the low-pressure pulmonary circulation. Only 2% is in the aorta, 8% in the arteries, 1% in the arterioles, and 5% in the capillaries (Table 31–9). When extra blood is administered by transfusion, less than 1% of it is distributed in the arterial system (the **"high-pressure system"**), and all the rest is found in the systemic veins, pulmonary circulation, and heart chambers other than the left ventricle (the **"low-pressure system"**).

ARTERIAL & ARTERIOLAR CIRCULATION

The pressure and velocities of the blood in the various parts of the systemic circulation are summarized in Figure 31–23. The general relationships in the pulmonary circulation are similar, but the pressure in the pulmonary artery is 25/10 mm Hg or less.

VELOCITY & FLOW OF BLOOD

Although the mean velocity of the blood in the proximal portion of the aorta is 40 cm/s, the flow is phasic, and velocity ranges from 120 cm/s during systole to a negative value at the time of the transient backflow before the aortic valve closes in diastole. In the distal portions of the aorta and in the large arteries, velocity is also much greater in systole than it is in diastole. However, the vessels are elastic, and forward flow is continuous because of the recoil during diastole of the vessel walls that have been stretched during systole (Figure 31–24). Pulsatile flow appears to maintain optimal function of the tissues, apparently via distinct effects on gene transcription. If an organ is perfused with a pump that delivers a nonpulsatile flow, inflammatory markers are produced, there is a gradual rise in vascular resistance, and ultimately tissue perfusion fails.

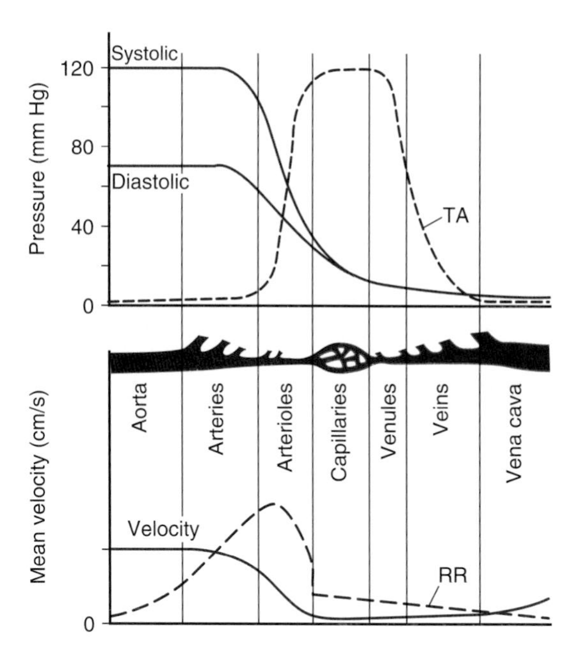

FIGURE 31–23 Diagram of the changes in pressure and velocity as blood flows through the systemic circulation. TA, total cross-sectional area of the vessels, which increases from 4.5 cm^2 in the aorta to 4500 cm^2 in the capillaries (Table 31–9). RR, relative resistance, which is highest in the arterioles.

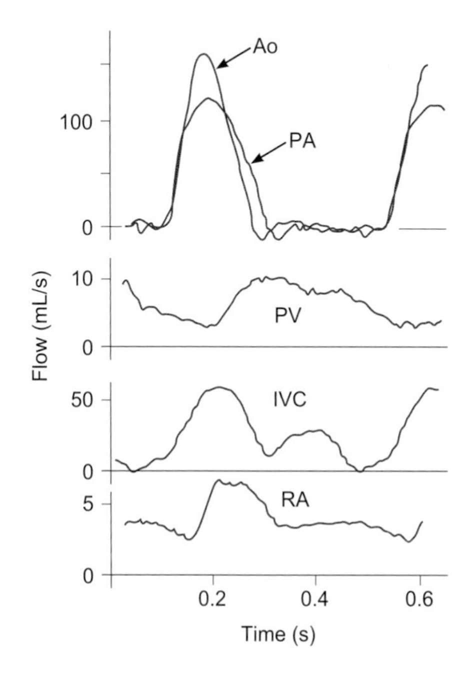

FIGURE 31–24 Changes in blood flow during the cardiac cycle in the dog. Diastole is followed by systole starting at 0.1 and again at 0.5 s. Flow patterns in humans are similar. Ao, aorta; PA, pulmonary artery; PV, pulmonary vein; IVC, inferior vena cava; RA, renal artery. (Reproduced with permission from Milnor WR: Pulsatile blood flow. N Engl J Med 1972;287:27.)

ARTERIAL PRESSURE

The pressure in the aorta and in the brachial and other large arteries in a young adult human rises to a peak value **(systolic pressure)** of about 120 mm Hg during each heart cycle and falls to a minimum **(diastolic pressure)** of about 70 mm Hg. The arterial pressure is conventionally written as systolic pressure over diastolic pressure, for example, 120/70 mm Hg. One millimeter of mercury equals 0.133 kPa, so in SI units this value is 16.0/9.3 kPa. The **pulse pressure,** the difference between the systolic and diastolic pressures, is normally about 50 mm Hg. The **mean pressure** is the average pressure throughout the cardiac cycle. Because systole is shorter than diastole, the mean pressure is slightly less than the value halfway between systolic and diastolic pressure. It can actually be determined only by integrating the area of the pressure curve (Figure 31–25); however, as an approximation, mean pressure equals the diastolic pressure plus one-third of the pulse pressure.

The pressure falls very slightly in the large- and medium sized arteries because their resistance to flow is small, but it falls rapidly in the small arteries and arterioles, which are the main sites of the peripheral resistance against which the heart pumps. The mean pressure at the end of the arterioles is 30–38 mm Hg. Pulse pressure also declines rapidly to about 5 mm Hg at the ends of the arterioles. The magnitude of the pressure drop along the arterioles varies considerably depending on whether they are constricted or dilated.

EFFECT OF GRAVITY

The pressures in Figure 31–24 are those in blood vessels at heart level. The pressure in any vessel below heart level is increased and that in any vessel above heart level is decreased by the effect of gravity. The magnitude of the gravitational effect is 0.77 mm Hg/cm of vertical distance above or below the heart at the density of normal blood. Thus, in an adult human in the upright position, when the mean arterial pressure at heart level is 100 mm Hg, the mean pressure in a large artery in the head (50 cm above the heart) is 62 mm Hg (100 – [0.77 × 50]) and the pressure in a large artery in the foot (105 cm below the heart) is 180 mm Hg (100 + [0.77 × 105]). The effect of gravity on venous pressure is similar (Figure 31–26).

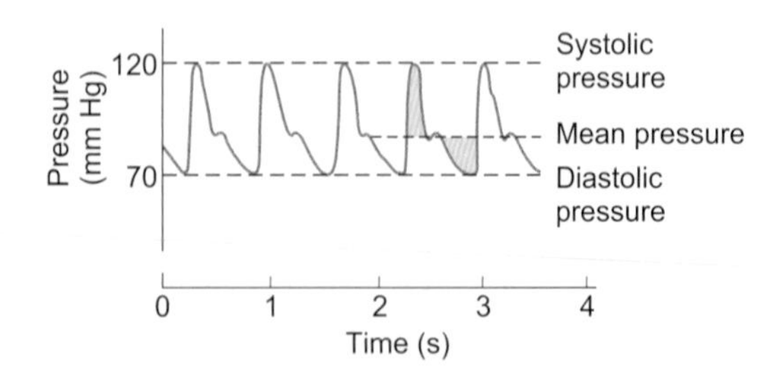

FIGURE 31–25 Brachial artery pressure curve of a normal young human, showing the relation of systolic and diastolic pressure to mean pressure. The shaded area above the mean pressure line is equal to the shaded area below it.

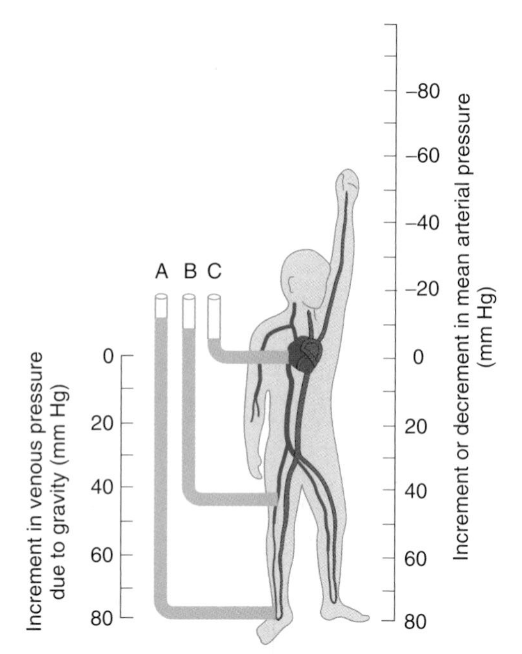

FIGURE 31–26 Effects of gravity on arterial and venous pressure. The scale on the right indicates the increment (or decrement) in mean pressure in a large artery at each level. The mean pressure in all large arteries is approximately 100 mm Hg when they are at the level of the left ventricle. The scale on the left indicates the increment in venous pressure at each level due to gravity. The manometers on the left of the figure indicate the height to which a column of blood in a tube would rise if connected to an ankle vein (A), the femoral vein (B), or the right atrium (C), with the subject in the standing position. The approximate pressures in these locations in the recumbent position; that is, when the ankle, thigh, and right atrium are at the same level, are A, 10 mm Hg; B, 7.5 mm Hg; and C, 4.6 mm Hg.

METHODS OF MEASURING BLOOD PRESSURE

If a cannula is inserted into an artery, the arterial pressure can be measured directly with a mercury manometer or a suitably calibrated strain gauge. When an artery is tied off beyond the point at which the cannula is inserted, an **end pressure** is recorded, flow in the artery is interrupted, and all the kinetic energy of flow is converted into pressure energy. If, alternatively, a T tube is inserted into a vessel and the pressure is measured in the side arm of the tube, the recorded **side pressure,** under conditions where pressure drop due to resistance is negligible, is lower than the end pressure by the kinetic energy of flow. This is because in a tube or a blood vessel the total energy—the sum of the kinetic energy of flow and the potential energy—is constant **(Bernoulli's principle).**

It is worth noting that the pressure drop in any segment of the arterial system is due both to resistance and to conversion of potential into kinetic energy. The pressure drop due to energy lost in overcoming resistance is irreversible, since the energy is dissipated as heat; but the pressure drop due to conversion of potential to kinetic energy as a vessel narrows is reversed when the vessel widens out again (Figure 31–27).

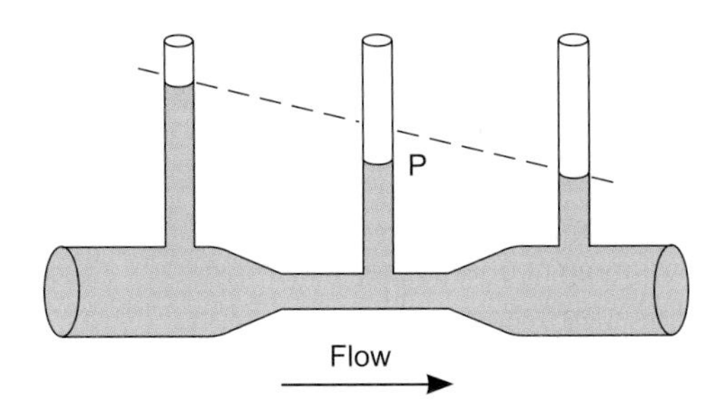

FIGURE 31–27 Bernoulli's principle. When fluid flows through the narrow portion of the tube, the kinetic energy of flow is increased as the velocity increases, and the potential energy is reduced. Consequently, the measured pressure (P) is lower than it would have been at that point if the tube had not been narrowed. The dashed line indicates what the pressure drop due to frictional forces would have been if the tube had been of uniform diameter.

Bernoulli's principle also has a significant application in pathophysiology. According to the principle, the greater the velocity of flow in a vessel, the lower the lateral pressure distending its walls. When a vessel is narrowed, the velocity of flow in the narrowed portion increases and the distending pressure decreases. Therefore, when a vessel is narrowed by a pathologic process such as an atherosclerotic plaque, the lateral pressure at the constriction is decreased and the narrowing tends to maintain itself.

AUSCULTATORY METHOD

The arterial blood pressure in humans is routinely measured by the **auscultatory method.** An inflatable cuff **(Riva–Roccicuff)** attached to a mercury manometer **(sphygmomanometer)** is wrapped around the arm and a stethoscope is placed over the brachial artery at the elbow. The cuff is rapidly inflated until the pressure is well above the expected systolic pressure in the brachial artery. The artery is occluded by the cuff, and no sound is heard with the stethoscope. The pressure in the cuff is then lowered slowly. At the point at which systolic pressure in the artery just exceeds the cuff pressure, a spurt of blood passes through with each heartbeat and, synchronously with each beat, a tapping sound is heard below the cuff. The cuff pressure at which the sounds are first heard is the systolic pressure. As the cuff pressure is lowered further, the sounds become louder, then dull and muffled. These are the **sounds of Korotkoff.** Finally, in most individuals, they disappear. When direct and indirect blood pressure measurements are made simultaneously, the diastolic pressure in resting adults correlates best with the pressure at which the sound disappears. However, in adults after exercise and in children, the diastolic pressure correlates best with the pressure at which the sounds become muffled. This is also true in diseases such as hyperthyroidism and aortic insufficiency.

The sounds of Korotkoff are produced by turbulent flow in the brachial artery. When the artery is narrowed by the cuff, the velocity of flow through the constriction exceeds the **critical velocity** and turbulent flow results (Figure 31–19). At cuff pressures just below the systolic pressure, flow through the artery occurs only at the peak of systole, and the intermittent turbulence produces a tapping sound. As long as the pressure in the cuff is above the diastolic pressure in the artery, flow is interrupted at least during part of diastole, and the intermittent sounds have a staccato quality. When the cuff pressure is near the arterial diastolic pressure, the vessel is still constricted, but the turbulent flow is continuous. Continuous sounds have a muffled rather than a staccato quality.

NORMAL ARTERIAL BLOOD PRESSURE

The blood pressure in the brachial artery in young adults in the sitting position at rest is approximately 120/70 mm Hg. Because the arterial pressure is the product of the cardiac output and the peripheral resistance, it is affected by conditions that affect either or both of these factors. Emotion increases the cardiac output and peripheral resistance, and about 20% of hypertensive patients have blood pressures that are higher in the doctor's office than at home, going about their regular daily activities ("white coat hypertension"). Blood pressure normally falls up to 20 mm Hg during sleep. This fall is reduced or absent in hypertension.

There is general agreement that blood pressure rises with advancing age, but the magnitude of this rise is uncertain because hypertension is a common disease and its incidence increases with advancing age (see Clinical Box 31–4). Individuals who have systolic blood pressures < 120 mm Hg at age 50–60 and never develop clinical hypertension still have systolic pressures that rise throughout life (Figure 31–28). This rise may be the closest approximation to the rise in normal individuals. Individuals with mild hypertension that is untreated show a significantly more rapid rise in systolic pressure. In both groups, diastolic pressure also rises, but then starts to fall in middle age as the stiffness of arteries increases. Consequently, pulse pressure rises with advancing age.

It is interesting that systolic and diastolic blood pressures are lower in young women than in young men until age 55–65, after which they become comparable. Because there is a positive correlation between blood pressure and the incidence of heart attacks and strokes (see below), the lower blood pressure before menopause in women may be one reason that, on average, they live longer than men.

CAPILLARY CIRCULATION

At any one time, only 5% of the circulating blood is in the capillaries, but this 5% is in a sense the most important part of the blood volume because it is the only pool from which O_2 and nutrients can enter the interstitial fluid and into which CO_2 and waste products can enter the bloodstream. Exchange across the capillary walls is essential to the survival of the tissues.

CLINICAL BOX 31–4

Hypertension

Hypertension is a sustained elevation of the systemic arterial pressure. It is most commonly due to increased peripheral resistance and is a very common abnormality in humans. It can be produced by many diseases (Table 31–12) and causes a number of serious disorders. When the resistance against which the left ventricle must pump (afterload) is elevated for a long period, the cardiac muscle hypertrophies. The initial response is activation of immediate-early genes in the ventricular muscle, followed by activation of a series of genes involved in growth during fetal life. Left ventricular hypertrophy is associated with a poor prognosis. The total O_2 consumption of the heart, already increased by the work of expelling blood against a raised pressure (see Chapter 30), is increased further because there is more muscle. Therefore, any decrease in coronary blood flow has more serious consequences in hypertensive patients than it does in normal individuals, and degrees of coronary vessel narrowing that do not produce symptoms when the size of the heart is normal may produce myocardial infarction when the heart is enlarged.

The incidence of atherosclerosis increases in hypertension, and myocardial infarcts are common even when the heart is not enlarged. Eventually, the ability to compensate for the high peripheral resistance is exceeded, and the heart fails. Hypertensive individuals are also predisposed to thromboses of cerebral vessels and cerebral hemorrhage. An additional complication is renal failure. However, the incidence of heart failure, strokes, and renal failure can be markedly reduced by active treatment of hypertension, even when the hypertension is relatively mild. In the vast majority of patients with elevated blood pressure, the cause of the hypertension is unknown, and they are said to have **essential hypertension** (Table 31–12). At present, essential hypertension is treatable but not curable. It is probably polygenic in origin, and environmental factors are also involved.

In other, less common forms of hypertension, the cause is known. A review of these is helpful because it emphasizes ways disordered physiology can lead to disease. Pathology that compromises the renal blood supply leads to renal hypertension, as does narrowing (coarctation) of the thoracic aorta, which both increases renin secretion and increases peripheral resistance. Pheochromocytomas, adrenal medullary tumors that secrete norepinephrine and epinephrine, can cause sporadic or sustained hypertension (see Chapter 20). Estrogens increase angiotensinogen secretion, and contraceptive pills containing large amounts of estrogen cause hypertension (pill hypertension) on this basis (see Chapter 22). Increased secretion of aldosterone or other mineralocorticoids causes renal Na^+ retention, which leads to hypertension. A primary increase in plasma mineralocorticoids inhibits renin secretion. For unknown reasons, plasma renin is also low in 10–15% of patients with essential hypertension and normal circulating mineralocortical levels (low renin hypertension). Mutations in a number of single genes are also known to cause hypertension. These cases of monogenic hypertension are rare but informative. One of these is glucocorticoid-remediable aldosteronism (GRA), in which a hybrid gene encodes an adrenocorticotropic hormone (ACTH)-sensitive aldosterone synthase, with resulting hyperaldosteronism (see Chapter 20). 11-β hydroxylase deficiency also causes hypertension by increasing the secretion of deoxycorticosterone (see Chapter 20). Normal blood pressure is restored when ACTH secretion is inhibited by administering a glucocorticoid. Mutations that decrease 11-β hydroxysteroid dehydrogenase cause loss of specificity of the mineralocorticoid receptors (see Chapter 20) with stimulation of them by cortisol and, in pregnancy, by the elevated circulating levels of progesterone. Finally, mutations of the genes for ENaCs that reduce degradation of the β or γ subunits increase ENaC activity and lead to excess renal Na^+ retention and hypertension (Liddle syndrome; see Chapter 38).

THERAPEUTIC HIGHLIGHTS

Effective lowering of the blood pressure can be produced by drugs that block α-adrenergic receptors, either in the periphery or in the central nervous system; drugs that block β-adrenergic receptors; drugs that inhibit the activity of angiotensin-converting enzyme; and calcium channel blockers that relax vascular smooth muscle.

METHODS OF STUDY

It is difficult to obtain accurate measurements of capillary pressures and flows. Capillary pressure has been estimated by determining the amount of external pressure necessary to occlude the capillaries or the amount of pressure necessary to make saline start to flow through a micropipette inserted so that its tip faces the arteriolar end of the capillary.

CAPILLARY PRESSURE & FLOW

Capillary pressures vary considerably, but typical values in human nail bed capillaries are 32 mm Hg at the arteriolar end and 15 mm Hg at the venous end. The pulse pressure is approximately 5 mm Hg at the arteriolar end and zero at the venous end. The capillaries are short, but blood moves slowly (about 0.07 cm/s) because the total cross-sectional area of the

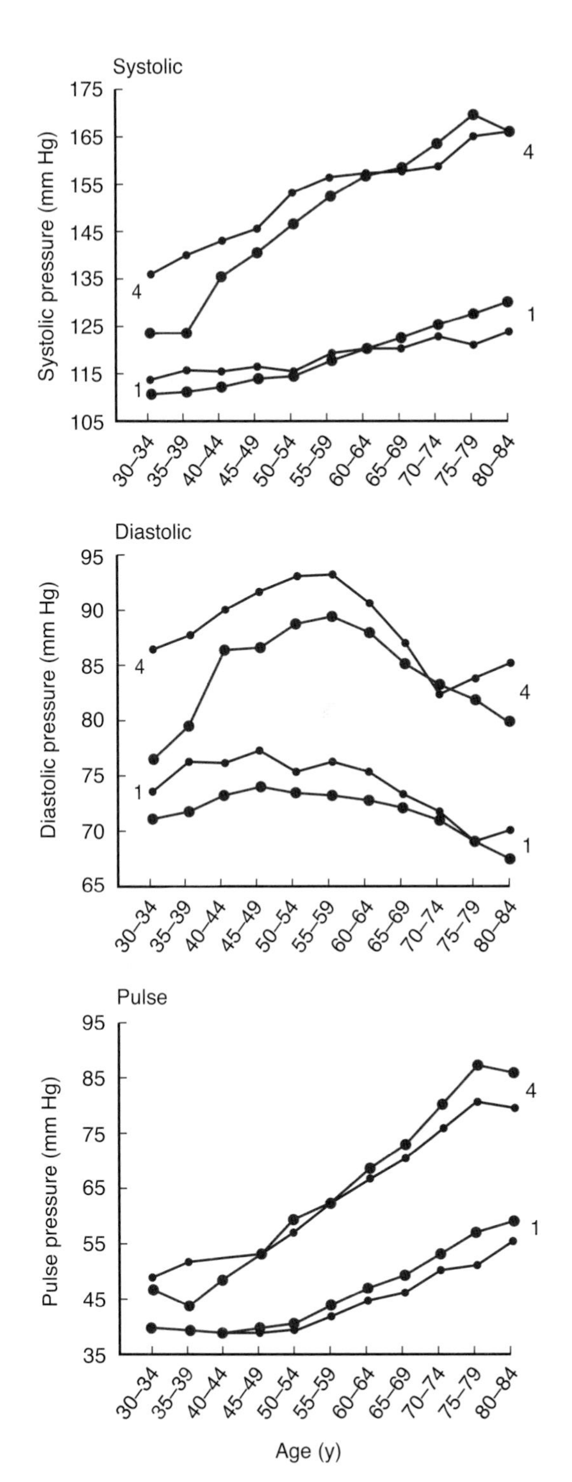

FIGURE 31–28 Effects of age and sex on arterial pressure components in humans. Data are from a large group of individuals who were studied every 2 years throughout their adult lives. Group 1: Individuals who had systolic blood pressures < 120 mm Hg at age 50–60. Group 4: Individuals who had systolic blood pressure ≥ 160 mm Hg at age 50–60, that is, individuals with mild, untreated hypertension. The red line shows the values for women, and the blue line shows the values for men. (Modified and reproduced with permission from Franklin SS, et al: Hemodynamic patterns of age-related changes in blood pressure: The Framingham Heart Study. Circulation 1997;96:308.)

TABLE 31–12 Estimated frequency of various forms of hypertension in the general hypertensive population.

	Percentage of Population
Essential hypertension	88
Renal hypertension	
Renovascular	2
Parenchymal	3
Endocrine hypertension	
Primary aldosteronism	5
Cushing syndrome	0.1
Pheochromocytoma	0.1
Other adrenal forms	0.2
Estrogen treatment ("pill hypertension")	1
Miscellaneous (Liddle syndrome, coarctation of the aorta, etc)	0.6

Data from Williams GH: Hypertensive vascular disease. In Braunwald E, et al (editors): *Harrison's Principles of Internal Medicine,* 15th ed. McGraw-Hill, 2001.

capillary bed is large. Transit time from the arteriolar to the venular end of an average-sized capillary is 1–2 s.

EQUILIBRATION WITH INTERSTITIAL FLUID

As noted above, the capillary wall is a thin membrane made up of endothelial cells. Substances pass through the junctions between endothelial cells and through fenestrations when they are present. Some also pass through the cells by vesicular transport.

The factors other than vesicular transport that are responsible for transport across the capillary wall are diffusion and filtration (see Chapter 1). Diffusion is quantitatively much more important. O_2 and glucose are in higher concentration in the bloodstream than in the interstitial fluid and diffuse into the interstitial fluid, whereas CO_2 diffuses in the opposite direction.

The rate of filtration at any point along a capillary depends on a balance of forces sometimes called the **Starling forces,** after the physiologist who first described their operation in detail. One of these forces is the **hydrostatic pressure gradient** (the hydrostatic pressure in the capillary minus the hydrostatic pressure of the interstitial fluid) at that point. The interstitial fluid pressure varies from one organ to another, and there is considerable evidence that it is subatmospheric (about –2 mm Hg) in subcutaneous tissue. It is, however, positive in the liver and kidneys and as high as 6 mm Hg in the brain. The other force is the **osmotic pressure gradient** across the capillary wall (colloid osmotic pressure of plasma minus colloid osmotic pressure of interstitial fluid). This component is directed inward.

Thus:

$$\text{Fluid movement} = k\,[(P_c - P_i) - (\pi_c - \pi_i)]$$

where

k = capillary filtration coefficient
P_c = capillary hydrostatic pressure
P_i = interstitial hydrostatic pressure
π_c = capillary colloid osmotic pressure
π_i = interstitial colloid osmotic pressure

π_i is usually negligible, so the osmotic pressure gradient ($\pi_c - \pi_i$) usually equals the oncotic pressure. The capillary filtration coefficient takes into account, and is proportional to, the permeability of the capillary wall and the area available for filtration. The magnitude of the Starling forces along a typical muscle capillary is shown in Figure 31–29. Fluid moves into the interstitial space at the arteriolar end of the capillary and into the capillary at the venular end. In other capillaries, the balance of Starling forces may be different. For example, fluid moves out of almost the entire length of the capillaries in the renal glomeruli. On the other hand, fluid moves into the capillaries through almost their entire length in the intestines. About 24 L of fluid is filtered through the capillaries per day. This is about 0.3% of the cardiac output. About 85% of the filtered fluid is reabsorbed into the capillaries, and the remainder returns to the circulation via the lymphatics.

It is worth noting that small molecules often equilibrate with the tissues near the arteriolar end of each capillary. In this situation, total diffusion can be increased by increasing blood flow; that is, exchange is **flow-limited** (Figure 31–30). Conversely, transfer of substances that do not reach equilibrium with the tissues during their passage through the capillaries is said to be **diffusion-limited.**

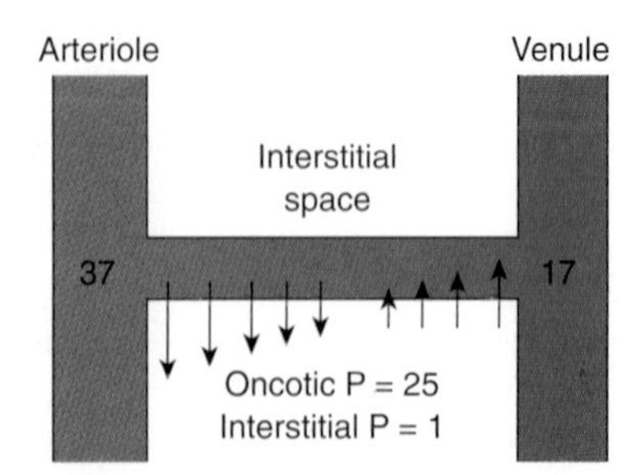

FIGURE 31–29 Schematic representation of pressure gradients across the wall of a muscle capillary. The numbers at the arteriolar and venular ends of the capillary are the hydrostatic pressures in mm Hg at these locations. The arrows indicate the approximate magnitude and direction of fluid movement. In this example, the pressure differential at the arteriolar end of the capillary is 11 mm Hg ([37 – 1] –25) outward; at the opposite end, it is 9 mm Hg (25 – [17 – 1]) inward.

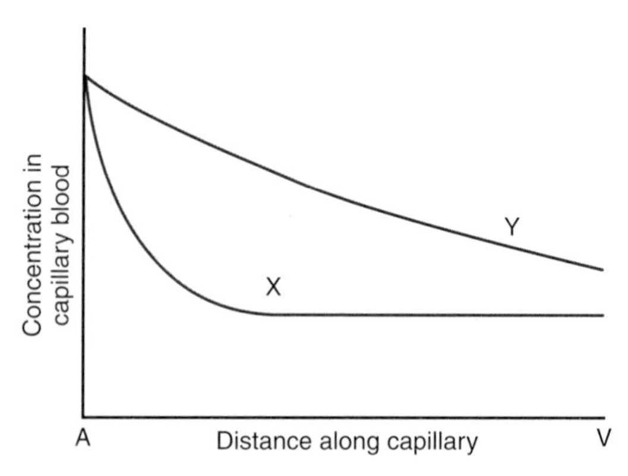

FIGURE 31–30 Flow-limited and diffusion-limited exchange across capillary walls. A and V indicate the arteriolar and venular ends of the capillary. Substance X equilibrates with the tissues (movement into the tissues equals movement out) well before the blood leaves the capillary, whereas substance Y does not equilibrate. If other factors stay constant, the amount of X entering the tissues can be increased only by increasing blood flow; that is, it is flow-limited. The movement of Y is diffusion-limited.

ACTIVE & INACTIVE CAPILLARIES

In resting tissues, most of the capillaries are collapsed. In active tissues, the metarterioles and the precapillary sphincters dilate. The intracapillary pressure rises, overcoming the critical closing pressure of the vessels, and blood flows through all of the capillaries. Relaxation of the smooth muscle of the metarterioles and precapillary sphincters is due to the action of vasodilator metabolites formed in active tissue (see Chapter 32).

After noxious stimulation, substance P released by the axon reflex (see Chapter 33) increases capillary permeability. Bradykinin and histamine also increase capillary permeability. When capillaries are stimulated mechanically, they empty (white reaction; see Chapter 33), probably due to contraction of the precapillary sphincters.

VENOUS CIRCULATION

Blood flows through the blood vessels, including the veins, primarily because of the pumping action of the heart. However, venous flow is aided by the heartbeat, the increase in the negative intrathoracic pressure during each inspiration, and contractions of skeletal muscles that compress the veins **(muscle pump).**

VENOUS PRESSURE & FLOW

The pressure in the venules is 12–18 mm Hg. It falls steadily in the larger veins to about 5.5 mm Hg in the great veins outside the thorax. The pressure in the great veins at their entrance into the right atrium **(central venous pressure)** averages 4.6 mm Hg, but fluctuates with respiration and heart action.

Peripheral venous pressure, like arterial pressure, is affected by gravity. It is increased by 0.77 mm Hg for each centimeter below the right atrium and decreased by a like amount for each centimeter above the right atrium the pressure is measured (Figure 31–26). Thus, on a proportional basis, gravity has a greater effect on venous than on arterial pressures.

When blood flows from the venules to the large veins, its average velocity increases as the total cross-sectional area of the vessels decreases. In the great veins, the velocity of blood is about one fourth that in the aorta, averaging about 10 cm/s.

THORACIC PUMP

During inspiration, the intrapleural pressure falls from −2.5 to −6 mm Hg. This negative pressure is transmitted to the great veins and, to a lesser extent, the atria, so that central venous pressure fluctuates from about 6 mm Hg during expiration to approximately 2 mm Hg during quiet inspiration. The drop in venous pressure during inspiration aids venous return. When the diaphragm descends during inspiration, intra-abdominal pressure rises, and this also squeezes blood toward the heart because backflow into the leg veins is prevented by the venous valves.

EFFECTS OF HEARTBEAT

The variations in atrial pressure are transmitted to the great veins, producing the **a, c,** and **v waves** of the venous pressure-pulse curve (see Chapter 30). Atrial pressure drops sharply during the ejection phase of ventricular systole because the atrioventricular valves are pulled downward, increasing the capacity of the atria. This action sucks blood into the atria from the great veins. The sucking of the blood into the atria during systole contributes appreciably to the venous return, especially at rapid heart rates.

Close to the heart, venous flow becomes pulsatile. When the heart rate is slow, two periods of peak flow are detectable, one during ventricular systole, due to pulling down of the atrioventricular valves, and one in early diastole, during the period of rapid ventricular filling (Figure 31–24).

MUSCLE PUMP

In the limbs, the veins are surrounded by skeletal muscles, and contraction of these muscles during activity compresses the veins. Pulsations of nearby arteries may also compress veins. Because the venous valves prevent reverse flow, the blood moves toward the heart. During quiet standing, when the full effect of gravity is manifest, venous pressure at the ankle is 85–90 mm Hg (Figure 31–26). Pooling of blood in the leg veins reduces venous return, with the result that cardiac output is reduced, sometimes to the point where fainting occurs. Rhythmic contractions of the leg muscles while the person is standing serve to lower the venous pressure in the legs to less than 30 mm Hg by propelling blood toward the heart. This heartward movement of the blood is decreased in patients with **varicose veins** because their valves are incompetent. These patients may develop stasis and ankle edema. However, even when the valves are incompetent, muscle contractions continue to produce a basic heartward movement of the blood because the resistance of the larger veins in the direction of the heart is less than the resistance of the small vessels away from the heart.

VENOUS PRESSURE IN THE HEAD

In the upright position, the venous pressure in the parts of the body above the heart is decreased by the force of gravity. The neck veins collapse above the point where the venous pressure is close to zero. However, the dural sinuses have rigid walls and cannot collapse. The pressure in them in the standing or sitting position is therefore subatmospheric. The magnitude of the negative pressure is proportional to the vertical distance above the top of the collapsed neck veins, and in the superior sagittal sinus may be as much as −10 mm Hg. This fact must be kept in mind by neurosurgeons. Neurosurgical procedures are sometimes performed with the patient seated. If one of the sinuses is opened during such a procedure it sucks air, causing **air embolism.**

AIR EMBOLISM

Because air, unlike fluid, is compressible, its presence in the circulation has serious consequences. The forward movement of the blood depends on the fact that blood is incompressible. Large amounts of air fill the heart and effectively stop the circulation, causing sudden death because most of the air is compressed by the contracting ventricles rather than propelled into the arteries. Small amounts of air are swept through the heart with the blood, but the bubbles lodge in the small blood vessels. The surface capillarity of the bubbles markedly increases the resistance to blood flow, and flow is reduced or abolished. Blockage of small vessels in the brain leads to serious and even fatal neurologic abnormalities. Treatment with hyperbaric oxygen (see Chapter 36) is of value because the pressure reduces the size of the gas emboli. In experimental animals, the amount of air that produces fatal air embolism varies considerably, depending in part on the rate at which it enters the veins. Sometimes as much as 100 mL can be injected without ill effects, whereas at other times as little as 5 mL is lethal.

MEASURING VENOUS PRESSURE

Central venous pressure can be measured directly by inserting a catheter into the thoracic great veins. **Peripheral venous pressure** correlates well with central venous pressure in most conditions. To measure peripheral venous pressure, a needle attached to a manometer containing sterile saline is inserted into an arm vein. The peripheral vein should be at the level of the right atrium (a point half the chest diameter from the back in the supine position). The values obtained in millimeters of saline can be converted into millimeters of mercury (mm Hg) by dividing by 13.6 (the density of mercury). The amount by which peripheral venous pressure exceeds central venous pressure increases with the distance from the heart along the veins. The mean pressure in the antecubital vein is normally 7.1 mm Hg, compared with a mean pressure of 4.6 mm Hg in the central veins.

A fairly accurate estimate of central venous pressure can be made without any equipment by simply noting the height to which the external jugular veins are distended when the subject lies with the head slightly above the heart. The vertical distance between the right atrium and the place the vein collapses (the place where the pressure in it is zero) is the venous pressure in mm of blood.

Central venous pressure is decreased during negative pressure breathing and shock. It is increased by positive pressure breathing, straining, expansion of the blood volume, and heart failure. In advanced congestive heart failure or obstruction of the superior vena cava, the pressure in the antecubital vein may reach values of 20 mm Hg or more.

LYMPHATIC CIRCULATION & INTERSTITIAL FLUID VOLUME

LYMPHATIC CIRCULATION

Fluid efflux normally exceeds influx across the capillary walls, but the extra fluid enters the lymphatics and drains through them back into the blood. This keeps the interstitial fluid pressure from rising and promotes the turnover of tissue fluid. The normal 24-h lymph flow is 2–4 L.

Lymphatic vessels can be divided into two types: initial lymphatics and collecting lymphatics (Figure 31–31). The former lack valves and smooth muscle in their walls, and they are found in regions such as the intestine or skeletal muscle. Tissue fluid appears to enter them through loose junctions between the endothelial cells that form their walls. The fluid in them apparently is massaged by muscle contractions of the organs and contraction of arterioles and venules, with which they are often associated. They drain into the collecting lymphatics, which have valves and smooth muscle in their walls and contract in a peristaltic fashion, propelling the lymph along the vessels. Flow in the collecting lymphatics is further aided by movements of skeletal muscle, the negative intrathoracic pressure during inspiration, and the suction effect of high velocity flow of blood in the veins in which the lymphatics terminate. However, the contractions are the principal factor propelling the lymph.

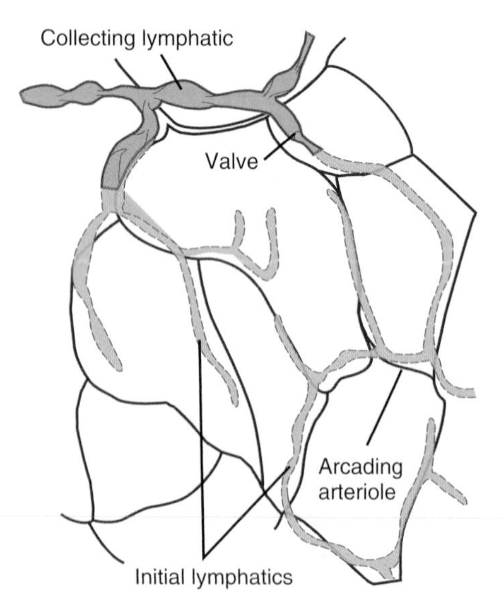

FIGURE 31–31 Initial lymphatics draining into collecting lymphatics in the mesentery. Note the close association with arcading arterioles, indicated by the single red lines. (Reproduced with permission from Schmid Schönbein GW, Zeifach BW: Fluid pump mechanisms in initial lymphatics. News Physiol Sci 1994;9:67.)

OTHER FUNCTIONS OF THE LYMPHATIC SYSTEM

Appreciable quantities of protein enter the interstitial fluid in the liver and intestine, and smaller quantities enter from the blood in other tissues. The macromolecules enter the lymphatics, presumably at the junctions between the endothelial cells, and the proteins are returned to the bloodstream via the lymphatics. The amount of protein returned in this fashion in 1 day is equal to 25–50% of the total circulating plasma protein. The transport of absorbed long-chain fatty acids and cholesterol from the intestine via the lymphatics has been discussed in Chapter 26.

INTERSTITIAL FLUID VOLUME

The amount of fluid in the interstitial spaces depends on the capillary pressure, the interstitial fluid pressure, the oncotic pressure, the capillary filtration coefficient, the number of active capillaries, the lymph flow, and the total extracellular fluid (ECF) volume. The ratio of precapillary to postcapillary venular resistance is also important. Precapillary constriction lowers filtration pressure, whereas postcapillary constriction raises it. Changes in any of these variables lead to changes in the volume of interstitial fluid. Factors promoting an increase in this volume are summarized in Table 31–13. **Edema** is the accumulation of interstitial fluid in abnormally large amounts.

In active tissues, capillary pressure rises, often to the point where it exceeds the oncotic pressure throughout the length of the capillary. In addition, osmotically active metabolites may temporarily accumulate in the interstitial fluid because they cannot be washed away as rapidly as they are formed. To the extent that they accumulate, they exert an osmotic effect that decreases the magnitude of the osmotic gradient due to the oncotic pressure. The amount of fluid leaving the capillaries is therefore markedly increased and the amount entering them reduced. Lymph flow is increased, decreasing the degree to which the fluid would otherwise accumulate, but exercising muscle, for example, still increases in volume by as much as 25%.

TABLE 31–13 Causes of increased interstitial fluid volume and edema.

Increased filtration pressure
Venular constriction
Increased venous pressure (heart failure, incompetent valves, venous obstruction, increased total ECF volume, effect of gravity, etc)
Decreased osmotic pressure gradient across capillary
Decreased plasma protein level
Accumulation of osmotically active substances in interstitial space
Increased capillary permeability
Substance P
Histamine and related substances
Kinins, etc
Inadequate lymph flow

Interstitial fluid tends to accumulate in dependent parts because of the effect of gravity. In the upright position, the capillaries in the legs are protected from the high arterial pressure by the arterioles, but the high venous pressure is transmitted to them through the venules. Skeletal muscle contractions keep the venous pressure low by pumping blood toward the heart (see above) when the individual moves about; however, if one stands still for long periods, fluid accumulates and edema eventually develops. The ankles also swell during long trips when travelers sit for prolonged periods with their feet in a dependent position. Venous obstruction may contribute to the edema in these situations.

Whenever there is abnormal retention of salt in the body, water is also retained. The salt and water are distributed throughout the ECF, and since the interstitial fluid volume is therefore increased, there is a predisposition to edema. Salt and water retention is a factor in the edema seen in heart failure, nephrosis, and cirrhosis, but there are also variations in the mechanisms that govern fluid movement across the capillary walls in these diseases. In congestive heart failure, for example, venous pressure is usually elevated, with a consequent elevation in capillary pressure. In cirrhosis of the liver, oncotic pressure is low because hepatic synthesis of plasma proteins is depressed; and in nephrosis, oncotic pressure is low because large amounts of protein are lost in the urine.

Another cause of edema is inadequate lymphatic drainage. Edema caused by lymphatic obstruction is called **lymphedema,** and the edema fluid has a high protein content. If it persists, it causes a chronic inflammatory condition that leads to fibrosis of the interstitial tissue. One cause of lymphedema is radical mastectomy, during which removal of the axillary lymph nodes leads to reduced lymph drainage. In filariasis, parasitic worms migrate into the lymphatics and obstruct them. Fluid accumulation plus tissue reaction lead in time to massive swelling, usually of the legs or scrotum **(elephantiasis).**

CHAPTER SUMMARY

- Blood consists of a suspension of red blood cells (erythrocytes), white blood cells, and platelets in a protein-rich fluid known as plasma.
- Blood cells arise in the bone marrow and are subject to regular renewal; the majority of plasma proteins are synthesized by the liver.
- Hemoglobin, stored in red blood cells, transports oxygen to peripheral tissues. Fetal hemoglobin is specialized to facilitate diffusion of oxygen from mother to fetus during development. Mutated forms of hemoglobin lead to red cell abnormalities and anemia.
- Complex oligosaccharide structures, specific to groups of individuals, form the basis of the ABO blood group system. AB blood group oligosaccharides, as well as other blood group molecules, can trigger the production of antibodies in naïve individuals following inappropriate transfusions, with potentially serious consequences due to erythrocyte agglutination.
- Blood flows from the heart to arteries and arterioles, thence to capillaries, and eventually to venules and veins and back to the heart. Each segment of the vasculature has specific contractile properties and regulatory mechanisms that subserve physiologic function. Physical principles of pressure, wall tension, and vessel caliber govern the flow of blood through each segment of the circulation.
- Transfer of oxygen and nutrients from the blood to tissues, as well as collection of metabolic wastes, occurs exclusively in the capillary beds.
- Fluid also leaves the circulation across the walls of capillaries. Some is reabsorbed; the remainder enters the lymphatic system, which eventually drains into the subclavian veins to return fluid to the bloodstream.
- Hypertension is an increase in mean blood pressure that is usually chronic and is common in humans. Hypertension can result in serious health consequences if left untreated. The majority of hypertension is of unknown cause, but several gene mutations underlie rare forms of the disease and are informative about mechanisms that control the dynamics of the circulatory system and its integration with other organs.

MULTIPLE-CHOICE QUESTIONS

For all questions, select the single best answer unless otherwise directed.

1. Which of the following has the highest *total* cross-sectional area in the body?
   A. Arteries
   B. Arterioles
   C. Capillaries
   D. Venules
   E. Veins
2. Lymph flow from the foot is
   A. increased when an individual rises from the supine to the standing position.
   B. increased by massaging the foot.
   C. increased when capillary permeability is decreased.
   D. decreased when the valves of the leg veins are incompetent.
   E. decreased by exercise.

3. The pressure in a capillary in skeletal muscle is 35 mm Hg at the arteriolar end and 14 mm Hg at the venular end. The interstitial pressure is 0 mm Hg. The colloid osmotic pressure is 25 mm Hg in the capillary and 1 mm Hg in the interstitium. The net force producing fluid movement across the capillary wall at its arteriolar end is
   A. 3 mm Hg out of the capillary.
   B. 3 mm Hg into the capillary.
   C. 10 mm Hg out of the capillary.
   D. 11 mm Hg out of the capillary.
   E. 11 mm Hg into the capillary.
4. The velocity of blood flow
   A. is higher in the capillaries than the arterioles.
   B. is higher in the veins than in the venules.
   C. is higher in the veins than the arteries.
   D. falls to zero in the descending aorta during diastole.
   E. is reduced in a constricted area of a blood vessel.
5. When the radius of the resistance vessels is increased, which of the following is increased?
   A. Systolic blood pressure
   B. Diastolic blood pressure
   C. Viscosity of the blood
   D. Hematocrit
   E. Capillary blood flow
6. A 30-year-old patient comes to her primary care physician complaining of headaches and vertigo. A blood test reveals a hematocrit of 55%, and a diagnosis of polycythemia is made. Which of the following would also be increased?
   A. Mean blood pressure
   B. Radius of the resistance vessels
   C. Radius of the capacitance vessels
   D. Central venous pressure
   E. Capillary blood flow
7. A pharmacologist discovers a drug that stimulates the production of VEGF receptors. He is excited because his drug might be of value in the treatment of
   A. coronary artery disease.
   B. cancer.
   C. emphysema.
   D. diabetes insipidus.
   E. dysmenorrhea.
8. Why is the dilator response to injected acetylcholine changed to a constrictor response when the endothelium is damaged?
   A. More Na^+ is generated.
   B. More bradykinin is generated.
   C. The damage lowers the pH of the remaining layers of the artery.
   D. The damage augments the production of endothelin by the endothelium.
   E. The damage interferes with the production of NO by the endothelium.

CHAPTER RESOURCES

Curry, RE, Adamson RH: Vascular permeability modulation at the cell, microvessel, or whole organ level: towards closing gaps in our knowledge. Cardiovasc Res 2010;87:218.

de Montalembert M: Management of sickle cell disease. Br Med J 2008;337:626.

Miller JL: Signaled expression of fetal hemoglobin during development. Transfusion 2005;45:1229.

Perrotta S, Gallagher PG, Mohandas N: Hereditary spherocytosis. Lancet 2008;372:1411.

Semenza GL: Vasculogenesis, angiogenesis, and arteriogenesis: Mechanisms of blood vessel formation and remodeling. J Cell Biochem 2007;102:840.

CHAPTER 32

Cardiovascular Regulatory Mechanisms

OBJECTIVES

After studying this chapter, you should be able to:

- Outline the neural mechanisms that control arterial blood pressure and heart rate, including the receptors, afferent and efferent pathways, central integrating pathways, and effector mechanisms involved.
- Describe the direct effects of CO_2 and hypoxia on the rostral ventrolateral medulla.
- Describe how the process of autoregulation contributes to control of vascular caliber.
- Identify the paracrine factors and hormones that regulate vascular tone, their sources, and their mechanisms of action.

INTRODUCTION

In humans and other mammals, multiple cardiovascular regulatory mechanisms have evolved. These mechanisms increase the blood supply to active tissues and increase or decrease heat loss from the body by redistributing the blood. In the face of challenges such as hemorrhage, they maintain the blood flow to the heart and brain. When the challenge faced is severe, flow to these vital organs is maintained at the expense of the circulation to the rest of the body.

Circulatory adjustments are effected by altering the output of the pump (the heart), changing the diameter of the resistance vessels (primarily the arterioles), or altering the amount of blood pooled in the capacitance vessels (the veins). Regulation of cardiac output is discussed in Chapter 30. The caliber of the arterioles is adjusted in part by autoregulation (Table 32–1). It is also increased in active tissues by locally produced vasodilator metabolites, is affected by substances secreted by the endothelium, and is regulated systemically by circulating vasoactive substances and the nerves that innervate the arterioles. The caliber of the capacitance vessels is also affected by circulating vasoactive substances and by vasomotor nerves. The systemic regulatory mechanisms synergize with the local mechanisms and adjust vascular responses throughout the body.

The terms **vasoconstriction** and **vasodilation** are generally used to refer to constriction and dilation of the resistance vessels. Changes in the caliber of the veins are referred to as **venoconstriction** or **venodilation.**

NEURAL CONTROL OF THE CARDIOVASCULAR SYSTEM

INNERVATION OF THE BLOOD VESSELS

Most of the vasculature is an example of an autonomic effector organ that receives innervation from the sympathetic but not the parasympathetic division of the autonomic nervous system. Sympathetic noradrenergic fibers terminate on vascular smooth muscle in all parts of the body to mediate vasoconstriction. In some species, resistance vessels in skeletal muscles of the limbs are also innervated by vasodilator fibers, which, although they travel with the sympathetic nerves, are cholinergic **(sympathetic cholinergic vasodilator system).** These nerve are inactive at rest but can be activated during stress or exercise. Evidence for a sympathetic cholinergic vasodilator system in humans is lacking. It is more likely that vasodilation of skeletal muscle vasculature in response to activation of the sympathetic nervous system is due to the actions of epinephrine released from the adrenal medulla. Activation of β_2-adrenoceptors on skeletal muscle blood vessels promotes vasodilation.

There are a few exceptions to the rule that only the sympathetic nervous system controls the vascular smooth muscle. The arteries in the erectile tissue of the reproductive organs, uterine and some facial blood vessels, and blood vessels in salivary glands, may also be controlled by parasympathetic nerves.

Although the arterioles and the other resistance vessels are most densely innervated, all blood vessels except capillaries and venules contain smooth muscle and receive motor nerve fibers from the sympathetic division of the autonomic nervous system. The fibers to the resistance vessels regulate tissue blood flow and arterial pressure. The fibers to the venous capacitance vessels vary the volume of blood "stored" in the veins. The innervation of most veins is sparse, but the splanchnic veins are well innervated. Venoconstriction is produced by stimuli that also activate the vasoconstrictor nerves to the arterioles. The resultant decrease in venous capacity increases venous return, shifting blood to the arterial side of the circulation.

When the sympathetic nerves are sectioned **(sympathectomy),** the blood vessels dilate. A change in the level of activity (increase or decrease) in sympathetic nerves is just one of the many factors that mediate vasoconstriction or vasodilation (Table 32–1).

TABLE 32–1 Summary of factors affecting the caliber of the arterioles.

Vasoconstriction	Vasodilation
Local factors	
Decreased local temperature	Increased CO_2 and decreased O_2
Autoregulation	Increased K^+, adenosine, lactate
	Decreased local pH
	Increased local temperature
Endothelial products	
Endothelin-1	Nitric oxide
Locally released platelet serotonin	Kinins
Thromboxane A_2	Prostacyclin
Circulating neurohumoral agents	
Epinephrine (except in skeletal muscle and liver)	Epinephrine in skeletal muscle and liver
Norepinephrine	Calcitonin G-related protein
Arginine vasopressin	Substance P
Angiotensin II	Histamine
Endogenous digitalis-like substance	Atrial natriuretic peptide
Neuropeptide Y	Vasoactive intestinal polypeptide
Neural factors	
Increased discharge of sympathetic nerves	Decreased discharge of sympathetic nerves
	Activation of sympathetic cholinergic vasodilator nerves to vasculature of skeletal muscles of the limbs

INNERVATION OF THE HEART

The heart is one example of an effector organ that receives opposing influences from the sympathetic and parasympathetic divisions of the autonomic nervous system. Release of norepinephrine from postganglionic sympathetic nerves activates β_1-adrenoceptors in the heart, notably on the sinoatrial (SA) node, atrioventricular (AV) node, His-Purkinje conductive tissue, and atrial and ventricular contractile tissue. In response to stimulation of sympathetic nerves, the heart rate **(chronotropy),** rate of transmission in the cardiac conductive tissue **(dromotropy),** and the force of ventricular contraction **(inotropy)** are increased. On the other hand, release of acetylcholine from postganglionic parasympathetic (vagus) nerves activates nicotinic receptors in the heart, notably on the SA and AV nodes and atrial muscle. In response to stimulation of the vagus nerve, the heart rate, the rate of transmission through the AV node, and atrial contractility are reduced.

The above description presents an oversimplified explanation of autonomic control of cardiac function. There are adrenergic and cholinergic receptors on autonomic nerve terminals that modulate transmitter release from nerve endings. For example, release of acetylcholine from vagal nerve terminals inhibits the release of norepinephrine from sympathetic nerve terminals, so this can enhance the effects of vagal nerve activation on the heart.

There is a moderate amount of tonic discharge in the cardiac sympathetic nerves at rest, but there is considerable tonic vagal discharge **(vagal tone)** in humans and other large animals. After the administration of nicotinic cholinergic receptor antagonists such as atropine, the heart rate in humans increases from 70, its normal resting value, to 150–180 beats/min because the sympathetic tone is unopposed. In humans in whom both noradrenergic and cholinergic systems are blocked, the heart rate is approximately 100 beats/min.

CARDIOVASCULAR CONTROL

The cardiovascular system is under neural influences coming from several parts of the brain stem, forebrain, and insular cortex. The brain stem receives feedback from sensory receptors in the vasculature (eg, baroreceptors and chemoreceptors). A simplified model of the feedback control circuit is shown

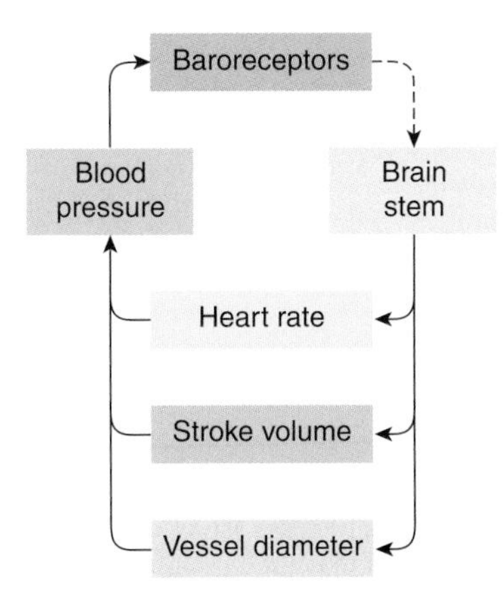

FIGURE 32–1 Feedback control of blood pressure. Brain stem excitatory input to sympathetic nerves to the heart and vasculature increases heart rate and stroke volume and reduces vessel diameter. Together these increase blood pressure, which activates the baroreceptor reflex to reduce the activity in the brain stem.

in Figure 32–1. An increase in neural output from the brain stem to sympathetic nerves leads to a decrease in blood vessel diameter (arteriolar vasoconstriction) and increases in stroke volume and heart rate, which contribute to a rise in blood pressure. This in turn causes an increase in baroreceptor activity, which signals the brain stem to reduce the neural output to sympathetic nerves.

Venoconstriction and a decrease in the stores of blood in the venous reservoirs usually accompany increases in arteriolar constriction, although changes in the capacitance vessels do not always parallel changes in the resistance vessels. In the presence of an increase in sympathetic nerve activity to the heart and vasculature, there is usually an associated decrease in the activity of vagal fibers to the heart. Conversely, a decrease in sympathetic activity causes vasodilation, a fall in blood pressure, and an increase in the storage of blood in the venous reservoirs. There is usually a concomitant decrease in heart rate, but this is mostly due to stimulation of the vagal innervation of the heart.

MEDULLARY CONTROL OF THE CARDIOVASCULAR SYSTEM

One of the major sources of excitatory input to sympathetic nerves controlling the vasculature is a group of neurons located near the pial surface of the medulla in the **rostral ventrolateral medulla** (RVLM; Figure 32–2). This region is sometimes called a vasomotor area. The axons of RVLM neurons course dorsally and medially and then descend in the lateral column of the spinal cord to the thoracolumbar intermediolateral gray column (IML). They contain phenylethanolamine-N-methyltransferase (PNMT; see Chapter 7), but it appears that the excitatory transmitter they secrete is glutamate rather than epinephrine. Neurovascular compression of the RVLM has been linked to some cases of **essential hypertension** in humans (see Clinical Box 32–1).

The activity of RVLM neurons is determined by many factors (see Table 32–2). They include not only the very important fibers from arterial baroreceptors, but also fibers from other parts of the nervous system and from the carotid and aortic chemoreceptors. In addition, some stimuli act directly on the vasomotor area.

There are descending tracts to the vasomotor area from the cerebral cortex (particularly the limbic cortex) that relay in the hypothalamus. These fibers are responsible for the blood pressure rise and tachycardia produced by emotions such as stress, sexual excitement, and anger. The connections between the hypothalamus and the vasomotor area are reciprocal, with afferents from the brain stem closing the loop.

Inflation of the lungs causes vasodilation and a decrease in blood pressure. This response is mediated via vagal afferents from the lungs that inhibit vasomotor discharge. Pain usually causes a rise in blood pressure via afferent impulses in the reticular formation converging in the RVLM. However, prolonged severe pain may cause vasodilation and fainting. The activity in afferents from exercising muscles probably exerts a similar pressor effect via a pathway to the RVLM. The pressor response to stimulation of somatic afferent nerves is called the **somatosympathetic reflex.**

The medulla is also a major site of origin of excitatory input to cardiac vagal motor neurons in the nucleus ambiguus (Figure 32–3). Table 32–3 is a summary of factors that affect the heart rate. In general, stimuli that increase the heart rate also increase blood pressure, whereas those that decrease the heart rate lower blood pressure. However, there are exceptions, such as the production of hypotension and tachycardia by stimulation of atrial stretch receptors and the production of hypertension and bradycardia by increased intracranial pressure.

BARORECEPTORS

The **baroreceptors** are stretch receptors in the walls of the heart and blood vessels. The **carotid sinus** and **aortic arch** receptors monitor the arterial circulation. Receptors are also located in the walls of the right and left atria at the entrance of the superior and inferior venae cavae and the pulmonary veins, as well as in the pulmonary circulation. These receptors in the low-pressure part of the circulation are referred to collectively as the **cardiopulmonary receptors.**

The carotid sinus is a small dilation of the internal carotid artery just above the bifurcation of the common carotid into external and internal carotid branches (Figure 32–4). Baroreceptors are located in this dilation. They are also found in the wall of the arch of the aorta. The receptors are located in the adventitia of the vessels. The afferent nerve fibers from the carotid sinus form a distinct branch of the glossopharyngeal nerve, the **carotid sinus nerve.** The fibers from the aortic arch form a branch of the vagus nerve, the **aortic depressor nerve.**

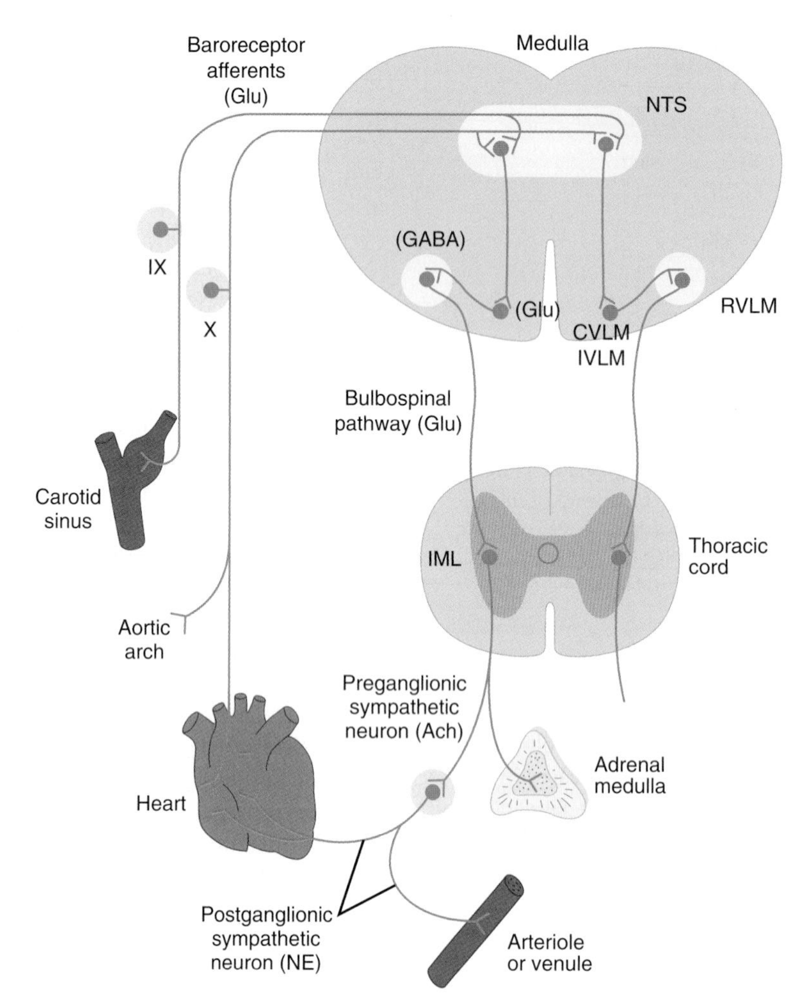

FIGURE 32–2 Basic pathways involved in the medullary control of blood pressure. The rostral ventrolateral medulla (RVLM) is one of the major sources of excitatory input to sympathetic nerves controlling the vasculature. These neurons receive inhibitory input from the baroreceptors via an inhibitory neuron in the caudal ventrolateral medulla (CVLM). The nucleus of the tractus solitarius (NTS) is the site of termination of baroreceptor afferent fibers. The putative neurotransmitters in the pathways are indicated in parentheses. Ach, acetylcholine; GABA, γ-aminobutyric acid; Glu, glutamate; IML, intermediolateral gray column; IVLM, intermediate ventrolateral medulla; NE, norepinephrine; NTS, nucleus of the tractus solitarius; IX and X, glossopharyngeal and vagus nerves.

The baroreceptors are stimulated by distention of the structures in which they are located, and so they discharge at an increased rate when the pressure in these structures rises. Their afferent fibers pass via the glossopharyngeal and vagus nerves to the medulla. Most of them end in the **nucleus of the tractus solitarius** (NTS), and the excitatory transmitter they secrete is glutamate (Figure 32–2). Excitatory (glutamate) projections extend from the NTS to the **caudal ventrolateral medulla** (CVLM), where they stimulate γ-aminobutyrate (GABA)-secreting inhibitory neurons that project to the RVLM. Excitatory projections also extend from the NTS to the vagal motor neurons in the nucleus ambiguus and dorsal motor nucleus (Figure 32–3). Thus, increased baroreceptor discharge *inhibits* the tonic discharge of sympathetic nerves and *excites* the vagal innervation of the heart. These neural changes produce vasodilation, venodilation, hypotension, bradycardia, and a decrease in cardiac output.

BARORECEPTOR NERVE ACTIVITY

Baroreceptors are more sensitive to pulsatile pressure than to constant pressure. A decline in pulse pressure without any change in mean pressure decreases the rate of baroreceptor discharge and provokes a rise in systemic blood pressure and tachycardia. At normal blood pressure levels (about 100 mm Hg mean pressure), a burst of action potentials appears in a single baroreceptor fiber during systole, but there are few action potentials in early diastole (Figure 32–5). At lower mean pressures, this phasic change in firing is even more dramatic with activity only occurring during systole. At these lower pressures, the overall firing rate is considerably reduced. The threshold for eliciting activity in the carotid sinus nerve is approximately 50 mm Hg; maximal activity occurs at approximately 200 mm Hg, with activity throughout the cardiac cycle.

When one carotid sinus is isolated and perfused and the other baroreceptors are denervated, there is no discharge in the

CLINICAL BOX 32-1

Essential Hypertension & Neurovascular Compression of the RVLM

In about 88% of patients with elevated blood pressure, the cause of the hypertension is unknown, and they are said to have **essential hypertension** (see Chapter 31). There are data available to support the view that **neurovascular compression** of the RVLM is associated with essential hypertension in some subjects. For example, patients with a schwannoma (acoustic neuroma) or meningioma lying close to the RVLM also have hypertension. Magnetic resonance angiography (MRA) has been used to compare the incidence of neurovascular compression in hypertensive and normotensive individuals and to correlate indices of sympathetic nerve activity with the presence or absence of compression. Some of these studies showed a higher incidence of coexistence of neurovascular compression with essential hypertension than in other forms of hypertension or normotension, but others showed the presence of a compression in normotensive subjects. On the other hand, there was a strong positive relationship between the presence of neurovascular compression and increased sympathetic activity.

THERAPEUTIC HIGHLIGHTS

In the 1970s, Dr. Peter Jannetta, a neurosurgeon in Pittsburgh, PA, developed a technique for "microvascular decompression" of the medulla to treat trigeminal neuralgia and hemifacial spasm, which he attributed to pulsatile compression of the vertebral and posterior inferior cerebellar arteries impinging on the fifth and seventh cranial nerves. Moving the arteries away from the nerves led to reversal of the neurologic symptoms in many cases. Some of these patients were also hypertensive, and they showed reductions in blood pressure postoperatively. Later, a few human studies claimed that surgical decompression of the RVLM could sometimes relieve hypertension. There are also reports that hypertension is relieved after surgical decompression in patients with a schwannoma or meningioma in the vicinity of the RVLM.

afferent fibers from the perfused sinus and no drop in the animal's arterial pressure or heart rate when the perfusion pressure is below 30 mm Hg (Figure 32–6). At carotid sinus perfusion pressures of 70–110 mm Hg, there is a near linear relationship between perfusion pressure and the fall in systemic blood pressure and heart rate. At perfusion pressures above 150 mm Hg there is no further increase in response, presumably because the rate of baroreceptor discharge and the degree of inhibition of sympathetic nerve activity are maximal.

TABLE 32–2 Factors affecting the activity of the RVLM.

Direct stimulation
CO_2
Hypoxia
Excitatory inputs
Cortex via hypothalamus
Mesencephalic periaqueductal gray
Brain stem reticular formation
Pain pathways
Somatic afferents (somatosympathetic reflex)
Carotid and aortic chemoreceptors
Inhibitory inputs
Cortex via hypothalamus
Caudal ventrolateral medulla
Caudal medullary raphé nuclei
Lung inflation afferents
Carotid, aortic, and cardiopulmonary baroreceptors

From the foregoing discussion, it is apparent that the baroreceptors on the arterial side of the circulation, their afferent connections to the medullary cardiovascular areas, and the efferent pathways from these areas constitute a reflex feedback mechanism that operates to stabilize blood pressure and

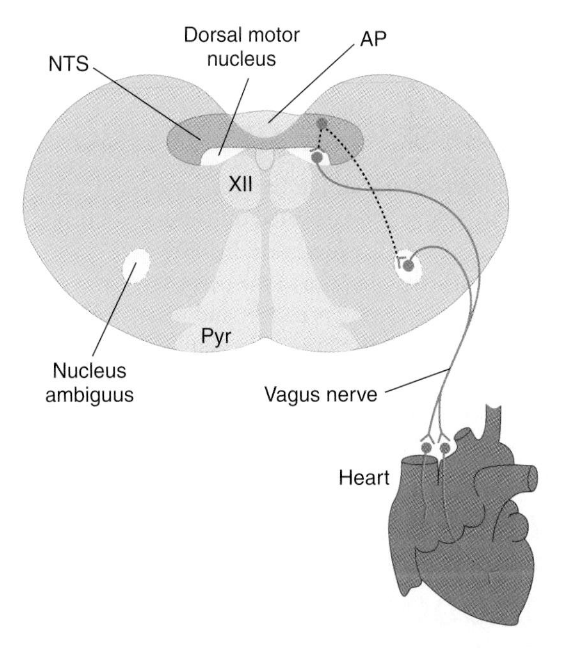

FIGURE 32–3 Basic pathways involved in the medullary control of heart rate by the vagus nerves. Neurons in the nucleus of the tractus solitarius (NTS) project to and excite cardiac preganglionic parasympathetic neurons primarily in the nucleus ambiguus. Some are also located in the dorsal motor nucleus of the vagus; however, this nucleus primarily contains vagal motor neurons that project to the gastrointestinal tract. AP, area postrema; Pyr, pyramid; XII, hypoglossal nucleus.

TABLE 33–3 Factors affecting heart rate.

Heart rate accelerated by:
Decreased activity of arterial baroreceptors
Increased activity of atrial stretch receptors
Inspiration
Excitement
Anger
Most painful stimuli
Hypoxia
Exercise
Thyroid hormones
Fever
Heart rate slowed by:
Increased activity of arterial baroreceptors
Expiration
Fear
Grief
Stimulation of pain fibers in trigeminal nerve
Increased intracranial pressure

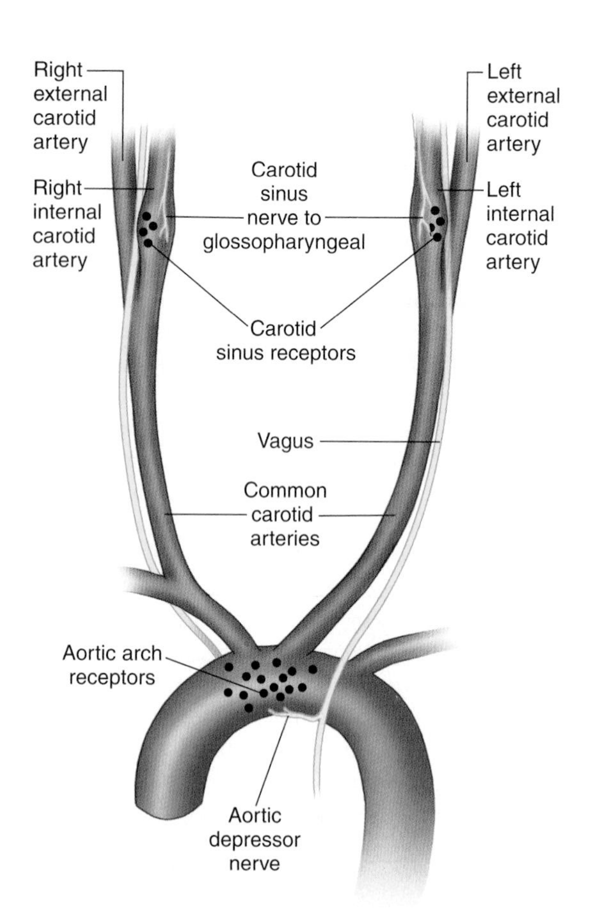

FIGURE 32–4 Baroreceptor areas in the carotid sinus and aortic arch. One set of baroreceptors (stretch receptors) is located in the carotid sinus, a small dilation of the internal carotid artery just above the bifurcation of the common carotid into external and internal carotid branches. These receptors are innervated by a branch of the glossopharyngeal nerve, the carotid sinus nerve. A second set of baroreceptors is located in the wall of the arch of the aorta. These receptors are innervated by a branch of the vagus nerve, the aortic depressor nerve.

heart rate. Any drop in systemic arterial pressure decreases the discharge in the buffer nerves, and there is a compensatory rise in blood pressure and cardiac output. Any rise in pressure produces dilation of the arterioles and decreases cardiac output until the blood pressure returns to its previous baseline level.

BARORECEPTOR RESETTING

In chronic hypertension, the baroreceptor reflex mechanism is "reset" to maintain an elevated rather than a normal blood pressure. In perfusion studies on hypertensive experimental animals, raising the pressure in the isolated carotid sinus lowers the elevated systemic pressure, and decreasing the perfusion pressure raises the elevated pressure (Figure 32–6). Little is known about how and why this occurs, but resetting occurs rapidly in experimental animals. It is also rapidly reversible, both in experimental animals and in clinical situations.

ROLE OF BARORECEPTORS IN SHORT-TERM CONTROL OF BLOOD PRESSURE

The changes in pulse rate and blood pressure that occur in humans on standing up or lying down are due for the most part to baroreceptor reflexes. The function of the receptors can be tested by monitoring changes in heart rate as a function of increasing arterial pressure during infusion of the α-adrenoceptor agonist phenylephrine. A normal response is shown in Figure 32–7; from a systolic pressure of about 120–150 mm Hg, there is a linear relation between pressure and lowering of the heart rate (longer RR interval). Baroreceptors are very important in short-term control of arterial pressure. Activation of the reflex allows for rapid adjustments in blood pressure in response to abrupt changes in posture, blood volume, cardiac output, or peripheral resistance during exercise.

Blood pressure initially rises dramatically after bilateral section of baroreceptor nerves or bilateral lesions of the NTS. However, after a period of time, mean blood pressure returns to near control levels, but there are large fluctuations in pressure during the course of a day. Removal of the baroreceptor reflex prevents an individual from adjusting their blood pressure in response to stimuli that cause abrupt changes in blood volume, cardiac output, or peripheral resistance, including exercise and postural changes. A long-term change in blood pressure resulting from loss of baroreceptor reflex control is called **neurogenic hypertension.**

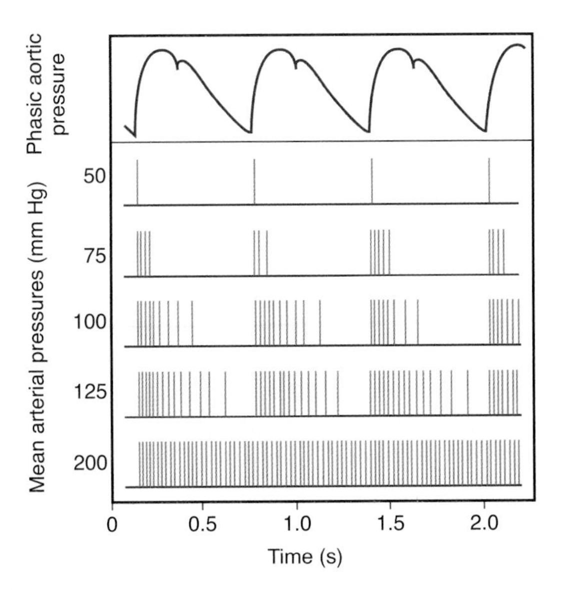

FIGURE 32–5 Discharges (vertical lines) in a single afferent nerve fiber from the carotid sinus at various levels of mean arterial pressures, plotted against changes in aortic pressure with time. Baroreceptors are very sensitive to changes in pulse pressure as shown by the record of phasic aortic pressure. (Reproduced with permission from Levy MN & Pappano AJ: *Cardiovascular Physiology*, 9th ed. Mosby, 2007.)

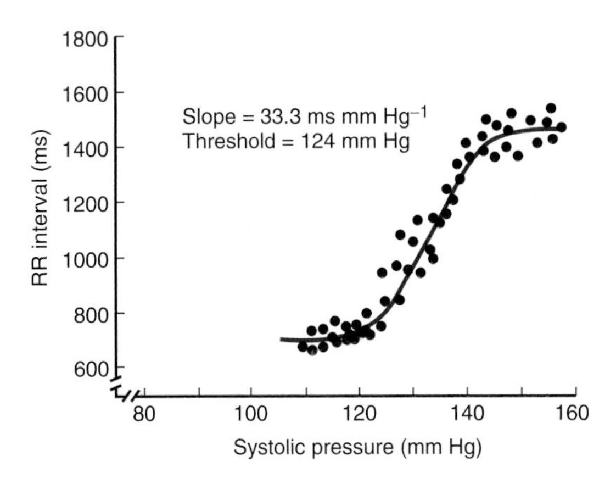

FIGURE 32–7 Baroreflex-mediated lowering of the heart rate during infusion of phenylephrine in a human subject. Note that the values for the RR interval of the electrocardiogram, which are plotted on the vertical axis, are inversely proportional to the heart rate. (Reproduced with permission from Kotrly K, et al: Effects of fentanyl-diazepam-nitrous oxide anaesthesia on arterial baroreflex control of heart rate in man. Br J Anaesth 1986;58:406.)

ATRIAL STRETCH AND CARDIOPULMONARY RECEPTORS

The stretch receptors in the atria are of two types: those that discharge primarily during atrial systole (type A), and those that discharge primarily late in diastole, at the time of peak atrial filling (type B). The discharge of type B baroreceptors is increased when venous return is increased and decreased by positive-pressure breathing, indicating that these baroreceptors respond primarily to distention of the atrial walls. The reflex circulatory adjustments initiated by increased discharge from most if not all of these receptors include vasodilation and a fall in blood pressure. However, the heart rate is increased rather than decreased.

Receptors in the endocardial surfaces of the ventricles are activated during ventricular distention. The response is a vagal bradycardia and hypotension, comparable to a baroreceptor reflex. Left ventricular stretch receptors may play a role in the maintenance of vagal tone that keeps the heart rate low at rest. Various chemicals are known to elicit reflexes due to activation of cardiopulmonary chemoreceptors and may play a role in various cardiovascular disorders (see Clinical Box 32–2).

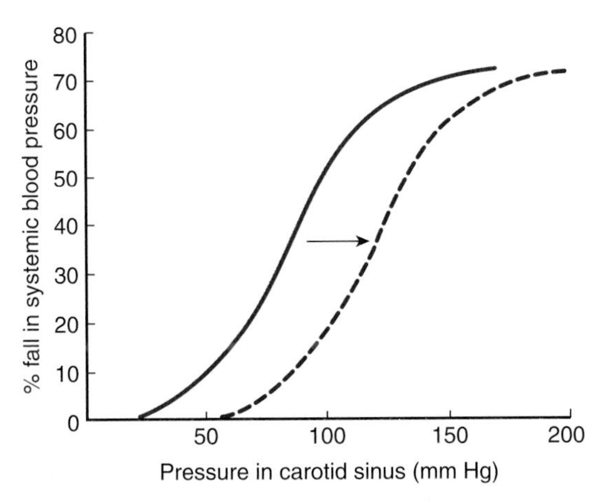

FIGURE 32–6 Fall in systemic blood pressure produced by raising the pressure in the isolated carotid sinus to various values. Solid line: Response in a normal monkey. **Dashed line:** Response in a hypertensive monkey, demonstrating baroreceptor resetting (arrow).

VALSALVA MANEUVER

The function of the receptors can also be tested by monitoring the changes in pulse and blood pressure that occur in response to brief periods of straining (forced expiration against a closed glottis: the **Valsalva maneuver**). Valsalva maneuvers occur regularly during coughing, defecation, and heavy lifting. The blood pressure rises at the onset of straining (Figure 32–8) because the increase in intrathoracic pressure is added to the pressure of the blood in the aorta. It then falls because the high intrathoracic pressure compresses the veins, decreasing venous return and cardiac output. The decreases in arterial pressure and pulse pressure inhibit the baroreceptors, causing tachycardia and a rise in peripheral resistance. When the glottis is opened and the intrathoracic pressure returns to normal, cardiac output is restored but the peripheral vessels are constricted. The blood pressure therefore rises above normal, and this stimulates the baroreceptors, causing bradycardia and a drop in pressure to normal levels.

CLINICAL BOX 32–2

Cardiopulmonary Chemosensitive Receptors

For nearly 150 years, it has been known that activation of chemosensitive vagal C fibers in the cardiopulmonary region (eg, juxtacapillary region of alveoli, ventricles, atria, great veins, and pulmonary artery) causes profound bradycardia, hypotension, and a brief period of apnea followed by rapid shallow breathing. This response pattern is called the **Bezold-Jarisch reflex** and was named after the individuals who first reported these findings. This reflex can be elicited by a variety of substances including capsaicin, serotonin, phenylbiguanide, and veratridine. Although originally viewed as a pharmacologic curiosity, there is a growing body of evidence supporting the view that the Bezold-Jarisch reflex is activated during certain pathophysiologic conditions. For example, this reflex may be activated during myocardial ischemia and reperfusion as a result of increased production of oxygen radicals and by agents used as radiocontrast for coronary angiography. This can contribute to the hypotension that is frequently a stubborn complication of heart disease. Activation of cardiopulmonary chemosensitive receptors may also be part of a defense mechanism protecting individuals from toxic chemical hazards. Activation of cardiopulmonary reflexes may help reduce the amount of inspired pollutants that gets absorbed into the blood, protecting vital organs from potential toxicity of these pollutants, and facilitating their elimination. Finally, the syndrome of cardiac slowing with hypotension **(vasovagal syncope)** has also been attributed to activation of the Bezold-Jarisch reflex. Vasovagal syncope can occur after prolonged upright posture that results in pooling of blood in the lower extremities and diminished intracardiac blood volume (also called **postural syncope**). This phenomenon is exaggerated if combined with dehydration. The resultant arterial hypotension is sensed in the carotid sinus baroreceptors, and afferent fibers from these receptors trigger autonomic signals that increase cardiac rate and contractility. However, pressure receptors in the wall of the left ventricle respond by sending signals that trigger paradoxical bradycardia and decreased contractility, resulting in sudden marked hypotension. The individual also feels lightheaded and may experience a brief episode of loss of consciousness.

THERAPEUTIC HIGHLIGHTS

The most critical intervention for individuals who experience episodes of neurogenic syncope is to avoid dehydration and to avoid situations that trigger the adverse event. Episodes of syncope may be reduced or prevented by an increased dietary salt intake or administration of mineralocorticoids. Vasovagal syncope has been treated with the use of **β-adrenoceptor antagonists** and **disopyramide,** an antiarrhythmic agent that blocks Na^+ channels. Cardiac pacemakers have also been used to stabilize the heart rate during episodes that normally trigger bradycardia.

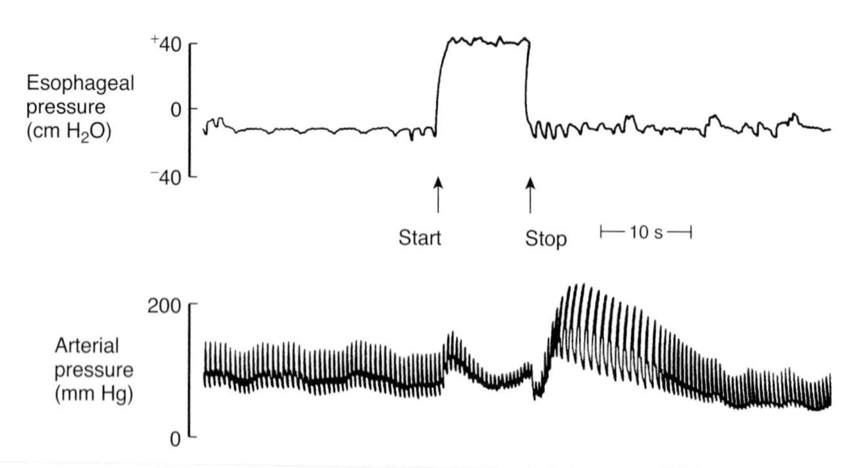

FIGURE 32–8 Diagram of the response to straining (the Valsalva maneuver) in a normal man, recorded with a needle in the brachial artery. Blood pressure rises at the onset of straining because increased intrathoracic pressure is added to the pressure of the blood in the aorta. It then falls because the high intrathoracic pressure compresses veins, decreasing venous return and cardiac output. (Courtesy of M McIlroy.)

In patients whose sympathetic nervous system is not functional, heart rate changes still occur because the baroreceptors and the vagi are intact. However, in patients with autonomic insufficiency, a syndrome in which autonomic function is widely disrupted, the heart rate changes are absent. For reasons that are still obscure, patients with primary hyperaldosteronism also fail to show the heart rate changes and the blood pressure rise when the intrathoracic pressure returns to normal. Their response to the Valsalva maneuver returns to normal after removal of the aldosterone-secreting tumor.

PERIPHERAL CHEMORECEPTOR REFLEX

Peripheral arterial chemoreceptors in the **carotid and aortic bodies** have very high rates of blood flow. These receptors are primarily activated by a reduction in partial pressure of oxygen (PaO_2), but they also respond to an increase in the partial pressure of carbon dioxide ($PaCO_2$) and pH. Chemoreceptors exert their main effects on respiration; however, their activation also leads to vasoconstriction. Heart rate changes are variable and depend on various factors, including changes in respiration. A direct effect of chemoreceptor activation is to increase vagal nerve activity. However, hypoxia also produces hyperpnea and increased catecholamine secretion from the adrenal medulla, both of which produce tachycardia and an increase in cardiac output. Hemorrhage that produces hypotension leads to chemoreceptor stimulation due to decreased blood flow to the chemoreceptors and consequent stagnant anoxia of these organs. Chemoreceptor discharge may also contribute to the production of **Mayer waves.** These should not be confused with **Traube–Hering waves,** which are fluctuations in blood pressure synchronized with respiration. The Mayer waves are slow, regular oscillations in arterial pressure that occur at the rate of about one per 20–40 s during hypotension. Under these conditions, hypoxia stimulates the chemoreceptors. The stimulation raises the blood pressure, which improves the blood flow in the receptor organs and eliminates the stimulus to the chemoreceptors, so that the pressure falls and a new cycle is initiated.

CENTRAL CHEMORECEPTORS

When intracranial pressure is increased, the blood supply to RVLM neurons is compromised, and the local hypoxia and hypercapnia increase their discharge. This activates a central chemoreceptors located on the ventrolateral surface of the medulla. The resultant rise in systemic arterial pressure **(Cushing reflex)** tends to restore the blood flow to the medulla. Over a considerable range, the blood pressure rise is proportional to the increase in intracranial pressure. The rise in blood pressure causes a reflex decrease in heart rate via the arterial baroreceptors. This is why bradycardia rather than tachycardia is characteristically seen in patients with increased intracranial pressure.

A rise in arterial P_{CO_2} stimulates the RVLM, but the direct peripheral effect of hypercapnia is vasodilation. Therefore, the peripheral and central actions tend to cancel each other out. Moderate hyperventilation, which significantly lowers the CO_2 tension of the blood, causes cutaneous and cerebral vasoconstriction in humans, but there is little change in blood pressure. Exposure to high concentrations of CO_2 is associated with marked cutaneous and cerebral vasodilation, but vasoconstriction occurs elsewhere and usually there is a slow rise in blood pressure.

LOCAL REGULATION

AUTOREGULATION

The capacity of tissues to regulate their own blood flow is referred to as **autoregulation.** Most vascular beds have an intrinsic capacity to compensate for moderate changes in perfusion pressure by changes in vascular resistance, so that blood flow remains relatively constant. This capacity is well developed in the kidneys (see Chapter 37), but it has also been observed in the mesentery, skeletal muscle, brain, liver, and myocardium. It is probably due in part to the intrinsic contractile response of smooth muscle to stretch **(myogenic theory of autoregulation).** As the pressure rises, the blood vessels are distended and the vascular smooth muscle fibers that surround the vessels contract. If it is postulated that the muscle responds to the tension in the vessel wall, this theory could explain the greater degree of contraction at higher pressures; the wall tension is proportional to the distending pressure times the radius of the vessel (law of Laplace; see Chapter 31), and the maintenance of a given wall tension as the pressure rises would require a decrease in radius. Vasodilator substances tend to accumulate in active tissues, and these "metabolites" also contribute to auto-regulation **(metabolic theory of autoregulation).** When blood flow decreases, they accumulate and the vessels dilate; when blood flow increases, they tend to be washed away.

VASODILATOR METABOLITES

The metabolic changes that produce vasodilation include, in most tissues, decreases in O_2 tension and pH. These changes cause relaxation of the arterioles and precapillary sphincters. A local fall in O_2 tension, in particular, can initiate a program of vasodilatory gene expression secondary to production of hypoxia-inducible factor-1α (HIF-1α), a transcription factor with multiple targets. Increases in CO_2 tension and osmolality also dilate the vessels. The direct dilator action of CO_2 is most pronounced in the skin and brain. The neurally mediated vasoconstrictor effects of systemic as opposed to local hypoxia and hypercapnia have been discussed above. A rise in temperature exerts a direct vasodilator effect, and the temperature rise in active tissues (due to the heat of metabolism) may contribute to the vasodilation. K^+ is another substance that accumulates locally, and has demonstrated dilator activity secondary to the hyperpolarization of vascular smooth muscle cells. Lactate may also contribute to the dilation. In injured tissues, histamine released from damaged cells increases capillary permeability. Thus, it is probably responsible for some of the swelling in areas of inflammation. Adenosine may play a vasodilator role in cardiac muscle but not in skeletal muscle. It also inhibits the release of norepinephrine.

LOCALIZED VASOCONSTRICTION

Injured arteries and arterioles constrict strongly. The constriction appears to be due in part to the local liberation of serotonin from platelets that stick to the vessel wall in the injured area. Injured veins also constrict.

A drop in tissue temperature causes vasoconstriction, and this local response to cold plays a part in temperature regulation (see Chapter 17).

SUBSTANCES SECRETED BY THE ENDOTHELIUM

ENDOTHELIAL CELLS

As noted in Chapter 31, the endothelial cells constitute a large and important tissue. They secrete many growth factors and vasoactive substances. The vasoactive substances include prostaglandins and thromboxanes, nitric oxide (NO), and endothelins.

PROSTACYCLIN & THROMBOXANE A_2

Prostacyclin is produced by endothelial cells and thromboxane A_2 by platelets from their common precursor arachidonic acid via the cyclooxygenase pathway. Thromboxane A_2 promotes platelet aggregation and vasoconstriction, whereas prostacyclin inhibits platelet aggregation and promotes vasodilation. The balance between platelet thromboxane A_2 and prostacyclin fosters localized platelet aggregation and consequent clot formation (see Chapter 31) while preventing excessive extension of the clot and maintaining blood flow around it.

The thromboxane A_2–prostacyclin balance can be shifted toward prostacyclin by administration of low doses of aspirin. Aspirin produces irreversible inhibition of cyclooxygenase by acetylating a serine residue in its active site. Obviously, this reduces production of both thromboxane A_2 and prostacyclin. However, endothelial cells produce new cyclooxygenase in a matter of hours, whereas platelets cannot manufacture the enzyme, and the level rises only as new platelets enter the circulation. This is a slow process because platelets have a half-life of about 4 days. Therefore, administration of small amounts of aspirin for prolonged periods reduces clot formation and has been shown to be of value in preventing myocardial infarctions, unstable angina, transient ischemic attacks, and stroke.

NITRIC OXIDE

A chance observation two decades ago led to the discovery that the endothelium plays a key role in vasodilation. Many different stimuli act on the endothelial cells to produce **endothelium-derived relaxing factor (EDRF),** a substance that is now known to be **nitric oxide (NO).** NO is synthesized from arginine (Figure 32–9) in a reaction catalyzed by nitric oxide synthase (NO synthase, NOS). Three isoforms of NOS have been identified: NOS 1, found in the nervous system; NOS 2, found in macrophages and other immune cells; and NOS 3, found in endothelial cells. NOS 1 and NOS 3 are activated by agents that increase intracellular Ca^{2+} concentrations, including the vasodilators acetylcholine and bradykinin. The NOS in immune cells is not activated by Ca^{2+} but is induced by cytokines. The NO that is formed in the endothelium diffuses to smooth muscle cells, where it activates soluble guanylyl cyclase, producing cyclic 3,5-guanosine monophosphate (cGMP; see Figure 32–9), which in turn mediates the relaxation of vascular smooth muscle. NO is inactivated by hemoglobin.

Adenosine, atrial natriuretic peptide (ANP), and histamine via H_2 receptors produce relaxation of vascular smooth muscle that is independent of the endothelium. However,

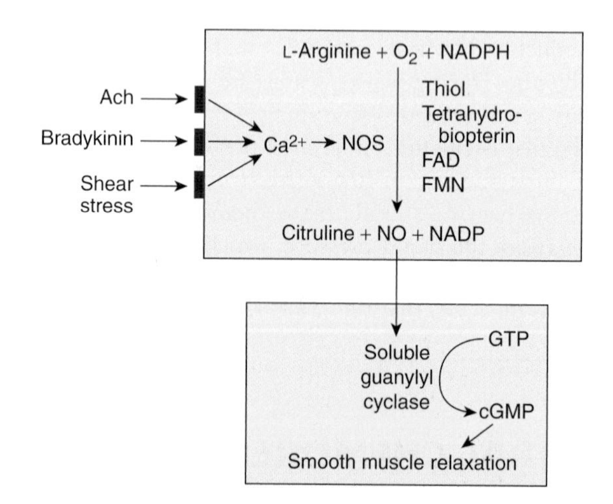

FIGURE 32–9 Synthesis of NO from arginine in endothelial cells and its action via stimulation of soluble guanylyl cyclase and generation of cGMP to produce relaxation in vascular smooth muscle cells. The endothelial form of nitric oxide synthase (NOS) is activated by increased intracellular Ca^{2+} concentration (top), and an increase is produced by acetylcholine (Ach), bradykinin, or shear stress acting on the cell membrane. Thiol, tetrahydrobiopterin, FAD, and FMN are requisite cofactors. NO then diffuses to adjacent smooth muscle cells in the wall of the vessel (bottom), diffuses across the plasma membrane, and activates soluble guanylyl cyclase to evoke an increase in cellular cGMP and smooth muscle relaxation.

acetylcholine, histamine via H_1 receptors, bradykinin, vasoactive intestinal peptide (VIP), substance P, and some other polypeptides act via the endothelium, and various vasoconstrictors that act directly on vascular smooth muscle would produce much greater constriction if their effects were not limited by their ability simultaneously to cause release of NO. When flow to a tissue is suddenly increased by arteriolar dilation, the large arteries to the tissue also dilate. This flow-induced dilation is due to local release of NO. Products of platelet aggregation also cause release of NO, and the resulting vasodilation helps keep blood vessels with an intact endothelium patent. This is in contrast to injured blood vessels, where the endothelium is damaged at the site of injury and platelets therefore aggregate and produce vasoconstriction (see Chapter 31).

Further evidence for a physiologic role of NO is the observation that mice lacking NOS 3 are hypertensive. This suggests that tonic release of NO is necessary to maintain normal blood pressure.

NO is also involved in vascular remodeling and angiogenesis, and NO may be involved in the pathogenesis of atherosclerosis. It is interesting in this regard that some patients with heart transplants develop an accelerated form of atherosclerosis in the vessels of the transplant, and there is reason to believe that this is triggered by endothelial damage. Nitroglycerin and other nitrovasodilators that are of great value in the treatment of angina act by stimulating guanylyl cyclase in the same manner as NO.

Penile erection is also produced by release of NO, with consequent vasodilation and engorgement of the corpora cavernosa (see Chapter 23). This accounts for the efficacy of drugs such as Viagra, which slow the breakdown of cGMP.

OTHER FUNCTIONS OF NO

NO is present in the brain and, acting via cGMP, it is important in brain function (see Chapter 7). NO is also necessary for the antimicrobial and cytotoxic activity of various inflammatory cells, although the net effect of NO in inflammation and tissue injury depends on the amount and kinetics of release, which in turn may depend on the specific NOS isoform involved. In the gastrointestinal tract, NO is important in the relaxation of smooth muscle. Other functions of NO are mentioned in other parts of this book.

CARBON MONOXIDE

The production of carbon monoxide (CO) from heme is shown in Figure 28–4. HO_2, the enzyme that catalyzes the reaction, is also present in cardiovascular tissues, and there is growing evidence that CO as well as NO produces local dilation in blood vessels. Interestingly, hydrogen sulfide is likewise emerging as a third gaseotransmitter that regulates vascular tone, although the relative roles of NO, CO, and H_2S have yet to be established.

ENDOTHELINS

Endothelial cells also produce **endothelin-1,** one of the most potent vasoconstrictor agents yet isolated. Endothelin-1 (ET-1), endothelin-2 (ET-2), and endothelin-3 (ET-3) are the members of a family of three similar 21-amino-acid polypeptides. Each is encoded by a different gene. The unique structure of the endothelins resembles that of the sarafotoxins, polypeptides found in the venom of a snake, the Israeli burrowing asp.

ENDOTHELIN-1

In endothelial cells, the product of the endothelin-1 gene is processed to a 39-amino-acid prohormone, **big endothelin-1,** which has about 1% of the activity of endothelin-1. The prohormone is cleaved at a tryptophan-valine (Trp-Val) bond to form endothelin-1 by **endothelin-converting enzyme.** Small amounts of big endothelin-1 and endothelin-1 are secreted into the blood, but for the most part, they are secreted locally and act in a paracrine fashion.

Two different endothelin receptors have been cloned, both of which are coupled via G proteins to phospholipase C (see Chapter 2). The ET_A receptor, which is specific for endothelin-1, is found in many tissues and mediates the vasoconstriction produced by endothelin-1. The ET_B receptor responds to all three endothelins, and is coupled to G_i. It may mediate vasodilation, and it appears to mediate the developmental effects of the endothelins (see below).

REGULATION OF SECRETION

Endothelin-1 is not stored in secretory granules, and most regulatory factors alter the transcription of its gene, with changes in secretion occurring promptly thereafter. Factors activating and inhibiting the gene are summarized in Table 32–4.

CARDIOVASCULAR FUNCTIONS

As noted above, endothelin-1 appears to be primarily a paracrine regulator of vascular tone. However, endothelin-1 is not increased in hypertension, and in mice in which one allele of the endothelin-1 gene is knocked out, blood pressure is actually elevated rather than reduced. The concentration of circulating endothelin-1 is, however, elevated in congestive heart failure and after myocardial infarction, so it may play a role in the pathophysiology of these diseases.

OTHER FUNCTIONS OF ENDOTHELINS

Endothelin-1 is found in the brain and kidneys as well as the endothelial cells. Endothelin-2 is produced primarily in the kidneys and intestine. Endothelin-3 is present in the

TABLE 32–4 Regulation of endothelin-1 secretion via transcription of its gene.

Stimulators
Angiotensin II
Catecholamines
Growth factors
Hypoxia
Insulin
Oxidized LDL
HDL
Shear stress
Thrombin
Inhibitors
NO
ANP
PGE_2
Prostacyclin

ANP, atrial natriuretic peptide; HDL, high density lipoprotein; LDL, low density lipoprotein; NO, nitric oxide; PGE_2, prostaglandin E_2; VIP, vasoactive intestinal polypeptide.

blood and is found in high concentrations in the brain. It is also found in the kidneys and gastrointestinal tract. In the brain, endothelins are abundant and, in early life, are produced by both astrocytes and neurons. They are found in the dorsal root ganglia, ventral horn cells, the cortex, the hypothalamus, and cerebellar Purkinje cells. They also play a role in regulating transport across the blood–brain barrier. There are endothelin receptors on mesangial cells (see Chapter 37), and the polypeptide participates in tubuloglomerular feedback.

Mice that have both alleles of the endothelin-1 gene deleted have severe craniofacial abnormalities and die of respiratory failure at birth. They also have megacolon (Hirschsprung disease), apparently because the cells that normally form the myenteric plexus fail to migrate to the distal colon (see Chapter 27). In addition, endothelins play a role in closing the ductus arteriosus at birth.

SYSTEMIC REGULATION BY NUEROHUMORAL AGENTS

Many circulating substances affect the vascular system. The vasodilator regulators include kinins, VIP, and ANP. Circulating vasoconstrictor hormones include vasopressin, norepinephrine, epinephrine, and angiotensin II.

KININS

Two related vasodilator peptides called **kinins** are found in the body. One is the nonapeptide **bradykinin,** and the other is the decapeptide **lysylbradykinin,** also known as **kallidin**

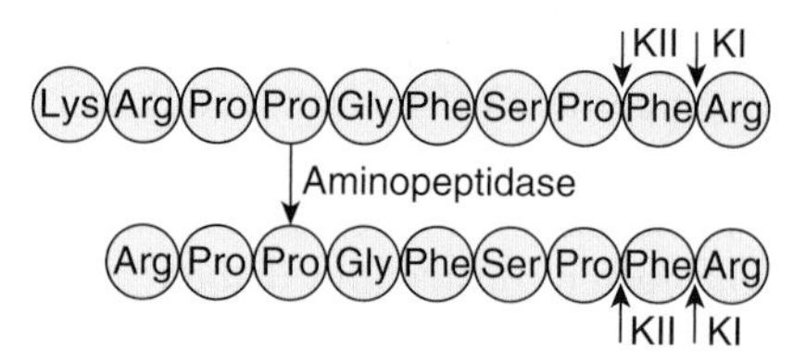

FIGURE 32–10 Kinins. Lysylbradykinin **(top)** can be converted to bradykinin **(bottom)** by aminopeptidase. The peptides are inactivated by kininase I (KI) or kininase II (KII) at the sites indicated by the short arrows.

(Figure 32–10). Lysylbradykinin can be converted to bradykinin by aminopeptidase. Both peptides are metabolized to inactive fragments by **kininase I,** a carboxypeptidase that removes the carboxyl terminal arginine (Arg). In addition, the dipeptidylcarboxypeptidase **kininase II** inactivates bradykinin and lysylbradykinin by removing phenylalanine-arginine (Phe-Arg) from the carboxyl terminal. Kininase II is the same enzyme as **angiotensin-converting enzyme,** which removes histidine-leucine (His-Leu) from the carboxyl terminal end of angiotensin I.

Bradykinin and lysylbradykinin are formed from two precursor proteins: **high-molecular-weight kininogen** and **low-molecular-weight kininogen** (Figure 32–11). They are formed by alternative splicing of a single gene located on chromosome 3. Proteases called **kallikreins** release the peptides from their precursors. They are produced in humans by a family of three genes located on chromosome 19. There are two types of kallikreins: **plasma kallikrein,** which circulates in an inactive form, and **tissue kallikrein,** which appears to be located primarily on the apical membranes of cells concerned with transcellular electrolyte transport. Tissue kallikrein is found in many tissues, including sweat and salivary glands, the pancreas, the prostate, the intestine, and the kidneys. Tissue kallikrein acts on high-molecular-weight kininogen to form bradykinin and low-molecular-weight kininogen to form lysylbradykinin. When activated, plasma kallikrein acts on high-molecular-weight kininogen to form bradykinin.

Inactive plasma kallikrein **(prekallikrein)** is converted to the active form, kallikrein, by active factor XII, the factor

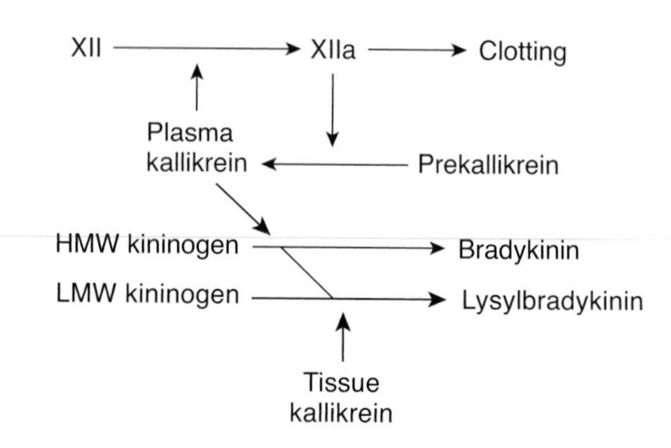

FIGURE 32–11 Formation of kinins from high-molecular-weight (HMW) and low-molecular-weight (LMW) kininogens.

that initiates the intrinsic blood clotting cascade. Kallikrein also activates factor XII in a positive feedback loop, and high-molecular-weight kininogen has a factor XII-activating action (see Figure 31–12).

The actions of both kinins resemble those of histamine. They are primarily paracrines, although small amounts are also found in the circulating blood. They cause contraction of visceral smooth muscle, but they relax vascular smooth muscle via NO, lowering blood pressure. They also increase capillary permeability, attract leukocytes, and cause pain upon injection under the skin. They are formed during active secretion in sweat glands, salivary glands, and the exocrine portion of the pancreas, and they are probably responsible for the increase in blood flow when these tissues are actively secreting their products.

Two bradykinin receptors, B_1 and B_2, have been identified. Their amino acid residues are 36% identical, and both are coupled to G proteins. The B_1 receptor may mediate the pain-producing effects of the kinins, but little is known about its distribution and function. The B_2 receptor has strong homology to the H_2 receptor and is found in many different tissues.

NATRIURETIC HORMONES

There is a family of natriuretic peptides involved in vascular regulation, including ANP secreted by the heart, brain natriuretic peptide (BNP), and C-type natriuretic peptide (CNP). They are released in response to hypervolemia. ANP and BNP circulate, whereas CNP acts predominantly in a paracrine fashion. In general, these peptides antagonize the action of various vasoconstrictor agents and lower blood pressure. ANP and BNP also serve to coordinate the control of vascular tone with fluid and electrolyte homeostasis via actions on the kidney.

CIRCULATING VASOCONSTRICTORS

Vasopressin is a potent vasoconstrictor, but when it is injected in normal individuals, there is a compensating decrease in cardiac output, so that there is little change in blood pressure. Its role in blood pressure regulation is discussed in Chapter 17.

Norepinephrine has a generalized vasoconstrictor action, whereas epinephrine dilates the vessels in skeletal muscle and the liver. The relative unimportance of circulating norepinephrine, as opposed to norepinephrine released from vasomotor nerves, is pointed out in Chapter 20, where the cardiovascular actions of catecholamines are discussed in detail.

Angiotensin II has a generalized vasoconstrictor action. It is formed by the action of angiotensin converting enzyme (ACE) on angiotensin I, which itself is liberated by the action of renin from the kidney on circulating angiotensinogen (see Chapter 38). Renin secretion, in turn, is increased when the blood pressure falls or extracellular fluid (ECF) volume is reduced, and angiotensin II therefore helps to maintain blood pressure. Angiotensin II also increases water intake and stimulates aldosterone secretion, and increased formation of angiotensin II is part of a homeostatic mechanism that operates to maintain ECF volume (see Chapter 20). In addition, there are renin–angiotensin systems in many different organs, and there may be one in the walls of blood vessels. Angiotensin II produced in blood vessel walls could be important in some forms of clinical hypertension. The role of angiotensin II in cardiovascular regulation is also amply demonstrated in the widespread use of ACE inhibitors as antihypertensive medications.

Urotensin-II, a polypeptide first isolated from the spinal cord of fish, is present in human cardiac and vascular tissue. It is one of the most potent mammalian vasoconstrictors known, and is being explored for its role in a large range of different human disease states. For example, levels of both urotensin-II and its receptor have been shown to be elevated in hypertension and heart failure, and may be markers of disease in these and other conditions.

CHAPTER SUMMARY

- RVLM neurons project to the thoracolumbar IML and release glutamate on preganglionic sympathetic neurons that innervate the heart and vasculature.
- The NTS is the major excitatory input to cardiac vagal motor neurons in the nucleus ambiguus.
- Carotid sinus and aortic depressor baroreceptors are innervated by branches of the ninth and tenth cranial nerves, respectively (glossopharyngeal and aortic depressor nerves). These receptors are most sensitive to changes in pulse pressure but also respond to changes in mean arterial pressure.
- Baroreceptor nerves terminate in the NTS and release glutamate. NTS neurons project to the CVLM and nucleus ambiguus and release glutamate. CVLM neurons project to RVLM and release GABA. This leads to a reduction in sympathetic activity and an increase in vagal activity (ie, the baroreceptor reflex).
- Activation of peripheral chemoreceptors in the carotid and aortic bodies by a reduction in PaO_2 or an increase in $PaCO_2$ leads to an increase in vasoconstriction. Heart rate changes are variable and depend on a number of factors including changes in respiration.
- In addition to various neural inputs, RVLM neurons are directly activated by hypoxia and hypercapnia.
- Most vascular beds have an intrinsic capacity to respond to changes in blood pressure within a certain range by altering vascular resistance to maintain stable blood flow. This property is known as autoregulation.
- Local factors such as oxygen tension, pH, temperature, and metabolic products contribute to vascular regulation; many produce vasodilation to restore blood flow.
- The endothelium is an important source of vasoactive mediators that act to either contract or relax vascular smooth muscle.

- Three gaseous mediators—NO, CO, and H_2S—are important regulators of vasodilation.
- Endothelins and angiotensin II induce vasoconstriction and may be involved in the pathogenesis of some forms of hypertension.

MULTIPLE-CHOICE QUESTIONS

For all questions, select the single best answer unless otherwise directed.

1. When a pheochromocytoma (tumor of the adrenal medulla) suddenly discharges a large amount of epinephrine into the circulation, the patient's heart rate would be expected to
   A. increase, because the increase in blood pressure stimulates the carotid and aortic baroreceptors.
   B. increase, because epinephrine has a direct chronotropic effect on the heart.
   C. increase, because of increased tonic parasympathetic discharge to the heart.
   D. decrease, because the increase in blood pressure stimulates the carotid and aortic chemoreceptors.
   E. decrease, because of increased tonic parasympathetic discharge to the heart.
2. A 65-year-old male had been experiencing frequent episodes of syncope as he got out of bed in the mornings. He was diagnosed with orthostatic hypotension due to a malfunction in his baroreceptor reflex. Activation of the baroreceptor reflex
   A. is primarily involved in short-term regulation of systemic blood pressure.
   B. leads to an increase in heart rate because of inhibition of the vagal cardiac motor neurons.
   C. inhibits neurons in the CVLM.
   D. excites neurons in the RVLM.
   E. occurs only under situations in which blood pressure is markedly elevated.
3. A 45-year-old female had a blood pressure of 155/95 when she was at her physician's office for a physical. It was her first time to see this physician and her first physical in over 10 years. The doctor suggested that she begin monitoring her pressure at home. Sympathetic nerve activity would be expected to increase
   A. if glutamate receptors were activated in the NTS.
   B. if GABA receptors were activated in the RVLM.
   C. if glutamate receptors were activated in the CVLM.
   D. during stress.
   E. when one transitions from an erect to a supine posture.
4. Which of the following neurotransmitters are correctly matched with an autonomic pathway?
   A. GABA is released by NTS neurons projecting to the RVLM.
   B. Glutamate is released by CVLM neurons projecting to the IML.
   C. GABA is released by NTS neurons projecting to the nucleus ambiguus.
   D. GABA is released by CVLM neurons projecting to the RVLM.
   E. Glutamate is released by CVLM neurons projecting to the NTS.
5. A 53-year-old woman with chronic lung disease was experiencing difficulty breathing. Her arterial Po_2 and Pco_2 were 50 and 60 mm Hg, respectively. Which one of the following statements about chemoreceptors is correct?
   A. Peripheral chemoreceptors are very sensitive to small increases in arterial Pco_2.
   B. Activation of arterial chemoreceptors leads to a fall in arterial pressure.
   C. Peripheral chemoreceptors are located in the NTS.
   D. Central chemoreceptors can be activated by an increase in intracranial pressure that compromises blood flow in the medulla.
   E. Central chemoreceptors are activated by increases in tissue pH.
6. A 55-year-old man comes to his primary care physician complaining of erectile dysfunction. He is given a prescription for Viagra, and on follow-up, reports that his ability to sustain an erection has been improved markedly by this treatment. The action of which of the following vasoactive mediators would primarily be increased in this patient?
   A. Histamine
   B. Endothelin-1
   C. Prostacyclin
   D. Nitric oxide
   E. Atrial natriuretic peptide

CHAPTER RESOURCES

Ahluwalia A, MacAllister RJ, Hobbs AJ: Vascular actions of natriuretic peptides. Cyclic GMP-dependent and -independent mechanisms. Basic Res Cardiol 2004;99:83.

Benarroch EE: *Central Autonomic Network. Functional Organization and Clinical Correlations.* Futura Publishing, 1997.

Chapleau MW, Abboud F (editors): *Neuro-cardiovascular regulation: From molecules to man.* Ann NY Acad Sci 2001;940.

Charkoudian N, Rabbitts JA: Sympathetic neural mechanisms in human cardiovascular health and disease. Mayo Clinic Proc 2009;84:822.

de Burgh Daly M: *Peripheral Arterial Chemoreceptors and Respiratory-Cardiovascular Integration.* Clarendon Press, 1997.

Haddy FJ, Vanhouttee PM, Feletou M: Role of potassium in regulating blood flow and blood pressure. Am J Physiol Regul Integr Comp Physiol 2006;290:R546.

Loewy AD, Spyer KM (editors): *Central Regulation of Autonomic Function.* Oxford University Press, 1990.

Marshall JM: Peripheral chemoreceptors and cardiovascular regulation. Physiol Rev 1994;74:543.

Paffett ML, Walker BR: Vascular adaptations to hypoxia: Molecular and cellular mechanisms regulating vascular tone. Essays Biochem 2007;43:105.

Ross B, McKendy K, Giaid A: Role of urotensin II in health and disease. Am J Physiol Regul Integr Comp Physiol 2010;298:R1156.

Trouth CO, Millis RM, Kiwull-Schöne HF, Schläfke ME: *Ventral Brainstem Mechanisms and Control of Respiration and Blood Pressure.* Marcel Dekker, 1995.

CHAPTER

33 Circulation Through Special Regions

OBJECTIVES

After studying this chapter, you should be able to:

- Define the special features of the circulation in the brain, coronary vessels, skin, and fetus, and how these are regulated.
- Describe how cerebrospinal fluid (CSF) is formed and reabsorbed, and its role in protecting the brain from injury.
- Understand how the blood–brain barrier impedes the entry of specific substances into the brain.
- Delineate how the oxygen needs of the contracting myocardium are met by the coronary arteries, and the consequences of their occlusion.
- List the vascular reactions of the skin and the reflexes that mediate them.
- Understand how the fetus is supplied with oxygen and nutrients in utero, and the circulatory events required for a transition to independent life after birth.

INTRODUCTION

The distribution of the cardiac output to various parts of the body at rest in a normal man is shown in Table 33–1. The general principles described in preceding chapters apply to the circulation of all these regions, but the vascular supplies of many organs have additional special features that are important to their physiology. The portal circulation of the anterior pituitary is discussed in Chapter 18, the pulmonary circulation in Chapter 35, the renal circulation in Chapter 37, and the circulation of the splanchnic area, particularly the intestines and liver, in Chapters 25 and 28. This chapter is concerned with the special circulations of the brain, the heart, and the skin, as well as the placenta and fetus.

CEREBRAL CIRCULATION: ANATOMIC CONSIDERATIONS

VESSELS

The principal arterial inflow to the brain in humans is via four arteries: two internal carotids and two vertebrals. In humans, the carotid arteries are quantitatively the most significant. The vertebral arteries unite to form the basilar artery, and the basilar artery and the carotids form the **circle of Willis** below the hypothalamus. The circle of Willis is the origin of the six large vessels supplying the cerebral cortex. Substances injected into one carotid artery are distributed almost exclusively to the cerebral hemisphere on that side. Normally no crossing over occurs, probably because the pressure is equal on both sides. Even when it is not, the anastomotic channels in the circle do not permit a very large flow. Occlusion of one carotid artery, particularly in older patients, often causes serious symptoms of cerebral ischemia. There are precapillary anastomoses between the cerebral vessels, but flow through these channels is generally insufficient to maintain the circulation and prevent infarction when a cerebral artery is occluded.

Venous drainage from the brain by way of the deep veins and dural sinuses empties principally into the internal jugular veins in humans, although a small amount of venous blood drains through the ophthalmic and pterygoid venous plexuses, through emissary veins to the scalp, and down the system of paravertebral veins in the spinal canal.

TABLE 33–1 Resting blood flow and O_2 consumption of various organs in a 63-kg adult man with a mean arterial blood pressure of 90 mm Hg and an O_2 consumption of 250 mL/min.

		Blood Flow		Arteriovenous Oxygen Difference	Oxygen Consumption		Resistance (R units)[a]		Percentage of Total	
Region	Mass (kg)	mL/min	mL/100 g/min	(mL/L)	mL/min	mL/100 g/min	Absolute	per kg	Cardiac Output	Oxygen Consumption
Liver	2.6	1500	57.7	34	51	2.0	3.6	9.4	27.8	20.4
Kidneys	0.3	1260	420.0	14	18	6.0	4.3	1.3	23.3	7.2
Brain	1.4	750	54.0	62	46	3.3	7.2	10.1	13.9	18.4
Skin	3.6	462	12.8	25	12	0.3	11.7	42.1	8.6	4.8
Skeletal muscle	31.0	840	2.7	60	50	0.2	6.4	198.4	15.6	20.0
Heart muscle	0.3	250	84.0	114	29	9.7	21.4	6.4	4.7	11.6
Rest of body	23.8	336	1.4	129	44	0.2	16.1	383.2	6.2	17.6
Whole body	63.0	5400	8.6	46	250	0.4	1.0	63.0	100.0	100.0

[a]R units are pressure (mm Hg) divided by blood flow (mL/s).

Reproduced with permission from Bard P (editor): *Medical Physiology*, 11th ed. Mosby, 1961.

The cerebral vessels have a number of unique anatomic features. In the choroid plexuses, there are gaps between the endothelial cells of the capillary wall, but the choroid epithelial cells that separate them from the cerebrospinal fluid (CSF) are connected to one another by tight junctions. The capillaries in the brain substance resemble nonfenestrated capillaries in muscle (see Chapter 31), but there are tight junctions between the endothelial cells that limit the passage of substances via the paracellular route. In addition, there are relatively few vesicles in the endothelial cytoplasm, and presumably little vesicular transport. However, multiple transport systems are present in the capillary cells. The brain capillaries are surrounded by the endfeet of astrocytes (Figure 33–1). These endfeet are closely applied to the basal lamina of the capillaries, but they do not cover the entire capillary wall, and gaps of about 20 nm occur between endfeet (Figure 33–2). However, the endfeet induce the tight junctions in the capillaries (see Chapter 31). The protoplasm of astrocytes is also found around synapses, where it appears to isolate the synapses in the brain from one another.

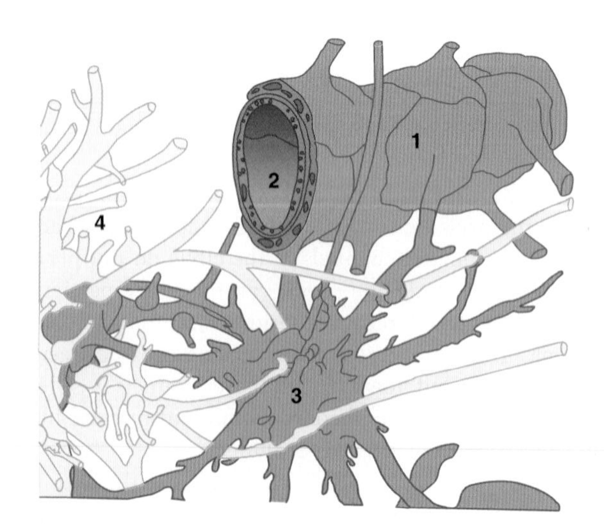

FIGURE 33–1 Relation of fibrous astrocyte (3) to a capillary (2) and neuron (4) in the brain. The endfeet of the astrocyte processes form a discontinuous membrane around the capillary (1). Astrocyte processes also envelop the neuron. (Adapted from Krstic RV: *Die Gewebe des Menschen und der Säugetiere*. Springer, 1978.)

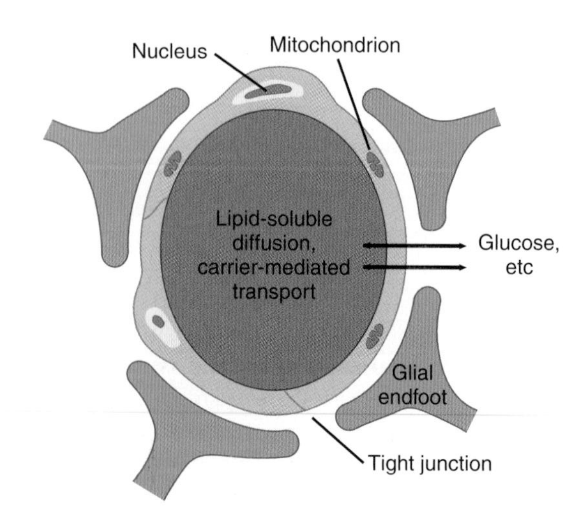

FIGURE 33–2 Transport across cerebral capillaries. Only free lipid-soluble substances can move passively across the endothelial cells. Water-soluble solutes, such as glucose, require active transport mechanisms. Proteins and protein-bound lipids are excluded.

INNERVATION

Three systems of nerves innervate the cerebral blood vessels. Postganglionic sympathetic neurons have their cell bodies in the superior cervical ganglia, and their endings contain norepinephrine. Many also contain neuropeptide Y. Cholinergic neurons that probably originate in the sphenopalatine ganglia also innervate the cerebral vessels, and the postganglionic cholinergic neurons on the blood vessels contain acetylcholine. Many also contain vasoactive intestinal peptide (VIP) and peptide histidyl methionine (PHM-27) (see Chapter 7). These nerves end primarily on large arteries. Sensory nerves are found on more distal arteries. They have their cell bodies in the trigeminal ganglia and contain substance P, neurokinin A, and calcitonin gene-related peptide (CGRP). Substance P, CGRP, VIP, and PHM-27 cause vasodilation, whereas neuropeptide Y is a vasoconstrictor. Touching or pulling on the cerebral vessels causes pain.

CEREBROSPINAL FLUID

FORMATION & ABSORPTION

CSF fills the ventricles and subarachnoid space. In humans, the volume of CSF is about 150 mL and the rate of CSF production is about 550 mL/d. Thus the CSF turns over about 3.7 times a day. In experiments on animals, it has been estimated that 50–70% of the CSF is formed in the choroid plexuses and the remainder is formed around blood vessels and along ventricular walls. Presumably, the situation in humans is similar. The CSF in the ventricles flows through the foramens of Magendie and Luschka to the subarachnoid space and is absorbed through the **arachnoid villi** into veins, primarily the cerebral venous sinuses. The villi consist of projections of the fused arachnoid membrane and endothelium of the sinuses into the venous sinuses. Similar, smaller villi project into veins around spinal nerve routes. These projections may contribute to the outflow of CSF into venous blood by a process known as **bulk flow,** which is unidirectional. However, recent studies suggest that, at least in animals, a more important route for CSF reabsorption into the bloodstream in health is via the cribriform plate above the nose and thence into the cervical lymphatics. However, reabsorption via one-way valves (of uncertain structural basis) in the arachnoid villi may assume a greater role if CSF pressure is elevated. Likewise, when CSF builds up abnormally, aquaporin water channels may be expressed in the choroid plexus and brain microvessels as a compensatory adaptation.

CSF is formed continuously by the choroid plexus in two stages. First, plasma is passively filtered across the choroidal capillary endothelium. Next, secretion of water and ions across the choroidal epithelium provides for active control of CSF composition and quantity. Bicarbonate, chloride, and potassium ions enter the CSF via channels in the epithelial cell apical membranes. Aquaporins provide for water movement to balance osmotic gradients. The composition of CSF (Table 33–2) is essentially the same as that of brain extracellular fluid (ECF), which in living humans makes up 15% of the brain volume. In adults, free communication appears to take place between the brain interstitial fluid and CSF, although the diffusion distances from some parts of the brain to the CSF are appreciable. Consequently, equilibration may take some time to occur, and local areas of the brain may have extracellular microenvironments that are transiently different from CSF.

TABLE 33–2 Concentration of various substances in human CSF and plasma.

	Units	CSF	Plasma	Ratio CSF/ Plasma
Na^+	(meq/kg H_2O)	147.0	150.0	0.98
K^+	(meq/kg H_2O)	2.9	4.6	0.62
Mg^{2+}	(meq/kg H_2O)	2.2	1.6	1.39
Ca^{2+}	(meq/kg H_2O)	2.3	4.7	0.49
Cl^-	(meq/kg H_2O)	113.0	99.0	1.14
HCO_3^-	(meq/L)	25.1	24.8	1.01
Pco_2	(mm Hg)	50.2	39.5	1.28
pH		7.33	7.40	...
Osmolality	(mosm/kg H_2O)	289.0	289.0	1.00
Protein	(mg/dL)	20.0	6000.0	0.003
Glucose	(mg/dL)	64.0	100.0	0.64
Inorganic P	(mg/dL)	3.4	4.7	0.73
Urea	(mg/dL)	12.0	15.0	0.80
Creatinine	(mg/dL)	1.5	1.2	1.25
Uric acid	(mg/dL)	1.5	5.0	0.30
Cholesterol	(mg/dL)	0.2	175.0	0.001

Lumbar CSF pressure is normally 70–180 mm H_2O. Up to pressures well above this range, the rate of CSF formation is independent of intraventricular pressure. However, absorption is proportional to the pressure (Figure 33–3). At a pressure of 112 mm H_2O, which is the average normal CSF pressure, filtration and absorption are equal. Below a pressure of approximately 68 mm H_2O, absorption stops. Large amounts of fluid accumulate when the capacity for CSF reabsorption is decreased **(external hydrocephalus, communicating hydrocephalus).** Fluid also accumulates proximal to the block and distends the ventricles when the foramens of Luschka and Magendie are blocked or there is obstruction within the ventricular system **(internal hydrocephalus, noncommunicating hydrocephalus).**

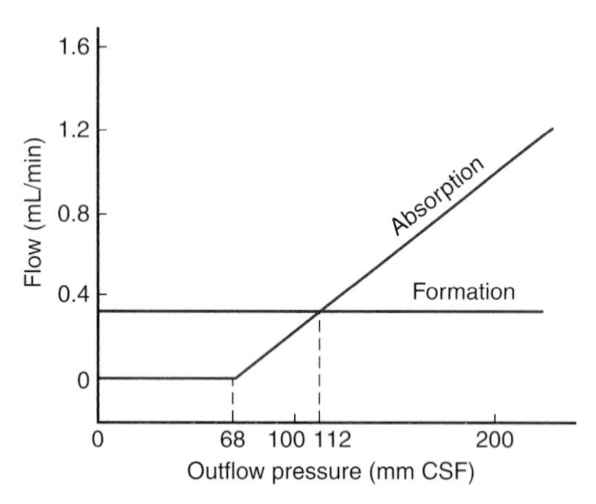

FIGURE 33–3 CSF formation and absorption in humans at various CSF pressures. Note that at 112 mm CSF, formation and absorption are equal, and at 68 mm CSF, absorption is zero. (Modified and reproduced with permission from Cutler RWP, et al: Formation and absorption of cerebrospinal fluid in man. Brain 1968;91:707.)

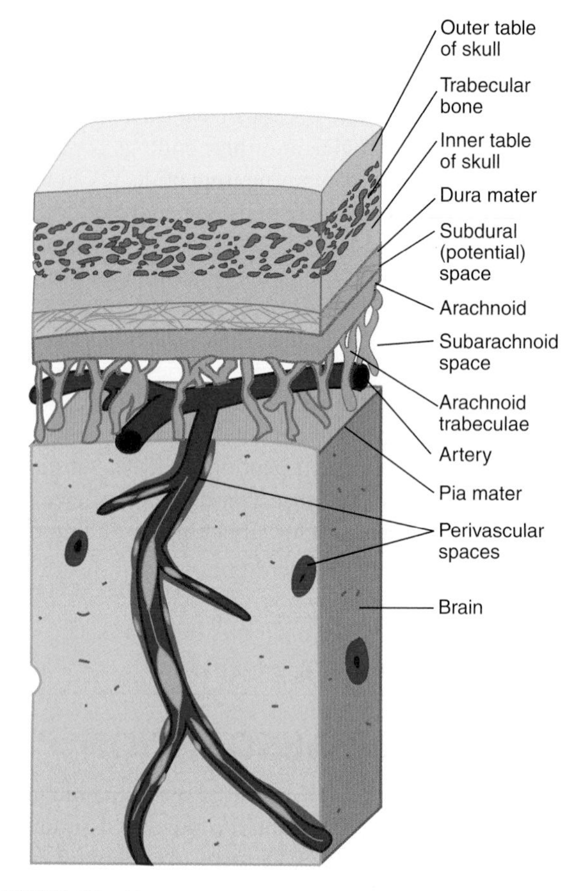

FIGURE 33–4 Investing membranes of the brain, showing their relation to the skull and to brain tissue. (Reproduced with permission from Young B, Heath JW: *Wheater's Functional Histology,* 4th ed. Churchill Livingstone, 2000.)

PROTECTIVE FUNCTION

The most critical role for CSF (and the meninges) is to protect the brain. The dura is attached firmly to bone. Normally, there is no "subdural space," with the arachnoid being held to the dura by the surface tension of the thin layer of fluid between the two membranes. As shown in Figure 33–4, the brain itself is supported within the arachnoid by the blood vessels and nerve roots and by the multiple fine fibrous **arachnoid trabeculae.** The brain weighs about 1400 g in air, but in its "water bath" of CSF it has a net weight of only 50 g. The buoyancy of the brain in the CSF permits its relatively flimsy attachments to suspend it very effectively. When the head receives a blow, the arachnoid slides on the dura and the brain moves, but its motion is gently checked by the CSF cushion and by the arachnoid trabeculae.

The pain produced by spinal fluid deficiency illustrates the importance of CSF in supporting the brain. Removal of CSF during lumbar puncture can cause a severe headache after the fluid is removed, because the brain hangs on the vessels and nerve roots, and traction on them stimulates pain fibers. The pain can be relieved by intrathecal injection of sterile isotonic saline.

HEAD INJURIES

Without the protection of the spinal fluid and the meninges, the brain would probably be unable to withstand even the minor traumas of everyday living; but with the protection afforded, it takes a fairly severe blow to produce cerebral damage. The brain is damaged most commonly when the skull is fractured and bone is driven into neural tissue (depressed skull fracture), when the brain moves far enough to tear the delicate bridging veins from the cortex to the bone, or when the brain is accelerated by a blow on the head and is driven against the skull or the tentorium at a point opposite where the blow was struck **(contrecoup injury).**

THE BLOOD–BRAIN BARRIER

The tight junctions between capillary endothelial cells in the brain and between the epithelial cells in the choroid plexus effectively prevent proteins from entering the brain in adults and slow the penetration of some smaller molecules as well. An example is the slow penetration of urea (Figure 33–5). This uniquely limited exchange of substances into the brain is referred to as the **blood–brain barrier,** a term most commonly used to encompass this barrier overall and more specifically the barrier in the choroid epithelium between blood and CSF.

Passive diffusion across the tight cerebral capillaries is very limited, and little vesicular transport takes place. However, there are numerous carrier-mediated and active transport systems in the cerebral capillaries. These move substances out of as well as into the brain, though movement out of the brain is generally more free than movement into it.

PENETRATION OF SUBSTANCES INTO THE BRAIN

Water, CO_2, and O_2 penetrate the brain with ease, as do the lipid-soluble free forms of steroid hormones, whereas their

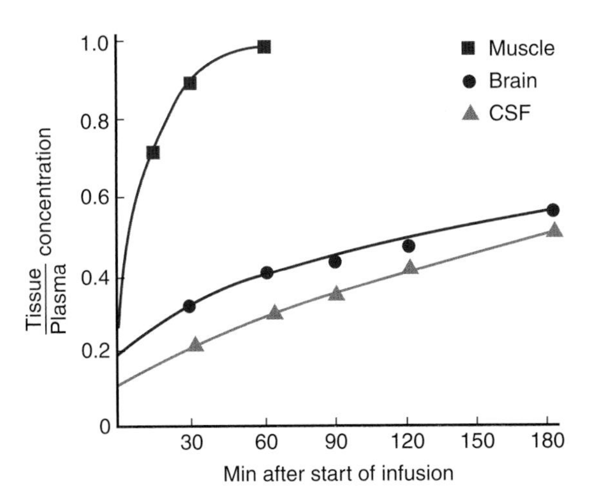

FIGURE 33–5 Penetration of urea into muscle, brain, spinal cord, and CSF. Urea was administered by constant infusion.

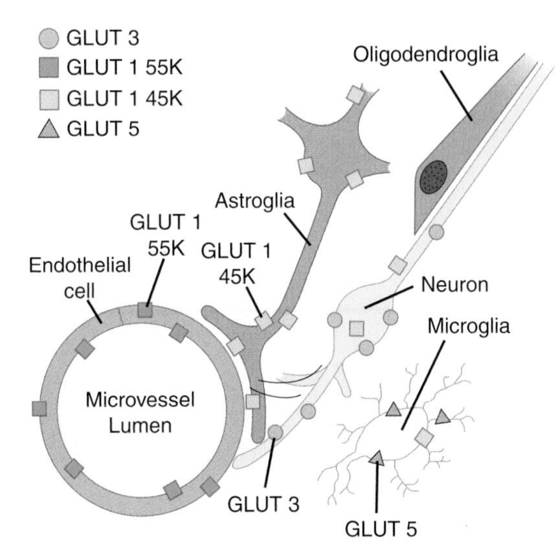

FIGURE 33–6 Localization of the various GLUT transporters in the brain. (Adapted from Maher F, Vannucci SJ, Simpson IA: Glucose transporter proteins in brain. FASEB J 1994;8:1003.)

protein-bound forms and, in general, all proteins and polypeptides do not. The rapid passive penetration of CO_2 contrasts with the regulated transcellular penetration of H^+ and HCO_3^- and has physiologic significance in the regulation of respiration (see Chapter 35).

Glucose is the major ultimate source of energy for nerve cells. Its diffusion across the blood–brain barrier would be very slow, but the rate of transport into the CSF is markedly enhanced by the presence of specific transporters, including the glucose transporter 1 (GLUT 1). The brain contains two forms of GLUT 1: GLUT 1 55K and GLUT 1 45K. Both are encoded by the same gene, but they differ in the extent to which they are glycosylated. GLUT 1 55K is present in high concentration in brain capillaries (Figure 33–6). Infants with congenital GLUT 1 deficiency develop low CSF glucose concentrations in the presence of normal plasma glucose, and they have seizures and delayed development. In addition, transporters for thyroid hormones; several organic acids; choline; nucleic acid precursors; and neutral, basic, and acidic amino acids are present at the blood–brain barrier.

A variety of drugs and peptides actually cross the cerebral capillaries but are promptly transported back into the blood by a multidrug nonspecific transporter in the apical membranes of the endothelial cells. This **P-glycoprotein** is a member of the family of adenosine triphosphate (ATP) binding cassettes that transport various proteins and lipids across cell membranes (see Chapter 2). In the absence of this transporter in mice, much larger proportions of systemically administered doses of various chemotherapeutic drugs, analgesics, and opioid peptides are found in the brain than in controls. If pharmacologic agents that inhibit this transporter can be developed, they could be of value in the treatment of brain tumors and other central nervous system (CNS) diseases in which it is difficult to introduce adequate amounts of therapeutic agents into the brain.

CIRCUMVENTRICULAR ORGANS

When dyes that bind to proteins in the plasma are injected, they stain many tissues but spare most of the brain. However, four small areas in or near the brain stem do take up the stain. These areas are (1) the **posterior pituitary** (neurohypophysis) and the adjacent ventral part of the **median eminence** of the hypothalamus, (2) the **area postrema,** (3) the **organum vasculosum of the lamina terminalis (OVLT,** supraoptic crest), and (4) the **subfornical organ (SFO).**

These areas are referred to collectively as the **circumventricular organs** (Figure 33–7). All have fenestrated capillaries, and because of their permeability they are said to be "outside the blood–brain barrier." Some of them function as **neurohemal organs;** that is, areas in which polypeptides secreted by neurons enter the circulation. Others contain receptors for many different peptides and other substances, and function as chemoreceptor zones in which substances in the circulating blood can act to trigger changes in brain function without penetrating the blood–brain barrier. For example, the area postrema is a chemoreceptor trigger zone that initiates vomiting in response to chemical changes in the plasma (see Chapter 27). It is also concerned with cardiovascular control, and in many species circulating angiotensin II acts on the area postrema to produce a neurally mediated increase in blood pressure. Angiotensin II also acts on the SFO and possibly on the OVLT to increase water intake. In addition, it appears that the OVLT is the site of the osmoreceptor controlling vasopressin secretion (see Chapter 38), and evidence suggests that circulating interleukin-1 (IL-1) produces fever by acting here too.

The subcommissural organ (Figure 33–7) is closely associated with the pineal gland and histologically resembles the circumventricular organs. However, it does not have fenestrated

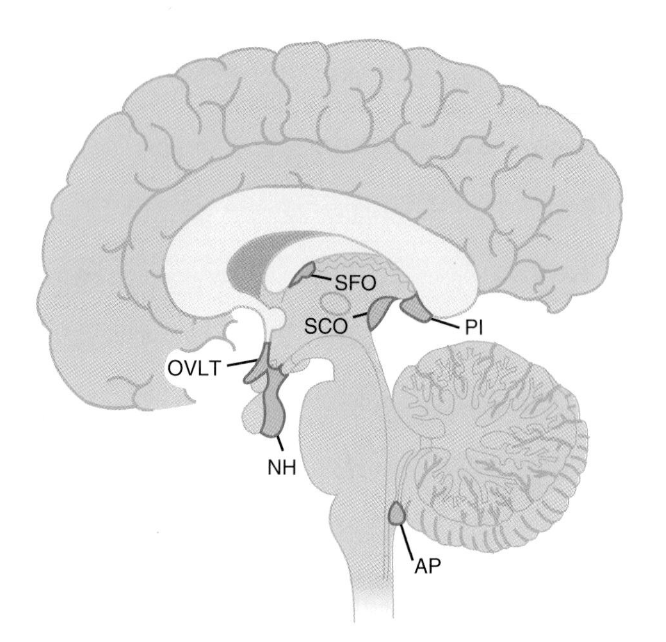

FIGURE 33–7 Circumventricular organs. The neurohypophysis (NH), organum vasculosum of the lamina terminalis (OVLT, organum vasculosum of the lamina terminalis), subfornical organ (SFO), and area postrema (AP) are shown projected on a sagittal section of the human brain. PI, pineal; SCO, subcommissural organ.

capillaries, is not highly permeable, and has no established function. Conversely, the pineal and the anterior pituitary do have fenestrated capillaries and are outside the blood–brain barrier, but both are endocrine glands and are not part of the brain.

FUNCTION OF THE BLOOD–BRAIN BARRIER

The blood–brain barrier strives to maintain the constancy of the environment of the neurons in the CNS (see Clinical Box 33–1). Even minor variations in the concentrations of K^+, Ca^{2+}, Mg^{2+}, H^+, and other ions can have far-reaching consequences. The constancy of the composition of the ECF in all parts of the body is maintained by multiple homeostatic mechanisms (see Chapters 1 and 38), but because of the sensitivity of the cortical neurons to ionic change, it is not surprising that an additional defense has evolved to protect them. Other functions of the blood–brain barrier include protection of the brain from endogenous and exogenous toxins in the blood and prevention of the escape of neurotransmitters into the general circulation.

DEVELOPMENT OF THE BLOOD–BRAIN BARRIER

In experimental animals, many small molecules penetrate the brain more readily during the fetal and neonatal period than they do in the adult. On this basis, it is often stated that the blood–brain barrier is immature at birth. Humans are more mature at birth than rats and various other experimental animals, and detailed data on passive permeability of the human blood–brain barrier are not available. However, in severely jaundiced infants with high plasma levels of free bilirubin and an immature hepatic bilirubin-conjugating system, free bilirubin enters the brain and, in the presence of asphyxia, damages the basal ganglia **(kernicterus).** The counterpart of this situation in later life is the Crigler–Najjar syndrome in which there is a congenital deficiency of glucuronyl transferase. These individuals can have very high free bilirubin levels in the blood and develop encephalopathy. In other conditions, free bilirubin levels are generally not high enough to produce brain damage.

CLINICAL BOX 33–1

Clinical Implications of the Blood–Brain Barrier

Physicians must know the degree to which drugs penetrate the brain in order to treat diseases of the nervous system intelligently. For example, it is clinically relevant that the amines dopamine and serotonin penetrate brain tissue to a very limited degree but their corresponding acid precursors, L-dopa and 5-hydroxytryptophan, respectively, enter with relative ease (see Chapters 7 and 12). Another important clinical consideration is the fact that the blood–brain barrier tends to break down in areas of infection or injury. Tumors develop new blood vessels, and the capillaries that are formed lack contact with normal astrocytes. Therefore, there are no tight junctions, and the vessels may even be fenestrated. The lack of a barrier helps in identifying the location of tumors; substances such as radioactive iodine-labeled albumin penetrate normal brain tissue very slowly, but they enter tumor tissue, making the tumor stand out as an island of radioactivity in the surrounding normal brain. The blood–brain barrier can also be temporarily disrupted by sudden marked increases in blood pressure or by intravenous injection of hypertonic fluids.

CEREBRAL BLOOD FLOW & ITS REGULATION

KETY METHOD

According to the **Fick principle** (see Chapter 30), the blood flow of any organ can be measured by determining the amount of a given substance (Q_x) removed from the bloodstream by the organ per unit of time and dividing that value by the difference between the concentration of the substance in arterial

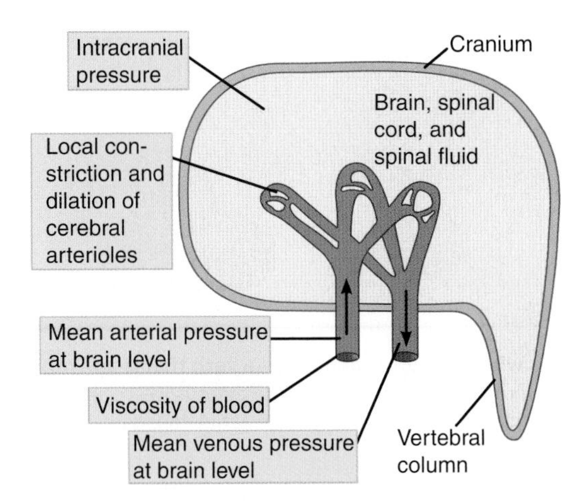

FIGURE 33–8 Diagrammatic summary of the factors affecting overall cerebral blood flow.

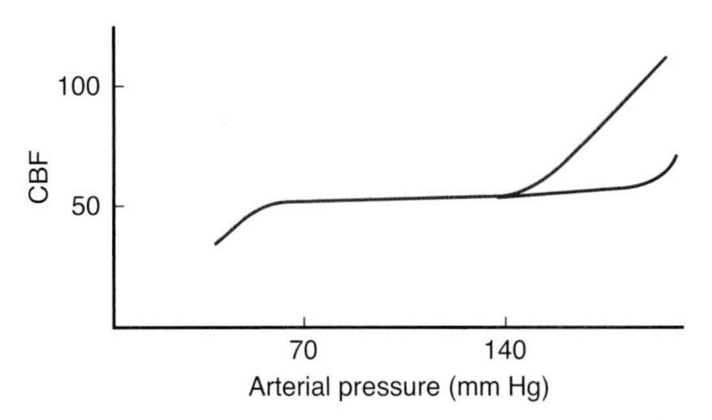

FIGURE 33–9 Autoregulation of cerebral blood flow (CBF) during steady-state conditions. The blue line shows the alteration produced by sympathetic stimulation during autoregulation.

blood and the concentration in the venous blood from the organ ($[A_x] - [V_x]$). Thus:

$$\text{Cerebral blood flow (CBF)} = \frac{Q_x}{[A_x] - [V_x]}$$

This can be applied clinically using inhaled nitrous oxide (N_2O) **(Kety method).** The average cerebral blood flow in young adults is 54 mL/100 g/min. The average adult brain weighs about 1400 g, so the flow for the whole brain is about 756 mL/min. Note that the Kety method provides an average value for perfused areas of brain because it gives no information about regional differences in blood flow. It also can only measure flow to perfused parts of the brain. If the blood flow to a portion of the brain is occluded, the measured flow does not change because the nonperfused area does not take up any N_2O.

In spite of the marked local fluctuations in brain blood flow with neural activity, the cerebral circulation is regulated in such a way that total blood flow remains relatively constant. The factors involved in regulating the flow are summarized in Figure 33–8.

ROLE OF INTRACRANIAL PRESSURE

In adults, the brain, spinal cord, and spinal fluid are encased, along with the cerebral vessels, in a rigid bony enclosure. The cranial cavity normally contains a brain weighing approximately 1400 g, 75 mL of blood, and 75 mL of spinal fluid. Because brain tissue and spinal fluid are essentially incompressible, the volume of blood, spinal fluid, and brain in the cranium at any time must be relatively constant **(Monro–Kellie doctrine).** More importantly, the cerebral vessels are compressed whenever the intracranial pressure rises. Any change in venous pressure promptly causes a similar change in intracranial pressure. Thus, a rise in venous pressure decreases cerebral blood flow both by decreasing the effective perfusion pressure and by compressing the cerebral vessels. This relationship helps to compensate for changes in arterial blood pressure at the level of the head. For example, if the body is accelerated upward (positive *g*), blood moves toward the feet and arterial pressure at the level of the head decreases. However, venous pressure also falls and intracranial pressure falls, so that the pressure on the vessels decreases and blood flow is much less severely compromised than it would otherwise be. Conversely, during acceleration downward, force acting toward the head (negative *g*) increases arterial pressure at head level, but intracranial pressure also rises, so that the vessels are supported and do not rupture. The cerebral vessels are protected during the straining associated with defecation or delivery in the same way.

AUTOREGULATION

As seen in other vascular beds, autoregulation is prominent in the brain (Figure 33–9). This process, by which the flow to many tissues is maintained at relatively constant levels despite variations in perfusion pressure, is discussed in Chapter 31. In the brain, autoregulation maintains a normal cerebral blood flow at arterial pressures of 65–140 mm Hg.

ROLE OF VASOMOTOR & SENSORY NERVES

The innervation of large cerebral blood vessels by postganglionic sympathetic and parasympathetic nerves and the additional distal innervation by sensory nerves have been described above. The nerves may also modulate tone indirectly, via the release of paracine substances from astrocytes. The precise role of these nerves, however, remains a matter of debate. It has been argued that noradrenergic discharge occurs when the blood pressure is markedly elevated. This reduces the resultant passive increase in blood flow and helps protect the blood–brain barrier from the disruption that could otherwise occur (see above). Thus, vasomotor discharges affect autoregulation. With sympathetic stimulation, the constant-flow, or plateau, part of the pressure-flow curve is extended to the right (Figure 33–9); that is, greater increases in pressure can occur

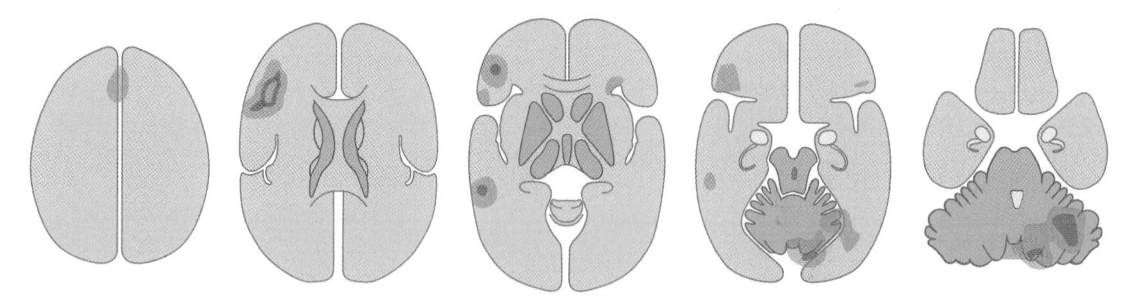

FIGURE 33–10 Activity in the human brain at five different horizontal levels while a subject generates a verb that is appropriate for each noun presented by an examiner. This mental task activates the frontal cortex (slices 1–4), anterior cingulate gyrus (slice 1), and posterior temporal lobe (slice 3) on the left side and the cerebellum (slices 4 and 5) on the right side. Light purple, moderate activation; dark purple, marked activation. (Based on PET scans in Posner MI, Raichle ME: *Images of Mind.* Scientific American Library, 1994.)

without an increase in flow. On the other hand, the vasodilator hydralazine and the angiotensin-converting enzyme (ACE) inhibitor captopril reduce the length of the plateau. Finally, neurovascular coupling may adjust local perfusion in response to changes in brain activity (see below).

BLOOD FLOW IN VARIOUS PARTS OF THE BRAIN

A major advance in recent decades has been the development of techniques for monitoring regional blood flow in living, conscious humans. Among the most valuable methods are **positron emission tomography (PET)** and related techniques in which a short-lived radioisotope is used to label a compound and the compound is injected. The arrival and clearance of the tracer are monitored by scintillation detectors placed over the head. Because blood flow is tightly coupled to brain metabolism, local uptake of 2-deoxyglucose is also a good index of blood flow (see below and Chapter 1). If the 2-deoxyglucose is labeled with a short-half-life positron emitter such as ^{18}F, ^{11}O, or ^{15}O, its concentration in any part of the brain can be monitored.

Another valuable technique involves magnetic resonance imaging (MRI). MRI is based on detecting resonant signals from different tissues in a magnetic field. **Functional magnetic resonance imaging (fMRI)** measures the amount of blood in a tissue area. When neurons become active, their increased discharge alters the local field potential. A still unsettled mechanism triggers an increase in local blood flow and oxygen. The increase in oxygenated blood is detected by fMRI. PET scanning can be used to measure not only blood flow but the concentration of molecules, such as dopamine, in various regions of the living brain. On the other hand, fMRI does not involve the use of radioactivity. Consequently, it can be used at frequent intervals to measure changes in regional blood flow in a single individual.

In resting humans, the average blood flow in gray matter is 69 mL/100 g/min compared with 28 mL/100 g/min in white matter. A striking feature of cerebral function is the marked variation in local blood flow with changes in brain activity. An example is shown in Figure 33–10. In subjects who are awake but at rest, blood flow is greatest in the premotor and frontal regions. This is the part of the brain that is believed to be concerned with decoding and analyzing afferent input and with intellectual activity. During voluntary clenching of the right hand, flow is increased in the hand area of the left motor cortex and the corresponding sensory areas in the postcentral gyrus. Especially when the movements being performed are sequential, the flow is also increased in the supplementary motor area. When subjects talk, there is a bilateral increase in blood flow in the face, tongue, and mouth-sensory and motor areas and the upper premotor cortex in the categorical (usually the left) hemisphere. When the speech is stereotyped, Broca's and Wernicke's areas do not show increased flow, but when the speech is creative—that is, when it involves ideas—flow increases in both these areas. Reading produces widespread increases in blood flow. Problem solving, reasoning, and motor ideation without movement produce increases in selected areas of the premotor and frontal cortex. In anticipation of a cognitive task, many of the brain areas that will be activated during the task are activated beforehand, as if the brain produces an internal model of the expected task. In right-handed individuals, blood flow to the left hemisphere is greater when a verbal task is being performed and blood flow to the right hemisphere is greater when a spatial task is being performed (see Clinical Box 33–2).

BRAIN METABOLISM & OXYGEN REQUIREMENTS

UPTAKE & RELEASE OF SUBSTANCES BY THE BRAIN

If the cerebral blood flow is known, it is possible to calculate the consumption or production by the brain of O_2, CO_2, glucose, or any other substance present in the bloodstream by multiplying the cerebral blood flow by the difference between the concentration of the substance in arterial blood and its concentration in cerebral venous blood (Table 33–3). When

CLINICAL BOX 33–2

Changes in Cerebral Blood Flow in Disease

Several disease states are now known to be associated with localized or general changes in cerebral blood flow, as revealed by PET scanning and fMRI techniques. For example, epileptic foci are hyperemic during seizures, whereas flow is reduced in other parts of the brain. Between seizures, flow is sometimes reduced in the foci that generate the seizures. Parietooccipital flow is decreased in patients with symptoms of agnosia (see Chapter 11). In Alzheimer disease, the earliest change is decreased metabolism and blood flow in the superior parietal cortex, with later spread to the temporal and finally the frontal cortex. The pre- and postcentral gyri, basal ganglia, thalamus, brain stem, and cerebellum are relatively spared. In Huntington disease, blood flow is reduced bilaterally in the caudate nucleus, and this alteration in flow occurs early in the disease. In manic depressives (but interestingly, not in patients with unipolar depression), there is a general decrease in cortical blood flow when the patients are depressed. In schizophrenia, some evidence suggests decreased blood flow in the frontal lobes, temporal lobes, and basal ganglia. Finally, during the aura in patients with migraine, a bilateral decrease in blood flow starts in the occipital cortex and spreads anteriorly to the temporal and parietal lobes.

calculated in this fashion, a negative value indicates that the brain is producing the substance.

OXYGEN CONSUMPTION

O_2 consumption by the human brain (**cerebral metabolic rate for O_2, $CMRO_2$**) averages approximately 20% of the total body resting O_2 consumption (Table 33–1). The brain is extremely sensitive to hypoxia, and occlusion of its blood supply produces unconsciousness in a period as short as 10 s. The vegetative structures in the brain stem are more resistant to hypoxia than the cerebral cortex, and patients may recover from accidents such as cardiac arrest and other conditions causing fairly prolonged hypoxia with normal vegetative functions but severe, permanent intellectual deficiencies. The basal ganglia use O_2 at a very high rate, and symptoms of Parkinson disease as well as intellectual deficits can be produced by chronic hypoxia. The thalamus and the inferior colliculus are also very susceptible to hypoxic damage (see Clinical Box 33–3).

ENERGY SOURCES

Glucose is the major ultimate source of energy for the brain; under normal conditions, 90% of the energy needed to maintain ion gradients across cell membranes and transmit electrical impulses comes from this source. Glucose enters the brain via GLUT 1 in cerebral capillaries (see above). Other transporters then distribute it to neurons and glial cells.

TABLE 33–3 Utilization and production of substances by the adult human brain in vivo.

Substance	Uptake (+) or Output (–) per 100 g of Brain/min	Total/min
Substances utilized		
Oxygen	+3.5 mL	+49 mL
Glucose	+5.5 mg	+77 mg
Glutamate	+0.4 mg	+5.6 mg
Substances produced		
Carbon dioxide	–3.5 mL	–49 mL
Glutamine	–0.6 mL	–8.4 mg

Substances not used or produced in the fed state: lactate, pyruvate, total ketones, and α-ketoglutarate.

Glucose is taken up from the blood in large amounts, and the RQ (respiratory quotient; see Chapter 24) of cerebral tissue is 0.95–0.99 in normal individuals. Importantly, insulin is not required for most cerebral cells to utilize glucose. In general, glucose utilization at rest parallels blood flow and O_2 consumption. This does not mean that the total source of energy is always glucose. During prolonged starvation, appreciable utilization of other substances occurs. Indeed, evidence indicates that as much as 30% of the glucose taken up under normal conditions is converted to amino acids, lipids, and proteins, and that substances other than glucose are metabolized for energy during convulsions. Some utilization of amino acids from the circulation may also take place even though the amino acid arteriovenous difference across the brain is normally minute.

The consequences of hypoglycemia in terms of neural function are discussed in Chapter 24.

GLUTAMATE & AMMONIA REMOVAL

The brain's uptake of glutamate is approximately balanced by its output of glutamine. Glutamate entering the brain associates with ammonia and leaves as glutamine. The glutamate–glutamine conversion in the brain—the opposite of the reaction in the kidney that produces some of the ammonia entering the tubules—serves as a detoxifying mechanism to keep the brain free of ammonia. Ammonia is very toxic to nerve cells, and ammonia intoxication is believed to be a major cause of the bizarre neurologic symptoms in hepatic coma (see Chapter 28).

CLINICAL BOX 33-3

Stroke

When the blood supply to a part of the brain is interrupted, ischemia damages or kills the cells in the area, producing the signs and symptoms of a stroke. There are two general types of strokes: hemorrhagic and ischemic. Hemorrhagic stroke occurs when a cerebral artery or arteriole ruptures, sometimes but not always at the site of a small aneurysm. Ischemic stroke occurs when flow in a vessel is compromised by atherosclerotic plaques on which thrombi form. Thrombi may also be produced elsewhere (eg, in the atria in patients with atrial fibrillation) and pass to the brain as emboli where they then lodge and interrupt flow. In the past, little could be done to modify the course of a stroke and its consequences. However, it has now become clear that in the penumbra, the area surrounding the most severe brain damage, ischemia reduces glutamate uptake by astrocytes, and the increase in local glutamate causes excitotoxic damage and death to neurons (see Chapter 7).

THERAPEUTIC HIGHLIGHTS

The clot-lysing drug, tissue-type plasminogen activator (t-PA) (see Chapter 31) is of great benefit in ischemic strokes. In experimental animals, drugs that prevent excitotoxic damage can also significantly reduce the effects of strokes, and drugs that would produce this effect in humans are currently undergoing clinical trials. However, t-PA and presumably antiexcitotoxic treatment must be given early in the course of a stroke to be of maximum benefit. This is why stroke has become a condition in which rapid diagnosis and treatment are extremely important. In addition, of course, it is important to determine if a stroke is thrombotic or hemorrhagic, since clot lysis is contraindicated in the latter.

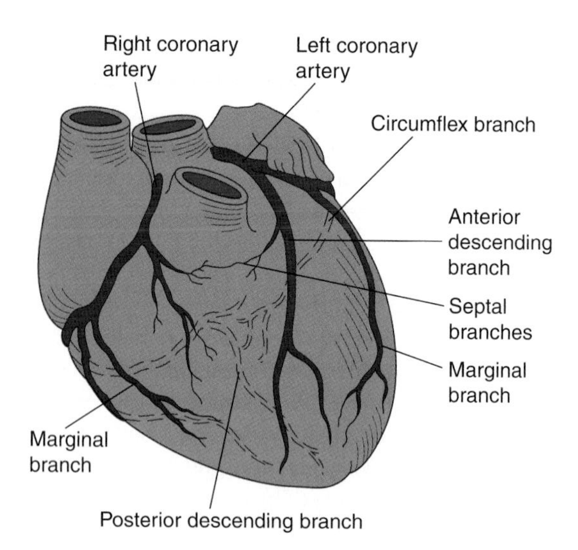

FIGURE 33–11 Coronary arteries and their principal branches in humans. (Reproduced with permission from Ross G: The cardiovascular system. In: *Essentials of Human Physiology*. Ross G [editor]. Copyright © 1978 by Year Book Medical Publishers.)

CORONARY CIRCULATION

ANATOMIC CONSIDERATIONS

The two coronary arteries that supply the myocardium arise from the sinuses behind two of the cusps of the aortic valve at the root of the aorta (Figure 33–11). Eddy currents keep the valves away from the orifices of the arteries, and they are patent throughout the cardiac cycle. Most of the venous blood returns to the heart through the coronary sinus and anterior cardiac veins (Figure 33–12), which drain into the right atrium. In addition, there are other vessels that empty directly into the heart chambers. These include **arteriosinusoidal vessels,** sinusoidal capillary-like vessels that connect arterioles to the chambers; **thebesian veins** that connect capillaries to the chambers; and a few **arterioluminal vessels** that are small arteries draining directly into the chambers. A few anastomoses occur between the coronary arterioles and extracardiac arterioles, especially around the mouths of the great veins. Anastomoses between coronary arterioles in humans only pass particles less than 40 μm in diameter, but evidence indicates that these channels enlarge and increase in number in patients with coronary artery disease.

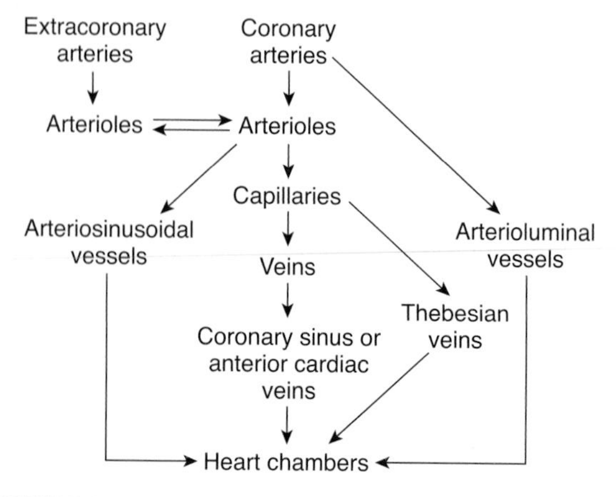

FIGURE 33–12 Diagram of the coronary circulation.

PRESSURE GRADIENTS & FLOW IN THE CORONARY VESSELS

The heart is a muscle that, like skeletal muscle, compresses its blood vessels when it contracts. The pressure inside the left ventricle is slightly higher than in the aorta during systole (Table 33–4). Consequently, flow occurs in the arteries supplying the subendocardial portion of the left ventricle only during diastole, although the force is sufficiently dissipated in the more superficial portions of the left ventricular myocardium to permit some flow in this region throughout the cardiac cycle. Because diastole is shorter when the heart rate is high, left ventricular coronary flow is reduced during tachycardia. On the other hand, the pressure differential between the aorta and the right ventricle, and the differential between the aorta and the atria, are somewhat greater during systole than during diastole. Consequently, coronary flow in those parts of the heart is not appreciably reduced during systole. Flow in the right and left coronary arteries is shown in Figure 33–13. Because no blood flow occurs during systole in the subendocardial portion of the left ventricle, this region is prone to ischemic damage and is the most common site of myocardial infarction. Blood flow to the left ventricle is decreased in patients with stenotic aortic valves because the pressure in the left ventricle must be much higher than that in the aorta to eject the blood. Consequently, the coronary vessels are severely compressed during systole. Patients with aortic stenosis are particularly prone to develop symptoms of myocardial ischemia, in part because of this compression and in part because the myocardium requires more O_2 to expel blood through the stenotic aortic valve. Coronary flow is also decreased when the aortic diastolic pressure is low. The rise in venous pressure in conditions such as congestive heart failure reduces coronary flow because it decreases effective coronary perfusion pressure (see Clinical Box 33–4).

Coronary blood flow has been measured by inserting a catheter into the coronary sinus and applying the Kety method to the heart on the assumption that the N_2O content of coronary venous blood is typical of the entire myocardial effluent. Coronary flow at rest in humans is about 250 mL/min (5% of the cardiac output). A number of techniques utilizing **radionuclides,** radioactive tracers that can be detected with radiation detectors over the chest, have been used to study regional blood flow in the heart and to detect areas of ischemia and infarct as well as to evaluate ventricular function. Radionuclides such as thallium-201 (^{201}Tl) are pumped into cardiac muscle cells by Na, K ATPase and equilibrate with the intracellular K^+ pool. For the first 10–15 min after intravenous injection, ^{201}Tl distribution is directly proportional to myocardial blood flow, and areas of ischemia can be detected by their low uptake. The uptake of this isotope is often determined soon after exercise and again several hours later to bring out areas in which exertion leads to compromised flow. Conversely, radiopharmaceuticals such as technetium-99m stannous pyrophosphate (^{99m}Tc-PYP) are selectively taken up by infarcted tissue by an incompletely understood mechanism and make infarcts stand out as "hot spots" on scintigrams of the chest. Coronary angiography can be combined with measurement of ^{133}Xe washout (see above) to provide detailed analysis of coronary blood flow. Radiopaque contrast medium is first injected into the coronary arteries, and X-rays are used to outline their distribution. The angiographic camera is then replaced with a scintillation camera, and ^{133}Xe washout is measured.

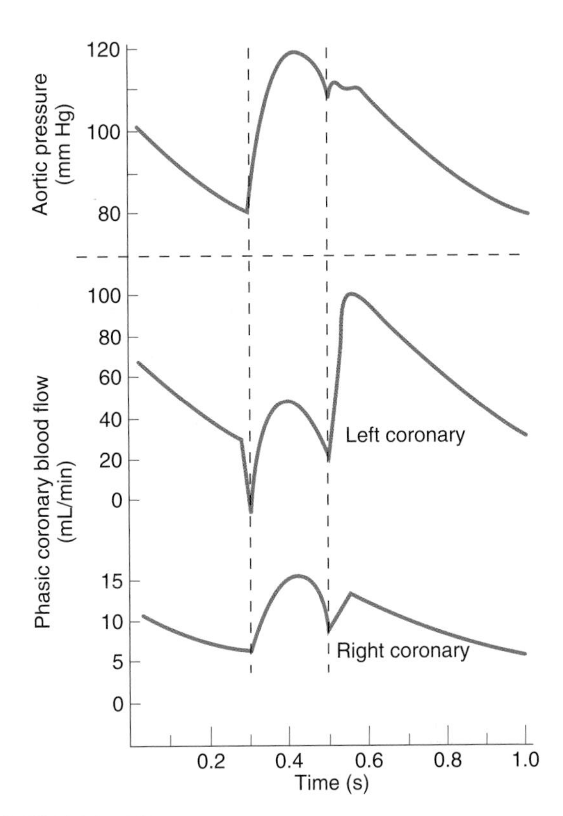

FIGURE 33–13 Blood flow in the left and right coronary arteries during various phases of the cardiac cycle. Systole occurs between the two vertical dashed lines. (Reproduced with permission from Berne RM, Levy MN: *Physiology,* 2nd ed. Mosby, 1988.)

TABLE 33–4 Pressure in aorta and left and right ventricles (vent) in systole and diastole.

	Pressure (mm Hg) in			Pressure Differential (mm Hg) between Aorta and	
	Aorta	Left Vent	Right Vent	Left Vent	Right Vent
Systole	120	121	25	–1	95
Diastole	80	0	0	80	80

VARIATIONS IN CORONARY FLOW

At rest, the heart extracts 70–80% of the O_2 from each unit of blood delivered to it (Table 33–1). O_2 consumption can only be increased significantly by increasing blood flow. Therefore, it is

CLINICAL BOX 33–4

Coronary Artery Disease

When flow through a coronary artery is reduced to the point that the myocardium it supplies becomes hypoxic, **angina pectoris** develops (see Chapter 30). If the myocardial ischemia is severe and prolonged, irreversible changes occur in the muscle, and the result is **myocardial infarction.** Many individuals have angina only on exertion, and blood flow is normal at rest. Others have more severe restriction of blood flow and have anginal pain at rest as well. Partially occluded coronary arteries can be constricted further by vasospasm, producing myocardial infarction. However, it is now clear that the most common cause of myocardial infarction is rupture of an **atherosclerotic plaque,** or hemorrhage into it, which triggers the formation of a coronary-occluding clot at the site of the plaque. The electrocardiographic changes in myocardial infarction are discussed in Chapter 29. When myocardial cells actually die, they leak enzymes into the circulation, and measuring the rise in serum enzymes and isoenzymes produced by infarcted myocardial cells also plays an important role in the diagnosis of myocardial infarction. The enzymes most commonly measured are the MB isomer of creatine kinase (CK-MB), troponin T, and troponin I. Myocardial infarction is a very common cause of death in developed countries because of the widespread occurrence of atherosclerosis. In addition, there is a relation between atherosclerosis and circulating levels of **lipoprotein(a) (Lp(a)).** Lp(a) has an outer coat count of apo(a). It interferes with fibrinolysis by down-regulating plasmin generation (see Chapter 31). It now appears that atherosclerosis has an important inflammatory component as well. The lesions of the disease contain inflammatory cells, and there is a positive correlation between increased levels of C-reactive protein and other **inflammatory markers** in the circulation with subsequent myocardial infarction.

THERAPEUTIC HIGHLIGHTS

Treatment of myocardial infarction aims to restore flow to the affected area as rapidly as possible while minimizing reperfusion injury. Needless to say, it should be started as promptly as possible to avoid irreversible changes in heart function. In acute disease, antithrombotic agents are often given, but these can be problematic, leading to increased mortality due to bleeding if cardiac surgery is subsequently needed. Mechanical/surgical approaches to coronary artery disease include balloon angioplasty and/or implantation of stents to hold the vessels open, or grafting of coronary vessels to bypass blocked segments (CABG).

not surprising that blood flow increases when the metabolism of the myocardium is increased. The caliber of the coronary vessels, and consequently the rate of coronary blood flow, is influenced not only by pressure changes in the aorta but also by chemical and neural factors. The coronary circulation also shows considerable autoregulation.

CHEMICAL FACTORS

The close relationship between coronary blood flow and myocardial O_2 consumption indicates that one or more of the products of metabolism cause coronary vasodilation. Factors suspected of playing this role include a lack of O_2 and increased local concentrations of CO_2, H^+, K^+, lactate, prostaglandins, adenine nucleotides and adenosine. Likely several or all of these vasodilator metabolites act in an integrated fashion, redundant fashion, or both. Asphyxia, hypoxia, and intracoronary injections of cyanide all increase coronary blood flow 200–300% in denervated as well as intact hearts, and the feature common to these three stimuli is hypoxia of the myocardial fibers. A similar increase in flow is produced in the area supplied by a coronary artery if the artery is occluded and then released. This **reactive hyperemia** is similar to that seen in the skin (see below). Evidence suggests that in the heart it is due to release of adenosine.

NEURAL FACTORS

The coronary arterioles contain α-adrenergic receptors, which mediate vasoconstriction, and β-adrenergic receptors, which mediate vasodilation. Activity in the noradrenergic nerves to the heart and injections of norepinephrine cause coronary vasodilation. However, norepinephrine increases the heart rate and the force of cardiac contraction, and the vasodilation is due to production of vasodilator metabolites in the myocardium secondary to the increase in its activity. When the inotropic and chronotropic effects of noradrenergic discharge are blocked by a β-adrenergic blocking drug, stimulation of the noradrenergic nerves or injection of norepinephrine in unanesthetized animals elicits coronary vasoconstriction. Thus, the direct effect of noradrenergic stimulation is constriction rather than dilation of the coronary vessels. On the other hand, stimulation of vagal fibers to the heart dilates the coronaries.

When the systemic blood pressure falls, the overall effect of the reflex increase in noradrenergic discharge is increased coronary blood flow secondary to the metabolic changes in the myocardium at a time when the cutaneous, renal, and splanchnic vessels are constricted. In this way the circulation of the heart, like that of the brain, is preserved when flow to other organs is compromised.

CUTANEOUS CIRCULATION

The amount of heat lost from the body is regulated to a large extent by varying the amount of blood flowing through the skin. The fingers, toes, palms, and earlobes contain well-innervated anastomotic connections between arterioles and venules (arteriovenous anastomoses; see Chapter 31). Blood flow in response to thermoregulatory stimuli can vary from 1 to as much as 150 mL/100 g of skin/min, and it has been postulated that these variations are possible because blood can be shunted through the anastomoses. The subdermal capillary and venous plexus is a blood reservoir of some importance, and the skin is one of the few places where the reactions of blood vessels can be observed visually.

WHITE REACTION

When a pointed object is drawn lightly over the skin, the stroke lines become pale **(white reaction).** The mechanical stimulus apparently initiates contraction of the precapillary sphincters, and blood drains out of the capillaries and small veins. The response appears in about 15 s.

TRIPLE RESPONSE

When the skin is stroked more firmly with a pointed instrument, instead of the white reaction there is reddening at the site that appears in about 10 s **(red reaction).** This is followed in a few minutes by local swelling and diffuse, mottled reddening around the injury. The initial redness is due to capillary dilation, a direct response of the capillaries to pressure. The swelling **(wheal)** is local edema due to increased permeability of the capillaries and postcapillary venules, with consequent extravasation of fluid. The redness spreading out from the injury **(flare)** is due to arteriolar dilation. This three-part response—the red reaction, wheal, and flare—is called the **triple response** and is part of the normal reaction to injury (see Chapter 3). It persists after total sympathectomy. On the other hand, the flare is absent in locally anesthetized skin and in denervated skin after the sensory nerves have degenerated, but it is present immediately after nerve block or section above the site of the injury. This, plus other evidence, indicates that it is due to an **axon reflex,** a response in which impulses initiated in sensory nerves by the injury are relayed antidromically down other branches of the sensory nerve fibers (Figure 33–14). This is the one situation in the body in which there is substantial evidence for a physiologic effect due to antidromic conduction. The transmitter released at the central termination of the sensory C fiber neurons is substance P (see Chapter 7), and substance P and CGRP are present in all parts of the neurons. Both dilate arterioles and, in addition, substance P causes extravasation of fluid. Effective nonpeptide antagonists to substance P have now been developed, and they reduce the extravasation. Thus, it appears that these peptides produce the wheal.

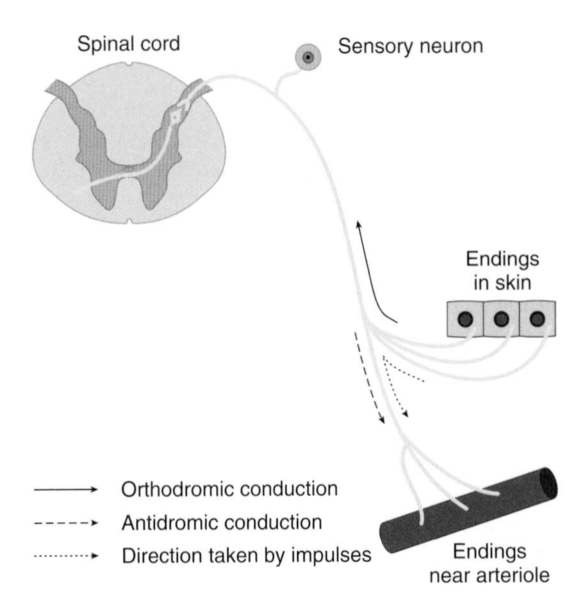

FIGURE 33–14 Axon reflex.

REACTIVE HYPEREMIA

A response of the blood vessels that occurs in many organs but is visible in the skin is **reactive hyperemia,** an increase in the amount of blood in a region when its circulation is reestablished after a period of occlusion. When the blood supply to a limb is occluded, the cutaneous arterioles below the occlusion dilate. When the circulation is reestablished, blood flowing into the dilated vessels makes the skin become fiery red. O_2 in the atmosphere can diffuse a short distance through the skin, and reactive hyperemia is prevented if the circulation of the limb is occluded in an atmosphere of 100% O_2. Therefore, the arteriolar dilation is apparently due to a local effect of hypoxia.

GENERALIZED RESPONSES

Noradrenergic nerve stimulation and circulating epinephrine and norepinephrine constrict cutaneous blood vessels. No known vasodilator nerve fibers extend to the cutaneous vessels, and thus vasodilation is brought about by a decrease in constrictor tone as well as the local production of vasodilator metabolites. Skin color and temperature also depend on the state of the capillaries and venules. A cold blue or gray skin is one in which the arterioles are constricted and the capillaries dilated; a warm red skin is one in which both are dilated.

Because painful stimuli cause diffuse noradrenergic discharge, a painful injury causes generalized cutaneous vasoconstriction in addition to the local triple response. When the body temperature rises during exercise, the cutaneous blood vessels dilate in spite of continuing noradrenergic discharge in other parts of the body. Dilation of cutaneous vessels in response to a rise in hypothalamic temperature overcomes other reflex activity. Cold causes cutaneous vasoconstriction;

however, with severe cold, superficial vasodilation may supervene. This vasodilation is the cause of the ruddy complexion seen on a cold day.

Shock is more profound in patients with elevated temperatures because of cutaneous vasodilation, and patients in shock should not be warmed to the point that their body temperature rises. This is sometimes a problem because well-meaning laymen have read in first-aid books that "injured patients should be kept warm," and they pile blankets on accident victims who are in shock.

PLACENTAL & FETAL CIRCULATION

UTERINE CIRCULATION

The blood flow of the uterus parallels the metabolic activity of the myometrium and endometrium and undergoes cyclic fluctuations that correlate with the menstrual cycle in nonpregnant women. The function of the spiral and basilar arteries of the endometrium in menstruation is discussed in Chapter 22. During pregnancy, blood flow increases rapidly as the uterus increases in size (Figure 33–15). Vasodilator metabolites are undoubtedly produced in the uterus, as they are in other active tissues. In early pregnancy, the arteriovenous O_2 difference across the uterus is small, and it has been suggested that estrogens act on the blood vessels to increase uterine blood flow in excess of tissue O_2 needs. However, even though uterine blood flow increases 20-fold during pregnancy, the size of the conceptus increases much more, changing from a single cell to a fetus plus a placenta that weighs 4–5 kg at term in humans. Consequently, more O_2 is extracted from the uterine blood during the latter part of pregnancy, and the O_2 saturation of uterine blood falls. Corticotrophin-releasing hormone appears to play an important role in up-regulating uterine blood flow, as well as in the eventual timing of birth.

PLACENTA

The placenta is the "fetal lung" (Figures 33–16 and 33–17). Its maternal portion is in effect a large blood sinus. Into this "lake" project the villi of the fetal portion containing the small branches of the fetal umbilical arteries and vein (Figure 33–16). O_2 is taken up by the fetal blood and CO_2 is discharged into the maternal circulation across the walls of the villi in a fashion analogous to O_2 and CO_2 exchange in the lungs (see Chapter 35). However, the cellular layers covering the villi are thicker and less permeable than the alveolar membranes in the lungs, and exchange is much less efficient. The placenta is also the route by which all nutritive materials enter the fetus and by which fetal wastes are discharged to the maternal blood.

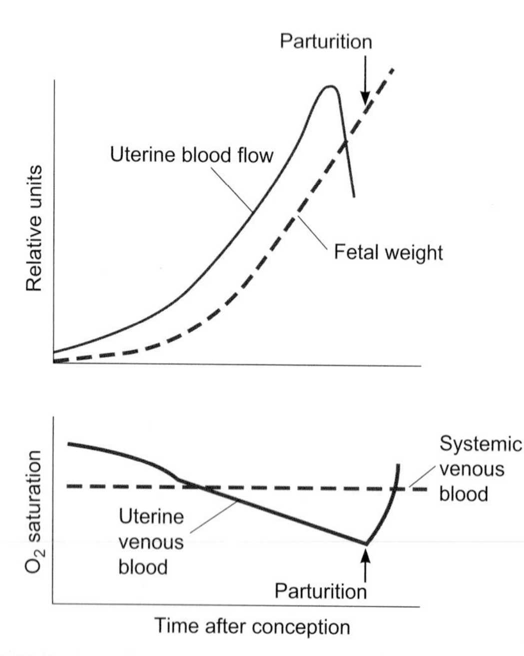

FIGURE 33–15 Changes in uterine blood flow and the amount of O_2 in uterine venous blood during pregnancy. (After Barcroft H. Modified and redrawn with permission from Keele CA, Neil E: *Samson Wright's Applied Physiology,* 12th ed. Oxford University Press, 1971.)

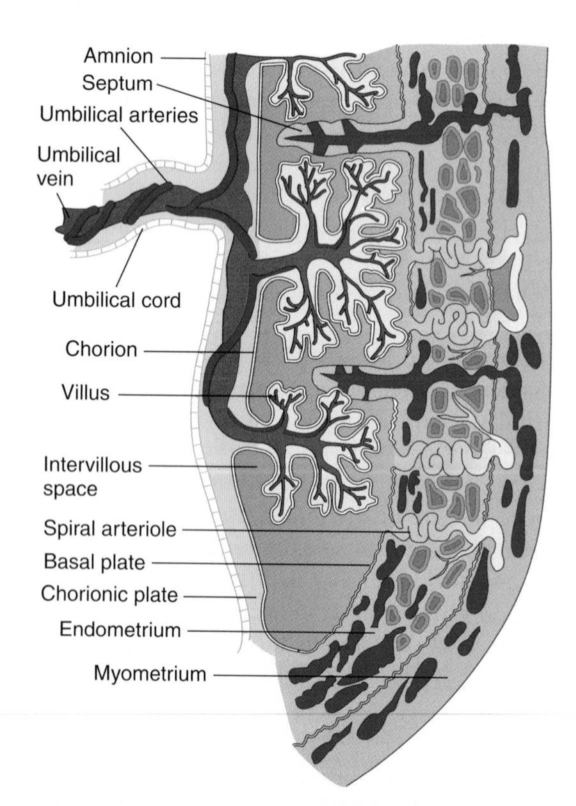

FIGURE 33–16 Diagram of a section through the human placenta, showing the way the fetal villi project into the maternal sinuses. (Reproduced with permission from Benson RC: *Handbook of Obstetrics and Gynecology,* 8th ed, and modified after Netter. Originally published by Appleton & Lange. Copyright © 1983 McGraw-Hill.)

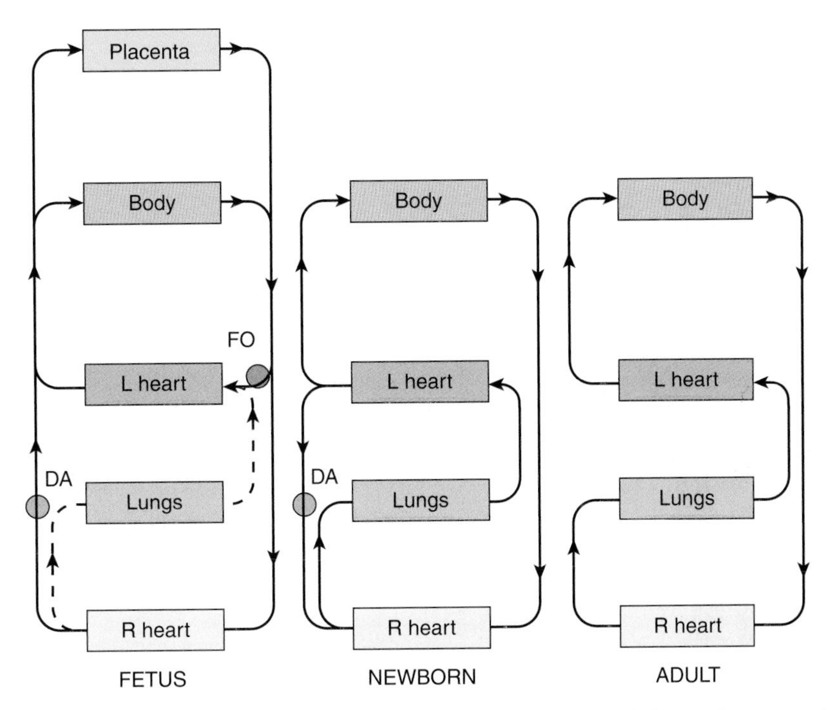

FIGURE 33–17 Diagram of the circulation in the fetus, the newborn infant, and the adult. DA, ductus arteriosus; FO, foramen ovale. (Redrawn and reproduced with permission from Born GVR, et al: Changes in the heart and lungs at birth. Cold Spring Harbor Symp Quant Biol 1954;19:102.)

FETAL CIRCULATION

The arrangement of the circulation in the fetus is shown diagrammatically in Figure 33–17. Fifty-five per cent of the fetal cardiac output goes through the placenta. The blood in the umbilical vein in humans is believed to be about 80% saturated with O_2, compared with 98% saturation in the arterial circulation of the adult. The **ductus venosus** (Figure 33–18) diverts some of this blood directly to the inferior vena cava, and the remainder mixes with the portal blood of the fetus. The portal and systemic venous blood of the fetus is only 26% saturated, and the saturation of the mixed blood in the inferior vena cava is approximately 67%. Most of the blood entering the heart through the inferior vena cava is diverted directly to the left atrium via the patent foramen ovale. Most of the blood from the superior vena cava enters the right ventricle and is expelled into the pulmonary artery. The resistance of the collapsed lungs is high, and the pressure in the pulmonary artery is several mm Hg higher than it is in the aorta, so that most of the blood in the pulmonary artery passes through the **ductus arteriosus** to the aorta. In this fashion, the relatively unsaturated blood from the right ventricle is diverted to the trunk and lower body of the fetus, while the head of the fetus receives the better-oxygenated blood from the left ventricle. From the aorta, some of the blood is pumped into the umbilical arteries and back to the placenta. The O_2 saturation of the blood in the lower aorta and umbilical arteries of the fetus is approximately 60%.

FETAL RESPIRATION

The tissues of fetal and newborn mammals have a remarkable but poorly understood resistance to hypoxia. However, the O_2 saturation of the maternal blood in the placenta is so low that the fetus might suffer hypoxic damage if fetal red cells did not have a greater O_2 affinity than adult red cells (Figure 33–19). The fetal red cells contain fetal hemoglobin (hemoglobin F), whereas the adult cells contain adult hemoglobin (hemoglobin A). The cause of the difference in O_2 affinity between the two is that hemoglobin F binds 2, 3-DPG less effectively than hemoglobin A does. The decrease in O_2 affinity due to the binding of 2, 3-DPG is discussed in Chapter 31.

CHANGES IN FETAL CIRCULATION & RESPIRATION AT BIRTH

Because of the patent ductus arteriosus and foramen ovale (Figure 33–18), the left heart and right heart pump in parallel in the fetus rather than in series as they do in the adult. At birth, the placental circulation is cut off and the peripheral resistance suddenly rises. Meanwhile, the infant becomes increasingly asphyxial. Finally, the infant gasps several times, and the lungs expand. The markedly negative intrapleural pressure (–30 to –50 mm Hg) during the gasps contributes to the expansion of the lungs, but other factors are likely also involved. The sucking action of the first breath plus constriction of the umbilical

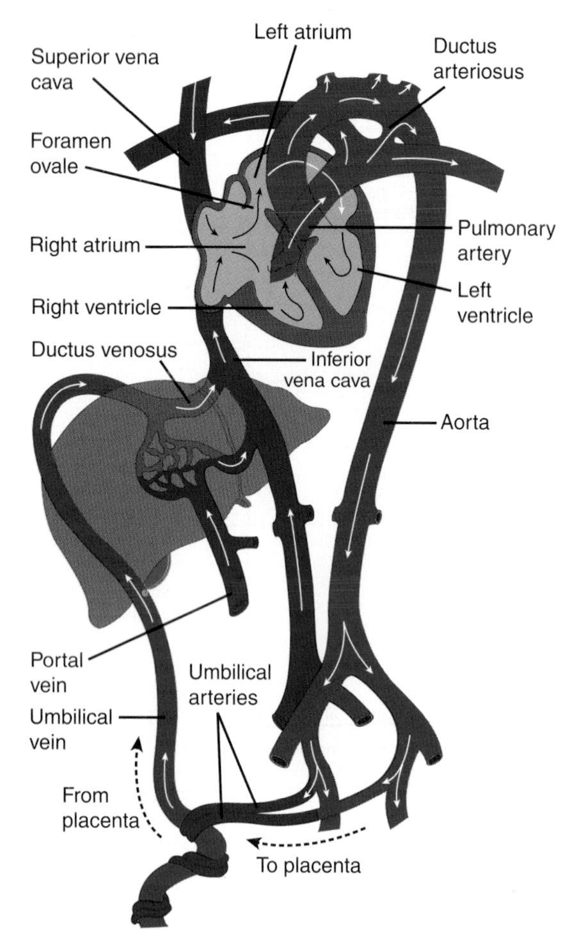

FIGURE 33–18 Circulation in the fetus. Most of the oxygenated blood reaching the heart via the umbilical vein and inferior vena cava is diverted through the foramen ovale and pumped out the aorta to the head, while the deoxygenated blood returned via the superior vena cava is mostly pumped through the pulmonary artery and ductus arteriosus to the feet and the umbilical arteries.

veins squeezes as much as 100 mL of blood from the placenta (the "placental transfusion").

Once the lungs are expanded, the pulmonary vascular resistance falls to less than 20% of the value in utero, and pulmonary blood flow increases markedly. Blood returning from the lungs raises the pressure in the left atrium, closing the foramen ovale by pushing the valve that guards it against the interatrial septum. The ductus arteriosus constricts within a few hours after birth, producing functional closure, and permanent anatomic closure follows in the next 24–48 h due to extensive intimal thickening. The mechanism producing the initial constriction involves the increase in arterial O_2 tension and bradykinin, which is released from the lungs during their initial inflation. In addition, relatively high concentrations of vasodilators are present in the ductus in utero—especially prostaglandin F_{2a}—and synthesis of these prostaglandins is blocked by inhibition of cyclooxygenase at birth. In many

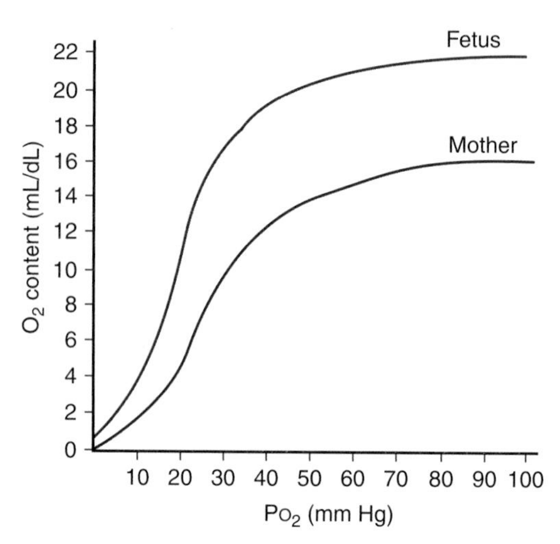

FIGURE 33–19 Dissociation curves of hemoglobin in human maternal and fetal blood.

premature infants the ductus fails to close spontaneously, but closure can be produced by infusion of drugs that inhibit cyclooxygenase. NO may also be involved in maintaining ductal patency in this setting.

CHAPTER SUMMARY

- Cerebrospinal fluid is produced predominantly in the choroid plexus of the brain, in part via active transport mechanisms in the choroid epithelial cells. Fluid is reabsorbed into the bloodstream to maintain appropriate pressure in the setting of continuous production.
- The permeation of circulating substances into the brain is tightly controlled. Water, CO_2, and O_2 permeate freely. Other substances (such as glucose) require specific transport mechanisms, whereas entry of macromolecules is negligible. The effectiveness of the blood–brain barrier in preventing entry of xenobiotics is bolstered by active efflux mediated by P-glycoprotein.
- The coronary circulation supplies oxygen to the contracting myocardium. Metabolic products and neural input induce vasodilation as needed for oxygen demand. Blockage of coronary arteries may lead to irreversible injury to heart tissue.
- Control of cutaneous blood flow is a key facet of temperature regulation, and is underpinned by varying levels of shunting through arteriovenous anastomoses. Hypoxia, axon reflexes, and sympathetic input are all important determinants of flow through the cutaneous vasculature.
- The fetal circulation cooperates with that of the placenta and uterus to deliver oxygen and nutrients to the growing fetus, as well as carrying away waste products. Unique anatomic features of the fetal circulation as well as biochemical properties of fetal hemoglobin serve to ensure adequate O_2 supply, particularly to the head. At birth, the foramen ovale and the ductus arteriosus close such that the neonatal lungs now serve as the site for oxygen exchange.

MULTIPLE-CHOICE QUESTIONS

For all questions, select the single best answer unless otherwise directed.

1. Blood in which of the following vessels normally has the lowest PO_2?
   A. Maternal artery
   B. Maternal uterine vein
   C. Maternal femoral vein
   D. Umbilical artery
   E. Umbilical vein
2. The pressure differential between the heart and the aorta is least in the
   A. left ventricle during systole.
   B. left ventricle during diastole.
   C. right ventricle during systole.
   D. right ventricle during diastole.
   E. left atrium during systole.
3. Injection of tissue plasminogen activator (t-PA) would probably be most beneficial
   A. after at least 1 year of uncomplicated recovery following occlusion of a coronary artery.
   B. after at least 2 months of rest and recuperation following occlusion of a coronary artery.
   C. During the second week after occlusion of a coronary artery.
   D. During the second day after occlusion of a coronary artery.
   E. During the second hour after occlusion of a coronary artery.
4. Which of the following organs has the greatest blood flow per 100 g of tissue?
   A. Brain
   B. Heart muscle
   C. Skin
   D. Liver
   E. Kidneys
5. Which of the following does *not* dilate arterioles in the skin?
   A. Increased body temperature
   B. Epinephrine
   C. Bradykinin
   D. Substance P
   E. Vasopressin
6. A baby boy is brought to the hospital because of convulsions. In the course of a workup, his body temperature and plasma glucose are found to be normal, but his cerebrospinal fluid glucose is 12 mg/dL (normal, 65 mg/dL). A possible explanation of his condition is
   A. constitutive activation of GLUT 3 in neurons.
   B. SGLT 1 deficiency in astrocytes.
   C. GLUT 5 deficiency in cerebral capillaries.
   D. GLUT 1 55K deficiency in cerebral capillaries.
   E. GLUT 1 45K deficiency in microglia.

CHAPTER RESOURCES

Begley DJ, Bradbury MW, Kreater J (editors): *The Blood–Brain Barrier and Drug Delivery to the CNS.* Marcel Dekker, 2000.

Birmingham K (editor): The heart. Nature 2002;415:197.

Duncker DJ, Bache RJ: Regulation of coronary blood flow during exercise. Physiol Rev 2008;88:1009.

Hamel E: Perivascular nerves and the regulation of cerebrovascular tone. J Appl Physiol 2006;100:1059.

Johanson CE, Duncan JA 3rd, Klinge PM, et al. Multiplicity of cerebrospinal fluid functions: New challenges in health and disease. Cerebrospinal Fluid Res 2008;5:10.

Ward JPT: Oxygen sensing in context. Biochim Biophys Acta 2008;1777:1.

SECTION VI Respiratory Physiology

Respiration, or the uptake of O_2 and removal of CO_2 from the body as a whole, is the primary goal of the lung. At rest, a normal human breathes 12–15 times a minute. With each breath containing ~500 mL of air, this translates to 6–8 L of air that is inspired and expired every minute. Once the air reaches the depths of the lung in the alveoli, simple diffusion allows O_2 to enter the blood in the pulmonary capillaries and CO_2 to enter the alveoli, from where it can be expired. Using some basic math, on average, 250 mL of O_2 enters the body per minute and 200 mL of CO_2 is excreted. In addition to the O_2 that enters the respiratory system, inspired air also contains a variety of particulates that must be properly filtered and/or removed to maintain lung health. Finally, although we have a certain amount of control over breathing, most of the minute to minute function, including the fine adjustments necessary for proper lung function, are accomplished independent of voluntary control. The goal of this section is to review basic concepts that underlie important aspects of the control and outcome of breathing as well as to highlight other important functions in respiratory physiology.

The respiratory system is connected to the outside world by the upper airway that leads down a set of conduits before reaching the gas-exchanging areas (the alveoli). The function of the lungs is supported by a variety of anatomical features that serve to inflate/deflate the lung, thereby allowing the movement of gases to and from the rest of the body. Supporting features include the chest wall; the respiratory muscles (which increase and decrease the size of the thoracic cavity); the areas in the brain that control the muscles; and the tracts and nerves that connect the brain to the muscles. Finally, the lung supports the pulmonary circulation, which allows for movement of gases to other organs and tissues of the body. In the first chapter of this section we will explore the unique anatomical and cellular makeup of the respiratory system and how the intricate structure of the lung contributes to respiratory physiology. This examination will lead into basic lung measurements that both define and allow for lung inflation/deflation, as well as some of the nonrespiratory functions essential for lung health.

Our discussion will continue with an overview of the primary function of the respiratory system—the capture of O_2 from the outside environment and its delivery to tissues, as well as the simultaneous removal of CO_2 from the tissues to the outside environment. During this discussion, the critical role of pH in gas exchange as well as the ability of the lung to contribute to pH regulation of the blood is examined. A discussion of respiratory responses to altered O_2 or CO_2 concentrations, caused by environmental and/or physiological changes, is used to better understand the overall control of coordinated uptake of O_2 and excretion of CO_2.

Control of breathing is quite complex, and includes not only the repetitive neuronal firing that controls muscle movements that inflate/deflate the lung, but also a series of feedback loops that increase/decrease deflation depending on the gas content of the blood. The final chapter in this section begins with an overview of some of the key factors that aid in this regulation of respiration. Specific examples of common respiratory abnormalities and how they relate to altered regulation of breathing are also discussed to better understand the intricate feedback loops that help to regulate breathing.

Due to the complexity of the lung and thus the variety of working parts that can be compromised, there is a wide-ranging list of diseases that impact its function. Such diseases include common (and uncommon) respiratory infections, asthma, chronic obstructive pulmonary disease (COPD), acute respiratory distress syndrome, pulmonary hypertension, lung cancer and many more. The health burden from such a diverse collection of disorders cannot be overstated. Using COPD as an example, current conservative estimates hold that over 12 million adults in the United States suffer from the condition. In fact, COPD is the fourth leading cause of death (and rising) and a contributory factor in an equal number of non-COPD deaths. Although treatment strategies for COPD, largely based on continuing research efforts and understanding, have contributed to an improved lifestyle, the underlying causes are, as of yet, untreatable. The continued and improved understanding of respiratory physiology and lung function (and its dysfunction) will provide opportunities to develop new strategies for treatment of COPD, as well as the myriad of other lung diseases.

CHAPTER

34 Introduction to Pulmonary Structure and Mechanics

OBJECTIVES

After studying this chapter, you should be able to:

- List the passages through which air passes from the exterior to the alveoli, and describe the cells that line each of them.
- List the major muscles involved in respiration, and state the role of each.
- Define the basic measures of lung volume and give approximate values for each in a normal adult.
- Define lung compliance and airway resistance.
- Compare the pulmonary and systemic circulations, and list some major differences between them.
- Describe basic lung defense and metabolic functions
- Define partial pressure and calculate the partial pressure of each of the important gases in the atmosphere at sea level.

INTRODUCTION

The structure of the respiratory system is uniquely suited to its primary function, the transport of gases in and out of the body. In addition, the respiratory system provides a large volume of tissue that is constantly exposed to the outside environment, and thus, to potential infection and injury. Finally, the pulmonary system includes a unique circulation that must handle the blood flow. This chapter begins with the basic anatomy and cellular physiology that contribute to the respiratory system and some of their unique features. The chapter also includes discussion of how the anatomical features contribute to the basic mechanics of breathing, as well as some highlights of nonrespiratory physiology in the pulmonary system.

ANATOMY OF THE LUNGS

Regions of the Respiratory Tract

Airflow through the respiratory system can be broken down into three interconnected regions: the **upper airway;** the **conducting airway**; and the **alveolar airway** (also known as the **lung parenchyma** or **acinar tissue**). The upper airway consists of the entry systems, the nose/nasal cavity and mouth that lead into the pharynx. The larynx extends from the lower part of the pharynx to complete the upper airway. The nose is the primary point of entry for inhaled air; therefore, the mucosal epithelium lining the nasopharyngeal airways is exposed to the highest concentration of inhaled allergens, toxicants, and particulate matter. With this in mind, it is easy to understand that in addition to olfaction, the nose and upper airway provides two additional crucial functions in airflow—(1) filtering out large particulates to prevent them from reaching the conducting and alveolar airways and (2) serving to warm and humidify air as it enters the body. Particulates larger than 30–50 μm in size tend to not to be inhaled through the nose whereas particulates on the order of 5–10 μm impact on the nasopharynx and do enter the conducting airway. Most of these latter particles settle on mucous membranes in the nose and pharynx. Because of their momentum, they do not follow the airstream as it curves downward into the lungs, and they impact on or near the **tonsils** and **adenoids,** large collections of immunologically active lymphoid tissue in the back of the pharynx.

Conducting Airway

The conducting airway begins at the trachea and branches dichotomously to greatly expand the surface area of the tissue in the lung. The first 16 generations of passages form the

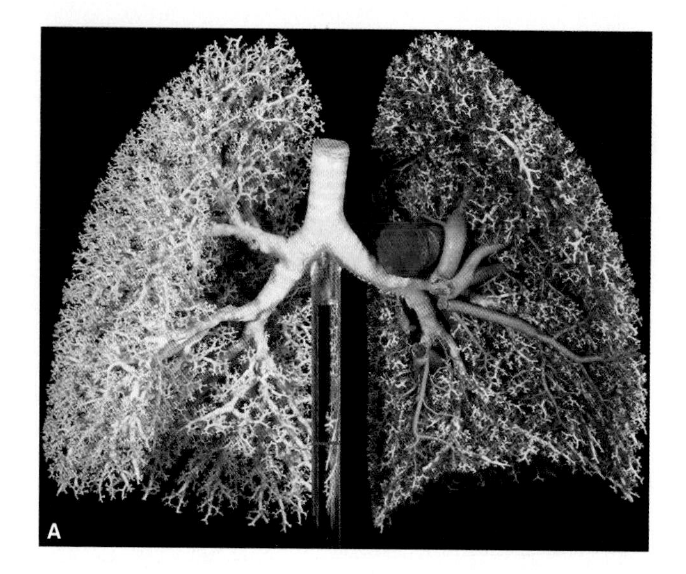

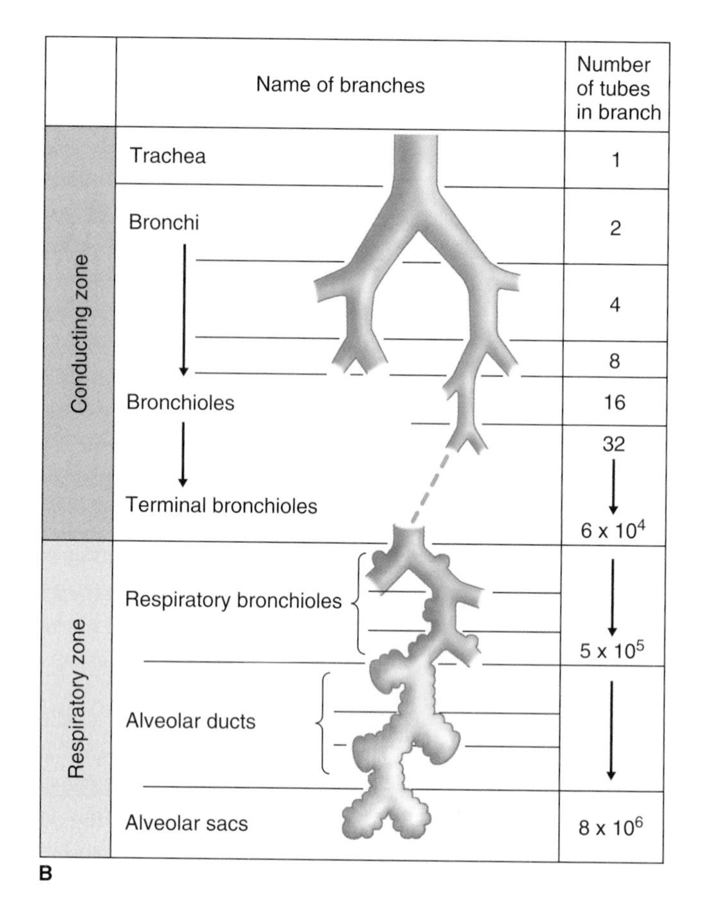

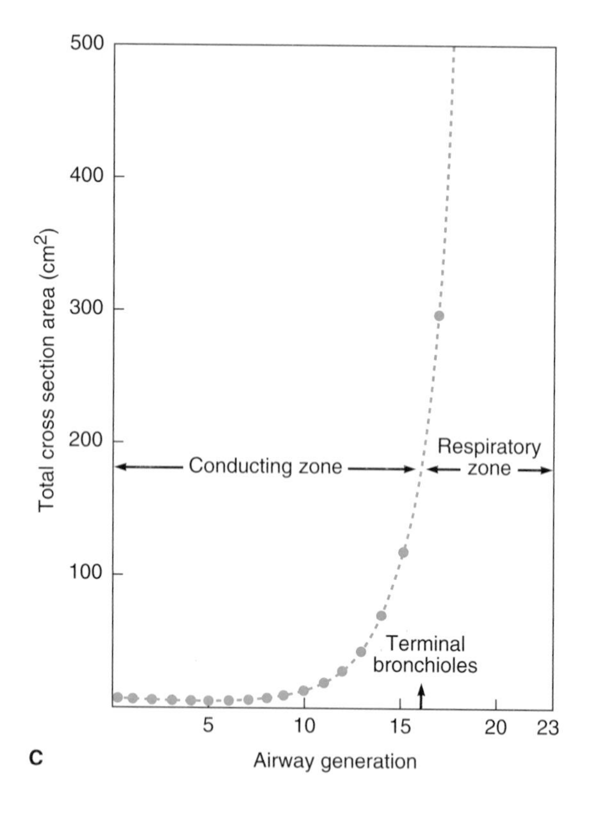

FIGURE 34–1 Conducting and respiratory zones in the airway. A) Resin cast of the human airway tree shows dichotomous branching beginning at the trachea. Note the added pulmonary arteries (red) and veins (blue) displayed in the left lung. **B)** The branching patterns of the airway from the conducting to the transitional and respiratory airway zones are drawn (not all divisions are drawn, and drawings are not to scale). Numbers indicate approximate airway passages following generational branching. **C)** The total airway cross-sectional area rapidly increases as the airways transition from the conducting to the respiratory zone. (A Reproduced with permission from Fishman AP: *Fishman's Pulmonary Diseases and Disorders*, 4th ed. McGraw Hill Medical, 2008; B and C Reproduced with permission from West JB: *Respiratory Physiology: The Essentials*, 7th ed. Williams & Wilkins, 2005.)

conducting zone of the airways that transports gas from and to the upper airway described above (Figure 34–1). These branches are made up of bronchi, bronchioles, and terminal bronchioles. The conducting airway is made up of a variety of specialized cells that provide more than simply a conduit for air to reach the lung (Figure 34–2). The mucosal epithelium is attached to a thin basement membrane, and beneath this, the lamina propria. Collectively these are referred to as the "airway mucosa." Smooth muscle cells are found beneath the epithelium and an enveloping connective tissue is likewise interspersed with cartilage that is more predominant in the portions of the conducting airway of greater caliber. The epithelium is organized as a pseudostratified epithelium and contains several cell types, including ciliated and secretory cells (eg, goblet cells and glandular acini) that provide key components for airway innate immunity, and basal cells that can serve

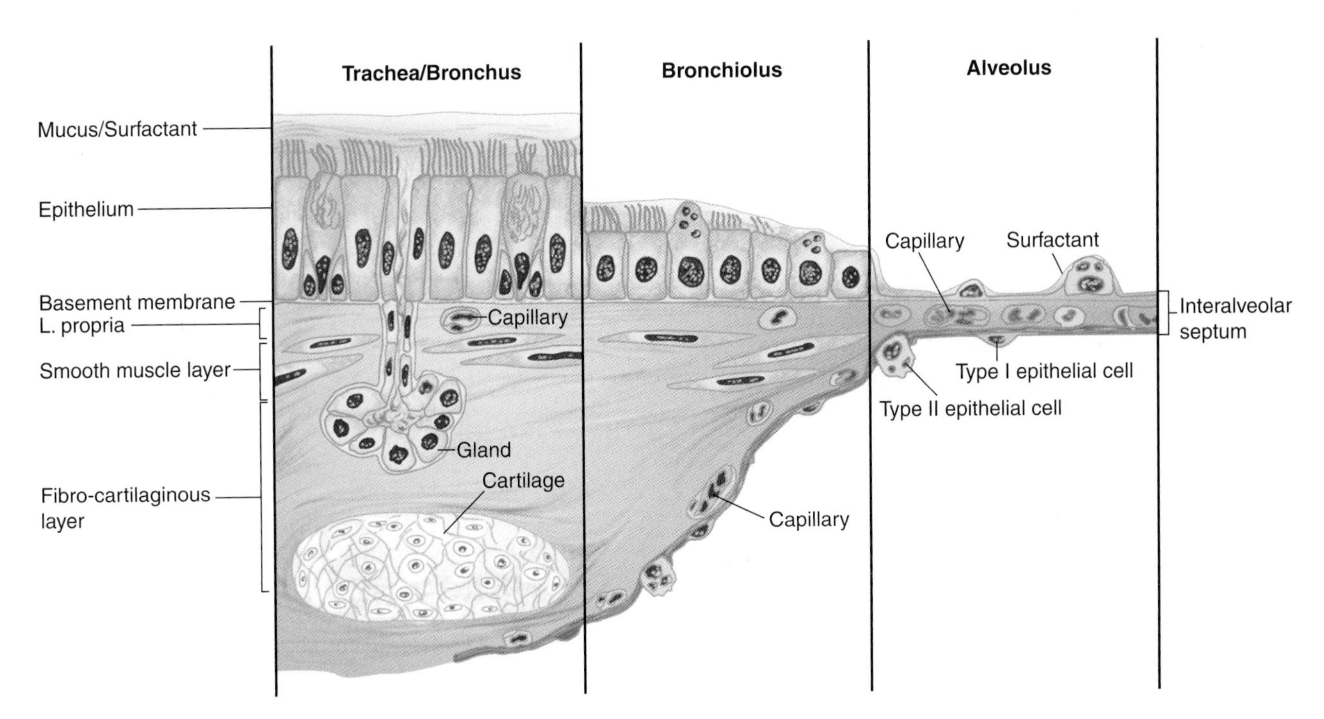

FIGURE 34–2 Cellular transition from conducting airway to the alveolus. The epithelial layer transitions from pseudostratified layer with submucosal glands to a cuboidal and the to a squamous epithelium. The underlying mesynchyme tissue and capillary structure also changes with the airway transition (Reproduced from Fishman AP: *Fishman's Pulmonary Diseases and Disorders,* 4th ed. McGraw Hill Medical, 2008).

as progenitor cells during injury. As the conducting airway transitions to terminal and transitional bronchioles, the histological appearance of the conducting tubes change. Secretory glands are absent from the epithelium of the bronchioles and terminal bronchioles, smooth muscle plays a more prominent role and cartilage is largely absent from the underlying tissue. Clara cells, nonciliated cuboidal epithelial cells that secrete important defense markers and serve as progenitor cells after injury, make up a large portion of the epithelial lining in the latter portions of the conducting airway.

Epithelial cells in the conducting airway can secrete a variety of molecules that aid in lung defense. Secretory immunoglobulins (IgA), collectins (including surfactant protein (SP) -A and SP-D), defensins and other peptides and proteases, reactive oxygen species, and reactive nitrogen species are all generated by airway epithelial cells. These secretions can act directly as antimicrobials to help keep the airway free of infection. Airway epithelial cells also secrete a variety of chemokines and cytokines that recruit traditional immune cells and other immune effector cells to site of infections. The smaller particles that make it through the upper airway, ~2–5 μm in diameter, generally fall on the walls of the bronchi as the airflow slows in the smaller passages. There they can initiate reflex bronchial constriction and coughing. Alternatively, they can be moved away from the lungs by the "mucociliary escalator." The epithelium of the respiratory passages from the anterior third of the nose to the beginning of the respiratory bronchioles is ciliated (Figure 34–2). The cilia are bathed in a periciliary fluid where they typically beat at rates of 10–15 Hz. On top of the periciliary layer and the beating cilia rests a mucus layer, a complex mixture of proteins and polysaccharides secreted from specialized cells, glands, or both in the conducting airway. This combination allows for the trapping of foreign particles (in the mucus) and their transport out of the airway (powered by ciliary beat). The ciliary mechanism is capable of moving particles away from the lungs at a rate of at least 16 mm/min. When ciliary motility is defective, as can occur in smokers, or as a result of other environmental conditions or genetic deficiencies, mucus transport is virtually absent. This can lead to chronic sinusitis, recurrent lung infections, and bronchiectasis. Some of these symptoms are evident in cystic fibrosis (Clinical Box 34–1).

The walls of the bronchi and bronchioles are innervated by the autonomic nervous system. Nerve cells in the airways sense mechanical stimuli or the presence of unwanted substances in the airways such as inhaled dusts, cold air, noxious gases and cigarette smoke. These neurons can signal the respiratory centers to contract the respiratory muscles and initiate sneeze or cough reflexes. The receptors show rapid adaptation when they are continuously stimulated to limit sneeze and cough under normal conditions. The β_2 receptors mediate bronchodilation. They also increase bronchial secretions (eg, mucus), while α_1 adrenergic receptors inhibit secretions.

CLINICAL BOX 34–1

Cystic Fibrosis

Among Caucasians, cystic fibrosis is one of the most common genetic disorders: greater than 3% of the United States population are carriers for this autosomal recessive disease.

The gene that is abnormal in cystic fibrosis is located on the long arm of chromosome 7 and encodes the **cystic fibrosis transmembrane conductance regulator (CFTR),** a regulated Cl^- channel located on the apical membrane of various secretory and absorptive epithelia. The number of reported mutations in the *CFTR* gene that cause cystic fibrosis is large (> 1000) and the mutations are now grouped into five classes (I–V) based on their cellular function. Class I mutations do not allow for synthesis of the protein. Class II mutations have protein processing defects. Class III mutations have a block in their channel regulation. Class IV mutations display altered conductance of the ion channel. Class V mutations display reduced synthesis of the protein. The severity of the defect varies with the class and the individual mutation. The most common mutation causing cystic fibrosis is loss of the phenylalanine residue at amino acid position 508 of the protein (ΔF508), a Class II mutation that limits the amount of protein that gets to the plasma membrane.

One outcome of cystic fibrosis is repeated pulmonary infections, particularly with *Pseudomonas aeruginosa,* and progressive, eventually fatal destruction of the lungs. There is also suppressed chloride secretion across the wall of the airways. One would expect Na^+ reabsorption to be depressed as well, and indeed in sweat glands it is. However, in the lungs, it is enhanced, so that the Na^+ and water move out of airways, leaving their other secretions inspissated and sticky. This results in a reduced periciliary layer that inhibits function of the mucociliary escalator, and alters the local environment to reduce the effectiveness of antimicrobial secretions.

THERAPEUTIC HIGHLIGHTS

Traditional treatments of cystic fibrosis address the various symptoms. Chest physiotherapy and mucolytics are used to loosen thick mucus and aid lung clearance. Antibiotics are used to prevent new infections and keep chronic infections in check. Bronchodilators and anti-inflammatory medications are used to help expand and clear air passages. Pancreatic enzymes and nutritive supplements are used to increase nutrient absorption and promote weight gain. Because of the "single gene" mutation of this disease, gene therapy has been closely examined; however, results have not been successful. More recently, drugs that target the molecular defects have been advancing in clinical trials and are showing great promise for better treatments.

Alveolar Airway

Between the trachea and the alveolar sacs, the airways divide 23 times. The last seven generations form the transitional and respiratory zones where gas exchange occurs are made up of transitional and respiratory bronchioles, alveolar ducts, and alveoli (Figure 34–1a,b). These multiple divisions greatly increase the total cross-sectional area of the airways, from 2.5 cm^2 in the trachea to 11,800 cm^2 in the alveoli (Figure 34–1c). Consequently, the velocity of airflow in the small airways declines to very low values. The transition from the conducting to the respiratory region that ends in the alveoli also includes a change in cellular arrangements (Figure 34–2; and Figure 34–3). Humans have 300 million alveoli, and the total area of the alveolar walls in contact with capillaries in both lungs is about 70 m^2.

The alveoli are lined by two types of epithelial cells. **Type I cells** are flat cells with large cytoplasmic extensions and are the primary lining cells of the alveoli, covering approximately 95% of the alveolar epithelial surface area. **Type II cells (granular pneumocytes)** are thicker and contain numerous lamellar inclusion bodies. Although these cells make up only 5% of the surface area, they represent approximately 60% of the epithelial cells in the alveoli. Type II cells are important in alveolar repair as well as other cellular physiology. One prime function of the type II cell is the production of **surfactant** (Figure 34–3d). Typical **lamellar bodies,** membrane-bound organelles containing whorls of phospholipid, are formed in these cells and secreted into the alveolar lumen by exocytosis. Tubes of lipid called **tubular myelin** form from the extruded bodies, and the tubular myelin in turn forms a phospholipid film. Following secretion, the phospholipids of surfactant line up in the alveoli with their hydrophobic fatty acid tails facing the alveolar lumen. This surfactant layer plays an important role in maintaining alveolar structure by reducing surface tension (see below). Surface tension is inversely proportional to the surfactant concentration per unit area. The surfactant molecules move further apart as the alveoli enlarge during inspiration, and surface tension increases, whereas it decreases when they move closer together during expiration. Some of the protein–lipid complexes in surfactant are taken up by endocytosis in type II alveolar cells and recycled.

The alveoli are surrounded by pulmonary capillaries. In most areas, air and blood are separated only by the alveolar epithelium and the capillary endothelium, so they are about

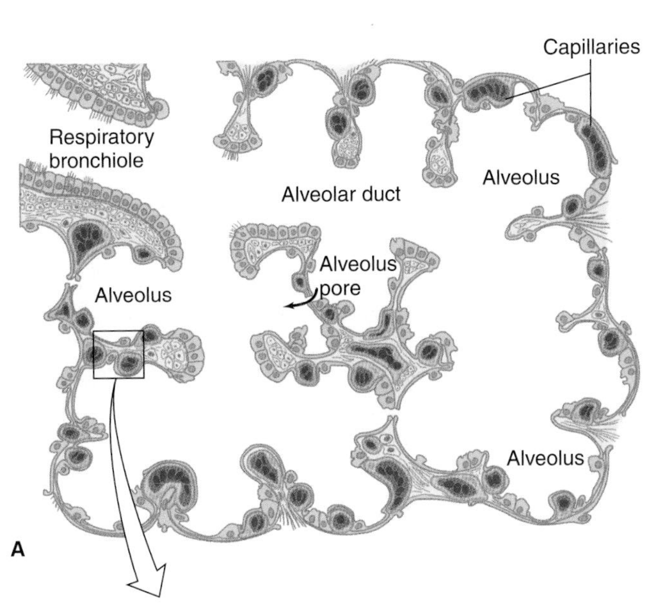

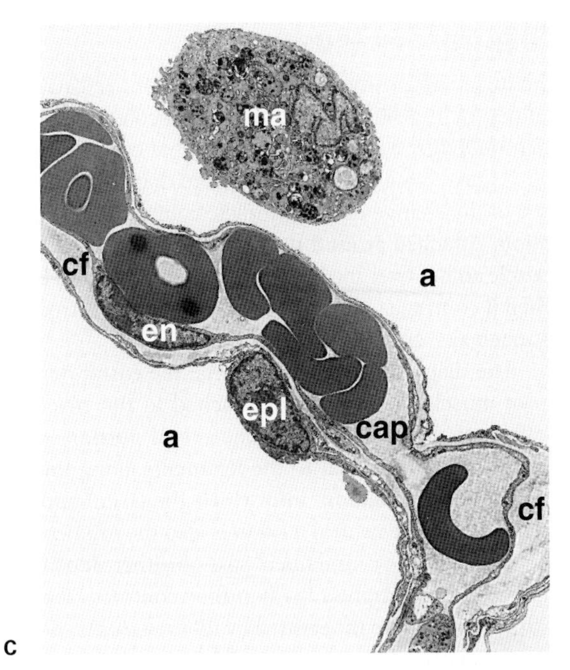

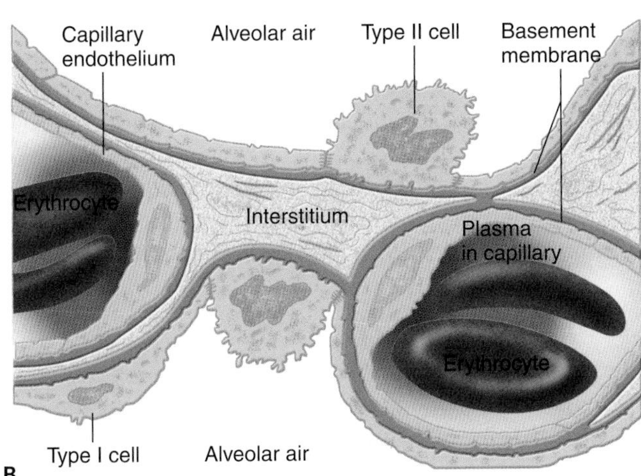

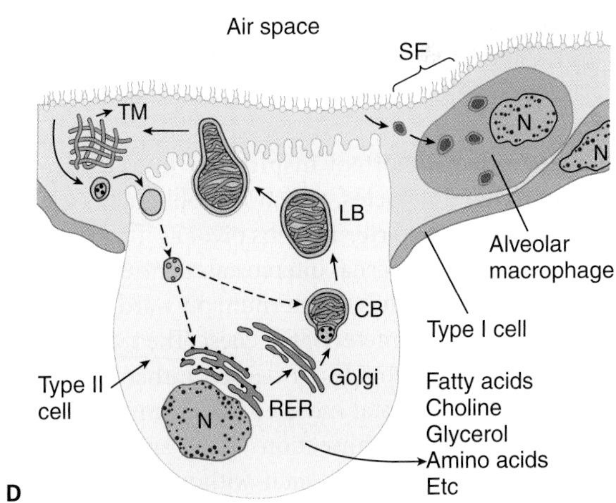

FIGURE 34–3 Prominent cells in the adult human alveolus. **A)** A cross-section of the respiratory zone shows the relationship between capillaries and the airway epithelium. Only 4 of the 18 alveoli are labeled. **B)** Enlargement of the boxed area from (A) displaying intimate relationship between capillaries, the interstitium, and the alveolar epithelium. **C)** Electron micrograph displaying a typical area depicted in (B). The pulmonary capillary (cap) in the septum contains plasma with red blood cells. Note the closely apposed endothelial and pulmonary epithelial cell membranes separated at places by additional connective tissue fibers (cf); en, nucleus of endothelial cell; epl, nucleus of type I alveolar epithelial cell; a, alveolar space; ma, alveolar macrophage. **D)** Type II cell formation and metabolism of surfactant. Lamellar bodies (LB) are formed in type II alveolar epithelial cells and secreted by exocytosis into the fluid lining the alveoli. The released lamellar body material is converted to tubular myelin (TM), and the TM is the source of the phospholipid surface film (SF). Surfactant is taken up by endocytosis into alveolar macrophages and type II epithelial cells. N, nucleus; RER, rough endoplasmic reticulum; CB, composite body. (For (A) From Greep RO, Weiss L: *Histology,* 3rd ed. New York: McGraw-Hill, 1973; (B) Reproduced with permission from Widmaier EP, Raff H, Strang KT: *Vander's Human Physiology: The Mechanisms of Body Function,* 11th ed. McGraw-Hill, 2008; (C) Burri PA: Development and growth of the human lung. In: *Handbook of Physiology,* Section 3, *The Respiratory System.* Fishman AP, Fisher AB [editors]. American Physiological Society, 1985; and (D) Wright JR: Metabolism and turnover of lung surfactant. Am Rev Respir Dis 136:426, 1987.)

0.5 (μm apart (Figure 34–3). The alveoli also contain other specialized cells, including pulmonary alveolar macrophages (PAMs, or AMs), lymphocytes, plasma cells, neuroendocrine cells, and mast cells. PAMs are an important component of the pulmonary defense system. Like other macrophages, these cells come originally from the bone marrow. PAMs are actively phagocytic and ingest small particles that evade the mucociliary escalator and reach the alveoli. They also help process inhaled antigens for immunologic attack, and they secrete substances that attract granulocytes to the lungs as well as substances that stimulate granulocyte and monocyte formation in the bone marrow. PAM function can also be detrimental—when they ingest large amounts of the substances in cigarette smoke or other irritants, they may release lysosomal products into the extracellular space to cause inflammation.

Respiratory Muscles

The lungs are positioned within the thoracic cavity, which is defined by the rib cage and the spinal column. The lungs are surrounded by a variety of muscles that contribute to breathing (Figure 34–4). Movement of the diaphragm accounts for 75% of the change in intrathoracic volume during quiet inspiration. Attached around the bottom of the thoracic cage, this muscle arches over the liver and moves downward like a piston when it contracts. The distance it moves ranges from 1.5 cm to as much as 7 cm with deep inspiration.

The diaphragm has three parts: the costal portion, made up of muscle fibers that are attached to the ribs around the bottom of the thoracic cage; the crural portion, made up of fibers that are attached to the ligaments along the vertebrae; and the central tendon, into which the costal and the crural fibers insert. The central tendon is also the inferior part of the pericardium. The crural fibers pass on either side of the esophagus and can compress it when they contract. The costal and crural portions are innervated by different parts of the phrenic nerve and can contract separately. For example, during vomiting and eructation, intra-abdominal pressure is increased by contraction of the costal fibers but the crural fibers remain relaxed, allowing material to pass from the stomach into the esophagus.

The other important **inspiratory muscles** are the **external intercostal muscles,** which run obliquely downward and forward from rib to rib. The ribs pivot as if hinged at the back, so that when the external intercostals contract they elevate the lower ribs. This pushes the sternum outward and increases the anteroposterior diameter of the chest. The transverse diameter also increases, but to a lesser degree. Either the diaphragm or the external intercostal muscles alone can maintain adequate ventilation at rest. Transection of the spinal cord above the third cervical segment is fatal without artificial respiration, but transection below the fifth cervical segment is not, because it leaves the phrenic nerves that innervate the diaphragm intact; the phrenic nerves arise from cervical segments 3–5. Conversely, in patients with bilateral phrenic nerve palsy but intact innervation of their intercostal muscles, respiration is somewhat labored but adequate to maintain life. The scalene and sternocleidomastoid muscles in the neck are accessory inspiratory muscles that help to elevate the thoracic cage during deep labored respiration.

A decrease in intrathoracic volume and forced expiration result when the **expiratory muscles** contract. The internal intercostals have this action because they pass obliquely downward and posteriorly from rib to rib and therefore pull the rib cage downward when they contract. Contractions of the muscles of the anterior abdominal wall also aid expiration by pulling the rib cage downward and inward and by increasing the intra-abdominal pressure, which pushes the diaphragm upward.

In order for air to get into the conducting airway it must pass through the **glottis,** defined as the area including and between the vocal folds within the larynx. The abductor muscles in the larynx contract early in inspiration, pulling the vocal cords apart and opening the glottis. During swallowing or gagging, a reflex contraction of the adductor muscles closes the glottis and prevents aspiration of food, fluid, or vomitus into the lungs. In unconscious or anesthetized patients, glottic closure may be incomplete and vomitus may enter the trachea, causing an inflammatory reaction in the lung **(aspiration pneumonia).**

Lung Pleura

The **pleural cavity** and its infoldings serve as a lubricating fluid/area that allows for lung movement within the thoracic cavity (Figure 34–5A). There are two layers that contribute to the pleural cavity: the **parietal pleura** and the **visceral pleura.**

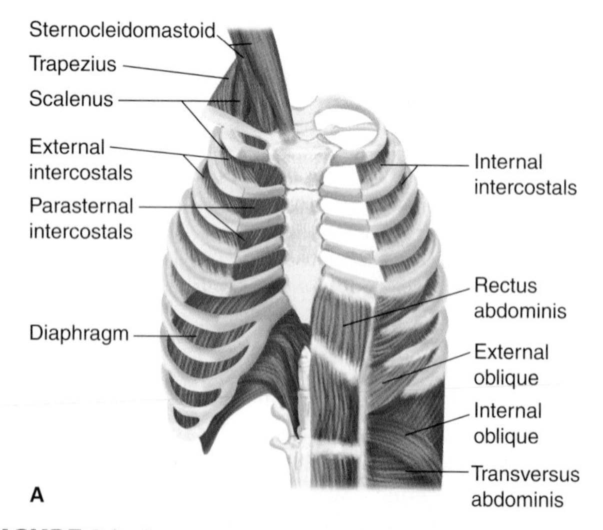

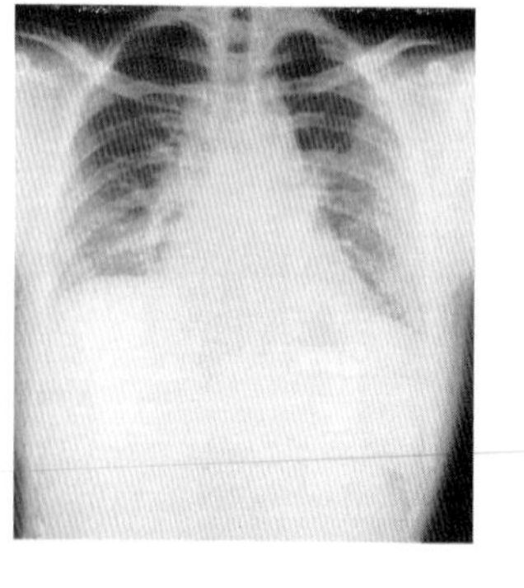

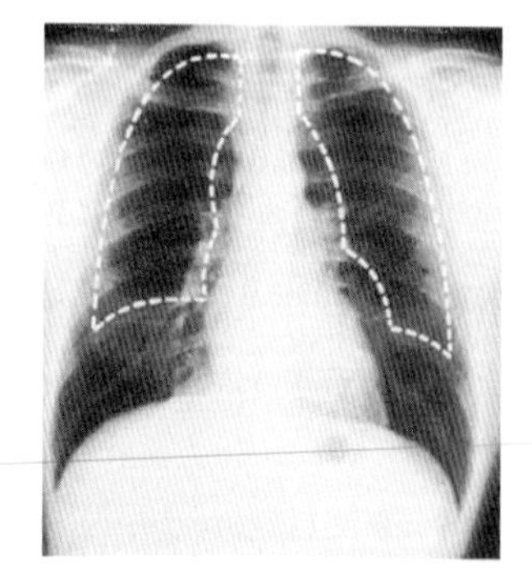

FIGURE 34–4 Muscles and movement in respiration. **A)** An idealized diagram of respiratory muscles surrounding the rib cage. The diaphragm and intercostals play prominent roles in respiration. **B)** and **C)** X-ray of chest in full expiration (B) and full inspiration (C). The dashed white line on in C is an outline of the lungs in full expiration. Note the difference in intrathoracic volume. (Reproduced with permission A) from Fishman AP: *Fishman's Pulmonary Diseases and Disorders,* 4th ed. McGraw Hill Medical, 2008; B, C from Comroe JH Jr: *Physiology of Respiration,* 2nd ed., Year Book, 1974.)

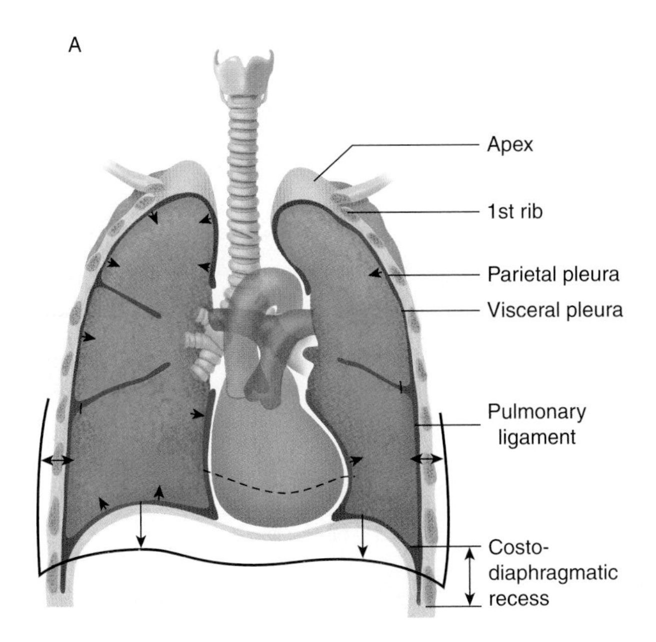

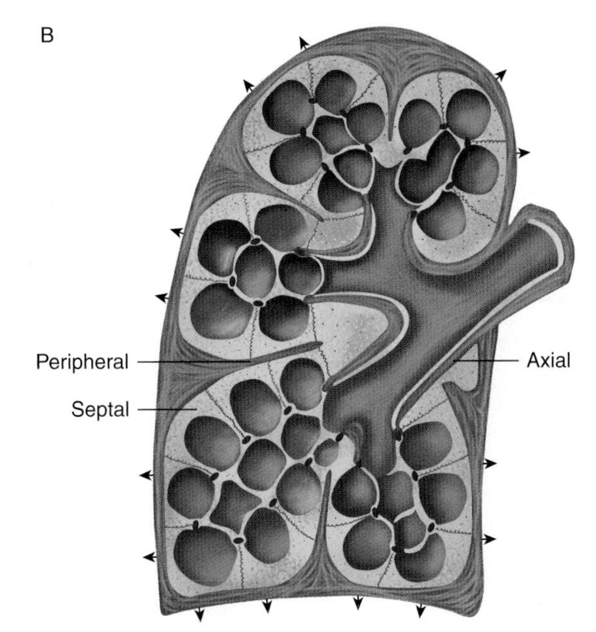

FIGURE 34–5 Pleural space and connective fibers. **A)** Front sectional drawing of lung within the rib cage. Note the parietal and visceral pleura and the infoldings around the lung lobes that include pleural space. **B)** Connective fiber tracts of the lung are highlighted. Note the axial fibers along the airways, the peripheral fibers in the pleura, and the septal fibers. (Reproduced with permission from Fishman AP: *Fishman's Pulmonary Diseases and Disorders*, 4th ed. McGraw Hill Medical, 2008)

The parietal pleura is a membrane which lines the chest cavity containing the lungs. The **visceral pleura** is a membrane which lines the lung surface. The pleural fluid (~15–20 ml) forms a thin layer between the pleural membranes and prevents friction between surfaces during inspiration and expiration.

The lung itself contains a vast amount of free space—it is ~80% air. Although this maximizes surface area for gas exchange, it also requires an extensive support network to maintain lung shape and function. The **connective tissue** within the visceral pleura contains three layers that help to support the lung. Elastic fibers that follow the mesothelium effectively wrap the three lobes of the right lung and the two lobes of the left lung (Figure 34–5B). A deep sheet of fine fibers that follow the outline of the alveoli provide support to individual air sacks. Between these two separate sheets lies connective tissue that is interspersed with individual cells for support and lung maintenance/function.

Blood and Lymph in the Lung

Both the **pulmonary circulation** and the **bronchial circulation** contribute to blood flow in the lung. In the pulmonary circulation (Figure 34–6), almost all the blood in the body passes via the pulmonary artery to the pulmonary capillary bed, where it is oxygenated and returned to the left atrium via the pulmonary veins. The pulmonary arteries strictly follow the branching of the bronchi down to the respiratory bronchioles. The pulmonary veins, however, are spaced between the bronchi on their return to the heart. The separate and much smaller bronchial circulation includes the bronchial arteries that come from systemic arteries. They form capillaries, which drain into bronchial veins or anastomose with pulmonary capillaries or veins. The bronchial veins drain into the azygos vein. The bronchial circulation nourishes the trachea down to the terminal bronchioles and also supplies the pleura and hilar lymph nodes. It should be noted that lymphatic channels are more abundant in the lungs than in any other organ. Lymph nodes are arranged along the bronchial tree and extend down until the bronchi are ~5 mm in diameter. Lymph node sizes can range from 1 mm at the bronchial periphery to 10 mm along the trachea. The nodes are connected by lymph vessels and allow for unidirectional flow of lymph to the subclavian veins.

MECHANICS OF RESPIRATION

INSPIRATION & EXPIRATION

The lungs and the chest wall are elastic structures. Normally, no more than a thin layer of fluid is present between the lungs and the chest wall (intrapleural space). The lungs slide easily on the chest wall, but resist being pulled away from it in the same way that two moist pieces of glass slide on each other but resist separation. The pressure in the "space" between the lungs and chest wall (intrapleural pressure) is subatmospheric (Figure 34–7). The lungs are stretched when they expand at birth, and at the end of quiet expiration their tendency to recoil from the chest wall is just balanced by the tendency of the chest wall to recoil in the opposite direction. If the chest wall is opened, the lungs collapse; and if the lungs lose their elasticity, the chest expands and becomes barrel-shaped.

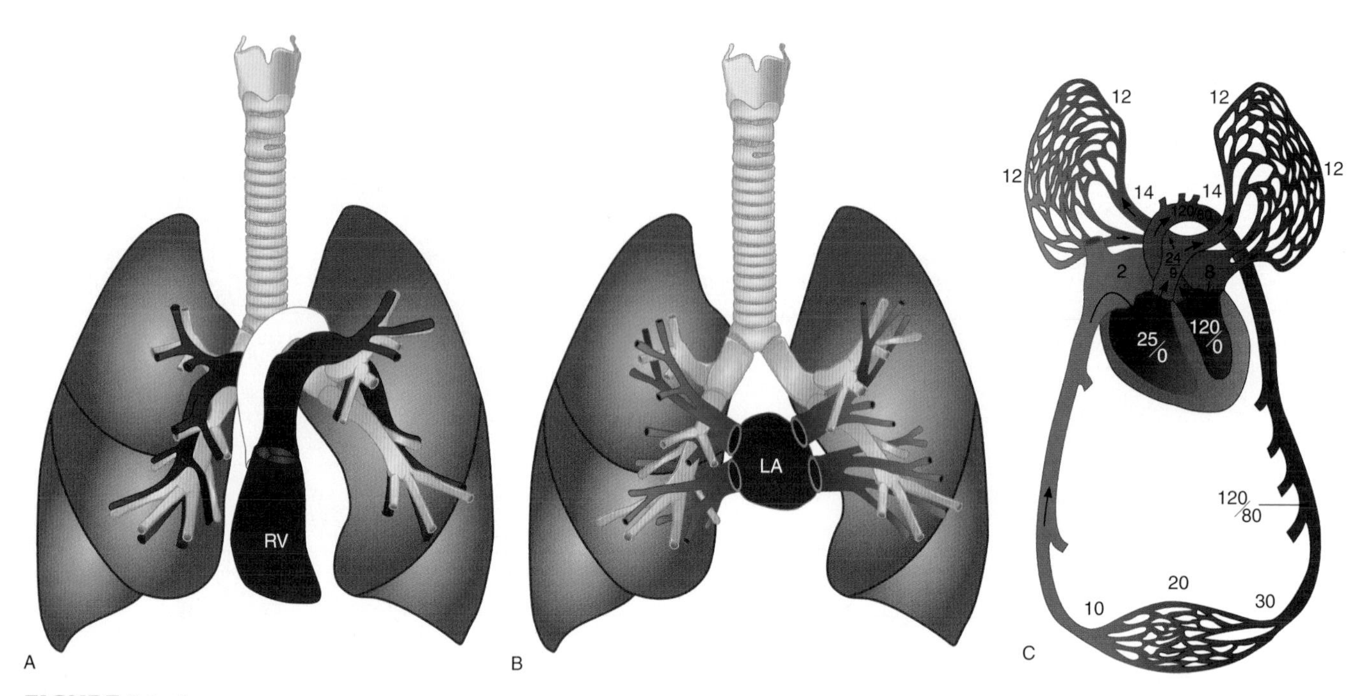

FIGURE 34–6 Pulmonary circulation. **A, B)** Schematic diagrams of the relation of the main branches of pulmonary arteries (A) and pulmonary veins (B) to the bronchial tree. LA = left atrium; RV = right ventricle. **C)** Representative areas of blood flow are labeled with corresponding blood pressure (mm Hg). (A, B Reproduced with permission from Fishman AP: *Fishman's Pulmonary Diseases and Disorders,* 4th ed. McGraw Hill Medical, 2008; C Modified from Comroe JH Jr: *Physiology of Respiration,* 2nd ed. Year Book, 1974.)

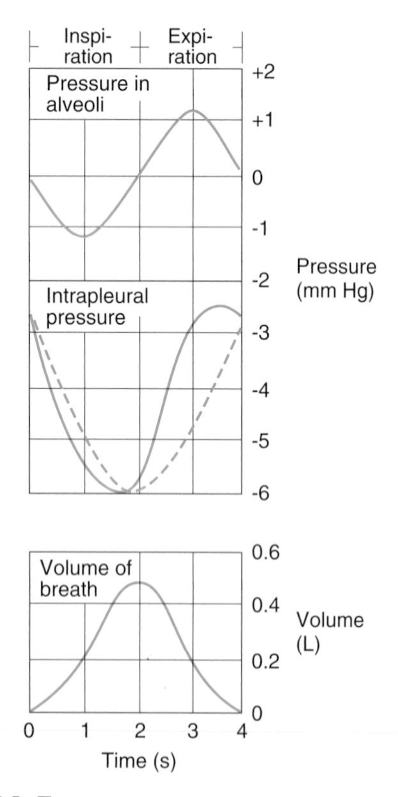

FIGURE 34–7 Pressure in the alveoli and the pleural space relative to atmospheric pressure during inspiration and expiration. The dashed line indicates what the intrapleural pressure would be in the absence of airway and tissue resistance; the actual curve (solid line) is skewed to the left by the resistance. Volume of breath during inspiration/expiration is graphed for comparison.

Inspiration is an active process. The contraction of the inspiratory muscles increases intrathoracic volume. The intrapleural pressure at the base of the lungs, which is normally about –2.5 mm Hg (relative to atmospheric) at the start of inspiration, decreases to about –6 mm Hg. The lungs are pulled into a more expanded position. The pressure in the airway becomes slightly negative, and air flows into the lungs. At the end of inspiration, the lung recoil begins to pull the chest back to the expiratory position, where the recoil pressures of the lungs and chest wall balance (see below). The pressure in the airway becomes slightly positive, and air flows out of the lungs. Expiration during quiet breathing is passive in the sense that no muscles that decrease intrathoracic volume contract. However, some contraction of the inspiratory muscles occurs in the early part of expiration. This contraction exerts a braking action on the recoil forces and slows expiration. Strong inspiratory efforts reduce intrapleural pressure to values as low as –30 mm Hg, producing correspondingly greater degrees of lung inflation. When ventilation is increased, the extent of lung deflation is also increased by active contraction of expiratory muscles that decrease intrathoracic volume.

QUANTITATING RESPIRATORY PHENOMENA

Modern spirometers permit direct measurement of gas intake and output. Since gas volumes vary with temperature and pressure and since the amount of water vapor in them varies, these devices have the ability to correct respiratory measurements involving volume to a stated set of standard conditions. It should be noted that correct measurements are highly dependent on the ability for the practitioner to properly encourage the patient to fully utilize the device. Modern techniques for gas analysis make possible rapid, reliable measurements of the composition of gas mixtures and the gas content of body fluids. For example, O_2 and CO_2 electrodes, small probes sensitive to O_2 or CO_2, can be inserted into the airway or into blood vessels or tissues and the Po_2 and Pco_2 recorded continuously. Chronic assessment of oxygenation is carried out noninvasively with a **pulse oximeter,** which can be easily placed on a fingertip or earlobe.

Lung Volumes and Capacities

Important quantitation of lung function can be gleaned from the displacement of air volume during inspiration and/or expiration. Lung capacities refer to subdivisions that contain two or more volumes. Volumes and capacities recorded on a spirometer from a healthy individual are shown in Figure 34–8. Diagnositic spirometry is used to assess a patient's lung function for purposes of comparison with a normal population, or with previous measures from the same patient. The amount of air that moves into the lungs with each inspiration (or the amount that moves out with each expiration) during quiet breathing is called the **tidal volume** (TV). Typical values for TV are on the order of 500–750 mL. The air inspired with a maximal inspiratory effort in excess of the TV is the **inspiratory reserve volume** (IRV; typically ~2 L). The volume expelled by an active expiratory effort after passive expiration is the **expiratory reserve volume** (ERV; ~1 L), and the air left in the lungs after a maximal expiratory effort is the **residual volume** (RV; ~1.3 L). When all four of the above components are taken together, they make up the **total lung capacity** (~5 L). The total lung capacity can be broken down into alternative capacities

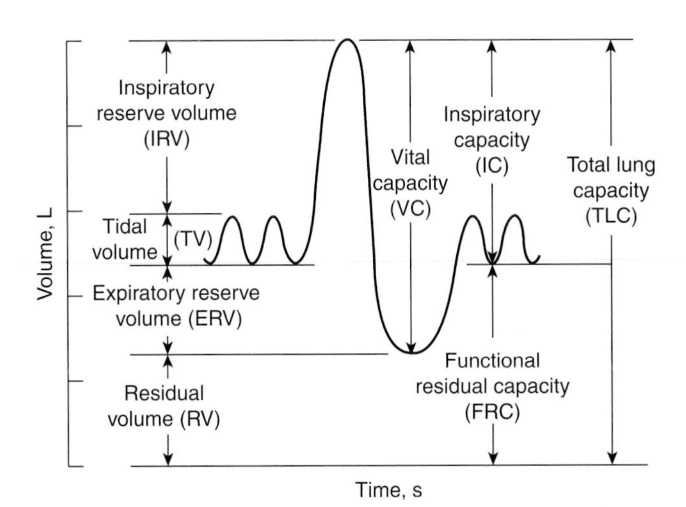

FIGURE 34–8 Lung volumes and capacity measurements. Lung volumes recorded by a spirometer. Lung capacities are determined from volume recordings. See text for definitions. (Reproduced with permission from Fishman AP: *Fishman's Pulmonary Diseases and Disorders*, 4th ed. McGraw Hill Medical, 2008).

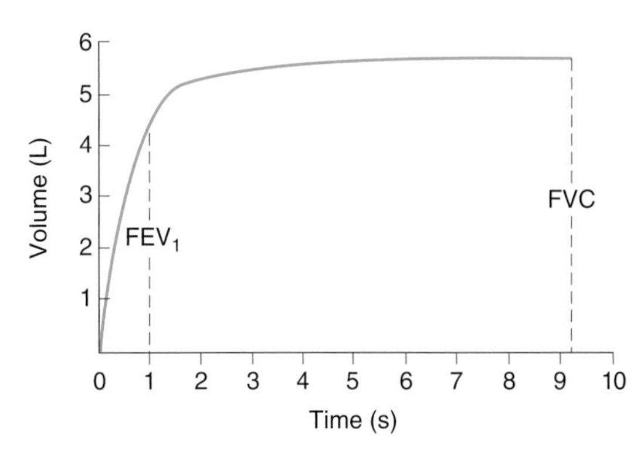

FIGURE 34–9 Volume of gas expired by a normal adult man during a forced expiration, demonstrating the FEV_1 and the forced vital capacity (FVC). From the graph the forced expiratory volume in 1 s (FEV_1) to FVC ration (FEV_1/FVC) can be calculated (4L/5L = 80%). (Reproduced, with permission, from Crapo RO: Pulmonary-function testing. N Engl J Med 1994; 331:25. Copyright © 1994, Massachusetts Medical Society.)

that help to define functioning lungs. The **vital lung capacity** (~3.5 L) refers to the maximum amount of air expired from the fully inflated lung, or maximum inspiratory level (this represents TV + IRV + ERV). The **inspiratory capacity** (~2.5 L) is the maximum amount of air inspired from the end-expiratory level (IRV + TV). The **functional residual capacity** (FRC; ~2.5 L) represents the volume of the air remaining in the lungs after expiration of a normal breath (RV + ERV).

Dynamic measurements of lung volumes and capacities have been used to help determine lung dysfunction. The **forced vital capacity (FVC),** the largest amount of air that can be expired after a maximal inspiratory effort, is frequently measured clinically as an index of pulmonary function. It gives useful information about the strength of the respiratory muscles and other aspects of pulmonary function. The fraction of the vital capacity expired during the first second of a forced expiration is referred to as **FEV_1** (forced expiratory volume in 1 sec; Figure 34–9). The FEV_1 to FVC ratio (FEV_1/FVC) is a useful tool in the recognizing classes of airway disease (Clinical Box 34–2). Other dynamic measurements include the **respiratory minute volume (RMV)** and the **maximal voluntary ventilation (MVV).** RMV is normally ~6 L (500 mL/ breath × 12 breaths/min). The **MVV** is the largest volume of gas that can be moved into and out of the lungs in 1 min by voluntary effort. Typically this is measured over a 15 s period and prorated to a minute; normal values range from 140 to 180 L/min for healthy adult males. Changes in RMV and MVV in a patient can be indicative of lung dysfunction.

COMPLIANCE OF THE LUNGS & CHEST WALL

Compliance is developed due to the tendency for tissue to resume its original position after an applied force has been removed. After an inspiration during quiet breathing (eg, at the FRC), the lungs have a tendency to collapse and the chest wall

CLINICAL BOX 34–2

Altered Airflow in Disease:

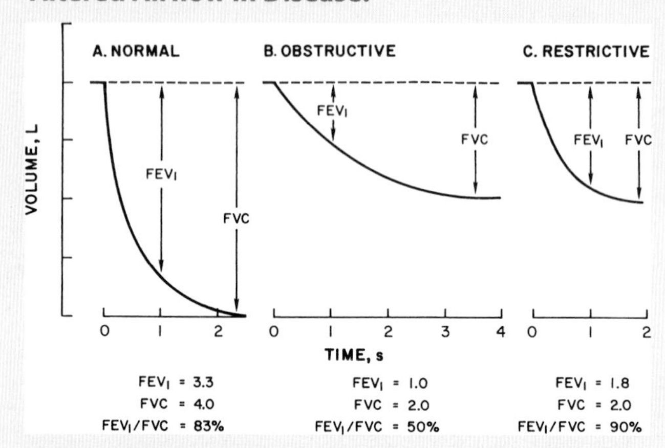

Representative spirograms measuring volume over time (sec) for subjects displaying normal (A), obstructive (B) or restrictive (C) patterns. Note the differences in the FEV_1, FVC and FEV_1/FVC (shown at bottom). Fishman's Pulmonary Diseases and Disorders, Chapter 34, Figure 34–16.

Airflow Measurements of Obstructive & Restrictive Disease

In the example above, a healthy FVC is ~4.0 L and a healthy FEV_1 is ~3.3 L. The calculated FEV_1/FVC is ~80%. Patients with obstructive or restrictive diseases can display reduced FVC, on the order of 2.0 L in the example above. Measurement of FEV_1, however, tends to vary significantly between the two diseases. In obstructive disorders, patients tend to show a slow, steady slope to the FVC, resulting in a small FEV_1, on the order of 1.0 L in the example. However, in the restrictive disorder patients, airflow tends to be fast at first, and then quickly level out to approach FVC. The resultant FEV_1 is much greater, on the order of 1.8 L in the example, even though FVC is equivalent (compare B, C above). A quick calculation of FEV_1/FVC for obstructive (50%) versus restrictive (90%) patients defines the hallmark measurements in evaluating these two diseases. Obstructive disorders result in a marked decrease in both FVC and FEV_1/FVC, whereas restrictive disorders result in a loss of FVC without loss in FEV_1/FVC. It should be noted that these examples are idealized and several disorders can show mixed readings.

Obstructive Disease—Asthma

Asthma is characterized by episodic or chronic wheezing, cough, and a feeling of tightness in the chest as a result of bronchoconstriction. Although the disease is not fully understood, three airway abnormalities are present: **airway obstruction** that is at least partially reversible, airway inflammation, and airway hyperresponsiveness to a variety of stimuli. A link to allergy has long been recognized, and plasma IgE levels are often elevated. Proteins released from eosinophils in the inflammatory reaction may damage the airway epithelium and contribute to the hyperresponsiveness. Leukotrienes are released from eosinophils and mast cells, and can enhance bronchoconstriction. Numerous other amines, neuropeptides, chemokines, and interleukins have effects on bronchial smooth muscle or produce inflammation, and they may be involved in asthma.

THERAPEUTIC HIGHLIGHTS

Because β_2-adrenergic receptors mediate bronchodilation, β_2-adrenergic agonists have long been the mainstay of treatment for mild to moderate asthma attacks. Inhaled and systemic steroids are used even in mild to moderate cases to reduce inflammation; they are very effective, but their side effects can be a problem. Agents that block synthesis of leukotrienes or their $CysLT_1$ receptor have also proved useful in some cases.

has a tendency to expand. The interaction between the recoil of the lungs and recoil of the chest can be demonstrated in living subjects through a spirometer that has a valve just beyond the mouthpiece. The mouthpiece contains a pressure-measuring device. After the subject inhales a given amount, the valve is shut, closing off the airway. The respiratory muscles are then relaxed while the pressure in the airway is recorded. The procedure is repeated after inhaling or actively exhaling various volumes. The curve of airway pressure obtained in this way, plotted against volume, is the **pressure–volume curve** of the total respiratory system (P_{TR} in Figure 34–10). The pressure is zero at a lung volume that corresponds to the volume of gas in the lungs at **FRC (relaxation volume).** As can be noted from Figure 34–10, this relaxation pressure is the sum of slightly negative pressure component from the chest wall (P_W) and a slightly positive pressure from the lungs (P_L). P_{TR} is positive at greater volumes and negative at smaller volumes. **Compliance** of the lung and chest wall is measured as the slope of the P_{TR} curve, or, as a change in lung volume per unit change in airway pressure ($\Delta V/\Delta P$). It is normally measured in the pressure range where the relaxation pressure curve is steepest, and normal values are ~0.2 L/cm H_2O in a healthy adult male. However, compliance depends on lung volume and thus can vary. In an extreme example, an individual with only one lung has approximately half the ΔV for a given ΔP. Compliance is also slightly greater when measured during deflation than when measured during inflation. Consequently, it is more informative to examine the whole pressure–volume curve.

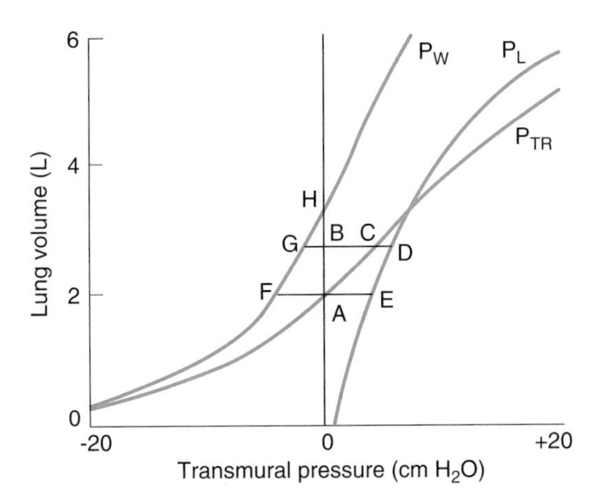

FIGURE 34–10 Pressure–volume curves in the lung. The pressure–volume curves of the total respiratory system (P_{TR}), the lungs (P_L), and the chest (P_W) are plotted together with standard volumes for functional residual capacity and tidal volume. The transmural pressure is intrapulmonary pressure minus intrapleural pressure in the case of the lungs, intrapleural pressure minus outside (barometric) pressure in the case of the chest wall, and intrapulmonary pressure minus barometric pressure in the case of the total respiratory system. From these curves, the total and actual elastic work associated with breathing can be derived (see text). (Modified from Mines AH: *Respiratory Physiology*, 3rd ed. Raven Press, 1993.)

The curve is shifted downward and to the right (compliance is decreased) by pulmonary edema and interstitial pulmonary fibrosis (Figure 34–11). Pulmonary fibrosis is a progressive restrictive airway disease in which there is stiffening and scarring of the lung. The curve is shifted upward and to the left (compliance is increased) in emphysema.

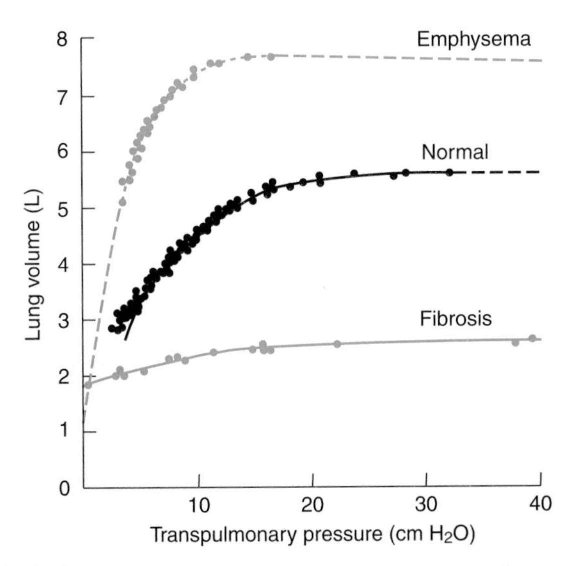

FIGURE 34–11 Static expiratory pressure–volume curves of lungs in normal subjects and subjects with severe emphysema and pulmonary fibrosis. (Modified and reproduced with permission from Pride NB, Macklem PT: Lung mechanics in disease. In: *Handbook of Physiology*. Section 3, *The Respiratory System*. Vol III, part 2. Fishman AP [editor]. American Physiological Society, 1986.)

Airway Resistance

Airway resistance is defined as the change of pressure (ΔP) from the alveoli to the mouth divided by the change in flow rate ($\dot{V}$). Because of the structure of the bronchial tree, and thus the pathway for air that contributes to its resistance, it is difficult to apply mathematical estimates of the movement through the bronchial tree. However, measurements where alveolar and intrapleural pressure can be compared to actual pressure (eg, Figure 34–7 middle panel) show the contribution of airway resistance. Airway resistance is significantly increased as lung volume is reduced. Also, bronchi and bronchioles significantly contribute to airway resistance. Thus, contraction of the smooth muscle that lines the bronchial airways will increase airway resistance, and make breathing more difficult.

Role of Surfactant in Alveolar Surface Tension

An important factor affecting the compliance of the lungs is the surface tension of the film of fluid that lines the alveoli. The magnitude of this component at various lung volumes can be measured by removing the lungs from the body of an experimental animal and distending them alternately with saline and with air while measuring the intrapulmonary pressure. Because saline reduces the surface tension to nearly zero, the pressure–volume curve obtained with saline measures only the tissue elasticity (Figure 34–12), whereas the curve obtained with air measures both tissue elasticity and surface tension. The difference between the saline and air curves is much smaller when lung volumes are small. Differences are also obvious in the curves generated during inflation and deflation. This difference is termed **hysteresis** and notably is not present in the saline generated curves. The alveolar

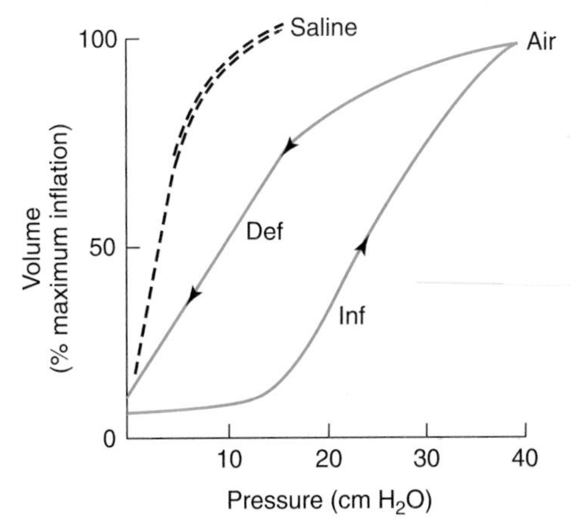

FIGURE 34–12 Pressure–volume curves in the lungs of a cat after removal from the body. Saline: lungs inflated and deflated with saline to reduce surface tension, resulting in a measurement of tissue elasticity. **Air:** lungs inflated (Inf) and deflated (Def) with air results in a measure of both tissue elasticity and surface tension. (Reproduced with permission from Morgan TE: Pulmonary surfactant. N Engl J Med 1971;284:1185.)

environment, and specifically the secreted factors that help to reduce surface tension and keep alveoli from collapsing, contribute to hysteresis.

The low surface tension when the alveoli are small is due to the presence of **surfactant** in the fluid lining the alveoli. Surfactant is a mixture of dipalmitoylphosphatidylcholine (DPPC), other lipids, and proteins. If the surface tension is not kept low when the alveoli become smaller during expiration, they collapse in accordance with the law of Laplace. In spherical structures like an alveolus, the distending pressure equals two times the tension divided by the radius ($P = 2T/r$); if T is not reduced as r is reduced, the tension overcomes the distending pressure. Surfactant also helps to prevent pulmonary edema. It has been calculated that if it were not present, the unopposed surface tension in the alveoli would produce a 20 mm Hg force favoring transudation of fluid from the blood into the alveoli.

Formation of the phospholipid film is greatly facilitated by the proteins in surfactant. This material contains four unique proteins: surfactant protein (SP)-A, SP-B, SP-C, and SP-D. SP-A is a large glycoprotein and has a collagen-like domain within its structure. It has multiple functions, including regulation of the feedback uptake of surfactant by the type II alveolar epithelial cells that secrete it. SP-B and SP-C are smaller proteins, which are the key protein members of the monomolecular film of surfactant. Like SP-A, SP-D is a glycoprotein. Its full function is uncertain, however it plays an important role in the organization of SP-B and SP-C into the surfactant layer. Both SP-A and SP-D are members of the collectin family of proteins that are involved in innate immunity in the conducting airway as well as in the alveoli. Some clinical aspects of surfactant are discussed in Clinical Box 34–3.

CLINICAL BOX 34–3

Surfactant

Surfactant is important at birth. The fetus makes respiratory movements in utero, but the lungs remain collapsed until birth. After birth, the infant makes several strong inspiratory movements and the lungs expand. Surfactant keeps them from collapsing again. Surfactant deficiency is an important cause of **infant respiratory distress syndrome** (IRDS, also known as **hyaline membrane disease**), the serious pulmonary disease that develops in infants born before their surfactant system is functional. Surface tension in the lungs of these infants is high, and the alveoli are collapsed in many areas **(atelectasis).** An additional factor in IRDS is retention of fluid in the lungs. During fetal life, Cl^- is secreted with fluid by the pulmonary epithelial cells. At birth, there is a shift to Na^+ absorption by these cells via the epithelial Na^+ channels (ENaCs), and fluid is absorbed with the Na^+. Prolonged immaturity of the ENaCs contributes to the pulmonary abnormalities in IRDS.

Overproduction/dysregulation of surfactant proteins can also lead to respiratory distress and is the cause of Pulmonary Alveolar Proteinosis (PAP).

THERAPEUTIC HIGHLIGHTS

Treatment of IRDS is commonly done with surfactant replacement therapy. Interestingly, such surfactant replacement therapy has not been as successful in clinical trials for adults experiencing respiratory distress due to surfactant dysfunction.

WORK OF BREATHING

Work is performed by the respiratory muscles in stretching the elastic tissues of the chest wall and lungs (elastic work; approximately 65% of the total work), moving inelastic tissues (viscous resistance; 7% of total), and moving air through the respiratory passages (airway resistance; 28% of total). Because pressure times volume ($g/cm^2 \times cm^3 = g \times cm$) has the same dimensions as work (force × distance), the work of breathing can be calculated from the previously presented pressure–volume curve (Figure 34–10). The total elastic work required for inspiration is represented by the area ABCA. Note that the relaxation pressure curve of the total respiratory system differs from that of the lungs alone. The actual elastic work required to increase the volume of the lungs alone is area ABDEA. The amount of elastic work required to inflate the whole respiratory system is less than the amount required to inflate the lungs alone because part of the work comes from elastic energy stored in the thorax. The elastic energy lost from the thorax (area AFGBA) is equal to that gained by the lungs (area AEDCA).

Estimates of the total work of quiet breathing range from 0.3 up to 0.8 kg-m/min. The value rises markedly during exercise, but the energy cost of breathing in normal individuals represents less than 3% of the total energy expenditure during exercise. The work of breathing is greatly increased in diseases such as emphysema, asthma, and congestive heart failure with dyspnea and orthopnea. The respiratory muscles have length–tension relations like those of other skeletal and cardiac muscles, and when they are severely stretched, they contract with less strength. They can also become fatigued and fail (pump failure), leading to inadequate ventilation.

DIFFERENCES IN VENTILATION & BLOOD FLOW IN DIFFERENT PARTS OF THE LUNG

In the upright position, ventilation per unit lung volume is greater at the base of the lung than at the apex. The reason for this is that at the start of inspiration, intrapleural pressure is less negative at the base than at the apex (Figure 34–13), and since the intrapulmonary intrapleural pressure difference

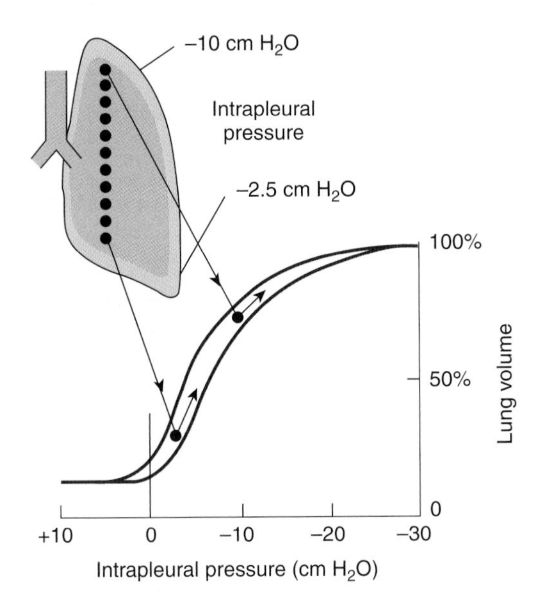

FIGURE 34–13 Intrapleural pressures in the upright position and their effect on ventilation. Note that because intrapulmonary pressure is atmospheric, the more negative intrapleural pressure at the apex holds the lung in a more expanded position at the start of inspiration. Further increases in volume per unit increase in intrapleural pressure are smaller than at the base because the expanded lung is stiffer. (Reproduced with permission from West JB: *Ventilation/Blood Flow and Gas Exchange,* 5th ed. Blackwell, 1990.)

is less than at the apex, the lung is less expanded. Conversely, at the apex, the lung is more expanded; that is, the percentage of maximum lung volume is greater. Because of the stiffness of the lung, the increase in lung volume per unit increase in pressure is smaller when the lung is initially more expanded, and ventilation is consequently greater at the base. Blood flow is also greater at the base than the apex. The relative change in blood flow from the apex to the base is greater than the relative change in ventilation, so the ventilation/perfusion ratio is low at the base and high at the apex.

The ventilation and perfusion differences from the apex to the base of the lung have usually been attributed to gravity: they tend to disappear in the supine position and; the weight of the lung would be expected to create pressure at the base in the upright position. However, the inequalities of ventilation and blood flow in humans were found to persist to a remarkable degree in the weightlessness of space. Therefore, other factors also play a role in producing the inequalities.

DEAD SPACE & UNEVEN VENTILATION

Because gaseous exchange in the respiratory system occurs only in the terminal portions of the airways, the gas that occupies the rest of the respiratory system is not available for gas exchange with pulmonary capillary blood. Normally, the volume (in mL) of this **anatomic dead space** is approximately equal to the body weight in pounds. As an example, in a man

TABLE 34–1 Effect of variations in respiratory rate and depth on alveolar ventilation.

Respiratory rate	30/min	10/min
Tidal volume	200 mL	600 mL
Minute volume	6 L	6 L
Alveolar ventilation	(200 – 150) × 30 = 1500 mL	(600 – 150) × 10 = 4500 mL

who weighs 150 lb (68 kg), only the first 350 mL of the 500 mL inspired with each breath at rest mixes with the air in the alveoli. Conversely, with each expiration, the first 150 mL expired is gas that occupied the dead space, and only the last 350 mL is gas from the alveoli. Consequently, the **alveolar ventilation,** ie, the amount of air reaching the alveoli per minute, is less than the RMV. Note that because of the dead space, rapid shallow breathing produces much less alveolar ventilation than slow deep breathing at the same RMV (Table 34–1).

It is important to distinguish between the **anatomic dead space** (respiratory system volume exclusive of alveoli) and the **total (physiologic) dead space** (volume of gas not equilibrating with blood; ie, wasted ventilation). In healthy individuals, the two dead spaces are identical and can be estimated by body weight. However, in disease states, no exchange may take place between the gas in some of the alveoli and the blood, and some of the alveoli may be overventilated. The volume of gas in nonperfused alveoli and any volume of air in the alveoli in excess of that necessary to arterialize the blood in the alveolar capillaries is part of the dead space (nonequilibrating) gas volume. The anatomic dead space can be measured by analysis of the single-breath N_2 curves (Figure 34–14). From mid-inspiration, the subject takes as deep a breath as possible of pure O_2, then exhales steadily while the N_2 content of the expired gas is continuously measured. The initial gas exhaled (phase I) is the gas that filled the dead space and that consequently contains no N_2. This is followed by a mixture of dead space and alveolar

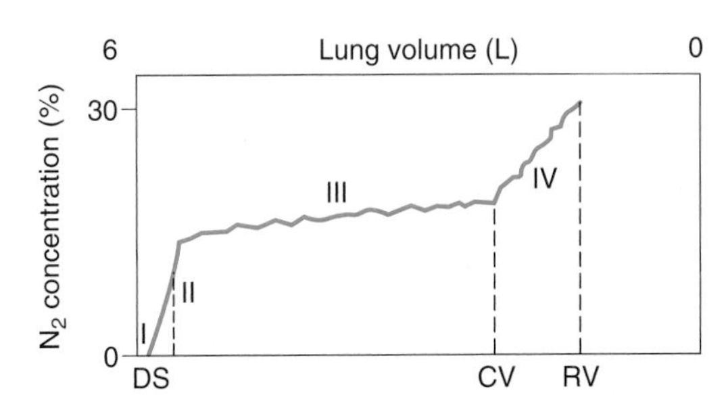

FIGURE 34–14 Single-breath N_2 curve. From mid-inspiration, the subject takes a deep breath of pure O_2 then exhales steadily. The changes in the N_2 concentration of expired gas during expiration are shown, with the various phases of the curve indicated by roman numerals. Notably, region I is representative of the dead space (DS); from I–II is a mixture of DS and alveolar gas; the transition form III–IV is the closing volume (CV), and the end of IV is the residual volume (RV).

gas (phase II) and then by alveolar gas (phase III). The volume of the dead space is the volume of the gas expired from peak inspiration to the midportion of phase II.

Phase III of the single-breath N_2 curve terminates at the **closing volume (CV)** and is followed by phase IV, during which the N_2 content of the expired gas is increased. The CV is the lung volume above RV at which airways in the lower, dependent parts of the lungs begin to close off because of the lesser transmural pressure in these areas. The gas in the upper portions of the lungs is richer in N_2 than the gas in the lower, dependent portions because the alveoli in the upper portions are more distended at the start of the inspiration of O_2 and, consequently, the N_2 in them is less diluted with O_2. It is also worth noting that in most normal individuals, phase III has a slight positive slope even before phase IV is reached. This indicates that even during phase III there is a gradual increase in the proportion of the expired gas coming from the relatively N_2-rich upper portions of the lungs.

The total dead space can be calculated from the P_{CO_2} of expired air, the P_{CO_2} of arterial blood, and the TV. The tidal volume (V_T) times the P_{CO_2} of the expired gas (P_{ECO_2}) equals the arterial P_{CO_2} (P_{aCO_2}) times the difference between the TV and the dead space (V_D) plus the P_{CO_2} of inspired air (P_{ICO_2}) times V_D **(Bohr's equation):**

$$P_{ECO_2} \times V_T = P_{aCO_2} \times (V_T - V_D) + P_{ICO_2} \times V_D$$

The term $P_{ICO_2} \times V_D$ is so small that it can be ignored and the equation solved for V_D, where $V_D = V_T - (P_{ECO_2} \times V_T)/(P_{aCO_2})$

If, for example: P_{ECO_2} = 28 mm Hg; P_{aCO_2} = 40 mm Hg and V_T = 500 mL, then V_D = 150 mL

The equation can also be used to measure the anatomic dead space if one replaces P_{aCO_2} with alveolar P_{CO_2} (P_{ACO_2}), which is the P_{CO_2} of the last 10 mL of expired gas. P_{CO_2} is an average of gas from different alveoli in proportion to their ventilation regardless of whether they are perfused. This is in contrast to P_{aCO_2}, which is gas equilibrated only with perfused alveoli, and consequently, in individuals with under-perfused alveoli, is greater than P_{CO_2}.

GAS EXCHANGE IN THE LUNGS

PARTIAL PRESSURES

Unlike liquids, gases expand to fill the volume available to them, and the volume occupied by a given number of gas molecules at a given temperature and pressure is (ideally) the same regardless of the composition of the gas. **Partial pressures** are frequently used to describe gases in respiration. The pressure of a gas is proportional to its temperature and number of moles occupying a certain volume (Table 34–2). The pressure exerted by any one gas in a mixture of gases (its partial pressure) is equal to the total pressure times the fraction of the total amount of gas it represents.

TABLE 34–2 Properties of Gases.

$P = \frac{nRT}{V}$ (from equation of state of ideal gas)

The pressure of a gas is proportional to its temperature and the number of moles per volume; P, Pressure; n, Number of moles; R, Gas constant; T, Absolute temperature; V, Volume.

The composition of dry air is 20.98% O_2, 0.04% CO_2, 78.06% N_2, and 0.92% other inert constituents such as argon and helium. The barometric pressure (PB) at sea level is 760 mm Hg (1 atmosphere). The partial pressure (indicated by the symbol P) of O_2 in dry air is therefore 0.21 × 760, or 160 mm Hg at sea level. The P_{N_2} and the other inert gases is 0.79 × 760, or 600 mm Hg; and the P_{CO_2} is 0.0004 × 760, or 0.3 mm Hg. The water vapor in the air in most climates reduces these percentages, and therefore the partial pressures, to a slight degree. Air equilibrated with water is saturated with water vapor, and inspired air is saturated by the time it reaches the lungs. The P_{H_2O} at body temperature (37°C) is 47 mm Hg. Therefore, the partial pressures at sea level of the other gases in the air reaching the lungs are P_{O_2}, 150 mm Hg; P_{CO_2}, 0.3 mm Hg; and P_{N_2} (including the other inert gases), 563 mm Hg.

Gas diffuses from areas of high pressure to areas of low pressure, with the rate of diffusion depending on the concentration gradient and the nature of the barrier between the two areas. When a mixture of gases is in contact with and permitted to equilibrate with a liquid, each gas in the mixture dissolves in the liquid to an extent determined by its partial pressure and its solubility in the fluid. The partial pressure of a gas in a liquid is the pressure that, in the gaseous phase in equilibrium with the liquid, would produce the concentration of gas molecules found in the liquid.

SAMPLING ALVEOLAR AIR

Theoretically, all but the first 150 mL expired from a healthy 150 lb man (ie, the dead space) with each expiration is the gas that was in the alveoli **(alveolar air),** but some mixing always occurs at the interface between the dead-space gas and the alveolar air (Figure 34–14). A later portion of expired air is therefore the portion taken for analysis. Using modern apparatus with a suitable automatic valve, it is possible to collect the last 10 mL expired during quiet breathing. The composition of alveolar gas is compared with that of inspired and expired air in Figure 34–15.

P_{AO_2} can also be calculated from the **alveolar gas equation:**

$$P_{AO_2} = P_{IO_2} - P_{ACO_2}\left(F_{IO_2} + \frac{1 - F_{IO_2}}{R}\right)$$

where F_{IO_2} is the fraction of O_2 molecules in the dry gas, P_{IO_2} is the inspired P_{O_2}, and R is the respiratory exchange ratio; that is, the flow of CO_2 molecules across the alveolar membrane

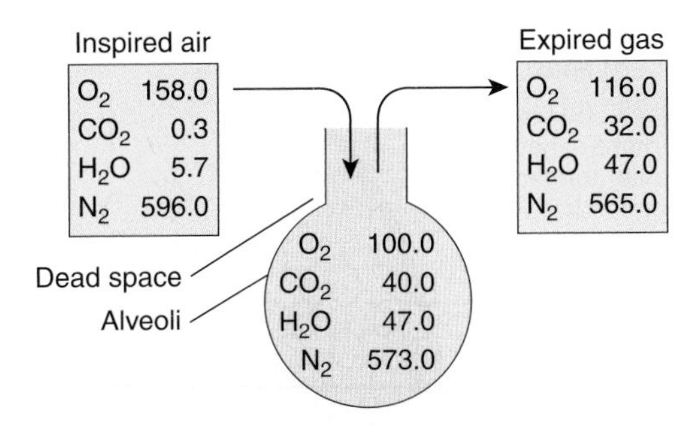

FIGURE 34–15 Partial pressures of gases (mm Hg) in various parts of the respiratory system. Typical partial pressures for inspired air, alveolar air, and expired air are given. See text for additional details.

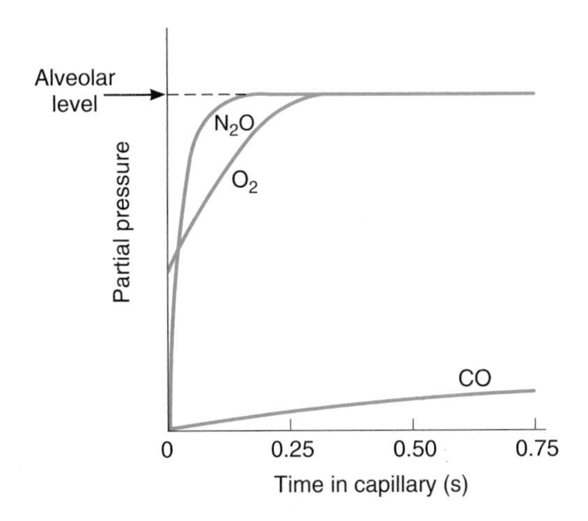

FIGURE 34–16 Uptake of various substances during the 0.75 s they are in transit through a pulmonary capillary. N_2O is not bound in blood, so its partial pressure in blood rises rapidly to its partial pressure in the alveoli. Conversely, CO is avidly taken up by red blood cells, so its partial pressure reaches only a fraction of its partial pressure in the alveoli. O_2 is intermediate between the two.

per minute divided by the flow of O_2 molecules across the membrane per minute.

COMPOSITION OF ALVEOLAR AIR

Oxygen continuously diffuses out of the gas in the alveoli into the bloodstream, and CO_2 continuously diffuses into the alveoli from the blood. In the steady state, inspired air mixes with the alveolar gas, replacing the O_2 that has entered the blood and diluting the CO_2 that has entered the alveoli. Part of this mixture is expired. The O_2 content of the alveolar gas then falls and its CO_2 content rises until the next inspiration. Because the volume of gas in the alveoli is about 2 L at the end of expiration (FRC), each 350 mL increment of inspired and expired air has relatively little effect on P_{O_2} and P_{CO_2}. Indeed, the composition of alveolar gas remains remarkably constant, not only at rest but also under a variety of other conditions.

DIFFUSION ACROSS THE ALVEOLOCAPILLARY MEMBRANE

Gases diffuse from the alveoli to the blood in the pulmonary capillaries or vice versa across the thin alveolocapillary membrane made up of the pulmonary epithelium, the capillary endothelium, and their fused basement membranes (Figure 34–3). Whether or not substances passing from the alveoli to the capillary blood reach equilibrium in the 0.75 s that blood takes to traverse the pulmonary capillaries at rest depends on their reaction with substances in the blood. Thus, for example, the anesthetic gas nitrous oxide (N_2O) does not react and reaches equilibrium in about 0.1 s (Figure 34–16). In this situation, the amount of N_2O taken up is not limited by diffusion but by the amount of blood flowing through the pulmonary capillaries; that is, it is **flow-limited.** On the other hand, carbon monoxide (CO) is taken up by hemoglobin in the red blood cells at such a high rate that the partial pressure of CO in the capillaries stays very low and equilibrium is not reached in the 0.75 s the blood is in the pulmonary capillaries. Therefore, the transfer of CO is not limited by perfusion at rest and instead is **diffusion-limited.** O_2 is intermediate between N_2O and CO; it is taken up by hemoglobin, but much less avidly than CO, and it reaches equilibrium with capillary blood in about 0.3 s. Thus, its uptake is **perfusion-limited.**

The **diffusing capacity** of the lung for a given gas is directly proportional to the surface area of the alveolocapillary membrane and inversely proportional to its thickness. The diffusing capacity for CO (DLCO) is measured as an index of diffusing capacity because its uptake is diffusion-limited. DLCO is proportional to the amount of CO entering the blood (VCO) divided by the partial pressure of CO in the alveoli minus the partial pressure of CO in the blood entering the pulmonary capillaries. Except in habitual cigarette smokers, this latter term is close to zero, so it can be ignored and the equation becomes:

$$D_{LCO} = \frac{\dot{V}_{CO}}{P_{ACO}}$$

The normal value of DLCO at rest is about 25 mL/min/mm Hg. It increases up to threefold during exercise because of capillary dilation and an increase in the number of active capillaries. The P_{O_2} of alveolar air is normally 100 mm Hg, and the P_{O_2} of the blood entering the pulmonary capillaries is 40 mm Hg. The diffusing capacity for O_2, like that for CO at rest, is about 25 mL/min/mm Hg, and the P_{O_2} of blood is raised to 97 mm Hg, a value just under the alveolar P_{O_2}.

The P_{CO_2} of venous blood is 46 mm Hg, whereas that of alveolar air is 40 mm Hg, and CO_2 diffuses from the blood into the alveoli along this gradient. The P_{CO_2} of blood leaving the lungs is 40 mm Hg. CO_2 passes through all biological membranes with ease, and the diffusing capacity of the lung for CO_2 is much greater than the capacity for O_2. It is for this reason that CO_2 retention is rarely a problem in patients with alveolar fibrosis even when the reduction in diffusing capacity for O_2 is severe.

PULMONARY CIRCULATION

PULMONARY BLOOD VESSELS

The pulmonary vascular bed resembles the systemic one, except that the walls of the pulmonary artery and its large branches are about 30% as thick as the wall of the aorta, and the small arterial vessels, unlike the systemic arterioles, are endothelial tubes with relatively little muscle in their walls. The walls of the postcapillary vessels also contain some smooth muscle. The pulmonary capillaries are large, and there are multiple anastomoses, so that each alveolus sits in a capillary basket.

PRESSURE, VOLUME, & FLOW

With two quantitatively minor exceptions, the blood put out by the left ventricle returns to the right atrium and is ejected by the right ventricle, making the pulmonary vasculature unique in that it accommodates a blood flow that is almost equal to that of all the other organs in the body. One of the exceptions is part of the bronchial blood flow. There are anastomoses between the bronchial capillaries and the pulmonary capillaries and veins, and although some of the bronchial blood enters the bronchial veins, some enters the pulmonary capillaries and veins, bypassing the right ventricle. The other exception is blood that flows from the coronary arteries into the chambers of the left side of the heart. Because of the small **physiologic shunt** created by those two exceptions, the blood in systemic arteries has a Po_2 about 2 mm Hg lower than that of blood that has equilibrated with alveolar air, and the saturation of hemoglobin is 0.5% less.

The pressure in the various parts of the pulmonary portion of the pulmonary circulation is shown in Figure 34–6c. The pressure gradient in the pulmonary system is about 7 mm Hg, compared with a gradient of about 90 mm Hg in the systemic circulation. Pulmonary capillary pressure is about 10 mm Hg, whereas the oncotic pressure is 25 mm Hg, so that an inward-directed pressure gradient of about 15 mm Hg keeps the alveoli free of all but a thin film of fluid. When the pulmonary capillary pressure is more than 25 mm Hg, pulmonary congestion and edema result.

The volume of blood in the pulmonary vessels at any one time is about 1 L, of which less than 100 mL is in the capillaries. The mean velocity of the blood in the root of the pulmonary artery is the same as that in the aorta (about 40 cm/s). It falls off rapidly, then rises slightly again in the larger pulmonary veins. It takes a red cell about 0.75 s to traverse the pulmonary capillaries at rest and 0.3 s or less during exercise.

EFFECT OF GRAVITY

Gravity has a relatively marked effect on the pulmonary circulation. In the upright position, the upper portions of the lungs are well above the level of the heart, and the bases are at or below it. Consequently, in the upper part of the lungs, the blood flow is less, the alveoli are larger, and ventilation is less than at the base (Figure 34–17). The pressure in the capillaries at the top of the lungs is close to the atmospheric pressure in the alveoli. Pulmonary arterial pressure is normally just sufficient to maintain perfusion, but if it is reduced or if alveolar pressure is increased, some of the capillaries collapse. Under these circumstances, no gas exchange takes place in the affected alveoli and they become part of the physiologic dead space.

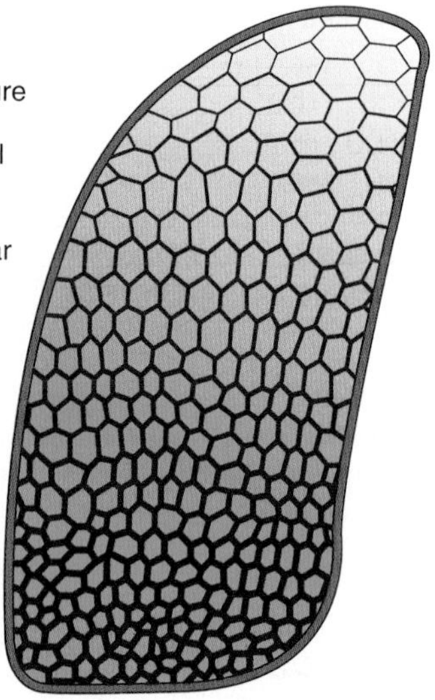

FIGURE 34–17 Diagram of normal differences in ventilation and perfusion of the lung in the upright position. Outlined areas are representative of changes in alveolar size (not actual size). Note the gradual change in alveolar size from top (apex) to bottom. Characteristic differences of alveoli at the apex of the lung are stated. (Modified from Levitzky MG: *Pulmonary Physiology,* 6th ed. McGraw-Hill, 2003).

In the middle portions of the lungs, the pulmonary arterial and capillary pressure exceeds alveolar pressure, but the pressure in the pulmonary venules may be lower than alveolar pressure during normal expiration, so they are collapsed. Under these circumstances, blood flow is determined by the pulmonary artery–alveolar pressure difference rather than the pulmonary artery–pulmonary vein difference. Beyond the constriction, blood "falls" into the pulmonary veins, which are compliant and take whatever amount of blood the constriction lets flow into them. This has been called the **waterfall effect.** Obviously, the compression of vessels produced by alveolar pressure decreases and pulmonary blood flow increases as the arterial pressure increases toward the base of the lung. In the lower portions of the lungs, alveolar pressure is lower than the pressure in all parts of the pulmonary circulation and blood flow is determined by the arterial–venous pressure difference. Examples of diseases affecting the pulmonary circulation are given in Clinical Box 34–4.

VENTILATION/PERFUSION RATIOS

The ratio of pulmonary ventilation to pulmonary blood flow for the whole lung at rest is about 0.8 (4.2 L/min ventilation divided by 5.5 L/min blood flow). However, relatively marked differences occur in this **ventilation/perfusion ratio** in various parts of the normal lung as a result of the effect of gravity, and local changes in the ventilation/perfusion ratio are common in disease. If the

CLINICAL BOX 34–4

Diseases Affecting the Pulmonary Circulation

Pulmonary Hypertension

Sustained idiopathic pulmonary hypertension can occur at any age. Like systemic arterial hypertension, it is a syndrome with multiple causes. However, the causes are different from those causing systemic hypertension. They include hypoxia, inhalation of cocaine, treatment with dexfenfluramine and related appetite-suppressing drugs that increase extracellular serotonin, and systemic lupus erythematosus. Some cases are familial and appear to be related to mutations that increase the sensitivity of pulmonary vessels to growth factors or cause deformations in the pulmonary vascular system.

All these conditions lead to increased pulmonary vascular resistance. If appropriate therapy is not initiated, the increased right ventricular afterload can lead eventually to right heart failure and death. Treatment with vasodilators such as prostacyclin and prostacyclin analogs is effective. Until recently, these had to be administered by continuous intravenous infusion, but aerosolized preparations that appear to be effective are now available.

ventilation to an alveolus is reduced relative to its perfusion, the Po_2 in the alveolus falls because less O_2 is delivered to it and the Pco_2 rises because less CO_2 is expired. Conversely, if perfusion is reduced relative to ventilation, the Pco_2 falls because less CO_2 is delivered and the Po_2 rises because less O_2 enters the blood. These effects are summarized in Figure 34–18.

As noted above, ventilation, as well as perfusion in the upright position, declines in a linear fashion from the bases to the apices of the lungs. However, the ventilation/perfusion ratios are high in the upper portions of the lungs. When widespread, nonuniformity of ventilation and perfusion in the lungs can cause CO_2 retention and lowers systemic arterial Po_2.

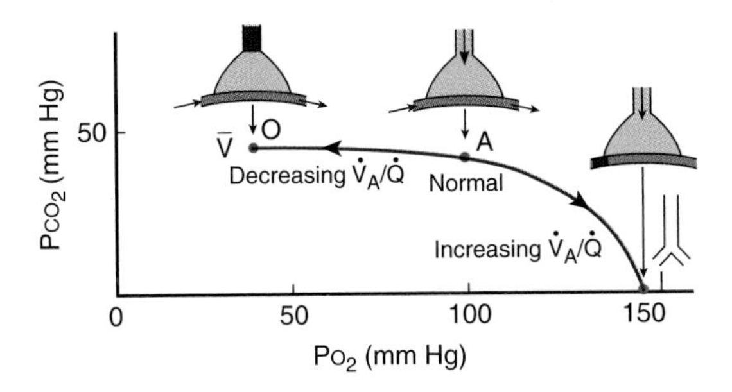

FIGURE 34–18 Effects of decreasing or increasing the ventilation/perfusion ratio ($\dot{V}_A/\dot{Q}$) on the Pco_2 and Po_2 in an alveolus. The drawings above the curve represent an alveolus and a pulmonary capillary, and the dark red areas indicate sites of blockage. With complete obstruction of the airway to the alveolus, Pco_2 and Po_2 approximate the values in mixed venous ($\bar{V}$) blood . With complete block of perfusion, Pco_2 and Po_2 approximate the values in inspired air. (Reproduced with permission from West JB: *Ventilation/Blood Flow and Gas Exchange*, 5th ed. Blackwell, 1990.)

REGULATION OF PULMONARY BLOOD FLOW

It is unsettled whether pulmonary veins and pulmonary arteries are regulated separately, although constriction of the veins increases pulmonary capillary pressure and constriction of pulmonary arteries increases the load on the right side of the heart.

Pulmonary blood flow is affected by both active and passive factors. There is an extensive autonomic innervation of the pulmonary vessels, and stimulation of the cervical sympathetic ganglia reduces pulmonary blood flow by as much as 30%. The vessels also respond to circulating humoral agents. A diversity of some the receptors involved and their effect on pulmonary smooth muscle are summarized in Table 34–3.

TABLE 34–3 Receptors affecting smooth muscle in pulmonary arteries and veins.

Receptor	Subtype	Response	Endothelium Dependency
Autonomic			
Adrenergic	α_1	Contraction	No
	α_2	Relaxation	Yes
	β_2	Relaxation	Yes
Muscarinic	M_3	Relaxation	Yes
Purinergic	P_{2x}	Contraction	No
	P_{2y}	Relaxation	Yes
Tachykinin	NK_1	Relaxation	Yes
	NK_2	Contraction	No
VIP	?	Relaxation	?
CGRP	?	Relaxation	No
Humoral			
Adenosine	A_1	Contraction	No
	A_2	Relaxation	No
Angiotensin II	AT_1	Contraction	No
ANP	ANP_A	Relaxation	No
	ANP_B	Relaxation	No
Bradykinin	B_1?	Relaxation	Yes
	B_2	Relaxation	Yes
Endothelin	ET_A	Contraction	No
	ET_B	Relaxation	Yes
Histamine	H_1	Relaxation	Yes
	H_2	Relaxation	No
5-HT	5-HT_1	Contraction	No
	5-HT_1C	Relaxation	Yes
Thromboxane	TP	Contraction	No
Vasopressin	V_1	Relaxation	Yes

Modified and reproduced with permission from Barnes PJ, Lin SF: Regulation of pulmonary vascular tone. Pharmacol Rev 1995;47:88.

Many of the dilator responses are endothelium-dependent and presumably operate via release of nitric oxide (NO).

Passive factors such as cardiac output and gravitational forces also have significant effects on pulmonary blood flow. Local adjustments of perfusion to ventilation occur with local changes in O_2. With exercise, cardiac output increases and pulmonary arterial pressure rises. More red cells move through the lungs without any reduction in the O_2 saturation of the hemoglobin in them, and consequently, the total amount of O_2 delivered to the systemic circulation is increased. Capillaries dilate, and previously underperfused capillaries are "recruited" to carry blood. The net effect is a marked increase in pulmonary blood flow with few, if any, alterations in autonomic outflow to the pulmonary vessels.

When a bronchus or a bronchiole is obstructed, hypoxia develops in the underventilated alveoli beyond the obstruction. The O_2 deficiency apparently acts directly on vascular smooth muscle in the area to produce constriction, shunting blood away from the hypoxic area. Accumulation of CO_2 leads to a drop in pH in the area, and a decline in pH also produces vasoconstriction in the lungs, as opposed to the vasodilation it produces in other tissues. Conversely, reduction of the blood flow to a portion of the lung lowers the alveolar P_{CO_2} in that area, and this leads to constriction of the bronchi supplying it, shifting ventilation away from the poorly perfused area. Systemic hypoxia also causes the pulmonary arterioles to constrict, with a resultant increase in pulmonary arterial pressure.

METABOLIC & ENDOCRINE FUNCTIONS OF THE LUNGS

In addition to their functions in gas exchange, the lungs have a number of metabolic functions. They manufacture surfactant for local use, as noted above. They also contain a fibrinolytic system that lyses clots in the pulmonary vessels. They release a variety of substances that enter the systemic arterial blood (Table 34–4), and they remove other substances from the systemic venous blood that reach them via the pulmonary artery. Prostaglandins are removed from the circulation, but they are also synthesized in the lungs and released into the blood when lung tissue is stretched.

The lungs play an important role in activating angiotensin. The physiologically inactive decapeptide angiotensin I is converted to the pressor, aldosterone-stimulating octapeptide angiotensin II in the pulmonary circulation. The reaction occurs in other tissues as well, but it is particularly prominent in the lungs. Large amounts of the angiotensin-converting enzyme responsible for this activation are located on the surface of the endothelial cells of the pulmonary capillaries. The converting enzyme also inactivates bradykinin. Circulation time through the pulmonary capillaries is less than 1 s, yet 70% of the angiotensin I reaching the lungs is converted to angiotensin II in a single trip through the capillaries. Four other peptidases have been identified on the surface of the pulmonary endothelial cells, but their full physiologic role is unsettled.

TABLE 34–4 Biologically active substances metabolized by the lungs.

Synthesized and used in the lungs
Surfactant
Synthesized or stored and released into the blood
Prostaglandins
Histamine
Kallikrein
Partially removed from the blood
Prostaglandins
Bradykinin
Adenine nucleotides
Serotonin
Norepinephrine
Acetylcholine
Activated in the lungs
Angiotensin I → angiotensin II

Removal of serotonin and norepinephrine reduces the amounts of these vasoactive substances reaching the systemic circulation. However, many other vasoactive hormones pass through the lungs without being metabolized. These include epinephrine, dopamine, oxytocin, vasopressin, and angiotensin II. In addition, various amines and polypeptides are secreted by neuroendocrine cells in the lungs.

CHAPTER SUMMARY

- Air enters the respiratory system in the upper airway, proceeds to the conducting airway and then on to the respiratory airway that ends in the alveoli. The cross-sectional area of the airway gradually increases through the conducting zone, and then rapidly increases during the transition from conducting to respiratory zones.
- The mucociliary escalator in the conducting airway helps to keep particulates out of the respiratory zone.
- There are several important measures of lung volume, including: tidal volume; inspiratory volume; expiratory reserve volume; forced vital capacity (FVC); the forced expiratory volume in one second (FEV_1); respiratory minute volume and maximal voluntary ventilation.
- Net "driving pressure" for air movement into the lung includes the force of muscle contraction, lung compliance ($\Delta P/\Delta V$) and airway resistance ($\Delta P/\Delta \dot{V}$).
- Surfactant decreases surface tension in the alveoli and helps to keep them from deflating.
- Not all air that enters the airway is available for gas exchange. The regions where gas is not exchanged in the airway are

termed "dead space." The conducting airway represents anatomical dead space. Increased dead space can occur in response to disease that affects air exchange in the respiratory zone (physiological dead space).

- The pressure gradient in the pulmonary circulation system is much less than that in the systemic circulation.
- There are a variety of biologically activated substances that are metabolized in the lung. These include substances that are made and function in the lung (eg, surfactant), substances that are released or removed from the blood (eg, prostaglandins), and substances that are activated as they pass through the lung (eg, angiotensin II).

MULTIPLE-CHOICE QUESTIONS

For all questions, select the single best answer unless otherwise directed.

1. On the summit of Mt. Everest, where the barometric pressure is about 250 mm Hg, the partial pressure of O_2 in mm Hg is about
   A. 0.1
   B. 0.5
   C. 5
   D. 50
   E. 100
2. The forced vital capacity is
   A. the amount of air that normally moves into (or out of) the lung with each respiration.
   B. the amount of air that enters the lung but does not participate in gas exchange.
   C. the amount of air expired after maximal expiratory effort.
   D. the largest amount of gas that can be moved into and out of the lungs in 1 min.
3. The tidal volume is
   A. the amount of air that normally moves into (or out of) the lung with each respiration.
   B. the amount of air that enters the lung but does not participate in gas exchange.
   C. the amount of air expired after maximal expiratory effort.
   D. the amount of gas that can be moved into and out of the lungs in 1 min.
4. Which of the following is responsible for the movement of O_2 from the alveoli into the blood in the pulmonary capillaries?
   A. Active transport
   B. Filtration
   C. Secondary active transport
   D. Facilitated diffusion
   E. Passive diffusion
5. Airway resistance
   A. is increased if the lungs are removed and inflated with saline.
   B. does not affect the work of breathing.
   C. is increased in paraplegic patients.
   D. is increased in following bronchial smooth muscle contraction.
   E. makes up 80% of the work of breathing.
6. Surfactant lining the alveoli
   A. helps prevent alveolar collapse.
   B. is produced in alveolar type I cells and secreted into the alveolus.
   C. is increased in the lungs of heavy smokers.
   D. is a glycolipid complex.

CHAPTER RESOURCES

Barnes PJ: Chronic obstructive pulmonary disease. N Engl J Med 2000;343:269.

Crystal RG, West JB (editors): *The Lung: Scientific Foundations,* 2nd ed. Raven Press, 1997.

Fishman AP, et al (editors): *Fishman's Pulmonary Diseases and Disorders,* 4th ed. McGraw-Hill, 2008.

Prisk GK, Paiva M, West JB (editors): *Gravity and the Lung: Lessons from Micrography.* Marcel Dekker, 2001.

West JB: *Pulmonary Pathophysiology,* 5th ed. McGraw-Hill, 1995.

Wright JR: Immunoregulatory functions of surfactant proteins. Nat Rev Immunol 2005;5:58.

CHAPTER

35

Gas Transport & pH

OBJECTIVES

After studying this chapter, you should be able to:

- Describe the manner in which O_2 flows "downhill" from the lungs to the tissues and CO_2 flows "downhill" from the tissues to the lungs.
- List the important factors affecting the affinity of hemoglobin for O_2 and the physiologic significance of each.
- List the reactions that increase the amount of CO_2 in the blood, and draw the CO_2 dissociation curve for arterial and venous blood.
- Define alkalosis and acidosis and list typical causes and compensatory responses to each.
- Define hypoxia and describe differences in subtypes of hypoxia.
- Describe the effects of hypercapnia and hypocapnia, and give examples of conditions that can cause them.

INTRODUCTION

The partial pressure gradients for O_2 and CO_2, plotted in graphical form in Figure 35–1, emphasize that they are the key to gas movement and that O_2 "flows downhill" from the air through the alveoli and blood into the tissues, whereas CO_2 "flows downhill" from the tissues to the alveoli. However, the amount of both of these gases transported to and from the tissues would be grossly inadequate if it were not for the fact that about 99% of the O_2 that dissolves in the blood combines with the O_2-carrying protein hemoglobin and that about 94.5% of the CO_2 that dissolves enters into a series of reversible chemical reactions that convert it into other compounds. Thus, the presence of hemoglobin increases the O_2-carrying capacity of the blood 70-fold, and the reactions of CO_2 increase the blood CO_2 content 17-fold. In this chapter, physiologic details that underlie O_2 and CO_2 movement under various conditions are discussed.

OXYGEN TRANSPORT

OXYGEN DELIVERY TO THE TISSUES

Oxygen delivery, or by definition, the volume of oxygen delivered to the systemic vascular bed per minute, is the product of the cardiac output and the arterial oxygen concentration. The ability to deliver O_2 in the body depends on both the respiratory and the cardiovascular systems. O_2 delivery to a particular tissue depends on the amount of O_2 entering the lungs, the adequacy of pulmonary gas exchange, the blood flow to the tissue, and the capacity of the blood to carry O_2. Blood flow to an individual tissue depends on cardiac output and the degree of constriction of the vascular bed in the tissue. The amount of O_2 in the blood is determined by the amount of dissolved O_2, the amount of hemoglobin in the blood, and the affinity of the hemoglobin for O_2.

REACTION OF HEMOGLOBIN & OXYGEN

The dynamics of the reaction of hemoglobin with O_2 make it a particularly suitable O_2 carrier. Hemoglobin is a protein made up of four subunits, each of which contains a **heme** moiety attached to a polypeptide chain. In normal adults, most of

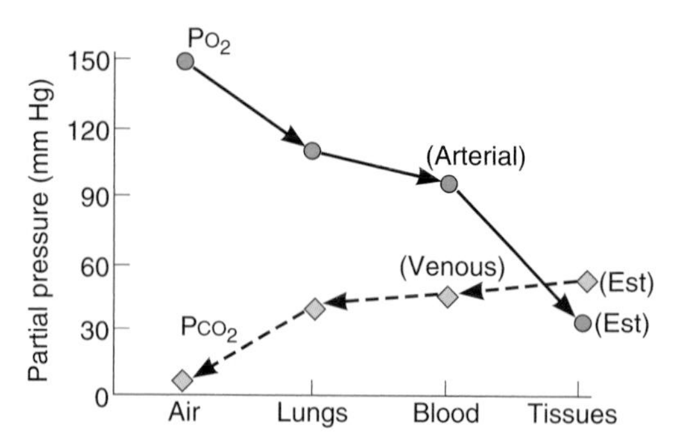

FIGURE 35–1 Po_2 and Pco_2 values in air, lungs, blood, and tissues. Note that both O_2 and CO_2 diffuse "downhill" along gradients of decreasing partial pressure. Est, estimated. (Redrawn and reproduced with permission from Kinney JM: Transport of carbon dioxide in blood. Anesthesiology 1960;21:615.)

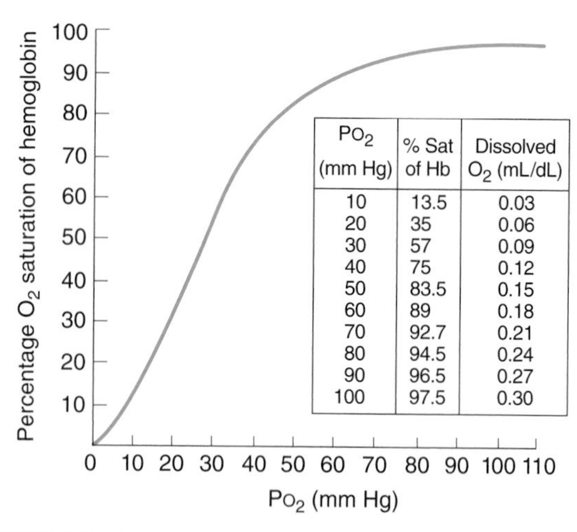

Po_2 (mm Hg)	% Sat of Hb	Dissolved O_2 (mL/dL)
10	13.5	0.03
20	35	0.06
30	57	0.09
40	75	0.12
50	83.5	0.15
60	89	0.18
70	92.7	0.21
80	94.5	0.24
90	96.5	0.27
100	97.5	0.30

FIGURE 35–2 Oxygen–hemoglobin dissociation curve. pH 7.40, temperature 38°C. Inset table relates the percentage of saturated hemoglobin (SaO_2) to Po_2 and dissolved O_2. (Redrawn and reproduced with permission from Comroe JH Jr, et al: *The Lung: Clinical Physiology and Pulmonary Function Tests, 2nd ed. Year Book, 1962.*)

the hemoglobin molecules contain two α and two β chains. Heme (see Figure 31–7) is a porphyrin ring complex that includes one atom of ferrous iron. Each of the four iron atoms in hemoglobin can reversibly bind one O_2 molecule. The iron stays in the ferrous state, so that the reaction is **oxygenation** (not oxidation). It has been customary to write the reaction of hemoglobin with O_2 as $Hb + O_2 \rightleftarrows HbO_2$. Because it contains four deoxyhemoglobin (Hb) units, the hemoglobin molecule can also be represented as Hb_4, and it actually reacts with four molecules of O_2 to form Hb_4O_8.

$$Hb_4 + O_2 \rightleftarrows Hb_4O_2$$
$$Hb_4O_2 + O_2 \rightleftarrows Hb_4O_4$$
$$Hb_4O_4 + O_2 \rightleftarrows Hb_4O_6$$
$$Hb_4O_6 + O_2 \rightleftarrows Hb_4O_8$$

The reaction is rapid, requiring less than 0.01 s. The deoxygenation of Hb_4O_8 is also very rapid.

The quaternary structure of hemoglobin determines its affinity for O_2. In deoxyhemoglobin, the globin units are tightly bound in a **tense (T) configuration,** which reduces the affinity of the molecule for O_2. When O_2 is first bound, the bonds holding the globin units are released, producing a **relaxed (R) configuration,** which exposes more O_2 binding sites. The net result is a 500-fold increase in O_2 affinity. In tissues, these reactions are reversed, resulting in O_2 release. The transition from one state to another has been calculated to occur about 10^8 times in the life of a red blood cell.

The **oxygen–hemoglobin dissociation curve** relates percentage saturation of the O_2 carrying power of hemoglobin (abbreviated as SaO_2) to the Po_2 (Figure 35–2). This curve has a characteristic sigmoid shape due to the T–R interconversion. Combination of the first heme in the Hb molecule with O_2 increases the affinity of the second heme for O_2, and oxygenation of the second increases the affinity of the third, and so on, so that the affinity of Hb for the fourth O_2 molecule is many times that for the first. Especially note that small changes at low Po_2 lead to large changes in SaO_2.

When blood is equilibrated with 100% O_2, the normal hemoglobin becomes 100% saturated. When fully saturated, each gram of normal hemoglobin contains 1.39 mL of O_2. However, blood normally contains small quantities of inactive hemoglobin derivatives, and the measured value in vivo is thus slightly lower. Using the traditional estimate of saturated hemoglobin in vivo, 1.34 mL of O_2, the hemoglobin concentration in normal blood is about 15 g/dL (14 g/dL in women and 16 g/dL in men). Therefore, 1 dL of blood contains 20.1 mL (1.34 mL × 15) of O_2 bound to hemoglobin when the hemoglobin is 100% saturated. The amount of dissolved O_2 is a linear function of the Po_2 (0.003 mL/dL blood/mm Hg Po_2).

In vivo, the hemoglobin in the blood at the ends of the pulmonary capillaries is about 97.5% saturated with O_2 (Po_2 = 100 mm Hg). Because of a slight admixture with venous blood that bypasses the pulmonary capillaries (ie, physiologic shunt), the hemoglobin in systemic arterial blood is only 97% saturated. The arterial blood therefore contains a total of about 19.8 mL of O_2 per dL: 0.29 mL in solution and 19.5 mL bound to hemoglobin. In venous blood at rest, the hemoglobin is 75% saturated and the total O_2 content is about 15.2 mL/dL: 0.12 mL in solution and 15.1 mL bound to hemoglobin. Thus, at rest the tissues remove about 4.6 mL of O_2 from each deciliter of blood passing through them (Table 35–1); 0.17 mL of

TABLE 35–1 Gas content of blood.

	mL/dL of Blood Containing 15 g of Hemoglobin			
	Arterial Blood (Po_2 95 mm Hg; Pco_2 40 mm Hg; Hb 97% Saturated)		Venous Blood (Po_2 40 mm Hg; Pco_2 46 mm Hg; Hb 75% Saturated)	
Gas	Dissolved	Combined	Dissolved	Combined
O_2	0.29	19.5	0.12	15.1
CO_2	2.62	46.4	2.98	49.7
N_2	0.98	0	0.98	0

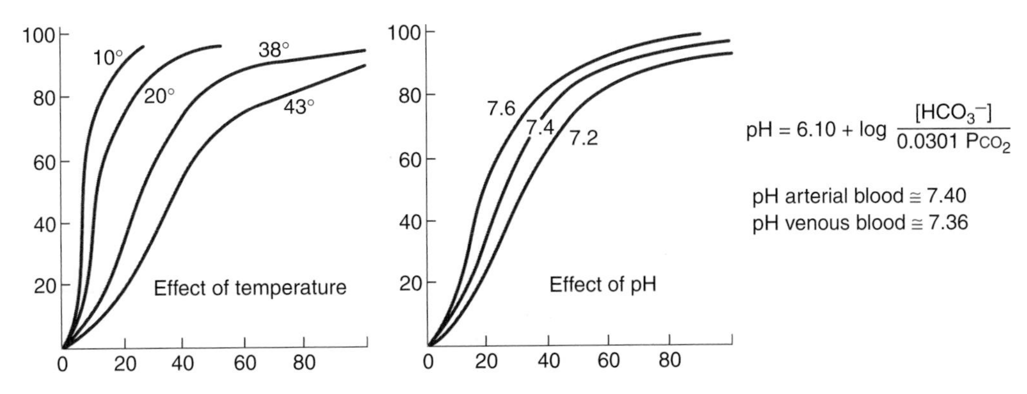

FIGURE 35–3 Effects of temperature and pH on the oxygen–hemoglobin dissociation curve. Both changes in temperature **(left)** and pH **(right)** can alter the affinity of hemoglobin for O_2. Plasma pH can be estimated using the modified Henderson–Hasselbalch equation, as shown. (Redrawn and reproduced with permission from Comroe JH Jr, et al: *The Lung: Clinical Physiology and Pulmonary Function Tests,* 2nd ed. Year Book, 1962.)

this total represents O_2 that was in solution in the blood, and the remainder represents O_2 that was liberated from hemoglobin. In this way, 250 mL of O_2 per minute is transported from the blood to the tissues at rest.

FACTORS AFFECTING THE AFFINITY OF HEMOGLOBIN FOR OXYGEN

Three important conditions affect the oxygen–hemoglobin dissociation curve: the **pH,** the **temperature,** and the concentration of **2,3-diphosphoglycerate (DPG; 2,3-DPG).** A rise in temperature or a fall in pH shifts the curve to the right (Figure 35–3). When the curve is shifted in this direction, a higher Po_2 is required for hemoglobin to bind a given amount of O_2. Conversely, a fall in temperature or a rise in pH shifts the curve to the left, and a lower Po_2 is required to bind a given amount of O_2. A convenient index for comparison of such shifts is the P_{50}, the Po_2 at which hemoglobin is half saturated with O_2. The higher the P_{50}, the lower the affinity of hemoglobin for O_2.

The decrease in O_2 affinity of hemoglobin when the pH of blood falls is called the **Bohr effect** and is closely related to the fact that deoxygenated hemoglobin (deoxyhemoglobin) binds H^+ more actively than does oxygenated hemoglobin (oxyhemoglobin). The pH of blood falls as its CO_2 content increases, so that when the Pco_2 rises, the curve shifts to the right and the P_{50} rises. Most of the unsaturation of hemoglobin that occurs in the tissues is secondary to the decline in the Po_2, but an extra 1–2% unsaturation is due to the rise in Pco_2 and consequent shift of the dissociation curve to the right.

2,3-DPG is very plentiful in red cells. It is formed from 3-phosphoglyceraldehyde, which is a product of glycolysis via the Embden–Meyerhof pathway. It is a highly charged anion that binds to the β chains of deoxyhemoglobin. One mole of deoxyhemoglobin binds 1 mol of 2,3-DPG. In effect,

$$HbO_2 + 2,3\text{-BPG} \rightleftarrows Hb - 2,3\text{-BPG} + O_2$$

In this equilibrium, an increase in the concentration of 2,3-DPG shifts the reaction to the right, causing more O_2 to be liberated.

Because acidosis inhibits red cell glycolysis, the 2,3-DPG concentration falls when the pH is low. Conversely, thyroid hormones, growth hormones, and androgens can all increase the concentration of 2,3-DPG and the P_{50}.

Exercise has been reported to produce an increase in 2,3-DPG within 60 min (although the rise may not occur in trained athletes). The P_{50} is also increased during exercise, because the temperature rises in active tissues and CO_2 and metabolites accumulate, lowering the pH. In addition, much more O_2 is removed from each unit of blood flowing through active tissues because the tissue Po_2 declines. Finally, at low Po_2 values, the oxygen–hemoglobin dissociation curve is steep, and large amounts of O_2 are liberated per unit drop in Po_2. Some clinical features of hemoglobin are discussed in Clinical Box 35–1.

An interesting contrast to hemoglobin is **myoglobin,** an iron-containing pigment found in skeletal muscle. Myoglobin resembles hemoglobin but binds 1 rather than 4 mol of O_2 per mole protein. The lack of cooperative binding is reflected in the myoglobin dissociation curve, a rectangular hyperbola rather than the sigmoid curve observed for hemoglobin (Figure 35–4). Additionally, the leftward shift of the myoglobin O_2 binding curve when compared with hemoglobin demonstrates a higher affinity for O_2, and thus promotes a favorable transfer of O_2 from hemoglobin in the blood. The steepness of the myoglobin curve also shows that O_2 is released only at low Po_2 values (eg, during exercise). The myoglobin content is greatest in muscles specialized for sustained contraction. The muscle blood supply is compressed during such contractions, and myoglobin can continue to provide O_2 under reduced blood flow and/or reduced Po_2 in the blood.

CLINICAL BOX 35–1

Hemoglobin & O_2 Binding In Vivo

Cyanosis

Reduced hemoglobin has a dark color, and a dusky bluish discoloration of the tissues, called **cyanosis,** appears when the reduced hemoglobin concentration of the blood in the capillaries is more than 5 g/dL. Its occurrence depends on the total amount of hemoglobin in the blood, the degree of hemoglobin unsaturation, and the state of the capillary circulation. Cyanosis is most easily seen in the nail beds and mucous membranes and in the earlobes, lips, and fingers, where the skin is thin. Although visible observation is indicative of cyanosis, it is not fully reliable. Further tests of arterial oxygen tension and saturation, blood and hemoglobin counts can provide more reliable diagnoses.

Effects of 2,3-DPG on Fetal & Stored Blood

The affinity of fetal hemoglobin (hemoglobin F) for O_2, which is greater than that for adult hemoglobin (hemoglobin A), facilitates the movement of O_2 from the mother to the fetus. The cause of this greater affinity is the poor binding of 2,3-DPG by the γ polypeptide chains that replace β chains in fetal hemoglobin. Some abnormal hemoglobins in adults have low P_{50} values, and the resulting high O_2 affinity of the hemoglobin causes enough tissue hypoxia to stimulate increased red cell formation, with resulting polycythemia. It is interesting to speculate that these hemoglobins may not bind 2,3-DPG.

Red cell 2,3-DPG concentration is increased in anemia and in a variety of diseases in which there is chronic hypoxia. This facilitates the delivery of O_2 to the tissues by raising the Po_2 at which O_2 is released in peripheral capillaries. In banked blood that is stored, the 2,3-DPG level falls and the ability of this blood to release O_2 to the tissues is reduced. This decrease, which obviously limits the benefit of the blood if it is transfused into a hypoxic patient, is less if the blood is stored in citrate–phosphate–dextrose solution rather than the usual acid–citrate–dextrose solution.

THERAPEUTIC HIGHLIGHTS

Cyanosis is an indication of poorly oxygenated hemoglobin rather than a disease, and thus can have many causes, from cold exposure to drug overdose to chronic lung disease. As such, proper treatment depends upon the underlying cause. For cyanosis caused by exposure to cold, maintaining a warm environment can be effective, whereas supplemental oxygen administration may be required under conditions of chronic disease.

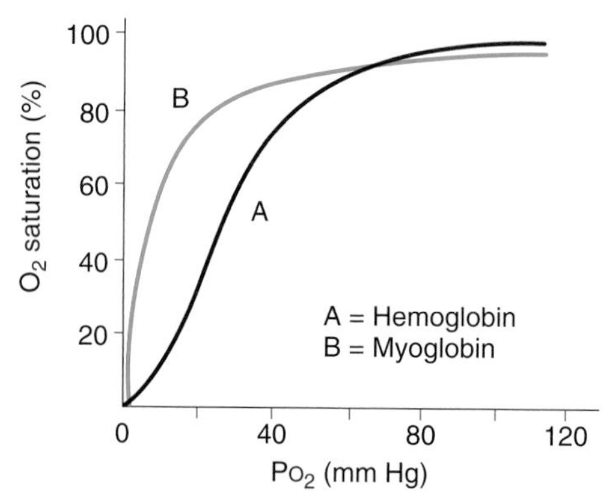

FIGURE 35–4 Comparison of dissociation curves for hemoglobin and myoglobin. The myoglobin binding curve (B) lacks the sigmoidal shape of the hemoglobin binding curve (A) because of the single O_2 binding site in each molecule. Myoglobin also has greater affinity for O_2 than hemoglobin (curve shifted left) and thus can release O_2 in muscle when Po_2 in blood is low (eg, during exercise).

CARBON DIOXIDE TRANSPORT

MOLECULAR FATE OF CARBON DIOXIDE IN BLOOD

The solubility of CO_2 in blood is about 20 times that of O_2; therefore, considerably more CO_2 than O_2 is present in simple solution at equal partial pressures. The CO_2 that diffuses into red blood cells is rapidly hydrated to H_2CO_3 because of the presence of carbonic anhydrase (Figure 35–5). The H_2CO_3 dissociates to H^+ and HCO_3^-, and the H^+ is buffered, primarily by hemoglobin, while the HCO_3^- enters the plasma. Some of the CO_2 in the red cells reacts with the amino groups of hemoglobin and other proteins (R), forming **carbamino compounds.**

$$CO_2 + R{-}N(H)(H) \rightleftarrows R{-}N(H)(COOH)$$

Because deoxyhemoglobin binds more H^+ than oxyhemoglobin and forms carbamino compounds more readily, binding of O_2 to hemoglobin reduces its affinity for CO_2. The **Haldane effect** refers to the increased capacity of deoxygenated hemoglobin to bind and carry CO_2. Consequently, venous blood carries more CO_2 than arterial blood, CO_2 uptake is facilitated in the tissues, and CO_2 release is facilitated in the lungs. About 11% of the CO_2 added to the blood in the systemic capillaries is carried to the lungs as carbamino-CO_2.

CHLORIDE SHIFT

Because the rise in the HCO_3^- content of red cells is much greater than that in plasma as the blood passes through the capillaries, about 70% of the HCO_3^- formed in the red cells

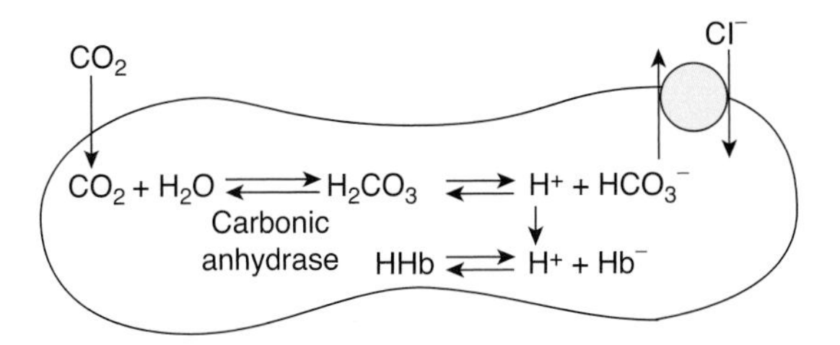

FIGURE 35–5 Fate of CO_2 in the red blood cell. Upon entering the red blood cell, CO_2 is rapidly hydrated to H_2CO_3 by carbonic anhydrase. H_2CO_3 is in equilibrium with H^+ and its conjugate base, HCO_3^-. H^+ can interact with deoxyhemoglobin, whereas HCO_3^- can be transported outside of the cell via anion exchanger 1 (AE1 or Band 3). In effect, for each CO_2 molecule that enters the red cell, there is an additional HCO_3^- or Cl^- in the cell.

TABLE 35–2 Fate of CO_2 in blood.

In plasma
1. Dissolved
2. Formation of carbamino compounds with plasma protein
3. Hydration, H^+ buffered, HCO_3^- in plasma
In red blood cells
1. Dissolved
2. Formation of carbamino-Hb
3. Hydration, H^+ buffered, 70% of HCO_3^- enters the plasma
4. Cl^- shifts into cells; mOsm in cells increases

enters the plasma. The excess HCO_3^- leaves the red cells in exchange for Cl^- (Figure 35–5). This process is mediated by **anion exchanger 1 (AE1**; also called Band 3), a major membrane protein in the red blood cell. Because of this **chloride shift**, the Cl^- content of the red cells in venous blood is significantly greater than that in arterial blood. The chloride shift occurs rapidly and is essentially complete within 1 s.

Note that for each CO_2 molecule added to a red cell, there is an increase of one osmotically active particle in the cell—either an HCO_3^- or a Cl^- (Figure 35–6). Consequently, the red cells take up water and increase in size. For this reason, plus the fact that a small amount of fluid in the arterial blood returns via the lymphatics rather than the veins, the hematocrit of venous blood is normally 3% greater than that of the arterial blood. In the lungs, the Cl^- moves back out of the cells and they shrink.

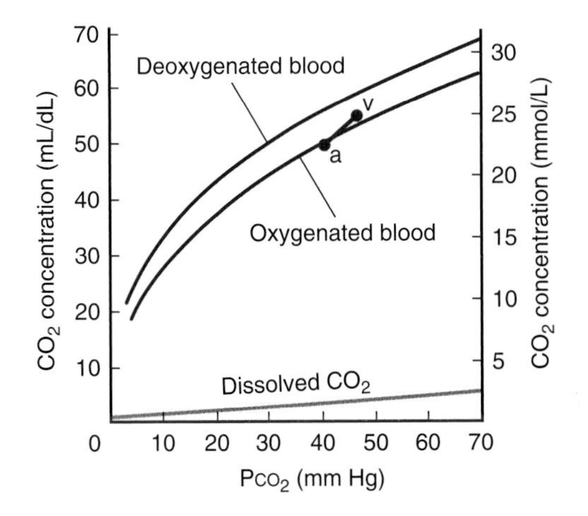

FIGURE 35–6 CO_2 dissociation curves. The arterial point (a) and the venous point (v) indicate the total CO_2 content found in arterial blood and venous blood of normal resting humans. Note the low amount of CO_2 that is dissolved (orange trace) compared to that which can be carried by other means (Table 35–2). (Modified and reproduced with permission from Schmidt RF, Thews G [editors]: *Human Physiology.* Springer, 1983.)

SPATIAL DISTRIBUTION OF CARBON DIOXIDE IN BLOOD

For convenience, the various fates of CO_2 in the plasma and red cells are summarized in Table 35–2. The extent to which they increase the capacity of the blood to carry CO_2 is indicated by the difference between the lines indicating the dissolved CO_2 and the total CO_2 in the dissociation curves for CO_2 shown in Figure 35–6.

Of the approximately 49 mL of CO_2 in each deciliter of arterial blood (Table 35–1), 2.6 mL is dissolved, 2.6 mL is in carbamino compounds, and 43.8 mL is in HCO_3^-. In the tissues, 3.7 mL of CO_2 per deciliter of blood is added; 0.4 mL stays in solution, 0.8 mL forms carbamino compounds, and 2.5 mL forms HCO_3^-. The pH of the blood drops from 7.40 to 7.36. In the lungs, the processes are reversed, and the 3.7 mL of CO_2 is discharged into the alveoli. In this fashion, 200 mL of CO_2 per minute at rest and much larger amounts during exercise are transported from the tissues to the lungs and excreted. It is worth noting that this amount of CO_2 is equivalent in 24 h to over 12,500 mEq of H^+.

ACID–BASE BALANCE & GAS TRANSPORT

The major source of acids in the blood under normal conditions is through cellular metabolism. The CO_2 formed by metabolism in the tissues is in large part hydrated to H_2CO_3, resulting in the large total H^+ load noted above (> 12,500 mEq/d). However, most of the CO_2 is excreted in the lungs, and the small quantities of the remaining H^+ are excreted by the kidneys.

BUFFERING IN THE BLOOD

Acid and base shifts in the blood are largely controlled by three main buffers in blood: (1) proteins, (2) hemoglobin, and (3) the carbonic acid–bicarbonate system. Plasma **proteins** are effective buffers because both their free carboxyl and their free amino groups dissociate:

$$RCOOH \rightleftarrows RCOO^- + H^+$$

$$pH = pK'_{RCOOH} + \log \frac{[RCOO^-]}{[RCOOH]}$$

$$RNH_3^+ \rightleftarrows RNH_2 + H^+$$

$$pH = pK'_{RNH_3} + \log \frac{[RNH_2]}{[RNH_3^+]}$$

The second buffer system is provided by the dissociation of the imidazole groups of the histidine residues in **hemoglobin.**

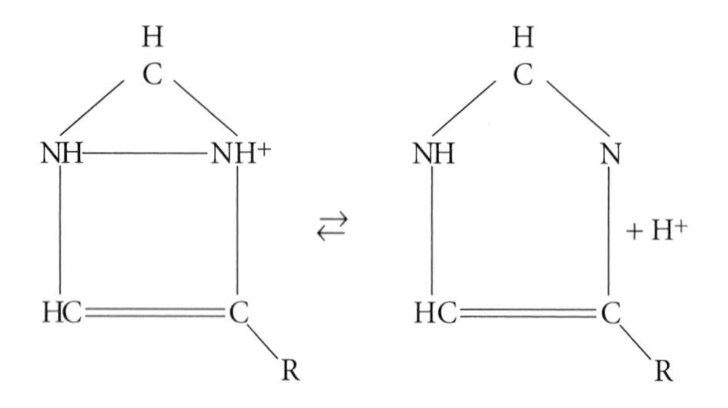

In the pH 7.0–7.7 range, the free carboxyl and amino groups of hemoglobin contribute relatively little to its buffering capacity. However, the hemoglobin molecule contains 38 histidine residues, and on this basis—plus the fact that hemoglobin is present in large amounts—the hemoglobin in blood has six times the buffering capacity of the plasma proteins. In addition, the action of hemoglobin is unique because the imidazole groups of deoxyhemoglobin (Hb) dissociate less than those of oxyhemoglobin (HbO_2), making Hb a weaker acid and therefore a better buffer than HbO_2. The titration curves for Hb and HbO_2 (Figure 35–7) illustrate the differences in H^+ buffering capacity.

The third and major buffer system in blood is the **carbonic acid–bicarbonate system:**

$$H_2CO_3 \rightleftarrows H^+ + HCO_3^-$$

The Henderson–Hasselbalch equation for this system is

$$pH = pK + \log \frac{[HCO_3^-]}{[H_2CO_3]}$$

The pK for this system in an ideal solution is low (about 3), and the amount of H_2CO_3 is small and hard to measure accurately. However, in the body, H_2CO_3 is in equilibrium with CO_2:

$$H_2CO_3 \rightleftarrows CO_2 + H_2O$$

If the pK is changed to pK′ (apparent ionization constant; distinguished from the true pK due to less than ideal

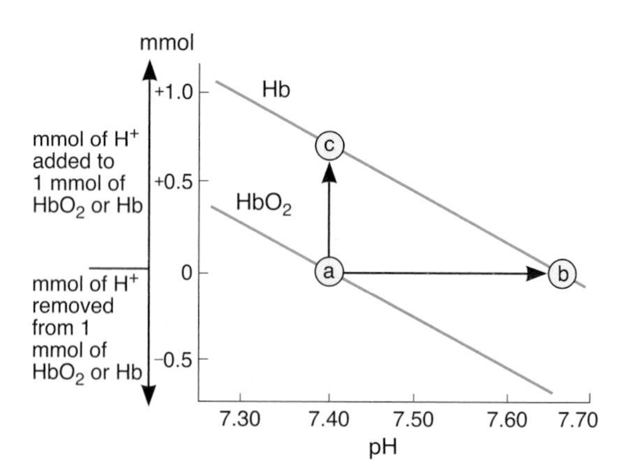

FIGURE 35–7 Comparative titration curves for oxygenated hemoglobin (HbO_2) and deoxyhemoglobin (Hb). The arrow from a to c indicates the number of additional millimoles of H^+ that Hb can buffer compared with a similar concentration of HbO_2 (ie, no shift in pH). The arrow from a to b indicates the pH shift that would occur on deoxygenation of HbO_2 without additional H^+.

conditions for the solution) and $[CO_2]$ is substituted for $[H_2CO_3]$, the pK′ is 6.1:

$$pH = 6.10 + \log \frac{[HCO_3^-]}{[CO_2]}$$

The clinically relevant form of this equation is:

$$pH = 6.10 + \log \frac{[HCO_3^-]}{0.0301\ P_{CO_2}}$$

since the amount of dissolved CO_2 is proportional to the partial pressure of CO_2 and the solubility coefficient of CO_2 in mmol/L/mm Hg is 0.0301. $[HCO_3^-]$ cannot be measured directly, but pH and P_{CO_2} can be measured with suitable accuracy with pH and P_{CO_2} glass electrodes, and $[HCO_3^-]$ can then be calculated.

The pK′ of this system is still low relative to the pH of the blood, but the system is one of the most effective buffer systems in the body because the amount of dissolved CO_2 is controlled by respiration (ie, it is an "open" system). Additional control of the plasma concentration of HCO_3^- is provided by the kidneys. When H^+ is added to the blood, HCO_3^- declines as more H_2CO_3 is formed. If the extra H_2CO_3 were not converted to CO_2 and H_2O and the CO_2 excreted in the lungs, the H_2CO_3 concentration would rise. Without CO_2 removal to reduce H_2CO_3, sufficient H^+ addition that would halve the plasma HCO_3^- would alter the pH 7.4 to 6.0. However, such a H^+ concentration increase is tolerated because: (1) extra H_2CO_3 that is formed is removed and (2) the H^+ rise stimulates respiration and therefore produces a drop in P_{CO_2}, so that some additional H_2CO_3 is removed. The net pH after such an increase in H^+ concentration is actually 7.2 or 7.3.

There are two additional factors that make the carbonic-acid-bicarbonate system such a good biological buffer. First, the reaction $CO_2 + H_2O \rightleftarrows H_2CO_3$ proceeds slowly in either direction unless the enzyme **carbonic anhydrase** is present. There is no carbonic anhydrase in plasma, but there is an abundant supply in red blood cells, spatially confining and controlling the reaction. Second, the presence of hemoglobin in the blood increases the buffering of the system by binding free H^+ produced by the hydration of CO_2 and allowing for movement of the HCO_3^- into the plasma.

ACIDOSIS & ALKALOSIS

The pH of the arterial plasma is normally 7.40 and that of venous plasma slightly lower. A decrease in pH below the norm **(acidosis)** is technically present whenever the arterial pH is below 7.40 and an increase in pH **(alkalosis)** is technically present whenever pH is above 7.40. In practice, variations of up to 0.05 pH unit occur without untoward effects. Acid–base disorders are split into four categories: respiratory acidosis, respiratory alkalosis, metabolic acidosis, and metabolic alkalosis. In addition, these disorders can occur in combination. Some examples of acid–base disturbances are shown in Table 35–3.

RESPIRATORY ACIDOSIS

Any short-term rise in arterial P_{CO_2} (ie, above 40 mm Hg, due to hypoventilation) results in **respiratory acidosis**. Recall that CO_2 that is retained is in equilibrium with H_2CO_3, which in turn is in equilibrium with HCO_3^-. The effective rise in plasma HCO_3^- means that a new equilibrium is reached at a lower pH. This can be indicated graphically on a plot of plasma HCO_3^- concentration versus pH (Figure 35–8). The pH change observed at any increase in P_{CO_2} during respiratory acidosis is dependent on the buffering capacity of the blood. The initial changes shown in Figure 35–8 are those that occur independently of any compensatory mechanism; that is, they are those of **uncompensated respiratory acidosis**.

TABLE 35–3 Plasma pH, HCO_3^-, and P_{CO_2} values in various typical disturbances of acid–base balance.[a]

	Arterial Plasma			
Condition	**pH**	**HCO_3^- (mEq/L)**	**P_{CO_2} (mm Hg)**	**Cause**
Normal	7.40	24.1	40	
Metabolic acidosis	7.28	18.1	40	NH_4Cl ingestion
	6.96	5.0	23	Diabetic acidosis
Metabolic alkalosis	7.50	30.1	40	$NaHCO_3$ ingestion
	7.56	49.8	58	Prolonged vomiting
Respiratory acidosis	7.34	25.0	48	Breathing 7% CO_2
	7.34	33.5	64	Emphysema
Respiratory alkalosis	7.53	22.0	27	Voluntary hyperventilation
	7.48	18.7	26	Three-week residence at 4000-m altitude

[a]In the diabetic acidosis and prolonged vomiting examples, respiratory compensation for primary metabolic acidosis and alkalosis has occurred, and the P_{CO_2} has shifted from 40 mm Hg. In the emphysema and high-altitude examples, renal compensation for primary respiratory acidosis and alkalosis has occurred and has made the deviations from normal of the plasma HCO_3^- larger than they would otherwise be.

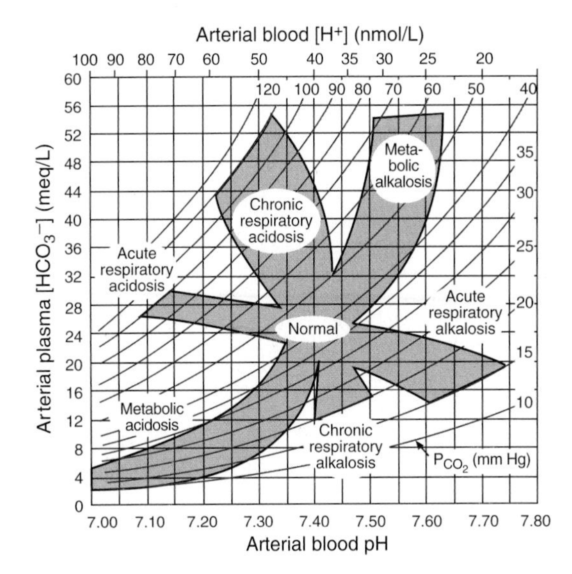

FIGURE 35–8 Acid–base nomogram. Changes in the P_{CO_2} (curved lines), plasma HCO_3^-, and pH (or $[H^+]$) of arterial blood in respiratory and metabolic acidosis are shown. Note the shifts in HCO_3^- and pH as acute respiratory acidosis and alkalosis are compensated, producing their chronic counterparts. (Reproduced with permission from Brenner BM, Rector FC Jr. (editors): *Brenner & Rector's The Kidney*, 7th ed. Saunders, 2004.)

RESPIRATORY ALKALOSIS

Any short-term lowering of P_{CO_2} below what is needed for proper CO_2 exchange (ie, below 35 mm Hg, as can occur during hyperventilation) results in **respiratory alkalosis**. The decreased CO_2 shifts the equilibrium of the carbonic acid–bicarbonate system to effectively lower $[H^+]$ and increase pH. As in respiratory acidosis, initial pH changes corresponding to respiratory alkalosis (Figure 35–8) are those that occur independently of any compensatory mechanism and are thus **uncompensated respiratory alkalosis**.

METABOLIC ACIDOSIS & ALKALOSIS

Blood pH changes can also arise by nonrespiratory mechanism. **Metabolic acidosis** (or nonrespiratory acidosis) occurs when strong acids are added to blood. If, for example, a large amount of acid is ingested (eg, aspirin overdose), acids in the blood are quickly increased. The H_2CO_3 that is formed is

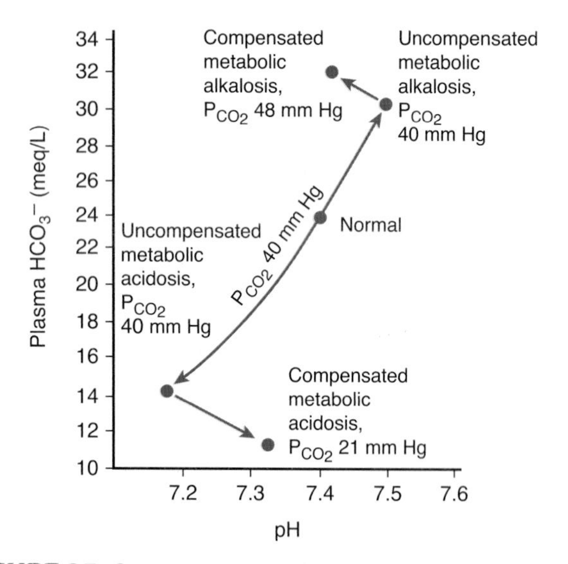

FIGURE 35–9 Acid–base paths during metabolic acidosis. Changes in true plasma pH, HCO_3^-, and Pco_2 at rest, during metabolic acidosis and alkalosis, and following respiratory compensation are plotted. Metabolic acidosis or alkalosis causes changes in pH along the Pco_2 isobar line (middle line). Respiratory compensation moves pH towards normal by altering Pco_2 (top and bottom arrows). (This is called a Davenport diagram and is based on Davenport HW: *The ABC of Acid–Base Chemistry*, 6th ed. University of Chicago Press, 1974.)

converted to H_2O and CO_2, and the CO_2 is rapidly excreted via the lungs. This is the situation in **uncompensated metabolic acidosis** (Figure 35–8). Note that in contrast to respiratory acidosis, metabolic acidosis does not include a change in Pco_2; the shift toward metabolic acidosis occurs along an isobar line (Figure 35–9). When the free [H^+] level falls as a result of addition of alkali, or more commonly, the removal of large amounts of acid (eg, following vomiting), **metabolic alkalosis** results. In uncompensated metabolic alkalosis the pH rises along the isobar line (Figures 35–8 and 35–9).

RESPIRATORY & RENAL COMPENSATION

Uncompensated acidosis and alkalosis as described above are seldom seen because of compensation systems. The two main compensatory systems are **respiratory compensation** and **renal compensation.**

The respiratory system compensates for metabolic acidosis or alkalosis by altering ventilation, and consequently, the Pco_2, which can directly change blood pH. Respiratory mechanisms are fast. In response to metabolic acidosis, ventilation is increased, resulting in a decrease of Pco_2 (eg, from 40 mm Hg to 20 mm Hg) and a subsequent increase in pH toward normal (Figure 35–9). In response to metabolic alkalosis, ventilation is decreased, Pco_2 is increased, and a subsequent decrease in pH occurs. Because respiratory compensation is a quick response, the graphical representation in Figure 35–9 overstates the two-step adjustment in blood pH. In actuality, as soon as metabolic acidosis begins, respiratory compensation is invoked and the large shifts in pH depicted do not occur.

For complete compensation from respiratory or metabolic acidosis/alkalosis, renal compensatory mechanisms are invoked. The kidney responds to acidosis by actively secreting fixed acids while retaining filtered HCO_3^-. In contrast, the kidney responds to alkalosis by decreasing H^+ secretion and by decreasing the retention of filtered HCO_3^-.

Renal tubule cells in the kidney have active carbonic anhydrase and thus can produce H^+ and HCO_3^- from CO_2. In response to acidosis, these cells secrete H^+ into the tubular fluid in exchange for Na^+ while the HCO_3^- is actively reabsorbed into the peritubular capillary; for each H^+ secreted, one Na^+ and one HCO_3^- are added to the blood. The result of this renal compensation for respiratory acidosis is shown graphically in the shift from acute to chronic respiratory acidosis in Figure 35–8. Conversely, in response to alkalosis, the kidney decreases H^+ secretion and depresses HCO_3^- reabsorption. The result of this renal compensation for respiratory alkalosis is shown graphically in the shift from acute to chronic respiratory alkalosis in Figure 35–8. Clinical evaluations of acid–base status are discussed in Clinical Box 35–2 and the role of the

CLINICAL BOX 35–2

Clinical Evaluation of Acid–Base Status

In evaluating disturbances of acid–base balance, it is important to know the pH and HCO_3^- content of arterial plasma. Reliable pH determinations can be made with a pH meter and a glass pH electrode. Using pH and a direct measurement of the Pco_2 with a CO_2 electrode, HCO_3^- concentration can be calculated. The Pco_2 is ~8 mm Hg higher and the pH 0.03–0.04 unit lower in venous than arterial plasma because venous blood contains the CO_2 being carried from the tissues to the lungs. Therefore, the calculated HCO_3^- concentration is about 2 mmol/L higher. However, if this is kept in mind, free-flowing venous blood can be substituted for arterial blood in most clinical situations.

A measurement that is of some value in the differential diagnosis of metabolic acidosis is the **anion gap.** This gap, which is something of a misnomer, refers to the difference between the concentration of cations other than Na^+ and the concentration of anions other than Cl^- and HCO_3^- in the plasma. It consists for the most part of proteins in anionic form, HPO_4^{2-}, SO_4^{2-}, and organic acids; a normal value is about 12 mEq/L. It is increased when the plasma concentration of K^+, Ca^{2+}, or Mg^+ is decreased; when the concentration of (or the charge on) plasma proteins is increased; or when organic anions such as lactate or foreign anions accumulate in blood. It is decreased when cations are increased or when plasma albumin is decreased. The anion gap is increased in metabolic acidosis due to ketoacidosis, lactic acidosis, and other forms of acidosis in which organic anions are increased.

kidneys in acid–base homeostasis is discussed in more detail in Chapter 38.

HYPOXIA

Hypoxia is O_2 deficiency at the tissue level. It is a more correct term than **anoxia** (lack of O_2), since there is rarely no O_2 at all left in the tissues.

Numerous classifications for hypoxia have been used, but the more traditional four-type system still has considerable utility if the definitions of the terms are kept clearly in mind. The four categories are (1) **hypoxemia** (sometimes termed **hypoxic hypoxia**), in which the Po_2 of the arterial blood is reduced; (2) **anemic hypoxia,** in which the arterial Po_2 is normal but the amount of hemoglobin available to carry O_2 is reduced; (3) **ischemic** or **stagnant hypoxia,** in which the blood flow to a tissue is so low that adequate O_2 is not delivered to it despite a normal Po_2 and hemoglobin concentration; and (4) **histotoxic hypoxia,** in which the amount of O_2 delivered to a tissue is adequate but, because of the action of a toxic agent, the tissue cells cannot make use of the O_2 supplied to them. Some specific effects of hypoxia on cells and tissues are discussed in Clinical Box 35–3.

HYPOXEMIA

By definition, hypoxemia is a condition of reduced arterial Po_2. Hypoxemia is a problem in normal individuals at high altitudes and is a complication of pneumonia and a variety of other diseases of the respiratory system.

EFFECTS OF DECREASED BAROMETRIC PRESSURE

The composition of air stays the same, but the total barometric pressure falls with increasing altitude (Figure 35–10), and thus, the Po_2 also falls. At 3000 m (~10,000 ft) above sea level, the alveolar Po_2 is about 60 mm Hg and there is enough hypoxic stimulation of the chemoreceptors under normal breathing to cause increased ventilation. As one ascends higher, the alveolar Po_2 falls less rapidly and the alveolar Pco_2 declines because of the hyperventilation. The resulting fall in arterial Pco_2 produces respiratory alkalosis. A number of compensatory mechanisms operate over a period of time to increase altitude tolerance **(acclimatization),** but in unacclimatized subjects, mental symptoms such as irritability appear at about 3700 m. At 5500 m, the hypoxic symptoms are severe; and at altitudes above 6100 m (20,000 ft), consciousness is usually lost.

HYPOXIC SYMPTOMS AND BREATHING OXYGEN

Some of the effects of high altitude can be offset by breathing 100% O_2. Under these conditions, the total atmospheric pressure becomes the limiting factor in altitude tolerance.

CLINICAL BOX 35–3

Effects of Hypoxia on Cells and Selected Tissues

Effects on Cells

Hypoxia causes the production of transcription factors (hypoxia-inducible factors; HIFs). These are made up of α and β subunits. In normally oxygenated tissues, the α sub-units are rapidly ubiquitinated and destroyed. However, in hypoxic cells, the α subunits dimerize with β subunits, and the dimers activate genes that produce several proteins including angiogenic factors and erythropoietin, among others.

Effects on the Brain

In hypoxymia and the other generalized forms of hypoxia, the brain is affected first. A sudden drop in the inspired Po_2 to less than 20 mm Hg, which occurs, for example, when cabin pressure is suddenly lost in a plane flying above 16,000 m, causes loss of consciousness in 10–20 s and death in 4–5 min. Less severe hypoxia causes a variety of mental aberrations not unlike those produced by alcohol: impaired judgment, drowsiness, dulled pain sensibility, excitement, disorientation, loss of time sense, and headache. Other symptoms include anorexia, nausea, vomiting, tachycardia, and, when the hypoxia is severe, hypertension. The rate of ventilation is increased in proportion to the severity of the hypoxia of the carotid chemoreceptor cells.

Respiratory Stimulation

Dyspnea is by definition difficult or labored breathing in which the subject is conscious of shortness of breath; **hyperpnea** is the general term for an increase in the rate or depth of breathing regardless of the patient's subjective sensations. **Tachypnea** is rapid, shallow breathing. In general, a normal individual is not conscious of respiration until ventilation is doubled, and breathing is not uncomfortable until ventilation is tripled or quadrupled. Whether or not a given level of ventilation is uncomfortable also appears to depend on a variety of other factors. Hypercapnia and, to a lesser extent, hypoxia cause dyspnea. An additional factor is the effort involved in moving the air in and out of the lungs (the work of breathing).

The partial pressure of water vapor in the alveolar air is constant at 47 mm Hg, and that of CO_2 is normally 40 mm Hg, so that the lowest barometric pressure at which a normal alveolar Po_2 of 100 mm Hg is possible is 187 mm Hg, the pressure at about 10,400 m (34,000 ft). At greater altitudes, the increased ventilation due to the decline in alveolar Po_2 lowers the alveolar Pco_2 somewhat, but the maximum alveolar Po_2 that can be attained when breathing 100% O_2 at the ambient barometric pressure of 100 mm Hg at 13,700 m is

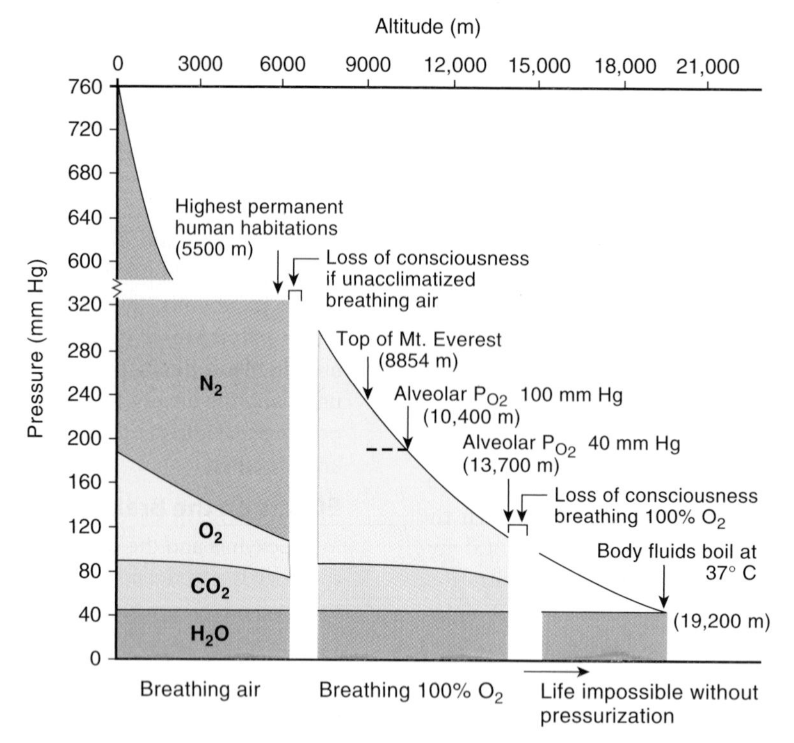

FIGURE 35–10 Composition of alveolar air in individuals breathing air (0–6100 m) and 100% O_2 (6100–13,700 m). The minimal alveolar Po_2 that an unacclimatized subject can tolerate without loss of consciousness is about 35–40 mm Hg. Note that with increasing altitude, the alveolar Pco_2 drops because of the hyperventilation due to hypoxic stimulation of the carotid and aortic chemoreceptors. The fall in barometric pressure with increasing altitude is not linear, because air is compressible.

~40 mm Hg. At ~14,000 m, consciousness is lost in spite of the administration of 100% O_2. At 19,200 m, the barometric pressure is 47 mm Hg, and at or below this pressure the body fluids boil at body temperature. The point is largely academic, however, because any individual exposed to such a low pressure would be dead of hypoxia before the bubbles of body fluids could cause death.

Of course, an artificial atmosphere can be created around an individual; in a pressurized suit or cabin supplied with O_2 and a system to remove CO_2, it is possible to ascend to any altitude and to live in the vacuum of interplanetary space. Some delayed effects of high altitude are discussed in Clinical Box 35–4.

ACCLIMATIZATION

Acclimatization to altitude is due to the operation of a variety of compensatory mechanisms. The respiratory alkalosis produced by the hyperventilation shifts the oxygen–hemoglobin dissociation curve to the left, but a concomitant increase in red blood cell 2,3-DPG tends to decrease the O_2 affinity of hemoglobin. The net effect is a small increase in P_{50}. The decrease in O_2 affinity makes more O_2 available to the tissues. However, the value of the increase in P_{50} is limited because when the arterial Po_2 is markedly reduced, the decreased O_2 affinity also interferes with O_2 uptake by hemoglobin in the lungs.

The initial ventilatory response to increased altitude is relatively small, because the alkalosis tends to counteract the stimulating effect of hypoxia. However, ventilation steadily increases over the next 4 days (Figure 35–11) because the active transport of H^+ into cerebrospinal fluid (CSF), or possibly a developing lactic acidosis in the brain, causes a fall in CSF pH that increases the response to hypoxia. After 4 days, the ventilatory response begins to decline slowly, but it takes years of residence at higher altitudes for it to decline to the initial level, if it is reached at all.

Erythropoietin secretion increases promptly on ascent to high altitude and then falls somewhat over the following 4 days as the ventilatory response increases and the arterial Po_2 rises. The increase in circulating red blood cells triggered by the erythropoietin begins in 2–3 days and is sustained as long as the individual remains at high altitude. Compensatory changes also occur in the tissues. The mitochondria, which are the site of oxidative reactions, increase in number, and myoglobin increases, which facilitates the movement of O_2 into the tissues. The tissue content of cytochrome oxidase also increases.

The effectiveness of the acclimatization process is indicated by the fact that permanent human habitations exist in the Andes and Himalayas at elevations above 5500 m (18,000 ft). The natives who live in these villages are barrel-chested and markedly polycythemic. They have low alveolar Po_2 values, but in most other ways they are remarkably normal.

CLINICAL BOX 35–4

Delayed Effects of High Altitude

When they first arrive at a high altitude, many individuals develop transient "mountain sickness." This syndrome develops 8–24 h after arrival at altitude and lasts 4–8 days. It is characterized by headache, irritability, insomnia, breathlessness, and nausea and vomiting. Its cause is unsettled, but it appears to be associated with cerebral edema. The low P_{O_2} at high altitude causes arteriolar dilation, and if cerebral autoregulation does not compensate, there is an increase in capillary pressure that favors increased transudation of fluid into brain tissue.

Two more serious syndromes that are associated with high-altitude illness: **high-altitude cerebral edema and high-altitude pulmonary edema.** In high-altitude cerebral edema, the capillary leakage in mountain sickness progresses to frank brain swelling, with ataxia, disorientation, and in some cases coma and death due to herniation of the brain through the tentorium. High-altitude pulmonary edema is a patchy edema of the lungs that is related to the marked pulmonary hypertension that develops at high altitude. It has been argued that it occurs because not all pulmonary arteries have enough smooth muscle to constrict in response to hypoxia, and in the capillaries supplied by those arteries, the general rise in pulmonary arterial pressure causes a capillary pressure increase that disrupts their walls (stress failure).

THERAPEUTIC HIGHLIGHTS

All forms of high-altitude illness are benefited by descent to lower altitude and by treatment with the diuretic acetazolamide. This drug inhibits carbonic anhydrase, and results in stimulated respiration, increased $PaCO_2$, and reduced formation of CSF. When cerebral edema is marked, large doses of glucocorticoids are often administered as well. Their mechanism of action is unsettled. In high-altitude pulmonary edema, prompt treatment with O_2 is essential—and, if available, use of a hyperbaric chamber. Portable hyperbaric chambers are now available in a number of mountain areas. Nifedipine, a Ca^{2+} channel blocker that lowers pulmonary artery pressure, can also be useful.

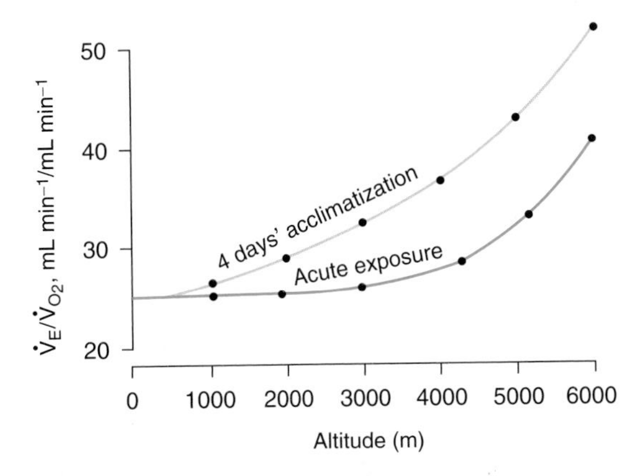

FIGURE 35–11 Effect of acclimatization on the ventilatory response at various altitudes. $\dot{V}_E/\dot{V}_{O_2}$ is the ventilatory equivalent, the ratio of expired minute volume ($\dot{V}_E$) to the O_2 consumption ($\dot{V}_{O_2}$). (Reproduced with permission from Lenfant C, Sullivan K: Adaptation to high altitude. N Engl J Med 1971;284:1298.)

DISEASES CAUSING HYPOXEMIA

Hypoxemia is the most common form of hypoxia seen clinically. The diseases that cause it can be roughly divided into those in which the gas exchange apparatus fails, those such as congenital heart disease in which large amounts of blood are shunted from the venous to the arterial side of the circulation, and those in which the respiratory pump fails. Lung failure occurs when conditions such as pulmonary fibrosis produce alveolar–capillary block, or there is ventilation–perfusion imbalance. Pump failure can be due to fatigue of the respiratory muscles in conditions in which the work of breathing is increased or to a variety of mechanical defects such as pneumothorax or bronchial obstruction that limit ventilation. It can also be caused by abnormalities of the neural mechanisms that control ventilation, such as depression of the respiratory neurons in the medulla by morphine and other drugs. Some specific causes of hypoxemia are discussed in the following text.

VENOUS-TO-ARTERIAL SHUNTS

When a cardiovascular abnormality such as an interatrial septal defect permits large amounts of unoxygenated venous blood to bypass the pulmonary capillaries and dilute the oxygenated blood in the systemic arteries ("right-to-left shunt"), chronic hypoxemia and cyanosis **(cyanotic congenital heart disease)** result. Administration of 100% O_2 raises the O_2 content of alveolar air but has little effect on hypoxia due to venous-to-arterial shunts. This is because the deoxygenated venous blood does not have the opportunity to get to the lung to be oxygenated.

VENTILATION–PERFUSION IMBALANCE

Patchy ventilation–perfusion imbalance is by far the most common cause of hypoxemia in clinical situations. In disease processes that prevent ventilation of some of the alveoli, the ventilation–blood flow ratios in different parts of

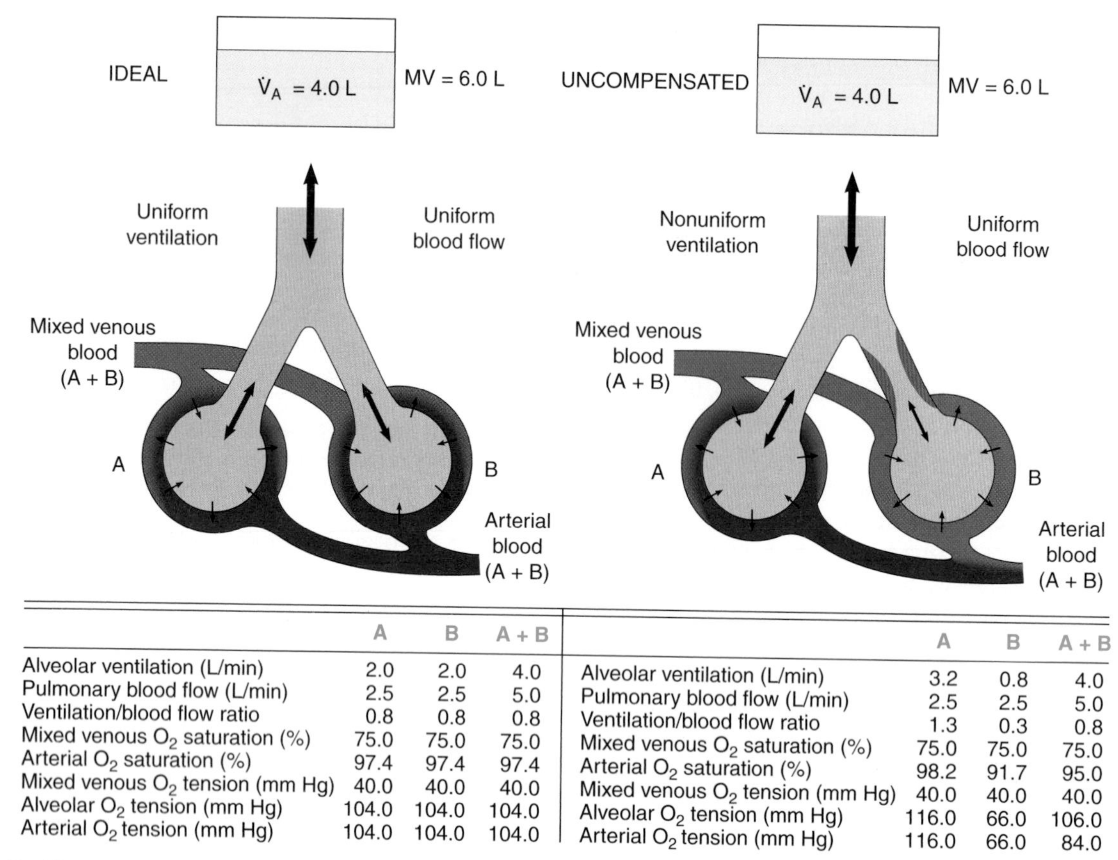

	A	B	A + B
Alveolar ventilation (L/min)	2.0	2.0	4.0
Pulmonary blood flow (L/min)	2.5	2.5	5.0
Ventilation/blood flow ratio	0.8	0.8	0.8
Mixed venous O_2 saturation (%)	75.0	75.0	75.0
Arterial O_2 saturation (%)	97.4	97.4	97.4
Mixed venous O_2 tension (mm Hg)	40.0	40.0	40.0
Alveolar O_2 tension (mm Hg)	104.0	104.0	104.0
Arterial O_2 tension (mm Hg)	104.0	104.0	104.0

	A	B	A + B
Alveolar ventilation (L/min)	3.2	0.8	4.0
Pulmonary blood flow (L/min)	2.5	2.5	5.0
Ventilation/blood flow ratio	1.3	0.3	0.8
Mixed venous O_2 saturation (%)	75.0	75.0	75.0
Arterial O_2 saturation (%)	98.2	91.7	95.0
Mixed venous O_2 tension (mm Hg)	40.0	40.0	40.0
Alveolar O_2 tension (mm Hg)	116.0	66.0	106.0
Arterial O_2 tension (mm Hg)	116.0	66.0	84.0

FIGURE 35–12 Comparison of ventilation/blood flow relationships in health and disease. Left: "Ideal" ventilation/blood flow relationship. Right: Nonuniform ventilation and uniform blood flow, uncompensated. $\dot{V}_A$, alveolar ventilation; MV, respiratory minute volume. See text for details. (Reproduced with permission from Comroe JH Jr, et al: *The Lung: Clinical Physiology and Pulmonary Function Tests,* 2nd ed. Year Book, 1962.)

the lung determine the extent to which systemic arterial Po_2 declines. If nonventilated alveoli are perfused, the nonventilated but perfused portion of the lung is in effect a right-to-left shunt, dumping unoxygenated blood into the left side of the heart. Lesser degrees of ventilation–perfusion imbalance are more common. In the example illustrated in Figure 35–12, the balanced ventilation-perfusion example on the left illustrates a uniform distribution throughout gas exchange. However, when ventilation is not in balance with perfusion, O_2 exchange is compromised. Note that the underventilated alveoli (B) have a low alveolar Po_2, whereas the overventilated alveoli (A) have a high alveolar Po_2 while both have the same blood flow. The unsaturation of the hemoglobin of the blood coming from B is not completely compensated by the slightly greater saturation of the blood coming from A, because hemoglobin is normally nearly saturated in the lungs and the higher alveolar Po_2 adds only a little more O_2 to the hemoglobin than it normally carries. Consequently, the arterial blood is unsaturated. The CO_2 content of the arterial blood is generally normal in such situations, since extra loss of CO_2 in overventilated regions can balance diminished loss in underventilated areas.

OTHER FORMS OF HYPOXIA

ANEMIC HYPOXIA

Hypoxia due to anemia is not severe at rest unless the hemoglobin deficiency is marked, because 2,3-DPG increases in the red blood cells. However, anemic patients may have considerable difficulty during exercise because of a limited ability to increase O_2 delivery to the active tissues (Figure 35–13).

CARBON MONOXIDE POISONING

Small amounts of carbon monoxide (CO) are formed in the body, and this gas may function as a chemical messenger in the brain and elsewhere. In larger amounts, it is poisonous. Outside the body, it is formed by incomplete combustion of carbon. It was used by the Greeks and Romans to execute criminals, and today it causes more deaths than any other gas. CO poisoning has become less common in the United States, since natural gas replaced other gases such as coal gas, which contain large amounts of CO. CO is, however, still readily available, as the exhaust of gasoline engines is 6% or more CO.

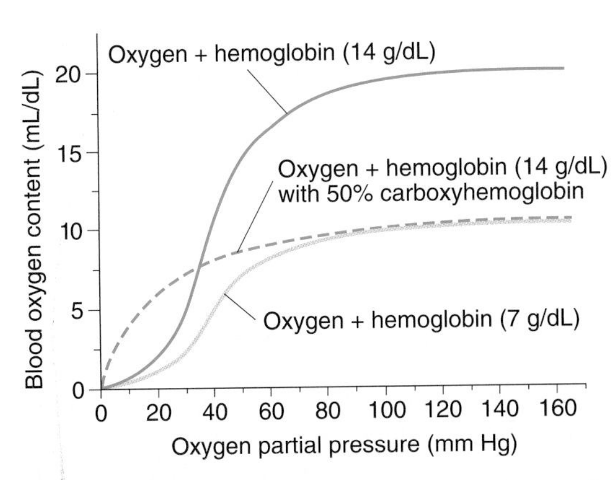

FIGURE 35–13 Effects of anemia and CO on hemoglobin binding of O_2. Normal oxyhemoglobin (14 g/dL hemoglobin) dissociation curve compared with anemia (7 g/dL hemoglobin) and with oxyhemoglobin dissociation curves in CO poisoning (50% carboxy-hemoglobin). Note that the CO-poisoning curve is shifted to the left of the anemia curve. (Reproduced with permission from Leff AR, Schumacker PT: *Respiratory Physiology: Basics and Applications.* Saunders, 1993.)

CO is toxic because it reacts with hemoglobin to form **carbon monoxyhemoglobin (carboxyhemoglobin, COHb),** and COHb does not take up O_2 (Figure 35–13). CO poisoning is often listed as a form of anemic hypoxia because the amount of hemoglobin that can carry O_2 is reduced, but the total hemoglobin content of the blood is unaffected by CO. The affinity of hemoglobin for CO is 210 times its affinity for O_2, and COHb liberates CO very slowly. An additional difficulty is that when COHb is present, the dissociation curve of the remaining HbO_2 shifts to the left, decreasing the amount of O_2 released. This is why an anemic individual who has 50% of the normal amount of HbO_2 may be able to perform moderate work, whereas an individual with HbO_2 reduced to the same level because of the formation of COHb is seriously incapacitated.

Because of the affinity of CO for hemoglobin, progressive COHb formation occurs when the alveolar PCO is greater than 0.4 mm Hg. However, the amount of COHb formed depends on the duration of exposure to CO as well as the concentration of CO in the inspired air and the alveolar ventilation.

CO is also toxic to the cytochromes in the tissues, but the amount of CO required to poison the cytochromes is 1000 times the lethal dose; tissue toxicity thus plays no role in clinical CO poisoning.

The symptoms of CO poisoning are those of any type of hypoxia, especially headache and nausea, but there is little stimulation of respiration, since in the arterial blood, P_{O_2} remains normal and the carotid and aortic chemoreceptors are not stimulated. The cherry-red color of COHb is visible in the skin, nail beds, and mucous membranes. Death results when about 70–80% of the circulating hemoglobin is converted to COHb. The symptoms produced by chronic exposure to sublethal concentrations of CO are those of progressive brain damage, including mental changes and, sometimes, a parkinsonism-like state.

Treatment of CO poisoning consists of immediate termination of the exposure and adequate ventilation, by artificial respiration if necessary. Ventilation with O_2 is preferable to ventilation with fresh air, since O_2 hastens the dissociation of COHb. Hyperbaric oxygenation (see below) is useful in this condition.

ISCHEMIC HYPOXIA

Ischemic hypoxia, or stagnant hypoxia, is due to slow circulation and is a problem in organs such as the kidneys and heart during shock. The liver and possibly the brain are damaged by ischemic hypoxia in congestive heart failure. The blood flow to the lung is normally very large, and it takes prolonged hypotension to produce significant damage. However, acute respiratory distress syndrome (ARDS) can develop when there is prolonged circulatory collapse.

HISTOTOXIC HYPOXIA

Hypoxia due to inhibition of tissue oxidative processes is most commonly the result of cyanide poisoning. Cyanide inhibits cytochrome oxidase and possibly other enzymes. Methylene blue or nitrites are used to treat cyanide poisoning. They act by forming **methemoglobin,** which then reacts with cyanide to form **cyanmethemoglobin,** a nontoxic compound. The extent of treatment with these compounds is, of course, limited by the amount of methemoglobin that can be safely formed. Hyperbaric oxygenation may also be useful.

OXYGEN TREATMENT OF HYPOXIA

Administration of oxygen-rich gas mixtures is of very limited value in hypoperfusion, anemic, and histotoxic hypoxia because all that can be accomplished in this way is an increase in the amount of dissolved O_2 in the arterial blood. This is also true in hypoxemia when it is due to shunting of unoxygenated venous blood past the lungs. In other forms of hypoxemia, O_2 is of great benefit. Treatment regimens that deliver less than 100% O_2 are of value both acutely and chronically, and administration of O_2 24 h/d for 2 years in this fashion has been shown to significantly decrease the mortality of chronic obstructive pulmonary disease. O_2 toxicity and therapy are discussed in Clinical Box 35–5.

HYPERCAPNIA & HYPOCAPNIA

HYPERCAPNIA

Retention of CO_2 in the body **(hypercapnia)** initially stimulates respiration. Retention of larger amounts produces symptoms due to depression of the central nervous system: confusion, diminished sensory acuity, and, eventually, coma

CLINICAL BOX 35–5

Administration of Oxygen & Its Potential Toxicity

It is interesting that while O_2 is necessary for life in aerobic organisms, it is also toxic. Indeed, 100% O_2 has been demonstrated to exert toxic effects not only in animals but also in bacteria, fungi, cultured animal cells, and plants. The toxicity seems to be due to the production of reactive oxygen species including superoxide anion (O_2^-) and H_2O_2. When 80–100% O_2 is administered to humans for periods of 8 h or more, the respiratory passages become irritated, causing substernal distress, nasal congestion, sore throat, and coughing.

Some infants treated with O_2 for respiratory distress syndrome develop a chronic condition characterized by lung cysts and densities **(bronchopulmonary dysplasia).** This syndrome may be a manifestation of O_2 toxicity. Another complication in these infants is **retinopathy of prematurity (retrolental fibroplasia),** the formation of opaque vascular tissue in the eyes, which can lead to serious visual defects. The retinal receptors mature from the center to the periphery of the retina, and they use considerable O_2. This causes the retina to become vascularized in an orderly fashion. Oxygen treatment before maturation is complete provides the needed O_2 to the photoreceptors, and consequently the normal vascular pattern fails to develop. Evidence indicates that this condition can be prevented or ameliorated by treatment with vitamin E, which exerts an anti-oxidant effect, and, in animals, by growth hormone inhibitors.

Administration of 100% O_2 at increased pressure accelerates the onset of O_2 toxicity, with the production not only of tracheobronchial irritation but also of muscle twitching, ringing in the ears, dizziness, convulsions, and coma. The speed with which these symptoms develop is proportional to the pressure at which the O_2 is administered; for example, at 4 atm, symptoms develop in half the subjects in 30 min, whereas at 6 atm, convulsions develop in a few minutes.

On the other hand, exposure to 100% O_2 at 2–3 atm can increase dissolved O_2 in arterial blood to the point that arterial O_2 tension is greater than 2000 mm Hg and tissue O_2 tension is 400 mm Hg. If exposure is limited to 5 h or less at these pressures, O_2 toxicity is not a problem. Therefore, **hyperbaric O_2** therapy in closed tanks is used to treat diseases in which improved oxygenation of tissues cannot be achieved in other ways. It is of demonstrated value in carbon monoxide poisoning, radiation-induced tissue injury, gas gangrene, very severe blood loss anemia, diabetic leg ulcers, and other wounds that are slow to heal, and rescue of skin flaps and grafts in which the circulation is marginal. It is also the primary treatment for decompression sickness and air embolism.

In hypercapnic patients in severe pulmonary failure, the CO_2 level may be so high that it depresses rather than stimulates respiration. Some of these patients keep breathing only because the carotid and aortic chemoreceptors drive the respiratory center. If the hypoxic drive is withdrawn by administering O_2, breathing may stop. During the resultant apnea, the arterial P_{O_2} drops but breathing may not start again, as P_{CO_2} further depresses the respiratory center. Therefore, O_2 therapy in this situation must be started with care.

with respiratory depression and death. In patients with these symptoms, the P_{CO_2} is markedly elevated and severe respiratory acidosis is present. Large amounts of HCO_3^- are excreted, but more HCO_3^- is reabsorbed, raising the plasma HCO_3^- and partially compensating for the acidosis.

CO_2 is so much more soluble than O_2 that hypercapnia is rarely a problem in patients with pulmonary fibrosis. However, it does occur in ventilation–perfusion inequality and when for any reason alveolar ventilation is inadequate in the various forms of pump failure. It is exacerbated when CO_2 production is increased. For example, in febrile patients there is a 13% increase in CO_2 production for each 1°C rise in temperature, and a high carbohydrate intake increases CO_2 production because of the increase in the respiratory quotient. Normally, alveolar ventilation increases and the extra CO_2 is expired, but it accumulates when ventilation is compromised.

HYPOCAPNIA

Hypocapnia is the result of hyperventilation. During voluntary hyperventilation, the arterial P_{CO_2} falls from 40 to as low as 15 mm Hg while the alveolar P_{O_2} rises to 120–140 mm Hg.

The more chronic effects of hypocapnia are seen in neurotic patients who chronically hyperventilate. Cerebral blood flow may be reduced 30% or more because of the direct constrictor effect of hypocapnia on the cerebral vessels. The cerebral ischemia causes light-headedness, dizziness, and paresthesias. Hypocapnia also increases cardiac output. It has a direct constrictor effect on many peripheral vessels, but it depresses the vasomotor center, so that the blood pressure is usually unchanged or only slightly elevated.

Other consequences of hypocapnia are due to the associated respiratory alkalosis, the blood pH being increased to

7.5 or 7.6. The plasma HCO_3^- level is low, but HCO_3^- reabsorption is decreased because of the inhibition of renal acid secretion by the low P_{CO_2}. The plasma total calcium level does not change, but the plasma Ca^{2+} level falls and hypocapnic individuals develop carpopedal spasm, a positive Chvostek sign, and other signs of tetany.

CHAPTER SUMMARY

- Partial pressure differences between air and blood for O_2 and CO_2 dictate a net flow of O_2 into the blood and CO_2 out of the blood in the pulmonary system.
- The amount of O_2 in the blood is determined by the amount dissolved (minor) and the amount bound (major) to hemoglobin. Each hemoglobin molecule contains four subunits that each can bind O_2. Hemoglobin O_2 binding is cooperative and also affected by pH, temperature, and the concentration of 2,3-diphosphoglycerate (2,3-DPG).
- CO_2 in blood is rapidly converted into H_2CO_3 due to the activity of carbonic anhydrase. CO_2 also readily forms carbamino compounds with blood proteins (including hemoglobin). The rapid net loss of CO_2 allows more CO_2 to dissolve in blood.
- The pH of plasma is 7.4. A decrease in plasma pH is termed acidosis and an increase of plasma pH is termed alkalosis. A short-term change in arterial P_{CO_2} due to decreased ventilation results in respiratory acidosis. A short-term change in arterial P_{CO_2} due to increased ventilation results in respiratory alkalosis. Metabolic acidosis occurs when strong acids are added to the blood, and metabolic alkalosis occurs when strong bases are added to (or strong acids are removed from) the blood.
- Respiratory compensation to acidosis or alkalosis involves quick changes in ventilation. Such changes effectively change the P_{CO_2} in the blood plasma. Renal compensation mechanisms are much slower and involve H^+ secretion or HCO_3^- reabsorption.
- Hypoxia is a deficiency of O_2 at the tissue level. Hypoxia has powerful consequences at the cellular, tissue, and organ level: It can alter cellular transcription factors and thus protein expression; it can quickly alter brain function and produce symptoms similar to alcohol (eg, dizziness, impaired mental function, drowsiness, headache); and it can affect ventilation. Long-term hypoxia results in cell and tissue death.

MULTIPLE-CHOICE QUESTIONS

For all questions, select the single best answer unless otherwise directed.

1. Most of the CO_2 transported in the blood is
   A. dissolved in plasma.
   B. in carbamino compounds formed from plasma proteins.
   C. in carbamino compounds formed from hemoglobin.
   D. bound to Cl^-.
   E. in HCO_3^-.
2. Which of the following has the greatest effect on the ability of blood to transport oxygen?
   A. Capacity of the blood to dissolve oxygen
   B. Amount of hemoglobin in the blood
   C. pH of plasma
   D. CO_2 content of red blood cells
   E. Temperature of the blood
3. Which of the following is true of the system?
   $CO_2 + H_2O \overset{1}{\rightleftarrows} H_2CO_3 \overset{2}{\rightleftarrows} H^+ + HCO_3^-$
   A. Reaction 2 is catalyzed by carbonic anhydrase.
   B. Because of reaction 2, the pH of blood declines during hyperventilation.
   C. Reaction 1 occurs in the red blood cell.
   D. Reaction 1 occurs primarily in plasma.
   E. The reactions move to the right when there is excess H^+ in the tissues.
4. In comparing uncompensated respiratory acidosis and uncompensated metabolic acidosis which one of the following is true?
   A. Plasma pH change is always greater in uncompensated respiratory acidosis compared to uncompensated metabolic acidosis.
   B. There are no compensation mechanisms for respiratory acidosis, whereas there is respiratory compensation for metabolic acidosis.
   C. Uncompensated respiratory acidosis involves changes in plasma $[HCO_3^-]$, whereas plasma $[HCO_3^-]$ is unchanged in uncompensated metabolic acidosis.
   D. Uncompensated respiratory acidosis is associated with a change in P_{CO_2}, whereas in uncompensated metabolic acidosis P_{CO_2} is constant.

CHAPTER RESOURCES

Crystal RG, West JB (editors): *The Lung: Scientific Foundations*, 2nd ed. Raven Press, 1997.

Fishman AP, et al (editors): *Fishman's Pulmonary Diseases and Disorders*, 4th ed. McGraw-Hill, 2008.

Hackett PH, Roach RC: High-altitude illness. N Engl J Med 2001;345:107.

Laffey JG, Kavanagh BP: Hypocapnia. N Engl J Med 2002;347:43.

Voelkel NF: High-altitude pulmonary edema. N Engl J Med 2002;346:1607.

West JB: *Pulmonary Pathophysiology*, 7th ed. Wolters Kluwer/ Lippincott Williams & Wilkins, 2008.

West JB: *Respiratory Physiology*, 8th ed. Wolters Kluwer/Lippincott Williams & Wilkins, 2008.

CHAPTER

36

Regulation of Respiration

OBJECTIVES

After studying this chapter, you should be able to:

- Locate the pre-Bötzinger complex and describe its role in producing spontaneous respiration.
- Identify the location and probable functions of the dorsal and ventral groups of respiratory neurons, the pneumotaxic center, and the apneustic center in the brain stem.
- List the specific respiratory functions of the vagus nerves and the respiratory receptors in the carotid body, the aortic body, and the ventral surface of the medulla oblongata.
- Describe and explain the ventilatory responses to increased CO_2 concentrations in the inspired air.
- Describe and explain the ventilatory responses to decreased O_2 concentrations in the inspired air.
- Describe the effects of each of the main nonchemical factors that influence respiration.
- Describe the effects of exercise on ventilation and O_2 exchange in the tissues.
- Define periodic breathing and explain its occurrence in various disease states.

INTRODUCTION

Spontaneous respiration is produced by rhythmic discharge of motor neurons that innervate the respiratory muscles. This discharge is totally dependent on nerve impulses from the brain; breathing stops if the spinal cord is transected above the origin of the phrenic nerves. The rhythmic discharges from the brain that produce spontaneous respiration are regulated by alterations in arterial P_{O_2}, P_{CO_2}, and H^+ concentration, and this chemical control of breathing is supplemented by a number of nonchemical influences. The physiological bases for these phenomena are discussed in this chapter.

NEURAL CONTROL OF BREATHING

CONTROL SYSTEMS

Two separate neural mechanisms regulate respiration. One is responsible for voluntary control and the other for automatic control. The voluntary system is located in the cerebral cortex and sends impulses to the respiratory motor neurons via the corticospinal tracts. The automatic system is driven by a group of pacemaker cells in the medulla. Impulses from these cells activate motor neurons in the cervical and thoracic spinal cord that innervate inspiratory muscles. Those in the cervical cord activate the diaphragm via the phrenic nerves, and those in the thoracic spinal cord activate the external intercostal muscles. However, the impulses also reach the innervation of the internal intercostal muscles and other expiratory muscles.

The motor neurons to the expiratory muscles are inhibited when those supplying the inspiratory muscles are active, and vice versa. Although spinal reflexes contribute to this **reciprocal innervation,** it is due primarily to activity in descending pathways. Impulses in these descending pathways excite agonists and inhibit antagonists. The one exception to the reciprocal inhibition is a small amount of activity in phrenic axons for a short period after inspiration. The function of this postinspiratory output appears to be to brake the lung's elastic recoil and make respiration smooth.

MEDULLARY SYSTEMS

The main components of the **respiratory control pattern generator** responsible for automatic respiration are located in the medulla. Rhythmic respiration is initiated by a small group of synaptically coupled pacemaker cells in the **pre-Bötzinger complex** (pre-BÖTC) on either side of the medulla between the nucleus ambiguus and the lateral reticular nucleus (Figure 36–1). These neurons discharge rhythmically, and they produce rhythmic discharges in phrenic motor neurons that are abolished by sections between the pre-Bötzinger complex and these motor neurons. They also contact the hypoglossal nuclei, and the tongue is involved in the regulation of airway resistance.

Neurons in the pre-Bötzinger complex discharge rhythmically in brain slice preparations in vitro, and if the slices become hypoxic, discharge changes to one associated with gasping. Addition of cadmium to the slices causes occasional sigh-like discharge patterns. There are NK1 receptors and μ-opioid receptors on these neurons, and, in vivo, substance P stimulates and opioids inhibit respiration. Depression of respiration is a side effect that limits the use of opioids in the treatment of pain. However, it is now known that $5HT_4$ receptors are present in the pre-Bötzinger complex and treatment with $5HT_4$ agonists blocks the inhibitory effect of opiates on respiration in experimental animals, without inhibiting their analgesic effect.

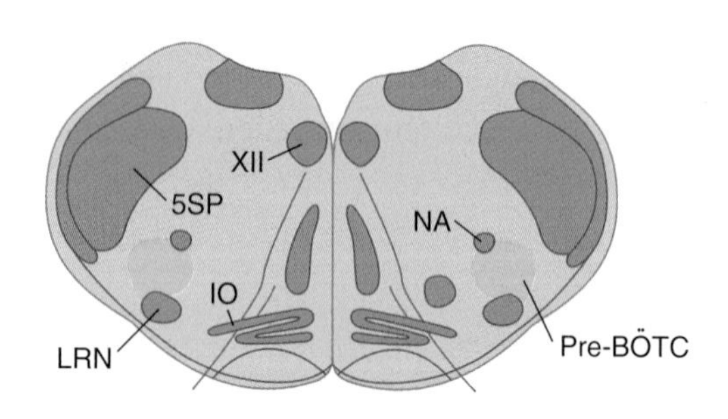

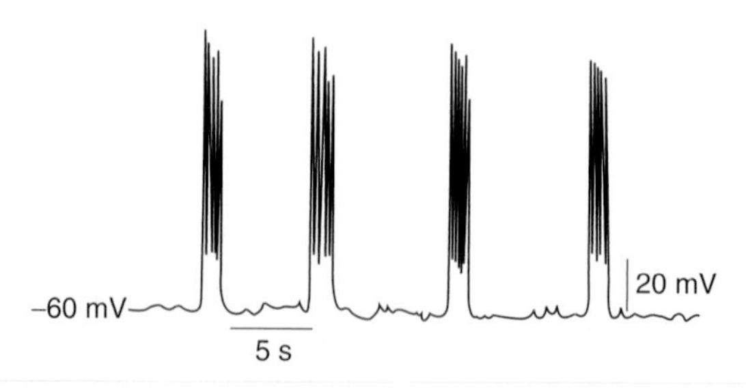

FIGURE 36–1 Pacemaker cells in the pre-Bötzinger complex (pre-BÖTC). Top: Anatomical diagram of the pre-BÖTC from a neonatal rat. **Bottom:** Sample rhythmic discharge tracing of neurons in the pre-BÖTC complex from a brain slice of a neonatal rat. IO, inferior olive; LRN, lateral reticular nucleus; NA, nucleus ambiguus; XII, nucleus of 12th cranial nerve; 5SP, spinal nucleus of trigeminal nerve. (Modified from Feldman JC, Gray PA: Sighs and gasps in a dish. Nat Neurosci 2000;3:531.)

In addition, dorsal and ventral groups of respiratory neurons are present in the medulla (Figure 36–2). However, lesions of these neurons do not abolish respiratory activity, and they apparently project to the pre-Bötzinger pacemaker neurons.

PONTINE & VAGAL INFLUENCES

Although the rhythmic discharge of medullary neurons concerned with respiration is spontaneous, it is modified by neurons in the pons and afferents in the vagus from receptors in the airways and lungs. An area known as the **pneumotaxic center** in the medial parabrachial and Kölliker–Fuse nuclei of the dorsolateral pons contains neurons active during inspiration and neurons active during expiration. When this area is damaged, respiration becomes slower and tidal volume greater, and when the vagi are also cut in anesthetized animals, there are prolonged inspiratory spasms that resemble breath holding (**apneusis;** section B in Figure 36–2). The normal function of the pneumotaxic center is unknown, but it may play a role in switching between inspiration and expiration.

Stretching of the lungs during inspiration initiates impulses in afferent pulmonary vagal fibers. These impulses inhibit inspiratory discharge. This is why the depth of inspiration is increased after vagotomy (Figure 36–2) and apneusis develops if the vagi are cut after damage to the pneumotaxic center. Vagal feedback activity does not alter the rate of rise of the neural activity in respiratory motor neurons (Figure 36–3).

When the activity of the inspiratory neurons is increased in intact animals, the rate and the depth of breathing are increased. The depth of respiration is increased because the lungs are stretched to a greater degree before the amount of vagal and pneumotaxic center inhibitory activity is sufficient to overcome the more intense inspiratory neuron discharge. The respiratory rate is increased because the after-discharge in the vagal and possibly the pneumotaxic afferents to the medulla is rapidly overcome.

REGULATION OF RESPIRATORY ACTIVITY

A rise in the P_{CO_2} or H^+ concentration of arterial blood or a drop in its P_{O_2} increases the level of respiratory neuron activity in the medulla, and changes in the opposite direction have a slight inhibitory effect. The effects of variations in

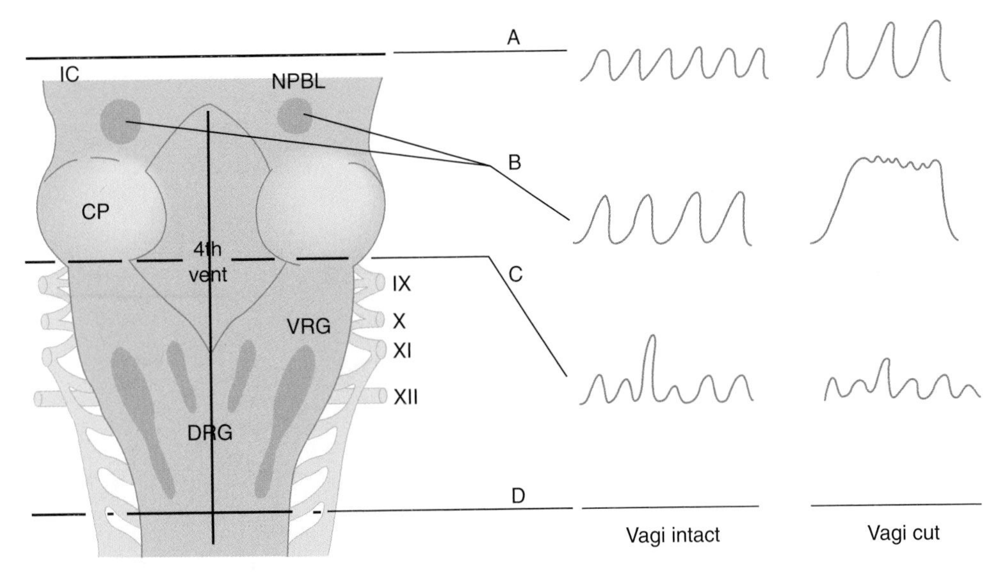

FIGURE 36–2 Respiratory neurons in the brain stem. Dorsal view of brain stem; cerebellum removed. The effects of various lesions and brain stem transections are shown; the spirometer tracings at the right indicate the depth and rate of breathing. If a lesion is introduced at D, breathing ceases. The effects of higher transections, with and without vagus nerves transection, are shown (see text for details). CP, middle cerebellar peduncle; DRG, dorsal group of respiratory neurons; IC, inferior colliculus; NPBL, nucleus parabrachialis (pneumotaxic center); VRG, ventral group of respiratory neurons; 4th vent, fourth ventricle. The roman numerals identify cranial nerves. (Modified from Mitchell RA, Berger A: State of the art: Review of neural regulation of respiration. Am Rev Respir Dis 1975;111:206.)

blood chemistry on ventilation are mediated via respiratory **chemoreceptors**—the carotid and aortic bodies and collections of cells in the medulla and elsewhere that are sensitive to changes in the chemistry of the blood. They initiate impulses that stimulate the respiratory center. Superimposed on this basic **chemical control of respiration,** other afferents provide nonchemical controls that affect breathing in particular situations (Table 36–1).

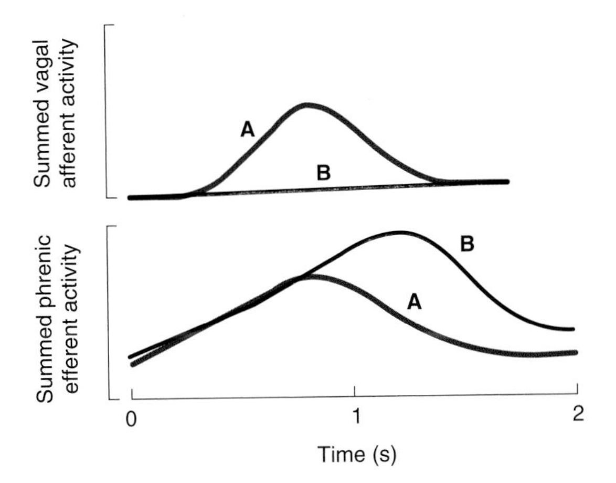

FIGURE 36–3 Afferent vagal fibers inhibit inspiratory discharge. Superimposed records of two breaths: **A)** with and **B)** without feedback vagal afferent activity from stretch receptors in the lungs. Note that the rate of rise in phrenic nerve activity to the diaphragm is unaffected but the discharge is prolonged in the absence of vagal input.

CHEMICAL CONTROL OF BREATHING

The chemical regulatory mechanisms adjust ventilation in such a way that the alveolar P_{CO_2} is normally held constant, the effects of excess H^+ in the blood are combated, and the P_{O_2} is raised when it falls to a potentially dangerous level. The respiratory minute volume is proportional to the metabolic rate, but the link between metabolism and ventilation is CO_2, not O_2. The receptors in the carotid and aortic bodies are stimulated by a rise in the P_{CO_2} or H^+ concentration of arterial blood or a decline in its P_{O_2}. After denervation of the carotid chemoreceptors, the response to a drop in P_{O_2} is abolished; the predominant effect of hypoxia after denervation of the carotid

TABLE 36–1 Stimuli affecting the respiratory center.

Chemical control
CO_2 (via CSF and brain interstitial fluid H^+ concentration)
O_2, H^+ } (via carotid and aortic bodies)
Nonchemical control
Vagal afferents from receptors in the airways and lungs
Afferents from the pons, hypothalamus, and limbic system
Afferents from proprioceptors
Afferents from baroreceptors: arterial, atrial, ventricular, pulmonary

bodies is a direct depression of the respiratory center. The response to changes in arterial blood H^+ concentration in the pH 7.3–7.5 range is also abolished, although larger changes exert some effect. The response to changes in arterial Pco_2, on the other hand, is affected only slightly; it is reduced no more than 30–35%.

CAROTID & AORTIC BODIES

There is a carotid body near the carotid bifurcation on each side, and there are usually two or more aortic bodies near the arch of the aorta (Figure 36–4). Each carotid and aortic body **(glomus)** contains islands of two types of cells, type I and type II cells, surrounded by fenestrated sinusoidal capillaries. The type I or **glomus cells** are closely associated with cuplike endings of the afferent nerves (Figure 36–5). The glomus cells resemble adrenal chromaffin cells and have dense-core granules containing catecholamines that are released upon exposure to hypoxia and cyanide. The cells are excited by hypoxia, and the principal transmitter appears to be dopamine, which excites the nerve endings by way of D_2 receptors. The type II cells are glia-like, and each surrounds four to six type I cells. The function of type II cells is not fully defined.

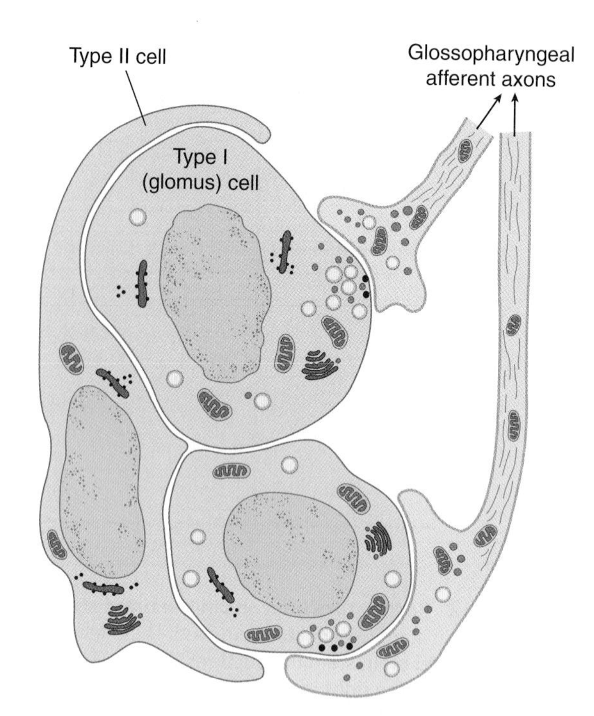

FIGURE 36–5 Organization of the carotid body. Type I (glomus) cells contain catecholamines. When exposed to hypoxia, they release their catecholamines, which stimulate the cuplike endings of the carotid sinus nerve fibers in the glossopharyngeal nerve. The glia-like type II cells surround the type I cells and probably have a sustentacular function.

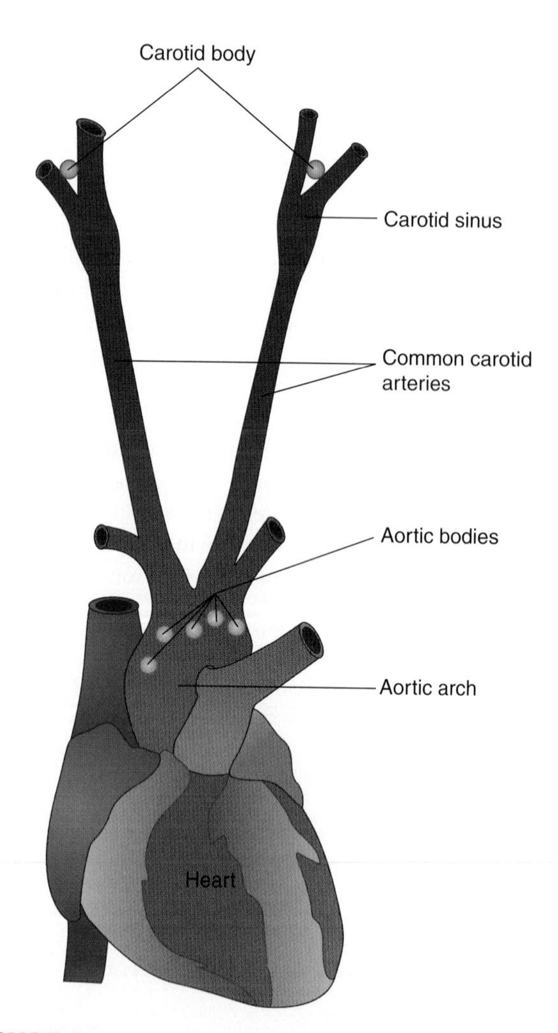

FIGURE 36–4 Location of carotid and aortic bodies. Carotid bodies are positioned near a major arterial baroreceptor, the carotid sinus. Aortic bodies are shown near the aortic arch.

Outside the capsule of each body, the nerve fibers acquire a myelin sheath; however, they are only 2–5 μm in diameter and conduct at the relatively low rate of 7–12 m/s. Afferents from the carotid bodies ascend to the medulla via the carotid sinus and glossopharyngeal nerves, and fibers from the aortic bodies ascend in the vagi. Studies in which one carotid body has been isolated and perfused while recordings are being taken from its afferent nerve fibers show that there is a graded increase in impulse traffic in these afferent fibers as the Po_2 of the perfusing blood is lowered (Figure 36–6) or the Pco_2 is raised.

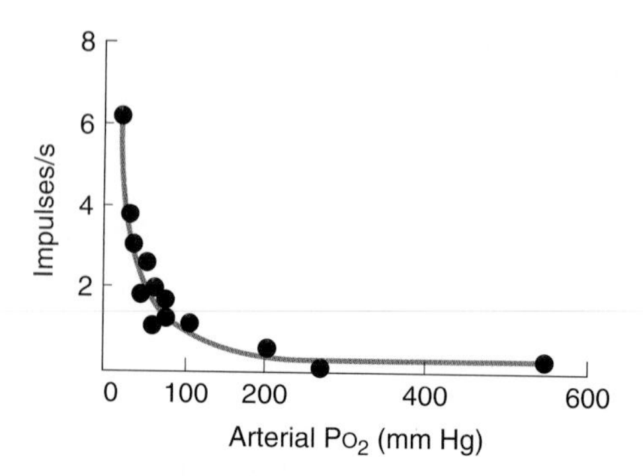

FIGURE 36–6 Effect of Pco_2 on afferent nerve firing. The rate of discharge of a single afferent fiber from the carotid body (circles) is plotted at several Po_2 values and fitted to a line. A sharp increase in firing rate is observed as Po_2 falls below normal resting levels (ie, near 100 mm Hg). (Courtesy of S Sampson.)

Type I glomus cells have O_2-sensitive K^+ channels, whose conductance is reduced in proportion to the degree of hypoxia to which they are exposed. This reduces the K^+ efflux, depolarizing the cell and causing Ca^{2+} influx, primarily via L-type Ca^{2+} channels. The Ca^{2+} influx triggers action potentials and transmitter release, with consequent excitation of the afferent nerve endings. The smooth muscle of pulmonary arteries contains similar O_2-sensitive K^+ channels, which mediate the vasoconstriction caused by hypoxia. This is in contrast to systemic arteries, which contain adenosine triphosphate-(ATP) dependent K^+ channels that permit more K^+ efflux with hypoxia and consequently cause vasodilation instead of vasoconstriction.

The blood flow in each 2 mg carotid body is about 0.04 mL/min, or 2000 mL/100 g of tissue/min compared with a blood flow of 54 mL or 420 mL per 100 g/min in the brain and kidneys, respectively. Because the blood flow per unit of tissue is so enormous, the O_2 needs of the cells can be met largely by dissolved O_2 alone. Therefore, the receptors are not stimulated in conditions such as anemia or carbon monoxide poisoning, in which the amount of dissolved O_2 in the blood reaching the receptors is generally normal, even though the combined O_2 in the blood is markedly decreased. The receptors are stimulated when the arterial P_{O_2} is low or when, because of vascular stasis, the amount of O_2 delivered to the receptors per unit time is decreased. Powerful stimulation is also produced by cyanide, which prevents O_2 utilization at the tissue level. In sufficient doses, nicotine and lobeline activate the chemoreceptors. It has also been reported that infusion of K^+ increases the discharge rate in chemoreceptor afferents, and because the plasma K^+ level is increased during exercise, the increase may contribute to exercise-induced hyperpnea.

Because of their anatomic location, the aortic bodies have not been studied in as great detail as the carotid bodies. Their responses are probably similar but of lesser magnitude. In humans in whom both carotid bodies have been removed but the aortic bodies left intact, the responses are essentially the same as those following denervation of both carotid and aortic bodies in animals: little change in ventilation at rest, but the ventilatory response to hypoxia is lost and the ventilatory response to CO_2 is reduced by 30%.

Neuroepithelial bodies composed of innervated clusters of amine-containing cells are found in the airways. These cells have an outward K^+ current that is reduced by hypoxia, and this would be expected to produce depolarization. However, the function of these hypoxia-sensitive cells is uncertain because, as noted above, removal of the carotid bodies alone abolishes the respiratory response to hypoxia.

CHEMORECEPTORS IN THE BRAIN STEM

The chemoreceptors that mediate the hyperventilation produced by increases in arterial P_{CO_2} after the carotid and aortic bodies are denervated are located in the medulla oblongata and consequently are called **medullary chemoreceptors.** They are separate from the dorsal and ventral respiratory neurons and are located on the ventral surface of the medulla (Figure 36–7). Recent evidence indicates that additional chemoreceptors are located in the vicinity of the solitary tract nuclei, the locus ceruleus, and the hypothalamus.

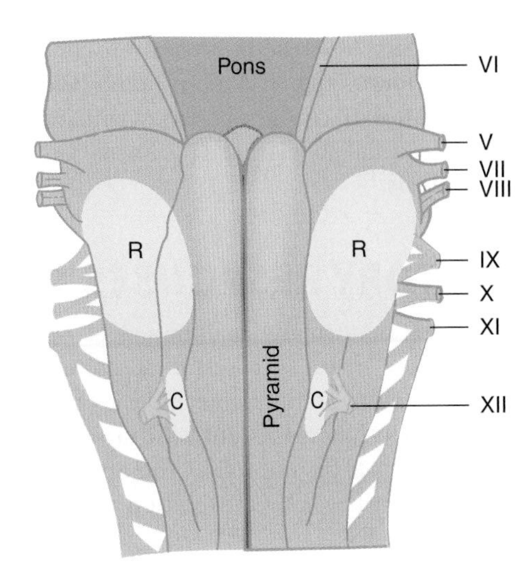

FIGURE 36–7 Rostral (R) and caudal (C) chemosensitive areas on the ventral surface of the medulla. Cranial nerves, pyramid and pons are labeled for reference.

The chemoreceptors monitor the H^+ concentration of cerebrospinal fluid (CSF), including the brain interstitial fluid. CO_2 readily penetrates membranes, including the blood–brain barrier, whereas H^+ and HCO_3^- penetrate slowly. The CO_2 that enters the brain and CSF is promptly hydrated. The H_2CO_3 dissociates, so that the local H^+ concentration rises. The H^+ concentration in brain interstitial fluid parallels the arterial P_{CO_2}. Experimentally produced changes in the P_{CO_2} of CSF have minor, variable effects on respiration as long as the H^+ concentration is held constant, but any increase in spinal fluid H^+ concentration stimulates respiration. The magnitude of the stimulation is proportional to the rise in H^+ concentration. Thus, the effects of CO_2 on respiration are mainly due to its movement into the CSF and brain interstitial fluid, where it increases the H^+ concentration and stimulates receptors sensitive to H^+.

VENTILATORY RESPONSES TO CHANGES IN ACID–BASE BALANCE

In metabolic acidosis due, for example, to the accumulation of acid ketone bodies in the circulation in diabetes mellitus, there is pronounced respiratory stimulation (Kussmaul breathing). The hyperventilation decreases alveolar P_{CO_2} ("blows off CO_2") and thus produces a compensatory fall in blood H^+ concentration. Conversely, in metabolic alkalosis due, for example, to protracted vomiting with loss of HCl from the body, ventilation is depressed and the arterial P_{CO_2} rises, raising the H^+ concentration toward normal. If there is an increase in ventilation that is not secondary to a rise in arterial H^+ concentration, the drop in P_{CO_2} lowers the H^+ concentration below normal **(respiratory alkalosis);** conversely, hypoventilation that is not secondary to a fall in plasma H^+ concentration causes **respiratory acidosis.**

VENTILATORY RESPONSES TO CO_2

The arterial P_{CO_2} is normally maintained at 40 mm Hg. When arterial P_{CO_2} rises as a result of increased tissue metabolism, ventilation is stimulated and the rate of pulmonary excretion of CO_2 increases until the arterial P_{CO_2} falls to normal, shutting off the stimulus. The operation of this feedback mechanism keeps CO_2 excretion and production in balance.

When a gas mixture containing CO_2 is inhaled, the alveolar P_{CO_2} rises, elevating the arterial P_{CO_2} and stimulating ventilation as soon as the blood that contains more CO_2 reaches the medulla. CO_2 elimination is increased, and the alveolar P_{CO_2} drops toward normal. This is why relatively large increments in the P_{CO_2} of inspired air (eg, 15 mm Hg) produce relatively slight increments in alveolar P_{CO_2} (eg, 3 mm Hg). However, the P_{CO_2} does not drop to normal, and a new equilibrium is reached at which the alveolar P_{CO_2} is slightly elevated and the hyperventilation persists as long as CO_2 is inhaled. The essentially linear relationship between respiratory minute volume and the alveolar P_{CO_2} is shown in Figure 36–8.

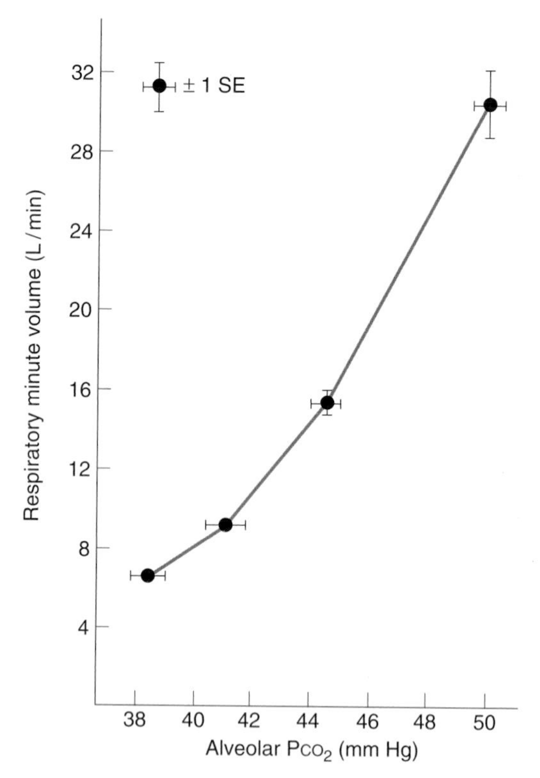

FIGURE 36–8 Responses of normal subjects to inhaling O_2 and approximately 2, 4, and 6% CO_2. The relatively linear increase in respiratory minute volume in response to increased CO_2 is due to an increase in both the depth and rate of respiration. (Reproduced with permission from Lambertsen CJ in: *Medical Physiology,* 13th ed. Mountcastle VB [editor]. Mosby, 1974.)

Of course, this linearity has an upper limit. When the P_{CO_2} of the inspired gas is close to the alveolar P_{CO_2}, elimination of CO_2 becomes difficult. When the CO_2 content of the inspired gas is more than 7%, the alveolar and arterial P_{CO_2} begin to rise abruptly in spite of hyperventilation. The resultant accumulation of CO_2 in the body **(hypercapnia)** depresses the central nervous system, including the respiratory center, and produces headache, confusion, and eventually coma **(CO_2 narcosis).**

VENTILATORY RESPONSE TO OXYGEN DEFICIENCY

When the O_2 content of the inspired air is decreased, respiratory minute volume is increased. The stimulation is slight when the P_{O_2} of the inspired air is more than 60 mm Hg, and marked stimulation of respiration occurs only at lower P_{O_2} values (Figure 36–9). However, any decline in arterial P_{O_2} below 100 mm Hg produces increased discharge in the nerves from the carotid and aortic chemoreceptors. There are two reasons why this increase in impulse traffic does not increase ventilation to any extent in normal individuals until the P_{O_2} is less than 60 mm Hg. First, because Hb is a weaker acid than HbO_2, there is a slight decrease in the H^+ concentration of arterial blood when the arterial P_{O_2} falls and hemoglobin becomes less saturated with O_2. The fall in H^+ concentration tends to inhibit respiration. In addition, any increase in ventilation that does occur lowers the alveolar P_{CO_2}, and this also tends to inhibit respiration. Therefore, the stimulatory effects of hypoxia on ventilation are not clearly manifest until they become strong enough to override the counterbalancing inhibitory effects of a decline in arterial H^+ concentration and P_{CO_2}.

The effects on ventilation of decreasing the alveolar P_{O_2} while holding the alveolar P_{CO_2} constant are shown in Figure 36–10. When the alveolar P_{CO_2} is stabilized at a level 2–3 mm Hg above normal, there is an inverse relationship between ventilation and the alveolar P_{O_2} even in the 90–110 mm Hg range; but when the alveolar P_{CO_2} is fixed at lower than normal values, there is no stimulation of ventilation by hypoxia until the alveolar P_{O_2} falls below 60 mm Hg.

EFFECTS OF HYPOXIA ON THE CO_2 RESPONSE CURVE

When the converse experiment is performed—that is, when the alveolar P_{O_2} is held constant while the response to varying amounts of inspired CO_2 is tested—a linear response is obtained (Figure 36–11). When the CO_2 response is tested at different fixed P_{O_2} values, the slope of the response curve changes, with the slope increased when alveolar P_{O_2} is decreased. In other words, hypoxia makes the individual more sensitive to increases in arterial P_{CO_2}. However, the alveolar P_{CO_2} level at

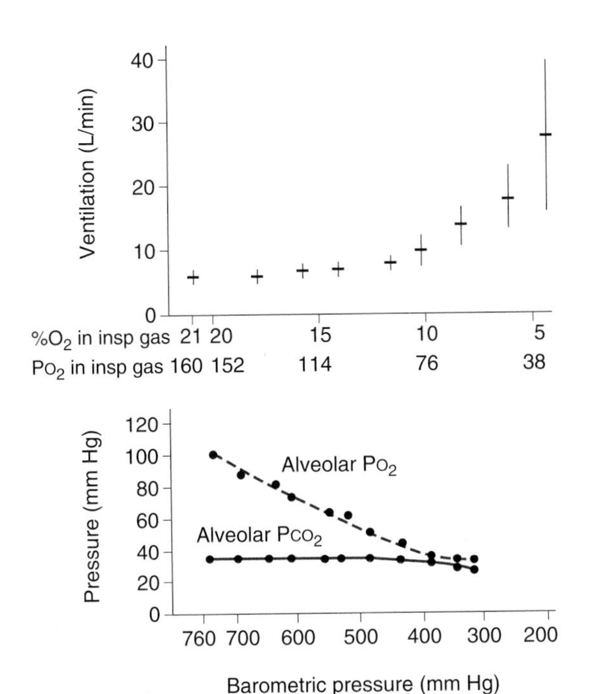

FIGURE 36–9 Top: Average respiratory minute volume during the first half hour of exposure to gases containing various amounts of O_2. Marked changes in ventilation occur at Po_2 values lower than 60 mm Hg. The horizontal line in each case indicates the mean; the vertical bar indicates one standard deviation. **Bottom:** Alveolar Po_2 and Pco_2 values when breathing air at various barometric pressures. The two graphs are aligned so that the Po_2 of the inspired gas mixtures in the upper graph correspond to the Po_2 at the various barometric pressures in the lower graph. (Courtesy of RH Kellogg.)

which the curves in Figure 36–11 intersect is unaffected. In the normal individual, this threshold value is just below the normal alveolar Pco_2, indicating that normally there is a very slight but definite "CO_2 drive" of the respiratory area.

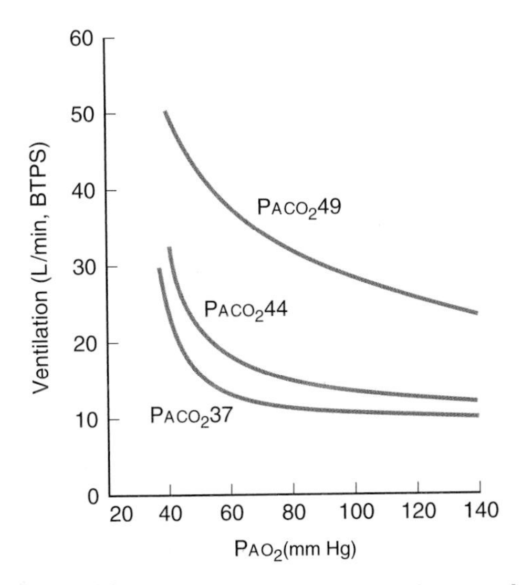

FIGURE 36–10 Ventilation at various alveolar Po_2 values when Pco_2 is held constant at 49, 44, or 37 mm Hg. Note the dramatic effect on the ventilatory response to P_{AO_2} when P_{ACO_2} is increased above normal. (Data from Loeschke HH and Gertz KH).

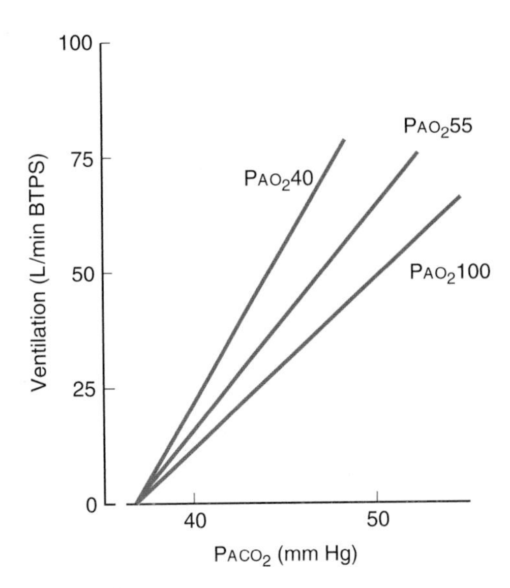

FIGURE 36–11 Fan of lines showing CO_2 response curves at various fixed values of alveolar Po_2. Decreased P_{AO_2} results in a more sensitive response to P_{ACO_2}.

EFFECT OF H^+ ON THE CO_2 RESPONSE

The stimulatory effects of H^+ and CO_2 on respiration appear to be additive and not, like those of CO_2 and O_2, complexly interrelated. In metabolic acidosis, the CO_2 response curves are similar to those in Figure 36–11, except that they are shifted to the left. In other words, the same amount of respiratory stimulation is produced by lower arterial Pco_2 levels. It has been calculated that the CO_2 response curve shifts 0.8 mm Hg to the left for each nanomole rise in arterial H^+. About 40% of the ventilatory response to CO_2 is removed if the increase in arterial H^+ produced by CO_2 is prevented. As noted above, the remaining 60% is probably due to the effect of CO_2 on spinal fluid or brain interstitial fluid H^+ concentration.

BREATH HOLDING

Respiration can be voluntarily inhibited for some time, but eventually the voluntary control is overridden. The point at which breathing can no longer be voluntarily inhibited is called the **breaking point.** Breaking is due to the rise in arterial Pco_2 and the fall in Po_2. Individuals can hold their breath longer after removal of the carotid bodies. Breathing 100% oxygen before breath holding raises alveolar Po_2 initially, so that the breaking point is delayed. The same is true of hyperventilating room air, because CO_2 is blown off and arterial Pco_2 is lower at the start. Reflex or mechanical factors appear to influence the breaking point, since subjects who hold their breath as long as possible and then breathe a gas mixture low in O_2 and high in CO_2 can hold their breath for an additional 20 s or more. Psychological factors also play a role, and subjects can hold their breath longer when they are told their performance is very good than when they are not.

NONCHEMICAL INFLUENCES ON RESPIRATION

RESPONSES MEDIATED BY RECEPTORS IN THE AIRWAYS & LUNGS

Receptors in the airways and lungs are innervated by myelinated and unmyelinated vagal fibers. The unmyelinated fibers are C fibers. The receptors innervated by myelinated fibers are commonly divided into **slowly adapting receptors** and **rapidly adapting receptors** on the basis of whether sustained stimulation leads to prolonged or transient discharge in their afferent nerve fibers (Table 36–2). The other group of receptors presumably consists of the endings of C fibers, and they are divided into pulmonary and bronchial subgroups on the basis of their location.

The shortening of inspiration produced by vagal afferent activity (Figure 36–3) is mediated by slowly adapting receptors, as are the **Hering–Breuer reflexes.** The Hering–Breuer inflation reflex is an increase in the duration of expiration produced by steady lung inflation, and the Hering–Breuer deflation reflex is a decrease in the duration of expiration produced by marked deflation of the lung. Because the rapidly adapting receptors are stimulated by chemicals such as histamine, they have been called **irritant receptors.** Activation of rapidly adapting receptors in the trachea causes coughing, bronchoconstriction, and mucus secretion, and activation of rapidly adapting receptors in the lung may produce hyperpnea.

Because the C fiber endings are close to pulmonary vessels, they have been called J (juxtacapillary) receptors. They are stimulated by hyperinflation of the lung, but they respond as well to intravenous or intracardiac administration of chemicals such as capsaicin. The reflex response that is produced is apnea followed by rapid breathing, bradycardia, and hypotension **(pulmonary chemoreflex).** A similar response is produced by receptors in the heart (**Bezold–Jarisch reflex** or the **coronary chemoreflex**). The physiologic role of this reflex is uncertain, but it probably occurs in pathologic states such as pulmonary congestion or embolization, in which it is produced by endogenously released substances.

COUGHING & SNEEZING

Coughing begins with a deep inspiration followed by forced expiration against a closed glottis. This increases the intrapleural pressure to 100 mm Hg or more. The glottis is then suddenly opened, producing an explosive outflow of air at velocities up to 965 km (600 mi) per hour. Sneezing is a similar expiratory effort with a continuously open glottis. These reflexes help expel irritants and keep airways clear. Other aspects of innervation are considered in a special case (Clinical Box 36–1).

AFFERENTS FROM PROPRIOCEPTORS

Carefully controlled experiments have shown that active and passive movements of joints stimulate respiration, presumably because impulses in afferent pathways from proprioceptors in muscles, tendons, and joints stimulate the inspiratory neurons. This effect probably helps increase ventilation during exercise. Other afferents are considered in Clinical Box 36–2.

TABLE 36–2 Airway and lung receptors.

Vagal Innervation	Type	Location in Interstitium	Stimulus	Response
Myelinated	Slowly adapting	Among airway smooth muscle cells (?)	Lung inflation	Inspiratory time shortening Hering–Breuer inflation and deflation reflexes Bronchodilation Tachycardia Hyperpnea
	Rapidly adapting	Among airway epithelial cells	Lung hyperinflation Exogenous and endogenous substances (eg, histamine, prostaglandins)	Cough Bronchoconstriction Mucus secretion
Unmyelinated C fibers	Pulmonary C fibers Bronchial C fibers	Close to blood vessels	Lung hyperinflation Exogenous and endogenous substances (eg, capsaicin, bradykinin, serotonin)	Apnea followed by rapid breathing Bronchoconstriction Bradycardia Hypotension Mucus secretion

Modified and reproduced with permission from Berger AJ, Hornbein TF: Control of respiration. In: *Textbook of Physiology,* 21st ed, Vol 2. Patton HD, et al (editors). Saunders, 1989.

CLINICAL BOX 36–1

Lung Innervation & Patients with Heart–Lung Transplants

Transplantation of the heart and lungs is now an established treatment for severe pulmonary disease and other conditions. In individuals with transplants, the recipient's right atrium is sutured to the donor heart, and the donor heart does not reinnervate, so the resting heart rate is elevated. The donor trachea is sutured to the recipient's just above the carina, and afferent fibers from the lungs do not regrow. Consequently, healthy patients with heart–lung transplants provide an opportunity to evaluate the role of lung innervation in normal physiology. Their cough responses to stimulation of the trachea are normal because the trachea remains innervated, but their cough responses to stimulation of the smaller airways are absent. Their bronchi tend to be dilated to a greater degree than normal. In addition, they have the normal number of yawns and sighs, indicating that these do not depend on innervation of the lungs. Finally, they lack Hering–Breuer reflexes, but their pattern of breathing at rest is normal, indicating that these reflexes do not play an important role in the regulation of resting respiration in humans.

CLINICAL BOX 36–2

Afferents from "Higher Centers"

Pain and emotional stimuli affect respiration, suggesting that afferents from the limbic system and hypothalamus signal to the respiratory neurons in the brain stem. In addition, even though breathing is not usually a conscious event, both inspiration and expiration are under voluntary control. The pathways for voluntary control pass from the neo-cortex to the motor neurons innervating the respiratory muscles, bypassing the medullary neurons.

Because voluntary and automatic control of respiration are separate, automatic control is sometimes disrupted without loss of voluntary control. The clinical condition that results has been called **Ondine's curse.** In German legend, Ondine was a water nymph who had an unfaithful mortal lover. The king of the water nymphs punished the lover by casting a curse on him that took away all his automatic functions. In this state, he could stay alive only by staying awake and remembering to breathe. He eventually fell asleep from sheer exhaustion, and his respiration stopped. Patients with this intriguing condition generally have bulbar poliomyelitis or disease processes that compress the medulla.

RESPIRATORY COMPONENTS OF VISCERAL REFLEXES

Inhibition of respiration and closure of the glottis during vomiting, swallowing, and sneezing not only prevent the aspiration of food or vomitus into the trachea but, in the case of vomiting, fix the chest so that contraction of the abdominal muscles increases the intra-abdominal pressure. Similar glottic closure and inhibition of respiration occur during voluntary and involuntary straining.

Hiccup is a spasmodic contraction of the diaphragm and other inspiratory muscles that produces an inspiration during which the glottis suddenly closes. The glottic closure is responsible for the characteristic sensation and sound. Hiccups occur in the fetus in utero as well as throughout extrauterine life. Their function is unknown. Most attacks of hiccups are usually of short duration, and they often respond to breath holding or other measures that increase arterial P_{CO_2}. Intractable hiccups, which can be debilitating, sometimes respond to dopamine antagonists and perhaps to some centrally acting analgesic compounds.

Yawning is a peculiar "infectious" respiratory act whose physiologic basis and significance are uncertain. Like hiccuping, it occurs in utero, and it occurs in fish and tortoises as well as mammals. The view that it is needed to increase O_2 intake has been discredited. Underventilated alveoli have a tendency to collapse, and it has been suggested that the deep inspiration and stretching them open prevents the development of atelectasis. However, in actual experiments, no atelectasis-preventing effect of yawning could be demonstrated. Yawning increases venous return to the heart, which may benefit the circulation. It has been suggested that yawning is a nonverbal signal used for communication between monkeys in a group, and one could argue that on a different level, the same thing is true in humans.

RESPIRATORY EFFECTS OF BARORECEPTOR STIMULATION

Afferent fibers from the baroreceptors in the carotid sinuses, aortic arch, atria, and ventricles relay to the respiratory neurons, as well as the vasomotor and cardioinhibitory neurons in the medulla. Impulses in them inhibit respiration, but the inhibitory effect is slight and of little physiologic importance. The hyperventilation in shock is due to chemoreceptor stimulation caused by acidosis and hypoxia secondary to local stagnation of blood flow, and is not baroreceptor-mediated. The activity of inspiratory neurons affects blood pressure and heart rate, and activity in the vasomotor and cardiac areas in the medulla may have minor effects on respiration.

EFFECTS OF SLEEP

Respiration is less rigorously controlled during sleep than in the waking state, and brief periods of apnea occur in normal sleeping adults. Changes in the ventilatory response to hypoxia vary. If the Pco_2 falls during the waking state, various stimuli from proprioceptors and the environment maintain respiration, but during sleep, these stimuli are decreased and a decrease in Pco_2 can cause apnea. During rapid eye movement (REM) sleep, breathing is irregular and the CO_2 response is highly variable.

RESPIRATORY ABNORMALITIES

ASPHYXIA

In asphyxia produced by occlusion of the airway, acute hypercapnia and hypoxia develop together. Stimulation of respiration is pronounced, with violent respiratory efforts. Blood pressure and heart rate rise sharply, catecholamine secretion is increased, and blood pH drops. Eventually the respiratory efforts cease, the blood pressure falls, and the heart slows. Asphyxiated animals can still be revived at this point by artificial respiration, although they are prone to ventricular fibrillation, probably because of the combination of hypoxic myocardial damage and high circulating catecholamine levels. If artificial respiration is not started, cardiac arrest occurs in 4–5 min.

DROWNING

Drowning is asphyxia caused by immersion, usually in water. In about 10% of drownings, the first gasp of water after the losing struggle not to breathe triggers laryngospasm, and death results from asphyxia without any water in the lungs. In the remaining cases, the glottic muscles eventually relax and fluid enters the lungs. Fresh water is rapidly absorbed, diluting the plasma and causing intravascular hemolysis. Ocean water is markedly hypertonic and draws fluid from the vascular system into the lungs, decreasing plasma volume. The immediate goal in the treatment of drowning is, of course, resuscitation, but long-term treatment must also take into account the circulatory effects of the water in the lungs.

PERIODIC BREATHING

The acute effects of voluntary hyperventilation demonstrate the interaction of the chemical mechanisms regulating respiration. When a normal individual hyperventilates for 2–3 min, then stops and permits respiration to continue without exerting any voluntary control over it, a period of apnea occurs. This is followed by a few shallow breaths and then by another period of apnea, followed again by a few breaths **(periodic breathing).** The cycles may last for some time before normal breathing is resumed (Figure 36–12). The apnea apparently is due to a lack of CO_2 because it does not occur following hyperventilation with gas mixtures containing 5% CO_2. During the apnea, the alveolar Po_2 falls and the Pco_2 rises. Breathing resumes because of hypoxic stimulation of the carotid and aortic chemoreceptors before the CO_2 level has returned to normal. A few breaths eliminate the hypoxic stimulus, and breathing stops until the alveolar Po_2 falls again. Gradually, however, the Pco_2 returns to normal, and normal breathing resumes. Changes in breathing patterns can be symptomatic of disease (Clinical Box 36–3).

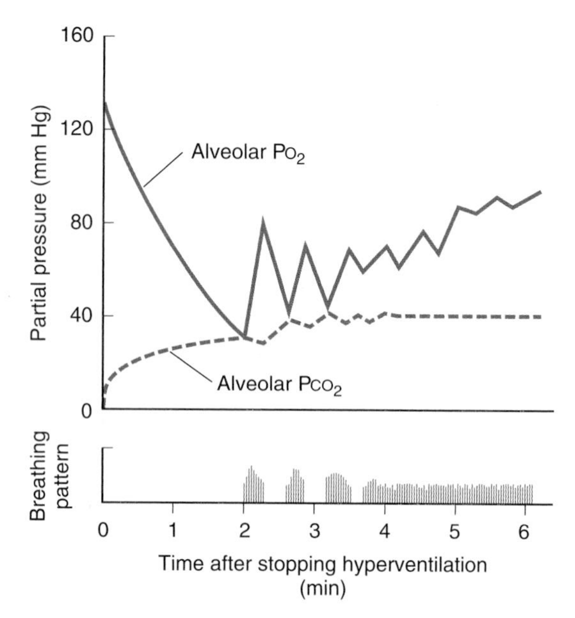

FIGURE 36–12 Changes in breathing and composition of alveolar air after forced hyperventilation for 2 min. Bars in bottom indicate breathing, whereas blank spaces are indicative of apnea.

EFFECTS OF EXERCISE

Exercise provides a physiological example to explore many of the control systems discussed above. Of course, many cardiovascular and respiratory mechanisms must operate in an integrated fashion if the O_2 needs of the active tissue are to be met and the extra CO_2 and heat removed from the body during exercise. Circulatory changes increase muscle blood flow while maintaining adequate circulation in the rest of the body. In addition, there is an increase in the extraction of O_2 from the blood in exercising muscles and an increase in ventilation. This provides extra O_2, eliminates some of the heat, and excretes extra CO_2. A focus on regulation of ventilation and tissue O_2 is presented below, as many other aspects of regulation have been presented in previous chapters.

CHANGES IN VENTILATION

During exercise, the amount of O_2 entering the blood in the lungs is increased because the amount of O_2 added to each unit of blood and the pulmonary blood flow per minute are increased. The Po_2 of blood flowing into the pulmonary

CLINICAL BOX 36–3

Periodic Breathing in Disease

Cheyne–Stokes Respiration

Periodic breathing occurs in various disease states and is often called Cheyne–Stokes respiration. It is seen most commonly in patients with congestive heart failure and uremia, but it occurs also in patients with brain disease and during sleep in some normal individuals. Some of the patients with Cheyne–Stokes respiration have increased sensitivity to CO_2. The increased response is apparently due to disruption of neural pathways that normally inhibit respiration. In these individuals, CO_2 causes relative hyperventilation, lowering the arterial P_{CO_2}. During the resultant apnea, the arterial P_{CO_2} again rises to normal, but the respiratory mechanism again overresponds to CO_2. Breathing ceases, and the cycle repeats.

Another cause of periodic breathing in patients with cardiac disease is prolongation of the lung-to-brain circulation time, so that it takes longer for changes in arterial gas tensions to affect the respiratory area in the medulla. When individuals with a slower circulation hyperventilate, they lower the P_{CO_2} of the blood in their lungs, but it takes longer than normal for the blood with a low P_{CO_2} to reach the brain. During this time, the P_{CO_2} in the pulmonary capillary blood continues to be lowered, and when this blood reaches the brain, the low P_{CO_2} inhibits the respiratory area, producing apnea. In other words, the respiratory control system oscillates because the negative feedback loop from lungs to brain is abnormally long.

Sleep Apnea

Episodes of apnea during sleep can be central in origin (ie, due to failure of discharge in the nerves producing respiration) or they can be due to airway obstruction **(obstructive sleep apnea).** Apnea can occur at any age and is produced when the pharyngeal muscles relax during sleep. In some cases, failure of the genioglossus muscles to contract during inspiration contributes to the blockage. The genioglossus muscles pull the tongue forward, and without (or with weak) contraction the tongue can obstruct the airway. After several increasingly strong respiratory efforts, the patient wakes up, takes a few normal breaths, and falls back to sleep. Apneic episodes are most common during REM sleep, when the muscles are most hypotonic. The symptoms are loud snoring, morning headaches, fatigue, and daytime sleepiness. When severe and prolonged, the condition can lead to hypertension and its complications. Frequent apneas can lead to numerous brief awakenings during sleep and to sleepiness during waking hours. With this in mind, it is not surprising to find that the incidence of motor vehicle accidents in sleep apnea patients is seven times greater than it is in the general driving population.

THERAPEUTIC HIGHLIGHTS

Treatment of sleep apnea is dependent on the patient and on the cause (if known). Treatments range from mild to moderate interventions to surgery. Interventions including positional therapy, dental appliances that rearrange the architecture of the airway, avoidance of muscle relaxants (eg, alcohol) or drugs that reduce respiratory drive, or continuous positive airway pressure. Because sleep apnea is increased in overweight or obese individuals, weight loss can also be effective.

capillaries falls from 40 to 25 mm Hg or less, so that the alveolar–capillary P_{O_2} gradient is increased and more O_2 enters the blood. Blood flow per minute is increased from 5.5 L/min to as much as 20–35 L/min. The total amount of O_2 entering the blood therefore increases from 250 mL/min at rest to values as high as 4000 mL/min. The amount of CO_2 removed from each unit of blood is increased, and CO_2 excretion increases from 200 mL/min to as much as 8000 mL/min. The increase in O_2 uptake is proportional to work load, up to a maximum. Above this maximum, O_2 consumption levels off and the blood lactate level continues to rise (Figure 36–13). The lactate comes from muscles in which aerobic resynthesis of energy stores cannot keep pace with their utilization, and an **oxygen debt** is being incurred.

Ventilation increases abruptly with the onset of exercise, which is followed after a brief pause by a further, more gradual increase (Figure 36–14). With moderate exercise, the increase is due mostly to an increase in the depth of respiration; this is accompanied by an increase in the respiratory rate when the exercise is more strenuous. Ventilation abruptly decreases when exercise ceases, which is followed after a brief pause by a more gradual decline to preexercise values. The abrupt increase at the start of exercise is presumably due to psychic stimuli and afferent impulses from proprioceptors in muscles, tendons, and joints. The more gradual increase is presumably humoral, even though arterial pH, P_{CO_2}, and P_{O_2} remain constant during moderate exercise. The increase in ventilation is proportional to the increase in O_2 consumption, but the mechanisms responsible for the stimulation of respiration are still the subject of much debate. The increase in body temperature may play a role. Exercise increases the plasma K^+ level, and this increase may stimulate the peripheral chemoreceptors. In addition, it may be that the sensitivity of the neurons controlling the response to CO_2 is increased or that the respiratory fluctuations in arterial P_{CO_2} increase so that, even though the

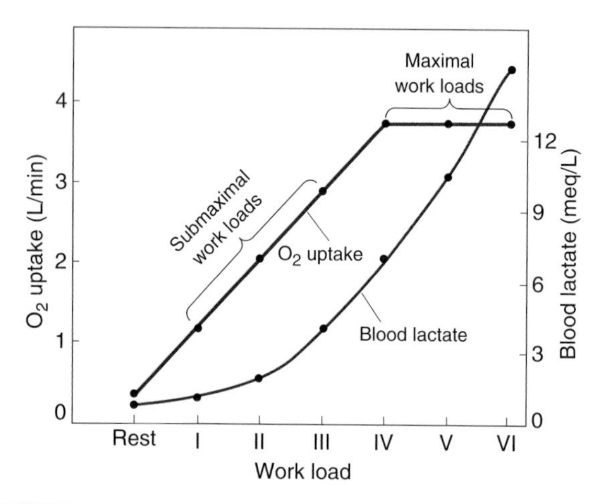

FIGURE 36–13 Relation between work load, blood lactate level, and O_2 uptake. I–VI, increasing work loads produced by increasing the speed and grade of a treadmill on which the subjects worked. (Reproduced with permission from Mitchell JH, Blomqvist G: Maximal oxygen uptake. N Engl J Med 1971;284:1018.)

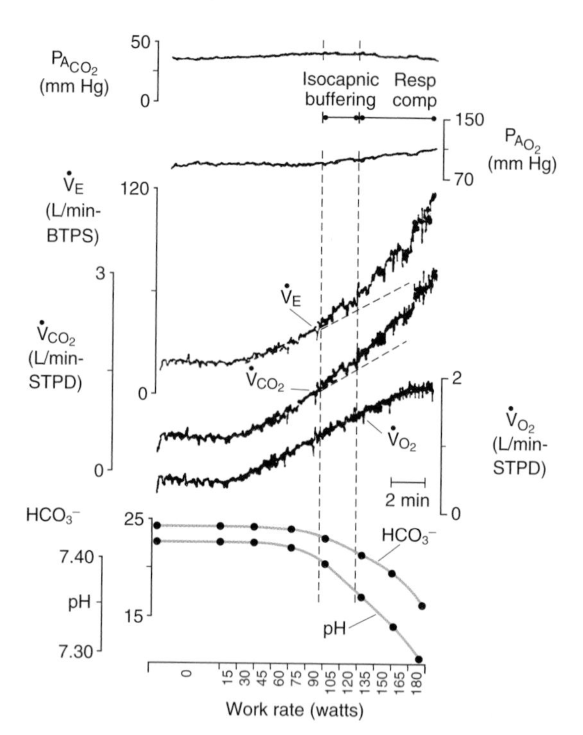

FIGURE 36–15 Physiologic responses to work rate during exercise. Changes in alveolar P_{CO_2}, alveolar Po, ventilation ($\dot{V}_E$) consumption production ($\dot{V}_{CO_2}$) consumption ($\dot{V}_{O_2}$) arterial HCO_3^-, and arterial pH with graded increases in work by an adult male on a bicycle ergometer. Resp comp, respiratory compensation. STPD, standard temperature (0°C) and pressure (760 mm Hg), dry. Dashed lines emphasize deviation from linear response. See text for additional details. (Reproduced with permission from Wasserman K: Breathing during exercise. NEJM 1978 Apr 6;298(14):780–785.)

mean arterial P_{CO_2} does not rise, it is CO_2 that is responsible for the increase in ventilation. O_2 also seems to play some role, despite the lack of a decrease in arterial P_{O_2}, since during the performance of a given amount of work, the increase in ventilation while breathing 100% O_2 is 10–20% less than the increase while breathing air. Thus, it currently appears that a number of different factors combine to produce the increase in ventilation seen during moderate exercise.

When exercise becomes more vigorous, buffering of the increased amounts of lactic acid that are produced liberates more CO_2, and this further increases ventilation. The response to graded exercise is shown in Figure 36–15. With increased production of acid, the increases in ventilation and CO_2 production remain proportional, so alveolar and arterial CO_2 change relatively little **(isocapnic buffering).** Because of the hyperventilation, alveolar P_{O_2} increases. With further accumulation of lactic acid, the increase in ventilation outstrips CO_2 production and alveolar P_{CO_2} falls, as does arterial P_{CO_2}. The decline in arterial P_{CO_2} provides respiratory compensation for the metabolic acidosis produced by the additional lactic acid. The additional increase in ventilation produced by the acidosis is dependent on the carotid bodies and does not occur if they are removed.

The respiratory rate after exercise does not reach basal levels until the O_2 debt is repaid. This may take as long as 90 min. The stimulus to ventilation after exercise is not the arterial P_{CO_2}, which is normal or low, or the arterial P_{O_2}, which is normal or high, but the elevated arterial H^+ concentration due to the lactic acidemia. The magnitude of the O_2 debt is the amount by which O_2 consumption exceeds basal consumption from the end of exertion until the O_2 consumption has returned to preexercise basal levels. During repayment of the O_2 debt, the O_2 concentration in muscle myoglobin rises slightly. ATP and phosphorylcreatine are resynthesized, and lactic acid is removed. Eighty per cent of the lactic acid is converted to glycogen and 20% is metabolized to CO_2 and H_2O.

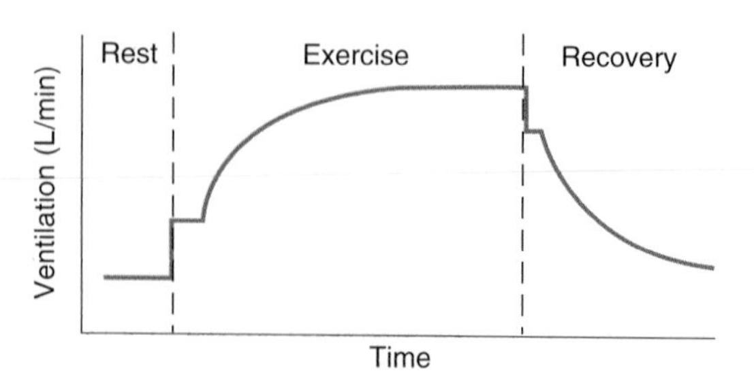

FIGURE 36–14 Diagrammatic representation of changes in ventilation during exercise. See text for details.

CHANGES IN THE TISSUES

Maximum O_2 uptake during exercise is limited by the maximum rate at which O_2 is transported to the mitochondria in the exercising muscle. However, this limitation is not normally due to deficient O_2 uptake in the lungs, and hemoglobin in arterial blood is saturated even during the most severe exercise.

During exercise, the contracting muscles use more O_2, and the tissue Po_2 and the Po_2 in venous blood from exercising muscle fall nearly to zero. More O_2 diffuses from the blood, the blood Po_2 of the blood in the muscles drops, and more O_2 is removed from hemoglobin. Because the capillary bed of contracting muscle is dilated and many previously closed capillaries are open, the mean distance from the blood to the tissue cells is greatly decreased; this facilitates the movement of O_2 from blood to cells. The oxygen–hemoglobin dissociation curve is steep in the Po_2 range below 60 mm Hg, and a relatively large amount of O_2 is supplied for each drop of 1 mm Hg in Po_2 (see Figure 35–2). Additional O_2 is supplied because, as a result of the accumulation of CO_2 and the rise in temperature in active tissues—and perhaps because of a rise in red blood cell 2,3-diphosphoglycerate (2,3-DPG)—the dissociation curve shifts to the right. The net effect is a threefold increase in O_2 extraction from each unit of blood (see Figure 35–3). Because this increase is accompanied by a 30-fold or greater increase in blood flow, it permits the metabolic rate of muscle to rise as much as 100-fold during exercise.

EXERCISE TOLERANCE & FATIGUE

What determines the maximum amount of exercise that can be performed by an individual? Obviously, exercise tolerance has a time as well as an intensity dimension. For example, a fit young man can produce a power output on a bicycle of about 700 W for 1 min, 300 W for 5 min, and 200 W for 40 min. It used to be argued that the limiting factors in exercise performance were the rate at which O_2 could be delivered to the tissues or the rate at which O_2 could enter the body in the lungs. These factors play a role, but it is clear that other factors also contribute and that exercise stops when the sensation of **fatigue** progresses to the sensation of exhaustion. Fatigue is produced in part by bombardment of the brain by neural impulses from muscles, and the decline in blood pH produced by lactic acidosis also makes one feel tired, as does the rise in body temperature, dyspnea, and, perhaps, the uncomfortable sensations produced by activation of the J receptors in the lungs.

CHAPTER SUMMARY

- Breathing is under both voluntary control (located in the cerebral cortex) and automatic control (driven by pacemaker cells in the medulla). There is a reciprocal innervation to expiratory and inspiratory muscles in that motor neurons supplying expiratory muscles are inactive when motor neurons supplying inspiratory muscles are active, and vice versa.
- The pre-Bötzinger complex on either side of the medulla contains synaptically coupled pacemaker cells that allow for rhythmic generation of breathing. The spontaneous activity of these neurons can be altered by neurons in the pneumotaxic center, although the full regulatory function of these neurons on normal breathing is not understood.
- Breathing patterns are sensitive to chemicals in the blood through activation of respiratory chemoreceptors. There are chemoreceptors in the carotid and aortic bodies and in collections of cells in the medulla. These chemoreceptors respond to changes in Po_2 and Pco_2 as well as H^+ to regulate breathing.
- Receptors in the airway are additionally innervated by slowly adapting and rapidly adapting myelinated vagal fibers. Slowly adapting receptors can be activated by lung inflation. Rapidly adapting receptors, or irritant receptors, can be activated by chemicals such as histamine and result in cough or even hyperpnea.
- Receptors in the airway are also innervated by unmyelinated vagal fibers (C fibers) that are typically found next to pulmonary vessels. They are stimulated by hyperinflation (or exogenous substances including capsaicin) and lead to the pulmonary chemoreflex. The physiologic role for this response is not fully understood.

MULTIPLE-CHOICE QUESTIONS

For all questions, select the single best answer unless otherwise directed.

1. The main respiratory control neurons
   A. send out regular bursts of impulses to expiratory muscles during quiet respiration.
   B. are unaffected by stimulation of pain receptors.
   C. are located in the pons.
   D. send out regular bursts of impulses to inspiratory muscles during quiet respiration.
   E. are unaffected by impulses from the cerebral cortex.
2. Intravenous lactic acid increases ventilation. The receptors responsible for this effect are located in the
   A. medulla oblongata.
   B. carotid bodies.
   C. lung parenchyma.
   D. aortic baroreceptors.
   E. trachea and large bronchi.
3. Spontaneous respiration ceases after
   A. transection of the brain stem above the pons.
   B. transection of the brain stem at the caudal end of the medulla.
   C. bilateral vagotomy.
   D. bilateral vagotomy combined with transection of the brain stem at the superior border of the pons.
   E. transection of the spinal cord at the level of the first thoracic segment.
4. The following physiologic events that occur in vivo are listed in random order: **(1)** decreased CSF pH; **(2)** increased arterial Pco_2; **(3)** increased CSF Pco_2; **(4)** stimulation of medullary chemoreceptors; **(5)** increased alveolar Pco_2.

   What is the usual sequence in which they occur when they affect respiration?
   A. 1, 2, 3, 4, 5
   B. 4, 1, 3, 2, 5
   C. 3, 4, 5, 1, 2
   D. 5, 2, 3, 1, 4
   E. 5, 3, 2, 4, 1

5. The following events that occur in the carotid bodies when they are exposed to hypoxia are listed in random order: **(1)** depolarization of type I glomus cells; **(2)** excitation of afferent nerve endings; **(3)** reduced conductance of hypoxia-sensitive K^+ channels in type I glomus cells; **(4)** Ca^{2+} entry into type I glomus cells; **(5)** decreased K^+ efflux.

   What is the usual sequence in which they occur on exposure to hypoxia?

   A. 1, 3, 4, 5, 2
   B. 1, 4, 2, 5, 3
   C. 3, 4, 5, 1, 2
   D. 3, 1, 4, 5, 2
   E. 3, 5, 1, 4, 2

6. Injection of a drug that stimulates the carotid bodies would be expected to cause

   A. a decrease in the pH of arterial blood.
   B. a decrease in the P_{CO_2} of arterial blood.
   C. an increase in the HCO_3^- concentration of arterial blood.
   D. an increase in urinary Na^+ excretion.
   E. an increase in plasma Cl^-.

7. Variations in which of the following components of blood or CSF do *not* affect respiration?

   A. Arterial HCO_3^- concentration
   B. Arterial H^+ concentration
   C. Arterial Na^+ concentration
   D. CSF CO_2 concentration
   E. CSF H^+ concentration

CHAPTER RESOURCES

Barnes PJ: Chronic obstructive pulmonary disease. N Engl J Med 2000;343:269.

Crystal RG, West JB (editors): *The Lung: Scientific Foundations,* 2nd ed. Lippincott-Raven, 1997.

Fishman AP, Elias JA, Fishman JA, et al. (editors): *Fishman's Pulmonary Diseases and Disorders,* 4th ed. McGraw-Hill, 2008.

Jones NL, Killian KJ: Exercise limitation in health and disease. N Engl J Med 2000;343:632.

Laffey JG, Kavanagh BP: Hypocapnia. N Engl J Med 2002;347:43.

Putnam RW, Dean JB, Ballantyne D (editors): Central chemosensitivity. Respir Physiol 2001;129:1.

Rekling JC, Feldman JL: Pre-Bötzinger complex and pacemaker neurons: Hypothesized site and kernel for respiratory rhythm generation. Annu Rev Physiol 1998;60:385.

West, JB: *Respiratory Physiology: The Essentials,* 8th ed. Wolers Kluwer/Lippincott Williams & Wilkins, 2008.

SECTION VII Renal Physiology

The kidneys, the bladder, and the ureters make up the urinary system. Within the kidney, the functional unit is the nephron and each human kidney has approximately 1 million nephrons. The kidneys play an essential role in the regulation of water homeostasis, electrolyte composition (eg, Na, Cl, K, HCO_3), regulation of extracellular volume (thus blood pressure), and acid–base homeostasis (Chapter 39). The kidneys filter the plasma of the blood and produce urine which allows the kidneys to excrete metabolic waste products from the body such as urea, ammonium, and foreign chemicals such as drug metabolites. The kidneys are responsible for the reabsorption of glucose and amino acids from the plasma filtrate in addition to regulated calcium and phosphate uptake (high in children). The kidneys play a role in gluconeogenesis and during fasting can synthesize and release glucose into the blood, producing almost 20% of the liver's glucose capacity. The kidneys are also endocrine organs, making kinins (see Chapter 32), 1, 25-dihydroxycholecalciferol (see Chapter 21), erythropoietin (see Chapter 37) and making and secreting renin (see Chapter 38).

Specifically, in the kidneys, a fluid that resembles plasma is filtered through the glomerular capillaries into the renal tubules **(glomerular filtration).** As this glomerular filtrate passes down the tubules, its volume is reduced and its composition altered by the processes of **tubular reabsorption** (removal of water and solutes from the tubular fluid) and **tubular secretion** (secretion of solutes into the tubular fluid) to form the urine that enters the renal pelvis. From the renal pelvis, the urine passes to the bladder and is expelled to the exterior by the process of urination, or **micturition.**

Diseases of the kidney are numerous. Commonly occurring clinical conditions include acute kidney injury, chronic kidney disease, diabetic kidney disease, nephritic and nephrotic syndromes, polycystic kidney disease, urinary tract obstruction, urinary tract infection, renal cancer. When renal function decreases such that the kidneys are no longer functioning to maintain health, patients sometimes undergo dialysis and eventually kidney transplantation.

The prevalence of kidney disease is increasing across the world as is the cost of treating the diseases which cause kidney damage; thus kidney disease represents a large threat to healthcare resources worldwide. The two diseases that cause the largest prevalence of kidney dysfunction are diabetes and hypertension. Nearly one billion people worldwide have high blood pressure and that number is expected to increase to 1.56 billion by 2025. There are currently over 240 million people with diabetes worldwide and this figure is anticipated to rise to 380 million by 2025. Of those with diabetes, about 40% will develop chronic kidney disease (CKD), which means that their risk of cardiovascular complications also increases. The cumulative global cost for dialysis and transplantation over the next decade is predicted to exceed US$ 1 trillion.

CHAPTER 37

Renal Function & Micturition

OBJECTIVES

After reading this chapter, you should be able to:

- Describe the morphology of a typical nephron and its blood supply.
- Define autoregulation and list the major theories advanced to explain autoregulation in the kidneys.
- Define glomerular filtration rate, describe how it can be measured, and list the major factors affecting it.
- Outline tubular handling of Na^+ and water.
- Discuss tubular reabsorption and secretion of glucose and K^+.
- Describe how the countercurrent mechanism in the kidney operates to produce hypertonic or hypotonic urine.
- List the major classes of diuretics; understand how each operates to increase urine flow.
- Describe the voiding reflex and draw a cystometrogram.

FUNCTIONAL ANATOMY

THE NEPHRON

Each individual renal tubule and its glomerulus is a unit **(nephron).** The size of the kidneys between species varies, as does the number of nephrons they contain. Each human kidney has approximately 1 million nephrons. The specific structures of the nephron are shown in diagrammatic fashion in Figure 37–1.

The glomerulus, which is about 200 μm in diameter, is formed by the invagination of a tuft of capillaries into the dilated, blind end of the nephron **(Bowman's capsule).** The capillaries are supplied by an **afferent arteriole** and drained by the **efferent arteriole** (Figure 37–2), and it is from the glomerulus that the filtrate is formed. The diameter of the afferent arteriole is larger than the efferent arteriole. Two cellular layers separate the blood from the glomerular filtrate in Bowman's capsule: the capillary endothelium and the specialized epithelium of the capsule. The endothelium of the glomerular capillaries is fenestrated, with pores that are 70–90 nm in diameter. The endothelium of the glomerular capillaries is completely surrounded by the glomerular basement membrane along with specialized cells called podocytes. **Podocytes** have numerous pseudopodia that interdigitate (Figure 37–2) to form **filtration slits** along the capillary wall. The slits are approximately 25 nm wide, and each is closed by a thin membrane. The glomerular basement membrane, the basal lamina, does not contain visible gaps or pores. Stellate cells called **mesangial cells** are located between the basal lamina and the endothelium. They are similar to cells called **pericytes,** which are found in the walls of capillaries elsewhere in the body. Mesangial cells are especially common between two neighboring capillaries, and in these locations the basal membrane forms a sheath shared by both capillaries (Figure 37–2). The mesangial cells are contractile and play a role in the regulation of glomerular filtration. Mesangial cells secrete the extracellular matrix, take up immune complexes, and are involved in the progression of glomerular disease.

Functionally, the glomerular membrane permits the free passage of neutral substances up to 4 nm in diameter and almost totally excludes those with diameters greater than 8 nm. However, the charge on molecules as well as their diameters affects their passage into Bowman's capsule. The total area of glomerular capillary endothelium across which filtration occurs in humans is about 0.8 m^2.

The general features of the cells that make up the walls of the tubules are shown in Figure 37–1; however, there are cell sub-types in all segments, and the anatomic differences between them correlate with differences in function.

The human **proximal convoluted tubule** is about 15 mm long and 55 μm in diameter. Its wall is made up of a single layer of cells that interdigitate with one another and are united by apical tight junctions. Between the cells are extensions of

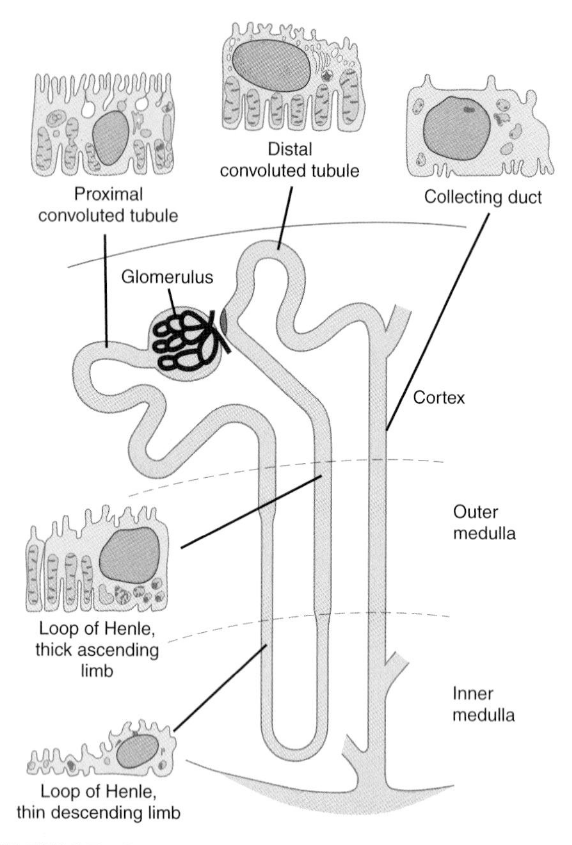

FIGURE 37–1 Diagram of a nephron. The main histologic features of the cells that make up each portion of the tubule are also shown.

the extracellular space called the **lateral intercellular spaces.** The luminal edges of the cells have a striated **brush border** due to the presence of many microvilli.

The convoluted proximal tubule straightens and the next portion of each nephron is the **loop of Henle.** The descending portion of the loop and the proximal portion of the ascending limb are made up of thin, permeable cells. On the other hand, the thick portion of the ascending limb (Figure 37–1) is made up of thick cells containing many mitochondria. The nephrons with glomeruli in the outer portions of the renal cortex have short loops of Henle **(cortical nephrons),** whereas those with glomeruli in the juxtamedullary region of the cortex **(juxtamedullary nephrons)** have long loops extending down into the medullary pyramids. In humans, only 15% of the nephrons have long loops.

The thick end of the ascending limb of the loop of Henle reaches the glomerulus of the nephron from which the tubule arose and nestles between its afferent and efferent arterioles. Specialized cells at the end form the **macula densa,** which is close to the efferent and particularly the afferent arteriole (Figure 37–2). The macula, the neighboring **lacis cells,** and the renin-secreting **granular cells** in the afferent arteriole form the **juxtaglomerular apparatus** (see Figure 38–8).

The **distal convoluted tubule,** which starts at the macula densa, is about 5 mm long. Its epithelium is lower than that of the proximal tubule, and although a few microvilli are present, there is no distinct brush border. The distal tubules coalesce to form **collecting ducts** that are about 20 mm long and pass through the renal cortex and medulla to empty into the pelvis of the kidney at the apexes of the medullary pyramids. The epithelium of the collecting ducts is made up of **principal cells (P cells)** and **intercalated cells (I cells).** The P cells, which predominate, are relatively tall and have few organelles. They are involved in Na^+ reabsorption and vasopressin-stimulated water reabsorption. The I cells, which are present in smaller numbers and are also found in the distal tubules, have more microvilli, cytoplasmic vesicles, and mitochondria. They are concerned with acid secretion and HCO_3^- transport. The total length of the nephrons, including the collecting ducts, ranges from 45 to 65 mm.

Cells in the kidneys that appear to have a secretory function include not only the granular cells in the juxtaglomerular apparatus but also some of the cells in the interstitial tissue of the medulla. These cells are called **renal medullary interstitial cells (RMICs)** and are specialized fibroblast-like cells. They contain lipid droplets and are a major site of cyclooxygenase 2 (COX-2) and prostaglandin synthase (PGES) expression. PGE_2 is the major prostanoid synthesized in the kidney and is an important paracrine regulator of salt and water homeostasis. PGE_2 is secreted by the RMICs, by the macula densa, and by cells in the collecting ducts; prostacyclin (PGI_2) and other prostaglandins are secreted by the arterioles and glomeruli.

BLOOD VESSELS

The renal circulation is diagrammed in Figure 37–3. The **afferent arterioles** are short, straight branches of the interlobular arteries. Each divides into multiple capillary branches to form the tuft of vessels in the glomerulus. The capillaries coalesce to form the **efferent arteriole,** which in turn breaks up into capillaries that supply the tubules **(peritubular capillaries)** before draining into the interlobular veins. The arterial segments between glomeruli and tubules are thus technically a portal system, and the glomerular capillaries are the only capillaries in the body that drain into arterioles. However, there is relatively little smooth muscle in the efferent arterioles.

The capillaries draining the tubules of the cortical nephrons form a peritubular network, whereas the efferent arterioles from the juxtamedullary glomeruli drain not only into a peritubular network, but also into vessels that form hairpin loops (the **vasa recta**). These loops dip into the medullary pyramids alongside the loops of Henle (Figure 37–3). The descending vasa recta have a nonfenestrated endothelium that contains a facilitated transporter for urea, and the ascending vasa recta have a fenestrated endothelium, consistent with their function in conserving solutes.

The efferent arteriole from each glomerulus breaks up into capillaries that supply a number of different nephrons. Thus, the tubule of each nephron does not necessarily receive blood solely from the efferent arteriole of the same nephron. In humans, the total surface of the renal capillaries is approximately equal to the total surface area of the

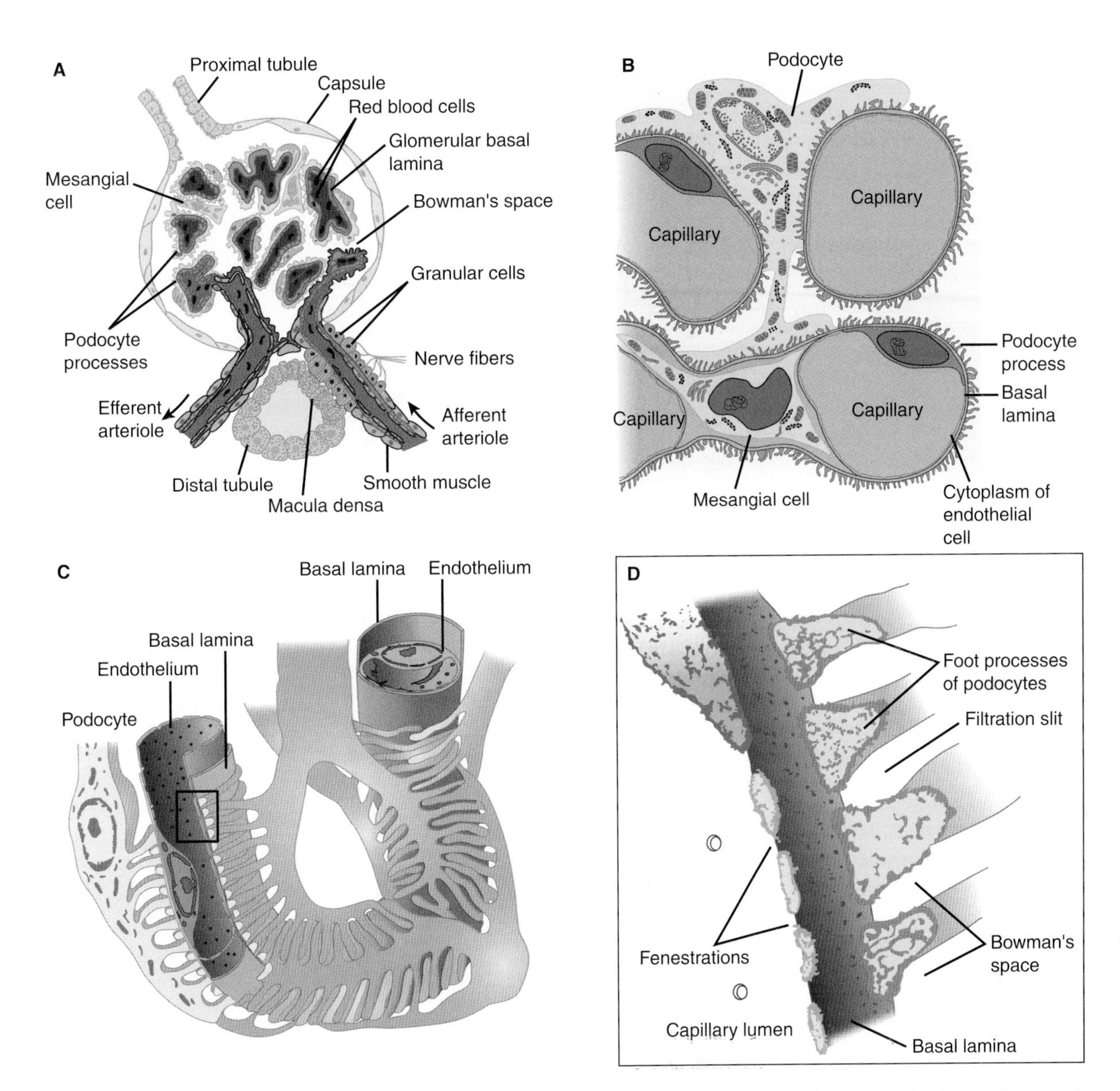

FIGURE 37–2 Structural details of glomerulus. A) Section through vascular pole, showing capillary loops. **B)** Relation of mesangial cells and podocytes to glomerular capillaries. **C)** Detail of the way podocytes form filtration slits on the basal lamina, and the relation of the lamina to the capillary endothelium. **D)** Enlargement of the rectangle in C to show the podocyte processes. The fuzzy material on their surfaces is glomerular polyanion.

tubules, both being about 12 m^2. The volume of blood in the renal capillaries at any given time is 30–40 mL.

LYMPHATICS

The kidneys have an abundant lymphatic supply that drains via the thoracic duct into the venous circulation in the thorax.

CAPSULE

The renal capsule is thin but tough. If the kidney becomes edematous, the capsule limits the swelling, and the tissue pressure **(renal interstitial pressure)** rises. This decreases the glomerular filtration rate (GFR) and is claimed to enhance and prolong anuria in acute renal failure.

INNERVATION OF THE RENAL VESSELS

The renal nerves travel along the renal blood vessels as they enter the kidney. They contain many postganglionic sympathetic efferent fibers and a few afferent fibers. There also appears to be a cholinergic innervation via the vagus nerve, but its function is uncertain. The sympathetic preganglionic innervation comes primarily from the lower thoracic and upper lumbar segments of the spinal cord, and the cell bodies of the postganglionic neurons are in the sympathetic ganglion chain, in the superior mesenteric ganglion, and along the renal artery. The sympathetic fibers are distributed primarily to the afferent and efferent arterioles, the proximal and distal

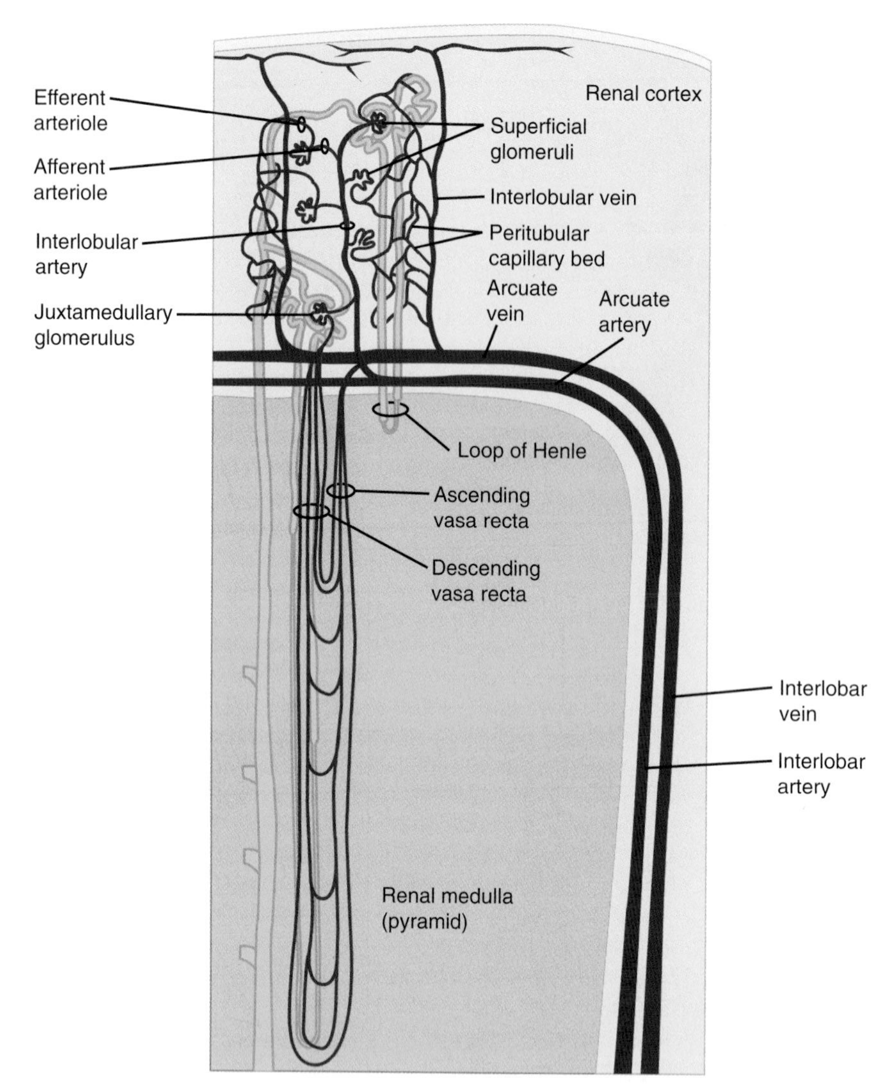

FIGURE 37–3 Renal circulation. Interlobar arteries divide into arcuate arteries, which give off interlobular arteries in the cortex. The inter-lobular arteries provide an afferent arteriole to each glomerulus. The efferent arteriole from each glomerulus breaks up into capillaries that supply blood to the renal tubules. Venous blood enters interlobular veins, which in turn flow via arcuate veins to the interlobar veins. (Modified from Boron WF, Boulpaep EL: *Medical Physiology*. Saunders, 2009.)

tubules, and the juxtaglomerular apparatus (see Chapter 38). In addition, there is a dense noradrenergic innervation of the thick ascending limb of the loop of Henle.

Nociceptive afferents that mediate pain in kidney disease parallel the sympathetic efferents and enter the spinal cord in the thoracic and upper lumbar dorsal roots. Other renal afferents presumably mediate a **renorenal reflex** by which an increase in ureteral pressure in one kidney leads to a decrease in efferent nerve activity to the contralateral kidney. This decrease permits an increase in its excretion of Na^+ and water.

RENAL CIRCULATION

BLOOD FLOW

In a resting adult, the kidneys receive 1.2–1.3 L of blood per minute, or just under 25% of the cardiac output. Renal blood flow can be measured with electromagnetic or other types of flow meters, or it can be determined by applying the Fick principle (see Chapter 30) to the kidney; that is, by measuring the amount of a given substance taken up per unit of time and dividing this value by the arteriovenous difference for the substance across the kidney. Because the kidney filters plasma, the **renal plasma flow** (RPF) equals the amount of a substance excreted per unit of time divided by the renal arteriovenous difference as long as the amount in the red cells is unaltered during passage through the kidney. Any excreted substance can be used if its concentration in arterial and renal venous plasma can be measured and if it is not metabolized, stored, or produced by the kidney and does not itself affect blood flow.

RPF can be measured by infusing p-aminohippuric acid (PAH) and determining its urine and plasma concentrations. PAH is filtered by the glomeruli and secreted by the tubular cells, so that its **extraction ratio** (arterial concentration minus renal venous concentration divided by arterial concentration) is high. For example, when PAH is infused at low doses, 90% of the PAH in arterial blood is removed in a single circulation through the kidney. It has therefore become commonplace to calculate the "RPF" by dividing the amount of PAH in the urine by the plasma PAH level, ignoring the level in renal venous blood. Peripheral venous plasma can be used because its PAH concentration is essentially identical to that in the arterial plasma reaching the kidney. The value obtained should be called the **effective renal plasma flow (ERPF)** to indicate that the level in renal venous plasma was not measured. In humans, ERPF averages about 625 mL/min.

$$\text{ERPF} = \frac{U_{PAH}\dot{V}}{P_{PAH}} = \text{Clearance of PAH } (C_{PAH})$$

Example:

Concentration of PAH in urine (U_{PAH}): 14 mg/mL
Urine flow ($\dot{V}$): 0.9 mL/min
Concentration of PAH in plasma (P_{PAH}): 0.02 mg/mL

$$\text{ERPF} = \frac{14 \times 0.9}{0.02} = 630\text{mL/min}$$

It should be noted that the ERPF determined in this way is the **clearance** of PAH. The concept of clearance is discussed in detail below.

ERPF can be converted to actual renal plasma flow (RPF):

Average PAH extraction ratio: 0.9

$$\frac{\text{ERP}}{\text{Extraction ration}} = \frac{630}{0.9} = \text{Actual RPF} = 700\text{mL/min}$$

From the renal plasma flow, the renal blood flow can be calculated by dividing by 1 minus the hematocrit:

Hematocrit (Hct): 45%

$$\text{Renal blood flow} = \text{RPF} \times \frac{1}{1-\text{Hct}} = 700 \times \frac{1}{0.55} = 1273 \text{ mL/min}$$

PRESSURE IN RENAL VESSELS

The pressure in the glomerular capillaries has been measured directly in rats and has been found to be considerably lower than predicted on the basis of indirect measurements. When the mean systemic arterial pressure is 100 mm Hg, the glomerular capillary pressure is about 45 mm Hg. The pressure drop across the glomerulus is only 1–3 mm Hg, but a further drop occurs in the efferent arteriole so that the pressure in the peritubular capillaries is about 8 mm Hg. The pressure in the renal vein is about 4 mm Hg. Pressure gradients are similar in squirrel monkeys and presumably in humans, with a glomerular capillary pressure that is about 40% of systemic arterial pressure.

REGULATION OF THE RENAL BLOOD FLOW

Norepinephrine (noradrenaline) constricts the renal vessels, with the greatest effect of injected norepinephrine being exerted on the interlobular arteries and the afferent arterioles. Dopamine is made in the kidney and causes renal vasodilation and natriuresis. Angiotensin II exerts a constrictor effect on both the afferent and efferent arterioles. Prostaglandins increase blood flow in the renal cortex and decrease blood flow in the renal medulla. Acetylcholine also produces renal vasodilation. A high-protein diet raises glomerular capillary pressure and increases renal blood flow.

FUNCTIONS OF THE RENAL NERVES

Stimulation of the renal nerves increases renin secretion by a direct action of released norepinephrine on β_1-adrenergic receptors on the juxtaglomerular cells (see Chapter 38) and it increases Na^+ reabsorption, probably by a direct action of norepinephrine on renal tubular cells. The proximal and distal tubules and the thick ascending limb of the loop of Henle are richly innervated. When the renal nerves are stimulated to increasing extents in experimental animals, the first response is an increase in the sensitivity of the granular cells in the juxtaglomerular apparatus (Table 37–1), followed by increased renin secretion, then increased Na^+ reabsorption, and finally, at the highest threshold, renal vasoconstriction with decreased

TABLE 37–1 Renal responses to graded renal nerve stimulation.

Renal Nerve Stimulation Frequency (Hz)	RSR	$U_{Na}V$	GFR	RBF
0.25	No effect on basal values; augments RSR mediated by nonneural stimuli	0	0	0
0.50	Increased without changing $U_{Na}V$, GFR, or RBF	0	0	0
1.0	Increased with decreased $U_{Na}V$, without changing GFR or RBF	↓	0	0
2.50	Increased with decreased $U_{Na}V$, GFR, and RBF	↓	↓	↓

RSR, renin secretion rate; $U_{Na}V$, urinary sodium excretion; GFR, glomerular filtration rate; RBF, renal blood flow.

Reproduced from DiBona GF: Neural control of renal function: Cardiovascular implications. Hypertension 1989;13:539. By permission of the American Heart Association.

glomerular filtration and renal blood flow. It is still unsettled whether the effect on Na^+ reabsorption is mediated via α- or β-adrenergic receptors, and it may be mediated by both. The physiologic role of the renal nerves in Na^+ homeostasis is also unsettled, in part because most renal functions appear to be normal in patients with transplanted kidneys, and it takes some time for transplanted kidneys to acquire a functional innervation.

Strong stimulation of the sympathetic noradrenergic nerves to the kidneys causes a marked decrease in renal blood flow. This effect is mediated by α_1-adrenergic receptors and to a lesser extent by postsynaptic α_2-adrenergic receptors. Some tonic discharge takes place in the renal nerves at rest in animals and humans. When systemic blood pressure falls, the vasoconstrictor response produced by decreased discharge in the baroreceptor nerves includes renal vasoconstriction. Renal blood flow is decreased during exercise and, to a lesser extent, on rising from the supine position.

AUTOREGULATION OF RENAL BLOOD FLOW

When the kidney is perfused at moderate pressures (90–220 mm Hg in the dog), the renal vascular resistance varies with the pressure so that renal blood flow is relatively constant (Figure 37–4). Autoregulation of this type occurs in other organs, and several factors contribute to it (see Chapter 32). Renal autoregulation is present in denervated and in isolated, perfused kidneys, but is prevented by the administration of drugs that paralyze vascular smooth muscle. It is probably produced in part by a direct contractile response to stretch of the smooth muscle of the afferent arteriole. NO may also be involved. At low perfusion pressures, angiotensin II also appears to play a role by constricting the efferent arterioles, thus maintaining the GFR. This is believed to be the explanation of the renal failure that sometimes develops in patients with poor renal perfusion who are treated with drugs that inhibit angiotensin-converting enzyme.

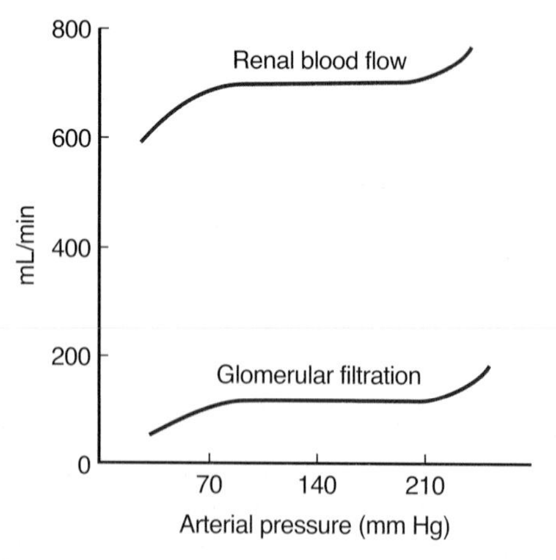

FIGURE 37–4 Autoregulation in the kidneys.

REGIONAL BLOOD FLOW & OXYGEN CONSUMPTION

The main function of the renal cortex is filtration of large volumes of blood through the glomeruli, so it is not surprising that the renal cortical blood flow is relatively great and little oxygen is extracted from the blood. Cortical blood flow is about 5 mL/g of kidney tissue/min (compared with 0.5 mL/g/min in the brain), and the arteriovenous oxygen difference for the whole kidney is only 14 mL/L of blood, compared with 62 mL/L for the brain and 114 mL/L for the heart (see Table 33–1). The Po_2 of the cortex is about 50 mm Hg. On the other hand, maintenance of the osmotic gradient in the medulla requires a relatively low blood flow. It is not surprising, therefore, that the blood flow is about 2.5 mL/g/min in the outer medulla and 0.6 mL/g/min in the inner medulla. However, metabolic work is being done, particularly to reabsorb Na^+ in the thick ascending limb of Henle, so relatively large amounts of O_2 are extracted from the blood in the medulla. The PO_2 of the medulla is about 15 mm Hg. This makes the medulla vulnerable to hypoxia if flow is reduced further. NO, prostaglandins, and many cardiovascular peptides in this region function in a paracrine fashion to maintain the balance between low blood flow and metabolic needs.

GLOMERULAR FILTRATION

MEASURING GFR

Glomerular filtration rate (GFR) is the amount of plasma ultrafiltrate formed each minute and can be measured in intact experimental animals and humans by measuring the plasma level of a substance and the amount of that substance that is excreted. A substance to be used to measure GFR must be freely filtered through the glomeruli and must be neither secreted nor reabsorbed by the tubules.

In addition to the requirement that it be freely filtered and neither reabsorbed nor secreted in the tubules, a substance suitable for measuring the GFR should be nontoxic and not metabolized by the body. Inulin, a polymer of fructose with a molecular weight of 5200, meets these criteria in humans and most animals and can be used to measure GFR.

Renal plasma clearance is the volume of plasma from which a substance is completely removed by the kidney in a given amount of time (usually minutes). The amount of that substance that appears in the urine per unit of time is the result of the renal filtering of a certain number of milliliters of plasma that contained this amount. GFR and clearance are measured in mL/min.

Therefore, if the substance is designated by the letter X, the GFR is equal to the concentration of X in urine (U_X) times the **urine flow** per unit of time ($\dot{V}$) divided by the **arterial plasma level** of X (P_X), or $U_X\dot{V}/P_X$. **This value is called the clearance of X (C_X).**

In practice, a loading dose of inulin is administered intravenously, followed by a sustaining infusion to keep the arterial plasma level constant. After the inulin has equilibrated with body fluids, an accurately timed urine specimen is collected and a plasma sample obtained halfway through the collection. Plasma and urinary inulin concentrations are determined and the clearance is calculated:

$$U_{IN} = 35 \text{ mg/mL}$$
$$\dot{V} = 0.9 \text{ mL/min}$$
$$P_{IN} = 0.25 \text{ mg/mL}$$
$$C_{IN} = \frac{U_{IN}\dot{V}}{P_{IN}} = \frac{35 \times 0.9}{0.25}$$
$$C_{IN} = 126 \text{ mL/min}$$

Clearance of creatinine (C_{Cr}) can also be used to determine GFR, however some creatinine is secreted by the tubules thus the clearance of creatinine will be slightly higher than inulin. In spite of this, the clearance of endogenous creatinine is a reasonable estimate of GFR as the values agree quite well with the GFR values measured with inulin (see Table 37–2). More common though is the use of **P_{Cr} values as an index of renal function (normal=1 mg/dL).**

NORMAL GFR

The GFR in a healthy adult of average size is approximately 125 mL/min. Its magnitude correlates fairly well with surface area, but values in women are 10% lower than those in men even after correction for surface area. A rate of 125 mL/min is 7.5 L/h, or 180 L/d, whereas the normal urine volume is about 1 L/d. Thus, 99% or more of the filtrate is normally reabsorbed. At the rate of 125 mL/min, in 1 day the kidneys filter an amount of fluid equal to four times the total body water, 15 times the ECF volume, and 60 times the plasma volume.

TABLE 37–2 Normal clearance values of different solutes.

Substance	Clearance (mL/min)
Glucose	0
Sodium	0.9
Chloride	1.3
Potassium	12
Phosphate	25
Urea	75
Inulin	125
Creatinine	140
PAH	560

CONTROL OF GFR

The factors governing filtration across the glomerular capillaries are the same as those governing filtration across all other capillaries (see Chapter 31), that is, the size of the capillary bed, the permeability of the capillaries, and the hydrostatic and osmotic pressure gradients across the capillary wall. For each nephron:

$$GFR = K_f [(P_{GC} - P_T) - (\pi_{GC} - \pi_T)]$$

where K_f, the glomerular ultrafiltration coefficient, is the product of the glomerular capillary wall hydraulic conductivity (ie, its permeability) and the effective filtration surface area. P_{GC} is the mean hydrostatic pressure in the glomerular capillaries, P_T the mean hydrostatic pressure in the tubule (Bowman's space), π_{GC} the oncotic pressure of the plasma in the glomerular capillaries, and π_T the oncotic pressure of the filtrate in the tubule (Bowman's space).

PERMEABILITY

The permeability of the glomerular capillaries is about 50 times that of the capillaries in skeletal muscle. Neutral substances with effective molecular diameters of less than 4 nm are freely filtered, and the filtration of neutral substances with diameters of more than 8 nm approaches zero. Between these values, filtration is inversely proportional to diameter. However, sialoproteins in the glomerular capillary wall are negatively charged, and studies with negatively and positively charged dextrans indicate that the negative charges repel negatively charged substances in blood, with the result that filtration of anionic substances 4 nm in diameter is less than half that of neutral substances of the same size. This probably explains why albumin, with an effective molecular diameter of approximately 7 nm, normally has a glomerular concentration only 0.2% of its plasma concentration rather than the higher concentration that would be expected on the basis of diameter alone; circulating albumin is negatively charged. Conversely, filtration of cationic substances is greater than that of neutral substances.

The amount of protein in the urine is normally less than 100 mg/d, and most of this is not filtered but comes from shed tubular cells. The presence of significant amounts of albumin in the urine is called **albuminuria.** In nephritis, the negative charges in the glomerular wall are dissipated, and albuminuria can occur for this reason without an increase in the size of the "pores" in the membrane.

SIZE OF THE CAPILLARY BED

K_f can be altered by the mesangial cells, with contraction of these cells producing a decrease in K_f that is largely due to a reduction in the area available for filtration. Contraction of points where the capillary loops bifurcate probably shifts flow away from some of the loops, and elsewhere, contracted

TABLE 37–3 Agents causing contraction or relaxation of mesangial cells.

Contraction	Relaxation
Endothelins	ANP
Angiotensin II	Dopamine
Vasopressin	PGE_2
Norepinephrine	cAMP
Platelet-activating factor	
Platelet-derived growth factor	
Thromboxane A_2	
PGF_2	
Leukotrienes C_4 and D_4	
Histamine	

mesangial cells distort and encroach on the capillary lumen. Agents that have been shown to affect the mesangial cells are listed in Table 37–3. Angiotensin II is an important regulator of mesangial contraction, and there are angiotensin II receptors in the glomeruli. In addition, some evidence suggests that mesangial cells make renin.

HYDROSTATIC & OSMOTIC PRESSURE

The pressure in the glomerular capillaries is higher than that in other capillary beds because the afferent arterioles are short, straight branches of the interlobular arteries. Furthermore, the vessels "downstream" from the glomeruli, the efferent arterioles, have a relatively high resistance. The capillary hydrostatic pressure is opposed by the hydrostatic pressure in Bowman's capsule. It is also opposed by the oncotic pressure gradient across the glomerular capillaries ($\pi_{GC} - \pi_T$). π_T is normally negligible, and the gradient is essentially equal to the oncotic pressure of the plasma proteins.

The actual pressures in one strain of rats are shown in Figure 37–5. The net filtration pressure (P_{UF}) is 15 mm Hg at the afferent end of the glomerular capillaries, but it falls to zero—that is, filtration equilibrium is reached—proximal to the efferent end of the glomerular capillaries. This is because fluid leaves the plasma and the oncotic pressure rises as blood passes through the glomerular capillaries. The calculated change in $\Delta\pi$ along an idealized glomerular capillary is also shown in Figure 37–5. It is apparent that portions of the glomerular capillaries do not normally contribute to the formation of the glomerular ultrafiltrate; that is, exchange across the glomerular capillaries is flow-limited rather than diffusion-limited. It is also apparent that a decrease in the rate of rise of the Δ curve produced by an increase in RPF would

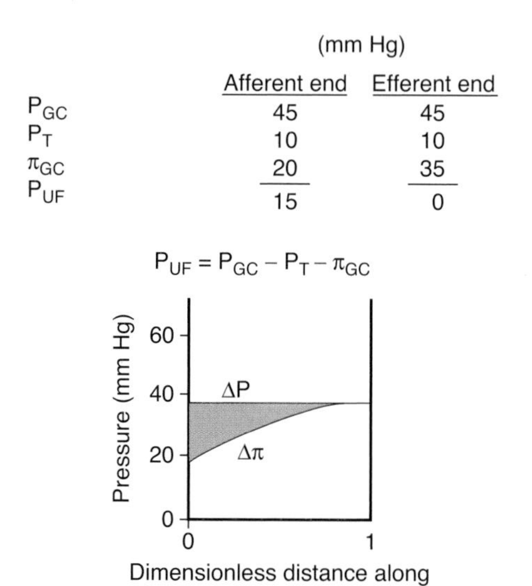

FIGURE 37–5 Hydrostatic pressure (P_{GC}) and osmotic pressure (π_{GC}) in a glomerular capillary in the rat. P_T, pressure in Bowman's capsule; P_{UF}, net filtration pressure. π_T is normally negligible, so $\Delta\pi = \pi_{GC}$. $\Delta P = P_{GC} - P_T$. (Reproduced with permission from Mercer PF, Maddox DA, Brenner BM: Current concepts of sodium chloride and water transport by the mammalian nephron. West J Med 1974;120:33.)

increase filtration because it would increase the distance along the capillary in which filtration was taking place.

There is considerable species variation in whether filtration equilibrium is reached, and some uncertainties are inherent in the measurement of K_f. It is uncertain whether filtration equilibrium is reached in humans.

CHANGES IN GFR

Variations in the factors discussed in the preceding paragraphs and listed in Table 37–4 have predictable effects on the GFR. Changes in renal vascular resistance as a result of

TABLE 37–4 Factors affecting the GFR.

Changes in renal blood flow
Changes in glomerular capillary hydrostatic pressure
Changes in systemic blood pressure
Afferent or efferent arteriolar constriction
Changes in hydrostatic pressure in Bowman's capsule
Ureteral obstruction
Edema of kidney inside tight renal capsule
Changes in concentration of plasma proteins: dehydration, hypoproteinemia, etc (minor factors)
Changes in K_f
Changes in glomerular capillary permeability
Changes in effective filtration surface area

autoregulation tend to stabilize filtration pressure, but when the mean systemic arterial pressure drops below the autoregulatory range (Figure 37–4), GFR drops sharply. The GFR tends to be maintained when efferent arteriolar constriction is greater than afferent constriction, but either type of constriction decreases blood flow to the tubules.

FILTRATION FRACTION

The ratio of the GFR to the RPF, the **filtration fraction,** is normally 0.16–0.20. The GFR varies less than the RPF. When there is a fall in systemic blood pressure, the GFR falls less than the RPF because of efferent arteriolar constriction, and consequently the filtration fraction rises.

TUBULAR FUNCTION

GENERAL CONSIDERATIONS

The amount of any substance (X) that is filtered is the product of the GFR and the plasma level of the substance ($C_{In}P_X$). The tubular cells may add more of the substance to the filtrate (tubular secretion), may remove some or all of the substance from the filtrate (tubular reabsorption), or may do both. The amount of the substance excreted per unit of ($V_X\dot{V}$) time equals the amount filtered plus the **net amount transferred** by the tubules. This latter quantity is conveniently indicated by the symbol T_X (Figure 37–6). The clearance of the substance equals the GFR if there is no net tubular secretion or reabsorption, exceeds the GFR if there is net tubular secretion, and is less than the GFR if there is net tubular reabsorption.

Much of our knowledge about glomerular filtration and tubular function has been obtained by using micropuncture techniques. Micropipettes can be inserted into the tubules of the living kidney and the composition of aspirated tubular fluid determined by the use of microchemical techniques. In addition, two pipettes can be inserted in a tubule and the tubule perfused in vivo. Alternatively, isolated perfused segments of tubules can be studied in vitro, and tubular cells can be grown and studied in culture.

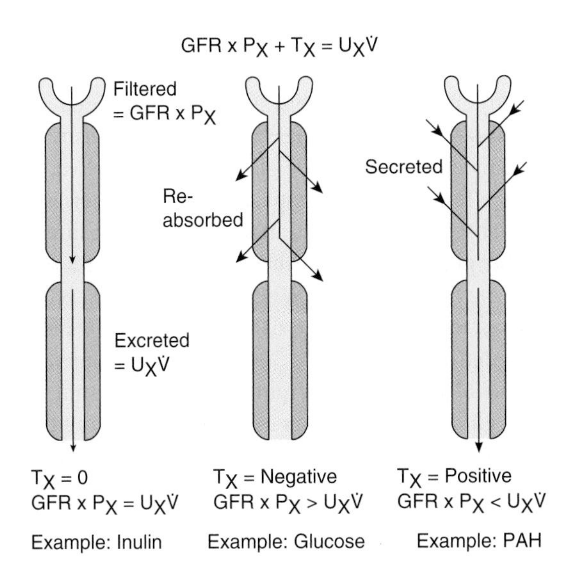

FIGURE 37–6 Tubular function. For explanation of symbols, see text.

MECHANISMS OF TUBULAR REABSORPTION & SECRETION

Small proteins and some peptide hormones are reabsorbed in the proximal tubules by endocytosis. Other substances are secreted or reabsorbed in the tubules by passive diffusion between cells and through cells by facilitated diffusion down chemical or electrical gradients or active transport against such gradients. Movement is by way of ion channels, exchangers, cotransporters, and pumps. Many of these have now been cloned, and their regulation is being studied.

It is important to note that the pumps and other transporters in the luminal membrane are different from those in the basolateral membrane. As was discussed for the gastrointestinal epithelium, it is this polarized distribution that makes possible net movement of solutes across the epithelia.

Like transport systems elsewhere, renal active transport systems have a maximal rate, or **transport maximum (Tm),** at which they can transport a particular solute. Thus, the amount of a particular solute transported is proportional to the amount present up to the Tm for the solute, but at higher concentrations, the transport mechanism is **saturated** and there is no appreciable increment in the amount transported. However, the Tms for some systems are high, and it is difficult to saturate them.

It should also be noted that the tubular epithelium, like that of the small intestine, is a **leaky epithelium** in that the tight junctions between cells permit the passage of some water and electrolytes. The degree to which leakage by this **paracellular pathway** contributes to the net flux of fluid and solute into and out of the tubules is controversial since it is difficult to measure, but current evidence seems to suggest that it is a significant factor in the proximal tubule. One indication of this is that paracellin-1, a protein localized to tight junctions, is related to Mg^{2+} reabsorption, and a loss-of-function mutation of its gene causes severe Mg^{2+} and Ca^{2+} loss in the urine.

Na^+ REABSORPTION

The reabsorption of Na^+ and Cl^- plays a major role in body electrolyte and water homeostasis. In addition, Na^+ transport is coupled to the movement of H^+, glucose, amino acids, organic acids, phosphate, and other electrolytes and substances across the tubule walls. The principal cotransporters and exchangers in the various parts of the nephron are listed in Table 37–5. In the proximal tubules, the thick portion of the ascending limb of the loop of Henle, the distal

TABLE 37–5 Transport proteins involved in the movement of Na^+ and Cl^- across the apical membranes of renal tubular cells.[a]

Site	Apical Transporter	Function
Proximal tubule	Na/glucose CT	Na^+ uptake, glucose uptake
	Na^+/P_i CT	Na^+ uptake, P_i uptake
	Na^+ amino acid CT	Na^+ uptake, amino acid uptake
	Na/lactate CT	Na^+ uptake, lactate uptake
	Na/H exchanger	Na^+ uptake, H^+ extrusion
	Cl/base exchanger	Cl^- uptake
Thick ascending limb	Na–K–2Cl CT	Na^+ uptake, Cl^- uptake, K^+ uptake
	Na/H exchanger	Na^+ uptake, H^+ extrusion
	K^+ channels	K^+ extrusion (recycling)
Distal convoluted tubule	NaCl CT	Na^+ uptake, Cl^- uptake
Collecting duct	Na^+ channel (ENaC)	Na^+ uptake

[a]Uptake indicates movement from tubular lumen to cell interior, extrusion is movement from cell interior to tubular lumen. CT, cotransporter; P_i, inorganic phosphate.

Data from Schnermann JB, Sayegh EI: *Kidney Physiology*. Lippincott-Raven, 1998.

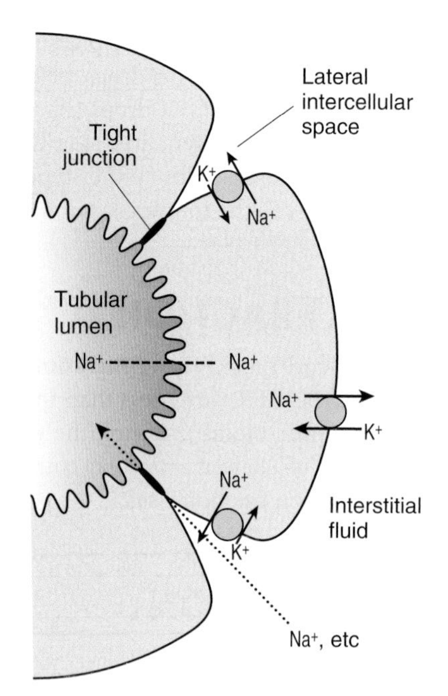

FIGURE 37–7 Mechanism of Na^+ reabsorption in the proximal tubule. Na^+ moves out of the tubular lumen by cotransport and exchange mechanism through the apical membrane of the tubule (dashed line). The Na^+ is then actively transported into the interstitial fluid by Na, K ATPase in the basolateral membrane (solid lines). K^+ enters the interstitial fluid via K^+ channels. A small amount of Na^+, other solutes, and H_2O re-enter the tubular lumen by passive transport through the tight junctions (dotted lines).

tubules, and the collecting ducts, Na^+ moves by cotransport or exchange from the tubular lumen into the tubular epithelial cells down its concentration and electrical gradients, and is then actively pumped from these cells into the interstitial space. Na^+ is pumped into the interstitium by Na, K ATPase in the basolateral membrane. Thus, Na^+ is actively transported out of all parts of the renal tubule except the thin portions of the loop of Henle. The operation of the ubiquitous Na^+ pump is considered in detail in Chapter 2. It extrudes three Na^+ in exchange for two K^+ that are pumped into the cell.

The tubular cells along the nephron are connected by tight junctions at their luminal edges, but there is space between the cells along the rest of their lateral borders. Much of the Na^+ is actively transported into these **lateral intercellular spaces** (Figure 37–7).

Normally about 60% of the filtered Na^+ is reabsorbed in the proximal tubule, primarily by Na–H exchange. Another 30% is absorbed via the Na–2Cl–K cotransporter in the thick ascending limb of the loop of Henle. In both of these segments of the nephron, passive paracellular movement of Na^+ also contributes to overall Na^+ reabsorption. In the distal convoluted tubule 7% of the filtered Na^+ is absorbed by the Na–Cl cotransporter. The remainder of the filtered Na^+, about 3%, is absorbed via ENaC channels in the collecting ducts, and this is the portion that is regulated by aldosterone to permit homeostatic adjustments in Na^+ balance.

GLUCOSE REABSORPTION

Glucose, amino acids, and bicarbonate are reabsorbed along with Na^+ in the early portion of the proximal tubule (Figure 37–8). Glucose is typical of substances removed from the urine by secondary active transport. It is filtered at a rate of approximately 100 mg/min (80 mg/dL of plasma × 125 mL/min). Essentially all of the glucose is reabsorbed, and no more than a few milligrams appear in the urine per 24 h. The amount reabsorbed is proportional to the amount filtered and hence to the plasma glucose level (P_G) times the GFR up to the transport maximum (Tm_G). When the Tm_G is exceeded, the amount of glucose in the urine rises (Figure 37–9). The Tm_G is about 375 mg/min in men and 300 mg/min in women.

The **renal threshold** for glucose is the plasma level at which the glucose first appears in the urine in more than the normal minute amounts. One would predict that the renal threshold would be about 300 mg/dL, that is, 375 mg/min (Tm_G) divided by 125 mL/min (GFR). However, the actual renal threshold is about 200 mg/dL of arterial plasma, which corresponds to a venous level of about 180 mg/dL. Figure 37–9 shows why the actual renal threshold is less than the predicted

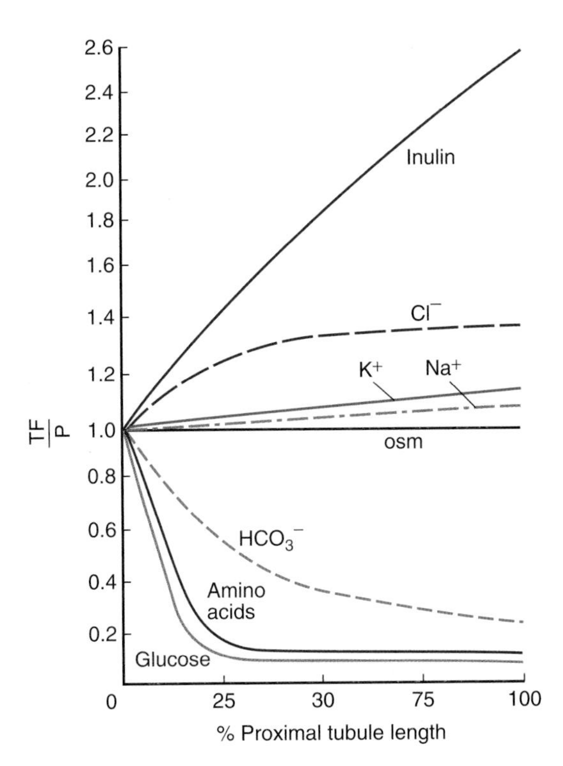

FIGURE 37–8 Reabsorption of various solutes in the proximal tubule. TF/P, tubular fluid:plasma concentration ratio. (Courtesy of FC Rector Jr)

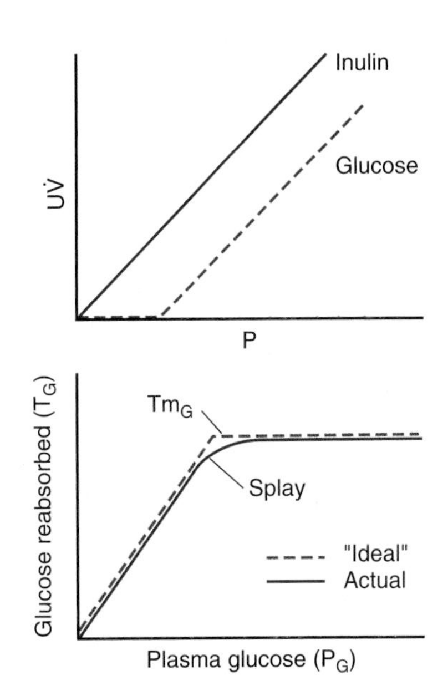

FIGURE 37–9 Renal glucose transport. Top: Relation between the plasma level (P) and excretion (UV) of glucose and inulin. **Bottom:** Relation between the plasma glucose level (P_G) and amount of glucose reabsorbed (T_G).

threshold. The "ideal" curve shown in this diagram would be obtained if the Tm_G in all the tubules was identical and if all the glucose were removed from each tubule when the amount filtered was below the Tm_G. This is not the case, and in humans, for example, the actual curve is rounded and deviates considerably from the "ideal" curve. This deviation is called **splay.** The magnitude of the splay is inversely proportional to the avidity with which the transport mechanism binds the substance it transports.

GLUCOSE TRANSPORT MECHANISM

Glucose reabsorption in the kidneys is similar to glucose reabsorption in the intestine (see Chapter 26). Glucose and Na^+ bind to the sodium-dependent glucose transporter (SGLT) 2 in the apical membrane, and glucose is carried into the cell as Na^+ moves down its electrical and chemical gradient. The Na^+ is then pumped out of the cell into the interstitium, and the glucose exits by facilitated diffusion via glucose transporter (GLUT) 2 into the interstitial fluid. At least in the rat, there is some transport by SGLT 1 and GLUT 1 as well.

SGLT 2 specifically binds the D isomer of glucose, and the rate of transport of D-glucose is many times greater than that of L-glucose. Glucose transport in the kidneys is inhibited, as it is in the intestine, by the plant glucoside **phlorhizin,** which competes with D-glucose for binding to the carrier.

ADDITIONAL EXAMPLES OF SECONDARY ACTIVE TRANSPORT

Like glucose reabsorption, amino acid reabsorption is most marked in the early portion of the proximal convoluted tubule. Absorption in this location resembles absorption in the intestine (see Chapter 26). The main carriers in the apical membrane cotransport Na^+, whereas the carriers in the basolateral membranes are not Na^+-dependent. Na^+ is pumped out of the cells by Na, K ATPase and the amino acids leave by passive or facilitated diffusion to the interstitial fluid.

Some Cl^- is reabsorbed with Na^+ and K^+ in the thick ascending limb of the loop of Henle. In addition, two members of the family of **Cl channels** have been identified in the kidney. Mutations in the gene for one of the renal channels is associated with Ca^{2+}-containing kidney stones and hypercalciuria **(Dent disease),** but how tubular transport of Ca^{2+} and Cl^- are linked is still unsettled.

PAH TRANSPORT

The dynamics of PAH transport illustrate the operation of the active transport mechanisms that secrete substances into the tubular fluid (see Clinical Box 37–1). The filtered load of PAH is a linear function of the plasma level, but PAH secretion increases as P_{PAH} rises only until a maximal secretion rate (Tm_{PAH}) is reached (Figure 37–10). When P_{PAH} is low, C_{PAH} is high; but as P_{PAH} rises above Tm_{PAH}, C_{PAH} falls progressively. It eventually approaches the clearance of inulin (C_{In}) (Figure 37–11), because the amount of PAH secreted becomes a smaller and smaller fraction of the total amount excreted.

CLINICAL BOX 37–1

Substances Secreted by the Tubules

Derivatives of hippuric acid in addition to PAH, phenol red and other sulfonphthalein dyes, penicillin, and a variety of iodinated dyes are actively secreted into the tubular fluid. Substances that are normally produced in the body and secreted by the tubules include various ethereal sulfates, steroid and other glucuronides, and 5-hydroxyindoleacetic acid, the principal metabolite of serotonin.

THERAPEUTIC HIGHLIGHTS

The loop diuretic, furosemide, and the thiazide diuretics are organic anions which gain access to their tubular sites of action (thick ascending limb and distal convoluted tubule, respectively) when they are secreted into the urine by the proximal tubule.

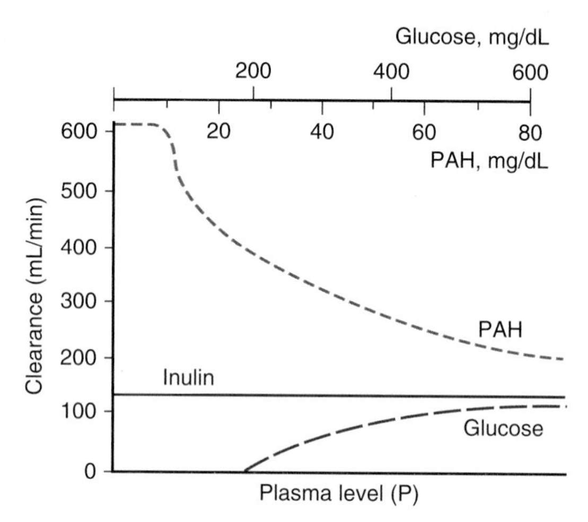

FIGURE 37–11 Clearance of inulin, glucose, and PAH at various plasma levels of each substance in humans.

Conversely, the clearance of glucose is essentially zero at P_G levels below the renal threshold; but above the threshold, C_G rises to approach C_{In} as P_G is raised.

The use of C_{PAH} to measure ERPF is discussed above.

TUBULOGLOMERULAR FEEDBACK & GLOMERULOTUBULAR BALANCE

Signals from the renal tubule in each nephron feed back to affect filtration in its glomerulus. As the rate of flow through the ascending limb of the loop of Henle and first part of the distal tubule increases, glomerular filtration in the same nephron decreases, and, conversely, a decrease in flow increases the GFR (Figure 37–12). This process, which is called **tubuloglomerular feedback,** tends to maintain the constancy of the load delivered to the distal tubule.

The sensor for this response is the **macula densa.** The amount of fluid entering the distal tubule at the end of the thick ascending limb of the loop of Henle depends on the amount of Na^+ and Cl^- in it. The Na^+ and Cl^- enter the macula densa cells via the Na–K–2Cl cotransporter in their apical membranes. The increased Na^+ causes increased Na, K ATPase activity and the resultant increased ATP hydrolysis causes more adenosine to be formed. Presumably, adenosine is secreted from the basal membrane of the cells. It acts via adenosine A_1 receptors on the macula densa cells to increase their release of Ca^{2+} to the vascular smooth muscle in the afferent arterioles. This causes afferent vasoconstriction and a resultant decrease in GFR. Presumably, a similar mechanism generates a signal that decreases renin secretion by the adjacent juxtaglomerular cells in the afferent arteriole (see Chapter 38), but this remains unsettled.

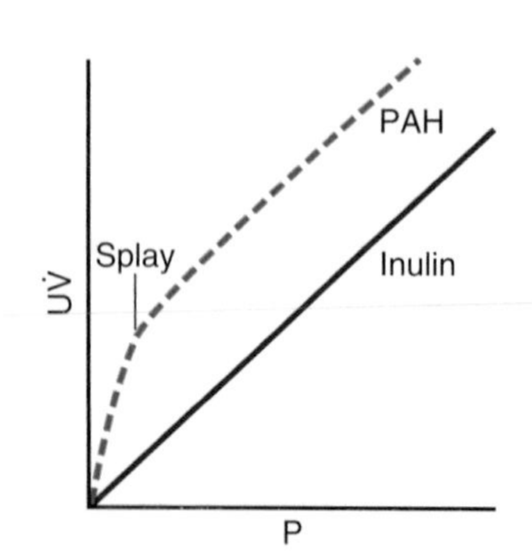

FIGURE 37–10 Relation between plasma levels (P) and excretion (UV) of PAH and inulin.

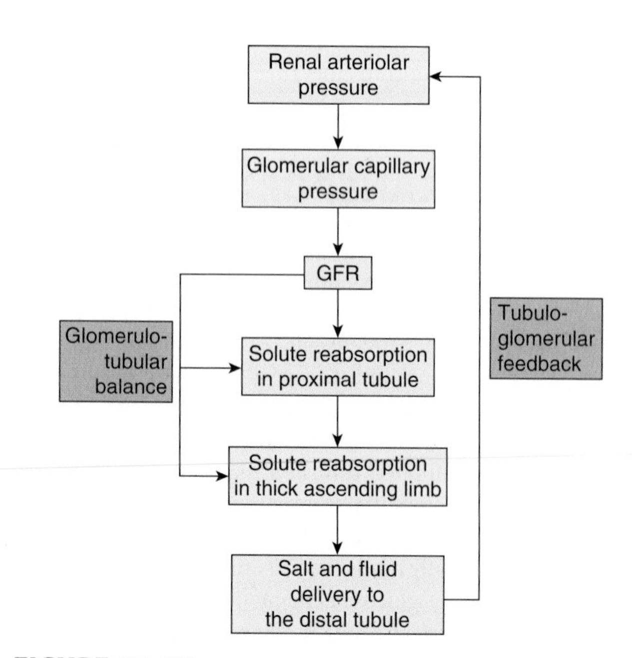

FIGURE 37–12 Mechanisms of glomerulotubular balance and tubuloglomerular feedback.

TABLE 37–6 Alterations in water metabolism produced by vasopressin in humans. In each case, the osmotic load excreted is 700 mOsm/d.

	GFR (mL/min)	Percentage of Filtered Water Reabsorbed	Urine Volume (L/d)	Urine Concentration (mOsm/kg H_2O)	Gain or Loss of Water in Excess of Solute (L/d)
Urine isotonic to plasma	125	98.7	2.4	290	...
Vasopressin (maximal antidiuresis)	125	99.7	0.5	1400	1.9 gain
No vasopressin ("complete" diabetes insipidus)	125	87.1	23.3	30	20.9 loss

Conversely, an increase in GFR causes an increase in the reabsorption of solutes, and consequently of water, primarily in the proximal tubule, so that in general the percentage of the solute reabsorbed is held constant. This process is called **glomerulotubular balance,** and it is particularly prominent for Na^+. The change in Na^+ reabsorption occurs within seconds after a change in filtration, so it seems unlikely that an extrarenal humoral factor is involved. Alternatively, one mediating factor is the oncotic pressure in the peritubular capillaries. When the GFR is high, there is a relatively large increase in the oncotic pressure of the plasma leaving the glomeruli via the efferent arterioles and hence in their capillary branches. This increases the reabsorption of Na^+ from the tubule. However, other as yet unidentified intrarenal mechanisms are likely also involved.

WATER TRANSPORT

Normally, 180 L of fluid is filtered through the glomeruli each day, while the average daily urine volume is about 1 L. The same load of solute can be excreted per 24 h in a urine volume of 500 mL with a concentration of 1400 mOsm/kg or in a volume of 23.3 L with a concentration of 30 mOsm/kg (Table 37–6). These figures demonstrate two important facts: First, at least 87% of the filtered water is reabsorbed, even when the urine volume is 23 L; and second, the reabsorption of the remainder of the filtered water can be varied without affecting total solute excretion. Therefore, when the urine is concentrated, water is retained in excess of solute; and when it is dilute, water is lost from the body in excess of solute. Both facts have great importance in the regulation of the osmolality of the body fluids. A key regulator of water output is vasopressin acting on the collecting ducts.

AQUAPORINS

Rapid diffusion of water across cell membranes depends on the presence of water channels, integral membrane proteins called **aquaporins.** To date, 13 aquaporins have been cloned; however, only four aquaporins (aquaporin-1, -2, -3, and -4) play a key role in the kidney. The roles played by aquaporin-1 and aquaporin-2 in renal water transport are discussed below.

PROXIMAL TUBULE

Active transport of many substances occurs from the fluid in the proximal tubule, but micropuncture studies have shown that the fluid remains essentially iso-osmotic until the end of the proximal tubule (Figure 37–8). **Aquaporin-1** is localized to both the basolateral and apical membrane of the proximal tubules and its presence allows water to move rapidly out of the tubule along the osmotic gradients set up by active transport of solutes, and isotonicity is maintained. Because the ratio of the concentration in tubular fluid to the concentration in plasma (TF/P) of the nonreabsorbable substance inulin is 2.5 to 3.3 at the end of the proximal tubule, it follows that 60–70% of the filtered solute and 60–70% of the filtered water have been removed by the time the filtrate reaches this point (Figure 37–13).

When aquaporin-1 was knocked out in mice, proximal tubular water permeability was reduced by 80%. When the mice were subjected to dehydration, their urine osmolality did not increase (< 700 mOsm/kg), even though other renal aquaporins were present. In humans with mutations that eliminate aquaporin-1 activity, the defect in water homeostasis is not as severe, though their response to dehydration is defective.

LOOP OF HENLE

As noted above, the loops of Henle of the juxtamedullary nephrons dip deeply into the medullary pyramids before draining into the distal convoluted tubules in the cortex, and all the collecting ducts descend back through the medullary pyramids to drain at the tips of the pyramids into the renal pelvis. There is a graded increase in the osmolality of the interstitium of the pyramids in humans: The osmolality at the tips of the papillae can reach about 1200 mOsm/kg of H_2O, approximately four times that of plasma. The descending limb of the loop of Henle is permeable to water, due to the presence of **aquaporin-1** in both the apical and basolateral membranes, but the ascending limb is impermeable to water. Na^+,

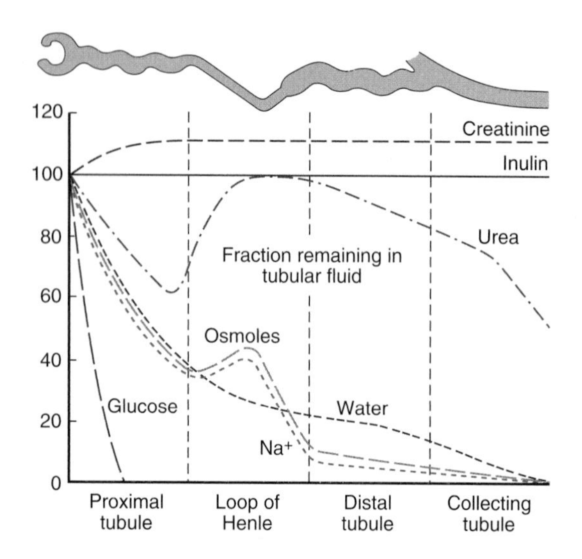

FIGURE 37–13 Changes in the percentage of the filtered amount of substances remaining in the tubular fluid along the length of the nephron in the presence of vasopressin. (Modified from Sullivan LP, Grantham JJ: *Physiology of the Kidney*, 2nd ed. Lea & Febiger, 1982.)

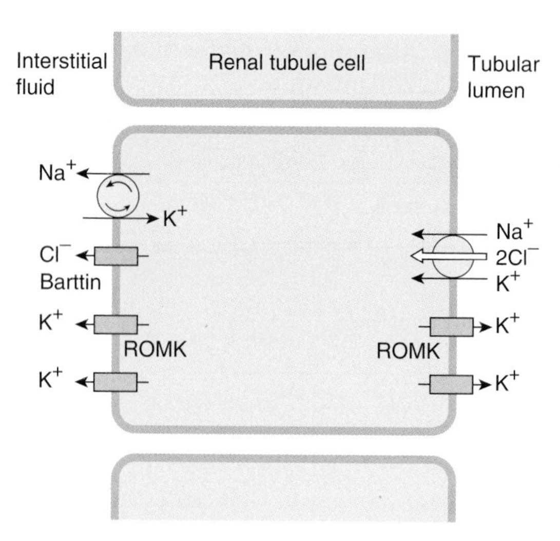

FIGURE 37–14 NaCl transport in the thick ascending limb of the loop of Henle. The Na–K–2Cl cotransporter moves these ions into the tubular cell by secondary active transport. Na^+ is transported out of the cell into the interstitium by Na, K ATPase in the basolateral membrane of the cell. Cl^- exits in basolateral ClC-Kb Cl^- channels. Barttin, a protein in the cell membrane, is essential for normal ClC-Kb function. K^+ moves from the cell to the interstitium and the tubular lumen by ROMK and other K^+ channels (see Clinical Box 37–2).

K^+, and Cl^- are cotransported out of the thick segment of the ascending limb. Therefore, the fluid in the descending limb of the loop of Henle becomes **hypertonic** as water moves out of the tubule into the hypertonic interstitium. In the ascending limb it becomes more dilute because of the movement of Na^+ and Cl^- out of the tubular lumen, and when fluid reaches the top of the ascending limb (called the **diluting segment**) it is now **hypotonic** to plasma. In passing through the descending loop of Henle, another 15% of the filtered water is removed, so approximately 20% of the filtered water enters the distal tubule, and the TF/P of inulin at this point is about 5.

In the thick ascending limb, a carrier cotransports one Na^+, one K^+, and $2Cl^-$ from the tubular lumen into the tubular cells. This is another example of secondary active transport; the Na^+ is actively transported from the cells into the interstitium by Na, K ATPase in the basolateral membranes of the cells, keeping the intracellular Na^+ low. The Na–K–2Cl cotransporter has 12 transmembrane domains with intracellular amino and carboxyl terminals. It is a member of a family of transporters found in many other locations, including salivary glands, the gastrointestinal tract, and the airways.

The K^+ diffuses back into the tubular lumen and back into the interstitium via ROMK and other K^+ channels. The Cl^- moves into the interstitium via ClC-Kb channels (Figure 37–14).

DISTAL TUBULE

The distal tubule, particularly its first part, is in effect an extension of the thick segment of the ascending limb. It is relatively impermeable to water, and continued removal of the solute in excess of solvent further dilutes the tubular fluid.

COLLECTING DUCTS

The collecting ducts have two portions: a cortical portion and a medullary portion. The changes in osmolality and volume in the collecting ducts depend on the amount of vasopressin acting on the ducts. This antidiuretic hormone from the posterior pituitary gland increases the permeability of the collecting ducts to water. The key to the action of vasopressin on the collecting ducts is aquaporin-2. Unlike the other aquaporins, this aquaporin is stored in vesicles in the cytoplasm of principal cells. Vasopressin causes rapid insertion of these vesicles into the apical membrane of cells. The effect is mediated via the vasopressin V_2 receptor, cyclic adenosine 5-monophosphate (cAMP), and protein kinase A. Cytoskeletal elements are involved, including microtubule-based motor proteins (dynein and dynactin) as well as actin filament-binding proteins such as myosin-1.

In the presence of enough vasopressin to produce maximal antidiuresis, water moves out of the hypotonic fluid entering the cortical collecting ducts into the interstitium of the cortex, and the tubular fluid becomes isotonic. In this fashion, as much as 10% of the filtered water is removed. The isotonic fluid then enters the medullary collecting ducts with a TF/P inulin of about 20. An additional 4.7% or more of the filtrate is reabsorbed into the hypertonic interstitium of the medulla, producing a concentrated urine with a TF/P inulin of over 300. In humans, the osmolality of urine may reach 1400 mOsm/kg of H_2O, almost five times the osmolality of plasma, with a total of 99.7% of the filtered water being reabsorbed (Table 37–6). In other species, the ability

CLINICAL BOX 37–2

Genetic Mutations in Renal Transporters

Mutations of individual genes for many renal sodium transporters and channels cause specific syndromes such as Bartter syndrome, Liddle syndrome, and Dent disease. A large number of mutations have been described.

Bartter syndrome is a rare but interesting condition that is due to defective transport in the thick ascending limb. It is characterized by chronic Na^+ loss in the urine, with resultant hypovolemia causing stimulation of renin and aldosterone secretion without hypertension, plus hyperkalemia and alkalosis. The condition can be caused by loss-of-function mutations in the gene for any of four key proteins: the Na–K–2Cl cotransporter, the ROMK K^+ channel, the ClC–Kb Cl^- channel, or **barttin,** a recently described integral membrane protein that is necessary for the normal function of ClC–Kb Cl^- channels.

The stria vascularis in the inner ear is responsible for maintaining the high K^+ concentration in the scala media that is essential for normal hearing. It contains both ClC–Kb and ClC–Ka Cl^- channels. Bartter syndrome associated with mutated ClC–Kb channels is not associated with deafness because the Clc–Ka channels can carry the load. However, both types of Cl^- channels are barttin-dependent, so patients with Bartter syndrome due to mutated barttin are also deaf.

Another interesting example involves the proteins polycystin-1 (PKD-1) and polycystin-2 (PKD-2). PKD-1 appears to be a Ca^{2+} receptor that activates a nonspecific ion channel associated with PKD-2. The normal function of this apparent ion channel is unknown, but both proteins are abnormal in **autosomal dominant polycystic kidney disease,** in which the renal parenchyma is progressively replaced by fluid-filled cysts until there is complete renal failure.

to concentrate urine is even greater. Maximal urine osmolality is about 2500 mOsm/kg in dogs, about 3200 mOsm/kg in laboratory rats, and as high as 5000 mOsm/kg in certain desert rodents.

When vasopressin is absent, the collecting duct epithelium is relatively impermeable to water. The fluid therefore remains hypotonic, and large amounts flow into the renal pelvis. In humans, the urine osmolality may be as low as 30 mOsm/kg of H_2O. The impermeability of the distal portions of the nephron is not absolute; along with the salt that is pumped out of the collecting duct fluid, about 2% of the filtered water is reabsorbed in the absence of vasopressin. However, as much as 13% of the filtered water may be excreted, and urine flow may reach 15 mL/min or more.

THE COUNTERCURRENT MECHANISM

The concentrating mechanism depends upon the maintenance of a gradient of **increasing osmolality** along the medullary pyramids. This gradient is produced by the operation of the loops of Henle as **countercurrent multipliers** and maintained by the operation of the vasa recta as **countercurrent exchangers.** A countercurrent system is a system in which the inflow runs parallel to, counter to, and in close proximity to the outflow for some distance. This occurs for both the loops of Henle and the vasa recta in the renal medulla (Figure 37–3).

The operation of each loop of Henle as a countercurrent multiplier depends on the high permeability of the thin descending limb to water (via aquaporin-1), the active transport of Na^+ and Cl^- out of the thick ascending limb, and the inflow of tubular fluid from the proximal tubule, with outflow into the distal tubule. The process can be explained using hypothetical steps leading to the normal equilibrium condition, although the steps do not occur in vivo. It is also important to remember that the equilibrium is maintained unless the osmotic gradient is washed out. These steps are summarized in Figure 37–15 for a cortical nephron with no thin ascending limb. Assume first a condition in which osmolality is 300 mOsm/kg of H_2O throughout the descending and ascending limbs and the medullary interstitium (Figure 37–15A). Assume in addition that the pumps in the thick ascending limb can pump 100 mOsm/kg of Na^+ and Cl^- from the tubular fluid to the interstitium, increasing interstitial osmolality to 400 mOsm/kg of H_2O. Water then moves out of the thin descending limb, and its contents equilibrate with the interstitium (Figure 37–15B). However, fluid containing 300 mOsm/kg of H_2O is continuously entering this limb from the proximal tubule (Figure 37–15C), so the gradient against which the Na^+ and Cl^- are pumped is reduced and more enters the interstitium (Figure 37–15D). Meanwhile, hypotonic fluid flows into the distal tubule, and isotonic and subsequently hypertonic fluid flows into the ascending thick limb. The process keeps repeating, and the final result is a gradient of osmolality from the top to the bottom of the loop.

In juxtamedullary nephrons with longer loops and thin ascending limbs, the osmotic gradient is spread over a greater distance and the osmolality at the tip of the loop is greater. This is because the thin ascending limb is relatively impermeable to water but permeable to Na^+ and Cl^-. Therefore, Na^+ and Cl^- move down their concentration gradients into the interstitium, and there is additional passive countercurrent multiplication. The greater the length of the loop of Henle, the greater the osmolality that can be reached at the tip of the medulla.

The osmotic gradient in the medullary pyramids would not last long if the Na^+ and urea in the interstitial spaces were removed by the circulation. These solutes remain in the pyramids primarily because the vasa recta operate as countercurrent

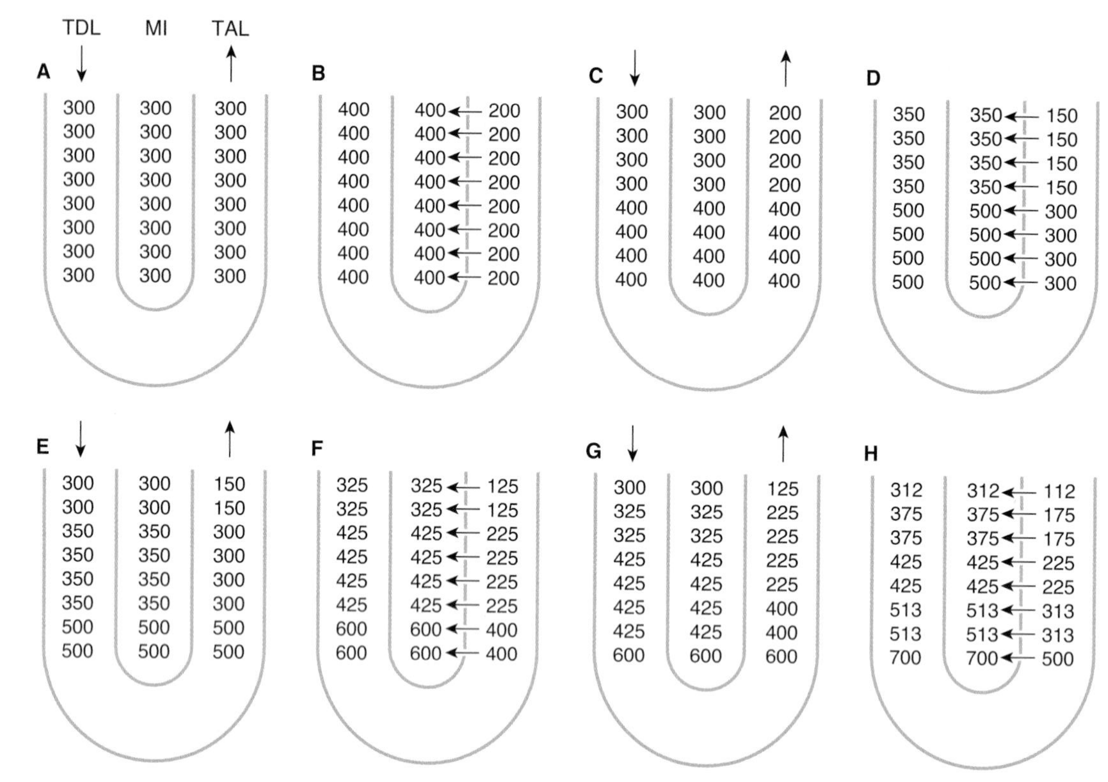

FIGURE 37–15 Operation of the loop of Henle as a countercurrent multiplier producing a gradient of hyperosmolarity in the medullary interstitium (MI). TDL, thin descending limb; TAL, thick ascending limb. The process of generation of the gradient is illustrated as occurring in hypothetical steps, starting at A, where osmolality in both limbs and the interstitium is 300 mOsm/kg of water. The pumps in the thick ascending limb move Na^+ and Cl^- into the interstitium, increasing its osmolality to 400 mOsm/kg, and this equilibrates with the fluid in the thin descending limb. However, isotonic fluid continues to flow into the thin descending limb and hypotonic fluid out of the thick ascending limb. Continued operation of the pumps makes the fluid leaving the thick ascending limb even more hypotonic, while hypertonicity accumulates at the apex of the loop. (Modified and reproduced with permission from Johnson LR [editor]: *Essential Medical Physiology*. Raven Press, 1992.)

exchangers (Figure 37–16). The solutes diffuse out of the vessels conducting blood toward the cortex and into the vessels descending into the pyramid. Conversely, water diffuses out of the descending vessels and into the fenestrated ascending vessels. Therefore, the solutes tend to recirculate in the medulla and water tends to bypass it, so that hypertonicity is maintained. The water removed from the collecting ducts in the pyramids is also removed by the vasa recta and enters the general circulation. Countercurrent exchange is a passive process; it depends on movement of water and could not maintain the osmotic gradient along the pyramids if the process of counter-current multiplication in the loops of Henle were to cease.

It is worth noting that there is a very large osmotic gradient in the loop of Henle and, in the presence of vasopressin, in the collecting ducts. It is the countercurrent system that makes this gradient possible by spreading it along a system of tubules 1 cm or more in length, rather than across a single layer of cells that is only a few micrometers thick. There are other examples of the operation of countercurrent exchangers in animals. One is the heat exchange between the arteries and venae comitantes of the limbs. To a minor degree in humans, but to a major degree in mammals living in cold water, heat is transferred from the arterial blood flowing into the limbs to the adjacent veins draining blood back into the body, making the tips of the limbs cold while conserving body heat (see Chapter 33).

ROLE OF UREA

Urea contributes to the establishment of the osmotic gradient in the medullary pyramids and to the ability to form a concentrated urine in the collecting ducts. Urea transport is mediated by urea transporters, presumably by facilitated diffusion. There are at least four isoforms of the transport protein UT-A in the kidneys (UT-A1 to UT-A4); UT-B is found in erythrocytes and in the descending limbs of the vasa recta. Urea transport in the collecting duct is mediated by UT-A1 and UT-A3, and both are regulated by vasopressin. During antidiuresis, when vasopressin is high, the amount of urea deposited in the medullary interstitium increases, thus increasing the concentrating capacity of the kidney. In addition, the amount of urea in the medullary interstitium and, consequently, in the urine varies with the amount of urea filtered, and this in turn varies with the dietary intake of protein. Therefore, a high-protein diet increases the ability of the kidneys to concentrate

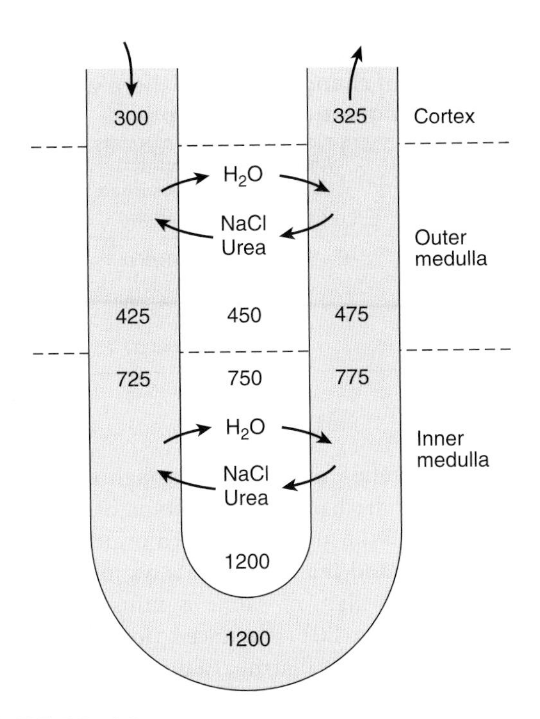

FIGURE 37–16 Operation of the vasa recta as countercurrent exchangers in the kidney. NaCl and urea diffuse out of the ascending limb of the vessel and into the descending limb, whereas water diffuses out of the descending and into the ascending limb of the vascular loop. (Modified and reproduced with permission from Pitts RF: *Physiology of the Kidney and Body Fluid,* 3rd ed. Chicago: Yearbook Medical Publications, 1974.)

the urine and a low-protein diet reduces the kidneys' ability to concentrate the urine.

OSMOTIC DIURESIS

The presence of large quantities of unreabsorbed solutes in the renal tubules causes an increase in urine volume called **osmotic diuresis.** Solutes that are not reabsorbed in the proximal tubules exert an appreciable osmotic effect as the volume of tubular fluid decreases and their concentration rises. Therefore, they "hold water in the tubules." In addition, the concentration gradient against which Na^+ can be pumped out of the proximal tubules is limited. Normally, the movement of water out of the proximal tubule prevents any appreciable gradient from developing, but Na^+ concentration in the fluid falls when water reabsorption is decreased because of the presence in the tubular fluid of increased amounts of unreabsorbable solutes. The limiting concentration gradient is reached, and further proximal reabsorption of Na^+ is prevented; more Na^+ remains in the tubule, and water stays with it. The result is that the loop of Henle is presented with a greatly increased volume of isotonic fluid. This fluid has a decreased Na^+ concentration, but the total amount of Na^+ reaching the loop per unit time is increased. In the loop, reabsorption of water and Na^+ is decreased because the medullary hypertonicity is decreased. The decrease is due primarily to decreased reabsorption of Na^+, K^+, and Cl^- in the ascending limb of the loop because the limiting concentration gradient for Na^+ reabsorption is reached. More fluid passes through the distal tubule, and because of the decrease in the osmotic gradient along the medullary pyramids, less water is reabsorbed in the collecting ducts. The result is a marked increase in urine volume and excretion of Na^+ and other electrolytes.

Osmotic diuresis is produced by the administration of compounds such as mannitol and related polysaccharides that are filtered but not reabsorbed. It is also produced by naturally occurring substances when they are present in amounts exceeding the capacity of the tubules to reabsorb them. For example, in **diabetes mellitus,** if blood glucose is high, glucose in the glomerular filtrate is high, thus the filtered load will exceed the Tm_G and glucose will remain in the tubules causing polyuria. Osmotic diuresis can also be produced by the infusion of large amounts of sodium chloride or urea.

It is important to recognize the difference between osmotic diuresis and water diuresis. In water diuresis, the amount of water reabsorbed in the proximal portions of the nephron is normal, and the maximal urine flow that can be produced is about 16 mL/min. In osmotic diuresis, increased urine flow is due to decreased water reabsorption in the proximal tubules and loops and very large urine flows can be produced. As the load of excreted solute is increased, the concentration of the urine approaches that of plasma (Figure 37–17) in spite of

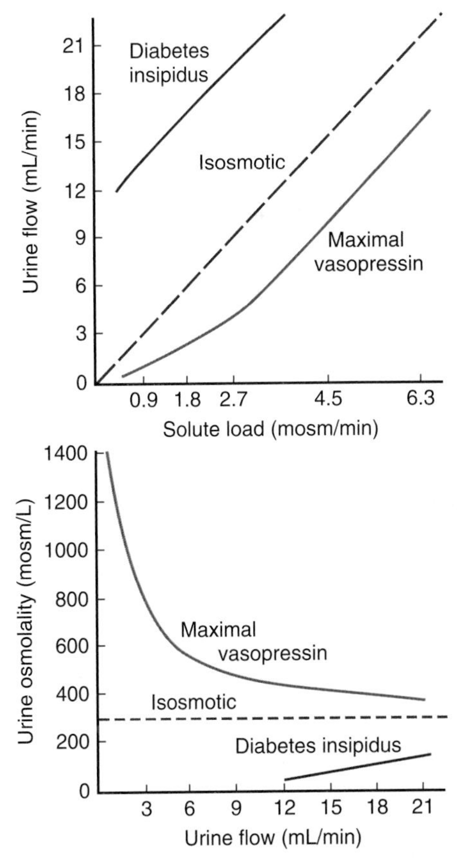

FIGURE 37–17 Approximate relationship between urine concentration and urine flow in osmotic diuresis in humans. The dashed line in the lower diagram indicates the concentration at which the urine is isosmotic with plasma. (Reproduced with permission from Berliner RW, Giebisch G: In *Best and Taylor's Physiological Basis of Medical Practice,* 9th ed. Brobeck JR [editor]. Williams & Wilkins, 1979.)

maximal vasopressin secretion, because an increasingly large fraction of the excreted urine is isotonic proximal tubular fluid. If osmotic diuresis is produced in an animal with diabetes insipidus, the urine concentration rises for the same reason.

RELATION OF URINE CONCENTRATION TO GFR

The magnitude of the osmotic gradient along the medullary pyramids is increased when the rate of flow of fluid through the loops of Henle is decreased. A reduction in GFR such as that caused by dehydration produces a decrease in the volume of fluid presented to the countercurrent mechanism, so that the rate of flow in the loops declines and the urine becomes more concentrated. When the GFR is low, the urine can become quite concentrated in the absence of vasopressin. If one renal artery is constricted in an animal with diabetes insipidus, the urine excreted on the side of the constriction becomes hypertonic because of the reduction in GFR, whereas that excreted on the opposite side remains hypotonic.

"FREE WATER CLEARANCE"

In order to quantitate the gain or loss of water by excretion of a concentrated or dilute urine, the "free water clearance" (C_{H_2O}) is sometimes calculated. This is the difference between the urine volume and the clearance of osmoles (C_{Osm}):

$$C_{H_2O} = \dot{V} - \frac{U_{Osm}\dot{V}}{P_{Osm}}$$

where $\dot{V}$ is the urine flow rate and U_{Osm} and P_{Osm} the urine and plasma osmolality, respectively. C_{Osm} is the amount of water necessary to excrete the osmotic load in a urine that is isotonic with plasma. Therefore, C_{H_2O} is negative when the urine is hypertonic and positive when the urine is hypotonic. For example, using the data in Table 37–6, the values for C_{H_2O} are −1.3 mL/min (−1.9 L/d) during maximal antidiuresis and 14.5 mL/min (20.9 L/d) in the absence of vasopressin.

REGULATION OF Na^+ EXCRETION

Na^+ is filtered in large amounts, but it is actively transported out of all portions of the tubule except the descending thin limb of Henle's loop. Normally about 99% of the filtered Na^+ is reabsorbed. Because Na^+ is the most abundant cation in ECF and because Na^+ salts account for over 90% of the osmotically active solute in the plasma and interstitial fluid, the amount of Na^+ in the body is a prime determinant of the ECF volume. Therefore, it is not surprising that multiple regulatory mechanisms have evolved in terrestrial animals to control the excretion of this ion. Through the operation of these regulatory mechanisms, the amount of Na^+ excreted is adjusted to equal the amount ingested over a wide range of dietary intakes, and the individual stays in Na^+ balance. When Na intake is high, or saline is infused, natriuresis occurs, whereas when ECF is reduced (for example, fluid loss following vomiting or diarrhea) a decrease in Na^+ excretion occurs. Thus, urinary Na^+ output ranges from less than 1 mEq/d on a low-salt diet to 400 mEq/d or more when the dietary Na^+ intake is high.

TABLE 37–7 Changes in Na^+ excretion that would occur as a result of changes in GFR if there were no concomitant changes in Na^+ reabsorption.

GFR (mL/min)	Plasma Na^+ (μEq/mL)	Amount Filtered (μEq/min)	Amount Reabsorbed (μEq/min)	Amount Excreted (μEq/min)
125	145	18,125	18,000	125
127	145	18,415	18,000	415
124.1	145	18,000	18,000	0

MECHANISMS

Variations in Na^+ excretion are brought about by changes in GFR (Table 37–7) and changes in tubular reabsorption, primarily in the 3% of filtered Na^+ that reaches the collecting ducts. The factors affecting GFR, including tubuloglomerular feedback, have been discussed previously. Factors affecting Na^+ reabsorption include the circulating level of aldosterone and other adrenocortical hormones, the circulating level of ANP and other natriuretic hormones, and the rate of tubular secretion of H^+ and K^+.

EFFECTS OF ADRENOCORTICAL STEROIDS

Adrenal mineralocorticoids such as aldosterone increase tubular reabsorption of Na^+ in association with secretion of K^+ and H^+ and also Na^+ reabsorption with Cl^-. When these hormones are injected into adrenalectomized animals, a latent period of 10–30 min occurs before their effects on Na^+ reabsorption become manifest, because of the time required for the steroids to alter protein synthesis via their action on DNA. Mineralocorticoids may also have more rapid membrane-mediated effects, but these are not apparent in terms of Na^+ excretion in the whole animal. The mineralocorticoids act primarily in the collecting ducts to increase the number of active epithelial sodium channels (ENaCs) in this part of the nephron. The molecular mechanisms believed to be involved are discussed in Chapter 20 and summarized in Figure 37–18.

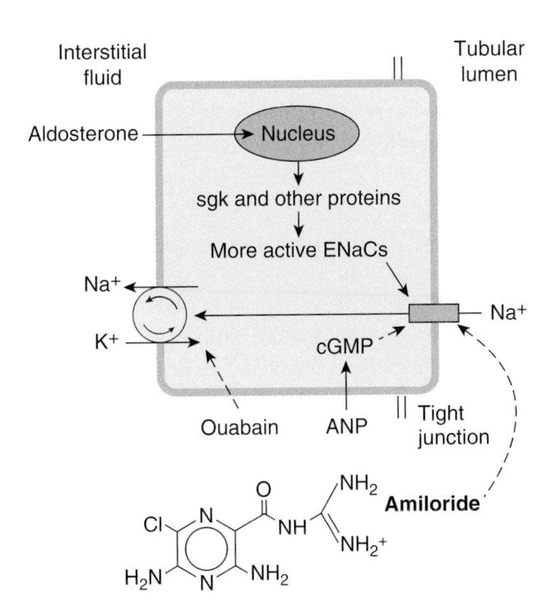

FIGURE 37–18 Renal principal cell. Na^+ enters via the ENaCs in the apical membrane and is pumped into the interstitial fluid by Na, K ATPases in the basolateral membrane. Aldosterone activates the genome to produce serum- and glucocorticoid-regulated kinase (sgk) and other proteins, and the number of active ENaCs is increased.

In Liddle syndrome, mutations in the genes that code for the β subunit and less commonly the γ subunit of ENaC cause the channels to become constitutively active in the kidney. This leads to Na^+ retention and hypertension.

OTHER HUMORAL EFFECTS

Reduction of dietary intake of salt increases aldosterone secretion (see Figure 20–24), producing marked but slowly developing decreases in Na^+ excretion. A variety of other humoral factors affect Na^+ reabsorption. PGE_2 causes a natriuresis, possibly by inhibiting Na, K ATPase and possibly by increasing intracellular Ca^{2+}, which in turn inhibits Na^+ transport via ENaCs. Endothelin and IL-1 cause natriuresis, probably by increasing the formation of PGE_2. ANP and related molecules increase intracellular cyclic 3',5'-guanosine monophosphate (cGMP), and this inhibits transport via ENaC. Inhibition of Na, K ATPase by another natriuretic hormone, which appears to be endogenously produced ouabain, also increases Na^+ excretion. Angiotensin II increases reabsorption of Na^+ and HCO_3^- by an action on the proximal tubules. There is an appreciable amount of angiotensin-converting enzyme in the kidneys, and the kidneys convert 20% of the circulating angiotensin I to angiotensin II. In addition, angiotensin I is generated in the kidneys.

Prolonged exposure to high levels of circulating mineralocorticoids does not cause edema in otherwise normal individuals because eventually the kidneys escape from the effects of the steroids. This **escape phenomenon,** which may be due to increased secretion of ANP, is discussed in Chapter 20. It appears to be reduced or absent in nephrosis, cirrhosis, and heart failure, and patients with these diseases continue to retain Na^+ and become edematous when exposed to high levels of mineralocorticoids.

REGULATION OF WATER EXCRETION

WATER DIURESIS

The feedback mechanism controlling vasopressin secretion and the way vasopressin secretion is stimulated by a rise and inhibited by a drop in the effective osmotic pressure of the plasma are discussed in Chapter 17. The **water diuresis** produced by drinking large amounts of hypotonic fluid begins about 15 min after ingestion of a water load and reaches its maximum in about 40 min. The act of drinking produces a small decrease in vasopressin secretion before the water is absorbed, but most of the inhibition is produced by the decrease in plasma osmolality after the water is absorbed.

WATER INTOXICATION

During excretion of an average osmotic load, the maximal urine flow that can be produced during a water diuresis is about 16 mL/min. If water is ingested at a higher rate than this for any length of time, swelling of the cells because of the uptake of water from the hypotonic ECF becomes severe and, rarely, the symptoms of **water intoxication** may develop. Swelling of the cells in the brain causes convulsions and coma and leads eventually to death. Water intoxication can also occur when water intake is not reduced after administration of exogenous vasopressin or when secretion of endogenous vasopressin occurs in response to nonosmotic stimuli such as surgical trauma. Administration of oxytocin after parturition (to contract the uterus) can also lead to water intoxication if water intake is not monitored carefully.

REGULATION OF K^+ EXCRETION

Much of the filtered K^+ is removed from the tubular fluid by active reabsorption in the proximal tubules, and K^+ is then secreted into the fluid by the distal tubular cells. The rate of K^+ secretion is proportional to the rate of flow of the tubular fluid through the distal portions of the nephron, because with rapid flow there is less opportunity for the tubular K^+ concentration to rise to a value that stops further secretion. In the absence of complicating factors, the amount secreted is approximately equal to the K^+ intake, and K^+ balance is maintained. In the collecting ducts, Na^+ is generally reabsorbed and K^+ is secreted. There is no rigid one-for-one exchange, and much of the movement of K^+ is passive. However, there is electrical coupling in the sense that intracellular migration of Na^+ from the lumen tends to lower the potential difference across the tubular cell, and this favors movement of K^+ into the tubular lumen. K^+ excretion is decreased when the amount of Na^+ reaching the distal tubule is small. In addition, if H^+ secretion is increased, K^+ excretion will decrease as K^+ is reabsorbed in collecting duct cells in exchange for H^+, via the action of the H,K-ATPase.

DIURETICS

Although a detailed discussion of diuretic agents is beyond the scope of this book, consideration of their mechanisms of action constitutes an informative review of the factors affecting urine volume and electrolyte excretion. These mechanisms are summarized in Table 37–8. Water, alcohol, osmotic diuretics, xanthines, and acidifying salts have limited clinical usefulness, and the vasopressin antagonists are currently undergoing clinical trials. However, many of the other agents on the list are used extensively in medical practice.

The carbonic anhydrase-inhibiting drugs are only moderately effective as diuretic agents, but because they inhibit acid secretion by decreasing the supply of carbonic acid, they have far-reaching effects. Not only is Na^+ excretion increased because H^+ secretion is decreased, but HCO_3^- reabsorption is also depressed; and because H^+ and K^+ compete with each other and with Na^+, the decrease in H^+ secretion facilitates the secretion and excretion of K^+.

Furosemide and the other loop diuretics inhibit the Na–K–2Cl cotransporter in the thick ascending limb of Henle's loop. They cause a marked natriuresis and kaliuresis. Thiazides act by inhibiting Na–Cl cotransport in the distal tubule. The diuresis they cause is less marked, but both loop diuretics and thiazides cause increased delivery of Na^+ (and fluid) to the collecting ducts, facilitating K^+ excretion. Thus, over time, K^+ depletion and hypokalemia are common complications in those who use them if they do not supplement their K^+ intake. On the other hand, the so-called K^+-sparing diuretics act in the collecting duct by inhibiting the action of aldosterone or blocking ENaCs.

TABLE 37–8 Mechanism of action of various diuretics.

Agent	Mechanism of Action
Water	Inhibits vasopressin secretion
Ethanol	Inhibits vasopressin secretion
Antagonists of V_2 vasopressin receptors such as astolvaptan	Inhibit action of vasopressin on collecting duct
Large quantities of osmotically active substances such as mannitol and glucose	Produce osmotic diuresis
Xanthines such as caffeine and theophylline	Decrease tubular reabsorption of Na^+ and increase GFR
Acidifying salts such as $CaCl_2$ and NH_4Cl	Supply acid load; H^+ is buffered, but an anion is excreted with Na^+ when the ability of the kidneys to replace Na^+ with H^+ is exceeded
Carbonic anhydrase inhibitors such as acetazolamide (Diamox)	Decrease H^+ secretion, with resultant increase in Na^+ and K^+ excretion
Metolazone (Zaroxolyn), thiazides such as chlorothiazide (Diuril)	Inhibit the Na–Cl cotransporter in the early portion of the distal tubule
Loop diuretics such as furosemide (Lasix), ethacrynic acid (Edecrin), and bumetanide	Inhibit the Na–K–2Cl cotransporter in the medullary thick ascending limb of the loop of Henle
K^+-retaining natriuretics such as spironolactone (Aldactone), triamterene (Dyrenium), and amiloride (Midamor)	Inhibit Na^+–K^+ "exchange" in the collecting ducts by inhibiting the action of aldosterone (spironolactone) or by inhibiting the ENaCs (amiloride)

EFFECTS OF DISORDERED RENAL FUNCTION

A number of abnormalities are common to many different types of renal disease. The secretion of renin by the kidneys and the relation of the kidneys to hypertension are discussed in Chapter 38. A frequent finding in various forms of renal disease is the presence in the urine of protein, leukocytes, red cells, and **casts,** which are proteinaceous material precipitated in the tubules and washed into the bladder. Other important consequences of renal disease are loss of the ability to concentrate or dilute the urine, uremia, acidosis, and abnormal retention of Na^+ (see Clinical Box 37–3).

LOSS OF CONCENTRATING & DILUTING ABILITY

In renal disease, the urine becomes less concentrated and urine volume is often increased, producing the symptoms of **polyuria** and **nocturia** (waking up at night to void). The ability to form a dilute urine is often retained, but in advanced renal disease, the osmolality of the urine becomes fixed at about that of plasma, indicating that the diluting and concentrating functions of the kidney have both been lost. The loss is due in part to disruption of the countercurrent mechanism, but a more important cause is a loss of functioning nephrons. When one kidney is removed surgically, the number of functioning nephrons is halved. The number of osmoles to be excreted is not reduced to this extent, and so the remaining nephrons must each be filtering and excreting more osmotically active substances, producing what is in effect an osmotic diuresis. In osmotic diuresis, the osmolality of the urine approaches that of plasma.

CLINICAL BOX 37–3

Proteinuria

In many renal diseases and in one benign condition, the permeability of the glomerular capillaries is increased, and protein is found in the urine in more than the usual trace amounts **(proteinuria).** Most of this protein is **albumin,** and the defect is commonly called albuminuria. The relation of charges on the glomerular membrane to albuminuria has been discussed above. The amount of protein in the urine may be very large, and especially in nephrosis, the urinary protein loss may exceed the rate at which the liver can synthesize plasma proteins. The resulting hypoproteinemia reduces the oncotic pressure, and the plasma volume declines, sometimes to dangerously low levels, while edema fluid accumulates in the tissues.

A benign condition that causes proteinuria is a poorly understood change in renal hemodynamics, which in some otherwise normal individuals, causes protein to appear in urine when they are in the standing position **(orthostatic albuminuria).** Urine formed when these individuals are lying down is protein-free.

The same thing happens when the number of functioning nephrons is reduced by disease. The increased filtration in the remaining nephrons eventually damages them, and thus more nephrons are lost. The damage resulting from increased filtration may be due to progressive fibrosis in the proximal tubule cells, but this is unsettled. However, the eventual result of this positive feedback is loss of so many nephrons that complete renal failure with **oliguria,** or even **anuria,** results.

UREMIA

When the breakdown products of protein metabolism accumulate in the blood, the syndrome known as **uremia** develops. The symptoms of uremia include lethargy, anorexia, nausea and vomiting, mental deterioration and confusion, muscle twitching, convulsions, and coma. The blood urea nitrogen (BUN) and creatinine levels are high, and the blood levels of these substances are used as an index of the severity of the uremia. It probably is not the accumulation of urea and creatinine per se but rather the accumulation of other toxic substances—possibly organic acids or phenols—that produces the symptoms of uremia.

The toxic substances that cause the symptoms of uremia can be removed by dialyzing the blood of uremic patients against a bath of suitable composition in an artificial kidney **(hemodialysis).** Patients can be kept alive and in reasonable health for many months on dialysis, even when they are completely anuric or have had both kidneys removed. However, the treatment of choice today is certainly transplantation of a kidney from a suitable donor.

Other features of chronic renal failure include anemia, which is caused primarily by failure to produce erythropoietin, and secondary hyperparathyroidism due to 1,25-dihydroxycholecalciferol deficiency (see Chapter 21).

ACIDOSIS

Acidosis is common in chronic renal disease because of failure to excrete the acid products of digestion and metabolism (see Chapter 39). In the rare syndrome of **renal tubular acidosis,** there is specific impairment of the ability to make the urine acidic, and other renal functions are usually normal. However, in most cases of chronic renal disease the urine is maximally acidified, and acidosis develops because the total amount of H^+ that can be secreted is reduced because of impaired renal tubular production of NH_4^+.

ABNORMAL Na^+ HANDLING

Many patients with renal disease retain excessive amounts of Na^+ and become edematous. Na^+ retention in renal disease has at least three causes. In acute glomerulonephritis, a disease that affects primarily the glomeruli, the amount of Na^+ filtered is decreased markedly. In the nephrotic syndrome, an increase in aldosterone secretion contributes to salt retention. The plasma protein level is low in this condition, and so fluid moves from the plasma into the interstitial spaces and the plasma volume falls. The decline in plasma volume triggers an increase in aldosterone secretion via the renin–angiotensin system. A third cause of Na^+ retention and edema in renal disease is **heart failure.** Renal disease predisposes to heart failure, partly because of the hypertension it frequently produces.

THE BLADDER

FILLING

The walls of the ureters contain smooth muscle arranged in spiral, longitudinal, and circular bundles, but distinct layers of muscle are not seen. Regular peristaltic contractions occurring one to five times per minute move the urine from the renal pelvis to the bladder, where it enters in spurts synchronous with each peristaltic wave. The ureters pass obliquely through the bladder wall and, although there are no ureteral sphincters as such, the oblique passage tends to keep the ureters closed except during peristaltic waves, preventing reflux of urine from the bladder.

EMPTYING

The smooth muscle of the bladder, like that of the ureters, is arranged in spiral, longitudinal, and circular bundles. Contraction of the circular muscle, which is called the

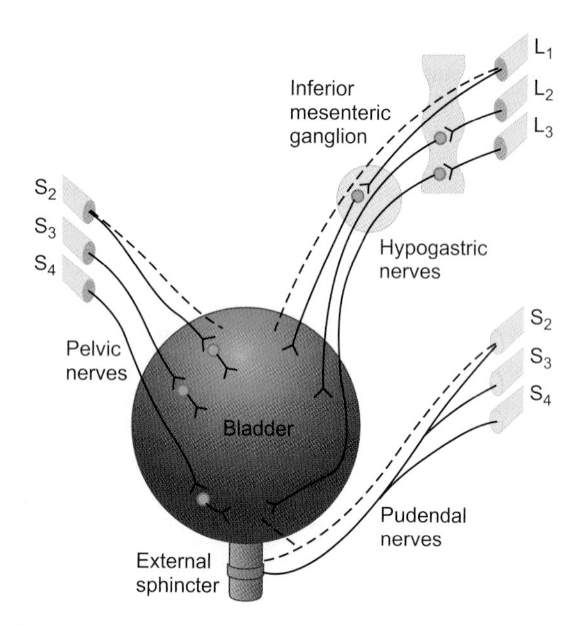

FIGURE 37–19 Innervation of the bladder. Dashed lines indicate sensory nerves. Parasympathetic innervation is shown at the left, sympathetic at the upper right, and somatic at the lower right.

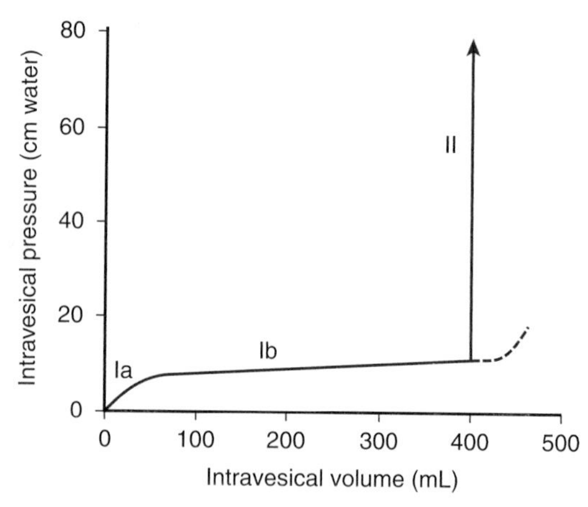

FIGURE 37–20 Cystometrogram in a normal human. The numerals identify the three components of the curve described in the text. The dashed line indicates the pressure–volume relations that would have been found had micturition not occurred and produced component II. (Modified and reproduced with permission from Tanagho EA, McAninch JW: *Smith's General Urology,* 15th ed. McGraw-Hill, 2000.)

detrusor muscle, is mainly responsible for emptying the bladder during urination (micturition). Muscle bundles pass on either side of the urethra, and these fibers are sometimes called the **internal urethral sphincter,** although they do not encircle the urethra. Farther along the urethra is a sphincter of skeletal muscle, the sphincter of the membranous urethra **(external urethral sphincter).** The bladder epithelium is made up of a superficial layer of flat cells and a deep layer of cuboidal cells. The innervation of the bladder is summarized in Figure 37–19.

The physiology of bladder emptying and the physiologic basis of its disorders are subjects about which there is much confusion. Micturition is fundamentally a spinal reflex facilitated and inhibited by higher brain centers and, like defecation, subject to voluntary facilitation and inhibition. Urine enters the bladder without producing much increase in intravesical pressure until the viscus is well filled. In addition, like other types of smooth muscle, the bladder muscle has the property of plasticity; when it is stretched, the tension initially produced is not maintained. The relation between intravesical pressure and volume can be studied by inserting a catheter and emptying the bladder, then recording the pressure while the bladder is filled with 50-mL increments of water or air **(cystometry).** A plot of intravesical pressure against the volume of fluid in the bladder is called a **cystometrogram** (Figure 37–20). The curve shows an initial slight rise in pressure when the first increments in volume are produced; a long, nearly flat segment as further increments are produced; and a sudden, sharp rise in pressure as the micturition reflex is triggered. These three components are sometimes called segments Ia, Ib, and II. The first urge to void is felt at a bladder volume of about 150 mL, and a marked sense of fullness at about 400 mL. The flatness of segment Ib is a manifestation of the law of Laplace. This law states that the pressure in a spherical viscus is equal to twice the wall tension divided by the radius. In the case of the bladder, the tension increases as the organ fills, but so does the radius. Therefore, the pressure increase is slight until the organ is relatively full.

During micturition, the perineal muscles and external urethral sphincter are relaxed, the detrusor muscle contracts, and urine passes out through the urethra. The bands of smooth muscle on either side of the urethra apparently play no role in micturition, and their main function in males is believed to be the prevention of reflux of semen into the bladder during ejaculation.

The mechanism by which voluntary urination is initiated remains unsettled. One of the initial events is relaxation of the muscles of the pelvic floor, and this may cause a sufficient downward tug on the detrusor muscle to initiate its contraction. The perineal muscles and external sphincter can be contracted voluntarily, preventing urine from passing down the urethra or interrupting the flow once urination has begun. It is through the learned ability to maintain the external sphincter in a contracted state that adults are able to delay urination until the opportunity to void presents itself. After urination, the female urethra empties by gravity. Urine remaining in the urethra of the male is expelled by several contractions of the bulbocavernosus muscle.

REFLEX CONTROL

The bladder smooth muscle has some inherent contractile activity; however, when its nerve supply is intact, stretch receptors in the bladder wall initiate a reflex contraction that has a lower threshold than the inherent contractile response of the muscle. Fibers in the pelvic nerves are the afferent limb of the voiding reflex, and the parasympathetic fibers to the bladder that constitute the efferent limb also travel in these nerves. The reflex is integrated in the sacral portion of the spinal cord. In the adult, the volume of urine in the bladder that normally initiates a reflex contraction is about 300–400 mL. The sympathetic nerves to the bladder play no part in micturition, but in males they do mediate the contraction of the bladder muscle that prevents semen from entering the bladder during ejaculation.

The stretch receptors in the bladder wall have no small motor nerve system. However, the threshold for the voiding reflex, like the stretch reflexes, is adjusted by the activity of facilitatory and inhibitory centers in the brainstem. There is a facilitatory area in the pontine region and an inhibitory area in the midbrain. After transection of the brain stem just above the pons, the threshold is lowered and less bladder filling is required to trigger it, whereas after transection at the top of the midbrain, the threshold for the reflex is essentially normal. There is another facilitatory area in the posterior hypothalamus. Humans with lesions in the superior frontal gyrus have a reduced desire to urinate and difficulty in stopping micturition once it has commenced. However, stimulation experiments in animals indicate that other cortical areas also affect the process. The bladder can be made to contract by voluntary facilitation of the spinal voiding reflex when it contains only a few milliliters of urine. Voluntary contraction of the abdominal muscles aids the expulsion of urine by increasing the intra-abdominal pressure, but voiding can be initiated without straining even when the bladder is nearly empty.

EFFECTS OF DEAFFERENTATION

When the sacral dorsal roots are cut in experimental animals or interrupted by diseases of the dorsal roots, such as **tabes dorsalis** in humans, all reflex contractions of the bladder are abolished. The bladder becomes distended, thin-walled, and hypotonic, but some contractions occur because of the intrinsic response of the smooth muscle to stretch.

EFFECTS OF DENERVATION

When the afferent and efferent nerves are both destroyed, as they may be by tumors of the cauda equina or filum terminale, the bladder is flaccid and distended for a while. Gradually, however, the muscle of the "decentralized bladder" becomes active, with many contraction waves that expel dribbles of urine out of the urethra. The bladder becomes shrunken and the bladder wall hypertrophied. The reason for the difference between the small, hypertrophic bladder seen in this condition and the distended, hypotonic bladder seen when only the afferent nerves are interrupted is not known. The hyperactive state in the former condition suggests the development of denervation hypersensitization even though the neurons interrupted are preganglionic rather than postganglionic (see Clinical Box 37–4).

CLINICAL BOX 37–4

Abnormalities of Micturition

Three major types of bladder dysfunction are due to neural lesions: (1) due to interruption of the afferent nerves from the bladder, (2) due to interruption of both afferent and efferent nerves, and (3) due to interruption of facilitatory and inhibitory pathways descending from the brain. In all three types the bladder contracts, but the contractions are generally not sufficient to empty the viscus completely, and residual urine is left in the bladder.

EFFECTS OF SPINAL CORD TRANSECTION

During spinal shock, the bladder is flaccid and unresponsive. It becomes overfilled, and urine dribbles through the sphincters **(overflow incontinence).** After spinal shock has passed, the voiding reflex returns, although there is, of course, no voluntary control and no inhibition or facilitation from higher centers when the spinal cord is transected. Some paraplegic patients train themselves to initiate voiding by pinching or stroking their thighs, provoking a mild mass reflex (see Chapter 12). In some instances, the voiding reflex becomes hyperactive, bladder capacity is reduced, and the wall becomes hypertrophied. This type of bladder is sometimes called the **spastic neurogenic bladder.** The reflex hyperactivity is made worse by, and may be caused by, infection in the bladder wall.

CHAPTER SUMMARY

- Plasma enters the kidneys and is filtered in the glomerulus. As the filtrate passes down the nephron and through the tubules its volume is reduced and water and solutes are removed (tubular reabsorption) and waste products are secreted (tubular secretion).
- A nephron consists of an individual renal tubule and its glomerulus. Each tubule has several segments, beginning with the proximal tubule, followed by the loop of Henle (descending and ascending limbs), the distal convoluted tubule, the connecting tubule, and the collecting duct.
- The kidneys receive just under 25% of the cardiac output and renal plasma flow can be measured by infusing p-aminohippuric acid (PAH) and determining its urine and plasma concentrations.
- Renal blood flow enters the glomerulus via the afferent arteriole and leaves via the efferent arteriole (whose diameter is smaller). Renal blood flow is regulated by norepinephrine (constriction, reduction of flow), dopamine (vasodilation, increases flow), angiotensin II (constricts), prostaglandins (dilation in the renal cortex and constriction in the renal medulla), and acetylcholine (vasodilation).

- Glomerular filtration rate can be measured by a substance that is freely filtered and neither reabsorbed nor secreted in the tubules, is nontoxic, and is not metabolized by the body. Inulin meets these criteria and is extensively used to measure GFR.
- Urine is stored in the bladder before voiding (micturition). The micturition response involves reflex pathways, but is under voluntary control.

MULTIPLE-CHOICE QUESTIONS

For all questions, select the single best answer unless otherwise directed.

1. In the presence of vasopressin, the greatest fraction of filtered water is absorbed in the
   A. proximal tubule.
   B. loop of Henle.
   C. distal tubule.
   D. cortical collecting duct.
   E. medullary collecting duct.
2. In the absence of vasopressin, the greatest fraction of filtered water is absorbed in the
   A. proximal tubule.
   B. loop of Henle.
   C. distal tubule.
   D. cortical collecting duct.
   E. medullary collecting duct.
3. If the clearance of a substance which is freely filtered is less than that of inulin,
   A. there is net reabsorption of the substance in the tubules.
   B. there is net secretion of the substance in the tubules.
   C. the substance is neither secreted nor reabsorbed in the tubules.
   D. the substance becomes bound to protein in the tubules.
   E. the substance is secreted in the proximal tubule to a greater degree than in the distal tubule.
4. Glucose reabsorption occurs in the
   A. proximal tubule.
   B. loop of Henle.
   C. distal tubule.
   D. cortical collecting duct.
   E. medullary collecting duct.
5. On which of the following does aldosterone exert its greatest effect?
   A. Glomerulus
   B. Proximal tubule
   C. Thin portion of the loop of Henle
   D. Thick portion of the loop of Henle
   E. Cortical collecting duct
6. What is the clearance of a substance when its concentration in the plasma is 10 mg/dL, its concentration in the urine is 100 mg/ dL, and urine flow is 2 mL/min?
   A. 2 mL/min
   B. 10 mL/min
   C. 20 mL/min
   D. 200 mL/min
   E. Clearance cannot be determined from the information given.
7. As urine flow increases during osmotic diuresis
   A. the osmolality of urine falls below that of plasma.
   B. the osmolality of urine increases because of the increased amounts of nonreabsorbable solute in the urine.
   C. the osmolality of urine approaches that of plasma because plasma leaks into the tubules.
   D. the osmolality of urine approaches that of plasma because an increasingly large fraction of the excreted urine is isotonic proximal tubular fluid.
   E. the action of vasopressin on the renal tubules is inhibited.

CHAPTER RESOURCES

Anderson K-E: Pharmacology of lower urinary tract smooth muscles and penile erectile tissue. Pharmacol Rev 1993;45:253.

Brenner BM, Rector FC Jr (editors): *The Kidney*, 6th ed, 2 Vols, Saunders, 1999.

Brown D: The ins and outs of aquaporin-2 trafficking. Am J Physiol Renal Physiol 2003;284:F893.

Brown D, Stow JL: Protein trafficking and polarity in kidney epithelium: From cell biology to physiology. Physiol Rev 1996;76:245.

DiBona GF, Kopp UC: Neural control of renal function. Physiol Rev 1997; 77:75.

Garcia NH, Ramsey CR, Knox FG: Understanding the role of paracellular transport in the proximal tubule. News Physiol Sci 1998;13:38.

Nielsen S, et al: Aquaporins in the kidney: From molecules to medicine. Physiol Rev 2002;82:205.

Spring KR: Epithelial fluid transport: A century of investigation. News Physiol Sci 1999;14:92.

Valten V: Tubuloglomerular feedback and the control of glomerular filtration rate. News Physiol Sci 2003;18:169.

Regulation of Extracellular Fluid Composition & Volume

CHAPTER 38

OBJECTIVES

After reading this chapter, you should be able to:

- Describe how the tonicity (osmolality) of the extracellular fluid is maintained by alterations in water intake and vasopressin secretion.
- Discuss the effects of vasopressin, the receptors on which it acts, and how its secretion is regulated.
- Describe how the volume of the extracellular fluid is maintained by alterations in renin and aldosterone secretion.
- Outline the cascade of reactions that lead to the formation of angiotensin II and its metabolites in the circulation.
- List the functions of angiotensin II and the receptors on which it acts to carry out these functions.
- Describe the structure and functions of ANP, BNP, and CNP and the receptors on which they act.
- Describe the site and mechanism of action of erythropoietin, and the feedback regulation of its secretion.

INTRODUCTION

This chapter is a review of the major homeostatic mechanisms that operate, primarily through the kidneys and the lungs, to maintain the **tonicity,** the **volume,** and the **specific ionic composition** of the extracellular fluid (ECF). The interstitial portion of this fluid is the fluid environment of the cells, and life depends upon the constancy of this "internal sea" (see Chapter 1).

DEFENSE OF TONICITY

The defense of the tonicity of the ECF is primarily the function of the vasopressin-secreting and thirst mechanisms. The total body osmolality is directly proportional to the total body sodium plus the total body potassium divided by the total body water, so that changes in the osmolality of the body fluids occur when a mismatch exists between the amount of these electrolytes and the amount of water ingested or lost from the body. When the effective osmotic pressure of the plasma rises, vasopressin secretion is increased and the thirst mechanism is stimulated; water is retained in the body, diluting the hypertonic plasma; and water intake is increased (Figure 38–1). Conversely, when the plasma becomes hypotonic, vasopressin secretion is decreased and "solute-free water" (water in excess of solute) is excreted. In this way, the tonicity of the body fluids is maintained within a narrow normal range. In health, plasma osmolality ranges from 280 to 295 mOsm/kg of H_2O, with vasopressin secretion maximally inhibited at 285 mOsm/kg and stimulated at higher values (Figure 38–2).

VASOPRESSIN RECEPTORS

There are at least three kinds of vasopressin receptors: V_{1A}, V_{1B}, and V_2. All are G protein-coupled. The V_{1A} and V_{1B} receptors act through phosphatidylinositol hydrolysis to increase the intracellular Ca^{2+} concentration. The V_2 receptors act through G_s to increase cyclic adenosine 3',5'-monophosphate (cAMP) levels.

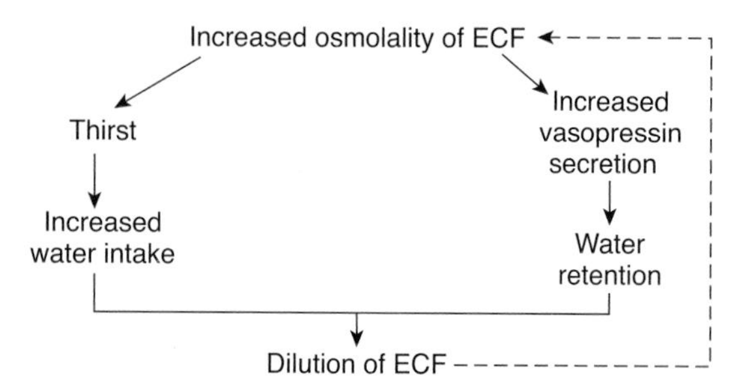

FIGURE 38–1 Mechanisms for defending ECF tonicity. The dashed arrow indicates inhibition. (Courtesy of J Fitzsimmons.)

EFFECTS OF VASOPRESSIN

Because one of its principal physiologic effects is the retention of water by the kidney, vasopressin is often called the **antidiuretic hormone (ADH).** It increases the permeability of the collecting ducts of the kidney, so that water enters the hypertonic interstitium of the renal pyramids. The urine becomes concentrated, and its volume decreases. The overall effect is therefore retention of water in excess of solute; consequently, the effective osmotic pressure of the body fluids is decreased. In the absence of vasopressin, the urine is hypotonic to plasma, urine volume is increased, and there is a net water loss. Consequently, the osmolality of the body fluid rises.

The mechanism by which vasopressin exerts its antidiuretic effect is activated by **V_2 receptors** and involves the insertion of aquaporin 2 into the apical (luminal) membranes of the principal cells of the collecting ducts. Movement of water across membranes by simple diffusion is now known to be augmented by movement through these water channels. These channels are stored in endosomes inside the cells, and vasopressin causes their rapid translocation to the luminal membranes.

V_{1A} receptors mediate the vasoconstrictor effect of vasopressin, and vasopressin is a potent stimulator of vascular smooth muscle in vitro. However, relatively large amounts of vasopressin are needed to raise blood pressure in vivo, because vasopressin also acts on the brain to decrease in cardiac output. The site of this action is the **area postrema,** one of the circumventricular organs (see Chapter 33). Hemorrhage is a potent stimulus for vasopressin secretion, and the blood pressure fall after hemorrhage is more marked in animals that have been treated with synthetic peptides that block the pressor action of vasopressin. Consequently, it appears that vasopressin does play a role in blood pressure homeostasis.

V_{1A} receptors are also found in the liver and the brain. Vasopressin causes glycogenolysis in the liver, and, as noted above, it is a neurotransmitter in the brain and spinal cord.

The V_{1B} receptors (also called V_3 receptors) appear to be unique to the anterior pituitary, where they mediate increased secretion of adrenocorticotropic hormone (ACTH) from the corticotropes.

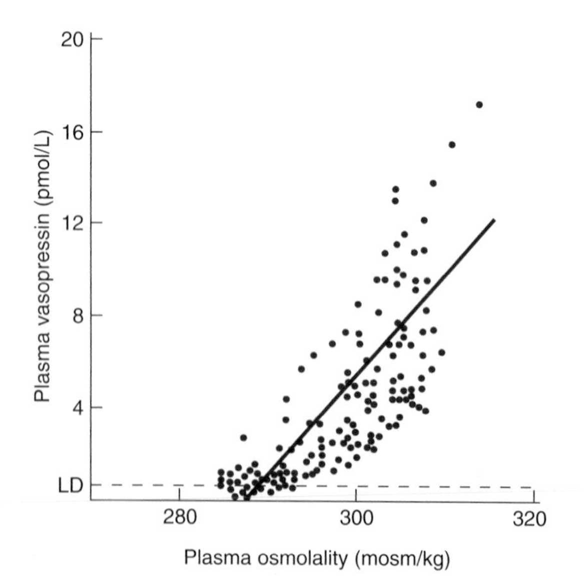

FIGURE 38–2 Relation between plasma osmolality and plasma vasopressin in healthy adult humans during infusion of hypertonic saline. LD, limit of detection. (Reproduced with permission from Thompson CJ, et al: The osmotic thresholds for thirst and vasopressin are similar in healthy humans. Clin Sci [Colch] 1986;71:651.)

METABOLISM

Circulating vasopressin is rapidly inactivated, principally in the liver and kidneys. It has a **biologic half-life** of approximately 18 min in humans.

CONTROL OF VASOPRESSIN SECRETION: OSMOTIC STIMULI

Vasopressin is stored in the posterior pituitary and released into the bloodstream in response to impulses in the nerve fibers that contain the hormone. The factors affecting its secretion are summarized in Table 38–1. When the effective

TABLE 38–1 Summary of stimuli affecting vasopressin secretion.

Vasopressin Secretion Increased	Vasopressin Secretion Decreased
Increased effective osmotic pressure of plasma	Decreased effective osmotic pressure of plasma
Decreased ECF volume	Increased ECF volume
Pain, emotion, "stress," exercise	Alcohol
Nausea and vomiting	
Standing	
Clofibrate, carbamazepine	
Angiotensin II	

osmotic pressure of the plasma is increased above 285 mOsm/kg, the rate of discharge of neurons containing vasopressin increases and vasopressin secretion occurs (Figure 38–2). At 285 mOsm/kg, plasma vasopressin is at or near the limits of detection by available assays, but its levels probably decrease when plasma osmolality is below this level. Vasopressin secretion is regulated by osmoreceptors located in the anterior hypothalamus. They are outside the blood–brain barrier and appear to be located in the circumventricular organs, primarily the organum vasculosum of the lamina terminalis (OVLT) (see Chapter 33). The osmotic threshold for thirst (Figure 38–1) is the same as or slightly greater than the threshold for increased vasopressin secretion (Figure 38–2), and it is still uncertain whether the same osmoreceptors mediate both effects.

Vasopressin secretion is thus controlled by a delicate feedback mechanism that operates continuously to defend the osmolality of the plasma. Significant changes in secretion occur when osmolality is changed as little as 1%. In this way, the osmolality of the plasma in normal individuals is maintained very close to 285 mOsm/L.

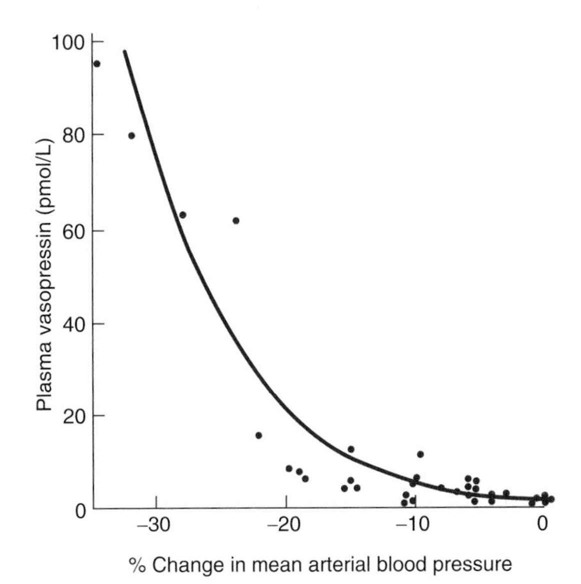

FIGURE 38–3 Relation of mean arterial blood pressure to plasma vasopressin in healthy adult humans in whom a progressive decline in blood pressure was induced by infusion of graded doses of the ganglionic blocking drug trimethaphan. The relation is exponential rather than linear. (Drawn from data in Baylis PH: Osmoregulation and control of vasopressin secretion in healthy humans. Am J Physiol 1987;253:R671.)

VOLUME EFFECTS ON VASOPRESSIN SECRETION

ECF volume also affects vasopressin secretion. Vasopressin secretion is increased when ECF volume is low and decreased when ECF volume is high (Table 38–1). There is an inverse relationship between the rate of vasopressin secretion and the rate of discharge in afferents from stretch receptors in the low- and high-pressure portions of the vascular system. The low-pressure receptors are those in the great veins, right and left atria, and pulmonary vessels; the high-pressure receptors are those in the carotid sinuses and aortic arch (see Chapter 32). The exponential increases in plasma vasopressin produced by decreases in blood pressure are documented in Figure 38–3. However, the low-pressure receptors monitor the fullness of the vascular system, and moderate decreases in blood volume that reduce central venous pressure without lowering arterial pressure can also increase plasma vasopressin.

Thus, the low-pressure receptors are the primary mediators of volume effects on vasopressin secretion. Impulses pass from them via the vagi to the nucleus of the tractus solitarius (NTS). An inhibitory pathway projects from the NTS to the caudal ventrolateral medulla (CVLM), and there is a direct excitatory pathway from the CVLM to the hypothalamus. Angiotensin II reinforces the response to hypovolemia and hypotension by acting on the circumventricular organs to increase vasopressin secretion (see Chapter 33).

Hypovolemia and hypotension produced by conditions such as hemorrhage release large amounts of vasopressin, and in the presence of hypovolemia, the osmotic response curve is shifted to the left (Figure 38–4). Its slope is also increased. The result is water retention and reduced plasma osmolality.

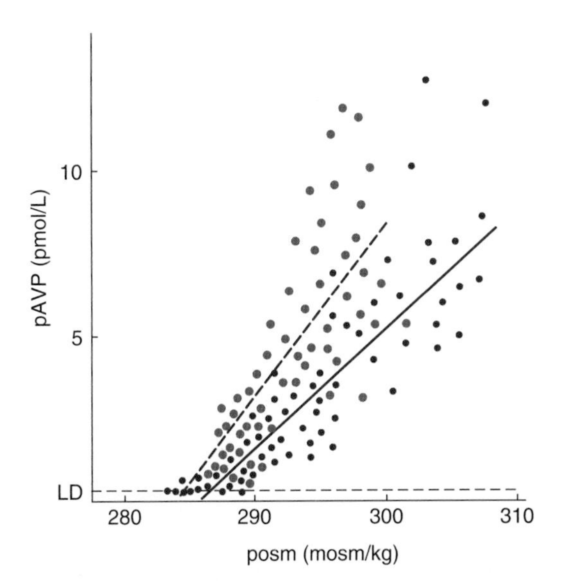

FIGURE 38–4 Effect of hypovolemia and hypervolemia on the relation between plasma vasopressin (pAVP) and plasma osmolality (posm). Seven blood samples were drawn at various times from 10 normal men when hypovolemia was induced by water deprivation (green circles, dashed line) and again when hypervolemia was induced by infusion of hypertonic saline (red circles, solid line). Linear regression analysis defined the relationship pAVP = 0.52 (posm − 283.5) for water deprivation and pAVP = 0.38 (posm − 285.6) for hypertonic saline. LD, limit of detection. Note the steeper curve as well as the shift of the intercept to the left during hypovolemia. (Courtesy of CJ Thompson.)

This includes hyponatremia, since Na^+ is the most abundant osmotically active component of the plasma.

OTHER STIMULI AFFECTING VASOPRESSIN SECRETION

A variety of stimuli in addition to osmotic pressure changes and ECF volume aberrations increase vasopressin secretion. These include pain, nausea, surgical stress, and some emotions (Table 38–1). Nausea is associated with particularly large increases in vasopressin secretion. Alcohol decreases vasopressin secretion.

CLINICAL IMPLICATIONS

In various clinical conditions, volume and other nonosmotic stimuli bias the osmotic control of vasopressin secretion. For example, patients who have had surgery may have elevated levels of plasma vasopressin because of pain and hypovolemia, and this may cause them to develop a low plasma osmolality and dilutional hyponatremia (see Clinical Box 38–1).

Diabetes insipidus is the syndrome that results when there is a vasopressin deficiency **(central diabetes insipidus)** or when the kidneys fail to respond to the hormone **(nephrogenic diabetes insipidus).**

Causes of vasopressin deficiency include disease processes in the supraoptic and paraventricular nuclei, the hypothalamohypophysial tract, or the posterior pituitary gland. It has been estimated that 30% of clinical cases are due to neoplastic lesions of the hypothalamus, either primary or metastatic; 30% are posttraumatic; 30% are idiopathic; and the remainder are due to vascular lesions, infections, systemic diseases such as sarcoidosis that affect the hypothalamus, or mutations in the gene for prepropressophysin. Disease that develops after surgical removal of the posterior lobe of the pituitary may be temporary if the distal ends of the supraoptic and paraventricular fibers are only damaged, because the fibers recover, make new vascular connections, and begin to secrete vasopressin again.

The symptoms of diabetes insipidus are passage of large amounts of dilute urine **(polyuria)** and the drinking of large amounts of fluid **(polydipsia),** provided the thirst mechanism is intact. It is the polydipsia that keeps these patients healthy. If their sense of thirst is depressed for any reason and their intake of dilute fluid decreases, they develop dehydration that can be fatal.

Another cause of diabetes insipidus is inability of the kidneys to respond to vasopressin **(nephrogenic diabetes insipidus).** Two forms of this disease have been described. In one form, the gene for the V_2 receptor is mutated, making the receptor unresponsive. The V_2 receptor gene is on the X chromosome, thus this condition is X-linked and inheritance is sex-linked recessive. In the other form of the condition, mutations occur in the autosomal gene for aquaporin-2 and produce nonfunctional versions of this water channel, many of which do not reach the apical membrane of the collecting duct but are trapped in intracellular locations.

CLINICAL BOX 38–1

Syndrome of Inappropriate Antidiurectic Hormone

The **syndrome of "inappropriate" hypersecretion of antidiuretic hormone (SIADH)** occurs when vasopressin is inappropriately high relative to serum osmolality. Vasopressin is responsible not only for dilutional **hyponatremia** (serum sodium < 135 mmol/L) but also for loss of salt in the urine when water retention is sufficient to expand the ECF volume, reducing aldosterone secretion (see Chapter 20). This occurs in patients with cerebral disease ("cerebral salt wasting") and pulmonary disease ("pulmonary salt wasting"). Hypersecretion of vasopressin in patients with pulmonary diseases such as lung cancer may be due in part to the interruption of inhibitory impulses in vagal afferents from the stretch receptors in the atria and great veins.

A significant number of lung tumors and some other cancers secrete vasopressin. There is a process called **"vasopressin escape"** that counteracts the water-retaining action of vasopressin to limit the degree of hyponatremia in SIADH. Studies in rats have demonstrated that prolonged exposure to elevated levels of vasopressin can lead eventually to down-regulation of the production of aquaporin-2. This permits urine flow suddenly to increase and plasma osmolality to fall despite exposure of the collecting ducts to elevated levels of the hormone; that is, the individual escapes from the renal effects of vasopressin.

THERAPEUTIC HIGHLIGHTS

Patients with inappropriate hypersecretion of vasopressin have been successfully treated with demeclocycline, an antibiotic that reduces the renal response to vasopressin.

SYNTHETIC AGONISTS & ANTAGONISTS

Synthetic peptides that have selective actions and are more active than naturally occurring vasopressin have been produced by altering the amino acid residues. For example, 1-deamino-8-D-arginine vasopressin (desmopressin; dDAVP) has very high antidiuretic activity with little pressor activity, making it valuable in the treatment of vasopressin deficiency.

DEFENSE OF VOLUME

The volume of the ECF is determined primarily by the total amount of osmotically active solute in the ECF. The composition of the ECF is discussed in Chapter 1. Because Na^+ and Cl^- are by far the most abundant osmotically active solutes in ECF, and because changes in Cl^- are to a great extent secondary to changes in Na^+, the amount of Na^+ in the ECF is the most important determinant of ECF volume. Therefore, the mechanisms that control Na^+ balance are the major mechanisms defending ECF volume. However, there is volume control of water excretion as well; a rise in ECF volume inhibits vasopressin secretion, and a decline in ECF volume produces an increase in the secretion of this hormone. Volume stimuli override the osmotic regulation of vasopressin secretion. Angiotensin II stimulates aldosterone and vasopressin secretion. It also causes thirst and constricts blood vessels, which help to maintain blood pressure. Thus, angiotensin II plays a key role in the body's response to hypovolemia (Figure 38–5). In addition, expansion of the ECF volume increases the secretion of

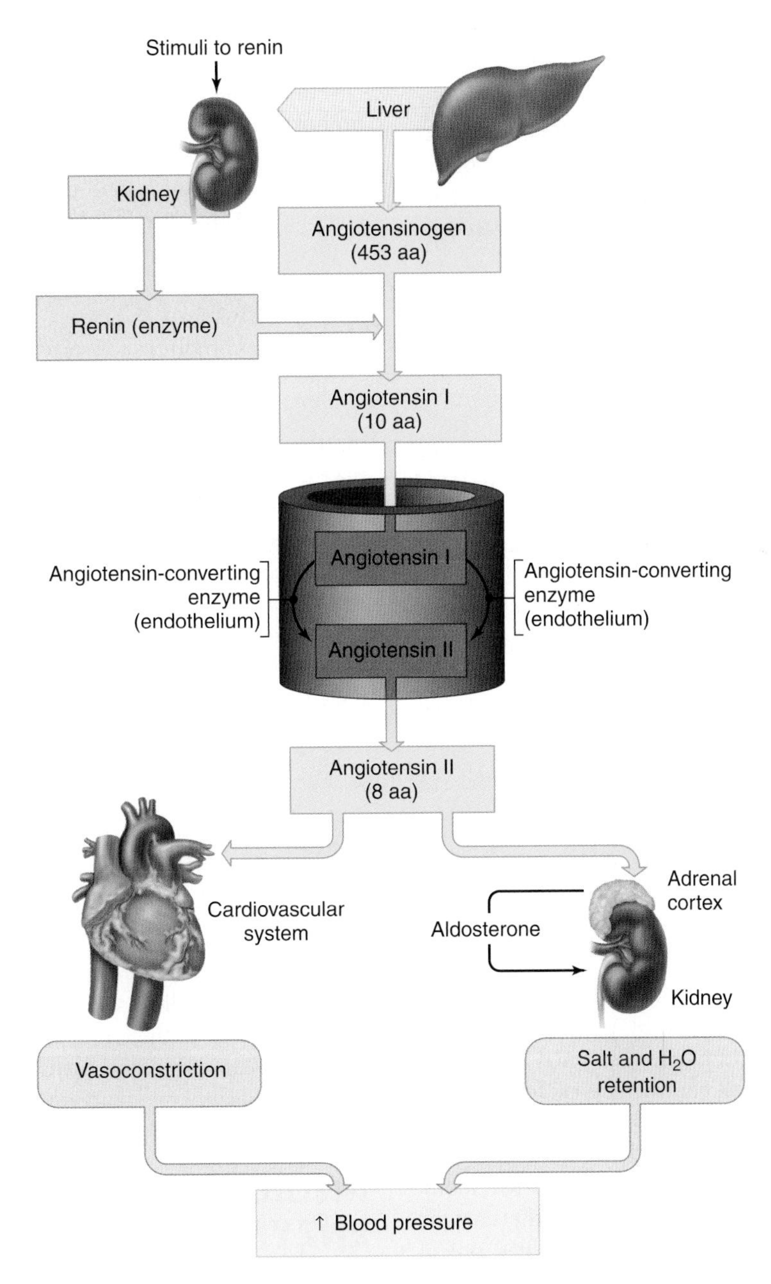

FIGURE 38–5 Summary of the renin–angiotensin system and the stimulation of aldosterone secretion by angiotensin II. The plasma concentration of renin is the rate-limiting step in the renin–angiotensin system; therefore, it is the major determinant of plasma angiotensin II concentration.

atrial natriuretic peptide (ANP) and brain natriuretic peptide (BNP) by the heart, and this causes natriuresis and diuresis.

In disease states, loss of water from the body **(dehydration)** causes a moderate decrease in ECF volume, because water is lost from both the intracellular and ECF compartments; but excessive loss of Na^+ in the stools (diarrhea), urine (severe acidosis, adrenal insufficiency), or sweat (heat prostration) decreases ECF volume markedly and eventually leads to shock. The immediate compensations in shock operate principally to maintain intravascular volume, but they also affect Na^+ balance. In adrenal insufficiency, the decline in ECF volume is not only due to loss of Na^+ in the urine but also to its movement into cells. Because of the key position of Na^+ in volume homeostasis, it is not surprising that more than one mechanism has evolved to control the excretion of this ion.

The filtration and reabsorption of Na^+ in the kidneys and the effects of these processes on Na^+ excretion are discussed in Chapter 37. When ECF volume is decreased, blood pressure falls, glomerular capillary pressure declines, and the glomerular filtration rate (GFR) therefore falls, reducing the amount of Na^+ filtered. Tubular reabsorption of Na^+ is increased, in part because the secretion of aldosterone is increased. Aldosterone secretion is controlled in part by a feedback system in which the change that initiates increased secretion is a decline in mean intravascular pressure. Other changes in Na^+ excretion occur too rapidly to be solely due to changes in aldosterone secretion. For example, rising from the supine to the standing position increases aldosterone secretion. However, Na^+ excretion is decreased within a few minutes, and this rapid change in Na^+ excretion occurs in adrenalectomized subjects. It is probably due to hemodynamic changes and possibly to decreased ANP secretion.

The kidneys produce three hormones: 1,25-dihydroxycholecalciferol (see Chapter 21), renin, and erythropoietin. Natriuretic peptides, substances secreted by the heart and other tissues, increase excretion of sodium by the kidneys, and an additional natriuretic hormone (endogenous ouabain) inhibits Na, K ATPase.

THE RENIN–ANGIOTENSIN SYSTEM

RENIN

The rise in blood pressure produced by injection of kidney extracts is due to **renin,** an acid protease secreted by the kidneys into the bloodstream. This enzyme acts in concert with angiotensin-converting enzyme (ACE) to form angiotensin II (Figure 38–6). It is a glycoprotein with a molecular weight of 37,326 in humans. The molecule is made up of two lobes, or domains, between which the active site of the enzyme is located in a deep cleft. Two aspartic acid residues, one at position 104 and one at position 292 (residue numbers from human preprorenin), are juxtaposed in the cleft and are essential for activity. Thus, renin is an aspartyl protease.

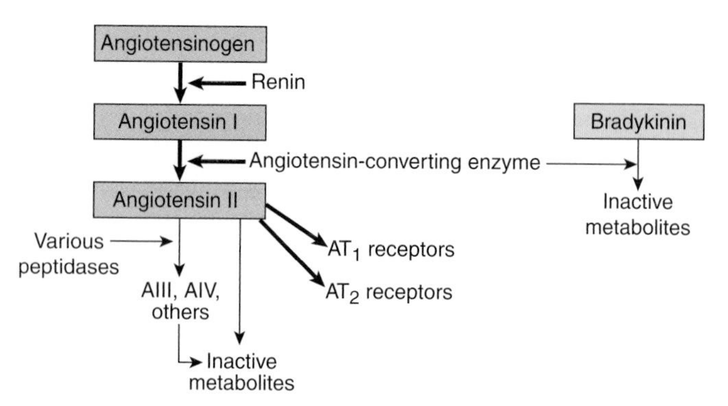

FIGURE 38–6 Formation and metabolism of circulating angiotensins.

Like other hormones, renin is synthesized as a large preprohormone. Human **preprorenin** contains 406 amino acid residues. The **prorenin** that remains after removal of a leader sequence of 23 amino acid residues from the amino terminal contains 383 amino acid residues, and after removal of the pro sequence from the amino terminal of prorenin, active **renin** contains 340 amino acid residues. Prorenin has little if any biologic activity.

Some prorenin is converted to renin in the kidneys, and some is secreted. Prorenin is also secreted by other organs, including the ovaries. After nephrectomy, the prorenin level in the circulation is usually only moderately reduced and may actually rise, but the level of active renin falls to essentially zero. Thus, very little prorenin is converted to renin in the circulation, and active renin is a product primarily, if not exclusively, of the kidneys. Prorenin is secreted constitutively, whereas active renin is formed in the secretory granules of the granular cells in the juxtaglomerular apparatus, the same cells that produce renin (see below). Active renin has a half-life in the circulation of 80 min or less. Its only known function is to cleave the decapeptide **angiotensin I** from the amino terminal end of **angiotensinogen (renin substrate)** (Figure 38–7).

ANGIOTENSINOGEN

Circulating angiotensinogen is found in the α_2-globulin fraction of the plasma (Figure 38–6). It contains about 13% carbohydrate and is made up of 453 amino acid residues. It is synthesized in the liver with a 32-amino-acid signal sequence that is removed in the endoplasmic reticulum. Its circulating level is increased by glucocorticoids, thyroid hormones, estrogens, several cytokines, and angiotensin II.

ANGIOTENSIN-CONVERTING ENZYME & ANGIOTENSIN II

ACE is a dipeptidyl carboxypeptidase that splits off histidylleucine from the physiologically inactive angiotensin I, forming the octapeptide **angiotensin II** (Figure 38–7). The same

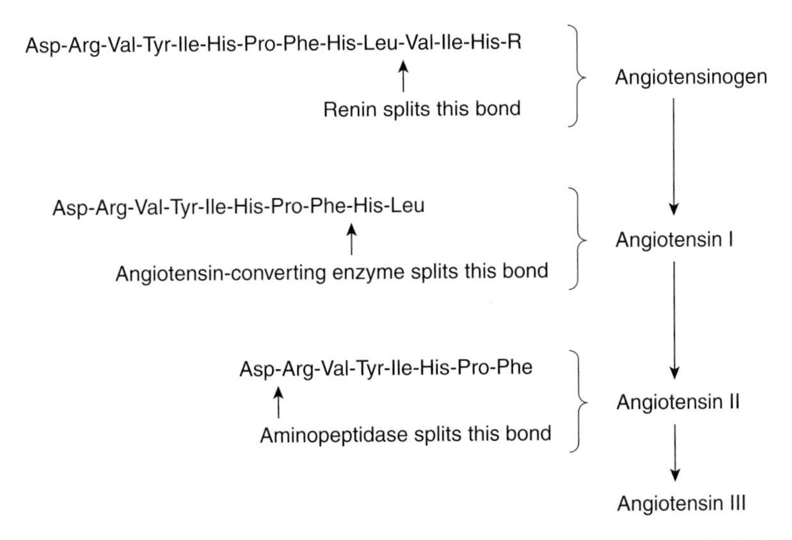

FIGURE 38–7 Structure of the amino terminal end of angiotensinogen and angiotensins I, II, and III in humans. R, remainder of protein. After removal of a 24-amino-acid leader sequence, angiotensinogen contains 453 amino acid residues. The structure of angiotensin II in dogs, rats, and many other mammals is the same as that in humans. Bovine and ovine angiotensin II have valine instead of isoleucine at position 5.

enzyme inactivates bradykinin (Figure 38–6). Increased tissue bradykinin produced when ACE is inhibited acts on B_2 receptors to produce the cough that is an annoying side effect in up to 20% of patients treated with ACE inhibitors (see Clinical Box 38–2). Most of the converting enzyme that forms angiotensin II in the circulation is located in endothelial cells. Much of the conversion occurs as the blood passes through the lungs, but conversion also occurs in many other parts of the body.

ACE is an ectoenzyme that exists in two forms: a **somatic** form found throughout the body and a **germinal** form found solely in postmeiotic spermatogenic cells and spermatozoa (see Chapter 23). Both ACEs have a single transmembrane domain and a short cytoplasmic tail. However, somatic ACE is a 170-kDa protein with two homologous extracellular domains, each containing an active site (Figure 38–8). Germinal ACE is a 90-kDa protein that has only one extracellular domain and active site. Both enzymes are formed from a single gene. However, the gene has two different promoters, producing two different mRNAs. In male mice in which the ACE gene has been knocked out, blood pressure is lower than normal, but in females it is normal. In addition, fertility is reduced in males but not in females.

CLINICAL BOX 38–2

Pharmacologic Manipulation of the Renin–Angiotensin System

It is now possible to inhibit the secretion or the effects of renin in a variety of ways. Inhibitors of prostaglandin synthesis such as **indomethacin** and β-adrenergic blocking drugs such as **propranolol** reduce renin secretion. The peptide **pepstatin** and newly developed renin inhibitors such as **enalkiren** prevent renin from generating angiotensin I. Angiotensin-converting enzyme inhibitors (ACE inhibitors) such as **captopril** and **enalapril** prevent conversion of angiotensin I to angiotensin II. **Saralasin** and several other analogs of angiotensin II are competitive inhibitors of the action of angiotensin II on both AT_1 and AT_2 receptors. **Losartan** (DuP-753) selectively blocks AT_1 receptors, and PD-123177 and several other drugs selectively block AT_2 receptors.

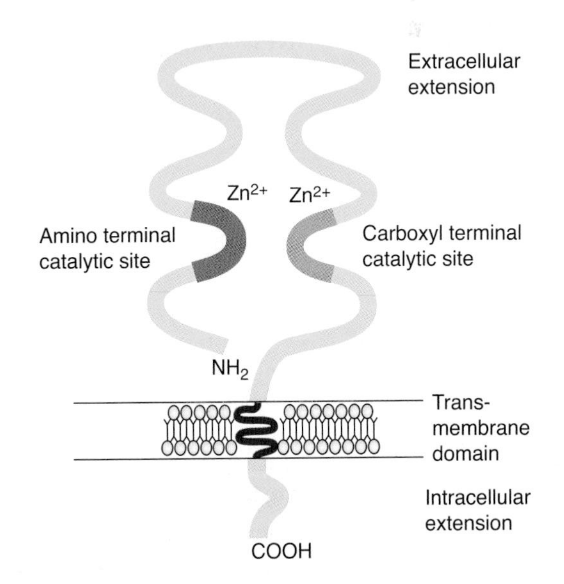

FIGURE 38–8 Diagrammatic representation of the structure of the somatic form of angiotensin-converting enzyme. Note the short cytoplasmic tail of the molecule and the two extracellular catalytic sites, each of which binds a zinc ion (Zn^{2+}). (Reproduced with permission from Johnston CI: Tissue angiotensin-converting enzyme in cardiac and vascular hypertrophy, repair, and remodeling. Hypertension 1994;23:258. Copyright © 1994 by The American Heart Association.)

METABOLISM OF ANGIOTENSIN II

Angiotensin II is metabolized rapidly; its half-life in the circulation in humans is 1–2 min. It is metabolized by various peptidases. An aminopeptidase removes the aspartic acid (Asp) residue from the amino terminal of the peptide (Figure 38–7). The resulting heptapeptide has physiologic activity and is sometimes called **angiotensin III.** Removal of a second amino terminal residue from angiotensin III produces the hexapeptide sometimes called angiotensin IV, which is also said to have some activity. Most, if not all, of the other peptide fragments that are formed are inactive. In addition, aminopeptidase can act on angiotensin I to produce (des-Asp^1) angiotensin I, and this compound can be converted directly to angiotensin III by the action of ACE. Angiotensin-metabolizing activity is found in red blood cells and many tissues. In addition, angiotensin II appears to be removed from the circulation by some sort of trapping mechanism in the vascular beds of tissues other than the lungs.

Renin is usually measured by incubating the sample to be assayed and measuring by immunoassay the amount of angiotensin I generated. This measures the **plasma renin activity (PRA)** of the sample. Deficiency of angiotensinogen as well as renin can cause low PRA values, and to avoid this problem, exogenous angiotensinogen is often added, so that **plasma renin concentration (PRC)** rather than PRA is measured. The normal PRA in supine subjects eating a normal amount of sodium is approximately 1 ng of angiotensin I generated per milliliter per hour. The plasma angiotensin II concentration in such subjects is about 25 pg/mL (approximately 25 pmol/L).

ACTIONS OF ANGIOTENSINS

Angiotensin I appears to function solely as the precursor of angiotensin II and does not have any other established action.

Angiotensin II produces arteriolar constriction and a rise in systolic and diastolic blood pressure. It is one of the most potent vasoconstrictors known, being four to eight times as active as norepinephrine on a weight basis in normal individuals. However, its pressor activity is decreased in Na^+-depleted individuals and in patients with cirrhosis and some other diseases. In these conditions, circulating angiotensin II is increased, and this down-regulates the angiotensin receptors in vascular smooth muscle. Consequently, there is less response to injected angiotensin II.

Angiotensin II also acts directly on the adrenal cortex to increase the secretion of aldosterone, and the renin–angiotensin system is a major regulator of aldosterone secretion. Additional actions of angiotensin II include facilitation of the release of norepinephrine by a direct action on postganglionic sympathetic neurons, contraction of mesangial cells with a resultant decrease in GFR (see Chapter 37), and a direct effect on the renal tubules to increase Na^+ reabsorption.

Angiotensin II also acts on the brain to decrease the sensitivity of the baroreflex, and this potentiates the pressor effect of angiotensin II. In addition, it acts on the brain to increase water intake and increase the secretion of vasopressin and ACTH. It does not penetrate the blood–brain barrier, but it triggers these responses by acting on the circumventricular organs, four small structures in the brain that are outside the blood–brain barrier (see Chapter 33). One of these structures, the area postrema, is primarily responsible for the pressor potentiation, whereas two of the others, the subfornical organ (SFO) and the OVLT, are responsible for the increase in water intake (dipsogenic effect). It is not certain which of the circumventricular organs are responsible for the increases in vasopressin and ACTH secretion.

Angiotensin III [(des-Asp^1) angiotensin II] has about 40% of the pressor activity of angiotensin II, but 100% of the aldosterone-stimulating activity. It has been suggested that angiotensin III is the natural aldosterone-stimulating peptide, whereas angiotensin II is the blood-pressure-regulating peptide. However, this appears not to be the case, and instead angiotensin III is simply a breakdown product with some biologic activity. The same is probably true of angiotensin IV, though some researchers have argued that it has unique effects in the brain.

TISSUE RENIN–ANGIOTENSIN SYSTEMS

In addition to the system that generates circulating angiotensin II, many different tissues contain independent renin–angiotensin systems that generate angiotensin II, apparently for local use. Components of the renin–angiotensin system are found in the walls of blood vessels and in the uterus, the placenta, and the fetal membranes. Amniotic fluid has a high concentration of prorenin. In addition, tissue renin–angiotensin systems, or at least several components of the renin–angiotensin system, are present in the eyes, exocrine portion of the pancreas, heart, fat, adrenal cortex, testis, ovary, anterior and intermediate lobes of the pituitary, pineal, and brain. Tissue renin contributes very little to the circulating renin pool, because PRA falls to undetectable levels after the kidneys are removed. The functions of these tissue renin–angiotensin systems are unsettled, though evidence is accumulating that angiotensin II is a significant growth factor in the heart and blood vessels. ACE inhibitors or AT_1 receptor blockers are now the treatment of choice for congestive heart failure, and part of their value may be due to inhibition of the growth effects of angiotensin II.

ANGIOTENSIN II RECEPTORS

There are at least two classes of angiotensin II receptors. AT_1 receptors are serpentine receptors coupled by a G protein (G_q) to phospholipase C, and angiotensin II increases the cytosolic free Ca^{2+} level. It also activates numerous tyrosine kinases. In vascular smooth muscle, AT_1 receptors are associated with caveolae (see Chapter 2), and AII increases production of caveolin-1, one of the three isoforms of the protein

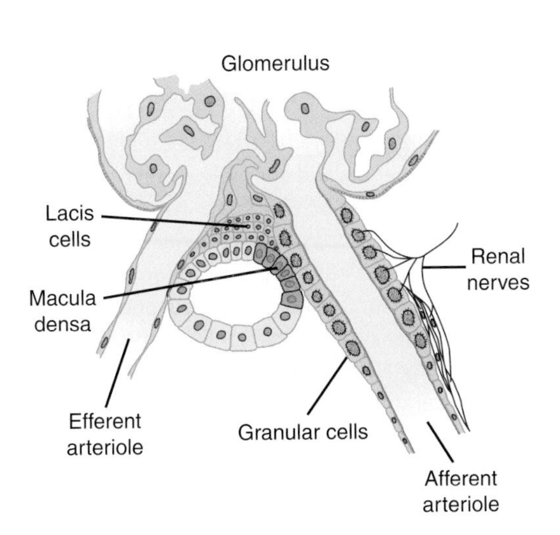

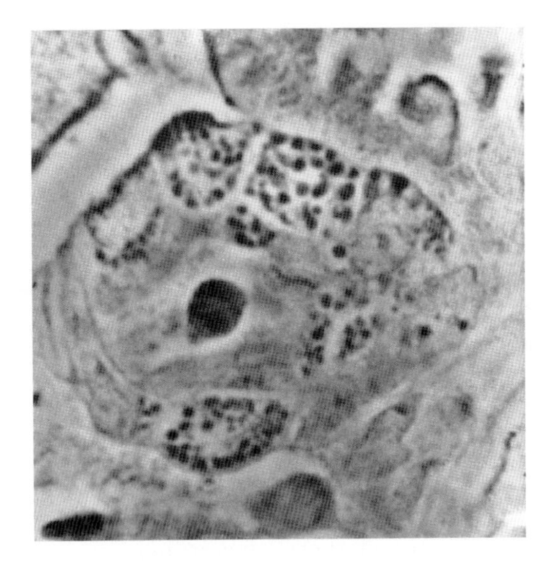

FIGURE 38–9 Left: Diagram of glomerulus, showing the juxtaglomerular apparatus. **Right:** Phase contrast photomicrograph of afferent arteriole in an unstained, freeze-dried preparation of the kidney of a mouse. Note the red blood cell in the lumen of the arteriole and the granular cells in the wall. (Courtesy of C Peil.)

that is characteristic of caveolae. In rodents, two different but closely related AT_1 subtypes, AT_{1A} and AT_{1B}, are coded by two separate genes. The AT_{1A} subtype is found in blood vessel walls, the brain, and many other organs. It mediates most of the known effects of angiotensin II. The AT_{1B} subtype is found in the anterior pituitary and the adrenal cortex. In humans, an AT_1 receptor gene is present on chromosome 3. There may be a second AT_1 type, but it is still unsettled whether distinct AT_{1A} and AT_{1B} subtypes occur.

There are also AT_2 receptors, which are coded in humans by a gene on the X chromosome. Like the AT_1 receptors, they have seven transmembrane domains, but their actions are different. They act via a G protein to activate various phosphatases which in turn antagonize growth effects and open K^+ channels. In addition, AT_2 receptor activation increases the production of NO and therefore increases intracellular cyclic 3,5-guanosine monophosphate (cGMP). The overall physiologic consequences of these second-messenger effects are unsettled. AT_2 receptors are more plentiful in fetal and neonatal life, but they persist in the brain and other organs in adults.

The AT_1 receptors in the arterioles and the AT_1 receptors in the adrenal cortex are regulated in opposite ways: an excess of angiotensin II down-regulates the vascular receptors, but it up-regulates the adrenocortical receptors, making the gland more sensitive to the aldosterone-stimulating effect of the peptide.

THE JUXTAGLOMERULAR APPARATUS

The renin in kidney extracts and the bloodstream is produced by the **juxtaglomerular cells (JG cells).** These epitheloid cells are located in the media of the afferent arterioles as they enter the glomeruli (Figure 38–9). The membrane-lined secretory granules in them have been shown to contain renin. Renin is also found in agranular **lacis cells** that are located in the junction between the afferent and efferent arterioles, but its significance in this location is unknown.

At the point where the afferent arteriole enters the glomerulus and the efferent arteriole leaves it, the tubule of the nephron touches the arterioles of the glomerulus from which it arose. At this location, which marks the start of the distal convolution, there is a modified region of tubular epithelium called the **macula densa** (Figure 38–9). The macula densa is in close proximity to the JG cells. The lacis cells, the JG cells, and the macula densa constitute the **juxtaglomerular apparatus.**

REGULATION OF RENIN SECRETION

Several different factors regulate renin secretion (Table 38–2), and the rate of renin secretion at any given time is determined by the summed activity of these factors. One factor is an intra-renal baroreceptor mechanism that causes renin secretion to decrease when arteriolar pressure at the level of the JG cells increases and to increase when arteriolar pressure at this level falls. Another renin-regulating sensor is in the macula densa. Renin secretion is inversely proportional to the amount of Na^+ and Cl^- entering the distal renal tubules from the loop of Henle. Presumably, these electrolytes enter the macula densa cells via the $Na–K–2Cl^-$ transporters in their apical membranes, and the increase in some fashion triggers a signal that decreases renin secretion in the juxtaglomerular cells in the adjacent afferent arterioles. A possible mediator is NO, but the identity of the signal remains unsettled. Renin secretion also varies inversely with the

TABLE 38–2 Factors that affect renin secretion.

Stimulatory
Increased sympathetic activity via renal nerves
Increased circulating catecholamines
Prostaglandins
Inhibitory
Increased Na^+ and Cl^- reabsorption across macula densa
Increased afferent arteriolar pressure
Angiotensin II
Vasopressin

plasma K^+ level, but the effect of K^+ appears to be mediated by the changes it produces in Na^+ and Cl^- delivery to the macula densa.

Angiotensin II feeds back to inhibit renin secretion by a direct action on the JG cells. Vasopressin also inhibits renin secretion in vitro and in vivo, although there is some debate about whether its in vivo effect is direct or indirect.

Finally, increased activity of the sympathetic nervous system increases renin secretion. The increase is mediated both by increased circulating catecholamines and by norepinephrine secreted by postganglionic renal sympathetic nerves. The catecholamines act mainly on β_1-adrenergic receptors on the JG cells and renin release is mediated by an increase in intracellular cAMP.

The principal conditions that increase renin secretion in humans are listed in Table 38–3. Most of the listed conditions decrease central venous pressure, which triggers an increase in sympathetic activity, and some also decrease renal arteriolar pressure (see Clinical Box 38–3). Renal artery constriction and constriction of the aorta proximal to the renal arteries produces a decrease in renal arteriolar pressure. Psychologic stimuli increase the activity of the renal nerves.

TABLE 38–3 Conditions that increase renin secretion.

Na^+ depletion
Diuretics
Hypotension
Hemorrhage
Upright posture
Dehydration
Cardiac failure
Cirrhosis
Constriction of renal artery or aorta
Various psychologic stimuli

CLINICAL BOX 38–3

Role of Renin in Clinical Hypertension

Constriction of one renal artery causes a prompt increase in renin secretion and the development of sustained hypertension **(renal** or **Goldblatt hypertension).** Removal of the ischemic kidney or the arterial constriction cures the hypertension if it has not persisted too long. In general, the hypertension produced by constricting one renal artery with the other kidney intact (one-clip, two-kidney Goldblatt hypertension) is associated with increased circulating renin. The clinical counterpart of this condition is **renal hypertension** due to atheromatous narrowing of one renal artery or other abnormalities of the renal circulation. However, plasma renin activity is usually normal in one-clip one-kidney Goldblatt hypertension. The explanation of the hypertension in this situation is unsettled. However, many patients with hypertension respond to treatment with ACE inhibitors or losartan even when their renal circulation appears to be normal and they have normal or even low plasma renin activity.

HORMONES OF THE HEART & OTHER NATRIURETIC FACTORS

STRUCTURE

The existence of various **natriuretic hormones** has been postulated for some time. Two of these are secreted by the heart. The muscle cells in the atria and, to a much lesser extent in the ventricles, contain secretory granules (Figure 38–10). The granules increase in number when

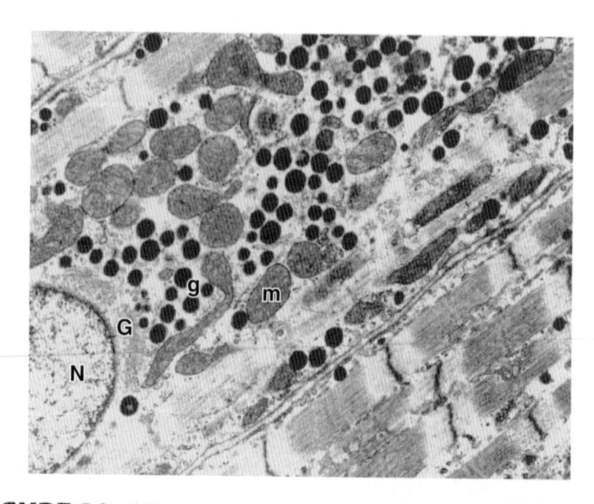

FIGURE 38–10 ANP granules (g) interspersed between mitochondria (m) in rat atrial muscle cell. G, Golgi complex; N, nucleus. The granules in human atrial cells are similar (× 17,640). (Courtesy of M Cantin.)

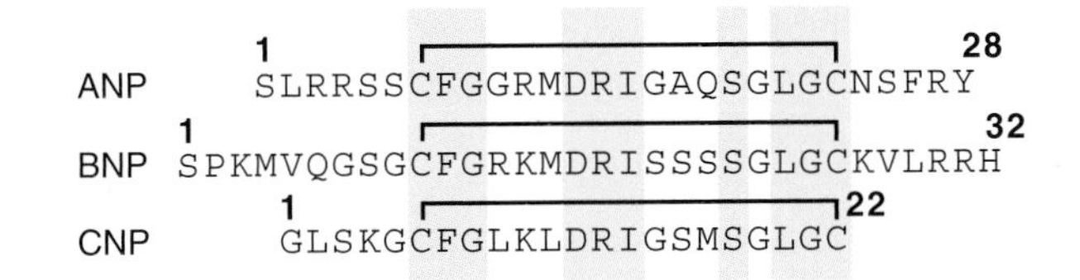

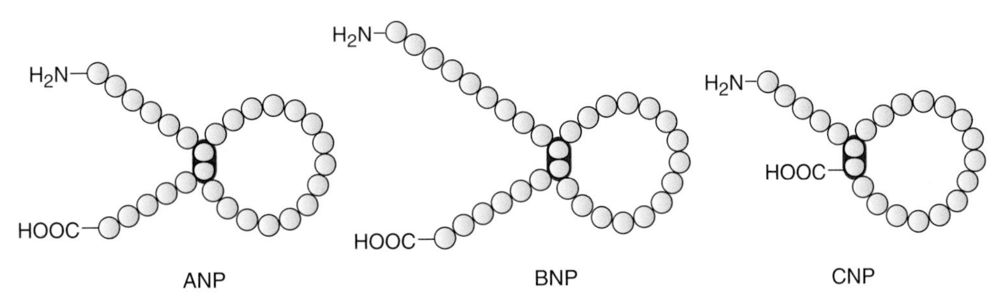

FIGURE 38–11 Human ANP, BNP, and CNP. Top: Single-letter codes for amino acid residues aligned to show common sequences (colored). Bottom: Shape of molecules. Note that one cysteine is the carboxyl terminal amino acid residue in CNP, so there is no carboxyl terminal extension from the 17-member ring. (Modified from Imura H, Nakao K, Itoh H: The natriuretic peptide system in the brain: Implication in the central control of cardiovascular and neuroendocrine functions. Front Neuroendocrinol 1992;13:217.)

NaCl intake is increased and ECF expanded, and extracts of atrial tissue cause natriuresis.

The first natriuretic hormone isolated from the heart was **ANP**, a polypeptide with a characteristic 17-amino-acid ring formed by a disulfide bond between two cysteines. The circulating form of this polypeptide has 28 amino acid residues (Figure 38–11). It is formed from a large precursor molecule that contains 151 amino acid residues, including a 24-amino-acid signal peptide. ANP was subsequently isolated from other tissues, including the brain, where it exists in two forms that are smaller than circulating ANP. A second natriuretic polypeptide was isolated from porcine brain and named **brain natriuretic peptide** (**BNP**; also known as **B-type natriuretic peptide**). It is also present in the brain in humans, but more is present in the human heart, including the ventricles. The circulating form of this hormone contains 32 amino acid residues. It has the same 17-member ring as ANP, though some of the amino acid residues in the ring are different (Figure 38–11). A third member of this family has been named **C-type natriuretic peptide (CNP)** because it was the third in the sequence to be isolated. It contains 22 amino acid residues (Figure 38–11), and there is also a larger 53-amino-acid form. CNP is present in the brain, the pituitary, the kidneys, and vascular endothelial cells. However, very little is present in the heart and the circulation, and it appears to be primarily a paracrine mediator.

ACTIONS

ANP and BNP in the circulation act on the kidneys to increase Na^+ excretion and injected CNP has a similar effect. They appear to produce this effect by dilating afferent arterioles and relaxing mesangial cells. Both of these actions increase glomerular filtration (see Chapter 37). In addition, they act on the renal tubules to inhibit Na^+ reabsorption (see Chapter 37). Other actions include an increase in capillary permeability, leading to extravasation of fluid and a decline in blood pressure. In addition, they relax vascular smooth muscle in arterioles and venules. CNP has a greater dilator effect on veins than ANP and BNP. These peptides also inhibit renin secretion and counteract the pressor effects of catecholamines and angiotensin II.

In the brain, ANP is present in neurons, and an ANP-containing neural pathway projects from the anteromedial part of the hypothalamus to the areas in the lower brain stem that are concerned with neural regulation of the cardiovascular system. In general, the effects of ANP in the brain are opposite to those of angiotensin II, and ANP-containing neural circuits appear to be involved in lowering blood pressure and promoting natriuresis. CNP and BNP in the brain probably have functions similar to those of ANP, but detailed information is not available.

NATRIURETIC PEPTIDE RECEPTORS

Three different natriuretic peptide receptors (NPR) have been isolated and characterized (Figure 38–12). The NPR-A and NPR-B receptors both span the cell membrane and have cytoplasmic domains that are guanylyl cyclases. ANP has the greatest affinity for the NPR-A receptor, and CNP has the greatest affinity for the NPR-B receptor. The third receptor, NPR-C, binds all three natriuretic peptides but has a markedly truncated cytoplasmic domain. Some evidence suggests that it acts via G proteins to activate phospholipase C and inhibit adenylyl cyclase. However, it has also been argued that

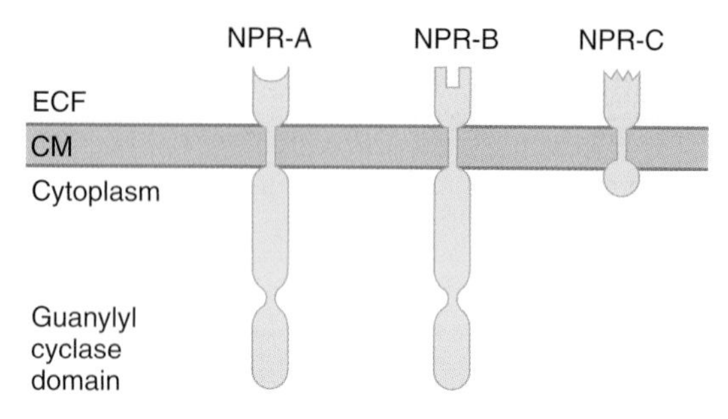

FIGURE 38–12 Diagrammatic representation of natriuretic peptide receptors. The NPR-A and NPR-B receptor molecules have intracellular guanylyl cyclase domains, whereas the putative clearance receptor, NPR-C, has only a small cytoplasmic domain. CM, cell membrane.

this receptor does not trigger any intracellular change and is instead a **clearance receptor** that removes natriuretic peptides from the bloodstream and then releases them later, helping to maintain a steady blood level of the hormones.

SECRETION & METABOLISM

The concentration of ANP in plasma is about 5 fmol/mL in normal humans ingesting moderate amounts of NaCl. ANP secretion is increased when the ECF volume is increased by infusion of isotonic saline and when the atria are stretched. BNP secretion is increased when the ventricles are stretched. ANP secretion is also increased by immersion in water up to the neck (Figure 38–13), a procedure that counteracts the effect of gravity on the circulation and increases central venous and consequently atrial pressure. Note that immersion also decreases the secretion of renin and aldosterone. Conversely, a small but measurable decrease in plasma ANP occurs in association with a decrease in central venous pressure on rising from the supine to the standing position. Thus, it seems clear that the atria respond directly to stretch in vivo and that the rate of ANP secretion is proportional to the degree to which the atria are stretched by increases in central venous pressure. Similarly, BNP secretion is proportional to the degree to which the ventricles are stretched. Plasma levels of both hormones are elevated in congestive heart failure, and their measurement is seeing increasing use in the diagnosis of this condition.

Circulating ANP has a short half-life. It is metabolized by neutral endopeptidase (NEP), which is inhibited by thiorphan. Therefore, administration of thiorphan increases circulating ANP.

Na, K ATPase-INHIBITING FACTOR

Another natriuretic factor is present in blood. This factor produces natriuresis by inhibiting Na, K ATPase and raises rather than lowers blood pressure. Current evidence indicates that it may well be the digitalis-like steroid **ouabain** and that it comes from the adrenal glands. However, its physiologic significance is not yet known.

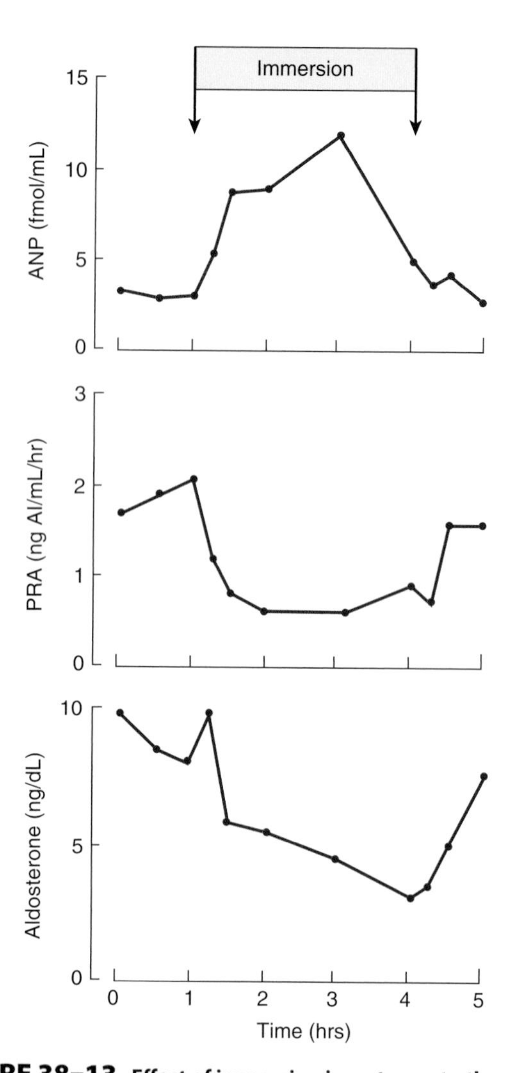

FIGURE 38–13 Effect of immersion in water up to the neck for 3 h on plasma concentrations of ANP, PRA, and aldosterone. (Modified and reproduced with permission from Epstein M, et al: Increases in circulating atrial natriuretic factor during immersion-induced central hypervolaemia in normal humans. J Hypertension Suppl 1986 June;4(2):S93–S99.)

DEFENSE OF SPECIFIC IONIC COMPOSITION

Special regulatory mechanisms maintain the levels of certain specific ions in the ECF as well as the levels of glucose and other nonionized substances important in metabolism (see Chapter 1). The feedback of Ca^{2+} on the parathyroids and the calcitonin-secreting cells to adjust their secretion maintains the ionized calcium level of the ECF (see Chapter 21). The Mg^{2+} concentration is subject to close regulation, but the mechanisms controlling Mg^{+} homeostasis are incompletely understood.

The mechanisms controlling Na^{+} and K^{+} content are part of those determining the volume and tonicity of ECF and have been discussed above. The levels of these ions are also dependent on the H^{+} concentration, and pH is one of the major factors affecting the anion composition of ECF. This will be discussed in Chapter 39.

ERYTHROPOIETIN

STRUCTURE & FUNCTION

When an individual bleeds or becomes hypoxic, hemoglobin synthesis is enhanced, and production and release of red blood cells from the bone marrow **(erythropoiesis)** are increased (see Chapter 31). Conversely, when the red cell volume is increased above normal levels by transfusion, the erythropoietic activity of the bone marrow decreases. These adjustments are brought about by changes in the circulating level of **erythropoietin,** a circulating glycoprotein that contains 165 amino acid residues and four oligosaccharide chains that are necessary for its activity in vivo. Its blood level is markedly increased in anemia (Figure 38–14).

Erythropoietin increases the number of erythropoietin-sensitive committed stem cells in the bone marrow that are converted to red blood cell precursors and subsequently to mature erythrocytes (see Chapter 31). The receptor for erythropoietin is a linear protein with a single transmembrane domain that is a member of the cytokine receptor superfamily (see Chapter 3). The receptor has tyrosine kinase activity, and it activates a cascade of serine and threonine kinases, resulting in inhibited apoptosis of red cells and their increased growth and development.

The principal site of inactivation of erythropoietin is the liver, and the hormone has a half-life in the circulation of about 5 h. However, the increase in circulating red cells that it triggers takes 2–3 days to appear, since red cell maturation is a relatively slow process.

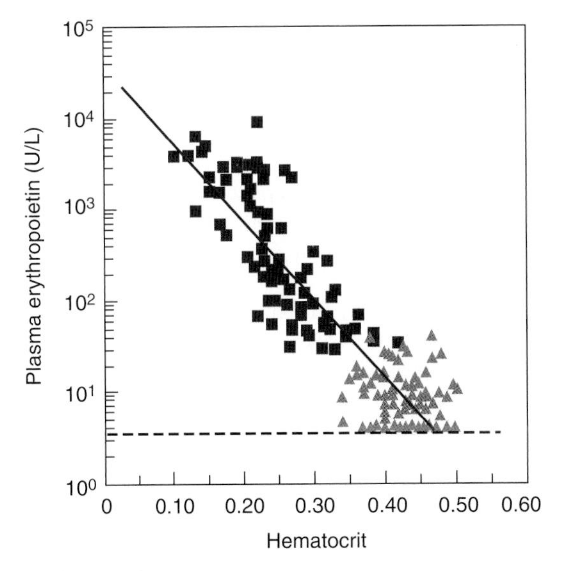

FIGURE 38–14 Plasma erythropoietin levels in normal blood donors (triangles) and patients with various forms of anemia (squares). (Reproduced with permission from Erslev AJ: Erythropoietin. N Engl J Med 1991;324:1339.)

SOURCES

In adults, about 85% of the erythropoietin comes from the kidneys and 15% from the liver. Both these organs contain the mRNA for erythropoietin. Erythropoietin can also be extracted from the spleen and salivary glands, but these tissues do not contain its mRNA and consequently do not appear to manufacture the hormone. When renal mass is reduced in adults by renal disease or nephrectomy, the liver cannot compensate and anemia develops.

Erythropoietin is produced by interstitial cells in the peritubular capillary bed of the kidneys and by perivenous hepatocytes in the liver. It is also produced in the brain, where it exerts a protective effect against excitotoxic damage triggered by hypoxia; and in the uterus and oviducts, where it is induced by estrogen and appears to mediate estrogen-dependent angiogenesis.

The gene for the hormone has been cloned, and recombinant erythropoietin produced in animal cells is available for clinical use as epoetin alfa. The recombinant erythropoietin is of value in the treatment of the anemia associated with renal failure; 90% of the patients with end-stage renal failure who are on dialysis are anemic as a result of erythropoietin deficiency. Erythropoietin is also used to stimulate red cell production in individuals who are banking a supply of their own blood in preparation for autologous transfusions during elective surgery (see Chapter 31).

REGULATION OF SECRETION

The usual stimulus for erythropoietin secretion is hypoxia, but secretion of the hormone can also be stimulated by cobalt salts and androgens. Recent evidence suggests that the O_2 sensor regulating erythropoietin secretion in the kidneys and the liver is a heme protein that in the deoxy form stimulates and in the oxy form inhibits transcription of the erythropoietin gene to form erythropoietin mRNA. Secretion of the hormone is also facilitated by the alkalosis that develops at high altitudes. Like renin secretion, erythropoietin secretion is facilitated by catecholamines via a β-adrenergic mechanism, although the renin–angiotensin system is totally separate from the erythropoietin system.

CHAPTER SUMMARY

- Total body osmolality is directly proportional to the total body sodium plus the total body potassium divided by the total body water. Changes in the osmolality of the body fluids occur when a disproportion exists between the amount of these electrolytes and the amount of water ingested or lost from the body.
- Vasopressin's main physiologic effect is the retention of water by the kidney by increasing the water permeability of the renal collecting ducts. Water is absorbed from the urine, the urine becomes concentrated, and its volume decreases.
- Vasopressin is stored in the posterior pituitary and released into the bloodstream in response to the stimulation of osmoreceptors or baroreceptors. Increases in secretion occur when osmolality is changed as little as 1%, thus keeping the osmolality of the plasma very close to 285 mOsm/L.
- The amount of Na^+ in the ECF is the most important determinant of ECF volume, and mechanisms that control Na^+ balance are the major mechanisms defending ECF volume. The main mechanism regulating sodium balance is the renin–angiotensin system, a hormone system that regulates blood pressure.
- The kidneys secrete the enzyme renin and renin acts in concert with angiotensin-converting enzyme to form angiotensin II. Angiotensin II acts directly on the adrenal cortex to increase the secretion of aldosterone. Aldosterone increases the retention of sodium from the urine via action on the renal collecting duct.

MULTIPLE-CHOICE QUESTIONS

For all questions, select the single best answer unless otherwise directed.

1. Dehydration increases the plasma concentration of all the following hormones *except*
   A. vasopressin.
   B. angiotensin II.
   C. aldosterone.
   D. norepinephrine.
   E. atrial natriuretic peptide.
2. In a patient who has become dehydrated, body water should be replaced by intravenous infusion of
   A. distilled water.
   B. 0.9% sodium chloride solution.
   C. 5% glucose solution.
   D. hyperoncotic albumin.
   E. 10% glucose solution.
3. Renin is secreted by
   A. cells in the macula densa.
   B. cells in the proximal tubules.
   C. cells in the distal tubules.
   D. granular cells in the juxtaglomerular apparatus.
   E. cells in the peritubular capillary bed.
4. Erythropoietin is secreted by
   A. cells in the macula densa.
   B. cells in the proximal tubules.
   C. cells in the distal tubules.
   D. granular cells in the juxtaglomerular apparatus.
   E. cells in the peritubular capillary bed.
5. When a woman who has been on a low-sodium diet for 8 days is given an intravenous injection of captopril, a drug that inhibits angiotensin-converting enzyme, one would expect
   A. her blood pressure to rise because her cardiac output would fall.
   B. her blood pressure to rise because her peripheral resistance would fall.
   C. her blood pressure to fall because her cardiac output would fall.
   D. her blood pressure to fall because her peripheral resistance would fall.
   E. her plasma renin activity to fall because her circulating angiotensin I level would rise.
6. Which of the following would *not* be expected to increase renin secretion?
   A. Administration of a drug that blocks angiotensin-converting enzyme
   B. Administration of a drug that blocks AT1 receptors
   C. Administration of a drug that blocks β-adrenergic receptors
   D. Constriction of the aorta between the celiac artery and the renal arteries
   E. Administration of a drug that reduces ECF volume
7. Which of the following is *least* likely to contribute to the beneficial effects of angiotensin-converting enzyme inhibitors in the treatment of congestive heart failure?
   A. Vasodilation
   B. Decreased cardiac growth
   C. Decreased cardiac afterload
   D. Increased plasma renin activity
   E. Decreased plasma aldosterone

CHAPTER RESOURCES

Adrogue HJ, Madias NE: Hypernatremia. N Engl J Med 2000;342:1493.
Adrogue HJ, Madias NE: Hyponatremia. N Engl J Med 2000;342:101.
Luft FC. Mendelian forms of human hypertension and mechanisms of disease. Clin Med Res 2003;1:291–300.
Morel F: Sites of hormone action in the mammalian nephron. Am J Physiol 1981;240:F159.
McKinley MS, Johnson AK: The physiologic regulation of thirst and fluid intake. News Physiol Sci 2004;19:1.
Robinson AG, Verbalis JG: Diabetes insipidus. Curr Ther Endocrinol Metab 1997;6:1.
Verkman AS: Mammalian aquaporins: Diverse physiological roles and potential clinical significance. Expert Rev Mol Med 2008;10:13.
Zeidel ML: Hormonal regulation of inner medullary collecting duct sodium transport. Am J Physiol 1993;265:F159.

CHAPTER

39

Acidification of the Urine & Bicarbonate Excretion

OBJECTIVES

After reading this chapter, you should be able to:

- Outline the processes involved in the secretion of H^+ into the tubules and discuss the significance of these processes in the regulation of acid–base balance.
- Define acidosis and alkalosis, and give (in mEq/L and pH) the normal mean and the range of H^+ concentrations in blood that are compatible with health.
- List the principal buffers in blood, interstitial fluid, and intracellular fluid, and, using the Henderson–Hasselbalch equation, describe what is unique about the bicarbonate buffer system.
- Describe the changes in blood chemistry that occur during the development of metabolic acidosis and metabolic alkalosis, and the respiratory and renal compensations for these conditions.
- Describe the changes in blood chemistry that occur during the development of respiratory acidosis and respiratory alkalosis, and the renal compensation for these conditions.

INTRODUCTION

The kidneys play a key role in the maintenance of acid–base balance and to do this they must excrete acid in the amount equivalent to the production of nonvolatile acids in the body. The production of nonvolatile acids will vary with diet, metabolism, and disease. The kidneys must also filter and reabsorb plasma bicarbonate, and thus prevent the loss of bicarbonate in the urine. Both processes are linked physiologically, due to the nephron's ability to secrete H^+ ions into the filtrate.

RENAL H^+ SECRETION

The cells of the proximal and distal tubules, like the cells of the gastric glands (see Chapter 25), secrete hydrogen ions. Hydrogen secretion also occurs in the collecting ducts. The transporter that is responsible for H^+ secretion in the proximal tubules is the Na–H exchanger (primarily NHE3) (Figure 39–1). This is an example of secondary active transport; Na^+ is moved from the inside of the cell to the interstitium by Na, K ATPase on the basolateral membrane, which keeps intracellular Na^+ low, thus establishing the drive for Na^+ to enter the cell, via the Na–H exchanger, from the tubular lumen. The Na–H exchanger secretes H^+ into the lumen in exchange for Na^+.

The secreted H^+ ion combines with filtered HCO_3^- to form H_2CO_3 and the presence of **carbonic anhydrase** on the apical membrane of the proximal tubule catalyzes the formation of H_2O and CO_2 from H_2CO_3. The apical membrane of epithelial cells lining the proximal tubule is permeable to CO_2 and H_2O, and they enter the tubule rapidly. 80% of the filtered load of HCO_3^- is reabsorbed in the proximal tubule.

Inside the cell, carbonic anhydrase is also present and can catalyze the formation of H_2CO_3 from CO_2 and H_2O. H_2CO_3 dissociates into H^+ ions and HCO_3^-; the H^+ is secreted into the tubular lumen, as mentioned above, and the HCO_3^- that is formed diffuses into the interstitial fluid. Thus, for each H^+ ion

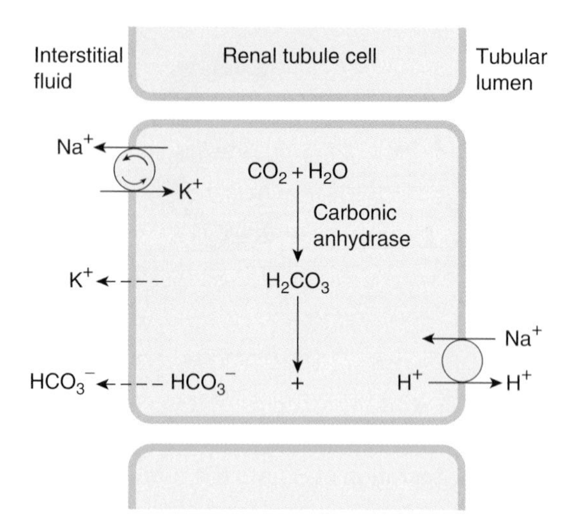

FIGURE 39–1 Secretion of acid by proximal tubular cells in the kidney. H^+ is transported into the tubular lumen by an antiport in exchange for Na^+. Active transport by Na, K ATPase is indicated by arrows in the circle. Dashed arrows indicate diffusion.

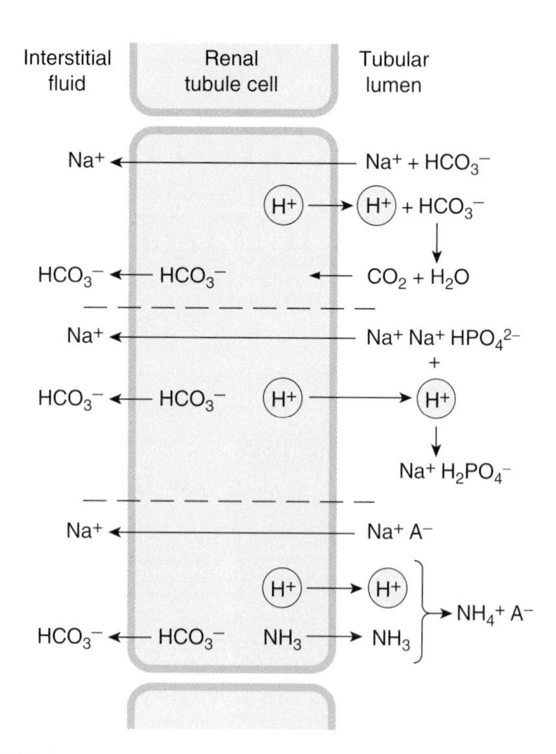

FIGURE 39–2 Fate of H^+ secreted into a tubule in exchange for Na^+. Top: Reabsorption of filtered bicarbonate via CO_2. Middle: Formation of monobasic phosphate. **Bottom:** Ammonium formation. Note that in each instance one Na^+ ion and one HCO_3^- ion enter the bloodstream for each H^+ ion secreted. A^-, anion.

secreted, one Na^+ ion and one HCO_3^- ion enter the interstitial fluid. Because carbonic anhydrase catalyzes the formation of H_2CO_3, drugs that inhibit carbonic anhydrase depress both secretion of acid by the proximal tubules and the reactions that depend on it.

Some evidence suggests that H^+ is secreted in the proximal tubules by other types of transporters, but the evidence for these additional transporters is controversial, and in any case, their contribution is small relative to that of the Na–H exchange mechanism.

This is in contrast to what occurs in the distal tubules and collecting ducts, where H^+ secretion is relatively independent of Na^+ in the tubular lumen. In this part of the tubule, most H^+ is secreted by an ATP-driven proton pump. Aldosterone acts on this pump to increase distal H^+ secretion. The I cells in this part of the renal tubule secrete acid and, like the parietal cells in the stomach, contain abundant carbonic anhydrase and numerous tubulovesicular structures. There is evidence that the H^+-translocating ATPase that produces H^+ secretion is located in these vesicles as well as in the apical cell membrane and that, in acidosis, the number of H^+ pumps is increased by insertion of these tubulovesicles into the apical cell membrane. Some of the H^+ is additionally secreted by $H–K^+$ ATPase. The I cells also contain **anion exchanger 1 (AE1,** formerly known as **Band 3)**, an anion exchange protein, in their basolateral cell membranes. This protein may function as a Cl/HCO_3 exchanger for the transport of HCO_3^- to the interstitial fluid.

FATE OF H⁺ IN THE URINE

The amount of acid secreted depends upon the subsequent events that modify the composition of the tubular urine. The maximal H^+ gradient against which the transport mechanisms can secrete in humans corresponds to a urine pH of about 4.5; that is, an H^+ concentration in the urine that is 1000 times the concentration in plasma. pH 4.5 is thus the **limiting pH.** This is normally reached in the collecting ducts. If there were no buffers that "tied up" H^+ in the urine, this pH would be reached rapidly, and H^+ secretion would stop. However, three important reactions in the tubular fluid remove free H^+, permitting more acid to be secreted (Figure 39–2). These are the reactions of H^+ with HCO_3^- to form CO_2 and H_2O (discussed above), with HPO_4^{2-} to form $H_2PO_4^-$ (titratable acids), and with NH_3 to form NH_4^+.

REACTION WITH BUFFERS

The three buffers of importance in the renal handling of acid and its secretion into the lumen are thus bicarbonate, dibasic phosphate, and ammonia. On an average diet, approximately 40% of nonvolatile acids (about 30 mEq/day), produced by the body in the course of various metabolic reactions is excreted as **titratable acid** (ie, phosphate system) and 60% of nonvolatile acid (about 50 mEq/day) is excreted as NH_4^+. The pK' of the bicarbonate system is 6.1, that of the dibasic phosphate system is 6.8, and that of the ammonia system is 9.0. The concentration of HCO_3^- in the plasma, and consequently in the glomerular filtrate, is normally about 24 mEq/L, whereas that of phosphate is only 1.5 mEq/L. Therefore, in the proximal tubule, most of the secreted H^+ reacts with HCO_3^- as described above, to form H_2CO_3 (Figure 39–2) and this enters the cell as CO_2 and H_2O following the action of carbonic anhydrase in the brush border of the proximal tubule cells. The CO_2 entering the tubular cells adds to the pool of CO_2 available to form H_2CO_3. Because most of the H^+ is removed from the tubule, the pH of the fluid is changed very little. This is the mechanism by which HCO_3^- is reabsorbed; for each mole of HCO_3^- removed from the tubular fluid, 1 mol of HCO_3^- diffuses from the tubular cells into the blood, although it is important to note that it is not the same mole that disappeared from the tubular fluid. About 4500 mEq of HCO_3^- are filtered and reabsorbed each day.

Secreted H^+ also reacts with dibasic phosphate (HPO_4^{2-}) to form monobasic phosphate ($H_2PO_4^-$). This happens to the greatest extent in the distal tubules and collecting ducts, because it is here that the phosphate that escapes proximal reabsorption is greatly concentrated by the reabsorption of water. H^+ ions are also known to combine to a minor degree with other buffer anions.

The ammonia buffering system allows secreted H^+ to combine with NH_3, and this occurs in the proximal tubule (where NH_3 is made, see below) and in the distal tubules. The pK' of the ammonia system is 9.0, and the ammonia system is titrated only from the pH of the urine to pH 7.4, so it contributes very little to the titratable acidity. Each H^+ ion that reacts with the buffers contributes to the urinary **titratable acidity,** which is measured by determining the amount of alkali that must be added to the urine to return its pH to 7.4, the pH of the glomerular filtrate. However, the titratable acidity obviously measures only a fraction of the acid secreted, since it does not account for the H_2CO_3 that has been converted to H_2O and CO_2.

The reabsorption of HCO_3^- is crucial to the maintenance of acid–base balance, as a loss of a single HCO_3^- ion in the urine would be the equivalent of adding a H^+ ion to the blood. However, the kidneys have the ability to replenish the body with new bicarbonate ions. This occurs when H^+ ions are removed from the body as NH_4^+ or titratable acid, as there is formation of new bicarbonate within the cells, and this enters the blood (ie, these bicarbonate ions are not those originally filtered, and yet they still enter the blood).

AMMONIA SECRETION

As mentioned above, reactions in the renal tubular cells produce NH_4^+ and HCO_3^-. NH_4^+ is in equilibrium with NH_3 and H^+ in the cells. Because the pK' of this reaction is 9.0, the ratio of NH_3 to NH_4^+ at pH 7.0 is 1:100 (Figure 39–3). However, NH_3 is lipid-soluble and diffuses across the cell membranes down its concentration gradient into the interstitial fluid and tubular urine. In the urine it reacts with H^+ to form NH_4^+, and the NH_4^+ remains "trapped" in the urine.

The principal reaction producing NH_4^+ in cells is conversion of glutamine to glutamate. This reaction is catalyzed by the enzyme **glutaminase,** which is abundant in renal tubular cells (Figure 39–3). **Glutamic dehydrogenase** catalyzes the conversion of glutamate to α-ketoglutarate, with the production of more NH_4^+. Subsequent metabolism of α-ketoglutarate utilizes $2H^+$, freeing $2HCO_3^-$.

$$NH_4^+ \rightleftharpoons NH_3 + H^+$$

$$pH = pK' + \log \frac{[NH_3]}{[NH_4^+]}$$

$$\text{Glutamine} \xrightarrow{\text{Glutaminase}} \text{Glutamate} + NH_4^+$$

$$\text{Glutamate} \xrightarrow{\text{Glutamic dehydrogenase}} \alpha\text{–Ketoglutarate} + NH_4^+$$

FIGURE 39–3 Major reactions involved in ammonia production in the kidneys.

In chronic acidosis, the amount of NH_4^+ excreted at any given urine pH also increases, because more NH_3 enters the tubular urine. The effect of this **adaptation** of NH_3 secretion, the cause of which is unsettled, is further removal of H^+ from the tubular fluid and consequently a further enhancement of H^+ secretion by the renal tubules and excretion in the urine. Because the amount of phosphate buffer filtered at the glomerulus cannot be increased, urinary excretion of acid via the phosphate buffer system is limited. The production of NH_4^+ by the renal tubules is the only way the kidneys can remove even the normal amount, much less an increased amount, of nonvolatile acid produced in the body.

In the inner medullary cells of the collecting duct, the main process by which NH_3 is secreted into the urine and then changed to NH_4^+, is called **nonionic diffusion** (see Chapter 2), thereby maintaining the concentration gradient for diffusion of NH_3. In the proximal tubule, nonionic diffusion of NH_4^+ is less important as NH_4^+ can be secreted into the lumen, often by replacing H^+ on the Na–H exchanger.

Salicylates and a number of other drugs that are weak bases or weak acids are also secreted by nonionic diffusion. They diffuse into the tubular fluid at a rate that depends on the pH of the urine, so the amount of each drug excreted varies with the pH of the urine.

pH CHANGES ALONG THE NEPHRON

A moderate drop in pH occurs in the proximal tubular fluid, but, as noted above, most of the secreted H^+ has little effect on luminal pH because of the formation of CO_2 and H_2O from H_2CO_3. In contrast, the distal tubule has less capacity to secrete H^+, but secretion in this segment has a greater effect on urinary pH.

FACTORS AFFECTING ACID SECRETION

Renal acid secretion is altered by changes in the intracellular P_{CO_2}, K^+ concentration, carbonic anhydrase level, and adrenocortical hormone concentration. When the P_{CO_2} is high **(respiratory acidosis),** more intracellular H_2CO_3 is available to buffer the hydroxyl ions and acid secretion is enhanced, whereas the reverse is true when the P_{CO_2} falls. K^+ depletion enhances acid secretion, apparently because the loss of K^+ causes intracellular acidosis even though the plasma pH may be elevated. Conversely, K^+ excess in the cells inhibits acid secretion. When carbonic anhydrase is inhibited, acid secretion is inhibited because the formation of H_2CO_3 is decreased. Aldosterone and the other adrenocortical steroids that enhance tubular reabsorption of Na^+ also increase the secretion of H^+ and K^+.

BICARBONATE EXCRETION

Although the process of HCO_3^- reabsorption does not involve actual transport of this ion into the tubular cells, HCO_3^- reabsorption is proportional to the amount filtered over a relatively wide range. There is no demonstrable Tm, but HCO_3^- reabsorption is decreased by an unknown mechanism when the extracellular fluid (ECF) volume is expanded (Figure 39–4). When the plasma HCO_3^- concentration is low, all the filtered HCO_3^- is reabsorbed; but when the plasma HCO_3^- concentration is high; that is, above 26–28 mEq/L (the renal threshold for HCO_3^-), HCO_3^- appears in the urine and the urine becomes alkaline. Conversely, when the plasma HCO_3^- falls below about 26 mEq/L, the value at which all the secreted H^+ is being used to reabsorb HCO_3^-, more H^+ becomes available to combine with other buffer anions. Therefore, the lower the plasma HCO_3^- concentration drops, the more acidic the urine becomes and the greater its NH_4^+ content (see Clinical Box 39–1).

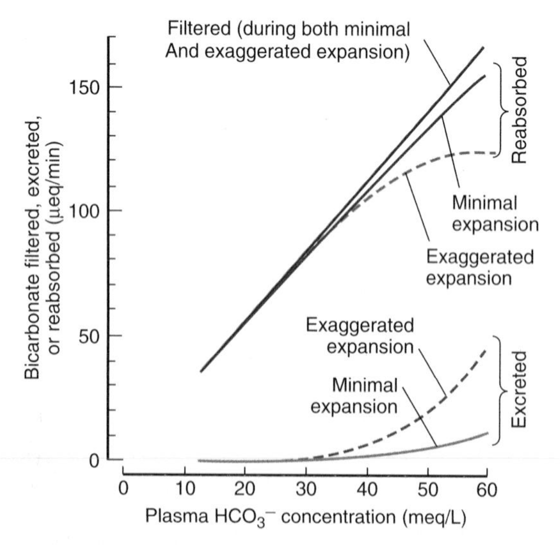

FIGURE 39–4 Effect of ECF volume on HCO_3^- filtration, reabsorption, and excretion in rats. The pattern of HCO_3^- excretion is similar in humans. The plasma HCO_3^- concentration is normally about 24 mEq/L. (Reproduced with permission from Valtin H: *Renal Function,* 2nd ed. Little, Brown, 1983.)

CLINICAL BOX 39–1

Implications of Urinary pH Changes

Depending on the rates of the interrelated processes of acid secretion, NH_4^+ production, and HCO_3^- excretion, the pH of the urine in humans varies from 4.5 to 8.0. Excretion of urine that is at a pH different from that of the body fluids has important implications for the body's electrolyte and acid–base economy. Acids are buffered in the plasma and cells, the overall reaction being $HA + NaHCO_3 \rightarrow NaA + H_2CO_3$. The H_2CO_3 forms CO_2 and H_2O, and the CO_2 is expired, while the NaA appears in the glomerular filtrate. To the extent that the Na^+ is replaced by H^+ in the urine, Na^+ is conserved in the body. Furthermore, for each H^+ ion excreted with phosphate or as NH_4^+, there is a net gain of one HCO_3^- ion in the blood, replenishing the supply of this important buffer anion. Conversely, when base is added to the body fluids, the OH^- ions are buffered, raising the plasma HCO_3^-. When the plasma level exceeds 28 mEq/L, the urine becomes alkaline and the extra HCO_3^- is excreted in the urine. Because the rate of maximal H^+ secretion by the tubules varies directly with the arterial P_{CO_2}, HCO_3^- reabsorption is also affected by the P_{CO_2}. This relationship has been discussed in more detail in the text.

THERAPEUTIC HIGHLIGHT

Sulfonamides inhibit carbonic anhydrase and sulfonamide derivatives have been used clinically as diuretics because of their inhibitory effects on carbonic anhydrase in the kidney (see Chapter 37).

DEFENSE OF H^+ CONCENTRATION

The mystique that envelopes the subject of acid–base balance makes it necessary to point out that the core of the problem is not "buffer base" or "fixed cation" or the like, but simply the maintenance of the H^+ concentration of the ECF. The mechanisms regulating the composition of the ECF are particularly important as far as this specific ion is concerned, because the machinery of the cells is very sensitive to changes in H^+ concentration. Intracellular H^+ concentration, which can be measured by using microelectrodes, pH-sensitive fluorescent dyes, and phosphorus magnetic resonance, is distinct from

TABLE 39–1 H^+ concentration and pH of body fluids.

	H^+ Concentration		
	mEq/L	mol/L	pH
Gastric HCl	150	0.15	0.8
Maximal urine acidity	0.03	3×10^{-5}	4.5
Plasma: Extreme acidocis	0.0001	1×10^{-7}	7.0
Plasma: Normal	0.00004	4×10^{-8}	7.4
Plasma: Extreme alkalosis	0.00002	2×10^{-8}	7.7
Pancreatic juice	0.00001	1×10^{-8}	8.0

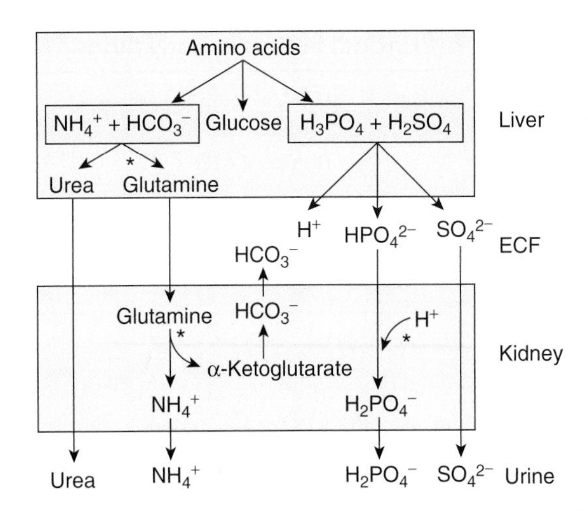

FIGURE 39–5 Role of the liver and kidneys in the handling of metabolically produced acid loads. Sites where regulation occurs are indicated by asterisks. (Modified and reproduced with permission from Knepper MA, et al: Ammonium, urea, and systemic pH regulation. Am J Physiol 1987;253:F199.)

extracellular pH and appears to be regulated by a variety of intracellular processes. However, it is sensitive to changes in ECF H^+ concentration.

The pH notation is a useful means of expressing H^+ concentrations in the body, because the H^+ concentrations are very low relative to those of other cations. Thus, the normal Na^+ concentration of arterial plasma that has been equilibrated with red blood cells is about 140 mEq/L, whereas the H^+ concentration is 0.00004 mEq/L (Table 39–1). The pH, the negative logarithm of 0.00004, is therefore 7.4. Of course, a decrease in pH of 1 unit, for example, from 7.0 to 6.0, represents a 10-fold increase in H^+ concentration. It is important to remember that the pH of blood is the pH of **true plasma**—plasma that has been in equilibrium with red cells—because the red cells contain hemoglobin, which is quantitatively one of the most important blood buffers (see Chapter 36).

H^+ BALANCE

The pH of the arterial plasma is normally 7.40 and that of venous plasma slightly lower. Technically, **acidosis** is present whenever the arterial pH is below 7.40, and **alkalosis** is present whenever it is above 7.40, although variations of up to 0.05 pH unit occur without untoward effects. The H^+ concentrations in the ECF that are compatible with life cover an approximately fivefold range, from 0.00002 mEq/L (pH 7.70) to 0.0001 mEq/L (pH 7.00).

Amino acids are utilized in the liver for gluconeogenesis, leaving NH_4^+ and HCO_3^- as products from their amino and carboxyl groups (Figure 39–5). The NH_4^+ is incorporated into urea (see Chapter 28) and the protons that are formed are buffered intracellularly by HCO_3^-, so little NH_4^+ and HCO_3^- escape into the circulation. However, metabolism of sulfur-containing amino acids produces H_2SO_4, and metabolism of phosphorylated amino acids such as phosphoserine produces H_3PO_4. These strong acids enter the circulation and present a major H^+ load to the buffers in the ECF. The H^+ load from amino acid metabolism is normally about 50 mEq/d. The CO_2 formed by metabolism in the tissues is in large part hydrated to H_2CO_3 (see Chapter 36), and the total H^+ load from this source is over 12,500 mEq/d. However, most of the CO_2 is excreted in the lungs, and only small quantities of the H^+ remain to be excreted by the kidneys. Common sources of extra acid loads are strenuous exercise (lactic acid), diabetic ketosis (acetoacetic acid and β-hydroxybutyric acid), and ingestion of acidifying salts such as NH_4Cl and $CaCl_2$, which in effect add HCl to the body. A failure of diseased kidneys to excrete normal amounts of acid is also a cause of acidosis. Fruits are the main dietary source of alkali. They contain Na^+ and K^+ salts of weak organic acids, and the anions of these salts are metabolized to CO_2, leaving $NaHCO_3$ and $KHCO_3$ in the body. $NaHCO_3$ and other alkalinizing salts are sometimes ingested in large amounts, but a more common cause of alkalosis is loss of acid from the body as a result of vomiting of gastric juice rich in HCl. This is, of course, equivalent to adding alkali to the body.

BUFFERING

Buffering is of key importance in maintaining H^+ homeostasis. It is defined in Chapter 1 and discussed in Chapter 36 in the context of gas transport, with an emphasis on roles for proteins, hemoglobin, and the carbonic anhydrase system in the blood. Carbonic anhydrase is also found in high concentration in gastric acid-secreting cells (see Chapter 25) and in renal tubular cells (see Chapter 37). Carbonic anhydrase is a protein with a molecular weight of 30,000 that contains an atom of zinc in each molecule. It is inhibited by cyanide, azide, and sulfide. In vivo, buffering is, of course, not limited to the blood. The principal buffers in the blood, interstitial fluid, and intracellular fluid are listed in Table 39–2. The principal buffers in cerebrospinal fluid (CSF) and urine are the bicarbonate and phosphate systems. In metabolic acidosis, only 15–20% of

TABLE 39–2 Principal buffers in body fluids.

Blood	$H_2CO_3 \rightleftarrows H^+ + HCO_3^-$ $HProt \rightleftarrows H^+ + Prot^-$ $HHb \rightleftarrows H^+ + Hb^-$
Interstitial fluid	$H_2CO_3 \rightleftarrows H^+ + HCO_3^-$
Intracellular fluid	$HProt \rightleftarrows H^+ + Prot^-$ $H_2PO_4^- \rightleftarrows H^+ + HPO_4^{2-}$

the acid load is buffered by the H_2CO_3-HCO_3^- system in the ECF, and most of the remainder is buffered in cells. In metabolic alkalosis, about 30–35% of the OH^- load is buffered in cells, whereas in respiratory acidosis and alkalosis, almost all of the buffering is intracellular.

In animal cells, the principal regulators of intracellular pH are HCO_3^- transporters. Those characterized to date include the Cl^--HCO_3^- exchanger **AE1,** three Na^+–HCO_3^- cotransporters, and a K^+–HCO_3^- cotransporter.

SUMMARY

When a strong acid is added to the blood, the major buffer reactions are driven to the left. The blood levels of the three "buffer anions" Hb^- (hemoglobin), $Prot^-$ (protein), and HCO_3^- consequently drop. The anions of the added acid are filtered into the renal tubules. They are accompanied ("covered") by cations, particularly Na^+, because electrochemical neutrality is maintained. By processes that have been discussed above, the tubules replace the Na^+ with H^+ and in so doing reabsorb equimolar amounts of Na^+ and HCO_3^-, thus conserving the cations, eliminating the acid, and restoring the supply of buffer anions to normal. When CO_2 is added to the blood, similar reactions occur, except that since it is H_2CO_3 that is formed, the plasma HCO_3^- rises rather than falls.

RENAL COMPENSATION TO RESPIRATORY ACIDOSIS AND ALKALOSIS

As noted in Chapter 36, a rise in arterial P_{CO_2} due to decreased ventilation causes **respiratory acidosis** and conversely, a decline in P_{CO_2} causes **respiratory alkalosis.** The initial changes shown in Figure 35-8 are those that occur independently of any compensatory mechanism; that is, they are those of **uncompensated** respiratory acidosis or alkalosis. In either situation, changes are produced in the kidneys, which then tend to **compensate** for the acidosis or alkalosis, adjusting the pH toward normal.

HCO_3^- reabsorption in the renal tubules depends not only on the filtered load of HCO_3^-, which is the product of the glomerular filtration rate (GFR) and the plasma HCO_3^- level, but also on the rate of H^+ secretion by the renal tubular cells, since HCO_3^- is reabsorbed by exchange for H^+. The rate of H^+ secretion—and hence the rate of HCO_3^- reabsorption—is proportional to the arterial P_{CO_2}, probably because the more CO_2 that is available to form H_2CO_3 in the tubular cells, the more H^+ that can be secreted. Furthermore, when the P_{CO_2} is high, the interior of most cells becomes more acidic. In respiratory acidosis, renal tubular H^+ secretion is therefore increased, removing H^+ from the body; and even though the plasma HCO_3^- is elevated, HCO_3^- reabsorption is increased, further raising the plasma HCO_3^-. This renal compensation for respiratory acidosis is shown graphically in the shift from acute to chronic respiratory acidosis in Figure 35–8. Cl^- excretion is increased, and plasma Cl^- falls as plasma HCO_3^- is increased. Conversely, in respiratory alkalosis, the low P_{CO_2} hinders renal H^+ secretion, HCO_3^- reabsorption is depressed, and HCO_3^- is excreted, further reducing the already low plasma HCO_3^- and lowering the pH toward normal.

METABOLIC ACIDOSIS

When acids stronger than Hb and the other buffer acids are added to blood, **metabolic acidosis** is produced; and when the free H^+ level falls as a result of addition of alkali or removal of acid, **metabolic alkalosis** results. Following the example from Chapter 35, if H_2SO_4 is added, the H^+ is buffered and the Hb^-, $Prot^-$, and HCO_3^- levels in plasma drop. The H_2CO_3 formed is converted to H_2O and CO_2, and the CO_2 is rapidly excreted via the lungs. This is the situation in **uncompensated** metabolic acidosis. Actually, the rise in plasma H^+ stimulates respiration, so that the P_{CO_2}, instead of rising or remaining constant, is reduced. This **respiratory compensation** raises the pH even further. The **renal** compensatory mechanisms then bring about the excretion of the extra H^+ and return the buffer systems to normal.

RENAL COMPENSATION

The anions that replace HCO_3^- in the plasma in metabolic acidosis are filtered, each with a cation (principally Na^+), thus maintaining electrical neutrality. The renal tubular cells secrete H^+ into the tubular fluid in exchange for Na^+; and for each H^+ secreted, one Na^+ and one HCO_3^- are added to the blood. The limiting urinary pH of 4.5 would be reached rapidly and the total amount of H^+ secreted would be small if no buffers were present in the urine to "tie up" H^+. However, secreted H^+ reacts with HCO_3^- to form CO_2 and H_2O (bicarbonate reabsorption); with HPO_4^{2-} to form $H_2PO_4^-$; and with NH_3 to form NH_4^+. In this way, large amounts of H^+ can be secreted, permitting correspondingly large amounts of HCO_3^- to be returned to (in the case of bicarbonate reabsorption) or added to the depleted body stores and large numbers of the cations to be reabsorbed. It is only when the acid load is very large that cations are lost with the anions, producing diuresis and depletion of body cation stores. In chronic acidosis, glutamine synthesis in the liver is increased, using some of the NH_4^+ that

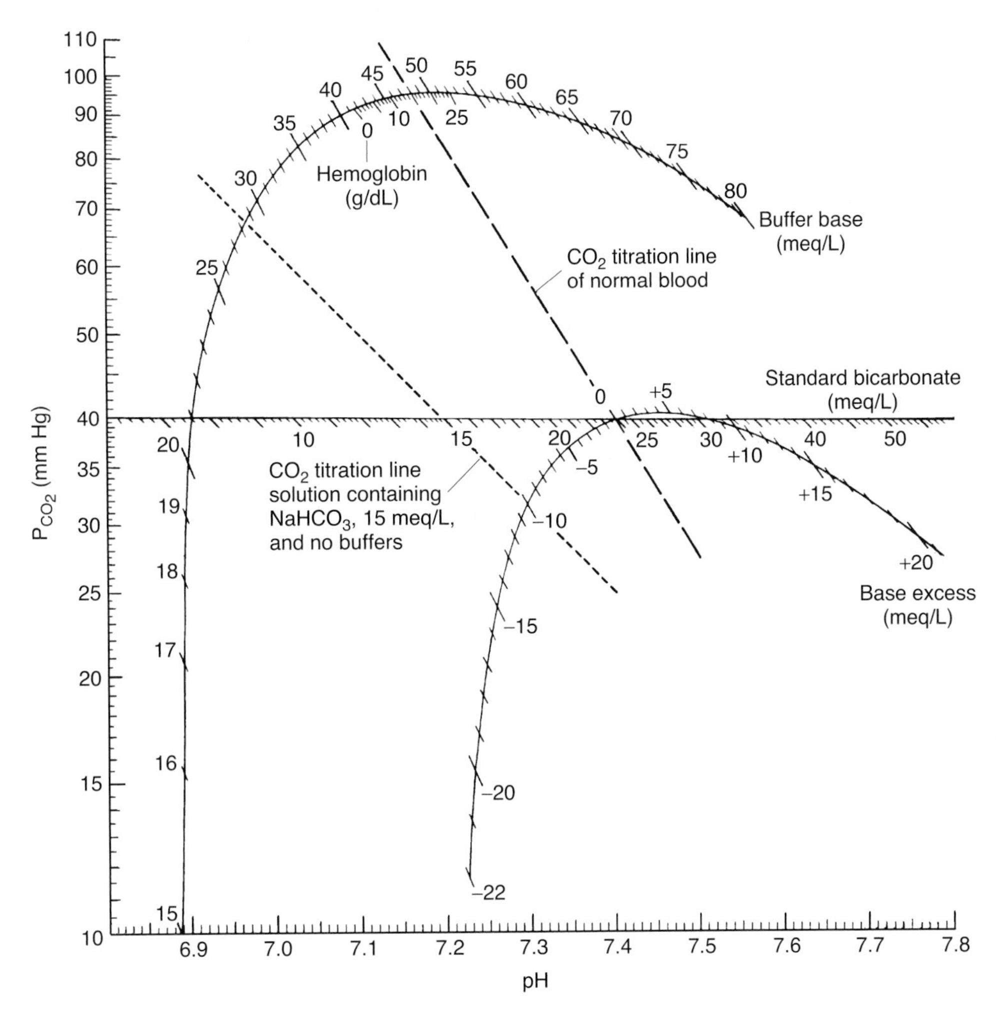

FIGURE 39–6 Siggaard-Andersen curve nomogram. (Courtesy of O Siggaard-Andersen and Radiometer, Copenhagen, Denmark.)

usually is converted to urea (Figure 39–5), and the glutamine provides the kidneys with an additional source of NH_4^+. NH_3 secretion increases over a period of days (adaptation of NH_3 secretion), further improving the renal compensation for acidosis. In addition, the metabolism of glutamine in the kidneys produces α-ketoglutarate, and this in turn is decarboxylated, producing HCO_3^-, which enters the bloodstream and helps buffer the acid load (Figure 39–5).

The overall reaction in blood when a strong acid such as H_2SO_4 is added is:

$$2NaHCO_3 + H_2SO_4 \rightarrow Na_2SO_4 + 2H_2CO_3$$

For each mole of H^+ added, 1 mole of $NaHCO_3$ is lost. The kidney in effect reverses the reaction:

$$Na_2SO_4 + 2\,H_2CO_3 \rightarrow 2NaHCO_3 + 2H^+ + SO_4^{2-}$$

and the H^+ and SO_4^{2-} are excreted. Of course, H_2SO_4 is not excreted as such, the H^+ appearing in the urine as titratable acidity and NH_4^+.

In metabolic acidosis, the respiratory compensation tends to inhibit the renal response in the sense that the induced drop in P_{CO_2} hinders acid secretion, but it also decreases the filtered load of HCO_3^- and so its net inhibitory effect is not great.

METABOLIC ALKALOSIS

In metabolic alkalosis, the plasma HCO_3^- level and pH rise (Figure 39–6). The respiratory compensation is a decrease in ventilation produced by the decline in H^+ concentration, and this elevates the P_{CO_2}. This brings the pH back toward normal while elevating the plasma HCO_3^- level still further. The magnitude of this compensation is limited by the carotid and aortic chemoreceptor mechanisms, which drive the respiratory center if any appreciable fall occurs in the arterial PO_2. In metabolic alkalosis, more renal H^+ secretion is expended in reabsorbing the increased filtered load of HCO_3^-; and if the HCO_3^- level in plasma exceeds 26–28 mEq/L, HCO_3^- appears in the urine. The rise in P_{CO_2} inhibits the renal compensation by facilitating acid secretion, but its effect is relatively slight.

THE SIGGAARD–ANDERSEN CURVE NOMOGRAM

Use of the Siggaard–Andersen curve nomogram (Figure 39–6) to plot the acid–base characteristics of arterial blood is helpful in clinical situations. This nomogram has P_{CO_2} plotted on a log scale on the vertical axis and pH on the horizontal axis. Thus, any point to the left of a vertical line through pH 7.40 indicates acidosis, and any point to the right indicates alkalosis. The position of the point above or below the horizontal line through a P_{CO_2} of 40 mm Hg defines the effective degree of hypoventilation or hyperventilation.

If a solution containing $NaHCO_3$ and no buffers were equilibrated with gas mixtures containing various amounts of CO_2, the pH and P_{CO_2} values at equilibrium would fall along the dashed line on the left in Figure 39–6 or a line parallel to it. If buffers were present, the slope of the line would be greater; and the greater the buffering capacity of the solution, the steeper the line. For normal blood containing 15 g of hemoglobin/dL, the CO_2 **titration line** passes through the 15-g/dL mark on the hemoglobin scale (on the underside of the upper curved scale) and the point where the P_{CO_2} = 40 mm Hg and pH = 7.40 lines intersect, as shown in Figure 39–6. When the hemoglobin content of the blood is low, there is significant loss of buffering capacity, and the slope of the CO_2 titration line diminishes. However, blood of course contains buffers in addition to hemoglobin, so that even the line drawn from the zero point on the hemoglobin scale through the normal P_{CO_2}–pH intercept is steeper than the curve for a solution containing no buffers.

For clinical use, arterial blood or arterialized capillary blood is drawn anaerobically and its pH measured. The pHs of the same blood after equilibration with each of two gas mixtures containing different known amounts of CO_2 are also determined. The pH values at the known P_{CO_2} levels are plotted and connected to provide the CO_2 titration line for the blood sample. The pH of the blood sample before equilibration is plotted on this line, and the P_{CO_2} of the sample is read off the vertical scale. The **standard bicarbonate** content of the sample is indicated by the point at which the CO_2 titration line intersects the bicarbonate scale on the P_{CO_2} = 40 mm Hg line. The standard bicarbonate is not the actual bicarbonate concentration of the sample but, rather, what the bicarbonate concentration would be after elimination of any respiratory component. It is a measure of the alkali reserve of the blood, except that it is measured by determining the pH rather than the total CO_2 content of the sample after equilibration. Like the alkali reserve, it is an index of the degree of metabolic acidosis or alkalosis present.

Additional graduations on the upper curved scale of the nomogram (Figure 39–6) are provided for measuring **buffer base** content; the point where the CO_2 calibration line of the arterial blood sample intersects this scale shows the mEq/L of buffer base in the sample. The buffer base is equal to the total number of buffer anions (principally $Prot^-$, HCO_3^-, and Hb^-) that can accept hydrogen ions in the blood. The normal value in an individual with 15 g of hemoglobin per deciliter of blood is 48 mEq/L.

The point at which the CO_2 calibration line intersects the lower curved scale on the nomogram indicates the **base excess.** This value, which is positive in alkalosis and negative in acidosis, is the amount of acid or base that would restore 1 L of blood to normal acid–base composition at a P_{CO_2} of 40 mm Hg. It should be noted that a base deficiency cannot be completely corrected simply by calculating the difference between the normal standard bicarbonate (24 mEq/L) and the actual standard bicarbonate and administering this amount of $NaHCO_3$ per liter of blood; some of the added HCO_3^- is converted to CO_2 and H_2O, and the CO_2 is lost in the lungs. The actual amount that must be added is roughly 1.2 times the standard bicarbonate deficit, but the lower curved scale on the nomogram, which has been developed empirically by analyzing many blood samples, is more accurate.

In treating acid–base disturbances, one must, of course, consider not only the blood but also all the body fluid compartments. The other fluid compartments have markedly different concentrations of buffers. It has been determined empirically that administration of an amount of acid (in alkalosis) or base (in acidosis) equal to 50% of the body weight in kilograms times the blood base excess per liter will correct the acid–base disturbance in the whole body. At least when the abnormality is severe, however, it is unwise to attempt such a large correction in a single step; instead, about half the indicated amount should be given and the arterial blood acid–base values determined again. The amount required for final correction can then be calculated and administered. It is also worth noting that, at least in lactic acidosis, $NaHCO_3$ decreases cardiac output and lowers blood pressure, so it should be used with caution.

CHAPTER SUMMARY

- The cells of the proximal and distal tubules secrete hydrogen ions. Acidification also occurs in the collecting ducts. The reaction that is primarily responsible for H^+ secretion in the proximal tubules is Na^+–H^+ exchange. Na is absorbed from the lumen of the tubule and H is excreted.
- The maximal H^+ gradient against which the transport mechanisms can secrete in humans corresponds to a urine pH of about 4.5. However, three important reactions in the tubular fluid remove free H^+, permitting more acid to be secreted. These are the reactions with HCO_3^- to form CO_2 and H_2O, with HPO_4^{2-} to form $H_2PO_4^-$, and with NH_3 to form NH_4^+.
- Carbonic anhydrase catalyzes the formation of H_2CO_3, and drugs that inhibit carbonic anhydrase depress secretion of acid by the proximal tubules.
- Renal acid secretion is altered by changes in the intracellular P_{CO_2}, K^+ concentration, carbonic anhydrase level, and adrenocortical hormone concentration.

MULTIPLE-CHOICE QUESTIONS

For all questions, select the single best answer unless otherwise directed.

1. Which of the following is the principal buffer in interstitial fluid?
   A. Hemoglobin
   B. Other proteins
   C. Carbonic acid
   D. H_2PO_4
   E. Compounds containing histidine
2. Increasing alveolar ventilation increases the blood pH because
   A. it activates neural mechanisms that remove acid from the blood.
   B. it makes hemoglobin a stronger acid.
   C. it increases the PO_2 of the blood.
   D. it decreases the Pco_2 in the alveoli.
   E. the increased muscle work of increased breathing generates more CO_2.
3. In uncompensated metabolic alkalosis
   A. the plasma pH, the plasma HCO_3^- concentration, and the arterial Pco_2 are all low.
   B. the plasma pH is high and the plasma HCO_3^- concentration and arterial Pco_2 are low.
   C. the plasma pH and the plasma HCO_3^- concentration are low and the arterial Pco_2 is normal.
   D. the plasma pH and the plasma HCO_3^- concentration are high and the arterial Pco_2 is normal.
   E. the plasma pH is low, the plasma HCO_3^- concentration is high, and the arterial Pco_2 is normal.
4. In a patient with a plasma pH of 7.10, the $[HCO_3^-]/[H_2CO_3]$ ratio in plasma is
   A. 20.
   B. 10.
   C. 2.
   D. 1.
   E. 0.1.

CHAPTER RESOURCES

Adrogué HJ, Madius NE: Management of life-threatening acid–base disorders. N Engl J Med 1998;338:26.

Brenner BM, Rector FC Jr (editors): *The Kidney,* 6th ed, 2 Vols. Saunders, 1999.

Davenport HW: *The ABC of Acid–Base Chemistry,* 6th ed. University of Chicago Press, 1974.

Halperin ML: *Fluid, Electrolyte, and Acid–Base Physiology,* 3rd ed. Saunders, 1998.

Lemann J Jr., Bushinsky DA, Hamm LL: Bone buffering of acid and base in humans. Am J Physiol Renal Physiol 2003;285:F811 (Review).

Vize PD, Wolff AS, Bard JBL (editors): *The Kidney: From Normal Development to Congenital Disease.* Academic Press, 2003.

Answers to Multiple Choice Questions

Chapter 1
1. B **2.** C **3.** B **4.** C **5.** C **6.** D **7.** E **8.** E

Chapter 2
1. A **2.** D **3.** D **4.** B **5.** C **6.** C **7.** B **8.** A

Chapter 3
1. B **2.** C **3.** E **4.** B **5.** B **6.** C **7.** D

Chapter 4
1. C **2.** E **3.** E **4.** A **5.** C **6.** B **7.** B **8.** C

Chapter 5
1.B **2.**D **3.** B **4.** C **5.** C

Chapter 6
1. C **2.** D **3.** E **4.** B **5.** D **6.** E

Chapter 7
1. D **2.** A **3.** C **4.** C **5.** B **6.** E

Chapter 8
1. D **2.** A **3.** B **4.** A **5.** D **6.** C **7.** B **8.** C
9. A **10.** E **11.** D

Chapter 9
1. D **2.** D **3.** C **4.** B **5.** E **6.** C **7.** D **8.** B
9. D **10.** D **11.** B

Chapter 10
1. A **2.** E **3.** E **4.** E **5.** B **6.** D **7.** D **8.** E
9. C **10.** A

Chapter 11
1. D **2.** C **3.** D **4.** D **5.** D **6.** C **7.** D **8.** E

Chapter 12
1. E **2.** C **3.** E **4.** C **5.** E **6.** B **7.** C **8.** A
9. E **10.** C **11.** D **12.** E

Chapter 13
1. A **2.** D **3.** C **4.** D **5.** C **6.** E

Chapter 14
1. C **2.** D **3.** C **4.** D **5.** A **6.** B

Chapter 15
1. C **2.** E **3.** C **4.** D **5.** B **6.** D **7.** D **8.** B

Chapter 16
No multiple choice questions

Chapter 17
1. B **2.** E **3.** B **4.** A **5.** A **6.** B **7.** D **8.** D

Chapter 18
1. E **2.** E **3.** A **4.** C **5.** B

Chapter 19
1. C **2.** B **3.** E **4.** C **5.** C **6.** A **7.** D **8.** A
9. D **10.** C

Chapter 20
1. D **2.** B **3.** E **4.** D **5.** C **6.** D **7.** D **8.** A **9.** A

Chapter 21
1. C **2.** E **3.** D **4.** A **5.** C **6.** D **7.** E

Chapter 22
1. C **2.** D **3.** C **4.** A

Chapter 23
1. E **2.** A **3.** C **4.** B

Chapter 24
1. E **2.** D **3.** D **4.** C **5.** E **6.** D **7.** C

Chapter 25
1. C **2.** E **3.** D **4.** C **5.** D

Chapter 26
1. E **2.** D **3.** E **4.** A **5.** C

Chapter 27
1. C **2.** D **3.** E **4.** A **5.** B

Chapter 28
1. E **2.** E **3.** C **4.** E **5.** E **6.** B

Chapter 29
1. C **2.** A **3.** A **4.** D **5.** D

Chapter 30
1. A **2.** C **3.** C **4.** C **5.** E **6.** D

Chapter 31
1. C **2.** B **3.** D **4.** B **5.** E **6.** A **7.** A **8.** E

Chapter 32
1. B **2.** A **3.** D **4.** D **5.** D

Chapter 33
1. D **2.** A **3.** E **4.** E **5.** E **6.** D

Chapter 34
1. D **2.** C **3.** A **4.** E **5.** D **6.** A

Chapter 35
1. E **2.** B **3.** D **4.** D

Chapter 36
1. D **2.** B **3.** B **4.** D **5.** E **6.** B **7.** C

Chapter 37
1. A **2.** A **3.** A **4.** A **5.** E **6.** C **7.** D

Chapter 38
1. E **2.** C **3.** D **4.** E **5.** D **6.** C **7.** D

Chapter 39
1. C **2.** D **3.** D **4.** B

Index

A
Aberrant sexual differentiation
  chromosomal abnormalities, 396
  hormonal abnormalities, 396–398
ABO system
  agglutination reactions, 561
  agglutinins, 561
  A and B antigens, 560–561
  H antigen, 560–561
  red cell agglutination, 561, 563
ABP. *See* Androgen binding protein
Absence seizures, 277
Absorption, 479–483
  of calcium, 485
  of iron, 485–486
  of vitamins, 485
Accelerated AV conduction, 533, 534
Acclimatization process, 650–651
Accommodation and aging, 188
ACE inhibitors. *See* Angiotensin-converting enzyme inhibitors
Acetylcholine, 144, 460, 461
  effect on intestinal smooth muscle, 115
  effect on unitary smooth muscle, 116
  functions of, 143
  removal from synapse, 144
  synthesis of, 143
  transmission at autonomic junctions, 259
  transport and release, 144
Acetylcholine receptors, 144–145
  pharmacologic properties, 144
Acetylcholinesterase, 144
Acetylcholinesterase inhibitors
  Alzheimer disease treatment by, 289
  myasthenia gravis treatment by, 129
ACh. *See* Acetylcholine
Achalasia, 501
Achondroplasia, 334
Achromatopsia, 193
Acid-base balance, 645
  HCO_3^- values, 647
  Pco_2 values, 647
  plasma pH values, 647
  ventilatory responses to changes in, 661–662
Acid–base disorders, 6
Acid–base nomogram, 647
Acid hydrolases
  in lysosomes, 39
Acidosis, 6, 647–649, 715
  uncompensated, 716
Acids, 6
Acquired immunity, 71
  diagrammatic representation of, 77
  T and B lymphocytes activation in, 71, 74–75
Acquired nystagmus, 212
Acromegaly, 305
  and gigantism, 328
Acrosomal reaction, 413
Acrosome, 420
ACTH. *See* Adrenocorticotropic hormone
Actinin in skeletal muscle, 100
Actin in skeletal muscle, 100
Action potentials
  all-or-none, 89–90
  in auditory nerve fibers, 207
  of cardiac muscle, 110–111
  changes in excitability during, 90–91
  conduction of, 87, 91
  generation in postsynaptic neuron, 124–125
  ionic fluxes during, 88–89
  recorded in dendrites, 125
  and twitch, 102
Active transport, 51
Activin receptors, 427
Acute pesticide poisoning, 262
Acute-phase proteins, 81, 512
Acuterespiratory distress syndrome, 653
Acyclovir, 47
Adaptation in olfactory system, 221
Addiction and motivation, 172
Addison disease, 375
Addisonian crisis, 375
Adenosine derivatives, 11
Adenosine triphosphate, 11–12, 149–150, 377
  formation process, 12
  generation in citric acid cycle, 23
  generation in glycolysis, 23–24
  as neurotransmitter, 150
  role in cell, 12
  turnover in muscle cells, 107
Adenylyl cyclase
  cAMP production by, 61–62
  regulatory properties, 61
ADH. *See* Antidiuretic hormone
Adipokines, 450
Adrenal cortex
  adrenocortical hormones. *See* Adrenocortical hormones
  constituents of, 354
  functions of, 353
  hormone biosynthesis in, 359
Adrenalectomy, 371
Adrenal gland, medulla and zones of cortex in, 355
Adrenal insufficiency, 365, 375
Adrenal medulla
  during fetal life, 354
  morphology of, 354–355
  secretions of inner, 353
  sympathetic ganglion, 353
Adrenal medullary hormones, 353
Adrenal medullary secretion regulation, 358
Adrenal responsiveness and ACTH, 368
Adrenal steroidogenic enzymes, nomenclature for, 362
Adrenergic receptors
  at junctions within ANS, 260
  subtypes of, 357
α–Adrenergic receptors, 146–147
Adrenoceptors, 146–147
  activation, 147
  epinephrine and norepinephrine as, 146–147
  subtypes of, 146
Adrenocortical hormones
  in adult humans, 360
  basic structure of, 358
  classification of, 359
  secreted steroids, 359
  species differences in, 359–360
  steroids. *See* Steroids
  types of, 358
Adrenocortical hyper-& hypofunction, effects of, 374–376
Adrenocortical secretion, 353–354
Adrenocorticotropic hormone
  and adrenal responsiveness, 368
  aldosterone secretion produced by, 372
  chemistry and metabolism, 368
  and circadian rhythm, 368–369
  deficiency, 336
  effect on adrenal, 368
  effect on aldosterone secretion, 372
  in fetus, 416
  functions, 368
  mechanism of action of, 361
  plasma concentrations of, 369
  response to stress, 369
  secretion of
    glucocorticoids effects on, 365
    and stress, 366
  stimulation, 362
Adrenogenital syndrome, 362, 374
Adrenomedullin, 357
Advanced glycosylation end products, 449
AE1. *See* Anion exchanger 1
Aerobic glycolysis and exercise, 107
Aerophagia, 502
Afferent and efferent neurons in spinal cord, polysynaptic connections between, 234
Afferent arterioles, 673, 674
Afferent fibers, central connections of, 230
Afferent nerve firing, Pco_2 effect, 660
Afferents, 319
Afferent vagal fibers, inspiratory discharge inhibition by, 659
2-AG, 151
Aganglionic megacolon, 505
Age-related macular degeneration, 180, 181
AGEs. *See* Advanced glycosylation end products
Ageusia, 225
Aging and accommodation, 188
Agnosia, 291
Agranulocytosis, 147
Air
  conduction, 206
  Po_2 and Pco_2 values in, 642

Airway conduction, 621–624
Airway obstruction, 630
Airway receptors, 664
Airway resistance, 631
Airways, responses mediated by receptors in, 664
Akinesia, 245
Albinos, 325
Albuminuria, 679
Aldosterone, 364
  effect on Na, K ATPase pump activity, 53
  in salt balance regulation, 374
  secretion, regulation of
    ACTH effect on, 372
    angiotensin II effect on, 372–373
    electrolytes effect on, 373–374
    renin effect on, 372–373
    second messengers, 375
    stimuli for, 371–372
Aldosterone deficiency, 375
Aldosterone-secreting mechanism, 374
Alerting response, 273
α-limit dextrins, 478
Alkalosis, 6, 647–649, 715
Allergic disease, actions of glucocorticoids in, 367
Allodynia and hyperalgesia, 164–165
Alpha rhythm, 273
  variations in, 273
ALS, 240
Alveolar air, 634–635
  composition of, 635, 650
  sampling of, 634–635
Alveolar airway, 621, 624–625
  acinar tissue, 621
  lung parenchyma, 621
Alveolar gas equation, 634
Alveolar surface tension, surfactant role, 631–632
Alveolar ventilation, 633
  variations effect in respiratory rate, 633
Alveoli
  adult human, prominent cells in, 625
  conducting airway, cellular transition from, 623
  pressure in, 628
  type I epithelial cells, 624
  type II epithelial cells. *See* Granular pneumocytes
Alveolocapillary membrane, diffusion across, 635
Alzheimer disease
  abnormalities associated with, 290
  cytopathologic hallmarks of, 290
  prevalence of, 289
  risk factors and pathogenic processes in, 290
  treatment of, 289
Amantadine, 140
Amatoxins, 263
Amblyopia, 187
AMD. *See* Age-related macular degeneration
AME, 371
α–Melanocyte-stimulating hormone, 415
Amenorrhea, 413
Amino acid pool, 17
  and common metabolic pool, interconversions between, 21
Amino acids, 16
  absorption of, 482
  activation in cytoplasm, 19
  in body, 19
  catabolism of, 21
  conditionally essential, 17
  found in proteins, 19
  metabolic functions of, 22
  nonessential, 17
  nutritionally essential, 17
Aminopyridines, Lambert–Eaton Syndrome treatment by, 130
Amiodarone, 181
Ammonia buffering system, 713
Ammonia processing to urea, 21–22
Amniocentesis, 397
AMPA receptors, 139–140
  in glia and neurons, 141
Amphetamine, 147
Amphipathic, 465
Ampullary responses to rotation, 211
Amyloid precursor protein, 290
Amylopectin, structure of, 478
Amylose, structure of, 478
Amyotrophic lateral sclerosis, 240
Anabolism, 488
Anaerobic glycolysis and exercise, 107
Anandamide, 151
Anatomic reserve, 491
Androgenbinding protein, 421
Androgen-dependent, 428
Androgen resistance, 397
Androgens, 364–365, 391, 419
Andropause, 401
Anemia, effects of, 653
Anemic hypoxia, 649, 652
Aneuploidy, 14
Angiogenesis
  vascular endothelial growth factor, 573
  vasculogenesis, 573
Angiotensin, 599
Angiotensin-converting enzyme
  diagrammatic representation of, 703
Angiotensin-converting enzyme inhibitors, 540
Angiotensin II
  action on subfornical organ, 310
  actions on adrenal cortex, 55
  effect on aldosterone secretion, 372–373
  mechanism of action of, 361
Angiotensin III, 704
Angiotensinogen. *See* Renin substrate
Anion exchanger 1, 645
Anion gap, 648
Anomic aphasia, 293
Anorectal area, sagittal view of, 506
Anorexin, 487
Anosmia, 221
Anovulatory cycles, 412
Anoxia, 649
ANP. *See* Atrial natriuretic peptide
ANP granules, 706
ANS. *See* Autonomic nervous system
Anterior pituitary gland
  hormone-secreting cells of, 324–325
  hormones secreted by, 323
    functions of, 323
    prolactin, 323
Anterior pituitary hormones, 313
  actions of, 314
  hypothalamus and, 314
  secretion, hypothalamic control of, 314
Anterograde amnesia, 285
Antibiotics, anosmia treatment by, 221
Antibodies against receptors, 64
Anticlotting mechanisms, 567–568
  antiaggregating effect, 567
  antithrombin III, 567
  clotting factors, 567
  fibrinolytic system, 567–568
  heparin, 567
  plasmin, 567–568
  plasminogen receptors, 568
  thrombomodulin, 567
Anticoagulants, 568–569
  chelating agents, 569
  coumarin derivatives, 569
  heparin, 568–569
Anticonvulsant drugs, 277
Antidepressants, Alzheimer disease treatment using, 289
Antidiuretic hormone, 313, 698
Antidromic and orthodromic conduction, 91–92
Antigen
  presentation of, 75
  recognition of, 75
Antigen-presenting cells
  and αβ T lymphocyte, interaction between, 76
  MHC protein–peptide complexes on, 75, 76
  types of, 75
Antihistamines, 160
Antimuscarinic syndrome, 263
Antipsychotic drugs
  for muscarinic poisoning treatment, 263
  for schizophrenia, 147
  thioridazine, 181
Antiviral drugs, 47
Antral systole, 502
Antrum
  formation, 401
  gastrin secretion inhibition by, 469
Aortic bodies, 660
  location of, 660
AP-1, 49
APCs. *See* Antigen-presenting cells
Aphasias, 293
Apneusis, 658
Apoptosis
  definition of, 47
  pathway bringing about, 47
APP, 290
Apparent mineralocorticoid excess, 371
Aquaporin-1, 685
Aquaporins, 685
Arachidonic acid, 31
2-Arachidonyl glycerol, 151
ARDS. *See* Acuterespiratory distress syndrome
Area postrema, 503
Arginine vasopressin, 311
Argyll Robertson pupil, 189
Arial muscle, 539
Arithmetic calculations, brain regions in, 294
Aromatase, 417
Aromatase inhibitors, 302

Arterial pressure
artery pressure curve, 578
diastolic pressure, 578
systolic pressure, 578
Ascending reticular activating system in brainstem, 272
Ascending sensory pathway, 167
Ascites, 511
Asphyxia, 666
Aspiration pneumonia, 626
Assembly protein 1, 49
Associative learning, 284, 285
Astereognosis, 291
Astigmatism, 188
Astrocytes, 83–84
protoplasmic, 84
Ataxia, 252
Atelectasis, 632
Atherosclerosis and cholesterol, 31
Athetosis, 245
ATP. *See* Adenosine triphosphate
Atretic follicles, 401
Atrial arrhythmias, 531
Atrial fibrillation, 531, 532
Atrial natriuretic peptide, 702, 707
effect of immersion, 708
Atrial systole, 521
Atrial tachycardia, 531
Atrioventricular node, 521
nodal block, 530
nodal delay, 524
Atrioventricular valves, 539
Atropine, 187, 262, 469
Atypical depression, 149
Auditory acuity, 209
Auditory nerve fibers, action potentials in, 207
Auditory ossicles, 199
schematic representation of, 206
Auditory pathways, 207–209
Auerbach's plexus, 473
Augmented limb leads, 526
Auras, 276
Auscultatory method
cuff pressure, 579
sounds of Korotkoff, 579
Autocrine communication, 54
Autonomic junctions, chemical transmission at
acetylcholine, 259
cholinergic neurotransmission, 259
nonadrenergic, noncholinergic transmitters, 264
noradrenergic neurotansmission, 260, 261, 264
norepinephrine, 259
Autonomic nerve activity, effector organs response to, 260
Autonomic nerve impulses, responses of effector organs to, 264–265
Autonomic nervous system, 473
divisions of, 255
dysfunction of, 266
features of, 256
functions of, 255, 264
parasympathetic division of, 257–259
peripheral motor portions of, 256
peripheral organization and transmitters released by, 257
and somatomotor nervous system, difference between, 264
sympathetic division of, 256–257
Autonomic neurotransmission, drugs affecting processes involved in, 261
Autonomic preganglionic neurons, 256
descending inputs to, 265–266
Autonomic responses
pathways controling, 265
triggered in hypothalamus, 309–310
Autoreceptor, 136
Autosomal dominant polycystic kidney disease, 687
AV node. *See* Atrioventricular node
AVP, 311
AV valves. *See* Atrioventricular valves
Axial and distal muscles, control of
corticobulbar tract, 239
corticospinal tracts, 238
movement and, 239
Axoaxonal synapses, 121
Axodendritic synapses, 121
Axonal conduction velocity, 92
Axonal regeneration, 94
Axonal transport, 86
along microtubules, 87
Axonemal dynein, 42
Axoneme, 42
Axon stump, degeneration of, 131
Axosomatic synapses, 121
Azathioprine, myasthenia gravis treatment using, 129
Azotemia, 548

B

Bachmann's bundle, 521
Baclofen
for ALS treatment, 240
for CP treatment, 236
Bacterial infections, actions of glucocorticoids in, 367
Bacterial toxins effect on cAMP, 62
β–Adrenergic receptors, 146–147
Ballism, 245
β-amino acids, 13
Barbiturates, 143
Barometric pressure, 634
effects of, 649
Baroreceptors
aortic arch, 589, 592
aortic depressor nerve, 589
cardiopulmonary receptors, 589
carotid sinus, 589, 592
carotid sinus nerve, 589
caudal ventrolateral medulla, 590
nerve activity of
cardiac output, 591
sympathetic nerve activity, 591
systemic blood pressure, 590
nucleus of tractus solitarius, 590
stimulation, respiratory effects of, 665
Barr body, 393
Bartter syndrome, 687
Barttin, 687
Basal ganglia
biochemical pathways in, 245
diseases of, 245, 246
functions of, 245
organization of, 243–244
Parkinson disease, 245, 247–248
principal connections of, 244
Basal lamina, 38. *See also* Cell membrane
Basal metabolic rate, 490
Basal plasma growth hormone level, 326
Bases, 6
Basic electrical rhythm, 498
of gastrointestinal smooth muscle, 499
spike potentials, 498
Basilar arteries, 403
Basket cells, 249, 270, 271
Basophils, 68
B cells, 69
activation in acquired immunity, 71, 75
exhaustion of, 443
maturation, sites of congenital blockade of, 80
responses, long-term changes, 443
role of cytokines in, 77
TH2 subtype, 77
BCR-ABL fusion gene, 57
BDNF, 94
Becker muscular dystrophy, 100
Benzodiazepines, 143, 262
BER. *See* Basic electrical rhythm
Beta rhythm, 273
Bezold–Jarisch reflex, 664
Bifascicular/trifascicular block, 530
Bile
cholesterol solubility, 515
human hepatic duct bile, 512
human hepatic duct, comparison of, 516
production of, 516
Bile acids, 465
Bile canaliculi, 510
Bile pigments, glucuronides of, 465, 513
Biliary secretion
bile, 464–466
neurohumoral control of, 516
Biliary system
bile formation, 514–516
biliary secretion, regulation of, 516
cholecystectomy, effects of, 516
gallbladder, functions of, 516
gallbladder, visualizing, 516–517
Bilirubin, 465, 513
handling of, 513
heme, conversion of, 513
uridine diphosphoglucuronic acid molecules, 513
Biliverdin, 465
Binocular vision, 195
Biofeedback, 288
Biologic oxidations, common form of, 11
Bitter taste, 224
Blastocyst, 414
Blind spot, 180
Blood
active and inactive capillaries, 582
angiogenesis
vascular endothelial growth factor, 573
vasculogenesis, 573
arterial blood pressure, 579, 581

Blood (*Cont'd.*)
  arterial pressure
    artery pressure curve, 578
    diastolic pressure, 578
    systolic pressure, 578
  arteries and arterioles, 570
  arteriovenous anastomoses, 572
  auscultatory method
    cuff pressure, 579
    sounds of Korotkoff, 579
  average velocity, 574–575
  blood flow
    cardiac cycle in dog, 577
    pressure and velocity, 577
  blood flow measurement, 573
  blood pressure measurement
    Bernoulli's principle, 578–579
    kinetic energy, 578–579
    and pathophysiology, 579
  blood type
    ABO system, 560–561
    agglutinogens, 564
    antigens inheritance, 562–564
    newborn, hemolytic disease, 564
    RH group, 564
    transfusion reactions, 561–562
  bone marrow
    blood cells, 556
    hematopoietic stem cells, 556
  buffering, 645–647
  capacitance vessels, 577
  capillaries, 570–572
  capillary circulation, 579
  capillary pressures, 580
    capillary filtration coefficient, 582
    flow-limited and diffusion-limited
      exchange, 582
    hydrostatic pressure gradient, 581
    interstitial fluid, 581–582
    osmotic pressure gradient, 581–582
    pressure gradients, 582
    starling forces, 581
  circulation time, 575
  critical closing pressure, 576
  Doppler flow meters, 573
  effective perfusion pressure, 573
  effect of gravity, 578
  endothelium, 569
  fate of CO_2 in, 645
  flow, 632–633, 676–677
  gas content of, 642
  genes in human, 574, 575
  hemoglobin
    carboxyhemoglobin, 559
    catabolism, 559–560
    in fetus, 559
    methemoglobin, 558–559
    oxyhemoglobin, 557–559
    synthesis, 559
  hemostasis
    anticlotting mechanisms, 567–568
    anticoagulants, 568–569
    clotting mechanism, 566–567
    response to injury, 565
  interstitial fluid volume
    edema, 584
    elephantiasis, 585
    lymphedema, 585
    precapillary constriction, 584
    promoting factors, 584, 585
  laminar flow
    effect of constriction, 574
    probability of turbulence, 574
    Reynolds number, 574
    velocity, 574
  law of Laplace
    curvature of viscus, 576
    dilated heart, 577
    surface tension, 577
    transmural pressure, 576
  lymph, 569
  lymphatic circulation
    functions of, 584
    lymphatics draining, 584
    lymphatic vessels, 584
  lymphatics, 572
  molecular fate in, 644
  Ohm's law, 573
  peripheral resistance, 573
  physical principles, 573–574
  plasma, 564–565
  plasma proteins
    afibrinogenemia, 565
    hypoproteinemia, 565
    origin of, 565
    physiologic functions, 565, 566
  platelets, 557, 558
  Po_2 and Pco_2 values in, 642
  Poiseuille–Hagen formula, 575
  red blood cells
    characteristics, 557, 559
    fibrin fibrils, 557, 559
    formation and destruction, 557, 559
  resistance vessels, 577
  shear stress, 574
  spatial distribution of, 645
  vascular smooth muscle
    contraction and relaxation, 569, 570
    latch-bridge mechanism, 569
  venous circulation, 582
  venous pressure and flow, 582
    air embolism, 583
    effects of heartbeat, 583
    muscle pump, 583
    thoracic pump, 583
    venous pressure measurement,
      583–584
  venules and veins, 572
  viscosity and resistance
    effect of changes, 576
    hematocrit, 575–576
  white blood cells
    cells grow, 556–558
    cellular elements, 556, 557
Blood–brain barrier, 514
  circumventricular organs
    angiotensin II, 605
    chemoreceptor zones, 605
    neurohemal organs, 605, 606
    subcommissural organ, 605–606
  development of, 606
  function of, 606
  penetration of substances
    adenosine triphosphate, 605
    localization of various GLUT
      transpoters, 605
    physiologic significance, 605
    specific transporters, 605
  penetration of urea, 604, 605
Blood cells, glucocorticoids effects on, 366
Blood osmolality in humans, 303
Blood–testis barrier, 419
Blood type
  ABO system
    agglutination reactions, 561
    agglutinins, 561
    A and B antigens, 560–561
    H antigen, 560–561
    red cell agglutination, 561, 563
  agglutinogens, 564
  antigens inheritance, 562–564
  newborn, hemolytic disease
    fetal–maternal hemorrhage, 564
    hydrops fetalis, 564
    kernicterus, 564
  RH group, 564
  transfusion reactions, 561–562
Blood vessels, innervation of
  splanchnic veins, 588
  sympathetic nerves, 587–588
  sympathetic noradrenergic fibers, 587–588
  venoconstriction, 588
BMI. *See* Body mass index
BMR. *See* Basal metabolic rate
BNP. *See* Brain natriuretic peptide
Body fluids
  buffering capacity of, 6
  inappropriate compartmentalization, 3–4
  organization of, 5
Body mass index, 488
Body mechanics, 110
Body temperature
  and fever, 319–320
  heat production and heat loss, 317–318
  mechanisms regulating, 318–319
    threshold core temperatures for, 319
  normal, 317
Body water, intracellular component of, 4
Body weight, 310
Bohr effect, 643
Bohr's equation, 634
Bombesin, 473. *See also* Gastrin releasing
    peptide
Bone conduction, 206
Bone marrow. *See* Erythropoiesis
Bone physiology, 385
  bone disease, 388
  bone formation & resorption, 386–388
  bone growth, 385–386
  structure, 385
Bony labyrinth, 200
Botulinum and tetanus toxins, 123
Botulinum toxin, 123
  for ALS treatment, 240
  for clonus treatment, 233
Bovine prepropressophysin, 312
Bowman's capsule, 673
Bradycardia, 529
Bradycardia-tachycardia, 529
Bradykinesia, 245
Brain and memory, link between, 285

Brain-derived neurotrophic factor, 94
Brain metabolism
  energy sources, 609
  glutamate, ammonia removal, 609
  oxygen consumption, 609
  uptake, release of substances, 608–609
Brain natriuretic peptide, 707
Brain regions in arithmetic calculations, 294
Brain stem
  ascending reticular activating system in, 272
  chemoreceptors in, 661
  respiratory neurons in, 659
Brain stem pathways, in voluntary movement
  lateral, 240–241
  medial, 239–240
Breaking point, 663
Breast cancer, 302
Breathing
  control and outcome, 619
  work of, 632
Breathing oxygen, 649–650
Bronchial circulation, 627
Bronchopulmonary dysplasia, 654
Brown fat, 27
Brown-Séquard syndrome, 170
Bruits, 544
Brush border hydrolases, 456
β-sheets, 18–19
Buffer, 6
Bundle branch block, 530
Bundle of Kent, 533

C
Ca^{2+}
  binding to troponin–tropomyosin complex, 102–103
  concentration gradient, 56, 57
  functions of, 56
  handling in mammalian cells, 57
  and phototransduction, 184
  as second messenger, 56–58
  smooth muscle contraction and, 114–115
  transport and muscle contraction, 102–104
Ca^{2+} channels, 51
  in cardiac myocytes, 111
Ca^{2+}-dependent process of exocytosis, 48
Cadherins, 42
Calbindin-D proteins, 380
Calcineurin, 58
Calcitonin, 377
  actions, 384–385
  calcium homeostatic mechanisms, 385
  origin, 384
  secretion & metabolism, 384
Calcitonin gene-related peptide, 151, 264
Calcium-binding proteins, 57–58
Calcium channel blockers, Raynaud disease treatment using, 264
Calcium metabolism, 377–378
  in adult human, 378
  distribution of, 378
  humoral agents on, 385
Calmodulin, 57
  secondary structure of, 58
Calmodulin-dependent kinases, 58
Calorie, 489
CaMKs, 58
cAMP. *See* Cyclic adenosine monophosphate
cAMP, intracellular, 464
CAMs. *See* Cell adhesion molecules
Cancer, genetic aspects of, 47
Capacitation, 422
Capillaries
  cross-sections of capillaries, 572
  endothelial glycocalyx, 571
  exocytosis, 571
  fenestrations, 571
  metarterioles, 570–571
  microcirculation, 570–571
  precapillary sphincters, 570–571
Capillary pressures, 580
  capillary filtration coefficient, 582
  flow-limited and diffusion-limited exchange, 582
  hydrostatic pressure gradient, 581
  interstitial fluid, 581–582
  osmotic pressure gradient, 581–582
  pressure gradients, 582
  starling forces, 581
Capillary wall
  structure of, 54
  transport across, 54
Capsaicin transdermal patches, chronic pain treatment using, 164
Carbamino compounds, 644
Carbidopa
  for MSA treatment, 256
  for Parkinson disease treatment, 247
Carbohydrate metabolism
  adrenal glucocorticoids, 447–448
  catecholamines, 447
  growth hormone, 448
  hormones/exercise, effects of, 447
  thyroid hormones, 447
Carbohydrate moieties, 484
Carbohydrates, 491
  breakdown during exercise, 106–107
  dietary, 22
  structural and functional roles, 22
  structures of, 22
Carbon dioxide transport, 644–647
  acid–base balance, 645
  blood, molecular fate in, 644
  blood, spatial distribution of, 645
  buffering in blood, 645–647
  chloride shift, 644–645
  dissociation curves, 645
  gas transport, 645
  response curves, 663
Carbonic acid, 7
Carbonic acid–bicarbonate system, 646
Carbonic anhydrase, 647
Carbonic anhydrase inhibitors, glaucoma treatment using, 179
Carbon monoxide poisoning, 652–653
Carbon monoxyhemoglobin. *See* Carboxyhemoglobin
Carboxyhemoglobin, 653
Carboxyl terminal tetrapeptide, 471
Cardiac arrhythmias, clinical applications
  abnormal pacemakers, 529–530
  accelerated AV conduction, 533–534
  arrhythmias, treatment of, 534
  atrial arrhythmias, 531–532
  atrial arrhythmias, consequences of, 532
  excitation, ectopic foci of, 530–531
  long QT syndrome, 533
  normal cardiac rate, 529
  reentry, 531
  ventricular arrhythmias, 532–533
Cardiac conduction system, 521
Cardiac cycle
  divisions of, 541
  events of, 543
Cardiac cycle, mechanical events
  arterial pulse, 542–544
  atrial pressure changes, 544
  atrial systole, 539–540
    length of, 542
  cardiac muscle
    length-tension relationship, 547
  cardiac output
    factors controlling, 546–547
    integrated control, 550
    measurement methods, 545–546
    in various conditions, 546
  diastole
    late, 539
    length of, 542
  diastole, early, 540–541
  echocardiography, 545
  end-diastolic volume, factors affecting, 547
  heart, oxygen consumption, 550–552
  heart sounds, 544
  murmurs/bruits, 544–545
  myocardial contractility, 547–550
  timing, 541–542
  ventricular systole, 540
Cardiac excitation, origin/spread of, 524
  anatomic considerations, 521–522
  cardiac muscle, properties of, 522–523
  pacemaker potentials, 523–524
Cardiac function with exercise, 550
Cardiac muscle
  action potential of, 110–111
  contractile response of, 111–112
  electrical responses of, 522
  electronmicrograph of, 111
  isoforms, 112
  length-tension relationship, 547
  length–tension relationship for, 112, 113
  metabolism, 113–114
  morphology, 110
  nerve endings in, 130
  resting membrane potential of, 110
  striations in, 110
Cardiac muscle cell, histology of, 522
Cardiac output, 546
  conditions, effect of, 546
  determination of, 546
  factors controlling, 546–547
  integrated control, 550
  interactions between, 546
  measurement methods, 545–546
  in various conditions, 546
Cardiac pacemaker, 521
Cardiac rate, 544
Cardiac remodeling, 540
Cardiac responses
  to moderate supine exercise, 552

Cardiac tissue
conduction speeds in, 524
depolarization of, 531
Cardiac vector, 528
Cardiovascular control
feedback control of blood pressure, 588–589
sensory receptors, 588–589
venoconstriction, 589
Cardiovascular regulatory mechanism
atrial stretch and cardiopulmonary receptors, 593
autoregulation, 587, 588, 595
baroreceptor nerve activity
cardiac output, 591
sympathetic nerve activity, 591
systemic blood pressure, 590
baroreceptor resetting, 592
baroreceptors
aortic arch, 589, 592
aortic depressor nerve, 589
cardiopulmonary receptors, 589
carotid sinus, 589, 592
carotid sinus nerve, 589
caudal ventrolateral medulla, 590
nucleus of tractus solitarius, 590
carbon monoxide, 597
cardiovascular control
feedback control of blood pressure, 588–589
sensory receptors, 588–589
venoconstriction, 589
central chemoreceptors, 595
circulating vasoconstrictors
angiotensin, 599
extracellular fluid, 599
norepinephrine, 599
urotensin-II, 599
endothelial cells, 596
endothelin-1
astrocytes and neurons, 598
cardiovascular functions, 597
craniofacial abnormalities, 598
megacolon, 598
regulation of secretion, 597, 598
endothelins, 597
heart innervation
adrenergic and cholinergic receptors, 588
cardiac conductive tissue, 588
heart rate, 588
postganglionic sympathetic nerves, 588
sinoatrial node, 588
vagal discharge, 588
ventricular contraction, 588
innervation of blood vessels
splanchnic veins, 588
sympathetic nerves, 587–588
sympathetic noradrenergic fibers, 587–588
venoconstriction, 588
kinins
active factor, 598–599
angiotensin-converting enzyme, 598
bradykinin receptors, 599
formation of, 598
kallikreins, 598
lysylbradykinin, 598
localized vasoconstriction, 596
medullary control
basic pathways, 589, 590
factors affecting activity of RVLM, 589, 591
factors affecting heart rate, 589, 592
heart rate by vagus nerves, 589, 591
intermediolateral gray column, 589
rostral ventrolateral medulla, 589
somatosympathetic reflex, 589
natriuretic hormones, 599
nitric oxide
adenosine, 596–597
arginine, 596
endothelium-derived relaxing factor, 596
functions, 597
physiologic role, 597
platelet aggregation, 597
synthesis of, 596
peripheral chemoreceptor reflex
hemorrhage, 595
Mayer waves, 595
Traube–Hering waves, 595
vasoconstriction, 595
prostacyclin, 596
role of baroreceptors
arterial pressure, 592
blood volume, 592
infusion of phenylephrine, 592, 593
neurogenic hypertension, 592
thromboxane A_2, 596
Valsalva maneuver
bradycardia, 593
heart rate, 595
hyperaldosteronism, 595
intrathoracic pressure, 593
response to straining, 593, 594
tachycardia, 593
vasodilator metabolites, 595–596
Cardioverter–defibrillator, 533
Carnitine, 26
deficiency, 28
Carotid body, 660
location of, 660
organization of, 660
CART. *See* Cocaine-and amphetamineregulated transcript
Caspases, activation of, 47
Catabolism, 488
Catch-up growth, 332
Catecholamines, 145–146, 301, 355, 442
biosynthesis and release, 145, 146
catabolism of, 146
half-life of, 356
metabolic effects of, 357–358
secretion regulation, 358
Categorical and representational hemispheres, lesions of, 291
Cathechol-O-methyltransferase (COMT) inhibitors, Parkinson disease treatment using, 247
CatSper, 422
Caudal ventrolateral medulla, 699
Caudate nucleus, 243–244
Causalgia, 164
Caveolae and rafts, 49
CBG. *See* Corticosteroid-binding globulin
CCK. *See* Cholecystokinin
CCK 4. *See* Carboxyl terminal tetrapeptide
CCK receptors, 151
CCK-releasing peptide, 471
CD8 and CD4 proteins
relation to MHC-I and MHC-II proteins, 76
on T cells, 76
Celiac disease, 491
Cell adhesion molecules, 36
classification of, 42
functions of, 42
nomenclature, 42
Cell-attached patch clamp, 49
Cell cycle
definition, 14
sequence of events during, 17
"Cell eating," 48
Cell membrane
composition of, 36
enzyme content of, 38
membrane potential across, 88
patch of, 49
permeability, 50
in prokaryotes and eukaryotes, 36
proteins embedded in, 36, 37
rafts and caveolae, 49
solubility properties of, 36
Cells
cytoskeletal elements of, 40
secretion from, 48
specialization of, 35
Cell signaling pathway
phosphorylation, 55–56
Cellular immunity, 75
Cellular lipids, types of, 27
Cellular membrane channels, alterations in, 285
Cellular phosphorylation, enzymes involved in, 55
Central diabetes insipidus, 700
Central herniation, 243
Central nervous system, 436, 473, 514
glycine excitatory and inhibitory effects in, 143
pathway linking skeletal muscles to, 256
Central neuromodulators, diffusely connected systems of, 1378
Central pathway of hearing, 207–210
Centrioles, 42
Centrosomes, 42
Cerebellar cortex
fundamental circuits of, 250
inputs to, 250
location and structure of neuronal types in, 249
Cerebellar disease, 251, 252
Cerebellar granule cells, 249
Cerebellar peduncles, 249
Cerebellum
afferent fibers into, 249
anatomical division of, 248
cerebellar cortex. *See* Cerebellar cortex
connection to brain stem, 248
damage to, 251
functional division of, 250–251
and learning, 251, 252
midsagittal section through, 248
organization of, 249–250
principal afferent systems to, 250
Cerebral and cerebellar cortex
synapses in, 120
synaptic knobs in, 120

Cerebral blood flow, regulation
  autoregulation, 607
  blood flow in brain
    activity in human brain, 608
    gray matter, 608
    hemispheres, 608
    magnetic resonance imaging, 608
    positron emission tomography, 608
  Kety method
    factors affecting cerebral blood flow, 607
    Fick principle, 606–607
  role of intracranial pressure, 607
  role of vasomotor, 607–608
Cerebral circulation
  blood-brain barrier
    circumventricular organs, 605–606
    development of, 606
    function of, 606
    penetration of substances, 604–605
  brain metabolism
    energy sources, 609
    glutamate, ammonia removal, 609
    oxygen consumption, 609
    uptake, release of substances, 608–609
  cerebral blood flow, regulation
    autoregulation, 607
    blood flow in brain, 608
    Kety method, 606–607
    role of intracranial pressure, 607
    role of vasomotor, 607–608
  cerebrospinal fluid
    formation and absorption, 603–604
    head injuries, 604
    protective function, 604
  innervation
    postganglionic sympathetic neurons, 603
    sphenopalatine ganglia, 603
    trigeminal ganglia, 603
    vasoactive intestinal peptide, 603
  vessels, 601–602
Cerebral cortex, structure of, 270
Cerebral dominance, 291–292
Cerebral hemisphere, areas concerned with face recognition, 294
Cerebral palsy, 236
Cerebrocerebellum, 251
Cerebrosides, 26
Cerebrospinal fluid, 70, 650, 715
  formation and absorption
    arachnoid villi, 603
    bulk flow, 603
    choroidal capillary endothelium, 603
    composition of, 603
    cribriform plate, 603
    extracellular fluid, 603
    pressure effect, 603, 604
  head injuries, 604
  protective function
    arachnoid trabeculae, 604
    membranes of brain, 604
    spinal fluid deficiency, 604
Cervical mucus, 404
CFF, 194–195
CFTR. *See* Cystic fibrosis transmembrane conductance regulator
CGRP. *See* Calcitonin gene-related peptide
Chandelier cell, 271
Channelopathies, 51, 53
Chaperones, 20
Chelating agents, 246
Chemical gradient, 7
Chemically sensitive nociceptors, 158
Chemical mediators
  action on receptors, 136
  as neurotransmitters and neuromodulators, 136
  in response to tissue damage, 165
Chemical messengers
  intercellular communication via, 54
  mechanism of action of, 55–56
  receptors for, 55
  recognition by cells, 55
  types of, 54
Chemical neurotransmission
  EPSP and, 259
  at synaptic junctions, 259
Chemical regulatory mechanisms, 659
Chemical synapses
  cell-to-cell communication via, 119
  synaptic cleft, 119
Chemical transmission at autonomic junctions
  acetylcholine, 259
  cholinergic neurotransmission, 259
  nonadrenergic, noncholinergic transmitters, 264
  noradrenergic neurotansmission, 260, 261, 264
  norepinephrine, 259
Chemokine receptors, 72
Chemokines, 72
Chemoreceptors, 157, 217, 659
  trigger zone of, 503
Chenodeoxycholic acid, 465
Chewing, 500
Cheyne–Stokes respiration, 667
Cheyne–Stokes respiratory pattern, 243
Chiari–Frommel syndrome, 417
Chloride/bicarbonate exchanger, 466
Chloride ions
  equilibrium potential, 9–10
  forces acting on, 9–10
  in mammalian spinal motor neurons, 10
Chloride shift, carbon dioxide transport, 644–645
Chloride transport, inhibitory postsynaptic potential and, 123
Cholecystectomized patients, 516
Cholecystokinin, 384, 443, 461, 502, 510
  secretion of, 471
Cholelithiasis, 517
Cholera, 467
Cholera toxin effect on cAMP, 62
Cholesterol, 360
  biosynthesis, 30
  interaction with caveolae and rafts, 49
  and vascular disease, relationship between, 31
Cholesterol desmolase, 360
Cholesterol esterase, 483
Cholesterol-lowering drugs, 30
Cholesterol solubility in bile, 515
Cholinergic agonists, glaucoma treatment using, 179
Cholinergic interneurons and the inhibitory dopaminergic input, excitatory discharge of, 248
Cholinergic nerve cells, 473
Cholinergic neurons, 497
Cholinergic neurotransmission in autonomic ganglia, 259
Cholinergic receptor, 145
Cholinergic receptors
  biochemical events at, 143
  at junctions within ANS, 260
Cholinesterase inhibitors, 262
Chorea, 245
Choroid, 177
Chromatin, 44
Chromosomal abnormalities, 396, 397
Chromosomal sex
  human chromosomes, 392
  sex chromatin, 392–393
  sex chromosomes, 392
Chromosomes
  composition of, 44
  Karyotype of, 393
  structure of, 44
Chronic myeloid leukemia, 57
Chronic pain, 164
Chronic sleep disorders, 279
Chronotropic action, 547
Chvostek's sign, 382
Chylomicron remnants, 29
Chylomicrons, 29
Chyme, 498
Cilia, 42
Ciliary disorders, 42
Ciliary neurotrophic factor, 94
Circadian rhythm, 278
  and ACTH levels, 368–369
  sleep disorders associated with disruption of, 279
Circhoral secretion, 401
Circulating angiotensins
  formation and metabolism of, 702
  metabolism of, 702
Circulating vasoconstrictors
  angiotensin, 599
  extracellular fluid, 599
  norepinephrine, 599
  urotensin-II, 599
Circulation, quantitative aspects of, 466
Circulatory changes
  with exercise, 551
Circulatory system
  blood vessels, 555
  bone marrow
    blood cells, 556
    hematopoietic stem cells, 556
    type, 556
  capillaries, 555
  systemic circulation, 555, 556
Circumventricular organs
  angiotensin II, 605
  chemoreceptor zones, 605
  neurohemal organs, 605, 606
  subcommissural organ, 605–606
Circus movement, 531
Citric acid cycle, 23
  in transamination and gluconeogenesis, 21
Clathrin-mediated endocytosis, 48–49
Cl^- channels, 51
CLD. *See* Chloride/bicarbonate exchanger
Clonus, 233

Closed-angle glaucoma, 179
Closing volume, 634
Clostridia, 123
Clotting mechanism, 566–567
  active factor, 567
  extrinsic system, 567
  fibrin formation, 566–567
  intrinsic system, 566–567
  tissue factor pathway inhibitor, 567
  tissue thromboplastin, 567
Clozapine, 246
  for schizophrenia, 147
CML, 57
CNP. *See* C-type natriuretic peptide
CNS. *See* Central nervous system
CNS lesions and somatosensory pathways, 170
CNTF, 94
CoA. *See* Coenzyme A
Coat complex and vesicle transport, 49
Cocaine-and amphetamineregulated transcript, 487
Cochlea
  chambers of, 200
  structure of, 200
Cochlear implants, 210
Coenzyme A, 11
  and its derivatives, 12
Coenzymes as hydrogen acceptors, 11
Colipase, 483
Collapsing, 542
Collecting ducts, 674
Colloid, 340
Colon, 504
  chloride secretion in, 467
  defecation, 506–507
  electrogenic sodium absorption in, 467
  human, 504
  motility of, 504–505
  transit time, 505–506
Colony-stimulating factors, 70
Color blindness, 193
Colors, 193
Color vision
  neural mechanisms of, 194
  retinal mechanisms of, 193–194
Colostrum and milk, composition, 417
Comedones, 408
Compensatory pause, 532
Complement system, 72–73
Compliance process, 629
Computer assisted therapies, 291
Concentration gradient, 7
Conditioned reflex
  biofeedback, 288
  definition of, 287
Conditioned stimulus, 287
Conduction, 318
Conduction aphasia, 293
Conductive deafness, 209
Cone photoreceptor, components of, 180
Cone pigments, 184
Cone receptor potential, 182–183
Cones
  density along horizontal meridian, 182
  schematic diagram of, 182
  sequence of events involved in phototransduction in, 184
Confabulation, 288
Congenital adrenal hyperplasia, 361–362
Congenital anosmia, 221
Congenital 5α-reductase deficiency, 426
Congenital hypothyroidism, 346
Congenital lipodystrophy, 450
Congenital lipoid adrenal hyperplasia, 362
Congenital myasthenia, 105
Congenital nystagmus, 212
Congestive heart failure, treatment of, 540
Conjunctiva, 177
Connective tissue, 627
Connexin, 43–44
  in disease, 45
  mutations, 45
Connexons, 43
Consciousness, brain stem and hypothalamic neurons influence on, 279–280
Consensual light response, 188
Constipation, 505
Constitutive pathway, 48
Continuous positive air-flow pressure, 276
Contraceptive methods, 412
Contractile mechanism, skeletal muscle, 97
Contraction, muscular
  fiber types involved in, 106
  flow of information leading to, 104
  heat production and, 108
  molecular basis of, 102–103
  muscle length, tension, and velocity of, 105
  muscle twitch, 102
  source of energy for
    carbohydrate and lipid breakdown, 106, 107
    oxygen debt mechanism, 107–108
    phosphorylcreatine, 106
  summation of, 104, 105
  types of, 103–104
Contraction of smooth muscle
  calmodulin-dependent myosin light chain kinase activity, 115
  Ca^{2+} role in, 114–115
  chemical mediators effect on, 115
  sequence of events in, 115
Contracture, 103
Convergence-projection theory for referred pain, 166
Convulsive generalized seizure, 277
Cool receptors, 161
COPD treatment strategies, 619
CO poisoning. *See* Carbon monoxide poisoning
Cornea, 177
Coronary chemoreflex, 664
Coronary circulation
  anatomic considerations
    arterioluminal vessels, 610
    arteriosinusoidal vessels, 610
    coronary arteries and branches, 610
    root of aorta, 610
    thebesian veins, 610
  chemical factors, 612
  diagram of, 610
  neural factors, 612
  pressure gradients
    blood flow in left and right coronary arteries, 611
    coronary angiography, 611
    radionuclides, 611
    systole and diastole, 611
    ventricular coronary flow, 611
    ventricular myocardium, 611
  variations in coronary flow, 611–612
Corpus albicans, 401
Corpus hemorrhagicum, 401
Corpus luteum, 401
Corrigan, 542
Cortex, electrical events in, 271
Cortical bone, 385
Cortical nephrons, 674
Cortical neuron, electrical responses of axon and dendrites of, 272
Cortical organization, 270–271
Cortical plasticity, 169–170
Corticobulbar tract, 239
  origins of, 239
  role in movement, 239
Corticospinal tracts
  lateral and ventral, 238
  origins of, 239
  role in movement, 239
  structure of, 239
Corticosteroid-binding globulin, 362
  rise in levels of, 363
Corticosteroids
  for anosmia treatment, 221
  for Brown-Séquard syndrome treatment, 170
  for MSA treatment, 256
  for MS treatment, 86
  vs. cortisol, relative potencies of, 361
Corticostriate pathway, 244
Corticotropes, 324
Corticotropin-releasing hormone, 314, 487
  functions of, 315
Cortisol
  half-life of, 362
  hepatic metabolism of, 363–364
  interrelationships of free and bound, 363
  and its binding protein, equilibrium between, 362
Cotransporter, 480
Coughing, 664
Counter-current exchangers, 687
Countercurrent multipliers, 687
Cowper's glands, 422
CP, 236
CPAP, 276
Creatine, phosphorylcreatine, and creatinine cycling in muscle, 107
Cretins, 346
CRH. *See* Corticotropin-releasing hormone
CRH receptors, 316
CRH-secreting neurons, 315
Crista ampullaris, 201
Critical fusion frequency, 194–195
Critical micelle concentration, 465
Crohn's disease, 75
Cryptorchidism, 428
Crypts, structure of, 457
CS, 287
CSF. *See* Cerebrospinal fluid
C-type natriuretic peptide, 707
Curling up, 319
Cushing syndrome, 366–367

Cutaneous circulation
  generalized responses, 613–614
  reactive hyperemia, 613
  triple response
    arteriolar dilation, 613
    axon reflex, 613
    local edema, 613
    physiologic effect, 613
    red reaction, 613
  white reaction, 613
Cutaneous mechanoreceptors
  generation of impulses in, 161
  sensory nerves from, 158
  types of, 158
CV. *See* Closing volume
CVLM. *See* Caudal ventrolateral medulla
C wave, 544
Cyanmethemoglobin, 653
Cyanosis, 644
Cyanotic congenital heart disease, 651
Cyclic adenosine monophosphate, 379
  activation of, 62
  bacterial toxins effect on, 62
  formation and metabolism of, 62
  as secondary messenger, 60–61
Cyclic adenosine 3',5'-monophosphate, 460, 473, 523, 705
Cyclic GMP/cGMP, 62
Cyclic guanosine monophosphate, 62
Cyclooxygenase 1 (COX1) and cyclooxygenase 2 (COX2), 30
Cyclosporine, myasthenia gravis treatment using, 129
Cystic fibrosis, 624
Cystic fibrosis transmembrane conductance regulator, 624
Cystic fibrosis transmembrane conductance regulator channel, 467
Cystinuria, 483
Cystometrogram, 694
Cystometry, 694
Cytokine receptor superfamilies, 74
Cytokines
  chemokines, 72
  receptors for, 72
  systemic and local paracrine effects, 72, 73
  systemic responses produced by, 80–81
Cytoplasmic dyneins, 41
Cytoskeleton
  intermediate filaments, 41
  microfilaments, 40, 41
  microtubules, 40
Cytotoxic T cells, 69
Cytotrophoblast, 414

D

Da, 4
DAG as second messenger, 60–61
Daily iodine intake and thyroid function, 340–341
Dalton, 4
Dantrolene
  for CP treatment, 236
Dark adaptation, 194
DBS. *See* Deep brain stimulation
Dead space, 633–634
Deafness, 209
  monogenic forms of, 210
Decerebrate rigidity, 241
Decerebration, midcollicular, 241, 242
Declarative memory, 285
Decomposition of movement, 252
Decortication, 243
Decreased peripheral utilization, 437
Deep and visceral pain, 165–166
Deep brain stimulation
  for ataxia treatment, 252
  for Parkinson disease treatment, 247
Deep tendon reflex, 229
Defects of image-forming mechanism, 186–188
  amblyopia, 187
  astigmatism, 188
  hyperopia, 186, 187
  myopia, 187, 188
  strabismus, 187
Defensins, 68
Dehydration, 702
Dehydroepiandrosterone, 398, 424
Dehydroepiandrosterone sulfate, 415
Deiodination, fluctuations in, 344–345
Demyelinating diseases, 86
Dendrites, functions of, 125–126
Denervation hypersensitivity, 130–131
Denervation supersensitivity, 130–131
Dent disease, 683
Deoxycholic acid, 465
Deoxycorticosterone, 359
Deoxyhemoglobin, comparative titration curves for, 646
Deoxyribonucleic acid
  constituents of, 14
  double helical structure of, 14
  double-helical structure of, 16
  fundamental unit of, 14
Depolarization
  of hair cells, 202
  in myelinated axons, 91
Depolarization process, 539
Depression, 149
Descending pathways, in pain control, 166
Desensitization, 55, 137
Desipramine, cataplexy treatment using, 276
Desmin in skeletal muscle, 100
Desmin-related myopathies, 100
Desmosomes, characteristics of, 43
Desynchronization, 273
Detoxification reactions, 512
Detrusor muscle, 694
DHEA. *See* Dehydroepiandrosterone
DHEAS. *See* Dehydroepiandrosterone sulfate
DHEA secretion, 362
DHPR, 103
DHT. *See* Dihydrotestosterone
Diabetes insipidus, 700
Diabetes mellitus, 431, 437, 449
  fat metabolism, 439
  juvenile, 449
  type 2, 449–450
  types of, 449
Diabetic nephropathy, 449
Diabetic neuropathy, 449
Diabetic retinopathy, 449
Diacylglycerol as second messenger, 60–61
Diapedesis, 68
Diaphragm, part of, 626
Diaschisis, 284
Diastole
  late, 539
  length of, 542
Diastolic dysfunction, 547
  effect of, 549
Diastolic failure, 540
Diastolic pressure, 539
Diazepam, CP treatment using, 236
Dichromats, 193, 194
Dicrotic notch, 542
Dietary lipid processing by pancreatic lipases, 29
Diffuse axonal injury, 284
Diffuse secondary response, 271
Diffusion
  across alveolocapillary membrane, 635
  definition, 7
  of diffusible anions, 9
  nonionic, 8
Digestion, 478–479
  protein digestion, 481–482
Digestive enzymes, 477
Digitalis, use of, 524
Dihydropyridine receptors, 103
Dihydrotestosterone, 425
  schematic diagram of, 426
1,25-Dihydroxycholecalciferol, 377, 380
Diluting segment, 686
Dimerized growth hormone receptor (GHR) signaling pathways activated by, 327
Direct Fick method, 545
Direct oxidative pathway, 22
Diseases causing hypoxemia, 651–652
  venous-to-arterial shunts, 651
  ventilation-perfusion imbalance, 651–652
Distal convoluted tubule, 674
Distal muscles, control of, 240
  corticobulbar tract, 239
  corticospinal tracts, 238
  movement and, 239
Distal stump, 94
Diuretics, mechanism of action, 692
Divalent metal transporter 1, 485
DMT, 149
DMT1. *See* Divalent metal transporter 1
DNA. *See* Deoxyribonucleic acid
Donnan effect, 8, 9
  on distribution of ions, 9
Dopamine, 147, 148
  metabolism of, 147
  physiologic function of, 358
  reuptake, 147
Dopamine receptors, categories of, 148
Dopaminergic neurons, 147
Dorsal column pathway, 167–169
Down syndrome, 397
2,3-DPG, 643
  effects on fetal & stored blood, 644
Drowning, 666
DTR, 229
Duchenne muscular dystrophy, 100
Dumping syndrome, 503
Dwarfism, 334
Dyneins, 41
Dysdiadochokinesia, 252

Dysequilibrium, 252
Dysgeusia, 225
Dyskinetic CP, 236
Dyslexia, 292
Dysmenorrhea, 413
Dysmetria, 252
Dysosmia, 221
Dyspnea, 649
Dysthymia, 149
Dystrophin–glycoprotein complex, 101

E

Ear
  electrical responses, 202–203
  external. *See* External ear
  inner. *See* Inner ear
  middle. *See* Middle ear
  sensory receptors in, 202
  threshold of, 205
Ear dust, 201
Ebner's glands, lingual lipase secretion, 483
ECF. *See* Extracellular fluid
ECG. *See* Electrocardiogram
Echocardiography, noninvasive technique of, 545
ECL cells. *See* Enterochromaffin-like cells
ECoG, 271
Ectopic focus, 531
Edema
  causes of, 3–4
  treatment for, 4
Edentulous patients, 500
EDRF, 116
EEG. *See* Electroencephalogram
Effective renal plasma flow, 677
Efferent arteriole, 673
Eicosanoids
  leukotrienes and lipoxins, 31, 32
  prostaglandins, 30–31
Einthoven's triangle, 524
Ejaculatory ducts, 419
Ejection fraction, 540
EJPs, 130
Electrical equivalence, 4
Electrical responses, hair cells of cochlea, 202–203
Electrical synapses, cell-to-cell communication via, 119
Electrical transmission at synaptic junctions, 123
Electroacupuncture, 172
Electrocardiogram, 539
  active/exploring electrode, 524
  bipolar leads, 526
  bipolar limb, 528
  body fluids, 524
  cardiac vector, 528
  heart
    electrical activity, spread of, 525
  with heart block, 530
  His bundle electrogram, 528–529
  intervals, 526
  monitoring, 529
  normal, 526–528
  normal ECG, 526–528
  unipolar (V) leads, 526
  waves of, 526
Electrocardiographic
  blood, ionic composition of, 537
  myocardial infarction, 534–537
Electrocorticogram, 271
Electroencephalogram
  alpha and beta rhythms, 273
  clinical uses of, 275
  recorded from scalp, 272
Electrogenic pump, 10
Electrolyte homeostasis and growth hormone, 327
Electrolytes
  definition of, 4
  effect on aldosterone secretion, 373–374
  in normal humans, 371
  organization of, 5
  in patients with adrenocortical diseases, 371
Electromyography, human extensor pollicis longus and flexor pollicis longus, 109
Electroneutral mechanism, 466
Electroneutral NaCl absorption, 466
Electrotonic potentials
  changes in excitability during, 90–91
  hyperpolarizing potential change in, 90
  and local response, 90–91
Embden–Meyerhof pathway, 22
Enchondral bone formation, 385
End-diastolic ventricular volume, 540
Endocardial lesions, 537
Endocrine cells, 469
Endocrine communication, 54
Endocrine disorders, 304
Endocrine glands, changes in, 335
Endocrine regulation, 468
Endocrine system, 303
Endocrine tumors, 305
Endocytosis, 48–49
  clathrin-mediated, 48–49
  phagocytosis, 48
  pinocytosis, 48
  types of, 48
Endogenous cannabinoids, 151–152
  stress-induced analgesia by, 173
Endogenous pyrogens, 319
Endometrium, spiral artery of, 403
Endopeptidases, 481
Endoplasmic reticulum, protein synthesis in, 20
Endoscopic retrograde cholangiopancreatography, 516, 517
Endothelial derived relaxing factor, 116
Endothelin-1
  astrocytes and neurons, 598
  cardiovascular functions, 597
  craniofacial abnormalities, 598
  megacolon, 598
  regulation of secretion, 597, 598
Energy balance, 490
Energy metabolism, 487–490
  calorie, 489
  energy balance, 490
  metabolic rate, 488
    factors affecting, 489–490
  respiratory quotient, 489
Energy production
  biologic oxidations, 11–12
  in Embden–Meyerhof pathway, 23–24
  energy transfer, 10–11
Energy production reactions, directional flow valves in, 24
Energy transfer, 10–11
Enophthalmos, 263
Enteric nervous system, 266, 468, 473
  extrinsic innervation, 473
Enterochromaffin-like cells, 459
Enteroendocrine cells, 468
Enterohepatic circulation, 465, 513
Enterokinase, 481
Enzymatic activity and protein structure sensitivity to pH, 6
Enzyme aromatase, 406
Enzyme glucuronyl transferase (UDP-glucuronosyltransferase, 513
Enzyme hydrolyzes, 483
Enzymes
  in cell membrane, 38
  in lysosomes, 39
Epicritic pain, 159
Epididymis, 419
Epidural anesthesia, 169
Epilepsy, 276
  genetic mutations in, 277
Epinephrine, 145, 447
  as adrenoceptors, 146–147
  catabolism of, 146
  effect on intestinal smooth muscle, 115
  metabolic effects of, 357–358
  and nonepinephrine levels in human venous blood, 356
  plasma levels, 355
  secretion of, 145
Epiphyses, 385
Epiphysial plate, 385, 387
Episodic memory, 284
Epithelia
  transport across, 53–54
Epithelial sodium channels, 51, 223, 466
EPSP. *See* Excitatory postsynaptic potential
Equilibrium potential
  for chloride ions, 9–10
  for potassium ions, 10
Equivalents, 4
ERCP. *See* Endoscopic retrograde cholangiopancreatography
Erection, 422
Eroding bone, 384
ERPF. *See* Effective renal plasma flow
ER-positive tumors, 302
ERV. *See* Expiratory reserve volume
Erythropoiesis, 709
  blood cells, 556
  hematopoietic stem cells, 556
Erythropoietin, 709
  regulation of secretion, 709
  sources, 709
  structure/function, 709
Escape phenomenon, 372, 691
Esophagogastric junction, 501
Esophagus, 501
  motor disorders of, 501
Essential fatty acids, 30
Estrogen-dependent, 417
Estrogen receptors, 302
Estrogens, 365, 385, 391
  biosynthesis and metabolism of, 406

Estrous cycle, 406
Estrus, 406
Ethosuximide, 277
Etiocholanolone, 364
Eukaryotic gene, basic structure of, 14, 16
Eunuchoidism, 428
Evoked cortical potentials, 271
Excitation-contraction coupling, 103
Excitation-contraction coupling in smooth muscle, 114
Excitatory and inhibitory amino acids
  acetylcholine, 144
  acetylcholine receptors, 144–145
  cholinergic receptor, 145
  GABA, 142–143
  glutamate, 138–142
  glycine, 143
Excitatory junction potentials, 130
Excitatory postsynaptic potential, 122, 123, 161, 259
  and inhibitory postsynaptic potentials, schematic of, 262
Excitotoxins, 140
Exercise effects, 666–669
  diagrammatic representation, 668
  tissues, changes in, 668–669
  tolerance & fatigue, 669
  ventilation, changes in, 666–668
Exocytosis, 48
Exopeptidases pancreas, carboxypeptidases of, 482
Expiration, 627–628
Expiratory muscles, 626
Expiratory reserve volume, 629
Explicit memories, areas concerned with encoding, 288
External ear
  functions of, 199
  structures of, 200
External intercostal muscles, 626
External urethral sphincter, 694
Exteroceptors, 217
Extracellular fluid, 5, 599
  characteristics of, 3
  classification of, 3
  defense of volume, 700–702
  effect of, 714
  ionic composition of, 697
  tonicity, defense of, 697
    mechanisms for, 698
Extracellular fluid volume, 548
Extraction ratio, 676
Extrahepatic biliary ducts, 510
Extranuclear receptors, 303
Extrinsic innervation, 468
Extrinsic sphincter, 501
Eye
  anatomy of, 178
  aqueous humor, 178
  choroid, 177
  conjunctiva, 177
  cornea, 177
  crystalline lens, 178
  fundus of, 180
  photoreceptor, 180, 181
  posterior chamber, 178
  principal structures of, 177
  pupillary light reflexes, 188
  retina
    blood vessels, 180
    layers, 178
    melanopsin, 185
    neural components of extrafoveal portion of, 178, 179
    pigment epithelium, 180
    potential changes initiating action potentials in, 182
    receptor layer of, 180
    visual information processing in, 185
  sclera, 177
Eye movements, 195–196
Eye muscle actions, 195

F

Face recognition, cerebral hemisphere areas concerned with, 294
Facilitated diffusion, 50
$FADH_2$, generation in citric acid cycle, 23
Familiarity and strangeness, 289
Farsightedness, 186, 187
Fasting state. *See* Gastric juice
Fast pain, causes of, 159
Fatigue, 669
Fat-soluble vitamins, 485
Fatty acids
  oxidation of, 26–27
    diseases associated with imbalance of, 28
  saturated/unsaturated, 26
  structure of, 26
  synthesis of, 27
Feedback control
  feedback inhibition of endocrine axes, 303
  principles of, 303–304. *See also* Hormones
Feedforward inhibit, 250
Female external genitalia *vs.* male, 395
Female pseudohermaphroditism, 396
Female reproductive system, 391
  aberrant sexual differentiation
    chromosomal abnormalities, 396
    hormonal abnormalities, 396–398
  chromosomal sex
    human chromosomes, 392
    sex chromatin, 392–393
    sex chromosomes, 392
  embryonic differentiation of, 394
  gametogenesis, 391
  hormones & cancer, 417
  human reproductive system, embryology of
    development of brain, 395–396
    embryology of genitalia, 393–395
    gonads, development of, 393
  lactation, 416–417
    breasts, development of, 416
    gynecomastia, 417
    initiation of, 416–417
    menstrual cycles, effect, 417
    milk, secretion/ejection of, 416
  menstrual cycle
    anovulatory cycles, 404
    breasts, cyclical changes, 404
    changes during intercourse, 404–405
    estrous cycle, 406
    indicators of ovulation, 405–406
    normal menstruation, 404
    ovarian cycle, 401–402
    uterine cervix, cyclical changes, 404
    uterine cycle, 402–404
    vaginal cycle, 404
  ovarian function, abnormalities of
    menstrual abnormalities, 412–413
  ovarian function, control of, 410–412
    contraception, 412
    control of cycle, 411–412
    feedback effects, 411
    hypothalamic components, 410–411
    reflex ovulation, 412
  ovarian hormones
    actions, 409
    breasts, effects on, 408
    central nervous system, effects on, 408
    chemistry, biosynthesis, 406–407
    endocrine organs, effects on, 407–408
    female genitalia, effects on, 407
    female secondary sex characteristics, 408
    mechanism of action, 408, 410
    progesterone, 409
    relaxin, 410
    secretion, 407, 409
    synthetic and environmental estrogens, 408–409
  precocious/delayed puberty
    delayed/absent puberty, 400
    menopause, 400–401
    sexual precocity, 399–400
  pregnancy
    endocrine changes, 414
    fertilization & implantation, 413–414
    fetal graft, 414
    fetoplacental unit, 415
    human chorionic gonadotropin, 414
    human chorionic somatomammotropin, 415
    infertility, 414
    parturition, 415–416
    placental hormones, 415
  puberty, 398
    control of onset, 398–399
  sex hormones, 391
Female reproductive tract, functional anatomy of, 402
Ferroportin, 486
Ferroportin 1, 485
Fertilization, 413
  in mammals, 413
  in vitro, 414
Fetal adrenal cortex, 354, 415
Fetal circulation, 614–616
Fetal development
  lymphocytes during, 69
Fetal respiration, 615, 616
Fever
  beneficial effect of, 320
  in homeothermic animals, 319
  pathogenesis of, 319–320
  produced by cytokines, 320
FGFR3, 334
Fibroblast growth factor receptor 3, 334
Fick principle, 545
Fick's law of diffusion, 7

Filtration, 54
Filtration fraction, 681
Final common pathway, 228
Firing level (threshold potential), 90
"First messengers," 55
Flatus, 502
Flavoprotein–cytochrome system, 11
Flexor responses, 234
Fluent aphasia, 293
Fluid intake. *See* Water intake
Fluoride stimulates osteoblasts, 388
Fluoxetine for MSA treatment, 256
Focusing point sources of light, 186
Follicle-stimulating hormone
  actions of, 333
  constituents of, 332
  half-life of, 332
  receptors for, 333
Follicle-stimulating hormone secretion, 421
Folliculostellate cells, 324
Food intake
  controlling, mechanisms, 487
  control of, 486–487
Forced vital capacity, 629
Forces acting on ions, 9–10
FP. *See* Ferroportin
Fractionation and occlusion, 234
Frank–Starling law, 547
  myocardial contractility effect of, 549
Free fatty acids, 439
  metabolism of, 29, 30
Free hormone and SBP-hormone complex, 302
Friedreich's ataxia, 252
Frontal lobe lesions, 294
Fructose
  metabolism of, 25, 26
  structure of, 22
Fructose 6-phosphate, 26
FSH. *See* Follicle-stimulating hormone
FSH receptor, mutations in, 332–333
FSH secretion. *See* Follicle-stimulating hormone secretion
Fundoplication, 501
Fundus of eye, 180
  in normal primate, 181
Fungiform papillae, 222
FVC. *See* Forced vital capacity

G

GABA, 142–143
  diagram of, 142
  formation of, 142
Gabapentin, 277
  for chronic pain, 164
GABA receptors
  low-level stimulation in CNS, 142
  pharmacological properties of, 142–143
  subtypes of, 142
gACE. *See* Germinal angiotensin converting enzyme
Galactorrhea, 417
Galactose
  metabolism of, 25
  structure of, 22
  transport of, 480
Galactose malabsorption, 480
Galactosemia, 26
Gallbladder, 510
  ducts of, 463
Gallbladder contraction, neurohumoral control of, 516
Gallstones, 517
Gametogenesis, 391, 419
  blood–testis barrier, 419–420
  ejaculation, 423
  erection, 422–423
  prostate specific antigen, 423
  semen, 422
  spermatogonia, 420–421
  spermatozoa, development of, 421–422
  temperature, effect of, 422
Gamma glutamyltranspeptidase, 514
Gamma hydroxybutyrate, cataplexy treatment using, 276
Gamma loop, 242
Gamma oscillations, 273
Ganciclovir, 47
Gap junctions
  connecting cytoplasm of cells, 44
  propagation of electrical activity from, 43
Gases, properties of, 634
Gas exchange in lung, 634–635
Gas transport, 645
Gastric acid
  regulation of, 460
  secretion, 469
Gastric acidity, 459
Gastric bypass surgery, consequences of, 503
Gastric gland, structure of, 459
Gastric inhibitory peptide, 443
Gastric juice, 459
Gastric parietal cells, 459
Gastric secretions, 457, 459, 460
  cephalic phase of, 459
  origin & regulation of, 457–461
  primary stimuli of, 458
Gastrin, 469, 472
  precursor for, 468
  release of, 460
Gastrin receptors, 471
Gastrin releasing peptide, 459, 460
Gastrin-releasing polypeptide, 469
Gastrin secretion stimuli, 471
Gastrocolic reflex, 507
Gastroesophageal reflux disease, 501
Gastroileal reflex, 505
Gastrointestinal circulation, 474
Gastrointestinal hormones, 151, 468
  cholecystokinin, 469–471
  enteroendocrine cells, 468
  gastrin, 468–469
  gastrointestinal peptides, 472–473
  GIP, 471–472
  integrated action of, 472
  motilin, 472
  production of, 469
  secretin, 471
  somatostatin, 472
  VIP, 472
Gastrointestinal hormones secretin, 461
Gastrointestinal immune system, 473–474
Gastrointestinal motility, 497
  basic electrical rhythm, 498
  colon, 504
    defecation, 506–507
    motility of, 504–505
    transit time, 505–506
  migrating motor complex, 498–500
  patterns of, 498
  peristalsis, 497–498
  segmentation/mixing, 498
  segment-specific patterns
    aerophagia/intestinal gas, 502
    lower esophageal sphincter, 500–501
    mastication, 500
    mouth/esophagus, 500
    swallowing, 500
  small intestine
    intestinal motility, 504
    transit time, 505–506
  stomach, 502–504
Gastrointestinal regulation, 468
Gastrointestinal secretions
  anatomic considerations, 457, 463
  biliary secretion, 464
    bile, 464–466
  gastric secretion, 457
    origin & regulation of, 457–461
  intestinal fluid/electrolyte transport, 466–468
  pancreatic juice
    composition of, 463
    regulation of, 464
  pancreatic secretion, 461–462
  salivary secretion, 456–457
Gastrointestinal smooth muscle, BER of, 499
Gastrointestinal system, 477
Gastrointestinal tract, 466, 472, 506
  function of, 455
  intestine wall organization, 456
  parts of, 455
  segments of, 498
  structural considerations, 455–456
  water turnover, 466
Gate-control mechanism of pain modulation, 170
Gating regulation in ion channels, 50
GBG. *See* Gonadal steroid-binding globulin
GDNF, 94–95
Gene
  definition, 14
  mutations, 14
  protein encoded by, 14
Generalized seizures, 276, 277
Generator potentials, 161
Genetic abnormalities, diseases caused by, 47
Genetic male, 392
Genetic sex determination, 392
Genital ducts, 394
Germinal, 703
Germinal angiotensin converting enzyme, 420
GFR. *See* Glomerular filtration rate
GGT. *See* Gamma glutamyltranspeptidase
GHB for cataplexy treatment, 276
Ghrelin, 472, 487
Gibbs–Donnan equation, 9
Gigantism, 305
  and acromegaly, 328
GIP. *See* Gastric inhibitory peptide

Glasgow Coma Scale, 284
Glatiramer acetate, for MS treatment, 86
Glaucoma, 179, 180
Glial cell line-derived neurotrophic factor, 94–95
Glial cells
  role in communication within CNS, 83–84
  types of, 83
Glicentin-related polypeptide, 444
Globus pallidus, 244
Glomerular filtration rate, 671, 675, 678, 702
  control of, 679
  factors affecting, 680
  normal, 679
Glomerulotubular balance mechanisms, 684, 685
Glomerulus
  diagram of, 705
  structural details, 675
Glomus, 660
Glomus cells, 660
Glottis, 626
GLP-1/2. *See* Glucagon-like polypeptides 1 and 2
Glucagon, 431, 444, 473
  action, 444
  chemistry, 443–444
  insulin-glucagon molar ratios, 445–446
  mechanisms, 444
  metabolism, 444
  regulation of secretion, 444–445
Glucagon-like polypeptides 1 and 2, 443
Glucagon secretion, factors affecting, 445
Glucocorticoid feedback, 369–370
Glucocorticoid-remediable aldosteronism, 372
Glucocorticoids, 385
  anti-inflammatory and anti-allergic effects, 367–368
  binding, 362–363
  metabolism and excretion of, 363–364
  pharmacologic and pathologic effects, 366–367
  physiologic effects, 365–366
  for SCI treatment, 235
  secretion regulation, 368–370
Gluconeogenesis, citric acid cycle in, 21
Glucose, 480
  breakdown during exercise, 107
  catabolism, 22
  fasting level of, 22
  infusions and plasma growth hormone levels, 331
  phosphorylation in cell, 22
  structure of, 22
  transport of, 480
    glucose transporter, 50
Glucose-dependent insulinotropic peptide, 468, 472
Glucose tolerance curves, 448
Glucose transport, 480
Glucose transporters, 434, 480
  in mammals, 435
Glucose uptake, 437
Glucuronyl transferase, deficiency of, 514
Glucuronyl transferase system, 514
Glutamate, 138–142
  action on ionotropic and metabotropic receptors, 139, 140
  binding to AMPA/kainate receptors, 140
  excessive levels of, 140
  excitotoxic damage due to, 140
  synthesis pathways, 138, 139
  uptake into neurons and glia, 137, 138
Glutamate receptors
  biochemical events at, 141
  pharmacology of, 142
Glutamic dehydrogenase, 713
Glutaminase, 713
Glutathione, 514
GLUT 1 deficiency, 446
GLUTs. *See* Glucose transporters
GLUT 4 transporters, cycling of, 435
Glycine, 143
  excitatory and inhibitory effects in CNS, 143
Glycogen, 490
  during fasting, 25
  synthesis and breakdown, 24, 25
Glycogenesis, 22
Glycogenin, 24
Glycogenolysis, 22
Glycogen storage, 512
Glycogen synthase, 24
Glycolysidic drugs and cardiac contractions, 113
Glycolysis, 22
Glycosylphosphatidylinositol (GPI) anchors, 37
γ-Motor neurons, 229, 230
  discharge, control of, 232
  discharge, effects of, 232
  stimulation of, 232
  tonic level of activity with, 229, 230
GnRH. *See* Gonadotropin-releasing hormone
GnRH-secreting neurons, 315
Goldblatt hypertension, 706
Golgi apparatus, 46, 484
Golgi cells, 249
Golgi tendon organ, inverse stretch reflex in, 232, 233
Gonadal dysgenesis, 334, 397
Gonadal steroid-binding globulin, 407, 424
Gonadal steroid, distribution of, 424
Gonadal steroids
  basic structure of, 358
  classification of, 359
  types of, 358
Gonadotropes, 324
Gonadotropin, carbohydrate in, 332
Gonadotropin-releasing hormone, 314, 398
  FSH-stimulating activity of, 315
Gout, 13
GPCRs. *See* G protein-coupled receptors
G protein-coupled oxytocin receptor, 313
G protein-coupled receptors, 136, 183, 223, 333
  abnormalities caused by loss-or gain-of-function mutations of, 64, 65
  drug targets, 60
  ligands for, 59
  structures of, 60
  subtypes of, 261
G proteins
  abnormalities caused by loss-or gain-of-function mutations of, 64, 65
  classification of, 58
  heterotrimeric, 58–59
  loss-of-function/gain-of-function mutations, 64
  regulation of, 58
G protein transducin, 184
GRA, 372
Graafian, 401
Graded renal nerve stimulation, renal responses to, 677
Granular cells, 674
Granular pneumocytes, 624
Granulocytes
  basophils, 68
  cytoplasmic granules, 67
  eosinophils, 68
  neutrophils, 67–68
  production of, 70
Granulosa cells, interactions, 407
Gray rami communicans, 257
GRH, 314
Growth factors
  gene activity alteration by, 63
  receptors for, 63
  types of, 62–63
Growth hormone
  action on carbohydrate metabolism, 327
  action on fat metabolism, 327
  binding of, 326
  biosynthesis and chemistry, 326
  direct and indirect actions of, 329
  effects on growth, 326
  effects on protein, 327
  electrolyte homeostasis and, 327
  insensitivity, 334
  metabolism of, 326
  physiology of growth, 331–332
  plasma levels, 326
  secretion
    diurnal variations in, 330
    feedback control of, 330
    hypothalamic & peripheral control of, 330–331
    stimuli affecting, 330–331
  and somatomedins, interaction between, 327–329
  species specificity, 326
Growth hormone receptor, 326
Growth hormone-releasing hormone, 314
Growth periods, 331, 332
Growth physiology, 331–332
  in boys and girls, 331
  catch-up growth, 332
  hormonal effects, 332
  normal and abnormal, 333
  role of nutrition in, 331
GRP. *See* Gastrin releasing peptide
GRPP. *See* Glicentin-related polypeptide
GTP generation in citric acid cycle, 23
Guanylin, 473
Guanylyl cyclases, 62
  diagrammatic representation of, 63
Gustation. *See* Taste
Gustducin, 223
Gynecomastia, 417

H

Habituation, 286
Hair cells of cochlea, 202
  arrangement of, 200
  structure of, 204
Hairline in children and adults, 425

Haldane effect, 644
Hallucinogenic agents, 149
Hartnup disease, 483
Haversian systems, 385
HBE. *See* His bundle electrogram
hCG. *See* Human chorionic gonadotropin
H^+ concentration, 714
  buffering, 715–716
  H^+ balance, 715
  metabolic acidosis, 716
  metabolic alkalosis, 717
  renal compensation, 716
  Siggaard–Andersen curve nomogram, 718
HDL, 29
Hearing
  sound transmission, 206
  sound waves, 203–206
Hearing loss, 210
Heart
  conducting system of, 522
  electrical activity, normal spread of, 525
  electrical activity of, 539
  Starling's law of, 547
Heart failure, 540
Heart innervation
  adrenergic and cholinergic receptors, 588
  cardiac conductive tissue, 588
  heart rate, 588
  postganglionic sympathetic nerves, 588
  sinoatrial node, 588
  vagal discharge, 588
  ventricular contraction, 588
Heart murmurs, 545
Heat loss, 317–318
Heat production
  causes of, 316
  by emotional excitement, 317
  endocrine mechanisms influence on, 317
  during exercise, 317
  and heat loss, 317, 318
    balance between, 316
Heat production in muscle, 108
Helicotrema, 200
Helper T cells, 69
Hematopoietic growth factors, 71
Hematopoietic stem cells, proliferation and self-renewal of, 70
Heme, 486, 641
Hemianopia, 190
Hemiblock/fascicular block, 530
Hemidesmosome and focal adhesions, 43
Hemidesmosomes, 43
Hemispheres
  anatomic differences between, 292
  complementary specialization of, 291–292
Hemoglobin, 646
  affinity for oxygen, factors affecting, 643–644
  carboxyhemoglobin, 559
  catabolism, 559–560
    biliverdin, 559–560
    iron deficiency anemia, 560
    phototherapy, 560
  diagrammatic representation of subunits, 557, 560
  dissociation curves, comparison of, 644
  in fetus, 559
  heme, 557
  methemoglobin, 558–559
  oxygenated/deoxygenated, comparative titration curves for, 646
  oxyhemoglobin, 557–559
  rate of synthesis, 559, 561
  reaction of, 641–643
  synthesis, 559
  in vivo binding, 644
Hemorrhage, 698
Hemosiderin, 486
Hemosiderosis, 486
Hemostasis
  anticlotting mechanisms, 567–568
    antiaggregating effect, 567
    antithrombin III, 567
    clotting factors, 567
    fibrinolytic system, 567–568
    heparin, 567
    plasmin, 567–568
    plasminogen receptors, 568
    thrombomodulin, 567
  anticoagulants, 568–569
    chelating agents, 569
    coumarin derivatives, 569
    heparin, 568–569
  clotting mechanism, 566–567
    active factor, 567
    extrinsic system, 567
    fibrin formation, 566–567
    intrinsic system, 566–567
    tissue factor pathway inhibitor, 567
    tissue thromboplastin, 567
  response to injury, 565
  summary of reactions, 565, 567
Henderson Hasselbalch equation, 6–7
Hepatic artery, 474
Hepatic encephalopathy, 512
Hepatic metabolism of cortisol, 363–364
Hepatic sinusoids, 510
Hepatic venous congestion, 511
Hepatocyte, 510
Hepatolenticular degeneration, 246
Hephaestin, 485
Hering–Breuer reflexes, 664
Herniation, 243
Herring bodies, 312
Heteroreceptor, 136
Heterotrimeric G proteins
  in cell signaling, 59
  composition of, 58
Hexose monophosphate shunt, 22
Hexoses, 479
Hiccup, 665
High-altitude cerebral edema, 651
High-altitude pulmonary edema, 651
High-density lipoproteins, 29
High-energy phosphate compounds, 11–12
High-output failure, 540
Hippocampus and medial temporal lobe, 288
Hirschsprung disease, 505
His bundle electrogram, 528
His, bundle of, 521
Histamine, 149–150
Histamine-1 receptor (H1-receptor) antagonists, 60
Histamine-2 receptor (H2-receptor) antagonists, 60
Histamine receptors as GPCR drug targets, 60
Histotoxic hypoxia, 649, 653
Holter monitor, 529
Homeostasis, hormonal contributors to, 303
Homeothermic animals, 316
  body temperature of, 317
Homometric regulation, 547
Hormone, 471
  biosynthesis in adrenal cortex, 359
Hormone excess, diseases associated with, 305–306
Hormone receptors, mutations in, 305
Hormone resistance, 305
Hormones. *See also* Specific hormones
  effects of, 385
  mechanism of action of, 302, 303
  peptide, 299
  radioimmunoassays for, 304
  receptors for, 303
  secretion of, 300
  significance of, 299
  synthesis and processing, 300
  transport in blood, 301–302
Hormone-secreting cells of anterior pituitary gland, 324–325
Hormone synthesis in zona glomerulosa, 360
Horner syndrome, 261, 263
Horripilation, 318
Hp. *See* Hephaestin
HSCs, proliferation and self-renewal of, 70
5-HT. *See* Serotonin
5-HT receptors. *See* Serotonergic receptors
Human audibility curve, 205
Human bile acids, 465
Human chorionic gonadotropin, 414
Human colon, 504
Human core temperature, circadian fluctuation of, 317
Human gastrocnemius muscle, electronmicrograph of, 99
Human gastrointestinal tract, hormonally active polypeptides secretion, 470
Human hepatic duct bile, 512
Human histocompatibility antigen, HLA-A2, structure of, 76
Human immunoglobulins, 78
Human luteal, structure of, 410
Human parathyroid, section of, 381
Human prolactin receptor, 333
Human reproductive system, embryology of
  development of brain, 395–396
  embryology of genitalia, 393–395
  gonads, development of, 393
Human right cerebral hemisphere, 191
Human semen, composition of, 422
Human spermatozoon, 421
Human stomach, effect of, 502
Human testis, 420
Humoral hypercalcemia of malignancy, 384
Humoral immunity, 74
Huntington disease, 245
  initial detectable damage in, 246
Hyaline membrane disease, 632
Hydrated ions, sizes of, 50

Hydrochloric acid secretion, 477
Hydrogen ion concentration. *See* H^+ concentration
Hydrostatic pressure, 680
Hydroxyapatites, 385
25-Hydroxycholecalciferol, 379
Hydroxycholecalciferols, 379–381
  chemistry, 379–380
  mechanism of action, 380
  regulation of synthesis, 380–381
5-Hydroxytryptamine. *See* Serotonin
Hyperaldosteronism, 374
Hyperalgesia and allodynia, 164–165
Hypercalcemia, 384
Hypercapnia, 653–654, 662
Hyperemia, 473
Hyperfunction, 428
Hyperglycemia, 438
Hypergonadotropic hypogonadism, 428
Hyperkinetic choreiform movements, 246
Hypernatremia, 310
Hyperopia, 186, 187
Hyperosmia, 221
Hyperosmolality, 9
Hyperosmolar coma, 439
Hyperpnea, 649
Hyperprolactinemia, 400
Hyperthermia, 320
Hyperthyroidism
  causes of, 347
  symptoms of, 347
Hypertonic solutions, 8
Hypervitaminosis A, 492
Hypervitaminosis D, 492
Hypervitaminosis K, 492
Hypesthesia, 221
Hypocalcemic tetany, 378
Hypocapnia, 654–655
Hypocretin, 276
Hypogeusia, 225
Hypoglycemia, 440, 441, 448–449
Hypogonadotropic hypogonadism, 316, 428
Hypokalemia, 435
Hypoketonemic hypoglycemia, 28
Hypomenorrhea, 413
Hyponatremia, 700
Hypoparathyroidism, 64
Hypophysiotropic hormones
  clinical implications of, 316
  effect on anterior pituitary hormone secretion, 314
  effect on secretion of anterior pituitary hormone, 315
  locations of cell bodies of neurons secreting, 315
  as neurotransmitters, 315
  prolactin-releasing hormone, 314
  receptors for, 316
  significance of, 316
  structure of, 315
  types of, 314
Hypophysis. *See* Pituitary gland
Hyposmia, 221
Hypospadias, 362
Hypothalamic disease, 316
Hypothalamic hormones, secretion of, 308
Hypothalamic regulatory mechanisms, 309
Hypothalamus
  afferent and efferent connections of, 307–308
  diagrammatic representation of, 308
  functions of, 309
  intrahypothalamic system, 308
  location of, 307
  postulated interrelationships, 427
  relation to autonomic function, 309–310
  relation to pituitary gland, 308, 309
  stimulation of, 309
Hypothermia, 320
Hypothyroidism, 346
  strategy for laboratory evaluation of, 305
Hypotonic CP, 236
Hypotonic solutions, 8
Hypovolemia, effect of, 699
Hypoxemia, 649–651
  acclimatization, 650–651
  breathing oxygen, 649–650
  decreased barometric pressure, effects of, 649
  definition, 649
  hypoxic symptoms, 649–650
Hypoxia, 649
  oxygen treatment of, 653
Hysteresis, 631

I

IC. *See* Intracellular canaliculi
IDDM. *See* Insulin-dependent diabetes mellitus
IDL, 29
IJPs, 130
Image-forming mechanism, 185
  common defects of, 186–188
    amblyopia, 187
    astigmatism, 188
    hyperopia, 186, 187
    myopia, 187, 188
    strabismus, 187
  principles of optics, 186
Imipramine for cataplexy treatment, 276
Immune effector cells
  colony-stimulating factors, 70
  granulocyte, 70
  granulocytes, 67–68
  lymphocytes, 69
  macrophage, 70
  mast cells, 68
  memory B and T cells, 70
  monocytes, 69
Immunity
  acquired, 74–75
  activated defenses, 70–72
  antigen presentation, 75
  antigen recognition, 75
  in apoptosis, 47
  B cells, 77
  complement system, 72, 73
  cytokines, 72
  genetic basis of, 78
  immunoglobulins, 77–78
  innate, 74
  T cell receptors and, 76–77
Immunoglobulins
  bacterial and viral antigens, 78
  basic component of, 77
  classes of, 77
  immunoglobulin G, 77
  polypeptide components, 77–78
Immunohistochemistry, 136
Immunosuppressants, clonus treatment using, 233
Immunosuppression, 449
Immunosuppressive drugs, myasthenia gravis treatment using, 129
Immunosympathectomy, 94
Immunotherapy, Lambert–Eaton Syndrome treatment using, 130
Implantation, 414
Increased intraocular pressure, 179
Indicator dilution method, 545
Infant respiratory distress syndrome, 632
Inferior cerebellar peduncle, 248
Inflammation
  sequence of reactions in, 80
  skin, 81
Inflammatory pain, 164
Inflammatory response, 68
  nuclear factor-κB role in, 80
Infranodal block, 530
Inhaler drugs, effects on smooth muscle, 116
Inhibitory interneurons, 270, 271
Inhibitory junction potentials, 130
Inhibitory postsynaptic potential, 123
  due to increased Cl^- influx, 125
  and excitatory postsynaptic potential, schematic of, 262
Inhibitory systems, 126–127
Innate immunity
  and acquired immunity, 71
  activated defenses, 70
  cells mediating, 74
  in *Drosophila,* 74
  first line of defense against infection, 72
Inner ear
  bony labyrinth, 200
  cochlea, 200
  membranous labyrinth, 200
  organ of Corti, 200
  otolith organs, 200
  schematic of, 202
  semicircular canals, 200
  structures of, 200
Inner hair cells, 207
Innervation of blood vessels
  splanchnic veins, 588
  sympathetic nerves, 587–588
  sympathetic noradrenergic fibers, 587–588
  venoconstriction, 588
Innumerable steroids, 359
Inositol trisphosphate
  as second messenger, 59–60
In situ hybridization histochemistry, 136
Insomnia, 279
Inspiration, 627–628
Inspiratory capacity, 629
Inspiratory muscles, 626
Inspiratory reserve volume, 629

Insulin, 431, 436, 481
 anabolic effect of, 436
 biosynthesis & secretion, 433
 deficiency
  acidosis, 439
  cholesterol metabolism, 440
  coma, 439–440
  fat metabolism in diabetes, 439
  glucose tolerance, 437–438
  hyperglycemia effects, 438
  intracellular glucose, effects, 438
  protein metabolism, changes, 438–439
 deficiency, effects of, 440
 effect on Na, K ATPase pump activity, 53
 effects of, 433, 434
  glucose transporters, 434
  hypoglycemic, 436
  preparations, 434–435
  relation to potassium, 435–436
 insulin-like activity in blood, 433
 and insulin-like growth factors
  comparison of, 329
  structure of, 328
 intracellular responses, 436
 mechanism of action
  receptors, 436–437
 metabolism, 433
 principal actions of, 434
 sensitivity, 335
 structure/species specificity, 432–433
Insulin-dependent diabetes mellitus, 449
Insulin excess
 compensatory mechanisms, 440–441
 symptoms, 440
Insulin–glucagon molar ratios, 445
Insulinoma, 448
Insulin receptors, 436
Insulin secretion, 442
 factors affecting, 441
 regulation of, 441
  autonomic nerves, effect of, 442–443
  B cell responses, long-term changes, 443
  cyclic amp, 442
  hypoglycemic agents, 442
  intestinal hormones, 443
  plasma glucose level, effects of, 441–442
  protein/fat derivatives, 442
Insulin-sensitive tissues, endosomes in, 435
Insulin sensitizers, 305
Integrins, 42
Intention tremor, 252
Intercalated cells, 674
Intercalated disks, 110
Intercellular communication
 by chemical mediators., 54–55
 types of, 54
Intercellular connections, 43
 in mucosa of small intestine, 43
Interleukins as multi-CSF, 70
Intermediary metabolism, actions of
  glucocorticoids on, 365
Intermediate-density lipo-proteins, 29
Intermediate filaments, 41
Internalization, 55
Internal urethral sphincter, 694
Internodal atrial pathways, 521
Interstitial cells of Cajal, 498
Interstitial fluid, 3, 5
Interstitial fluid volume
 edema, 584
 elephantiasis, 585
 lymphedema, 585
 precapillary constriction, 584
 promoting factors, 584, 585
Intestinal epithelial cells, disposition of, 482
Intestinal fluid/electrolyte transport, 466–468
Intestinal lumen, 477
Intestinal mucosa
 lipid digestion and passage in, 484
Intestinal smooth muscle
 effects of agents on membrane potential
  of, 115
Intestinal villi, 457
Intestine
 epithelium of, 456
 functional layers, 456
 substances, normal transport of, 459
Intracellular canaliculi, 460
Intracellular fluid, 3, 5
Intracellular H^+ concentration, defense of,
  714–718
Intracranial hematoma, 284
Intrafusal muscle fibers, 229
Intrahepatic portal vein radicles, 511
Intrahypothalamic system, 308
Intramembranous bone formation, 385
Intrauterine devices, 412
Intravenous immunoglobulin, Lambert–Eaton
  Syndrome treatment, 130
Intrinsic depression, 550
Inverse stretch reflex
 definition of, 232
 in Golgi tendon organ, 232, 233
 muscle tone and, 233
 pathways responsible for, 231
Invertebrates, body temperature of, 316
In vitro fertilization, 414
Iodide transport across thyrocytes, 341
Iodine homeostasis and thyroid hormones,
  340–341
Ion channels
 activation by chemical messengers, 55
 multiunit structure of, 51
 pore formation in, 51
 poreforming subunits of, 52
 regulation of gating in, 50
 spatial distribution of, 91
Ion distribution and fluxes, 102
Ionotropic glutamate receptors
 properties of, 140
 subtypes of, 139–140
Ionotropic receptors, 136
Ionotropic receptors, salt and sour tastes
  triggered by, 223
Ion transport proteins of parietal cells, 461
IOP, 179
IPSP. *See* Inhibitory postsynaptic potential
IRDS. *See* Infant respiratory distress syndrome
Iron
 absorption of, 486
 intestinal absorption of, 486
Iron uptake, disorders of, 486
Irritant receptors, 664
IRV. *See* Inspiratory reserve volume
Ischemic hypoxia, 653
Ischemic/stagnant hypoxia, 649
Ishihara charts, 193
Islet cell hormones
 effects of, 446
 pancreatic, organization of, 446–447
 pancreatic polypeptide, 446
 somatostatin, 446
Islet cell structure, 432
 human, cell types, 432
 human pancreatic islet
  electronmicrograph of, 432
 rat pancreas
  islet of Langerhans, 432
Isocapnic buffering, 668
Isohydric principle, 6
Isomaltase, 478
Isometric contractions, 104
Isotonic contractions, 104
 muscle preparation arranged for
  recording, 106
Isotonic muscle contractions, 488
Isotonic solutions, 8
Isovolumetric ventricular contraction, 540
Isovolumetric ventricular relaxation, 541
Itch, 159, 160
IUDs. *See* Intrauterine devices

J

JAK2–STAT pathways, 326
JAK–STAT pathways, signal transduction via,
  63, 64
Janus tyrosine kinases, 63
Jaundice
 bilirubin, 514
 obstructive, 517
JG cells. *See* Juxtaglomerular cells
Junctional potentials, 130
Juvenile diabetes, 449
Juxtacrine communication, 54
Juxtaglomerular apparatus, 674
Juxtaglomerular cells, 705
Juxtamedullary nephrons, 674

K

Kainate receptors, 141
Kallmann syndrome, 316
Karyotype, 392
Kaspar Hauser syndrome, 334
Kayser–Fleischer rings, 246
KCC1. *See* Potassium/chloride cotransporter
K^+ channels, 51
K^+ depletion, 443
Ketoacidosis, 28
Ketone bodies
 formation and metabolism of, 27, 28
 health problems due to, 28
Ketosis, 439
Ketosteroids, 364
17-Ketosteroids, 364
Kidneys
 acid–base balance maintenance, 711
 ammonia production, 713
 autoregulation, 678

countercurrent exchangers in, 689
distal tubule of, 466
metabolically produced acid loads, 715
proximal tubular cells
secretion of acid, 712
Kinesin, 41
Kinins
active factor, 598–599
angiotensin-converting enzyme, 598
bradykinin receptors, 599
formation of, 598
kallikreins, 598
lysylbradykinin, 598
Kinocilium, 202
Klinefelter syndrome, 397
Knee jerk reflex, 229
Kölliker–Fuse nuclei, 658
Krebs cycle. *See* Citric acid cycle
Kupffer cells, 510
Kupffer cells, 512
Kussmaul breathing, 439, 661

L

Labyrinth. *See* Inner ear
Lacis cells, 705
Lactase, 478
Lactation, 416–417
breasts, development of, 416
gynecomastia, 417
initiation of, 416–417
menstrual cycles, effect, 417
milk, secretion/ejection of, 416
Lactic acidosis, 440, 548
Lactose, 478
brush border digestion, 480
Lactose intolerance, treatment for, 479
Lactotropes, 324
Lambert–Eaton Syndrome, 130
Lamellar bodies, 624
Language
areas in categorical hemisphere concerned with, 293
physiology of, 292–293
Language-based activity
active areas of brain during, 284
Language disorders
aphasias, 293
stuttering, 294
Large-molecule transmitters, 136
calcitonin gene-related peptide, 151
CCK receptors, 151
gastrointestinal hormones, 151
neuropeptide Y, 151
opioid peptides, 150
oxytocin, 151
somatostatin, 150, 151
substance P, 150
vasopressin, 151
Laron dwarfism, 334
Lateral brain stem pathway, 240–241
Lateral intercellular spaces, 674, 682
L channels, 523
LDL, 29
L-Dopa, 334
Leader sequence, 20
Leaky epithelium, 681
Learning
associative, 284, 285
and cerebellum, 251, 252
definition of, 283
nonassociative, 285
and synaptic plasticity, 286
Left and right planum temporale in brain, 209
Left ventricular ejection time, 542
Length–tension relationship
in cardiac muscle, 112, 113
human triceps muscle, 107
sliding filament mechanism of muscle contraction and, 105
Leptin, 399
LES. *See* Lower esophageal sphincter
Lesions of representational and categorical hemispheres, 291
Leukemia inhibitory factor, 95
Levodopa, 140
for MSA treatment, 256
for Parkinson disease treatment, 247
Lewy bodies, 248
Leydig cells, 394, 426
interstitial cells of, 419
LH. *See* Luteinizing hormone
LHRH, 314
Lidocaine for chronic pain, 164
LIF, 95
Light adaptation, 194
Light microscope, hypothetical cell seen with, 36
Light therapy, 279
Linear acceleration, responses to, 211–212
Lipid breakdown during exercise, 106–107
Lipid digestion
intestinal mucosa, 484
intracellular handling of, 484
Lipids, 483–485
biologically important, 26
in cells, 27, 28
fat absorption, 484–485
fat digestion, 483–484
fatty acids. *See* Fatty acids
plasma, 29
short-chain fatty acids, 485
steatorrhea, 484
transport of, 29
Lipoproteins, 29
Liver
acinus, concept of, 511
bile synthesis in, 509
biliary system
bile formation, 514–516
biliary secretion, regulation of, 516
cholecystectomy, effects of, 516
gallbladder, functions of, 516
gallbladder, visualizing, 516–517
blood, detoxifies, 512
blood percolates, 509
excretion
ammonia metabolism, 514
substances, 514
functional anatomy, 509–510
functions of, 511
bile, 512–513
bilirubin metabolism/excretion, 513
glucuronyl transferase system, 514
jaundice, 514
metabolism/detoxification, 512
plasma proteins synthesis, 512
glucose buffer function of, 512
hepatic circulation, 510–511
metabolically produced acid loads, 715
principal functions of, 511
roles in, 512
schematic anatomy of, 510
transport/metabolic functions of, 509
Local anesthetics and nerve fibers, 93
Local-circuit interneurons, 171
Local current flow, in axon, 91
Local injury, 80
Localized cortical lesions, 294
Local response of membrane, 90
Long QT syndrome, 113
Long-term depression, 286
Long-term learned responses, 285
Long-term memory, 285
recalling of, 289
storage in neocortex, 288
Long-term potentiation
in NMDA receptor, 286
in Schaffer collaterals in hippocampus, 287
Loop of Henle, 674
operation, 688
Loops of Henle, 685–686
Loss-of-function receptor mutations, 64
Lou Gehrig disease, 240
Lovastatin, 30
Low-density lipoproteins, 29
Lower esophageal sphincter, 500
Lower motor neurons, 239
damage to, 240
Lown–Ganong–Levine syndrome, 534
LTP. *See* Long-term potentiation
Lung
airway conduction, 621–622, 624
alveolar air, 634–635
alveolar airway, 624–625
alveolocapillary membrane, diffusion across, 635
anatomy, 621–627
biologically active substances metabolism, 638
blood and lymph in, 627
blood flow, 632–633
bronchi, 623
capacities, 629
complexity, 619
compliance of, 629–632
diffusing capacity, 635
endocrine functions, 638
gas exchange in, 634–635
gas transport in, 641–655
metabolic functions, 638
parenchyma. *See also* Alveolar airway
partial pressures, 634
perfusion, 636
pleura, 626–627
Po_2 and Pco_2 values in, 642
pressure–volume curves in, 630, 631
receptors, 664
respiratory muscles, 626
respiratory system, 635
respiratory tract, regions of, 621
responses mediated by receptors in, 664
ventilation, 632–633, 636

Lung volume, 629
Luteal cells, 401
Luteinizing hormone, 304
  actions of, 333
  constituents of, 332
  episodic secretion of, 411
  half-life of, 332
  receptors for, 333
Luteinizing hormone-releasing hormone, 314
Luteolysis, 411
LVET. *See* Left ventricular ejection time
Lymphatic circulation
  functions of, 584
  lymphatics draining, 584
  lymphatic vessels, 584
Lymphatic organs, glucocorticoids effects on, 366
Lymph node, anatomy of, 69
Lymphocytes, 473
  in bloodstream, 69
  during fetal development, 69
Lysergic acid diethylamide, 149
Lysine vasopressin, 311
Lyso-PC damages cell membranes, 463
Lysosomal diseases, 39
Lysosomes
  definition of, 39
  enzymes found in, 39

M
Machado-Joseph disease, 252
Macroglia, 83–84
Macrophages, 69
Macrosomia, 446
Macula, 180
Macula densa, 674, 684, 705
Macular sparing, 191
Maculopathy, 181
Magnocellular neurons, electrical activity of, 312
Major depressive disorder, 149
Major histocompatibility complex, 491
Major histocompatibility complex (MHC) genes, 75
Major proglucagon fragment, 443
Malabsorption, 491
Malabsorption syndrome, 491
Male contraception, 423
Male menopause. *See* Andropause
Male pseudohermaphroditism, 397, 398
Male reproductive system, 419
  anatomical features of, 420
  embryonic differentiation of, 394
  gametogenesis or ejaculation
    blood–testis barrier, 419–420
    ejaculation, 423
    erection, 422–423
    prostate specific antigen (PSA), 423
    semen, 422
    spermatogonia, 420–421
    spermatozoa, development of, 421–422
    temperature, effect of, 422
  structure, 419
  testes, endocrine function of
    actions, 424–425
    anabolic effects, 425
    estrogens, testicular production of, 426–427
    inhibins, 427
    mechanism of action, 425–426
    secondary sex characteristics, 425
    secretion, 424
    steroid feedback, 427
    testosterone, chemistry/biosynthesis of, 423–424
    transport/metabolism, 424
  testicular function, abnormalities of
    androgen-secreting tumors, 428
    cryptorchidism, 428
    hormones and cancer, 428
    male hypogonadism, 428
Male secondary sex characteristics, 425
Malignancy, humoral hypercalcemia of, 384
Malignant hyperthermia, 105, 320
Maltase, 478
Maltose, 478
Maltotriose, 478
Mammalian nerve fibers, 92
Mammalian skeletal muscle, 98
  electrical and mechanical responses of, 102
Mammalian spinal motor neurons, ion concentration inside and outside, 10
Mammotropes, 324
Manubrium, 199
MAOIs, 149
Masculinization, 374
Masking, 205
Mass action contraction, 505
Mass discharge in stressful situations, 265
Mast cells, 68
Mastication, 500
Maximal voluntary ventilation, 629
Maximum metabolic rate, 490
McCune–Albright syndrome, 413
M cells. *See* Microfold cells
MDMA, 149
Mechanoreceptors, 157
  hair cells as, 202
Medial brain stem pathways, 239, 241
  medial tracts involved in, 240
Medial temporal lobe and hippocampus, 288
Median eminence, 309
Medullary chemoreceptors, 661
Medullary control
  basic pathways, 589, 590
  factors affecting activity of RVLM, 589, 591
  factors affecting heart rate, 589, 592
  heart rate by vagus nerves, 589, 591
  intermediolateral gray column, 589
  rostral ventrolateral medulla, 589
  somatosympathetic reflex, 589
Medullary hormones, structure and function of, 355–356
Medullary reticulospinal tracts and posture, 240
Meiosis, 14
Meissner's corpuscles, 158
Meissner's plexus, 473
Melanophore, 325
Melanopsin, 185
Melanotropins
  biosynthesis, 325
  physiological functions, 325
Melatonin
  secretion of, 278
  and sleep-wake state, 280
  synthesis, diurnal rhythms of compounds in, 280
Memantine, 140
  for Alzheimer disease treatment, 289
Membrane polarization, abnormalities of, 534
Membrane potential
  extracellular Ca^{2+} concentration and, 89
  genesis of, 10
  resting, 88
  from separation of positive and negative charges, 88
  sequential feedback control in, 88–89
  of smooth muscle, 114
  subthreshold stimuli effect on, 90
  upstroke in, 88
Membrane transport proteins
  aquaporins, 50
  carriers, 50
  uniports, 51
Membranous labyrinth, 200
Memory
  and brain, link between, 285
  episodic, 284
  explicit or declarative, 283, 284
  forms of, 285
  implicit, 284
  intercortical transfer of, 286–287
  long-term, 285
  neural basis of, 285–286
  procedural, 284
  semantic, 284
  short-term, 285
  working, 288
Memory B cells, 70
Memory T cells, 69–70
Menarche, 398
Menopause, 400
Menorrhagia, 413
Menstrual cycle
  anovulatory cycles, 404
  basal body temperature and plasma hormone concentrations, 405
  breasts, cyclical changes, 404
  changes during intercourse, 404–405
  estrous cycle, 406
  indicators of ovulation, 405–406
  normal menstruation, 404
  ovarian cycle, 401–402
  ovarian *vs.* uterine changes, 403
  uterine cervix, cyclical changes, 404
  uterine cycle, 402–404
  vaginal cycle, 404
Menstrual cycle ovulation, 405
Menstruation, 401
Mentation, 349
Merkel cells, 158
Mesangial cells, 673
  relaxation of, 680
Mesocortical system, 147
Metabolic acidosis, 647–648
  acid–base paths, 648
Metabolic alkalosis, 647–648
Metabolic myopathies, 100

Metabolic rate, 488
  and body weight, 490
  factors affecting, 489
Metabolic syndrome, 449–450
Metabotropic glutamate receptors
  activation of, 141
  subtypes of, 141
Metabotropic receptors, 136
Metahypophysial diabetes, 443
Metathyroid diabetes, 443
Methemoglobin, 653
3, 4-Methylenedioxymethamphetamine, 149
Mexiletine, 164
mGluR. *See* Metabotropic glutamate receptors
MHC. *See* Major histocompatibility complex
Micelles, 465, 483
Microfilaments, 40
  composition of, 41
  structure of, 41
Microfold cells, 483
Microglia, 83
microRNAs, 16
Microscopy techniques, cellular constituents examination by, 35
Microsomes, 36
Microtubule-organizing centers, 42
Microtubules
  composition of, 40
  drugs affecting, 40, 41
  structures of, 40
Microvilli, 460
Micturition, 671
Midcollicular decerebrate cats, decerebrate rigidity in, 242
Middle cerebellar peduncle, 248
Middle ear
  functions of, 199
  medial view of, 201
  structures of, 200
Mifepristone (RU 486), 410
Migrating motor complex, 498
Migrating motor complexes, 499
Milk ejection reflex, 313
Mineralocorticoids, 353
  aldosterone. *See* Aldosterone
  mechanism of action of, 370–371
  relation to glucocorticoid receptors, 371
  secondary effects of excess, 372
Miotics, 187
MIS. *See* Müllerian inhibiting substance
Mitochondria
  components involved in oxidative phosphorylation in, 38
  functions, 38
  genome, 38
Mitochondrial diseases, 39
Mitochondrial DNA, 38
  diseases caused by abnormalities in, 47
Mitochondrial genome
  and nuclear genome, interaction between, 38
Mitochondrial membrane
  proton transport across inner and outer lamellas of inner, 12
Mitogen activated protein (MAP) kinase cascade, 56
Mitosis, 14
Molecular building blocks
  deoxyribonucleic acid, 14
  mitosis and meiosis, 14
  nucleosides, nucleotides, and nucleic acids, 12–13
  ribonucleic acids, 14–16
Molecular medicine, 47
Molecular motors
  dyneins, 41
  kinesin, 41
  myosin, 41, 42
Molecular weight of substance, 4
Moles, 4
Mongolism. *See* Down syndrome
Monoamine oxidase inhibitors, 149
Monoamines
  adrenoceptors, 146–147
  ATP, 149–150
  catecholamines, 145–146
  dopamine, 147, 148
  epinephrine, 145
  histamine, 149–150
  noradrenergic synapses, 147
  norepinephrine, 145
  secreted at synaptic junctions, 140
  serotonergic receptors, 148, 149
  serotonergic synapses, 149
  serotonin, 148
Monocular and binocular visual fields, 195
Monocytes activated by cytokines, 69
Monogenic forms of deafness, 210
Monosynaptic reflexes, 229
Mood disorder, 149
Morphological changes, parietal cell receptors and schematic representation of, 461
Mosaicism, 396
Motilin, 472
Motivation and addiction, 172
Motor axons, branching of, 128
Motor cortex. *See also* Primary motor cortex
  axons of neurons from, 238
  somatotopic organization for, 238
  and voluntary movement
    plasticity, 238
    posterior parietal cortex, 238
    premotor cortex, 238
    primary motor cortex, 236–238
    supplementary motor area, 238
Motor homunculus, 237
Motor neurons, inputs converging on, 227
Motor neuron with myelinated axon, 84
Motor pathways, general principles of central organization of, 236
Motor system, division of, 239
Motor unit, 108, 109
Movement, corticospinal and corticobulbar system role in, 239
MPGF. *See* Major proglucagon fragment
MRF. *See* Müllerian regression factor
mRNA transcription, 16, 18
MRP-2. *See* Multidrug resistance protein 2
MSH. *See* α–Melanocyte-stimulating hormone
MTOCs, 42
Mucosa irritation, 502
Mucosal cells, 484
Mucosal immune system, 473
Müllerian inhibiting substance, 392
Müllerian regression factor, 394
Multidrug resistance protein 2, 513
Multiple sclerosis, 86
Multiple system atrophy, 256
Multiunit smooth muscle, chemical mediators effect on, 115
Murmurs, 544
Muscarinic cholinergic receptors, 144, 145
Muscarinic poisoning, 263
Muscarinic receptors, 259
Muscle channelopathies, 105
Muscle, contractile element of, 547
Muscle fibers, 97
  classification of, 106, 107
  depolarization of, 103
  electrical response to repeated stimulation, 104
  isometric tension of, 106
  length and tension, link between, 105
  in motor unit, 109
  sarcotubular system of. *See* Sarcotubular system
Muscle rigor, 108
Muscles, control of
  corticobulbar tract, 239
  corticospinal tracts, 238
  movement and, 239
Muscle spindle
  afferents, dynamic and static responses of, 231
  discharge, effect of conditions on, 231
  essential elements of, 229
  function of, 230–231
  loading, 230–231
  mammalian, 230
  motor nerve supply, 229, 230
  sensory endings in, 229
Muscle tone, 233
Muscle twitch, 102
Muscle weakness
  by autoimmune attack, 130
  myasthenia gravis, 129
Muscular dystrophy, 100
Mushroom poisoning, 263
MV. *See* Microvilli
MVV. *See* Maximal voluntary ventilation
Myasthenia, 105
Myasthenia gravis, 129
Mycetism, 263
Myelinated axons
  conduction in, 91
  depolarization in, 91
Myelin sheath, 85
  defects, adverse neurological consequences of, 86
  nodes of Ranvier and, 86
Myenteric plexus, 473
Myocardial fibers, 522
Myocardial hibernation, 550
Myocardial infarctions, 537
Myoepithelial cells, contraction of, 313
Myoglobin, 643
  dissociation curves, comparison of, 644
Myopia, 187, 188
Myosin, 41, 42
  power stroke in skeletal muscle, 102–103
  in skeletal muscle, 99
Myotatic reflex. *See* Inverse stretch reflex
Myotonia dystrophy, 105
Myxedema. *See* Hypothyroidism

Parathyroidectomy, effects of, 382
Parathyroid excess, diseases of, 383
Parathyroid glands
actions, 382
anatomy, 381
malignancy, hypercalcemia of, 384
mechanism of action, 382–383
parathyroid hormonerelated protein, 384
PTH, synthesis/metabolism of, 381–382
regulation of secretion, 383
Parathyroid hormone, 377, 382
signal transduction pathways, 378
Parathyroid hormonerelated protein, 384
effects of, 381
Paraventricular neurons, 308
Paravertebral ganglia, 257
Parietal cell, composite diagram of, 460
Parietal cell—gastrin, agonists of, 460
Parkinson disease
basal ganglia-thalamocortical circuitry in, 248
familial cases of, 248
hypokinetic features of, 245
lead pipe rigidity in, 248
pathogenesis of movement disorders in, 248
prevalence of, 247
in sporadic idiopathic form, 247
symptoms, 247
treatment of, 247
Paroxysmal atrial tachycardia with block, 531
Paroxysmal ventricular tachycardia, 532
Partial pressures, 634
Partial seizures, 276
Parturition, 416
Parvocellular pathway, 190
Patch clamping, 49
Patchy ventilation–perfusion imbalance, 651
Pattern recognition receptors, 74
pAVP. *See* Plasma vasopressin
PB. *See* Barometric pressure
PC. *See* Phosphatidylcholine
Pegaptanib sodium, 181
Pendred syndrome, 210
Penicillin G and silibinin
for muscarinic poisoning treatment, 263
Pepsins hydrolyze, 481
PepT1. *See* Peptide transporter 1
Peptic ulcer disease, 458
Peptide bonds, formation of, 19
Peptide hormones, 299
precursors for, 300
Peptides, 18
Peptide transporter 1, 482
Peptide YY structure, 472
Periaqueductal gray, 172
Pericytes, 673
Perilymph, ionic composition of, 205
Perinuclear cisterns, 45
Periodic breathing, 666
in disease, 667
Periodic limb movement disorder, 276
Periosteum, 386
Peripheral chemoreceptor reflex
hemorrhage, 595
Mayer waves, 595
Traube–Hering waves, 595
vasoconstriction, 595
Peripheral nerve damage, 94
Peripheral nerves
composition of, 92
sensitivity to hypoxia and anesthetics, 93
Peripheral proteins, 37
in cell membrane, 36, 37
Peristalsis, 504
Peritubular capillaries, 674
Permeability, 679
Permissive effects of glucocorticoids, 365
Peroxisome proliferator-activated receptor γ, 442
Peroxisome proliferator-activated receptors, 40
Peroxisomes, 40
Pertussis toxin effect on cAMP, 62
p53 gene mutation, 47
PGES expression. *See* Prostaglandin synthase expression
PGH_2, 30
pH, 643
definition, 6
effects on oxygen–hemoglobin dissociation curve, 643
proton concentration and, 6
Phagocytic function, disorders of, 71
Phagocytosis, 48
Phantom limb pain, 169
Pharynx, food movement, 500
Phasic bursting, 312, 313
Phenobarbital, 143
Phenylephrine, 263
Phenylketonuria, 146
Phenytoin, 277
Pheochromocytomas, 358
Phlorhizin, 683
"Phosphate timer," 56
Phosphatidylcholine, 463, 514
Phosphatidylinositol
metabolism in cell membranes, 61
Phosphodiesterase, 116
Phospholipid bilayer, organization of, 37
Phospholipids, 26
Phosphorus magnetic resonance, 714
Phosphorus metabolism, 379
Phosphorylcreatine, 106
Photoreceptor mechanism, 182
Photoreceptor potentials, ionic basis of, 183
Photoreceptors, 157
receptor potentials of, 182
rod and cone, 180, 181
sequence of events in, 184
Phototransduction in rods and cones, sequence of events involved in, 184
Physicochemical disturbances, 87
Physiological tremor, 231
Physostigmine, muscarinic poisoning treatment by, 263
Piebaldism, 325–326
Pigment abnormalities, 325–326
Pinocytosis, 48
Pituitary gland
anatomy of, 324
anterior. *See* Anterior pituitary gland
anterior and intermediate lobes of, 308
diagrammatic outline of formation of, 324
functions of, 323
histology of, 324
hypothalamus relation to, 308, 309
Pituitary insufficiency
causes of, 336
effects of, 335–336
PKU, 146
Placenta
circulation, 614
fetal adrenal cortex, interactions, 415
Planum temporale, 292
Plasma
carriers for hormones, 301
dopamine levels, 356
glucose homeostasis, 25
glucose level, 22
factors determining, 24–25
nonelectrolytes of, 8
osmolal concentration of, 8
Plasma erythropoietin levels, 709
Plasma glucose homeostasis, 438
Plasma glucose level, 440, 441
Plasma growth hormone, 332
Plasma hormone concentrations, changes, 399
Plasma K^+ level, correlation of, 536
Plasma lipids, 29
Plasma membrane. *See* Cell membrane
Plasma osmolality
and changes in ECF volume in thirst, 310
plasma vasopressin, relation between, 698
and thirst, link between, 310
Plasma osmolality and disease, 9
Plasma proteins
afibrinogenemia, 565
hypoproteinemia, 565
origin of, 565
physiologic functions, 565, 566
Plasma renin activity, 704
Plasma renin concentration, 704
Plasma testosterone, human male, 398
Plasma vasopressin, 699
relation between, 698
mean arterial blood pressure, 699
Plasticity of motor cortex, 238
Platelet-activating factor, 80
Platelets
ADP receptors in human, 79
aggregation, 80
clotting factors and PDGF in, 78, 79
cytoplasm of, 78
production, regulation of, 80
response to tissue injury, 78
wound healing role, 79
Pleural cavity, 626
connective fibers, 627
parietal pleura, 626
pleural space, 627
pressure in, 628
visceral pleura, 626, 627
PLMD, 276
PMS. *See* Premenstrual syndrome
Pneumotaxic center, 658
p^{75NTR} receptors, 93, 94
Podocytes, 673
Poikilothermic animals, 316
Point mutations, 14
Poly(A) tail, 16
Polydipsia, 700
Polymodal nociceptors, 158

Polypeptides, 18
Polypeptide YY, 446
Polysynaptic connections between afferent and efferent neurons in spinal cord, 234
Polysynaptic reflex, 234
Polysynaptic reflexes, 229
Polyuria, 692, 700
POMC. *See* Pro-opiomelanocortin
Pontine reticulospinal tracts and posture, 240
Pontogeniculo-occipital (PGO) spikes, 275
Portal hypophysial vessels, 309
Positive feedback loop, 88
Posterior lobe hormones, synthesis of, 311
Posterior parietal cortex, 238
Posterior pituitary glands, hormones of
  biosynthesis of, 311
  intraneuronal transport of, 311
  secretion of, 311–312
  vasopressin and oxytocin, 311
Postextrasystolic potentiation, 548
Postganglionic autonomic neurons, 131
Postganglionic neurons, 256
Postsynaptic density, 120, 141
Postsynaptic inhibition, 126
Postsynaptic membrane, receptors concentrated in clusters on, 136
Postsynaptic neurons
  action potential generation in, 124–125
  action potential in. *See* Postsynaptic potentials
  PSD on membrane of, 141
  synaptic endings, 120
Postsynaptic potentials
  excitatory and inhibitory, 121–123
  slow EPSPs and IPSPs, 123
  temporal and spatial summation of, 125
Posttetanic potentiation, 286
Posttranscriptional modification
  of polypeptide chain, 20
  of pre-mRNA, 16
  reactions in, 20
Posture maintenance
  brain stem pathways involved in
    lateral, 240–241
    medial, 239–240
Posture-regulating systems
  decerebration, 241–243
  decortication, 243
Potassium/chloride cotransporter, 466
Potassium ions
  active transport of, 53, 54
  changes in membrane conductance of, 88
  concentration difference and sodium ions, 88
  equilibrium potential for, 10
  in mammalian spinal motor neurons, 10
  and membrane potential, 10
Power stroke of myosin in skeletal muscle, 102–103
ΠΠΑΡγ. *See* Peroxisome proliferator-activated receptor γ
PPARs, 40
PRA. *See* Plasma renin activity
Pralidoxime, 262
PRC. *See* Plasma renin concentration
pre-Bötzinger complex, 658
  pacemaker cells in, 658
Precocious/delayed puberty
  delayed/absent puberty, 400
  menopause, 400–401
  sexual precocity, 399–400
Precocious pseudopuberty, 399
Precocious sexual development
  classification of, 399
Precursor proteins, 427
Prednisone
  for Lambert–Eaton Syndrome treatment, 130
  for MS treatment, 86
  for myasthenia gravis treatment, 129
Preejection period (PEP), 542
Preganglionic neurons, 256
Pregnancy
  endocrine changes, 414
  fertilization & implantation, 413–414
  fetal graft, 414
  fetoplacental unit, 415
  human chorionic gonadotropin, 414
  human chorionic somatomammotropin, 415
  infertility, 414
  parturition, 415–416
  placental hormones, 415
Pregnenolone, 360
Premature beat, heart, 531
Prematurity, retinopathy of, 654
Premenopausal women, 485
Premenstrual syndrome, 413
Premotor cortex, 238
pre-mRNA, posttranscriptional modification of, 16
Preproenkephalin, 357
Prepro-oxyphysin, 311, 312
Prepropressophysin, 311
PreproPTH, 381
Prestin, 207
Presynaptic inhibition
  comparison of neurons producing, 126
  effects on action potential and Ca^{2+} current, 127
  and facilitation, 126
Presynaptic nerve terminals, small synaptic vesicle cycle in, 122
Presynaptic receptor, 136
PRH, 314
Primary adrenal insufficiency, 375
Primary ciliary dyskinesia, 42
Primary colors, 193
Primary evoked potential, 271
Primary hyperaldosteronism, 374
Primary messengers, 56
Primary motor cortex, 236–238
  cell organization in, 237
  imaging techniques for mapping, 237–238
  location of, 236–237
  motor homunculus and, 237
Primary plexus, 309
Primary structure of proteins, 18
Primary visual cortex, 191–192
  connections to sensory areas, 192
  distribution in human brain, 192
  layers in, 191
  nerve cells in, 191
  ocular dominance columns in, 191–192
  orientation columns in, 191
  responses of neurons in, 191
  visual projections from, 192
Primordial follicles, 400
Principal cells, 674
Principal digestive enzymes, 462
Proarrhythmic, 534
Proenzymes, 481
Progesterone, biosynthesis of, 409
Programmed cell death. *See* Apoptosis
Progressive motility, 422
Prolactin
  actions of, 333, 334
  components of, 333
  functions of, 323
  half-life of, 333
  secretion, regulation of, 334–335
Prolactin-inhibiting hormone, 314
Prolactin-releasing hormone, 314
Proliferative phase, 403
Proopiomelanocortin
  biosynthesis, 325
Pro-opiomelanocortin, 487
  biosynthesis, 325
Proprioceptors, 157
Prorenin, 702
Prosopagnosia, 294
Prostacyclin, 596
Prostaglandin H_2, 30
Prostaglandins, 422
  pharmacology of, 31
  synthesis of, 30
Prostaglandin synthase expression, 674
Prostate, 419
Prostate specific antigen, 423
Protease-activated receptor-2 (PAR-2) activation, 160
Protein degradation
  and production, balance between, 20–21
  and ubiquitination, 20–21
Protein digestion, 481–482
Protein folding, 20
Protein kinases, 56
  in cancer, 57
  in mammalian cell signaling, 56
Protein linkages to membrane lipids, 37
Protein processing, cellular structures involved in, 46
Protein-rich food, 502
Proteins, 491, 645
  amino acids found in, 19
  composition of, 18
  embedded in cell membrane, 36, 37
  in skeletal muscle, 97
  structure of, 18–19
  ubiquitination of, 20
Protein synthesis
  activation of, 56
  definition of, 19
  in endoplasmic reticulum, 20
  initiation of, 20
  mechanism of, 19–20
  RNA role in, 16, 19–20
Protein transferrin, 485
Protein translation and rough endoplasmic reticulum, 45
Protodiastole, 540
Proton transport, 12
Proto-oncogenes, 47
Protopathic pain, 159

Protoplasmic astrocytes, 84
Proximal convoluted tubule, 673
Proximal tubule
  Na^+ reabsorption mechanism, 682
  solutes reabsorption, 683
PRRs, 74
PSA. *See* Prostate specific antigen
PSD. *See* Postsynaptic density
Pseudocholinesterase, 144
Pseudohypoparathyroidism, 64, 383
Psilocin, 149
Psychosocial dwarfism, 334
PTH. *See* Parathyroid hormone
PTHrP. *See* Parathyroid hormonerelated protein
Puberty, 398
  control of onset, 398–399
Pulmonary alveolar macrophages, 625
Pulmonary arteries/veins, 637
Pulmonary chemoreflex, 664
Pulmonary circulation, 627, 628, 636–638
  flow, 636
  gravity, 636
  pressure, 636
  pulmonary blood vessels, 636
  pulmonary reservoir, $603
  regulation of pulmonary blood flow, 637–638
  ventilation/perfusion ratios, 636–637
  volume, 636
Pulmonary fibrosis, 631
Pulmonary hypertension, 637
Pulse, 542
Pulse oximeter, 629
Pupillary light reflexes, 188
Purine adenosine triphosphate, 473
Purines
  compounds containing, 13
  ring structures of, 12
Purkinje cells, 249
  output of, 250
Purkinje fibers, 522, 524
Purkinje system, 521
Pyramidal neurons, 270–271
Pyridostigmine, 262
  for myasthenia gravis treatment, 129
Pyrimidines
  catabolism of, 13
  compounds containing, 13
  ring structures of, 12

Q
QRS vector, 528
Quality control, 46–47
Quantitating respiratory phenomena, 628–629
  lung volumes and capacities, 629
Quaternary structure, 19

R
Radiation, 318
Rafts and caveolae, 49
Ralfinamide, 164
Raloxifene, 409
Ramelteon, 279
Ranibizumab, 181
RANKL. *See* Receptor activator for nuclear factor kappa beta ligand
Rapid auditory processing theory, 292
Rapid eye movement sleep, 273, 666
  EEG waves during, 274
  PET scans of, 275
  phasic potentials, 275
  rapid movements of eyes, 275
Rapidly adapting receptors, 664
RAS, 271
Rasagiline for MSA treatment, 256
Rayleigh match, 194
Raynaud disease, 264
Raynaud phenomenon, 264
Rebound phenomenon, 252
Receptive relaxation, 502
Receptor activator for nuclear factor kappa beta ligand, 387
Receptor–ligand interaction, 55
Receptors. *See also* Specific receptors
  desensitization, 137
  G protein-coupled, 136
  and G protein diseases, 64
  ionotropic, 136
  metabotropic, 136
  for neurotransmitters and neuromodulators, 136, 139
  on nociceptive unmyelinated nerve terminals, 160
  on postsynaptic membrane, 136
  presynaptic, 136
  and transmitters, 136, 137
Reciprocal innervation, 232, 657
Recovery heat, 108
Recruitment of motor units, 234
Rectum
  distention of, 506
  responses to distention of, 506
Red blood cells
  characteristics, 557, 559
  fibrin fibrils, 557, 559
  formation and destruction, 557, 559
5α-Reductase deficiency, 397
Reduction, 11
Referred pain, 166
Reflex arc
  activity in, 228
  components of, 228
  monosynaptic reflexes, 229
  polysynaptic reflexes, 229
Reflexes
  general properties, 228
  and semireflex thermoregulatory responses, 318–319
  spinal integration of, 234–236
  stimulus for, 228
Reflex ovulation, 412
Reflex sympathetic dystrophy, 164
Refraction, 186
Refractive power, 186
Refractory period, 91
Regulatory elements, 14
Relaxation volume, 630
Relaxed (R) configuration, 642
Relaxin, 328
REM sleep. *See* Rapid eye movement sleep
Renal circulation, 676
Renal failure, acute, 548
Renal function, 673–693
  abnormal Na^+ handling, 693
  acidosis, 693
  adrenocortical steroids effects, 690–691
  aquaporins, 685
  bladder, 693–695
  bladder emptying, 693–694
  bladder filling, 693
  blood flow, 676–677
  blood vessels, 674–675
  capillary bed size, 679–680
  capsule, 675
  collecting ducts, 686–687
  countercurrent mechanism, 687–688
  deafferentation effects, 695
  denervation effects, 695
  diluting ability, 692–693
  disordered renal function effects, 692–693
  distal tubule, 686
  diuretics, 692
  filtration fraction, 681
  free water clearance, 690
  functional anatomy, 673–676
  glomerular filtration, 678–681
  glomerular filtration rate (GFR), 678–679
  glomerular filtration rate (GFR) changes in, 680–681
  glomerular filtration rate (GFR) control of, 679
  glomerulotubular balance, 684–685
  glucose reabsorption, 682–683
  glucose transport mechanism, 683
  humoral effects, 691
  hydrostatic pressure, 680
  K^+ excretion, regulation, 691–692
  loop of Henle, 685–686
  loss of concentrating, 692–693
  lymphatics, 675
  Na^+ excretion, regulation, 690–691
  Na^+ reabsorption, 681–682
  nephron, 673–674
  osmotic diuresis, 689–690
  osmotic pressure, 680
  oxygen consumption, 678
  PAH transport, 683–684
  permeability, 679
  proximal tubule, 685
  reflex control, 694–695
  regional blood flow, 678
  renal blood flow autoregulation, 678
  renal blood flow regulation, 677
  renal circulation, 676–678
  renal nerves functions, 677–678
  renal vessels innervation, 675–676
  renal vessels pressure, 677
  secondary active transport, additional examples, 683
  spinal cord transection effects, 695
  tubular function, 681–690
  tubular reabsorption & secretion mechanisms, 681
  tubuloglomerular feedback, 684–685
  urea role, 688–689
  uremia, 693
  urine concentration relation, 690
  water diuresis, 691

water excretion, regulation, 691
water intoxication, 691
water transport, 685
Renal glucose transport, 683
Renal H^+ secretion, 711–714
ammonia secretion, 713
bicarbonate excretion, 714
body fluids
principal buffers, 716
factors affecting acid secretion, 714
fate of, 712
fate of H^+ in urine, 712
Na–H exchanger, 711
pH changes, 713
pH of body fluids, 715
reaction with buffers, 712–713
Renal hypertension, 706
Renal interstitial pressure, 675
Renal medullary interstitial cells, 674
Renal physiology, 671
Renal plasma clearance, 678
Renal plasma flow, 676
Renal principal cell, 691
Renal threshold, 682
Renal tubular acidosis, 693
Renin and aldosterone secretion, 372–373
Renin–angiotensin system, 702–706
angiotensin-converting enzyme & angiotensin II, 702–703
metabolism of, 704
receptors, 704–705
angiotensinogen, 702
angiotensins, actions of, 704
pharmacologic manipulation of, 703
renin, 702
renin secretion, regulation of, 705–706
summary of, 701
tissue renin-angiotensin systems, 704
Renin secretion
conditions, 706
factors, 706
regulation of, 705–706
Renin substrate, 702
amino terminal end of, 703
Renorenal reflex, 676
Replication, 14
Representational and categorical hemispheres, lesions of, 291
Reproductive abnormalities, 413
Reserpine, 147
Residual volume, 629
Respiration, 619
acid-base balance, ventilatory responses to changes in, 661–662
airway resistance, 631
alveolar surface tension, surfactant role, 631–632
baroreceptor stimulation, respiratory effects of, 665
blood flow, 632–633
brain stem, chemoreceptors in, 661
breath holding, 663
breathing, work of, 632
carotid & aortic bodies, 660–661
chemical control, 659
chemical control of breathing, 659–663
chest wall, 629–632
compliance of lungs, chest wall, 629–632
control systems, 657–658
CO_2 response curve, hypoxia effects on, 662–663
CO_2 response, H^+ effect, 663
coughing & sneezing, 664
dead space, 633–634
exercise effects, 666–669
expiration, 627–628
inspiration, 627–628
lung capacities, 629
lungs, responses mediated by receptors in, 664
lung volumes, 629
medullary systems, 658
movement in, 626
muscles, 626
neural control of breathing, 657–658
nonchemical influences on, 664–666
oxygen deficiency, ventilatory response to, 662
pontine & vagal influences, 658
pressure–volume curves in, 631
proprioceptors, afferents from, 664–665
quantitating phenomena, 628–629
regulation of, 657–669
regulation of respiratory activity, 658–659
respiratory abnormalities, 666
respiratory muscles, 626
sleep, effects of, 665–666
uneven ventilation, 633–634
ventilation, 632–633
ventilatory responses to CO, 662
visceral reflexes, respiratory components of, 665
Respiration at birth, 615–616
Respiratory acidosis, 647, 662, 714, 716
Respiratory alkalosis, 647, 662
Respiratory burst, 68
Respiratory center, stimuli affecting, 659
Respiratory compensation, 648–649, 716
Respiratory control pattern generator, 658
Respiratory enzyme pathway, 107
Respiratory exchange ratio, 489
Respiratory minute volume, 629
Respiratory muscles, 626, 632
Respiratory quotient, 489
Respiratory system, 635
partial pressures of gases, 635
pressure-volume curve, 630, 631
Respiratory tract
alveolar airway, 621
conducting airway, 621
regions of, 621
upper airway, 621
Resting heat, 108
Resting membrane potential, 87–88
of cardiac muscle, 110
of hair cells, 203
Reticular activating system, 271
Reticuloendothelial system, 69
Reticulospinal tracts, 240
Retina
blood vessels, 180
layers, 178
melanopsin, 185
neural components of extrafoveal portion of, 178, 179
pigment epithelium, 180
potential changes initiating action potentials in, 182
receptor layer of, 180
visual information processing in, 185
Retinal, 183
Retinal ganglion cells
projections to right lateral geniculate body, 190
response to light focused on receptive fields, 185
types of, 190
Retinohypothalamic fibers, 278
Retinoid X receptor, 348
Retrograde amnesia, 286
Retrograde transport, 87
Retrolental fibroplasia, 654
Reuptake
catecholamines, 146
definition of, 137
of norepinephrine, 137
R-flurbiprofen, Alzheimer disease treatment by, 289
Rhodopsin, 183
structure of, 184
Ribonucleic acids
and DNA, difference between, 14
production from DNA, 16
role in protein synthesis, 16
types of, 16
Ribosomes, 46
Rickets, 380
Rigor mortis, 108
Riluzole, 140
for ALS treatment, 240
Rituximab
for MS treatment, 86
RMICs. *See* Renal medullary interstitial cells
RMV. *See* Respiratory minute volume
RNA. *See* Ribonucleic acids
Rod photoreceptor
components of, 180
Rod receptor potential, 182–183
Rods
density along horizontal meridian, 182
photosensitive pigment in, 183
schematic diagram of, 182
sequence of events involved in phototransduction in, 184
Rotational acceleration, responses to, 211
Rough endoplasmic reticulum and protein translation, 45
RPF. *See* Renal plasma flow
RQ. *See* Respiratory quotient
Rubrospinal tract, 240
Ruffini corpuscles, 158
RV. *See* Residual volume
Ryanodine receptor, 103

S

Saline cathartics, 468
Saliva, 456
Salivary α-amylase, 478
Salivary glands, 456
Salivary secretion, 456–457
regulation of, 458
Saltatory conduction, 91

Salt-sensitive taste, 223
SA node. *See* Sinoatrial node
Sarcomere, 99
Sarcotubular system
  components of, 100, 101
  T system and, 101
Satiety factor, 487
SBP, 302
Scala tympani, 200
Scala vestibuli, 200
SCFAs. *See* Short-chain fatty acids
Schizophrenia, 147
Schwann cells, 83–84
SCI, 235
Sclera, 177
SDA. *See* Specific dynamic action
Sealing zone, 387
Secondary active transport, 53–54
Secondary adrenal insufficiency, 375
Secondary hyperaldosteronism, 374
Secondary sex characteristics, 425
Secondary structure of proteins
  β-sheets, 18–19
  spatial arrangement, 18
Second messengers
  cyclic AMP, 60–61
  diacylglycerol as, 60–61
  effect on aldosterone secretion, 375
  effect on Na, K ATPase pump activity, 53
  inositol trisphosphate as, 59–60
  intracellular Ca^{2+} as, 56–58
  phosphorylation and, 55–56
  in regulation of aldosterone secretion, 375
  short-term changes in cell function by, 55
Secretory immunity, 78, 483
Segmentation contractions, 504
Seizures
  EEG activity during, 277, 278
  genetic mutations and, 277
  partial/generalized, 276
  treatment of, 277, 278
  types of, 276–277
Selectins, 42, 68
Selective dorsal rhizotomy, CP treatment by, 236
Selective estrogen receptor modulators, 409
Semantic memory, 284
Semen, 422
Seminiferous epithelium, 421
Seminiferous tubule dysgenesis, 397
Seminiferous tubules, 419
Senile dementia, 289
Sensitization and habituation, 286
Sensorineural and conduction deafness, tests for comparing, 211
Sensorineural hearing loss, 209
Sensory association area, 169
Sensory coding
  definition, 161
  intensity of sensation, 162, 163
  location, 162
  modality, 161–162
  neurological exam, 163
Sensory homunculus, 168
Sensory modalities, 157, 158
Sensory nerve fibers, numerical classification of, 92
Sensory receptors
  in ear, 202
  as transducers, 157
Serial electrocardiographic patterns, diagrammatic illustration of, 535
SERMs. *See* Selective estrogen receptor modulators
Serotonergic receptors, 148, 149
  classes of, 148
  functions of, 148, 149
  pharmacology of, 149
Serotonergic synapses, 149
Serotonin, 148, 504
Serotonin reuptake inhibitors, 149
Serpentine receptors. *See* G protein-coupled receptors
Sertoli cells, 419
Serum dehydroepiandrosterone sulfate (DHEAS), changes in, 364
Seven-helix receptors. *See* G protein-coupled receptors
Sex chromosomes, defects, 398
Sex chromosomes, nondisjunction of, 396
Sex determination, diagrammatic summary of, 396
Sex differentiation disorders, 397
Sex hormone-binding globulin, 302
Sex hormones, 391, 419
Sex steroid-binding globulin, 424
Sex steroids in women, 407
Sexual excitement, 405
SFO. *See* Subfornical organ
SGLT. *See* Sodium-dependent glucose transporter
SHBG, 302
Shock, 548
  hypovolemic, 548
  low-resistance, 548
  obstructive, 548
  treatment of, 548
Short-chain fatty acids, 485
"Short-circuit" conductance, 27
Short gut syndrome, 491
Short-term memory, 285
Shy–Drager syndrome, 256
SIADH. *See* Syndrome of "inappropriate" hypersecretion of antidiuretic hormone
Sick sinus syndrome, 529
Siggaard–Andersen curve nomogram, 717
Signal hypothesis, 20
Signal peptide, 20
Signal recognition particle, 20
Signal transduction
  in odorant receptor, 219–220
  in taste receptors, 223–224
  via JAK–STAT pathway, 63, 64
Sinemet for Parkinson disease treatment, 247
Single-breath N_2 curve, 633
Sinoatrial node, 521
  membrane potential of, 523
Sinus arrhythmia, 529
  in young/old man, 529
Sinus bradycardia, 529
Skeletal muscles
  actin and myosin arrangement in, 99
  body movements and, 110
  contractile mechanism in, 97
  contraction of. *See* Contraction, muscular
  cross-striations in, 98–100
  denervation of, 108
  electrical characteristics, 101, 102
  ion distribution and fluxes in, 102
  length-tension relation in, 105
  mammalian, 98
  mechanical efficiency of, 108
  in middle ear, 200
  motor unit, 108, 109
  muscle fibers in, 97
  pathway linking CNS to, 256
  proteins in, 97, 100
  relaxation of, 264–265
  strength of, 110
Skin coloration, control of, 325–326
Skin disorder, connexin mutations and, 45
Sleep
  effects of, 665–666
  importance of, 275
Sleep apnea, obstructive, 667
Sleep disorders, 276
Sleep stages
  distribution of, 275
  NREM sleep, 273–274
  REM sleep, 273
    EEG waves during, 274
    PET scans of, 275
    phasic potentials, 275
    rapid movements of eyes, 275
Sleep-wake cycle, 272
  alpha rhythm, 273
  beta rhythm, 273
  EEG and muscle activity during, 274
  sleep stages, 273
  thalamocortical loops and, 273
Sleep-wake state
  melatonin and, 280
  neurochemical mechanisms promoting, 278
    GABA, 280
    histamine, 280
    melatonin, 280
    midbrain reticular formation, 279
    RAS neurons, reciprocal activity, 279–280
  transitions between, 278
Sleepwalking, 276
Slowly adapting receptors, 664
Small-conductance calcium-activated potassium (SK) channels, 240
Small G proteins, 58
Small intestine
  chloride secretion in, 467
  intestinal motility, 504
  transit time, 505–506
Small-molecule neurotransmitters
  biosynthesis of, 137
  excitatory and inhibitory amino acids
    acetylcholine, 144
    acetylcholine receptors, 144–145
    cholinergic receptor, 145
    GABA, 142–143
    glutamate, 138–142
    glycine, 143
  monoamines
    adrenoceptors, 146–147
    ATP, 149–150

catecholamines, 145–146
dopamine, 147, 148
epinephrine, 145
histamine, 149–150
noradrenergic synapses, 147
norepinephrine, 145
serotonergic receptors, 148, 149
serotonergic synapses, 149
serotonin, 148
with neuropeptides, 136
Smell
adaptation and, 221
classification of, 217
odorant-binding proteins and, 221
odorant receptors and, 219–220
odor detection threshold, 220
olfactory cortex function in. *See* Olfactory cortex
olfactory epithelium. *See* Olfactory epithelium
signal transduction and, 219–220
Smooth muscle
contraction of
calmodulin-dependent myosin light chain kinase activity, 115
Ca^{2+} role in, 114–115
chemical mediators effect on, 115
sequence of events in, 115
drugs acting on, 116
effects of agents on membrane potential of, 115
electrical activity of, 114
force generation of, 116
mechanical activity of, 114
nerve endings in, 130
nerve supply to, 116
overexcitation in airways, 116
plasticity of, 116
postganglionic autonomic neurons on, 131
relaxation of
cellular mechanisms linked with, 115
skeletal and cardiac muscle, comparison of, 114
striations in, 114
types of, 114
of uterus, oxytocin effect on, 313
Sneezing, 664
SOCCs, 57
Sodium-dependent glucose transporter, 480
Sodium/hydrogen exchanger, 466
Sodium ions
active transport of, 53, 54
changes in membrane conductance of, 88
concentration difference and potassium ions, 88
in mammalian spinal motor neurons, 10
Sodium–potassium adenosine triphosphatase (Na, K ATPase)
ATP hydrolysis by, 51
α and β subunits of, 51–53
electrogenic pump, 51
heterodimer, 51, 52
ion binding sites, 53
regulation of, 53
Solutes, normal clearance values of, 679
Somatic cell division, 14
Somatic chromosomes, 392
Somatic motor activity, 227
Somatic nervous system, 257
Somatomedins
polypeptide growth factors, 327
principal circulating, 327, 328
relaxin isoforms, 328
Somatomotor nervous system
and ANS, difference between, 264
Somatosensory pathways, 166
cortical plasticity, 169, 170
dorsal column pathway, 167–169
effects of CNS lesions, 170
ventrolateral spinothalamic tract, 169
Somatostatin, 150, 151, 431, 472
for acromegaly, 328
Somatostatin inhibits, 472
Somatotopic organization, 167–169
Somatotropes, 324
Somnambulism, 276
Sound
definition of, 203
localization, 209
loudness and pitch of, 203
transmission of, 206
Sound frequencies audible to humans, 205
Sound waves
amplitude of, 205
characteristics of, 205
conduction of, 206
definition of, 203
SP. *See* Surfactant protein
Spastic CP, 236
Spastic neurogenic bladder, 695
Spatial summation, 125
Specific dynamic action, 489
Specific ionic composition, defense of, 708–709
Spermatogenesis, 419, 422
Spermatogonia, 420
Spermatozoon, ejaculation of, 422
Spherical follicle, 340
Sphincter of Oddi, 510
Sphincters, 455
Spinal cord injury, 235
Spinal cord stimulation, 169
Spinal cord, transection of, 235
Spinal motor neuron, 109
negative feedback inhibition of, 127
Spinal reflex, 228
Spinal shock, 235
Spinnbarkeit, 404
Spinocerebellum, 250
Spiny stellate cells, 271
Splanchnic circulation. *See also* Gastrointestinal circulation
schematic of, 474
SRP, 20
SSRIs, 149
Standard limb leads, 524, 526
StAR protein. *See* Steroidogenic acute regulatory protein
Stellate cells, 249
Stem cells, factors stimulating production of, 70
Stenosis, 544
Stereoagnosia, 162
Stereocilia, 202
Stereognosis, 162
Steroid binding proteins, 302
Steroidogenic acute regulatory protein
in adrenals and gonads, 362
functions of, 300
regulation of steroid biosynthesis by, 301
Steroids biosynthesis, 360–361
enzyme deficiencies effect on, 361–362
intracellular pathway of, 356
precursor of, 360
Steroid-secreting cells, structures of, 356
Steroids hormones, thyroid and, difference between, 300
Sterols, 26
Stokes–Adams syndrome, 530
Stomach, 502–504
acid-secreting cells of, 469
anatomy of, 459
gastrointestinal motility, 502
regulation of, 502
glandular secretions of, 458
small intestine
intestinal motility, 504
vomiting, 502–504
Store-operated Ca^{2+} channels, 57
Strabismus, 187
Strangeness and familiarity, 289
Stratum functionale, 403
Strength–duration curve, 89
Stress, glucocorticoids effects on, 366
Stress-induced analgesia, 173
Stretch reflexes, 229
pathways responsible for, 231
reciprocal innervation and, 232
Stretch reflex–inverse stretch reflex sequence, 233
Striations in cardiac muscle, 110
Striations in skeletal muscle
identification by letters, 98–99
thick filaments, 99
thin filaments, 99–100
Stroke volume, 546
Structural lipids, 27
Subcortical structures and navigation in humans, 294
Subfornical organ, 704
Submucous plexus, 473
Substance P, 150, 473
Substantia nigra, 244
Sucrose, 478
Sugars, intestinal transport of, 481
Superior cerebellar peduncle, 248
Superior colliculi, 196
Supplementary motor area, 238
Suppressor strip, 243
Suprachiasmatic nuclei (SCN), circadian activity of, 278
Surface tension, 624
Surfactant, 624, 632
Surfactant protein, 623, 632
Surgical approach
for acromegaly and gigantism, 328
for Parkinson disease treatment, 247
Swallowing, 500
Sweet-responsive receptors, 224
Sweet taste, 223–224
Sympathetic activation, 265
Sympathetic innervations, 265

Sympathetic nervous system
  sympathetic preganglionic and postganglionic fiber projection in, 257
  as thoracolumbar division of CNS, 256
Sympathetic noradrenergic discharge, 265
Sympathetic paravertebral ganglion, 257
Sympathetic preganglionic and postganglionic fibers
  projection of, 257
Sympathomimetic drugs, Horner syndrome treatment by, 263
Sympathomimetics, 147
Symport, 480
Synapses
  anatomic structure of, 120
  cell-to-cell communication via, 119
  in cerebral and cerebellar cortex, 120
  facilitation at, 126
  inhibition at
    in cerebellum, 127
    inhibitory systems for, 126–127
    postsynaptic and presynaptic, 126
  on motor neuron, 120
  transmission of action potential, 119
Synaptic delay, 122
Synaptic elements, functions of, 120–121
Synaptic junctions, electrical transmission at, 123
Synaptic knob, electronmicrograph of, 120
Synaptic physiology, 138
Synaptic plasticity and learning, 286
Synaptic transmission, 119
Synaptic vesicle docking and fusion in nerve endings, 122
Synaptic vesicles
  kinds of, 120
  transport along axon, 120–121
Synaptobrevin cleavage by botulinum toxin, 123
Synaptosome-associated protein (SNAP-25)
  cleavage by botulinum toxin, 123
Syncytiotrophoblast, 414
Syndrome of "inappropriate" hypersecretion of antidiuretic hormone, 700
Syndromic deafness, 210
Syntrophins, 101
Systemic response to injury, 80–81
System mediating acquired immunity, 69, 70
Systolic dysfunction, 547
  effect of, 549
Systolic failure, 540
Systolic pressure, 539

T

T_3. *See* Triiodothyronine
T_4. *See* Thyroxine
Tabes dorsalis, 695
Tachycardia, 529
Tachycardias, 529
Tachypnea, 649
Tactile agnosia, 162
Tamoxifen, 302, 409, 417
Tardive dyskinesia, 246
Taste
  classification of, 217
  reactions and contrast phenomena, 224
  sense organ for. *See* Taste buds
Taste buds
  basal cells, 221, 222
  fungiform papillae, 222
  innervation, 221
  in papillae of human tongue, 222
  types of, 221
Taste detection, abnormalities in, 225
Taste modalities
  receptors for, 223–224
  types of, 222–223
Taste pathways, 222, 223
Taste receptors
  signal transduction in, 223–224
  types of, 223
Taste threshold
  definition of, 224
  and intensity discrimination, 224
  of substances, 224
Taxol, 41
TBG concentration and thyroid hormones, 344
TBI, 284
T cell receptors
  heterodimers, 76
  MHC protein–peptide complexes and, 76
T cells
  activation in acquired immunity, 71, 74
  CD8 and CD4 proteins on, 76
  maturation, sites of congenital blockade of, 80
  polypeptides of circulating, 76
  types of, 69
T channels, 523
Temperature, 643
  effects on oxygen-hemoglobin dissociation curve, 643
Temperature regulation, 316–320
  mechanisms for, 318–319
Temporal lobe memory, 285
Temporal summation, 125
Tenotomy, CP treatment by, 236
Tense (T) configuration, 642
Tertiary adrenal insufficiency, 375
Tertiary structure of protein, 19
Testes, endocrine function of
  actions, 424–425
  anabolic effects, 425
  estrogens, testicular production of, 426–427
  inhibins, 427
  mechanism of action, 425–426
  secondary sex characteristics, 425
  secretion, 424
  steroid feedback, 427
  testosterone, chemistry/biosynthesis of, 423–424
  transport/metabolism, 424
Testicular descent, 428
Testicular feminizing syndrome, 397
Testicular function, abnormalities of
  androgen-secreting tumors, 428
  cryptorchidism, 428
  hormones and cancer, 428
  male hypogonadism, 428
Testosterone, 419, 425
  biosynthesis of, 424
  17-ketosteroid metabolites of, 425
  schematic diagram of, 426
Testosterone–receptor, 426
Testosterone secretion rate, 424
Testotoxicosis, 64
Tetanic contraction, 104–105
Tetanus, 106
Tetanus toxins and botulinum, 123
Tetanus toxoid vaccine, 123
Tetrabenazine, 246
TG. *See* Triglyceride
TGFα, 54
Thalamic pain syndrome, 170
Thalamocortical loops, 273
Thalamostriatal pathway, 244
Thalamus
  division of, 269
  functions of, 269
  sensory relay nuclei, 270
  ventral anterior and ventral lateral nuclei, 270
Theca interna, 401
Theca interna cells, interactions, 407
Thermal gradient, 318
Thermal nociceptors, 158
Thermodilution, 545
Thermoreceptors, 157
  threshold for activation of, 161
Thermoregulatory responses in humans, 318–319
Thiazolidinediones, 442
Thioridazine, 181
Thioureylenes, hyperthyroidism treatment by, 347
Thirst, 310
Thrombocytopenic purpura, 80
Thrombopoietin, 80
Thromboxane A_2, 596
Thymectomy, myasthenia gravis treatment by, 129
Thymidine–adenine–thymidine–adenine (TATA) sequence, 14
Thyrocytes
  basolateral membranes of, 341
  iodide transport across, 341
Thyroglossal duct, 339
Thyroid cell, 340
Thyroid gland
  anatomy of, 339–340
  histology of, 340
  lobes of, 339
  spherical follicle, 340
Thyroid growth, factors affecting, 345
Thyroid hormones, 385
  biological activity of, 340
  calorigenic action of, 348–349
  effect on Na, K ATPase pump activity, 53
  fluctuations in deiodination, 344–345
  iodine homeostasis and, 340–341
  mechanism of action of, 347–348
  metabolism of, 344–345
  physiologic effects on, 347
    brain, 349
    cardiovascular system, 349
    catecholamines, 350
    nervous system, 349–350
    normal growth, 350
    skeletal muscle, 350
  secretion, regulation of, 345–347
  steroids and, difference between, 300
  synthesis and secretion, 341–343
  transport of, 343–344

Thyroid-stimulating hormone
chemistry and metabolism of, 345
deficiency, 336
effect on thyroid, 345
secretion in cold, 319
Thyroid-stimulating immunoglobulins, 306
Thyrotropin-releasing hormone, 314
functions of, 315
Thyroxine, 340
calorigenic effect of, 348
mechanism of action of, 347–348
metabolism of, 344
plasma level in adults, 343
protein binding, 343–344
Tibialis muscular dystrophy, 100
Tickle, 159, 160
Tidal volume, 629
Tight junctions, 43
Tip links, 202
schematic representation of, 204
Tissue conductance, 318
Tissue macrophage system, 69
Tissues, Po_2 and Pco_2 values in, 642
Titin in skeletal muscle, 100
Titratable acid, 712
Tizanidine, ALS treatment by, 240
TLRs, 74
Toll-like receptors, 74
Tonic-clonic seizure, 277
Tonic contractions, 504
Tonicity, 8
Tonicity, defense of, 697
clinical implications, 700
metabolism, 698
synthetic agonists/antagonists, 700
vasopressin, effects of, 698
vasopressin receptors, 697–698
vasopressin secretion
control of, 698–699
stimuli, variety of, 700
volume effects on, 699–700
Topiramate, 277
for chronic pain, 164
Torsade de pointes, 532
Total blood volume, 3
Total body calcium, 388
Total lung capacity, 629
Trabecular bone, structure of, 386
Transamination reactions, citric acid cycle in, 21
Transcellular fluids, 5
Transcription
activation of, 56
definition, 16
diagrammatic outline of, 18
into pre-mRNA, 16
Transcytosis, 54
Transducin, 184
Transepithelial transport, 53
Transforming growth factor alpha, 54
Transient receptor potential channels, 159
Transient sleep disorders, 279
Translation. *See* Protein synthesis
Transmembrane proteins
in cell membrane, 36, 37
signal peptide, 20
in tight junctions, 43
Transmitter
chemistry of, 136–137
endogenous cannabinoids, 151–152
hypersensitivity of postsynaptic structure to, 131
nitric oxide, 151
quantal release of, 129
receptors and, 136, 137
Transport proteins, 682. *See also* Membrane transport proteins
Traumatic brain injury, 284
Traveling waves
movement in cochlea, 206
schematic representation of, 207
TR genes, 347–348
TRH. *See* Thyrotropin-releasing hormone
TRH-secreting neurons, 315
Triacylglycerols, 26
Tricarboxylic acid cycle. *See* Citric acid cycle
Trichromats, 193, 194
Tricyclic antidepressants, 147
Triglyceride, 484
Triglycerides, 26
Triiodothyronine, 340
calorigenic effect of, 348
mechanism of action of, 347–348
metabolism of, 344
plasma level in adults, 343
protein binding, 343–344
Trinucleotide repeat diseases, 47, 245
tRNA–amino acid–adenylate complex, 19
tRNA for amino acids, 19
Troglitazone (Rezulin), 442
Trophic action, 469
Tropic hormones, 323
Tropomyosin in skeletal muscle, 100
Troponin, 57
Trousseau's sign, 382
TRP channels, 159
True hermaphroditism, 396
Tryptophan hydroxylase in CNS, 148
TSH. *See* Thyroid-stimulating hormone
TSH receptor, 345
TSIs, 306
T tubule, 104
Tubular function, 681
Tubular myelin, 624
Tubular reabsorption, 671
Tubular secretion, 671
Tubuloglomerular feedback, 684
Tumor suppressor genes (p53 gene), 47
Turner syndrome. *See* Ovarian agenesis
Turnover rate of endogenous proteins, 17
Two-point threshold test, 162
Tympanic membrane, 199
movements of, 206
Type 1 diabetes mellitus, 305
Type 2 diabetes mellitus, 305
Typical depression, 149
Tyrosine kinase activity, 436
Tyrosine kinase associated (Trk) receptors, 93–94
Tyrosine kinases, diagrammatic representation of, 63
Tyrosine phosphatases, diagrammatic representation of, 63

U

Ubiquitination
definition, 20
and protein degradation, 20–21
of proteins, 20
Ultrasonography, 516
Umami taste, 224
Uncal herniation, 243
Uncompensated metabolic acidosis, 648
Uncompensated respiratory alkalosis, 647
Unconditioned stimulus, 287
Unipolar electrocardiographic leads, 526
Unipolar recording, 524
Uniports, 51
Units for measuring concentration of solutes, 4
Uper motor neuron lesion, 233
Upper motor neurons, 239, 240
damage to, 240
Urea, 688–689
Urea cycle, 515
Urea formation
enzymes involved in, 21
in liver, 21
precursor for, 21
Uremia, 548
Uric acid
excretion on purine-free diet, 13
synthesis and breakdown of, 13
Urinary pH changes
implications of, 714
Urine, 715
Urotensin-II, 599
US, 287
Uterine circulation, 614
Uterine musculature sensitivity to oxytocin, 313

V

Vagal outflow, 459
Valproate, 277
Valsalva maneuver
bradycardia, 593
diagram of response to straining, 593, 594
heart rate, 595
hyperaldosteronism, 595
intrathoracic pressure, 593
tachycardia, 593
Vasa recta, 674
Vascular endothelial growth factor, 401
Vascular link between hypothalamus and anterior pituitary, 308, 309
Vascular reactivity, 365
Vascular smooth muscle
contraction and relaxation, 569, 570
latch-bridge mechanism, 569
Vas deferens, 419
Vasectomy, 423
Vasoactive intestinal polypeptide, 264, 497, 516
Vasopressin, 151, 303, 685, 698
physiologic effects, 313
Vasopressin, effects of, 698
Vasopressin escape, 700
Vasopressin receptors, 313, 697–698
Vasopressin-secreting neurons
stimulation of, 312
in suprachiasmatic nuclei, 313

Vasopressin secretion
control of, 698–699
osmotic pressure of, 697
stimuli affecting, summary of, 698
stimuli, variety of, 700
volume effects on, 699–700
Vasospasm, 403
VEGF. *See* Vascular endothelial growth factor
Venous circulation, 582
Venous pressure and flow, 582
air embolism, 583
effects of heartbeat, 583
muscle pump, 583
thoracic pump, 583
venous pressure measurement, 583–584
Venous-to-arterial shunts, 651
Ventilation, 632–633
alveolar, variations effect in respiratory rate, 633
intrapleural pressures effect, 633
uneven, 633–634
Ventilation/blood flow comparison, 652
Ventilation-perfusion imbalance, 651–652
Ventilation/perfusion ratio, 636, 637
Ventricle, left, normal pressure–volume loop of, 542
Ventricular ejection, 540
Ventricular fibrillation, 532
Ventricular muscle, comparison of, 523
Ventricular premature beats, 532
Ventricular systole, 521, 539
start of, 540
Ventrolateral spinothalamic tract, 169
Venules and veins, 572
Vertebrates, body temperature of, 316
Very low density lipoprotein, 29
Very low density lipoproteins, 29, 492
Vesicle transport and coat complex, 49
Vesicular monoamine transporter, 137
Vesicular traffic, 46
small G proteins of Rab family and, 49
Vessels
brain capillaries, 602
cerebral ischemia, 601
cerebrospinal fluid, 602
circle of Willis, 601
paravertebral veins, 601
relation of fibrous astrocyte, 602
transport across cerebral capillaries, 602
vertebral arteries, 601
Vestibular labyrinth, 212
Vestibular system
ampullary responses to rotation, 211
division of, 209
responses to linear acceleration, 211–212
responses to rotational acceleration, 211
spatial orientation, 212
vestibular apparatus, 209
vestibular nuclei, 209
Vestibulocerebellar output, 250–251
Vestibulocerebellum, 250
Vestibulospinal tracts, medial and lateral, 240
Vibratory sensibility, 162
Villus sampling, chorionic, 397
VIP. *See* Vasoactive intestinal polypeptide
VIPomas. *See* VIP-secreting tumors
VIP-secreting tumors, 472
Virilization, 362
Visceral and deep pain, 165–166
Visceral reflexes, respiratory components of, 665
Visceral sensation, 169
Visual acuity, 181
Visual fields, 195
Visual function
binocular vision, 195
critical fusion frequency, 194–195
dark adaptation, 194
eye movements, 195–196
superior colliculi, 196
visual fields, 195
Visual information processing, 185
Visual pathways, 207–209
Visual pathways and cortex, responses in
cortical areas concerned with vision, 192–193
effect of lesions, 190–191
neural pathways, 189–190
primary visual cortex, 191–192
Visual projection areas in human brain, 193
Vital lung capacity, 629
Vitamin A deficiency, 183
Vitamin D
formation and hydroxylation of, 379
metabolite of, 379–381
Vitamin D-binding protein (DBP), 379
Vitamins, 492
human nutrition, 493–494
Vitamins, absorption of, 485
Vitiligo, 326
VLDL. *See* Very low density lipoprotein
VMAT, 137
Voltage-gated Ca^{2+} channels
Ca^{2+} influx by, 103
in cardiac myocytes, 110, 111
Voltage-gated K^+ channels, 51
feedback control in, 90
opening and closing of, 88
sequential feedback control in, action potential, 88–89
Voltage-gated Na^+ channels
feedback control in, 90
reponse to depolarizing stimulus, 88
sequential feedback control in, action potential, 88–89
spatial distribution of, 91
Volume conductor, 524
Voluntary movement
brain stem pathways involved in
lateral, 240–241
medial, 239–240
control of, 237
corticospinal and corticobulbar system role in, 239
Vomeronasal organ, 219
Vomiting, 502
neural pathways, 503
Von Willebrand factor, 79
VPB. *See* Ventricular premature beats
Vulnerable period, 533

W

Warmth receptors, 161
Water
dipole moment, 4
hydrogen bond network in, 4
Waterfall effect, 636
Water-hammer pulse, 542
Water intake
factors influencing, 310
plasma osmolality and changes in ECF volume effect on, 310
psychologic and social factors effect on, 311
Water metabolism, actions of glucocorticoids on, 365–366
Water metabolism and pituitary insufficiency, 335–336
Weak acid, buffering capacity of, 7
Weak base, buffering capacity of, 7
Wenckebach phenomenon, 530
Wernicke's area, 208
White blood cells
cells grow, 556–558
cellular elements, 556, 557
production of, 70
White fat depots, 27
White rami communicans, 257
Whole body ammonia homeostasis, 515
Wilson disease, 246
Withdrawal reflex, importance of, 234
Wolff–Chaikoff effect, 346
Wolff–Parkinson–White syndrome, 533, 534
Working memory, 285, 288
Wound healing, 81

X

X chromosomes, 392
Xerophthalmia, 183
Xerostomia, 457

Y

Yawning, 665
Y chromosomes, 392
Young–Helmholtz theory of color vision, 193

Z

Ziconotide, chronic pain by, 164
Zinc deficiency, 492
Zollinger–Ellison syndrome, 384
Zolpidem, 279
Zona fasciculata, 354
Zona glomerulosa, 354
Zona pellucida, 413
Zona reticularis, 354
Zonula adherens, 43
Zonula occludens, 43
Zymogen granules, 463